ANNUAL REVIEW OF MICROBIOLOGY

EDITORIAL COMMITTEE (1995)

ALBERT BALOWS
JOHN C. BOOTHROYD
JANET S. BUTEL
E. PETER GREENBERG
MARY E. LIDSTROM
BARRY L. MARRS
L. NICHOLAS ORNSTON
JOHN N. ROSAZZA
GRAHAM C. WALKER

Responsible for the organization of Volume 49
(Editorial Committee, 1993)

ALBERT BALOWS
JANET S. BUTEL
E. PETER GREENBERG
BARRY L. MARRS
L. NICHOLAS ORNSTON
JOHN P. N. ROSAZZA
H. JEAN SHADOMY (GUEST)
GRAHAM C. WALKER
C. C. WANG

Production Editor   AMANDA M. SUVER
Subject Indexer   STEVEN M. SORENSEN

ANNUAL REVIEW OF MICROBIOLOGY

VOLUME 49, 1995

L. NICHOLAS ORNSTON, *Editor*
Yale University

ALBERT BALOWS, *Associate Editor*
Centers for Disease Control, Atlanta

E. PETER GREENBERG, *Associate Editor*
University of Iowa, Iowa City

ANNUAL REVIEWS INC.   4139 EL CAMINO WAY   P.O. BOX 10139   PALO ALTO, CALIFORNIA 94303-0139

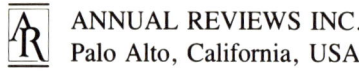 ANNUAL REVIEWS INC.
Palo Alto, California, USA

COPYRIGHT © 1995 BY ANNUAL REVIEWS INC., PALO ALTO, CALIFORNIA, USA. ALL RIGHTS RESERVED. The appearance of the code at the bottom of the first page of an article in this serial indicates the copyright owner's consent that copies of the article may be made for personal or internal use, or for the personal or internal use of specific clients. This consent is given on the condition, however, that the copier pay the stated per-copy fee of $5.00 per article through the Copyright Clearance Center, Inc. (222 Rosewood Drive, Danvers, MA 01923) for copying beyond that permitted by Sections 107 or 108 of the US Copyright Law. The per-copy fee of $5.00 per article also applies to the copying, under the stated conditions, of articles published in any *Annual Review* serial before January 1, 1978. Individual readers, and nonprofit libraries acting for them, are permitted to make a single copy of an article without charge for use in research or teaching. This consent does not extend to other kinds of copying, such as copying for general distribution, for advertising or promotional purposes, for creating new collective works, or for resale. For such uses, written permission is required. Write to Permissions Dept., Annual Reviews Inc., 4139 El Camino Way, P.O. Box 10139, Palo Alto, CA 94303-0139 USA.

International Standard Serial Number: 0066–4227
International Standard Book Number: 0–8243–1149-3
Library of Congress Catalog Card Number: 49-432

Annual Review and publication titles are registered trademarks of Annual Reviews Inc.

∞ The paper used in this publication meets the minimum requirements of American National Standard for Information Sciences—Permanence of Paper for Printed Library Materials, ANSI Z39.48-1984.

Annual Reviews Inc. and the Editors of its publications assume no responsibility for the statements expressed by the contributors to this *Review*.

Typesetting by Kachina Typesetting Inc., Tempe, Arizona; John Olson, President; Jeannie Kaarle, Typesetting Coordinator; and by the Annual Reviews Inc. Editorial Staff

PRINTED AND BOUND IN THE UNITED STATES OF AMERICA

PREFACE

Intentionally or unintentionally, we all practice the art of classification. There can be no order without categories for our information, and useful categories take on a force that guides our observations. In this sense, a useful category can become a trap. By developing an addiction to its convenience, we may let a valued category control our intellectual direction, and thus slave can become master. Such was the concern of CB van Niel years ago when he was confronted with a field that had taken on the label Objective Taxonomy. "I suppose it is all right," he commented, "as long as it is backed up by the intuition of an experienced taxonomist."

The experience of a taxonomist should foster fascination with observations that do not fit paradigms, as well as a thirst for shifts to paradigms that allow cleaner accommodation of our perceptions of reality. In this sense, we might consider it unfortunate that changing fashion phased out the term *taxonomy* in favor of the more chic expression *systematics*. Perhaps the change is for the better because, if systematics is the game, we all know that we all do it.

Survivors of the information explosion are required to challenge existing categories and to find new ones that will guide us into the unknown. The question is not what information is there, but what is most important. This is the question guiding the deliberations of the editorial committee of the *Annual Review of Microbiology* as it selects topics and authors for the next volume.

As with other things biological, the most constant feature of an editorial committee is constant change. Some changes deny even the illusion of constancy; one of these is the departure of Amanda Suver after 6 years as production editor with the *Review*. We shall miss her greatly, and we wish her good fortune in her new endeavors.

L. NICHOLAS ORNSTON
EDITOR

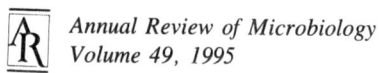

Annual Review of Microbiology
Volume 49, 1995

CONTENTS

THE ROAD TO YELLOWSTONE—AND BEYOND, *Thomas D. Brock*	1
MECHANISMS FOR THE PREVENTION OF DAMAGE TO DNA IN SPORES OF *BACILLUS* SPECIES, *Peter Setlow*	29
GENETICS, PHYSIOLOGY, AND EVOLUTIONARY RELATIONSHIPS OF THE GENUS *BUCHNERA*: Intracellular Symbionts of Aphids, *Paul Baumann, Linda Baumann, Chi-Yung Lai, Dadbeh Rouhbakhsh, Nancy A. Moran, and Marta A. Clark*	55
PHYSIOLOGICAL IMPLICATIONS OF STEROL BIOSYNTHESIS IN YEAST, *Leo W. Parks and Warren M. Casey*	95
THE STRUCTURE AND REPLICATION OF KINETOPLAST DNA, *Theresa A. Shapiro and Paul T. Englund*	117
HOW *SALMONELLA* SURVIVE AGAINST THE ODDS, *John W. Foster and Michael P. Spector*	145
THE MECHANISMS OF *TRYPANOSOMA CRUZI* INVASION OF MAMMALIAN CELLS, *Barbara A. Burleigh and Norma W. Andrews*	175
POLYKETIDE SYNTHASE GENE MANIPULATION: A Structure-Function Approach in Engineering Novel Antibiotics, *C. Richard Hutchinson and Isao Fujii*	201
NONOPSONIC PHAGOCYTOSIS OF MICROORGANISMS, *I. Ofek, J. Goldhar, Y. Keisari, and N. Sharon*	239
PEPTIDES AS WEAPONS AGAINST MICROORGANISMS IN THE CHEMICAL DEFENSE SYSTEM OF VERTEBRATES, *Pierre Nicolas and Amram Mor*	277
CO DEHYDROGENASE, *James G. Ferry*	305
NITROGENASE STRUCTURE AND FUNCTION: A Biochemical-Genetic Perspective, *John W. Peters, Karl Fisher, and Dennis R. Dean*	335
CONJUGATIVE TRANSPOSITION, *June R. Scott and Gordon G. Churchward*	367
CELLULOSE DEGRADATION IN ANAEROBIC ENVIRONMENTS, *Susan B. Leschine*	399

CONTENTS (*continued*)   vii

New Mechanisms of Drug Resistance in Parasitic Protozoa, *P. Borst and M. Ouellette*	427
Environmental Virology: From Detection of Virus in Sewage and Water by Isolation to Identification by Molecular Biology—A Trip of Over 50 Years, *T. G. Metcalf, J. L. Melnick, and M. K. Estes*	461
How Bacteria Sense and Swim, *David F. Blair*	489
Biodegradation of Nitroaromatic Compounds, *Jim C. Spain*	523
Biocatalytic Syntheses of Aromatics from D-Glucose: Renewable Microbial Sources of Aromatic Compounds, *J. W. Frost and K. M. Draths*	557
The Regulation of Methane Oxidation in Soil, *Rocco L. Mancinelli*	581
Discovery, Biosynthesis, and Mechanism of Action of the Zaragozic Acids: Potent Inhibitors of Squalene Synthase, *James D. Bergstrom, Claude Dufresne, Gerald F. Bills, Mary Nallin-Omstead, and Kevin Byrne*	607
Prospects for New Interventions in the Treatment and Prevention of Mycobacterial Disease, *Douglas B. Young and Kenneth Duncan*	641
Development and Application of Herpes Simplex Virus Vectors for Human Gene Therapy, *J. C. Glorioso, N. A. DeLuca, and D. J. Fink*	675
Microbial Biofilms, *J. William Costerton, Zbigniew Lewandowski, Douglas E. Caldwell, Darren R. Korber, and Hilary M. Lappin-Scott*	711
Leucine-Responsive Regulatory Protein: A Global Regulator of Gene Expression in *E. coli*, *E. B. Newman and Rongtuan Lin*	747
Microbiology to 10,500 Meters in the Deep Sea, *A. Aristides Yayanos*	777
Viral Vectors in Gene Therapy, *Alan E. Smith*	807

INDEXES

Subject Index	839
Cumulative Index of Contributing Authors, Volumes 45–49	855
Cumulative Index of Chapter Titles, Volumes 45–49	857

OTHER REVIEWS OF INTEREST TO MICROBIOLOGISTS

From the *Annual Review of Biochemistry*, Volume 64 (1995)

   The Envelope of Mycobacteria, P. J. Brennan and H. Nikaido
   DNA Processing Reactions in Bacterial Conjugation, E. Lanka and B. M. Wilkins
   The Molecular Biology of Hepatitis Delta Virus, M. M. C. Lai

From the *Annual Review of Cell and Developmental Biology*, Volume 11 (1995)

   Structure and Function of Kinetochores in Budding Yeast, A. A. Hyman and P. K. Sorger
   The Cell Biology of Infection by Intracellular Bacterial Pathogens, J. A. Theriot
   Receptor-Mediated Protein Sorting to the Vacuole in Yeast: Roles for Protein Kinase, Lipid Kinase, and GTP-Binding Proteins, J. H. Stack, B. Horazdovsky, and S. D. Emr

From the *Annual Review of Entomology*, Volume 40 (1995)

   Cellular and Molecular Interrelationships Between Ticks and Prokaryotic Tick-Borne Pathogens, U. G. Munderloh and T. J. Kurtti

From the *Annual Review of Genetics*, Volume 29 (1995)

   Chromosome Partitioning in Bacteria, R. G. Wake and J. Errington
   Genetic Networks Controlling the Initiation of Sporulation and the Development of Genetic Competence in Bacillus subtilis, A. D. Grossman
   Light-Harvesting Complexes in Oxygenic Photosynthesis: Diversity, Control, and Evolution, A. R. Grossman, D. Bhaya, K. E. Apt, and D. M. Kehoe
   Genetic Analysis of the Multidrug Transporter, M. M. Gottesman, C. A. Hrycyna, P. V. Schoenlein, U. A. Germann, and I. Pastan
   The Plant Response in Pathogenesis, Symbiosis, and Wounding: Variations on a Common Theme? C. Baron and P. C. Zambryski
   Chlamydomonas reinhardtii *as the Photosynthetic Yeast*, J.-D. Rochaix
   Genetics of Ustilago maydis, *a Fungal Pathogen That Induces Tumors in Maize*, F. Banuett

From the *Annual Review of Immunology*, Volume 13 (1995)

   Hepatitis B Virus Immunopathogenesis, F. V. Chisari and C. Ferrari
   The Regulation of Immunity to Leishmania major, S. L. Reiner and R. M. Locksley
   Prevention of AIDS Transmission Through Screening of the Blood Supply, S. A. Galel, J. D. Lifson, and E. G. Engleman

RELATED ARTICLES (*continued*)

From the *Annual Review of Medicine,* Volume 46 (1995)

Risks of HIV Infection in the Health Care Setting, V. J. Fraser and W. G. Powderly

From the *Annual Review of Plant Physiology and Plant Molecular Biology,* Volume 46 (1995)

Breaking the N≡N Bond, R. H. Burris

Molecular Genetics of Sexuality in Chlamydomonas, U. W. Goodenough, E. V. Armbrust, A. M. Campbell, and P. J. Ferris

Chemoperception of Microbial Signals in Plant Cells, T. Boller

Heterologous Expression of Genes in Bacterial, Fungal, Animal, and Plant Cells, W. B. Frommer and O. Ninnemann

ANNUAL REVIEWS INC. is a nonprofit scientific publisher established to promote the advancement of the sciences. Beginning in 1932 with the *Annual Review of Biochemistry*, the Company has pursued as its principal function the publication of high-quality, reasonably priced *Annual Review* volumes. The volumes are organized by Editors and Editorial Committees who invite qualified authors to contribute critical articles reviewing significant developments within each major discipline. The Editor-in-Chief invites those interested in serving as future Editorial Committee members to communicate directly with him. Annual Reviews Inc. is administered by a Board of Directors, whose members serve without compensation.

1995 Board of Directors, Annual Reviews Inc.

Richard N. Zare, Chairman of Annual Reviews Inc.
  Professor of Physical Chemistry, Stanford University
Winslow R. Briggs, Vice Chairman of Annual Reviews Inc.
  Director Emeritus, Carnegie Institution of Washington, Stanford
Peter F. Carpenter, *Founder, Mission and Values Institute*
W. Maxwell Cowan, *Vice President and Chief Scientific Officer, Howard Hughes Medical Institute, Bethesda*
Sidney D. Drell, *Deputy Director, Stanford Linear Accelerator Center*
Sandra M. Faber, *Professor of Astronomy, University of California, Santa Cruz*
Eugene Garfield, *Publisher,* The Scientist
Samuel Gubins, *Annual Reviews, Inc.*
Daniel E. Koshland, Jr., *Professor of Biochemistry, University of California, Berkeley*
Joshua Lederberg, *University Professor, The Rockefeller University*
Gardner Lindzey, *Director Emeritus, Center for Advanced Study in the Behavioral Sciences, Stanford*
Sharon R. Long, *Professor of Biological Sciences, Stanford University*
Harriet A. Zuckerman, *Vice President, The Andrew W. Mellon Foundation*

Management of Annual Reviews, Inc.

Samuel Gubins, President and Editor-in-Chief
John S. McNeil, Publisher and Secretary-Treasurer
Donald Svedeman, Business Manager
Richard L. Burke, Production and Technology Applications Manager
Richard A. Peterson, Advertising and Marketing Manager

ANNUAL REVIEWS OF
Anthropology
Astronomy and Astrophysics
Biochemistry
Biophysics and Biomolecular Structure
Cell Biology
Computer Science
Earth and Planetary Sciences
Ecology and Systematics
Energy and the Environment
Entomology
Fluid Mechanics
Genetics
Immunology

Materials Science
Medicine
Microbiology
Neuroscience
Nuclear and Particle Science
Nutrition
Pharmacology and Toxicology
Physical Chemistry
Physiology
Phytopathology
Plant Physiology and
  Plant Molecular Biology
Psychology

Public Health
Sociology

SPECIAL PUBLICATIONS

Excitement and Fascination of Science, Vols. 1, 2, and 3

Intelligence and Affectivity, by Jean Piaget

For the convenience of readers, a detachable order form/envelope is bound into the back of this volume.

Thomas D. Brock

THE ROAD TO YELLOWSTONE—AND BEYOND

Thomas D. Brock
1227 Dartmouth Road, Madison, Wisconsin 53705-2213

KEY WORDS: microbial ecology, thermophiles, biotechnology, history, limnology

CONTENTS

Introduction	2
Beginnings	2
First Professional Work	4
First Academic Appointment	5
Indiana University	7
Steps Toward Microbial Ecology	7
Friday Harbor Laboratories	8
From Leucothrix To Yellowstone	10
Discovery of Thermus aquaticus	13
Bacteria in Boiling Water	15
"Life at High Temperatures": the Science Paper	16
Travels in Search of Hot Springs	17
Deep Sea Vents	18
Microbial Prospecting in Yellowstone	18
Biology of Microorganisms	19
Beyond Yellowstone	20
History of Microbiology	22
Computers	23
Publishing	23
Final Words	24

ABSTRACT

This memoir describes the professional life and times of Thomas D. Brock, with an emphasis on those aspects of his career relating to research in microbial ecology, and how this work led to field research in Yellowstone. The first discovery of extremely thermophilic bacteria is described, followed by a discussion of some of the consequences of this discovery for biotechnology and microbiology. Also covered briefly in this memoir are Brock's activities in textbook writing, publishing, computers, and the history of science.

Introduction

The trajectory of my career provides an example of what can be done with a little knowledge, a fair bit of luck, and a lot of hard work. It also does not hurt to have had a good set of parents.

As I had only two microbiology courses during my university studies, one of which was a so-called nonmajors course, I am mostly a self-taught microbiologist. In fact, I am self-taught in almost everything. Yet the education I did receive laid the intellectual foundation for what I have learned.

I started out in botany, then specialized in mycology and yeast physiology, and after my PhD I worked in antibiotics research in the pharmaceutical industry, a move that brought me into microbiology and molecular biology. After leaving industry, I developed a brief, albeit successful, career in yeast genetics before becoming a medical microbiologist specializing on the streptococci. In 1963, I began work in marine microbiology, which brought me into microbial ecology. For seven or eight years I juggled research on antibiotics, yeast genetics, streptococci, and marine microbiology, before finally concentrating on microbial ecology with an emphasis on extreme environments. This work led me into geology and biogeochemistry. Later I became a specialist on lakes and limnology, before closing out my scholarly career in the history of science. Throughout most of my professional life I have also had a parallel career as a textbook author. Because of this, I taught myself how to use computers and became a pioneer in the use of microcomputers in scholarly work. This path led me into publishing and editing, and I established and ran my own publishing company.

I am a member of a favored generation, finishing high school in 1944. Too young to be killed in World War II, I was able to ride to the top in the post-war prosperity of the United States. My professional life encompassed the post-Sputnik years in the United States, when money for research was easy to obtain.

Although this is a memoir, it is not based just on memory, but on extensive documentation. For better or worse, I have been a "saver," and through the vagaries of life I have managed to keep my files more or less intact. I have gone through these files in detail in preparation for writing this article in order to keep my facts straight.

Beginnings

I was the only child of Helen Sophia Ringwald, of Chillicothe, Ohio, and Thomas Carter Brock, of Toronto, Canada. I was born (September 10, 1926) and raised in Cleveland, Ohio, and although I have traveled extensively, I have made my permanent residence in the midwestern states all my life.

Cleveland is predominantly an industrial city, but our house was on the top of a hill in a unique cul-de-sac adjacent to an errant farm and a forested park. There was a distant view of Lake Erie from the front porch. Although a short walk down a quite steep hill brought us to one of Cleveland's major thoroughfares (Euclid Avenue), behind our house were fields, cows, woods, and open spaces.

I grew up during the depths of the financial depression of the 1930s. My father had only an eighth-grade education but continued to educate himself through correspondence courses and self-study. Eventually, he became a power engineer, working in various industries in Cleveland. Although we were never well-off, my father did have a job throughout the depression (in the boiler room of St. Luke's Hospital) and my mother (who had been a registered nurse) was able to remain at home. In this working-class household, we had no music, no literature or art, and very few books. But I did have a stable home life in a pleasant neighborhood with lots of friendly playmates. In those days the schools in Cleveland were good and it was through them that I was introduced to music (the Cleveland Orchestra) and art (the Cleveland Institute of Art).

Unfortunately, when I was almost 15 my father got sick and we had to leave this idyllic environment and move to my mother's hometown, Chillicothe, Ohio. Within months my father was dead, and my mother and I lived through my high school years in what could best be called gentile poverty. I had to help financially and worked at a variety of clerical jobs in drug, clothing, and grocery stores. Because of the war-time labor shortages, I had no trouble finding work, but the pay was low ($0.25 per hour).

My father had always recognized his lack of formal education and encouraged me to think about college. He brought home discarded electrical equipment and showed me how to make coils, electromagnets, and radios. When I was 10 years old, I received a chemistry set for Christmas and he helped me set up a simple laboratory in the basement. After we moved to Chillicothe, I met David Thornburgh, who was also interested in chemistry, and he and I set up a small research laboratory in the loft of a barn behind my house. We did lots of crazy experiments (explosives, toxic gas), but Thornburgh had also heard about penicillin (this was 1943), and we made some fleeting attempts to enrich soil for antibiotic-producing microorganisms. Hence, I decided to go to college and become a chemist.

However, the war was on, so after graduation from high school I enlisted in an electronics program in the US Navy, spending about 18 months in various Navy schools in the Chicago area. I finished out my Navy career in Kodiak, Alaska, where I worked not in electronics but as a member of the Shore Patrol. One of my jobs in the town of Kodiak was to clear the bars of sailors at curfew and to make sure the brothels were empty. After my sheltered home life, the Navy was a riotous experience. Although I had always been a reader, I had never had any contact with great literature. However, in the Navy I began to read voraciously. Soon I had decided to become a writer.

As a military veteran I was eligible for the GI Bill, and in the fall of 1946, I enrolled at Ohio State University. Although I was a good student at OSU, I soon began to have doubts about a writing career and began to think again about chemistry and science. One reason may have been Sinclair Lewis's *Arrowsmith* (the protagonist of which was a research bacteriologist), a book I was required to read in a twentieth-century American literature course. However, for reasons too complex to go into, I did not major in bacteriology but in botany, graduating (with honors) in 1949.

I was offered a graduate assistantship in botany at OSU, and with my limited financial resources, I found it difficult to go to another school. However, the study of higher plants soon bored me and I switched to mycology, receiving my MS and PhD working on a mushroom (*Morchella esculenta*) and the yeast *Hansenula anomala* (3, 5).

Although I learned very little bacteriology as a student, I did learn a lot about plant ecology. The ecology group at OSU was very strong, and I had the pleasure of taking courses from John N Wolfe, an enthusiastic lecturer. Wolfe was above all a field-oriented ecologist, and every Saturday he took us on a field trip to an interesting habitat. Furthermore, a lot of my fellow graduate students were plant ecologists. One friend, Theodore Sudia, coaxed me to think about microbial ecology. Although years would pass before I would finally move in this direction, Sudia's sharp discussions along these lines stuck with me.

First Professional Work

I received my PhD in 1952. I had been hoping for a faculty position, but at this time academic jobs were very hard to find, and postdoctoral positions were virtually nonexistent. I obtained for the summer a position as a temporary research associate working on soil fungi at the Ohio Agricultural Experiment Station (Wooster, Ohio) and spent most of my spare moments looking for a job. Fortunately, a position in the Antibiotics Research Department at The Upjohn Company came along, and because of my work in soil mycology, I was offered the job. I abandoned hope of a teaching position and went off with my new wife, Louise, to Kalamazoo, Michigan.

I actually became a bacteriologist via on-the-job training during my Upjohn years, learning from my Upjohn colleagues, most of whom had received their degrees from midwestern bacteriology departments. Antibiotics research is microbial physiology under the simplest possible conditions, and it was quite easy to pick up. By the time I left Upjohn five years later, I had published papers in respectable journals (6), had become a member of the Society of

American Bacteriologists, and had attended the society's annual meetings in New York, Houston, and Detroit.

Two other accomplishments came out of my five years in Kalamazoo—an interest in historical research and a knowledge of German. Because industrial research was an eight-to-five job, I had lots of free time. Interest in an abandoned railroad near my home led me to begin doing research in local history. I had soon published a lengthy article on my railroad research in a scholarly journal (4). In my last two years in Kalamazoo, I also began studying German, using records and audio tapes. Although I never quite reached fluency, I did become adept at translating scientific papers, a skill I was eventually to use extensively in my work on the history of microbiology.

First Academic Appointment

I left Upjohn because I became bored with the routine research program and the lack of control I had over my own work. Being out of the academic world, I had quite a bit of difficulty finding an academic job, but I finally managed to obtain one in my old hometown, Cleveland, in the Biology Department of Western Reserve University (WRU, now Case Western Reserve University). It was a rather poor department; the teaching load was enormous; and the salary was miserable (about half my Upjohn salary), but it gave me the opportunity to develop my own research program and to teach microbiology rather than botany. Another attraction was that across the street in the medical school was a very fine Department of Microbiology, where I could attend seminars and borrow equipment for my research.

At WRU I taught general bacteriology to the undergraduates, and two separate courses in nursing microbiology, one for degree students in the School of Nursing, the other for nondegree students from three hospitals in Cleveland. In addition to giving all the lectures, I had to supervise the laboratories, unfortunately without any qualified teaching assistants. In subsequent semesters, I taught mycology, medical microbiology, and an advanced microbiology course that was predominantly microbial physiology and genetics.

Although I had learned a lot of microbiology at Upjohn, it was rather narrowly focused. Teaching is the best way to learn, and after two years with this heavy teaching load I was well on the way to becoming a real microbiologist. Surprisingly, I also found time for some significant research. I obtained two research grants, one from the NIH on the mode of action of antibiotics and one from the NSF on yeast mating. I had never used radioisotopes as tracers, but I ordered some and taught myself how to work with them. (In those days, one could order 50 µCi of any ^{14}C compound without a license!) Louise worked as a technician and collaborator on the NIH grant, but I did the yeast work myself. By the time I left the Biology Department, Louise and I had done enough research to publish 13 papers (for instance, 7, 9, 30).

I also continued my study of German, and during the summer of 1958 I did most of the translations for what would become my first book, *Milestones in Microbiology* (8). Doing *Milestones* combined two areas that had been hobbies during my Kalamazoo days, German and history. The book was finished just before I left Cleveland and initiated a long-term relationship with the publishing company Prentice-Hall that has continued to this day. However, after *Milestones* many years passed before I returned to serious work on the history of microbiology.

Now that I was out of industry I had more free time for vacations. Louise's father was member of a wilderness fishing camp on Lake Memesagemissing in northern Ontario, and I also joined, spending the months of August in 1958 and 1959 in veritable isolation. I brought along my *Milestones* project to work on, but mainly I learned about boating and fishing, which introduced me to the aquatic environment and got me interested in canoeing, kayaking, and the great outdoors. I had never really done things like this before, and these years were to provide important background for my later research in marine microbiology, limnology, and microbial ecology.

After the second year of teaching nurses and general bacteriology, I realized that the teaching load in the Biology Department at WRU was so heavy that I would never be able to develop a really significant research program. Then LO Krampitz, Chairman of the Department of Microbiology, offered me a postdoctoral position. I resigned my Assistant Professor position and moved across the street. The research project I was assigned, on the biosynthesis of the M protein of group A streptococci, turned out to be a dead end, but it taught me a lot of biochemistry, as well as a fair bit of immunology and clinical microbiology. Krampitz's department was packed with top-flight people, many of whom went on to distinguished positions elsewhere.

Before beginning my postdoctorate, I had taken the bacterial genetics course at Cold Spring Harbor, where I associated with many distinguished geneticists and molecular biologists. As usual, I learned as much from my fellow students as from the faculty. (Many of the students would become distinguished scientists: Julius Adler, Marshall Nirenberg, Gordon Tompkins, David C White, Solomon Bartnicki-Garcia.) Many years later I returned to Cold Spring Harbor when I wrote my book on the history of bacterial genetics (27).

During the year I was in Krampitz's laboratory, I continued to supervise the research that Louise was still conducting in the Biology Department for my NIH grant. Also, because they had been unable to find a replacement, I still taught the undergraduate bacteriology course. Although this was a traumatic year in many ways, at long last I was in a legitimate microbiology department and associating with real microbiologists. By the time I left WRU I could talk bacteriology without making a fool of myself.

Indiana University

As chance would have it, toward the end of my first year as a postdoctorate, a position opened up at Indiana University (IU) in Bloomington. By this time, I had had enough of a large industrial city, even if it had been my hometown, and the small university town of Bloomington seemed quite attractive. The chairman of the department, LS McClung, offered me a position as an Assistant Professor of Bacteriology, which I accepted enthusiastically, even though the teaching responsibility was medical microbiology. Joining the IU department made my shift from botany to microbiology complete.

Although I had received my PhD in 1952, it was just now, eight years later, that I finally could get down to long-term work. In a way, it was fortunate that I had waited to settle in, since the 1960s were a time when the US economy was flourishing. Also, it was a time of great university expansion, with a special focus on increasing research activity. Training grants were easy to come by, and most of my early students at IU were supported by a Microbiology Training Grant (NIH), which allowed me to reserve my own grants for the salaries of technicians. Because of the baby boomers, enrollment was increasing. Everyone was in an expansionist mood. It was quite easy to obtain money, support, students, space, etc. Also, prices were relatively low.

However, the position at Indiana University required that I teach and do research in the field of medical microbiology. I had taught a few medical students at WRU, and a lot of nurses, so I had a fair grasp of how to put such a program together. Based on my postdoctorate work at WRU, I obtained two grants on streptococci, one of which involved extensive work on genetics and phage. I also did a lot of work on amino acid and peptide transport in enterococci (38) and finished up the work on the mode of action of antibiotics that I had begun at Upjohn and continued at WRU. With several students and technicians, I became heavily involved in research on streptococcal genetics. Probably because of my medical emphasis, I had no difficulty obtaining one of the newly established Research Career Development Awards (RCDA) from the NIH. Nepotism rules prevented Louise from working in my laboratory (later she was permitted to work as an unpaid research associate), but I had no difficulty finding competent students (34, 49) and technicians (for instance, 36, 37).

Steps Toward Microbial Ecology

During my three years at WRU, my interests in ecology had been strongly suppressed. However, I continued to think about microbial ecology, and with the RCDA to pay my salary (and a consequent reduction in teaching load), I decided that now was the time to take some steps in this direction. I had always been interested in aquatic and marine microbiology. When I took the bacterial genetics course at Cold Spring Harbor in the summer of 1959, I had lived within view of Long Island Sound and had been fascinated by the marine life and changing tides. Canoeing and kayaking in Ontario and Florida also furthered my interest in freshwater habitats.

I once read that play in children is never just play, but the first steps toward building useful skills. Outdoor activities such as canoeing and backpacking certainly fit into this category for me, as once I became involved in microbial ecology research, I not only used the outdoor skills I had acquired, but I also used the equipment I had bought. One reason I became so firmly field-oriented in my research may have been because I had become so enamored of the outdoors.

In November 1962, just before attending the American Society for Cell Biology meeting in San Francisco, I visited the laboratory of CB van Niel at the Hopkins Marine Station, Pacific Grove, California. This was the first time I had ever been at a marine station, and although this was hardly a typical marine laboratory, I found the environment quite fascinating. I had always been a fan of John Steinbeck's books and was excited to see the area around Cannery Row.

Friday Harbor Laboratories

A major turning point in my career came in the spring of 1963 when I made plans to actually do research at a marine laboratory, Friday Harbor Laboratories of the University of Washington. I had learned about this laboratory from Brooks Church, a seminar visitor at Indiana who was at that time on the faculty at the University of Washington. Through Church, I made contact with the director of the laboratory and applied for research space for the summer of 1963. The research I proposed, on the presence of enterococci in marine animals, was excessively naive, but it was a logical progression from my streptococcus work. I arrived at Friday Harbor about July 1 and stayed through the middle of August. It was a complete change of pace from Indiana, not only because of the marine environment and the quaint laboratories, but because of the chance to associate with field-oriented biologists in a variety of disciplines. I had always had broad interests; Friday Harbor showed me how these interests could be channeled in new ways.

However, my work on streptococci in marine animals lasted not much more than two weeks. Through a circuitous but logical route, I moved from ostensibly marine streptococci to *Leucothrix mucor,* which I discovered to be a widespread marine microorganism. I taught myself how to isolate pure cultures directly from nature, instead of using the accepted but unecological enrichment culture technique. The resulting work could best be called autoecological. My first paper on knots in *L. mucor* made the cover of *Science* (11) (Figure 1) and was featured in the New York *Times* (May 15, 1964). My work on *Leucothrix*

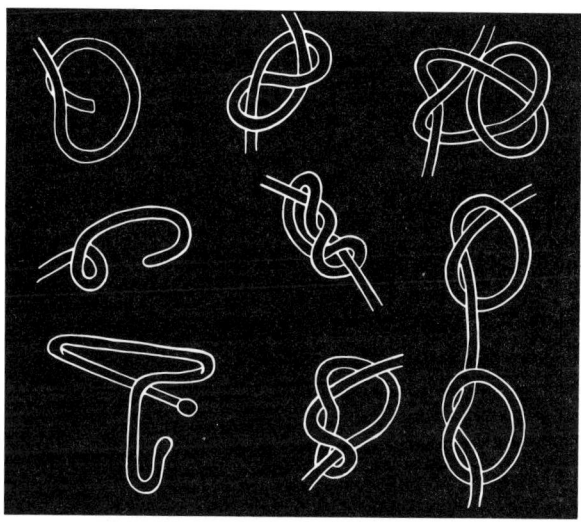

Figure 1 My first paper on knots in *L. mucor* made the cover of *Science* and was a feature article in the New York *Times*. Reprinted from Science, copyright © 1964 (11).

extended over the next eight or nine years and resulted in quite a few papers and two PhD theses (1, 16, 17, 41).

My *L. mucor* work is beyond the scope of the present article, but it is important to note that it not only pushed me toward research in general microbiology, but got me interested in sulfur springs, the habitat of members of the related genus *Thiothrix*. From cold sulfur springs I quickly moved to Yellowstone hot springs.

In addition, the Friday Harbor work led me to begin writing *Principles of Microbial Ecology* (14), which really focused my mind on what microbial ecology was all about. I began making contact with limnologists and macroecologists and understood that the proper approach to microbial ecology was through direct studies in the natural environment. My training in plant ecology at OSU certainly had a lot to do with my insistence on studying microorganisms directly in nature. Thus, even though my principal research continued in the areas of streptococcal genetics (10), yeast mating (I had just obtained a new NSF grant in this area) (12, 13, 39), and the mode of action of antibiotics [which led to the first of three invitations to the Society for General Microbiology annual symposium (15)], my reading over the period 1963–1964 concentrated on microbial ecology, as I worked on the first draft of my microbial ecology book (14).

From Leucothrix to Yellowstone

Although I had become interested in sulfur springs because of *Thiothrix,* hot springs were another matter. I had never seen any, nor had I any desire to go to Yellowstone because of its reputation as a heavily visited "amusement" park, rather than a natural area. My first visit to Yellowstone, on the way to Friday Harbor in July 1964 after a backpacking trip in Grand Teton National Park, was a revelation. I had not expected such enormous developments of microorganisms as were present in the runoff channels of the Yellowstone hot springs. Returning from Friday Harbor that summer, I stopped again in Yellowstone, and this time I sampled some hot springs, looking for possible habitats for *Thiothrix* (Figure 2).

Figure 2 Sampling a small hot spring along the Yellowstone River, August 22, 1964.

Soon after returning to Bloomington, my interests in the Yellowstone habitats broadened as a result of further work on my book. By this time, I was reading a lot of the ecosystem literature, with its focus on steady-state systems. I began to think of springs as steady-state ecosystems (19).

However, I did my first work in Yellowstone to get some field experience before research planned for the summer of 1965 on the new volcanic island of Surtsey, which had developed off the south coast of Iceland (28). Thus, Louise and I visited Yellowstone again in late June and early July 1965. We planned this trip, which was one of the more exciting two weeks of my career (even though it rained most of the time), as a working vacation. The focus was to be on the thermal algae (cyanobacteria, actually) that formed the colorful mats found in the outflow channels of many of the springs and geysers.

Casual observations the previous summer had indicated that the outflow channels of Yellowstone hot springs often had extensive developments of photosynthetic organisms. I reasoned that if these effluents were steady-state ecosystems, quantitative measurements of chlorophyll in the thermal gradients should be possible. However, as an additional twist, I planned to measure not only chlorophyll, but also macromolecules: RNA, DNA, and protein. Therefore, we gathered supplies and equipment that could be taken along for such assays (Figure 3).

While getting ready for the Yellowstone trip, I had decided that I should set up some culture studies to attempt to get growth of high-temperature bacteria. Thus, I came to Yellowstone in June 1965 with supplies of bacterial nutrient media, and I tried to obtain growth at high temperatures using hot spring water as the inoculum, and the springs themselves as incubators.

Although these culture experiments came to naught, they did lead to an important discovery. In the outflow channel of a spring in the White Creek area, which we called Pool A (Octopus Spring), I saw pink gelatinous masses of material, obviously biological, at surprisingly high temperatures. The notes on this, from page 62 of field book L-IV for June 20, 1965, read in part as follows: "Pool A. N.E. of The Diamond. White Creek drainage.... Effluent at 82°C has pink gelatinous stringy (organism?). In strong flow.... Micr. exam of pinkish material. Long thin *Vitreoscilla*-like filaments, wavy, attached more or less to a central core of similar fils. Very heavy growth. Definitely living...." [*Vitreoscilla* was a gliding organism that I had made some observations on at Friday Harbor.]

Using the assays we had worked out, we found that although the pink material had considerable protein, there was no chlorophyll, although chlorophyll could readily be found in samples from temperatures below 70°C. I became convinced that the pink material was definitely bacterial and that bacteria, but not phototrophs, were living at temperatures near boiling (see also below). These observations led me to commit early to the idea of what I later called extreme thermophiles (hyperthermophiles).

The state of research on thermophilic bacteria at that time is exemplified by the following quote: "By incubating enrichment media at very high temperatures (e.g. 55 or 60°C), cultures of thermophilic bacteria can be obtained" (47). Note what was meant here by high temperature. I have never been fond of the enrichment culture technique for research in microbial ecology, and this quotation provides one reason why. The study of microbes directly in the natural habitat led to the discovery of extreme thermophiles. A reliance on enrichment culture techniques and standard incubation temperatures of 55°C had caused investigators working up to that time to miss them.

In the fall of 1965 I wrote up a research proposal for the NSF on the Yellowstone work, emphasizing the cyanobacteria research and the possibility

Figure 3 Getting ready for the first Yellowstone research trip, summer 1965, in my Indiana University laboratory, Jordan Hall Room 361. Three women who were to make important contributions to the Yellowstone project: Sally Murphy (foreground), Pat Holleman (left rear), and Louise Brock.

of using thermal springs as model ecosystems. This proposal was funded, and I began serious work in Yellowstone in June 1966.

The early Yellowstone research was done in temporary laboratory facilities we set up in a rented cabin. Most of this work is described in detail (21). What does not come across well in this book is my gradual realization of the broad practical implications of what I was doing. Initially I had viewed the Yellowstone hot springs primarily as model ecosystems for studying basic questions of microbial ecology. However, as I began to publish papers, I received queries from other scientists, especially biochemists, who had different interests and agendas. Because of my broad interests, I always encouraged such inquiries and arranged for samples, cultures, and even housing accommodations and research space in our Yellowstone project. After we had established the laboratory at West Yellowstone, Montana (see Figure 14.1 in Reference 21), these arrangements became not only easier, but one of my goals, as I viewed the laboratory as a generalized research facility. Also, as the years went by, I involved more of my graduate students and postdoctoral workers in the Yellowstone work, as well as key laboratory technicians and undergraduate students. The first graduate student to work on the project, and one of the most productive, was Bill Doemel (Figure 4).

Discovery of Thermus aquaticus

Most of the early work in Yellowstone focused on measurements of photosynthesis in organisms from the thermal mats and on the temperature optima of the various populations, but from the beginning I attempted to study bacteria in the higher-temperature regions where the photosynthetic organisms were absent. Hudson Freeze was an honors undergraduate student who was with us in Yellowstone the summer of 1966, and he was interested in a project he could do for his honors thesis the following year. I suggested that he try to culture the pink bacteria from Pool A (Octopus Spring). Just before leaving at the end of the summer, we collected pink bacteria as well as mat samples from several other sources, including the outflow channel of Mushroom Spring, which had a temperature of about 69°C.

I knew from my work on marine and freshwater microbiology that culture media for aquatic microbes should not be too rich in organic constituents, so we developed a medium using synthetic salts to which we added small amounts of tryptone and yeast extract. After inoculation, the cultures were incubated at 70°C in a water bath. Although we did not succeed in culturing the pink bacteria [they are still uncultivated, although their molecular phylogeny is known (45)], extensive growth of yellow-pigmented bacteria occurred with the 69°C sample from Mushroom Spring. The mat here had been very thin, near the upper temperature limit for photosynthesis, so we used a small bit of the underlying siliceous sinter (to which the organisms adhered) for the inocu-

Figure 4 Two pioneer participants in the Yellowstone project, Nancy and Bill Doemel, summer of 1967. They were actually on their honeymoon. Nancy served as a field assistant for most of Bill's early work. She later worked on my textbook, *Biology of Microorganisms*.

lum. *Thermus aquaticus* itself presumably represented only a very small amount of the microbes in the inoculum. By October 1966, Freeze had his first culture, a strain he designated YT-1. Subsequently, I isolated quite a few more strains from other sources, both in Yellowstone and in thermal areas in other parts of the world (as well as from hot water heaters!).

The work on *T. aquaticus* continued sporadically over the next two years, much of it done by Pat Holleman, a technician in my laboratory who had done extensive work on a variety of my research problems. (Pat Holleman was one of the most important people in my laboratory for several years, as can be seen by credits in various papers. She worked on yeast mating, *L. mucor,* and various Yellowstone projects from about 1964–1969. When her husband was sent to Vietnam in 1968–1969, she lived with Louise and me for about a year, both in Bloomington and West Yellowstone.) Freeze's undergraduate course load meant that his work on *T. aquaticus* tended to come in fits and starts. However, he was responsible for all of the DNA base composition work, as well as the measurements of growth rates at various temperatures, and I ended up doing the bulk of the taxonomic work (35).

Most of the thermophilic bacteria that had been described by earlier researchers were spore-forming bacteria, whereas the new bacterium was definitely not a spore-former. In deciding on a name, I did an extensive survey of the literature of thermophilic bacteria. The name I first selected was *Caldobacter trichogenes,* but sometime after the first draft was typed I changed the name to *Thermus aquaticus.* I do not remember why. (What would Taq polymerase be called if the original name, *Caldobacter trichogenes,* had been used for this organism?)

At the time that the paper on *Thermus aquaticus* was being written, I also deposited representative cultures of the organism in the American Type Culture Collection, in Washington, DC. Among these cultures was YT-1 (ATCC 25104), which later became the source of Taq endonuclease (46) and Taq polymerase for the polymerase chain reaction (PCR). Cultures of YT-1 were passed around by Richard Roberts of Cold Spring Harbor as sources of endonuclease. Soon, *T. aquaticus* was eliciting considerable interest among microbiologists, biochemists, and molecular biologists. Throughout the 1970s, the number of papers on this bacterium continued to increase, and when PCR was reported in the mid 1980s, the reports became even more prolific. By the end of 1990 over 1000 papers on *T. aquaticus* had been published.

Bacteria in Boiling Water

Although I was convinced in 1965 that bacteria were living at much higher temperatures than had been previously suspected, I had still not realized that they were living in boiling water (i.e. at temperatures near 100°C). My ideas had been based only on observations in outflow channels with visible accumulations (such as the pink bacteria at temperatures above 80°C in Octopus Spring). However, near the end of the summer of 1967, I started using an immersion slide technique and realized that although there were no visible accumulations in the boiling pools themselves, bacteria were present in virtually every pool I examined. Although the immersion slide technique had been widely used in microbial ecology, it took me quite a while to get around to using it in boiling springs.

Because of the importance of the immersion slide technique for the discovery of bacteria in boiling water, the history of its use in Yellowstone boiling springs warrants some discussion. I had been invited by the Society of Applied Bacteriology (UK) to participate in a symposium on extreme environments that was to take place in Belfast in July 1967. I decided to present some material on high-temperature bacteria (32), so Louise and I attempted to obtain incorporation of ^{14}C glucose into the pink bacteria to determine their temperature optimum, as we had been doing with the phototrophs. However, incorporation of the radioisotope was virtually nil, leading me to hypothesize that the visible accumulations of pink bacteria might be moribund. If so, it seemed possible that good incorporation would be possible if actively growing populations were used. To get such populations, we immersed pieces of string directly in the outflow channel, as well as in the source pool itself. We got good growth on the string in the channel, but what was really exciting was that I also saw small amounts of pink material on the string from the source pool, organisms that had obviously grown on the string during the immersion period. These observations suggested for the first time that bacteria were growing in the source pool itself, which had a temperature of over 90°C. Obviously, organisms would have been seen microscopically on the string after a shorter incubation time, so I decided to immerse microscope slides directly in the pool. Microscopic observations of a slide one day after immersion in Octopus Spring showed large numbers of filaments throughout the slide. I took photomicrographs, one of which was subsequently used in my 1967 *Science* paper (see below).

Obviously, these organisms were growing rapidly in these high-temperature systems. The upper temperature limit for bacterial growth was obviously not in the 80°C range, as I had thought, and could conceivably be much higher.

The discovery of the wide distribution of bacteria in boiling springs came about more or less by chance. On August 11, 1967, I had a visit from Anne and Jerry Mosser, who were then graduate students at Rockefeller University and were on a honeymoon trip (Jerry had been an honors undergraduate student in my laboratory in the early 1960s). When I had visitors, I usually tried to show them some interesting areas, but I also liked to do some science. What better than to carry out a widespread slide study in boiling and superheated pools? The experiment was simple: Tie one or two slides to a piece of string, drop the slides in the pool, and tie the other end of the string to a log, rock, or nail. We spent a pleasant day wandering over the Lower Geyser Basin, dropping slides in boiling pools (Figure 5).

On Sunday, August 20, all the slides were removed and returned to the laboratory for microscopy. Virtually every slide, from every boiling or superheated pool, had heavy bacterial growth readily visible microscopically. In some cases, the density was so heavy that the slides had a film visible to the naked eye. The following summer (1968) Tom Bott did an outstanding job of proving not only that growth occurred, but of actually measuring growth rates (2).

"Life at High Temperatures": the Science Paper

In early 1967, the editors of *Science* asked me to write a lead article "in the general area of microbial ecology." By this time the Yellowstone work occupied most of my thoughts, so I decided to write the paper "Life at High Temperatures," with an emphasis on my own work in Yellowstone. In fact, I

Figure 5 The first study of bacterial growth in superheated pools, using the immersion slide technique. A Mosser and TD Brock at Steep Cone, August 11, 1967.

turned eagerly to this task, and although the editor's deadline was not until June 1, I completed the article and submitted it by April 24, 1967. I had not yet discovered bacteria in boiling water (described above), but by the time the edited version was returned to me in mid August, I had already begun the extensive slide study of bacteria living in boiling and superheated pools. Thus, I had much stronger evidence for life in boiling water and published in this paper photomicrographs of organisms from the immersion slides.

This article, published in the fall of 1967 (18), elicited a lot of interest and led to several fruitful collaborations, of which the most significant was that with Mercedes Edwards, an electron microscopist at the State Laboratory of Public Health, Albany, New York. Mercedes and I worked jointly on the fine structure of several of the high-temperature bacteria, and her stunning electron micrographs greatly enhanced several publications as well as lectures that I was now giving around the world.

Travels in Search of Hot Springs

I had always loved to travel, and now I found myself in the exciting position of having valid scientific reasons for going to exotic places. Most of the geothermal regions of the world are in interesting areas: Italy, Iceland, New Zealand, Japan, Central America, the Caribbean. Over the years 1966–1972 I visited all these areas, some of them more than once.

The initial rationale for visiting such regions was to find boiling springs at lower altitudes than Yellowstone. Because of its altitude, water boils at about

92°C there. I had found bacteria in many Yellowstone boiling springs but wondered, was 92°C the upper temperature limit, or would bacteria be present at even higher temperatures? Especially in Iceland and New Zealand, the temperatures in low-altitude boiling springs range up to 100°C or a little higher in superheated waters. I found bacteria in virtually every boiling spring of neutral or alkaline pH in these areas, thus extending to a somewhat higher value the upper temperature limit for life.

All of these trips were part of other travels. In the case of Iceland, I had begun studying the new volcanic island Surtsey (see above), and return visits there (four in all) enabled me to visit springs on the mainland. My New Zealand visit was part of an extended trip prompted by an invitation to attend an international symposium on cyanobacteria in India. My Italian visits were combined with marine work I was doing on *L. mucor*.

At the time of my visits to Iceland and New Zealand, I found that the local microbiologists had neither knowledge nor curiosity about the biology of their thermal habitats. In fact, I published the first work from these countries (31, 33). Later, after the biotechnology industry had discovered thermophiles, the local scientists got interested, and both countries now have active research groups (40, 44). However, little of the present work in these regions is ecological. Mainly, it involves the study of cultures obtained from natural samples.

Deep Sea Vents

For many years my Yellowstone work seemed somewhat exotic to many microbiologists, perhaps because of the presumed restricted distribution of hot springs. This attitude changed after the discovery of the deep sea vents, with their very high temperatures and the associated diverse and flourishing life forms. After the discovery of deep sea vents in the late 1970s, my Yellowstone work took on broader significance; it not only legitimized hypotheses that microbes might be present in some of these high-temperature systems, it also provided the essential foundation for studies on the microbiology of thermal vents. The techniques and principles that we had developed for proving that bacteria live and reproduce in boiling water could be applied to the deep sea habitats. Furthermore, working in Yellowstone was much easier and cheaper, so even those interested primarily in marine thermal vents also visited Yellowstone.

Microbial Prospecting in Yellowstone

Although it was clear when I ended my project in 1975 that the Yellowstone hot springs contained many interesting organisms, it was not until the advent of PCR that widespread attention really focused on thermophiles. Not only has the biotechnology industry discovered Yellowstone, but the National Park Service itself has finally realized that Yellowstone has more of biological interest than grizzly bears and lodgepole pines (42, 43). Dozens of research groups now have permits to collect microbial samples in the park. Perhaps some of these groups will find valuable organisms, just as we found *T. aquaticus,* which led to the discovery of Taq polymerase. Never before has industry profited directly from living creatures taken from a national park, and the Yellowstone administrators are concerned about whether the Park itself should participate in the largess. "When you see the money that's being made," says Yellowstone research chief John Varley, "that's hard for a starving bureaucrat to overlook" (43).[1]

Yellowstone, of course, has no monopoly on thermophiles, but it provides the most accessible location where a wide variety of thermal habitats are available.

Biology of Microorganisms

One of the most satisfying activities of my career has been the success of my textbook, *Biology of Microorganisms (BOM)* (20). I signed the contract for this book with Prentice-Hall (P-H) on January 11, 1967. Although I had already published two books (8, 14), they were both modest in scope, whereas the new book was to be a major effort, with a lot of money riding on the outcome. I had always been interested in writing, and with the broad knowledge that I had acquired since leaving graduate school, I felt capable of tackling the challenge of a major textbook. Another significant reason was the potential financial reward of a successful textbook. I had a mortgage on a house, and as a depression-era baby, I had a strong aversion to debt. A decent book would enable me to pay off this mortgage, the only debt I owed. (This proved to be true.)

From early 1967 until the fall of 1969, I spent many weekend and early-morning hours working on this text. I did the first outline on long winter evenings at Old Faithful at the end of January 1967, while I was participating in a winter field expedition. I wrote several important chapters early in the morning at West Yellowstone, while waiting for students or technicians to assemble for a day of field work.

---

[1]This issue, coupled with the use of Taq polymerase in DNA testing, has seemed to galvanize the news media. In addition to the articles cited, news reports have appeared in *Audubon Magazine* ("The Microbe Miners," December 1994), *Gannett News Service* ("Yellowstone's Living Lab," October 10, 1994), Los Angeles *Times* ("Simpson Case Boosts Microbe Conservation," August 31, 1994), *Genetic Engineering News* ("Biotech Finds Yellowstone National Park a Thermophilic Microbe Hotbed," March 15, 1994), *High Country News* ("Firms Milk Park's Wildlife," December 27, 1993), *Billings Gazette* ("Tiny Treasures; Yellowstone Secrets," December 5, 1993), *Wisconsin State Journal* ("Hot Water Bacterium; Fight over Tiny Critter Looms," May 24, 1993), London *Times* ("Going to Extremes," August 14, 1993), *Milwaukee Journal* ("Heat-Loving Bacterium Roils Two Worlds," May 9, 1993). There have also been news reports on national television.

Because P-H was investing a lot of money in the book, they were prepared to do a handsome production job, with state-of-the-art design, typography, printing, etc. The first edition of *BOM* was intended primarily for the so-called nonmajors market, which would have required lots of practical material, but I chose to emphasize general concepts instead. As P-H also published *The Microbial World* by Stanier, Doudoroff & Adelberg (46a), it was important that the two books not compete. Despite these limitations, my book proved successful, and I soon considered a new edition. However, by the time I began working on the second edition, in 1972, P-H had concluded (erroneously, as it turned out) that their other textbook was not going to be revised. Consequently, the editor, Chester Lucido, asked me to raise the level of my book so that it would serve the Stanier et al market. I happily took this opportunity to cover the material at a level I was more comfortable with. Three subsequent editions, as well as translations into Japanese and Spanish, all did very well throughout the 1970s and early 1980s. Then about 1986, I convinced P-H to permit me to completely redesign the book using new full-color art. Thus, the fifth edition of *BOM,* published in 1988, became the first full-color microbiology textbook on the market. By this time I had acquired an outstanding coauthor, Michael Madigan, who took over most of the writing for the revision for the sixth edition (1991), while I concentrated on the technical aspects of book production. The book, now in its seventh edition (1994), continues to be successful.

Beyond Yellowstone

At a cocktail party at the American Society for Microbiology in May 1970, someone asked me where I was from. "Indiana University," I answered. "Are you with Brock?" he queried. "No," I countered, "I *am* Brock." The conversation then ended in confusion, but it was at that point that I knew I had "arrived."

The title of my book *Principles of Microbial Ecology* was the first to address this field, and the book itself was published just before the big interest in environmental awareness. (The first US Earth Day was in the spring of 1969.) For me, the wave of enthusiasm for ecology culminated in the offer of a chair professorship (EB Fred Professor of Natural Sciences) at the University of Wisconsin-Madison (UW) in 1971. This professorship meant a lot to me, as it validated everything I had done up to that time. I had always been attracted to UW and had actually applied for admission as an undergraduate when I was discharged from the Navy in 1946. (I was denied admission because I was a nonresident.) The UW position was very attractive, and although I owed a lot to IU (which among other things had provided the seed money to start my West Yellowstone research laboratory), things had changed there since 1960, and there was no good reason for me to stay. As Jacob Henle once said, the

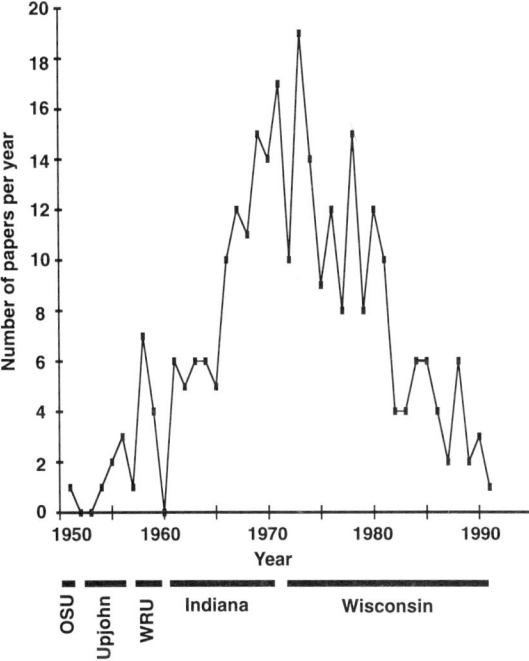

Figure 6 The trajectory of a career. The number of papers and/or books published by TD Brock each year.

only time a professor feels really independent is when he receives a call from another university. Fortunately, I was able to transfer my Yellowstone operation in its entirety. At about the same time, my personal life changed, as I married Katherine Middleton in February 1971. I never regretted the move to Madison, as I found myself in one of the top bacteriology departments in the country.

The Madison years were the most productive of my career (Figure 6). However, within a few years I started to think of other research programs besides Yellowstone. I had always been fascinated by lakes, and limnology in particular had appealed to me since my vacations in northern Ontario. The Yellowstone work continued at Wisconsin for five more years, and in fact many important studies were carried out in that time, including especially the work on the genera *Sulfolobus* and *Chloroflexus*. During this period, I had another excellent technician, Charlene Knaack. Nevertheless, I had always been fascinated with lakes, and the Wisconsin lakes kept beckoning. Also, I thought I should do something for the taxpayers of Wisconsin, who were, after all, paying part of my salary. Finally, in the fall of 1975 I called it quits in Yellowstone and initiated a major study on cyanobacterial populations of Lake Mendota. This was fun work, and I learned lots about limnology. I also learned lots about computer programming and began to develop computer models for natural microbial populations. The work dealt not only with Lake Mendota, but with a variety of lakes throughout Wisconsin. Several major PhD theses resulted, including those of Tim Parkin, Bob Fallon, and Carlos Pedros-Alio, and the excellent postdoctoral work of Al Konopka. Eventually, the Lake Mendota work was published in a book (25) and I moved on to other things.

The work I did on lakes never elicited the excitement in others that my Yellowstone research had aroused. In contrast, the Yellowstone work continued to find interest, partly because of the work of Carl Woese on microbial evolution (50). As it turned out, several of the organisms we had discovered in high-temperature systems fell into Woese's classification as *Archaebacteria* (*Archea*). The fact that we had cultured and described these organisms made it possible, I believe, for Woese to quickly extend the *Archaebacteria* concept. Soon, other laboratories, especially the German laboratories of Wolfgang Zillig and Karl Stetter, were isolating large numbers of new extreme thermophiles, most of which were *Archaebacteria* (48). Some of this work was motivated by the discovery of the deep sea vents.

In the subsequent years, especially because the archaebacterial concept and PCR have elicited wide interest in thermophiles, I have been called upon frequently to discuss my Yellowstone work. Now, I can do it only in a historical context, but as I have become primarily a historian, this has not been difficult.

History of Microbiology

I never had a history course in college, which is perhaps why I have found history fascinating. In the 1980s I returned to the old love from my Kalamazoo days, but this time in a more involved way. My book *Milestones of Microbiology* was an early effort, done when I knew very little of microbiology itself and almost no history of the field. In retrospect, it is amazing that the book is as good as it is. I had never, however, attempted to pass myself off as a historian of microbiology. (The Preface of *Milestones* begins: "This is not a history of microbiology.")

In the late 1970s, as I felt the years passing by, I became more interested in doing the "real" history of microbiology. Historical research is a respectable scholarly activity, and something I could do without research grants, students, or technicians. I was blessed with excellent library facilities at UW. My first book, a biography of Robert Koch, made use of the German-language skills I had acquired in Kalamazoo (26). Surprisingly, there had never been a real biography of Koch in English, and my book was well received. My second book, the history of bacterial genetics (27), was more of a stretch, but the weeks I spent at Cold Spring Harbor in the summer of 1959 paid off, and this book was also received favorably. I was surprised, though, that my department did not view this work as valid research, and I was pressured to retire.

In recent years, I have returned to one of my early loves, local history, combining my expertise in book publishing with my feel for history, producing on a pro bono basis several publications for Historic Madison, Inc. In addition, I have become the unofficial historian of the Village of Shorewood Hills, where I have made my home since moving to Madison.

Computers

My interest in computers arose initially from my writing activities, but when I became involved in lakes research I also used computers in the analysis of data. I initially began with the UW mainframe computers, but when the microcomputer became available in the late 1970s, I became heavily involved with these fascinating tools. Beginning with the Apple II, then moving on to CP/M-based systems and finally to the IBM-PC, I have spent most of the 1980s and early 1990s sitting in front of a computer screen. I published numerous articles in microcomputing magazines (22, 23), learned programming languages, and wrote extensive programs. At one time in the early 1980s, I thought that I might give up microbiology and become a computer scientist, but I decided that the computer is simply a tool, in my case useful primarily for writing and publishing scientific books.

Publishing

My publishing activities arose out of my interest in computers. When I realized that a word processor file created on a microcomputer could be transferred to a typesetter and used to set type, I realized that publishing had entered a revolution. Unable to convince Prentice-Hall to use computers (this was around 1980), I decided to go into publishing myself. Taking the bull by the horns, I not only wrote a book and set the type (using a program called T_EX that was available on the UW mainframe computer), but I actually printed and published the book (24), setting up a company for marketing and sales. I chose a topic from my microbial ecology work that was sufficiently broad to engender a substantial market for the book. During this period, I was also a full-time faculty member, so my publishing activities were carried out at nights and on weekends. In this endeavor I was greatly assisted by my wife Kathie, who actually became the principal operating officer of the company.

This first book was successful and led to more. Eventually, I found myself retired from the university and running a company with its own building and nine employees. In addition to publishing our own books, we also handled production for other publishers. Out of approximately 60 titles, the most extensive project was the design, copyediting, and production of the second edition of *The Prokaryotes,* a four-volume enormity that consumed all of 1990–1991. (I am the only person in existence who has read *every* chapter in this immense undertaking!)

However, the most important part of my publishing activities is the knowledge I have gained about book design and production, printing, graphic arts, and related areas. It was with this knowledge, along with the track record I had already established, that I convinced Prentice-Hall to allow me to redesign my textbook, *Biology of Microorganisms,* as a full-color book. The ensuing success was certainly one of the most satisfying experiences of my career and topped off a very fulfilling life.

Final Words

Modern scientific research is almost always a collective activity of many people. Unfortunately, space limitations have kept me from crediting all those who have made important contributions to my own research. I list them in Table 1 (students) and Table 2 (postdoctoral associates). Technicians, except for those specifically mentioned above, are credited in specific research papers.

In reviewing my life and career, I can see that my broad interests have carried me into quite a few fields. This has been both a strength and a weakness—a strength, as many of my innovations have come from applying ideas from one field to another; a weakness, because I have had to work hard to become accepted by the workers in each field. I have occasionally thought how nice it would have been to have specialized in one narrow area (e.g. mycoplasmatology or methanogenesis). I could have published in only one or a few journals, gone to one type of meeting, interacted with a just handful of laboratories around the world. Collegiality stands for a lot in science, and it takes quite a while to build up a reputation in a given field. As it was, every time I switched arenas I had to qualify myself with a new set of peers. Peer groups often operate like closed corporations, and working one's way in is difficult, sometimes impossible.

I made a list of the various groups to which at one time or another I contributed. The count was 32! Many of my colleagues were completely unaware of my associates or even activities in other fields.

Certain key choices have led me to where I am today. Some of them were made with serious deliberation, whereas some of the most important were made casually. However, the things I have not done account for where I am today as much as what I have done. These negative decisions do not end up in a narrative such as this, but I know that I came within a whisker, several times,

Table 1 Students receiving the PhD or whose work led to published papers

Student name	Dates in lab	Research topic
Allen, Stephen D	1966–1967	Temperature optima of intestinal microflora (Master's)
Bauld, John	1970–1972	*Chloroflexus* mats
Bland, Judith	1970–1971	*Leucothrix* as an algal epiphyte
Clyne, Jenny	1981–1982	Bacterial utilization of algal excretion products (Master's)
Crandall, Marjorie	1965–1967	Biochemical basis of mating in yeast
Davie, Joseph M	1963–1965	Bacteriocine/hemolysin of group D streptococci
Delmer (Pierson), Deborah	1962–1963	Bacteriocines of group D streptococci (undergraduate)
Doemel, William	1967–1970	Physiological ecology of *Cyanidium*
Entenmann (Cook), Susan	1974	*Sulfolobus* ecology (undergraduate)
Fallon, Robert	1976–1980	Cyanobacteria in Lake Mendota
Fliermans, Carl	1969–1971	Ecology of hot acid soils
Freeze, Hudson	1966–1969	*Thermus aquaticus*/thermostable aldolase (undergraduate)
Gustafson, John	1975–1976	Ferric iron as an electron acceptor by sulfur bacteria (undergraduate)
Herdrich, Gary	1974	Temperature optima of bacteria from cold habitats (undergraduate)
Hoffman, James	1973–1974	Thermal pollution of Madison lakes (undergraduate)
Kelly, Michael	1966–1968	Physiological ecology of *Leucothrix*
Lay, Bibiana	1982	Long-term changes in phytoplankton in Wisconsin lakes
Madigan, Michael	1973–1976	Ecology and physiology of *Chloroflexus*
Moo-Penn, Gloria	1961	Amino acid transport in group D streptococci (undergraduate)
Mosser, Jerry	1962–1964	*Streptococcus* phage/*Leucothrix* (undergraduate)
O'Dea, Katherine	1976	Ferrous sulfide as a redox agent for anaerobes (undergraduate)
Parkin, Timothy	1976–1980	Ecology of phototrophic bacteria in lakes
Passman, Fred	1969–1970	Ecology of cold springs (undergraduate)
Pedros-Alio, Carlos	1977–1980	Bacterioplankton ecology
Peterson, Sandra	1973–1975	*Chloroflexus*/halophilic bacteria (undergraduate)
Ray, Paul	1967–1970	Lipids and membranes of *Thermus aquaticus*
Remington, Patrick	1973–1975	Microbial mats in Yellowstone (undergraduate)
Smith, David	1970–1972	Water relations of *Cyanidium* in hot acid soils
Vidaver, Anne	1962–1966	Biochemistry of phage receptor of *Streptococcus faecium*
Ward, David	1972–1975	Hydrocarbon decomposition in lakes
Weiss, Richard	1970–1971	Fine structure of *Sulfolobus*
Weller, Donald	1974	Stromatolitic structures of *Phormidium* (undergraduate)
Zeikus, J Gregory	1969–1970	Protein synthesis at high temperature
Zinder, Stephen	1974–1977	Organic sulfur compounds in thermal and aquatic environments

Table 2 Postdoctoral associates

Name	Dates in lab	Research topic
Belly, Robert	1970–1973	*Thermoplasma* ecology; acidothermophiles; coal mining ecology
Bohlool, Ben	1972–1973	*Sulfolobus* and *Thermoplasma* ecology
Bott, Thomas	1968–1969	Growth rates of bacteria in boiling springs
Boylen, Charles	1971–1972	Firehole River ecology
Darland, Gary	1969–1970	Acidothermophile ecology; *Thermoplasma*
Doemel, William	1973–1975	Growth and decomposition of microbial mats
Ingvorsen, Kjeld	1979–1980	Sulfate reduction in Lake Mendota
Johnson, Roy	1963–1964	*Streptococcus* phage
Konopka, Alan	1975–1977	Cyanobacteria ecology in lakes
Leon, Shalom	1965–1966	Interaction of streptomycin with ribosomes
Lynn, Raymond	1968	*Zygogonium* in acidothermophilic habitats
Middleton (Brock), Katherine	1970–1971	*Sulfolobus* and *Thermoplasma* ecology
Mosser, Jerry	1970–1974	Yellowstone ecology; ecology of high-alpine habitats
Shivvers, Douglas	1971–1972	Metabolism of sulfur by *Sulfolobus*
Tansey, Michael	1970–1971	Thermophilic fungi
Watson, Vicki	1981–1983	Modeling cyanobacterial populations in Lake Mendota
Zehnder, Alexander	1978–1980	Anaerobic methane oxidation
Zeikus, J Gregory	1970	Firehole River ecology

of going off in what would have been a professionally fatal direction. As always, neither hard work nor brilliance can substitute for good luck.

The recent interest in Taq polymerase and *T. aquaticus* has brought my earlier work to the attention of a new generation of scientists and the public. The managers of Yellowstone National Park have called on me once more to provide new material for their interpretive programs, and in 1994 I wrote a work called *Life at High Temperatures,* a small book in which I could use some of the now historic color photographs that I had taken over the years (29). It was indeed a pleasure to revisit scenes of former research activities and to think again about those exciting days in the late 1960s when every study we carried out in Yellowstone led to new discoveries. I have also developed an interest in the history of Yellowstone Park itself and have begun a major collection of early reports, photographs, paper ephemera, and guidebooks relating to the history and development of the Park. Who knows—eventually this collection may lead to another book on Yellowstone!

The Road to Yellowstone was never straight, but it was almost never bumpy, and the view was marvelous all the way.

Any *Annual Review* chapter, as well as any article cited in an *Annual Review* chapter, may be purchased from the Annual Reviews Preprints and Reprints service.
1-800-347-8007; 415-259-5017; email: arpr@class.org

Literature Cited[2]

1. Bland JA, Brock TD. 1973. The marine bacterium *Leucothrix mucor* as an algal epiphyte. *Marine Biology* 23:283–92
2. Bott TL, Brock TD. 1969. Bacterial growth rates above 90°C in Yellowstone hot springs. *Science* 164:1411–12
3. Brock TD. 1951. Studies on the nutrition of *Morchella esculenta* Fries. *Mycologia* 43:402–22
4. Brock TD. 1955. Paw Paw versus the railroads. *Michigan History* 39(June):129–82 plus map; 1955. Addendum, Vol. 39(September)
5. Brock TD. 1956. Lipid synthesis in *Hansenula anomala*. *Mycologia* 48:337–44
6. Brock TD. 1956. Studies on the mode of action of novobiocin. *J. Bacteriol.* 72:320–23
7. Brock TD. 1959. Biochemical basis of mating in yeasts. *Science* 129:960–61
8. Brock TD. 1961. *Milestones in Microbiology*. Englewood Cliffs, NJ: Prentice-Hall
9. Brock TD. 1961. Physiology of the conjugation process in the yeast *Hansenula wingei*. *J. Gen. Microbiol.* 26:487–98
10. Brock TD. 1964. Host range of certain virulent and temperate bacteriophages attacking group D streptococci. *J. Bacteriol.* 88:165–71
11. Brock TD. 1964. Knots in *Leucothrix mucor*. *Science* 144:870–72
12. Brock TD. 1965. Biochemical and cellular changes occurring during conjugation in *Hansenula wingei*. *J. Bacteriol.* 90:1019–25
13. Brock TD. 1965. The purification and characterization of an intracellular sex-specific mannan protein from yeast. *Proc. Natl. Acad. Sci. USA* 54:1104–12
14. Brock TD. 1966. *Principles of Microbial Ecology*. Englewood Cliffs, NJ: Prentice-Hall
15. Brock TD. 1966. Streptomycin. In *Biochemical Studies of Antimicrobial Drugs. 16th Symposium Society for General Microbiology*, pp. 131–68. Cambridge: Cambridge Univ. Press
16. Brock TD. 1966. The habitat of *Leucothrix mucor*, a widespread marine microorganism. *Limnol. Oceanogr.* 11:303–7
17. Brock TD. 1967. Bacterial growth rate in the sea: direct analysis by thymidine autoradiography. *Science* 155:81–83
18. Brock TD. 1967. Life at high temperatures. *Science* 158:1012–19
19. Brock TD. 1967. The ecosystem and the steady state. *BioScience* 17:166–69
20. Brock TD. 1970. *Biology of Microorganisms*. Englewood Cliffs, NJ: Prentice-Hall
21. Brock TD. 1978. *Thermophilic Microorganisms and Life at High Temperatures*. New York: Springer-Verlag. 465 pp.
22. Brock TD. 1980. Hard copy for apple graphics. *Microcomputing* November:100–2
23. Brock TD. 1980. The microcomputer in microbiology. *ASM News* 46:171–73
24. Brock TD. 1983. *Membrane Filtration: a User's Guide and Reference Manual*. Madison, WI: Science Tech/Heidelberg: Springer-Verlag. 381 pp.
25. Brock TD. 1985. *A Eutrophic Lake: Lake Mendota, Wisconsin*. New York: Springer-Verlag
26. Brock TD. 1988. *Robert Koch. A Life in Bacteriology and Medicine*. Madison, WI: Science Tech/Heidelberg: Springer-Verlag
27. Brock TD. 1990. *The Emergence of Bacterial Genetics*. Cold Spring Harbor, NY: Cold Spring Harbor Lab.
28. Brock TD. 1992. Research on thermophiles: a memoir. In *Thermophiles: Science and Technology, Int. Conf., Reykjavik, Iceland, 23–26 August 1992*, pp. A3–A12. Reykjavik: IceTec
29. Brock TD. 1994. *Life at High Temperatures*. Yellowstone Natl. Park, WY: Yellowstone Assoc.
30. Brock TD, Brock ML. 1959. Similarity in mode of action of chloramphenicol and erythromycin. *Biochim. Biophys. Acta* 33:274–75
31. Brock TD, Brock ML. 1966. Temperature optima for algal development in Yellowstone and Iceland hot springs. *Nature* 209:733–34
32. Brock TD, Brock ML. 1968. Relationship between environmental temperature and optimum temperature of bacteria along a hot spring thermal gradient. *J. Appl. Bacteriol.* 31:54–58
33. Brock TD, Brock ML. 1971. Microbio-

---

[2] Because of space limitations, my own publications in this bibliography are incomplete. A copy of the complete one will be sent to any interested reader.

logical studies of thermal habitats on the central volcanic region, North Island, New Zealand. *NZ J. Mar. Freshwater Res.* 5:233–57
34. Brock TD, Davie JM. 1963. Probable identity of a group D hemolysin with a bacteriocine. *J. Bacteriol.* 86:708–12
35. Brock TD, Freeze H. 1969. *Thermus aquaticus* gen. n. and sp. n., a nonsporulating extreme thermophile. *J. Bacteriol.* 98:289–97
36. Brock TD, Peacher B, Pierson D. 1963. Survey of the bacteriocines of enterococci. *J. Bacteriol.* 86:702–7
37. Brock TD, Wooley SO. 1963. Streptomycin as an antiviral agent: mode of action. *Science* 141:1065–67
38. Brock TD, Wooley SO. 1964. Glycylglycine uptake in streptococci and a possible role of peptides in amino acid transport. *Arch. Biochem. Biophys.* 105:51–57
39. Crandall MA, Brock TD. 1968. Molecular aspects of specific cell contact. *Science* 161:473–75
40. Geirsdottir AM, Brown HP, Skjenstad T. 1992. *Thermophiles: Science and Technology. Int. Conf., Reykjavik, Iceland, 23–26 August 1992.* p. 112. Reykjavik: IceTech
41. Kelly MT, Brock TD. 1969. Physiological ecology of *Leucothrix mucor. J. Gen. Microbiol.* 59:153–62
42. Milstein M. 1994. There's gold in them thermophiles. *Outside* August:22
43. Milstein M. 1994. Yellowstone managers eye profits from hot microbes. *Science* 264:655
44. Reeve JN. 1994. Thermophiles in New Zealand. *ASM News* 60:541–45
45. Reysenbach A-L, Wickham GS, Pace NR. 1994. Phylogenetic analysis of the hyperthermophilic pink filament community in Octopus Spring, Yellowstone National Park. *Appl. Environ. Microbiol.* 60:2113–19
46. Sata S, Hutchison CA, Harris JI. 1977. A thermostable sequence-specific endo nuclease from *Thermus aquaticus. Proc. Natl. Acad. Sci. USA* 74:542–46
46a. Stanier RY, Doudoroff M, Adelberg EA. 1957. *The Microbial World.* Englewood Cliffs, NJ: Prentice-Hall
47. Stanier RY, Doudoroff M, Adelberg EA. 1957. See Ref. 46a, p. 294
48. Stetter KO. 1986. Diversity of extremely thermophilic archaebacteria. In *Thermophiles. General, Molecular, and Applied Microbiology,* ed. TD Brock, pp. 39–74. New York: Wiley
49. Vidaver AK, Brock TD. 1966. Purification and properties of a bacteriophage receptor material from *Streptococcus faecium. Biochim. Biophys. Acta* 121:298–314
50. Woese CR, Wolfe RS, eds. 1985. *Archaebacteria.* New York: Academic

MECHANISMS FOR THE PREVENTION OF DAMAGE TO DNA IN SPORES OF *BACILLUS* SPECIES

Peter Setlow

Department of Biochemistry, University of Connecticut Health Center, Farmington, Connecticut 06030

KEY WORDS: UV resistance, heat resistance, DNA-binding proteins, UV photochemistry, DNA depurination

CONTENTS

INTRODUCTION	30
ENVIRONMENT OF SPORE DNA	33
SPORE RESISTANCE TO OXIDIZING AGENTS, HEAT, AND DESICCATION	35
Mechanisms of Spore Killing by Heat and Oxidizing Agents	35
Resistance to Oxidizing Agents	37
Resistance to Heat	37
Resistance to Desiccation	38
SPORE RADIATION RESISTANCE	38
Resistance to UV Radiation	38
Resistance to γ Radiation	40
DNA REPAIR DURING SPORE GERMINATION	40
α/β-TYPE SASPs	41
DNA DAMAGE IN α⁻β⁻ SPORES	43
Resistance of α⁻β⁻ Spores to UV Radiation	44
Resistance of α⁻β⁻ Spores to Heat	45
Resistance of α⁻β⁻ Spores to Oxidizing Agents	46
Resistance of α⁻β⁻ Spores to Desiccation	46
INTERACTION OF α/β-TYPE SASPs WITH DNA IN VITRO	47
Structure of the α/β-Type SASP–DNA Complex	47
Effect of α/β-Type SASPs on DNA Reactivity	48
CONCLUSIONS	49

ABSTRACT

The DNA in dormant spores of *Bacillus subtilis* as well as other *Bacillus* species is extremely well protected against damage resulting from treatments such as desiccation, heat, oxidizing agents, and UV and γ radiation. This high degree of DNA protection is a major factor in the survival of spores of these species, not only when subjected to the treatments noted above, but also when incubated under common environmental conditions for many years. Factors that play major roles in overall spore resistance include the low permeability of spores to toxic chemicals and the decreased spore-core water content. However, although decreased spore permeability and water content appear to at least partially protect spore DNA from oxidative damage, these factors seem to play little or no role in protecting spore DNA from heat damage. The major factor preventing damage to spore DNA is the saturation of this DNA with a novel group of small, acid-soluble proteins of the α/β-type whose binding greatly alters DNA's chemical and enzymatic reactivity as well as its UV photochemistry. Binding of these proteins is also a key factor in spore DNA resistance to desiccation, heat, oxidizing agents, and UV radiation.

INTRODUCTION

When growing cells of *Bacillus* or *Clostridium* species are starved for certain nutrients, these cells initiate the process of sporulation in which a growing cell is converted into a dormant spore. At the end of a cell's growth cycle, at which point it contains two nucleoids (termed Stage 0 of sporulation) (Figure 1*a*), a landmark morphological event in sporulation begins: the formation of an asymmetric septum that divides the sporulating cell into large and small daughter cells, each of which contains a complete chromosome (Figure 1*b*). The small cell, termed the forespore, is then engulfed by the larger cell, termed the mother cell (Figure 1*c*), and is eventually enclosed within the mother cell (Figure 1*d*). After a series of biochemical events including addition of several external layers to the forespore (see below) the mother cell begins to lyse (Figure 1*e*), eventually releasing the free spore (Figure 1*f*) into the environment. The structure of the spore differs significantly from that of a growing cell (Figure 2), consisting of several novel layers. From the outside in, these layers include the exosporium, which has not been too well characterized; the proteinaceous spore coats; the cortex composed of peptidoglycan with a structure similar but not identical to that of cell wall peptidoglycan; and finally the central core, the site of most spore enzymes, ribosomes, and most importantly for this review, spore DNA (24, 25, 80).

Dormant spores of *Bacillus* and *Clostridium* species are specialized cells that have undergone these numerous changes during sporulation to ensure spore survival in the absence of exogenous nutrients for extremely long periods. Good data are available indicating that spores can remain viable for decades and possibly more than a century, and some findings have been interpreted indicating that spores can survive for thousands and possibly mil-

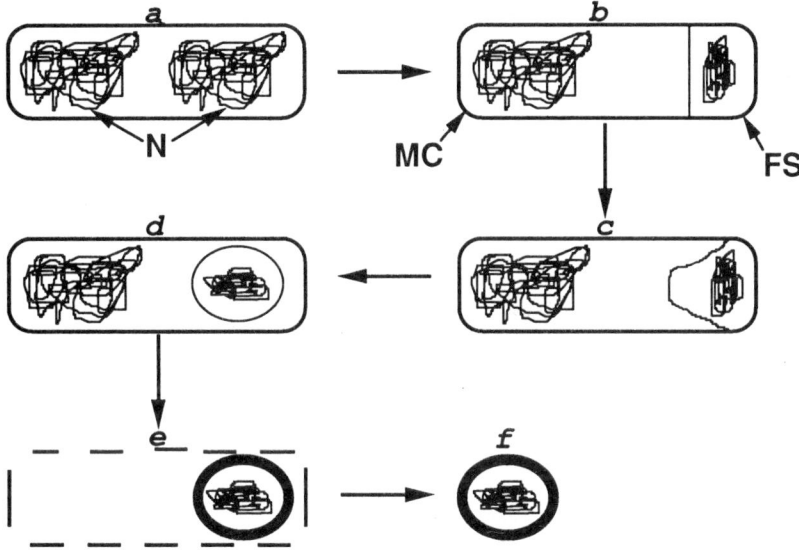

Figure 1 Changes in cell and nucleoid structure during sporulation of a *Bacillus* species. The cell, which has just entered sporulation (*a*), contains two completed nucleoids (N) or chromosomes. (*b*) An asymmetric septum has formed separating the larger mother cell (MC) and smaller forespore (FS)—note the more condensed nucleoid in the forespore. (*c*) The mother cell is beginning to engulf the forespore, and (*d*) engulfment is complete. (*e*) Several layers have been added to the outer surface of the forespore (hence the much thicker outline); the mother cell is beginning to lyse; and its nucleoid has been degraded. (*f*) The mature free spore has been released from the mother cell.

lions of years (36, 84). Obviously no controlled studies of spore survival over the latter long periods of time can be done, but spore populations from several species exhibit survival frequencies up to 75% during storage in water for periods of several years (18, 84). Clearly there must be many adaptations that allow spore survival over long periods of time. The most important effect of these adaptations must be to ensure the survival of spore DNA in undamaged form. A spore might easily survive the loss of 99% of any single enzyme, but spores usually have only one, and rarely two, genomes (7, 77). Consequently, significant DNA damage accumulated during dormancy could lead to potentially lethal mutations. Although DNA is a rather stable molecule chemically, it is not completely so and is subject to significant rates of both oxidative damage and depurination, either of which can result in mutagenic or lethal events (1, 2, 34, 37).

Normally growing cells minimize the effects of these types of DNA damage through the action of various DNA repair pathways. However, one important adaptation involved in long-term spore survival is that spores are metabolically

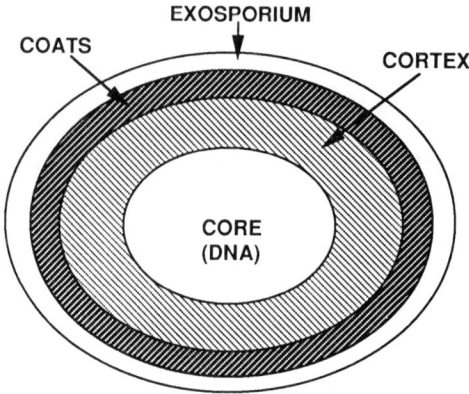

Figure 2 Schematic structure of the spore. The various layers of the mature spore are not drawn completely to scale. Note that the exosporium is hardly visible in spores of some species, that the spore coats contain at least two distinct layers that can be visualized in the electron microscope, and that the spore's DNA is in the core (24, 80). The spore's membranes are not shown. However, a complete membrane separates the cortex and the core, and a second membrane, which may or may not be complete, lies between the cortex and the coats (24, 25).

dormant and therefore carry out no detectable catabolism of exogenous or endogenous metabolites (82). Consequently, spores of both *Bacillus* and *Clostridium* species do not have significant levels of the common high-energy compounds found in growing cells, such as nucleoside triphosphates, reduced pyridine nucleotides, and acyl-coenzyme A molecules (82). Although metabolic dormancy may be essential for spore survival in the absence of exogenous nutrients, the absence of compounds such as nucleoside triphosphates precludes the possibility of rapid repair of most types of DNA damage in the dormant spore. Furthermore, there is good evidence that DNA repair in dormant spores is also precluded by the virtual inactivity of all enzymes present in the spore core, which is also the site of spore DNA (Figure 2) (82).

The precise mechanism(s) causing the lack of activity of spore-core enzymes is not completely clear, but a major contributor appears to be the extremely low water content in the spore core (25). Although the spore core is not anhydrous, estimates of its water content range from 0.3 to 0.7 g H_2O/g dry wt depending on the species, in contrast to values of 3 to 4 g H_2O/g dry wt in growing cells (25). Indeed, a recent theoretical analysis led to the suggestion that the spore core is in a glassy state (62).

Because of the apparent impossibility of DNA repair during their potentially long periods of dormancy, spores have only two possible methods for minimizing the deleterious effects of DNA damage: (*a*) ensure rapid repair of any

DNA damage when spores germinate (i.e. return to life) and resume metabolism, rapidly accumulate nucleoside triphosphates, and begin macromolecular biosynthesis, and/or (b) protect dormant spore DNA from damage in the first place. Although evidence supports the operation of at least one special DNA repair process in the first minutes of spore germination (47, 48, 89) (see below), the most important mechanism for ensuring long-term survival of spore DNA appears to be the prevention of DNA damage in the dormant spore (75, 79, 82). Consequently, the latter topic is the major focus of this review. Essentially all data on specific mechanisms of DNA protection in spores have come from studies utilizing *Bacillus subtilis*, which is therefore the main species discussed. However, the mechanisms ensuring DNA protection in dormant *B. subtilis* spores probably also operate in spores of other *Bacillus* species, as well as in spores of *Clostridium* species.

In addition to exhibiting long-term survival, dormant spores are extremely resistant to a variety of environmental stresses that can potentially damage DNA, including heat, radiation, oxidizing agents, and desiccation, which are also discussed in this review. For more detailed reviews on various aspects of spore resistance, including some with information directly applicable to spore DNA resistance, the reader is referred to earlier reviews (9, 22, 25, 43, 59, 75, 79, 80, 82).

ENVIRONMENT OF SPORE DNA

Some major factors involved in preventing DNA damage in dormant spores might well be found in the environment of the spore DNA, i.e. in the spore core (Figure 2). These include such general environmental features as: (a) pH, (b) water content, and (c) small molecules unique to the dormant spore. (a) Studies in several *Bacillus* species including *B. subtilis* have shown that the pH in the core of the dormant spore is 6.5–7.0 and is approximately one unit lower than the value in growing cells (41). Although this decreased spore pH may be important in bringing about spore dormancy late in sporulation (41), no evidence indicates that it plays any significant role in protection of spore DNA (82). Indeed, a lower pH in the spore core might actually increase the rate of DNA depurination, as this process is acid catalyzed (39). (b) As noted above, the spore core has a significantly lower water content than does a growing cell. This decreased spore-core water content plays a major role in spore heat resistance, as there is a good correlation across *Bacillus* species between lower spore-core water content and increased spore heat resistance (25) (see below). Although the spore core has less total water per gram dry weight than does a growing cell, some water certainly remains in the spore core—potentially more than enough to hydrate spore DNA. Unfortunately, no good data are available on either the amount of free water (if any) in the spore

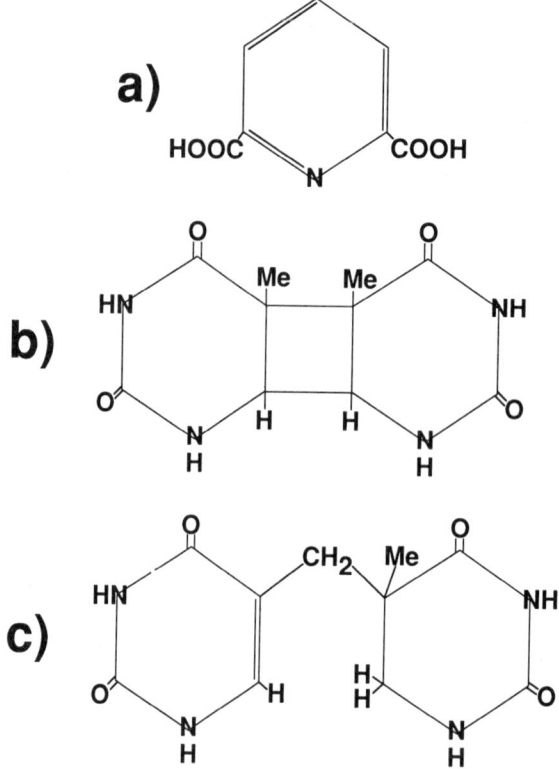

Figure 3 Structure of (*a*) pyridine-2,6-dicarboxylic acid (dipicolinic acid or DPA), (*b*) cyclobutane-type thymine dimer (TT), and (*c*) thyminyl-thymine adduct (spore photoproduct or SP). Me (*b, c*) denotes a methyl group.

core or the distribution of water among the various spore-core components. (*c*) A third general feature of the environment in the spore core is the presence of an enormous depot of pyridine-2,6-dicarboxylic acid [dipicolinic acid (DPA)] (Figure 3*a*), which makes up 5–10% of the dry weight of bacterial spores. DPA is not present in growing cells and is acquired by the developing spore very late in sporulation, at about the time of acquisition of spore heat resistance (82). DPA probably exists in spores as a 1:1 chelate with divalent cations, predominantly Ca^{2+}, and does not leak out significantly from spores stored in water (25).

Given the novel features of the dormant spore core's environment noted above, it is not surprising that the spore shows very different permeability characteristics than does a growing cell. In particular, the membrane surrounding the spore core (i.e. between the spore core and cortex, Figure 2) seems under normal conditions to be relatively impermeable to molecules >280

Daltons, to hydrophilic molecules, and to charged molecules (26). In addition, both the proteinaceous spore coats and the peptidoglycan cortex, which surround the spore core (Figure 2), also appear to play a role in restricting the access of potentially toxic molecules to the spore core (9). The detailed structural reasons for the greatly decreased permeation of most molecules into and out of the spore core are not clear, but this permeability barrier will protect spore DNA from many otherwise damaging chemicals.

In addition to novel general features of the dormant spore DNA's environment, several novel features are specific to spore DNA. Early in sporulation when the forespore has been generated, the forespore chromosome becomes significantly condensed compared with the chromosome in either the mother cell or growing cells (Figure 1a,b) (66, 77). The mechanism for the forespore chromosome condensation is not yet clear and may simply reflect the small size of the forespore relative to that of the mother cell. Approximately 2 h after forespore-chromosome condensation, the forespore DNA becomes saturated with a group of small, acid-soluble proteins (SASPs) of the α/β-type; these proteins are named for the two major proteins of this type in *B. subtilis* spores—termed SASP-α and β (23, 75, 76, 78, 80, 81). These novel DNA-binding proteins are synthesized only in the forespore, remain on the spore chromosome in the dormant spore, and play a major role in the protection of spore DNA from many types of damage (76). In the first minutes of spore germination, α/β-type SASPs are rapidly degraded by a sequence-specific protease (76).

SPORE RESISTANCE TO OXIDIZING AGENTS, HEAT, AND DESICCATION

As noted above, wild-type spores not only exhibit long-term survival in water under "normal" environmental conditions, they are also extremely resistant to killing by a variety of harsh conditions. Representative data for *B. subtilis* given in Table 1 indicate the tremendous heat resistance, the significant resistance to oxidizing agents and UV and γ radiation, and the essentially complete desiccation resistance provided by conversion of a cell to a dormant spore. Furthermore, *B. subtilis* is by no means among the most resistant of spore formers. Indeed, D values (times for killing of 90% of the population) for spores of species such as *Bacillus stearothermophilus* may be several orders of magnitude higher than for spores of *B. subtilis* upon both heat and hydrogen peroxide treatment (25, 83). Spores of some *Clostridium* species are even more heat resistant than are those of *B. stearothermophilus* (25, 59).

Mechanisms of Spore Killing by Heat and Oxidizing Agents

Whereas UV and γ radiation kill spores by causing DNA damage, the precise killing mechanism(s) of heat and oxidizing agents is less clear, although the

Table 1 Resistance of cells and spores of *Bacillus subtilis* to various treatments[a]

Treatment	Survival-%	
	Log-phase cells	Dormant spores
Heat, 85°C for 30 min[b]	$<10^{-4}$	79
H_2O_2, 4 M for 20 min	$<10^{-6}$	64
Freeze-drying, once	2	100[c]
UV radiation, 100 J/m^2	0.3	50
γ-radiation, 100 krads	1	28

[a] Data are for spores or cells treated in water or 25 mM KPO_4 (pH 7.4) and are from References 18, 19, 31, 71, and 79.
[b] Spores were heated in water.
[c] Spores also survived six cycles of freeze-drying and rehydration with no detectable killing.

weight of available evidence suggests that for spores in water this mechanism is not DNA damage. Analysis of the survivors of hydrogen peroxide treatment of *B. subtilis* spores revealed no increase in mutation frequency, and no DNA damage, as evidenced by a lack of single-strand breaks in spore DNA (71). In contrast, electron microscopic analyses of *B. subtilis* spores treated with hydrogen peroxide showed that extensive degradation of outer spore layers (i.e. coats and cortex, Figure 2) accompanied spore killing (83). Detailed analyses of spore populations suffering 90–99% heat-induced mortality in water also revealed neither increased mutations nor significant DNA damage indicated by either single-strand breaks or abasic sites (18, 73). Differential scanning calorimetry, used to study heat-induced spore killing, has led to the suggestion that the target for heat killing of spores in water is either a spore protein or spore membranes and not DNA (8). In contrast to these results, another report suggests that heat killing of *B. subtilis* spores in water results in significant mutagenesis; surprisingly, the resultant mutations appeared to be almost exclusively glycine auxotrophs (35). Although these results were not obtained in a second study (18), in some spores the gly region of the chromosome may be exquisitely sensitive to heat damage.

Several reports (15, 53, 91; B Setlow & P Setlow, unpublished data) have also shown that the killing of dry spores by heat causes a variety of mutations, although DNA damage under these conditions has not been directly examined. The damage to DNA caused by heating dry spores may be DNA depurination, and the rates of depurination of DNA in dry spores at high temperatures (145–155°C) were reported to be similar to those in purified DNA in vitro (28, 53, 91). Unfortunately, these suggestions have never been directly tested. However, for hydrated *B. subtilis* spores, the rate of DNA depurination is at least 20-fold slower than for DNA in solution (73). Because spores of some species are much more heat resistant than are *B. subtilis* spores, DNA in hydrated spores must be extremely well protected against heat damage, in particular against depurination.

Resistance to Oxidizing Agents

Apparently, two main environmental factors aid in protecting spore DNA from oxidative damage: (*a*) decreased spore permeability to oxidizing agents and (*b*) decreased spore water content. Certainly for many large oxidizing agents, their access to the DNA in the spore core is restricted, and these agents may cause spore killing by breaking down external spore layers long before they cause DNA damage (9, 83). However, evidence indicates that oxidizing agents such as hydrogen peroxide (or possibly the hydroxyl radical derived from hydrogen peroxide) can gain access to the core of spores of certain *B. subtilis* mutants (see below) and kill the spores, at least in part by DNA damage (71). Consequently, decreased spore permeability to oxidizing agents, while a factor in protecting spore DNA from such agents, cannot be the sole protective mechanism. The decreased spore water content may also decrease the rate of attack of oxidizing agents on DNA, and there are some, albeit limited, data suggesting that a decreased spore water content is associated not only with increased spore heat resistance, but also increased resistance to hydrogen peroxide (83). However, because the available evidence suggests that hydrogen peroxide does not kill wild-type spores by DNA damage (71), whether the low spore-core water content plays a direct role in the resistance of spore DNA to oxidative damage remains unknown. Data also suggest that the low spore-core water content does not play a significant role in protecting spore DNA from heat damage (18) (see below).

Another mechanism contributing to the protection of spore DNA from oxidizing agents such as hydrogen peroxide may be the absence of reduced pyridine nucleotides in the spore, as this would prevent continuous production of damaging free radicals such as the hydroxyl radical via the Fenton reaction (1, 34). Indeed, prevention of hydroxyl-radical formation from hydrogen peroxide with either metal ion chelators or thiols significantly protects spores against hydrogen peroxide (43, 83).

Resistance to Heat

Spore heat resistance is acquired during sporulation approximately in parallel with dehydration of the spore core and uptake of DPA (25, 80). The major, although certainly not the only, factor involved in spore heat resistance is the decreased spore-core water content; spores of different species that have decreasing spore-core water contents exhibit increasing levels of heat resistance (25). In addition, two otherwise isogenic strains of *B. subtilis* that differ in only one locus exhibit markedly different spore water contents; the more hydrated spores are significantly more heat sensitive (D Popham & P Setlow, unpublished data). Although the low spore-core water content is clearly a major factor in spore heat resistance, whether it is a major factor in protection of spore DNA from heat damage, primarily depurination, is not known. The amounts of spore-core water are low, yet a significant amount of water remains, and one would expect that this water could cause DNA depurination. In addition, as noted above, DNA is probably not a major target for the heat killing of spores in water. Thus, the decreased spore-core water content probably plays only a minor or an extremely indirect role in protection of the DNA of hydrated spores from damage caused by heat (see below).

Resistance to Desiccation

As noted in Table 1, *B. subtilis* spores are essentially completely resistant to repeated freeze-drying/rehydration cycles, in contrast to vegetative cells, which are killed by such processes (19). Similar results have been obtained with other *Bacillus* species (59). Because freeze-drying of bacterial cells can cause DNA damage and mutations (5, 10, 11, 90), freeze-drying can kill cells at least in part by means of DNA damage. The extreme resistance of spores to freeze-drying indicates that spore DNA must be protected against deleterious effects of this process, although how freeze-drying causes DNA damage in vivo is not clear. Unfortunately, there have been no thorough systematic studies of the effects of general spore environmental factors on spore desiccation resistance.

SPORE RADIATION RESISTANCE

Resistance to UV Radiation

Spores of *Bacillus* species are generally 7–50 times more resistant to UV radiation (260 nm) than are growing cells of the same species (75, 79, 82) (Table 1). As DNA is the target for UV radiation, UV damage to spore DNA must either be prevented in some fashion or be rapidly repaired in the first minutes of spore germination. The developing forespore acquires resistance to UV radiation about 1–2 h before acquisition of heat resistance (and spore-core dehydration) and much of the spore's resistance to hydrogen peroxide (82). In contrast to UV's effects on growing cells, in which the major photoproducts are cyclobutane-type pyrimidine dimers residing predominantly between adjacent thymine residues on the same DNA strand (TT) (Figure 3b), UV irradiation of dormant spores generates no detectable TT and probably no cyclobutane-type cytosine-thymine dimers (CT) (Table 2) (16). Rather, the major DNA photoproduct detected in UV-irradiated dormant spores is a 5-thyminyl-5,6-dihydrothymine adduct (Figure 3c), initially termed spore photoproduct (SP) (17, 88) (Table 2); SP is also formed between adjacent thymine residues on the same DNA strand. The amounts of SP formed in spores or TT

Table 2 UV photoproducts from DNA with or without α/β-type SASP irradiated in vivo or in vitro[a]

	% of total thymine in vivo[b]			
	Wild-type		α-β-	
Sample irradiated	TT	SP	TT	SP
Dormant *B. subtilis* spores	<0.1	21	2.9	8.8
Log-phase *B. subtilis* cells	4.8	<0.1	4.7	0.1
	% of total cytidine or thymine in vitro[c]			
	plus α/β-type SASP		without α/β-type SASP	
DNA irradiated	PyPy[d]	SP	PyPy[d]	SP
plasmid pUC19	<0.1	3.1	5.4	0.3
poly(dAdG)-poly(dCdT)	<0.2	e	3.8	e
poly(dG)-poly(dC)	<0.1	e	2.2	e

[a] Data are from References 20, 50, and 68.
[b] Irradiation was with 2.5 kJ/m^2.
[c] Irradiation was with 10 kJ/m^2, and DNA was irradiated in aqueous solution.
[d] PyPy denotes total cyclobutane-type pyrimidine dimers, including TT, CT, and dimers between adjacent cytosine residues.
[e] SP only forms between adjacent thymine residues in DNA.

formed in cells as a function of UV fluence are similar, and SP is a potentially lethal photoproduct (17, 75). However, spores have an SP-specific repair system that rapidly and efficiently repairs this lesion in the first minutes of spore germination (see below) (47, 48, 89). The explanation for the increased UV resistance of bacterial spores is thus to be found in the SP-specific repair system and in the mechanism(s) that changes the UV photochemistry of spore DNA from TT to SP formation. Indeed, spores in which mutations result in significant TT formation upon UV irradiation are actually less UV resistant than are growing cells (44, 79).

Although UV irradiation of purified DNA in solution generates predominantly TT and little SP (Table 2), SP formation is increased and TT formation decreased when UV irradiation is carried out with poorly hydrated or dry DNA (54, 57). This change in spore DNA photochemistry with hydration thus seems consistent with the decreased water content of the spore core (25). However, when poorly hydrated or dry DNA is irradiated with UV, significant TT is still generated (54, 57). Thus spore-core dehydration cannot be the sole cause of the change in spore DNA's UV photochemistry. In addition, because the acquisition of spore UV resistance and the change in spore DNA photochemistry take place 1–2 h before spore-core dehydration and DPA accumulation (which take place approximately in parallel and between *d* and *e* in Figure 1), as well as in mutants blocked in spore-core dehydration, spore-core dehydration and DPA appear to play no major role in spore UV resistance or the change in spore DNA photochemistry (6, 64, 75).

One feature of spore UV resistance that has often been somewhat puzzling is that when UV resistance is first acquired by forespores during sporulation, the forespores are significantly more UV resistant than are dormant spores (6, 27). Similarly, in the first minutes of spore germination, spore UV resistance rises dramatically until falling to the lower UV resistance of the growing cell (85). Studies have shown that in forespores or germinated spores with higher UV resistance than the dormant spore, the yield of SP decreases as a function of UV fluence compared with that in the dormant spore—as if DNA in dormant spores is sensitized to UV (27, 85). Studies in vivo and in vitro have strongly suggested that this photosensitizer is DPA, which is accumulated by developing forespores 1–2 h after acquisition of UV resistance, and whose excretion is one of the earliest events in spore germination (6, 72). DPA also sensitizes SP formation upon UV irradiation of DNA in vitro (72). Although DPA may sensitize DNA for SP formation in spores, as noted above DPA itself does not change spore DNA photochemistry from TT to SP formation, nor does this change result from spore-core dehydration. What then causes this change? The culprit appears to be α/β-type SASPs, which are synthesized in parallel with the acquisition of spore UV resistance and which saturate the spore chromosome and markedly change its UV photochemistry as well as its chemical reactivity (79, 82) (see below).

Resistance to γ Radiation

As is the case with UV radiation, spores are generally more resistant to γ radiation than are their vegetative cell counterparts, although these differences tend to be smaller than for UV radiation or heat resistance (Table 1) (22, 82). Generally γ-radiation resistance is acquired during sporulation just slightly before acquisition of spore heat resistance. However, to date no definitive correlation has been made between acquisition of spore γ-radiation resistance and any biochemical or physiological change in the spore. Although the decreased water content of the spore core ought to protect against γ radiation, no studies have been done to measure the γ-radiation resistance of spores with different core water contents. Similarly, the role of other spore components in γ-radiation resistance has not been systematically evaluated, although α/β-type SASPs are not involved in γ-radiation resistance (31). Given our almost complete lack of real knowledge of the mechanisms protecting spore DNA from γ radiation, this may be a fruitful area for future study.

DNA REPAIR DURING SPORE GERMINATION

As noted above, efficient SP repair is crucial for spore UV resistance, as SP is a potentially lethal photoproduct. SP repair is carried out during spore germination and outgrowth by either of two DNA repair systems: (*a*) excision repair, which works not only on SP but also on TT, and (*b*) SP-specific repair (75). Excision repair of SP during spore germination appears to be identical to excision repair operating in growing cells, with the same gene products involved in both stages of growth. However, spores of *uvrA* mutants of *B. subtilis* exhibit at most a twofold decrease in UV resistance compared to wild-type spores, indicating the presence of another pathway for SP repair (21, 70). This pathway is mediated by the *spl* locus, which encodes a protein that probably monomerizes SP in DNA back to two thymine residues (21, 89). This process does not require metabolic energy or light, although Spl has some sequence homology to DNA photolyases (21, 89). In contrast to the components of the excision-repair pathway that are present throughout all stages of growth, *spl* transcription appears restricted to the developing forespore late in sporulation, and Spl synthesis is coregulated with that of α/β-type SASP (55, 75). Strikingly, *B. subtilis* strains with mutations in the *spl* gene produce spores with a seven- to eightfold decrease in UV resistance (70). However, *uvrA spl* double mutants of *B. subtilis* produce spores that are ~40-fold less UV resistant than are wild-type spores, indicating the potential importance of the excision-repair pathway in SP repair (70).

The importance of DNA repair in spore UV resistance suggests that this mechanism could also be important in resistance to treatments such as heat or oxidizing agents. However, as DNA damage does not appear to be a major killing mechanism for heat- or oxidizing agent–treatment of spores in water, repair of DNA damage early in spore germination may not be important after all. Nevertheless, in some spores, particularly those in which mutations have rendered the spores sensitive to killing from DNA damage induced by heat or oxidizing agents (see below), DNA repair might become an important factor in spore survival. Although one study failed to detect significant DNA repair in spores that were heat killed in large part through DNA damage (B Setlow & P Setlow, unpublished data), another showed that spores with mutations in DNA repair genes are slightly more heat sensitive than are their wild-type counterparts (32). The differences in the *D* values between mutant and wild-type spores in this latter study were only approximately twofold, but reexamination of these mutants—particularly in combination with mutations rendering spore DNA sensitive to heat or oxidative damage—might be worthwhile.

α/β-TYPE SASPs

One of the most striking novel features of the DNA in dormant spores of *Bacillus* and *Clostridium* species is that the spore chromosome is saturated with a group of α/β-type SASPs (76, 77, 80, 81). These almost identical proteins are synthesized only in the developing forespore, in parallel with the

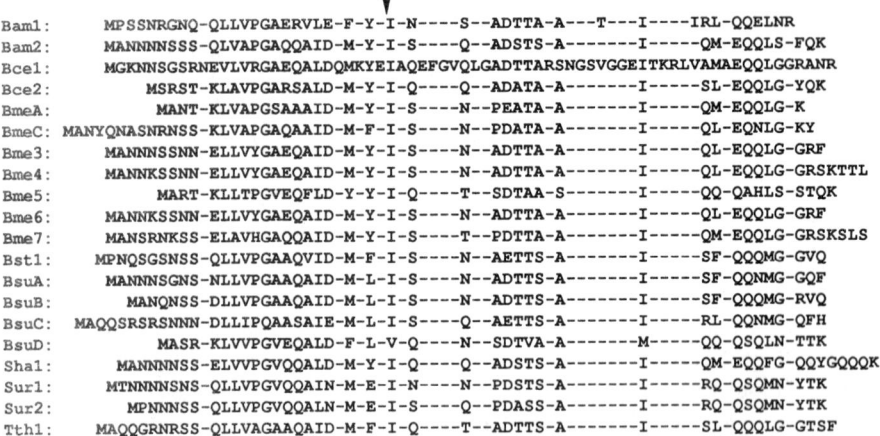

Figure 4 Comparison of amino acid sequences of α/β-type SASPs from different *Bacillus* species, as well as from closely related genera (40, 42, 76; C Loshon, EZ Grey, ML Santiago-Lara & P Setlow, unpublished data). Amino acids are given in the one letter code. At positions denoted by dashes the residue present is identical to that in Bce1. Note that the amino-terminal methionine residue is absent in the mature protein. The vertical arrow denotes the site of cleavage of α/β-type SASPs by the sequence-specific protease, GPR. The species represented are: Bam, *Bacillus aminovorans*; Bce, *Bacillus cereus*; Bme, *Bacillus megaterium*; Bst, *B. stearothermophilus*; Bsu, *B. subtilis*; Sha, *Sporosarcina halophila*; Sur, *Sporosarcina ureae*; and Tth, "*Thermoactinomyces thalpophilus.*"

acquisition of spore UV resistance and a significant component of spore resistance to hydrogen peroxide but 1–2 h before acquisition of spore heat resistance. They are coded for by a multigene family consisting of at least seven members in each *Bacillus* species, although two proteins generally are expressed at high levels while the rest are synthesized at much lower levels. The genes coding for α/β-type SASPs (termed *ssp*) are monocistronic, scattered on the chromosome, and transcribed by an RNA polymerase containing a forespore-specific σ-factor. The α/β-type SASPs are small proteins (60–75 residues) whose sequence is tremendously conserved both within a single species and across various *Bacillus* species (Figure 4). However, the amino acid sequences of α/β-type SASPs show no significant sequence similarity to any other protein in available databases.

Immunoelectron microscopy has localized α/β-type SASPs to the forespore nucleoid; when these proteins are expressed in *Escherichia coli,* they also appear to be associated essentially exclusively with the cell's nucleoid (23, 65). The α/β-type SASPs have been shown to be DNA-binding proteins in vitro (74), and the absence of the two major α/β-type SASPs results in *B.*

subtilis spores with dramatically altered DNA properties (see below); hence, these proteins almost certainly directly interact with DNA in the spore. Indeed, cross-linking studies with intact spores have demonstrated that α/β-type SASPs are directly associated with the spore's DNA (67). As the α/β-type SASPs in spores are sufficient to saturate the spore chromosome (based on in vitro binding stoichiometries), it seems likely that essentially the whole chromosome is covered with these proteins. However, the interesting question of the fate of other chromosomal DNA-binding proteins such as HU in spores has not been answered. The genes for several α/β-type SASPs, including those for the two major proteins, SASP-α and β, have been cloned from *B. subtilis*. This has allowed construction of strains with deletions in the genes coding for both SASP-α and β (termed α⁻β⁻ strains) or with deletions in only one of these genes. These strains sporulate essentially normally, and the α⁻β⁻ spores lack both SASP-α and β (~75% of the α/β-type SASP pool). These α⁻β⁻ spores exhibit greatly decreased survival and resistance properties as compared with wild-type spores (see below).

Although α/β-type SASPs are accumulated to high levels (~5% of total spore protein) late in *B. subtilis* sporulation and are stable in the dormant spore, they are rapidly degraded to amino acids upon initiation of spore germination. The amino acids produced in this process support much protein synthesis early in spore germination. Degradation of α/β-type SASPs is initiated by a sequence-specific protease termed GPR, which begins SASP degradation by cleaving at one site in these proteins (arrow, Figure 4). GPR is maintained in the spore in an inactive form and acts only upon initiation of germination (76). Knockout mutations in the *gpr* gene have been constructed; although spores of these *gpr* mutants trigger germination normally, degradation of α/β-type SASPs is extremely slow during their germination (60).

DNA DAMAGE IN α⁻β⁻ SPORES

As noted above, the cloning of the genes coding for SASP-α and β in *B. subtilis* has allowed the construction of strains with deletions in one or both of the *ssp* genes coding for major α/β-type SASP (44). Detailed analysis of these mutant spores has shown that loss of SASP-α results in increased spore levels of SASP-β, and vice versa. However, in spores of the α⁻β⁻ strain, levels of the normally minor α/β-type SASP are not increased. Strikingly, α⁻β⁻ spores are much more sensitive than wild-type spores to UV radiation, heat, oxidizing agents, and freeze-drying, although the spores' γ-radiation resistance is the same (Table 3). As expected, the α⁻β⁻ mutations have no effect on the phenotype of growing cells. The decreased resistance of α⁻β⁻ spores to these various treatments can be restored to wild-type or near wild-type levels if the α⁻β⁻ spores are provided with sufficient levels of any wild-type α/β-type SASP

Table 3 Resistance and mutagenesis of wild-type and $\alpha^-\beta^-$ spores of *B. subtilis* upon various treatments[a]

	Dose to kill 90% of spore population		Mutants upon killing to 0.01–10% survival (%)[b]	
Treatment	Wild-type	α-β	Wild-type	α-β
UV radiation	325 J/m^2	25 J/m^2	—	—
Heat, 85°C[c]	320 min	14 min	<0.2	13
Heat, 65°C[c]	105 h	10 h	—	—
Heat, 22°C[c]	2.5 years	2.8 mo	<0.4	18
H$_2$O$_2$, 4 M	40 min	11 min	<0.5	10–15
NaOCl, 1%	33 min	13 min	<0.5	4
Freeze-drying	>20 times	3–4 times	[d]	14
γ Radiation	180 krads	180 krads	—	—

[a] Data are from References 18, 19, 31, 65, 71, 79, and M-ZH Sabli & WM Waites, unpublished data.
[b] Mutants are auxotrophs, asporogenous, or have grossly altered colony morphology.
[c] Spores were heated in water.
[d] Note that freeze-drying gave no killing of wild-type spores.

(75). This protein can be either SASP-α or β, a normally minor *B. subtilis* α/β-type SASP, or an α/β-type SASP from another *Bacillus* species. However, α/β-type SASPs that no longer bind DNA in vitro because of changes in highly conserved amino acid residues cannot restore resistance to $\alpha^-\beta^-$ spores (14, 87). These results indicate that α/β-type SASPs play a key role in many aspects of spore resistance. This role appears to be a direct one because (*a*) the spore-core water content in $\alpha^-\beta^-$ spores is identical to that in wild-type spores (31), and (*b*) killing of $\alpha^-\beta^-$ spores by a variety of agents is accompanied by much DNA damage, as described below.

Resistance of $\alpha^-\beta^-$ Spores to UV Radiation

$\alpha^-\beta^-$ Spores are actually slightly more UV sensitive than are growing cells (Table 3), and their irradiation generates significant levels of TT, some CT, and reduced amounts of SP (79) (Table 2). The reason for the increased UV sensitivity of $\alpha^-\beta^-$ spores compared to growing cells is probably the spore's depot of DPA, which appears to act as a photosensitizer (see above). Several studies have shown that the generation of cyclobutane-type pyrimidine dimers is the reason for the loss in UV resistance of $\alpha^-\beta^-$ spores (70, 79). The acquisition of spore UV resistance during sporulation, in parallel with acquisition of α/β-type SASPs and 1–2 h before spore-core dehydration, as well as the UV sensitivity of and TT formation in $\alpha^-\beta^-$ spores, indicates that α/β-type SASP binding is the major, if not the sole, cause of the altered UV photochemistry of spore DNA. Indeed, synthesis in *E. coli* of levels of α/β-type SASP sufficient to half-saturate this organism's chromosome also results in reduced TT formation and significant SP formation upon UV irradiation (65). As would be expected, $\alpha^-\beta^-$ spores exhibit no transient increase in UV resistance during spore germination, presumably because TT formation as a function of UV fluence does not change during germination of these mutant spores (69). Similarly, *gpr* mutant spores that lack the protease that initiates α/β-type SASP degradation have a much longer-lasting transient UV-resistant period during spore germination, presumably because UV irradiation continues to generate only SP long after the photosensitizer DPA has been released (60).

Although the causative role for α/β-type binding in promoting SP formation has been conclusively demonstrated both in vivo and in vitro (see below), the precise mechanism whereby α/β-type SASP binding alters the UV photochemistry of DNA is not clear. It was suggested over 30 years ago that DNA in spores might be in an A-like helix and that this structure causes the altered photochemistry of the spore's DNA (85). Unfortunately, the postulated SP structure on which this early hypothesis was based turned out to be incorrect. In addition, further work with purified DNA in vitro showed that SP was readily generated by UV irradiation of DNA that was not in an A-like helix (54). More recent work has suggested that DNA saturated with α/β-type SASPs may indeed be in an A-like helix (see below); however, this structure is not a classical A-like helix and its precise nature has not yet been established.

Resistance of $\alpha^-\beta^-$ Spores to Heat

Although $\alpha^-\beta^-$ spores are much more heat-resistant than growing cells, they are significantly more heat sensitive than wild-type spores when incubated in water at a variety of temperatures (18) (Table 3). The effects at lower temperatures are particularly noteworthy, as they may have been a major selective force driving evolution of α/β-type SASPs. Analysis of the survivors of heat treatment of $\alpha^-\beta^-$ spores at several temperatures has disclosed an extremely high percentage of mutants among the survivors, as well as significant DNA damage (Table 3) (18, 72). The most obvious DNA damage is single-strand breaks (18). Although the precise mechanism generating these breaks has not been established, a significant amount of abasic sites in DNA of $\alpha^-\beta^-$ spores are also generated upon lethal heat treatment; secondary reactions at these abasic sites may produce the single-strand breaks (73). Strikingly, the rate of generation of the abasic sites in $\alpha^-\beta^-$ spores is very similar to the rate of DNA depurination predicted from in vitro measurements of heat treatment. Because $\alpha^-\beta^-$ spores have the same low spore-water content as do wild-type spores (31), these data suggest that the low spore-core water content itself has very little effect on the rate of spore DNA depurination in vivo. The data also suggest that a second major protective effect of α/β-type SASPs on spore DNA is to greatly slow DNA depurination, thus rendering spore DNA relatively resistant to damage by heat. Again, this effect has been directly demonstrated in vitro (18). In contrast, rates of DNA single-strand breakage and depurination in wild-type spores are at least 20-fold lower than in $\alpha^-\beta^-$ spores at comparable levels of killing by heat (18, 73). Again, as noted above, DNA in wild-type spores is extremely well protected by α/β-type SASPs against these types of DNA damage.

Resistance of $\alpha^-\beta^-$ Spores to Oxidizing Agents

Killing by oxidizing agents, either hydrogen peroxide or hypochlorite, is also faster with $\alpha^-\beta^-$ spores than with wild-type spores (Table 3). Normally, spore resistance to hydrogen peroxide is acquired during sporulation in at least two increments—one increment in parallel with α/β-type SASP synthesis, and a second increment 1–2 h later in parallel with the onset of spore-core dehydration, metabolic dormancy, and the dramatic changes in spore permeability (71). In $\alpha^-\beta^-$ strains the first increment of hydrogen peroxide resistance is not acquired, although the second is; consequently, $\alpha^-\beta^-$ spores have much greater hydrogen peroxide resistance than growing cells but significantly less than wild-type spores. Analysis of the wild-type– or $\alpha^-\beta^-$-spore survivors of treatment with hydrogen peroxide or hypochlorite reveals no significant number of mutants among the wild-type–spore survivors, but a high frequency of mutants among the $\alpha^-\beta^-$-spore survivors, including auxotrophic, asporogenous, and colony-morphology mutants (71) (Table 3). The DNA from hydrogen peroxide–killed $\alpha^-\beta^-$ spores also has a high frequency of single-strand breaks, although few if any abasic sites (71, 73). In contrast, DNA from wild-type spores killed to the same extent by hydrogen peroxide contains few if any single strand breaks (71). The mechanism whereby hydrogen peroxide damages DNA in spores remains unknown, but hydrogen peroxide can cause single-strand breaks in DNA in vivo as well as in vitro (34). Taken together, these data suggest that α/β-type SASPs also play a key role in the protection of spore DNA from oxidative damage, which has been substantiated by experiments carried out in vitro (see below).

Resistance of $\alpha^-\beta^-$ Spores to Desiccation

Although spores of wild-type *B. subtilis* are resistant to up to six cycles of freeze-drying and rehydration, $\alpha^-\beta^-$ spores undergo a 30–70% loss in viability during each cycle (19) (Table 3). The precise mechanism of spore killing by freeze-drying is not clear, yet killing of $\alpha^-\beta^-$ spores is accompanied by both DNA damage (single-strand breaks) and mutagenesis (Table 3). Consequently, as found for the other killing agents studied, α/β-type SASPs also provide a significant component of the protection of spore DNA against damage resulting from freeze-drying.

INTERACTION OF α/β-TYPE SASPs WITH DNA IN VITRO

The evidence that α/β-type SASPs significantly protect spore DNA against various treatments has prompted studies on the interaction of α/β-type SASPs with DNA in vitro. These studies have utilized proteins purified either from spores of *Bacillus* or *Clostridium* species, or from *E. coli* overexpressing the products of cloned *ssp* genes (49, 50, 71, 74). In all cases, the general effects of purified α/β-type SASPs on DNA have been essentially identical. As is possibly not surprising given the association of α/β-type SASPs with DNA in vivo, these proteins are nonspecific double-stranded DNA-binding proteins in vitro; they do not bind to single-stranded DNA or to single- or double-stranded RNA (74). The stoichiometry of DNA binding is ~1 protein/5 bp, which is close to the ratio of α/β-type SASP–DNA found in spores. While α/β-type SASPs bind to all natural double-stranded DNAs and saturate them completely, binding is weakest to A+T-rich regions, in particular to runs of As. The binding preference of α/β-type SASP to natural DNAs is reflected in the binding preference for synthetic polydeoxynucleotides: Poly(dG)•poly(dC) is bound most strongly, followed by poly(dGdC)•poly(dGdC) and then natural DNAs such as plasmid pUC19, and poly(dAdT)•poly(dAdT). Poly(dA)•poly(dT) is bound poorly if at all. However, oligo(dA) tracts up to 13 residues long in natural DNAs are covered by α/β-type SASPs. Binding to most DNAs appears rather cooperative, as seen to the greatest extent with poly(dAdT)•poly(dAdT) and to the least extent with poly(dG)•poly(dC) (71). In addition, oligo(dG)$_{12}$•oligo(dC)$_{12}$ is bound, while oligo(dAdT)$_{70}$•oligo(dAdT)$_{70}$ is not. The cooperative binding of α/β-type SASPs to DNA appears to result from protein-protein interactions on the DNA, as the proteins are monomers in solution (76; B Setlow & P Setlow, unpublished data). As might be expected given the extreme sequence conservation of α/β-type SASPs among *Bacillus* species, α/β-type SASPs with alterations in highly conserved residues generally bind poorly, if at all, to DNA (14, 61, 87).

Structure of the α/β-Type SASP–DNA Complex

Analysis of the structure of the α/β-type SASP–DNA complex has shown that both the protein and the DNA undergo significant conformational change upon complex formation. The protein becomes much more highly α-helical and is quite protease resistant in the complex (S Mohr, B Setlow & P Setlow, unpublished data), whereas both circular dichroism and Fourier transform infrared spectroscopy indicate that the DNA adopts an A-like helical structure in the complex (46), which is consistent with the binding preference of these proteins for synthetic polydeoxynucleotides: Poly(dG)•poly(dC) is in or close to an A-like helix in solution; poly(dGdC)•poly(dGdC) and poly(dAdT)

•poly(dAdT) adopt such a structure moderately readily and less readily, respectively, whereas poly(dA)•poly(dT) adopts this structure poorly if at all (3, 4, 30, 33, 45, 52, 63). However, electron microscopic analysis has shown that the length of the DNA helix (i.e. the rise per base pair) does not change appreciably upon α/β-type SASP binding (29). Thus the DNA cannot be in a classical DNA fiber A-like helix when complexed with an α/β-type SASP.

The electron microscopic analysis of α/β-type SASP–DNA complexes also indicated that DNA does not wrap around the protein (29). Instead this analysis suggested that the protein binds on the outside of the DNA helix, probably forming a helix of subunits around the backbone. This latter suggestion is strengthened by the essentially complete protection given to the DNA backbone against cleavage by both chemicals and enzymes upon α/β-type SASP binding (see below). The region of α/β-type SASP that interacts directly with the DNA is thought to be in the carboxyl-terminal half of the protein, including the 13-residue stretch conserved almost exactly throughout evolution (Figure 4) (58). Strikingly, binding of α/β-type SASP does not bend DNA, but rather straightens sequence-directed DNA bends and greatly increases the rigidity of the DNA, as evidenced by an increase in the DNA persistence length of >20-fold (29). The binding of α/β-type SASPs to DNA in vitro also has a striking effect on DNA topology; binding introduces numerous apparent negative supertwists per kilobase into covalently closed DNA (49). The mechanism for this introduction of apparent negative supertwists is not yet completely clear (29), but it operates in vivo as well as in vitro (49, 51).

Effect of α/β-Type SASPs on DNA Reactivity

As noted above, the binding of α/β-type SASPs to DNA in vitro has dramatic effects on the enzymatic and chemical reactivity, as well as the photoreactivity, of DNA. α/β-Type SASP binding blocks cleavage of the DNA backbone by either restriction endonucleases or nonspecific DNases (74). For plasmids from 2.7–4.5 kb in length, this backbone protection extends to the whole molecule, although A+T-rich regions are protected less well than other areas (74). Similar protection is afforded against free radical cleavage of the DNA backbone in vitro, either from hydrogen peroxide, orthophenanthroline-Cu^{2+}, or hydroxyl radicals generated using Fe^{2+}-ascorbate (71, 74). The protection against hydrogen peroxide cleavage of the DNA backbone by α/β-type SASP binding is most notable, as it duplicates in vitro the action of these proteins in vivo (71). Strikingly, purine methylation by dimethylsulfate is not affected by binding of α/β-type SASPs (74), which is also consistent with the majority of the protein's interactions with DNA occurring through the DNA backbone.

Binding of α/β-type SASP also dramatically affects DNA depurination rates in vitro. Studies with several different DNAs have shown that binding of α/β-type SASPs slows DNA depurination by a factor of at least 20 (18).

Because DNA depurination is a major potential cause of DNA damage in wild-type spores and appears to be a predominant form of DNA damage in $\alpha^-\beta^-$ spores (73), the in vitro results again strongly support a key role for α/β-type SASPs in the prevention of this type of DNA damage in vivo.

Finally, binding of α/β-type SASP completely changes the photochemistry of DNA in vitro. Whereas UV irradiation of purified DNA in aqueous solution gives predominantly TT and little if any SP, irradiation of an α/β-type SASP–DNA complex under the same conditions gives essentially no TT and much SP (50) (Table 2). However, the yield of SP from an α/β-type SASP–DNA complex in vitro as a function of UV fluence is approximately sevenfold lower than in spores. This difference appears to result from the action of DPA as a photosensitizer in spores, because addition of DPA to α/β-type SASP–DNA complexes in vitro greatly increases SP yield as a function of UV fluence (72).

In addition to suppression of TT formation, α/β-type SASP binding to DNA suppresses formation of CT dimers, as well as cyclobutane dimers between adjacent cytosine residues and pyrimidine-pyrimidone adduct formation between adjacent pyrimidines (Table 2) (20). However, α/β-type SASPs in which specific mutations have abolished the ability of the mutant protein to bind DNA have no effect on DNA photochemistry in vitro (87). These effects essentially mimic the changes in the UV photochemistry of DNA in dormant spores. As these effects in vivo are mostly abolished upon loss of ~75% of the spore's α/β-type SASPs (i.e. in $\alpha^-\beta^-$ spores), α/β-type SASP binding seems largely, if not completely, responsible for the change in the UV photochemistry of spore DNA. Consequently, α/β-type SASP binding, in conjunction with the SP-repair systems, is sufficient to completely explain the protection of spore DNA against UV, and thus spore UV resistance.

CONCLUSIONS

Clearly, for spores to survive months, years, decades, or even centuries of dormancy, spore DNA must be maintained in an undamaged state—either by prevention of DNA damage during dormancy or by ensuring rapid DNA repair in the first minutes of spore germination. Because spore death upon extended periods at low or high temperatures or upon treatment with oxidizing agents is normally not accompanied by significant DNA damage or mutagenesis, spore DNA must normally be extremely well protected against these types of DNA damage. Thus, heat and oxidizing agents probably kill spores through destruction of molecules other than DNA—presumably proteins and/or membranes. Some of the protection of spore DNA against oxidizing agents appears to result from the spore's relative impermeability to most chemicals as well as the spore's metabolic dormancy. The decreased water content of the spore core may also play a role in protection of spore DNA against oxidizing agents.

Although low spore-core water content plays a major role in overall spore heat resistance, water content does not seem particularly important in protecting spore DNA from damage.

In contrast, a major mechanism protecting spore DNA from oxidative and heat damage is the binding of α/β-type SASPs, which both in vivo and in vitro retards oxidative cleavage of the DNA backbone as well as DNA depurination caused by heat. Presumably in wild-type spores α/β-type SASPs protect DNA so well against damage by these agents that spores are killed only by damage to molecules other than DNA. However, in $\alpha^-\beta^-$ spores, DNA damage can accumulate at temperatures or levels of oxidizing agents that normally do not kill wild-type spores. We suspect that this is also true for damage resulting from freeze-drying; still, little detailed knowledge is available about the mechanism of killing by this treatment.

In UV resistance, the binding of α/β-type SASPs is clearly the major cause of spore UV resistance, as binding of these proteins alone alters spore DNA photochemistry from TT to SP formation. SP is a much less lethal photoproduct because it is rapidly repaired during spore germination, in large part by an SP-specific repair system present only in spores.

The work outlined above has clearly shown that α/β-type SASPs play a major role in the protection of spore DNA from potentially lethal damage by alteration of the chemical and photoreactivity of spore DNA. To fully understand spore DNA protection by these mechanisms, we must understand the structural bases of the alterations in DNA reactivity upon α/β-type SASP binding. The answers to these fundamental questions will most likely be obtained upon the elucidation of a detailed structure of an α/β-type SASP–DNA complex, and this seems an obvious goal for research in the immediate future.

While a detailed understanding of the mechanisms for protection of spore DNA against damage is clearly of interest in itself, the protection of spore DNA from damage over long periods of time may have implications not only for spore survival, but possibly for long-term DNA survival as well. Several recent studies have used the polymerase chain reaction to describe the amplification of DNA fragments from samples many millions of years old (12, 13, 86). Interestingly, in one of these reports the DNA amplified appeared to be from *Bacillus* species that had presumably lived as intestinal microflora in bees 23–40 million years old (12, 13). The definitive identification of this amplified DNA is still controversial; the chemical instability of DNA might preclude its survival in amplifiable form over millions of years (37, 38), although this suggestion has been disputed (56). While this controversy cannot be resolved in this review, it is striking that the DNA in spores of *Bacillus* species is so well protected by α/β-type SASPs against the major chemical actions that destroy DNA—oxidative damage and depurination. Might the DNA amplified from the *Bacillus* species noted above have come from spores originally in the bee's intestinal tract? The logical conclusion of this idea would suggest a truly feasible scenario for a Jurassic Park: one populated only by ancient *Bacillus* (and possibly *Clostridium*) species.

ACKNOWLEDGMENTS

The work in the author's laboratory has been supported by grants from the Army Research Office and the National Institutes of Health (GM-19698). It is a pleasure to acknowledge the many contributors to this work in the author's laboratory: Rosa Martha Cabrera-Martinez, Michael Connors, Everardo Curiel-Quesada, Heather Fairhead, Edward Fliss, Rebecca Hackett, Charles Loshon, Nancy Magill, Wayne Nicholson, David Popham, Jose-Luis Sanchez-Salas, Leticia Santiago-Lara, Barbara Setlow, Dongxu Sun, Michael Sussman, and Federico Tovar-Rojo, as well as collaborators at other institutions: Jack Griffith, Scott Mohr, and William Waites.

> Any *Annual Review* chapter, as well as any article cited in an *Annual Review* chapter, may be purchased from the Annual Reviews Preprints and Reprints service.
> 1-800-347-8007; 415-259-5017; email: arpr@class.org

Literature Cited

1. Ahern H. 1991. Cellular responses to oxidative stress. *Am. Soc. Microbiol. News* 57:627–29
2. Ames BN, Shigenaga MK, Hagen TM. 1993. Oxidants, antioxidants, and the degenerative diseases of aging. *Proc. Natl. Acad. Sci. USA* 90:7915–22
3. Arnott S, Selsing E. 1974. Structures for the polynucleotide complexes poly(dA)•poly(dT) and poly(dT)•poly(dA)•poly(dT). *J. Mol. Biol.* 88:509–21
4. Arnott S, Selsing E. 1974. The structure of polydeoxyguanylic acid•polydeoxycytidylic acid. *J. Mol. Biol.* 88:551–52
5. Ashwood-Smith MJ, Grant E. 1976. Mutation induction in bacteria by freeze-drying. *Cryobiology* 13:206–13
6. Baillie E, Germaine GR, Murrell WG, Ohye DF. 1974. Photoreactivation, photoproduct formation, and deoxyribonucleic acid state in ultraviolet-irradiated sporulating cultures of *Bacillus cereus. J. Bacteriol.* 120:516–23
7. Belliveau BH, Beaman TC, Gerhardt P. 1990. Heat resistance correlated with DNA content in *Bacillus megaterium* spores. *Appl. Environ. Microbiol.* 56:2919–21
8. Belliveau BH, Beaman TC, Pankratz S, Gerhardt P. 1992. Heat killing of bacterial spores analyzed by differential scanning calorimetry. *J. Bacteriol.* 174:4463–74
9. Bloomfield SF, Arthur M. 1994. Mechanisms of inactivation and resistance of spores to chemical biocides. *J. Appl. Bacteriol.* 76:91S–104S
10. Bousfield IJ, Mackenzie AR. 1976. Inactivation of bacteria by freeze drying. In *Inactivaton of Bacteria by Freeze Drying,* ed. FA Skinner, WB Hugo, pp. 329–44. New York: Academic
11. Calcott PH. 1986. Cryopreservation of microorganisms. *Crit. Rev. Biotechnol.* 4:279–97
12. Cano RJ. 1994. *Bacillus* DNA in amber: a window to ancient symbiotic relationships. *Am. Soc. Microbiol. News* 60:129–34
13. Cano RJ, Borucki MK, Higby-Schweitzer M, Poinar HN, Poinar GO Jr, Pollard KJ. 1994. *Bacillus* DNA in fossil bees: an ancient symbiosis? *Appl. Environ. Microbiol.* 60:2164–67
14. Carrillo-Martinez Y, Setlow P. 1994. Properties of *Bacillus subtilis* small, acid-soluble, spore proteins with changes in the sequence recognized by their specific protease. *J. Bacteriol.* 176:5357–63
15. Chiasson LP, Zamenhof S. 1966. Studies on induction of mutations by heat in spores of *Bacillus subtilis. Can. J. Microbiol.* 12:43–46

16. Donnellan JE Jr, Setlow RB. 1965. Thymine photoproducts but not thymine dimers are found in ultraviolet irradiated bacterial spores. *Science* 149:308–10
17. Donnellan JE Jr, Stafford RS. 1968. The ultraviolet photochemistry and photobiology of vegetative cells and spores of *Bacillus megaterium*. *Biophys. J.* 8:17–28
18. Fairhead H, Setlow B, Setlow P. 1993. Prevention of DNA damage in spores and in vitro by small, acid-soluble proteins from *Bacillus* species. *J. Bacteriol.* 175:1367–74
19. Fairhead H, Setlow B, Waites WM, Setlow P. 1994. Small, acid-soluble proteins bound to DNA protect *Bacillus subtilis* spores from killing by freeze-drying. *Appl. Environ. Microbiol.* 60: 2647–49
20. Fairhead H, Setlow P. 1992. Binding of DNA to α/β-type small, acid-soluble proteins from spores of *Bacillus* or *Clostridium* species prevents formation of cytosine dimers, cytosine-thymine dimers and bipyrimidine photoadducts upon ultraviolet irradiation. *J. Bacteriol.* 174:2874–80
21. Fajardo-Cavazos P, Salazar C, Nicholson WL. 1993. Molecular cloning and characterization of the *Bacillus subtilis* spore photoproduct lyase (spl) gene, which is involved in the repair of UV radiation–induced DNA damage during spore germination. *J. Bacteriol.* 175: 1735–44
22. Farkas J. 1994. Tolerance of spores to ionizing radiation: mechanisms of inactivation, injury and repair. *J. Appl. Bacteriol.* 76:81S–90S
23. Francesconi SC, MacAlister TJ, Setlow B, Setlow P. 1988. Immunoelectron microscopic localization of small, acid-soluble spore proteins in sporulating cells of *Bacillus subtilis*. *J. Bacteriol.* 170: 5963–67
24. Gauthier JJ, Tipper DJ. 1972. Structure of the bacterial endospore. See Ref. 31a, pp. 3–12
25. Gerhardt P, Marquis RE. 1989. Spore thermoresistance mechanisms. In *Regulation of Procaryotic Development*, ed. I Smith, R Slepecky, P Setlow, pp. 17–63. Washington DC: Am. Soc. Microbiol.
26. Gerhardt P, Scherrer R, Black SH. 1972. Molecular seiving by dormant spore structures. See Ref. 31a, pp. 68–74
27. Germaine GR, Murrell WG. 1973. Effect of dipicolinic acid on the ultraviolet radiation resistance of *Bacillus cereus* spores. *Photochem. Photobiol.* 17:145–54
28. Greer S, Zamenhof S. 1962. Studies on depurination of DNA by heat. *J. Mol. Biol.* 4:123–41
29. Griffith J, Makhov A, Santiago-Lara L, Setlow P. 1994. Electron microscopic studies of the interaction between a *Bacillus* α/β-type small, acid-soluble spore protein with DNA: protein binding is cooperative, stiffens the DNA and induces negative supercoiling. *Proc. Natl. Acad. Sci. USA* 91:8224–28
30. Gudibande SR, Jayasena SD, Behe MJ. 1988. CD studies of double-stranded polydeoxyribonucleotides composed of repeating units of contiguous homopurine residues. *Biopolymers* 27: 1905–15
31. Hackett RH, Setlow P. 1988. Properties of spores of *Bacillus subtilis* strains which lack the major small, acid-soluble protein. *J. Bacteriol.* 170:1403–4
31a. Halvorson HO, Hanson R, Campbell LL, eds. 1972. *Spores V*. Washington DC: Am. Soc. Microbiol.
32. Hanlin JH, Lombardi SJ, Slepecky RA. 1985. Heat and UV light resistance of vegetative cells and spores of *Bacillus subtilis* Rec⁻ mutants. *J. Bacteriol.* 163: 774–77
33. Heinemann V, Alings C, Lauble H. 1989. Structural features of G/C rich DNA going A or B. In *Structure and Methods: DNA and RNA*, ed. RH Sarma, MS Sarma, 3:39–53. Guilderland: Academic
34. Imlay JA, Linn S. 1988. DNA damage and oxygen radical toxicity. *Science* 240:1302–9
35. Kadota H, Uchida A, Sako Y, Harada K. 1978. Heat induced DNA injury in spores and vegetative cells of *Bacillus subtilis*. In *Spores VII*, ed. G Chambliss, JC Vary, pp. 27–30. Washington DC: Am. Soc. Microbiol.
36. Kennedy MJ, Reader SL, Swierczynski LM. 1994. Preservation records of micro-organisms: evidence of the tenacity of life. *Microbiology* 140:2513–29
37. Lindahl T. 1993. Instability and decay of the primary structure of DNA. *Nature* 362:709–15
38. Lindahl T. 1993. Recovery of antediluvian DNA. *Nature* 365:700
39. Lindahl T, Nyberg N. 1972. Rate of depurination of native deoxyribonucleic acid. *Biochemistry* 11:3610–18
40. Loshon CA, Fliss ER, Setlow B, Foerster HF, Setlow P. 1986. Cloning and nucleotide sequence of genes for small, acid-soluble, spore proteins of *Bacillus cereus, Bacillus stearothermophilus,* and "*Thermoactinomyces thalpophilus*." *J. Bacteriol.* 167:417–25
41. Magill NG, Cowan AE, Koppel DE,

Setlow P. 1994. The internal pH of the forespore compartment of *Bacillus megaterium* decreases by about 1 pH unit during sporulation. *J. Bacteriol.* 176:2252–58
42. Magill NG, Loshon CA, Setlow P. 1990. Small, acid-soluble proteins and their genes from two species of *Sporosarcina*. *FEMS Microbiol. Lett.* 72:293–98
43. Marquis RE, Sim J, Shin SY. 1994. Molecular mechanisms of resistance to heat and oxidative damage. *J. Appl. Bacteriol.* 70:40S–48S
44. Mason JM, Setlow P. 1986. Essential role for small, acid-soluble, spore proteins in the resistance of *Bacillus subtilis* spores to ultraviolet light. *J. Bacteriol.* 170:239–44
45. McCall M, Brown T, Kennard O. 1985. The crystal structure of d(GGGG-CCCC). A model for poly(dG)•poly(dC). *J. Mol. Biol.* 183:385–96
46. Mohr SC, Sokolov NVHA, He C, Setlow P. 1991. Binding of small acid-soluble spore proteins from *Bacillus subtilis* changes the conformation of DNA from B to A. *Proc. Natl. Acad. Sci. USA* 88:77–81
47. Munakata N, Rupert CS. 1972. Genetically controlled removal of "spore photoproduct" from deoxyribonucleic acid of ultraviolet irradiated *Bacillus subtilis* spores. *J. Bacteriol.* 111:192–98
48. Munakata N, Rupert CS. 1974. Dark repair of DNA containing "spore photoproduct" in *Bacillus subtilis*. *Mol. Gen. Genet.* 130:239–50
49. Nicholson WL, Setlow B, Setlow P. 1990. Binding of DNA in vitro by a small, acid-soluble spore protein and its effect on DNA topology. *J. Bacteriol.* 172:6900–6
50. Nicholson WL, Setlow B, Setlow P. 1991. Ultraviolet irradiation of DNA complexed with α/β-type small, acid-soluble proteins from spores of *Bacillus* or *Clostridium* species makes spore photoproduct but not thymine dimers. *Proc. Natl. Acad. Sci. USA* 88:8288–92
51. Nicholson WL, Setlow P. 1990. Dramatic increase in the negative superhelicity of plasmid DNA in the forespore compartment of sporulating cells of *Bacillus subtilis*. *J. Bacteriol.* 172:7–14
52. Nishimura Y, Torigoe C, Tsuboi M. 1985. An A-form poly(dG)•poly(dC) in H$_2$O solution. *Biopolymers* 24:1841–44
53. Northrop J, Slepecky RA. 1967. Sporulation mutations induced by heat in *Bacillus subtilis*. *Science* 155:838–39
54. Patrick MH, Gray DM. 1976. Independence of photoproduct formation on DNA conformation. *Photochem. Photobiol.* 24:507–13
55. Pedraza-Reyes M, Gutierrez-Corona F, Nicholson WL. 1994. Temporal regulation and forespore-specific expression of the spore photoproduct lyase gene by sigma-G RNA polymerase during *Bacillus subtilis* sporulation. *J. Bacteriol.* 176:3983–91
56. Poinar GO Jr. 1993. Recovery of antediluvian DNA. *Nature* 365:700
57. Rahn RO, Hosszu JL. 1969. Influence of relative humidity on the photochemistry of DNA films. *Biochim. Biophys. Acta* 190:126–31
58. Rao H, Mohr SC, Fairhead H, Setlow P. 1992. Synthesis and characterization of a 29-amino acid residue DNA-binding peptide derived from α/β-type small, acid-soluble spore proteins (SASP) of bacteria. *FEBS Lett.* 305:115–20
59. Roberts TA, Hitchins AD. 1969. Resistance of spores. In *The Bacterial Spore*, ed. GW Gould, A Hurst, pp. 611–70. New York: Academic
60. Sanchez-Salas J-L, Santiago-Lara ML, Setlow B, Sussman MD, Setlow P. 1992. Properties of mutants of *Bacillus megaterium* and *Bacillus subtilis* which lack the protease that degrades small, acid-soluble proteins during spore germination. *J. Bacteriol.* 174:807–14
61. Sanchez-Salas J-L, Sharon M, Setlow P. 1992. Effect of mutant small, acid-soluble spore proteins containing cysteine or tryptophan on DNA properties in vivo or in vitro. *Biochimie* 74:651–60
62. Sapru V, Labuza TP. 1993. Glassy state in bacterial spores predicted by polymer glass-transition theory. *J. Food Sci.* 58:445–48
63. Sarma MH, Gupta G, Sarma RH. 1986. 500-MHz ^{1}H NMR study of poly(dG)•poly(dC) in solution using one-dimensional nuclear Overhauser effect. *Biochemistry* 25:3659–65
64. Setlow B, Hackett RH, Setlow P. 1982. Non-involvement of the spore cortex in the acquisition of low molecular weight basic proteins and ultraviolet light resistance during sporulation of *Bacillus sphaericus*. *J. Bacteriol.* 149:494–98
65. Setlow B, Hand AR, Setlow P. 1991. Synthesis of a *Bacillus subtilis* small, acid-soluble spore protein in *Escherichia coli* causes cell DNA to assume some characteristics of spore DNA. *J. Bacteriol.* 173:1642–53
66. Setlow B, Magill N, Febroriello P, Nakhimovsky L, Koppel DE, Setlow P. 1991. Condensation of the forespore nucleoid early in sporulation of *Bacillus* species. *J. Bacteriol.* 173:6270–78

67. Setlow B, Setlow P. 1979. Localization of low-molecular-weight basic proteins in *Bacillus megaterium* spores by cross-linking with ultraviolet light. *J. Bacteriol.* 139:486–94
68. Setlow B, Setlow P. 1987. Thymine containing dimers as well as spore photoproducts are found in ultraviolet-irradiated *Bacillus subtilis* spores that lack small, acid-soluble proteins. *Proc. Natl. Acad. Sci. USA* 84:421–23
69. Setlow B, Setlow P. 1988. Absence of transient elevated UV resistance during germination of *Bacillus subtilis* spores lacking small, acid-soluble spore proteins α and β. *J. Bacteriol.* 170:2858–59
70. Setlow B, Setlow P. 1988. Decreased UV resistance of spores of *Bacillus subtilis* strains deficient in UV repair and small, acid-soluble spore proteins. *Appl. Environ. Microbiol.* 54:1275–76
71. Setlow B, Setlow P. 1993. Binding of small, acid-soluble spore proteins to DNA plays a significant role in the resistance of *Bacillus subtilis* spores to hydrogen peroxide. *Appl. Environ. Microbiol.* 59:3418–23
72. Setlow B, Setlow P. 1993. Dipicolinic acid greatly enhances the production of spore photoproduct in bacterial spores upon ultraviolet irradiation. *Appl. Environ. Microbiol.* 59:640–43
73. Setlow B, Setlow P. 1994. Heat inactivation of *Bacillus subtilis* spores lacking small, acid-soluble spore proteins is accompanied by generation of abasic sites in spore DNA. *J Bacteriol.* 176:2111–12
74. Setlow B, Sun D, Setlow P. 1992. Studies of the interaction between DNA and α/β-type small, acid-soluble spore proteins: a new class of DNA binding protein. *J. Bacteriol.* 174:2312–22
75. Setlow P. 1988. Resistance of bacterial spores to ultraviolet light. *Comments Mol. Cell. Biophys.* 5:253–64
76. Setlow P. 1988. Small acid-soluble, spore proteins of *Bacillus* species: structure, synthesis, genetics, function and degradation. *Annu. Rev. Microbiol.* 42:319–38
77. Setlow P. 1991. Changes in forespore chromosome structure during sporulation in *Bacillus* species. *Sem. Dev. Biol.* 2:55–62
78. Setlow P. 1992. DNA in dormant spores of *Bacillus* species is in an A-like conformation. *Mol. Microbiol.* 6:563–67
79. Setlow P. 1992. I will survive: protecting and repairing spore DNA. *J. Bacteriol.* 174:2737–41
80. Setlow P. 1993. DNA structure, spore formation and spore properties. In *Regulation of Bacterial Differentiation*, ed. PJ Piggot, P Youngman, CP Moran Jr, pp. 181–94. Washington DC: Am. Soc. Microbiol.
81. Setlow P. 1993. Spore structural proteins. In *Bacillus subtilis and Other Gram-Positive Bacteria: Biochemistry, Physiology, and Molecular Genetics*, ed. JA Hoch, R Losick, AL Sonenshein, pp. 801–9. Washington, DC: Am. Soc. Microbiol.
82. Setlow P. 1994. Mechanisms which contribute to the long-term survival of spores of *Bacillus* species. *J. Appl. Bacteriol.* 176:49S–60S
83. Shin S-Y, Calvisi EG, Beaman TC, Pankratz HS, Gerhardt P, Marquis RE. 1994. Microscopic and thermal characterization of hydrogen peroxide killing and lysis of spores and protection by transition metal ions, chelators, and antioxidants. *Appl. Environ. Microbiol.* 60:3192–97
84. Slepecky RA, Leadbetter ER. 1983. On the prevalence and roles of spore-forming bacteria and their spores in nature. In *The Bacterial Spore*, ed. A Hurst, GW Gould, 2:79–99. New York: Academic
85. Stafford RS, Donnellan JE Jr. 1968. Photochemical evidence for conformation changes in DNA during germination of bacterial spores. *Proc. Natl. Acad. Sci. USA* 59:822–29
86. Sykes B. 1993. Less cause for grave concern. *Nature* 366:513
87. Tovar-Rojo F, Setlow P. 1991. Analysis of the effects of mutant small, acid-soluble spore proteins from *Bacillus subtilis* on DNA in vivo and in vitro. *J. Bacteriol.* 173:4827–35
88. Varghese A. 1970. 5-Thyminyl-5,6-dihydrothymine from DNA irradiated with ultraviolet light. *Biochem. Biophys. Res. Commun.* 38:484–90
89. Wang T-C, Rupert CS. 1977. Evidence for the monomerization of spore photoproduct to two thymines by the light-independent "spore repair" process in *Bacillus subtilis*. *Photochem. Photobiol.* 25:123–27
90. Webb SJ. 1967. Mutation of bacterial cells by controlled dessication. *Nature* 213:1137–39
91. Zamenhof S. 1960. Effects of heating dry bacteria and spores on their phenotype and genotype. *Proc. Natl. Acad. Sci. USA* 46:101–5

GENETICS, PHYSIOLOGY, AND EVOLUTIONARY RELATIONSHIPS OF THE GENUS *BUCHNERA*: Intracellular Symbionts of Aphids

Paul Baumann, Linda Baumann, Chi-Yung Lai, and Dadbeh Rouhbakhsh

Microbiology Section, University of California, Davis, California 95616-8665

Nancy A. Moran

Department of Ecology and Evolutionary Biology, University of Arizona, Tucson, Arizona 85721

Marta A. Clark

Microbiology Section, University of California, Davis, California 95616-8665

KEY WORDS: endosymbionts, genome analysis, tryptophan biosynthesis, gene amplification, evolutionary relationships, coevolution, mutualism

CONTENTS

INTRODUCTION	57
GENERAL PROPERTIES OF APHIDS	57
ULTRASTRUCTURE, LOCATION, AND TRANSMISSION OF *BUCHNERA*	58
KINETICS OF GROWTH DURING PARTHENOGENETIC REPRODUCTION	60
EVOLUTIONARY RELATIONSHIPS	62
GENERAL GENETICS AND PHYSIOLOGY OF *BUCHNERA*	64
Buchnera Genome and Methods of Genetic Characterization	64
DNA Synthesis, Transcription, and Translation	65
Genes for Other Properties	68
Comparisons of Gene Order and Potential Regulatory Sequences	69
rRNA Operons	73
GroES and GroEL	75
Metabolic Studies	78

ROLE OF *BUCHNERA* IN THE SYMBIOTIC ASSOCIATION	78
Requirement of Buchnera by the Aphid Host	78
Nutritional Studies	79
Potential Modifications of Biosynthetic Pathways	80
Tryptophan Biosynthesis	81
Genes of the Common Pathway of Aromatic Amino Acid Biosynthesis	84
Sulfate Reduction and Cysteine and Methionine Biosynthesis	84
CONCLUSIONS AND PERSPECTIVES	85

ABSTRACT

Evolutionary studies suggest that 200–250 million years ago an aphid ancestor was infected with a free-living eubacterium. The latter became established within aphid cells. Host and endosymbiont (genus *Buchnera*) became interdependent and unable to survive without each other. The growth of *Buchnera* became integrated with that of the aphids, which acquired the endosymbionts from their mothers before birth. Speciation of host lineages was paralleled by divergence of associated endosymbiont lineages, resulting in parallel evolution of *Buchnera* and aphids. Present day *Buchnera* retains many of the properties of its free-living ancestor, containing genes for proteins involved in DNA replication, transcription, and translation, as well as chaperonins and proteins involved in secretion, energy-yielding metabolism, and amino acid biosynthesis. Some of these processes are also observed in isolated endosymbiont cells. Genetic and physiological studies indicate that *Buchnera* can synthesize methionine, cysteine, and tryptophan and supply these amino acids to the aphid host. In the case of some fast-growing species of aphids, the overproduction of tryptophan by *Buchnera* involves plasmid-amplification of the gene coding for anthranilate synthase, the first enzyme of the tryptophan biosynthetic pathway. These recent studies provide a beginning in our understanding of *Buchnera* and its role in the endosymbiosis with aphids.

* * *

Buchnera:
Son io, son io, son io
Che vi fa scaltri.
L'arguzia mia crea
L'arguzia degli altri.
Afide (ospite):
Ma bravo!

Falstaff, G Verdi[1]

[1]*Buchnera:*/It is I, it is I, it is I/Who makes you clever./My cleverness creates/The cleverness of the others./Aphid (host): Well bravo!

INTRODUCTION

Associations between insects and intracellular prokaryotes (endosymbionts) are widespread among members of the insect orders Homoptera (aphids, whiteflies, mealybugs, psyllids, cicadas), Blattaria (cockroaches), and Coleoptera (beetles) (13). Many of these insects utilize diets deficient in one or several classes of nutrients, and the endosymbionts, through their biosynthetic activities, are thought to provide the host with the missing essential nutrients (30). The classical treatise on these and other associations is Buchner's (13), published in 1965. Since then several reviews have dealt with the diversity of beneficial associations between insects and endosymbionts (17, 31, 38, 72, 80). This review concentrates on the more recent studies of the genetics, physiology, and evolutionary relationships of aphid endosymbionts. For a comprehensive review of the literature up to 1979 on endosymbionts of Homoptera, readers are referred to Houk & Griffiths (72); for a review dealing only with aphid endosymbionts, see Houk (71).

The endosymbionts that appear to be common to most aphids have been assigned recently to the genus *Buchnera* (see section on evolutionary relationships for a discussion of taxonomy) (120). The symbiotic association between *Buchnera* and aphids is obligate and mutualistic in that neither partner can reproduce in the absence of the other (72). This association differs from the extensively studied association between insects and *Wolbachia pipientis*, an intracellular prokaryote that causes reproductive alterations in insects (11, 130, 144, 156). *W. pipientis* is not essential for the host's survival and some aspects of its biology resemble parasitism and not mutualism.

GENERAL PROPERTIES OF APHIDS

Aphids are insects that feed on plant sap, a nutrient rich in carbohydrates but deficient in nitrogenous compounds (33, 107). Aphids penetrate plant tissue by means of flexible stylets that probe until they reach the sieve tubes in the phloem tissue. Penetration is primarily between cells and requires the dissolution of plant cell–cementing material by means of pectinases and other enzymes found in aphid saliva (18, 100). The aphid secretes a sheath that surrounds the stylets and makes them rigid. Due to the low amount of nitrogenous compounds in plant sap and because aphids do not fix nitrogen, these insects ingest a large amount of food and then excrete the excess sugar as honeydew (72). Aphid populations can reach enormous sizes (2×10^9 aphids/acre), and aphids are among the major pests of agriculturally important plants. Perhaps the main impact of aphids on plant health is the transmission of viral diseases (157). Additional effects are nutrient deprivation, leaf curling, and gall formation (105, 106).

Aphids vary greatly in their annual life cycles and host-plant preferences (33, 34, 107). In the simplest case, female aphids hatch in spring from overwintering eggs, feed and develop on their host plants, and reproduce by parthenogenesis. A succession of 3–15 parthenogenetic generations continues throughout the growing season. The large number of generations is possible because during parthenogenetic reproduction, the young undergo prenatal development and are born at an advanced stage.

Although genetically similar, members of a parthenogenetic clone may differ in form. Early in the season and during optimal growth conditions, wingless females with high fecundity are produced; adverse conditions such as overcrowding induce the production of winged females able to disperse to new host plants. In autumn, males and sexual females are produced; after mating, the sexual females deposit overwintering eggs, which hatch in spring to renew the cycle. In the studies considered in this review, the aphids are usually in their most active reproductive state—that is wingless parthenogenetic females. Three species have most commonly been the subjects of physiological and/or genetic studies: *Schizaphis graminum* (greenbug), *Acyrthosiphon pisum* (pea aphid), and *Myzus persicae* (green peach aphid).

ULTRASTRUCTURE, LOCATION, AND TRANSMISSION OF *BUCHNERA*

Members of the genus *Buchnera* are spherical or oval cells, 2–5 µm in diameter, with a cell wall resembling that of gram-negative eubacteria (Figure 1*a,b*) (60, 72, 103). In electron micrographs, a thin line indicative of the peptidoglycan layer has been detected between the two unit membranes (73). The presence of peptidoglycan is indicated by chemical evidence and by the major alterations of cell wall structure observed upon addition of penicillin to the aphid diet (61, 73). Septum formation has not been observed in endosymbionts undergoing division (60, 103). *Buchnera* reside within host-derived membrane vesicles designated as symbiosomes (Figure 1*a–c*) (141) located within specialized polyploid cells called bacteriocytes or mycetocytes (Figure 1*d,e*) (9, 13, 60, 71, 72, 103). The bacteriocytes are arranged as a bilobed structure within the body cavity of the aphid. This structure is known as the bacteriome or mycetome (Figure 1*e*) (134, 146).

In young aphids the bacteriome is surrounded by a sheath consisting of a thin layer of flattened and often syncytial cells (42, 72). In some aphids (e.g. *A. pisum*), the sheath cells may contain rod-shaped eubacteria (0.5–1.5 µm) that have a gram-negative cell wall and are also enclosed in vesicles (60, 61, 71, 72, 74). This rod-shaped organism frequently is designated as the secondary (S-) endosymbiont to differentiate it from *Buchnera*, which is often called the primary endosymbiont. The S-endosymbiont is never found in bacteriocytes

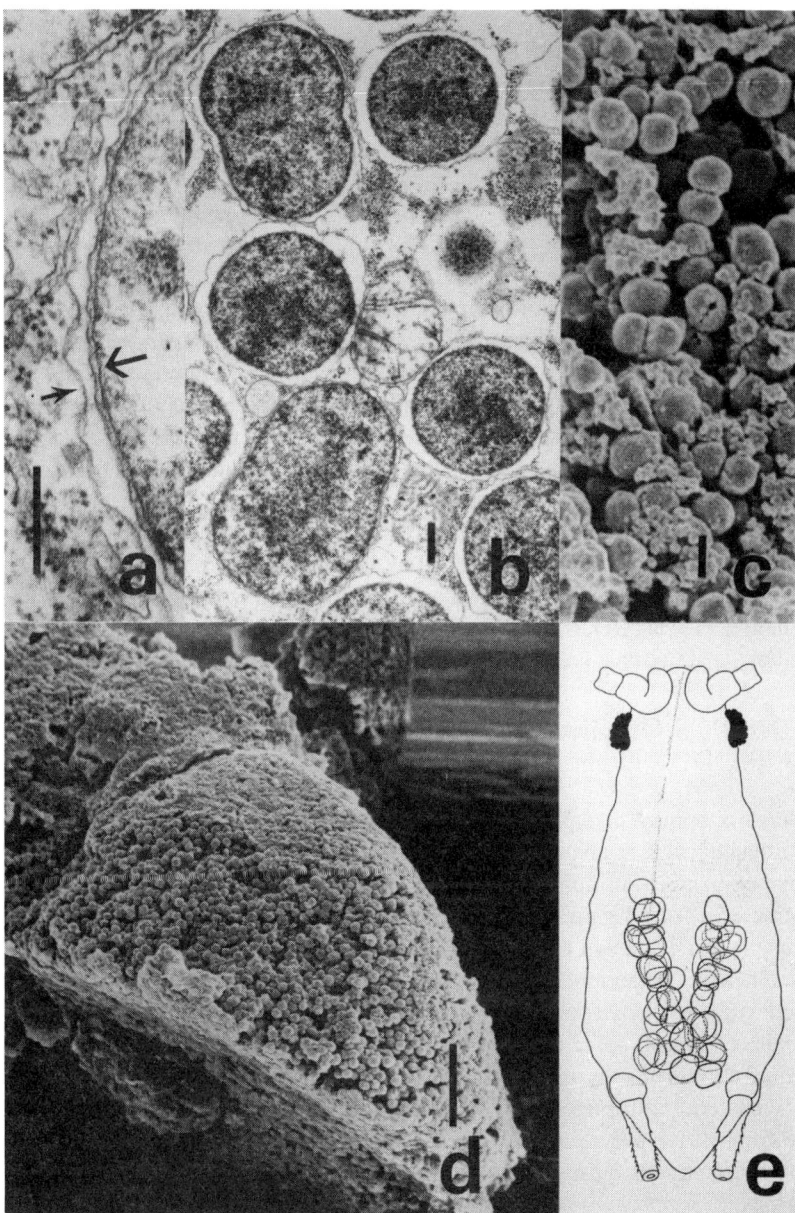

Figure 1 Photographs illustrating the morphology of *Buchnera* and its position in the aphid. Transmission electron micrographs: (*a*) gram-negative cell wall of *Buchnera* (*large arrow*) and symbiosome membrane (*small arrow*) (bar = 0.5 µm), (*b*) *Buchnera* within symbiosomes (bar = 1 µm). Scanning electron micrographs: (*c*) symbiosomes in bacteriocyte (bar = 10 µm), (*d*) bacteriocyte stripped of the cell membrane showing symbiosomes (bar = 100 µm). (*e*) Diagram of an aphid showing internal location of bacteriocytes and their arrangement as a bacteriome. All photographs reproduced by permission: (*a, b, c*) DL McLean & M Kinsey; (*d*) Reference 103; (*e*) Reference 134, with kind permission from author(s) and Elsevier Science Publishers.

and is present in much lower numbers relative to *B. aphidicola*. In older aphids as well as adults, the bacteriome breaks apart and groups of bacteriocytes become dispersed throughout the abdomen (12, 42). At this stage, the bacteriome sheath is not apparent and the S-endosymbiont may be found in the hemolymph. Other morphological types of S-endosymbionts were recently reported (54).

Both *Buchnera* and the S-endosymbionts are maternally transmitted to eggs and embryos. The mechanism of transmission is complex and has not been studied extensively (9). At an early stage of development during parthenogenetic reproduction, bacteriocytes are found in close proximity to the embryos; *Buchnera* enters the embryo; and the bacteriome then begins to develop (12, 70). *Buchnera* cells in the process of being transferred from the bacteriome to the embryo are not surrounded by their vesicular membranes. The transfer of the S-endosymbionts has not been studied.

Bacteriomes and *Buchnera* are nearly universal within the true aphids (Aphidoidea) (13), some species of the tribe Ceratophidini lack both (13, 50, 52). Instead, the body cavity of these aphids contains a yeast-like extracellular symbiont (13, 50, 51). In addition, some species of aphids produce dwarf males and/or sterile female soldiers that may lack endosymbionts (13, 50, 52).

KINETICS OF GROWTH DURING PARTHENOGENETIC REPRODUCTION

Wingless aphids that produce live young are in their most active state of reproduction. Such aphids contain embryos in the process of development, so that the increase in aphid size during the period of growth results from an increase in both the mass of the mother as well as the mass of her embryos. This process is known as the telescoping of generations because it allows the simultaneous development of two generations and accounts for the rapid reproduction of many species of aphids (35). Although there have been several studies of overall aphid growth (42, 150, 168), the numbers of *Buchnera* have been determined only recently (7). *Buchnera* contains only one copy of *rrs*, the gene coding for 16S rRNA (117, 119, 160, 161). Quantitation of *rrs* by the competitive polymerase chain reaction (PCR) has permitted the enumeration of *Buchnera* during the period from birth to maturity of the host (7). Aphids (*S. graminum*) born during a 16- to 18-h time period had an average wet weight of 24 µg and grew to 540 µg in 9–10 days, representing a 23-fold increase in mass. During this time, the number of *Buchnera* increased by approximately the same factor, from 0.2×10^6 to 5.6×10^6 endosymbionts per aphid. These values represent the sum of endosymbionts in the mother and her embryos. Observations on the morphology of *Buchnera* have indicated a somewhat higher number of cells undergoing division in embryonic bacteriocytes

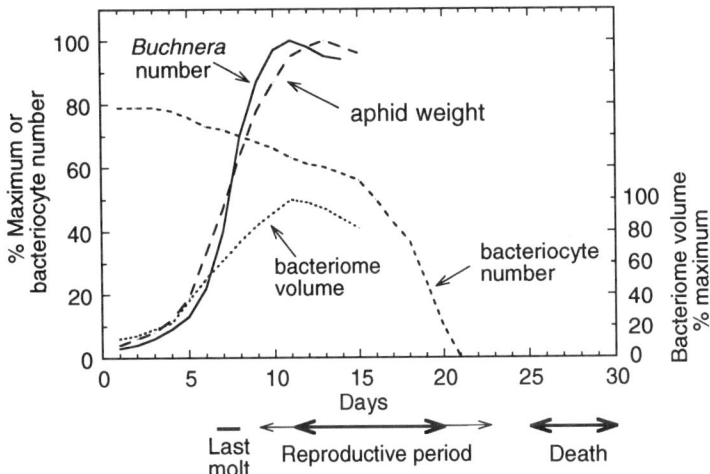

Figure 2 Kinetics of *Buchnera* and aphid growth during parthenogenetic reproduction. Composite graph derived from References 7, 42, and 150. Only maternal bacteriome volume and bacteriocyte number are given.

as compared to maternal bacteriocytes (168). The increase in the total DNA and protein per aphid approximately paralleled the increase in wet weight and *Buchnera* number (7). These results suggest that from birth to maturity there is an integration of aphid growth with an increase in the number of *Buchnera*. The increase in *Buchnera* numbers implies the presence of the biosynthetic machinery required for an orderly increase of all the endosymbiont's constituents as is characteristic of bacterial growth. Using the genome size of *Buchnera* from *A. pisum* (79), it was estimated that about 9% of the total DNA of *S. graminum* is from the endosymbiont (7).

Figure 2 presents a composite graph of aphid and endosymbiont growth, based on studies of *A. pisum* (150), *Megoura viciae* (42), and *S. graminum* (7). From birth to adulthood an aphid passes through four instars (the period between molts); the last molt occurs at 7–8 days. During the initial phase of growth, the wet weight of the aphid doubles approximately every 1.5–2.0 days. Birth of young begins at 9–11 days and finishes at 20–23 days. Each aphid can give rise to 50–60 live offspring. Aphids begin to die at 25 days, and few survive beyond 30 days. During growth, as measured by weight increase, and about halfway into the reproductive period, the number of maternal bacteriocytes gradually decreases. However, at the same time, there is a major increase in the volume of the maternal bacteriocytes, which suggests that they and their increasing endosymbiont population are essential for embryonic development. Toward the end of the reproductive period, when their function is no longer required, bacteriocyte numbers decrease greatly, and virtually none are detected at 25 days (42). Degradation of *Buchnera* at this stage has been noted by means of electron microscopy (60, 70).

EVOLUTIONARY RELATIONSHIPS

The genes coding for *rrs* of *Buchnera* from 12 species of aphids have been cloned and sequenced (112, 119). Figure 3 presents the resulting evolutionary tree. *Escherichia coli* and members of the family *Enterobacteriaceae* are close relatives of *Buchnera*. The recently characterized bacteriome endosymbionts of tsetse constitute a distinct lineage that appears to be more closely related to the *Enterobacteriaceae* than to *Buchnera* (2). All of these organisms, including *Ruminobacter amylophilus* (Figure 3), are members of the γ-3 subdivision of the class *Proteobacteria* (129, 152). The groupings obtained on the basis of *Buchnera rrs* relationships agree with those of classical aphid taxonomy (Figure 3). In addition, the branching order within the *Buchnera* phylogenetic tree is identical to the proposed phylogeny of aphids based primarily on morphology (Figure 3) (67, 111, 112). Approximately the same number of substitutions in *rrs* separate each of the 12 endosymbionts from

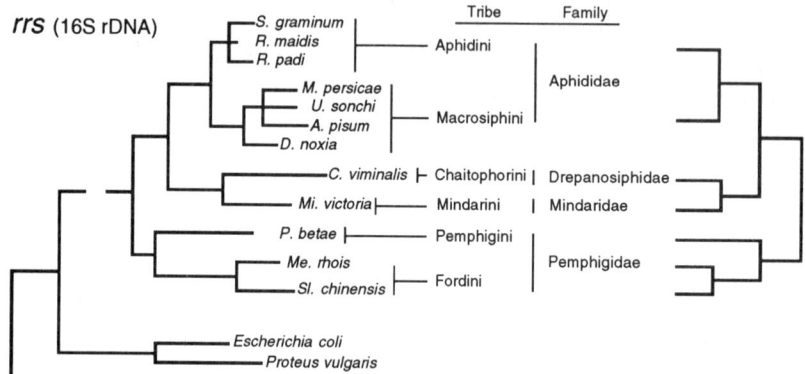

Figure 3 *Buchnera* phylogeny based on *rrs* and comparison with a proposed phylogeny of the aphid hosts (67, 111, 112, 119). Names designate the aphid host. The following abbreviations designate aphid genera: *S.*, *Schizaphis*; *R.*, *Rhopalosiphum*; *M.*, *Myzus*; *U.*, *Uroleucon*; *A.*, *Acyrthosiphon*; *D.*, *Diuraphis*; *C.*, *Chaitophorus*; *Mi.*, *Mindarus*; *P.*, *Pemphigus*; *Me.*, *Melaphis*; *Sl.*, *Schlechtendalia*.

E. coli, indicating similarity in the substitution rates among the symbiont lineages (112). Estimated divergence dates were assigned, based primarily on the aphid fossil record, to several nodes of the *Buchnera* tree. Evolutionary rates based on these dates fell within the range of 0.01–0.02 substitutions/*rrs* site per 50 million years (112).

The results presented in Figure 3 support vertical evolution of the endosymbionts, in that these organisms do not appear to be exchanged or lost and the host reinfected with other bacteria. The one exception is the absence of a *Buchnera* species and bacteriomes in certain species of Cerataphidini, which appears to be the result of their loss from a common, *Buchnera*-containing ancestor and their replacement by a yeast-like extracellular symbiont within the aphid body cavity (13, 50, 51). The congruence between the evolutionary trees of endosymbiont and host (Figure 3) argues for a monophyletic origin of *Buchnera* and suggests the following scenario (111, 112, 119): About 200–250 million years ago an ancestor of the present-day aphids was infected with a free-living bacterium, and an intracellular symbiotic association was established. Subsequent parallel divergence led to cospeciation of endosymbiont and host, resulting in the present species of aphids and strains of *Buchnera*. The parallel phylogenies of *Buchnera* and hosts contrast with results for the only other insect-bacterium association studied in some detail, namely that of *W. pipientis,* which causes cytoplasmic incompatibility or other reproductive alterations in a variety of insect orders (11, 130, 144, 156). Phylogenetic trees based on *W. pipientis rrs* show a lack of congruence with host phylogenies and imply multiple infections of the same as well as different insect species (horizontal evolution). *W. pipientis* is in the α-subdivision of the protobacteria, which also contains a variety of intracellular eubacteria that are associated with arthropods (11, 130, 144).

The lineage represented by the endosymbionts of aphids presented in Figure 3 has been given the generic and specific designation *B. aphidicola,* and the symbiont from *S. graminum* has been designated as the type strain of the species (120). Rules of bacterial nomenclature indicate that a genus cannot be named unless a type species is declared (97). An interim solution is the assignment of a species name to an entire lineage containing considerable internal diversity, which may accommodate several species or subspecies. Further studies are necessary to establish the range of diversity of *Buchnera* within the same aphid species as well as the relationships of endosymbionts from other aphid species. In addition, a gene that is not as highly conserved as *rrs* would be more useful for the delineation of close relationships. In this review we designate *Buchnera* from different species of aphids by the abbreviation B followed by parentheses enclosing the host genus and species designations listed in Figure 3.

Whiteflies (Aleyrodoidea) and mealybugs (Pseudococcidae) contain prokaryotic endosymbionts that, based on *rrs* sequences, are distinct from *Buchnera* (24, 121). The endosymbionts of whiteflies represent a separate lineage within the γ-subdivision, whereas mealybug endosymbionts are in the β-subdivision of the *Proteobacteria* (24, 121). *A. pisum* contains an S-endosymbiont that, on the basis of *rrs*, has been placed in the family *Enterobacteriaceae* (160, 161). This endosymbiont also has a single copy of *rrs* in its genome (160, 161). As is indicated in other studies (20, 24, 59), members of the *Enterobacteriaceae* appear able to enter into different associations with insects.

GENERAL GENETICS AND PHYSIOLOGY OF *BUCHNERA*

Buchnera Genome and Methods of Genetic Characterization

The DNA of B(*A. pisum*) has a guanine + cytosine (G+C) content of 28–30 mol% (79). Its genome size has been reported as 1.4×10^{10} Daltons (79) or 5.6 times larger than that of *E. coli* [2.5×10^9 Daltons (123)]. This is an unexpected finding because a sheltered, nutrient-rich, intracellular habitat is generally thought to select for a reduction in genome size, such as in the obligate intracellular pathogens *Rickettsia prowazekii* and *Chlamydia trachomatis* [genome sizes of 1.1×10^9 and 0.5×10^9 Daltons, respectively (115, 166)]. A major reduction in genome size is a characteristic of organelles such as chloroplasts and mitochondria (4, 131). In addition, the genomes of mycoplasmas, which are not intracellular organisms but live in a rich milieu in close association with animals and plants, are also substantially reduced in size (145).

The approach used for the genetic characterization of *Buchnera* was similar to that used for cloning genes from *C. trachomatis* (47). Sequences of proteins involved in essential functions, such as initiation of chromosome replication, transcription, and translation, were compared for a variety of organisms and the conserved regions identified so that oligonucleotide primers could be designed for use in PCR. Because the G+C content of *Buchnera* is low (79), codons containing A and T in the first and third positions were favored. This bias was justified by the success in using these primers and was borne out by the subsequently compiled tables of codon usage (23, 127). PCR products were cloned and then sequenced to confirm the identity of the genes prior to use as probes for their detection in restriction enzyme and Southern blot analysis and in λ bacteriophage recombinants (94, 95). To date, the sequence of over 54 kb of DNA has been determined for B(*S. graminum*), 11 kb for B(*Schlechtendalia chinensis*) (96), and 5.2 kb for B(*A. pisum*) (128). The 54-kb B(*S. graminum*) DNA had a G+C content of 28.0 mol% and represents only 0.2% of the endosymbiont genome (79). Figures 4, 5, and 6 (see below) present a genetic map of the cloned and sequenced B(*S. graminum*) DNA fragments. Individual genes were identified by the similarity of their deduced amino acid sequences to the proteins of *E. coli;* the *E. coli* genetic designations were retained. Tables 1 and 2 (see below) group most of the genes presented in the genetic maps according to function. In cases where the amino acid sequences of proteins from B(*S. graminum*) could be compared to those of several eubacteria, the greater similarity was always to the protein from *E. coli*, which is consistent with the evolutionary relationships shown in Figure 3.

DNA Synthesis, Transcription, and Translation

Methods developed for the isolation of *Buchnera* from the aphid host involve disruption of the aphid in solutions of high osmolarity and separation of *Buchnera* from host constituents by centrifugation through Percoll gradients (65, 75). These or similar preparations have been used in studies of incorporation of precursors into DNA, rRNA, and protein, as well as uptake of nutrients (76–78, 113, 167). It is not always clear whether these preparations included the symbiosome membrane.

Initiation of bidirectional, eubacterial chromosome replication involves the DnaA protein (125, 179). The deduced amino acid sequence of B(*S. graminum*) DnaA indicates that the protein from the endosymbiont shares the general characteristics of DnaA proteins from other organisms (175). The presence of genes for DNA primase, DNA polymerase III (β- and ϵ-subunits), DNA gyrase, and RNase II (Table 1) is in agreement with the incorporation of ^{3}H-thymidine into DNA by isolated B(*A. pisum*). Inhibition of the incorporation by nalidixic acid, which acts on subunit A of DNA gyrase (104), is also consistent with the presence and function of this protein. All the genes for the holoenzyme of the DNA-dependent RNA polymerase (*rpoABCD*) have been detected as have genes for 16S, 23S, and 5S rRNA (Table 1). These observations are consistent with the incorporation of ^{3}H-uridine into rRNA by B(*A. pisum*) (76) and the inhibition of this process by rifampicin, an antibiotic that acts on the β-subunit of RNA polymerase (82). Similarly, the capacity for protein synthesis is indicated by (*a*) the detection of genes for three tRNA synthases, RNase P, tRNA^{Glu}, seven ribosomal proteins, and initiation factor-3 (Table 1); (*b*) the incorporation of radioactive amino acids by the endosymbiont into at least 210 different proteins; and (*c*) the inhibition of this incorporation by rifampicin and chloramphenicol (77, 78). The totality of these genetic and physiological observations indicates that *Buchnera* possesses the genes and enzymatic machinery essential for growth and that it is eubacterial in nature, as indicated by sequence similarity to homologous *E. coli* proteins and inhibition by eubac-

Table 1 Genes involved in polymerization reactions that have been found in B(S. graminum) and some properties of the deduced products

General category	Gene	Protein or RNA	% identity to E. coli	Function or property	Figure; references
DNA synthesis	dnaA	DnaA	77	Initiation of chromosome replication	4a; 95, 104, 175
	dnaG	Primase	(48)[a]	Synthesis of RNA primers	4d; 94, 104, 175
	dnaN	DNA polymerase III β-subunit	41	Elongation of DNA	4a; 95, 104, 175
	dnaQ	DNA polymerase III ξ-subunit	(56)	Proofreading exonuclease	5a; 104, 117
	gyrB	DNA gyrase subunit B	(81)	Breaking and resealing of double-stranded DNA	4a; 95, 104, 175
	rnh	RNase H	57	Excision of RNA primer	5a; 104, 117
Transcription		RNA polymerase		Synthesis of mRNA, rRNA, tRNA	98
	rpoA	α-subunit	(84)		4b; 116
	rpoB	β-subunit	84	Rifampicin sensitivity	4c; 23, 82
	rpoC	β'-subunit	(91)		4c; 23
	rpoD	σ-subunit	80	Major σ factor, promoter recognition	4d; 94
Translation		Ribosomal RNA		Protein synthesis	98
	rrs	16S rRNA	90		5a; 117
	rrl	23S rRNA	87		5b; 142

ENDOSYMBIONTS OF APHIDS 67

rrf	5S rRNA		80	5b; 142
	tRNA synthases			98
argS	arginyl-tRNA synthase	Charging tRNA^{Arg}	(36)	5a; 117
cysS	cysteinyl-tRNA synthase	Charging tRNA^{Cys}	(54)	5b; 142
thrS	threonyl-tRNA synthase	Charging tRNA^{Thr}	70	4e; 87
	Small ribosomal subunits	Protein synthesis		98
rpsA	protein S1		75	4g; b
rpsD	protein S4		80	4b; 116
rpsK	protein S11		(100)	4b; 116
	Large ribosomal subunits	Protein synthesis		98
rplL	protein L7/L12		(71)	4c; 23
rplT	protein L20		(89)	4e; 87, 116
rpmH	protein L34		83	4a; 95
rpmI	protein L35		72	4e; 87, 116
infC	Initiation factor-3	Selection of initiator tRNA	93	4e; 87, 116
tRNA^{Glu}	glutamate tRNA	Protein synthesis	89	5b; 142
rnpA	RNase P	Processing of tRNA	47	4a; 95

[a] Numbers in parentheses indicate partial sequence.
[b] MA Clark, unpublished observations.

teria-specific antibiotics. Because the increase in B(*S. graminum*) numbers is coupled to aphid growth (7), all or most of the components of this essential enzymatic machinery must increase in a parallel fashion.

Genes for Other Properties

In gram-negative organisms, the secretion of proteins destined for either the periplasmic space or the outer membrane involves several multicomponent systems (136). One such system includes a cytoplasmic protein (SecB) that keeps the nascent peptide in a configuration that precludes tight folding and allows its export (90, 91, 136). The gene coding for this protein has been found in B(*S. graminum*) (Table 2), suggesting that the endosymbionts have a SecB-dependent system of translocation and the potential of exporting proteins to the host. A gene for SohB, a possible periplasmic protease (6), has been detected in B(*S. chinensis*) (Table 2). Two potentially membrane-associated proteins are ORF-A and ORF-C (Table 2). The hydrophobicity profile of ORF-A suggests the presence of a signal peptide and nine membrane-spanning segments; the properties of ORF-C suggest a membrane-associated ATP-binding protein (154).

A B(*S. graminum*) DNA fragment containing a portion of the gene coding for the ATP synthase β-subunit (*atpD*) has been sequenced (Table 2). The presence of *atpD* suggests that *Buchnera* may have the full ATP synthase (163). Because the endosymbiont resides in the host-derived symbiosome membrane, this enzyme could be used to synthesize ATP from the proton motive force generated by *Buchnera* or by the host bacteriocytes. B(*S. graminum*) contains the genes for glyceraldehyde-3-phosphate dehydrogenase (*gapA*) and triose phosphate isomerase (*tpiA*) (Table 2). Depending on the source of carbon and energy used by *Buchnera*, these enzymes could play a role in energy-yielding metabolism or in biosynthesis. A comparison of sequences of glyceraldehyde-3-phosphate dehydrogenases has indicated that the *E. coli* enzyme (GapA) more closely resembles the enzyme of eukaryotes than that of other eubacteria (36). This finding has been cited as an example of horizontal evolution, in which a bacterial ancestor obtained an enzyme from a eukaryote. The presence of *gapA* in B(*S. graminum*) suggests that this event occurred prior to the divergence of *Buchnera* and the Enterobacteriaceae (Figure 3). A fragment of the aphid host ATP synthase β-subunit was also sequenced and found to contain two introns (25).

Various eubacterial intracellular pathogens, including *R. prowazekii*, are ATP parasites in that they have a membrane-associated ATP/ADP translocase, which exchanges parasite ADP for host ATP (169). Because the G+C content of *R. prowazekii* is similar to that of *Buchnera*, a DNA fragment containing the gene coding for this enzyme was used to probe B(*S. graminum*) DNA

(169). No hybridization was detected, suggesting that this endosymbiont lacks an ATP/ADP translocase (25).

Comparisons of Gene Order and Potential Regulatory Sequences

In most organisms the region between *rpmH* and *dnaA* contains the origin of replication (Figure 4a) (175, 179). Consequently, this region has been the object of numerous comparative studies, and sequence data are available for many different organisms (125, 175). In *E. coli, Proteus mirabilis, Pseudomonas putida,* and *Bacillus subtilis,* the gene order is *rnpA-rpmH-dnaA-dnaN-recF-gyrB* (147, 175). B(*S. graminum*) differs from these organisms only in the absence of *recF* (Figure 4a). This finding is surprising because *recF*, which is involved in recombination and repair reactions including repair of UV damage (104, 165), is found between *dnaN* and *gyrB* in widely divergent organisms. *recF* might be located elsewhere on the *Buchnera* chromosome, or the protected niche occupied by *Buchnera* and their genetic isolation might make RecF function unnecessary. *Spiroplasma citri,* a plant-pathogenic, insect-transmitted organism occupying protected niches, also does not contain *recF* between *dnaN* and *gyrB* (174).

In *P. putida* and *B. subtilis,* a 15-kb DNA fragment bearing *dnaA* also contains 12 different homologous genes arranged in the same order (125). In *E. coli* and B(*S. graminum*) the genes in the conserved order are fewer, consisting only of *50 kDa* to *gyrB* (Figure 4a). The order of genes in *P. putida* and *B. subtilis* is believed to be the ancestral arrangement, and the *E. coli* order probably results from a translocation of a chromosomal segment (125). The genes to the left of *50 kDa* (Figure 4a) are different in *Buchnera* and *E. coli,* suggesting translocation events after their divergence from a common ancestor (Figure 3).

Members of the *Enterobacteriaceae* differ from other organisms studied in that their origin of replication is about 40 kb from *dnaA* (175). Origins of replication have several characteristic features, the most prominent of which are an AT-rich direct repeat and several copies of a sequence of nine nucleotides known as a DnaA box, to which the DnaA protein binds (57, 175, 179). In *E. coli* and other organisms, one DnaA box is present directly upstream of *dnaA* and is involved in the autoregulation of synthesis of DnaA (179). B(*S. graminum*) lacks sequences resembling DnaA boxes between *rpmH* and *dnaA* (Figure 4a) (95). This observation is striking because DnaA boxes have been detected in such diverse species as *E. coli, P. putida, Micrococcus luteus, B. subtilis, Streptomyces coelicolor, Mycoplasma capricolum,* and *S. citri* (15, 49, 174, 175). *M. capricolum* and *S. citri* have a G+C content slightly lower than *Buchnera* and contain DnaA boxes, indicating that a low G+C content

Table 2 Additional genes found in B(*S. graminum*) and some properties of the deduced products

General category	Gene	Protein	% Identity to *E. coli*	Function or property	Figure; references
ATP synthesis	*atpD*	ATP synthase β-subunit	(93)[a]	Generation of ATP from proton motive force	25
Sugar metabolism	*gapA*	Glyceraldehyde-3-phosphate dehydrogenase	84	Enzymatic activity	25, 86
	tpiA	triose phosphate isomerase	47	Enzymatic activity	4g; b
Chaperonins	*secB*	SecB	47	Protein folding and unfolding, secretion, stress response	94
	groEL[c]	GroEL	85		4a; d
	groES[c]	GroES	78		4a; d
DNA-binding and -bending proteins	*himD*	Integration host factor β-subunit	64	Multifunctional protein, alters gene expression	4g; b
Unknown	*50 kDa*	50 kDa	49	Essential for cell	4a; d
	60 kDa	60 kDa		Essential for cell	4a; d
	ORF-V	ORF-V	29	Function unknown	6; 118
	ORF-VI	ORF-VI	43	Function unknown	6; 118
	P14	P14	(65)	Function unknown	6; 118
	ORF-A	ORF-A	51[e]	Transmembrane protein?	6; 118

ENDOSYMBIONTS OF APHIDS 71

	ORF-C	ORF-C		
	sohB[f]	SohB[f]	57[f]	Membrane associated, ATP-binding protein	4f; 86

				Periplasmic protease?	6, 96
Amino acid biosynthesis	trpE[g]	Anthranilate synthase, α-subunit	61	Enzymatic activity	6; 93
	trpG[g]	Anthranilate synthase, β-subunit	59	Enzymatic activity	6; 93
	trpD[g]	Phosphoribosylanthranilate transferase	57	Enzymatic activity	6; 118
	trpC(F)[g,h]	Indoleglycerol phosphate synthase and N-phosphoribosylanthranilate isomerase	54	Enzymatic activity	6; 118
	trpB[g]	Tryptophan synthase, β-subunit	75	Enzymatic activity	6; 118
	trpA[g]	Tryptophan synthase, α-subunit	57	Enzymatic activity	6; 118
	aro[b]	5-Enolpyruvoylshikimate-3-phosphate synthase	(68)	Enzymatic activity	4g
	aroH	DAHP synthase (Trp-sensitive)	80	Enzymatic activity	4e; 87
	aroE	Shikimate dehydrogenase	48	Enzymatic activity	5b; 142
	cysE	Serine acetyltransferase	52	Enzymatic activity	4d; 94

[a] Numbers in parentheses indicate partial sequence.
[b] MA Clark, unpublished observations.
[c] Also cloned and sequenced from B(*A. pisum*) (127, 128).
[d] N Ogasawara & H Yoshikawa, unpublished observations.
[e] Comparisons with ORF-A of B(*S. chinensis*) (96).
[f] Gene or protein from B(*S. chinensis*) (96).
[g] Also cloned and sequenced from B(*S. chinensis*) (96).
[h] Gene fusion.

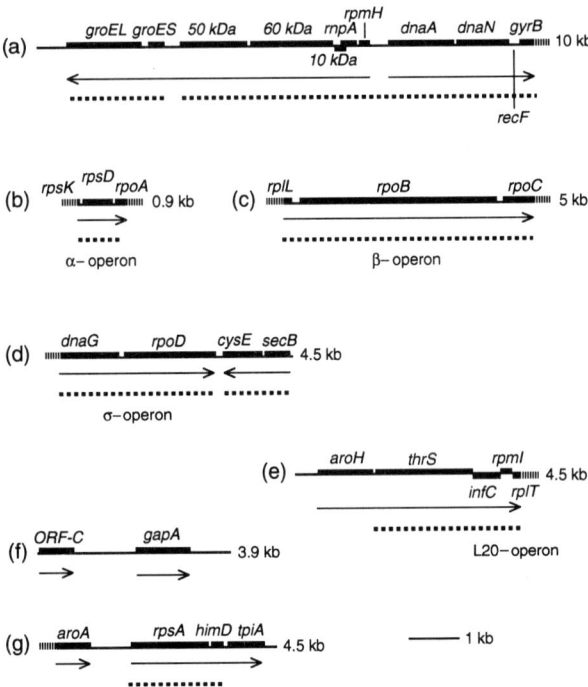

Figure 4 Genetic map of B(*S. graminum*) DNA fragments. Thin line, DNA; thick solid line, regions coding for protein; striped thick line, partial sequence of gene; arrow, direction of transcription; dashed line, linkage relationship found in *E. coli*. For explanation of gene designations, see Tables 1 and 2.

does not preclude their presence. Furthermore, a plasmid found in B(*S. graminum*) has DnaA boxes and an AT-rich direct repeat (see section on tryptophan biosynthesis). These results indicate that B(*S. graminum*), like *E. coli*, has its origin of replication elsewhere on the chromosome; they also suggest that DnaA synthesis is not autoregulated. In an endosymbiont, whose growth must be integrated with that of the host, an alternative host-controlled mechanism of regulation of DnaA synthesis may be in effect. *Borrelia burgdorferi* also lacks DnaA boxes upstream of *dnaA* (144a); this organism is in a separate evolutionary lineage far distant from the species considered above.

In B(*S. graminum*) and *E. coli*, *dnaG* and *rpoD* (Figure 4*d*) are next to each other; in *E. coli* both are part of the σ-operon (14). B(*S. graminum*) *cysE* and *secB* (Figure 4*d*) are also adjacent but are 81 min distant from *dnaG* and *rpoD* on the *E. coli* chromosome (5). Endosymbiont genes for RNA-polymerase core-enzyme, initiation factor-3, and ribosomal proteins (Figure 4*a–c,e*) have the same order as in *E. coli,* where they are part of three major operons (83). In general, comparisons of linkage relationships of B(*S. graminum*) with those of *E. coli* indicate conservation of some gene orders and considerable rearrangement of others.

In *E. coli* and other organisms some of the structural genes on the fragments shown in Figure 4 have neighboring nucleotide sequences that are involved in regulation of gene expression. It is usually not possible to use similarity to *E. coli* consensus sequences to designate potential −10 and −35 regions in *Buchnera* DNA, because the average G+C content of its intergenic regions is 17.6 mol%, and consequently, many sequences related to the *E. coli* −10 consensus sequence are found. *Rickettsia* spp., which also have a low G+C content, represent a similar situation (170). One type of identifiable and potentially significant regulatory sequence is an adjacent inverted repeat followed by a string of Ts, an arrangement characteristic of rho-independent terminators (98). In six different eubacterial species, *rpoD* is followed by such a sequence (94); no inverted repeat follows B(*S. graminum*) *rpoD* (94). In *E. coli, rpoB* is preceded by an inverted repeat corresponding to an RNase III processing site (44); a similar repeat precedes B(*S. graminum*) *rpoB* (Figure 4c) (23). Regulation of *thrS* in *E. coli* involves translational control and a readily recognizable nucleotide sequence with an extensive secondary structure resembling tRNA (8, 110). The lack of such a sequence upstream of B(*S. graminum*) *thrS* (Figure 4e) indicates the absence of translational control of the type found in *E. coli.* In general, nucleotide sequences known to have a regulatory function in *E. coli* were absent from B(*S. graminum*). The significance of these observations is not known.

rRNA Operons

In most bacteria, the genes for rRNA are organized as a single operon in the order *rrs-rrl-rrf* (16S-23S-5S rRNA) (83). Usually one or two tRNAs are found between *rrs* and *rrl;* in some operons tRNAs also follow *rrf*. The resulting messenger RNA is processed to the individual RNA species. Preceding the transcription start site are one or two tandem promoters, and at the end of the messenger RNA are direct nucleotide repeats that are rho-independent terminators. The sequences of *rrs, rrl,* and *rrf* are highly conserved among bacteria (171). Additional conserved sequences that precede *rrs* and *rrl* are designated *boxA* and *boxC* (151). *boxA* is involved in antitermination; the function of *boxC* is not known. Because ribosomes are major constituents of bacteria, the promoters for rRNA are among the strongest in the cell. Free-living bacteria such as *E. coli* and *B. subtilis* contain seven and ten rRNA operons, respectively. Slow-growing bacteria, including intracellular parasites such as *R. prowazekii,* contain one copy of the rRNA operon (132, 172).

Buchnera rRNA genes differ from most eubacteria in their organization into

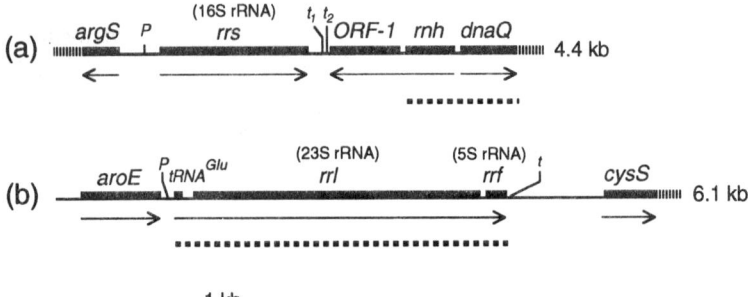

Figure 5 Genetic map of two B(*S. graminum*) DNA fragments containing rRNA genes. P, promoter; t, terminator. For explanation of gene designations, see Tables 1 and 2.

two transcription units consisting of (*a*) *rrs* and (*b*) tRNA^{Glu}-*rrl*-*rrf* (Figure 5). Each is present as a single copy in the *Buchnera* genome (117, 142). The nucleotide sequence of these two transcription units and adjacent genes was obtained initially for B(*S. graminum*). Because sequences of regulatory importance are usually conserved among closely related organisms, we obtained nucleotide sequences of regions upstream of *rrs* and *rrl* for *Buchnera* from six additional species of aphids. The results for *rrs* indicate conservation of sequences corresponding to the −35 (TTGACA) and −10 (T^G/$_A$TAAT) promoter regions as well as *boxA* and *boxC*. Similarly, for tRNA^{Glu}-*rrl*-*rrf*, the results indicated conservation of sequences corresponding to −35 (TTGACT) and −10 (TGTA^A/$_T$T), tRNA^{Glu}, *boxA*, and *boxC* (142). Both *rrs* and *rrf* are followed by one or two direct nucleotide repeats suggestive of rho-independent terminators. In the case of *rrs*, the putative promoters from the endosymbionts of seven species of aphids and the putative B(*S. graminum*) terminator function in *E. coli* (117). The organization of rRNA genes into two transcription units has so far been found only in *B. burgdorferi* (55), *Mycoplasma gallisepticum* (21), *Pirellula marina* (99), and *Thermus thermophilus* (66). None of these are members of the *Proteobacteria*, which includes *Buchnera* (119, 152).

In several *E. coli* rRNA operons, tRNA^{Glu} is located between *rrs* and *rrl* (83). Consequently, in the early *Buchnera* lineage, a translocation of a portion of the *rrs*-tRNA^{Glu}-*rrl*-*rrf* operon probably occurred in which *rrs* gained a terminator and tRNA^{Glu}-*rrl*-*rrf* a promoter. The resulting rearrangement has some convenient practical applications in that it is possible to identify *Buchnera* using PCR and oligonucleotides complementary to *aroE* and *rrl* (Figure 5*b*) (143). The potential diagnostic specificity of this approach is great because it tests for linkage between these two genes, an indication of an unusual rRNA operon. Other methods test for the linkage of *argS* to *rrs* (Figure 5*a*) and for the presence of *Buchnera*-specific sequences in *rrs* (143). The arrangement of rRNA genes of the S-endosymbiont of *A. pisum* resembles that of most other eubacteria in having the gene order *rrs-rrl* (160). The space between these genes contains tRNA^{Glu}. The bacteriocyte-associated endosymbionts of tsetse, which are also in the γ-3 subdivision of the *Proteobacteria*, differ from *Buchnera* in having the order *rrl-rrs* (1, 2). They resemble *Buchnera* and other slow-growing bacteria in having only one copy of the rRNA operon (1).

GroES and GroEL

Research over the past decade has shown that all cells contain proteins belonging to highly conserved families whose function is to control the higher-order structure of proteins. These proteins have been called chaperonins, and their major role is to prevent misfolding, which might arise during protein synthesis, translocation of proteins across membranes, and recovery from stress (58, 68, 164, 178). One such family of proteins, designated GroEL (or hsp60), is common to prokaryotes as well as to mitochondria and chloroplasts, which are prokaryote-derived organelles (68, 178). In organelles, GroEL-homologues may have become adapted to ensuring the proper folding of proteins imported from the cytosol (68). In *E. coli* and many of the prokaryotes examined, the gene for GroEL is part of the *groESL* operon (68, 178). Both GroES and GroEL are essential for cell survival (48, 178). GroEL has a M_r of 57,300, is composed of two rings of seven subunits, and has ATPase activity; GroES has a M_r of 10,300 (62, 178). Transcription of the *groESL* operon is regulated by σ^{32}, which recognizes −10 and −35 sequences that are different from those recognized by σ^{70}, the principal *E. coli* σ-factor (62, 178). This transcription confers on the operon the typical pattern of heat-shock expression (177). The characteristic −10 and −35 sequences recognized by σ^{32} have been found in different eubacteria (140). *E. coli* also exhibit a low level of *groESL* expression from a σ^{70}-directed promoter (177, 178).

A variety of stimuli, including heat, heavy metals, oxygen radicals, ethanol, antibiotics, starvation (for carbon, nitrogen, or phosphorous), DNA-damaging agents, alkali shift, bacteriophage infection, abnormal polypeptides, and anaerobiosis, can lead to a major increase in the amount of GroEL and other proteins (58, 62, 101, 162, 164, 178). An increase in the rate of synthesis of these proteins is usually accompanied by a decrease in the rate of synthesis of other proteins. Because the initial stimulus used was heat-shock, these proteins have been called heat-shock proteins. After the more general nature of the stimulus was established, they were often called stress proteins. These designations are unfortunate, because many or all of these proteins are essential to the cell and are overproduced in response to a variety of stimuli. For example, GroEL, one of the major proteins found in *E. coli*, constitutes 1% of the total cell protein at 30°C, whereas at 46°C it increases to 10% (58, 62).

More recent studies have shown that intracellular pathogenic bacteria have elevated levels of GroEL, a finding attributed to the hostile intracellular environment (56, 114, 140, 164, 176). Similarly, *Rhizobium* spp. in plant nodules as well as the intracellular X-symbiotic bacterium of *Amoeba proteus* have elevated levels of this protein (22). It is perhaps somewhat artificial to always equate increases of GroEL protein with stressful conditions imposed on the cell because this assumes that the amount of GroEL found in cells growing under laboratory conditions is the norm. Instead, elevated levels of GroEL and other stress proteins may simply reflect a homeostatic adjustment of the cell to different environments, including its usual environment or any it may encounter as part of its life cycle.

Work from Ishikawa's laboratory has demonstrated the heat-shock response in isolated cells of B(*A. pisum*) and the presence of *groEL* and *groES* (designated by the authors as *symL* and *symS*, respectively) (113, 127, 128). Isolated cells incubated at elevated temperatures contained increased levels of 45-, 63-, and 73-kDa proteins (113). These levels were also increased when the cells were treated with arsenate, ethanol, or $CdCl_2$. The 63-kDa protein reacted with antiserum to B(*A. pisum*) GroEL (64, 113). In electron micrographs, the purified GroEL had the characteristic double-ring appearance previously observed with the *E. coli* protein (178). B(*A. pisum*) GroEL had ATPase activity and could partially reconstitute denatured *Rhodospirillum rubrum* ribulose-1,5-biphosphate carboxylase in the presence of ATP and *E. coli* GroES, thereby indicating chaperonin activity. Immunohistochemistry with antibody to B(*A. pisum*) GroEL localized this protein exclusively in endosymbionts of maternal and embryonic bacteriocytes (53, 63).

The *groES* and *groEL* genes were first cloned and sequenced from B(*A. pisum*) (127, 128) and later from B(*S. graminum*) (N Ogasawara & H Yoshikawa, unpublished observations) (Table 2). In the latter endosymbiont, the two genes are located next to *50 kDa*, a gene in the vicinity of *dnaA* that codes for a protein of unknown function (Figure 4a) (125). B(*A. pisum*) DNA containing *groEL* and *groES* could complement *E. coli* temperature-sensitive *groEL* and *groES* mutants (128). The sequences of GroES and GroEL from B(*S. graminum*) and B(*A. pisum*) were 91 and 97% identical, respectively. A comparison of the extragenic regions of the DNAs of these endosymbionts showed that only the sequences upstream of *groES* resembling the −10 and −35 sequences recognized by *E. coli* σ^{32} (62, 128), and of a segment downstream of *groEL* that included an inverted repeat followed by a row of Ts characteristic of a rho-independent terminator, were conserved. These results suggest that *groES* and *groEL* of *Buchnera* are contained within an operon similar in organization to that of *E. coli* (68, 178). One possible difference noted in B(*A. pisum*) is that *groEL* appears to be more highly expressed than *groES* (84, 128). Evidence also indicates that B(*A. pisum*) produces high levels of GroEL (84, 113). The presence of elevated levels of GroEL in bacteriocyte endosymbionts of tsetse has been well documented (1). Endosymbionts within isolated tsetse bacteriocytes incorporated radioactive methionine into various proteins, of which GroEL appeared to be one of the more prominent (1).

Ishikawa and associates have performed extensive experiments on the effect of eukaryote- and prokaryote-specific antibiotics on endosymbiont protein synthesis in vivo and in vitro (77, 80). In aphids injected with cyclohexamide (to eliminate host protein synthesis) and radioactive amino acids, the endosymbionts incorporated virtually all of the radioactivity into a protein that in gels migrated at a position corresponding to 63 kDa (77, 78). In contrast, isolated endosymbionts incorporated radioactive amino acids into various proteins, and this process was not affected by cyclohexamide (77, 78). The 63-kDa protein detected in polyacrylamide gels was called symbionin, and the conclusions derived from these and other experiments involving antibiotics were the basis of a model of the symbiotic association between *Buchnera* and aphids (77, 80).

The following are the key characteristics of this model, which is in part based on an analogy between endosymbionts and organelles such as mitochondria and chloroplasts: (*a*) When inside the aphid, the endosymbiont preferentially synthesizes the 63-kDa protein to the near exclusion of other proteins. (*b*) Removal of the endosymbiont from the aphid releases the control that prevented the synthesis of proteins other than symbionin. (*c*) The 63-kDa protein was said to be the basis of the symbiotic association between *Buchnera* and aphids. (*d*) Antibiotic inhibition studies were interpreted to indicate that the gene for the 63-kDa protein resides in the nucleus of aphid cells and that the messenger RNA was imported into the endosymbiont and translated using the *Buchnera* protein synthesis machinery (77).

The identification of symbionin as GroEL as well as other recent work from Ishikawa's laboratory (84, 113, 128) and others (see previous sections) makes this model untenable. One point, namely the conclusion that GroEL is virtually the sole protein made by *Buchnera*, needs to be addressed (77). During aphid growth *Buchnera* numbers increase in parallel to aphid weight and consequently GroEL cannot be the only protein made by endosymbionts because there must be an orderly increase in the enzymatic machinery and other cellular constituents as is characteristic of bacterial growth (7). Although GroEL is overexpressed in *Buchnera*, the conclusion that the synthesis of other proteins is virtually nonexistent is probably derived from noncomparable experiments involving in vivo and in vitro incorporation of radioactive amino acids by endosymbionts. The autoradiograms indicate that much more radioactivity is incorporated by endosymbionts in vitro than in vivo. This observation is not surprising because the injected radioactive amino acids have to transverse the bacteriocytes and the symbiosome membrane to reach the endosymbionts. Furthermore, the injection of cyclohexamide (to eliminate host protein synthesis) into the aphid may itself trigger a stress response in the endosymbiont, leading to depression of synthesis of most proteins and increased synthesis of GroEL (58). Consequently, low incorporation of radioactive amino acids in the in vivo experiments and the possibly resulting activated stress response may prevent the detection of proteins other than GroEL. It should also be noted that in isolated tsetse bacteriomes, although there is considerable incorporation of radioactive amino acids into GroEL, there is also incorporation into other proteins in the presence of α-amanitin, an antibiotic that inhibits transcription of host genes (1).

Metabolic Studies

Symbiosomes containing B(*A. pisum*) were isolated from aphid embryos (167) and shown to take up the tricarboxylic acid cycle intermediates, aspartate, and glutamate and convert these substrates to CO_2. Oxygen was also consumed. These results indicate that *Buchnera* is capable of aerobic respiration and suggest the presence of the tricarboxylic acid cycle. Glucose was poorly metabolized; other carbohydrates were not tested.

ROLE OF *BUCHNERA* IN THE SYMBIOTIC ASSOCIATION

Requirement of Buchnera by the Aphid Host

The complex and elaborate mechanisms employed by aphids to assure the transmission of *Buchnera* to their offspring (12, 70) would arguably imply an essential role of the endosymbionts in the life of the aphid (72). A more direct demonstration of this dependence is an examination of the consequences of the elimination of *Buchnera* by a variety of treatments that presumably have little or no effect on the host. The treatments used include antibiotics (present in the diet or injected into the aphid) as well as heat (17, 38, 72, 80, 126). The most frequently used antibiotics are chlortetracycline and rifampicin, which at higher concentrations may have an effect on the host (17, 61). Some species of aphids can grow on synthetic media containing all the amino acids as well as a variety of other constituents (109); studies have included antibiotics in synthetic diets (reviewed in 30, 38, 109). Here, we describe a recent study examining the effect of chlortetracycline on aphid growth in both synthetic media and on plants (40). The results presented in Table 3 indicate that most of the first-generation aphids survive to adulthood no matter the source of food. Compared with growth on plants (without antibiotic), both the addition of antibiotic and the synthetic diet without antibiotic (*a*) increase the time required to reach adulthood, (*b*) decrease the weight of the adult, and (*c*) decrease the relative growth rate. Aphids grown on a synthetic diet also have fewer off-

Table 3 Comparison of the effect of chlortetracycline treatment on growth and survival of *A. pisum* grown on plants or a synthetic diet[a]

	Plant		Synthetic diet[b]	
Chlortetracycline	−	+	−	+
Survival to adulthood	100%	80%	80%	77%
Time to adulthood (days)	8.2	10.0	10.4	14.2
Weight (mg)/adult aphid	3.06	0.49	0.95	0.44
Relative growth rate	1.00	0.40	0.51	0.26
Viable offspring/aphid	59	None[c]	14	None[c]

[a] Compiled from Reference 40.
[b] *A. pisum* survived for at least two generations on this synthetic diet (without antibiotic).
[c] A few were stillborn, or if alive, died within a few days.

spring; the effect of the antibiotic in both plants and synthetic diets is the elimination of viable progeny. These results as well as others (reviewed in 17, 38, 72, 80, 126) indicate that *Buchnera* has a role in aphid growth and is essential for continued reproduction.

Nutritional Studies

There has been a great deal of speculation about the contribution(s) that *Buchnera* makes to the aphid host (17, 38, 72, 80); there is little if any direct experimental evidence. It has been suggested that the endosymbionts synthesize pectinases, which are involved in the degradation of the intracellular plant matrix, allowing stylet penetration and feeding of the aphid on plant sap (18). Some credence to this argument appeared to come from a genetic study that purported to show that the ability to overcome host-plant resistance was inherited as an extranuclear trait, presumably owing to modification of an endosymbiont-made pectinase (46). A more extensive genetic study did not support this conclusion and indicated that this trait is polygenic (137, 138). The previous conclusion that B(*S. graminum*) provides the aphid with sterols has also been disproved (19).

A number of studies suggest that endosymbionts may be involved in the synthesis of amino acids and, possibly, vitamins (reviewed in 30, 38, 72). Aphids, like other animals, are thought to require 10 essential amino acids (i.e. arginine, histidine, isoleucine, leucine, lysine, methionine, threonine, tryptophan, valine, and phenylalanine). The diet of aphids is rich in carbohydrates but poor in nitrogenous compounds (41, 149), and one role of the endosymbionts may be to synthesize the essential amino acids and make them available to the host. Experiments designed to test this possibility involve the use of synthetic diets with and without the essential amino acid and in the presence or absence of antibiotics (see above). Ideally, these experiments would show that growth and reproduction is unaffected by the omission of the amino acid because it is synthesized by the endosymbiont. Introduction of an antibiotic would result in dependence on the essential amino acid because the endosymbionts would be eliminated. In practice, such clear-cut results are not obtained, and interpretation is often difficult because of several complications: (*a*) Long-term maintenance of the aphid on any diet is not possible in the presence of an antibiotic (Table 3), and it may also not be possible on a synthetic diet in the absence of an essential amino acid. (*b*) The antibiotics used reduce aphid growth, and they may have some undetected effect on the host. (*c*) Microbial contamination of the synthetic diet could modify the results. (*d*) The feeding behavior of aphids is complex, and addition of antibiotics or nutrient omissions may modify the rate of food utilization. Because of these problems, most experiments involve comparisons of aphid growth and the number of first-generation progeny produced on synthetic diets. The inocula used are aphids, newly born to mothers that had been reared on plants. In synthetic media containing antibiotics, increased growth resulting from the presence of an amino acid, as compared to its omission, is taken as an indication of amino acid synthesis by the endosymbiont. In spite of these complications, the nutritional studies as a whole have been interpreted to indicate that one function of *Buchnera* is the synthesis of some or all of the essential amino acids and, in some aphids, additional amino acids as well (30, 108, 109, 148). Perhaps the nutritional study with the most clear-cut results is a recent one indicating that tryptophan is produced by the endosymbionts (43) (discussed in the section on tryptophan biosynthesis).

Potential Modifications of Biosynthetic Pathways

As previously indicated, *Buchnera* is found within symbiosomes in the aphid bacteriocytes (Figure 1*a,b*). Potentially, the host-derived symbiosome membrane could be the site regulating nutrient availability to the endosymbiont and the export of products of its metabolism. Compared to a free-living organism, *Buchnera* is in a sheltered and, probably, a relatively constant environment during aphid growth. If this is the case, many of the mechanisms designed to regulate the activity of biosynthetic pathways in response to the external availability of the end product need not be present. In fact, if the endosymbionts are to provide their host with amino acids, these mechanisms would have to be modified to allow amino acid overproduction.

The activity of many amino acid biosynthetic pathways is regulated at the level of enzyme synthesis and/or feedback inhibition of the initiating enzyme reaction by the end product of the pathway (123). For *Buchnera* to overproduce these amino acids for the host, changes in these regulatory mechanisms would be required. Two potential modifications are the constitutive synthesis of enzymes of a biosynthetic pathway and changes in the initiating enzyme, which would nullify feedback inhibition by the end product. Regulatory mutations resulting in constitutive synthesis frequently are observed in studies of directed evolution involving the utilization of new substrates (26). Because some end-product accumulation is expected regardless of the efficiency with which it is removed from the symbiont, some mechanism must ensure sufficient activity of the initiating enzyme in the presence of the end product. One such mechanism would be a mutation resulting in loss of allosteric inhibition. Under laboratory conditions, mutants that overproduce amino acids frequently are desensitized to the allosteric inhibition of a key regulatory enzyme (32). Because the activity of many allosteric enzymes is not fully inhibited by their feedback inhibitors, an alternative solution is to increase the total activity in the presence of the effector by overproduction of the enzyme. In fact, gene amplification is a frequent response of a population of cells subjected to growth limitation resulting from inadequate enzyme activity or antibiotics (3, 153). In prokaryotes, tandem duplication of chromosomal or plasmid-borne genes may occur at a frequency of 10^{-4} to 10^{-5}, which is considerably higher than the mutation frequency in structural genes altering feedback inhibition or enzyme specificity (3, 45). In most cases, duplications are unstable and are lost once the selective pressure is removed.

Tryptophan Biosynthesis

Evidence indicates that B(*A. pisum*) and B(*S. graminum*) can synthesize the essential amino acid tryptophan and supply it to the aphid host. Nutritional studies show that for the first generation, growth and survival of B(*A. pisum*) on a synthetic diet is unaffected by the absence of tryptophan (43). Addition of chlortetracycline to a tryptophan-deficient synthetic diet results in death of virtually all aphids, whereas most of the aphids survive to adulthood on a chlortetracycline-containing diet when this amino acid is supplied. The tryptophan content of honeydew from aphids treated with chlortetracycline also was greatly reduced (135). Tryptophan synthase (TrpBA) activity has been detected in extracts of whole aphids and isolated B(*A. pisum*) but not in chlortetracycline-treated aphids (43).

The comparative genetics and biochemistry of the tryptophan biosynthetic pathway have been studied extensively (28). The pathway from chorismate to tryptophan involves five reactions and seven proteins (some of which may be fused to make bifunctional enzymes). In almost all organisms, tryptophan biosynthesis is regulated by feedback inhibition of anthranilate synthase (TrpEG) by tryptophan (28, 29). In *E. coli*, tryptophan mediates the repression of an operon containing genes coding for all of the enzymes of tryptophan biosynthesis [*trpEG(D)C(F)BA*] (173). In this and some other organisms, levels of tryptophan biosynthetic enzymes are also regulated by attenuation (transcription termination) based on the availability of charged tRNA^{Trp} (28, 173). A property characteristic of this attenuation mechanism is the presence of DNA coding for a short tryptophan-containing peptide preceding the structural gene for TrpEG. All of these regulatory mechanisms, which are characteristic of free-living bacteria, respond to accumulation of tryptophan within the cell.

S. graminum is a fast-growing aphid that during parthenogenetic reproduction reaches maturity in about one week (7, 34). The tryptophan biosynthetic pathway of its endosymbiont is organized into two transcription units (93, 118), one chromosomal and the other plasmid-borne. *trpEG* is present as four tandem repeats of a 3.6-kb DNA fragment that makes up a plasmid of 14.3 kb (Figure 6) (93). Each 3.6-kb repeated unit contains two DnaA boxes and an AT-rich direct repeat, characteristics of an origin of replication (93, 175). Relative to the remaining genes of the pathway [*trpDC(F)BA*] (Figure 6), which are located as a single copy on the B(*S. graminum*) chromosome, *trpEG* is amplified 14- to 15-fold (93). The deduced amino acid sequence of TrpE contains all of the residues known to be involved in feedback inhibition (16, 102); there is no evidence for an attenuation mechanism of the type found in *E. coli*. These observations suggest the following adaptations to the symbiotic association: In order for B(*S. graminum*) to overproduce tryptophan for the aphid host, the enzymes of the tryptophan biosynthetic pathway have to be produced constitutively (28, 44). A subsequent translocation event resulted in the removal of *trpEG* from the B(*S. graminum*) chromosome onto a plasmid, followed by tandem duplications. Because the activity of TrpEG is not fully inhibited by tryptophan and as gene amplification provides an increase in the enzyme level, synthesis of tryptophan could take place even under conditions of end-product

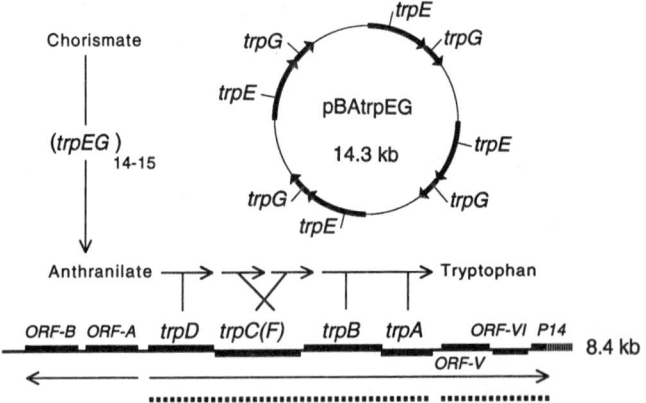

Figure 6 Genetics of tryptophan biosynthesis in B(*S. graminum*). *trpEG* is on a plasmid and is amplified 14- to 15-fold relative to the remaining genes of the pathway, which are present as one copy on the chromosome. For explanation of symbols, see Table 2.

accumulation. An analogous situation occurs in *Corynebacterium glutamicum*, in which increasing allosterically inhibitable TrpEG leads to a major accumulation of tryptophan in the medium (85).

Amplification of *trpEG* on plasmids may be widespread in rapidly growing members of the Aphididae (Figure 3). We have found that in B(*R. maidis*) the plasmid consists of a single 3.7-kb unit, whereas B(*A. pisum*) plasmids contain five, six, and ten tandem repeats of a 3.7-kb unit (C-Y Lai, D Rouhbakhsh & L Baumann, unpublished data). For *A. pisum,* whether different aphids have endosymbionts containing one-size plasmids or plasmids of all three sizes is unknown. The 3.7-kb units from B(*R. maidis*) and B(*A. pisum*) were cloned and sequenced. Comparisons indicated high sequence conservation of *trpEG* and a region upstream of *trpE*. This region has sequences resembling an origin of replication and presumably the promoter of *trpEG* (93, 175). A comparison of the intergenic region between *trpEG* and the putative origin of replication did not reveal any sequence conservation. A phylogenetic analysis of *trp* and the deduced protein products from B(*S. graminum*), B(*R. maidis*), B(*A. pisum*), and B(*S. chinensis*) gave the same order of branching as observed with 16S rDNA (Figure 3), indicating vertical evolution of the *trpEG* genes and 16S rDNA (112). In the case of B(*S. graminum*), the sequence of the intergenic region from two 3.6-kb tandem repeats is virtually identical (93). The lack of sequence conservation in the intergenic region of plasmids from the endosymbionts of closely related aphid species (e.g. *S. graminum* and *R. maidis*) indicates that the sequence change in this region is relatively rapid. The virtual identity of the same intergenic region in at least two tandem repeats in the same plasmid, and the presence of plasmids of five, six, and ten tandem repeats in the endosymbionts of a single aphid species, suggests that their number is subject to rapid variation, an increase possibly related to periods of high nutrient availability and rapid growth.

In contrast to the fast-growing members of the Aphididae, *S. chinensis* and *Melaphis rhois* (members of the Pemphigidae; Figure 3) have a slow development time, maturing within about a month (34, 158, 159). Curiously, when the genes for the tryptophan biosynthetic pathway of B(*S. chinensis*) were cloned and sequenced, a single chromosomal copy of *trpEG* was found unlinked to *trpDC(F)BA* (96). Preliminary results also indicated a chromosomal location for B(*M. rhois*) *trpEG* (C-Y Lai, unpublished data). The lack of amplification of *trpEG* in these endosymbionts suggests that the tryptophan demand of these slow-growing aphids may be satisfied by a lower rate of synthesis.

In *E. coli,* the *trp* operon is followed by six ORFs (designated ORF-I–VI) (155). Only ORF-V and ORF-VI have been detected in B(*S. graminum*) (Figure 6) (118), and their conservation suggests a function. Upstream of *trpD* of B(*S. graminum*) is ORF-A, which is also found in B(*S. chinensis*) (96). In the latter organism, ORF-B (Figure 6) is replaced by a gene with a high sequence similarity to *E. coli sohB,* which encodes a putative periplasmic protease (10, 96). This difference indicates a chromosomal rearrangement during the 90–180 million years that have elapsed since the hosts (*S. graminum* and *S. chinensis*) and their endosymbionts diverged (112).

Genes of the Common Pathway of Aromatic Amino Acid Biosynthesis

The synthesis of aromatic amino acids involves a set of common reactions up to the intermediate chorismate, which is the branch point for pathways leading to the synthesis of tryptophan, phenylalanine, and tyrosine (69, 133). In addition to the evidence that *Buchnera* can make tryptophan, there is nutritional evidence for the synthesis of phenylalanine (108); both observations imply that *Buchnera* contains the common portion of the pathway. The synthesis of aromatic amino acids is initiated by a reaction in which erythrose 4-phosphate and phosphoenol pyruvate are converted to 3-deoxy-D-*arabino*-heptulosonate 7-phosphate (DAHP). The rate of this reaction, catalyzed by DAHP synthase, is regulated in bacteria (69, 81, 133). *E. coli* has three isofunctional DAHP synthases (*aroF, aroG, aroH*) that are feedback inhibited by tyrosine, phenylalanine, and tryptophan, respectively (133). AroH is present in the least amount and its maximal inhibition by tryptophan is only 60% (139). The residual activity is thought to be required for the synthesis of vitamins and other compounds that branch off from the aromatic amino acid pathway (69, 133). In B(*S. graminum*) only one gene for DAHP synthase has been detected (Figure 4e, Table 2) and it codes for a protein with an amino acid sequence most closely resembling *E. coli* AroH (87). In addition, B(*S. graminum*) contains *aroA* and *aroE* (Figure 5b, Table 2), which encode 5-enolpyruvoylshikimate-3-phosphate synthase and shikimate dehydrogenase, respectively, enzymes that are also part of the common pathway of aromatic amino acid biosynthesis.

Sulfate Reduction and Cysteine and Methionine Biosynthesis

Insects cannot reduce sulfate to sulfide (30). There is good evidence that B(*M. persicae*) can perform this process, as well as synthesize cysteine and methionine (37, 39). The pathways of sulfate reduction and cysteine biosynthesis have been studied extensively in *Salmonella typhimurium* and *E. coli* (88, 89). In these organisms, an uptake system and four enzymatic steps result in the reduction of sulfate to sulfide. Cysteine biosynthesis involves the conversion of serine and acetyl CoA to O-acetyl serine by serine transacetylase (CysE); subsequently, sulfide and O-acetyl serine are converted to cysteine. In *E. coli* and *S. typhimurium* the enzymes of this pathway are constitutive and cysteine biosynthesis is regulated by feedback inhibition of CysE by this amino acid (88, 89). Cysteine is the source of sulfur for methionine (27).

Through the use of synthetic diets containing ^{35}S-sulfate, it was shown that the aphid *M. persicae* incorporates ^{35}S into proteins; cysteine and methionine were identified as the radioactive products (37, 39). Treatment of the aphid with chlortetracycline abolished incorporation. Isolated embryos (which contain *Buchnera*) incorporated ^{35}S-sulfate, while isolated aphid guts (which lack endosymbionts) did not. After the initiation of feeding, radioactivity was detected first at the posterior end of the aphid and subsequently at the anterior end, consistent with the synthesis of cysteine and methionine by B(*M. persicae*) in the bacteriome and its movement to other parts of the aphid.

During the genetic characterization of *rpoD* of B(*S. graminum*), an adjacent open reading frame was found that was 52% similar to *E. coli* CysE (94). A striking difference between the deduced amino acid sequences of B(*S. graminum*) and *E. coli* CysE was the complete dissimilarity of their C-terminal regions. An *E. coli* cysteine-excreting mutant contains a single amino acid substitution (Met→Ile) in the C-terminal portion of CysE (32). This mutation resulted in desensitization of the enzyme to feedback inhibition by cysteine and excretion of this amino acid into the medium. Because the location of this mutation falls within the C-terminal region of B(*S. graminum*) CysE, which is totally different from that of *E. coli,* the endosymbiont enzyme probably is not subject to feedback inhibition by cysteine. As with the *E. coli* mutant, this difference in the C-terminal region of B(*S. graminum*) *cysE* could lead to overproduction of cysteine and its utilization by the host. B(*S. graminum*) *cysE* could not complement *E. coli cysE* mutants, even when placed under the control of the *tac* promoter (92). Attempts to overproduce the endosymbiont CysE in *E. coli* were unsuccessful owing to extensive degradation of the protein (92). Consequently, the effect of cysteine on the activity of the enzyme from *Buchnera* could not be tested.

CONCLUSIONS AND PERSPECTIVES

In the past decade a considerable amount of new information has been obtained on the evolutionary relationships and the general genetics and physiology of aphid endosymbionts. This information is primarily descriptive and allows us to make some conclusions as to the nature of *Buchnera* and some of the contributions it makes to the aphid host. The following are some of the principal conclusions for which there is a reasonable experimental basis (references were cited in the preceding text):

1. *Buchnera* from at least 12 species of aphids (representing four families) have a monophyletic origin. The age of the symbiotic association is 200–250 million years.
2. *Buchnera* is a lineage within the γ-subdivision of the *Proteobacteria* and is distinct from lineages containing the endosymbionts of whiteflies and

mealybugs. The *Enterobacteriaceae* and the endosymbionts of tsetse are the closest known relatives of *Buchnera*.
3. A wide range of genetic properties shows that *Buchnera* is more similar to free-living bacteria than to organelles such as mitochondria or chloroplasts.
4. During parthenogenetic reproduction, there is a parallel increase in the weight of the aphid and the number of *Buchnera*, showing an integration of aphid and endosymbiont growth. A mature aphid has 5.6×10^6 endosymbionts.
5. The rRNA genes are arranged in *Buchnera* as two transcription units consisting of *rrs* and tRNA-*rrl*-*rrf*. This arrangement is distinctive and rare in eubacteria. Only one copy of each transcription unit is present per endosymbiont genome.
6. *Buchnera* is capable of synthesizing DNA, RNA, and proteins. The enzymatic machinery is eubacterial in nature, as judged by antibiotic sensitivity and the presence of genes coding for proteins involved in DNA biosynthesis, transcription, and translation.
7. The genes and proteins involved in the stress response are present in *Buchnera;* at least one stress protein (GroEL) is found in high amounts.
8. *Buchnera* can reduce sulfate and sulfide and synthesize methionine and cysteine, which are subsequently utilized by the aphid host.
9. *Buchnera* can synthesize tryptophan for the aphid host. All of the genes of the tryptophan biosynthetic pathway have been detected in the endosymbiont. The endosymbionts of several rapidly growing aphids exhibit plasmid amplification of the genes coding for anthranilate synthase, the first enzyme of the tryptophan biosynthetic pathway.

What are the specific adaptations of *Buchnera* to its endosymbiotic association? The few answers we have to this question contain a considerable amount of speculation. Perhaps the best example of an adaptation is the tandem duplication of genes coding for anthranilate synthase and their location on a plasmid, which represents an effective way of overcoming the endosymbiont's preexisting regulatory mechanisms and thereby overproducing tryptophan. Another example is the presence of one copy of the genes coding for rRNA, a characteristic of slow-growing bacteria. An additional rather speculative example is the modification of CysE so that it no longer responds to feedback inhibition by cysteine, resulting in the overproduction of this amino acid. There are also several cases in which nucleotide sequences with known regulatory functions in *E. coli* or other bacteria are absent from their expected locations in B(*S. graminum*). The significance of these latter observations is not known.

Aphid endosymbionts were originally thought to be akin to organelles such as mitochondria and chloroplasts, and this assumption has had considerable influence on the conceptual framework within which the endosymbionts were viewed. A characteristic of these organelles is the major reduction in genome size and coding capacity and the transfer of some of the genes to the host nucleus (4, 98, 131). The genome size of *Buchnera* (79), as well as the recent information on its genetic properties (23, 94, 95, 118), suggests that comparisons to free-living bacteria are more meaningful. An unanswered question is the reason for the major difference between the characteristics of insect endosymbionts and organelles such as mitochondria and chloroplasts. One possible explanation is the length of time over which the association has occurred. Chloroplast and mitochondria associations are over 1 billion years old (124), whereas the association of aphids and *Buchnera* is only 200–250 million years old (119). Consequently it is possible that insufficient time has elapsed for many of the alterations to occur. Perhaps a better explanation is that unlike the case of mitochondria and chloroplasts, the ancestor of *Buchnera* originally infected a complex multicellular host in which it was sequestered within cells that were separate from those of the germ line. This separation may have limited opportunities for the transfer of genes from the prokaryotic chromosome to the host nucleus and their incorporation into the aphid germ line.

Although these recent studies help elucidate the nature of the aphid endosymbionts, they are barely a beginning, and many questions remain unanswered. Perhaps the more interesting questions relate to additional functions that might be attributed to *Buchnera*. Also, to understand the endosymbiotic association, one must know the properties and role of the host-derived symbiosome membrane, because this structure could control the rates of entry and exit of nutrients and serve as a mediating barrier between the host and the endosymbiont. Other questions concern the nature of the mechanisms responsible for the integration of aphid and endosymbiont growth and the maternal transmission of endosymbionts. Because *Buchnera* cannot be cultivated on laboratory media, much of the information obtained in recent studies has depended on recombinant DNA methodologies that have also been instrumental in our understanding of *Wolbachia* spp. and intracellular pathogenic genera such as *Rickettsia* and *Chlamydia*.

ACKNOWLEDGMENTS

Work from the authors' laboratories was supported by the National Science Foundation IBN-9201285 (PB), MCB-9402813 (PB), DEB-9306495 (NAM), Entotech Inc, (Novo Nordisk) (PB), and the University of California Experiment Station (PB).

> Any *Annual Review* chapter, as well as any article cited in an *Annual Review* chapter, may be purchased from the Annual Reviews Preprints and Reprints service.
> 1-800-347-8007; 415-259-5017; email: arpr@class.org

Literature Cited

1. Aksoy S. 1994. Molecular analysis of the endosymbionts of tsetse fly; 16S rDNA locus and over-expression of molecular chaperonins. *Insect Mol. Biol.* 4:23–29
2. Aksoy S, Pourhosseini AA, Chow A. 1994. Mycetome endosymbionts of tsetse flies constitute a distinct lineage related to *Enterebacteriaceae*. *Insect Mol. Biol.* 4:15–22
3. Anderson RP, Roth JR. 1977. Tandem genetic duplications in phage and bacteria. *Annu. Rev. Microbiol.* 31:473–505
4. Attardi G. 1985. Animal mitochondrial DNA: an extreme example of genetic economy. *Int. Rev. Cytol.* 93:93–145
5. Bachman BJ. 1990. Linkage map of *Escherichia coli* K12, edition 8. *Microbiol. Rev.* 54:130–97
6. Baird L, Lipinska B, Raina S, Georgopoulos C. 1991. Identification of the *Escherichia coli sohB* gene, a multicopy suppressor of the HtrA (DegP) null phenotype. *J. Bacteriol.* 173: 5763–70
7. Baumann L, Baumann P. 1994. Growth kinetics of the endosymbiont, *Buchnera aphidicola*, in the aphid *Schizaphis graminum*. *Appl. Environ. Microbiol.* 60:3440–43
8. Bedouelle H. 1993. Symmetrical interactions between the translational operator of the *thrS* gene and dimeric threonyl transfer RNA synthetase. *J. Mol. Biol.* 230:704–8
9. Blackman RL. 1987. Reproduction, cytogenetics and development. See Ref. 107, 2A:163–95
10. Branlant G, Branlant C. 1985. Nucleotide sequence of the *Escherichia coli gap* gene. Different evolutionary behavior of the NAD$^+$-binding domain and of the catalytic domain of D-glyceraldehyde-3-phosphate dehydrogenase. *Eur. J. Biochem.* 150:61–66
11. Breeuwer JA, Stouthamer R, Barnes SM, Pelletier DA, Weisburg WG, Werren JH. 1992. Phylogeny of cytoplasmic incompatibility microorganisms in the parasitoid wasp genus *Nasonia* (Hymenoptera: Pteromalidae) based on 16S ribosomal DNA sequences. *Insect Mol. Biol.* 1:25–36
12. Brough CN, Dixon AFG. 1990. Ultrastructural features of egg development in oviparae of the vetch aphid, *Megura viciae* Buckton. *Tissue Cell* 22:51–63
13. Buchner P. 1965. In *Endosymbiosis of Animals with Plant Microorganisms*, pp. 210–332. New York: Interscience
14. Burton ZF, Gross CA, Watanabe KK, Burges RR. 1983. The operon that encodes sigma subunit of RNA polymerase also encodes ribosomal protein S21 and DNA primase in *E. coli* K12. *Cell* 32:335–49
15. Calcutt MJ, Schmidt FJ. 1992. Conserved gene arrangement in the origin region of the *Streptomyces coelicor* chromosome. *J. Bacteriol.* 174:3220–26
16. Caligiuri MG, Bauerle R. 1991. Identification of amino acid residues involved in feedback regulation of anthranilate synthase complex from *Salmonella typhimurium*. *J. Biol. Chem.* 266:8328–35
17. Campbell BC. 1990. On the role of microbial symbiotes in herbivorous insects. In *Insect Plant Interactions*, ed. EA Bernays, 1:1–44. Boca Raton, FL: CRC
18. Campbell BC, Dreyer DL. 1985. Host-plant resistance of sorghum: differential hydrolysis of sorghum pectic substances by polysaccharases of greenbug biotypes (*Schizaphis graminum*, Homoptera: Aphididae). *Arch. Insect Biochem. Physiol.* 2:203–15
19. Campbell BC, Nes WD. 1983. A reappraisal of sterol biosynthesis and metabolism in aphids. *J. Insect Physiol.* 29: 149–56
20. Campbell BC, Purcell AH. 1993. Phylogenetic affiliation of BEV, a bacterial parasite of the leafhopper *Euscelidius variegatus*, on the basis of 16S rDNA sequences. *Curr. Microbiol.* 26:37–41
20a. Campbell RK, Eikenbary RD, eds. 1990. *Aphid-Plant Genotype Interactions*. Amsterdam: Elsevier
21. Chen X, Finch LR. 1989. Novel arangement of rRNA genes in *Mycoplasma gallisepticum*: separation of the 16S gene of one set from the 23S and 5S genes. *J. Bacteriol.* 171:2876–78
22. Choi EY, Ahn GS, Jeon KW. 1991. Elevated levels of stress proteins associated with bacterial symbiosis in *Amoeba proteus* and soybean root nodule cells. *BioSystems* 25:205–12
23. Clark MA, Baumann L, Baumann P. 1992. Sequence analysis of an aphid endosymbiont DNA fragment containing *rpoB* (β-subunit of RNA polymerase) and portions of *rplL* and *rpoC*. *Curr. Microbiol.* 25:283–90
24. Clark MA, Baumann L, Munson MA, Baumann P, Campbell BC, et al. 1992. The eubacterial endosymbionts of whiteflies (Homoptera: Aleyrodoidea) constitute a lineage distinct from the

endosymbionts of aphids and mealybugs. *Curr. Microbiol.* 25:119–23
25. Clark MA, Baumann P. 1993. Aspects of energy-yielding metabolism in the aphid, *Schizaphis graminum*, and its endosymbiont: detection of gene fragments potentially coding for the ATP synthase β-subunit and glyceraldehyde-3-phosphate dehydrogenase. *Curr. Microbiol.* 26:233–37
26. Clarke PH. 1978. Experiments in microbial evolution. In *The Bacteria*, ed. LN Ornston, JR Sokatch, 6:137–218. New York: Academic
27. Cohen GN, Saint-Girons I. 1987. Biosynthesis of threonine, lysine, and methionine. See Ref. 122, 1:429–44
28. Crawford IP. 1989. Evolution of a biosynthetic pathway: the tryptophan paradigm. *Annu. Rev. Microbiol.* 43:567–600
29. DaCosta e Silva O, Kosuge T. 1991. Molecular characterization and expression analysis of the anthranilate synthase gene of *Pseudomonas syringae* subsp. *savastanoi*. *J. Bacteriol.* 173:463–71
30. Dadd RH. 1985. Nutrition: organisms. In *Comprehensive Insect Physiology, Biochemistry, and Pharmacology*, ed. GA Kerkut, Ll Gilbert, 4:315–19. Oxford: Pergamon
31. Dasch GA, Weiss E, Chang K-P. 1984. Endosymbionts of insects. See Ref. 89a, pp. 811–33
32. Denk D, Böck A. 1987. L-Cysteine biosynthesis in *Escherichia coli:* nucleotide sequence and expression of the serine acetyltransferase (*cysE*) gene from the wild-type and a cysteine-excreting mutant. *J. Gen. Microbiol.* 133:515–25
33. Dixon AFG. 1973. *Biology of Aphids.* London: Edward Arnold
34. Dixon AFG. 1985. *Aphid Ecology.* Glasgow: Blackie
35. Dixon AFG. 1992. Constraints on the rate of parthenogenetic reproduction and pest status of aphids. *Invertebr. Reprod. Dev.* 22:159–63
36. Doolittle RF, Feng DF, Anderson KL, Alberro MR. 1990. A naturally occurring horizontal gene transfer from a eukaryote to a prokaryote. *J. Mol. Evol.* 31:383–88
37. Douglas AE. 1988. Sulfate utilization in an aphid symbiosis. *Insect Biochem.* 18:599–605
38. Douglas AE. 1989. Mycetocyte symbiosis in insects. *Biol. Rev. Cambridge Philos. Soc.* 64:409–34
39. Douglas AE. 1990. Nutritional interactions between *Myzus persicae* and its symbionts. See Ref. 20a, pp. 319–27
40. Douglas AE. 1992. Requirement of pea aphids (*Acyrthosiphon pisum*) for their symbiotic bacteria. *Entomol. Exp. Appl.* 65:195–98
41. Douglas AE. 1993. The nutritional quality of phloem sap utilized by natural aphid populations. *Ecol. Entomol.* 18:31–38
42. Douglas AE, Dixon AFG. 1987. The mycetocyte symbiosis of aphids: variation with age and morph in virginoparae of *Megoura viciae* and *Acyrthosiphon pisum*. *J. Insect Physiol.* 33:109–13
43. Douglas AE, Prosser WA. 1992. Synthesis of the essential amino acid tryptophan in the pea aphid (*Acyrthosiphon pisum*) symbiosis. *J. Insect Physiol.* 38:565–68
44. Downing WL, Dennis PP. 1987. Transcription products from the *rplKAJL-rpoBC* gene cluster. *J. Mol. Biol.* 194:609–20
44a. Drlica K, Riley M, eds. 1990. *The Bacterial Chromosome.* Washington, DC: Am. Soc. Microbiol.
45. Edlund T, Normark S. 1981. Recombination between short DNA homologies causes tandem duplications. *Nature* 292:269–71
46. Eisenbach J, Mittler TE. 1987. Extranuclear inheritance in a sexually produced aphid: the ability to overcome host plant resistance by biotype hybrids of the greenbug, *Schizaphis graminum*. *Experientia* 43:332–34
47. Engel JN, Ganem D. 1990. A polymerase chain reaction–based approach to cloning sigma factors from eubacteria and its application to the isolation of a sigma-70 homolog from *Chlamydia trachomatis*. *J. Bacteriol.* 172:2447–55
48. Fayet O, Ziegelhoffer T, Georgopoulos C. 1989. The *groES* and *groEL* heat shock gene products of *Escherichia coli* are essential for bacterial growth at all temperatures. *J. Bacteriol.* 171:1379–85
49. Fujita Q, Yoshikawa H, Ogasawara N. 1992. Structure of *dnaA* and DnaA box region in *Mycoplasma capricolum* chromosome: conservation and variations in course of evolution. *Gene* 110:17–23
50. Fukatsu T, Aoki S, Kurosu U, Ishikawa H. 1994. Phylogeny of Ceratophidini aphids revealed by their symbiotic microorganisms and basic structure of their galls: implications for host-symbiont coevolution and evolution of sterile soldier castes. *Zool. Sci.* 11:613–23
51. Fukatsu T, Ishikawa H. 1992. A novel eukaryotic extracellular symbiont in an aphid, *Astegopteryx styraci* (Homoptera, Aphididae, Hormaphidinae). *J. Insect Physiol.* 38:765–73

52. Fukatsu T, Ishikawa H. 1992. Soldier and male of an eusocial aphid *Colophina arma* lack endosymbiont: implications for physiological and evolutionary interaction between host and symbiont. *J. Insect Physiol.* 38:1033–42
53. Fukatsu T, Ishikawa H. 1992. Synthesis and localization of symbionin, an aphid endosymbiont protein. *Insect Biochem. Mol. Biol.* 22:167–74
54. Fukatsu T, Ishikawa H. 1993. Occurence of chaperonin 60 and chaperonin 10 in primary and secondary bacterial symbionts of aphids: implications for the evolution of an endosymbiotic system in aphids. *J. Mol. Evol.* 36:568–77
55. Fukunaga M, Yanagihara Y, Sohnaka M. 1992. The 23S/5S ribosomal RNA genes (*rrl/rrf*) are separate from the 16S ribosomal RNA gene (*rrs*) in *Borellia burgdorferi*, the aetiological agent of Lyme disease. *J. Gen . Microbiol.* 138:871–77
56. Garbe TR. 1992. Heat shock proteins and infection: interactions of pathogen and host. *Experientia* 48:635–39
57. Georgopoulos C. 1989. The *E. coli dnaA* initiation protein: a protein for all seasons. *Trends Genet.* 5:319–21
58. Georgopoulos C, Ang D, Maddock A, Raina S, Lipinska B, Zylicz M. 1990. Heat shock response of *Escherichia coli*. See Ref. 44a, pp. 405–19
59. Gherna RL, Werren JH, Weisburg W, Cote R, Woese CR, et al. 1991. *Arsenophonus nasoniae* gen. nov., sp. nov., the causative agent of son-killer trait in the parasitic wasp *Nasonia vitripennis*. *Int. J. Syst. Bacteriol.* 41:563–65
60. Griffiths GW, Beck SD. 1973. Intracellular symbiotes of the pea aphid, *Acyrthosiphon pisum*. *J. Insect Physiol.* 19:75–84
61. Griffiths GW, Beck SD. 1974. Effects of antibiotics on intracellular symbiotes in the pea aphid, *Acyrthosiphon pisum*. *Cell Tissue Res.* 148:287–300
62. Gross CA, Straus DB, Erikson JW, Yura T. 1990. The function and regulation of heat shock proteins in *Escherichia coli*. In *Stress Proteins in Biology and Medicine*, ed. RI Morimoto, A Tissières, C Georgopoulos, pp. 167–89. Cold Spring Harbor, NY: Cold Spring Harbor Lab.
63. Hara E, Fukatsu T, Ishikawa H. 1990. Characterization of symbionin with anti-symbionin antiserum. *Insect Biochem.* 20:429–36
64. Hara E, Ishikawa H. 1990. Purification and partial characterization of symbionin, an aphid endosymbiont-specific protein. *Insect Biochem.* 20:421–27
65. Harrison CP, Douglas AE, Dixon AFG. 1989. A rapid method to isolate symbiotic bacteria from aphids. *J. Invertebr. Pathol.* 53:427–28
66. Hartmann RK, Toschka HY, Erdman VA. 1991. Processing and termination of 23S rRNA-5S rRNA-tRNA^{Gly} primary transcripts in *Thermus thermophilus* HB8. *J. Bacteriol.* 173:2681–90
67. Heie OE. 1987. Palaeontology and phylogeny. See Ref. 107, 2A:367–91
68. Hemmingsen SM, Woolford C, van der Vies SM, Tilly K, Dennis DT, et al. 1988. Homologous plant and bacterial proteins chaperone oligomeric protein assembly. *Nature* 333:330–34
69. Herrmann KM. 1983. The common aromatic biosynthetic pathway. In *Amino Acids: Biosynthesis and Genetic Regulation*, ed. KM Herrmann, RL Somerville, pp. 301–18. Reading, MA: Addison-Wesley
70. Hinde R. 1971. The control of mycetome symbiotes of the aphids *Brevicoryne brassicae, Myzus persicae,* and *Microsiphum rosae*. *J. Insect Physiol.* 17:1791–800
71. Houk EJ. 1987. Symbionts. See Ref. 107, 2A:123–29
72. Houk EJ, Griffiths GW. 1980. Intracellular symbiotes of the Homoptera. *Annu. Rev. Entomol.* 25:161–87
73. Houk EJ, Griffiths GW, Hadjokas NE, Beck SD. 1977. Peptidoglycan in the cell wall of the primary intracellular symbiote of the pea aphid. *Science* 198:401–3
74. Iaccarino FM, Tremblay E. 1973. Comparazione ultrastruttruale della disimbiosi di *Macrosiphum rosea* (L.) e *Dactynotus jaceae* (l.) *Boll. Ist. Entomol. Univ. Studi Napoli* 30:319–29
75. Ishikawa H. 1982. Isolation of the intracellular symbionts and partial characterization of their RNA species of the elder aphid, *Acyrthosiphon magnoliae*. *Comp. Biochem. Physiol.* 72B:239–47
76. Ishikawa H. 1982. DNA, RNA and protein synthesis in the isolated symbionts from the pea aphid, *Acyrthosiphon pisum*. *Insect Biochem.* 12:605–12
77. Ishikawa H. 1984. Molecular aspects of intracellular symbiosis in the aphid mycetocyte. *Zool. Sci.* 1:509–22
78. Ishikawa H. 1984. Characterization of the protein species synthesized *in vivo* and *in vitro* by an aphid endosymbiont. *Insect Biochem.* 14:417–25
79. Ishikawa H. 1987. Nucleotide composition and kinetic complexity of the genomic DNA of an intracellular symbiont of the pea aphid *Acyrthosiphon pisum*. *J. Mol. Evol.* 24:205–11

80. Ishikawa H. 1989. Biochemical and molecular aspects of endosymbiosis in insects. *Int. Rev. Cytol.* 116:1–45
81. Jensen RA. 1985. Biochemical pathways in prokaryotes can be traced backward through evolutionary time. *Mol. Biol. Evol.* 2:92–108
82. Jin DJ, Gross C. 1988. Mapping and sequencing of mutations in the *Escherichia coli rpoB* gene that lead to rifampicin resistance. *J. Mol. Biol.* 202:45–58
83. Jinks-Robertson S, Nomura M. 1987. Ribosomes and tRNA. See Ref. 122, 2:1358–85
84. Kakeda K, Ishikawa H. 1991. Molecular chaperon produced by an intracellular symbiont. *J. Biochem.* 110:583–87
85. Katsumata R, Ikeda M. 1993. Hyperproduction of tryptophan in *Corynebacterium glutamicum* by pathway engineering. *Biotechnology* 11:921–25
86. Kolibachuk D, Baumann P. 1994. *Buchnera aphidicola* (aphid-endosymbiont) glyceraldehyde-3-phosphate dehydrogenase: molecular cloning and sequence analysis. *Curr. Microbiol.* 30:133–38
87. Kolibachuk D, Rouhbakhsh D, Baumann P. 1995. Aromatic amino acid biosynthesis in *Buchnera aphidicola* (endosymbiont of aphids): cloning and sequencing of a DNA fragment containing *aroH-thrS-infC-rpml-rplT*. *Curr. Microbiol.* 30:313–16
88. Kredich NM. 1987. Biosynthesis of cysteine. See Ref. 122, 1:419–28
89. Kredich NM. 1992. The molecular basis for positive regulation of *cys* promoters in *Salmonella typhimurium* and *Escherichia coli*. *Mol. Microbiol.* 6:2747–53
89a. Krieg NR, Holt JG, eds. 1984. *Bergey's Manual of Systematic Bacteriology*, Vol. 1. Baltimore: Williams & Wilkins
90. Kumamoto CA. 1991. Molecular chaperones and protein translocation across the *Escherichia coli* inner membrane. *Mol. Microbiol.* 5:19–22
91. Kumamoto CA, Nault AK. 1989. Characterization of the *Escherichia coli* protein-export gene *secB*. *Gene* 75:167–75
92. Lai C-Y. 1994. *Comparative genetic and physiological characterization of the procaryotic endosymbionts of aphids*. PhD thesis. Univ. Calif., Davis
93. Lai C-Y, Baumann L, Baumann P. 1994. Amplification of *trpEG*; adaptation of *Buchnera aphidicola* to an endosymbiotic association with aphids. *Proc. Natl. Acad. Sci. USA* 91:3819–23
94. Lai C-Y, Baumann P. 1992. Sequence analysis of a DNA fragment from *Buchnera aphidicola* (an endosymbiont of aphids) containing genes homologous to *dnaG, rpoD, cysE*, and *secB*. *Gene* 119:113–18
95. Lai C-Y, Baumann P. 1992. Genetic analysis of an aphid endosymbiont DNA fragment homologous to the *rnpA-rpmH-dnaA-dnaN-gyrB* region of eubacteria. *Gene* 113:175–81
96. Lai C-Y, Baumann P, Moran NA. 1995. Genetics of the tryptophan biosynthetic pathway of the prokaryotic endosymbiont (*Buchnera*) of the aphid *Schlechtendalia chinensis*. *Insect Mol. Biol.* 4:47–59
97. Lapage SP, Sneath PHA, Lessel EF, Skerman VBD, Seeliger HPR, Clark WA, eds. 1975. *International Code of Nomenclature of Bacteria*. Washington, DC: Am. Soc. Microbiol.
98. Lewin B. 1994. In *Genes* V, pp. 377–489. Oxford: Oxford Univ. Press
99. Liesack W, Stackebrandt E. 1989. Evidence for unlinked *rrn* operons in Planctomycete *Pirellula marina*. *J. Bacteriol.* 171:5025–30
100. Ma R, Reese JC, Black WC IV, Bramel-Cox P. 1990. Detection of pectinesterase and polygalacturonase from salivary secretions of living greenbugs, *Schizaphis graminum* (Homoptera: Aphididae). *J. Insect Physiol.* 36:507–12
101. Matin A. 1991. The molecular basis of carbon-starvation-induced general resistance in *Escherichia coli*. *Mol. Microbiol.* 5:3–10
102. Matsui K, Miwa K, Sano K. 1987. Two single-base-pair substitutions causing desensitization to tryptophan feedback inhibition of anthranilate synthase and enhanced expression of tryptophan genes of *Brevibacterium lactofermentum*. *J. Bacteriol.* 169:5330–32
103. McLean DL, Houk EJ. 1973. Phase contrast and electron microscopy of the mycetocytes and symbiotes of the pea aphid, *Acyrthosiphon pisum*. *J. Insect Physiol.* 19:625–33
104. McMacken R, Silver L, Georgopolous C. 1987. DNA replication. See Ref. 122, 1:564–612
105. Miles PW. 1989. The responses of plants to the feeding of Aphidoidea: principles. See Ref. 107, 2C:1–21
106. Miles PW. 1989. Specific responses and damage caused by Aphidoidea. See Ref. 107, 2C:23–47
107. Minks AK, Harrewijn P, eds. 1987, 1988, 1989. *Aphids: Their Biology, Natural Enemies, and Control*, Vols. 2A, 2B, 2C. Amsterdam: Elsevier
108. Mittler TE. 1971. Dietary amino acid requirements of the aphid *Myzus persi-*

cae affected by antibiotic uptake. *J. Nutr.* 101:1023–28
109. Mittler TE. 1988. Applications of artificial feeding techniques for aphids. See Ref. 107, 2B:145–70
110. Moine H, Fayat G, Romby P, Springer M, Grunberg-Manago M, et al. 1983. Structural and transcriptional evidence for related *thrS* and *infC* expression. *Proc. Natl. Acad. Sci. USA* 80:6152–56
111. Moran NA, Baumann P. 1994. Phylogenetics of cytoplasmically inherited microorganisms of arthropods. *Trends Ecol. Evol.* 9:15–20
112. Moran NA, Munson MA, Baumann P, Ishikawa H. 1993. A molecular clock in endosymbiotic bacteria is calibrated using the insect hosts. *Proc. R. Soc. London Ser. B* 253:167–71
113. Morioka M, Ishikawa H. 1992. Mutualism based on stress: selective synthesis and phosphorylation of a stress protein by an intracellular symbiont. *J. Biochem.* 111:431–35
114. Moulder JW. 1979. The cell as an extreme environment. *Proc. R. Soc. London Ser. B* 204:199–210
115. Moulder JW, Hatch TP, Kuo C-C, Schachter J, Storz J. 1984. Chlamydia. See Ref. 89a, pp. 729–39
116. Munson MA, Baumann L, Baumann P. 1992. *Buchnera aphidicola*, the endosymbiont of aphids, contains genes for four ribosomal RNA proteins, initiation factor-3, and the α-subunit of RNA polymerase. *Curr. Microbiol.* 24:23–29
117. Munson MA, Baumann L, Baumann P. 1993. *Buchnera aphidicola* (a prokaryotic endosymbiont of aphids) contains a putative 16S rRNA operon unlinked to the 23S rRNA-encoding gene: sequence determination, and promoter and terminator analysis. *Gene* 137:171–78
118. Munson MA, Baumann P. 1993. Molecular cloning and nucleotide sequence of a putative *trpDC(F)BA* operon in *Buchnera aphidicola* (endosymbiont of the aphid *Schizaphis graminum*). *J. Bacteriol.* 175:6426–32
119. Munson MA, Baumann P, Clark MA, Baumann L, Moran NA, et al. 1991. Evidence for the establishment of aphideubacterium endosymbiosis in an ancestor of four aphid families. *J. Bacteriol.* 173:6321–24
120. Munson MA, Baumann P, Kinsey MG. 1991. *Buchnera* gen. nov. and *Buchnera aphidicola* sp. nov., a taxon consisting of the mycetocyte-associated, primary endosymbionts of aphids. *Int. J. Syst. Bacteriol.* 41:566–68
121. Munson MA, Baumann P, Moran NA. 1992. Phylogenetic relationships of the endosymbionts of mealybugs (Homoptera: Pseudococcidae) based on 16S rDNA sequences. *Mol. Phylogenet. Evol.* 1:26–30
122. Neidhardt FC, Ingraham JL, Low KB, Magasanik B, Schaechter M, Umbarger HE, eds. 1987. Escherichia coli *and* Salmonella typhimurium: *Cellular and Molecular Biology,* Vols. 1, 2. Washington, DC: Am. Soc. Microbiol.
123. Neidhardt FC, Ingraham JL, Schaechter M. 1990. *Physiology of the Bacterial Cell: A Molecular Approach.* Sunderland, MA: Sinauer
124. Ochman H, Wilson AC. 1987. Evolution of bacteria: evidence for a universal substitution rate in cellular genomes. *J. Mol. Evol.* 26:74–78
125. Ogasawara N, Yoshikawa H. 1992. Genes and their organization in the replication origin of the bacterial chromosome. *Mol. Microbiol.* 6:629–34
126. Ohtaka C, Ishikawa H. 1991. Effects of heat treatment on the symbiotic system of an aphid mycetocyte. *Symbiosis* 11:19–30
127. Ohtaka C, Ishikawa H. 1993. Accumulation of adenine and thymine in a *groE*-homologous operon of an intracellular symbiont. *J. Mol. Evol.* 36:121–26
128. Ohtaka C, Nakamura H, Ishikawa H. 1992. Structures of chaperonins from an intracellular symbiont and their functional expression in *Escherichia coli groE* mutants. *J. Bacteriol.* 174:1869–74
129. Olsen GJ, Woese CR, Overbeek R. 1994. The winds of (evolutionary) change: breathing new life into microbiology. *J. Bacteriol.* 176:1–6
130. O'Neill SL, Giordano R, Colbert AME, Karr TL, Robertson HM. 1992. 16S rRNA phylogenetic analysis of the bacterial endosymbionts associated with cytoplasmic incompatibility in insects. *Proc. Natl. Acad. Sci. USA* 89:2699–702
131. Palmer JD. 1991. Plastid chromosomes: structure and function. In *The Molecular Biology of Plastids,* ed. L Bogorad, IK Vasil, pp. 5–53. San Diego: Academic
132. Pang H, Winkler H. 1993. Copy number of the 16S rRNA gene in *Rickettsia prowazekii*. *J. Bacteriol.* 175:3893–96
133. Pittard AJ. 1985. Biosynthesis of the aromatic amino acids. See Ref. 122, 1:368–94
134. Ponsen MB. 1987. Alimentary tract. See Ref. 107, 2A:79–97.
135. Prosser WA, Douglas AE. 1991. The aposymbiotic aphid: an analysis of chlortetracycline-treated pea aphid,

Acyrthosiphon pisum. J. Insect Physiol. 37:713-19
136. Pugsley AP. 1993. The complete general secretory pathway in Gram-negative bacteria. *Microbiol. Rev.* 57:50-108
137. Puterka GJ, Peters DC. 1989. Inheritance of greenbug, *Schizaphis graminum* (Rondani), virulence to *Gb2* and *Gb3* resistance genes in wheat. *Genome* 32:109-14
138. Puterka GJ, Peters DC. 1990. Sexual reproduction and inheritance of virulence in the greenbug, *Schizaphis graminum* (Rondani). See Ref. 20a, pp. 289-318
139. Ray JM, Bauerle R. 1991. Purification and properties of tryptophan-sensitive 3-deoxy-D-*arabino*-heptulosonate-7-phosphate synthase from *Escherichia coli*. *J. Bacteriol.* 173:1894-901
140. Rinke de Wit TF, Bekelie S, Osland A, Miko TL, Hermans D, et al. 1992. Mycobacteria contain two *groEL* genes: the second *mycobacterium lepare groEL* gene is arranged in an operon with *groES*. *Mol. Microbiol.* 6:1995-2007
141. Roth E, Jeon K, Stacey G. 1988. Homology in endosymbiotic systems: the term "symbiosome." In *Molecular Genetics of Plant-Microbe Interactions*, ed. R Palacios, DPS Verma, pp. 220-25. St. Paul, MN: Am. Phytopathol. Soc.
142. Rouhbakhsh D, Baumann P. 1995. Characterization of a putative 23S-5S rRNA operon of *Buchnera aphidicola* (a prokaryotic endosymbiont of aphids) unlinked to the 16S rRNA-encoding gene. *Gene.* 155:107-12
143. Roubakhsh D, Moran NA, Baumann L, Voegtlin DJ, Baumann P. 1994. Polymerase chain reaction-based identification of *Buchnera:* the primary prokaryotic endosymbiont of aphids. *Insect Mol. Biol.* 3:213-17
144. Rousset F, Vautrin D, Solignac M. 1992. Molecular identification of *Wolbachia*, the agent of cytoplasmic incompatibility in *Drosophila simulans*, and variability in relation with host mitochondrial types. *Proc. R. Soc. London Ser. B* 247:163-68
144a. Saint Girons I, Old IG, Davidson BE. 1994. Molecular biology of the *Borrelia*, bacteria with linear replicons. *Microbiology* 140:1803-16
145. Samuelsson T, Borén T. 1990. Evolution of macromolecule synthesis. In *Mycoplasmas: Molecular Biology and Pathogenesis*, ed. J Maniloff, RN McElhaney, LR Finch, JB Baseman, pp. 575-91. Washington, DC: Am. Soc. Microbiol.
146. Schwemmler W, Müller H. 1986. Role of insect lysozymes in endocytobiosis and immunity of leafhoppers. In *Hemocytic and Humoral Immunity in Arthropods*, ed. AP Gupta, pp. 449-60. New York: Wiley
147. Skovgaard O. 1990. Nucleotide sequence of a *Proteus mirabilis* DNA fragment homologous to the *60K-rnpA-rpmH-dnaA-dnaN-recF-gyrB* region of *Escherichia coli*. *Gene* 93:27-34
148. Smith DC, Douglas AE. 1987. *The Biology of Symbiosis*. London: Arnold
149. Srivastava PN. 1987. Nutritional physiology. See Ref. 107, 2A:99-121
150. Srivastava PN, Auclair JL, Srivastava U. 1980. Nucleic acid, nucleotide and protein concentrations in the pea aphid, *Acyrthosiphon pisum*, during larval growth and development. *Insect Biochem.* 10:209-13
151. Squires CL, Greenblatt J, Li J, Condon C, Squires CL. 1993. Ribosomal RNA antitermination *in vitro:* requirement for Nus factors and one or more unidentified cellular components. *Proc. Natl. Acad. Sci. USA* 90:970-74
152. Stakebrandt E, Murray RGE, Trüper HG. 1988. *Proteobacteria* classis nov., a name for the phylogenetic taxon that includes the "purple bacteria and their relatives." *Int. J. Syst. Bacteriol.* 38:321-25
153. Stark GR, Wahl GM. 1984. Gene amplification. *Annu. Rev. Biochem.* 53:447-91
154. Stirling DA, Hulton CSJ, Waddell L, Park SF, Stewart GSAB. 1989. Molecular characterization of the *proU* loci of *Salmonella typhimurium* and *Escherichia coli* encoding osmoregulated glycine betaine transport systems. *Mol. Microbiol.* 3:1025-38
155. Stoltzfus A, Leslie JF, Milkman R. 1988. Molecular evolution of the *Escherichia coli* chromosome. I. Analysis of structure and natural variation in a previously uncharacterized region between *trp* and *tonB*. *Genetics* 120:345-58
156. Stouthamer R, Breeuwer JAJ, Luck RF, Werren JH. 1993. Molecular identification of microorganisms associated with parthenogenesis. *Nature* 361:66-68
157. Sylvester ES. 1985. Multiple acquisition of viruses and vector-dependent prokaryotes: consequences on transmission. *Annu. Rev. Entomol.* 30:71-88
158. Takada H. 1991. Gall development of *Schlechtendalia chinensis* (Bell) (Homoptera: Pemphigidae) on *Rhus javanica* L. and emergence of alatae from galls. *Jpn. J. Appl. Entomol. Zool.* 35:71-76 (In Japanese)
159. Takada H. 1991. Does the sexual female

of *Schlechtendalia chinensis* (Bell) (Homoptera: Pemphigidae) "viviparously" produce the fundatrix? *Appl. Entomol. Zool.* 26:117–21
160. Unterman BM, Baumann P. 1990. Partial characterization of ribosomal RNA operons of the pea-aphid endosymbionts: evolutionary and physiological implications. See Ref. 20a, pp. 329–50
161. Unterman BM, Baumann P, McLean DL. 1989. Pea aphid symbiont relationships established by analysis of 16S rRNAs. *J. Bacteriol.* 171:2970–74
162. Van Boglen RA, Neidhardt FC. 1990. Ribosomes as sensors of heat and cold shock in *Escherichia coli. Proc. Natl. Acad. Sci. USA* 87:5589–93
163. Walker JE, Saraste M, Gay NJ. 1984. The *unc* operon. Nucleotide sequence, regulation and structure of ATP-synthase. *Biochim. Biophys. Acta* 768:164–200
164. Watson K. 1990. Microbial stress proteins. *Adv. Microbiol. Physiol.* 31:183–223
165. Weinstock GM. 1987. General recombination in *Escherichia coli*. See Ref. 122, 2:1034–43
166. Weiss E, Moulder JW. 1984. *Rickettsia*. See Ref. 89a, pp. 688–98
167. Whitehead LF, Douglas AE. 1993. A metabolic study of *Buchnera*, the intracellular bacterial symbionts of the pea aphid *Acyrthosiphon pisum. J. Gen Microbiol.* 139:821–26
168. Whitehead LF, Douglas AE. 1993. Populations of symbiotic bacteria in the parthenogenetic pea aphid (*Acyrthosiphon pisum*) symbiosis. *Proc. R. Soc. London Ser. B* 254:29–32
169. Williamson LR, Plano GV, Winkler HH, Krause DC, Wood DO. 1989. Nucleotide sequence of the *Rickettsia prowazekii* ATP/ADP translocase-encoding gene. *Gene* 80:269–78
170. Winkler HH. 1990. *Rickettsia* species (as organisms). *Annu. Rev. Microbiol.* 44:131–53
171. Woese CR. 1987. Bacterial evolution. *Microbiol. Rev.* 51:221–71
172. Wolfe CJ, Haygood MG. 1993. Bioluminescent symbionts of the Caribbean flashlight fish (*Kryptophanaron alfredi*) have a single rRNA operon. *Mol. Mar. Microbiol.* 2:189–97
173. Yanofsky C, Crawford IP. 1987. The tryptophan operon. See Ref. 122, 2:1453–72
174. Ye F, Renaudin J, Bové J-M, Laigret F. 1994. Cloning and sequencing of the replication origin (*oriC*) of the *Spiroplasma citri* chromosome and construction of autonomously replicating artificial plasmids. *Curr. Microbiol.* 29:23–29
175. Yoshikawa H, Ogasawara N. 1991. Structure and function of DnaA and the DnaA-box in eubacteria: evolutionary relationships of bacterial replication origins. *Mol. Microbiol.* 5:2589–97
176. Young DB, Mehlert A, Smith DF. 1990. Stress proteins and infectious diseases. In *Stress Proteins in Biology and Medicine*, ed. RI Morimoto, A Tissières, C Georgopoulos, pp. 131–65. Cold Spring Harbor, NY: Cold Spring Harbor Lab.
177. Yura T, Nagai H, Mori H. 1993. Regulation of the heat-shock response in bacteria. *Annu. Rev. Microbiol.* 47:321–50
178. Zeilstra-Ryalls J, Fayet O, Georgopoulos C. 1991. The universally conserved GroE(Hsp60) chaperonins. *Annu. Rev. Microbiol.* 45:301–25
179. Zyskind JW. 1990. Priming and growth rate regulation: questions concerning initiation of DNA replication in *Escherichia coli*. See Ref. 44a, pp. 269–78

PHYSIOLOGICAL IMPLICATIONS OF STEROL BIOSYNTHESIS IN YEAST[1]

Leo W. Parks

Department of Microbiology, North Carolina State University, Raleigh, North Carolina 27695-7615

Warren M. Casey

Glaxo, Five Moore Drive, Research Triangle Park, North Carolina 27709

KEY WORDS: yeast, *Saccharomyces*, sterols, ergosterol

CONTENTS

INTRODUCTION	96
STEROL BIOSYNTHESIS AND REGULATION	96
Biosynthetic Pathways	96
Sites of Regulation	99
Ergosterol Features and Regulation	101
STEROL METABOLISM	101
Effect of Cultural Conditions on Sterol Metabolism	101
Interactions with Heme Compounds	102
Sterols and Respiration	102
Esterification	103
PHYSIOLOGICAL EFFECTS OF STEROL ALTERATIONS IN YEAST	107
STEROL TRAFFICKING	109
Sterol Uptake and Transport	109
Lipid-Transport Proteins	110
WHY ERGOSTEROL?	110

ABSTRACT

Fungi are among the most primitive organisms that synthesize sterols. The fungal sterol, ergosterol, is similar to animal sterol, cholesterol, but with

[1]The literature review for this manuscript was completed on September 30, 1994.

significant structural differences. The genetics and biochemistry for most of the steps in sterol biosynthesis have been studied in the yeast, *Saccharomyces cerevisiae*. Yet, little is known of the precise physiological roles that sterols play in the cell. Work with strains that are auxotrophic for ergosterol has led to the prediction of at least four growth-dependent functions for sterols. Most of the antifungal compounds in medical and agricultural use affect some aspect of sterol synthesis or function. Extensive studies on the modes of action of those substances and research on the effects of altering sterol metabolism by sterol mutants are providing new insights into sterol functions in the cells. In addition, questioning why fungi require ergosterol rather than the simpler cholesterol provides heuristic impetus for further experimentation.

INTRODUCTION

For many years sterols in primitive eukaryotes have been relegated to passive structural roles in which they modulate the biophysical properties of membranes. Indeed, sterols do contribute to that biophysical environment, but for the purpose of maintaining an appropriately fluid membrane environment in response to changes in temperature or culture medium, they are extremely metabolically expensive cellular components. For example, the 80-kDa enzyme, lanosterol synthase, which converts epoxysqualene to lanosterol, precisely alters 20 bonds, forms 4 rings, and establishes 7 stereocenters (18). The sterol methyltransferase enzyme mediates the transfer of a methyl group from S-adenosylmethionine to zymosterol, forming fecosterol, at a metabolic cost of 12–14 ATP equivalents (74). Because of the complexity and cost of sterol biosynthesis and the availability of other components for altering microviscosity, the sterols must play other critical roles; otherwise sterol biosynthesis in these organisms would not have evolved and been maintained. Hence, a more comprehensive picture is emerging in which sterols have multiple, essential cellular roles. In the past 20 years almost 3000 papers on sterols in fungi have appeared. This review presents a physiological view of the cell with regard to the roles of the sterols therein; an exhaustive survey of the extensive literature on fungal sterols is not intended. In general, research presented in a previous review (74) is not discussed further here.

STEROL BIOSYNTHESIS AND REGULATION

Biosynthetic Pathways

The pathways for sterol biosynthesis in *Saccharomyces* are shown in Figures 1 and 2. The synthesis and regulation of sterols in yeast have been reviewed (73) and are not presented in detail here. The literature contains some ambiguity regarding the genetic designations of the structural genes in ergosterol biosynthesis; however, those detailed in that recent review are used.

Many suspected precursors of ergosterol have been isolated from yeast (6), but some of these compounds were detected in very small amounts or were present in mutants defective in one or more steps of sterol biosynthesis. A divergent pathway originally proposed by Oehlschlager and his colleagues (23) and subsequently modified (74) reflected many of those unusual intermediates. The relative nonspecificity of the substrate requirements in the later enzymes in sterol synthesis contributed to this mix of suspected ergosterol precursors. Some of the normally minor compounds accumulate abundantly only in unusual circumstances. For example, while the methylation reaction in *Saccharomyces cerevisiae* preferentially utilizes zymosterol, aerobically adapting yeast also use small amounts of 4α-methyl-zymosterol as substrate (75). The utilization of lower-affinity substrates has been offered as evidence for alternative pathways for sterol biosynthesis in yeast (68, 117).

To date all of the principal structural genes in ergosterol biosynthesis have been identified by means of molecular biology except for *ERG5* (C22 desaturase) and those involved in the demethylations at carbon 4 (43, 73). Recent experiments demonstrating the reproducible isolation of *erg5* mutants portend an early cloning of that gene as well. Stable mutants, even conditional mutants, that are defective in the 4,4-demethylations have not been characterized. The removal of the C4 methyl groups is obligatory for satisfying the sterol bulk function for growth of the GL7 sterol auxotroph (66). An inhibitor for the 4,4-demethylations has been reported (41). If this inhibition proves to be growth restricting and is sufficiently specific, it may offer a means of defining the genetics and biochemistry for this key reaction in ergosterol formation.

The pathway to ergosterol in yeast has been of interest because of the complexity of the biochemistry involved and for its exploitation in the development of antifungal compounds. Many of the agents currently in use affect sterol biosynthesis or function (96). Several interesting aspects of sterol synthesis in yeast distinguish it from sterol synthesis in animals, and these have been considered for developing antifungal agents that discriminate between cholesterol biosynthesis and the sterols of fungi. The obvious differences in the ergosterol structure are the C5,7 double unsaturations of the B ring, the additional unsaturation at carbon 22, and the methylation at carbon 24. These reactions provide interesting starting points for assessing uniqueness in the functions of ergosterol.

Although the focus here is on ergosterol, this compound is only one of an interesting group of isoprenoid compounds produced by yeast. Ubiquinone, dolichols, farnesylpyrophosphate, geranylgeranylpyrophosphate, isopentenylpyrophosphate, and ergosterol are all important products of the isopentenoid pathway. In addition to having distinctive structural roles, these isoprenoid

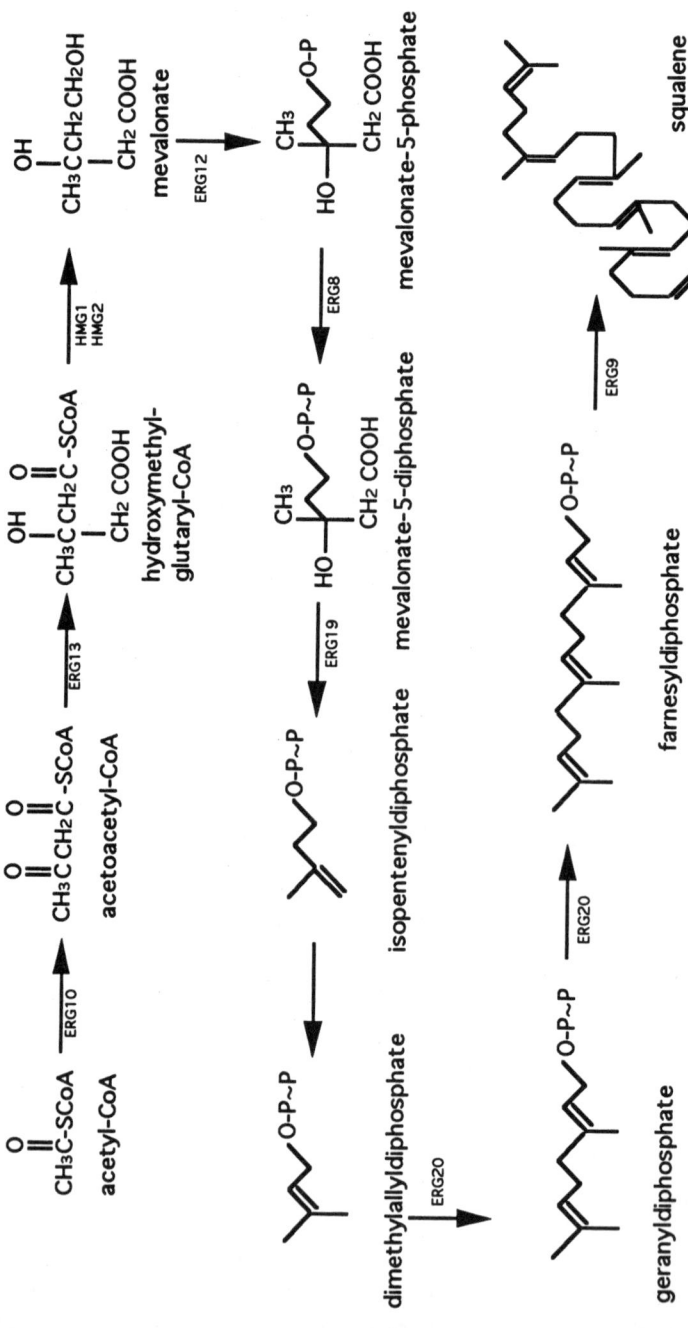

Figure 1 Biochemical transformations in the synthesis of squalene. The gene designations are as described previously (73).

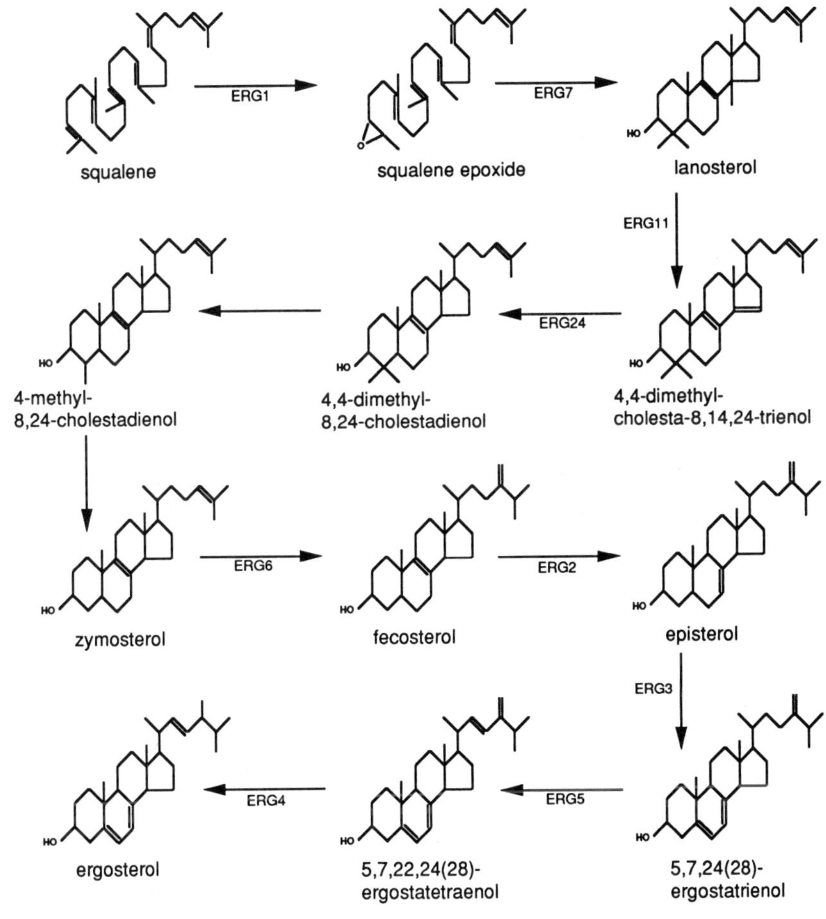

Figure 2 Biochemical transformations in the conversion of squalene to ergosterol. The gene designations are as described previously (73).

derivatives are involved in respiration and the covalent modifications of proteins and tRNA. The synthesis of 3-hydroxy-3-methylglutaryl coenzyme A (HMG-CoA) is the first committed step in isoprenoid biosynthesis.

Sites of Regulation

The regulation of sterol biosynthesis has proved to be very complex, and most studies have focused on the early reactions in the pathway. The isoprenoid pathway has been faithfully conserved in evolution. Biochemical comparisons of sterol synthesis between mammals and the fungi have been instructive, although yeast produce ergosterol while animal cells produce a more saturated and nonalkylated derivative, cholesterol. The 3-hydroxy-3-methylglutaryl coenzyme A reductase (HMGR) is a critical regulatory enzyme in sterol biosynthesis in animals (12, 25).

Yeast contain two genes, *HMG1* and *HMG2*, producing two isozymes of HMGR, Hmg1p and Hmg2p, that map on separate chromosomes (8). Strains with a null mutation in either of these genes are viable, but null mutations in both alleles render the cells inviable (7). Approximately 83% of the HMGR activity is attributable to Hmg1p. Single losses of the *HMG* alleles do reduce survival. When isogenic cultures were grown in complete liquid media for prolonged times with daily dilution to fresh media, the *hmg1,HMG2*-containing strain was the least fit for survival; the *HMG1,hmg2* construction was of intermediate fitness, and the wild-type (*HMG1,HMG2*) strain was most fit for survival (7). The persistence of certain strains could reflect advantageous growth-doubling times, lag periods for recovery from stationary phase, viability in stationary phase, etc.

Despite a huge disparity in the specific activities of the two HMGRs, each of the isozymes direct essentially equal amounts of carbon to the biosynthesis of sterols under heme-competent conditions (17). The monounsaturated C16 fatty acid, palmitoleic acid (C16:1,Δ^9), acts as a rate-limiting positive regulator, and ergosterol acts as an inhibitor of sterol synthesis, in strains that contain only the Hmg1p isozyme. Sterol synthesis is regulated by oleic acid (C18:1,Δ^9) and to a lessor degree by ergosterol in those strains with only Hmg2p. The specific activities of the two reductases are regulated differently by hemes. The differential regulation of sterol biosynthesis and discrepancies in the results of sterol feeding experiments led to speculation of physiologically compartmentalized pathways of sterol synthesis in yeast (17, 67).

Feed-back regulation of Hmg1p synthesis is independent of any change in the level of *HMG1* mRNA but is translationally regulated by levels of mevalonate (21). The Hmg2p isozyme is subject to regulated nonvacuolar degradation, whereas Hmg1p is stable (28). This degradation is modulated by mevalonate pathway flux, and a squalene precursor, possibly a farnesylated protein, is necessary for degradation. Squalene or another isoprenoid derivative might play a significant role in regulating HMGR (4).

The addition of aminolevulinic acid to a heme and sterol auxotrophic mutant, which makes the cells heme competent, results in the production of high endogenous concentrations of 2,3-oxidosqualene and 2,3;22,23-dioxidosqualene in the presence of glucose but not ethanol (50, 52). The increased synthesis of the oxidosqualenes is accompanied by a fivefold increase in the activity of HMGR. The failure of heme competency to stimulate lipid synthesis in the presence of ethanol may be partially a consequence of the sensitivity of HMGR activity to the presence of ethanol (113).

Squalene synthetase (farnesyl diphosphate:farnesyl diphosphate farnesyltransferase; *ERG9*) is a branch point of isoprenoid biosynthesis and has also been proposed as a regulatory site for sterol synthesis. The *ERG9*-encoded protein contains a predicted PEST (proline, glutamic acid, serine, threonine) consensus sequence that is present in many proteins that have a short cellular half-life (33). Inhibition of sterol synthesis with lovastatin at HMGR results in an elevation of the squalene synthetase mRNA (82), supporting the proposal that the mevalonic acid pool may regulate squalene synthetase levels (59). Anaerobic conditions and excess sterol repress squalene synthetase and squalene epoxidase (59). A coordination of sterol synthesis and the availability of unsaturated fatty acids is predicted from the observation that squalene epoxidase activity requires unsaturated fatty acids (14).

Ergosterol Features and Regulation

Sterols that contain both a C22 unsaturation and a C24 methyl group can reduce sterol biosynthesis by approximately 50%, regardless of the B-ring structure (16). Oxy-derivatives of cholesterol had no effect on sterol synthesis, but 24,25-epoxylanosterol reduced the endogenous synthesis of ergosterol by 49%. The use of the GL7 heme and sterol auxotroph led to the proposal that sterols with a C5 double bond were especially effective in controlling ergosterol synthesis (67).

Under some conditions of ergosterol deprivation, acetoacetyl-coenzyme A thiolase and HMG-CoA synthase are generally increased (87), but exogenously supplied sterols do not suppress enzyme activities under aerobic conditions. Under semianaerobic conditions, cholesterol and ergosterol are effective in lowering the accumulation of squalene, ergosterol being fourfold more effective than the nonalkylated sterol (78).

STEROL METABOLISM

Effect of Cultural Conditions on Sterol Metabolism

The synthesis of ergosterol requires molecular oxygen to provide the oxygen in the hydroxyl group at carbon 3. Cultures grown anaerobically exhibit extensive accumulation of squalene. On exposure to oxygen, a rapid conversion of squalene to sterols occurs (9, 40). The effect of oxygen, nutrition, and temperature on sterol accumulation in yeast was discussed extensively in a previous review (74).

The sterol content of yeast may be perturbed, depending on the media used for culturing the organism. For example, the supplementation of culture media with homocysteine results principally in the formation of sterols lacking the alkylation at C24 (32). Because of the facile enzymatic conversion of homocysteine to S-adenosylhomocysteine (S-AH), the low methylation can be explained by the inhibition of the sterol methyltransferase by S-AH (61). Other examples of aberrant sterol profiles as a consequence of cultural conditions have been described but are not understood as easily as the homocysteine effect.

Ergosterol synthesis appears to be one of the more temperature-sensitive processes in yeast. Although cells can grow at 40°C in defined medium, they must have exogenously supplied sterols and oleic acid (88, 89, 94). The yield of respiratory deficient cells is high at that temperature, but the proportion of respiratory-deficient cells is reduced upon the addition of ergosterol and oleic acid (77). Recently, human-pathogenic isolates of *Saccharomyces cerevisiae* were characterized (63). Most of those isolates can grow at 42°C, whereas the saprophyte forms do not. It will be interesting to see if the temperature limits of ergosterol and oleate synthesis are aberrantly high in the pathogenic strains, and if these lipids contribute to the yeast's pathogenicity and tolerance for elevated temperatures.

Interactions with Heme Compounds

Sterol biosynthesis and oxygen in yeast are intimately associated. Besides the requirement for molecular oxygen for the epoxidation of squalene, cytochrome hemoproteins are involved in demethylation and desaturation steps. The *HAP1* gene activates target genes in response to the presence of hemes. Features of the *HAP1*-encoded protein implicate it as a general sensor of the redox state of the cell. A comprehensive review of this area was published recently (121). In aerobic, heme-competent cells, *HAP1* activates the transcription of *ERG11* (110). Lanosterol 14α-demethylase (Erg11p) is an essential enzyme for ergosterol synthesis and is a major cytochrome P-450 component of yeast. Furthermore, it may be transcriptionally regulated (93). The *ERG11* mRNA levels increase during growth on glucose in the presence of heme and during semianaerobic and oxygen-limiting growth (108). Expression of the gene under anaerobic conditions is precluded because of the absence of heme. Because oxygen is not present for the epoxidation of squalene, the C30 hydrocarbon accumulates. Large amounts of squalene are deleterious to the yeast and have been proposed as the cause of action of some fungicidal compounds (85).

Although their role is less well understood than that of heme in lanosterol demethylase, cytochrome derivatives also participate in the desaturations at C5 and at C22 (31, 72).

Sterols and Respiration

Among the earliest observations with ergosterol were its associations with oxygen and respiration (58, 74). The formation of ergosterol coincides with the acquisition of oxidative activity of yeast (58), and ergosterol is required for the anaerobic growth of yeast (1). Restricting ergosterol formation by conditions of high temperature or mutations results in a high frequency of respiratory deficient (petites) mutants (34, 74). When given a choice of available sterols, yeast cells discriminate in favor of ergosterol for inclusion in the mitochondrial structure (76). A quantitative relationship between sterols and the functioning of the adenine nucleotide transporter and ATP generation in mitochondria has been established (30).

The multiple interactions of hemes appear to pervade the different functions of sterols in yeast (116). That the synthesis of ergosterol and unsaturated fatty acids, and the expression of heme formation, all coincide with the development of functional respiratory structures suggests interrelatedness. Numerous experiments support this notion. Qualitative changes in the sterol composition of cells result in corresponding physical and enzymatic alterations of the mitochondria (64). Extensive fractionation techniques of yeast subcellular fractions have localized sterols and steryl esters (120); both the inner and outer mitochondrial membranes contain ergosterol, with the inner membrane having a higher ergosterol-to-phospholipid ratio (92).

ERG3 encodes the C5-desaturase that introduces the C5=6 unsaturation in ergosterol biosynthesis. Although early work suggested that the C5=6 unsaturation was essential for growth (84), when *ERG3* was cloned and inactivated, the cells grew well (2). Another study, in which *ERG3* was independently cloned and inactivated, found that cells lacking the C5=6 do not grow if there is a heme deficiency or if the C5=6 deficient cultures are incubated on respiratory substrates such as ethanol or glycerol in a defined medium (91). The sensitivity to antifungal agents that are inhibitors of sterol biosynthesis is enhanced when the cultures are grown on respiratory substrates (51, 53, 54, 76). These studies provide an interesting entree into the biochemical and genetic study of the physiological interactions between heme compounds, ergosterol, and aerobic energy metabolism.

Esterification

ESTERIFICATION AND THE CULTURE CYCLE   Sterols in yeast are found with the 3β-OH group free or esterified to long chain fatty acids. The cellular content of steryl esters varies, depending on the nature of the sterol and the phase of the culture cycle in which the cells are harvested (3). A sharp increase in the rate and accumulation of steryl esters occurs upon entry of the culture to the stationary phase. The interconversion of free sterols and steryl esters was studied radioisotopically in cultures of yeast (103). These experiments demonstrated a culture-cycle dependent esterification of ergosterol and hydrolysis of ergosteryl esters. Steryl ester synthesis occurs principally in the microsomal fraction, but steryl ester hydrolysis is attributed to the plasma membrane. Taken together, these observations argue for a controlled mobilization and transfer of sterols and steryl esters. Moreover, small lipid droplets of low density are the major repository of steryl esters (119), and a role for low-density vesicles in the movement of sterol within the cell has been proposed (118).

Although yeast sterols are esterified principally to C16 and C18 long-chain fatty acids, over 20 types of steryl esters have been detected in these organisms (81). Selection of a specific fatty acid for esterification does not appear to be linked to the culture cycle (22). Ester accumulation in lovastatin-treated cultures also decreases as a consequence of diminished synthesis of sterols (51).

Cultures of Upc20 (*upc2,* described below) accumulate large amounts of endogenously synthesized steryl esters, taking sterols from the medium for that purpose. This mutant fails to show the culture cycle–dependent interconversions between free and esterified sterols (38). When *upc2* is included in a sterol auxotrophic background and the supply of sterol in the medium is exhausted, hydrolysis of the accumulated esters occurs (NV Shenvi & LW Parks, submitted). The flux of free sterols apparently has a significant effect on the esterification and hydrolysis of the steryl ester cycle. Independent extragenic suppressors of *upc2* return the mutant phenotype to that of the wild-type culture. Excess sterols are deposited principally in the ester fraction (47), and the membrane pool of free sterols is only marginally increased with sterol uptake.

ESTER SYNTHESIS AND HYDROLYSIS   The steryl ester synthase enzyme (SES) from yeast has been characterized biochemically (102) and is microsomally localized. Substrate specificities indicated that the yeast enzyme is a broadly specific acyl-coenzyme A–ergosterol acyl transferase (AEAT). SES activity is directly inhibited by the hypocholesterolemic drug, lovastatin (mevinolin). The inhibition is enhanced in heme-competent cells (37). However, repeated attempts have failed to show the sensitivity of this enzyme to most other inhibitors of the homologous cholesterol esterifying enzyme of animals, acyl-coenzyme A–cholesterol acyl transferase (ACAT).

Sterol and heme auxotrophs were used to determine the effect of heme depletion on esterification. SES activity was stimulated fourfold in cells that were auxotrophic for heme (*hem1*) and that were made heme competent by feeding the cells aminolevulinic acid. Whereas heme stimulation of sterol biosynthesis was observed only on glucose, the stimulation of esterification by heme was observed on either fermentative or respiratory carbon sources. SES activity increased as the cells entered stationary phase and was also enhanced by heme competency (37).

The selectivity for sterol esterification in whole cells of a sterol auxotroph of yeast is for sterols lacking double bonds at the C7 and C22 positions and lacking methylation at C24 (104). Cultures that are fed sterols with 5,7 double bonds have severely reduced steryl ester levels and are prevented from accumulating other sterols from the medium. Phosphatidylcholine or lysophosphatidylcholine stimulates the hydrolysis of steryl esters by the steryl ester hydrolase (100). Whether this phenomenon has physiological significance or is a consequence of the solubilization of the ester by the phospholipid remains to be determined.

FUNCTIONS FOR ESTERIFICATION The 3-methylether derivatives of ergosterol or cholesterol can sustain growth of yeast cultures both aerobically and anaerobically (44). However, substantial conversion of the ethers to ergosterol was observed. Because the ether linkages cannot be esterified, the esterification process is probably not essential for the growth of yeast.

Esterification is likely an important event in ergosterol metabolism; otherwise selection would have eliminated the process. Furthermore, numerous attempts to isolate mutants lacking the steryl ester synthase have failed, which enhances the notion that the maintenance of SES provides a selective advantage to the yeast.

The esterification of ergosterol may achieve at least two beneficial physiological ends: a proofing function whereby nonergosterol sterols could be incarcerated to lipid droplets and be isolated from participating in critical ergosterol-dependent functions, and the salvage of sterols or the storage of excess sterol that results from over-synthesis.

Proofing Proofing is defined here as the identification and removal of less desirable sterols from active metabolism. Preliminary support for esterification as a physiologically significant proofing process has been obtained from experiments in which nonergosterol sterols were preferentially esterified and then deposited and retained in the lipid droplets as esters. Labeling experiments and analyses of sterol pools showed that ergosterol was not the predominant sterol in the ester fraction while the culture was growing (74, 103). This observation suggests an active cellular qualitative control over the partitioning of sterols to the free and steryl ester fraction. In the removal of sterols from the steryl ester pool, cells may discriminate in favor of ergosterol from the various steryl esters found in the mixture (76). The ester-rich low-density lipid particles contain, in addition to ergosterol, large amounts of zymosterol, lanosterol, fecosterol, and episterol, all esterified to long-chain fatty acids (108, 119). Although zymosterol is the preferred substrate for the sterol methyltransferase, esters of zymosterol do not serve as substrates for transmethylation (60).

The accumulation of nonmethylated sterols substantially increases the total steryl ester pool in the mutants incapable of sterol transmethylation (*erg6*). These cells have sterol levels eight times higher than those of their parents (60), and the bulk of the excess sterols are found in the ester fraction. The C28 sterols regulate sterol biosynthesis (16), and their absence could enhance sterol biosynthesis in the mutant. Alternatively, the active removal of the nonmethylated sterols to the ester fraction as a consequence of proofing could elicit a perceived sterol deficiency, which the cell attempts to remediate by increasing sterol production. These observations suggest that during active growth, esterification may selectively remove nonergosterol sterols from active metabolism.

Sterol storage and salvage by esterification Esterification of surplus ergosterol may also serve a simple storage function. Because efficient feedback mechanisms are available for the cell to use in regulating biosynthetic activities, excessive production of almost all cellular constituents is avoided, except for authentic storage compounds. Major storage products usually serve as a ready sources for carbon and energy. Glycogen, trehalose, and neutral fat (triacylglycerols) fulfill those functions. Ergosterol is a very expensive compound to produce. Because it is not readily degraded, ergosterol (stored as a fatty acid ester) would not appear to be a very readily metabolized storage product.

No matter what reason sterol accumulates, if the cell contains an excess of free sterol, such excess may be detrimental to normal membrane functions. There appears to be an optimal concentration of free sterol in the cells, and further accumulation of free sterol results in near quantitative conversion of the excess sterol to ester (48, 55). Furthermore, although Upc20 (*upc2*) accumulates large amounts of sterols either by endogenous synthesis or uptake from the medium, the bulk of these compounds are deposited to the ester fraction. If excess free sterols are toxic, this could explain in part the failure to isolate SES mutants (47).

Esterification might also serve a transport function as in higher organisms. Nevertheless, this function would not account for the high accumulations of steryl esters in the late exponential phase of the growth of yeast cultures. An ergosterol-rich membranous component may be present that is obligatory but transient in a phase of the actively growing cell. When that part of the cell cycle has passed, the free ergosterol would be physiologically excessive and would be esterified to avoid adventitious association and possible disruption of membranes already containing the proper amount of sterol. When, during the next budding cycle, there is a high demand for sterol for the hypothetical ergosterol-rich membrane, the ester could be hydrolyzed to provide a ready source of ergosterol. Stationary phase could reflect an alignment of the bulk of the cells, when salvaged ergosterol is at its highest concentration. Esterification with fatty acids assures that fatty acids will also be available. A cholesterol-rich intracellular precursor of the plasma membrane has been described in tissue-culture cells (45).

If steryl esters are indeed true storage compounds, they could be critical to cells reinitiating growth following a senescent period. Triacylglycerols accumulate concomitant with steryl-ester accumulation. Glycerol, fatty acids, ergosterol, and possibly diacylglycerol would be available for the many new membranes that are required under these conditions.

PHYSIOLOGICAL EFFECTS OF STEROL ALTERATIONS IN YEAST

One technique to define the various physiological functions of sterols is to monitor which activities are perturbed when the sterol composition is altered. Numerous studies have been conducted on the effects of differing sterols on the physical and enzymic activities of membrane lipids (74). More recent experiments have shown a coordination of sterol formation and other membrane lipids. The isolation of stable sterol auxotrophs allowed the effects of specific sterols on the phospholipids of cells to be tested. Using these auxotrophs, an early study showed that yeast cells grown on ergosterol contained less unsaturated fatty acids and more medium-chain saturated fatty acids in their phospholipids than did cholesterol-grown cells (15).

Steady-state fluorescence anisotropy was used to detect changes in the physical properties of membranes in yeast sterol mutants and sterol auxotrophs that were grown on differing amounts of sterols and with sterols with different structural features. These experiments were interpreted to indicate that changes in the lipid properties of the auxotroph reflected the ability of the organisms to modify their phospholipid composition with respect to the sterol being provided (10, 11, 57). Quantitative analyses revealed that a sterol supplied exogenously to a sterol auxotroph coordinately regulated specific phospholipid species, fatty acid composition, and sterol-to-phospholipid ratios. In yeast cells subjected to thermal stress, membrane thermotropic transitions were 3°C to 6°C higher than in control cells (56). Alterations in the phospholipid composition of a mutant incapable of sterol transmethylation (*erg6*) were also observed (60).

The activity of some phospholipid biosynthetic enzymes is altered in a culture-cycle dependent manner in cholesterol-grown cells as opposed to yeast grown with ergosterol (36, 80). The activities of phosphatidylethanolamine:phosphatidylcholine-N-methyl transferase and acyl-coenzyme A α-glycerol-3-phosphate transacylase were greater in ergosterol-grown than in cholesterol-grown cells. Phosphatidylinositol kinase activity is also stimulated by ergosterol (19, 20). Sterols could affect the phospholipid composition by affecting the transfer of phospholipids by phospholipid-transfer proteins. Ergosterol reduces the transfer rate of phospholipids by a phosphatidylinositol-transfer protein (97). Additionally, phosphoinositide turnover in yeast has been proposed to be regulated by sterol (109).

The presence of coordinate regulatory effects attendant to changing sterol may serve to optimize the membrane for the function that is required. That is, if the sterol is changed, then phospholipids and other membrane components could be adjusted to provide an environment as near normal as possible for the activities of the membrane (62). Incorporation of the nonfungal sterol cholesterol into yeast membranes reduces membrane elasticity and increases susceptibility to osmotic lysis. In contrast, mutants that accumulate growth-permissive sterols do not affect the resistance to osmotically induced swelling (65).

The effect of sterol alterations has an impact in areas other than phospholipids as well. While the amount of dolichyl phosphate synthesized in wild-type and sterol mutants is equivalent, mannosylation and glucosylation of dolichyl phosphates is about four times lower in ergosterol mutants compared with parental strains (98). Ergosterol-containing lipid fractions inhibit lipid peroxidation when introduced into phospholipid liposomes (115).

Alterations to the sterol composition of cells caused by feeding of auxotrophs or by mutations can profoundly affect membrane permeability because of changes to the sterol itself or phospholipid changes mandated by the aberrant sterol. Effects on membrane properties have been observed most dramatically with the strains that have a defect in the gene for sterol methyl transferase, *erg6*. Early attempts to map *erg6* genetically employed a strain with a nutritional marker for tryptophan biosynthesis, *trp1* (60), and resulted in the erroneous mapping of *erg6* "close to *trp1*" (60). This location was corrected several years later when it was observed that *erg6* mutants are defective in tryptophan transport (24), and the gene was then mapped to chromosome XIII. The tryptophan effect is particularly interesting, because both tryptophan permeases must be affected. The mutants with *erg6* were also observed to be hypersensitive to the protein synthetic inhibitor cycloheximide and shown to be permeable to Ni^+ (5, 24). Greater membrane rigidity than wild-type (46) was also found. *ERG6* was recloned recently in a search for complementation to a mutation with enhanced sensitivity to the monovalent cations, Na^+ and Li^+ (114). Unlike the wild-type, mutants with *erg6* are sensitive to brefeldin A, which causes a block in protein transport through the secretory pathway (27, 112), again emphasizing the altered permeability of these mutants.

ERG6 in high copy number can act as a suppressor of the *erd2* gene of yeast (29). *ERD2* encodes a receptor that is required for retrieval from the Golgi apparatus of luminal proteins of the endoplasmic reticulum. How *ERG6* itself or its encoded transmethylase mediates the suppression of *erd2* remains unknown.

ERG3 (C5 desaturase) has been cloned on the basis of complementation of a gene that simultaneously leads to resistance to the phytotoxin, syringomycin, and sensitivity of growth to high calcium concentrations (99). A deficiency in C5=6 sterols results in greater sensitivity to calcium ions and an increased rate of calcium flux. Membrane changes accompanying the presence of C5=6 sterols appear to enhance the resistance of yeast to ethanol-induced death (70, 105).

Changing the sterol composition of the plasma membrane can reduce the ability of cells to complete conjugation. The mating efficiency of sterol auxotrophs was perturbed by the replacement of ergosterol with other sterols. This is manifested in reduced cytoplasmic fusion in populations of the mated pairs. Cells become adherent in the normal fashion of mating but remain prezygotic without membrane dissolution and cytoplasmic and nuclear mixing (106).

STEROL TRAFFICKING

Sterol trafficking is defined here as the orderly qualitative and quantitative discrimination, distribution, transport, and maintenance of sterol composition in cellular structures. A comprehensive review of intracellular lipid transport in eukaryotes has been written (111). An analysis of the sterol composition of different cellular fractions (60, 64, 71, 119) shows a wide discrepancy in the amount of sterol present in different membranes. Furthermore, the quantitative and qualitative requirements differ for separate sterol functions in the cell (11, 83), and the yeast cells have an active cellular control over what types of sterols are found in the free and esterified sterol pools (101, 103). Although sterol synthesis occurs in the endoplasmic reticulum (69), the plasma membrane and the secretory vesicles are most highly enriched for ergosterol (119, 120). Taken together, these disparities argue for mechanisms of maintaining distinctive sterol content of membranes that avoid randomization of the lipid.

Sterol Uptake and Transport

Despite the enormous metabolic expense for forming ergosterol, yeast cells under aerobic conditions cannot take substantial amounts of the sterol from the growth medium. One might expect that cells would use exogenously supplied sterols and thus save the metabolic costs of synthesizing sterols. In the presence of exogenously supplied sterols, aerobic yeast cells continue to produce ergosterol and do not take up a significant quantity of the supplied sterol (47, 107). This phenomenon has been called aerobic sterol exclusion and depends on heme biosynthesis (49). Under anaerobic conditions, heme and sterol synthesis are precluded and sterol uptake proceeds readily (79).

In aerobic cells incapable of heme biosynthesis, the amount of sterol taken from the medium is inversely proportional to the amount of cellular free sterol (51, 55), and when the cells are saturated with sterol, further accumulation depends on cell growth (86). A preference in uptake of sterols for those that are most structurally similar to ergosterol was found (104). Cells that were saturated with free ergosterol were prevented from accumulating additional ergosterol or any other sterols from the medium. In these same cells, the steryl ester fraction was low. This was not the case for cholesterol, in which the cells took up large amounts of cholesterol and deposited most of it in the ester fraction. The accumulation of sterols from the growth medium in sterol-depleted cells does not depend on cellular energy, and cell viability is not required (52).

Early attempts to isolate viable, nonleaky sterol auxotrophs in yeast were unsuccessful without a concurrent mutation early in the heme biosynthetic pathway (13, 26, 35). Aerobic sterol exclusion is so effective that most sterol auxotrophs without a heme deficiency cannot grow (50), even with sterols provided in the medium (37). Saturation of the cell membrane with ergosterol is not responsible for the heme-mediated sterol exclusion (90). Mutants defective in the aerobic sterol exclusion process, even in fully heme-competent cells, have been isolated (47). Called Upc$^-$ (upc, uptake control) strains, these isolates showed a broad range of sterol accumulation. Strains with *upc2*, the most widely studied of this group, have normal amounts of steryl ester synthase activity. Reversion of the *upc2* strains to the wild-type phenotype was accomplished by two nonlinked suppressor mutations whose effects were additive (GA Keesler & LW Parks, unpublished experiments).

Lipid-Transport Proteins

Some proteins capable of transporting various lipids in higher organisms, such as the phospholipid-transfer proteins, have homologous counterparts in yeast. A phosphatidylinositol-transfer protein described in yeast complexes with phosphatidylinositol and phosphatidylcholine but transfers the former much more readily than the latter. Ergosterol affects the phospholipid transfer rate (97). A phosphatidylserine-transfer protein has been characterized and shown to have a low transport rate for ergosterol (42).

Apolipoproteins mediate the transport of cholesterol and cholesteryl esters in higher organisms. Anti-rat apolipoprotein antibodies obtained from hyperimmunized rabbits cross-react with proteins in yeast. Anti-apolipoproteins A1, B, and E reacted both with proteins in a crude cell-free preparation of yeast and proteins of the low-density floating-lipid fraction (39). The Upc20 strain showed a different protein reactivity from the wild-type cells, but the banding pattern of the immunoblot was returned to the wild-type by extragenic suppressors (38). A cDNA to human apolipoprotein E has been expressed in yeast, but the protein is secreted only in those mutants that have the *upc2* gene (95). The presence of cholesterol increased the secretion of the protein.

WHY ERGOSTEROL?

Cholesterol appears to satisfy the complex sterolic needs of higher organisms, yet it is not the sterol of choice for yeast or other fungi. For the synthesis of present in mammalian cells, S-adenosylmethionine–dependent sterol methyltransferase, the C24=28 methylene reductase, and the C22 desaturase, the structural genes for which are *ERG6, ERG4,* and *ERG5,* respectively. The methyltransferase is a very metabolically expensive reaction for the cell (74). Furthermore, sterol biosynthesis or function is a principal target for compounds with antifungal activity. Therefore, we can assume that these reactions are important to the survival of the yeast; otherwise as a consequence of Darwinian selection, these cells would have been eliminated in favor of strains lacking these functions. Although all but two of the structural genes for ergosterol biosynthesis (those involved in the demethylations at C4 and *ERG5*) have been identified as of this writing, the genetics, biochemistry, and molecular biology have clearly far surpassed our understanding of the physiology of ergosterol in yeast.

It seems fruitless to offer an explanation for ergosterol in favor of other sterols on the basis of its ability (or inability) to pack in certain phospholipid bilayers, because other sterols appear to be accommodated adequately in these organisms. Although the intimate relationship between ergosterol and oxygen is obvious, this is likely not a unique aspect of ergosterol in fungi, as higher organisms have functional respiratory mechanisms that use cholesterol.

Ergosterol is probably the preferred sterol in yeast because it serves as a consensus sterol, being able to satisfy a variety of functions, some of which may not be critical in higher organisms. For instance, the sensitivity to cations of the ergosterol mutants is probably an important observation in our attempts to answer the question, "why ergosterol?" Yeast survive in an ecological niche that usually contains high concentrations of solutes. Ergosterol may provide the proper association with phospholipids to exclude solutes. It is interesting that all of the sterol mutants that have been tested have some cation sensitivity, and that the extremely effective polyene fungicidal compounds, which interact directly with membrane sterols, cause leaching of cations from the cells.

If ergosterol is a consensus sterol as we predict, the role of proofing becomes important to prevent nonergosterol sterols from participating in critical ergosterol-dependent functions. A low, but measurable, influx of sterols from the medium occurs even under the most stringent conditions of aerobic sterol exclusion. The different features of ergosterol are not equally significant in the separate sterol roles. However, nonergosterol sterols may interfere with ergosterol-dependent activities and diminish their effectiveness. Although this may not be lethal in the classic genetic sense, the reduced cellular functions could lead to lowered evolutionary fitness.

Whatever its unique functions, understanding the roles of ergosterol in yeast cells provides an interesting challenge. With this understanding should come a greater insight into physiological interactions, not only in the primitive eukaryotes, but in higher organisms as well.

> Any *Annual Review* chapter, as well as any article cited in an *Annual Review* chapter, may be purchased from the Annual Reviews Preprints and Reprints service.
> 1-800-347-8007; 415-259-5017; email: arpr@class.org

Literature Cited

1. Andreasen AA, Stier TJB. 1953. Anaerobic nutrition of *Saccharomyces cerevisiae*. I. Ergosterol requirements for growth in a defined medium. *J. Cell. Comp. Physiol.* 41:23–36
2. Arthington BA, Bennett LG, Skatrud PL, Guynn CJ, Barbuch RJ, et al. 1991. Cloning, disruption and sequence of the gene encoding yeast C-5 sterol desaturase. *Gene* 102:39–44
3. Bailey RB, Parks LW. 1975. Yeast sterol esters and their relationship to the growth of yeast. *J. Bacteriol.* 124:606–12
4. Bard M, Downing JF. 1981. Genetic and biochemical aspects of yeast sterol regulation involving 3-hydroxy-3-methylglutaryl coenzyme A reductase. *J. Gen. Microbiol.* 125:415–20
5. Bard M, Lees ND, Burrows LS, Kleinhaus FW. 1978. Differences in crystal violet uptake and cation-induced death among yeast sterol mutants. *J. Bacteriol.* 135:1146–48
6. Barton DHR, Corrie JET, Widdowson DA, Bard M, Woods RA. 1974. Biosynthesis of terpenes and steroids. IX. The sterols of some mutant yeasts and their relationship to the biosynthesis of ergosterol. *J. Chem. Soc. Perkin Trans.* 1:1326–33
7. Basson ME, Moore RL, O'Rear J, Rine J. 1987. Identifying mutations in duplicated functions in *Saccharomyces cerevisiae:* recessive mutations in HMG-CoA reductase genes. *Genetics* 117:645–55
8. Basson ME, Thorsness M, Rine J. 1986. *Saccharomyces cerevisiae* contains two functional genes encoding 3-hydroxy-3-methylglutaryl-coenzyme A reductase. *Proc. Natl. Acad. Sci. USA* 83:5563–67
9. Boll M, Lowel M, Berndt J. 1980. Effect of unsaturated fatty acids on sterol biosynthesis in yeast. *Biochim. Biophys. Acta* 620:429–39
10. Bottema CDK, McLean-Bowen CA, Parks LW. 1983. Role of sterol structure in the thermotropic behavior of plasma membranes of *Saccharomyces cerevisiae*. *Biochim. Biophys. Acta* 734:235–48
11. Bottema CD, Rodriguez RJ, Parks LW. 1985. Influence of sterol structure on yeast plasma membrane properties. *Biochim. Biophys. Acta* 813:313–20
12. Brown MS, Goldstein JL. 1980. A receptor-mediated pathway for cholesterol homeostasis. *Science* 232:34–47
13. Buttke TM, Bloch K. 1980. Comparative responses of the yeast mutant GL7 to lanosterol, cycloartenol, and cyclolaudenol. *Biochem. Biophys. Res. Commun.* 92:229–36
14. Buttke TM, Brint SL, Lowe MR. 1988. Regulation of squalene epoxidase activity by membrane fatty acid composition in yeast. *Lipids* 23:68–71
15. Buttke TM, Reynolds R, Pyle AL. 1982. Phospholipid synthesis in *S. cerevisiae* strain GL7 grown without unsaturated fatty acid supplements. *Lipids* 17:361–66
16. Casey WM, Burgess JP, Parks LW. 1991. Effect of sterol side-chain structure on the feed-back control of sterol biosynthesis in yeast. *Biochim. Biophys. Acta* 1081:279–84
17. Casey WM, Keesler GA, Parks LW. 1992. Regulation of partitioned sterol biosynthesis in *Saccharomyces cerevisiae*. *J. Bacteriol.* 174:7283–88
18. Corey EJ, Matsuda SP, Bartel B. 1994. Molecular cloning, characterization, and overexpression of ERG7, the *Saccharomyces cerevisiae* gene encoding lanosterol synthase. *Proc. Natl. Acad. Sci. USA* 91:2211–15
19. Dahl C, Biemann HP, Dahl J. 1987. A protein kinase antigenically related to pp60v-src possibly involved in yeast cell cycle control: positive *in vivo* regulation by sterol. *Proc. Natl. Acad. Sci. USA* 84:4012–16
20. Dahl JS, Dahl CE. 1985. Stimulation of cell proliferation and polyphosphoinositide metabolism in *Saccharomyces cere-*

visiae GL7 by ergosterol. *Biochem. Biophys. Res. Commun.* 133:844–50
21. Dimster-Denk D, Thorsness MK, Rine J. 1994. Feedback regulation of 3-hydroxy-3-methylglutaryl coenzyme A reductase in *Saccharomyces cerevisiae*. *Mol. Biol. Cell* 5:655–65
22. Fenner GP, Parks LW. 1989. Gas chromatographic analysis of intact steryl esters in wild-type *Saccharomyces cerevisiae* and in an ester accumulating mutant. *Lipids* 24:625–29
23. Fryberg M, Oehlschlager AC, Unrau AM. 1973. Biosynthesis of ergosterol in yeast. Evidence for multiple pathways. *J. Am. Chem. Soc.* 95:5747–57
24. Gaber RF, Copple DM, Kennedy BK, Vidal M, Bard M. 1989. The yeast gene ERG6 is required for normal membrane function but is not essential for biosynthesis of the cell-cycle-sparking sterol. *Mol. Cell. Biol.* 9:3447–56
25. Goldstein JL, Brown MS. 1990. Regulation of the mevalonate pathway. *Nature* 343:425–30
26. Gollub EG, Liu KP, Dayan J, Adlersberg M, Sprinson DB. 1977. Yeast mutants deficient in heme biosynthesis and a heme mutant additionally blocked in cyclization of 2,3-oxidosqualene. *J. Biol. Chem.* 252:2846–54
27. Graham TR, Scott PA, Emr SD. 1993. Brefeldin A reversibly blocks early but not late protein transport steps in the yeast secretory pathway. *EMBO J.* 12:869–77
28. Hampton RY, Rine J. 1994. Regulated degradation of HMG-CoA reductase, an integral membrane protein of the endoplasmic reticulum in yeast. *J. Cell Biol.* 125:299–312
29. Hardwick KG, Pelham HRB. 1994. SED6 is identical to ERG6, and encodes a putative methyltransferase required for ergosterol synthesis. *Yeast* 10:265–69
30. Haslam JM, Astin AM, Nichols WW. 1977. The effects of altered sterol composition on the mitochondrial adenine nucleotide transporter of *Saccharomyces cerevisiae*. *Biochem. J.* 166:559–63
31. Hata S, Nishino T, Komori M, Katsuki H. 1981. Involvement of cytochrome P-450 in Δ^{22}-desaturation in ergosterol biosynthesis of yeast. *Biochem. Biophys. Res. Commun.* 103:272–77
32. Hatanaka H, Ariga N, Nagai J, Katsuki H. 1974. Accumulation of sterol intermediate during reaction in the presence of homocysteine with cell-free extracts of yeast. *Biochem. Biophys. Res. Commun.* 60:787–93
33. Jennings SM, Tsay YH, Fisch TM, Robinson GW. 1991. Molecular cloning and characterization of the yeast gene for squalene synthetase. *Proc. Natl. Acad. Sci. USA* 88:6038–42
34. Jimenez J, Longo E, Benitez T. 1988. Induction of petite yeast mutants by membrane-active agents. *Appl. Environ. Microbiol.* 54:3126–32
35. Karst F, Lacroute F. 1973. Isolation of pleiotropic yeast mutants requiring ergosterol for growth. *Biochem. Biophys. Res. Commun.* 52:741–47
36. Kawasaki S, Ramgopal M, Chin J, Bloch K. 1985. Sterol control of the phosphatidylethanolamine-phosphatidylcholine conversion in the yeast mutant GL7. *Proc. Natl. Acad. Sci. USA* 82:5715–19
37. Keesler GA, Casey WM, Parks LW. 1992. Stimulation by heme of steryl ester synthase and aerobic sterol exclusion in the yeast *Saccharomyces cerevisiae*. *Arch. Biochem. Biophys.* 296:474–81
38. Keesler GA, Laster SM, Parks LW. 1992. A defect in the sterol:steryl ester interconversion in a mutant of the yeast, *Saccharomyces cerevisiae*. *Biochim. Biophys. Acta* 1123:127–32
39. Keesler GA, Moore S, Usher DC, Parks LW. 1991. Yeast proteins with reactivity to antibodies elicited against mammalian apolipoproteins. *Biochem. Biophys. Res. Commun.* 174:631–37
40. Klein HP. 1955. Synthesis of lipids in resting cells of *Saccharomyces cerevisiae*. *J. Bacteriol.* 69:620–27
41. Kuchta T, Bartkova K, Kubinec R. 1992. Ergosterol depletion and 4-methyl sterols accumulation in the yeast, *Saccharomyces cerevisiae*, treated with an antifungal, 6-amino-2-n-pentylthiobenzothiazole. *Biochem. Biophys. Res. Commun.* 189:85–91
42. Lafer G, Szolderits G, Paltauf F, Daum G. 1991. Isolation of a phosphatidylserine transfer protein from yeast cytosol. *Biochim. Biophys. Acta* 1069:139–44
43. Lai MH, Bard M, Pierson CA, Alexander JF, Goebl M, et al. 1994. The identification of a gene family in the *Saccharomyces cerevisiae* ergosterol biosynthetic pathway. *Gene* 140:41–49
44. Lala AK, Buttke TM, Bloch K. 1979. On the role of the sterol hydroxyl group in membranes. *J. Biol. Chem.* 254:10582–85
45. Lange Y, Steck TL. 1985. Cholesterol-rich intracellular membranes: a precursor of the plasma membrane. *J. Biol. Chem.* 260:15592–97
46. Lees ND, Bard M, Kemple MD, Hank RA, Kleinhaus FW. 1979. ESR determination of membrane order parameter in yeast sterol mutants. *Biochim. Biophys. Acta* 553:469–75

47. Lewis TL, Keesler GA, Fenner GP, Parks LW. 1988. Pleiotropic mutations in *Saccharomyces cerevisiae* affecting sterol uptake and metabolism. *Yeast* 4:93–106
48. Lewis TA, Rodriguez RJ, Parks LW. 1987. Relationship between intracellular sterol content and sterol esterification and hydrolysis in *Saccharomyces cerevisiae*. *Biochim. Biophys. Acta* 921:205–12
49. Lewis TA, Taylor FR, Parks LW. 1985. Involvement of heme biosynthesis in control of sterol uptake by *Saccharomyces cerevisiae*. *J. Bacteriol.* 163:199–207
50. Lorenz RT, Parks LW. 1987. Regulation of ergosterol biosynthesis and sterol uptake in a sterol-auxotrophic yeast. *J. Bacteriol.* 169:3707–11
51. Lorenz RT, Parks LW. 1990. Effects of lovastatin (mevinolin) on sterol levels and on activity of azoles in *Saccharomyces cerevisiae*. *Antimicrob. Agents Chemother.* 34:1660–65
52. Lorenz RT, Parks LW. 1991. Involvement of heme components in sterol metabolism of *Saccharomyces cerevisiae*. *Lipids* 26:598–603
53. Lorenz RT, Parks LW. 1991. Physiological effects of fenpropimorph on wild-type *Saccharomyces cerevisiae* and fenpropimorph-resistant mutants. *Antimicrob. Agents Chemother.* 35:1532–37
54. Lorenz RT, Parks LW. 1992. Cloning, sequencing, and disruption of the gene encoding sterol C-14 reductase in *Saccharomyces cerevisiae*. *DNA Cell Biol.* 11:685–92
55. Lorenz RT, Rodriguez RJ, Lewis TA, Parks LW. 1986. Characteristics of sterol uptake in *Saccharomyces cerevisiae*. *J. Bacteriol.* 167:981–85
56. Low C, Parks LW. 1987. Sterol and phospholipid acyl chain alterations in *Saccharomyces cerevisiae* secretion mutants as a function of temperature stress. *Lipids* 22:715–20
57. Low C, Rodriguez RJ, Parks LW. 1985. Modulation of yeast plasma membrane composition of a yeast sterol auxotroph as a function of exogenous sterol. *Arch. Biochem. Biophys.* 240:530–38
58. Maguigan WH, Walker E. 1940. Sterol metabolism of microorganisms. I. Yeast. *Biochem. J.* 34:804–13
59. M'Baya B, Fegueur M, Servouse M, Karst F. 1989. Regulation of squalene synthetase and squalene epoxidase activities in *Saccharomyces cerevisiae*. *Lipids* 24:1020–23
60. McCammon MT, Hartmann MA, Bottema CDK, Parks LW. 1984. Sterol methylation in *Saccharomyces cerevisiae*. *J. Bacteriol.* 157:475–83
61. McCammon MT, Parks LW. 1981. Inhibition of sterol transmethylation by analogs of S-adenosylhomocysteine. *J. Bacteriol.* 145:106–12
62. McCammon MT, Parks LW. 1982. Lipid synthesis in inositol-starved *Saccharomyces cerevisiae*. *Biochim. Biophys. Acta* 713:86–93
63. McCusker JH, Clemons KV, Stevens DA, Davis RW. 1994. Genetic characterization of pathogenic *Saccharomyces cerevisiae* isolates. *Genetics* 136:1261–69
64. McLean-Bowen CA, Parks LW. 1981. Corresponding changes in kynurenine hydroxylase activity, membrane fluidity, and sterol composition in *Saccharomyces cerevisiae* mitochondria. *J. Bacteriol.* 143:1325–33
65. McLean-Bowen CA, Parks LW. 1982. Effect of altered sterol composition on the osmotic behavior of spheroplasts and mitochondria of *Saccharomyces cerevisiae*. *Lipids* 17:662–65
66. Nes WD, Janssen GG, Crumley FG, Kalinowska M, Akihisa T. 1993. The structural requirements of sterols for membrane function in *Saccharomyces cerevisiae*. *Arch. Biochem. Biophys.* 300:724–33
67. Nes WR, Dhanuka IC. 1988. Inhibition of sterol synthesis by Δ^5-sterols in a sterol auxotroph of yeast defective in oxidosqualene cyclase and cytochrome P-450. *J. Biol. Chem.* 263:11844–50
68. Nishino T, Hata S, Osumi T, Katsuki H. 1980. Biosynthesis of ergosterol in cell-free system of yeast. *J. Biochem. Tokyo* 88:247–54
69. Nishino T, Hata S, Taketani S, Yabusaki Y, Katsuki H. 1981. Subcellular localization of the enzymes involved in the late stages of ergosterol biosynthesis in yeast. *J. Biochem. Tokyo* 89:1391–96
70. Novotny C, Flieger M, Panos J, Karst F. 1992. Effect of 5,7-unsaturated sterols on ethanol tolerance in *Saccharomyces cerevisiae*. *Biotechnol. Appl. Biochem.* 15:314–20
71. Nurminen T, Konttinen K, Suomalainen H. 1975. Neutral lipids in the cells and cell envelope fractions of aerobic baker's yeast and anaerobic brewer's yeast. *Chem. Phys. Lipids* 14:15–32
72. Osumi T, Nishino T, Katsuki H. 1979. Studies on the Δ^5-desaturation in ergosterol biosynthesis in yeast. *J. Biochem. Tokyo* 85:819–26
73. Paltauf F, Kohlwein SD, Henry SA. 1992. Regulation and compartmentali-

zation of lipid synthesis in yeast. In *The Molecular and Cellular Biology of the Yeast* Saccharomyces: *Gene Expression*, 2:415–500. Cold Spring Harbor, NY: Cold Spring Harbor Lab.
74. Parks LW. 1978. Metabolism of sterols in yeast. *CRC Crit. Rev. Microbiol.* 6:301–41
75. Parks LW, Anding C, Ourisson G. 1974. Sterol transmethylation during aerobic adaptation of yeast. *Eur. J. Biochem.* 43:451–58
76. Parks LW, McClean-Bowen CA, Taylor FR, Hough S. 1978. Sterols in yeast subcellular fractions. *Lipids* 13:730–35
77. Parks LW, Starr PR. 1963. A relationship between ergosterol and respiratory competency in yeast. *J. Cell. Comp. Physiol.* 61:61–65
78. Pinto WJ, Lozano R, Nes WR. 1985. Inhibition of sterol biosynthesis by ergosterol and cholesterol in *Saccharomyces cerevisiae*. *Biochim. Biophys. Acta* 836:89–95
79. Pinto WJ, Nes WR. 1983. Stereochemical specificity for sterols in *Saccharomyces cerevisiae*. *J. Biol. Chem.* 258:4472–76
80. Ramgopal M, Zundel M, Block K. 1990. Sterol effects on phospholipid biosynthesis in the yeast strain GL7. *J. Lipid Res.* 31:653–58
81. Rezanka T. 1992. Analysis of sterol esters from alga and yeast by high-performance liquid chromatography and capillary gas chromatography–mass spectrometry with chemical ionization. *J. Chromatogr.* 598:2196–26
82. Robinson GW, Tsay YH, Kienzle BK, Smith-Monroy CA, Bishop RW. 1993. Conservation between human and fungal squalene synthetases: similarities in structure, function, and regulation. *Mol. Cell. Biol.* 13:2706–17
83. Rodriguez RJ, Low C, Bottema CDK, Parks LW. 1985. Multiple functions for sterols in *Saccharomyces cerevisiae*. *Biochim. Biophys. Acta.* 837:336–43
84. Rodriguez RJ, Parks LW. 1983. Structural and physiological features of sterols necessary to satisfy bulk membrane and sparking requirements in yeast sterol auxotrophs. *Arch. Biochem. Biophys.* 225:861–71
85. Ryder NS. 1992. Terbinafine: mode of action and properties of the squalene epoxidase inhibition. *Br. J. Dermatol.* 126(Suppl.)39:2–7
86. Salerno LF, Parks LW. 1983. Sterol uptake in the yeast, *Saccharomyces cerevisiae*. *Biochim. Biophys. Acta* 752:240–43
87. Servouse M, Karst M. 1986. Regulation of early enzymes of ergosterol biosynthesis in *Saccharomyces cerevisiae*. *Biochem. J.* 240:541–47
88. Sherman F. 1959. The effects of elevated temperatures on yeast. I. Nutrient requirements for growth at elevated temperatures. *J. Cell. Comp. Physiol.* 54:29–35
89. Sherman F. 1959. The effects of elevated temperatures on yeast. II. Induction of respiratory-deficient mutants. *J. Cell. Comp. Physiol.* 54:37–52
90. Shinabarger DL, Keesler GA, Parks LW. 1989. Regulation by heme of sterol uptake in *Saccharomyces cerevisiae*. *Steroids* 53:607–23
91. Smith SJ, Parks LW. 1993. The ERG3 gene in *Saccharomyces cerevisiae* is required for the utilization of respiratory substrates and in heme-deficient cells. *Yeast* 9:1177–87
92. Sperka-Gottlieb CD, Hermetter A, Paltauf F, Daum G. 1988. Lipid topology and physical properties of the outer mitochondrial membrane of the yeast, *Saccharomyces cerevisiae*. *Biochim. Biophys. Acta* 946:227–34
93. Stansfield I, Cliffe KR, Kelly SL. 1991. Chemostat studies of microsomal enzyme induction in *Saccharomyces cerevisiae*. *Yeast* 7:147–56
94. Starr PR, Parks LW. 1962. Effect of temperature on sterol metabolism in yeast. *J. Cell. Comp. Physiol.* 59:107–10
95. Sturley SL, Culbertson MR, Attie AD. 1991. Secretion and lipid association of human apolipoprotein E in *Saccharomyces cerevisiae* requires a host mutation in sterol esterification and uptake. *J. Biol. Chem.* 266:16273–76
96. Sutcliffe J, Georgopapadakou NH. 1992. *Emerging Targets in Antibacterial and Antifungal Chemotherapy*. New York: Chapman & Hall
97. Szolderits G, Hermstetter A, Paltauf F, Daum G. 1989. Membrane properties modulate the activity of a phosphatidylinositol transfer protein from the yeast, *Saccharomyces cerevisiae*. *Biochim. Biophys. Acta* 986:301–9
98. Szkopinska A, Rytka J, Karst F, Palamarczyk G. 1993. The deficiency of sterol biosynthesis in *Saccharomyces cerevisiae* affects the synthesis of glycosyl derivatives of dolichyl phosphates. *FEMS Microbiol. Lett.* 112:325–28
99. Taguchi N, Takano Y, Julmanop C, Wang Y, Stock S, et al. 1994. Identification and analysis of the *Saccharomyces cerevisiae* SYR1 gene reveals that ergosterol is involved in the action of syringomycin. *Microbiology* 140:353–59

100. Taketani S, Nishino T, Katsuki H. 1981. Purification and properties of sterol-ester hydrolase from *Saccharomyces cerevisiae*. *J. Biochem. Tokyo* 89:1667–73
101. Taketani S, Nagai J, Katsuki H. 1978. Quantitative aspects of free and esterified sterols in *Saccharomyces cerevisiae* under various conditions. *Biochim. Biophys. Acta* 528:416–23
102. Taketani S, Nishino T, Katsuki H. 1979. Characterization of sterol-ester synthetase in *Saccharomyces cerevisiae*. *Biochim. Biophys. Acta* 575:148–55
103. Taylor FR, Parks LW. 1978. Metabolic interconversion of free sterols and steryl esters in *Saccharomyces cerevisiae* J. *Bacteriol.* 136:531–37
104. Taylor FR, Parks LW. 1981. An assessment of the specificity of sterol uptake and esterification in *Saccharomyces cerevisiae*. *J. Biol. Chem.* 256:13048–54
105. Thomas DS, Hossack JA, Rose AH. 1978. Plasma-membrane lipid composition and ethanol tolerance in *Saccharomyces cerevisiae*. *Arch. Microbiol.* 117:239–45
106. Tomeo ME, Fenner G, Tove SR, Parks LW. 1992. Effect of sterol alterations on conjugation in *Saccharomyces cerevisiae*. *Yeast* 8:1015–24
107. Trocha PJ, Sprinson DB. 1976. Location and regulation of early enzymes of sterol biosynthesis in yeast. *Arch. Biochem. Biophys.* 174:45–51
108. Turi TG, Loper JC. 1992. Multiple regulatory elements control expression of the gene encoding the *Saccharomyces cerevisiae* cytochrome P450, lanosterol 14α-demethylase (ERG11). *J. Biol. Chem.* 267:2046–56
109. Uno I, Fukami K, Kato H, Takenawa T, Ishikawa T. 1988. Essential role for phosphatidylinositol- 4,5-bisphosphate in yeast cell proliferation. *Nature* 333:188–90
110. Verdiere J, Gaisne M, Labbe-Bois R. 1991. CYP1 (HAP1) is a determinant effector of alternative expression of heme-dependent transcribed genes in yeast. *Mol. Gen. Genet.* 228:300–6
111. Voelker DR. 1991. Organelle biogenesis and intracellular lipid transport in eukaryotes. *Microbiol. Rev.* 55:543–60
112. Vogel JP, Lee JN, Kirsch DR, Rose MD, Sztul ES. 1993. Brefeldin A causes a defect in secretion in *Saccharomyces cerevisiae*. *J. Biol. Chem.* 268:3040–43
113. Walker-Caprioglio HM, Casey WM, Parks LW. 1990. *Saccharomyces cerevisiae* membrane sterol modifications in response to growth in the presence of ethanol. *Appl. Environ. Microbiol.* 56:2853–57
114. Welihinda AA, Beavis AD, Trumbly RJ. 1994. Mutations in LIS1 (ERG6) gene confer increased sodium and lithium uptake in *Saccharomyces cerevisiae*. *Biochim. Biophys. Acta* 1193:107–17
115. Wiseman H, Cannon M, Arnstein HR, Halliwell B. 1993. Enhancement by tamoxifen of the membrane antioxidant action of the yeast membrane sterol ergosterol: relevance to the antiyeast and anticancer action of tamoxifen. *Biochim. Biophys. Acta* 1181:201–6
116. Woods RA, Sanders HK, Briquet M, Foury F, Drysdale B, Mattoon JR. 1975. Regulation of mitochondrial biogenesis: enzymatic changes in cytochrome-deficient yeast mutants requiring δ-aminolevulinic acid. *J. Biol. Chem.* 250:9090–98
117. Yabusaki Y, Nishino T, Ariga N, Katsuki H. 1979. Studies on the Δ^8-Δ^9 isomerization and methyl transfer of sterols in ergosterol biosynthesis of yeast. *J. Biochem. Tokyo* 85:1513–17
118. Youings A, Rose AH. 1989. Sterol uptake by anaerobically grown *Saccharomyces cerevisiae*. *Yeast* 5:459–63
119. Zinser E, Paltauf F, Daum G. 1993. Sterol composition of yeast organelle membranes and subcellular distribution of enzymes involved in sterol metabolism. *J. Bacteriol.* 175:2853–58
120. Zinser E, Sperka-Gottlieb CD, Fasch EV, Kohlwein SD, Paltauf F, Daum G. 1991. Phospholipid synthesis and lipid composition of subcellular membranes in the unicellular eukaryote *Saccharomyces cerevisiae*. *J. Bacteriol.* 173:2026–34
121. Zitomer RS, Lowry CV. 1992. Regulation of gene expression by oxygen in *Saccharomyces cerevisiae*. *Microbiol. Rev.* 56:1–11

THE STRUCTURE AND REPLICATION OF KINETOPLAST DNA

Theresa A. Shapiro* and Paul T. Englund**

Division of Clinical Pharmacology, *Department of Medicine and **Department of Biological Chemistry, Johns Hopkins School of Medicine, Baltimore, Maryland 21205

KEY WORDS: kDNA, mitochondrial DNA, DNA network, *Trypanosoma*, *Crithidia*

CONTENTS

INTRODUCTION	118
THE MOLECULAR COMPONENTS OF kDNA	119
Maxicircles	119
Minicircles	120
STRUCTURE OF THE kDNA NETWORK	121
The Isolated Network	121
The kDNA Network In Vivo	122
What Conditions Are Required for Network Formation?	124
REPLICATION OF THE kDNA NETWORK	125
Network Replication in C. fasciculata	125
Network Replication in T. brucei	128
REPLICATION MECHANISMS FOR MINICIRCLES AND MAXICIRCLES	129
Minicircle Replication	129
Maxicircle Replication	130
IMPLICATIONS OF THE kDNA REPLICATION MODEL	131
Minicircle Inheritance	131
Minicircle Exchange During T. brucei Mating	132
Transkinetoplastidy	132
Dyskinetoplastidy	133
ENZYMES AND PROTEINS INVOLVED WITH kDNA STRUCTURE AND REPLICATION	133
DNA Topoisomerases	133
DNA Polymerase	135
Minicircle Origin–Binding Proteins	135
Kinetoplast-Associated Protein (KAP)	136
Proteins that Condense kDNA	136
Heat-Shock Proteins	136
CONCLUSION	137

Abstract

The mitochondrial DNA of trypanosomatid protozoa, termed kinetoplast DNA (kDNA), is unique in its structure, function, and mode of replication. kDNA is a massive network, composed of thousands of topologically interlocked DNA circles, which resembles the chain mail of medieval armor. Each cell contains one network condensed into a disk-shaped structure within the matrix of its single mitochondrion. The kDNA circles are of two types, maxicircles present in a few dozen copies and minicircles present in several thousand copies. The maxicircles, which encode ribosomal RNAs and a few mitochondrial proteins, are similar in structure and genetic function to the mitochondrial DNA of other eukaryotes. Many maxicircle transcripts undergo editing, a remarkable process involving the insertion or deletion of uridine residues at specific sites. The minicircles encode small guide RNAs that control the specificity of editing. During kDNA replication, covalently closed minicircles are released from the network by a topoisomerase II. The free minicircles replicate as θ-structures within one of two complexes of replication proteins that are positioned on opposite sides of the kinetoplast disk. The progeny minicircles, which contain nicks or gaps, are attached to the network periphery. Maxicircles also replicate as θ-structures, but they remain linked to the network. As replication proceeds, the number of minicircles and maxicircles increases. When the network has doubled in size, all of the minicircle nicks and gaps are repaired, and the network splits in two. The two progeny networks then segregate into the daughter cells.

INTRODUCTION

Kinetoplast DNA (kDNA) is the mitochondrial DNA of trypanosomatid protozoa. These ancient organisms are widespread in nature, and they are important because they cause disease in humans, livestock, and commercially important plants. This group includes the African trypanosome *Trypanosoma brucei*, the South American trypanosome *Trypanosoma cruzi*, and *Leishmania* spp., all of which are human pathogens. Trypanosomatids also include *Phytomonas* spp., which infect plants, and *Crithidia fasciculata*, which is a parasite of insects. The trypanosomatids have attracted the attention of the research community not only because they cause disease, but also because they have a fascinating biology.

kDNA is one of these parasites' most unusual features. It is a structure unique in nature, a giant DNA network consisting of several thousand circular DNAs that are topologically interlocked. Each cell has one network within the matrix of its single mitochondrion. The network contains two types of DNA molecules: maxicircles present in a few dozen copies and minicircles present

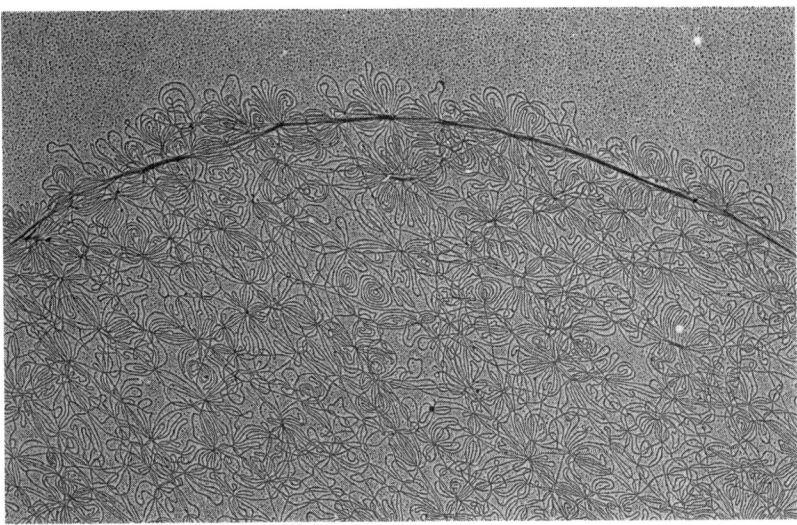

Figure 1 Part of a kDNA network from *C. fasciculata,* shown by electron microscopy (EM). Each small loop represents a 2.5-kb minicircle.

in several thousand copies. Figure 1 shows an electron micrograph of part of a network isolated from *Crithidia fasciculata.* (For previous reviews on kDNA, see 73, 86, 96, 102, 113.)

The first indications of the existence of kinetoplast DNA came nearly 100 years ago. Although trypanosomes were probably first described in 1841 by Valentin, and the dense structure that we now term the kinetoplast (2) was noted by subsequent investigators, it was in 1898 that Ziemann reported that the kinetoplast (like the nucleus) stained a brilliant red with the newly discovered Romanovsky counterstains (125). We now know that the basophilic character of the kinetoplast is attributable to the high local concentration of DNA, a finding confirmed by measurements of [^{3}H]thymidine incorporation and DNase sensitivity (109). The remarkable molecular configuration of the kDNA network was finally revealed by electron microscopy (EM) in the late 1960s and early 1970s (48, 76, 79, 105).

THE MOLECULAR COMPONENTS OF kDNA

Maxicircles

Maxicircles are the functional homologues of mitochondrial DNA in other eukaryotes, and in different species they range in size from about 20 to nearly

40 kb. They encode ribosomal RNAs and several proteins, most of which are involved in mitochondrial energy transduction (e.g. cytochrome b, subunits of cytochrome oxidase, and subunits of NADH dehydrogenase). They apparently do not encode tRNAs. A most remarkable feature of maxicircles is that most of their transcripts undergo RNA editing to form a functional mRNA (5). Editing involves addition or deletion of uridine residues at specific sites in the transcript, and in some mRNAs more than half of the nucleotides in the coding sequence are introduced by editing (25). We do not discuss the mechanism of editing in this article, but recent reviews on this subject are available (4, 34, 103, 113). Within each maxicircle resides a noncoding variable region [about 8 kb in *T. brucei* (63, 107) and about 12 kb in *Leishmania tarentolae* (62)], which contains many repetitive sequences. The sequence of this region has been determined only for *T. brucei*, and it contains two copies of a sequence associated with the minicircle replication origin (63, 107) (see below).

Minicircles

Minicircles make up over 90% of the mass of the network and in different species they range in size from about 0.5 to 2.9 kb. In most species, all the minicircles in a network are about the same size. A crucial genetic function of minicircles is to encode small guide RNAs (~70 nucleotides, with a short nonencoded oligo-U tail at the 3' end) that govern the specificity of RNA editing (114). Maxicircles also encode some guide RNAs (8). Most *T. brucei* minicircles encode three guide RNAs (69), and those from *L. tarentolae* encode one (115). In most trypanosomatid species, the minicircles in a network are heterogeneous in sequence, and the degree of heterogeneity appears to reflect the number of guide RNAs needed for editing. *T. brucei*, in which many maxicircle transcripts are extensively edited, contains over 250 different minicircle sequence classes, all differing in copy number (111). A laboratory strain of *L. tarentolae*, in which editing occurs at a more modest level, probably has no more than 17 different minicircle classes (57). In *C. fasciculata*, over 90% of the minicircles have about the same sequence (116), and in *Trypanosoma equiperdum* (3) and most stocks of *Trypanosoma evansi* (10), minicircles are virtually homogeneous. The latter two species are African trypanosomes related to *T. brucei*. Because of defective or missing maxicircle DNA, and the absence of most guide RNAs, *T. equiperdum* and *T. evansi* cannot produce the cytochromes and other mitochondrial proteins essential for life in the tsetse vector. These organisms propagate only in the host animal bloodstream where the mitochondrial functions controlled by maxicircle genes are normally inactive.

Aside from the region encoding guide RNAs, minicircles also have a conserved region (usually 100–180 nucleotides and virtually identical in all minicircles in a network) that contains the replication origin. Some minicircles, such as those in *T. brucei* and *L. tarentolae* (13, 41), contain a single conserved region. *C. fasciculata* minicircles contain two (116), and those in *T. cruzi* contain four (16). Multiple conserved regions, when present, are usually positioned 180° or 90° apart on the minicircle map. Minicircles from most trypanosomatids (*T. cruzi* is an exception) contain a single region of bent helix, consisting of short A tracts (4 to 6 residues each) spaced about every 10 base pairs (44, 53, 66). Small bends are associated with each A tract. Because these bends are spaced in phase with the 10.5–base pair helical repeat, there is an additive effect that results in substantial curvature of the helix. As viewed by EM, a 0.2-kb *C. fasciculata* minicircle fragment, containing 18 sequential A tracts, bends approximately 360° (32). The function of the bent helix is not known, but it may facilitate the compaction of the network in vivo (54).

STRUCTURE OF THE kDNA NETWORK

The Isolated Network

Because kDNA networks are massive in size ($\sim 10^{10}$ kDa) and compact in structure, they are easy to isolate from cell lysates by means of low-speed centrifugation. The following description applies to networks from *C. fasciculata*, which can be isolated in milligram quantities, but networks from other species probably have similar characteristics. *C. fasciculata* networks not undergoing replication contain about 5000 minicircles organized in a planar array (20); see the schematic diagram of a network in Figure 2a. As viewed with EM, the isolated network has an elliptical shape about 10 by 15 µm in size, and it appears to have a fairly regular structure (70). When visualized in solution by light microscopy, after staining with a fluorescent dye, the network appears cup shaped. Topoisomerase II can decatenate all the DNA rings in a network, and therefore all of the linkages between the rings must be topological. Inspection of the products of a topoisomerase II decatenation of highly purified networks indicates that minicircles and maxicircles are the only DNA components (55). EM of minicircle oligomers released from the network by means of partial digestion with a restriction enzyme demonstrates that linked minicircles are joined by a single interlock (72). Minicircles in a network not undergoing replication are all covalently closed, but they are also relaxed (72). This arrangement is unusual, as other circular DNAs in prokaryotes and eukaryotes are negatively supercoiled (see following section for discussion of the significance of this finding). Because neighboring circles share only a single interlock, no strain is associated with the topological linkages in a kDNA network.

Each minicircle in a *C. fasciculata* network is linked to an average of three neighbors (12). This value, termed the minicircle valence, was determined by analyzing the structure and relative abundance of catenated minicircle oligom-

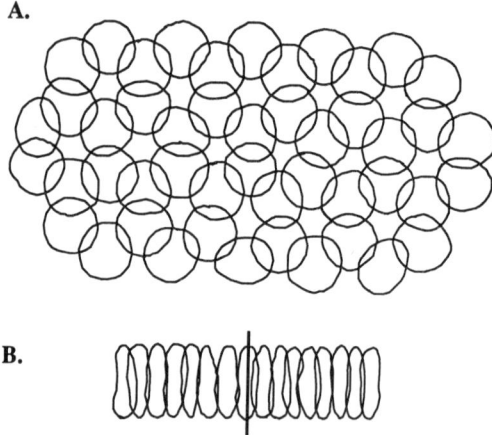

Figure 2 (A) Diagram of a segment of an isolated *C. fasciculata* kDNA network, in which each minicircle lies flat in a plane. Except for those on an edge, each minicircle is linked topologically to three neighboring minicircles (i.e. the minicircle valence is 3). (B) Diagram of a section through a kinetoplast disk in vivo. Instead of lying in a plane as in A, each minicircle stands perpendicular to the plane. The vertical line shows the disk axis. The model in B was first proposed in Reference 18.

ers released from the network by partial digestion with a restriction enzyme that cleaves minicircles. Each minicircle in the model network shown in Figure 2a (except for those on the edges) has a valence of 3.

How the 25 maxicircles are organized in *C. fasciculata* networks is not known. However, in networks from the African trypanosome *T. equiperdum*, cleavage of all of the identical minicircles by a restriction enzyme (which does not cleave maxicircles) revealed that the maxicircles are topologically linked to one another (90). Therefore, kDNA networks consist of two independent catenanes that are interlocked to each other, forming a network within a network. The significance of this finding is not yet clear.

The kDNA Network In Vivo

EM of thin sections or confocal fluorescence microscopy (using an acridine stain) indicates that within the mitochondrial matrix the network is condensed into a disk-like structure (26). For *C. fasciculata*, the disk is about 1 µm in diameter and roughly 0.4 µm thick. Compaction of the network in vivo does not involve folding of the network sheet. Instead, the minicircles stand perpendicular to the plane of the network and apparently stretch out so that the DNA fibers run parallel to the disk's axis [this arrangement can be seen in EMs of thin sections (e.g. 77)]. Therefore, in vivo the network remains a

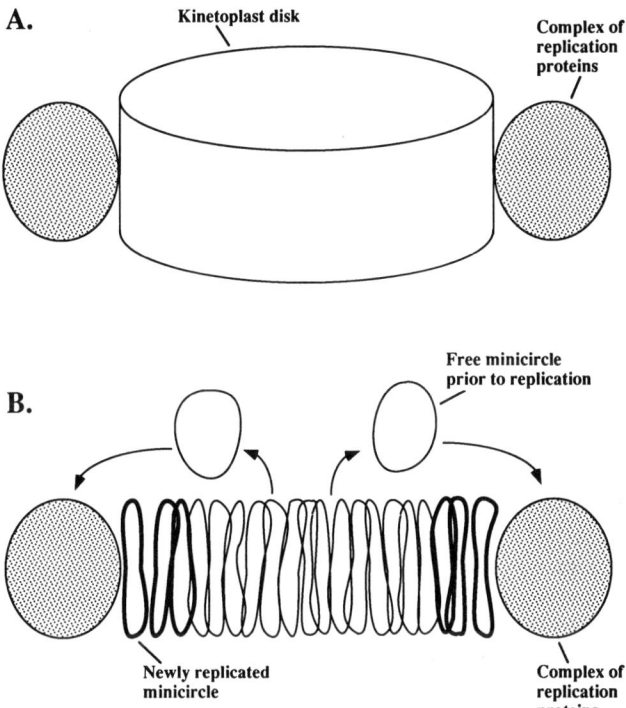

Figure 3 (A) Diagram of the kinetoplast disk, showing the two antipodal complexes of replication proteins. (B) Model for kDNA replication in vivo (redrawn with permission from Reference 26). The diagram shows a section through the kinetoplast disk. Minicircles in bold have undergone replication and are nicked or gapped. See text for details.

monolayer. Figure 2b diagrams a section through the kinetoplast disk. In support of this model, the thickness of the disk is roughly half the circumference of a minicircle (which varies in different trypanosomatid species) (18, 101). The DNA concentration inside the kinetoplast disk, about 50 mg/ml, is comparable to that in a bacterial nucleoid (about 20 mg/ml) but is far less than that inside the head of a T4 bacteriophage (800 mg/ml) (40). Therefore, in its compacted state, there must be considerable space between the DNA strands in the kinetoplast.

Fluorescence in situ hybridization using maxicircle probes indicates that *C. fasciculata* maxicircles are condensed in roughly 8–10 foci embedded within the kinetoplast disk (27). Because the network contains about 25 maxicircles (55), each focus may contain multiple maxicircles. The significance of this localization is not yet known. In *T. brucei,* maxicircles seem spread throughout the kinetoplast disk (27). If these maxicircles are also organized within foci, they are too closely packed to be resolved using available microscopic techniques.

In cells undergoing kDNA replication, the kinetoplast disk is flanked by two antipodal complexes of replication enzymes (26) (see diagram in Figure 3*a*). These complexes, roughly 0.4 μm in diameter, contain a topoisomerase II (60), a DNA polymerase (26), and minicircles that are probably replication intermediates (26). They probably also contain other proteins required for DNA replication. The function of these structures is discussed below.

What Conditions Are Required for Network Formation?

A network would probably form if relaxed minicircles, in the presence of a topoisomerase II, were aligned at high concentration in a monolayer, in a volume delimited by the 1×0.4 μm disk. Network formation almost certainly requires the minicircles to be relaxed, as much more space is available in the center of a relaxed circle than in a supercoiled circle. The additional space would facilitate interpenetration and catenation. In fact, if DNA rings in vitro are aggregated by spermidine and then treated with a topoisomerase II, they will form a large network only if the rings are not supercoiled (46). Because supercoiling provides critical energy for helix unwinding during processes such as replication or transcription, the trypanosomatids must have sacrificed the advantages of supercoiling to assemble the minicircles into a network.

What keeps the minicircles inside the volume of this small disk? We do not know what constrains them above or below the disk, but one candidate for confining them at the disk periphery is the mitochondrial membrane. Many electron micrographs of thin sections show that the kinetoplast disk directly abuts the mitochondrial membrane (e.g. see 77).

The minicircle valence (the number of neighbors linked to each minicircle) is probably determined by their concentration when the topological bonds are formed. The alignment of 5000 *C. fasciculata* minicircles within the 1×0.4 μm disk must result in an average valence of 3. If more minicircles occupied the same space, or the 5000 were constrained in a smaller space, then the valence would be higher. We discuss below the increase in the average minicircle valence as the minicircle number increases during network replication. Unlike *C. fasciculata* networks, those from *L. tarentolae* are characteristically fragile, full of holes, and contain many chains of minicircles. This suggests that in *L. tarentolae* the average minicircle valence must be less than 3 and that the topological linkages must form when the minicircle concentration is less than that in *C. fasciculata.*

REPLICATION OF THE kDNA NETWORK

kDNA synthesis involves replication of each individual minicircle and maxicircle and then distribution of the daughter circles into two progeny networks. The networks then partition into daughter cells at the time of cell division. kDNA replication occurs during a discrete S phase, nearly coincident with that of the nuclear DNA (15, 104, 123). In contrast, replication of mitochondrial DNA in mammalian cells occurs throughout the entire cell cycle (14). In the *C. fasciculata* cell cycle at 25°C, the G_1 phase is 200 min, the S phase is 80 min, the G_2 phase is 70 min, and the division phase is 22 min (15).

In most of this section, we focus on minicircle replication, which must overcome two significant problems. The first is topological. How does a DNA ring that is interlocked to several neighbors replicate? The parasite solves this problem by using a topoisomerase II to release the minicircle from the network (21). The free minicircles can replicate as would a DNA circle in any other cell, and the progeny are then attached to the network. At no time does the whole network decatenate completely into free minicircles. Instead, at any point during the S phase, there are several hundred free minicircles undergoing replication (22). The second problem for minicircle replication concerns bookkeeping. How can the parasite ensure that each minicircle replicates once and only once per cell cycle? It appears that nicks or gaps, introduced into progeny minicircles during replication (20), may distinguish replicated minicircles from those that have not yet replicated. Only when all the minicircles have replicated are the nicks repaired.

Network Replication in C. fasciculata

In the current model (Figure 3b), minicircles are released by a topoisomerase II from the center of the network, probably from random positions. The free minicircles then migrate (or are transported) to one of the two antipodal protein complexes, where they undergo replication as θ-structures (26) (the details of these reactions are summarized below). The progeny minicircles, which contain nicks or gaps, are attached to the network periphery. Autoradiographic analysis of networks isolated from cells that had been pulse-labeled with [³H]thymidine reveals that the newly synthesized minicircles are positioned at two peripheral sites on opposite sides of the network (71, 100). In the cell, these sites must have been adjacent to the two protein complexes.

Inspection of partly replicated networks, either by EM (after network isolation) or by fluorescence in situ hybridization, reveals that the newly synthesized nicked minicircles are eventually distributed uniformly around the network periphery (26, 70). The partly replicated network resembles a doughnut in which the nonreplicated covalently closed minicircles form the doughnut hole. If minicircles are attached to the network adjacent to only the two protein

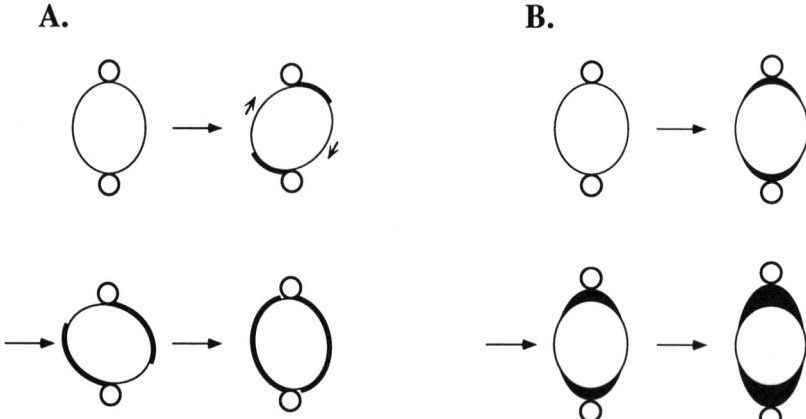

Figure 4 Comparison of network replication in *C. fasciculata* (A) and *T. brucei* (B). The diagram shows the kinetoplast disk flanked by the two complexes of replication proteins (*small circles*). The region containing newly synthesized and reattached minicircles appears in bold. In *C. fasciculata*, the kinetoplast may rotate during replication, causing distribution of newly synthesized minicircles around the network periphery (rotation indicated by small arrows). See text for details. (Reprinted with permission from Reference 27.)

complexes, how do they become distributed around the periphery? EM autoradiography of kDNA labeled in vivo for 6 min with [^{3}H]thymidine provided evidence that the kinetoplast disk and the two complexes of replication proteins may actually be in relative motion (71). As shown in the model in Figure 4a, the kinetoplast may rotate between the two fixed complexes, allowing orderly distribution of the newly synthesized minicircles around the network periphery. Based on the time needed to label the entire network periphery, the network must make one rotation every 12 min. Subsequent rotation would result in minicircle attachment in a spiral pattern. Although all available data are consistent with this model, more evidence for a spinning kinetoplast is certainly needed. However, we assume this model is correct for the rest of this review.

As kDNA replication proceeds, minicircles are continuously released from the network's central zone and their progeny are attached around the periphery. The central zone shrinks, the peripheral zone enlarges, and the total number of minicircles increases. Unexpectedly, the newly attached minicircles have a higher valence than those in a nonreplicating network. The average valence may be as high as 5 or 6, rather than 3 (J Chen, PT Englund & NR Cozzarelli, unpublished results). The probable reason for this difference is that the number of minicircles increases during replication, but the available space (perhaps limited by the mitochondrial membrane) does not. The continuing release of

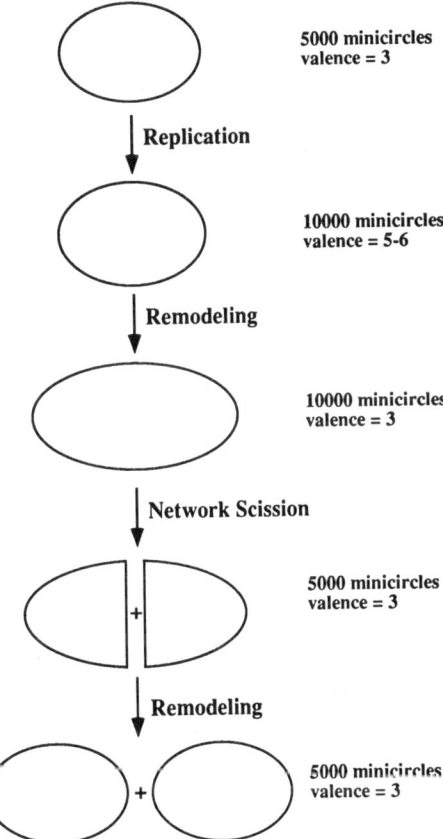

Figure 5 Remodeling of a *C. fasciculata* kDNA network. Scheme shows the changes in minicircle number and valence during replication as well as the changes in network size and shape resulting from remodeling.

covalently closed minicircles from the central region of the network causes holes to form, which are visible by EM of isolated networks (70). However, addition of minicircles to the network periphery must compress the central region, favoring eventual repair of the holes. Finally, at the end of replication, the network contains about 10,000 nicked or gapped minicircles. The minicircle valence is high, and the surface area of the isolated network has grown little from that of the prereplication network (see model in Figure 5).

When the replication of individual minicircles is complete, the network undergoes a dramatic remodeling (Figure 5). As the available space for the network increases, possibly because of enlargement of the mitochondrial membrane, the network expands and topoisomerase action results in a gradual reduction in minicircle valence. A corresponding increase in network surface area also occurs (as measured by EM of the isolated network) (J Chen, PT Englund & NR Cozzarelli, unpublished observation). During this time, minicircle nicks and gaps are gradually repaired (70). Finally, when the minicircle valence falls to 3 and the network has a surface area double that of a prereplication network, the structure splits in two. Splitting undoubtedly occurs by topoisomerase action, unlinking neighboring circles between the two daughter networks. EM of newly formed networks revealed that they have a characteristic flat edge (70), as if the splitting of the double-size elliptical network occurred by a straight cut along its short axis. Subsequent remodeling of the flat-edged daughter networks must reestablish the characteristic oval shape. The contour of the mitochondrial membrane may dictate the final shape.

Network Replication in T. brucei

Much less is known about the mode of kDNA replication in *T. brucei*, but studies on *T. equiperdum* (a closely related parasite) suggest that many features of the mechanism are similar to those in *C. fasciculata* (65, 67, 84, 85, 91). However, instead of the uniform peripheral distribution that produces the doughnut configuration of a replicating *C. fasciculata* network, the newly synthesized nicked minicircles of trypanosomes cluster at two antipodal sites (27, 82). This pattern, found at all stages of network replication, suggests that *T. brucei* also has two complexes of replication proteins flanking the kinetoplast. However, it also indicates that the *T. brucei* kinetoplast disk does not rotate during replication (see scheme in Figure 4*b*). Interestingly, partly replicated networks from *T. cruzi* and *Leishmania donovani*, as visualized by EM, resemble doughnuts like those of *C. fasciculata* (T Zimmers & PT Englund, unpublished observation). *T. brucei* is the most ancient of these parasites (28, 47, 56), which suggests that the rotating kinetoplast probably developed more recently in evolution. Below, we propose a possible advantage for kinetoplast rotation.

Another study on *T. brucei* has addressed the mechanism of segregation of daughter kDNA networks after their replication (81). The basal body of the flagellum is always positioned adjacent to the kinetoplast. Immunofluorescence with probes specific for the basal body and kinetoplast indicated that these two structures were in close proximity during the entire segregation process. Direct evidence for a molecular interaction between the kinetoplast and basal body was obtained by visualization of kDNA-flagellum complexes after cell disruption either in hypotonic buffer or by nonionic detergent. These facts suggest that kinetoplast segregation may be mediated by interaction with the basal body.

REPLICATION MECHANISMS FOR MINICIRCLES AND MAXICIRCLES

Minicircle Replication

The replication mechanism for free minicircles has been studied extensively in *C. fasciculata* and *T. equiperdum*, from which free minicircle replication intermediates are readily isolated and characterized. The sequence homogeneity of *T. equiperdum* minicircles and the near homogeneity of those from *C. fasciculata* have made possible the mapping of the origin and other important sites in minicircle replication. More details of free minicircle replication are presented in earlier reviews (73, 86).

Replication begins when a covalently closed minicircle is decatenated from the network. Like those bound to the network, the covalently closed free minicircles are topologically relaxed (42). Minicircle replication occurs via θ-intermediates and is unidirectional (6, 21, 84, 95). Leading strand synthesis initiates complementary to the universal minicircle sequence, GGGG-TTGGTGTA, a 12-mer found in the conserved region of minicircles from nearly all trypanosomatids that have been examined (67, 74). In *C. fasciculata* minicircles, which have two of these sequences in conserved regions positioned 180° apart, either can serve as the site for replication initiation (7). The leading strand has one or more ribonucleotides at its 5' end, indicating that DNA synthesis is initiated by a primase (7, 67). A second, less well-conserved sequence, ACGCCC, resides 70–90 nucleotides downstream from the 12-mer, and the first Okazaki fragment initiates complementary to this 6-mer (85). In both *C. fasciculata* and *T. equiperdum*, the leading strand is synthesized continuously around the molecule, and the lagging strand fragments are roughly 100 nucleotides in size (6, 7, 43, 85).

In the final stages of minicircle replication, topoisomerase II segregates the daughter circles from one another; if this activity is inhibited, multiple interlocked daughter circles accumulate (91). Knotted topoisomers of the daughter circle containing the newly synthesized leading strand have been described (87, 95, 121). Although knots can arise through several mechanisms, the fact that newly replicated minicircles seem to be the only knotted form suggests that knots may arise at the time of daughter-circle segregation. A knot could be formed if topoisomerase II catalyzes an intramolecular strand passage, rather than passing one daughter circle through the other. Minicircles are unusual compared with other circular DNAs in that their nascent strands are not covalently closed either before or shortly after the segregation of daughter circles. Instead, covalent closure occurs only after all the minicircles have replicated and their progeny have attached to the network.

After segregation, daughter minicircles containing the continuously synthesized strand retain a single gap that overlies the GGGGTTGGTGTA sequence (7, 65); those containing the discontinuously synthesized strand have a gap at the site of the ACGCCC sequence and nicks or small gaps throughout the nascent strand (7, 42, 85). The discontinuously synthesized strand gradually matures both before and after the minicircle is attached to the network (42, 85). However, once on the network, the gap in the continuously synthesized strand and those flanking the first Okazaki fragment are the last interruptions to be repaired (7, 84, 85). The minicircles are covalently closed just before the network splits in two (70).

Maxicircle Replication

Recent EM of *C. fasciculata* maxicircles released from networks by restriction enzyme cleavage revealed many branched replication intermediates. Analysis of these molecules indicates that maxicircle replication initiates in a unique region and proceeds unidirectionally as a θ-structure (LJ Rocco & PT Englund, unpublished observations). The maxicircle replication origin is in the variable region, which in *T. brucei* contains two copies of the GGGGTTGGTGT sequence found at the minicircle replication origin (this sequence is one base shorter than the minicircle 12-mer; see previous section) (63, 107). If the *C. fasciculata* maxicircle variable region also contains this sequence, and if it serves as a site for initiation of DNA synthesis, then the mechanism of maxicircle replication may be very similar to that of minicircles. Both molecules replicate simultaneously (35), and both replicate unidirectionally as θ-structures. If they initiate at similar sequences, the factors that control initiation could possibly be the same. The major difference appears to be that minicircles replicate free of the network whereas maxicircles replicate while still linked to it.

These data contrast with an earlier report that suggested that maxicircles replicate by a rolling circle mechanism (35). It was thought that when replication was complete, the branch of the rolling circle was cleaved to form a linearized maxicircle; this structure would subsequently circularize and attach to the network. The most plausible explanation for the early data is the presence of many topoisomerase II consensus binding sites near the replication origin, as has been reported for *T. brucei* (63). DNA bound to topoisomerase II may be cleaved during cell lysis by sodium dodecylsulfate (SDS) or sarkosyl (88), which would linearize the newly synthesized molecules at a site near the origin. Cleavage of θ-structures by this mechanism could result in the formation of molecules that resemble rolling circles.

As mentioned above, fluorescence in situ hybridization of prereplication *C. fasciculata* cells reveals that maxicircles are concentrated in 8–10 discrete foci distributed throughout the kinetoplast disk. During the S phase, when minicircles replicate, the number of maxicircle foci increases, but they remain scattered throughout the disk (27). This situation contrasts with that in *T. brucei*. As demonstrated by EM of isolated networks (38) or fluorescence in situ hybridization (27), *T. brucei* networks in the later stage of replication have all the maxicircles clustered in the central region near the point where network division will take place. Maxicircle replication in *C. fasciculata* and trypanosomes may differ in other ways. For example, *T. brucei* maxicircles may not replicate simultaneously with minicircles. EM of *T. brucei* kDNA labeled in vivo with bromodeoxyuridine revealed networks in which minicircles were heavily labeled (as determined with a gold-labeled antibody), but maxicircles were not (82). Other clues to maxicircle replication in *T. brucei* are based on studies of the closely related *T. equiperdum* (90). Maxicircle catenanes isolated from *T. equiperdum* networks have a characteristic size distribution, consistent with the idea that all maxicircles in a network are catenated to one another and that the maxicircle catenane doubles in size during replication. The topological organization of the maxicircle catenane changes with increasing size, and the largest structures have the distinctive rosette appearance seen at the division point of intact networks in the final stages of replication (24, 38). The significance of the differences between *C. fasciculata* and trypanosomes is not yet clear.

IMPLICATIONS OF THE kDNA REPLICATION MODEL

Minicircle Inheritance

A surprising prediction of the kDNA replication model concerns minicircle inheritance. According to this model, a covalently closed free minicircle migrates randomly to one of the two complexes of replication proteins. Once the minicircle has replicated, its daughters attach to the network, presumably at neighboring positions. Eventually, after all of the other minicircles have replicated, cleavage of the double-size network results in the two sister minicircles segregating into the same daughter cell. After a few generations, this process could lead to loss of minicircle sequences, which would be lethal because the minicircles encode crucial guide RNAs. Therefore, the parasite's survival would depend on its possession of multiple copies of each minicircle or of minicircles that encode redundant guide RNAs (78).

Selection for a complete set of minicircles would be strong, although the ratios of the different minicircle classes could probably vary. This variation could contribute to the changes in the pattern of restriction enzyme digests observed in different isolates of various trypanosomatid species (e.g. 9, 61). Even the passaging of *C. fasciculata* (37) or *L. tarentolae* (106) for long periods in the laboratory results in small changes in the pattern of minicircle restriction fragments. If there were no selective pressure to preserve a repertoire of minicircles, there would be rapid loss of minicircle sequences, leading ultimately to minicircle homogeneity, a situation observed in the African trypanosomes *T. equiperdum* (3) and *T. evansi* (10). These parasites survive because they live only in the animal host's bloodstream, where maxicircle gene products are not essential for viability.

The spinning kinetoplast may have evolved to facilitate distribution of sister minicircles into different daughter networks. In *C. fasciculata,* the sister minicircles are attached to the network at different times after replication (42). The sister with the continuously synthesized strand is attached immediately after synthesis. The other, with the discontinuously synthesized strand, is attached after a few minutes' delay. Rotation of the kinetoplast during the interval between attachment of the sister molecules could ultimately result in their segregation into different daughter networks.

Minicircle Exchange During T. brucei Mating

Trypanosomes undergo genetic exchange of nuclear DNA markers during passage through the insect vector (39). Exchange of kDNA sequences also occurs, in which the hybrid progeny inherit maxicircles from only one parent but minicircles from both (30). One possible mechanism for the minicircle exchange involves fusion of the mitochondria of the parent cells. If both parental networks were undergoing replication within the same compartment, the progeny of minicircles released from one network could attach to the other. Because maxicircles are not released from the network during replication, these molecules could not exchange.

Transkinetoplastidy

Over the past two years, an intriguing series of papers has described a drug-induced alteration in the kDNA of *Leishmania mexicana amazonensis*. This phenomenon, termed transkinetoplastidy (52), consists of changes in the sequence of maxicircles (51) accompanied by a dramatic shift in the population of minicircles (50). In these studies, insect forms of the parasite were rendered drug-resistant by acclimatization to escalating concentrations of drug (either sodium arsenite or tunicamycin). During transkinetoplastidy, which requires about 15 generations, the predominant sequence class of minicircles is drastically reduced and supplanted by a previously minor class (which differs in arsenite- and tunicamycin-resistant cells). In the intermediate stages of transkinetoplastidy, the total number of minicircles in the network declines, and each network appears to contain minicircles characteristic of both wild-type and resistant cells (49). Transkinetoplastidy is reversible; between eight and ten months after drug pressure is removed, the minicircle composition reverts to wild-type (50). It also seems to occur only in drug-resistant cells that have drug-induced amplification of chromosomal DNA; furthermore, the mitochondrial alterations may lag behind the nuclear events (50, 52).

What is the significance of transkinetoplastidy? One possibility is that the parasite is adapting its mitochondrial function (by changing the sequence or abundance of maxicircle gene products) to the stressful conditions caused by growth in the presence of drug. Nevertheless, it is difficult to explain the mechanism of transkinetoplastidy in terms of the current model of kDNA replication. The observations suggest that some sort of selection and amplification mechanism may be at work during kDNA replication. The problem is to explain how some classes of minicircles can be depleted from the network and subsequently replaced by a different class during the relatively short period when the cell undergoes transkinetoplastidy. More studies are needed to follow up on these important findings.

Dyskinetoplastidy

African trypanosomes lacking all or part of their kDNA have been known since 1910 (122). These dyskinetoplastic cells have mitochondrial membranes, and they survive because they can live in their host's bloodstream, where maxicircle gene products are not needed for viability (see 33, 101 for reviews on dyskinetoplastidy). Dyskinetoplastic cells may arise spontaneously in nature (dyskinetoplastic *T. equiperdum, T. evansi,* and *T. brucei* have been isolated from infected animals) or may be induced experimentally. Dyskinetoplastic bloodstream trypanosomes are readily generated by treatment with DNA-binding agents such as ethidium bromide, acriflavin, or hydroxystilbamidine. As described in the following section, these drugs inhibit mitochondrial topoisomerases in trypanosomes.

ENZYMES AND PROTEINS INVOLVED WITH kDNA STRUCTURE AND REPLICATION

To date, no cell-free kDNA replication system is available to facilitate the purification and study of enzymes mediating kDNA synthesis. Nevertheless, progress has been made in isolating and characterizing kDNA-metabolizing enzymes and kDNA-binding proteins. Much of this work has been done with *C. fasciculata,* which grows rapidly and reaches high cell densities in vitro. In this section, we focus on proteins that localize within the mitochondrion.

DNA Topoisomerases

These enzymes are classified as type I (which make single-stranded breaks in DNA) or as type II (which make double-stranded breaks) (88). Two apparently distinct mitochondrial type II topoisomerases have been isolated from *C. fasciculata.* One, containing four 60-kDa subunits, is an ATP-dependent topoisomerase II (98, 99). Nicking of network-bound minicircles by a *C. fasciculata* endonuclease inhibits decatenation by this topoisomerase but has no effect on the catenation of monomeric minicircles (97). Immunolocalization at the EM level indicates that this topoisomerase is distributed throughout the kinetoplast disk (96; J Shlomai, personal communication). Therefore, it is a candidate for the enzyme that releases covalently closed minicircles from the network prior to replication and that remodels the network after replication.

A second *C. fasciculata* mitochondrial ATP-dependent topoisomerase II is a dimer of 132-kDa subunits (59). Polyclonal antibodies to this enzyme do not recognize the enzyme with 60-kDa subunits. Immunolocalization studies show that this enzyme is in the replication complexes sited on opposite sides of the kinetoplast disk (60). No detectable antibody binding to the nucleus occurs. The single-copy gene for this enzyme encodes a predicted 138-kDa polypeptide that shares 30–70% homology with other eukaryotic type II topoisomerases, including those from *T. brucei* and *T. cruzi* (29, 68, 112). This enzyme may facilitate minicircle replication or may be involved in attachment of newly synthesized minicircles to the network.

Potent and highly specific inhibitors of topoisomerase II (e.g. etoposide) have yielded valuable insights into the normal function of these enzymes. In vivo, these inhibitors can stabilize cleavable complexes consisting of the inhibitor, the enzyme, and the DNA substrate (88). If a treated cell is lysed with strong detergent, the DNA is cleaved and each 5' end is covalently linked to a denatured topoisomerase II subunit. When trypanosomes are treated with etoposide and lysed with detergent, about 20% of minicircles and virtually all maxicircles are released from networks as protein-bound linearized molecules (75, 93, 94). This result indicates that in the cell each of these DNA circles is associated with topoisomerase II. In view of the structural complexity of kDNA and the myriad topological interconversions required for its metabolism, it is not surprising that the topoisomerase II:DNA ratio is very high (about 1 molecule:8 kb). However, the linearization of all maxicircles, including those from cells not in S phase, suggests that topoisomerase II may play a structural as well as an enzymatic role in the kinetoplast. Interestingly, in minicircles the most prominent topoisomerase II cleavage sites map to the 5' ends of the continuously synthesized strand and of the early Okazaki fragments, a result suggesting that the enzyme may operate at these sites during replication.

Several classical antitrypanosomal drugs (e.g. pentamidine, berenil, ethidium bromide, isometamidium chloride) bind to DNA. In trypanosomes, these compounds form cleavable complexes with mitochondrial, but not nuclear, topoisomerase II (92). This selective effect may explain why these drugs preferentially disrupt kDNA structure in trypanosomes (64), thereby generating dyskinetoplastic cells (17, 80).

Efforts to purify a mitochondrial topoisomerase I from *C. fasciculata* yielded an enzyme that immunolocalizes to the nucleus but not to the kinetoplast (58), raising the possibility that kinetoplast-specific topoisomerases are exclusively type II enzymes. However, when trypanosomes are treated with camptothecin (a specific inhibitor of topoisomerase I) and then lysed with detergent, about 15% of minicircles covalently bind to protein (126). This observation strongly supports the existence of a mitochondrial topoisomerase I in trypanosomes.

DNA Polymerase

Among the five classes of eukaryotic DNA polymerase, polymerase γ is the mitochondrial replication enzyme. Several DNA polymerase activities have been detected in trypanosomatids, but only one, from *C. fasciculata*, is known to be mitochondrial. This unusual protein is monomeric in solution, and with a molecular weight of 43 kDa, it is much smaller than typical γ polymerases (118). Moreover, whereas γ polymerases are highly processive, highly accurate, and have a proofreading exonuclease activity, the *C. fasciculata* mitochondrial enzyme is nonprocessive, has low fidelity, and has no detectable exonuclease activity (119). Together these characteristics suggest that the *C. fasciculata* mitochondrial enzyme is a β polymerase, which in other eukaryotes are gap-filling enzymes involved in DNA repair in the nucleus. The *C. fasciculata* enzyme's sequence has 31% amino acid identity with human DNA polymerase β (118a). Therefore, it is the first example of a mitochondrial polymerase β.

The protein has a cleaved N-terminal presequence similar to the mitochondrial import signals of other *C. fasciculata* mitochondrial proteins (124). The *C. fasciculata* polymerase immunolocalizes to the complex of replication proteins located at opposite sides of the kinetoplast disk (26). It may have been imported into the trypanosomatid mitochondrion for the specific purpose of repairing the many gaps in the discontinuously synthesized strand of newly synthesized minicircles. A replicative polymerase, perhaps related to DNA polymerase γ, is probably also present in trypanosomatid mitochondria.

Minicircle Origin–Binding Protein

A single-stranded DNA-binding protein that recognizes the universal minicircle sequence GGGGTTGGTGTA has been isolated from whole cell lysates of *C. fasciculata* (120). The protein is a homodimer consisting of two 13.5-kDa subunits (96), and its gene sequence predicts five CCHC-type zinc fingers (1). This protein may be involved in replication initiation, recognizing the GGGGTTGGTGTA sequence after the strands in this region have been separated. In addition, it could function after the minicircle has completed replication, binding to and preserving the unique gap in the continuously synthesized strand. In this way it would ensure that minicircles replicated only once per generation. Removal of this protein after all minicircles have replicated would then allow gap repair and covalent closure of these molecules.

Kinetoplast-Associated Protein (KAP)

This 118-kDa polypeptide is the product of a single-copy *T. cruzi* gene (31). Its amino acid sequence is highly repetitive, and its predicted secondary structure has substantial α-helical character, which suggests the protein may be structural. Immunolocalization experiments showed that KAP localizes in the kinetoplast of *T. cruzi* amastigotes and epimastigotes, two stages of the life cycle that are replicative; it is not found in the kinetoplast of *T. cruzi* trypomastigotes, which are nonreplicative.

Proteins that Condense kDNA

Structural proteins are likely required to organize the kDNA network into the condensed structure diagrammed in Figure 2b. Early histochemical studies suggested that trypanosomatids had histones in the nucleus but not in the kinetoplast (110), and EM of *Crithidia luciliae* DNA revealed typical bead-like nucleosome structures in nuclear DNA but not in kDNA (11). However, EM cytochemical methods have suggested that basic proteins are bound to the kinetoplast of *T. cruzi* (108).

Only recently has progress been made in the identification of proteins that are candidates for condensing and organizing kDNA in vivo. One family of five proteins, 15–21 kDa in size, was isolated by reversible cross-linking to *C. fasciculata* kDNA in vivo using formaldehyde (124). These proteins are highly basic, and two that have been sequenced have predicted isoelectric points of 11.1 and 10.5. Immunoelectron microscopy has revealed that one of these proteins, p16, localizes within the kinetoplast disk (DS Ray, personal communication). Three of these proteins have cleaved N-terminal presequences of nine amino acid residues, which are likely to be mitochondrial import signals. Similar *C. fasciculata* proteins, which are in the same size range and that also localize to the kinetoplast, have been isolated in another laboratory (117).

Heat-Shock Proteins

Heat-shock proteins control protein folding, facilitate translocation of proteins across membranes, and assist in the assembly and disassembly of protein complexes. hsp70 heat-shock proteins have been detected in or isolated from mitochondria of *T. cruzi* (23), *C. fasciculata* (19), and *Leishmania donovani* (89). The genes for all three are homologous to hsp70 proteins in other cells and exhibit substantial similarity to *E. coli dnaK*. All three have a 20-residue N-terminal presequence that is probably a mitochondrial import signal. These proteins are concentrated around the kinetoplast disk, but what restricts them to this location remains unknown. Comparison of kinetoplastic and dyskinetoplastic strains of *T. brucei* indicates that hsp70 is localized around the kDNA in the kinetoplastic strain but is dispersed throughout the mitochondrion in the dyskinetoplastic strain (45). These proteins probably have the same function as their homologue in yeast, facilitating the transport of proteins into mitochondria. However, their association with kDNA raises the possibility that they may also be involved in assembly and disassembly of protein complexes involved in processes such as kDNA replication or RNA editing.

CONCLUSION

Why is kDNA organized in a network? Despite extensive investigation of kDNA structure, replication, and gene expression, why the mitochondrial DNA in trypanosomatids is the only example of a DNA network remains a mystery. We conclude this review by offering some speculations on this subject (12).

Some of the most ancient kinetoplastid protozoa do not organize their mitochondrial DNA in a network. In *Bodo caudatus*, a free-living kinetoplastid protozoan, the mitochondrial DNA circles are not catenated (36), and in *Trypanoplasma borreli*, a parasite of fish, the minicircle-like sequences are tandemly repeated on long molecules (57a). There must be a compelling reason why more recently evolved trypanosomatids have organized their mitochondrial DNA in a network, especially since these parasites had to sacrifice the advantages of supercoiling their minicircles to form this network (see discussion above). One speculation is that catenation of kDNA into a network facilitates the interaction of maxicircle and minicircle transcripts, thereby expediting RNA editing. Another is that a network structure permits rapid change in the minicircle repertoire. As discussed above, the minicircle content can differ dramatically among isolates from the same trypanosomatid species (e.g. 9, 61, 83), and at least in the case of *T. brucei*, minicircle exchange occurs during mating (30). Assuming that exchange of minicircles is important, exchange between networks would be more efficient than recombining minicircle-like sequences in linear chromosomes. Rapid exchange could also be achieved with noncatenated minicircles. However, the network provides more efficient housekeeping, both in organizing the maxicircles and minicircles and in allowing for an orderly mechanism for replication of the kDNA genetic information.

What would be the advantage of exchanging minicircles? One possibility concerns mutations in maxicircles. A mutation in an edited region of a maxicircle gene could be lethal unless a corresponding change occurred in the minicircle that encodes the guide RNA sequence. Two simultaneous complementing mutations would be rare, and therefore a continuous change in the minicircle repertoire would increase the chance that one would encode a guide RNA that matched the mutated maxicircle sequence.

ACKNOWLEDGMENTS

Work in the authors' laboratories was supported by NIH grants AI28855 (to TAS) and GM27608 (to PTE) and by a grant from the MacArthur Foundation.

> Any *Annual Review* chapter, as well as any article cited in an *Annual Review* chapter, may be purchased from the Annual Reviews Preprints and Reprints service.
> 1-800-347-8007; 415-259-5017; email: arpr@class.org

Literature Cited

1. Abeliovich H, Tzfati Y, Shlomai J. 1993. A trypanosomal CCHC-type zinc finger protein which binds the conserved universal sequence of kinetoplast DNA minicircles: isolation and analysis of the complete cDNA from *Crithidia fasciculata*. *Mol. Cell Biol.* 13:7766–73
2. Alexeieff A. 1917. Sur la fonction glycoplastique du kinetoplaste (=kinetonucleus) chez les flagellés. *C. R. Soc. Biol.* 80:512–14
3. Barrois M, Riou G, Galibert F. 1981. Complete nucleotide sequence of minicircle kinetoplast DNA from *Trypanosoma equiperdum*. *Proc. Natl. Acad. Sci. USA* 78:3323–27
4. Benne R. 1994. RNA editing in trypanosomes. *Eur. J. Biochem.* 221:9–23
5. Benne R, Van den Burg J, Brakenhoff JPJ, Sloof P, Van Boom JH, et al. 1986. Major transcript of the frameshifted coxII gene from trypanosome mitochondria contains four nucleotides that are not encoded in the DNA. *Cell* 46:819–26
6. Birkenmeyer L, Ray DS. 1986. Replication of kinetoplast DNA in isolated kinetoplasts from *Crithidia fasciculata*. Identification of minicircle DNA replication intermediates. *J. Biol. Chem.* 261:2362–68
7. Birkenmeyer L, Sugisaki H, Ray DS. 1987. Structural characterization of site-specific discontinuities associated with replication origins of minicircle DNA from *Crithidia fasciculata*. *J. Biol. Chem.* 262:2384–92
8. Blum B, Bakalara N, Simpson L. 1990. A model for RNA editing in kinetoplastid mitochondria: "guide" RNA molecules transcribed from maxicircle DNA provide the edited information. *Cell* 60:189–98
9. Borst P, Fase-Fowler F, Frasch ACC, Hoeijmakers JHJ, Weijers PJ. 1980. Characterization of DNA from *Trypanosoma brucei* and related trypanosomes by restriction endonuclease digestion. *Mol. Biochem. Parasitol.* 1:221–46
10. Borst P, Fase-Fowler F, Gibson WC. 1987. Kinetoplast DNA of *Trypanosoma evansi*. *Mol. Biochem. Parasitol.* 23:31–38
11. Borst P, Hoeijmakers JHJ. 1979. Kinetoplast DNA. *Plasmid* 2:20–40
12. Chen J, Rauch CA, White JH, Englund PT, Cozzarelli NR. 1995. The topology of the kinetoplast DNA network. *Cell* 80:61–69
13. Chen KK, Donelson JE. 1980. Sequences of two kinetoplast DNA minicircles of *Trypanosoma brucei*. *Proc. Natl. Acad. Sci. USA* 77:2445–49
14. Clayton DA. 1991. Replication and transcription of vertebrate mitochondrial DNA. *Annu. Rev. Cell Biol.* 7:453–78
15. Cosgrove WB, Skeen MJ. 1970. The cell cycle in *Crithidia fasciculata*. Temporal relationships between synthesis of deoxyribonucleic acid in the nucleus and in the kinetoplast. *J. Protozool.* 17:172–77
16. Degrave W, Fragoso SP, Britto C, van Heuverswyn H, Kidane GZ, et al. 1988. Peculiar sequence organization of kinetoplast DNA minicircles from *Trypanosoma cruzi*. *Mol. Biochem. Parasitol.* 27:63–70
17. Delain E, Brack C, Riou G, Festy B. 1971. Ultrastructural alterations of *Trypanosoma cruzi* kinetoplast induced by the interaction of a trypanocidal drug

(hydroxystilbamidine) with the kinetoplast DNA. *J. Ultrastruct. Res.* 37:200–18
18. Delain E, Riou G. 1969. Ultrastructure du DNA du kinétoplaste de *Trypanosoma cruzi* cultivé in vitro. *C. R. Acad. Sci. Ser. D* 268:1225–27
19. Effron PN, Torri AF, Engman DM, Donelson JE, Englund PT. 1993. A mitochondrial heat shock protein from *Crithidia fasciculata*. *Mol. Biochem. Parasitol.* 59:191–200
20. Englund PT. 1978. The replication of kinetoplast DNA networks in *Crithidia fasciculata*. *Cell* 14:157–68
21. Englund PT. 1979. Free minicircles of kinetoplast DNA in *Crithidia fasciculata*. *J. Biol. Chem.* 254:4895–900
22. Englund, PT. 1981. Kinetoplast DNA. In *Biochemistry and Physiology of Protozoa*, ed. M Levandowsky, SH Hutner, 4:333–83. New York: Academic
23. Engman DM, Kirchhoff LV, Donelson JE. 1989. Molecular cloning of mtp70, a mitochondrial member of the hsp70 family. *Mol. Cell Biol.* 9:5163–68
24. Fairlamb AH, Weislogel PO, Hoeijmakers JHJ, Borst P. 1978. Isolation and characterization of kinetoplast DNA from bloodstream form of *Trypanosoma brucei*. *J. Cell Biol.* 76:293–309
25. Feagin JE, Abraham JM, Stuart K. 1988. Extensive editing of the cytochrome c oxidase III transcript in Trypanosoma brucei. *Cell* 53:413–22
26. Ferguson M, Torri AF, Ward DC, Englund PT. 1992. In situ hybridization to the *Crithidia fasciculata* kinetoplast reveals two antipodal sites involved in kinetoplast DNA replication. *Cell* 70:621–29
27. Ferguson MF, Torri AF, Pérez-Morga D, Ward DC, Englund PT. 1994. Kinetoplast DNA replication: mechanistic differences between *Trypanosoma brucei* and *Crithidia fasciculata*. *J. Cell Biol.* 126:631–39
28. Fernandes AP, Nelson K, Beverley SM. 1993. Evolution of nuclear ribosomal RNAs in kinetoplastid protozoa: perspectives on the age and origins of parasitism. *Proc. Natl. Acad. Sci. USA* 90:11608–12
29. Fragoso SP, Goldenberg S. 1992. Cloning and characterization of the gene encoding *Trypanosoma cruzi* DNA topoisomerase II. *Mol. Biochem. Parasitol.* 55:127–34
30. Gibson W, Garside L. 1990. Kinetoplast DNA minicircles are inherited from both parents in genetic hybrids of *Trypanosoma brucei*. *Mol. Biochem. Parasitol.* 42:45–54
31. González A, Rosales JL, Ley V, Diaz C. 1990. Cloning and characterization of a gene coding for a protein (KAP) associated with the kinetoplast of epimastigotes and amastigotes of *Trypanosoma cruzi*. *Mol. Biochem. Parasitol.* 40:233–44
32. Griffith J, Bleyman M, Rauch CA, Kitchin PA, Englund PT. 1986. Visualization of the bent helix in kinetoplast DNA by electron microscopy. *Cell* 46:717–24
33. Hajduk SL. 1978. Influence of DNA complexing compounds on the kinetoplast of trypanosomatids. *Prog. Mol. Subcell. Biol.* 6:158–200
34. Hajduk SL, Harris ME, Pollard VW. 1993. RNA editing in kinetoplastid mitochondria. *FASEB J.* 7:54–63
35. Hajduk SL, Klein VA, Englund PT. 1984. Replication of kinetoplast DNA maxicircles. *Cell* 36:483–92
36. Hajduk SL, Siqueira AM, Vickerman K. 1986. Kinetoplast DNA of *Bodo caudatus*: a noncatenated structure. *Mol. Cell Biol.* 6:4372–78
37. Hoeijmakers JHJ, Borst P. 1982. Kinetoplast DNA in the insect trypanosomes *Crithidia luciliae* and *Crithidia fasciculata*. II. Sequence evolution of the minicircles. *Plasmid* 7:210–20
38. Hoeijmakers JJ, Weijers PJ. 1980. The segregation of kinetoplast DNA networks in *Trypanosoma brucei*. *Plasmid* 4:97–116
39. Jenni L, Marti S, Schweizer J, Betschart B, Le Page RWF, et al. 1986. Hybrid formation between African trypanosomes during cyclical transmission. *Nature* 322:173–75
40. Kellenberger E, Carlemalm E, Sechaud J, Ryter A, De Haller G. 1986. Considerations on the condensation and the degree of compactness in non-eukaryotic DNA-containing plasmas. In *Bacterial Chromatin*, ed. CO Gualerzi, CL Pon, pp. 11–25. Berlin: Springer-Verlag
41. Kidane GZ, Hughes D, Simpson L. 1984. Sequence heterogeneity and anomalous electrophoretic mobility of kinetoplast minicircle DNA from *Leishmania tarentolae*. *Gene* 27:265–77
42. Kitchin PA, Klein VA, Englund PT. 1985. Intermediates in the replication of kinetoplast DNA minicircles. *J. Biol. Chem.* 260:3844–51
43. Kitchin PA, Klein VA, Fein BI, Englund PT. 1984. Gapped minicircles. A novel replication intermediate of kinetoplast DNA. *J. Biol. Chem.* 259:15532–39
44. Kitchin PA, Klein VA, Ryan KA, Gann KL, Rauch CA, et al. 1986. A highly bent fragment of *Crithidia fasciculata*

45. Klein KG, Olson CL, Engman DM. 1995. Mitochondrial heat shock protein 70 is found throughout the mitochondrion in a dyskinetoplastic mutant of *Trypanosoma brucei*. *Mol. Biochem. Parasitol.* In press
46. Kreuzer KN, Cozzarelli NR. 1980. Formation and resolution of DNA catenanes by DNA gyrase. *Cell* 20:245–54
47. Landweber LF, Gilbert W. 1994. Phylogenetic analysis of RNA editing: a primitive genetic phenomenon. *Proc. Natl. Acad. Sci. USA* 91:918–21
48. Laurent M, Steinert M. 1970. Electron microscopy of kinetoplastic DNA from *Trypanosoma mega*. *Proc. Natl. Acad. Sci. USA* 66:419–24
49. Lee S-T, Liu H-Y, Lee S-P, Tarn C. 1994. Selection for arsenite resistance causes reversible changes in minicircle composition and kinetoplast organization in *Leishmania mexicana*. *Mol. Cell Biol.* 14:587–96
50. Lee S-T, Tarn C, Chang K-P. 1993. Characterization of the switch of kinetoplast DNA minicircle dominance during development and reversion of drug resistance in *Leishmania*. *Mol. Biochem. Parasitol.* 58:187–203
51. Lee S-T, Tarn C, Wang C-Y. 1992. Characterization of sequence changes in kinetoplast DNA maxicircles of drug-resistant *Leishmania*. *Mol. Biochem. Parasitol.* 56:197–207
52. Lee S-Y, Lee S-T, Chang K-P. 1992. Transkinetoplastidy—a novel phenomenon involving bulk alterations of mitochondrion-kinetoplast DNA of a trypanosomatid protozoan. *J. Protozool.* 39:190–96
53. Marini JC, Levene SD, Crothers DM, Englund PT. 1982. Bent helical structure in kinetoplast DNA. *Proc. Natl. Acad. Sci. USA* 79:7664–68
54. Marini JC, Levene SD, Crothers DM, Englund PT. 1983. A bent helix in kinetoplast DNA. *Cold Spring Harbor Symp. Quant. Biol.* 47:279–83
55. Marini JC, Miller KG, Englund PT. 1980. Decatenation of kinetoplast DNA by topoisomerases. *J. Biol. Chem.* 255:4976–79
56. Maslov DA, Avila HA, Lake JA, Simpson L. 1994. Evolution of RNA editing in kinetoplastid protozoa. *Nature* 368:345–48
57. Maslov DA, Simpson L. 1992. The polarity of editing within a multiple gRNA-mediated domain is due to formation of anchors for upstream gRNAs by downstream editing. *Cell* 70:459–67
57a. Maslov DA, Simpson L. 1994. RNA editing and mitochondrial genomic organization in the cryptobiid kinetoplastid protozoan *Trypanoplasma borreli*. *Mol. Biochem. Parasitol.* 14: 8174–82
58. Melendy T, Ray DS. 1987. Purification and nuclear localization of a type I topoisomerase from *Crithidia fasciculata*. *Mol. Biochem. Parasitol.* 24:215–25
59. Melendy T, Ray DS. 1989. Novobiocin affinity purification of a mitochondrial type II topoisomerase from the trypanosomatid *Crithidia fasciculata*. *J. Biol. Chem.* 264:1870–76
60. Melendy T, Sheline C, Ray DS. 1988. Localization of a type II DNA topoisomerase to two sites at the periphery of the kinetoplast DNA of *Crithidia fasciculata*. *Cell* 55:1083–88
61. Morel C, Chiari E, Plessmann Camargo E, Mattei DM, Romanha AJ, et al. 1980. Strains and clones of *Trypanosoma cruzi* can be characterized by pattern of restriction endonuclease products of kinetoplast DNA minicircles. *Proc. Natl. Acad. Sci. USA* 77:6810–14
62. Muhich ML, Simpson L, Simpson AM. 1983. Comparison of maxicircle DNAs of *Leishmania tarentolae* and *Trypanosoma brucei*. *Proc. Natl. Acad. Sci. USA*. 80:4060–64
63. Myler PJ, Glick D, Feagin JE, Morales TH, Stuart KD. 1993. Structural organization of the maxicircle variable region of *Trypanosoma brucei*: identification of potential replication origins and topoisomerase II binding sites. *Nucleic Acids Res.* 21:687–94
64. Newton BA. 1974. The chemotherapy of trypanosomiasis and leishmaniasis: towards a more rational approach. In *Trypanosomiasis and Leishmaniasis with Special Reference to Chagas' Disease*, pp. 285–307. Amsterdam: Elsevier Excerpta Medica North-Holland
65. Ntambi JM, Englund PT. 1985. A gap at a unique location in newly replicated kinetoplast DNA minicircles from *Trypanosoma equiperdum*. *J. Biol. Chem.* 260:5574–79
66. Ntambi JM, Marini JC, Bangs JD, Hajduk SL, Jimenez HE, et al. 1984. Presence of a bent helix in fragments of kinetoplast DNA minicircles from several trypanosomatid species. *Mol. Biochem. Parasitol.* 12:273–86
67. Ntambi JM, Shapiro TA, Ryan KA, Englund PT. 1986. Ribonucleotides associated with a gap in newly replicated kinetoplast DNA minicircles from *Trypanosoma equiperdum*. *J. Biol. Chem.* 261:11890–95

68. Pasion SG, Hines JC, Aebersold R, Ray DS. 1992. Molecular cloning and expression of the gene encoding the kinetoplast-associated type II DNA topoisomerase of *Crithidia fasciculata. Mol. Biochem. Parasitol.* 50:57–68
69. Pollard VW, Rohrer SP, Michelotti EF, Hancock K, Hajduk SL. 1990. Organization of minicircle genes for guide RNAs in *Trypanosoma brucei. Cell* 63:783–90
70. Pérez-Morga D, Englund PT. 1993. The structure of replicating kinetoplast DNA networks. *J. Cell Biol.* 123:1069–79
71. Pérez-Morga D, Englund PT, 1993. The attachment of minicircles to kinetoplast DNA networks during replication. *Cell* 74:703–11
72. Rauch CA, Pérez-Morga D, Cozzarelli NR, Englund PT. 1993. The absence of supercoiling in kinetoplast DNA minicircles. *EMBO J.* 12:403–11
73. Ray DS. 1987. Kinetoplast DNA minicircles: high-copy-number mitochondrial plasmids. *Plasmid* 17:177–90
74. Ray DS. 1989. Conserved sequence blocks in kinetoplast minicircles from diverse species of trypanosomes. *Mol. Cell Biol.* 9:1365–67
75. Ray DS, Hines JC, Anderson M. 1992. Kinetoplast-associated DNA topoisomerase in *Crithidia fasciculata:* crosslinking of mitochondrial topoisomerase II to both minicircles and maxicircles in cells treated with the topoisomerase inhibitor VP16. *Nucleic Acids Res.* 20:3353–56
76. Renger HC, Wolstenholme DR. 1970. Kinetoplast deoxyribonucleic acid of the hemoflagellate *Trypanosoma lewisi. J. Cell Biol.* 47:689–702
77. Renger HC, Wolstenholme DR. 1972. The form and structure of kinetoplast DNA of *Crithidia. J. Cell Biol.* 54:346–64
78. Riley GR, Corell RA, Stuart K. 1994. Multiple guide RNAs for identical editing of *Trypanosoma brucei* apocytochrome b mRNA have an unusual minicircle location and are developmentally regulated. *J. Biol. Chem.* 269:6101–8
79. Riou G, Delain E. 1969. Electron microscopy of the circular kinetoplastic DNA from *Trypanosoma cruzi:* occurrence of catenated forms. *Proc. Natl. Acad. Sci. USA* 62:210–17
80. Riou GF, Belnat P, Benard J. 1980. Complete loss of kinetoplast DNA sequences induced by ethidium bromide or by acriflavine in *Trypanosoma equiperdum. J. Biol. Chem.* 255:5141–44
81. Robinson DR, Gull K. 1991. Basal body movements as a mechanism for mitochondrial genome segregation in the trypanosome cell cycle. *Nature* 352:731–33
82. Robinson DR, Gull K. 1994. The configuration of DNA replication sites within the *Trypanosoma brucei* kinetoplast. *J. Cell Biol.* 126:641–48
83. Rogers WO, Wirth DF. 1987. Kinetoplast DNA minicircles: regions of extensive sequence divergence. *Proc. Natl. Acad. Sci. USA* 84:565–69
84. Ryan KA, Englund PT. 1989. Synthesis and processing of kinetoplast DNA minicircles in *Trypanosoma equiperdum. Mol. Cell Biol.* 9:3212–17
85. Ryan KA, Englund PT. 1989. Replication of kinetoplast DNA in *Trypanosoma equiperdum.* Minicircle H strand fragments which map at specific locations. *J. Biol. Chem.* 264:823–30
86. Ryan KA, Shapiro TA, Rauch CA, Englund PT. 1988. Replication of kinetoplast DNA in trypanosomes. *Annu. Rev. Microbiol.* 42:339–58
87. Ryan KA, Shapiro TA, Rauch CA, Griffith JD, Englund PT. 1988. A knotted free minicircle in kinetoplast DNA. *Proc. Natl. Acad. Sci. USA* 85:5844–48
88. Schneider E, Hsiang Y-H, Liu LF. 1990. DNA topoisomerases as anticancer drug targets. *Adv. Pharmacol.* 21:149–83
89. Searle S, McCrossan MV, Smith DF. 1993. Expression of a mitochondrial stress protein in the protozoan parasite *Leishmania major. J. Cell Sci.* 104:1091–100
90. Shapiro TA. 1993. Kinetoplast DNA maxicircles: networks within networks. *Proc. Natl. Acad. Sci. USA* 90:7809–13
91. Shapiro TA. 1994. Mitochondrial topoisomerase II activity is essential for kinetoplast DNA minicircle segregation. *Mol. Cell Biol.* 14:3660–67
92. Shapiro TA, Englund PT. 1989. Selective cleavage of kinetoplast DNA minicircles promoted by antitrypanosomal drugs. *Proc. Natl. Acad. Sci. USA* 87:950–54
93. Shapiro TA, Klein VA, Englund PT. 1989. Drug promoted cleavage of kinetoplast DNA minicircles: evidence for type II topoisomerase activity in trypanosome mitochondria. *J. Biol. Chem.* 264:4173–78
94. Shapiro TA, Showalter AF. 1994. In vivo inhibition of trypanosome mitochondrial topoisomerase II: effects on kinetoplast DNA maxicircles. *Mol. Cell Biol.* 14:5891–97
95. Sheline C, Melendy T, Ray DS. 1989. Replication of DNA minicircles in kinetoplasts isolated from *Crithidia fasc-*

iculata: structure of nascent minicircles. *Mol. Cell Biol.* 9:169–76
96. Shlomai J. 1994. The assembly of kinetoplast DNA. *Parasitol. Today* 10:341–46
97. Shlomai J, Linial M. 1986. A nicking enzyme from trypanosomatids which specifically affects the topological linking of duplex DNA circles. Purification and characterization. *J. Biol. Chem.* 261:16219–25
98. Shlomai J, Zadok A. 1983. Reversible decatenation of kinetoplast DNA by a DNA topoisomerase from trypanosomatids. *Nucleic Acids Res.* 11:4019–34
99. Shlomai J, Zadok A, Frank D. 1984. A unique ATP-dependent DNA topoisomerase from trypanosomatids. *Adv. Exp. Med. Biol.* 179:409–22
100. Simpson AM, Simpson L. 1976. Pulse-labeling of kinetoplast DNA: localization of 2 sites of synthesis within the networks and kinetics of labeling of closed minicircles. *J. Protozool.* 23:583–87
101. Simpson L. 1972. The kinetoplast of the hemoflagellates. *Int. Rev. Cytol.* 32:139–207
102. Simpson L. 1987. The mitochondrial genome of kinetoplastid protozoa: genomic organization, transcription, replication, and evolution. *Annu. Rev. Microbiol.* 41:363–82
103. Simpson L. 1990. RNA editing—a novel genetic phenomenon? *Science* 250:512–13
104. Simpson L, Braly P. 1970. Synchronization of *Leishmania tarentolae* by hydroxyurea. *J. Protozool.* 17:511–17
105. Simpson L, Da Silva A. 1971. Isolation and characterization of kinetoplast DNA from *Leishmania tarentolae.* *J. Mol. Biol.* 56:443–73
106. Simpson L, Simpson AM, Kidane G, Livingston L, Spithill TW. 1980. The kinetoplast DNA of the hemoflagellate protozoa. *Am. J. Trop. Med. Hyg.* 29:1053–63
107. Sloof P, de Haan A, Eier W, van Iersel M, Boel E, et al. 1992. The nucleotide sequence of the variable region in *Trypanosoma brucei* completes the sequence analysis of the maxicircle component of mitochondrial kinetoplast DNA. *Mol. Biochem. Parasitol.* 56:289–99
108. Souto-Padrón T, De Souza W. 1978. Ultrastructural localization of basic proteins in *Trypanosoma cruzi.* *J. Histochem. Cytochem.* 26:349–58
109. Steinert G, Firket H, Steinert M. 1958. Synthèse d'acide désoxyribonucléique dans le corps parabasal de *Trypanosoma mega.* *Exp. Cell Res.* 15:632–35
110. Steinert M. 1965. L'absence d'histone dans le kinétonucléus des trypanosomes. *Exp. Cell Res.* 39:69–73
111. Steinert M, Van Assel S. 1980. Sequence heterogeneity in kinetoplast DNA: reassociation kinetics. *Plasmid* 3:7–17
112. Strauss PR, Wang JC. 1990. The *TOP2* gene of *Trypanosoma brucei:* a single-copy gene that shares extensive homology with other *TOP2* genes encoding eukaryotic DNA topoisomerase II. *Mol. Biochem. Parasitol.* 38:141–50
113. Stuart K, Feagin JE. 1992. Mitochondrial DNA of kinetoplastids. *Int. Rev. Cytol.* 141:65–88
114. Sturm NR, Simpson L. 1990. Kinetoplast DNA minicircles encode guide RNAs for editing of cytochrome oxidase subunit III mRNA. *Cell* 61:879–84
115. Sturm NR, Simpson L. 1991. *Leishmania tarentolae* minicircles of different sequence classes encode single guide RNAs located in the variable region approximately 150 bp from the conserved region. *Nucleic Acids Res.* 19:6277–81
116. Sugisaki H, Ray DS. 1987. DNA sequence of *Crithidia fasciculata* kinetoplast minicircles. *Mol. Biochem. Parasitol.* 23:253–63
117. Tittawella I, Carlsson L, Thornell L-E. 1993. Two proteins involved in kinetoplast compaction. *FEBS Lett.* 333:5–9
118. Torri AF, Englund PT. 1992. Purification of a mitochondrial DNA polymerase from *Crithidia fasciculata.* *J. Biol. Chem.* 267:4786–92
118a. Tori AF, Englund PT. 1995. A DNA polymerase β in the mitochondrion of the trypanosomatid *Crithidia fasciculata.* *J. Biol. Chem.* 270:3495–97
119. Torri AF, Kunkel TA, Englund PT. 1994. A β-like DNA polymerase from the mitochondrion of the trypanosomatid *Crithidia fasciculata.* *J. Biol. Chem.* 269:8165–71
120. Tzfati Y, Abeliovich H, Kapeller I, Shlomai J. 1992. A single-stranded DNA-binding protein from *Crithidia fasciculata* recognizes the nucleotide sequence at the origin of replication of kinetoplast DNA minicircles. *Proc. Natl. Acad. Sci. USA* 89:6891–95
121. Valenzuela ME, Bardhan S, Krishnamani MRS, Siddiqui KAI. 1991. Catenated dimers and knotted DNA structures: putative intermediates in the replication of *T. cruzi* kinetoplast mini-

circle DNA. *Biochem. Biophys. Res. Commun.* 174:958–68
122. Werbitzki FW. 1910. Über blepharoblastlose Trypanosomen. *Zentralbl. Bakteriol. Parasitenkd. Infektionskr.* 53:303–15
123. Woodward R, Gull K. 1990. Timing of nuclear and kinetoplast DNA replication and early morphological events in the cell cycle of *Trypanosoma brucei*. *J. Cell Sci.* 95:49–57
124. Xu C, Ray DS. 1993. Isolation of proteins associated with kinetoplast DNA networks in vivo. *Proc. Natl. Acad. Sci. USA* 90:1786–89
125. Ziemann H. 1898. Eine Methode der Doppelfärbung bei Flagellaten, Pilzen, Spirillen und Bakterien, sowie bei einigen Amöben. *Zentralbl. Bakteriol. Parasitenkd. Infektionskr.* 24:945–55
126. Bodley AL, Shapiro TA. 1995. Molecular and cytotoxic effects of camptothecin, a topoisomerase I inhibitor, on trypanosomes and Leishmania. *Proc. Natl. Acad. Sci. USA* 92:3726–30

HOW *SALMONELLA* SURVIVE AGAINST THE ODDS

John W. Foster* and Michael P. Spector**
*Department of Microbiology and Immunology, College of Medicine and
**Department of Biomedical Sciences, College of Allied Health, University of South Alabama, Mobile, Alabama 36688

KEY WORDS: environmental stress, stress and virulence, stress responses

CONTENTS

INTRODUCTION	146
VOYAGES THROUGH THE NATURAL ENVIRONMENT AND THROUGH AN ANIMAL HOST	146
SALMONELLA STRESS SURVIVAL STRATEGIES	147
Starvation Stress Response	147
Iron Stress	150
Acid-Tolerance Response	152
Oxidative-Stress Response	154
Heat-Shock Response and Thermotolerance	156
Osmotic-Shock Response	156
Resistance to Cationic Peptides	158
STRESS MANAGEMENT AND VIRULENCE	159
Escape from the Extracellular Environment	159
Macrophage Survival	161
In Vivo Environmental Cues	162
Salmonella Plasmid Virulence (spv) Genes	164
Salmonella Serum Resistance	164
The Salmonella mvi Mouse Virulence Genes	165
CONCLUSIONS	165

ABSTRACT

The enteric pathogen *Salmonella typhimurium* faces daunting odds during its voyages in the natural environment and through an infected host. It must manage stresses ranging from feast to famine, acid to base, and high to low osmolarity, among others, as well as counter various types of oxidative stress and a variety of antimicrobial peptides. The defenses used to survive these encounters can be specific or can provide cross protection to a variety of hostile conditions. Once inside a host, *Salmonella* spp. escape the extracellular environment and thus humoral immunity by invading professional and nonprofessional phagocytes in which a new set of challenges await. Some of these stresses are similar to those encountered in the natural environment (e.g. acid, starvation) but the bacterial response is complicated by the simultaneous occurrence of multiple stresses. *S. typhimurium* appears to sense various in vivo cues and responds by seducing the host signal-transduction pathways that are required to phagocytize the bacterial cell. The pathogen then calls upon components of its stress-response arsenal to survive the intracellular environment. These survival strategies enable the organism to persist in nature, where conditions are usually suboptimal, and equip the bacterium with pathogenic properties that, if successful, will provide it with a very rich and stress-free growth environment, a dead host.

INTRODUCTION

How microorganisms survive during periods of environmental stress is an exciting area of modern biology. The ability of bacteria to sense and respond to unscheduled changes in the environment is crucial to their survival. This is especially true for pathogens that not only encounter potentially lethal environmental extremes in the natural environment but that must also resist a battery of antimicrobial maneuvers initiated by an embattled host. The inability to cope with either one or the other of these stressful milieus (natural vs host environments) in part contributes to the fastidious nature of obligate intracellular parasites and the nonpathogenic nature of saprophytes. Few, if any, organisms cope as well as *Salmonella typhimurium* with the breadth of stresses offered by both niches (29, 189).

VOYAGES THROUGH THE NATURAL ENVIRONMENT AND THROUGH AN ANIMAL HOST

After evacuating a host, *S. typhimurium* usually enters a relatively nutrient-rich aquatic environment that may rapidly deteriorate through dilution. The organism will experience a temperature downshift (if expelled from a mammalian or avian host) followed by a variety of progressively serious nutrient limitations. In addition, low osmolarity and fluctuations in pH are common. Eventually, the organism may enter what is termed a viable-but-nonculturable state in which it can survive for long periods of time (171, 197). The organism may also encounter certain predatory protozoa in which it can persist (11).

Any successful pathogen must overcome or circumvent many levels of mammalian host defenses. Upon reentering a host through the ingestion of contaminated food or water, *Salmonella* spp. must first survive the low pH of the stomach. Once in the intestine, the organism encounters low oxygen levels, intestinal bile salts, weak acids (owing to fermentation), increased osmolarity, and competition with resident microorganisms for food. *Salmonella* spp. next invade the intestinal mucosa via columnar epithelial and/or M cells, entering a vacuolar environment low in nutrient. Upon exiting the basal side of intestinal epithelial cells, the organism enters the reticuloendothelial system and escapes humoral defenses by invading macrophages. Within the macrophage, *Salmonella* spp. reside within phagosomes or phagolysosomes, which also present many hazards to the invading microbe, including low pH, oxidative stress, nutrient limitations, and various antimicrobial peptides called defensins. *S. typhimurium* is considered an intracellular parasite, but it can live free within the bloodstream by virtue of its ability to survive the bactericidal effects of serum complement.

This review aims to summarize the strategies used by *S. typhimurium* to overcome these seemingly insurmountable odds working against its survival. Figure 1 presents an overview of the genes and regulatory circuitry that comprise the *S. typhimurium* stress-response systems.

SALMONELLA STRESS SURVIVAL STRATEGIES

Starvation Stress Response

One of the most common stresses encountered by *Salmonella* spp. is starvation for phosphate (P), carbon (C), and nitrogen (N) sources (88). Because of their lifestyle, salmonellae must tolerate sometimes long periods of nutrient starvation, both in the natural aquatic and terrestrial microcosms as well as within host microenvironments (105). The starvation stress response (SSR) of *S. typhimurium* encompasses the genetic and physiologic changes that occur upon starvation for a P, C, or N source. The SSR can be divided into individual P-, C-, or N-starvation stress responses, each of which involves both unique and overlapping sets of proteins (59, 181, 182, 185). Studies using Mud-*lac* gene fusions have been used to identify genes induced by various nutrient limitations. Six regulatory classes of nutrient starvation–inducible loci have been characterized (59, 184; A Turk, P Gulig & M Spector, unpublished results). Genes induced by two or more starvation conditions are broadly referred to as starvation-inducible (*sti*) loci, whereas those induced by only a single starvation condition are given specific designations, e.g. carbon starvation inducible (*csi*). Some of the genes induced by starvation are clearly required for starvation survival (*stiA, B, C* and two *csi* genes)(183).

In addition to their physiologic regulation, genes induced during C-limitation exhibit different induction-kinetics. Phase 0 genes (*stiK, csiG,* and *csiM*) are induced in late log phase just prior to starvation-induced stationary phase.

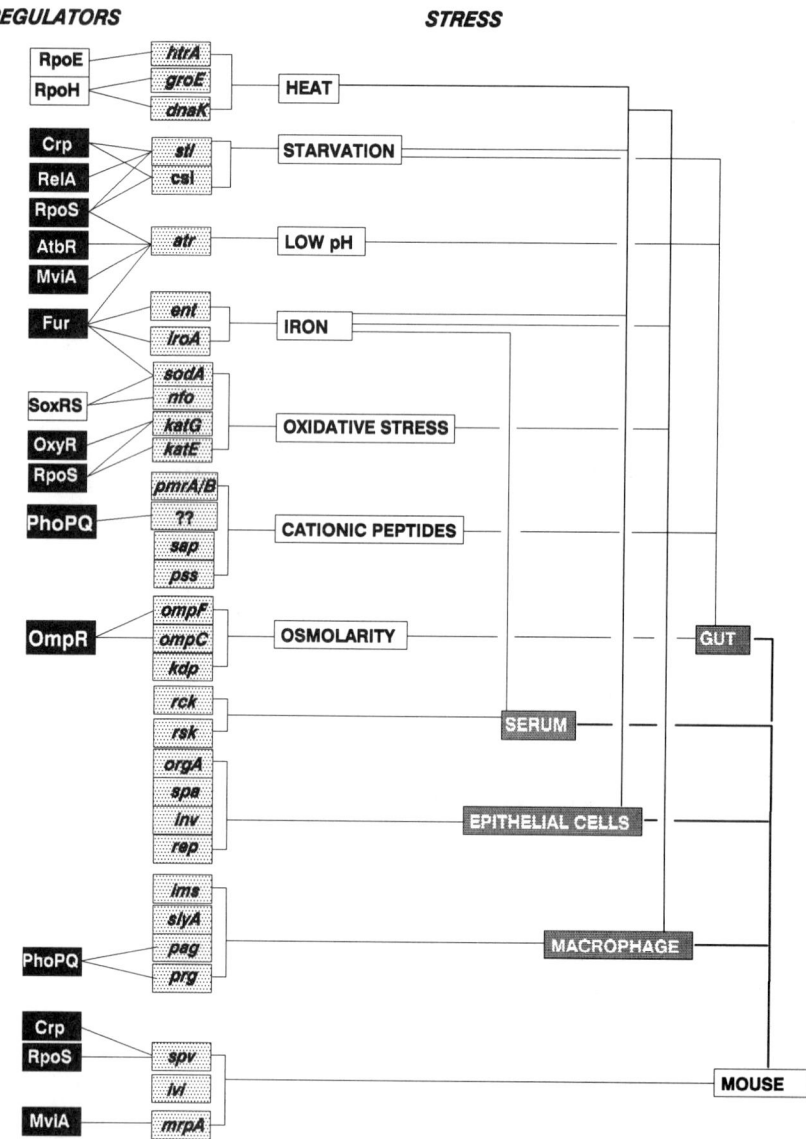

Figure 1 Stress response systems of *S. typhimurium*. Various environmental stresses encountered by *S. typhimurium* appear on the right. Intersecting lines show where stresses found in the natural environment also contribute to stress imposed by host locales. For example, starvation is encountered by the microbe in epithelial and macrophage vacuoles as well as in the gut. Regulatory genes and their target loci appear on the left with interconnections defining regulatory circuits. Regulators not yet identified in *S. typhimurium* are indicated with open boxes.

Phase 1 genes are induced during the transition into starvation-induced stationary phase, whereas phase 2 and 3 genes are induced at 1–2 h and 4–5 h after entry into stationary phase, respectively. The overwhelming majority of genes identified thus far via *lac* fusions are phase 1 genes, many of which continue to be expressed at relatively high levels for 24 h or more of C-starvation (184, 185).

The genetic regulation of starvation-inducible genes (both *sti* and *csi*) is very complex, involving cyclic AMP receptor protein (CRP) and the alternate sigma factor σ^{38} (RpoS), both of which have positive and negative effects on gene expression (111, 150, 183). In nonstarved *S. typhimurium*, CRP negatively regulates several genes, including three starvation survival genes, *stiA*, *stiB*, and *stiC* (183). CRP-mediated negative control of *stiB* is cAMP independent (183). In contrast, this complex positively controls induction of other phase 1 genes (W Gabriel & M Spector, in preparation). The alternative sigma factor, RpoS (σ^{38} or σ^S) (141, 192), positively regulates *stiA* and *stiC* as well as the virulence operon *spvABCD* during P-, C-, and N-starvation (43, 146, 152; A Turk, P Gulig & M Spector, unpublished results); but negatively regulates two phase 0 genes (*csiG* and *csiM*) and one phase 1 gene (*stiB*). Negative regulation by RpoS is most likely indirect, through σ^S-dependent synthesis of a repressor protein(s) that down-regulates *csiG* and *M* expression upon entry into starvation-induced stationary phase (W Gabriel & M Spector, in preparation) or that limits *stiB* induction during P- and C-starvation (152). In addition, most of the *sti* and *csi* genes identified require *relA*-dependent production of guanosine tetraphosphate (ppGpp) during C-starvation and N-starvation (if N-starvation-inducible) but not P-starvation (183; M Spector & R Hill, unpublished results). The role of ppGpp may be related to its involvement in σ^{38} expression (68).

As suggested above, survival during starvation requires protein synthesis. Some of the starvation-inducible proteins exhibiting increased expression during starvation are required for survival during simultaneous P-, C-, and N-starvation (PCN starvation) or C-starvation. Four genes are required for starvation survival in *S. typhimurium*: *rpoS*, *stiA*, *stiB*, and *stiC* (152, 183). Mutations in any of these four genes reduce starvation survival some 50- to 100-fold depending on the mutation. Even though σ^{38} is required for the induction of the *sti* loci, combining a *rpoS* mutation with any of the three *sti* mutations only reduces survival slightly more than a *sti* mutation alone. However, loss of any two of the *sti* mutations reduces survival 500- to 2000-fold, which suggests that even the basal levels of StiA, B, and C remaining in a *rpoS* mutant may aid in starvation survival (152, 183).

In addition to protecting cells against the effects of starvation, C-limitation can also induce cross-protection to several other environmental stresses (135). This function is in addition to the various specific stress responses described below. Starving *S. typhimurium* for 5–24 h can induce cross-protection to thermal (55°C for 20 min), oxidative (15 mM hydrogen peroxide for 60 min), and osmotic challenge (2.5 M NaCl for 55 h) (P Mishra & M Spector, in preparation). The development of C-starvation–induced cross-protection in *S. typhimurium* and *Escherichia coli* depends upon the *rpoS* gene product, whose expression is regulated transcriptionally and post-transcriptionally during starvation (110, 126, 134). The *S. typhimurium* starvation survival genes, *stiA, B,* and *C,* all appear necessary for C-starvation–induced cross-protection to thermal challenge. Furthermore, the *stiC* gene product is needed for the development of C-starvation–induced cross-protection against 15 mM hydrogen peroxide (P Mishra & M Spector, in preparation). Recent findings suggest C-starvation will also induce cross-protection against the membrane-permeabilizing, defensin-like small cationic peptide polymyxin B. Interestingly, unlike the cross-protection generated against the other stresses, cross-protection to polymyxin B is *rpoS* independent (M Spector & G McLeod, submitted).

In summary, starvation (especially C-starvation) can significantly affect the survival capabilities of *S. typhimurium* in natural microcosms and host microenvironments. C-starvation would appear to be a perfect signal to induce cross-protection against many host defenses until specific protective mechanisms to each stress can be generated.

Iron Stress

Numerous studies suggest that the ability to efficiently compete for iron is an important virulence characteristic of successful pathogens (125, 153, 161). Free iron is extremely limited in the mammalian host. The trace quantities of extracellular iron are bound by the high-affinity iron-binding glycoproteins transferrin and lactoferrin. In normal human serum, the amount of free iron in equilibrium with transferrin-bound iron is approximately 10^{-18} M, far below the level required for bacterial growth (10^{-6} M). During the acute-phase response to infection, iron availability is further limited by a cytokine-mediated increase in transferrin (21). This strategy enhances host resistance to various infectious agents, including *Salmonella* spp. (130, 164). In addition to surviving iron-poor environments in a host, *Salmonella* spp. must also survive in nature where most iron is insoluble at neutral or alkaline pH in the form of ferric hydroxide and thus is not readily accessible.

Microorganisms have solved the problem of iron acquisition by secreting low-molecular-weight, high-affinity iron chelators termed siderophores in response to iron stress. These compounds effectively compete with the host iron-binding compounds transferrin and lactoferrin and can solubilize iron from mineral complexes in the environment (8, 143). *S. typhimurium* utilizes enterobactin (enterochelin), a catechol type of siderophore (162), although some species can also synthesize the hydroxamate siderophore aerobactin (202).

Enterochelin is synthesized by the products of the *ent* gene cluster at 13 min on the *S. typhimurium* linkage map. Once secreted by the cell, this chelator encounters environmental iron and forms a ferric-enterochelin complex, which is then transported into the cell by a transport system specific for this complex (24). The ferric-enterochelin complex is too large to diffuse through the outer membrane pores, so a series of proteins (FepA–G), some outer membrane, some inner membrane, and some periplasmic in location, are required for transport into the cell. Three additional proteins—TonB, ExbB, and ExbD—are also required. TonB probably connects the outer and inner membranes, thereby serving as an energy-transducing system coupling the uptake process at the outer membrane to the energy-yielding elements of the cytoplasmic membrane (16, 23). ExbB and ExbD probably form a complex with TonB (50, 100, 180). A popular model is that the outer-membrane receptor (FepA) might form a gated pore that opens upon interaction with TonB. The periplasmic protein FepB then passes the ferric-enterochelin complex to the inner-membrane proteins FepGDC. Iron is removed from enterobactin within the cell by the *fes* product (enterochelin esterase), which cleaves enterochelin (reducing the binding constant from 10^{-52} M to 10^{-8} M) and reduces Fe(III) to Fe(II) (179).

Another bacterial strategy for iron acquisition is the production of hemolysins, which release iron complexed to intracellular heme and hemoglobin (i.e. HylA) (205). Whereas many members of the *Enterobacteriaceae* are hemolytic, *S. typhimurium* is considered nonhemolytic. However, low-passage clinical isolates of *S. typhimurium* are occasionally hemolytic on blood agar (120). The responsible gene, *slyA*, has been sequenced but lacks significant homology to other hemolysins. Purified salmolysin also behaves very differently from HylA. Thus, salmolysin may represent a new family of hemolysins. This newly described cytolysin is considered essential for *S. typhimurium* survival within certain macrophages. One of its functions may be the release of host-sequestered iron. The *sly* locus is not controlled by the *phoP/phoQ* or the *rpoS* regulators but has not been tested for regulation by the major iron regulator.

Regulation of iron uptake is mediated by a small 14-kDa protein called Fur. Mutants defective in the *fur* gene were identified originally in *S. typhimurium*, then in *E. coli* (20, 41, 86, 87). Fur protein from *E. coli* affixes Fe(II) as a cofactor and binds to specific DNA sequences called iron boxes located upstream of regulated genes (10, 87, 176). Fur has traditionally been considered a negative regulator that senses increasing intracellular Fe(II) concentrations and represses genes involved with iron uptake. Several studies have challenged this concept and suggest Fur has an even greater impact on global regulation than previously realized (54, 85, 124). Two-dimensional SDS-PAGE comparisons of polypeptides produced by *S. typhimurium* in response to conditions of iron excess and limitation reveal several distinct classes of Fur-regulated proteins. Nineteen proteins were the classical type, iron-repressed by Fur.

However, six iron-repressed proteins actually required Fur for their expression during iron limitation. Even more surprising was a set of nine iron-induced proteins that also required Fur for their expression. Thus, Fur acts both as a negative and a positive regulator of gene expression. Similar results were reported for *Vibrio cholerae* (124). It is not clear as yet whether Fur activates genes directly or indirectly by a repressor cascade. Considering the broad impact iron can have on the cell (for example, see sections on acid tolerance and oxidative stress below), it should not be surprising that the major iron sensor would influence genes that can buffer the myriad negative effects of iron excess or limitation.

The importance of enterobactin in the virulence of *S. typhimurium* is a subject of debate. One study using the virulent strain SR-11 revealed that an *ent* mutant of ill-defined nature (probably *entA, B,* or *C*) was less virulent when injected intraperitoneally into mice and grew slower in complement-inactivated serum (207). A contradictory study using virulent strain SL1344 derivatives showed that *ent* mutations had no effect on virulence in mice, although they confirmed *ent* mutants grew more slowly in serum (18). Thus, enterochelin is not essential for intracellular replication of *S. typhimurium*. However, the intracellular environment faced by *S. typhimurium* in the phagosomes of epithelial cells is apparently low in available iron. This conclusion is based on the induction of an iron-regulated *S. typhimurium* gene (*iroA*) observed following invasion of epithelial cells (66). So although enterochelin may not be essential, intracellular *S. typhimurium* must have some means for obtaining iron.

Acid-Tolerance Response

Acidic pH is one of the most frequent stress conditions encountered by microbial systems. Acid mine drainage, acid rain, and weak acids produced by microorganisms themselves all contribute to acid stress. The ability to survive and even to flourish during these encounters is crucial to perpetuation of the species. Some microorganisms (acidophiles) have evolved to the point where the preferred ecological niche is an extreme acid environment. However, other organisms with optimal pH values for growth in the neutral range (neutralophiles, formerly called neutrophiles) nevertheless must cope with frequent, although often transient, encounters with potentially lethal levels of acid. *S. typhimurium* appears to have developed several systems for surviving exposures to low pH. These include inducible systems for acid growth as well as log-phase and stationary-phase acid tolerance responses (ATRs). Minimum growth pH for *S. typhimurium* in minimal glucose medium is pH 4.3 (starting from a small inoculum, 37°C, aerobic). This value compares to pH 4.6 for *E. coli*. Huttanen (92) found that *S. typhimurium* could adapt to grow at lower pH on solid media. Cells struck across a gradient pH plate will grow a specific distance toward the most acid part of the medium. Huttanen found that cells taken from the acidic-growth edge were more acid resistant and could grow farther into the acidic end of a fresh gradient plate. This was not the result of selecting more acid resistant mutants because these cells lost their acid resistance after passage on a neutral pH medium. The molecular features of this phenomenon have not been pursued.

In addition to its ability to grow at pH 4.3, *S. typhimurium* can survive more severe acid conditions (down to pH 3.0) if it is allowed to induce one of two acid tolerance response systems (56, 114). One system involves log-phase cells, the other can be demonstrated in stationary-phase cells. If log-phase growth at neutral pH is interrupted by a shift to pH 4.0–4.5 (called acid shock), the induction of 50 acid-shock proteins (ASPs), of which ~20 are induced only by acid, leads to the development of profound acid tolerance to extremely low pH (51, 52). In contrast, stationary-phase cells shifted to pH 4.3 only induce 15 ASPs, four of which are also log-phase ASPs, that can still provide resistance to extreme acid. The two systems are thought to be different because mutations that dramatically affect one system have only small effects upon the other.

The log-phase ATR can be demonstrated in minimal or complex medium and not only protects against low pH but also ameliorates the effects of weak acids on viability (JW Foster & H Baik, unpublished observation). Both ATR systems require protein synthesis during the acid-shock adaptive period but not during acid challenge (52, 114). Several genes have been associated with acid tolerance (52, 53, 55). The genes identified affect iron metabolism (*fur, ent, atrD*), increase internal buffer (*icd*), interfere with proton pumping (*atp*), or impact DNA-repair functions (*polA*). A regulon involved with acid survival was identified through the isolation of an insertion mutant with decreased tolerance to dinitrophenol and low pH (53). The locus, *atrB*, is regulated by a gene designated *atbR*. Two-dimensional SDS-PAGE analysis of the *atbR* mutant revealed the overexpression of 10 proteins including AtrB. The *atbR* regulatory mutant proved to be constitutively acid tolerant, a finding that suggests that one or more members of the AtbR regulon contributes to acid tolerance.

The alternate sigma factor σ^S has an important role in log-phase ATR. Mutants defective in *rpoS*, although capable of transiently inducing a modest ATR, cannot produce the level of acid tolerance achieved by an *rpoS*$^+$ strain (IS Lee, J Lin, HK Hall, B Bearson & J Foster, submitted). The reason appears to involve the production of seven σ^S-dependent acid shock proteins.

Acid-shocked cells also develop significant cross-protection to heat, oxidative stress, and osmotic stress. This cross protection also depends upon RpoS. Conversely, neither heat shock nor osmotic shock will induce acid tolerance (118). Thus, acid exposure seems fairly unique among environmental stresses in terms of inducing protection against several different types of environmental challenge. It may, in fact, be another way of inducing the general stress resistance system previously only associated with starvation. The fact that acid shock induces the synthesis of RpoS supports this idea (IS Lee, J Lin, HK Hall, B Bearson & J Foster, submitted).

Transitions to acid pH environments also lead to dramatic changes in outer-membrane porin synthesis. At pH 5.8, OmpF synthesis decreases while OmpC expression increases (56, 57). Even more dramatic is the complete repression of OmpC and OmpF during acid shock (52). Decreased levels of these porins were previously associated with antibiotic resistance (7, 34, 72). One consequence of the acid shock–mediated reduction in porin synthesis could be increased antibiotic resistance in the stressed organisms.

Oxidative-Stress Response

Reactive oxygen species such as superoxide and hydrogen peroxide can damage DNA, proteins, and membranes. *S. typhimurium* can encounter potentially lethal levels of these intermediates from normal aerobic metabolism (62), many environmental agents (e.g. paraquot) (101), and the oxidative burst of phagocytes (9). *S. typhimurium* will induce a series of gene products in response to this stress. For example, *S. typhimurium* exposed to low concentrations of H_2O_2 (60 μM) will induce a set of proteins that enable survival in high concentrations of H_2O_2 (15 mM). The adaptive resistances to oxidative stress in *E. coli* and *S. typhimurium* involve two stimulons—one for hydrogen peroxide and one for superoxide. Each stimulon, produced in response to sublethal doses of the oxidative agent, consists of between 30 to 40 proteins with some overlap observed between the stimulons (71, 140). Most of these proteins are of unknown function. However, two sets of the inducible proteins constitute regulons. The *oxyR* gene mediates hydrogen peroxide-inducible expression of nine proteins (33), whereas the two-gene locus *soxRS* controls the induction of ten other proteins in response to superoxide generators but not hydrogen peroxide (72, 195).

OxyR is the transcriptional activator of genes for the peroxide-destroying enzymes catalase (*katG*), glutathione reductase (*gorA*) and NADPH-dependent alkyl hydroperoxidase (*ahpFC*) (140, 194). *E. coli* and *S. typhimurium* have two isozymes of catalase, KatE and KatG, that convert hydrogen peroxide to oxygen and water; however, only *katG* is regulated by OxyR. Glutathione reductase probably protects against oxidative stress by maintaining a pool of reduced glutathione that in turn can maintain the reduced state of cellular proteins. Alkyl hydroperoxide reductase converts lipid hydroperoxides, a consequence of hydrogen peroxide damage, into the corresponding nontoxic alcohols. In addition to these known enzymes, evidence indicates that the OxyR regulon is involved in DNA repair (83, 190).

The mechanism utilized by OxyR to sense oxidative damage and subsequently regulate gene expression is quite elegant. Purified OxyR binds specific sites (of approximately 45 bases) near target genes. The protein senses oxidative stress by a direct, reversible oxidation. When oxidized, OxyR undergoes a conformational change that activates transcription from the *katG* and *ahp* promoters by σ^{70} RNA polymerase (191). The conformational change presumably exposes or repositions a region of the protein that stimulates RNA polymerase. Curiously, the nature of the oxidative change remains unknown. An intramolecular disulfide bond is not the critical oxidation product because replacement of all but one of the OxyR cysteines with alanine will not affect the reversible activation (191). Loosely bound metals do not seem to be involved because EDTA-treated OxyR preparations retain redox-sensitive activity. However, OxyR may contain a tightly bound redox-sensitive metal.

The *soxRS* regulon provides a multilevel defense by increasing the synthesis of manganese-containing superoxide dismutase (*sodA*); the DNA-repair enzyme endonuclease IV (*nfo*); glucose-6-phosphate dehydrogenase (*zwf*); the *micF* antisense RNA, which suppresses synthesis of the OmpF outer-membrane protein; and at least six other proteins of unknown function (71, 72, 195). Superoxide-generating agents and nitric oxide induce the *soxRS* regulon (147), which is particularly important because macrophages, in which salmonellae can survive, produce nitric oxide as an essential killing mechanism. Although all of the published studies have used *E. coli*, *S. typhimurium* likely possesses a similar if not identical *soxRS* system.

SoxR senses an intracellular redox signal and activates transcription of *soxS*. Increased levels of SoxS will in turn activate transcription of the various *soxRS* regulon genes (119, 148, 206). The SoxS protein represses its own promoter, damping the system (149). Interestingly, activation of the *soxRS* system also increases resistance to many structurally unrelated antibiotics not implicated in redox stress. Resistance probably results from the decreased synthesis of the outer-membrane protein OmpF via *micF* antisense RNA destabilization of the *ompF* message (72).

The iron-regulatory protein Fur imposes another layer of control over *sodA*. In the presence of iron, H_2O_2 can give rise to the extremely reactive hydroxyl radical (OH•) through a Haber-Weiss Fenton type reaction (84). As a sensor of intracellular iron, Fur protein helps maintain a balance between intracellular iron concentration, superoxide flux, and H_2O_2 formation. Fur repression minimizes iron uptake during conditions of iron excess and will damp H_2O_2 production (and thus OH•) by repressing *sodA*. Conversely, during iron starvation when levels of FeSOD are low, derepression of MnSOD protects the cell from oxidative damage (142, 178, 193). Although these studies were carried out with *E. coli*, a similar situation is expected for *S. typhimurium*.

Heat-Shock Response and Thermotolerance

Cells exposed to temperatures above their optimum for growth synthesize increased amounts of heat-shock proteins. At the same time, the cell develops increased resistance to a higher heat challenge, a phenomenon called thermotolerance. Thermotolerance has been demonstrated in *S. typhimurium* (28, 128, 129), and two-dimensional analysis of heat-shock proteins has been reported (182). *S. typhimurium* produces heat-shock proteins in response to many stresses, but adaptations to adverse conditions other than starvation and acid shock do not produce cross-protection to heat. In *E. coli*, a 17- to 20-member subset of heat-shock proteins, including the chaperonins GroEL, GroES, DnaK, DnaJ, and GrpE as well as the housekeeping RNA polymerase sigma factor σ^{70} (RpoD), comprises a regulon whose temperature-dependent induction requires an alternate sigma factor, σ^{32} (RpoH). σ^{32} confers upon RNA polymerase (E) a promoter specificity for heat-shock promoters (208). Some of the resulting HSPs bind proteins denatured as a result of elevated temperature, protecting them from further degradation or facilitating their refolding (160). Other HSPs are involved in proteolysis (Lon and Clp proteases). Several genes involved in high-temperature resistance (*htr*) have been identified, although not all of them are thermally regulated (e.g. *htrB*, *htrD*, and *htrP*) (38).

Thermal induction of the Htp regulon requires the release of preformed σ^{32} from interactions with DnaKJ and GrpE so that more σ^{32} is available to express the target genes. Binding of DnaK/DnaJ to abnormally folded proteins that result from heat denaturation lowers the concentration of free DnaK/DnaJ available to bind and inactivate σ^{32}. The *rpoH* gene is also subject to control at the transcriptional and post-transcriptional levels. $E\sigma^{70}$ transcribes three *rpoH* promoters, P1, P4, and P5, whereas another promoter, P3, requires the sigma factor σ^{24}. σ^{24} is also required for heat-shock induction of *htrA* (*degP*), a periplasmic protease (121, 122) and for survival at extremely high temperatures. The *rpoH* transcript is also susceptible to secondary structure capable of blocking its translation. Upon temperature up-shift, this secondary structure may be transiently disrupted, which would permit higher rates of translation initiation.

Although many of the genes associated with thermotolerance in *S. typhimurium* have not been identified, the *S. typhimurium* system is probably analogous to the *E. coli* system. Genes that have been identified include *dnaJ*, *dnaK*, *lon*, and *htrA* (31, 174). Furthermore, increases in DnaK and GroEL have been documented (56; JW Foster, unpublished data). We discuss the association of some of these proteins with *S. typhimurium* virulence below.

Osmotic-Shock Response

S. typhimurium can experience osmotic stress in a variety of situations ranging from natural to host environments. When bacteria are subjected to changes in medium osmolarity, they will experience osmotic stress in which either the inward pressure on the cell membrane (high external osmolarity) or the outward pressure (low external osmolarity) are very high. To maintain its shape, the bacterium must maintain a minimum outward pressure called turgor (osmotic pressure 0.6 Mpa, 250 mosmol/kg). If external osmolarity increases, the cell possesses homeostatic mechanisms to transport compatible solutes (K^+, proline, glycine-betaine, glutamate, or trehalose) into the cell that will counterbalance the increase in external pressure. As a result of this adaptive response, cells maintain a constant volume over a wide range of external osmolarity.

Glycine-betaine and proline serve as osmoprotectants when added extracellularly, but glutamate and trehalose must be synthesized internally (35). Osmotically stressed *S. typhimurium* may synthesize glutamate by a route other than by glutamate synthase or glutamate dehydrogenase, the normal routes of glutamate synthesis (22). The *kdp* operon associated with turgor-induced uptake of potassium in *E. coli* has been identified by hybridization in *S. typhimurium* but is not yet characterized (204).

Regulation of the *proU* system for glycine-betaine and proline uptake identified in *S. typhimurium* has been linked to osmolarity (36, 188). The operon includes three genes, *proV*, *proW*, and *proX*, but its regulatory features have proven somewhat elusive. A *cis*-acting negative regulator identified within the first structural gene of the operon, *proV*, is required for the osmotic control (36, 37, 155). Several studies have found that *proU* is regulated by other environmental stresses (e.g. pH) and suggest a relationship with DNA supercoiling, although this remains to be proven (70, 91, 102).

The genes encoding the major outer membrane porins OmpC and OmpF are inversely regulated in response to changes in osmolarity (144). OmpC levels increase and OmpF levels decrease when cells are grown in high osmolarity. Regulation of this system is mediated in part by a two-component regulatory system involving a membrane sensor, EnvZ, and a signal-transducing regulatory protein, OmpR (35). EnvZ senses some consequence of high osmolarity and phosphorylates OmpR. Osmotic control of the *ompCF* genes appears to be based on the concentration of OmpR-P and the presence of multiple binding sites for this protein. Shifts in porin balance reflect an adaptation by organisms such as *S. typhimurium* to a transition from life in the animal gut to a free-living state. The high osmolarity of the gut would seem to favor *ompC* expression while *ompF* expression should predominate when outside the host. The smaller pore size of OmpC could aid in the exclusion of harmful molecules, such as bile salts, present in the gut. Studies showing that mutations in *ompR*, *ompC*, and *ompF* attenuate *S. typhimurium* point to a role for the products of these genes in *S. typhimurium* virulence (30, 39). Mutants defective in both OmpC and OmpF proved to be attenuated by the oral route but not by intravenous injection, which suggests a protective role for these proteins in the gut.

Resistance to Cationic Peptides

Various animal species synthesize small cationic peptides that have antimicrobial properties against gram-negative and gram-positive bacteria. Many of these peptides form voltage-gated channels in lipid bilayers, suggesting they kill bacteria by depolarizing the cytoplasmic membrane (32, 99). Following their ingestion, *S. typhimurium* encounter a class of antimicrobial peptides called cryptdins that are secreted by cells of the intestinal epithelium into the lumen (154). Later in infection, *S. typhimurium* reside in macrophage vacuolar compartments rich in another group of cationic peptides, defensins, that can make up to 5% of the protein (96, 115). Because *S. typhimurium* can successfully survive within these environments, resistance to antimicrobial peptides is required for pathogenesis.

Fields and coworkers (45) identified a locus that affected virulence and survival in macrophages but did not exhibit a nutritional requirement. The locus, *phoP*, originally described as necessary for phosphate starvation–induced expression of a nonspecific acid phosphatase (104), consists of an operon composed of two genes, *phoP* and *phoQ* (74, 136, 138). PhoP and PhoQ belong to the family of two-component regulatory systems; PhoP acts as a regulator and receiver and PhoQ as a sensor and transmitter (157). PhoPQ controls the expression of at least 40 polypeptides (25, 137). One role for the PhoPQ regulon in *S. typhimurium* virulence involves resistance to the antimicrobial peptides described above. Null mutations in *phoP* or *phoQ* increase susceptibility to defensins 1000-fold (44, 75). The PhoPQ-regulated proteins responsible for resistance are not known.

A second two-component regulatory system that controls resistance to antimicrobial peptides is *pmrA/pmrB* (173). Originally isolated as polymyxin E–resistant mutants (133), *pmrA* mutants have increased substitutions of lipid A phosphates that reduce the net negative charge on the LPS (198). This decreases binding of polymyxin or cationic peptides to the outer membrane, a first step in gaining access to the inner membrane. Another gene, designated *pmrD*, whose product likely interacts with PmrA or a PmrA-regulated gene product, confers polymyxin-resistance when present in multicopy (172). The nature of the resistance is still unclear.

Hypersensitivity to the antimicrobial peptide protamine has also been used to identify genetic loci involved in resistance to defensins (77). The *sap* (sensitivity to antimicrobial peptides) genes map to eight different chromosomal locations, and some of their functions have been deduced. For example, the *sapABCDF* operon encodes a new member of the ABC transporter family (158). Homology to the oligopeptide permease of *S. typhimurium* suggests the

SapABCDF transporter could transport cationic peptides into the cell for degradation by cellular proteases. Alternatively, the transporter could sense the presence of toxic peptides and set off a regulatory cascade that activates the relevant peptide-resistance determinants.

Another antimicrobial peptide-resistance mechanism may involve counteracting physiological imbalances produced as a consequence of the antimicrobial peptides. Affected processes include disruption of free-energy metabolism and potassium loss. Interestingly, SapG and SapJ are almost identical to TrkA and TrkH of *E. coli*, components of a low affinity K^+-transport system (159). Furthermore, the *sapABCDF* operon is identical to the *trkE* locus of *E. coli*, another component of the K^+-transport system (159). SapG, SapJ, and the SapABCDF transporter may function as a complex to mediate both peptide and K^+ transport. Defensin resistance could also be mediated by proteases at the cell surface. The isolation of a periplasmic peptidase (magaininase) that cleaves magainin provides evidence supporting this mechanism (73).

STRESS MANAGEMENT AND VIRULENCE

Escape from the Extracellular Environment

Salmonellae cause numerous diseases in animals and humans, including gastroenteritis, enteric (typhoid) fever, and bacteremia. Regardless of the disease manifestation, all *Salmonella* infections start with the invasion of columnar epithelial cells and/or M cells overlaying the Peyer's patches in the distal ileum and proximal colon of the bowel. This step makes *Salmonella* spp. inaccessible to the humoral response and qualifies as a survival strategy. Because epithelial cells are not normally phagocytic, *Salmonella* spp. initiate a novel endocytic/phagocytic event to potentiate uptake. They apparently do so by subverting existing host-signal-transduction pathways (65, 169).

Initial association between *Salmonella* spp. and the apical surface of the intestinal epithelial cell results in degeneration of the microvilli. Contact between bacterium and host cell triggers the synthesis of thin appendages on *Salmonella* spp. that may signal the host cell to begin engulfment and facilitate invasion (69). The host membrane begins to bleb and swell as a result of local reorganization of the cytoskeleton (49), and the resultant membrane ruffle encloses and internalizes the invading bacterium (60). Several different invasion genes (*inv*) and loci participate in this process (5, 63, 64). The *invA, E, G,* and *C* loci are involved in appendage formation. Mutants defective in these loci cannot induce ruffling and will not invade.

The region of the *S. typhimurium* chromosome between 57 and 60 min is rich with genes required for invasion. In addition to the *inv* cluster, the region contains a set of genes called *spa* (surface presentation of antigens) consisting of over 12 genes (76). The *spa* locus appears to encode a *sec*-independent protein secretory pathway required to process invasion proteins to the cell surface. Two of the Spa proteins, S and L, have domains typical of ATPases and may account for the energy-dependent requirement for invasion.

Whether or not new protein synthesis is required to efficiently invade host cells is controversial (48, 109, 127). However, growth state has a definite influence on invasion efficiency. Stationary-phase cells and aerobically grown cells are poorly invasive. Oxygen-limited growth conditions appear to be best for inducing invasiveness (42, 112, 177). An invasion locus called *hil*, mapping at 58 to 60 min, is involved with the anaerobic induction of invasion (113). Two genes within the *hil* locus have been identified, *orgA* (induced by anaerobiosis) (98) and *prgH* (a member of the PhoPQ regulon) (14). Iron limitation also plays a role in preparing the cell for invasion. One or more membrane-associated protein(s) required for invasion are induced in iron-rich conditions in a *fur*-independent manner (II Amin, L Burns-Keliher & R Curtiss III, submitted).

Invasion also involves subverting the host cell signaling pathways. *S. typhimurium* activates the epidermal growth factor (EGF) receptor tyrosine kinase activity, which is essential for cell responses, such as membrane ruffling, reorganization of actin microfilaments, and increases in calcium concentration, to EGF (65). Invasion stimulates a rapid increase in the levels of free intracellular calcium and causes profound rearrangements in the cytoskeleton of the host cell (49, 69, 156). However, a conflicting report could not reproduce the EGFR activation results. This study found that mitogenic activating protein kinase (MAPK) was activated by *S. typhimurium*, but the activation was independent of invasion (170). Consequently, the signals used by *S. typhimurium* to invade epithelial cells remain a mystery.

Following invasion, intracellular replication is essential for the virulence of *S. typhimurium*. Leung & Finlay (117) have identified several prototrophic mutants (designated *rep*) defective in intracellular replication, not invasion, that are attenuated in mice. Once within epithelial cells, *S. typhimurium* remains within a phagocytic vacuole that fuses with lysosomal vacuoles (47, 67). Intracellular replication depends on the formation of filamentous structures emanating from the vacuole and sometimes connecting different groups of bacteria (67). The *rep* mutants did not form these structures. Acidification of the vacuole is one signal necessary for activating the bacterial genes required to induce filamentation. These filaments may establish interactions with other organelles capable of providing nutrients to the bacterium.

By an unknown mechanism, *S. typhimurium* eventually moves to the basolateral side of epithelial cells and exits. From this point *S. typhimurium* travels to its principal target of infection, the macrophage.

Macrophage Survival

The ability of microorganisms to survive within phagocytic cells is a potent method of preventing clearance from the host. Survival within the macrophage is essential for *S. typhimurium* virulence, although little is known of the mechanisms by which *S. typhimurium* survive after phagocytosis (45, 168). *S. typhimurium* are thought to be sequestered within macrophages at all stages of the infection. However, one study found that 4 h after intraperitoneal injection, most of bacteria found in the spleen and liver were in polymorphonuclear leucocytes (PMNs) (40). Most studies, however, have focused upon survival of *S. typhimurium* within macrophages.

S. typhimurium can use two pathways for internalization into macrophages, normal phagocytosis and the *S. typhimurium*–induced pathway described for epithelial cells above. When these bacteria enter cells, they are enclosed within spacious phagosomes (2–6 µm diameter) large enough for the organism to swim freely (3). This situation contrasts with that of other organisms such as *Yersinia enterocholitica*, which enter macrophages within more tightly opposed phagosomes. Phagosomes containing *S. typhimurium* are delayed in acidification relative to phagosomes containing dead bacteria, but the eventual phagosome acidification results in increased transcription of some virulence (*pag*) genes (4).

Many genes associated with survival within macrophages have been identified (13, 15, 43, 45, 120). These include known genes such as *htrA, purD, phoPQ, nagA,* and *fliD*. Others, designated *ims* (impaired macrophage survival), are not well characterized. The best-characterized system to date is the *phoPQ* system. One reason why *phoPQ* mutants survive macrophage invasion poorly is a diminished resistance to defensins. Several genes regulated by PhoPQ have been identified using *lacZ* or *phoA* fusion studies (14, 15, 136, 139). PhoP activated genes (*pag*) and PhoP repressed genes (*prg*) include *pagC, pagD, pagK, pagM,* and *prgH*, which are required for virulence in mice (14, 136, 139). Other genes regulated by PhoPQ but not implicated in macrophage survival include *pagA* and *pagB*, which along with *pagC* are induced by intramacrophage acid pH (4). Most if not all of the Pag and Prg proteins are membrane proteins. However, none of these genes have been implicated in defensin resistance, suggesting other roles for PhoPQ regulated genes in surviving macrophages.

Considerable debate focuses on whether *S. typhimurium* survive within macrophages by preventing phagosome-lysosome fusion, as does *Legionella pneumophila*, thereby preventing exposure to toxic lysosomal contents (4, 26, 93). A mutant of *S. typhimurium* lacking the inhibitory function for phagosome-lysosome fusion has been reported (94). However, a study that tracked lysosomal membrane glycoproteins (lgps) found that *S. typhimurium* resides in lgp-containing vacuoles, suggesting fusion (67). This point still awaits consensus.

One stress *S. typhimurium* predictably encounters within macrophage phagolysosomes is oxidative damage. A light-emitting *lux* fusion to a H_2O_2-inducible *S. typhimurium* gene showed that *S. typhimurium* experiences oxidative stress in the macrophage (61). The macrophage-sensitive nature of *recA* mutants defective in DNA repair has been attributed to increased susceptibility to oxidative damage (27). Indeed, elevated levels of MnSOD (*sodA*) produced in response to iron limitation or a Fur⁻ phenotype will enhance macrophage survival (196). However, *sodA* mutants are only slightly attenuated in terms of virulence in mice. A possible explanation for this apparent discrepancy is that *S. typhimurium* reportedly elicits little to no oxidative burst upon entry into phagocytic cells (203). *S. typhimurium* might prevent the oxidative burst or selectively invade a set of phagocytic cells with diminished capacity for activating oxidative killing, as was found for resident macrophages in the liver and spleen (116). A separate locus involved in resistance to killing by polymorphonuclear leukocytes and hydrogen peroxide has been identified and designated *pss* (96 min). It encodes a 59-kDa outer membrane protein of unknown function that is strongly expressed in stationary phase (187).

Abshire & Neidhardt (2) suggest the presence of two populations of intracellular *S. typhimurium* in macrophages, one static and one growing rapidly. Rapidly growing cells of *S. typhimurium* contain roughly equal amounts of L7 and L12 ribosomal proteins, but slow-growing cells have much more L7 (the N-terminal acylated form of L12) than L12. Labeled bacteria prepared from macrophage-infected cells displayed an L7/L12 ratio characteristic of rapidly growing cells (40 min generation time), but when viable counts were measured, the organisms appeared to be growing very slowly (perhaps a 2-h generation time). The addition of a bactericidal antibiotic (ampicillin) to macrophage-infected cells did not dramatically affect cell numbers, suggesting most cells are not growing (but are viable) while a small proportion of cells does grow rapidly and accounts for the labeling results. One of the keys to intracellular survival may be to differentiate into this nongrowing state, which may be more resistant to a variety of environmental stresses.

In Vivo Environmental Cues

The communication between pathogen and host is a complex series of sense and respond maneuvers each designed to gain advantage in the struggle. The difficulties encountered when studying host-parasite relations of any organism include a basic ignorance of the host intravacuole composition and which aspects of that environment are threatening to the microorganism. One way to explore both of those questions is to identify what bacterial genes respond as the organism grows in vivo. Based upon the observation that *S. typhimurium*–containing vacuoles can induce *iroA-lac*, *mgtB-lac*, and *cadA-lac* gene fusions of *S. typhimurium*, Garcia-del Portillo et al (66) conclude the epithelial vacuolar environment is acidic and low in iron and magnesium but contains lysine. Alpuche-Aranda et al (4) have shown that the PhoP/PhoQ system is activated upon entry of *S. typhimurium* into macrophages but not epithelial cells and that a potential signal for that activation is the acid environment of the phagolysosome. Other conditions that may be sensed by the PhoPQ system include phosphate, carbon, and nitrogen limitations. Defensins apparently do not serve to induce *pag* loci (201).

Two-dimensional SDS-PAGE analyses have shown that *S. typhimurium* changes its pattern of gene expression in response to conditions found within the macrophage (1, 25). Approximately 40 proteins are produced and 100 proteins disappear while the bacteria grows in this environment. The classes of stress-induced proteins that are synthesized confirm that *S. typhimurium* experiences a variety of stresses while residing within macrophages, including low pH, oxidative stress, nutrient limitations (carbon, phosphate, nitrogen, sulfur), high osmolarity, and high temperature (1). However, no set of stress-induced proteins was induced in its entirety within the macrophage environment. The microbial response to the macrophage was not a simple sum of stress responses displayed during extracellular growth. These results could be explained by the use of different growth media to examine extracellular (minimal glucose) vs intracellular (tissue culture medium) stress responses. Alternatively, a special set of circumstances may occur within the macrophage that cannot be reproduced in vitro.

Mahan et al (132) have devised a very clever way to reveal *S. typhimurium* genes required for in vivo growth. Called in vivo expression technology (IVET), the strategy is based on the use of a *pur* mutant of *S. typhimurium* that grows poorly in mice because of the auxotrophic requirement for purine. In vivo induced (*ivi*) genes were identified by cloning fragments of *S. typhimurium* chromosomal DNA in front of promoterless *purA*$^+$ and *lacZY*$^+$ genes situated in tandem on a suicide plasmid. Properly positioned *S. typhimurium* promoters would drive expression of the tandem *purA-lacZY* genes and thus restore virulence to the mutant strain. Eighty-six percent of the bacterial survivors extracted from the livers of mice injected with a pool of integrated *purA-lac* fusion strains were Pur$^+$ and Lac$^+$. However, 5% of the survivors were Lac$^-$ on MacConkey lactose indicator medium. This suggests the fusions were induced in vivo (allowing for survival) but were not induced in vitro in rich medium. Two of the *ivi* genes were previously unrecognized and may represent genes induced specifically in response to in vivo signals difficult to reproduce in vitro. One potential criticism of this procedure is that for a cloned promoter to enhance survival of the organism in mice, the promoter may need to be expressed throughout the entire voyage of the organism through the mouse. Thus, genes specifically induced within macrophages but not epithelial cells could be missed. This criticism has been addressed recently. An IVET vector with the promoterless *purA* gene replaced by a promoterless *cat* gene was used to select *S. typhimurium* genes induced specifically in macrophages (131).

Salmonella Plasmid Virulence (spv) Genes

Several *Salmonella* species possess a high-molecular-weight plasmid (ranging from 50 to 90 kb) that is required for these serovars to cause systemic disease (79, 80). Although these virulence plasmids differ in size, they possess a highly conserved 8-kb segment that restores virulence to plasmid-cured strains. This region encodes five genes, *spvR* and the *spvABCD* operon (12, 78, 107, 108, 145). The *spv* genes are induced within macrophages, but whether a specific macrophage induction signal is required or if activation occurs in response to the numerous stresses that an intracellular compartment presents to the bacterium is not clear (46, 166). As stated above, the *spvABCD* operon is induced during C- (43, 199), P-, and N-starvation (A Turk, P Gulig & M Spector, unpublished results) in an *rpoS*-dependent manner. RpoS appears to control *spvR* expression, which in turn regulates *spvABCD* (106, 108). In addition, *spv* gene expression also appears to be negatively controlled by cAMP-CRP (150). Along with starvation, iron limitation and lowered pH induce synthesis of SpvR (186) as well as SpvA, SpvB, and SpvC (199). A member of the LysR family of regulators, SpvR is a positive regulator of the *spvABCD* operon as well as *spvR* itself (80). SpvA negatively regulates *spvR* expression and, therefore, *spvABCD* expression in a negative-feedback mechanism (186).

Although all the *spv* genes have been cloned and sequenced and their amino acid sequences deduced, the exact functions of these proteins in virulence have yet to be determined. However, plasmids carrying the *spvR spvABCD* genes increase the growth rate of *S. typhimurium* in mice as compared with plasmid-cured strains (81, 168). Regulation of *spv* genes by environmental stresses (e.g. starvation, low pH, iron limitation, etc) may reflect an important host-pathogen relationship (80, 163). That is, as the host tries to prevent the growth of *S. typhimurium* within the phagolysosome, the bacteria counter by expressing proteins that increase growth rate and enhance survival within the host.

Salmonella Serum Resistance

S. typhimurium isolated from blood cultures are almost always serum resistant. Components of the bacterial cell surface, including capsules, LPS, and outer membrane proteins, affect the interaction between serum complement and various gram-negative bacteria. One determinant of serum resistance in *S. typhimurium* is the length of the LPS. Because the C3 component of complement deposits on the long LPS O-antigen side chain, the terminal complement C5b-9 membrane attack complex (MAC) forms too far from the hydrophobic regions of the outer membrane. The MAC is then shed without disrupting membrane integrity (97). The virulence plasmid of *S. typhimurium* contains several genes whose products also contribute to serum resistance. They include a *traT*-like outer membrane protein (167), the *rsk* locus composed of a series of imperfectly repeated 10-bp sequences (200), and a third gene that restores complete O-antigen synthesis in certain rough strains of *S. typhimurium* (103). A fourth locus, *rck,* encodes a 17-kDa outer membrane protein (82) that interferes with MAC formation by altering C9 polymerization with C5b-9 (82, 89, 90). As an outer membrane protein, Rck presumably handles C5b-9 complexes that are not bound by the LPS and manage to insert into the membrane.

The Salmonella mvi Mouse Virulence Genes

An interesting locus called *mviA* regulates *S. typhimurium* virulence during infection of Ity^s mice (19). The major in vivo effect of the *Ity* locus is on the growth of *S. typhimurium,* although the mechanism is not known (17, 123). The *mviA* product is a 38-kDa protein with significant homology to the response regulator family of bacterial transcriptional regulatory proteins (WH Benjamin & E Swords, personal communication; 157). Regulation of virulence appears to involve collaboration between MviA and one or more other regulatory proteins. The Mvi system is predicted to influence virulence by controlling the rate of *S. typhimurium* growth in vivo. Bacteria that can grow more rapidly in vivo would be more virulent than slower growing cells. A 55-kDa periplasmic protein whose expression is MviA-dependent has been identified and designated MrpA (MviA regulated protein). Although virulence has been associated with the presence of this protein, its function is unknown (E Swords & WH Benjamin, personal communication). A correlation between the *mviA* system and stress responses was recently noted in that at least one of the RpoS-dependent acid-shock proteins described above is also controlled by *mviA* (JW Foster, S Duncan, E Swords & WH Benjamin, unpublished observation). This observation is significant because MviA may influence in vivo stress resistance or regulate genes required for in vivo growth by sensing acid conditions.

CONCLUSIONS

S. typhimurium encounters many stresses during its voyages between natural and host environments, as summarized in Figure 1. Yet the organism has proven remarkably versatile in its resilience to these stresses. For its defense, this organism has evolved a complex interconnected series of stress management response systems. An array of regulatory proteins provides overlapping control that links many of these systems. The organism then incorporates the individual stress response systems into a multidefense strategy for invading and surviving within a host. A successfully fought campaign by the microbe ultimately provides an extremely rich, almost stress-free environment in the form of a dead host. Much remains to be learned about how many of these survival strategies work.

ACKNOWLEDGMENTS

Studies by the authors presented in this review were supported by grants from the National Institutes of Health (GM48017 to JWF and GM47628 to MPS) and the National Science Foundation (DCB-89-04839 to JWF).

> Any *Annual Review* chapter, as well as any article cited in an *Annual Review* chapter, may be purchased from the Annual Reviews Preprints and Reprints service.
> 1-800-347-8007; 415-259-5017; email: arpr@class.org

Literature Cited

1. Abshire KZ, Neidhardt FC. 1993. Analysis of proteins synthesized by *Salmonella typhimurium* during growth within a host macrophage. *J. Bacteriol.* 175:3734–43
2. Abshire KZ, Neidhardt FC. 1993. Growth rate paradox of *Salmonella typhimurium* within host macrophage. *J. Bacteriol.* 175:3744–48
3. Alpuche-Aranda CM, Racoosin EL, Swanson JA, Miller SI. 1994. *Salmonella* stimulate macrophage macropinocytosis and persist within spacious phagosomes. *J. Exp. Med.* 179:601–8
4. Alpuche-Aranda CM, Swanson JA, Loomis WP, Miller SI. 1992. *Salmonella typhimurium* activates virulence gene transcription within acidified macrophage phagosomes. *Proc. Natl. Acad. Sci. USA* 89:10079–83
5. Altmeyer RM, McNern JK, Bossio JC, Rossenshine I, Finlay BB, et al. 1993. Cloning and molecular characterization of a gene involved in *Salmonella* adherence and invasion of cultured epithelial cells. *Mol. Microbiol.* 7:89–98
6. Deleted in proof
7. Anderson J, Delihas N. 1990. micF RNA binds to the 5' end of ompF mRNA and to a protein from *Escherichia coli*. *Biochemistry* 29:9249–56
8. Aznar R, Amaro C, Alcaide E, Lemos ML. 1989. Siderophore production by environmental strains of *Salmonella* species. *FEMS Microbiol. Lett.* 57:7–12
9. Babior BM. 1992. The respiratory burst oxidase. *Adv. Enzymol. Relat. Areas Mol. Biol.* 65:49–95
10. Bagg A, Neilands JB. 1987. Ferric uptake regulation protein acts as a repressor, employing iron (II) as a cofactor to bind the operator of an iron transport operon in *Escherichia coli*. *Biochemistry* 26:5471–77
11. Barker J, Brown MRW. 1994. Trojan horses of the microbial world: protozoa and survival of bacterial pathogens in the environment. *Microbiology* 140:1253–59
12. Barrow PA, Lovell MA. 1989. Functional homolog of virulence plasmids in *Salmonella gallinarum, S. pullorum,* and *S typhimurium*. *Infect. Immun.* 57:3136–41
13. Baumler AJ, Kusters JG, Stojiljkovic I, Heffron F. 1994. *Salmonella typhimurium* loci involved in survival within macrophages. *Infect. Immun.* 62:1623–30
14. Behlau I, Miller SI. 1993. A PhoP-repressed gene promotes *Salmonella typhimurium* invasion of epithelial cells. *J. Bacteriol.* 175:4475–84
15. Belden WJ, Miller SI. 1994. Further characterization of the PhoP regulon: identification of new PhoP-activated virulence loci. *Infect. Immun.* 62:5095–101
16. Bell PE, Nau CD, Brown JT, Konisky J, Kadner R. 1990. Genetic suppression demonstrates interaction of TonB protein with outer membrane transport proteins in *Escherichia coli*. *J. Bacteriol.* 172:3826–29
17. Benjamin WH, Hall P, Roberts SJ, Briles DE. 1990. The primary effect of the *Ity* locus is on the growth rate of *Salmonella typhimurium* that are relatively protected from killing. *J. Immunol.* 144:3143–51

18. Benjamin WH, Turnbough CLJ, Posey BSJ, Briles DE. 1985. The ability of *Salmonella typhimurium* to produce the siderophore enterobactin is not a virulence factor in mouse typhoid. *Infect. Immun.* 50:392–97
19. Benjamin WH, Yother J, Hall P, Briles DE. 1991. The *Salmonella typhimurium* locus *mviA* regulates virulence in *Ity*^r but not *Ity*^s mice: functional *mviA* results in avirulence; mutant (nonfunctional) *mviA* results in virulence. *J. Exp. Med.* 174:1073–83
20. Bennett RL, Rothfield LI. 1976. Genetic and physiological regulation of intrinsic proteins of the outer membrane of *Salmonella typhimurium*. *J. Bacteriol.* 127:498–504
21. Beutler B, Cerami A. 1987. Catchetin: more than a tumor necrosis factor. *N. Engl. J. Med.* 316:379–85
22. Botsford JL, Alvarez M, Hernandez R, Nichols R. 1994. Accumulation of glutamate by *Salmonella typhimurium* in response to osmotic stress. *Appl. Environ. Microbiol.* 60:2568–74
23. Braun V, Gunter K, Hantke K. 1991. Transport of iron across the outer membrane. *Biometals* 4:14–22
24. Briat J-F. 1992. Iron assimilation and storage in procaryotes. *Microbiology* 138:2475–83
25. Buchmeier NA, Heffron F. 1990. Induction of *Salmonella* stress proteins upon infection of macrophages. *Science* 248:730–32
26. Buchmeier NA, Heffron F. 1991. Inhibition of macrophage phagosome–lysosome fusion by *Salmonella typhimurium*. *Infect. Immun.* 59:2232–38
27. Buchmeier NA, Nathan C, Lipps CJ, So MH, Heffron F. 1993. Recombination-deficient mutants of *Salmonella typhimurium* are avirulent and sensitive to the oxidative burst of macrophages. *Mol. Microbiol.* 7:933–36
28. Bunning VK, Crawford RG, Tierney JT, Peeler JT. 1990. Thermotolerance of *Listeria monocytogenes* and *Salmonella typhimurium* after sublethal heat shock. *Appl. Environ. Microbiol.* 56:3216–19
29. Cabello F, Hormaeche C, Mastroeni P, Bonina L, eds. 1993. *Biology of Salmonella*. New York: Plenum
30. Chatfield SN, Dorman CJ, Hayward C, Dougan G. 1991. Role of *ompR*-dependent genes in *Salmonella typhimurium* virulence: Mutants deficient in both OmpC and OmpF are attenuated in vivo. *Infect. Immun.* 59:449–52
31. Chatfield SN, Strahan K, Pickard D, Charles IG, Hormaeche CE, et al. 1992. Evaluation of *Salmonella typhimurium* strain harboring defined mutations in *htrA* and *aroA* in the murine salmonellosis model. *Microb. Pathog.* 12:145–51
32. Christensen B, Fink J, Merrifield RB, Mauzerall D. 1988. Channel-forming properties of cecropins and related model compounds incorporated into planar lipid membranes. *Proc. Natl. Acad. Sci. USA* 85:5072–76
33. Christman M, Morgan R, Jacobson F, Ames B. 1985. Positive control of a regulon for defenses against oxidative stress and some heat shock proteins in Salmonella typhimurium. *Cell* 41:753–62
34. Cohen SP, McMurry LM, Levy SB. 1988. *marA* locus causes decreased expression of OmpF porin in multiple-antibiotic-resistant (Mar) mutants of *Escherichia coli*. *J. Bacteriol.* 170:5416–22
35. Csonka LN, Hanson AD. 1991. Procaryotic osmoregulation: genetics and physiology. *Annu. Rev. Microbiol.* 45:569–81
36. Csonka LN, Ikeda TP, Fletcher SA, Kustu S. 1994. The accumulation of glutamate is necessary for optimal growth of *Salmonella typhimurium* in media of high osmolality but not induction of the *proU* operon. *J. Bacteriol.* 176:6324–33
37. Dattananda CS, Rajkumari K, Gowrishankar J. 1991. Multiple mechanisms contribute to osmotic inducibility of *proU* operon expression in *Escherichia coli*: demonstration of two osmoresponsive promoters and of a negative regulatory element within the first structural gene. *J. Bacteriol.* 173:7481–90
38. Delaney JM, Wall D, Georgopolous C. 1993. Molecular characterization of the *Escherichia coli htrD* gene: cloning, sequence, regulation and involvement with cytochrome *d* oxidase. *J. Bacteriol.* 175:166–75
39. Dorman CJ, Chatfield S, Higgens CF, Hayward C, Dougan G. 1989. Characterization of porin and *ompR* mutants of a virulent strain of *Salmonella typhimurium*: *ompR* mutants are attenuated in vivo. *Infect. Immun.* 57:2136–40
40. Dunlap NE, Benjamin WB, Berry AK, Eldridge JH, Briles DE. 1992. A 'safe-site' for *Salmonella typhimurium* is within splenic polymorphonuclear cells. *Microb. Pathog.* 13:181–90
41. Ernst JF, Bennett RL, Rothfield LI. 1978. Constitutive expression of the iron enterochelin and ferrichrome uptake

systems in a mutant strain of *Salmonella typhimurium*. *J. Bacteriol.* 135:928–34
42. Ernst RK, Dombroski DM, Merrick J. 1990. Anaerobiosis, type 1 fimbriae and growth phase are factors that affect invasion of HEp-2 cells by *Salmonella typhimurium*. *Infect. Immun.* 58:2014–16
43. Fang FC, Libby SJ, Buchmeier NA, Loewen PC, Switala J, et al. 1992. The alternative sigma factor (RpoS) regulates *Salmonella* virulence. *Proc. Natl. Acad. Sci. USA* 89:11978–82
44. Fields PI, Groisman EA, Heffron F. 1989. A *Salmonella* locus that controls resistance to microbicidal proteins from phagocytic cells. *Science* 243:1059–62
45. Fields PI, Swanson RV, Haidaris CG, Heffron R. 1986. Mutants of *Salmonella typhimurium* that cannot survive within the macrophage are avirulent. *Proc. Natl. Acad. Sci. USA* 83:5189–93
46. Fierer J, Eckmann L, Fang R, Pfeiffer C, Finlay BB, et al. 1993. Expression of the *Salmonella* virulence plasmid gene *spvB* in cultured macrophages and nonphagocytic cells. *Infect. Immun.* 61:5231–36
47. Finlay BB, Falkow S. 1989. *Salmonella* as an intracellular parasite. *Mol. Microbiol.* 3:1833–41
48. Finlay BB, Heffron F, Falkow S. 1989. Epithelial cell surfaces induce *Salmonella* proteins required for bacterial adherence and invasion. *Science* 243:940–43
49. Finlay BB, Ruschkowski S, Dedhar S. 1991. Cytoskeletal rearrangements accompanying *Salmonella* entry into epithelial cells. *J. Cell. Sci.* 99:283–96
50. Fischer B, Gunter K, Braun V. 1989. Involvement of ExbB and TonB in transport across the outer membrane of *Escherichia coli*: phenotypic complementation of *exb* mutants by overexpressed *tonB* and physical stabilization of TonB by ExbB. *J. Bacteriol.* 171:5127–34
51. Foster JW. 1993. The acid tolerance response of *Salmonella typhimurium* involves transient synthesis of key acid shock proteins. *J. Bacteriol.* 175:1981–87
52. Foster JW. 1991. *Salmonella* acid shock proteins are required for the adaptive acid tolerance response. *J. Bacteriol.* 173:6896–902
53. Foster JW, Bearson B. 1994. Acid sensitive mutants of *Salmonella typhimurium* identified through a dinitrophenol selection strategy. *J. Bacteriol.* 176:2596–602
54. Foster JW, Hall HK. 1992. Effect of *Salmonella typhimurium* ferric uptake regulator *(fur)* mutations on iron and pH-regulated protein synthesis. *J. Bacteriol.* 174:4317–23
55. Foster JW, Hall HK. 1991. Inducible pH homeostasis and the acid tolerance response of *Salmonella typhimurium*. *J. Bacteriol.* 173:5129–35
56. Foster JW, Hall HK. 1990. Adaptive acidification tolerance response of *Salmonella typhimurium*. *J. Bacteriol.* 172:771–78
57. Foster JW, Park YK, Bang IS, Karem K, Betts H, et al. 1994. Regulatory circuits involved with pH-regulated gene expression in *Salmonella typhimurium*. *Microbiology* 140:341–52
58. Deleted in proof
59. Foster JW, Spector MP. 1986. Phosphate-starvation regulon of *Salmonella typhimurium*. *J. Bacteriol.* 166:666–69
60. Francis CL, Ryan TA, Jones BD, Smith SJ, Falkow S. 1993. Ruffles induced by *Salmonella* and other stimuli direct macropinocytosis of bacteria. *Nature* 364:639–42
61. Francis KP, Gallagher MP. 1993. Light emission from a mud-lux transcriptional fusion in *Salmonella typhimurium* is stimulated by hydrogen-peroxide and by interaction with the mouse macrophage cell-line J774.2. *Infect. Immun.* 61:640–49
62. Fridovich I. 1983. Superoxide radical: an endogenous toxicant. *Annu. Rev. Pharmacol. Toxicol.* 23:239–57
63. Galan JE, Curtiss R. 1990. Expression of *Salmonella typhimurium* genes required for invasion is regulated by changes in DNA supercoiling. *Infect. Immun.* 58:1876–85
64. Galan JE, Curtiss R. 1989. Cloning and molecular characterization of genes whose products allow *Salmonella typhimurium* to penetrate tissue culture cells. *Proc. Natl. Acad. Sci. USA* 86:6383–87
65. Galan JE, Pace J, Hayman MJ. 1992. Involvement of the epidermal growth factor receptor in the invasion of cultured mammalian cells by *Salmonella typhimurium*. *Nature* 357:588–89
66. Garcia-del Portillo F, Foster JW, Maguire ME, Finlay BB. 1992. Characterization of *Salmonella typhimurium*-containing vacuoles within MDCK epithelial cells. *Mol. Microbiol.* 6:3289–97
67. Garcia-del Portillo F, Zwick MB, Leung KY, Finlay BB. 1993. *Salmonella* induces the formation of filamentous structures containing lysosomal membrane glycoproteins in epithelial cells. *Proc. Natl. Acad. Sci. USA* 90:10544–48

68. Gentry DR, Hernandez VJ, Nguyen LH, Jensen DB, Cashel M. 1993. Synthesis of stationary phase sigma factor σ^s is positively regulated by ppGpp. *J. Bacteriol.* 175:7982–89
69. Ginnochio C, Pace J, Galan JE. 1992. Identification and molecular characterization of a *Salmonella typhimurium* gene involved in triggering the internalization of *Salmonella* into cultured epithelial cells. *Proc. Natl. Acad. Sci. USA* 89:5976–80
70. Graeme-Cook KA, May G, Bremer E, Higgins CF. 1989. Osmotic regulation of protein expression: a role for DNA supercoiling. *Mol. Microbiol.* 3:1287–94
71. Greenberg JT, Demple B. 1989. A global response induced in *Escherichia coli* by redox cycling agents overlaps with that induced by peroxide stress. *J. Bacteriol.* 171:3933–39
72. Greenberg JT, Monach P, Chou JH, Josephy PD, Demple B. 1990. Positive control of global antioxidant defense regulon activated by superoxide-generating agents in *Escherichia coli. Proc. Natl. Acad. Sci. USA* 87:6181–85
73. Groisman EA. 1994. How bacteria resist killing by host-defense peptides. *Trends Microbiol.* 2:444–49
74. Groisman EA, Chiao E, Lipps CJ, Heffron F. 1989. *Salmonella typhimurium phoP* virulence gene is a transcriptional regulator. *Proc. Natl. Acad. Sci. USA* 86:7077–81
75. Groisman EA, Heffron F, Solomon F. 1992. Molecular genetic analysis of the *Escherichia coli phoP* locus. *J. Bacteriol.* 174:486–91
76. Groisman EA, Ochman H. 1993. Cognate gene clusters govern invasion of host epithelial cells by *Salmonella typhimurium* and *Shigella flexneri. EMBO J.* 12:3779–87
77. Groisman EA, Parra-Lopez CA, Salcedo M, Lipps CJ, Heffron F. 1992. Resistance to host antimicrobial peptides is necessary for *Salmonella* virulence. *Proc. Natl. Acad. Sci. USA* 89:11939–43
78. Gulig PA, Caldwell AL, Chiodo V. 1992. Identification, genetic analysis and DNA sequence of a 7.8-kb virulence region of the *Salmonella typhimurium* virulence plasmid. *Mol. Microbiol.* 6:1395–11
79. Gulig PA, Curtiss R III. 1987. Plasmid-associated virulence of *Salmonella typhimurium. J. Bacteriol.* 55:2891–901
80. Gulig PA, Danbara H, Guiney DG, Lax AJ, Norel F, et al. 1993. Molecular analysis of *spv* virulence genes of the *Salmonella* virulence plasmids. *Mol. Microbiol.* 7:825–30
81. Gulig PA, Doyle TJ. 1993. The *Salmonella typhimurium* virulence plasmid increases the growth rate of salmonellae in mice. *Infect. Immun.* 61:504–11
82. Hackett J, Wyk P, Reeves P, Mathan V. 1987. Mediation of serum resistance in *Salmonella typhimurium* by an 11 kilodalton polypeptide encoded by the cryptic plasmid. *J. Inf. Dis.* 155:540–49
83. Hagensee ME, Moses RB. 1989. Multiple pathways for repair of hydrogen peroxide–induced DNA damage in *Escherichia coli. J. Bacteriol.* 171:991–95
84. Halliwell B, Gutteridge JMC. 1984. Oxygen toxicity, oxygen radicals, transition metals and disease. *Biochemistry* 219:1–14
85. Hantke K. 1987. Selection procedure for deregulated iron transport mutants (*fur*) in *Escherichia coli: fur* not only affects iron metabolism. *Mol. Gen. Genet.* 210:135–39
86. Hantke K. 1982. Negative control of iron uptake in *Escherichia coli. FEMS Microbiol. Lett.* 15:83–86
87. Hantke K. 1981. Regulation of ferric iron transport in *Escherichia coli* K12: isolation of a constitutive mutant. *Mol. Gen. Genet.* 182:288–92
88. Harder W, Dijkhuizen L. 1983. Physiological responses to nutrient limitation. *Annu. Rev. Microbiol.* 37:1–23
89. Heffernan EJ, Harwood J, Fierer J, Guiney D. 1992. The *Salmonella typhimurium* virulence plasmid complement resistance gene *rck* is homologous to a family of virulence related outer membrane protein genes, including *pagC* and *ail. J. Bacteriol.* 174:84–91
90. Heffernan EJ, Reed S, Hackett J, Fierer J, Roudier C, et al. 1992. Mechanism of resistance to complement-mediated killing of bacteria encoded by the *Salmonella typhimurium* virulence plasmid gene *rck. J. Clin. Invest.* 90:953–64
91. Higgins CF, Dorman CJ, Stirling DA, Waddel L, Booth LR, et al. 1988. A physiological role for DNA supercoiling in the osmotic regulation of gene expression in Salmonella typhimurium and E. coli. *Cell* 52:569–84
92. Huttanen CN. 1975. Use of pH gradient plates for increasing the acid tolerance of *Salmonella. Appl. Environ. Microbiol.* 29:309–12
93. Ishibashi Y, Arai T. 1990. Specific inhibition of phagosome-lysosome fusion in murine macrophages mediated by *Salmonella typhimurium* infection. *FEMS Immunol. Med. Microbiol.* 64:35–44

94. Ishibashi Y, Nobuta K, Toshihiko A. 1992. Mutant of *Salmonella typhimurium* lacking the inhibitory function for phagosome-lysosome fusion in murine macrophages. *Microb. Pathog.* 13:317–23
95. Deleted in proof
96. Joiner KA, Ganz T, Albert T, Rothosen D. 1989. The opsonizing ligand on *Salmonella typhimurium* influences the incorporation of specific, but not azurophil, granule constituent into neutrophil phagosomes. *J. Cell Biol.* 109:2771–82
97. Joiner KA, Hammer CH, Brown EJ, Frank MM. 1982. Studies on the mechanism of bacterial resistance to complement-mediated killing. II. C8 and C9 release C5b67 from the surface of *Salmonella minnesota* S218 because the terminal complex does not insert into the bacterial membrane. *J. Exp. Med.* 155:808–19
98. Jones BD, Falkow S. 1994. Identification and characterization of a *Salmonella typhimurium* oxygen regulated gene required for bacterial internalization. *Infect. Immun.* 62:3745–52
99. Kagan BL, Selsted ME, Ganz T, Albert J, Lehrer RI. 1990. Antimicrobial defensin peptides form voltage dependent ion permeable channels in planar lipid bilayer membranes. *Proc. Natl. Acad. Sci. USA* 87:210–14
100. Kampfenkel K, Braun V. 1993. Topology of the ExbB protein in the cytoplasmic membrane of *Escherichia coli*. *J. Biol. Chem.* 268:6050–57
101. Kappus H, Sies H. 1981. Toxic drug effects associated with oxygen metabolism: redox cycling and lipid peroxidation. *Experimentia* 37:1233–58
102. Karem K, Foster JW. 1993. The influence of DNA topology on the environmental regulation of a pH-regulated locus in *Salmonella typhimurium*. *Mol. Microbiol.* 10:75–86
103. Kawahara K, Hamaoka T, Suzuki S, Nakamura M, Murayama SY, et al. 1989. Lipopolysaccharide alteration mediated by the virulence plasmid of *Salmonella*. *Microb. Pathog.* 7:195–202
104. Kier LD, Weppelman RM, Ames BN. 1979. Regulation of a nonspecific acid phosphatase in *Salmonella*: *phoN* and *phoP* genes. *J. Bacteriol.* 138:155–61
105. Koch AL. 1971. The adaptive response of *Escherichia coli* to a feast or famine existence. *Adv. Microb. Physiol.* 6:147–217
106. Kowarz L, Coynault C, Robbe-Saule V, Norel F. 1994. The *Salmonella typhimurium katF* (*rpoS*) gene: cloning, nucleotide sequence and regulation of *spvR* and *spvABCD* virulence plasmid genes. *J. Bacteriol.* 176:6853–60
107. Krause M, Fang FC, Guiney DG. 1992. Regulation of plasmid virulence gene expression in *Salmonella dublin* involves an unusual operon structure. *J. Bacteriol.* 174:4482–89
108. Krause M, Harwood J, Fierer J, Guiney D. 1991. Molecular analysis of the virulence locus of *Salmonella dublin* plasmid pSDL2. *Mol. Microbiol.* 5:307–16
109. Kusters JG, Mulders-Kremers GAWM, van Doornik CEM, van der Zeijst BAM. 1993. Effect of multiplicity of infection, bacterial protein synthesis, and growth phase on adhesion to an invasion of human cell lines by *Salmonella typhimurium*. *Infect. Immunology* 61:5013–20
110. Lange R, Hengge-Aronis R. 1994. The cellular concentration of the σ^s subunit of RNA polymerase in *Escherichia coli* is controlled at the levels of transcription, translation, and protein stability. *Genes Dev.* 8:1600–12
111. Lange R, Hengge-Aronis R. 1991. Identification of a central regulator of stationary phase gene expression in *Escherichia coli*. *Mol. Microbiol.* 5:49–59
112. Lee CA, Falkow S. 1990. The ability of *Salmonella* to enter mammalian cells is affected by bacterial growth state. *Proc. Natl. Acad. Sci. USA* 87:4304–8
113. Lee CA, Jones BD, Falkow S. 1992. Identification of a *Salmonella typhimurium* invasion locus by selection for hyperinvasive mutants. *Proc. Natl. Acad. Sci. USA* 89:1847–51
114. Lee IS, Slonezewski JL, Foster JW. 1994. A low-pH inducible stationary phase acid tolerance response in *Salmonella typhimurium*. *J. Bacteriol.* 176:1422–26
115. Lehrer RL, Lichtenstein AK, Ganz T. 1993. Defensins: antimicrobial and cytotoxic peptides of mammalian cells. *Annu. Rev. Immunol.* 11:105–28
116. Lepay DA, Nathan CF, Steinman RM, Murray HW, Cohn ZA. 1985. Murine Kupffer cells. Mononuclear phagocytes deficient in the generation of reactive oxygen intermediates. *J. Exp. Med.* 161:1079–96
117. Leung KY, Finlay BB. 1991. Intracellular replication is essential for the virulence of *Salmonella typhimurium*. *Proc. Natl. Acad. Sci. USA* 88:11470–74
118. Leyer GJ, Johnson EA. 1993. Acid adaptation induces cross protection against environmental stresses in *Salmonella typhimurium*. *Appl. Environ. Microbiol.* 59:1842
119. Li Z, Demple B. 1994. SoxS, an acti-

vator of superoxide stress genes in *Escherichia coli*. Purification and interaction with DNA. *J. Biol. Chem.* 269: 18371–77
120. Libby SJ, Goebel W, Ludwig A, Buchmeier N, Bowe F, et al. 1994. A cytolysin encoded by *Salmonella* is required for survival within macrophages. *Proc. Natl. Acad. Sci. USA* 91:489–93
121. Lipinska B, Fayet O, Baird L, Georgopoulos C. 1989. Identification, characterization and mapping of the *Escherichia coli htr* gene, whose product is essential for bacterial growth only at elevated temperatures. *J. Bacteriol.* 171:1574–84
122. Lipinska B, Sharma S, Georgopoulos C. 1989. Sequence analysis and regulation of the *htrA* gene of *Escherichia coli:* a σ^{32}-independent mechanism of heat-inducible transcription. *Nucleic Acids Res.* 16:10053–67
123. Lissner CR, Weinstein DL, O'Brian AD. 1985. Mouse chromosome 1 *Ity* locus regulates microbicidal activity of isolated peritoneal macrophages against a diverse group of intracellular and extracellular bacteria. *J. Immunol.* 135: 544–47
124. Litwin CM, Calderwood SB. 1994. Analysis of the complexity of gene regulation by Fur in *Vibrio cholera*. *J. Bacteriol.* 176:240–48
125. Litwin CM, Calderwood SB. 1993. Role of iron in regulation of virulence genes. *Clin. Microbiol. Rev.* 6:137–49
126. Loewen PC, von Ossowski I, Switala J, Mulvey MR. 1993. KatF (σ^s) synthesis in *Escherichia coli* is subject to posttranscriptional regulation. *J. Bacteriol.* 175:2105–53
127. MacBeth KJ, Lee CA. 1993. Prolonged inhibition of bacterial protein synthesis abolishes *Salmonella* invasion. *Infect. Immun.* 61:1544–46
128. Mackey BM, Derrick C. 1990. Heat shock protein synthesis and thermotolerance in *Salmonella typhimurium*. *J. Appl. Bacteriol.* 69:373–83
129. Mackey BM, Derrick CM. 1986. Elevation of the heat resistance of *Salmonella typhimurium* by sublethal heat shock. *J. Appl. Bacteriol.* 61:389–93
130. Mackness GB. 1971. Resistance to intracellular infection. *J. Infect. Dis.* 123:439–45
131. Mahan MJ, Slauch JM, Hanna PC, Camilli A, Tobias JW, et al. 1994. Selection for bacterial genes that are specifically induced in host tissue: the hunt for virulence factors. *Infect. Agents Dis.* 2:263–68
132. Mahan MJ, Slauch JM, Mekalanos JJ. 1993. Selection of bacterial virulence genes that are specifically induced in host tissues. *Science* 259:686–88
133. Makelä PH, Sarvas M, Calcagno S, Lounatmaa K. 1978. Isolation and characterization of polymyxin-resistant mutants of *Salmonella*. *FEMS Microbiol. Rev.* 3:323–26
134. McCann MP, Fraley CD, Matin A. 1993. The putative σ factor KatF is regulated posttranscriptionally during carbon starvation. *J. Bacteriol.* 175: 2143–49
135. McCann MP, Kidwell JP, Matin A. 1991. The putative sigma factor KatF has a central role in development of starvation-mediated general resistance in *Escherichia coli*. *J. Bacteriol.* 173: 4188–94
136. Miller SI, Kukral AM, Mekalanos JJ. 1989. A two component regulatory system (*phoP phoQ*) controls *Salmonella typhimurium* virulence. *Proc. Natl. Acad. Sci. USA* 86:5054–58
137. Miller SI, Mekalanos JJ. 1990. Constitutive expression of the *phoP* regulon attenuates *Salmonella* virulence and survival within macrophages. *J. Bacteriol.* 172:2485–90
138. Miller SI, Pulkkinen WS, Selsted ME, Mekalanos JJ. 1990. Characterization of defense resistance phenotypes associated with mutations in the phoP virulence regulon of *Salmonella typhimurium*. *Infect. Immun.* 58:3706–10
139. Miller VL, Beer KB, Loomis WP, Olson JA, Miller SI. 1992. An unusual *pagC::TnphoA* mutation leads to an invasion- and virulence-defective phenotype in *Salmonella*. *Infect. Immun.* 60: 3763–70
140. Morgan RW, Christman MF, Jacobson FS, Sturz G, Ames BN. 1986. Hydrogen peroxide–inducible proteins in *Salmonella typhimurium* overlap with heat shock and other stress proteins. *Proc. Natl. Acad. Sci. USA* 83:8059–63
141. Mulvey MR, Loewen PC. 1993. Nucleotide sequence of *katF* of *Escherichia coli* suggests KatF protein is a novel σ transcription factor. *Nucleic Acids Res.* 17:9979–91
142. Neiderhoffer EC, Naranjo CM, Bradley KL, Fee JA. 1990. Control of *Escherichia coli* superoxide dismutase (*sodA* and *sodB*) genes by the ferric uptake regulation (*fur*) locus. *J. Bacteriol.* 172:1930–38
143. Neilands JB. 1981. Microbial iron compounds. *Annu. Rev. Biochem.* 50:715–31
144. Nikaido H, Vaara M. 1987. Outer membrane. In Escherichia coli *and* Salmo-

nella typhimurium: *Cellular and Molecular Biology*, pp. 7–27. Washington, DC: Am. Soc. Microbiol.
145. Norel F, Coynault C, Miras I, Hermant D, Popoff MY. 1989. Cloning and expression of plasmid DNA sequences involved in *Salmonella* serotype *typhimurium* virulence. *Mol. Microbiol.* 3:733–43
146. Norel F, Robbe-Saule V, Popoff MY, Coynault C. 1992. The putative sigma factor KatF (RpoS) is required for the transcription of the *Salmonella typhimurium* virulence gene *spvB* in *Escherichia coli*. *FEMS Microbiol. Lett.* 99:271–76
147. Nunoshiba T, de Rojas-Walker T, Wishnok JS, Tannebaum SR, Demple B. 1993. Activation by nitric oxide of an oxidative-stress response that defends *Escherichia coli* against activated macrophages. *Proc. Natl. Acad. Sci. USA* 90:9993–97
148. Nunoshiba T, Hidalgo E, Cuevas CFA, Demple B. 1992. Two stage control of an oxidative stress regulon: the *Escherichia coli* SoxR protein triggers redox-inducible expression of the *soxS* regulatory gene. *J. Bacteriol.* 174:6054–60
149. Nunoshiba T, Hidalgo E, Li ZY, Demple B. 1993. Negative autoregulation by the *Escherichia coli* SoxS protein: a dampening mechanism for the *soxRS* stress response. *J. Bacteriol.* 175:7492–94
150. O'Byrne CP, Dorman CJ. 1994. The *spv* virulence operon of *Salmonella typhimurium* LT-2 is regulated negatively by cyclic AMP (cAMP)–cAMP receptor protein system. *J. Bacteriol.* 176:905–12
151. Deleted in proof
152. O'Neal CR, Gabriel WM, Turk AM, Libby SJ, Fang FC, et al. 1994. RpoS is necessary for both the positive and negative regulation of starvation survival genes during phosphate, carbon, and nitrogen starvation in *Salmonella typhimurium*. *J. Bacteriol.* 176:4610–16
153. Otto BR, Verweij-van Vught AMJJ, MacLaren DM. 1992. Transferrins and heme-compounds as iron sources for pathogenic bacteria. *Crit. Rev. Microbiol.* 18:217–33
154. Ouellette AJ, Lualdi JC. 1990. A novel mouse gene family coding for cationic, cysteine-rich peptide regulation in small intestine and cells of myeloid origin. *J. Bacteriol.* 265:9831–37
155. Overdier DG, Csonka LN. 1992. A transcriptional silencer downstream of the promoter in the osmotically controlled *proU* operon of *Salmonella typhimurium*. *Proc. Natl. Acad. Sci. USA* 89:3140–44
156. Pace J, Hayman MJ, Galan JE. 1993. Signal transduction and invasion of epithelial cells by S. typhimurium. *Cell* 72:505–14
157. Parkinson JS, Kofoid EC. 1992. Communication modules in bacterial signaling proteins. *Annu. Rev. Genet.* 26:71–112
158. Parra-Lopez C, Baer MT, Groisman E. 1993. Molecular genetic analysis of a locus required for resistance to antimicrobial peptides in *Salmonella typhimurium*. *EMBO J.* 12:4053–62
159. Parra-Lopez C, Lin R, Aspedon A, Groisman EA. 1994. A *Salmonella* protein that is required for resistance to antimicrobial peptides and transport of potassium. *EMBO J.* 13:3964–72
160. Parsell DA, Lindquist S. 1993. The functions of heat shock proteins in stress tolerance: degradation and reactivation of damaged proteins. *Annu. Rev. Genet.* 27:437–96
161. Payne SM. 1988. Iron and virulence in the family *Enterobacteriaceae*. *Crit. Rev. Microbiol.* 16:81–111
162. Pollack JR, Neilands JB. 1970. Enterobactin, an iron transport compound from *Salmonella typhimurium*. *Biochem. Biophys. Res. Commun.* 38:982–89
163. Pullinger GD, Lax AJ. 1992. A *Salmonella dublin* virulence plasmid locus that affects bacterial growth under nutrient-limited conditions. *Mol. Microbiol.* 6:1631–43
164. Puschmann M, Ganzoni AM. 1977. Increased resistance of iron-deficient mice to *Salmonella* infection. *Infect. Immun.* 17:663–64
165. Deleted in proof
166. Rhen M, Riikonen P, Taira S. 1993. Transcriptional regulation of *Salmonella enterica* virulence plasmid genes in cultured macrophages. *Mol. Microbiol.* 10:45–56
167. Rhen M, Sukupolvi S. 1988. The role of the *traT* gene of the *Salmonella typhimurium* virulence plasmid for serum resistance and growth within liver macrophages. *Microb. Pathog.* 5:275–85
168. Riikonen P, Makela PH, Saarilahti J, Sukopolui S, Taira S, et al. 1992. The virulence plasmid does not contribute to growth of *Salmonella* in cultured murine macrophages. *Microb. Pathog.* 13:281–91
169. Rosenshine I, Finlay BB. 1993. Exploitation of host signal transduction pathways and cytoskeletal functions by invasive bacteria. *BioEssays* 15:17–24
170. Rosenshine I, Ruschkowski S, Foubister

V, Finlay BB. 1994. *Salmonella typhimurium* invasion of epithelial cells: role of induced host cell tyrosine protein phosphorylation. *Infect. Immun.* 62: 4969–74
171. Roszak DB, Grimes DJ, Colwell RR. 1984. Viable but nonrecoverable stage of *Salmonella enteritidis* in aquatic systems. *Can. J. Microbiol.* 30:334–38
172. Rowland KL, Esther CR, Spitznagel JK. 1994. Isolation and characterization of a gene, *pmrD*, that confers resistance to polymyxin when expressed in multiple copies. *J. Bacteriol.* 176:3589–97
173. Rowland KL, Martin LE, Esther CR, Spitznagel JK. 1993. Spontaneous *pmrA* mutants of *Salmonella typhimurium* LT2 define a new two component regulatory system with a possible role in virulence. *J. Bacteriol.* 175:4154–64
174. Rutz JM, Liu J, Lyons JA, Goranson J, Armstrong SK, et al. 1992. Formation of a gated channel by a ligand-specific transport protein in the bacterial outer membrane. *Science* 258:471–75
175. Deleted in proof
176. Schäffer S, Hantke K, Braun V. 1985. Nucleotide sequence of the iron regulatory gene *fur*. *Mol. Gen. Genet.* 200: 110–13
177. Schiemann DA, Shope SR. 1991. Anaerobic growth of *Salmonella typhimurium* results in increased uptake by Henle 407 epithelial and mouse peritoneal cells in vitro and repression of a major outer membrane protein. *Infect. Immun.* 59:437–40
178. Schrum LW, Hassan HM. 1994. The effects of *fur* on the transcriptional and post-transcriptional regulation of MnSOD gene (*sodA*) in *Escherichia coli*. *Arch. Biochem. Biophys.* 309:288–92
179. Silver S, Walderhaug M. 1992. Gene regulation of plasmid- and chromosome-determined inorganic ion transport in bacteria. *Microbiol. Rev.* 56:195–228
180. Skare JT, Postle K. 1991. Evidence for a TonB-dependent energy transduction complex in *Escherichia coli*. *Mol. Microbiol.* 5:2883–90
181. Spector MP. 1990. Gene expression in response to multiple nutrient-starvation conditions in *Salmonella typhimurium*. *FEMS Microbiol. Ecol.* 74:175–84
182. Spector MP, Aliabadi Z, Gonzalez T, Foster JW. 1986. Global control in *Salmonella typhimurium*: two-dimensional electrophoretic analysis of starvation-, anaerobiosis-, and heat shock–inducible proteins. *J. Bacteriol.* 168:420–24
183. Spector MP, Cubitt CL. 1992. Starvation-inducible loci of *Salmonella typhimurium*: regulation and roles in starvation survival. *Mol. Microbiol.* 6: 1467–76
184. Spector MP, Foster JW. 1993. Starvation-stress response (SSR) of *Salmonella typhimurium*: gene expression and survival during nutrient starvation. In *Starvation in Bacteria*, ed. S Kjelleberg, pp. 201–24. New York: Plenum
185. Spector MP, Park YK, Tirgari S, Gonzalez T, Foster JW. 1988. Identification and characterization of starvation-regulated genetic loci in *Salmonella typhimurium* by using Mud-directed *lacZ* operon fusions. *J. Bacteriol.* 170:345–51
186. Spink JM, Pullinger GC, Wood MW, Lax AJ. 1994. Regulation of spvR, the positive regulatory gene of *Salmonella* plasmid virulence genes. *FEMS Microbiol. Lett.* 116:113–22
187. Stinavage PS, Martin LE, Spitznagel JK. 1990. A 59 kilodalton outer membrane protein of *Salmonella typhimurium* protects against oxidative intraleukocytic killing due to human neutrophils. *Mol. Microbiol.* 4:283–93
188. Stirling DA, Hulton CSJ, Waddell L, Park SF, Stewart GSAB, et al. 1989. Molecular characterization of the *proU* loci of *Salmonella typhimurium* and *Escherichia coli* encoding osmoregulated glycine betaine transport systems. *Mol. Microbiol.* 3:1025–38
189. Stocker BAD, Makela PH. 1986. Genetic determinance of bacterial virulence with special reference to *Salmonella*. *Microbiol. Immunol.* 124:149–72
190. Storz G, Christman MF, Sies H, Ames BN. 1987. Spontaneous and oxidative damage to DNA in *Salmonella typhimurium*. *Proc. Natl. Acad. Sci. USA* 84:8917–21
191. Storz G, Tartaglia LA, Ames BN. 1990. Transcriptional regulator of oxidative stress-inducible genes: direct activation by oxidation. *Science* 248:189–94
192. Tanaka K, Takayanagi N, Fujita N, Ishihama A, Takahashi H. 1993. Heterogeneity of the principal σ factor in *Escherichia coli*: the *rpoS* gene product, σ^{38}, is second σ factor of RNA polymerase in stationary-phase *Escherichia coli*. *Proc. Natl. Acad. Sci. USA* 90: 3511–15
193. Tardat B, Touati D. 1991. Two global regulators repress the anaerobic expression of Mn SOD in *Escherichia coli*::Fur (ferric uptake regulation) and Arc (aerobic reparation control). *Mol. Microbiol.* 5:455–65
194. Tartaglia LA, Storz G, Brodsky MH,

Lai A, Ames BN. 1990. Alkyl hydroperoxide reductase from *Salmonella typhimurium*. Sequence and homology to thioredoxin reductase and other flavoprotein disulfide oxidoreductases. *J. Biol. Chem.* 265:10535–40

195. Tsaneva IR, Weiss B. 1990. *soxR*, a locus governing a superoxide response regulon in *Escherichia coli* K-12. *J. Bacteriol.* 172:4197–205

196. Tsolis RM, Baumler AJ, Heffron F. 1995. Role of *Salmonella typhimurium* Mn-superoxide dismutase (SodA) in protection against early killing by J774 macrophages. *Infect. Immun.* In press

197. Turpin PE, Maycroft KA, Rowlands CL, Wellington EMH. 1993. Viable but nonculturable salmonellas in soil. *J. Appl. Bacteriol.* 74:421–27

198. Vaara M, Vaara T, Jenson M, Helander I, Nurminen M, et al. 1981. Characterization of the lipopolysaccharide from the polymyxin resistant *pmrA* mutants of *Salmonella typhimurium*. *FEBS Lett.* 129:145–49

199. Valone SE, Chikami GK, Miller VL. 1993. Stress induction of virulence proteins (SpvA, -B, and -C) from native plasmid pSDL2 of *Salmonella dublin*. *Infect. Immun.* 61:705–13

200. Vandenbosch JL, Kurlandsky DR, Urdangaray R, Jones GW. 1989. Evidence of coordinate regulation of virulence in *Salmonella typhimurium* involving the *rsk* element of the 95 kilobase plasmid. *Infect. Immun.* 57:2566–68

201. Vescovi EG, Soncini FC, Groisman EA. 1994. The role of PhoP/PhoQ regulon in *Salmonella* virulence. *Res. Microbiol.* 145:473–80

202. Visca P, Filetici E, Anastasio MP, Vetriani C, Fantasia M, et al. 1991. Siderophore production by species isolated from different sources. *FEMS Microbiol. Lett.* 79:225–32

203. Vladoianu I-R, Chang HR, Pechere J-C. 1990. Expression of host resistance to *Salmonella typhi* and *Salmonella typhimurium:* bacterial survival within macrophages of murine and human origin. *Microb. Pathog.* 8:83–90

204. Walderhaug NO, Litwack ED, Epstein W. 1989. Wide distribution of *Escherichia coli* Kdp K^+-ATPase among gram negative bacteria. *J. Bacteriol.* 171:1192–95

205. Welch R. 1991. Pore-forming cytolysins of gram negative bacteria. *Mol. Microbiol.* 5:521–28

206. Wu J, Weiss B. 1992. Two-stage induction of the *soxRS* (superoxide response) regulon of *Escherichia coli*. *J. Bacteriol.* 174:3915–20

207. Yancey RJ, Breeding SAL, Lankford CE. 1979. Enterochelin (enterobactin): virulence factor for *Salmonella typhimurium*. *Infect. Immun.* 24:174–80

208. Yura T, Nagai H, Mori H. 1993. Regulation of the heat-shock response in bacteria. *Annu. Rev. Microbiol.* 47:321–50

THE MECHANISMS OF *TRYPANOSOMA CRUZI* INVASION OF MAMMALIAN CELLS

Barbara A. Burleigh and Norma W. Andrews

Department of Cell Biology, Yale University School of Medicine, 333 Cedar Street, New Haven, Connecticut 06510

KEY WORDS: protozoan, parasite, intracellular, antigen, attachment, signaling

CONTENTS

INTRODUCTION	176
LIFE-CYCLE STAGES	177
INVASION OF PHAGOCYTIC AND NONPHAGOCYTIC HOST CELLS	179
Amastigote Invasion of Macrophages	179
Lysosome Recruitment as a Mechanism for Invasion of Nonphagocytic Cells	180
Calcium Signaling in Host Cells as an Early Event in Invasion	182
THE ROLE OF PARASITE SURFACE MOLECULES IN HOST-CELL ATTACHMENT AND INVASION	183
Metacyclic Surface Antigens	184
Bloodstream Trypomastigote Surface Antigens	184
Trans-Sialidase and Related Surface Glycoproteins	186
Role of Sialic Acid Acceptor Molecules in Invasion	187
Amastigote Surface Antigens	188
THE ROLE OF HOST-CELL SURFACE MOLECULES	189
The Role of Cell Surface Carbohydrate	190
INHIBITORS OF INVASION	191
Antibodies	191
Chemical Inhibitors of Invasion	192
ESCAPE FROM THE VACUOLE	192
CONCLUDING REMARKS	193

ABSTRACT

The protozoan parasite *Trypanosoma cruzi* must enter cells of its vertebrate host in order to replicate. Once this is accomplished, the infective trypomastigotes can invade many different cell types from several host species. This observation is in agreement with the parasite's wide natural host range. Studies performed with cultured mammalian cells in vitro have shown that *T. cruzi* invasion is an unusual process, distinct from phagocytosis, that depends on parasite energy and on negatively charged surface molecules of the host cell. Several surface glycoproteins and mucin-like molecules of trypomastigotes have been implicated, mainly by inhibition studies with antibodies, in interactions with host cells. Recently, several of the trypomastigote surface glycoproteins were shown to be related members of a large family that includes the *T. cruzi trans*-sialidase. The mucin-like molecules are beginning to emerge as a separate family of threonine-rich, *O*-glycosylated molecules that function as acceptors of sialic acid in the infective stages. Several lines of evidence suggest that parasite surface molecules mediate binding to host cells, whereas invasion of nonphagocytic cells involves recruitment of host-cell lysosomes, an unusual event apparently triggered by signal transduction.

INTRODUCTION

The American trypanosome *Trypanosoma cruzi* is the causative agent of Chagas' disease in humans. With no immunoprophylactic agents available and the widespread distribution of insect vectors (hematophagous triatomids), Chagas' disease has become a prominent health problem in South and Central America. Although insect transmission has been efficiently controlled in the most developed areas, blood transfusion and congenital infections remain important sources of human disease.

The vertebrate stages of *T. cruzi* are obligate intracellular pathogens, because they must enter host cells in order to replicate. The parasites released from infected cells are circulated in the bloodstream before they attach and enter new cells for another cycle of differentiation and replication. The use of mammalian cells as cultured monolayers to support the growth and propagation of *T. cruzi* in vitro greatly facilitated the study of invasion, differentiation, and replication of these parasites. These experiments demonstrated that many different cell types from several species could serve as hosts for the intracellular cycle of *T. cruzi*, in agreement with the wide range of vertebrate hosts observed in nature (18). We focus here exclusively on studies performed with cultured cells, assuming that the basic mechanisms for parasite invasion in vivo are similar. However, complement and antibody opsonization, among other factors [see review by Ofek et al in this volume (77a)], must play an important role in natural infections. Tropism of *T. cruzi* trypomastigotes for specific vertebrate tissues has been observed (18), and its basis is still poorly understood. Although the parasite seems to preferentially invade cells capable of attaching and spreading on substrates, the intracellular cycle has been successfully reproduced in a myeloma line growing in suspension (25).

Many years of effort have been dedicated to the search for candidate receptors on the host cell as well as to the identification of surface membrane proteins of parasite infective stages involved in the host-parasite recognition process. The concept emerging from these studies is that parasite attachment and invasion do not appear to result from a simple process of receptor-ligand binding. Instead, the data implicate several host and parasite molecules in the recognition process as well as the involvement of enzymatic reactions and a requirement for energy on the part of the parasite. As *T. cruzi* can infect many cell types, it may utilize different receptors on the host cells that have common properties such as negative charge. In addition to the mechanisms that mediate parasite binding to host cells, we discuss recent evidence indicating that signal-transduction events are involved in triggering the internalization process.

LIFE-CYCLE STAGES

T. cruzi is a kinetoplastid protozoan parasite whose life cycle alternates between invertebrate and vertebrate hosts (Figure 1). Transmission is initiated by insect vectors belonging to the family Reduviidae, which after a blood meal defecate and release infective stages of the parasite near the bite wound. These infective stages, called metacyclic trypomastigotes, differentiate from epimastigotes, the noninfective forms that inhabit the insect digestive tract. When metacyclic trypomastigotes enter the vertebrate host, they invade cells through the formation of a membrane-bounded vacuole (discussed in more detail below) and initiate the intracellular cycle, which lasts an average of 4–5 days. The first step of the cycle is disruption of the vacuolar membrane, which occurs within the first 1–2 h after invasion. Upon reaching the cytoplasm, the parasites differentiate into amastigotes, the intracellular replicative stages. After a lag period of approximately 20 h, the amastigotes start dividing by binary fission with a doubling time of about 12 h; thus, 500 parasites are generated from each one originally internalized (38). After completion of the replicative period, amastigotes differentiate into trypomastigotes; the host cell ruptures; and the parasites are released into the bloodstream, where they can be ingested by the insect vector and complete the life cycle. These forms are therefore called bloodstream trypomastigotes and are equivalent to the ones released to the supernatant of infected cell culture monolayers. They have the same properties of metacyclic trypomastigotes, in the sense that they can invade vertebrate cells and repeat the intracellular cycle (see Figure 1).

Both free-living forms of *T. cruzi*—the noninfective epimastigotes and the infective trypomastigotes—are flagellated and highly motile. Amastigotes, the intracellular forms, have a very short flagellum and a more spherical body.

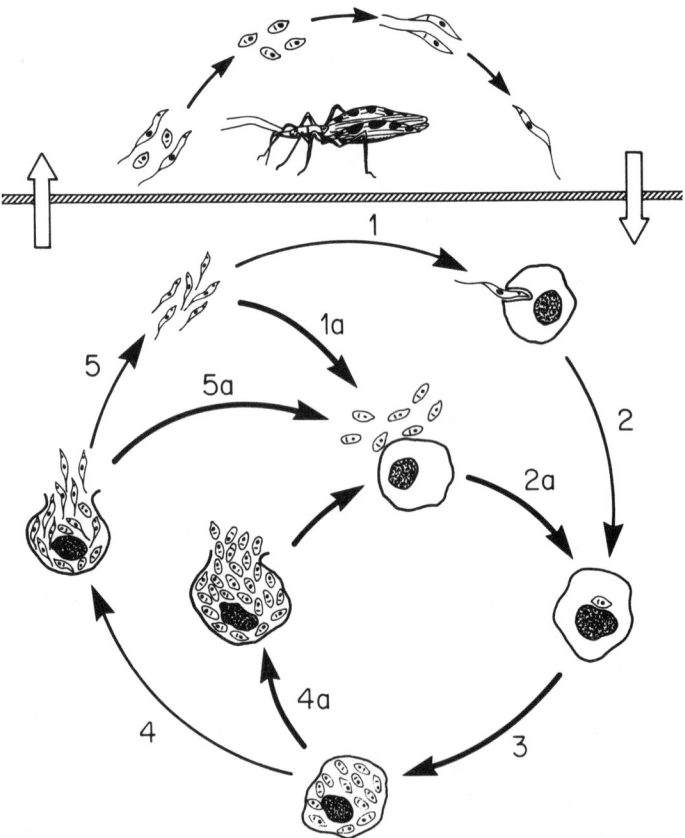

Figure 1 Life cycle of *Trypanosoma cruzi*. With a blood meal the insect vector ingests bloodstream trypomastigotes (*left arrow, top*), which in the lumen of the insect digestive tract transform into replicative, noninfective epimastigotes. Epimastigotes differentiate into infective metacyclic trypomastigotes, which are released with the feces of the insect and enter the vertebrate host through the bite wound (*right arrow, top*). Metacyclic trypomastigotes invade vertebrate cells (step 1); escape from the vacuole and transform into amastigotes (step 2); replicate in the cytoplasm (step 3); and differentiate into bloodstream trypomastigotes (step 4), which are released by rupture of the host cell (step 5). An alternative subcycle in the vertebrate can occur when amastigotes, either derived from premature rupture of host cells (steps 4a, 5a) or through extracellular differentiation of trypomastigotes (step 1a), are ingested by macrophages, where they can survive and complete the intracellular cycle.

Whether trypomastigote motility plays a role in the invasion process is unknown, because it has not been possible to inhibit parasite movement without affecting the host cells as well. This issue is not addressed here, but one should bear in mind that trypomastigotes are not inert particles that settle by gravity on host cells, but are actively moving flagellates that may be able to change their location according to stimuli.

Ultrastructural studies have shown that trypomastigotes and epimastigotes have distinct plasma membrane characteristics. Bloodstream forms have a surface coat threefold thicker than epimastigotes, and freeze-fracture reveals fewer intramembrane particles in trypomastigotes as compared to epimastigotes (35). Because of the assumption that host-cell invasion is mediated by specific properties of trypomastigotes, most studies have compared both forms and searched for differences. Although useful, as discussed below, this strategy has not yet led to the identification of a trypomastigote factor unequivocally involved in promoting cell entry.

INVASION OF PHAGOCYTIC AND NONPHAGOCYTIC HOST CELLS

Resident tissue macrophages are believed to play an important role in vivo as one of the first host cells to be invaded by *T. cruzi*. For this reason, many of the early studies of *T. cruzi*–host cell interactions focused on this cell type. Initial studies demonstrated that both trypomastigotes and epimastigotes were efficiently internalized by macrophages, and later experiments revealed their presence inside phagolysosomes (22, 69, 70). However, only trypomastigotes could escape from the phagolysosome and multiply in the cytosol, whereas epimastigotes were destroyed (70, 77). Subsequently, investigators demonstrated that macrophages activated by cytokines, particularly interferon-γ, could efficiently kill intracellular trypomastigotes as well (96). The killing mechanism has been attributed to activation of the respiratory burst and the generation of toxic oxygen intermediates (53, 76), although a recent report challenges this view and proposes a role for nitric oxide (65).

The mechanism by which trypomastigotes enter macrophages has been a subject of controversy. While several investigators reported that entry was blocked in the presence of inhibitors of actin polymerization such as cytochalasin B (68, 77), others found no inhibition and concluded that invasion was active on part of the parasites (59). The recent discovery of a cytochalasin-independent entry mechanism for trypomastigotes in nonphagocytic cells (see below) has clarified this issue. In macrophages, both mechanisms—active entry and passive phagocytosis—can probably occur simultaneously, which would explain the apparently contradictory early observations.

Amastigote Invasion of Macrophages

The intracellular stage of *T. cruzi* may play a role in maintaining the replicative cycle of the parasite in macrophages (23, 61, 66). Although poorly infective for nonprofessional phagocytes, these forms are readily ingested by macrophages, wherein they escape from the phagolysosome and replicate normally in the cytoplasm (61). Prior findings that amastigotes are present in the circulation of mice during the acute phase of the disease (8) raise the intriguing possibility that macrophage uptake of amastigotes may initiate an alternative subcycle of this parasite in the mammalian host (Figure 1). This alternative mode of propagation may be significant for survival of the parasite in the presence of a host cytotoxic response, in which lysis of infected cells would lead to the release of viable intracellular amastigotes.

Amastigote infection of nonphagocytic cells has also been reported, and the characteristics of this interaction were studied recently (74). Unlike trypomastigotes, which show a marked preference for invading at the host-cell margins, amastigotes attach seemingly at random to surface microvilli and become internalized at the dorsal surface of cells (74). Interestingly, cytochalasin D blocks the entry of amastigotes into these cells, which is another marked difference from trypomastigotes (see below). Even in nonphagocytic cells, the entry mechanism of amastigotes can be characterized as phagocytosis. Clustering of actin filaments is detected at the site of amastigote attachment, and the process is inhibited by the protein kinase inhibitors staurosporine and genistein (93).

Lysosome Recruitment as a Mechanism for Invasion of Nonphagocytic Cells

The first indication of an unusual mechanism for invasion of nonphagocytic cells by *T. cruzi* came from attempts to interfere with the process using inhibitors of actin polymerization. Unexpectedly, invasion assays performed in the presence of cytochalasin D showed no inhibition of parasite entry in HeLa and Madin-Darby canine kidney (MDCK) cells (111). Further studies demonstrated that cytochalasin D actually significantly facilitated trypomastigote invasion of normal rat kidney (NRK) fibroblasts and polarized MDCK cells. A 10-fold enhancement in parasite entry through the apical domain of MDCK cells was observed when the brush border cytoskeleton was disrupted by treatment with 10 µM cytochalasin D (115). Furthermore, staining of host-cell microfilaments with phalloidin revealed no accumulation of F-actin around recently internalized trypanosomes. These findings suggested that the entry mechanism of *T. cruzi* was independent of actin polymerization and that removal of the cortical actin cytoskeleton somehow facilitated the process. These highly unusual observations differ from those made during host-cell invasion by many bacterial pathogens (41). Moreover, trypomastigotes show a strong preference for invading flat lamellipodia close to the cell margins (102). These are regions known to be highly dynamic with respect to assembly and disassembly of the cortical actin cytoskeleton (49).

Morphological studies provided additional evidence to clarify the mechanism of *T. cruzi* invasion of nonphagocytic cells. Entry has long been known to involve the formation of a tight membrane-bounded vacuole around trypomastigotes, but the initial assumption was that the vacuole originated from the host-cell plasma membrane. As mentioned above, recent experiments showing that disruption of the F-actin cytoskeleton enhances invasion instead of inhibiting it challenged this view. It is difficult to imagine what would be the driving force for enveloping a particle as large as a trypanosome (10–15 µm long), in the absence of actin polymerization. Earlier studies of the morphology of the invasion process in epithelial cells and fibroblasts added to this puzzle: Scanning electron micrographs showed trypomastigotes entering cells with no pseudopod formation or any other alteration of the host-cell plasma membrane (102). The parasites appeared to slide gradually into the host cell, with no evidence of phagocytic cup formation or zippering, the classical features of phagocytosis. Observations of the invasion process using live video microscopy confirm this view (38; M Rioult & NW Andrews, unpublished data). One report describes pseudopod extension and phalloidin staining of F-actin at the site of *T. cruzi* invasion of HeLa cells in the presence of 10 µM cytochalasin D (110). The same authors made no similar observations in fibroblasts or epithelial cells and suggested that the pseudopod extension and staining might be restricted to the HeLa cell isolate used. Taken together, the morphology of the entry process in nonphagocytic cells and the apparent lack of involvement of actin raised the possibility that the membrane required to form the *T. cruzi* intracellular vacuole might originate from intracellular organelles.

Indeed, an unusual association of host-cell lysosomes with trypomastigotes was detected at very early stages of the invasion process. Using two different labeling methods—a monoclonal antibody specific for lgp 120, a major lysosomal membrane glycoprotein, or horseradish peroxidase internalized by fluid-phase endocytosis—investigators demonstrated that clusters of lysosomes gathered in close proximity to the host-cell plasma membrane at regions of close contact with trypomastigotes. These clusters, detected before the onset of internalization, were predominantly associated with the posterior end of the parasites, the extremity that initiates invasion. Lysosomal markers could be detected in partially formed intracellular vacuoles surrounding parasites, suggesting that gradual fusion of lysosomes provided the membrane required for vacuole formation (115).

Experimental manipulation of the cellular distribution and fusion capacity of lysosomes reinforced this hypothesis. Acidification of the cytosol, a treatment known to induce migration of lysosomes to the cell periphery (52), enhanced *T. cruzi* invasion of NRK cells, whereas alkalinization, which induces perinuclear clustering of lysosomes, had the opposite effect (115). Other drugs that change the distribution of lysosomes from the perinuclear area to the cell periphery, such as dBcAMP and brefeldin A, also enhanced trypomastigote entry (115). Loading of lysosomes with sucrose, which interferes with the fusion capacity of lysosomes (72), reduced *T. cruzi* invasion of NRK cells by 50% (115). This finding indicates that the low infection rate by *T. cruzi* in mitotic cells (39, 115) probably results from the block in membrane fusion events observed at this stage of the cell cycle (123) and not from loss of surface receptors as previously suggested (39, 129).

Calcium Signaling in Host Cells as an Early Event in Invasion

The discovery that trypomastigotes could recruit host-cell lysosomes to the invasion site (115) implied that *T. cruzi* must trigger a signal transduction mechanism in host cells. An initial investigation of the cytosolic pH of NRK cells showed that it remains stable at neutral pH during the entire process of attachment and invasion by trypomastigotes (115). However, subsequent studies of the intracellular free calcium concentration ($[Ca^{2+}]_i$) of host cells revealed a striking response. Upon contact with infective trypomastigotes (but not the noninfective epimastigotes), rapid and repetitive elevations in the $[Ca^{2+}]_i$ of NRK cells were observed (114). The Ca^{2+}-signaling activity proved to be associated with the soluble fraction of disrupted trypomastigotes (19a) and to be abolished by trypsin treatment (114). Distinct cell types from different species were responsive to this soluble factor, with the $[Ca^{2+}]_i$ reaching levels comparable to that induced by other Ca^{2+}-agonists. Recent evidence indicates that the trypomastigote Ca^{2+}-signaling factor originates from proteolytic processing of an inactive precursor, in a manner analogous to several mammalian hormones and growth factors (19a).

Buffering or depletion of intracellular Ca^{2+} in host cells resulted in significant inhibition of *T. cruzi* entry, suggesting a role for the $[Ca^{2+}]_i$-transients in invasion (79, 114). These findings, similar to the discovery of lysosome recruitment, explain previously puzzling observations. For example, the infection of a mammalian cell monolayer by *T. cruzi* does not follow a random, Poisson distribution. The distribution of intracellular parasites among a population of bovine embryo skeletal muscle cells corresponds more closely to a negative binomial prediction (38). The signaling events triggered by the parasites in host cells may make them more susceptible to invasion, thus explaining the occurrence of multiple infections. In this respect, it is tempting to speculate that the tropism observed in vivo for muscle and neuronal cells may also be based on the active Ca^{2+}-signaling capacity of these cell types (94).

$[Ca^{2+}]_i$-transients are involved in events involving rearrangement of the actin cytoskeleton such as exocytosis (19, 37, 60). Therefore, in light of the results discussed in the previous section, this recently identified trypomastigote Ca^{2+}-signaling mechanism may be involved in the recruitment of host-cell lysosomes that leads to invasion. A recent report has described detection of elevated $[Ca^{2+}]_i$ in both trypomastigotes and host cells (L6E9 myoblasts) during early stages of their interaction. The $[Ca^{2+}]_i$ of trypomastigotes increased from 20–30 nM to 100 nM upon host-cell invasion, and an area of elevated cytosolic free Ca^{2+} was detected in host cells around internalized parasites (73). Preincubation of trypomastigotes with the intracellular Ca^{2+} chelators Quin 2 or BAPTA resulted in decreased association with myoblasts, perhaps suggesting a role for Ca^{2+}-induced exocytosis of parasite molecules in the invasion process. In support of this view, pretreatment of trypomastigotes with the Ca^{2+} ionophore ionomycin reportedly enhanced their invasive capacity (124).

THE ROLE OF PARASITE SURFACE MOLECULES IN HOST-CELL ATTACHMENT AND INVASION

Attachment of *T. cruzi* to mammalian cells and subsequent invasion are active processes that appear to be subject to parasite controlling factors more than host-cell influences. The attachment and invasion events can be experimentally uncoupled by aldehyde fixation of mammalian cells, which has been useful for examining factors affecting parasite binding. Using this strategy, Schenkman et al (111) demonstrated that *T. cruzi* does not attach well to mammalian cells at 4°C, indicating that surface molecules that participate in host-cell recognition are not sufficient to promote strong binding interactions. In addition, parasite attachment is sensitive to inhibitors of oxidative phosphorylation, which suggests that attachment is an energy-dependent process. This finding does not rule out a requirement for parasite surface molecules in attachment and invasion, because their removal results in a drastic reduction in invasive capacity (9), and several of the monoclonal antibodies directed to parasite surface molecules inhibit invasion (5, 11, 95, 126). Thus, attachment of *T. cruzi* to host cells appears to require not only an immediate availability of specific surface molecules, but ATP-dependent processes as well. Why energy is needed for efficient binding is not clear, but perhaps, as mentioned above, some of the molecules that participate in binding are secreted by the parasite upon contact with the host cell. Alternatively, energy-dependent rearrangements or clustering of trypomastigote surface molecules may be required for generating the active ligand.

The concept of invasion that evolved over the past 10 years was that *T. cruzi*–host cell interactions were very complex; several different, predominantly stage-specific, molecules appeared to be involved in the recognition and invasion steps. Many of these molecules are members of a large multigene family that includes the surface *trans*-sialidase/sialidases (20, 32, 56), and although these proteins are highly heterogenous, they are related on a structural level and may perform similar functions during invasion. Therefore, attachment and invasion by *T. cruzi* may be less complicated than originally conceived. Nonetheless, in hopes of providing a clear understanding of the various surface molecules that have been implicated in invasion, we take a more historical approach to the description of these antigens.

Metacyclic Surface Antigens

As discussed above, the metacyclic trypomastigote stage of *T. cruzi* comprises nondividing, invasive forms of the parasite that originate in the insect vector by the transformation of noninfective epimastigote forms (see Figure 1). Metacyclics are released onto the skin of a potential host in the feces and urine of the feeding insect and gain access to the host through contamination of the bite. Once inside, *T. cruzi* invades professional phagocytic cells as well as nonphagocytic cells. Several metacyclic surface molecules have been implicated in invasion. Monoclonal antibodies (mAbs) raised to heat-killed metacyclic forms specifically precipitate the major surface proteins, which were identified by iodination to be 90, 82, and 75 kDa (116). Further characterization of the 90-kDa antigen revealed that it is a glycoprotein (gp90) (125) anchored to the membrane by a glycophosphatidylinositol (GPI) linkage (112). A mAb specific for gp90 neutralized infections in mice inoculated with metacyclic trypomastigotes (11) and partially inhibited invasion of cultured mammalian cells (125), therefore suggesting a role for this protein in invasion by metacyclic trypanosomes. The 82-kDa surface glycoprotein (gp82) (95) is also thought to be involved in parasite attachment and invasion. Isolated protein exhibited saturable binding to the surface of Vero cells, suggesting the presence of a specific receptor (95). In addition, parasite invasion was inhibited by mAbs and Fab fragments specific for gp82. gp90 and gp82 are now recognized as related molecules, belonging to the gp85/sialidase family (14, 43; see below).

A metacyclic surface 35/50 kDa mucin-like protein (98, 126) has been identified as the major metacyclic acceptor for exogenous sialic acid in a reaction involving the *T. cruzi trans*-sialidase (27, 32, 106, 107; see below). This heavily glycosylated protein contains many *O*-linked glycans (107), cannot be labeled with ^{125}I (116) nor removed from the surface by trypsinization (75), and is anchored to the membrane with a GPI linkage (107). Consistent with a possible role in invasion, monoclonal antibodies to the 35/50 kDa protein inhibit invasion in vitro (98) and neutralize infections of mice inoculated with metacyclic trypomastigotes (126). The 35/50 kDa mucin-like proteins of different strains of *T. cruzi* exhibit a high degree of structural polymorphism (75). However, based on biochemical differences, this molecule does not appear to be related to the gp90 and gp82 metacyclic proteins. A recently identified unusually small gene coding for a *T. cruzi* threonine-rich mucin-like protein supports this view (97).

Bloodstream Trypomastigote Surface Antigens

The bloodstream trypomastigote stages of *T. cruzi* are similar to the metacyclic stages in terms of morphology and their ability to invade a wide variety of mammalian cells in vitro, as previously described (Figure 1). Polypeptides on the surface of trypomastigote forms of *T. cruzi* are essential for attachment and invasion of mammalian cells, because trypsin treatment of parasites, which remove most of the iodinated surface glycoproteins, results in a 90% reduction in invasive capacity. The ability of trypsinized parasites to invade cells was restored following an incubation period in which the iodinatable surface proteins were regenerated (9). Even without prior removal of surface polypeptides, the invasive capacity of trypomastigotes was sensitive to inhibitors of protein synthesis (63) and N-linked glycosylation (130). Other evidence to support the involvement of trypomastigote surface molecules in invasion comes from mAbs to the heterogenous surface glycoprotein (Tc-85), which could inhibit invasion of cultured mammalian cells by 50–70% (1, 5).

Cumulative evidence from several investigations argues that after trypomastigotes are released from mammalian cells, remodeling of the parasite surface occurs, which leads to a severalfold increase in infective capacity. First, trypomastigotes become more infective upon extracellular incubation in the culture medium (88, 129). Then, when trypomastigotes emerge from ruptured cells, they are poorly sialylated. Subsequent incubation in medium or with sialic acid–containing macromolecules such as fetuin results in the sialylation of parasite surface proteins and an increase in infective capacity (87, 108). Finally, Piras et al (86) found that mild trypsinization of fresh trypomastigotes could mimic the apparent maturation process and that protease inhibitors blocked it.

Monoclonal antibodies to stage-specific trypomastigote surface epitopes, defined as Ssp-1, Ssp-2, and Ssp-3, have served as useful markers to follow changes at the parasite surface over time (8). Trypomastigotes that had recently emerged from mammalian cells expressed Ssp-1 and Ssp-2, but during extracellular incubation of parasites, the epitopes gradually disappeared from the parasite surface (8). In contrast, the amount of Ssp-3 increased during the first 6 h of extracellular incubation and then gradually decreased, but this epitope was still detectable after 48 h, whereas Ssp-1 and Ssp-2 were not (8). Molecules carrying the Ssp-1 epitope were glycoproteins of approximately 100, 120, and 150 kDa, and Ssp-2 corresponded to a surface membrane protein of 70 kDa.

The nature of the Ssp-3–containing molecule(s) was not known until several years later, because the mAb to Ssp-3 failed to precipitate surface iodinated proteins. Ssp-3 eventually turned out to be a sialic acid–dependent epitope (108) created by transfer of exogenous sialic acid residues to parasite acceptors by the *T. cruzi* surface *trans*-sialidase (105, 106; see below). On the basis of this finding and on biochemical properties, Ssp-3 was proposed to be similar to the 35/50 kDa mucin-like acceptor molecule described for metacyclic stages (3). A monoclonal antibody recognizing the sialic acid–containing Ssp-3 epitope inhibited trypomastigote attachment to host cells (104, 108). The creation of the Ssp-3 epitope through the acquisition of sialic acid is a clear example of the surface remodeling that occurs in trypomastigotes. However, the full extent of changes that occur at the parasite surface after release from infected cells is presently unknown, and whether proteolytic processing events are included has not been fully elucidated.

Penetrin, a 60-kDa trypomastigote-specific heparin-binding protein, has been implicated in the attachment of parasites to host cells (78). The saturable binding of purified 60-kDa protein to Vero cells was inhibited by heparin, heparan sulfate, and collagen, of which heparin and heparan sulfate also inhibited binding and invasion by trypomastigotes. Expression of the 60-kDa trypomastigote protein in a strain of *Escherichia coli* that normally does not invade mammalian cells appeared to confer the ability to bind and penetrate Vero cells (78). Sequence information from the cloned penetrin cDNA is not yet available.

Host-derived fibronectin has been implicated in invasion of mammalian cells by trypomastigotes (80–82). Fibronectin binds to parasites in a specific saturable manner; antifibronectin antibodies inhibit invasion; and soluble human fibronectin augments invasion (80). In addition, peptides that mimic the fibronectin attachment domain competitively inhibit parasite invasion (82). Therefore, receptors on the parasite surface may mediate attachment to fibronectin on the target cell membrane. A trypomastigote fibronectin-binding protein (85 kDa) was isolated by ligand affinity chromatography (81). This protein, which also binds collagen (119), was identified as one of the major excretory-secretory products of trypomastigotes (83) and has been detected as a circulating antigen in the blood of patients with chronic Chagas' disease (83).

Trans-Sialidase and Related Surface Glycoproteins

The structure, function, and the relationship of *trans*-sialidase to invasion of mammalian cells by *T. cruzi* has proven to be a complex and interesting story. As several excellent reviews have recently appeared in the literature describing the structural and functional aspects of *trans*-sialidase (20, 24, 27, 32, 105, 106), we limit our discussion of this enzyme to a possible role in invasion and its relationship to other surface glycoproteins. *Trans*-sialidase belongs to a large multigene family (20) that encodes a heterogeneous group of enzymatically active glycoproteins ranging in size from 60–250 kDa (92). This enzyme is expressed at relatively high levels on the surface of metacyclic and culture-derived trypomastigotes, where it is anchored by GPI and functions in the transfer of exogenous sialic residues to acceptor molecules on the parasite surface (108). mAbs directed to the parasite sialic acid acceptor molecules block parasite entry into host cells (108, 126); therefore *trans*-sialidase was thought to perform an essential function in invasion. However, the requirement for *trans*-sialidase remains uncertain since antibodies to the catalytic domain, the only ones available that inhibit enzyme activity in vitro (103), are not inhibitory in the vicinity of cell surfaces, presumably because of the high concentration of sialic acid donors at that site (106).

The *trans*-sialidase superfamily also includes glycoproteins (82–90 kDa) that lack enzymatic activity, although these proteins contain the eight–amino acid Asp boxes characteristic of bacterial sialidases (20, 32). This subfamily is referred to as gp85 (106) and contains the metacyclic surface proteins gp90 (43) and gp82 (14). The gp85 family includes the Tc-85 group of trypomastigote-specific surface glycoproteins (1, 5, 44, 56, 85) that display heterogeneity in size, isoelectric point, and glycosylation. Tc-85 was originally identified with an antitrypomastigote antiserum, which after removing antibodies that cross-reacted with epimastigotes, reacted with an 85-kDa surface protein in trypomastigotes (128). This protein was then purified using lectin affinity chromatography in which it was bound to wheatgerm agglutinin (WGA) and specifically eluted with *N*-acetylglucosamine (57). Neuraminidase treatment did not affect binding of Tc-85 to WGA (57), although sialic acid was later shown to be a component of this glycoprotein (29). When *N*-linked glycosylation was inhibited, the protein appeared as a molecule of 75 kDa that still resolved as several spots on two-dimensional gels (1).

The GPI anchor of Tc-85 (28) contains a modified inositol group (MJM Alves, personal communication), which explains earlier observations of poor reactivity with antibodies to the cross-reacting determinant (CRD), which is inositol-dependent and normally exposed when GPI-linked molecules are released in soluble form after cleavage with PI-PLC (21, 31). Antibodies to Tc-85 partially block entry of mammalian cells by trypomastigotes (5), suggesting that Tc-85 molecules are involved in invasion. A recent study (45) showed that an acidic component of the Tc-85 family binds the extracellular matrix protein, laminin, indicating a possible mechanism for parasite attachment to host cells. The related metacyclic gp82 protein exhibited saturable binding to the surface of mammalian cells (95), which suggests that it too has a role in host-cell recognition.

The putative fibronectin-binding protein identified in trypomastigotes is also a molecule of 85 kDa (81), and although it has yet to be included in the gp85 gene family, it is likely to be related. Cross & Takle (32) pointed out that several members of the gp85/*trans*-sialidase family contain a Ser-x-Asp-x-Gly-x-Thr-Trp sequence motif that is also found in lectin-like proteins. Therefore, as diverse as the members of the *trans*-sialidase multigene family are, they may contain specific domains that exert the common function of host-cell recognition. Ongoing studies of the functional domains of the diverse members of this family may soon clarify this issue.

Role of Sialic Acid Acceptor Molecules in Invasion

Sialic acid residues on the surface of the host cells (26, 36, 71, 101) and infective forms of *T. cruzi* (87, 108) appear to influence the invasion process.

T. cruzi cannot synthesize sialic acids (90, 91); these sugars are acquired from external donors through the enzymatic action of the *trans*-sialidase, as mentioned above. Metacyclic and bloodstream-form trypomastigotes express highly glycosylated GPI-linked mucin-like proteins on their surfaces (3, 107) that function as acceptors of exogenous sialic acid by the *T. cruzi* surface *trans*-sialidase (106). As already discussed, antibodies to the metacyclic 35/50 kDa acceptor molecule inhibit invasion (98) and neutralize infections in mice (126), and antibodies to the bloodstream-form trypomastigote acceptor, Ssp-3, block attachment (108, 109). These results and the observations that an increase in parasite-surface associated sialic acid is correlated with increased infectivity (87, 108) indicate that the sialic acid acceptor molecules perform an important function in the recognition or invasion of host cells by parasites. Schenkman et al (109) have used nonphagocytic cell lines transfected with Fc receptors to obtain additional evidence for the participation of the Ssp-3 epitope in invasion. Opsonization of trypomastigotes with antibodies to Ssp-3 resulted in more efficient binding of parasites to cells expressing Fc receptors, but invasion was inhibited. This effect was specific for Ssp-3, because opsonization of parasites with polyclonal antibodies that reacted with other determinants on trypomastigotes resulted in efficient binding and invasion, unless antibodies specific for Ssp-3 were added to the polyclonal antiserum (109).

Amastigote Surface Antigens

As the transformation from trypomastigotes to amastigotes occurs, significant changes in morphology are accompanied by the acquisition of new surface molecules in the molecular weight range of 70,000–92,000 (8, 55, 84, 117). The complete sequence of a new family of amastigote-specific surface proteins, called amastins, was recently published (117). The function of the amastins is not known, and no significant homologies were found when the sequence was compared to known proteins. Although the deduced amino acid sequence of amastin corresponds to a protein of 19.5 kDa, Western blot analysis revealed that this ConA-binding protein is expressed as a higher-molecular-weight glycoconjugate (117). No consensus sequences for *N*-linked glycosylation were observed, but many possible sites for *O*-linked glycosylation were present, which is reminiscent of the mucin-like proteins described above. The relationship of amastins to the previously described major amastigote surface marker Ssp-4 (8, 10) is not clear. The C-terminal sequence of the amastin clone (117) is compatible with the addition of a GPI anchor predicted to be present in Ssp-4 by studies showing myristic acid labeling, sensitivity to PI-PLC, and reaction with anti-CRD antibodies (10).

As mentioned earlier, although the amastigote stage is considered essentially intracellular, this form of the parasite occurs in the circulation of infected hosts (8) and can infect both phagocytic and nonphagocytic mammalian cells (61,

66, 74). The role of amastigote surface molecules in invasion is not clear, but because they interact with specific structures on the surface of host cells (74), the interaction appears to be mediated by surface molecules. Antibodies in serum from patients with Chagas' disease react with amastigote surface antigens (7, 84), but the effect of these sera on invasion has not been analyzed. However, a recently described monoclonal antibody to Ssp-4, which recognizes an epitope distinct from that recognized by the original antibody (8), can inhibit amastigote invasion (16).

THE ROLE OF HOST-CELL SURFACE MOLECULES

The specificity of the parasite–host cell interaction must be dictated by both parasite and host-cell receptor molecules. The fact that *T. cruzi* can invade many different cell types in vitro suggests that either a relatively ubiquitous host-cell receptor(s) is utilized or that the heterogeneity in parasite surface glycoproteins imparts the ability to bind many different specific receptors. No distinct mammalian cell receptor has yet been identified, so *T. cruzi* may interact with several different molecules with similar biochemical characteristics. The concept of specificity in the interaction of *T. cruzi* with mammalian cells is well supported. First, trypomastigotes prefer to invade at the exposed cell margins of adherent cells and to enter polarized cells at the basolateral surfaces (102). These observations actually explain earlier findings of reduced invasion rates when host-cell monolayers reached confluence (89). Second, trypomastigotes and amastigotes do not compete for the same structures on host cells during invasion (74). The fact that treatment of host cells with proteases (4, 50) or glycosidases (36, 121, 122) prevents attachment and invasion by *T. cruzi*, suggests that surface glycoproteins are required.

The role of β_1-integrins as host-cell receptors for *T. cruzi* attachment was recently investigated. Antibodies specific for β_1-integrins had a significant inhibitory effect on the invasion of macrophages by trypomastigotes (42). The monoclonal antibody found to be effective in inhibiting parasite binding also inhibited the binding of fibronectin to β_1-integrin, and other antibodies that were directed to different epitopes, not involved in binding fibronectin, did not affect invasion (42). These results suggest that, at least in macrophages, parasites may bind to β_1-integrins, either directly or through the binding of fibronectin. β_1-Integrins have also been detected in the membrane of the vacuole that surrounds the parasite for a short time after invasion (46). As integrins are a ubiquitous family of integral membrane proteins linking the extracellular matrix to the cortical cytoskeleton and involved in signal transduction (54), they are good candidates for mediating the cellular changes that accompany *T. cruzi* invasion, such as the recruitment of lysosomes discussed above.

The Role of Cell Surface Carbohydrates

The concept that carbohydrates are involved in the early events of the *T. cruzi* interaction with host cells, including parasite attachment and internalization, is generally accepted. However, the precise role of carbohydrate moieties in these processes remains largely undefined, because the results from studies that address this issue conflict (12, 13, 121, 122). These contradictions may result from variations in glycosylation observed between parasite strains (33, 118) and among the different mammalian cell types used. To clarify this issue, three basic approaches have been taken: (*a*) inhibition with soluble monosaccharides and lectins, (*b*) glycosidase treatment of the parasite or host cell, and (*c*) invasion of mammalian cell lines with mutations that render them defective in glycosylation or deficient in surface proteoglycans.

Parasite surface glycoconjugates have been implicated in invasion by experiments demonstrating an inhibitory effect when *N*-linked glycosylation was blocked by tunicamycin (130) and those showing a correlation between sialylation of surface molecules and invasive capacity (108). Other studies indicate that host-cell glycoconjugates are required. For example, removal of mannose (121), galactose (122), or sialic acid (36) from the surface of mammalian cells with specific glycosidases resulted in a significant reduction in invasion by *T. cruzi*, whereas similar treatment of parasites either had no effect (36) or resulted in more efficient invasion (12, 121, 122). Invasion was inhibited when the surface carbohydrate moieties of fibroblasts and primary chick muscle cells were masked with lectins that bind mannose (ConA), *N*-acetylglucosamine and sialic acid (WGA), galactose (ricin I), or *N*-acetylgalactosamine (PHA) residues (50). Crane & Dvorak (30) found that only the exogenous addition of *N*-acetylglucosamine, out of nine monosaccharides tested, inhibited invasion of nonphagocytic cells by trypomastigotes. In contrast, de Arruda et al (34) observed a 50% inhibition of parasite entry into epithelial cells in the presence of the synthetic disaccharide β-D-galactofuranose. Zenian & Kierszenbaum (127) found that preincubation of the host cells with ConA inhibited *T. cruzi* invasion of macrophages, whereas Araujo-Jorge & de Souza (12, 13) reported that the effect of exogenous competing sugars and lectins on the invasion of macrophages varied depending on the strain of *T. cruzi* tested.

Recently, a more direct and productive approach demonstrated a role for host-cell surface glycoconjugates in invasion by *T. cruzi*. This approach was based on invasion assays utilizing cell lines deficient in surface glycoconjugates. The level of invasion observed in the sialic acid–deficient mutant cell line Lec2 (113) was about 50% lower than that observed in wild-type cells (26, 71, 101). Sialylation of Lec2 cells with the purified *T. cruzi trans*-sialidase and an exogenous source of sialic acid restored invasion to control levels. These results strongly suggest that host-cell sialic acid moieties are involved in the parasite–host cell recognition and/or invasion process. A similar study, using mutant Chinese hamster ovary (CHO) cells lacking proteoglycans (40), indicated a role for host-cell heparin and heparan sulfate glycosaminoglycans in invasion (51). Invasion of these mutant CHO or parental cell lines pretreated with lyases that selectively depolymerize heparin and heparan sulfate, or with inhibitors of glycosaminoglycan biosynthesis, was less efficient (around 50%) than invasion of untreated parental controls. Interestingly, removal of sialic acid from the proteoglycan-deficient mutants resulted in a further decrease in invasion, reaching 80% inhibition over parental controls. Conversely, removal of heparin and heparan sulfate from the surface of the sialic acid–deficient Lec2 cells produced even lower invasion levels (51). Therefore, binding of *T. cruzi* to host cells appears to involve both sialic acid–containing glycoconjugates and heparin and heparan sulfate glycosaminoglycans, which presumably interact with parasite surface molecules to promote attachment and invasion, perhaps in a cooperative manner. These results suggest that *T. cruzi* may utilize ubiquitous negatively charged host-cell molecules as receptors, perhaps explaining its capacity to invade a wide variety of vertebrate cells.

INHIBITORS OF INVASION

Antibodies

The first indication that invasion-blocking antibodies could be generated against *T. cruzi* came from studies using sera from patients with chronic Chagas' disease, or rabbits immunized with intact trypomastigotes. These antisera would partially block the entry of trypomastigotes into cultured mammalian cells (128). Subsequently, experiments using both tissue culture and metacyclic trypomastigotes showed that polyclonal antibodies generated to proteins of 80–90 kDa excised from preparative gels of trypomastigote lysates partially inhibited invasion of LLC-MK$_2$ cells. These antibodies were reported to be mainly directed against β-D-galactofuranosyl carbohydrate epitopes (34).

More specific information about surface epitopes involved in invasion came from experiments using monoclonal antibodies. As described above in more detail, several of the monoclonal antibodies raised to *T. cruzi* surface antigens inhibit to some degree parasite invasion of cultured mammalian cells (1, 5, 95, 125, 126), and some of these antibodies have a neutralizing effect in animal infections (11, 95, 126). Although a complete block of invasion was not achieved in these studies, they provided valuable information regarding the participation of metacyclic and trypomastigote surface molecules in host-cell recognition and invasion. Clearly, most of the glycoproteins recognized by these antibodies are encoded by the large sialidase-related multigene family (20, 32, 56, 106). Unfortunately, the presence of multiple simultaneously expressed gene copies, in many cases scattered throughout the parasite's genome, constitutes a major obstacle to direct functional studies, based on the generation of null mutants, of these proteins.

Chemical Inhibitors of Invasion

Several chemical reagents with specific biochemical targets are effective inhibitors of parasite attachment and/or invasion. Inhibition of ATP synthesis in *T. cruzi* with drugs that interfere with glycolysis (2-deoxyglucose) or mitochondrial electron transport and oxidative phosphorylation (antimycin A, oligomycin, sodium azide) profoundly affects the level of invasion of host cells. The decrease in invasive capacity results from the severe inhibition of parasite attachment to host cells (111). In addition, inhibition of parasite protein kinase activity blocked attachment of *T. cruzi* to host cells (111), and inhibition of host-cell protein kinases resulted in decreased invasion of macrophages (120).

Inhibitors of polyamine oxidase (64) and arginine decarboxylase (58) reduce the invasive capacity of trypomastigotes and inhibit intracellular development of these parasites. The expected target enzymes have not yet been found in *T. cruzi*; therefore, these drugs may have alternative mechanisms of action.

Peptidyl diazomethane derivatives and peptide fluoromethyl ketones, which are irreversible inhibitors of cysteine proteases, significantly impair the establishment of *T. cruzi* infections in cultured cells (48, 67). Preincubation of trypomastigotes with low concentrations of Z-(S-Bzl)Cys-Phe-CHN$_2$ significantly reduced invasion of heart muscle cells (67). In addition, when this and related inhibitors were added to cultures that had been previously infected, the parasite life cycle was arrested, because amastigote replication and transformation to trypomastigotes was blocked. Similarly, peptidyl fluoromethyl ketones such as Z-Phe-Ala-FMK and Z-Phe-Arg-FMK effectively blocked intracellular replication of amastigotes and exerted their strongest inhibitory effect during differentiation (48). Although this study revealed no effect of the fluoromethyl ketones on trypomastigote invasion, recent results demonstrate that arginine-containing peptidyl fluoromethyl and chloromethyl ketone inhibitors of cysteine proteases block invasion in a dose-dependent manner (19a). A target of these synthetic peptide inhibitors appears to be the major cysteine protease, cruzain, since it could be affinity labeled with an iodinated peptidyl diazomethane inhibitor (67). However, Z-Phe-Arg-FMK is a potent inhibitor of at least one protease expressed by invasive forms of *T. cruzi* that has properties distinct from cruzain (15, 19a, 100).

ESCAPE FROM THE VACUOLE

For *T. cruzi* to successfully perpetuate its life cycle, it must escape from the membranous vacuole that surrounds it after invasion and replicate as amastigotes in the cytosol (77). The parasitophorous vacuole membrane is gradually disrupted in the first few hours immediately following invasion and eventually disintegrates completely. This process requires an acidic environment, because artificial increases in the pH inhibit escape (62). Evidence suggests that a secreted amastigote pore-forming protein, TcTox, participates in vacuole disruption, as this hemolysin is active at low pH and is immunologically related to the terminal component of complement, C9 (6). Pore-forming proteins mediate the release of other intracellular pathogens such as *Listeria monocytogenes* (17) and *Shigella flexneri* (99) from membranous vacuoles after invasion of mammalian cells.

A possible role for the *T. cruzi trans*-sialidase in lysis of the vacuolar membrane was also investigated, because the fusion of lysosomes with the plasma membrane results in a vacuole that is rich in sialic acid–containing lysosomal membrane glycoproteins (lgps). Hall et al (47) showed that *T. cruzi* escaped from vacuoles in Lec2 cells, the mutant CHO cell line defective in sialylation (113), significantly faster than from vacuoles formed in the parental cell line. Since lgp molecules are a substrate for the *trans*-sialidase, which maintained enzymatic activity under acidic conditions, and membranes were more susceptible to lysis with TcTox after desialylation, these authors proposed that *trans*-sialidase activity would increase the efficiency of lysis of the parasitophorous vacuole membrane by TcTox (47).

CONCLUDING REMARKS

Although the mechanisms by which *Trypanosoma cruzi* invades mammalian cells are still not fully understood, significant progress has been made in the past few years. A decade ago, the available evidence only suggested the involvement of antigenic glycosylated molecules. One of the major recent advancements has been the finding that many of the glycoproteins present on the surface of the infective trypomastigotes, derived from infected cells or from differentiation of epimastigotes inside the insect, have similar structures and belong to a large family, including the unique *T. cruzi trans*-sialidase (20, 32). Although whether these glycoproteins share a common motif responsible for association with mammalian cells is not yet clear, in the next few years this possibility will certainly be investigated.

Another recent breakthrough has been the realization that the mechanism of *T. cruzi* entry into nonphagocytic cells (and perhaps phagocytes as well, under some circumstances) is clearly distinct from phagocytosis. Trypomastigotes attach to many different cell types in an energy-dependent manner (111), and this interaction appears to require negatively charged molecules of the host cell (51). Invasion occurs slowly (video microscopy observations reveal that 5–10 min are required for complete parasite internalization) and at specific sites of the host cell, i.e. the leading edges of lamellipodia (102). No pseudopod extension or any other obvious alteration in the host-cell plasma membrane occurs during invasion, and exposure to cytochalasin D enhances the process (115). This finding is the opposite of what is observed in phagocytosis, which is totally blocked by inhibitors of actin polymerization. Recent observations of clustering of host-cell lysosomes at the invasion site, and gradual fusion with the parasitophorous vacuole during its formation (115), suggest an intriguing mechanism that involves recruitment of intracellular organelles and that is so far unique to *T. cruzi*. Future studies should clarify the role of signal transduction mechanisms in this unusual process, and may lead to the discovery of novel, hormone-like parasite molecules that interact with a new class of vertebrate receptors.

ACKNOWLEDGMENTS

We thank Victor Nussenzweig and Victoria Ley for permission to include Figure 1. Work in the author's laboratory was supported by National Institutes of Health grants RO1AI32056 and RO1AI34867. BB was supported by a Training Award from the Medical Research Council of Canada.

> Any *Annual Review* chapter, as well as any article cited in an *Annual Review* chapter, may be purchased from the Annual Reviews Preprints and Reprints service.
> 1-800-347-8007; 415-259-5017; email: arpr@class.org

Literature Cited

1. Abuin G, Colli W, de Souza W, Alves MJM. 1989. A surface antigen of *Trypanosoma cruzi* involved in cell invasion (Tc-85) is heterogenous in expression and molecular constitution. *Mol. Biochem. Parasitol.* 35:229–38
2. Deleted in proof
3. Acosta A, Schenkman RP, Schenkman S. 1994. Sialic acid acceptors of different stages of *Trypanosoma cruzi* are mucin-like glycoproteins linked to the parasite membrane by GPI anchors. *Brazil. J. Med. Biol. Res.* 27:439–42
4. Alcantara A, Brener Z. 1978. *Trypanosoma cruzi*: role of macrophage membrane components in the phagocytosis of bloodstream forms. *Exp. Parasitol.* 50:1–6
5. Alves MJM, Abuin G, Kuwajima VY, Colli W. 1986. Partial inhibition of trypomastigote entry into cultured mammalian cells by monoclonal antibodies against a surface glycoprotein of *Trypanosoma cruzi*. *Mol. Biochem. Parasitol.* 21:75–82
6. Andrews NW, Abrams CK, Slatin SL, Griffiths G. 1990. A *T. cruzi*–secreted protein immunologically related to the complement component C9: evidence for membrane pore-forming activity at low pH. *Cell* 61:1277–87
7. Andrews NW, Einstein M, Nussenzweig V. 1989. Presence of antibodies to the major surface glycoprotein of *Trypanosoma cruzi* amastigotes in sera from Chagasic patients. *Am. J. Trop. Med. Hyg.* 40:46–49
8. Andrews NW, Hong KS, Robbins ES, Nussenzweig V. 1987. Stage-specific surface antigens expressed during the morphogenesis of vertebrate forms of *Trypanosoma cruzi*. *Exp. Parasitol.* 64:474–84
9. Andrews NW, Katzin AM, Colli W. 1984. Mapping of surface glycoproteins of *Trypanosoma cruzi* by two-dimensional electrophoresis. *Eur. J. Biochem.* 140:599–604
10. Andrews NW, Robbins ES, Ley V, Hong KS, Nussenzweig V. 1988. Developmentally regulated, phospholipase C–mediated release of the major surface

glycoprotein of amastigotes of *Trypanosoma cruzi. J. Exp. Med.* 167:300–14

11. Araguth MF, Rodrigues MM, Yoshida N. 1988. *Trypanosoma cruzi* metacyclic trypanosomes: neutralization by the stage-specific monoclonal antibody 1G7 and immunogenicity of 90 kDa surface antigen. *Parasitol. Immunol.* 10:707–17
12. Araujo-Jorge TC, de Souza W. 1984. Effect of carbohydrates, periodate and enzymes in the process of endocytosis of *Trypanosoma cruzi* by macrophages. *Acta Trop.* 41:17–28
13. Araujo-Jorge TC, de Souza W. 1988. Interaction of *Trypanosoma cruzi* with macrophages. Involvement of surface galactose and N-acetyl-D-galactosamine residues in the recognition process. *Acta Trop.* 45:127–36
14. Araya JE, Cano MI, Yoshida N, da Silveira JF. 1994. Cloning and characterization of a gene for the stage-specific 82-kDa surface antigen of metacyclic trypomastigotes of *Trypanosoma cruzi. Mol. Biochem. Parasitol.* 65:161–69
15. Ashall F, Harris D, Roberts H, Healy N, Shaw E. 1990. Substrate specificity and inhibitor sensitivity of a trypanosomatid alkaline peptidase. *Biochim. Biophys. Acta* 1053:293–99
16. Barros HC, Mortara RA. 1994. Ssp-4, the major surface glycoprotein of *Trypanosoma cruzi* amastigotes, has two carbohydrate epitopes, 2C2 and 1D9, but only 1D9 appears to be involved in cell invasion. *Mem. Inst. Oswaldo Cruz* 89:72 (Abstr.)
17. Bielecki J, Youngman P, Connelly P, Portnoy DA. 1990. *Bacillus subtilis* expressing a hemolysin gene from *Listeria monocytogenes* can grow in mammalian cells. *Nature* 345:175–76
18. Brener Z. 1973. Biology of *Trypanosoma cruzi. Annu. Rev. Microbiol.* 27:347–83
19. Burgoyne RD, Cheek TR. 1987. Reorganization of peripheral actin filaments as a prelude to exocytosis. *Biosci. Rep.* 7:281–88
19a. Burleigh BA, Andrews NW. 1995. A 120 kDa alkaline peptidase from *Trypanosoma cruzi* is involved in the generation of a novel Ca^{2+} signaling factor for mammalian cells. *J. Biol. Chem.* 270:51782–80
20. Campetella O, Sanchez D, Cazzulo JJ, Frasch ACC. 1992. A superfamily of *Trypanosoma cruzi* surface antigens. *Parasitol. Today* 8:378–81
21. Cardoso de Almeida ML, Turner M. 1983. The membrane form of variant surface glycoproteins of *Trypanosoma brucei. Nature* 302:349–51
22. Carvalho TMV, de Souza W. 1989. Early events related with the behavior of *Trypanosoma cruzi* within an endocytic vacuole in mouse peritoneal macrophages. *Cell Struct. Funct.* 14:383–92
23. Carvalho TU, de Souza W. 1986. Infectivity of amastigotes of *Trypanosoma cruzi. Rev. Inst. Med. Trop. Sao Paulo* 28:205–10
24. Cazzulo JJ, Frasch ACC. 1992. SAPA/*trans*-sialidase and cruzipain: two antigens from *Trypanosoma cruzi* contain immunodominant but enzymatically inactive domains. *FASEB J.* 6:3259–64
25. Chao D, Lotz JM, Dusanic DG. 1984. Infection of rodent myeloma Y3-Ag 1.2.3 cells by *Trypanosoma cruzi* metacyclic trypomastigotes. *J. Parasitol.* 70:1005–7
26. Ciavaglia M, de Carvalho TU, de Souza W. 1993. Interaction of *Trypanosoma cruzi* with cells with altered glycosylation patterns. *Biochem. Biophys. Res. Commun.* 193:718–21
27. Colli W. 1993. Trans-sialidase: a unique enzyme activity discovered in the protozoan *Trypanosoma cruzi. FASEB J.* 7:1257–64
28. Couto AS, de Lederkremer RM, Colli W, Alves MJM. 1993. The glycosylphosphatidylinositol anchor of the trypomastigote-specific Tc-85 glycoprotein from *Trypanosoma cruzi*. Metabolic labeling and structural studies. *Eur. J. Biochem.* 217:597–602
29. Couto AS, Goncalves MF, Colli W, Lederkremer RM. 1990. The N-linked carbohydrate chain of the 85-kilodalton glycoprotein from *Trypanosoma cruzi* trypomastigotes contains sialyl, fucosyl and galactosyl (α1–3) galactose units. *Mol. Biochem. Parasitol.* 39:101–8
30. Crane MS, Dvorak JA. 1982. Influence of monosaccharides on the infection of vertebrate cells by *Trypanosoma cruzi* and *Toxoplasma gondii. Mol. Biochem. Parasitol.* 5:333–41
31. Cross GAM. 1979. Crossreacting determinants in the C-terminal region of trypanosome variant surface antigens. *Nature* 277:310–12
32. Cross GAM, Takle GB. 1993. The surface *trans*-sialidase family of *Trypanosoma cruzi. Annu. Rev. Microbiol.* 46:385–411
33. de Andrade AF, Esteves MJ, Angluster J, Gonzalez-Perdamo M. 1991. Changes in cell-surface carbohydrates of *Trypanosoma cruzi* during metacyclogene-

sis under chemically defined conditions. *J. Gen. Microbiol.* 137:2845–49
34. de Arruda MV, Colli W, Zingales B. 1989. Terminal beta-D-galactofuranosyl epitopes recognized by antibodies that inhibit *Trypanosoma cruzi* internalization into mammalian cells. *Eur. J. Biochem.* 182:413–21
35. de Souza W, Martinez Palomo A, Gonzales-Robles A. 1978. The cell surface of *Trypanosoma cruzi*. *J. Cell Sci.* 33: 285–99
36. de Titto EH, Araujo FG. 1987. Mechanism of cell invasion by *Trypanosoma cruzi*: importance of sialidase activity. *Acta Trop.* 44:273–82
37. Downey GP, Chan CK, Trudel S, Grinstein S. 1990. Actin assembly in electropermeabilized neutrophils: role of intracellular calcium. *J. Cell Biol.* 110: 1975–82
38. Dvorak JA. 1975. New in vitro approach to quantitation of *Trypanosoma cruzi* vertebrate cell interactions. *New Approaches Am. Trypanosom. Res. Sci. Pub.* 318:109–20
39. Dvorak JA, Crane MS. 1981. Vertebrate cell cycle modulates infection by protozoan parasites. *Science* 214:1034–36
40. Esko JD, Weinke JL, Taylor WH, Ekborg G, Roden L, et al. 1987. Inhibition of chondroitin and heparan sulphate biosynthesis in Chinese hamster ovary cell mutants defective in galactosyltransferase I. *J. Biol. Chem.* 262:12189–95
41. Falkow S, Isberg RR, Portnoy DA. 1992. The interaction of bacteria with mammalian cells. *Annu. Rev. Cell Biol.* 8:333–63
42. Fernandez MA, Munoz-Fernandez MA, Fresno M. 1993. Involvement of B_1-integrins in the binding and entry of *Trypanosoma cruzi* into human macrophages. *Eur. J. Immunol.* 23:552–57
43. Franco FRS, Paranhos-Bacalla GS, Yamamuchi LM, Yoshida N, da Silveira JF. 1993. Characterization of a cDNA clone encoding the carboxy-terminal domain of a 90-kilodalton surface antigen of *Trypanosoma cruzi* metacyclic trypomastigotes. *Infect. Immun.* 61:4196–201
44. Giordano R, Alves MJM, Colli W, Tewari D, Fouts DL, Manning JE. 1994. The epitope recognized by neutralizing antibody H1A10 is encoded by a member of the *Trypanosoma cruzi* TSA/trans-sialidase supergene family. *Mem. Inst. Oswaldo Cruz* 89:97 (Abstr.)
45. Giordano R, Chammas R, Veiga SS, Colli W, Alves MJM. 1994. An acidic component of the heterogenous Tc-85 protein family from the surface of *Trypanosoma cruzi* is a laminin binding glycoprotein. *Mol. Biochem. Parasitol.* 65:85–94
46. Hall BF, Furtado GC, Joiner KA. 1991. Characterization of host cell–derived membrane proteins of the vacuole surrounding different intracellular forms of *Trypanosoma cruzi* in J774 cells. Evidence for phagocyte receptor sorting during the early stages of parasite entry. *J. Immunol.* 147:4313–21
47. Hall BF, Webster P, Ma AK, Joiner KA, Andrews NW. 1992. Desialylation of lysosomal membrane glycoproteins by *Trypanosoma cruzi*: a role for the surface neuraminidase in facilitating parasite entry into the host cell cytoplasm. *J. Exp. Med.* 176:313–25
48. Harth G, Andrews NW, Mills AA, Engel JC, Smith R, McKerrow JH. 1993. Peptide-fluoromethyl ketones arrest intracellular replication and intercellular transmission of *Trypanosoma cruzi*. *Mol. Biochem. Parasitol.* 58:17–24
49. Heath JP, Holifield BF. 1991. Cell locomotion. Actin alone in lamellipodia. *Nature* 352:107–8
50. Henriquez D, Piras R, Piras MM. 1981. The effect of surface membrane modifications of fibroblastic cells on the entry process of *Trypanosoma cruzi* trypomastigotes. *Mol. Biochem. Parasitol.* 2: 359–66
51. Herrera EM, Ming M, Ortega-Barria E, Pereira MEA. 1994. Mediation of *Trypanosoma cruzi* invasion by heparan sulphate on host cells and penetrin counter-receptors on the trypanosomes. *Mol. Biochem. Parasitol.* 65:73–83
52. Heuser JE. 1989. Changes in lysosome shape and distribution correlated with changes in cytoplasmic pH. *J. Cell Biol.* 108:855–64
53. Ho JL, Reed SG, Sobel J, Arruda S, He SH, et al. 1992. Interleukin-3 induces antimicrobial activity against *Leishmania amazonensis* and *Trypanosoma cruzi* and tumoricidal activity in human peripheral blood-derived macrophages. *Infect. Immun.* 60:1984–93
54. Hynes RO. 1992. Integrins: versatility, modulation, and signaling in cell adhesion. *Cell* 69:11–25
55. Iida K, Ley V. 1991. Isolation and characterization of a 92-kD surface molecule of *Trypanosoma cruzi* amastigotes recognized by a monoclonal antibody that induces complement-mediated killing. *Am. J. Trop. Med. Hyg.* 45:619–28
56. Kahn S, Van VW, Eisen H. 1990. The major 85-kD surface antigen of the mammalian form of *Trypanosoma cruzi* is encoded by a large heterogeneous

family of simultaneously expressed genes. *J. Exp. Med.* 172:589–97
57. Katzin AM, Colli W. 1983. Lectin receptors in *Trypanosoma cruzi*. An N-acetyl-D-glucosamine–containing surface glycoprotein specific for the trypomastigote stage. *Biochim. Biophys. Acta* 727:403–11
58. Kierszenbaum F, Wirth JJ, McCann PP, Sjoerdsma A. 1987. Arginine decarboxylase inhibitors reduce the capacity of *Trypanosoma cruzi* to infect and multiply in mammalian host cells. *Proc. Natl. Acad. Sci. USA* 84:4278–82
59. Kipnis TL, Calich VLG, Dias da Silva W. 1979. Active entry of bloodstream forms of *Trypanosoma cruzi* into macrophages. *Parasitology* 78:89–99
60. Koffer A, Tatham PE, Gomperts BD. 1990. Changes in the state of actin during the exocytotic reaction of permeabilized rat mast cells. *J. Cell Biol.* 111:919–27
61. Ley V, Andrews NW, Robbins ES, Nussenzweig V. 1988. Amastigotes of *Trypanosoma cruzi* sustain an infective cycle in mammalian cells. *J. Exp. Med.* 168:649–59
62. Ley V, Robbins ES, Nessenzweig V, Andrews NW. 1990. The exit of *Trypanosoma cruzi* from the phagosome is inhibited by raising the pH of acidic compartments. *J. Exp. Med.* 171:401–13
63. Lima MF, Kierszenbaum F. 1982. Biochemical requirements for intracellular invasion by *Trypanosoma cruzi:* protein synthesis. *J. Protozool.* 29:566–70
64. Majumder S, Kierszenbaum F. 1993. N,N'-thiophene-substituted polyamine analogs inhibit mammalian host cell invasion and intracellular multiplication of *Trypanosoma cruzi*. *Mol. Biochem. Parasitol.* 60:231–40
65. McCabe R, Mullins BT. 1990. Failure of *Trypanosoma cruzi* to trigger the respiratory burst of activated macrophages. *J. Immunol.* 144:2384–88
66. McCabe RE, Remington JS, Araujo FG. 1984. Mechanisms of invasion and replication of the intracellular stage of *Trypanosoma cruzi*. *Infect. Immun.* 46:372–78
67. Meirelles MN, Juliano L, Carmona E, Silva SG, Costa EM, et al. 1992. Inhibitors of the major cysteinyl proteinase (GP57/51) impair host cell invasion and arrest the intracellular development of *Trypanosoma cruzi* in vitro. *Mol. Biochem. Parasitol.* 52:175–84
68. Meirelles MNL, Araujo Jorge TC, de Souza W. 1982. Interaction of *Trypanosoma cruzi* with macrophages in vitro: dissociation of the attachment and internalization phases by low temperature and cytochalasin B. *Z. Parasitenkd.* 68:7–14
69. Meirelles MNL, de Souza W. 1983. Interaction of lysosomes with endocytic vacuoles in macrophages simultaneously infected with *Trypanosoma cruzi* and *Toxoplasma gondii*. *J. Submicrosc. Cytol. Pathol.* 17:327–34
70. Milder R, Kloetzel J. 1980. The development of *Trypanosoma cruzi* in macrophages in vitro. Interaction with host cell lysosomes and host cell fate. *Parasitology* 80:139–45
71. Ming M, Chuenkova M, Ortega-Barria E, Pereira MEA. 1993. Mediation of *Trypanosoma cruzi* invasion by sialic acid on the host cell and *trans*-sialidase on the trypanosome. *Mol. Biochem. Parasitol.* 59:243–52
72. Montgomery R, Webster P, Mellman I. 1991. Accumulation of undigestible substances reduces fusion competence of macrophage lysosomes. *J. Immunol.* 147:3087–93
73. Moreno SNJ, Silva J, Vercesi AE, Docampo R. 1994. Cytosolic free calcium elevation in *Trypanosoma cruzi* is required for cell invasion. *J. Exp. Med.* 180:1535–40
74. Mortara RA. 1991. *Trypanosoma cruzi* amastigotes and trypomastigotes interact with different structures on the surface of HeLa cells. *Exp. Parasitol.* 73:1–14
75. Mortara RA, da Silva S, Araguth MF, Blanco SA, Yoshida N. 1992. Polymorphism of the 35- and 50-kilodalton surface glycoconjugates of *Trypanosoma cruzi* metacyclic trypomastigotes. *Infect. Immun.* 60:4673–78
76. Nathan CF, Nogueira N, Juangbhanich C, Ellis J, Cohn Z. 1979. Activation of macrophages in vivo and in vitro. Correlation between hydrogen peroxide release and killing of *Trypanosoma cruzi*. *J. Exp. Med.* 149:1056–68
77. Nogueira N, Cohn Z. 1976. *Trypanosoma cruzi:* mechanism of entry and intracellular fate in mammalian cells. *J. Exp. Med.* 143:1402–20
77a. Ofek I, Goldhar J, Keisari Y. 1995. Nonopsonic phagocytosis of microorganisms. *Annu. Rev. Microbiol.* 49:239–76
78. Ortega-Barria E, Pereira MEA. 1991. A novel *Trypanosoma cruzi* heparin-binding protein promoted fibroblast adhesion and penetration of engineered bacteria and trypanosomes into mammalian cells. *Cell* 67:411–21
79. Osuna A, Castanys S, Rodriguez-Cabezas MN, Gamarro F. 1990. *Try-*

80. Ouaissi MA, Afchain D, Capron A, Grimaud JA. 1984. Fibronectin receptors on *Trypanosoma cruzi* trypomastigotes and their biological function. *Nature* 308:380–82
81. Ouaissi MA, Cornette J, Capron A. 1986. Identification and isolation of *Trypanosoma cruzi* trypomastigote cell surface protein with properties expected of a fibronectin receptor. *Mol. Biochem. Parasitol.* 19:201–11
82. Ouaissi MA, Cornette J, Gohain D, Capron A, Grasmasse H, Tartar A. 1986. *Trypanosoma cruzi* infection inhibited by peptides modeled from fibronectin cell attachment domain. *Science* 234:603–7
83. Ouaissi MA, Taibi A, Cornette J, Velge P, Marty B, et al. 1990. Characterization of major surface and excretory-secretory immunogens of *Trypanosoma cruzi* trypomastigotes and identification of potential protective antigen. *Parasitology* 100:115–24
84. Pan AA, McMahon-Pratt D. 1989. Amastigote and epimastigote stage-specific components of *Trypanosoma cruzi* characterized by using monoclonal antibodies. *J. Immunol.* 143:1001–8
85. Peterson DS, Wrightsman RA, Manning JE. 1986. Cloning of a major surface antigen gene of *Trypanosoma cruzi* and identification of a nonapeptide repeat. *Nature* 322:566–68
86. Piras MM, Henriquez D, Piras R. 1985. The effect of proteolytic enzymes and protease inhibitors on the interaction with *Trypanosoma cruzi* fibroblasts. *Mol. Biochem. Parasitol.* 14:151–63
87. Piras MM, Henriquez D, Piras R. 1987. The effect of fetuin and other sialoglycoproteins on the in vitro penetration of *Trypanosoma cruzi* trypomastigotes into fibroblastic cells. *Mol. Biochem. Parasitol.* 22:135–43
88. Piras MM, Piras R, Henriquez D. 1982. Changes in morphology and infectivity of cell culture–derived trypomastigotes of *Trypanosoma cruzi*. *Mol. Biochem. Parasitol.* 6:67–81
89. Piras R, Henriquez D, Piras MM. 1980. Studies on host-parasite interactions: role of fibroblastic cell surface functions and *Trypanosoma cruzi* forms in the invasion process. In *Host-Invader Interplay*, ed. H van der Bosche, pp 131–34. Amsterdam: Elsevier/North Holland Biomedical
90. Previato JO, Andrade AF, Pessolani MC, Mendonca-Previato L. 1985. Incorporation of sialic acid into *Trypanosoma cruzi* macromolecules. A proposal for a new metabolic route. *Mol. Biochem. Parasitol.* 16:85–96
91. Previato JO, Andrade AFB, Vermelho A, Firmino JC, Mendonca-Previato L. 1990. Evidence for *N*-glycolylneuraminic acid incorporation by *Trypanosoma cruzi* from infected animals. *Mem. Inst. Oswaldo Cruz* 85:38 (Abstr.)
92. Prioli RP, Mejia JS, Pereira ME. 1990. Monoclonal antibodies against *Trypanosoma cruzi* neuraminidase reveal enzyme polymorphism, recognize a subset of trypomastigotes, and enhance infection in vitro. *J. Immunol.* 144:4384–91
93. Procopio DO, Mortara RA. 1994. The mechanisms used by two developmental stages of *Trypanosoma cruzi* to invade mammalian cells are different and may be dependent on the host cells. *Mem. Inst. Oswaldo Cruz* 89:61 (Abstr.)
94. Putney JW Jr. 1993. Excitement about calcium signaling in inexcitable cells. *Science* 262:676–78
95. Ramirez MI, Ruiz R, Araya JE, da Silveira JF, Yoshida N. 1993. Involvement of the stage-specific 82-kilodalton adhesion molecule of *Trypanosoma cruzi* metacyclic trypomastigotes in host cell invasion. *Infect. Immun.* 61:3636–41
96. Reed SG. 1988. In vivo administration of recombinant IFN-gamma induces macrophage activation, and prevents acute disease, immune suppression, and death in experimental *Trypanosoma cruzi* infections. *J. Immunol.* 140:4342–47
97. Reyes MB, Pollevick GD, Frasch ACC. 1994. An unusually small gene encodes a putative mucin-like glycoprotein in *Trypanosoma cruzi*. *Gene* 40:139–40
98. Ruiz R, Rigoni VL, Gonzalez J, Yoshida N. 1993. The 35/50 kDa surface antigen of *Trypanosoma cruzi* metacyclic trypomastigotes, an adhesion molecule involved in host cell invasion. *Parasitol. Immunol.* 15:121–25
99. Sansonetti PJ, Ryter A, Clerc P, Maurelli AT, Mounier J. 1986. Multiplication of *Shigella flexneri* within HeLa cells: lysis of the phagocytic vacuole and plasmid-mediated contact hemolysis. *Infect. Immun.* 51:461–69
100. Santana JM, Grellier P, Rodier M-H, Schrevel J, Teixeira A. 1992. Purification and characterization of a new 120 kDa alkaline proteinase of *Trypanosoma cruzi*. *Biochem. Biophys. Res. Commun.* 187:1466–73
101. Schenkman RPS, Vandekerckhove F,

Schenkman S. 1993. Mammalian cell sialic acid enhances *Trypanosoma cruzi* invasion. *Infect. Immun.* 61:898–902
102. Schenkman S, Andrews NW, Nussenzweig V, Robbins ES. 1988. *Trypanosoma cruzi* invade a mammalian epithelial cell in a polarized manner. *Cell* 55:157–65
103. Schenkman S, Chaves LB, Pontes de Carvalho L, Eichinger D. 1994. A proteolytic fragment of *Trypanosoma cruzi trans*-sialidase lacking the carboxyl-terminal domain is active, monomeric, and generates antibodies that inhibit enzymatic activity. *J. Biol. Chem.* 269:7970–75
104. Schenkman S, Diaz C, Nussenzweig V. 1991. Attachment of *Trypanosoma cruzi* trypomastigotes to receptors at restricted cell surface domains. *Exp. Parasitol.* 72:76–86
105. Schenkman S, Eichinger D. 1993. *Trypanosoma cruzi trans*-sialidase and cell invasion. *Parasitol. Today* 9:218–22
106. Schenkman S, Eichinger D, Pereira MEA, Nussenzweig V. 1994. Structural and functional properties of *Trypanosoma trans*-sialidase. *Annu. Rev. Microbiol.* 48:499–523
107. Schenkman S, Ferguson MAJ, Heise N, Cardoso de Almeida ML, Mortara R, Yoshida N. 1993. Mucin-like glycoproteins linked to the membrane by glycosylphosphatidylinositol are the major acceptors of sialic acid in a reaction catalysed by *trans*-sialidase in metacyclic forms of *Trypanosoma cruzi*. *Mol. Biochem. Parasitol.* 59:293–304
108. Schenkman S, Jiang MS, Hart GW, Nussenzweig V. 1991. A novel cell surface *trans*-sialidase of *Trypanosoma cruzi* generates a stage-specific epitope required for invasion of mammalian cells. *Cell* 65:1117–25
109. Schenkman S, Kurosaki T, Ravetch JV, Nussenzweig V. 1992. Evidence for the participation of the Ssp-3 antigen in the invasion of non-phagocytic mammalian cells by *Trypanosoma cruzi*. *J. Exp. Med.* 175:1635–41
110. Schenkman S, Mortara R. 1992. Hela cells extend and internalize pseudopodia during active invasion by *Trypanosoma cruzi* trypomastigotes. *J. Cell Sci.* 101:895–905
111. Schenkman S, Robbins ES, Nussenzweig V. 1991. Attachment of *Trypanosoma cruzi* to mammalian cells requires parasite energy, and invasion can be independent of the target cell cytoskeleton. *Infect. Immun.* 59:645–54
112. Schenkman S, Yoshida N, Cardoso de Almeida ML. 1988. Glycophosphatidylinositol-anchored proteins in metacyclic trypomastigotes of *Trypanosoma cruzi*. *Mol. Biochem. Parasitol.* 29:141–52
113. Stanley P, Siminovitch L. 1977. Complementation between mutants of CHO cells resistant to a variety of plant lectins. *Som. Cell. Genet.* 3:391–405
114. Tardieux I, Nathanson MH, Andrews NW. 1994. Role in host cell invasion of *Trypanosoma cruzi*–induced cytosolic-free Ca^{2+}-transients. *J. Exp. Med.* 179:1017–22
115. Tardieux I, Webster P, Ravesloot J, Boron W, Lunn JA, et al. 1992. Lysosome recruitment and fusion are early events required for trypanosome invasion of mammalian cells. *Cell* 71:1117–30
116. Teixeira MMG, Yoshida N. 1986. Stage-specific surface antigens of metacyclic trypomastigotes of *Trypanosoma cruzi* identified by monoclonal antibodies. *Mol. Biochem. Parasitol.* 18:271–82
117. Teixeira SMR, Russell DG, Kirchhoff LV, Donelson JE. 1994. A differentially expressed gene family encoding "amastin", a surface protein of *Trypanosoma cruzi* amastigotes. *J. Biol. Chem.* 269:20509–16
118. Toma HK, Romanha AJ. 1994. Effect of monosaccharides, lectins and serum from mice on the in vitro interaction of *Trypanosoma cruzi* strains, from the state of Santa Catarina, with Vero cells. *Mem. Inst. Oswaldo Cruz* 89:60 (Abstr.)
119. Velge P, Ouaissi MA, Cornette J, Ajchain D, Capron A. 1988. Identification and isolation of *Trypanosoma cruzi* trypomastigote collagen-binding proteins: possible role in cell-parasite interaction. *Parasitology* 97:2–6
120. Vieira MC, de Carvalho TU, de Souza W. 1994. Effect of protein kinase inhibitors on the invasion process of macrophages by *Trypanosoma cruzi*. *Biochem. Biophys. Res. Commun.* 203:967–71
121. Villalta F, Kierszenbaum F. 1983. Role of cell surface mannose residues in host cell invasion by *Trypanosoma cruzi*. *Biochim. Biophys. Acta* 736:39–44
122. Villalta F, Kierszenbaum F. 1984. Host cell invasion by *Trypanosoma cruzi*: role of cell surface galactose residues. *Biochem. Biophys. Res. Commun.* 119:228–35
123. Warren G. 1985. Membrane traffic and organelle division trends. *Biochem. Sci.* 10:438–43
124. Yakubu M, Majumder S, Kierzembaum F. 1994. Changes in *Trypanosoma cruzi*

infectivity by treatments that affect calcium ion levels. *Mol. Biochem. Parasitol.* 66:119–25
125. Yoshida N, Blanco SA, Araguth MF, Russo M, Gonzalez J. 1990. The stage-specific 90-kilodalton surface antigen of metacyclic trypomastigotes of *Trypanosoma cruzi*. *Mol. Biochem. Parasitol.* 39:39–46
126. Yoshida N, Mortara RN, Araguth MF, Gonzalez JC, Russo M. 1989. Metacyclic neutralizing effect of monoclonal antibody 10D8 directed to the 35- and 50-kilodalton surface glycoconjugates of *Trypanosoma cruzi*. *Infect. Immun.* 57:1663–67
127. Zenian A, Kierszenbaum F. 1982. Inhibition of macrophage–*Trypanosoma cruzi* interaction by concanavalin A and differential binding of bloodstream and culture forms to the macrophage surface. *J. Parasitol.* 68:408–15
128. Zingales B, Andrews NW, Kuwajima VY, Colli W. 1982. Cell surface antigens of *Trypanosoma cruzi:* possible correlation with the interiorization process in mammalian cells. *Mol. Biochem. Parasitol.* 6:111–24
129. Zingales B, Colli W. 1985. *Trypanosoma cruzi:* interaction with host cells. *Curr. Top. Microbiol. Immunol.* 117:129–52
130. Zingales B, Katzin AM, Arruda MV, Colli W. 1985. Correlation of tunicamycin-sensitive surface glycoproteins from *Trypanosoma cruzi* with parasite interiorization into mammalian cells. *Mol. Biochem. Parasitol.* 16:21–34

POLYKETIDE SYNTHASE GENE MANIPULATION: A Structure-Function Approach in Engineering Novel Antibiotics

C. Richard Hutchinson

School of Pharmacy and Department of Bacteriology, University of Wisconsin, Madison, Wisconsin 53706

Isao Fujii

Faculty of Pharmaceutical Sciences, The University of Tokyo, Bunkyo-ku, Tokyo 113, Japan

KEY WORDS: biosynthesis, enzymes, polyketide cyclases, secondary metabolites, *Streptomyces*

CONTENTS

INTRODUCTION	202
POLYKETIDE SYNTHASES: STRUCTURE AND FUNCTION	205
Type I Multifunctional Enzyme Systems	205
Type II Multienzyme Systems	210
Intermediate Attachment and Channeling	212
APPROACHES TO STUDYING THE MECHANISMS OF POLYKETIDE SYNTHASES	213
In Bacteria	213
In Fungi	225
ENZYMES THAT ACT ON THE PRODUCTS OF POLYKETIDE SYNTHASES	226
Reduction	227
Oxidation	228
Methylation	229
POLYKETIDE SYNTHASES AS A SOURCE OF CHEMICAL DIVERSITY AND DRUGS	231
Targeted Modification of Type I Enzymes	231
Combinatorial Studies of Type II Systems	231

Abstract

Polyketides are produced primarily in microorganisms through a specialized metabolism that is a variation of fatty acid biosynthesis. A strong sequence and mechanistic similarity among many of the fatty acid and polyketide synthase enzymes has led to two paradigms for explaining polyketide biochemistry. In one, polyketides are formed by enzyme complexes consisting of four to seven monofunctional proteins in which the β-carbonyl groups of the intermediates resulting from the condensation of acetate residues are largely not reduced and cyclization of the intermediates typically produces aromatic compounds. The intermediates in the other model are formed by multifunctional enzymes in which each of the initial condensation products is processed through reduction; reduction and dehydration; or reduction, dehydration, and further reduction cycles to produce highly reduced compounds from acetate, propionate, and butyrate residues. Expression of the genes encoding each type of polyketide synthase, or their mutant forms, has provided much information about the underlying biochemistry and, in some cases, resulted in the formation of novel natural products.

INTRODUCTION

The idea that naturally occurring materials can be made by directed polymerization of simpler substances may have originated with the study of polyketides as well as of proteins; the chemistry of both types of natural products was under examination at the turn of this century, when the idea arose. Collie (21) coined the term *polyketide* to represent natural products containing multiple carbonyl or hydroxyl groups, each separated by one carbon atom, as in the structural element $-CH_2C(=O)CH_2CH(OH)CH_2C(=O)-$. He imagined that this characteristic feature could result from the polymerization of acetic acid or its chemical equivalent via the formation of poly-β-carbonyl–containing intermediates. Fatty acid biosynthesis is the biochemical counterpart of this concept, and experiments in the 1950s and early 1960s showed that both fatty acids and polyketides are derived from acetyl-coenzyme A (CoA) or related acyl-CoAs. However, as knowledge about the biochemistry of these two processes developed over the ensuing decades, investigators still could not explain how the diversity of chemical structures found in typical aromatic or reduced polyketides [the latter are also called complex polyketides (60)] (Figure 1) could emerge from the comparatively simple process of fatty acid biosynthesis, in which two-carbon units are polymerized to form saturated fatty acids consisting of 16–20 carbons by the reiteration of four reactions (Figure 2a). Nonetheless, the results of isotope labeling experiments (98) strongly suggest that the mechanisms of polyketide biosynthesis fall within the four types of

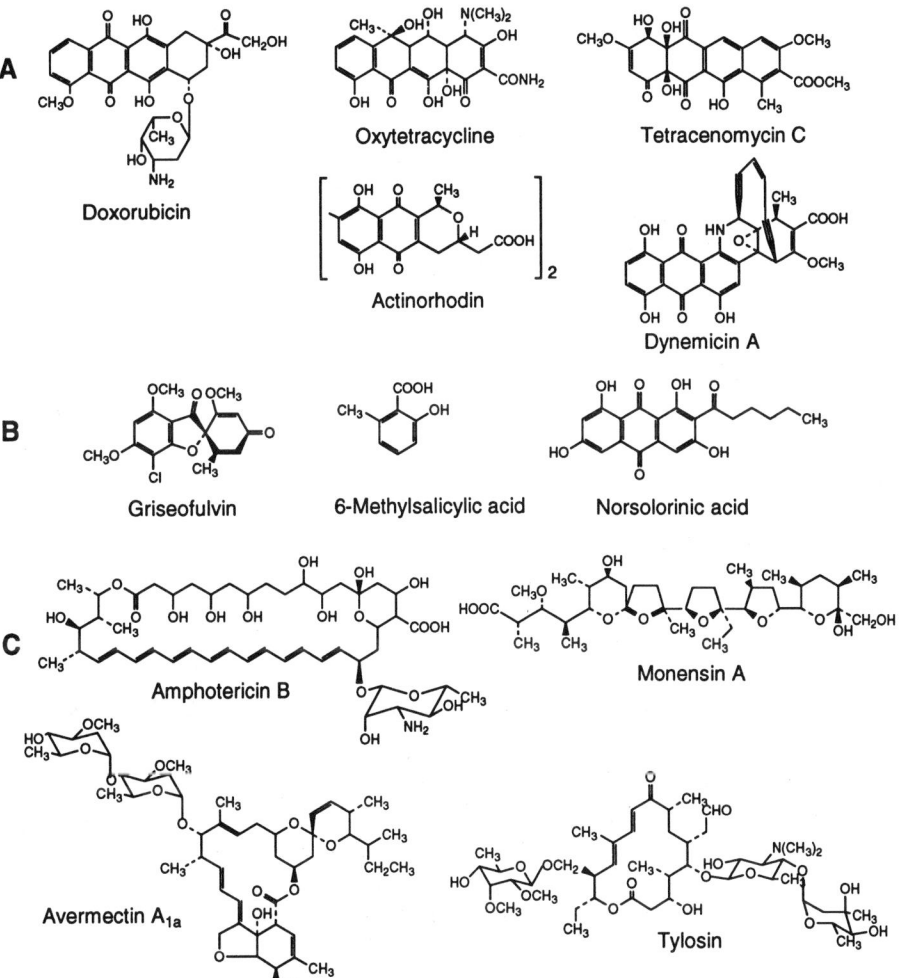

Figure 1 Representative aromatic polyketides from (*A*) actinomycete bacteria and (*B*) fungi and reduced (or complex) polyketides from (*C*) actinomycete bacteria.

reactions used to make fatty acids and that the monomeric building blocks are largely the same in the two systems. The difference is in how these steps and monomers are combined: Polyketides can be built from more than one type of monomer by the almost limitless permutations of the steps of fatty acid biosynthesis. Figure 2*b* illustrates this versatility for a hypothetical polyketide

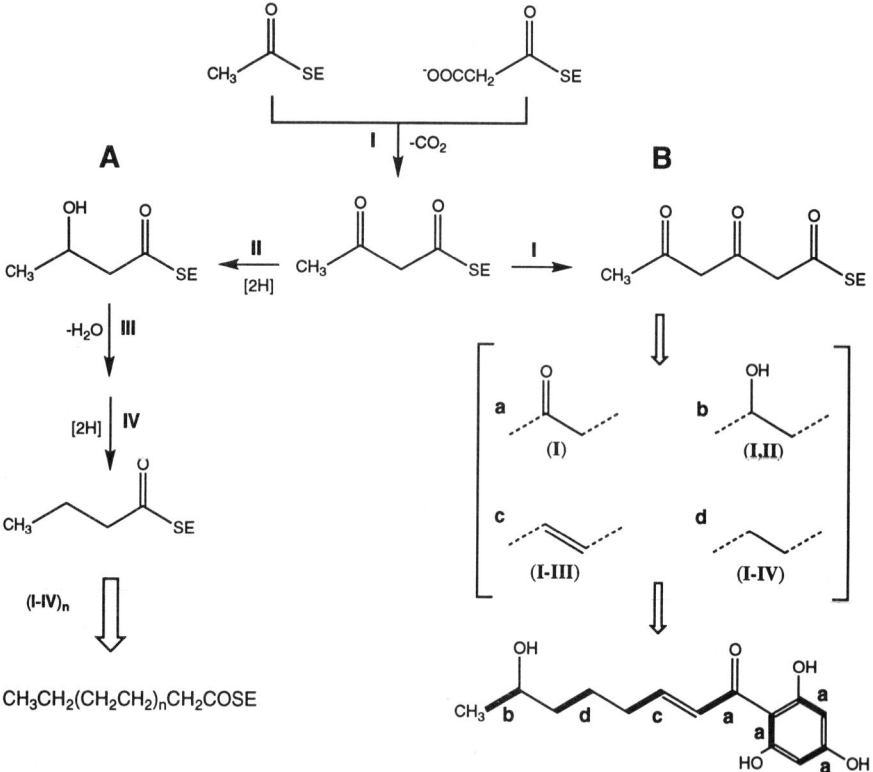

Figure 2 Comparison of the steps in (*A*) fatty acid and (*B*) polyketide synthesis. I. Condensation of acetyl-CoA (the starter unit) with malonyl-CoA (the carbon-chain extender unit) to produce a 3-oxobutyrylthioester. II. Reduction of the latter's carbonyl group at position 3 to a hydroxyl group. III. Dehydration of this hydroxyl to produce a double bond conjugated to the thioester carbonyl. IV. Reduction of this double bond to form the butyrylthioester. E, enzyme; [2H], reduction. (I-IV)$_n$ indicates repetition of steps I to IV n times; (I), (I,II), (I-III), and (I-IV) indicate that the structural segments are made through operation of the indicated steps, each of which is analogous to those shown in *A*.

constructed from the four structural segments, a–d, in the sequence b + d + c + 4 × a.

The complexity of polyketide biosynthesis might have remained largely unexplained were it not for advances in microbial genetics made since 1980, initially through the pioneering research on *Streptomyces* spp. of David Hopwood, his colleagues, and followers. Because the fruits of these efforts have provided several sets of bacterial and fungal genes encoding polyketide synthases (PKSs), interest in exploring the mechanisms of these enzymes has developed rapidly. The following review covers recent advances in this subject, highlighting the importance of microbial genetics in such work and showing how PKS-encoding genes can be used to make novel natural products. This survey aims only to illustrate approaches to answering the main questions currently being addressed and to present sufficient historical perspective to understand the reason for these lines of investigation. Readers interested in a more comprehensive coverage are encouraged to consult the excellent reviews by Hopwood & Sherman (53), Katz & Donadio (60), Hopwood & Khosla (52), and Hopwood (51); the monograph by O'Hagen (98); and the recent book edited by Vining & Stuttard (132).

POLYKETIDE SYNTHASES: STRUCTURE AND FUNCTION

Type I Multifunctional Enzyme Systems

As dictated by the convention for fatty acid synthases (FASs) (2), microbial PKSs have been classified into two types. The type I bacterial enzymes consist of giant multifunctional proteins, exemplified by 6-deoxyerythronolide B synthase (DEBS), which catalyzes the formation of the macrolactone portion of erythromycin A (Figure 3), a medically important antibiotic produced by *Saccharopolyspora erythraea*. Based on sequence data for the *eryAI, -II* and *-III* genes, Leadlay (10, 22) and Katz (30, 32) and their coworkers reported in 1990–1991 that the DEBS consists of three proteins of approximately 300 kDa, each of which is constructed of two modules containing the set of active-site domains needed to assemble 6-deoxyerythronolide B from propionyl-CoA and 2-methylmalonyl-CoA (Figure 3). The linear order of the catalytic sites in each module mirrors the arrangement of the seven active sites in animal FASs, although a given module may not have all of the FAS activities. Intriguingly, the order of the six modules in relation to the 5'-to-3' ends of the *eryA* genes is the same as the steps in 6-deoxyerythronolide B assembly (see Figure 6, below). The nature and number of catalytic sites in each module, together with the modules' ordering, suggest that the solution to the construction of 6-deoxyerythronolide B was to evolve a nonreiterative use of FAS-like activities.

Gene sequence data show analogous structure between other type I PKSs that catalyze the assembly of (*a*) macrolide antibiotics, such as the avermectins (77), FK506 (94), and rapamycin (64), whose structures resemble those of the macrolides, and (*b*) polyether antibiotics such as tetronasin (76), although the arrangement of the modules and enzyme activities can be different in individual type I PKSs (77). A set of genes encoding a type I–like PKS has been identified in a *Bacillus* sp. (109), but their role and the product made remain a mystery. Similarly, a recent entry in the DNA sequence databases (U00023) shows that

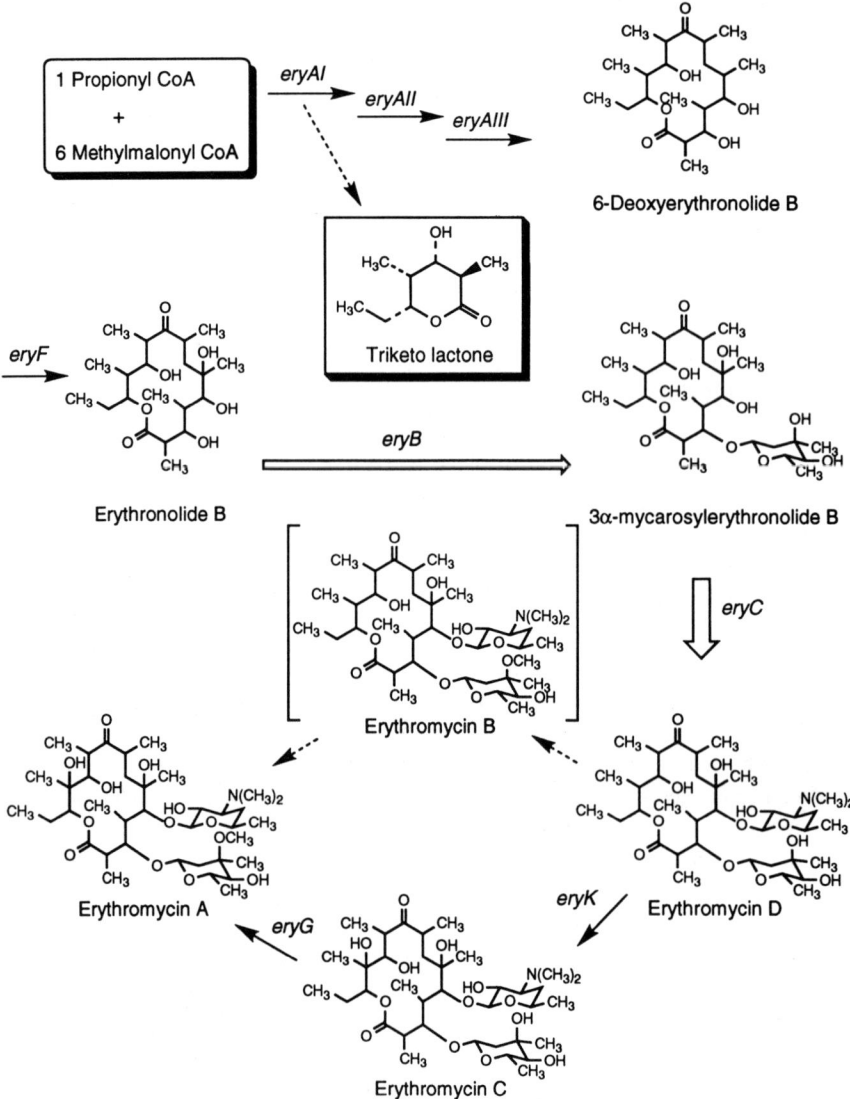

Figure 3 The erythromycin biosynthetic pathway in *Saccharopolyspora erythraea*. The names of the gene or classes of genes governing each step are indicated above or by the thin solid arrows for single genes or the open arrows for sets of genes. An alternate, minor pathway from erythromycin D to erythromycin A is indicated by dashed arrows and brackets. The dashed arrow beneath *eryAI* indicates that the triketo lactone can be made by the product of this gene alone (59).

Mycobacterium leprae contains a set of open reading frames that may encode six components of a putative PKS along with three proteins similar to the DrrAB daunorubicin resistance determinants (46). The function of these putative genes is unknown.

Fungal type I PKSs and some of their genes have also been discovered. In fact, 6-methylsalicylic acid synthase (MSAS) from the fungus *Penicillium patulum* was the first microbial PKS purified. Lynen and coworkers reported the purification of MSAS in a landmark 1971 paper (28) and showed that this protein of 1300±200 kDa [recent data support a size of 750 kDa (121)] contained all of the 11 catalytic activities needed for the synthesis of 6-methylsalicylic acid from acetyl- and malonyl-CoA. The sequence data for the MSAS-encoding gene isolated in 1990 (9) [and from *Penicillium urticae* in 1991 (134)] indicate that MSAS consists of four 190,731-Dalton subunits, each of which contains motifs resembling the acyl carrier protein (ACP), β-ketoacyl:ACP synthase (KS), and acetyl-CoA/malonyl-CoA:ACP acyltransferase (AT) domains of animal FASs (3). Synthesis of 6-methylsalicylic acid requires all of these activities for each condensation step, beginning with the reaction between the enzyme-bound forms of acetyl-CoA and malonyl-CoA. Therefore, unlike DEBS, MSAS uses partial reiteration of three FAS-like steps in the synthesis of 6-methylsalicylic acid. Ketoreductase (KR) and dehydratase (DH) activities are needed only once, when the 3-oxohexanoylthioester is reduced and dehydrated presumably to a *cis*-5-oxo-3-hexenoylthioester, but where the DH motif is located in MSAS is not yet clear. The combination of reiterated and nonreiterated catalytic activities suggests that MSAS can precisely recognize the structures of intermediates and that it can consequently determine when (and when not) to carry out the reduction-dehydration steps. This is clearly the case because MSAS does not perform the third condensation step in the absence of NADPH, the coenzyme used in reduction of the 3-oxohexanoylthioester. DEBS and the other bacterial type I PKSs, on the other hand, may not have such tight structure-recognition abilities: They exhibit a specific set of activities for the addition of each two-, three-, or four-carbon unit as the polyketide grows from the starting acyl-CoA. Furthermore, DEBS can skip different reduction steps and still continue synthesis of the polyketide chain (31, 32).

Other type I PKSs from fungi include the 130-kDa PKS from *Penicillium cyclopium* for the synthesis of orsellinic acid (57), the 4-hydroxy form of 6-methylsalicylic acid (whose synthesis would not require KR and DH activities), and the protein encoded by a 7.6-kb transcript from a putative PKS-encoding gene involved in hydroxynaphthalene and melanin biosynthesis in *Alternaria alternata* (68), as well as a similar protein involved in melanin biosynthesis in *Colletortrichum lagenarium* (70, 128a). Gene sequencing and disruption have proven that a type I PKS with subunits somewhat larger than those of MSAS is involved in sterigmatocystin synthesis in *Aspergillus nidulans* (J-H Yu & T Leonard, personal communication) and aflatoxin synthesis in *Aspergillus parasiticus* (J Linz, personal communication). The sterigmatocystin and aflatoxin clusters also contain two FAS-like genes that may encode the enzyme subunits responsible for assembling the hexanoyl-CoA starter for the synthesis of 10-deoxynorsolorinic acid, the earliest intermediate of sterigmatocystin and aflatoxin biosynthesis (34). *A. nidulans* also contains a gene at the *wA* locus that is predicted to encode a PKS-like type I protein of 216,633 Daltons governing the synthesis of a yellow conidiospore pigment (85). Although the structure of the latter polyketide is unknown, the ascospore pigment of *A. nidulans* is a dimeric hydroxylated anthraquinone, ascoquinone A (16), and *A. parasiticus* produces a hydroxymethylnaphthopyranone spore pigment called parasperone A (15).

Mevinolin (syn, lovastatin), an important cholesterol biosynthesis inhibitor from *Aspergillus terreus* (1), and the closely related compactin from *Monascus ruber* (36), are products of a pathway that begins with an almost fully reduced polyketide made from acetyl- and malonyl-CoA (Figure 4*a*). Although the exact identity of the putative mevinolin diol carboxylate (MDC) is unknown, dihydromonacolin L is a likely candidate on the basis of the suggestion (138) that the bicyclic ring system forms from cyclization of 5 to 6 (Figure 4*b*). This structure would be consistent with the processive formation of 1 to 5 as occurs in the macrolide antibiotics (18, 141) and other types of reduced fungal polyketides (75).

Sequence analysis of a gene encoding a 269-kDa protein essential for mevinolin biosynthesis has revealed that the putative PKS contains all of the active sites found in animal FAS, except for the absence of a recognizable thioesterase (TE) domain. It also has an additional motif between the DH and enoyl reductase (ER) domains that is quite similar to the sequences of methyltransferases from bacteria and eukaryotes (86). The latter finding is consistent with the fact that S-adenosylmethionine–dependent methylation takes place at C-6 of mevinolin during its biosynthesis. The remaining functional domains of the putative mevinolin PKS are ordered identically to those of most other type I PKS and FAS enzymes; however, at the level of protein domain and subunit organization, this enzyme bears little resemblance to the FAS of *P. patulum* (136).

How does the mevinolin PKS tell when to introduce the double bonds, hydroxyls, and the C-methyl group of the dihydromonacolin L precursor? The answer is not yet available, but clearly molecular recognition features must be involved that are different from those operating in the biosynthesis of macrolides and aromatic compounds made by other type I PKSs, because the mevinolin PKS lacks the reiterated domain structure of DEBS and must perform a greater variety of reductions and dehydrations than MSAS.

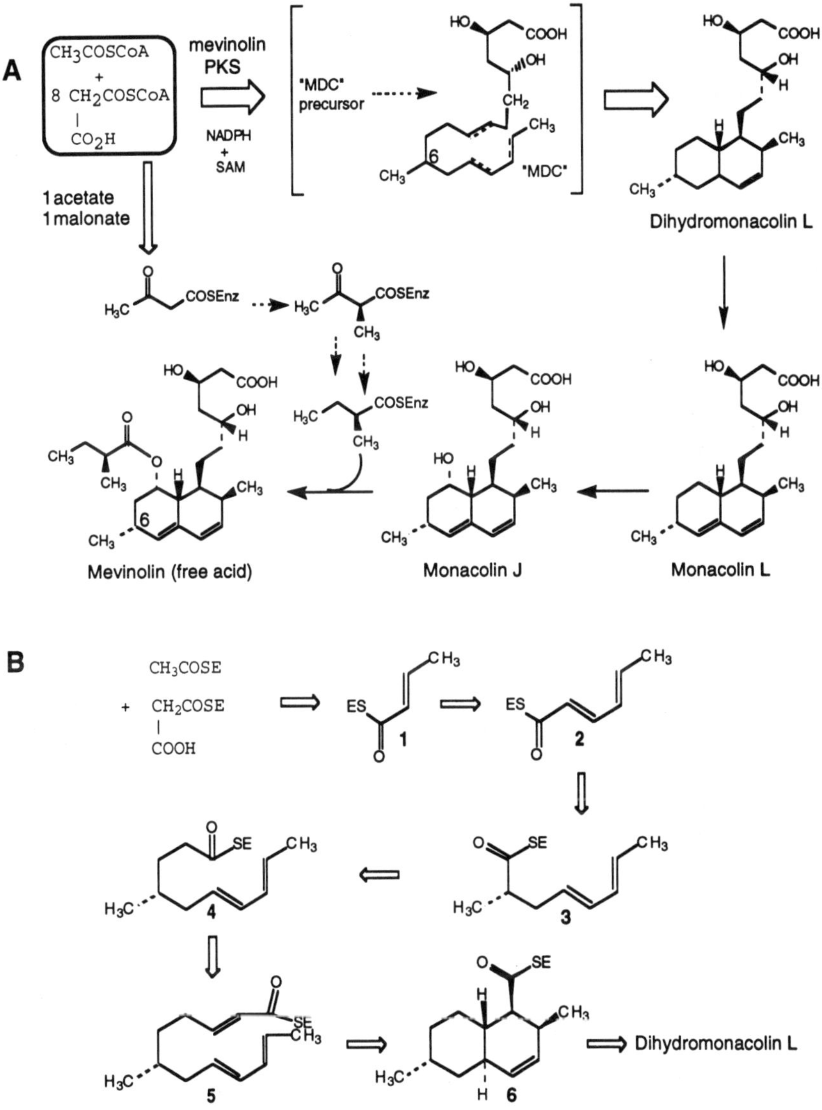

Figure 4 (A) The biosynthetic pathway to mevinolin free acid in *Aspergillus terreus*. The structure of the nonaketide product made by a type I PKS that requires NADPH and S-adenosylmethionine (SAM) is unknown: "MDC" with three double bonds (*dashed line*) is one possibility. The diketide made by a different PKS of unknown structure is probably acetoacetate or its 2-methyl derivative. Dashed arrows indicate hypothetical steps; thin solid and thick open arrows indicate single and multiple steps, respectively. (*B*) The hypothesis for the synthesis of dihydromonacolin L (138). Structures 1 to 6 would be formed processively by the mevinolin PKS; then 6 would be extended by malonyl-CoA to produce the enzyme-bound form of dihydromonacolin L. E, enzyme.

Type II Multienzyme Systems

PKSs consisting of several separate, largely monofunctional proteins are called type II systems. Information derived from gene sequences about type II PKSs predated the discovery of type I systems by two years, although purification of a complete set of type II PKS enzymes has not been achieved. Studies of the genetics of actinorhodin (act) biosynthesis in *Streptomyces coelicolor* (81) and tetracenomycin (tcm) biosynthesis in *Streptomyces glaucescens* (95) led to the isolation of the first genes encoding type II PKSs and the discovery of extensive cross-hybridization among some of these genes in different actinomycetes (80). This feature greatly facilitated the isolation of additional clusters of such genes so that 13 sets from different *Streptomyces* and *Saccharopolyspora* species have now been identified and characterized (Figure 5). Sequence analysis of the PKS-encoding DNA from the granaticin (*gra*) biosynthesis genes in *Streptomyces violaceoruber* (118) (the initial steps in actinorhodin and granaticin biosynthesis are the same), the *tcm* genes (12, 124, 125), and the *act* genes (37, 48) established the basic architecture of a type II PKS. It consists of individual enzymes with extensive sequence similarity to the KS, ACP, and KR enzymes of bacterial fatty acid biosynthesis, plus unique proteins called cyclases or aromatases that catalyze the formation of aromatic ring–containing compounds from the poly-β-ketone intermediates typically made by the KS and ACP from acetyl- and malonyl-CoA. The KS enzymes always occur with a highly homologous protein, recently named the chain length factor (CLF) (90), whose sequence is quite similar to that of the KS but which lacks the highly conserved Cys at the presumed site of the condensation reaction. Genes encoding KR enzymes are not present in all type II PKSs, and other systems may contain additional enzymes such as an AT in daunorubicin (45, 137) or ATs plus apparent acyl-CoA ligases and asparaginases in the oxytetracycline (otc) PKS (I Hunter, personal communication). These extra kinds of enzymes are probably involved in specifying the unusual starter units: propionyl-CoA for daunorubicin and possibly malonamyl-CoA for oxytetracycline.

The mechanisms of type II PKS enzymes are quite different from those of the type I class, because these enzyme complexes catalyze the formation of cyclic, aromatic compounds that do not require extensive reduction or reduction and dehydration cycles. Consequently, the principal challenges faced by type II PKSs are to determine how many times chain extension occurs, how to fold the poly-β-ketone intermediates into the correct orientation for cyclization, and how to select which carbonyl and methylene groups react to form the six-membered rings of the final cyclic compounds.

In spite of the obvious structural and mechanistic differences between type I and type II PKSs, the type II class does not reflect a unique solution to

MICROBIAL POLYKETIDE SYNTHASES 211

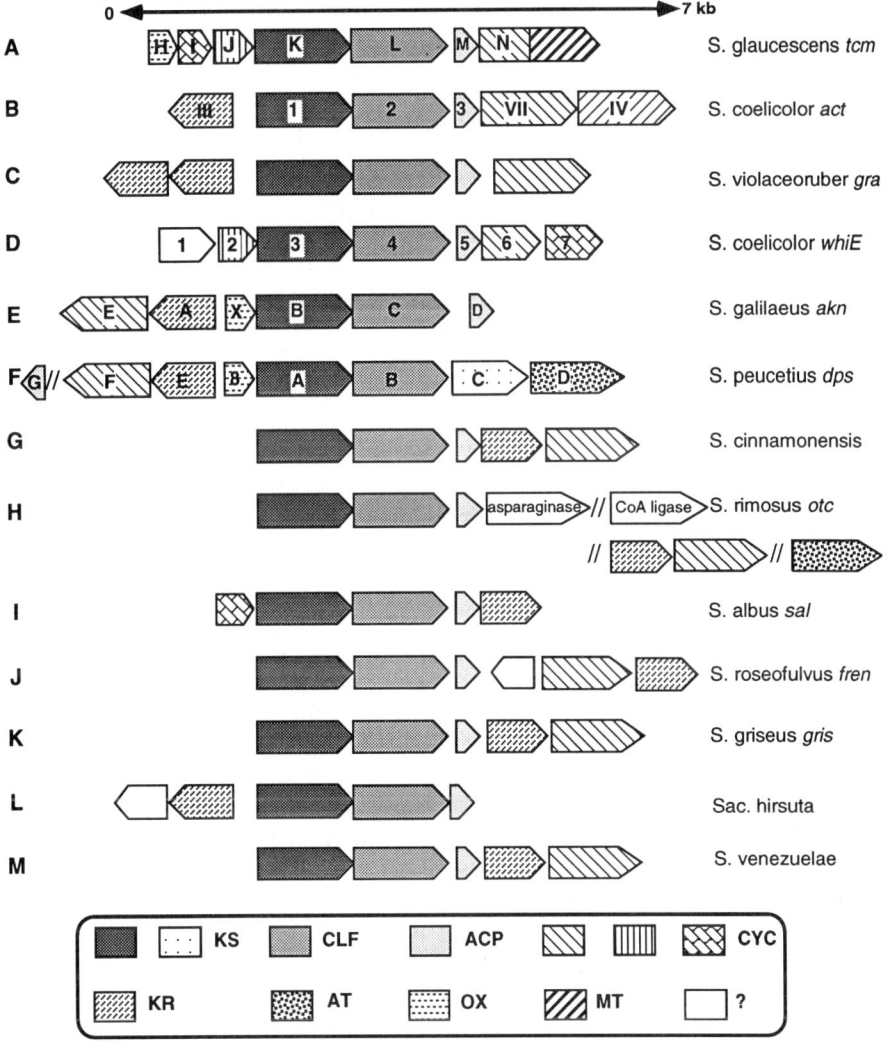

Figure 5 Comparison of the organization of the known gene clusters encoding type II PKSs. Wedges depict the relative sizes of each gene and are shaded to represent known functions (slashes indicate that open reading frames intervene between the ones shown). KS, CLF, ACP, KR, and AT are defined in the text; CYC, polyketide cyclase; OX, oxidation; MT, methyltransferase; ?, unknown function. References: *A* (12, 124, 125), *B* (37, 48), *C* (118), *D* (24), *E* (129a,b), *F* (45, 137), *G* (7), *H* (66a; I Hunter, personal communication), *I* (JW Suh & CR Hutchinson, unpublished results), *J* (13), *K* (140), *L* (73), *M* (48a).

creating cyclic, aromatic molecules, because these molecules are also formed by fungal type I PKSs. Hence, it is not clear whether the genes encoding type II enzymes simply evolved first (with the corollary that the type I enzymes with equivalent catalytic properties arose by fusion of genes encoding type II FASs or PKSs) or, as indicated by their apparent ubiquity, actually represent a better solution to the mechanistic problem.

Intermediate Attachment and Channeling

Regardless of the type of PKS used to synthesize a polyketide, the intermediates of carbon-chain assembly are thought to remain enzyme bound from the point of the condensation between the starter and the first extender unit until the carbon chain reaches its full length. This belief follows from the principles of fatty acid biosynthesis that have emerged from work with purified FASs from many organisms (2, 133) and is supported by the observed properties of MSAS (27, 28, 108, 121) and DEBS (6, 74, 82, 104). The intermediates are presumably attached via thioester linkages to a highly conserved Cys at the active site of an individual KS protein or domain or to the free SH group of the 4'-phosphopantetheine attached to a Ser in the holoenzyme form of an ACP or ACP domain. Removal of the attachment site therefore leads to enzyme inactivity, as observed for the S42A mutant of the actinorhodin PKS ACP (65), the C173A and C173S mutants of TcmK (92), the KS subunit of the tetracenomycin PKS, and the DEBS3 subunit isolated from a recombinant *Escherichia coli* that expressed the *eryAIII* gene but did not 4'-phosphopantethenylate the ACP domain of the resulting DEBS3 protein (104). The inhibition of polyketide synthesis by thiol-specific reagents such as cerulenin (97) also supports this idea. Covalent attachment explains why the carbon-chain assembly intermediates rarely, if ever, appear outside the cells unless the strains contain mutant forms of the PKS, such as those that produce tylosin (54), mycinamicin (69), and erythromycin (32, 33). On the other hand, because these intermediates can be incorporated into the final product if they are added to the culture medium in the form of their N-acetylcysteamine thioesters (14, 18, 47, 75, 100, 141), exchange of free and enzyme-bound intermediates must be possible. This exchange most likely takes place through an aberration of the normal transfer of substrates between the –SH groups of the KS and ACP active sites that occurs during each condensation step.

An important consequence of covalent attachment is the channeling of intermediates along the preferred pathway. For instance, attachment may enforce conformational constraints favoring the desired stereoselectivity in the condensation between $RCH_2COSEnz$ (R = rest of the polyketide) and 2-methyl- or 2-ethylmalonyl-SEnz intermediates, or those favoring the reduction of the resulting β-keto products of condensation in macrolide or polyether biosynthesis. Alternatively, it could ensure the correct folding of poly-β-ketone intermediates so that only the desired methylene and carbonyl groups undergo the intramolecular aldol reaction leading to a particular aromatic ring. Preventing undesired aldol reactions might also be important, because studies of chemical models have shown that the poly-β-ketone intermediates are quite reactive when in solution (49). Tethering also might be helpful in determining the size (length) of the poly-β-ketone intermediates, if this technique allowed, for example, a type II PKS to measure size with respect to the point of attachment and use the principle of cavity filling as the size determinant instead of kinetic effects on the condensation reaction or the rate of release of the polyketide intermediate. Of course, these considerations are not meant to minimize the well-recognized value of substrate channeling for achieving high efficiency in cellular metabolism (83).

APPROACHES TO STUDYING THE MECHANISMS OF POLYKETIDE SYNTHASES

In Bacteria

The mechanisms by which PKSs assemble the carbon skeletons of polyketides will eventually have to be studied in vitro using purified enzymes. Because purification and reconstitution of a complete type I or type II PKS from a bacterium has not been reported, most of the mechanistic studies have been done in vivo using recombinant organisms to express the native or mutant PKS genes or to study the effect of mutations in the chromosomal copy of the PKS-encoding genes. Although these studies have led to valuable mechanistic insights, many of the deductions are largely inferential.

TYPE I PKSs Among the type I systems known at the gene level, only DEBS has been studied mechanistically to any significant extent. Donadio et al (32) reported that an in-frame deletion of a 271–amino acid region containing the KR domain closest to the N-terminus of the DEBS3 subunit resulted in the formation of 5-deoxy-5-oxo-6-deoxyerythronolide B and its 3α-mycarosyl derivative as the major products. The lack of reduction of the C-5 carbonyl was predicted on the basis of the hypothesis, derived from the sequence data for *eryAI–III*, that the order of the active sites in the DEBS1, -2, and -3 subunits mirrors the sequence of the steps in the synthesis of 6-deoxyerythronolide B. The experimental results validated this idea and also showed that the mutant DEBS3 subunit could form a catalytically competent complex with DEBS1 and DEBS2. Additional support for the thought that mutant forms of a type I PKS can retain significant enzyme activity has come from studies of a DEBS2 mutant in which two amino acids in the normally functional ER active site were altered to inactivate this function, resulting in the synthesis of a small amount of 6,7-anhydroerythromycin C through metabolism of the 6,7-anhydro-6-deoxyerythronolide B produced by the mutant PKS (31). Moreover, the 2-norerythromycins formed upon introduction of a library of DNA from the oleandomycin-producing *Streptomyces antibioticus* into a *S. erythraea* mutant (87) containing an in-frame deletion of the AT domain in module 6 of DEBS3; this formation may have resulted from suppression of the *eryAIII* mutation either by the starter AT domain from the type I PKS of *S. antibioticus* (127) or by the FAS malonyl-CoA:ACP acyltransferase.

Besides causing the formation of modified full-length polyketides, DEBS can make products containing fewer carbons than 6-deoxyerythronolide B. 15-Norerythromycins are known (135) and assumed to result from the occasional use of acetyl-CoA as the starter unit instead of propionyl-CoA. The *eryAIII* mutant that produced 5-deoxy-5-oxo-6-deoxyerythronolide B also made a small amount of 3,5-dihydroxy-2,4-dimethyl-*n*-heptanoic acid-δ-lactone (triketide lactone) (33), the expected product of the DEBS1 subunit (Figure 3). The triketide lactone also results when the *eryAI* gene encoding DEBS1 is expressed alone in vivo either with (P Leadlay, personal communication) or without (59) the addition of the TE domain from the C-terminus of DEBS3. The TE domain has been postulated (32) to release and cyclize the linear form of 6-deoxyerythronolide B from DEBS. These results, together with the formation of various intermediates of macrolactone formation by mutants of other macrolide-producing bacteria, show that subunits of a type I PKS can turn over in the absence of the complete complex of multifunctional enzymes. Thus, engineered type I PKSs resulting from unusual combinations of the *eryA* modules, or their homologues from other bacteria, may be able to biosynthesize novel reduced polyketides, a feature that will be useful in the search for new natural products with potential importance as drugs.

Another question that has been studied with DEBS is how type I PKSs can create the enantiomeric stereochemistry at the carbons bearing the methyl or ethyl groups. The stereochemistry of these carbons varies as a function of their position in the growing polyketide chain (Figure 6), which is true of all 14- and 16-membered macrolide antibiotics (19). Similar stereochemical patterns are found in polyether antibiotics (17). Marsden et al (82) have reported that (2S)-2-methylmalonyl-CoA, the chain-extending unit in 6-deoxyerythronolide B synthesis (Figure 6), is loaded onto the DEBS subunits in vitro but the (2R) enantiomer is not. This result suggests that the substrate, or the newly created RCH(CH$_3$)C(=O)– center resulting from the condensation of (2S)-2-methylmalonyl-SEnz with the RC(=O)SEnz intermediate, undergoes C-2 epimerization during some chain extension (condensation) steps but not others. Epimerization immediately before the condensation reaction is supported by the confirmation that an analogous event takes place during the biosynthesis of the oligopeptide antibiotics actinomycin (123) and gramicidin (M Marahiel,

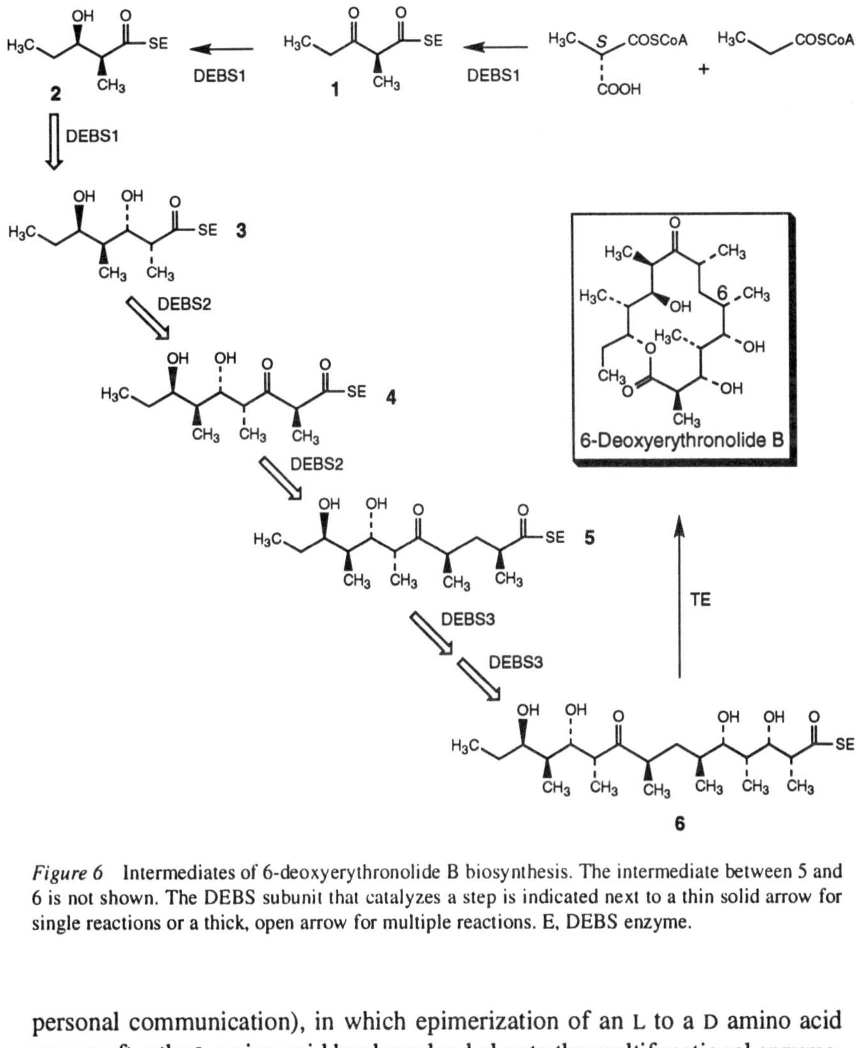

Figure 6 Intermediates of 6-deoxyerythronolide B biosynthesis. The intermediate between 5 and 6 is not shown. The DEBS subunit that catalyzes a step is indicated next to a thin solid arrow for single reactions or a thick, open arrow for multiple reactions. E, DEBS enzyme.

personal communication), in which epimerization of an L to a D amino acid occurs after the L amino acid has been loaded onto the multifunctional enzyme.

Purification of bacterial type I PKSs and reconstitution studies are needed for an understanding of other facets of polyketide assembly; for instance, how is the chain-extending substrate chosen for 16-membered macrolides, such as tylosin, that use a combination of malonyl-, 2-methylmalonyl-, and 2-ethylmalonyl-CoA, and how is the stereochemistry of the $RCH(OH)C(CH_3)-$ centers established by reduction of the β-carbonyl groups created in each condensation reaction? Advances in the purification of DEBS have been described by Aparicio et al (6), who reported that the native DEBS subunits exist as dimers. An N-terminal segment recovered from limited proteolysis of DEBS1 and containing the AT and ACP domains was labeled by [1-^{14}C] propionyl-CoA in vitro, which proves that this is the site for initiation of chain synthesis (6). Sherman and coworkers have expressed some of the tylactone synthase genes in insect cells (D Sherman, personal communication) to try to overcome the lack of adequate pantethenylation of the ACP domains observed for DEBS2 upon expression of the *eryAII* gene in *E. coli* (104). The same ploy has been successful for animal FAS (58).

TYPE II PKSs Mechanistic investigations of bacterial type II PKSs have also relied heavily on in vivo studies, although the purified components have been used to investigate some aspects of these systems. Expression of the *tcmM* gene, which encodes the tetracenomycin PKS ACP (12), in *E. coli* gave mostly apoprotein, whereas expression in *Streptomyces lividans* resulted in exclusive formation of the 4′-phosphopantethenylated holoenzyme, which accepted malonate from malonyl-CoA, catalyzed by a crude malonyl-CoA:ACP acyltransferase preparation from *S. glaucescens* (115). In contrast, the holo-ACP was the exclusive product when the *S. glaucescens fabC* gene encoding a putative FAS ACP was expressed in *E. coli* (RG Summers, A Ali, B Shen, WA Wessel & CR Hutchinson, submitted). Because these results imply that an [ACP]synthase (35) committed to fatty acid synthesis in *E. coli* functions poorly with a heterologous PKS apo-ACP but not its FAS counterpart, one wonders whether *S. glaucescens* has two [ACP]synthases: one for FabC and another for TcmM.

The following results show that a type II PKS has stricter requirements for some of its component enzymes than for others. Heterologous ACPs appear to function nearly as well as the actinorhodin PKS ACP in the synthesis of actinorhodin and its biosynthetic intermediates. Replacement of the *actI-ORF3* gene encoding the Act ACP with *tcmM,* the genes encoding PKS ACPs from the granaticin and oxytetracycline producers, or even the *S. erythraea* gene encoding a putative FAS ACP allowed the pigmentation typical of Act$^+$ strains (65, 66). As the structures of the decaketide intermediates (i.e. a poly-β-ketone made from 10 two-carbon units) of tetracenomycin and oxytetracycline biosynthesis are quite different from the octaketide precursor of actinorhodin and granaticin, the ACP component of a type II PKS must not have a major influence on the type of polyketide produced. Mutations in the *actI-ORF1* KS component were complemented *in trans* by the *graI-ORF1, whiE-ORF1,* and *otc-ORF1* genes (67, 117), which again resulted in pigmentation typical of an Act$^+$ strain. In contrast, mutations in the *actI-ORF2* gene encoding the CLF component were largely not complemented by their heterologous counterparts (*graI-ORF2* was the exception among the three genes tested) (67, 117).

A possibly better approach to investigating the role of each component and determining how well heterologous type II PKSs function is to express cassettes of their genes introduced on a suitable vector into a background devoid of the normal type II PKS. This technique should minimize the background activities and uncertainties associated with different levels of enzyme coming from chromosomal and plasmid-borne genes. Furthermore, it can also provide for regulated or higher-level expression of the PKS-encoding genes as a function of the vector components. Khosla, Hopwood, and their coworkers have described a system for the regulated expression of *act* genes in combination with type II PKS-encoding genes from *S. glaucescens* and several other bacteria (90). They used these gene cassettes to determine what factor controls the size of the polyketide made by the KS and ACP components and what controls its apparent folding pattern in concert with cyclization, or reduction and cyclization, to produce aromatic rings. Certain combinations of *actI-ORF1* and *-ORF2* with their respective *tcmK* and *tcmL* homologues, as well as the *actIII, actIV,* and *actVII* genes (which are required for the synthesis of an early intermediate of actinorhodin biosynthesis), were expressed in the *S. coelicolor* CH999 mutant from which all of the *act* cluster had been deleted (90). Substitution of *tcmKL* for *actI-ORF1* and *-ORF2* in a cassette of *actIII/actI/actVII/actIV* genes altered the nature of the metabolites produced: Instead of aloesaponarin II and its 2-carboxy derivative (Figure 7A, path a), which are normal shunt products of the actinorhodin pathway, RM20 was synthesized (Figure 7B, path b). Because the RM20 contains 20 carbons, it must arise from a decaketide, as does tetracenomycin C, and not from the octaketide that leads to 2-carboxyaloesaponarin II. The presence of *tcmL* presumably accounts for this change in metabolites, even though *tcmK* was also present (90). The versions of this cassette that contained *actI-ORF1* and *tcmL,* or *tcmK* and *actI-ORF2,* instead of *actI-ORF1* and *-ORF2* reportedly did not produce any metabolite, which shows that a heterologous KS/CLF system can be nonfunctional. This conclusion rests on the assumption, deduced from sequence similarities, that a type II KS consists of two different subunits (e.g. ActI-Orf1/ActI-Orf2 or TcmK/TcmL) each of which has a unique function, yet of which only one (ActI-Orf1 or TcmK) contains an active site(s) that is recognizable on the basis of our current understanding. Presumably, the second subunit provides a needed activity (chain-length specificity) in a way we do not currently understand. The suggestion that the *actI-ORF2* or *tcmL* genes influence the number of chain extender units added to the growing poly-β-ketone intermediate led McDaniel et al (90) to designate the products of genes such as *actI-ORF2* and *tcmL* as the CLF. Further work from this group has substantiated this assignment (39, 88) and shown that a PKS with a relaxed chain-length specificity can produce both a 16- and an 18-carbon polyketide (89).

These results do not prove whether, in addition to the CLF, kinetic effects help determine chain length by hydrolysis of the $R(C(=O)CH_2)_n COSEnz$ tether once the polyketide has reached the correct size. [Competition between the

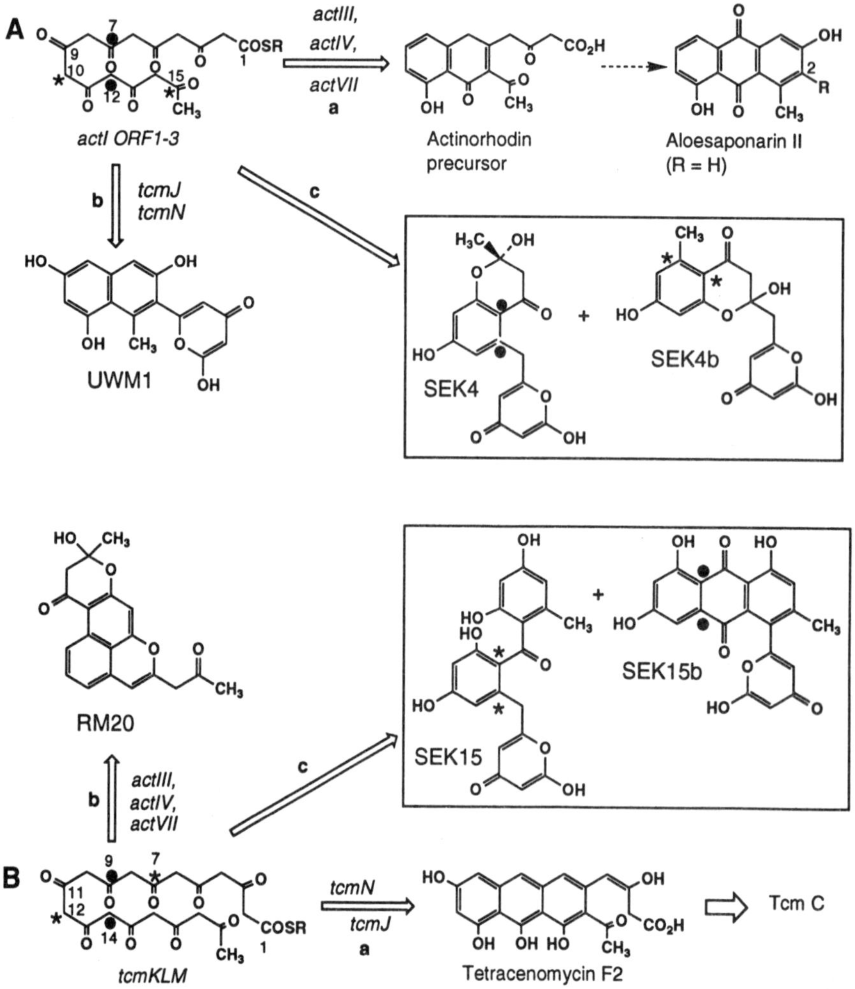

Figure 7 Poly-β-ketone and early pathway intermediates of (A) actinorhodin and (B) tetracenomycin biosynthesis. The genes that govern the steps of each process are shown beneath the poly-β-ketone or near the arrows. Products of minimal or heterologous PKSs are bracketed or boxed, respectively. The dashed arrow indicates a shunt process.

rate of elongation by the KS, the relative amount of each type of KS, the supply of malonyl-CoA, and the utilization of acyl-ACPs by glycerophosphate acyltransferase are the most significant factors determining fatty acid chain length in *E. coli* (79).] In fact, the lack of an obvious TE component in the type I fungal and type II bacterial PKSs implies that release of the poly-β-ketone intermediate does not always require a dedicated TE. This enzyme might be supplied by fatty acid biosynthesis, but the two known *E. coli* TEs are not required for fatty acid synthesis (79), unlike in animals (2, 133). Moreover, Spenser & Jordan (121) have noted in their study of MSAS that spontaneous hydrolysis of the $R(C(=O)CH_2)_n$COSEnz intermediate should be possible, even though such activity may be a special case for MSAS, in which aromatic ring formation can facilitate hydrolysis.

Hutchinson and coworkers have used a complementary approach to assess the roles of *tcmL* and the *tcmN* and *tcmJ* polyketide cyclase genes. Cassettes of the *tcmJKLMN* genes, known to produce Tcm F2 (Figure 7B, path a), the first intermediate following decaketide formation in Tcm C biosynthesis (124, 125), were cloned on a high-copy vector under the control of a strong, constitutive promoter. The cassettes were then expressed in the *S. glaucescens tcmIc* null mutant, in which the *tcmGHIJKLMNO* genes are not expressed at a significant level (25), or in a *tcmL* in-frame deletion mutant. Replacement of *tcmK* with *actI-ORF1* did not always lead to lack of Tcm F2 biosynthesis, although a metabolite was sometimes not observed, presumably because of a nonfunctional hybrid PKS. However, substitution of *tcmL* with *actI-ORF2* resulted in formation of both Tcm F2 (or Tcm C in the Δ*tcmL* background) and a new, 16-carbon compound, UWM1 (Figure 7A, path b) (B Shen, RG Summers, E Wendt-Pienkowski & CR Hutchinson, submitted). This compound predominated when *tcmKL* was replaced with *actI-ORF1/ORF2*, which reinforces the idea that the KS requires two subunits to determine, by their joint activity, the length of the poly-β-ketone made from acetate and malonate. Small amounts of SEK4 and SEK4b, shunt products of the *actI-ORF1/ORF2/ORF3* genes (38, 39), were also produced (Figure 7A, path c); their production decreased significantly in the absence of either *tcmJ* or *tcmN* (B Shen, RG Summers, E Wendt-Pienkowski & CR Hutchinson, submitted). Consequently, the fidelity of a type II PKS depends heavily on the genes encoding the KS and cyclase components: The presence of a heterologous CLF can subvert the chain-assembly process but not totally redirect it; the KS subunit, such as TcmK, that most closely resembles its FAS equivalent [*fabB* in *E. coli* (61)] has much less influence on the size of the poly-β-ketone; and the cyclase genes, such as *tcmN* and *tcmJ*, guide this intermediate in the proper direction for the synthesis of a given aromatic compound.

The collaborative efforts of the Khosla and Hopwood laboratories have established the importance of the last conclusion, and these investigators, together with the Hutchinson group, have shown that the polyketide cyclases can even influence the nature of the first cyclization step. Studies of the metabolites produced by a minimal PKS, defined as only the two KS subunits (KS + CLF) plus the ACP, have shown that the *actI-ORF1/ORF2/ORF3* genes produce a mixture of SEK4 and SEK4b through competing cyclizations between C-7 and C-12 vs C-10 and C-15 of the linear octaketide (38, 39) (Figure 7A, path c), whereas the *tcmKLactI-ORF3* (88) or *tcmJKLM* (B Shen, RG Summers, E Wendt-Pienkowski & CR Hutchinson, submitted) genes produce an approximately 1:1 mixture of SEK15 and SEK15b through competing C-7/12 vs C-9/14 cyclizations (Figure 7B, path c). Addition of *actVII* shifted the products to an approximately 16:1 mixture of SEK15 vs SEK15b (88). Addition of *tcmN* or *tcmJN* to the *actI-ORF1/ORF2/tcmM* cassette caused a marked reduction in the amounts of SEK4 and SEK4b and a commensurate increase in UWM1 formation (B Shen, RG Summers, E Wendt-Pienkowski & CR Hutchinson, submitted). The *tcmJactI-ORF1/ORF2/tcmMN* cassette produces UWM1 as the major metabolite along with traces of SEK4 and SEK4b (B Shen, RG Summers, E Wendt-Pienkowski & CR Hutchinson, submitted). These data suggest that the polyketide cyclases can govern the folding and cyclization(s) of the nascent poly-β-ketone intermediate: *actVII* forces a decaketide to favor the cyclization characteristic of an actinorhodin-like polyketide (SEK15), and *tcmJ* or *tcmN* favor cyclization of an octaketide to a tetracenomycin-like polyketide (UWM1). Yet in the absence of a cyclase, these intermediates seem to fold and cyclize in the normal manner (e.g. to SEK4 or SEK15b) as well as aberrantly (e.g. to SEK4b or SEK15). SEK4b, in fact, may result from formation of the aromatic ring by abnormal cyclization of the $CH_3C(=O)CH_2R$ end of the poly-β-ketone, possibly before chain synthesis has been completed and the octaketide released from the PKS (38). Formation of a six-membered aromatic ring by a spontaneous aldol reaction at the methyl-bearing end of a polyketide is consistent with the chemistry of polyketide model systems in which this type of ring formation is thought to be the least sterically hindered process compared with cyclizations at internal carbonyl groups (49). Consequently, Khosla, Hopwood, and their coworkers have speculated that the KS and CLF enzymes have some influence on the initial cyclization, and they consider SEK4 and SEK15b products of the normal cyclization route. But the influence of these enzymes seems to be less than that of the polyketide cyclases, which clearly can govern both the initial and subsequent cyclizations leading, for instance, to the compound preceding 2-carboxyaloesponarin II in Figure 7A, path a, and to Tcm F1 in Figure 8a.

Polyketide cyclases such as ActIV, ActVII, TcmJ, and TcmN are difficult enzymes to study because the presumed substrates, whether they are linear or monocyclic poly-β-ketones, are not directly available. [Octaketides and decaketides have been synthesized in a form that masks their inherent reactivity (49), but the unprotected molecules, even if they could be made, would be too unstable for use in the enzyme assays.] However, the *S. glaucescens tcmI* gene encodes the cyclase that converts Tcm F2 to Tcm F1 (125) (Figure 8a), a cyclization that closely resembles the intramolecular aldol and Claisen reactions involved in the synthesis of mono-, di-, and tricyclic aromatic polyketides.

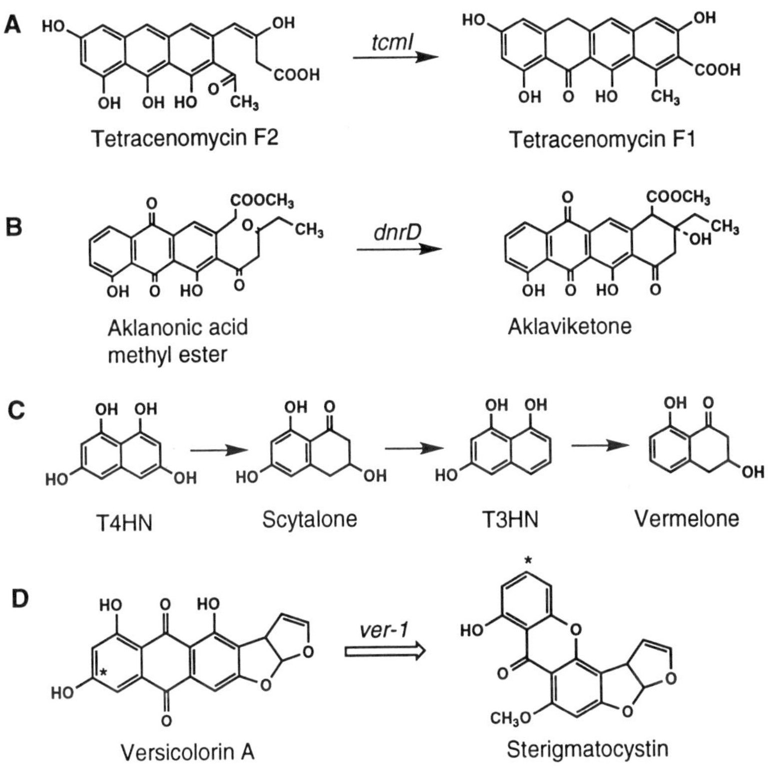

Figure 8 (*A, B*) Reactions analogous to steps catalyzed by polyketide cyclases in the biosynthesis of bacterial polyketides. (*C, D*) Reduction steps in the biosynthesis of fungal polyketides. The conversion of versicolorin A to sterigmatocystin involves more than reduction of the asterisked position. The genes governing certain steps are indicated above the arrows.

Because Tcm F2 is stable and can be isolated from a *S. glaucescens* mutant (142) or a recombinant strain carrying the *tcmJKLMN* genes (124, 125), the mechanism of TcmI catalysis is open to study. TcmI has been purified from *S. glaucescens* (114) and an overproducing *E. coli* strain (U Roos & CR Hutchinson, unpublished results) and shown to catalyze the conversion of Tcm F2 to Tcm F1 at pH ≥ 8 [9-decarboxy-Tcm F1 was the product at pH ≤ 6.5; this compound has also been isolated from *S. glaucescens* Tcm C$^-$ mutants (142)]. A related enzyme encoded by the *S. peucetius dnrD* gene (78) catalyzes the conversion of the tricyclic aklanonic acid methyl ester (Figure 8*b*) to aklaviketone in daunorubicin biosynthesis. Even though the substrates for these two enzymes differ in oxidation state and esterification, their cyclization

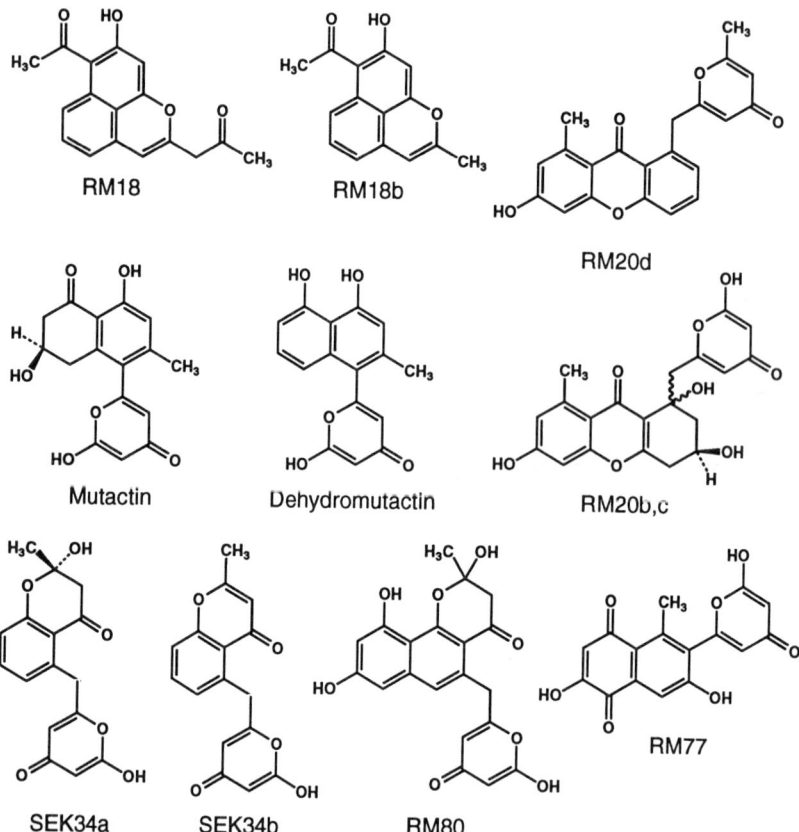

Figure 9 Novel polyketides made by heterologous type II PKSs. Dehydromutactin (88), RM20d, and SEK34b are the dehydration products of mutactin, RM20b and c (40a), and SEK34a, respectively. RM80 and RM77 are from R McDaniel, CR Hutchinson & C Khosla (submitted).

mechanisms, though similar in some respects, certainly differ in others. TcmI produces an aromatic acid resulting from dehydration of the intermediate aldol product, whereas DnrD yields an ester that does not undergo aromatization by loss of water.

In contrast, the exact function of ActIV, ActVII, TcmJ, and TcmN has not been established. Because *actVII* mutants accumulate the shunt product mutactin (8, 143) (Figure 9) and the addition of *actVII* to the *actI-ORF1/ORF2/ORF3/actIII* genes resulted in the formation of SEK34a,b (Figure 9) (91), the ActVII enzyme has been named an aromatase based on its apparent ability to aromatize the A-ring of an intermediate of actinorhodin biosynthesis (91), from which the SEK34a and -34b shunt products result. This may be a misnomer because mutactin is a shunt product of this pathway, and therefore, the lack of A-ring dehydration in mutactin could be an aberration. Furthermore, ActVII clearly influences the regiochemistry of polyketide cyclization, as noted above. The *actIV* product then provides the means for cyclization of the B-ring in the actinorhodin pathway (8, 91), although the ActIV substrate and exact product are still conjectural. TcmJ and TcmN seem to act independently in Tcm F2 biosynthesis because the addition of *tcmN* to the *tcmKLM* genes results in Tcm F2 formation only, whereas addition of *tcmJ* causes the formation of not only Tcm F2 but also SEK15 and SEK15b, along with other minor products (124, 125; B Shen, RG Summers, E Wendt-Pienkowski & CR Hutchinson, submitted). Interestingly, addition of *tcmJ* to the *tcmKLMN* genes simply increases the amount of Tcm F2 made (125). Moreover, a *S. glaucescens* mutant containing an in-frame deletion in *tcmJ* still produces all of the intermediates of Tcm C biosynthesis but accumulates more than the normal amount of Tcm F2 (125). Perhaps TcmN acts in part on a pool of spontaneously cyclized intermediates and forms Tcm F2, whereas TcmJ normally catalyzes only one cyclization [forming the first or second aromatic ring (125)] or acts imperfectly in the absence of TcmN.

It is difficult to reconcile the facts about the type II PKSs and cyclases just summarized and deduce a model that explains all of the results. For instance, if the three-component PKS determines the size (length) of the poly-β-ketone intermediate by cavity filling and the cyclase that forms the first ring acts within that cavity, then the cyclase must be able to distort this cavity so as to influence which methylene and carbonyl groups are properly juxtaposed for the initial intramolecular aldol reaction. Formation of aberrant compounds like SEK4b or SEK15b implies that the absence of a cyclase results in a lack of distortion that increases the degrees of freedom experienced by the poly-β-ketone intermediate. Perhaps some of the resulting sloppiness also results from premature release of the poly-β-ketone intermediate from the PKS in the absence of the cyclase(s). Further insight into these and other matters will have to await reconstitution of functional type II PKSs in vitro so that the mechanistic questions can be studied directly.

Besides determining the size and shape (cyclization pattern) of an aromatic polyketide, a type II PKS must also specify the starter unit. Acetate is most often used, as in the case of actinorhodin and tetracenomycin, while daunorubicin and oxytetracycline require propionate and presumed malonamide units, respectively. Sequence data have revealed that the clusters of genes governing the biosynthesis of the latter two compounds could produce enzymes not found in the *act* and *tcm* clusters. For instance, the *S. peucetius dpsCD* genes (45) appear to encode, respectively, a homologue of the *E. coli* FabH (KSIII) enzyme (129) and a propionyl-CoA acyltransferase (45). Grimm et al (45) have proposed that the DpsC and DpsD enzymes together determine the choice of starter unit by means of a PKS consisting of the products of the *dpsABCDEFG* genes (Figure 5f), because the absence of only *dpsD* did not alter the ability of the *S. lividans* (*dpsABCEFG*) transformant to make aklanonic acid. Whatever means actually determine the choice of starter unit must involve more than just the KS, CLF, and ACP enzymes, because the *otc* genes encoding only these enzymes produce a decaketide (RM20b,c in Figure 9) that begins with acetate in the CH999 host (40) and Tcm F2 is still produced when the *dpsAB* genes are substituted for *tcmKL* in a *tcmJKLMN* cassette in *S. glaucescens* (G Meurer & CR Hutchinson, submitted).

Why then do the *act* and *tcm* clusters not have genes for KS and AT enzymes similar to DpsC and DpsD? TcmK and its homologues from other type II PKSs contain a putative AT domain centered on the 345KSMXGHSLGA354 motif of TcmK that is highly conserved in many acyltransferases and esterases (GHSXG is the minimal requirement, the Ser residue being the point of substrate attachment during catalysis). This observation has led to the suggestion that these KS enzymes could also have AT activity (37, 113) and would thus not require a separate AT like DpsD. However, because the TcmK S351A mutant does not exhibit a diminished ability to make Tcm F2 in vivo (92), a separate AT appears to be required to provide malonyl-SACP for Tcm F2 biosynthesis or to catalyze the loading of acetyl-CoA onto TcmK or the transfer of $RC(C=O)CH_2C(=O)SE$ from TcmM to TcmK. Perhaps these enzymes can be subverted from fatty acid biosynthesis. In fact, a set of genes encoding four components of a putative FAS has been identified in *S. glaucescens* (RG Summers, A Ali, B Shen, WA Wessel & CR Hutchinson, submitted); one of these, *fabD*, encodes a strong homologue of the FabD malonyl-CoA:ACP acyltransferase component of the *E. coli* FAS and could complement the *fabD89* mutation in an *E. coli* FAS⁻ strain (RG Summers, A Ali, B Shen, WA Wessel & CR Hutchinson, submitted). Because a crude FabD preparation obtained from a recombinant *E. coli* catalyzed the transfer of malonate to TcmM as efficiently as to the FabC ACP in vitro (RG Summers, A Ali, B Shen, WA Wessel & CR Hutchinson, submitted), FabD may provide the AT activity required for the Tcm PKS. Clearly, the best way to answer this question is to determine whether a *S. glaucescens fabD* mutant can make both Tcm C and fatty acids. In addition to *S. glaucescens* (E Wendt-Pienkowski & CR Hutchinson, work in progress) *S. coelicolor* is also being used to address this problem. The latter organism contains a homologue of *fabD* that may be required for fatty acid and actinorhodin synthesis because its disruption was a lethal event (WP Revill, MJ Mibb & DA Hopwood, submitted).

All of the questions highlighted above can be answered best through direct studies of the type I and type II PKSs in vitro. Undertaking such analyses will not be a simple task, because the lack of function of any domain or monofunctional enzyme will usually result in either a nonfunctional or dysfunctional PKS. Thus, each subunit of a type I PKS or each component of a type II PKS must be purified in a fully active form before reconstitution experiments will be successful. If the activity of the protein cannot be assayed until a minimum level of reconstitution is reached, or if additional proteins are needed that lie outside the cluster of metabolite biosynthesis genes, then purification and reconstitution will be difficult work. Therefore, cell-free systems prepared from recombinant strains that overproduce a PKS are an attractive means to study the enzyme mechanisms.

Shen & Hutchinson have described such a system and used it to explore some of the properties of the tetracenomycin PKS (113). Cell-free preparations were made from *S. glaucescens tcmIc* transformants bearing different combinations of *tcm* genes carried on a high-copy vector and expressed from a strong, constitutive promoter. PKS activity was assayed through the synthesis of radioactive Tcm F2 from [^{14}C]malonyl-CoA and acetyl-CoA. It was inhibited by cerulenin, iodoacetamide, and *p*-methylphenylsulfonyl fluoride, reagents known to alkylate Cys and Ser residues in enzyme active sites, as well as by polyclonal antisera raised to the TcmK, -L, -M, and -N enzymes. Activity was stimulated by addition of purified TcmM but not the *E. coli* FAS or the *S. glaucescens* FabC (RG Summers, A Ali, B Shen, WA Wessel & CR Hutchinson, submitted) ACPs. Furthermore, Decker & Hutchinson (26) have reported that extra copies of the *tcmM* gene significantly increased the levels of Tcm C biosynthesis intermediates in vivo. Because Tcm F2 was synthesized from [^{14}C]acetyl-CoA but not from [^{14}C]propionyl- or [^{14}C]isobutyryl-CoA plus unlabeled malonyl-CoA, the tetracenomycin PKS appears to have a strict starter unit specificity, unlike the daunorubicin PKS. Based on the isolation of the feudomycins (96) and 13-methylaclacinomycins (120) from anthracycline-producing bacteria, daunorubicin PKS appears able to accept acetate, acetoacetate, and isobutyrate as alternative starter units. Similar evidence indicates that the oxytetracycline PKS can accept acetate instead of malonamate (40, 50, 93).

The results of the cell-free and in vivo experiments were highly correlated with respect to the synthesis of Tcm F2 as a function of different combinations of *tcm* genes (113), except that no radioactive metabolites were produced by the *tcmKLM* genes, which cause the synthesis of Tcm F2 and other unidentified metabolites in vivo (124, 125; B Shen, RG Summers, E Wendt-Pienkowski & CR Hutchinson, submitted). Moreover, Tcm F2 was synthesized upon admixture of cell-free extracts prepared from strains bearing the *tcmKMN* and *tcmKLM* genes. This particularly interesting result indicates that the activity of an individual component of the tetracenomycin PKS could be assayed by adding fractions obtained during enzyme purification to a cell-free system deficient in that enzyme but containing the remaining components.

In Fungi

Of the fungal PKS enzymes, only MSAS has been studied mechanistically. Lynen & Scott and their coworkers reported more than 20 years ago that MSAS and a FAS of *P. patulum* were copurified through the early steps, although the MSAS activity was generally 10 times higher than that exhibited by FAS (28, 107, 108). Triacetic acid lactone was formed at 10% of the rate of 6-methylsalicylic acid and became the exclusive product in the absence of NADPH, indicating that reduction of the C-3 carbonyl takes place at the triketide stage. This result also shows that, like DEBS and the triketo lactone (Figure 3), chain growth can be terminated by release of product before the normal length is reached. Schweizer and colleagues (106a) reported recently that the triketide intermediate is a better substrate for reduction by MSAS than by FAS (which prefers acetoacetate) and that the product is not dehydrated until the tetraketide stage, prior to cyclization and hydrolytic release of 6-methylsalicylic acid. They concluded that the FAS and MSAS produce different products as a consequence of differing substrate specificities of their reductase, dehydratase, and terminal acyl transferase enzymes. Other work has shown that MSAS accepts propionyl-CoA as the starter unit in place of acetyl-CoA (the rate of 6-ethylsalicylic acid synthesis is only 13% of that of 6-methylsalicylic acid) and that inhibition with iodoacetamide converted MSAS into a malonyl-CoA decarboxylase able to synthesize 6-methylsalicylic acid in the absence of acetyl-CoA (27). These findings have been extended by workers in PM Jordon's laboratory, who improved the methods for purifying MSAS (121) and investigated the mechanism of the condensation reaction between acetyl-CoA and malonyl-CoA using chirally labeled samples of the malonate (122). Their data showed that MSAS accepts acetoacetyl-CoA as an alternative starter unit and, as noted by Lynen, forms a small amount of triacetic acid lactone as an obligatory by-product (122). We agree with their conclusions from the results of the experiments with chiral malonate (122) that "a remarkable degree of steric control exists during manipulation of the enzyme-bound polyketide intermediates..." (p. 9114) and that "the overall stereochemical course [of MSAS] is likely to have significant implications for the understanding of conformational regulation and its influence on the programming of this and other polyketide synthase reactions" (p. 9114). Similar mechanistic studies of DEBS and other type I PKSs are eagerly awaited with the anticipation that the stereochemical features will illuminate how such enzymes control the outcome of the reductions and epimerizations that establish the new chiral centers in their macrolide products.

ENZYMES THAT ACT ON THE PRODUCTS OF POLYKETIDE SYNTHASES

In most cases, the initial products formed by a PKS are modified by oxidation, reduction, methylation, and other types of reactions that tailor the polyketide to the species-specific spectrum of metabolites. Because these reactions often imbue the polyketide with its characteristic biological activity, the specificity of the tailoring enzymes must be investigated to assess the potential of transforming novel polyketides made by a genetically modified PKS into potential drugs. The little information uncovered so far hints at a possible looseness in substrate specificity but also shows that some of the tailoring enzymes are too fastidious to act on compounds different from their normal substrates.

Reduction

In fungal melanin biosynthesis, 1,3,6,8-tetrahydroxynaphthalene (T4HN), the direct product of a PKS, is reduced to scytalone (Figure 8c) (105, 106). The enzyme responsible for this NADPH-dependent reduction step was purified to homogeneity from *Magnaporthe grisea* (130), the fungus responsible for a rice blast disease, and found to catalyze a second reduction to form vermelone from 1,3,8-trihydroxynaphthalene (T3HN) (Figure 8c). The cDNA coding for T4HN reductase was cloned from a cDNA expression library by using anti-T4HN reductase (130). Sequence analysis revealed a strong similarity with the product of *ver-1*, a gene involved in aflatoxin biosynthesis in *A. parasiticus* (119), and with oxidoreductases of the short-chain alcohol dehydrogenase class (101, 131). Scytalone dehydratase has been purified from *Cochliobolus miyabeanus* and catalyzes the formation of T3HN from scytalone (128). This 23-kDa polypeptide does not require a NADPH cofactor.

Aflatoxin biosynthesis (34) involves several reduction steps, and the genes encoding some of these enzymes were cloned recently. The *ver-1* gene (119) and its homologue from *A. nidulans* (63) are associated with conversion of versicolorin A to sterigmatocystin (Figure 8d), during which the asterisked carbonyl is reduced and dehydrated. The deduced *ver-1* gene product showed extensive identity with the *S. coelicolor actIII*-encoded ketoreductase described below. Two reductases acting on versiconal hemiacetal acetate have also been purified, but their genes have yet to be identified (84).

Ketoreductases interact with the bacterial PKSs and often reduce the carbonyl that is nine carbons from the carboxyl end of the poly-β-ketone intermediates (8), which results in the absence of a hydroxyl at the corresponding position of the polyketide. The ActIII enzyme of actinorhodin biosynthesis (48), for example, has this function. Fu et al (39) have studied the catalytic specificity of this enzyme in vivo as a function of the polyketides produced by heterologous PKSs. They reported that ActIII maintained its fidelity for

C-9 reduction and, as a consequence of this reduction, the regiochemistry of some of the cyclization reactions that take place before or after reduction was altered. For instance, RM20 or mutactin were formed when *actIII* was present, but SEK15 or SEK4 were made in its absence. If this outcome is not just coincidental, it suggests that the second cyclization step but not the first can be altered by the effect of ActIII, even though this conclusion is difficult to rationalize (perhaps reduction causes the monocyclic intermediates to adopt a different conformation). Although reduction probably takes place before aromatization in these bacterial examples, hydroxy groups are removed after formation of aromatic rings in fungi by enzymes such as T4HN reductase. Similar reductions occur in the biosynthesis of chrysophanol from emodin in *Pyrenochaete terrestris* (5, 56).

Oxidation

Oxidation of polyketides typically involves hydroxylation of aromatic or reduced polyketides, conversion of hydroquinols to quinones, oxidative cleavage of double bonds or aromatic rings, and oxidative coupling reactions. Examples of such purified enzymes are given below.

Cytochrome P450 hydroxylases are widely distributed among microorganisms, and several P450-encoding genes have now been found in clusters of secondary metabolism genes. In erythromycin biosynthesis, 6-deoxyerythronolide B hydroxylase, the product of the *eryF* gene (135), acts on the macrolactone produced by DEBS to produce erythronolide B (Figure 3). This P450 hydroxylase has been purified from *S. erythraea* (110) and a recombinant *E. coli* strain that overexpressed *eryF* (4). Like most other P450 hydroxylases, 6-deoxyerythronolide B hydroxylase requires two proteins to transfer electrons from NADPH to the heme-bound Fe^{3+} of the terminal hydroxylase (111). Studies of substrate-inducible P450 hydroxylases (99) suggest that the genes for the electron-transport proteins should be adjacent to *eryF*, but neither of the genes encoding the NAD(P)H:ferredoxin oxidoreductase (ForA) or ferredoxin (FdxA) used in vitro (111) resides in the cluster of erythromycin production genes (29, 144). Such a divergence appears to be typical of the P450 hydroxylases dedicated to secondary metabolism and may indicate that expression of electron-transport proteins need not be coordinated with that of the hydroxylase (144). The substrate specificity of 6-deoxyerythronolide B hydroxylase (4) reveals that this enzyme displays marked differences in turnover rates and substrate dissociation values between the normal substrate and its 5-oxo, 9-deoxo-9-hydroxy, and 15-demethyl derivatives. Some of the differences were explained on the basis of how the substrate's conformation might affect binding to the enzyme. These conformational effects were determined using molecular modeling techniques in which the active site was derived from the crystal structure of the camphor CamC P450 hydroxylase (102). A better understanding of these factors should now be possible because the three-dimensional (3D) structure of 6-deoxyerythronolide B hydroxylase with bound substrate has been solved with X-ray crystallography (23). The 3D structure shows that the enzyme has a larger substrate-binding pocket than CamC and contains several ordered water molecules. Hydrophobic interactions occur between certain carbons of the substrate and the side chains: C-15 and C-16 interact with Ile174 and Leu175, respectively, and C-14, C-19, C-20, and C-21 interact with Ala74, Val237, Thr92, and Leu391, respectively. Key hydrogen bonds possibly important for determining the specificity of hydroxylation occur between the C-5 hydroxyl, the C-9 ketone, and C-11 of the substrate and three solvent molecules in the binding pocket.

Erythromycin biosynthesis uses another cytochrome P450 enzyme, EryK, for the C-12 hydroxylation of erythromycin D. Studies of purified EryK have recently established that the k_{cat} for erythromycin D is 1200- to 1900-fold greater than that for erythromycin B (72), which explains why the erythromycin D-to-erythromycin C conversion is the preferred pathway (Figure 3).

In anthra- and naphthacenequinone biosynthesis, the direct product of the PKS is an anthrone or naphthacenone. Emodinanthrone is oxidized to the anthraquinone, emodin (Figure 10); this oxygenase activity was first found in *A. terreus* (41). Emodinanthrone oxygenase has been purified to near homogeneity and suggested to be a nonheme Fe^{3+} monooxygenase (20). An analogous enzyme may be involved in anthracycline biosynthesis at the step where 12-deoxyaklanonic acid is oxidized to aklanonic acid (Figure 10); *S. peucetius dnr-ORF8* (45) and *Streptomyces* sp. C5 *dauA-ORFE* (137) are the first-discovered examples of genes encoding this kind of enzyme. In tetracenomycin biosynthesis, the naphthacenequinone, Tcm F1, is oxidized to Tcm D3 (Figure 10) by the product of the *tcmH* gene (125). TcmH, purified from *S. glaucescens*, is a 37,500-Dalton trimer of 12,500-Dalton subunits and can be classified as an internal monooxygenase that requires only O_2 for the enzymatic reaction (112). Interestingly, TcmH does not possess any of the prosthetic groups of known monooxygenases nor does it utilize metal ions. Because TcmH is inactivated by N-ethylmaleimide and diethyl pyrocarbonate, sulfhydryl groups and histidine residues may be essential for its activity.

The key reaction in the conversion of an anthraquinone to a benzophenone is the Baeyer-Villiger type oxidative ring cleavage of one of the aromatic rings. Questin oxygenase, an enzyme that catalyzes this type of reaction, was detected in the cell-free extract of *A. terreus*, but further characterization has been hindered by its lability (43).

An interesting reaction involved in geodin and asterric acid biosynthesis is the stereospecific intramolecular phenol oxidative coupling reaction to form a unique spiro structure from the polyketide-derived benzophenone (like griseofulvin in Figure 1). Dihydrogeodin oxidase, a fungal enzyme catalyzing this

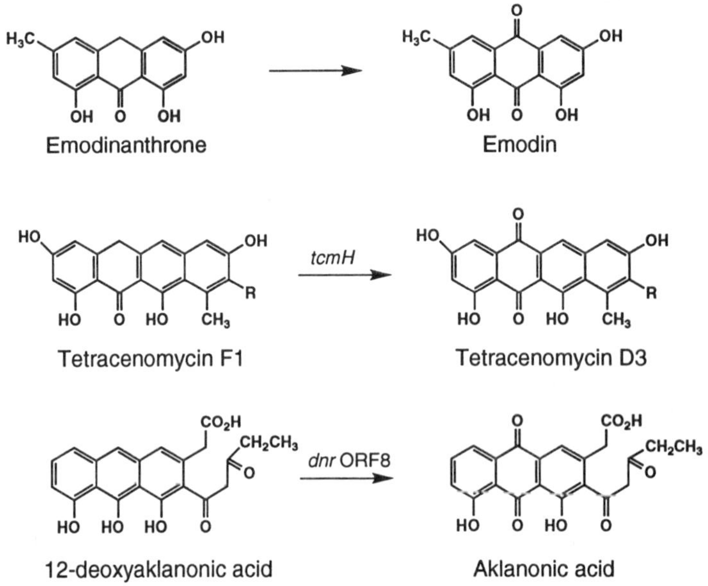

Figure 10 Reactions catalyzed by monooxygenases in fungal and bacterial polyketide biosynthesis.

reaction purified from *A. terreus*, is a blue copper protein (44). The gene encoding this enzyme was recently cloned (42).

Methylation

Methylation of hydroxyl and carboxyl groups is a common reaction in secondary microbial metabolism. Of the 16 enzymes estimated to be involved in the biosynthesis of the aflatoxins from 10-deoxynorsolorinic acid (34), two different methyltransferases have been purified to homogeneity that convert sterigmatocystin or its dihydro derivative to *O*-methylsterigmatocystin and its dihydro derivative (11, 62). The cDNA for one of these 42-kDa *O*-methyltransferases has been cloned from *A. parasiticus* and expressed in *E. coli* (139).

Polyketides often undergo more than one methylation during tailoring of the initial products made by type I and type II PKSs; therefore, the substrate specificity of the methyltransferases should be determined to assess the extent to which novel products of modified PKSs can be biotransformed. Structural changes distant from the site of methylation can have a negative effect: 6,7-anhydroerythromycin C, produced by targeted mutation of the *eryAIII* gene (31), is not converted to the erythromycin A analogue because it is a poor substrate for erythromycin C *O*-methyltransferase, the last enzyme in the erythromycin pathway (Figure 3). On the other hand, mutations in the *tcmN* or *tcmO* genes that encode two of the three methyltransferases of Tcm C biosynthesis still permit the remaining enzymes to O-methylate the normal position in substrates that have not undergone prior methylation by TcmN or TcmO (26). In fact, recent observations (G Meurer, E Wendt-Pienkowski & CR Hutchinson, unpublished results) hint at the possibility that overproduction of TcmN results in multiple methylations of positions other than the normal C-3 hydroxyl of Tcm D3 (124). Such sloppiness of secondary metabolism enzymes is an endearing trait responsible for the network of pathways by which a collection of closely related products is produced in one species.

POLYKETIDE SYNTHASES AS A SOURCE OF CHEMICAL DIVERSITY AND DRUGS

Targeted Modification of Type I Enzymes

The modular organization of type I PKSs offers an attractive way to determine whether novel metabolites could result from different combinations of the modules, using genes from the same or different organisms. If successful, this technique would augment the use of microorganisms in screening programs that require an ever-expanding source of chemical diversity in the search for new drugs (55). As noted above, novel erythromycins can be made by targeted mutation of the *eryAI, -II*, and *-III* genes, as reviewed elsewhere (60); additional types of metabolites will probably result from further work with these genes and others that encode different type I PKSs. This forecast is supported by the fact that a good yield of the triketo lactone (Figure 3) was achieved with expression of only *eryAI* in *S. coelicolor* (59). Thus, if modified forms of DEBS1 produced by *eryAI* mutants also turn over in vivo, or if analogues of the triketo lactone can be substrates for heterologous type I PKS subunits, a broad spectrum of novel products may be available by the genetic engineering of such genes. Nevertheless, our enthusiasm must be tempered with the knowledge that many attempts to produce new metabolites through targeted mutation of the *eryA* and similar genes have been largely unsuccessful (L Katz, D MacNeil, and B Schoner, personal communications).

Combinatorial Studies of Type II Systems

Recent studies of the properties of recombinant type II PKSs, especially the research efforts of Khosla, Hopwood, and their collaborators, have uncovered numerous new metabolites resulting from combinations of heterologous bacterial genes. Figure 9 lists the ones reported to date in addition to those mentioned above in the discussion of the *actI, actIV, actVII,* and *tcmJKLMN* genes. Our knowledge of the chemical diversity of bacteria harboring the ubiquitous type II PKS genes is expanding with our growing understanding of how the starter unit is specified, which genes lead to differences in chain length, how carbonyl reduction or its lack influences the nature of the intramolecular cyclizations, and what differences in these cyclizations are associated with each combination of PKS and polyketide cyclase-encoding genes. Consequently, the time is ripe to find out whether any of the newly formed aromatic compounds have significant biological activity and whether the tailoring enzymes of particular pathways will be able to catalyze the conversion of such compounds into new pharmacological agents.

ACKNOWLEDGMENTS

We thank Chaitan Khosla, Leonard Katz, David Hopwood, Guido Meurer, Ben Shen, Richard Summers, and Evelyn Wendt-Pienkowski for critical reading of the manuscript. Unpublished results described here were from research supported by the National Institutes of Health (CA 35381) and the American Cancer Society (CII-524).

> Any *Annual Review* chapter, as well as any article cited in an *Annual Review* chapter, may be purchased from the Annual Reviews Preprints and Reprints service.
> 1-800-347-8007; 415-259-5017; email: arpr@class.org

Literature Cited

1. Alberts AW, Chen J, Kuron G, Hunt V, Huff J, et al. 1980. Mevinolin: a highly potent competitive inhibitor of hydroxymethylglutaryl-coenzyme A reductase and a cholesterol-lowering agent. *Proc. Natl. Acad. Sci. USA* 77:3957–61
2. Alberts AW, Greenspan MD. 1984. Animal and bacterial fatty acid synthetase: structure, function and regulation. In *Fatty Acid Metabolism and Its Regulation,* ed. S Numa 2:29–58. Amsterdam: Elsevier
3. Amy CM, Witkowski A, Naggert J, Williams B, Randhawa Z, Smith S. 1989. Molecular cloning and sequencing of cDNAs encoding the entire rat fatty acid synthase. *Proc. Natl. Acad. Sci. USA* 86:3114–18
4. Andersen JF, Tatsuta K, Gunji H, Ishiyama T, Hutchinson CR. 1993. Substrate specificity of 6-deoxyerythronolide B hydroxylase, a bacterial cytochrome P450 of erythromycin A biosynthesis. *Biochemistry* 32:1905–13
5. Anderson JA, Lind B-K, Wang SS. 1990. Purification and properties of emodin deoxygenase from *Pyrenochaeta terrestris. Phytochemistry* 29:2415–18
6. Aparicio JF, Caffrey P, Marsden AFA, Staunton J, Leadlay PF. 1994. Limited proteolysis and active-site studies of the first multienzyme component of the erythromycin-producing polyketide synthase. *J. Biol. Chem.* 269:8524–28
7. Arrowsmith TJ, Malpartida F, Sherman DH, Birch AW, Hopwood DA, Robinson JA. 1992. Characterization of *actI*-homologous DNA encoding polyketide synthase genes from the monensin producer *Streptomyces cinnamonensis. Mol. Gen. Genet.* 234:254–64
8. Bartel PL, Zhu C-B, Lampel JS, Dosch DC, Conners NC, et al. 1990. Biosynthesis of anthraquinones by interspecies cloning of actinorhodin biosynthesis genes in streptomycetes: clarification of actinorhodin gene functions. *J. Bacteriol.* 172:4816–26
9. Beck J, Ripka S, Siegner A, Schiltz E, Schweizer E. 1990. The multifunctional 6-methylsalicylic acid synthase gene of *Penicillium patulum*: its gene structure relative to that of other polyketide synthases. *Eur. J. Biochem.* 192:487–98

10. Bevitt DJ, Cortes J, Haydock SF, Leadlay PF. 1992. 6-Deoxyerythronolide B synthase from *Saccharopolyspora erythraea*. Cloning of the structural gene, sequence analysis and inferred domain structure of the multifunctional enzyme. *Eur. J. Biochem.* 204:39–49
11. Bhatnagar DA, Ullah AHJ, Cleveland TE. 1988. Purification and characterization of a methyltransferase from *Aspergillus parasiticus* SRRC 163 involved in the aflatoxin biosynthetic pathway. *Prep. Biochem.* 18:321–49
12. Bibb MJ, Biró S, Motamedi H, Collins JF, Hutchinson CR. 1989. Analysis of the nucleotide sequence of the *Streptomyces glaucescens tcmI* genes provides key information about the enzymology of polyketide tetracenomycin C antibiotic biosynthesis. *EMBO J.* 8:2727–36
13. Bibb MJ, Sherman DH, Ōmura S, Hopwood DA. 1994. Cloning, sequencing and deduced functions of a cluster of *Streptomyces* genes probably encoding biosynthesis of the polyketide antibiotic frenolicin. *Gene* 142:31–39
14. Brobst S, Townsend CA. 1994. The potential role of fatty acid initiation in the biosynthesis of the fungal aromatic polyketide aflatoxin B1. *Can. J. Chem.* 72:200–7
15. Brown DW, Hauser FM, Tommasi R, Corlett S, Salvo JJ. 1993. Structural elucidation of a putative conidial pigment intermediate in *Aspergillus parasiticus*. *Tetrahedron Lett.* 34:419–22
16. Brown DW, Salvo JJ. 1994. Isolation and characterization of sexual spore pigments from *Aspergillus nidulans*. *Appl. Environ. Microbiol.* 60:979–83
17. Cane DE, Celmer WD, Westley JW. 1983. Unified stereochemical model of polyether antibiotic structure and biogenesis. *J. Am. Chem. Soc.* 105:3594–600
18. Cane DE, Lambalot RH, Prabhakaran PC, Ott WR. 1993. Macrolide biosynthesis. 7. Incorporation of polyketide chain elongation intermediates in methymycin. *J. Am. Chem. Soc.* 115:522–26
19. Celmer WD. 1965. Macrolide stereochemistry. III. A configurational model for macrolide antibiotics. *J. Am. Chem. Soc.* 87:1801–2
20. Chen ZG, Fujii I, Ebizuka Y, Sankawa U. 1995. Purification and characterization of emodinanthrone oxygenase from *Aspergillus terreus*. *Phytochemistry*. In press
21. Collie N. 1893. The formation of orcinol and other condensation products from dehydracetic acid. *J. Chem. Soc.* 1893:122–26
22. Cortes J, Haydock SF, Roberts GA, Bevitt DJ, Leadlay PF. 1990. An unusually large multifunctional polypeptide in the erythromycin-polyketide synthase of *Saccharopolyspora erythraea*. *Nature* 346:176–78
23. Cupp-Vickery JR, Poulos TL. 1995. Crystal structure of cytochrome P450eryF: a macrolide hydroxylase involved in erythromycin biosynthesis. *Nat. Struct. Biol.* 2:144–53
24. Davis NK, Chater KF. 1990. Spore colour in *Streptomyces coelicolor* A3(2) involves the developmentally regulated synthesis of a compound biosynthetically related to polyketide antibiotics. *Mol. Microbiol.* 4:1679–92
25. Decker H, Hutchinson CR. 1993. Transcriptional analysis of the *Streptomyces glaucescens* tetracenomycin C biosynthesis gene cluster. *J. Bacteriol.* 175:3887–92
26. Decker H, Hutchinson CR. 1994. Overproduction of the acyl carrier protein component of a type II polyketide synthase stimulated production of tetracenomycin biosynthetic intermediates in *Streptomyces glaucescens*. *J. Antibiot.* 47:54–63
27. Dimroth P, Ringelmann E, Lynen F. 1976. 6-Methylsalicylic acid synthetase from *Penicillium patulum*. *Eur. J. Biochem.* 68:591–96
28. Dimroth P, Walter H, Lynen F. 1971. Biosynthese von 6-methylsalicylsäure. *Eur. J. Biochem.* 13:98–110
29. Donadio S, Hutchinson CR. 1991. Cloning and characterization of the *Saccharopolyspora erythraea fdxA* gene encoding ferredoxin. *Gene* 100:231–35
30. Donadio S, Katz L. 1992. Organization of the enzymatic domains in the multifunctional polyketide synthase involved in erythromycin biosynthesis in *Saccharopolyspora erythraea*. *Gene* 111:51–60
31. Donadio S, McAlpine JB, Sheldon PA, Jackson MA, Katz L. 1993. An erythromycin analog produced by reprogramming of polyketide synthesis. *Proc. Natl. Acad. Sci. USA*. 90:7119–23
32. Donadio S, Staver MJ, McAlpine JB, Swanson SJ, Katz L. 1991. Modular organization of genes required for complex polyketide biosynthesis. *Science* 252:675–79
33. Donadio S, Staver MJ, McAlpine JB, Swanson SJ, Katz L. 1992. Biosynthesis of the erythromycin macrolactone and a rational approach for producing hybrid macrolides. *Gene* 115:97–103

34. Dutton MF. 1988. Enzymes and aflatoxin biosynthesis. *Microbiol. Rev.* 52:274–95
35. Elovson J, Vagelos PR. 1968. Acyl carrier protein. X. Acyl carrier protein synthetase. *J. Biol. Chem.* 243:3603–11
36. Endo A. 1979. Monacolin K, a new hypocholesterolemic agent produced by a *Monascus* species. *J. Antibiot.* 32:852–54
37. Fernández-Moreno MA, Martínez E, Boto L, Hopwood DA, Malpartida F. 1992. Nucleotide sequence and deduced functions of a set of co-transcribed genes of *Streptomyces coelicolor* A3(2) including the polyketide synthase for the antibiotic actinorhodin. *J. Biol. Chem.* 267:19278–90
38. Fu H, Hopwood DA, Khosla C. 1995. Engineered biosynthesis of novel polyketides: evidence for temporal, but not regiospecific, control of cyclization of an aromatic polyketide precursor. *Chem. Biol.* 1:205–10
39. Fu H, Ebert-Khosla S, Hopwood DA, Khosla C. 1994. Engineered biosynthesis of novel polyketides: dissection of the catalytic specificity of the act ketoreductase. *J. Am. Chem. Soc.* 116:4166–70
40. Fu H, Ebert-Khosla S, Hopwood DA, Khosla C. 1994. Relaxed specificity of the oxytetracycline polyketide synthase for an acetate primer in the absence of a malonamyl primer. *J. Am. Chem. Soc.* 116:6443–44
40a. Fu H, McDaniel R, Hopwood DA, Khosla C. 1994. Engineered biosynthesis of novel polyketides: stereochemical course of two reactions catalyzed by a polyketide synthase. *Biochemistry* 33:9321–26
41. Fujii I, Chen ZG, Ebizuka Y, Sankawa U. 1991. Identification of emodinanthrone oxygenase in fungus *Asperegillus terreus. Biochem. Intern.* 25:1043–49
42. Fujii I, Chen ZG, Tsukamoto N, Tada H, Huang KX, et al. 1993. *Fungal polyketide biosynthesis. Enzymological and molecular genetic approach.* Presented at Int. Botanical Congr., 15th, Yokohama
43. Fujii I, Ebizuka Y, Sankawa U. 1988. A novel anthraquinone ring cleavage enzyme from *Aspergillus terreus. J. Biochem.* 103:878–83
44. Fujii I, Iijima H, Tsukita S, Ebizuka Y, Sankawa U. 1987. Purification and properties of dihydrogeodin oxidase from *Aspergillus terreus. J. Biochem.* 101:11–18
45. Grimm A, Madduri K, Ali A, Hutchinson CR. 1994. Characterization of the *Streptomyces peucetius* ATCC 29050 genes encoding doxorubicin polyketide synthase. *Gene* 151:1–10
46. Guilfoile P, Hutchinson CR. 1991. A bacterial analog of the *mdr* gene of mammalian tumor cells is present in *Streptomyces peucetius*, the producer of daunorubicin and doxorubicin. *Proc. Natl. Acad. Sci. USA* 88:8553–57
47. Hailes HC, Jackson CM, Leadlay PF, Ley SV, Staunton J. 1994. Biosynthesis of tetronasin. Part 1. Introduction and investigation of the diketide and triketide intermediates bound to polyketide synthase. *Tetrahedron Lett.* 35:307–10
48. Hallam SE, Malpartida F, Hopwood DA. 1988. Nucleotide sequence, transcription and deduced function of a gene involved in polyketide antibiotic synthesis in *Streptomyces coelicolor. Gene* 74:305–20
48a. Han L, Yang K, Ramalingam E, Mosher RH, Vining LC. 1994. Cloning and characterization of polyketide synthase genes for jadomycin B synthesis in *Streptomyces venezuelae* ISP5230. *Microbiology* 140:3379–89
49. Harris TM, Harris CM. 1986. Biomimetic syntheses of aromatic polyketide metabolites. *Pure Appl. Chem.* 58:283–94
50. Hochstein FA, Schach von Wittenau M, Tanner FW, Maurai K. 1960. 2-Acetyl-2-decarboxamidooxytetracycline. *J. Am. Chem. Soc.* 82:5934–37
51. Hopwood DA. 1994. Genetic engineering of *Streptomyces* to create hybrid antibiotics. *Curr. Opin. Biotechnol.* 4:531–37
52. Hopwood DA, Khosla C. 1992. Genes for polyketide secondary metabolic pathways in microorganisms and plants. *Ciba Found. Symp.* 17:88–112
53. Hopwood DA, Sherman DH. 1990. Molecular genetics of polyketides and its comparison to fatty acid biosynthesis. *Annu. Rev. Genet.* 24:37–66
54. Huber MLB, Paschal JW, Leeds JP, Kirst HA, Wind JA, et al. 1990. Branched chain fatty acids produced by mutants of *Streptomyces fradiae*, putative precursors of the lactone ring of tylosin. *Antimicrob. Agents Chemother.* 34:1535–41
55. Hutchinson CR. 1994. Drug synthesis by genetically engineered microorganisms. *Bio/Technology* 12:375–80
56. Ichinose K, Kiyono J, Ebizuka Y, Sankawa U. 1993. Post-aromatic deoxygenation in polyketide biosynthesis: reduction of aromatic rings in the biosynthesis

of fungal melanin and anthraquinone. *Chem. Pharm. Bull.* 41:2015–21
57. Jordan PM, Spencer JB. 1993. The biosynthesis of tetraketides: enzymology, mechanism and molecular programming. *Biochem. Soc. Trans.* 21:222–28
58. Joshi AK, Smith S. 1993. Construction of a cDNA encoding the multifunctional animal fatty acid synthase and expression in *Spodoptera frugiperda* cells using baculoviral vectors. *Biochem. J.* 296:143–49
59. Kao CM, Luo G, Katz L, Cane DE, Khosla C. 1994. Engineered biosynthesis of a triketide lactone from an incomplete modular polyketide synthase. *J. Am. Chem. Soc.* 116:11612–13
60. Katz L, Donadio S. 1993. Polyketide synthesis: prospects for hybrid antibiotics. *Annu. Rev. Microbiol.* 47:875–912
61. Kauppinen S, Siggaard-Andersen M, Wettstein-Knowles P. 1988. β-Ketoacyl-ACP synthase I of *Escherichia coli*: nucleotide sequence of the *fabB* gene and identification of the cerulenin binding residue. *Carlsberg Res.* 53:357–70
62. Keller NP, Dischinger HC Jr, Bhatnagar D, Cleveland TE, Ullah AHJ. 1993. Purification of a 40-kilodalton methyltransferase active in the aflatoxin biosynthetic pathway. *Appl. Environ. Microbiol.* 59:479–84
63. Keller NP, Kantz NJ, Adams TH. 1994. *Aspergillus nidulans verA* is required for production of the mycotoxin sterigmatocystin. *Appl. Environ. Microbiol.* 60:1444–50
64. Khaw LE, Haydock SF, Leadlay PF. 1994. *Cloning and expression of an O-methyltransferase gene from Streptomyces hygroscopicus possibly involved in polyketide immunosuppressant biosynthesis*. Presented at Int. Symp. Biology of Actinomycetes, 9th, Moscow
65. Khosla C, Ebert-Khosla S, Hopwood DA. 1992. Targeted gene replacements in a *Streptomyces* polyketide synthase gene cluster: role for the acyl carrier protein. *Mol. Microbiol.* 6:3237–49
66. Khosla C, McDaniel R, Ebert-Khosla S, Torres R, Sherman D, et al. 1993. Genetic construction and functional analysis of hybrid polyketide synthases containing heterologous acyl carrier proteins. *J. Bacteriol.* 175:2197–204
66a. Kim E-S, Bibb MJ, Butler MJ, Hopwood DA, Sherman DH. 1994. Sequences of the oxytetracycline polyketide synthase-encoding *otc* genes from *Streptomyces rimosus*. *Gene* 141:141–42
67. Kim E-S, Hopwood DA, Sherman DH. 1994. Analysis of type II polyketide β-ketoacyl synthase specificity in *Streptomyces coelicolor* A3(2) by *trans* complementation of actinorhodin synthase mutants. *J. Bacteriol.* 176:1801–4
68. Kimura N, Tsuge T. 1993. Gene cluster in melanin biosynthesis of the filamentous fungus *Alternaria alternata*. *J. Bacteriol.* 175:4427–35
69. Kinoshita K, Takenaka S, Hayashi M. 1988. Isolation of proposed intermediates in the biosynthesis of mycinamicins. *J. Chem. Soc. Chem. Commun.* 1988:943–49
70. Kubo Y, Nakamoto H, Kobayashi K, Okuno T, Furusawa I. 1991. Cloning of a melanin biosynthesis gene essential for appressorial penetration of *Colletotrichum lagenarium*. *Mol. Plant-Microb. Interact.* 4:440–45
71. Deleted in proof
72. Lambalot RH, Cane DE, Aparicio JJ, Katz L. 1995. Overproduction and characterization of the erythromycin C-12 hydroxylase, EryK. *Biochemistry* 34:1858–66
73. Le Gouill C, Desmarais D, Dery CV. 1993. *Saccharopolyspora hirsuta* 367 encodes clustered polyketide synthases, polyketide reductase, acyl carrier protein, and biotin carboxyl carrier protein homologous genes. *Mol. Gen. Genet.* 240:146–50
74. Leadlay PF, Staunton J, Aparicio JF, Bevitt DJ, Caffrey P, et al. 1993. The erythromycin-producing polyketide synthase. *Biochem. Soc. Trans.* 21:217–20
75. Li Z, Martin FM, Vederas JC. 1992. Biosynthetic incorporation of labeled tetraketide intermediates into dehydrocurvularin, a phytotoxin from *Alternaria cineraria*, with assistance of β-oxidation inhibitors. *J. Am. Chem. Soc.* 114:1531–33
76. Linton KJ, Cooper HN, Hunter IS, Leadlay PF. 1994. An ABC-transporter from *Streptomyces longisporoflavus* confers resistance to the polyether-ionophore antibiotic tetronasin. *Mol. Microbiol.* 11:777–85
77. MacNeil DJ, Occi JL, Gewain KM, MacNeil T. 1994. Correlation of the avermectin polyketide synthase genes to the avermectin structure. *Ann. NY Acad. Sci.* 721:123–32
78. Madduri K, Hutchinson CR. 1995. Functional characterization and transcriptional analysis of a gene cluster governing early and late steps in daunorubicin biosynthesis in *Streptomyces peucetius*. *J. Bacteriol.* In press
79. Magnuson K, Jackowski S, Rock CO,

80. Cronan JE Jr. 1993. Regulation of fatty acid biosynthesis in *Escherichia coli*. *Microbiol. Rev.* 57:522–42
80. Malpartida F, Hallam SE, Kieser HM, Motamedi H, Hopwood DA, et al. 1987. Homology between *Streptomyces* genes coding for synthesis of different polyketides used to clone antibiotic biosynthetic genes. *Nature* 325:818–21
81. Malpartida F, Hopwood DA. 1984. Molecular cloning of the whole biosynthetic pathway of a *Streptomyces* antibiotic and its expression in a heterologous host. *Nature* 309:462–64
82. Marsden AFA, Caffrey P, Aparicio JF, Loughran MS, Staunton J, et al. 1994. Stereospecific acyl transfers on the erythromycin-producing polyketide synthase. *Science* 263:378–80
83. Mathews CK. 1993. The cell-bag of enzymes of network of channels? *J. Bacteriol.* 175:6377–81
84. Matsushima K, Ando Y, Hamasaki T, Yabe K. 1994. Purification and characterization of two versiconal hemiacetal acetate reductases involved in aflatoxin biosynthesis. *Appl. Environ. Microbiol.* 60:2561–67
85. Mayorga MA, Timberlake WE. 1992. The developmentally regulated *Aspergillus nidulans wA* gene encodes a polypeptide homologous to polyketide and fatty acid synthases. *Mol. Gen. Genet.* 235:205–12
86. McAda P. 1994. *Fungal polyketide synthases: lovastatin biosynthesis in* Aspergillus terreus. Presented at Int. Mycological Congr., 5th, Vancouver
87. McAlpine JB, Tuan JS, Brown DP, Grebner KD, Whittern DN, et al. 1987. New antibiotics from genetically engineered actinomycetes. I. 2-Norerythromycins, isolation and structural determinations. *J. Antibiot.* 40:1115–22
88. McDaniel R, Ebert-Khosla S, Fu H, Hopwood DA, Khosla C. 1994. Engineered biosynthesis of novel polyketides: influence of a downstream enzyme on the catalytic specificity of a minimal aromatic polyketide synthase. *Proc. Natl. Acad. Sci. USA* 91:11542–46
89. McDaniel R, Ebert-Khosla S, Hopwood DA, Khosla C. 1993. Engineered biosynthesis of novel polyketides: manipulation and analysis of an aromatic polyketide synthase with unproven catalytic specificities. *J. Am. Chem. Soc.* 115:11671–75
90. McDaniel R, Ebert-Khosla S, Hopwood DA, Khosla C. 1993. Engineered biosynthesis of novel polyketides. *Science* 262:1546–50
91. McDaniel R, Ebert-Khosla S, Hopwood DA, Khosla C. 1994. Engineered biosynthesis of novel polyketides: the *actVII* and *actIV* genes encode aromatase and cyclase enzymes, respectively. *J. Am. Chem. Soc.* 116:10855–59
92. Meurer G, Hutchinson CR. 1995. Functional analysis of putative β-ketoacyl:acyl carrier protein synthase and acyltransferase active-site-motifs in a type II polyketide synthase of *Streptomyces glaucescens*. *J. Bacteriol.* 177: 477–81
93. Miller MW, Hochstein FA. 1962. Isolation and characterization of two new tetracycline antibiotics. *J. Org. Chem.* 27:2525–28
94. Motamedi H, Cai SJ, Streicher SS, Shafiee A. 1994. *FK506 polyketide synthase is a large multifunctional polypeptide with 19 FAS-like domains*. Presented at Int. Symp. Genetics of Industrial Microorganisms, 7th, Montreal
95. Motamedi H, Hutchinson CR. 1987. Cloning and heterologous expression of a gene cluster for the biosynthesis of tetracenomycin C, the anthracycline antitumor antibiotic of *Streptomyces glaucescens*. *Proc. Natl. Acad. Sci. USA* 84:4445–49
96. Oki T, Matsuzawa Y, Kiyoshima K, Yoshimoto A. 1981. New anthracyclines, feudomycins, produced by the mutant from *Streptomyces coeruleorubidus* ME130-A4. *J. Antibiot.* 34:783–90
97. Ōmura S. 1981. Cerulenin. *Methods Enzymol.* 72:520–32
98. O'Hagen D. 1991. *The Polyketide Metabolites*. New York: Ellis Horwood
99. O'Keefe DP, Harder PA. 1991. Occurrence and biological function of cytochrome P450 monooxygenases in the actinomycetes. *Mol. Microbiol.* 5:2099–105
100. Patzelt H, Robinson JA. 1993. Biosynthesis of the polyether antibiotic monensin A: incorporation of a polyketide chain elongation intermediate. *J. Chem. Soc. Chem. Commun.* 1993:1258–60
101. Pernsson B, Krook M, Jörnvall H. 1991. Characteristics of short-chain alcohol dehydrogenases and related enzymes. *Eur. J. Biochem.* 200:537–43
102. Poulos TL, Finzel BC, Howard AJ. 1987. High-resolution crystal structure of cytochrome P450cam. *J. Mol. Biol.* 195:687–700
103. Deleted in proof
104. Roberts GA, Staunton J, Leadlay PF. 1993. Heterologous expression in *Escherichia coli* of an intact multienzyme component of the erythromycin-producing polyketide synthase. *Eur. J. Biochem.* 214:305–11
105. Sankawa U, Shimada H, Sato T, Ki-

noshita K, Yamasaki K. 1977. Biosynthesis of scytalone. *Tetrahedron Lett.* 483–86
106. Sankawa U, Shimada H, Sato T, Kinoshita K, Yamasaki K. 1981. Biosynthesis of scytalone. *Chem. Pharm. Bull.* 29:3536–42
106a. Schorr R, Mittag M, Mueller G, Schweizer E. 1994. Differential activities and intramolecular location of fatty-acid synthase and 6-methylsalicylic acid synthase component enzymes. *J. Plant Physiol.* 143:407–15
107. Scott AI, Beadling LC, Georgopapadakou NH, Subbarayan CR. 1974. Biosynthesis of polyketides. Purification and inhibition studies of 6-methylsalicylic acid synthase. *Bioorg. Chem.* 3:238–48
108. Scott AI, Phillips GT, Kircheis U. 1971. Biosynthesis of polyketides. The synthesis of 6-methylsalicylic acid and triacetic acid lactone in *Penicillium patulum. Bioorg. Chem.* 1:380–99
109. Scotti C, Piatti M, Cuzzoni A, Perani P, Tognoni A, et al. 1993. A *Bacillus subtilis* ORF coding for a polypeptide highly similar to polyketide synthases. *Gene* 130:65–71
110. Shafiee A, Hutchinson CR. 1987. Macrolide antibiotic biosynthesis: isolation and properties of two forms of 6-deoxyerythronolide B hydroxylase from *Saccharopolyspora erythraea* (*Streptomyces erythreus*). *Biochemistry* 26:6204–10
111. Shafiee A, Hutchinson CR. 1988. Purification and reconstitution of the electron transport components for 6-deoxyerythronolide B hydroxylase, a cytochrome P450 enzyme of macrolide antibiotic (erythromycin) biosynthesis. *J. Bacteriol.* 170:1548–53
112. Shen B, Hutchinson CR. 1993. Tetracenomycin F1 monooxygenase: oxidation of a naphthacenone to a naphthacenequinone in the biosynthesis of tetracenomycin C in *Streptomyces glaucescens. Biochemistry* 32:6656–63
113. Shen B, Hutchinson CR. 1993. Enzymatic synthesis of a bacterial polyketide from acetyl and malonyl coenzyme A. *Science* 262:1535–40
114. Shen B, Hutchinson CR. 1993. Tetracenomycin F2 cyclase: intramolecular aldol condensation in the biosynthesis of tetracenomycin C in *Streptomyces glaucescens. Biochemistry* 32:11149–54
115. Shen B, Summers RG, Gramajo HC, Bibb MJ, Hutchinson CR. 1992. Purification and characterization of the acyl carrier protein of the *Streptomyces glaucescens* tetracenomycin C polyketide synthase. *J. Bacteriol.* 174:3818–21
116. Deleted in proof
117. Sherman DH, Kim E-S, Bibb MJ, Hopwood DA. 1992. Functional replacement of genes for individual polyketide synthase components in *Streptomyces coelicolor* A3(2) by heterologous genes from a different polyketide pathway. *J. Bacteriol.* 174:6184–90
118. Sherman DH, Malpartida F, Bibb MJ, Kieser HM, Bibb MJ, Hopwood DA. 1989. Structure and deduced function of the granaticin-producing polyketide synthase gene cluster of *Streptomyces violaceoruber* TÜ22. *EMBO J.* 8:2717–25
119. Skory CD, Chang PK, Cary J, Linz JE. 1992. Isolation and characterization of a gene from *Aspergillus parasiticus* associated with the conversion of versicolorin A to sterigmatocystin in aflatoxin biosynthesis. *Appl. Environ. Microbiol.* 58:3524–37
120. Soga K, Furusho H, Mori S, Oki T. 1981. New antitumor antibiotics: 13-methylaclacinomycin A and its derivatives. *J. Antibiot.* 34:770–73
121. Spencer JB, Jordan PM. 1992. Purification and properties of 6-methylsalicylic acid synthase from *Penicillium patulum. Biochem. J.* 288:839–46
122. Spencer JB, Jordan PM. 1992. Investigation of the mechanism and steric course of the reaction catalyzed by 6-methylsalicylic acid synthase from *Penicillium patulum* using (R) [1-^{13}C; 2-^{2}H]- and (S)-[1-^{13}C; 2-^{2}H]malonates. *Biochemistry* 31:9107–16
123. Stindl A, Keller U. 1994. Epimerization of the D-valine portion in the biosynthesis of actinomycin D. *Biochemistry* 33:9358–64
124. Summers RG, Wendt-Pienkowski E, Motamedi H, Hutchinson CR. 1992. Nucleotide sequence of the *tcmII-tcmIV* region of the tetracenomycin biosynthetic gene cluster of *Streptomyces glaucescens* and evidence that the *tcmN* gene encodes a multifunctional cyclase-dehydratase-O-methyltransferase. *J. Bacteriol.* 174:1810–20
125. Summers RG, Wendt-Pienkowski E, Motamedi H, Hutchinson CR. 1993. The *tcmVI* region of the tetracenomycin C biosynthetic gene cluster of *Streptomyces glaucescens* encodes the tetracenomycin F1 monooxygenase, tetracenomycin F2 cyclase, and, most likely, a second cyclase. *J. Bacteriol.* 175:7571–80
126. Deleted in proof
127. Swan DG, Rodriguez AM, Vilches C,

Mendez C, Salas JA. 1994. Characterisation of a *Streptomyces antibioticus* gene encoding a type I polyketide synthase which has an unusual coding sequence. *Mol. Gen. Genet.* 242:358–62

128. Tajima S, Kubo Y, Furusawa I, Shishiyama J. 1989. Purification of a melanin biosynthetic enzyme converting scytalone to 1,3,8-trihydroxynaphthalene from *Cochliobolus miyabeanus*. *Exp. Mycol.* 13:69–76

128a. Takano Y, Kubo Y, Shimizu K, Mise K, Okuno T, Furusawa I. 1995. Structural analysis of PKS1, a polyketide synthase gene involved in melanin biosynthesis of *Colletotricum lagenarium*. *Mol. Gen. Genet.* In press

129. Tsay J-T, Oh W, Larson TJ, Jackowski S, Rock CO. 1992. Isolation and characterization of the β-ketoacyl-acyl carrier protein synthase III gene (*fabH*) from *Escherichia coli* K-12. *J. Biol. Chem.* 267:6807–14

129a. Tsukamoto N, Fujii I, Ebizua Y, Sankawa U. 1994. Nucleotide sequence of the *aknA* region of the alkavinone biosynthetic gene cluster of *Streptomyces galilaeus*. *J. Bacteriol.* 176:2473–75

129b. Tsukamoto N. 1994. *Cloning and analysis of biosynthetic genes of polyketide aklavinone from Streptomyces galilaeus*. PhD thesis. Univ. Tokyo, Tokyo, Japan

130. Vidal-Cros A, Viviani F, Laesse G, Boccara M, Gaudry M. 1994. Polyhydroxynaphthalene reductase involved in melanin biosynthesis in *Magnaporthe grisea*. Purification, cDNA cloning and sequencing. *Eur. J. Biochem.* 219:985–92

131. Villaroya A, Juan E, Egestd B, Jörnvall H. 1989. The primary structure of alcohol dehydrogenase from *Drosophila lebanonensis*. Extensive variation within insect short-chain alcohol dehydrogenase lacking zinc. *Eur. J. Biochem.* 180:191–97

132. Vining LC, Stuttard C, eds. 1995. *Genetics and Biochemistry of Antibiotic Production*. Boston: Butterworth-Heinemann

133. Wakil SA. 1989. Fatty acid synthase, a proficient multifunctional enzyme. *Biochemistry* 28:4523–30

134. Wang I-K, Reeves C, Gaucher GM. 1991. Isolation and sequencing of a genomic DNA clone containing the 3′ terminus of the 6-methylsalicylic acid polyketide synthase gene of *Penicillium urticae*. *Can. J. Microbiol.* 37:86–95

135. Weber JM, Leung JO, Swanson SJ, Idler KB, McAlpine JB. 1991. An erythromycin derivative produced by targeted gene disruption in *Saccharopolyspora erythraea*. *Science* 252:114–17

136. Weisner P, Beck J, Beck K-F, Ripka S, Muller G, et al. 1988. Isolation and sequence analysis of the fatty acid synthase FAS2 gene from *Penicillium patulum*. *Eur. J. Biochem.* 177:69–79

137. Ye J, Dickens ML, Plater R, Li Y, Lawrence J, et al. 1994. Isolation and sequence analysis of polyketide synthase genes from the daunomycin-producing *Streptomyces* sp. strain C5. *J. Bacteriol.* 176:6270–80

138. Yoshizawa Y, Witter DJ, Liu Y, Vederas JC. 1994. Revision of the biosynthetic origin of oxygens in mevinolin (lovastatin), a hypocholesterolemic drug from *Aspergillus terreus* MF 4845. *J. Am. Chem. Soc.* 116:2693–94

139. Yu J, Cary JW, Bhatnagar D, Cleveland TE, Keller NP, Chu FS. 1993. Cloning and characterization of a cDNA from *Aspergillus parasiticus* encoding an O-methyltransferase involved in aflatoxin biosynthesis. *Appl. Environ. Microbiol.* 59:3564–71

140. Yu T-W, Bibb MJ, Revill PW, Hopwood DA. 1994. Cloning, sequencing and analysis of the griseusin polyketide synthase gene cluster from *Streptomyces griseus*. *J. Bacteriol.* 176:2627–34

141. Yue S, Duncan JS, Yamamoto Y, Hutchinson CR. 1987. Macrolide biosynthesis. Tylactone formation involves the processive addition of three carbon units. *J. Am. Chem. Soc.* 109:1253–55

142. Yue S, Motamedi H, Wendt-Pienkowski E, Hutchinson CR. 1986. Anthracycline metabolites of tetracenomycin C-nonproducing *Streptomyces glaucescens* mutants. *J. Bacteriol.* 167:581–86

143. Zhang H-L, He X-G, Adefarati A, Gallucci J, Cole SP, et al. 1990. Mutactin, a novel polyketide from *Streptomyces coelicolor*. Structure and biosynthetic relationship to actinorhodin. *J. Org. Chem.* 55:1682–84

144. Zotchev S, Hutchinson CR. 1995. Cloning and expression of the genes encoding nonspecific electron transport components for a cytochrome P450 system of *Saccharopolyspora erythraea* involved in erythromycin production. *Gene*. In press

NONOPSONIC PHAGOCYTOSIS OF MICROORGANISMS

I. Ofek, J. Goldhar, and Y. Keisari

Department of Human Microbiology, Sackler Faculty of Medicine, Tel-Aviv University, Tel-Aviv, 69978 Israel

N. Sharon

Department of Membrane Research and Biophysics, The Weizmann Institute of Science, Rehovot, 76100, Israel

KEY WORDS: phagocytic receptors, opsonins, adhesins, lectins

CONTENTS

INTRODUCTION	240
LECTINOPHAGOCYTOSIS MEDIATED BY BACTERIAL SURFACE LECTINS	243
Type 1 Fimbriae of Enterobacteriaceae	244
Other Fimbrial and Nonfimbrial Adhesins of Escherichia coli	249
Outer Membrane Proteins (Opa) of Neisseria gonorrhoeae	250
Fimbriae and Other Adhesins of Pseudomonas aeruginosa	251
Type 2 Fimbriae of Actinomyces Spp.	252
MACROPHAGE LECTINS AS RECEPTORS IN LECTINOPHAGOCYTOSIS	253
The Gal/GalNAc-Specific Lectin	253
The Man/GlcNAc-Specific Lectin (Mannose Receptor)	254
Macrophage Lectins with Other Sugar Specificities	260
PHAGOCYTOSIS MEDIATED BY MACROPHAGE INTEGRINS	260
RGD-Containing Proteins of Bordetella pertussis	261
Interaction of CR3 with Leishmania Promastigotes	261
Interaction of Integrins with Other Microbial Constituents	262
HYDROPHOBIC INTERACTIONS	262
NONOPSONIC PHAGOCYTOSIS IN INFLAMMATION AND TISSUE INJURY	263
Type 1 Fimbriae and Renal Scarring	264
Stimulation of Phagocytic Cells by Lipoteichoic Acid	265
Stimulation of Neutrophils by Oral Bacteria	265
CONCLUSIONS	265

Abstract

Nonopsonic phagocytosis mediated by phagocyte receptors that recognize corresponding adhesins on microbial surfaces has attracted increasing interest as a potential host defense mechanism against extracellular pathogens and as a means of survival in the host for intracellular pathogens. Three types of nonopsonic phagocytosis involving carbohydrate-protein interactions (also termed lectinophagocytosis), protein-protein interactions, and hydrophobic interactions are discussed. A prominent receptor on phagocytic cells involved in recognizing pathogens belongs to the CD11/CD18 integrins. It mediates both opsonophagocytosis and nonopsonic phagocytosis and exhibits multiple specificity for different microbial adhesins. In other cases, similar specificity toward a microbial ligand (e.g. the *Klebsiella pneumoniae* capsule) is shared by dual molecules, one of which (e.g. the mannose-binding protein in serum) mediates opsonophagocytosis and the other (e.g. the macrophage mannose receptor) mediates nonopsonic phagocytosis of the microorganisms. In addition, we discuss how nonopsonic phagocytosis can trigger the phagocytes to release inflammatory agents and cause tissue injury. Further studies of the molecular mechanisms of nonopsonic phagocytosis, in particular those underlying the up-regulation of the phagocytic receptors by various agents, should lead to the development of new approaches for the prevention of infectious diseases.

INTRODUCTION

During their lifetimes, animals encounter numerous microbial species, but only a few of these microbes can grow and cause serious diseases in otherwise healthy hosts. Other microorganisms, often referred to as opportunistic pathogens, may cause disease only in immunocompromised hosts. Phagocytic cells such as neutrophils and macrophages serve as an important early defense against the invading microorganisms. These phagocytic cells recognize and subsequently ingest and kill many extracellular microbial species. Intracellular microorganisms are those species, fewer in number, that bind to and enter the host cells, including phagocytes, and resist intracellular killing and thus multiply within the cells; they thus escape antimicrobial agents of immune and nonimmune origin in the surrounding milieu.

Recognition of microorganisms by neutrophils and macrophages plays a crucial role in the process of phagocytosis and killing of extracellular pathogens as well as in the entry of intracellular pathogens into the phagocytes. The survival of extracellular pathogens in deep tissues required the evolution of mechanisms that allow the microorganisms to evade recognition by the phagocytes; whereas survival of intracellular bacteria depends on the ability of phagocytes to recognize the pathogen. Failure of the phagocytes to recognize the invading microorganism may render the host susceptible to infection by extracellular pathogens, but may also prevent the development of infection by intracellular pathogens. Elucidation of the molecular mechanisms underlying the recognition of microorganisms by phagocytes is essential for understanding the pathogenesis of the infectious process. It may also provide a basis for improved methods for treatment and prevention of some microbial infections.

Investigations over the past decade have revealed two basic molecular mechanisms of recognition of microorganisms by phagocytic cells: opsonin-dependent, or opsonic, and opsonin-independent, or nonopsonic (106, 167). The former mechanism requires serum components, opsonins, that function as a bridge between the microorganisms and the phagocytes. They act by binding to the surface of the microorganisms at one end and to specific receptors on the phagocyte surface at the other. The best-known opsonins are IgG, which binds via its Fc domain to the Fc receptor (FcR) on the phagocytes, and the iC3b fragment of the C3 component of complement, which binds to the complement receptor CR3 on the phagocyte surface.

Recently, other serum components, collectively termed collectins, that function as opsonins were described (50, 133). These lectins include the surfactant proteins SpA and SpD, the C-reactive protein, and the mannose-binding protein (MBP). They combine with complementary carbohydrates on microorganisms and with special collectin receptors on phagocytes. For example, the human mannose-binding protein, which appears to distinguish bacterial from host oligosaccharides, binds to the C1q receptor on macrophages and to mannose residues on the surfaces of microorganisms and thus enhances phagocytosis of the latter (Figure 1) (73). The indisputable role of opsonin-dependent phagocytosis that requires the participation of complement and antibody to protect against infection is termed opsonophagocytosis. In fact, several currently available vaccines, particularly active against encapsulated pathogenic bacteria, trigger the formation of protective opsonic antibodies aided by the complement system (80).

The immunoglobulin and complement receptors that participate in opsonophagocytosis were the first receptors identified on phagocytic cells. Research during the past two decades has established that phagocytic cells express specific receptors for other types of ligand (43, 92), some of which have recognition domains for structures expressed on the surface of different microorganisms. Recognition of such domains by the phagocytic receptors does not require opsonins and often leads to nonopsonic phagocytosis. Although we know that many different types of bacteria interact with phagocytic cells in serum-free media in vitro (106, 107, 180), our knowledge of nonopsonic phagocytosis, by which many microbial species are recognized, ingested, and killed, remains in its infancy. The molecules participating in

Table 1 Nonopsonic modes of recognition in phagocytosis of microorganisms

Type of interaction	Bacterial ligand (and example)	Phagocytic receptor (and example)	Reference
Lectin-carbohydrate	Lectin (Type 1 fimbriae)	Glycoprotein (Integrins)	33
	Polysaccharide (Capsule)	Lectin (Man/GlcNAc receptors)	7
Protein-protein	RGD-containing proteins[a] (Filamentous hemagglutinin)	RGD receptor (Integrins)	135
Hydrophobin-protein	Glycolipid (Lipoteichoic acid)	Lipid receptors? (Integrins)	115
	(Lipid A of lipopolysaccharides)	(Integrins)	185

[a] RGD, arginine-glycine-aspartic acid sequence.

nonopsonic recognition of only few species have been identified, and we are just beginning to understand the in vivo role of this mode of phagocytosis. In this review, we summarize knowledge on nonopsonic phagocytosis, referring specifically to the systems that have been investigated in considerable detail. We describe the properties and specificities of the various molecules and structures involved in the phagocytic recognition of microorganisms other than viruses. We also describe the role of nonopsonic phagocytosis in provoking inflammatory responses and in providing vital support to intracellular pathogens. In particular, we marshal the evidence showing that the nonopsonic mode of recognition of bacteria by phagocytes may act as a major defense mechanism against infections caused by many pathogens. A more detailed description of the biology of the interaction of microorganisms with phagocytic cells may be found in several excellent reviews and books (92, 154, 187).

Three main forms of recognition in nonopsonic phagocytosis have been described (Table 1). One mode, termed lectinophagocytosis, is based on recognition between surface lectins on one cell and surface carbohydrates on the opposing cell. The second mode is the result of protein-protein interactions via the Arg-Gly-Asp (RGD) peptide sequence of microorganisms and macrophage integrins. In the third mode, which is the least studied at the molecular level, hydrophobic interactions between bacteria and phagocytic cells produce phagocytosis. Furthermore, a particular microbial species can express multiple adhesins, each recognized by a distinct receptor present on phagocytic cells (Table 2).

Table 2 Examples of multiple mechanisms of molecular recognition in nonopsonic phagocytosis of microorganisms[a]

Microbial species	Microbial ligand	Phagocytic receptor	Reference
Extracellular			
K. pneumoniae	Type 1 fimbriae	CD11/CD18	33
	Capsule[b]	Mannose receptor	7
P. aeruginosa	Fimbriae	ND (expression stimulated by FN)	91
	NP, rpoN regulated	ND (expression inhibited by PTX)	91
	ND	ND (regulated by LPS)	91
	Mannose units in LPS	Mannose receptor	158
Intracellular			
B. pertussis	FHA	Galactose	173
	RGD sequence in FHA	CR3	121
Leishmania spp.	Lipophosphoglycan	Mannose receptor	183
	RDG sequence in gp63	LPS-binding site of integrin	183
	ND	Advanced glycosylation end product receptor	93
	ND	Mannose-6-phosphate receptor	132

[a] ND, not defined; FN, fibronectin; PTX, pertussis toxin; FHA, filamentous hemagglutinin.
[b] Capsules containing Manα2/3-Man or L-Rhaα2/3-L-Rha sequences.

LECTINOPHAGOCYTOSIS MEDIATED BY BACTERIAL SURFACE LECTINS

Virtually all bacterial species that colonize mammalian tissues express on their surfaces adhesins by which the organisms can bind to animal cells (104c). Many of these adhesins are sugar specific and are classified as lectins. Frequently they are associated with fimbriae or pili, hair-like appendages that protrude from the surface of the bacteria and are comprised of an assembly of protein subunits of different kinds. Although purified fimbriae also bind to cells in a sugar-specific manner, in most cases they do not act as agglutinins unless they have been cross-linked, so that presumably they are monovalent. Bacterial surface lectins may recognize a particular sugar(s) on the phagocytic cell surface and mediate phagocytosis of the bacteria (106, 107). The phagocytic cells, with which bacteria interact via their surface lectins, include human, rat, and mice peritoneal macrophages and human polymorphonuclear leukocytes (PMNs). Frequently, lectin-mediated binding to phagocytes leads to stimulation, ingestion, and killing of the bacteria. Whenever tested, specific sugars that inhibit binding of the bacteria to the phagocytes also inhibit the subsequent phagocytic activities. In contrast, such sugars do not affect the interaction of opsonized bacteria with the phagocytes, which reemphasizes that the activities just mentioned result from the lectin-mediated binding. Table 3 lists bacterial species reported to interact with phagocytic cells via surface lectins. The following sections discuss the lectins that have been studied in more detail during the past decade.

Type 1 Fimbriae of Enterobacteriaceae

Lectinophagocytosis of type 1 fimbriated *Escherichia coli* and some other members of the *Enterobacteriaceae* is the best-characterized system of phagocytosis. Type 1 fimbriae mediate the binding of bacteria to the mannose-containing receptors present on the surface of various animal and human cells, including all types of phagocytic cells tested (104a, 106, 107). These fimbriae are composed of a major 17-kDa subunit that is aligned in a right-handed array forming a filament with a diameter of 7 nm and is 100–400 nm long. In addition to the major subunit, a cassette of three minor subunits, FimG, FimF, and FimH (of mol wt 14,000, 16,000, and 29,000, respectively), is present at the fimbrial tips as well as at long intervals throughout the fimbrial length. FimH is the mannose-binding subunit of the fimbriae; its fine sugar specificity is influenced by the fimbrial shaft (82), and allelic variations in its gene give rise to strains that exhibit distinct receptor specificity (152). For example, allelic variants of FimH differing by only one amino acid substitution in their primary structure confer mannose-sensitive peptide-binding activity upon the fimbriated strains of *E. coli*.

LECTINOPHAGOCYTOSIS IN VITRO The detailed features of lectinophagocytosis mediated by type 1 fimbriae (reviewed in 104a, 106, 107) include attachment followed by ingestion and killing of the organisms by the phagocytic cells. Stimulation of the phagocytes induced by the mannose-specific binding of the bacteria was repeatedly demonstrated by measuring the various parameters of the oxidative burst or granule degranulation (Table 4).

Lectinophagocytosis mediated by type 1 fimbriae is quantitatively comparable with opsonophagocytosis in the attachment and stimulation phases (104a). Goetz et al (38) and Ohman et al (108) found that the rate of nonopsonic killing of type 1 fimbriated *E. coli* varies from strain to strain and is lower by 30–80% than that mediated by opsonizing antibodies. This decreased killing may result from inadequate stimulation of the intracellular killing mechanisms (38). Table 4 gives examples of phagocytic activities mediated by type 1 fimbriae and its FimH adhesin. Inter- and intraspecies strain variations were noted in lectinophagocytosis of enterobacteria expressing type 1 fimbriae. Such variations may result from differences in the fine sugar specificities of FimH (82, 152) or from other bacterial surface components, e.g. surface lipopolysaccharides (LPS), that affect the mannose-specific interaction of bacteria with phagocytic cells (108). Intraspecies strain variations in the ability of the bacteria to stimulate antimicrobial systems may explain the results of one study reporting that type 1 fimbriated *E. coli* containing FimH were not killed by

Table 3 Lectinophagocytosis mediated by bacterial surface lectins[a]

Lectin	Sugar[b]	Bacterium	Cells tested[c]
Type 1 fimbriae	Mannose	E. coli S. typhimurium K. pneumoniae	Human G and MØ, Mouse peritoneal MØ Rat peritoneal MØ
Type 2 fimbriae	Galβ3GalNAc	A. viscosus A. naeslundii	Human G Human G
Prs (F165₁) fimbriae	GalNAcβ3GalNAc	E. coli	Porcine G
Type P fimbriae	Gall,4Gal	E. coli	Human G coated by GalαGal4 containing glycolipid
Fimbriae	GalNAc	P. gingivalis	Mouse peritoneal MØ
Filamentous hemagglutinin and pertussis toxin	Galactose	B. pertussis	Human and rat alveolar MØ
Outer membrane	GalNAc Galactose	Eikenella corrodens Fusobacterium nucleatum	Guinea pig MØ Human G
Cell surface	GalNAc, NeuAc	Staphylocaecus saprophyticus	Human G

[a] With all the cells examined, the sugar-specific attachment, as well as stimulation of the phagocytes (oxidative burst and/or degranulation), have been demonstrated. In some of the systems, uptake of the bacteria (e.g. B. pertussis) and their killing (E. coli, A viscosus) were also observed. Adapted from 104a except for fimbriae of P. gingivalis (97) and E. coli Prs fimbriae (101).
[b] GalNAc, N-acetylgalactosamine; NeuAc, N-acetylneuraminic acid.
[c] MØ, macrophages; G, granulocytes (usually PMN).

Table 4 Phagocytic activities induced by intact type 1 fimbriae or by the FimH subunit[a]

Mode of presentation of FimH adhesin	Type of phagocytic cell[b]	Relative activity[a]				Reference
		Attachment	Stimulation	Ingestion	Killing	
On bacteria						
E. coli wild-type (Fim+)						
ABU2 strain	Human G	**100**	**100**[d]	**100**	**100**	81, 108
PN 7 strain		100	27	41	100	
E. coli recombinant						
Fim+ FimH+	Human, G	**100**	**100**[e]	ND	**100**	34, 170
Fim− FimH−		15	10	ND	0	
Fim+ FimH−		1.9	ND	ND	ND	
Shigella flexneri: recombinant						
Fim+ FimH+	Mouse, PM	**100**	**100**[e]	ND	ND	34, 170
Fim− FimH−		12	30	ND	ND	
Fim+ FimH−		1.9	ND	ND	ND	
On surfaces and particles						
Latex, fimbriae	Human, G	**100**	+[f]	+	ND	39
BSA[g]		0	−	−	ND	
Microsphere, FimH		**100**	+	+	ND	170
BSA		8.2	−	−	ND	
Nitrocellulose, FimH		**100**	ND	ND	ND	170
BSA		0				

[a] In each study the relative activity is determined by comparing the results to data from a reference strain or control [arbitrarily 100 or + (in bold face)]. All activities were inhibited by mannose or its methyl α-glycoside (see text for details).
[b] G, granulocytes (usually neutrophils); PM, peritoneal macrophages.
[c] ND, not done.
[d] Chemiluminescence.
[e] degranulation.
[f] oxygen uptake.
[g] bovine serum albumin control.

mouse macrophages and that growth of the organisms was enhanced compared to growth of isogenic strains of *E. coli* expressing fimbriae lacking FimH (59).

A growth advantage of *E. coli* bound via type 1 fimbriae to nonphagocytic cells over nonfimbriated bacteria has been documented (104e). The fate of such fimbriated *E. coli* depends on the structure of LPS on the bacterial surface, as well as on the relative hydrophobicity of bacterial and phagocytic surfaces. For example, a rough hydrophobic bacterial surface facilitates phagocytosis, whereas a smooth hydrophilic surface counteracts ingestion (e.g. strain PN 7 in Table 4) (81). Specific attachment, stimulation, ingestion, and killing occurred when type 1 fimbriae were expressed by recombinant strains of *E. coli* (170) or by heterologous species such as a recombinant strain of *Shigella flexneri* (34). In contrast, isogenic strains that do not express the fimbriae or that express fimbriae lacking the FimH subunit lack these activities. Moreover, inert particles coated by isolated type 1 fimbriae or purified FimH subunits stimulate phagocytic cells and are ingested by the latter (39, 170). Neutrophils also bind to the FimH subunit immobilized on nitrocellulose (170).

A recent study identified the leukocyte integrins (CD11/CD18) as a major receptor on human granulocytes (33). The bacteria bound in a dose-dependent manner to purified CD11/CD18 immobilized in microwells, whereas nonfimbriated *E. coli* failed to bind. The fimbriated bacteria also bound to the CD11a, CD11b, CD11c, and CD18 proteins, which were separated by polyacrylamide gel electrophoresis. The authors concluded that type 1 fimbriated *E. coli* bind to human granulocytes via the oligomannose and hybrid N-linked units of CD11/CD18. Because CD11b/CD18 and CD11c/CD18 function as receptors for iC3b, these findings provide a link between lectinophagocytosis and opsonophagocytosis (43, 92) (Figure 1).

Salmon et al (130) suggested that Fc receptors of human granulocytes are involved in ingestion, but not attachment, of type 1 fimbriated *E. coli* by the phagocytes and that the process is mediated by specific interaction between the oligosaccharide side chain of the receptors and the fimbriae. After attachment of the fimbriated bacteria to granulocytes via the carbohydrates of CD11/CD18, the fimbrial lectin may become concentrated in the proximity of the Fc receptor, which interacts with the lectin to trigger ingestion of the bacteria.

Leusch et al (78) and Sauter et al (136) found that *E. coli* and *Salmonella* strains carrying type 1 fimbriae bind to glycoproteins on the surface of different animal cells, including granulocytes. These glycoproteins belong to the family of carcinoembryonic antigens (CEA) and to the related nonspecific cross-reacting antigens (NCA). The latter consist of highly glycosylated, mannose-rich proteins that belong to the immunoglobulin supergene family and are expressed by granulocytes (137). The relationship of NCA to CD11/CD18 is, however, not known.

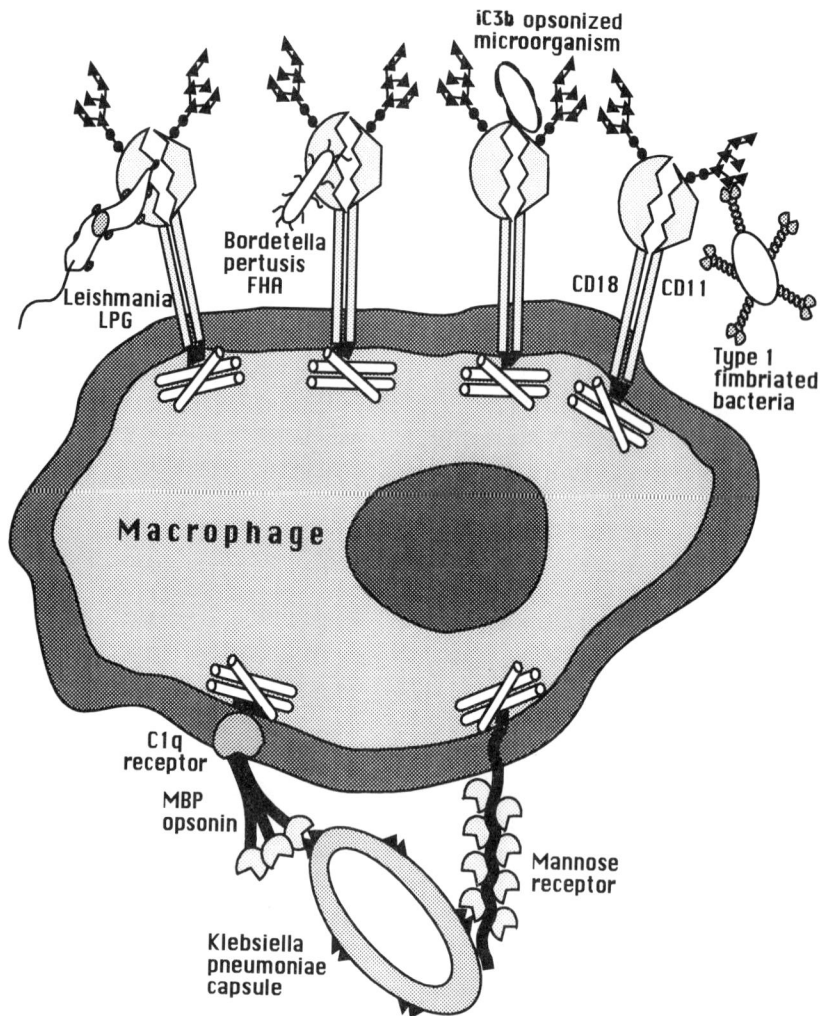

Figure 1 Diagram showing receptors on phagocytic cells and the corresponding adhesins on microbial surfaces participating in opsonic and nonopsonic phagocytosis. The diagram illustrates the CD18/CD11 integrins that contain attachment sites for different types of phagocytosis. Binding sites for nonopsonic phagocytosis in the integrins are illustrated for the RGD sequences in filamentous hemagglutinin of *Bordetella pertussis* (121), the LPS binding site for lipophosphoglycan of *Leishmania* spp. (183), and the mannose-containing oligosaccharide side chain for lectinophagocytosis mediated by type 1 fimbriated bacteria (33). Binding sites in the integrins for opsonic phagocytosis are illustrated for iC3b on microbial surfaces after complement activation. Also shown is the participation of a capsular polysaccharide of *Klebsiella pneumoniae* in nonopsonic phagocytosis mediated by the mannose receptor on macrophages and in opsonophagocytosis mediated by the mannose-binding protein (MBP). The carbohydrate-binding domains of the mannose receptor (adapted from 22) and of MBP (adapted from 50) share common homologous regions (26), which recognize mannoside units in the capsular polysaccharide of *Klebsiella pneumoniae*.

LECTINOPHAGOCYTOSIS IN VIVO   Although the occurrence of lectinophagocytosis mediated by bacterial lectins in vitro is well established, little is known about its occurrence and role in vivo. Because type 1 fimbriae mediate (*a*) binding of the bacteria to epithelial cells to promote colonization of mucosal surfaces and (*b*) binding to phagocytic cells, thereby causing clearing in deep tissues, bacteria expressing type 1 fimbriae should be more susceptible to phagocytosis and thus less virulent than those lacking the fimbriae. Indeed, when animals were infected with mixed bacterial strains or isogens, one of which was type 1 fimbriated (Fim$^+$) and the other nonfimbriated (Fim$^-$), the Fim$^-$ always survived at phagocyte-rich sites, whereas the Fim$^+$ phenotype always survived at phagocyte-poor sites (104a, 107). The selective survival of the fimbriated phenotype was attributed to its binding and colonization of epithelial surfaces, whereas that of the Fim$^-$ phenotype in phagocyte-rich sites likely resulted from elimination of the Fim$^+$ phenotype by phagocytes. Because *E. coli* cause intestinal and extraintestinal infections (164), phase variation, i.e. the ability of the bacteria to switch from phase Fim$^+$ to Fim$^-$ and back, is probably crucial for the survival of the bacteria in different niches during the natural course of infection (104c).

Lectinophagocytosis mediated by type 1 fimbriae may be responsible for blood clearance of the fimbriated bacteria (77, 126). A recent study confirmed the notion that Fim$^-$ variants are more virulent than Fim$^+$ variants in phagocyte-rich environments (90). Interaction of Fim$^+$ bacteria with host cells in vivo may be affected by the presence of soluble glycoproteins containing mannose. The best-known inhibitor of lectinophagocytosis mediated by type 1 fimbriae is Tamm-Horsfall protein, a glycoprotein synthesized by the kidney that is abundant in the normal urine (75, 104e). In an attempt to provide direct evidence of the occurrence of lectinophagocytosis in vivo, Bernhard et al (11) showed that type 1 fimbriated *E. coli* stimulate macrophages in the mouse peritoneum by measuring the release of lysosomal β-*N*-acetylglucosaminidase into peritoneal fluid (i.e. degranulation).

Other Fimbrial and Nonfimbrial Adhesins of Escherichia coli

Most pathogenic *E. coli* carry genetic information for expression of the other (besides type 1) fimbrial and nonfimbrial adhesins that cause hemagglutination of various animal erythrocytes (23). Among these adhesins, traditionally named mannose-resistant (MR), the P and S fimbriae (rigid 7-nm fimbriae) bind poorly or not at all to human granulocytes (12, 165). The diminished nonopsonic phagocytosis of P fimbriated strains was attributed to repulsive forces between the surfaces of the bacteria and the phagocytes caused by the negatively charged PapG subunit (tip adhesin) of the P fimbriae (171). In contrast to P fimbriae, *E. coli* bearing Prs fimbriae bind to porcine neutrophils, but only a small percentage of the bound bacteria was killed (101). Prs fimbriae are related to P fimbriae but are specific for the GalNAcα3GalNAc sequence of the Forssman antigen rather than for the Galα4Gal sequence alone (104d).

Many *E. coli* strains (and other enterobacterial strains) produce an MR hemagglutinin in the form of thin fimbriae (3 nm). These fimbriae participate in nonopsonic phagocytosis in at least one case (45). *Salmonella typhimurium* expressing such fimbrial hemagglutinins adhere to and are ingested by mouse peritoneal macrophages, but the specificity of the fimbriae is not known (45).

Several nonfimbrial (or fibrillar) adhesins (hemagglutinins) expressed by various *E. coli* isolates have been described (104c). In some cases, these adhesins are organized in a capsule-like form on the bacterial surface (40). NFA-1 and NFA-3 mediate binding of the bacteria to human granulocytes (41, 44). This conclusion is based on the findings showing: (a) The isolated adhesins bind to granulocytes, and they inhibited the attachment to granulocytes of bacteria carrying the homologous adhesins. (b) The recombinant *E. coli* LE 392 carrying either NFA-1 or NFA-3 bound to granulocytes, whereas the parent strain lacking the adhesin did not. NFA-3 recognizes the blood-group N-antigen on glycophorin A, but the specificity of NFA-1 was not defined. Although the expression of N-blood-group antigen on the surface of granulocytes was not reported before, evidence indicates that N-like determinants on the granulocyte surface serve as attachment sites for *E. coli* via its NFA-3 adhesin (44). Thus, donor granulocytes of blood group NN bound, ingested, and killed significantly more *E. coli* NFA-3 than donor granulocytes of blood group MM (41, 44). In addition, glycophorin A^{NN}, which inhibits the interaction between the bacteria and NN-erythrocytes, also inhibited binding of *E. coli*-NFA-3 to donor granulocytes of blood group NN donors.

Outer Membrane Proteins (Opa) of Neisseria gonorrhoeae

During gonococcal inflammatory infections, the bacteria are frequently found inside granulocytes. Internalization of the gonococci results from their adhesion to the granulocytes, which is mediated by a gonococcal outer membrane protein termed Opa or PII (146). Variants lacking this protein (Opa$^-$) were resistant to phagocytosis, and Opa-specific monoclonal antibodies (Fab fragments) inhibited gonococcal adhesion to and stimulation (i.e. chemiluminescence) of human granulocytes by the bacteria (24, 146). Binding of the organisms expressing Opa protein (Opa$^+$) to human granulocytes in the absence of serum was followed by ingestion and killing of the bound bacteria with concomitant stimulation of oxygen burst. The role of Opa protein as a mediator of the gonococcal attachment to and stimulation of granulocytes was further confirmed by experiments showing that purified Opa, Opa inserted into liposomes, and synthetic peptides composing the HV2 region of Opa bound to granulocytes and inhibited binding of Opa$^+$ bacteria to the phagocytic cells (98). The rate of killing and the ability to induce chemiluminescence of nonopsonized Opa⁺ gonococci were comparable to those of the opsonized bacteria (99, 122). Opa can be considered a lectin-like component because carbohydrates such as glucosamine, sialic acid, mannose, methyl α-mannoside, *N*-acetylglucosamine, and lactose inhibited the stimulation of oxidative burst of human neutrophils, as measured by chemiluminescence (146). Judged by the pattern of inhibition, the specificity of lectin is rather complex.

Like type 1 fimbriated bacteria, gonococci have underlying surface structures that influence the Opa-mediated interaction with the phagocytes. One such surface structure is the lipooligosaccharide. This constituent becomes sialylated during growth of the gonococci in media supplemented with N-acetylneuraminic acid, especially under anaerobic conditions (122). The presence of sialylated lipooligosaccharides on the surface of Opa⁺ gonococci confers resistance to phagocytosis.

The role of Opa-mediated recognition of gonococci by phagocytic cells in vivo is not clear, but several findings may be relevant to this issue. Expression of Opa proteins occurs under conditions of complex phase and antigenic variations, and the shift from one phase to another occurs rapidly, indicating a capacity for adaptation to the changing conditions during natural infection (104d). Urethral inoculation of volunteers with Opa⁻ variants resulted in strong selection of Opa⁺ variants that were in association with neutrophils (53). Another study showed that Opa⁺ variants are mostly isolated from urethral infections, whereas blood isolates are often Opa⁻ (146). These results may be explained by postulating that in blood Opa⁺ variants are eliminated by nonopsonic phagocytosis, while in the urethra adhesion to epithelial cells mediated by Opa proteins confers an advantage over nonadhering Opa⁻ variants.

Fimbriae and Other Adhesins of Pseudomonas aeruginosa

P. aeruginosa is responsible for various opportunistic infections, characteristically for chronic infections of patients with cystic fibrosis (CF) or with burns. The dynamic process of the infection results in phenotypic changes in the bacterial strains. For example, infection is usually initiated by nonmucoid strains, which are later replaced by mucoid ones. Speert and coworkers (155) first observed that clinical isolates of *P. aeruginosa* can be phagocytosed by granulocytes in serum-free media. By means of microscopic examination, reduction of nitroblue tetrazolium dye, enhancement of chemiluminescence, and bacterial killing, the authors showed that the magnitude of each of these activities varies within strains. One factor that may influence such variation is the production of a mucoid exopolysaccharide by CF isolates. This polysaccharide, an alginate, provides a survival advantage for the bacteria: in addition to promoting adhesion of the bacteria to the tracheal epithelium (104d), it increases the resistance of the bacteria to nonopsonic phagocytosis by human granulocytes and macrophages (15). This resistance depends on the negative charge of the well-hydrated polysaccharide causing repulsion between the bacteria and the phagocytes. Krieg et al (72) reported that a nonmucoid revertant of a clinical CF isolate and noncystic fibrosis isolates induced chemiluminescence stimulation and release of H_2O_2 by a rat alveolar macrophage cell line and by human granulocytes. The phagocyte response to the opsonized mucoid strain was lower than to the nonmucoid strain, suggesting that alginate might inhibit phagocytosis, even in the presence of specific antibodies. Purified alginate inhibited the granulocyte-respiratory response induced by either opsonized or nonopsonized *P. aeruginosa*. Alginate seems to represent another example of a surface structure that influences microbe-phagocytic cell interactions.

The susceptibility to phagocytosis of *P. aeruginosa* by granulocytes and monocyte-derived macrophages correlates with the presence of fimbriae and with hydrophobic interactions (156). The fimbriae may bind by hydrophobic interactions to putative structures on the phagocyte's surface and may mediate the initial attachment of *P. aeruginosa* to macrophages and neutrophils (91, 156). Fibronectin enhanced phagocytosis of fimbriated organisms but not of those lacking fimbriae (61, 65). The enhanced phagocytosis of a wild-type strain was inhibited by isolated fimbriae and pertussis toxin, but not by cholera toxin. The two latter toxins are carbohydrate-binding proteins (36), which suggests that distinct carbohydrates may also be involved in this phagocytic process. Because the fimbriae recognize *N*-acetylneuraminic, *N*-acetylglucosamine, and the GalNAcβ4Gal sequence present in asialo GM1 glycolipid (104f, 129), this process may be considered lectinophagocytosis.

Intracellular lectins produced by *P. aeruginosa* may also be involved in binding of the bacteria to phagocytic cells. This organism produces two intracellular lectins, one specific for galactose (PA-I) and one for fucose and mannose (PA-II) (36, 104f). Isolated PA-I lectin can serve as a bridge between heterologous cells: This protein bound to *E. coli* O86:B7 containing galactose in its LPS, and the lectin-treated bacteria were readily phagocytosed by human granulocytes. Release of intracellular PA-I and PA-II during infection may influence the interaction of *P. aeruginosa* with both epithelial and phagocytic cells (104c).

Type 2 Fimbriae of Actinomyces Spp.

Some species of the genus *Actinomyces*, which are gram-positive anaerobic bacteria involved in infections in the oral cavity, express on their surface type 2 fimbriae, a lectin specific for Galβ3GalNAc (104f). Binding of *Actinomyces* spp. to human granulocytes via this lectin was followed by ingestion, stimulation of superoxide and lactoferrin release, and killing of the bacteria (131). The attachment and phagocytosis were enhanced by pretreatment of the granulocytes by sialidase, which unmasks galactose residues on glycoproteins and glycolipids. The phagocytic activity was inhibited by lactose and methyl β-galactoside. Mutants of *Actinomyces viscosus* and *Actinomyces naeslundii* lacking type 2 fimbriae did not bind well to the granulocytes nor were they ingested or killed by these cells. Lectinophagocytosis by sialidase-treated granulocytes of *A. viscosus* increased when the bacteria were preincubated with IgM and complement, suggesting cooperative action of both lectin-mediated and complement-mediated killing (74).

MACROPHAGE LECTINS AS RECEPTORS IN LECTINOPHAGOCYTOSIS

After the discovery of hepatocyte lectins by Ashwell & Morell in the early 1970s, the presence of lectins on macrophages was demonstrated, and their role in clearance and endocytosis of glycoproteins and in cell-cell recognition is now well established (6, 22). They have a role in several macrophage functions (43, 159) including recognition of carbohydrates frequently present on the surfaces of different microorganisms, which results in the binding of the latter and their endocytosis (107). This process has been well documented, in particular for a macrophage surface lectin specific for galactose and N-acetylgalactosamine (Gal/GalNAc receptor) and for one specific for mannose, L-fucose, and N-acetylgalactosamine (Man/GlcNAc or Man/Fuc receptor, or mannose receptor). Both of these lectins belong to the family of Ca^{2+}-dependent, or C-type, lectins (22).

The Gal/GalNAc-Specific Lectin

The Gal/GalNAc-specific lectin of rat liver (the asialoglycoprotein receptor) is a type II transmembrane protein with an extracellular carboxy end and an intracellular amino terminus. It is made up of three types of subunits: a major one, RHL1, of 22.5 kDa, and two minor ones, RHL2 and RHL3, of 49.0 kDa and 54.0 kDa, respectively (22). Galactose on the β1-4 linked branch of complex oligosaccharide units of glycoproteins, whether free or peptide linked, binds solely to the minor subunits, whereas the galactose residues on the α1-6- and α1-3-linked branches bind specifically to the major subunit (123). Similar Gal/GalNAc-specific lectins present in rat Kupffer cells and mouse peritoneal macrophages have been studied in detail (68, 69). The lectins mediate phagocytosis of particles bearing surface galactose residues as well as of desialylated erythrocytes and thrombocytes in which such residues become exposed (6, 66, 69, 140). Experiments with both rats (6, 94) and mice (116) suggest that blood clearance of asialoglycoproteins and desialylated erythrocytes from the circulation by phagocytic cells of the liver and spleen are among the major physiological functions of the Gal/GalNAc-specific macrophage lectins in mammals (reviewed in 2).

The Gal/GalNAc-specific lectin of mouse macrophages recognizes galactosyl residues on bacterial surfaces, as demonstrated in experiments with *Streptococcus agalactiae* (104a). Removal of sialic acid from the capsular polysaccharide of these bacteria exposes galactose residues. Attachment in vitro of the desialylated streptococci to thioglycolate-elicited peritoneal macrophages was inhibited by galactosylated bovine serum albumin. Furthermore, the desialylated bacteria were cleared much faster than the untreated bacteria from the circulation of mice, and this clearance was inhibited by the galactosylated but not mannosylated neoglycoproteins. Similarly, desialylated *Trypanosoma cruzi* are also recognized by the Gal/GalNAc-specific lectin of macrophages (5). Because the Gal/GalNAc-specific lectin appears to be confined to hepatic and peritoneal macrophages, it may act in blood clearance and liver sequestration of microorganisms expressing surface galactosyl residues, as it acts in sequestration of desialylated erythrocytes, especially of aged ones.

The Man/GlcNAc-Specific Lectin (Mannose Receptor)

Among the various macrophage lectins, the Man/GlcNAc-specific lectin or mannose receptor is the most thoroughly studied with respect to its role in lectinophagocytosis of microorganisms. This lectin is expressed by all tissue macrophages examined, including alveolar, peritoneal, and hepatic ones. Its affinity to different monosaccharides is as follows: Man = L-Fuc > GlcNAc > glucose (6, 42, 117). It is a type I transmembrane glycoprotein consisting of a single subunit (165–170 kDa), with its carboxyl terminus on the cytoplasmic side of the membrane. The mannose receptor contains as many as eight adjoining carbohydrate-recognition domains (CRD), four of which are required to exhibit receptor activity (168). Mannose-containing glycoconjugates are cleared from body fluids as a result of the ability of the lectin to recognize the corresponding mannooligosaccharides in these macromolecules; they are then internalized (pinocytosis) and digested by the tissue macrophages (27, 42, 117). The mannose receptor also recognizes carbohydrates expressed on the surface of a wide spectrum of microorganisms (bacteria, fungi, and protozoa) and mediates their ingestion by tissue macrophages (Table 5).

Studies have shown that transfection of nonphagocytic COS (a line of monkey epithelial) cells with cDNA coding for the human macrophage mannose receptor confers the ability to bind and ingest *Pneumocystis carinii* and *Candida albicans,* two opportunistic pathogens that infect especially AIDS patients (26, 28). Thus, the macrophage lectin can endow nonphagocytic cells with the ability to bind and ingest microorganisms and can be considered a professional receptor for phagocytosis. Although opsonin-mediated phagocytosis often requires cooperation between more than one receptor on the macrophage surface (154), in Man/GlcNAc-dependent lectinophagocytosis, a single

Table 5 Phagocytosis of microorganisms mediated by the Man/GlcNAc-specific lectin (mannose receptor) of macrophages

Organism	Macrophages	Binding	Ingestion	Killing	Reference
Bacteria					
E. coli[a]	Mouse peritoneal	+	+	NT[b]	104a, 112
K. pneumoniae	Human monocyte derived	+	NT	NT	7
	Guinea pig alveolar	+	+	+	7
M. avium	Human monocyte derived	+	+	NT	141
M. tuberculosis	Human monocyte derived	+	+	NT	10
P. aeruginosa	Human monocyte derived	+	+	NT	158
S. agalactiae	Mouse peritoneal	+	NT	NT	104a
Fungi					
Aspergillus fumigatus	Mouse alveolar	+	+	NT	55
C. albicans	Mouse Kupffer	NT	NT	+	30
	Human monocyte derived	+	+	NT	86
Candida krusei	Mouse alveolar	+	NT	NT	177
Cryptococcus neoformans	Rat alveolar	NT	NT	+	13
Protozoa					
Leishmania donovani	Mouse peritoneal	+	+	NT	184
P. carinii	Human and rat alveolar	+	+	NT	28
	cos I transfected cells	+	+	NT	28

[a] Enteropathogenic (112) and uropathogenic (104a) *E. coli* strains.
[b] NT, not tested.

receptor on the macrophages is sufficient to bind and ingest the wide spectrum of microorganisms expressing the appropriate oligosaccharides.

MICROBIAL POLYSACCHARIDES RECOGNIZED BY THE MANNOSE RECEPTOR The identity of the surface structures involved in binding has been defined for only a few of the microorganisms bound specifically by the mannose receptor (Table 3). Lectinophagocytosis of yeasts (*Saccharomyces cerevisiae* or *C. albicans*) results from macrophage binding to the mannans coating the microorganisms. Thus, yeast mannan complexed with ferritin binds to human and mouse macrophages and inhibits phagocytosis of the yeast cells by the macrophages (70, 163, 177). Furthermore, mannan-coated particles were readily ingested by human monocyte-derived macrophages and rat-liver macrophages (58, 62). Recently, the mannosyl oligosaccharide of lipoarabinomannan on the surface of *Mycobacterium tuberculosis* was found to mediate the binding of the bacteria to human macrophages (142).

In another example, certain *K. pneumoniae* serotypes (e.g. K21a) express capsular polysaccharides that contain Manα2/3Man or L-Rhaα2/3L-Rha. Recognition of such sequences by the mannose receptor of macrophages results

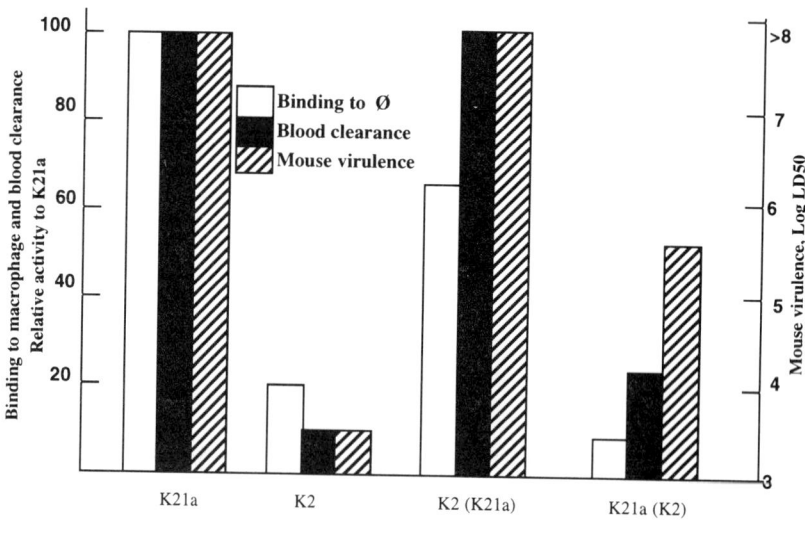

Figure 2 Relative binding to rat alveolar macrophages, blood clearance, and virulence in mice of *K. pneumoniae* capsular serotypes K2 and K21a parent strains and their capsule-switched derivatives K2(K21a) and K21a(K2), respectively. The binding to macrophages (based on the number of bacteria bound per phagocytic cell) and the blood clearance (survival of bacteria in blood, 15 min after intravenous inoculation) for each strain are shown relative to those of K21a, which is considered 100%. The virulence is expressed as LD$_{50}$ after intraperitoneal inoculation of mice for each strain. (Adapted from 54, 105.)

in phagocytosis (7). On the other hand, serotypes that lack such sequences (e.g. K2) are not recognized by the macrophage lectin, nor are they phagocytized. Isolated and purified capsular polysaccharides containing the repeating sequence Manα2/3Man or L-Rhaα2/3L-Rha bound to guinea pig alveolar macrophages, whereas those lacking these disaccharides did not. Moreover, the proximity between the capsular polysaccharide genes and the *his* locus made it possible to monitor the transfer of these genes using a conjugative plasmid from a donor serotype to a suitable recipient to allow interserotype switching of the capsular polysaccharide genes by reciprocal recombination (105). Such genetic manipulations produced the capsule-switched recombinant strains K2(K21a) and K21a(K2), which retained their respective recipient K2 and K21 strain background, as evidenced by a genetic marker, but inherited genes encoding the capsular polysaccharides of the donor strains. The capsule-switched recombinants K2(K21a) inherited the macrophage-binding phenotype of the K21a donor, whereas the K21a(K2) derivative bound poorly to

Table 6 The regulation of mannose receptor expression by various agents that affect macrophage activity

Effect on expression of receptor and agent	Modality affecting macrophage function	Reference
Increase		
PGE	Inhibition of production of pro inflammatory cytokines	14, 143
IL-4	Inhibition of inflammatory reactions	95, 161, 169
CSF-1	Production of monocyte-differentiating cytokine	9, 56
GM-CSF	Production of monocyte-differentiating cytokine	100, 124
Decrease		
BCG	Activation of macrophage in vivo	25, 43, 51a
IFN-γ	Production of macrophage-activating lymphokine	43, 100
LPS	Stimulation of production of inflammatory cytokine	88, 148
Phorbol ester	Stimulation of inflammatory reactions	4, 148
Glucocorticoids	Antiinflammatory agents	148

macrophages because they inherited the capsule genes of the donor K2 strain, which is not recognized by the macrophage's lectin (Figure 2).

MODULATION OF THE MANNOSE RECEPTOR Because the mannose receptor is involved in recognition of pathogens by macrophages (Table 5), it is important to determine the conditions and factors that modulate its expression and may thereby affect its ability to mediate lectinophagocytosis of microorganisms. Cytokines and drugs that change the activation status of macrophages affect the expression of the receptor. In general, decreased inflammatory capacity is associated with increased expression of the receptor, whereas activation reduced the expression of the receptor or its activity (Table 6).

Cytokines that activate macrophages down-regulate the expression of the mannose receptor (for review see 100). Lectinophagocytosis and killing of *P. aeruginosa* by human monocyte-derived macrophages was suppressed by IFN-γ (157). However, another study showed that although IFN-γ reduced the expression of the mannose receptor in human monocyte-derived macrophages, it concomitantly increased the receptor-mediated killing of *C. albicans* (87). The down-regulation of the mannose receptor by IFN-γ was attributed to inhibition of gene transcription (48). Activation of macrophages by other reagents was also accompanied by diminished expression of the mannose receptor, which in turn was attributed to receptor inactivation rather than increased turnover or decreased biosynthesis of the receptor (148).

Unlike the situation in macrophage activation, cytokines and agents that reduce the inflammatory potency of macrophages enhance mannose-receptor expression. The eicosanoid prostaglandin that is generated during inflammatory reactions and inhibits macrophage IL-1 and TNF-α production (14) augmented mannose-receptor expression and synthesis (143). This prostaglandin also reversed the IFN-γ induced diminution in mannose-receptor expression (144). Antiinflammatory agents such as glucocorticoids increased total mannose-receptor activity (150) and prevented the decrease in this activity that occurs after cell activation (148). IL-4, which is produced by Th-2 lymphocytes, may exhibit antiinflammatory activities (95, 169). Exposure of resident mouse peritoneal macrophages to recombinant IL-4 increased mannose receptor expression resulting from increased mRNA levels of the receptor (161).

Cytokines that promote monocyte and macrophage differentiation also affect mannose-receptor expression. Treatment of murine macrophages with macrophage colony-stimulating factors (M-CSF or CSF-1) augmented the expression of the receptor, as well as mannose-inhibitable killing of *C. albicans* (56). Granulocyte-macrophage colony-stimulating factor (GM-CSF), which promotes the differentiation of monocytes to macrophages, enhances the expression of mannose receptor in human monocyte-derived macrophages, as evidenced by the increased binding of yeast cells, and of *K. pneumoniae* expressing the complementary ligand (124).

Peripheral blood monocytes endowed with a high capacity to produce inflammatory cytokines (IL-1, TNF-α) and oxidative burst products lack the mannose receptor. In contrast, differentiated macrophages that are less reactive exhibit high mannose receptor levels (149).

ROLE OF THE MANNOSE RECEPTOR IN VIVO   Attachment of bacteria to macrophages via the mannose receptor may be a normal defense mechanism against bacterial infections in the lung and may be responsible for rapid clearance of the bacteria from the blood. Evidence to support this view comes from studies with the encapsulated strains of *K. pneumoniae*. Polysaccharide capsules are generally considered antiphagocytic (80). However, as discussed above, certain polysaccharide sequences may facilitate phagocytosis when they contain carbohydrates complementary to the macrophage lectin. Recently, the relative contribution of lectinophagocytosis mediated by the mannose receptor to the virulence of *K. pneumoniae* in mice was examined (54) using the serotypes K2 and K21a and their respective capsule-switched derivatives K2(K21a) and K21a(K2) (see above; Figure 2). The results suggest that switching of *cps* genes in *K. pneumoniae* serotypes markedly affects the interaction of the bacteria with macrophages and blood clearance and thus their virulence. The virulence, therefore, depends on the chemical structure of the capsule. Capsule types such as K21a are recognized by the macrophage lectin and as a result decrease virulence of the bacteria by enhancing the host cells' lectinophagocytosis and killing.

Studies of experimental infections in animals suggest that the Man/GlcNAc-

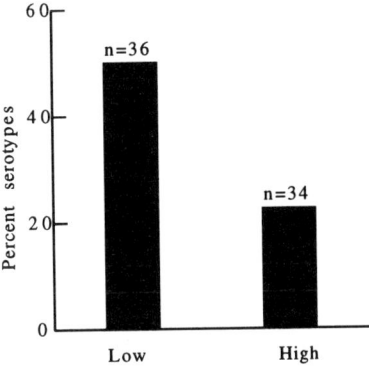

Figure 3 Distribution of serotypes expressing capsular polysaccharides that contain Manα2/3Man or Rhaα2/3Rha sequences between the group of serotypes isolated at high and low frequency from blood of patients. Fisher's exact test with two-tail analysis reveals a significant ($p < 0.05$) difference in the distribution of serotypes recognized by the macrophage Man/GlcNAc-specific lectin (mannose receptor) between the two groups of isolates. A high and low frequency of isolation represent ≥1% and <1%, respectively, of all isolates described (7, 19). The sequence of the repeating units of the capsular polysaccharides is based on data of Kenne & Lindberg (63) and Karunarante (57).

specific lectin may be responsible for blood clearance and liver sequestration not only of *K. pneumoniae* strains, but also of other bacterial species such as *E. coli* and *S. agalactiae* expressing the corresponding sugars on their surfaces (107, 106). This lectin may also be responsible for liver sequestration of *C. albicans* because mannose or yeast mannans inhibited trapping of the fungi by Kupffer cells (and other types of hepatocytes) in a perfused mouse liver system (139).

The possible role of macrophage lectin–capsular polysaccharide interactions in human infections of the lung is supported by epidemiological observations on the distribution of the various serogroups of isolates of *K. pneumoniae*. This organism is a frequently encountered gram-negative pathogen that causes nosocomial infections of the lower respiratory tract, often leading to fatal bacteremia (182). *Klebsiella* species have at least 77 different capsular serotypes (111), but strains belonging to only 13 of these serotypes account for about 70% of the *K. pneumoniae* blood isolates (19). The majority of these isolates do not express polysaccharides capable of binding to the macrophage lectin (Figure 3). The rapid clearance of small infecting doses of *Klebsiella* serotypes, which express the sequences Manα2/3-Man or L-Rhaα2/3-L-Rha, from the serum-poor environment of the lung may be attributed to lectinophagocytosis mediated by mannose receptors on alveolar macrophages.

Macrophage Lectins with Other Sugar Specificities

Macrophage lectins other than the mannose and galactose receptors probably also mediate lectinophagocytosis (43). This notion is supported by the early observations showing the inhibitory effect of monosaccharides on the binding of various bacteria to macrophages (178). For example, binding of *Staphylococcus albus* to macrophages via Ia antigens on the latter cells may be a mode of lectinophagocytosis that may lead to the development of the immune response (162). Other examples include the β-glucan receptor and the sialic acid–specific lectin on macrophages. The β-glucan receptor consists of two polypeptides of 150 and 170 kDa that recognize corresponding polymers contained in yeast cell walls (20). The sialic acid–specific lectin of macrophages binds specifically to sheep erythrocytes, but its role in recognizing sialyl-containing glycoconjugates on the microbial surface has not been described (18).

PHAGOCYTOSIS MEDIATED BY MACROPHAGE INTEGRINS

The β integrins, like other members of the integrin superfamily, consist of membrane heterodimeric glycoproteins, each composed of different α subunits and the same $β_1$ subunit to form unique heterodimers (e.g. $α_1β_1$, $α_2β_1$, etc) on the surface of various types of animal cells (51). The various subunits of the integrin molecules belong to the group of cluster-of-differentiation (CD) molecules and are identified by their interaction with specific monoclonal antibodies (e.g. CD11a/CD18, CD11b/CD18 etc). They are especially suited to the mediation of endocytosis because they are bound to cytoskeletal components inside the animal cells and extend through the membrane to serve as receptors for various ligands, such as fibronectin or complement receptors. In addition, they are involved in internalization of bacteria by both professional and nonprofessional phagocytes. Certain integrins also serve as receptors for microbial surface ligands in nonopsonic phagocytosis (Figure 1).

Integrins bind in a specific manner to accessible Arg-Gly-Asp (RGD)–containing peptides present in several mammalian proteins, in which the RGD sequence functions as a cell-binding site (127). They also contain domains that recognize the lipid A portion of bacterial lipopolysaccharides known as LPS-binding sites (also known as the alternate or lectin-like sites) (125, 185). Because integrins are glycoproteins, their carbohydrate units can be recognized by appropriate bacterial lectins. Finally, certain integrins serve as receptors in opsonophagocytosis of microorganisms coated with the C3b fragment of complement. For example, the complement receptor for the iC3b fragment is the integrin CD11b/CD18 dimer also known as CR3, which is a membrane receptor for the recognition of diverse ligands (125) (Figure 1). The following sections summarize the evidence showing that integrins are involved in nonopsonic phagocytosis of several microorganisms.

RGD-Containing Proteins of Bordetella pertussis

Recently, several microorganisms were found to express on their surfaces RGD-containing proteins that mediate adhesion of the microbe to mammalian cells, including phagocytic cells. A well-studied example is the filamentous hemagglutinin expressed by *B. pertussis,* which contains the RGD triplet that promotes adherence to the macrophage integrin CR3 (121). *B. pertussis* express another RGD-containing protein on their surfaces called pertactin. Whereas the filamentous hemagglutinins mediate the binding of the organisms to macrophages, pertactin mediates their binding to integrins on HeLa cells in tissue culture (76).

Although long-term survival of *B. pertussis* within human macrophages has been reported, how the RGD-mediated binding of the pathogen contributes to inactivation of the oxidative burst and other antimicrobial systems in the phagocytes is unclear (31).

The interaction of *B. pertussis* with macrophages demonstrates that pathogens can recognize, and modulate the expression of, receptors that exist both in inactive and active forms. In this system, binding of pertussis toxin to carbohydrates on the surface of macrophages up-regulates CR3 (134). The activated CR3 then binds the filamentous hemagglutinin via its RGD sequence, which leads to bacterial uptake by the macrophages. Binding by the toxin or by filamentous hemagglutinin alone results in significantly less ingestion than that observed in cells primed by preligation to the toxin. This cooperative (or coordinate) process is analogous to the interaction between selectins and integrins in leukocyte transmigration across endothelia (49).

Interaction of CR3 with Leishmania Promastigotes

Leishmania spp. are intracellular parasites transmitted to mammals by certain flies. Once in the mammalian host, the extracellular promastigotes enter and survive intracellularly in tissue macrophages. In vitro, promastigotes of different *Leishmania* species bind to macrophages in the absence of serum (183). Binding is to two distinct sites on the integrin CD11b/CD18 (CR3). One site is the lipopolysaccharide-binding site, which binds lipophosphoglycan, a major constituent of *Leishmania mexicana* promastigotes (128). The other attachment site is the iC3b-binding site, which recognizes gp63, a major surface glycoprotein on many *Leishmania* species (183). A recent study showed that gp63 is a fibronectin-like molecule that contains a conserved Ser-Arg-Tyr-Asp sequence that mimics antigenically and functionally the Arg-Gly-Asp sequence of fibronectin (153). Arg-Gly-Asp– and Ser-Arg-Tyr-Asp–containing peptides efficiently inhibited parasite attachment to mouse macrophages. CR3 may bind various *Leishmania* species via its RGD-binding site, which recognizes a similar, but not identical, sequence in the gp63 glycoprotein on the promastigote surface. Hence, *Leishmania* species are apparently capable of expressing multiple adhesins (Table 2). Expression of each type of adhesin is most likely under distinct genetic regulation, and the microbial population represents a mixture of variants or phenotypes that each express a different adhesin. This variation may explain the increased inhibition of *Leishmania* uptake by the macrophages when the recognition mediated by two receptors (e.g. mannose receptor and CR3) is specifically blocked, compared with that obtained when recognition mediated by only one receptor is blocked (113, 183).

Interaction of Integrins with Other Microbial Constituents

As discussed above, the mannose-specific type 1 fimbriae of enterobacteria recognize the oligomannoside and hybrid units of the integrin CD11/CD18 (Figure 1). In addition, CD11/CD18 can bind the phosphosugars of the sphingoglycolipids of *Histoplasma capsulatum* to mediate phagocytosis of these yeasts (8). These integrins also allow nonopsonic phagocytosis of group B streptococci by the mouse macrophage-like cell line, PU5-1.8, and by resident peritoneal macrophages of mice (3, 115). The integrin $\alpha_v\beta_3$ (vitronectin receptor) on human monocytes and monocyte-derived macrophages mediates the binding of the *Mycobacterium avium–Mycobacterium intracellulare* complex (119). Because CR3 also serves as a receptor in opsonophagocytosis, the participation of integrins as receptors for both nonopsonic and opsonic phagocytosis of microorganisms represents a link between the specialized and innate defense systems of animals.

The nonopsonic binding of *T. cruzi* and *C. albicans* to phagocytes is mediated by integrin-like molecules. Such molecules are present on the surface of these eukaryotic microorganisms, and they bind to RGD sequences of fibronectin on phagocyte surfaces (118, 138).

HYDROPHOBIC INTERACTIONS

The ability of the bacteria to bind to phagocytic cells frequently correlates to the magnitude of their surface hydrophobicity (1, 176). In this regard the phagocytic cell is not unique, because microbial surface hydrophobicity also promotes adhesion of the microorganisms to nonphagocytic cells (21). Surface hydrophobicity and susceptibility to nonopsonic, as well as opsonic, phagocytosis are influenced by bacterial surface constituents, especially by negatively charged residues of surface polysaccharides (181). Removal of such charged residues renders the microorganisms more hydrophobic and more susceptible to nonopsonic phagocytosis (181). Relevant to the situation in vivo are the findings showing that bacteria growing in media supplemented with sublethal concentrations of antibiotics exhibit hydrophobic surfaces and that their binding to human neutrophils is increased (67).

Apolar molecules on microbial surfaces involved in promoting adhesion to animal cells have been termed hydrophobins (104b). However, increased hydrophobicity is not always associated with increased nonopsonic phagocytosis (64). Surface hydrophobins can therefore serve to overcome repulsive forces between the microorganisms and the phagocyte surfaces to allow specific interactions of high avidity to take place. In only a few cases was a specific hydrophobin shown to mediate the binding of microorganisms to phagocytes. *Streptococcus pyogenes* binds to mouse peritoneal macrophages and human granulocytes in the absence of serum. Lipoteichoic acid inhibits this binding, which suggests that this polymer, which is present on the streptococcal surface, might serve as a ligand that is recognized by specific receptors on the phagocytic cells (17, 103). Lipoteichoic acid also mediates the binding of *S. agalactiae* (group B streptococci) to mouse peritoneal phagocytes (115). This compound may therefore be considered a hydrophobin that promotes the interaction of streptococci with integrins on phagocytic cells (see above).

Binding of *P. aeruginosa* to phagocytic cells was inhibited by hydrophobic compounds such as *p*-nitrophenol, and strains that were hydrophobic and piliated were taken up more readily than hydrophilic, nonpiliated strains (156). The lipophilic constituents were not defined, but experimental evidence suggests that they promote adhesion of *P. aeruginosa* to erythrocytes as well (32).

NONOPSONIC PHAGOCYTOSIS IN INFLAMMATION AND TISSUE INJURY

The nonopsonic recognition of intracellular microorganisms such as *B. pertussis*, mycobacteria, and *Leishmania* spp. enhances their persistence in the host and as a result their ability to cause tissue damage (154). Otherwise, the major role of the phagocytic process is the destruction of invading microorganisms that would normally proliferate only extracellularly. In most cases this process includes internalization within a phagocytic vacuole and the targeting of microbicidal agents toward the ingested microbe (16). During the course of these events, the phagocytes release into the extracellular milieu toxic antimicrobial agents such as reactive oxygen intermediates, neutral proteases, arachidonic acid, and cytokines (e.g. platelet activating factor, IL-1, IL-6, and TNF-α). Such agents cause tissue injury and inflammation (179). A major event occurring during the inflammatory response is the migration of leukocytes, initially neutrophils and macrophages and later lymphocytes and mast cells, from the circulation and the surrounding tissues into the site of microbial invasion. The interaction of these cells with the invading microbes can potentially result in the release of inflammation mediators and cause tissue injury. For example, nonopsonic activation of neutrophils by *Helicobacter pylori* appears to be responsible for the pathogenic mechanisms underlying acute or chronic gastritis and peptic ulcer disease (120). Tissue injury may be more pronounced with organisms that are ingested at a relatively reduced rate by the phagocytes but are capable of stimulating the phagocytes. For example, nonopsonized *Staphylococcus aureus* is recognized by granulocytes, but its ingestion is much slower than that of the opsonized bacteria (174). Nevertheless, the nonopsonized staphylococci induced degranulation in the phagocytes to the same extent as the opsonized ones.

In a few cases, clear evidence shows that ligands, which mediate nonopsonic phagocytosis, directly stimulate the phagocytes and sometimes cause tissue damage in vivo. Three of these are of particular interest and are discussed below.

Type 1 Fimbriae and Renal Scarring

Normally, the inflammatory response constitutes a rapid onset usually of short duration to eradicate an infectious agent. In some cases, especially when this response is chronic, it can have pathological consequences. An example is lectinophagocytosis mediated by type 1 fimbriae of uropathogenic *E. coli*. Renal scarring, which is closely associated with chronic pyelonephritis and inflammatory processes, often accompanies infections caused by type 1 fimbriated *E. coli* (47, 89, 172). Scarring may result from a unique pattern of degranulation stimulated by type 1 fimbriated bacteria in human granulocytes (52, 160) and from the bacteria's interaction with mast cells that internalize the bacteria with a resultant release of potent inflammatory agents (83, 84). In addition, strains isolated from patients with renal scarring triggered formation of primarily extracellular oxygen radicals, whereas no such activity was observed with strains isolated from urine of patients without renal scarring (96). This is the first demonstration that a bacterial ligand that mediates lectinophagocytosis is responsible for the chronic inflammatory process leading to renal scarring. Moreover, the neutrophil stimulation by *E. coli* expressing cross-linked type 1 fimbriae was enhanced compared with that by bacteria with intact fimbriae (114, 151). How such cross-linking of the fimbrial lectin on the bacterial surface potentiates a chronic inflammatory response in vivo remains to be seen.

Nonopsonic interaction of *E. coli* with neutrophils stimulated release of leukotrienes, which are potent inflammatory agents, in a manner dependent on the presence of fimbriae other than type 1 on the bacterial surface (71, 175). The molecular mechanisms underlying this stimulation have not been defined.

Stimulation of Phagocytic Cells by Lipoteichoic Acid

Another case is the nonopsonic phagocytosis of groups A and B streptococci mediated by lipoteichoic acid (see above) (Table 1). Isolated lipoteichoic acid can stimulate oxidative burst in human monocytes and neutrophils (37, 79, 110) as well as the release of cytokines from these cells (60, 85). These effects require cross-linking of the lipoteichoic acid receptors on the phagocytic cells (79, 85). Stimulation of monocytes and release of mediators of inflammation caused by lipoteichoic acids may be responsible for septic shock following gram-positive septicemias.

Stimulation of Neutrophils by Oral Bacteria

Periodontal disease is a chronic inflammatory process characterized by a massive infiltration of neutrophils and macrophages into periodontal tissues (35). Numerous bacterial species inhabiting the oral cavity have been implicated as playing a role in this tissue-destructive process. The current view is that the direct interaction between plaque bacteria and neutrophils in the microenvironment of the gingival niches may initiate an inflammatory response during periodontitis (145, 166). Several periodontopathogens cause activation and stimulation of lysosomal enzyme release from human and rabbit neutrophils (166). The N-acetylgalactosamine–specific fimbriae obtained from the periodontopathogen *Porphyromonas gingivalis* induced gene expression and production of inflammatory cytokines such as interleukin-1 and TNF-α, as well as the chemoattractant KC in mouse peritoneal macrophages (46, 97). These organisms also cause depolarization of human neutrophils, but an outer membrane constituent(s) of the bacteria is responsible for this activity (102). However, no data are available on the ability of *P. gingivalis* fimbriae to mediate binding of the intact bacteria to, and their ingestion by, the phagocytic cells.

CONCLUSIONS

The foregoing summarized the evidence demonstrating the occurrence of nonopsonic phagocytosis mediated by lectin-carbohydrate, protein-protein, or hydrophobic interactions. Opsonophagocytosis is mediated by C3b or iC3b generated by activation of complement via the alternative pathway that probably developed early in evolution, before the appearance of antibodies and Fc receptors (29, 167). Because CR3, which serves a receptor for C3b or iC3b, also participates in phagocytosis mediated by type 1 fimbriae or by RGD-containing proteins (Figure 1), these modes of nonopsonic phagocytosis may have developed even earlier, because they do not require complement. The requirement for antibodies to function as opsonins may have developed later to cope with microbial mutations that either rendered the microbes unable to activate complement or that resulted in constituents not recognized by phagocytes.

To cause infections in deep tissues of the nonimmune host, an invading pathogen must not only evade opsonophagocytosis but also nonopsonic phagocytosis. Microorganisms use several strategies to evade nonopsonic phagocytosis. Thus, certain bacteria express adhesins by which they bind to receptors present on both epithelial and phagocytic cells; examples are the Opa proteins of *N. gonorrhoeae,* the type 1 fimbriae of *E. coli,* and lipoteichoic acid of *S. pyogenes.* In the first stage of infection, these proteins mediate adhesion of the bacteria to epithelial cells, allowing the former to survive and colonize mucosal surfaces (104d). However, in deep tissue the same organisms run the risk of nonopsonic phagocytosis leading to their killing. To survive and continue the infectious process, some of these bacteria synthesize in vivo surface polymers (e.g. capsule) that interfere with the adhesin-receptor interaction by altering the net surface charge, the surface hydrophobicity, or the presentation and orientation of the adhesins on the microbial surfaces (17, 109). Another strategy is phase variation, in which the organisms oscillate at a relatively high frequency from a variant that expresses the adhesin to a variant that does not (104d). Variants that do not produce the adhesin usually predominate in phagocyte-rich sites, whereas those that do express the adhesin predominate at phagocyte-poor sites (104a, 107) (see also section on role in vivo). However, some microorganisms produce adhesins that can bind to both phagocytic and mucosal cells, but these organisms cannot alter their adhesiveness. It is tempting to speculate that such organisms are likely to colonize mucosal surfaces and less likely to successfully produce infections in deep tissues. Many of the normal microbial flora of the human body may belong to this category of microorganisms.

The dual role of microbial adhesins in promoting infection at mucosal surfaces and enhancing phagocytosis and killing of the organisms in deep tissues may jeopardize approaches to prevent infections by antiadhesive agents. For instance, carbohydrates that prevent colonization and infection (104e) could reach phagocyte-rich sites and block lectinophagocytosis mediated by either bacterial or phagocyte lectins, and thus prolong the duration of the infection.

Microorganisms produce multiple adhesins in order to colonize diverse surfaces (104e). These proteins are also involved in nonopsonic phagocytosis of many microbial pathogens (Table 2). Because expression of each adhesin requires complex genetic regulation, extracellular microorganisms capable of expressing such multiple adhesins are less capable of avoiding nonopsonic phagocytosis than those that express a single adhesin. The multiple molecular mechanisms involved in nonopsonic phagocytosis of extracellular pathogens (e.g. *P. aeruginosa*) may explain why the latter rarely cause infections in deep tissues, emphasizing the role of nonopsonic phagocytosis in host defense (91).

Although nonopsonic phagocytosis may play a critical role in protecting the host against symptomatic infections, under certain circumstances it may be damaging. It may result in release of inflammatory agents that cause tissue injury and be responsible for specific diseases (see section on type 1 fimbriated *E. coli* and scarring).

It is not clear what factors or which phagocyte receptors are more important for intracellular killing and which are needed for triggering release of mediators of inflammation. Particles phagocytosed via CR3 may not engender an oxidative burst (186), which suggests that this route of entry might permit survival of intracellular pathogens (e.g. *B. pertussis*) within the macrophage. However, examination of the different modes of nonopsonic phagocytosis of pathogens reveals that the outcome of the pathogen-phagocyte interaction does not depend on the receptor utilized for uptake of the pathogen. For example, the mannose receptor of macrophages recognizes some microorganisms that enter and survive within the phagocyte (e.g. *Leishmania* or *Mycobacteria* spp.) and others that are ingested and killed by the phagocytes (e.g. *K. pneumoniae*) (Table 4). In addition, the integrins serve as receptors for both intracellular (e.g. *Leishmania* spp. and *B. pertussis*) and extracellular (e.g. type 1 fimbriated *E. coli*) pathogens. The complex biochemical events (187) involved in killing of microorganisms ingested by phagocytes and the strategies used by intracellular pathogens to escape intracellular killing are outside the scope of this review.

Nonopsonic phagocytosis and opsonophagocytosis appear to be connected (Figure 1). The molecular region of the receptor on the macrophage surface involved in nonopsonophagocytosis may be either the same or distinct from that involved in opsonophagocytosis. For example, the galactose-specific lectins on liver macrophages recognize fibronectin-opsonized particles as well as Gal/GalNAc residues on bacterial surfaces (68).

The receptors involved in nonopsonic recognition of microorganisms not only participate in opsonophagocytosis but also in many other biological activities of the phagocytic cells, especially macrophages. The lectin-carbohydrate recognition system is a notable example (147). The evolutionary pressure that selected for such recognition systems to provide innate immunity against infections will become clearer as more systems of host defense employing nonopsonic mechanisms of recognition between microorganisms and phagocytic cells are elucidated.

A major goal in the study of nonopsonic phagocytosis is to develop means to manipulate the nonopsonic modes of recognition to enhance eradication of potential pathogens. One approach may involve manipulation of receptors present on phagocytes, e.g. up-regulation (Table 6), with minimal damage to other vital functions of these cells. It is hoped that further studies of nonopsonic phagocytosis will lead not only to a deeper understanding of this phenomenon and its role in infection in general, but may also result in development of new approaches for the prevention of infectious diseases.

ACKNOWLEDGMENTS

Research on the mannose receptor has been partially supported by the Kurt Leon Foundation, Konstanz, Germany, and by Tel-Aviv University Funds for Basic Research. We are grateful to Ms. R Perry for the computer-generated drawings.

> Any *Annual Review* chapter, as well as any article cited in an *Annual Review* chapter, may be purchased from the Annual Reviews Preprints and Reprints service.
> 1-800-347-8007; 415-259-5017; email: arpr@class.org

Literature Cited

1. Absolom DR. 1988. The role of bacterial hydrophobicity in infection: bacterial adhesion and phagocytic ingestion. *Can. J. Microbiol.* 34:287–98
2. Aminoff D. 1988. The role of sialoglycoconjugates in the aging and sequestration of red cells from circulation. *Blood Cells* 14:229–47
3. Antal JM, Cunningham JV, Goodrum KJ. 1992. Opsonin-independent phagocytosis of group B streptococci: role of complement receptor type 3. *Infect. Immun.* 60:1114–21
4. Apte RN, Keisari Y. 1987. Differential stimulation of mononuclear phagocyte IL-1 production and oxidative burst by tumor-promoting and non-tumor-promoting agents. *Immunobiology* 175:470–81
5. Araujo-Jorge TC, De Souza W. 1988. Interaction of *Trypanosoma cruzi* with macrophages. *Acta Trop.* 45:127–36
6. Ashwell C, Hartford H. 1982. Carbohydrate-specific receptors of the liver. *Annu. Rev. Biochem.* 51:531–54
7. Athamna A, Ofek I, Keisari Y, Markowitz S, Dutton GGS, Sharon N. 1991. Lectinophagocytosis of encapsulated *Klebsiella pneumoniae* mediated by surface lectins of guinea pig alveolar macrophages and human monocyte-derived macrophages. *Infect. Immun.* 59:1673–82
8. Barr K, Lester RL. 1984. Occurrence of novel antigenic phosphoinositol-containing sphingolipids in the pathogenic yeast *Histoplasma capsulatum*. *Biochemistry* 23:5581–89
9. Becker S, Warren MK, Haskill S. 1987. Colony-stimulating factor-induced monocyte survival and differentiation into macrophages in serum-free cultures. *J. Immunol.* 139:3703–9
10. Bermudez LE, Young LS, Enkel H. 1991. Interaction of *Mycobacterium avium* complex with human macrophages: roles of membrane receptors and serum proteins. *Infect. Immun.* 59:1697–702
11. Bernhard W, Gbarah A, Sharon N. 1992. Lectinophagocytosis of type 1 fimbriated (mannose-specific) *Escherichia coli* in the mouse peritoneum. *J. Leukocyte Biol.* 52:343–48
12. Blumenstock E, Jann K. 1982. Adhesion of piliated *Escherichia coli* strains to phagocytes: differences between bacteria with mannose-sensitive pili and those with mannose-resistant pili. *Infect. Immun.* 35:264–69
13. Bolanos B, Mitchell TG. 1989. Phagocytosis and killing of *Cryptococcus neoformans* by rat alveolar macrophages in the absence of serum. *J. Leukocyte Biol.* 46:521–28
14. Bonta IL, Ben-Efraim S. 1993. Involvement of inflammatory mediators in macrophage antitumor activity. *J. Leukocyte Biol.* 54:613–26
15. Cabral DA, Loh BA, Speert DP. 1987. Mucoid *Pseudomonas aeruginosa* resists nonopsonic phagocytosis by human neutrophils and macrophages. *Pediatr. Res.* 22:429–31
16. Cohen MS. 1994. Molecular events in the activation of human neutrophils for microbial killing. *Clin. Infect. Dis. Suppl. 2* 18:170–79
17. Courtney H, Hasty D, Ofek I. 1990. Hydrophobic characteristic of pyogenic

streptococci. In *Microbial Cell Surface Hydrophobicity*, ed. RJ Doyle, M Rosenberg, pp. 361–86. Washington, DC: Am. Soc. Microbiol. 425 pp.
18. Crocker PR, Gordon S. 1989. Mouse macrophage hemagglutinin (sheep erythrocyte receptor) with specificity for sialylated glycoconjugates characterized by monoclonal antibodies. *J. Exp. Med.* 169:1337–46
19. Cryz SJ Jr, Mortimer PM, Mansfield V, Germanier R. 1986. Seroepidemiology of *Klebsiella* bacteremic isolates and implications for vaccine development. *J. Clin. Microbiol.* 23:687–90
20. Czop JK, Kay J. 1991. Isolation and characterization of β-glucan receptors on human mononuclear phagocytes. *J. Exp. Med.* 173:1511–20
21. Doyle RJ, Rosenberg M. 1990. Microbial cell surface hydrophobicity: history, measurement, and significance. See Ref. 17, pp. 1–37
22. Drickamer K, Taylor ME. 1993. Biology of animal lectins. *Annu. Rev. Cell Biol.* 9:237–64
23. Duguid JP, Old DC. 1980. Adhesive properties of Enterobacteriaceae. In *Bacterial Adherence: Receptor and Recognition*, Ser. B, ed. EH Beachey, 6: 186–217. London: Chapman & Hall. 466 pp.
24. Elkins C, Rest RF. 1990. Monoclonal antibodies to outer membrane protein PII block interactions of *Neisseria gonorhoeae* with human neutrophils. *Infect. Immun.* 58:1078–84
25. Ezekowitz RAB, Gordon S. 1982. Down-regulation of mannosyl receptor-mediated endocytosis and antigen F4/80 in bacillus calmette-guerin activated mouse macrophages. *J. Exp. Med.* 155: 1623–28
26. Ezekowitz RAB, Sastry K, Baily P, Warner A. 1990. Molecular characterization of the human macrophage mannose receptor: demonstration of multiple carbohydrate recognition-like domains and phagocytosis of yeast in Cos-1 cells. *J. Exp. Med.* 172: 1785–94
27. Ezekowitz RAB, Stahl PD. 1988. The structure and function of vertebrate mannose-lectin like proteins. *J. Cell Sci. Suppl.* 9:121–33
28. Ezekowitz RAB, Williams DJ, Koziel H, Armstrong MYK, Warner A, et al. 1991. Uptake of *Pneumocystis carinii* mediated by the macrophage mannose receptor. *Nature* 351:155–58
29. Farries TC, Steuer KLK, Atkinson JP. 1990. Evolutionary implications of a new bypass activation pathway of the complement system. *Immunol. Today* 11:78–80
30. Felipe I, Bim S, Loyola W. 1989. Participation of mannose receptor on the surface of stimulated macrophages in the phagocytosis of gluteraldehyde-fixed *Candida albicans in vitro*. *Brazil. J. Med. Biol. Res.* 22:1251–54
31. Friedman RL, Nordensson K, Wilson L, Akporiaye ET, Yocum DE. 1992. Uptake and intracellular survival of *Bordetella pertussis* in human macrophages. *Infect. Immun.* 60:4578–85
32. Garber N, Sharon N, Shohet D, Lam JS, Doyle RJ. 1985. Contribution of hydrophobicity to hemagglutination reactions of *Pseudomonas aeruginosa*. *Infect. Immun.* 50:336–37
33. Gbarah A, Gahmberg CG, Ofek I, Jacobi U, Sharon N. 1991. Identification of the leukocyte adhesion molecules CD11/CD18 as receptors for type 1 fimbriated (mannose specific) *Escherichia coli*. *Infect. Immun.* 59:4524–30
34. Gbarah A, Mirelman D, Sansonetti PJ, Verdon R, Bernhard W, Sharon N. 1993. *Shigella flexneri* transformants expressing type 1 (mannose-specific) fimbriae bind to, activate, and are killed by phagocytic cells. *Infect. Immun.* 61: 1687–93
35. Genco RJ, Slots J. 1984. Host responses: host responses in periodontal diseases. *J. Dent. Res.* 63:441–51
36. Gilboa-Garber N, Garber N. 1992. Microbial lectins. In *Glycoconjugates: Composition, Structure and Function*, ed. HJ Allen, EC Kisailus, pp. 541–91. New York: Decker. 685 pp.
37. Ginsburg I, Fligiel SEG, Ward PA, Varani J. 1988. Lipoteichoic acid-antilipoteichoic acid complexes induce superoxide generation by human neutrophils. *Inflammation* 12:525–48
38. Goetz MB, Kuriyama SM, Silverblatt FT. 1987. Phagolysosome formation by polymorphonuclear leukocytes after ingestion of *Escherichia coli* that express type 1 pili. *J. Infect. Dis.* 156:229–33
39. Goetz MB, Silverblatt FJ. 1987. Stimulation of human polymorphonuclear leukocyte oxidative metabolism by type 1 pili from *Escherichia coli*. *Infect. Immun.* 55:534–40
40. Goldhar J, Perry R, Golecki JR, Hoschutzky H, Jann B, Jann K. 1987. Nonfimbrial mannose-resistant adhesins from uropathogenic *Escherichia coli* O83-K1-H4 and O14:K?:H:11. *Infect. Immun.* 55:1837–42
41. Goldhar J, Yavzori M, Keisari Y, Ofek I. 1991. Phagocytosis of *Escherichia coli* mediated by mannose resistant non-

fimbrial haemagglutinin (NFA-1). *Microb. Pathogen.* 11:171–78
42. Gordon S, Makena T. 1989. Receptors for mannosyl structures on mononuclear phagocytes. In *Mononuclear Phagocytes,* ed. M Zembda, GL Asherson, pp. 141–50. San Diego: Academic. 348 pp.
43. Gordon S, Perry VH, Rabinowitz S, Chung L-P, Rosen H. 1988. Plasma membrane receptors of the mononuclear phagocyte system. *J. Cell Sci. Suppl.* 9:1–26
44. Grünberg J, Ofek I, Perry R, Wiselka M, Boulnois G, Goldhar J. 1994. Blood group NN dependent phagocytosis mediated by NFA-3 haemagglutinin of *Escherichia coli. Immunol. Infect. Dis.* 4:28–32
45. Grund S, Seiler A. 1993. Fimbriae and lectinophagocytosis of *Salmonella typhimurium* variatio copenhagen (STMVC)—an electron microscopic study. *J. Vet. Med. B* 40:105–12
46. Hanazawa K, Murakami Y, Hirose SK, Amano S, Ohmori Y, et al. 1991. *Porphyromonas (Bacteroides) gingivalis* fimbriae activate mouse peritoneal macrophages and induce gene expression and production of interleukin-1. *Infect. Immun.* 59:1972–77
47. Harber MJ. 1986. Virulence factors of urinary pathogens in relation to kidney scarring. In *Microbial Disease in Nephrology,* ed. AW Asscher, W Brumfitt, pp. 69–78. Chichester: Wiley
48. Harris N, Super M, Rits M, Chang G, Ezekowitz RAB. 1992. Characterization of the murine macrophage mannose receptor: demonstration that the down-regulation of receptor expression mediated by interferon γ occurs at the level of transcription. *Blood* 80:2363–73
49. Hoepelman AIM, Tuomanen EI. 1992. Consequences of microbial attachment: directing host cell functions with adhesins. *Infect. Immun.* 60:1729–33
50. Holmskov U, Malhotra R, Sim RB, Jensenius JC. 1994. Collectins: collagenous C-type lectins of the innate immune defense system. *Immunol. Today* 15:67–74
51. Hynes RO. 1987. Integrins, a family of cell surface receptors. *Cell* 48:549–55
51a. Imber MJ, Pizzo SV, Johnson WJ, Adams DO. 1982. Selective diminution of the binding of mannose by murine macrophages in the late stages of activation. *J. Biol. Chem.* 257:5129–35
52. Jenner DE, Harber MJ, Davis M, Asscher AW. 1985. Polymorphonuclear leucocyte degranulation after stimulation by serum-treated and unopsonized bacteria. *Biochem. Soc. Trans.* 13:1196–97
53. Jerse AE, Cohen MS, Drown PM, Whicker LG, Isbey SF, et al. 1994. Multiple gonococcal opacity proteins are expressed during experimental urethral infection in the male. *J. Exp. Med.* 179:911–20
54. Kabha K, Athamna A, Keisari Y, Parolis H, Parolis LAS, et al. 1994. Analysis of the relationship between capsule type and virulence in *Klebsiella pneumoniae. Public Health Rev.* 21:95–96
55. Kan VL, Bennett JE. 1988. Lectin-like attachment sites on murine pulmonar alveolar macrophages bind *Aspargillus fumigatus* conidia. *J. Infect. Dis.* 158:407–14
56. Karbassi A, Becker JM, Foster JS, Moore RN. 1987. Enhanced killing of *Candida albicans* by murine macrophages treated with macrophage colony-stimulating factor: evidence for augmented expression of mannose receptors. *J. Immunol.* 139:417–21
57. Karunarante DN. 1985. *Structural investigation of the capsular polysaccharides of* K. pneumoniae. PhD thesis. Univ. BC, Vancouver, Canada
58. Kataoka M, Tavassoli M. 1985. Development of specific surface receptors recognizing mannose-terminal glycoconjugates in cultured monocytes: a possible early marker for differentiation of monocytes into macrophages. *Exp. Hematol.* 13:44–50
59. Keith BR, Harris SL, Russell PW, Orndorff PE. 1990. Effect of type 1 piliation on in vitro killing of *Escherichia coli* by mouse peritoneal macrophages. *Infect. Immun.* 58:3448–54
60. Keller R, Fisher W, Keist R, Bassetti S. 1992. Macrophage response to bacteria: induction of marked secretory and cellular activities by lipoteichoic acids. *Infect. Immun.* 60:3664–72
61. Kelly NM, Kluftinger JN, Pasloske BL, Paranchych W, Hancock REW. 1989. *Pseudomonas aeruginosa* pili as ligands for nonopsonic phagocytosis by fibronectin-stimulated macrophages. *Infect. Immun.* 57:3841–45
62. Kempka G, Kolb-Bachofen V. 1988. Binding, uptake, and transcytosis of ligands for mannose-specific receptors in rat liver: an electron microscopic study. *Exp. Cell. Res.* 176:38–48
63. Kenne L, Lindberg B. 1983. Bacterial polysaccharides. In *The Polysaccharides,* ed. GO Aspinall, 2:287–363. New York: Academic
64. Kerosuo E, Haapasalo M, Lounatmaa K. 1993. *Eubacterium yurii* subspecies

margaretiae is resistant to nonopsonic phagocytic ingestion. *Dent. Res.* 101: 304–10

65. Kluftinger JL, Kelly NM, Hancock REW. 1989. Stimulation by fibronectin of macrophage-mediated phagocytosis of *Pseudomonas aeruginosa. Infect. Immun.* 57:817–22
66. Kluge A, Reuter G, Lee H, Ruch-Heeger B, Schaur R. 1992. Interaction of rat peritoneal macrophages with homologous sialidase-treated thrombocytes in vitro: biochemical and morphological studies. Detection of N-(O-acetyl) glycolylneuraminic acid. *Eur. J. Cell Biol.* 59:12–20
67. Kohada A, Miyake Y, Sugai M, Tsuru H, Suginaka H. 1991. Effects of antibiotics on nonopsonized adherence of *Staphylococcus aureus* to human polymorphonuclear leukocytes. *Chemotherapy* 37:50–56
68. Kolb-Bachofen V, Abel F. 1991. Participation of D-galactose-specific receptors of liver macrophages in recognition of fibronectin-opsonized particles. *Carbohydr. Res.* 213:201–13
69. Kolb-Bachofen V, Schlepper-Schafer J, Roos P, Hulsmann D, Kolb H. 1984. GalNAc/Gal-specific rat liver lectins: their role in cellular recognition. *Biol. Cell* 51:219–26
70. Kolotila MP, Rogers AL, Beneke ES, Smith CW. 1987. The effects of soluble *Saccharomyces cerevisiae* mannan on the phagocytosis of *Candida albicans* by mouse peritoneal macrophages *in vivo. J. Med. Vet. Mycol.* 25:85–95
71. Konig B, Konig W. 1991. Roles of human peripheral blood leukocyte protein kinase C and G proteins in inflammatory mediator release by isogenic *Escherichia coli* strains. *Infect. Immun.* 59:3801–10
72. Krieg DP, Helmke RJ, German VF, Mangos JA. 1988. Resistance of mucoid *Pseudomonas aeruginosa* to nonopsonic phagocytosis by alveolar macrophages in vitro. *Infect. Immun.* 56:3173–79
73. Kuhlman M, Joiner K, Ezekowitz RAB. 1989. The human mannose-binding protein functions as an opsonin. *J. Exp. Med.* 169:1733–45
74. Kurashima C, Sandberg AL, Cisar JO, Mudrick LL. 1991. Cooperative complement—and bacterial—lectin-initiated bactericidal activity of polymorphonuclear leukocytes. *Infect. Immun.* 59:216–21
75. Kuriyma S, Silverblatt FJ. 1986. Effect of Tamm-Horsfall urinary glycoprotein on phagocytosis and killing of type 1– fimbriated *Escherichia coli. Infect. Immun.* 51:193–98
76. Leininger E, Ewanowich CA, Bhargawa A, Peppler MS, Kenimer JG, Brennan MJ. 1992. Comparative roles of the Arg-Gly-Asp sequence present in the *Bordetella pertussis* adhesins pertactin and filamentous hemagglutinin. *Infect. Immun.* 60:2380–85
77. Leunk RD, Moon RJ. 1982. Association of type 1 pili with the ability of livers to clear *Salmonella typhimurium. Infect. Immun.* 36:1168–74
78. Leusch H-G, Drzeniek Z, Markos-Pusztaj Z, Wagener C. 1991. Binding of *Escherichia coli* and *Salmonella* strains to members of carcinoembryonic antigen family: differential binding inhibition by aromatic alpha-glycosides of mannose. *Infect. Immun.* 59:2051–57
79. Levy R, Kotb M, Nagauker O, Majumdar G, Alkan M, et al. 1990. Stimulation of oxidative burst in human monocytes by lipoteichoic acids. *Infect. Immun.* 58:566–68
80. Lindberg AA. 1990. Polysaccharide vaccines: vaccines needed for the 1990s. In *New Antibacterial Strategies*, ed. HC Neu, pp. 69–85. New York: Churchill Livingstone. 312 pp.
81. Lock R, Dahlgren C, Linden M, Stendhal O, Svensbergh A, Ohman L. 1990. Neutrophil killing of two type 1 fimbriae-bearing *Escherichia coli* strains: dependence on respiratory burst activation. *Infect. Immun.* 58:37–42
82. Madison B, Ofek I, Clegg S, Abraham SN. 1994. Type 1 fimbrial shafts of *Escherichia coli* and *Klebsiella pneumoniae* influence sugar-binding specificities of their FimH adhesins. *Infect. Immun.* 62:834–48
83. Malaviya R, Ross EA, Jakschik BA, Abraham S. 1994. Mast cell degranulation induced by type 1 fimbriated *Escherichia coli* in mice. *J. Clin. Invest.* 93:1645–53
84. Malaviya R, Ross EA, MacGregor JI, Ikeda T, Little JR, et al. 1994. Mast cell phagocytosis of FimH-expressing enterobacteria. *J. Immunol.* 152:1907–14
85. Mancuso G, Tomasello F, Ofek I, Teti G. 1994. Anti-lipoteichoic acid antibodies enhance the release of cytokines by monocytes sensitized with lipoteichoic acid. *Infect. Immun.* 62:1470–73
86. Marodi L, Korchak HM, Johnston RB Jr. 1991. Mechanisms of host defense against *Candida* species. I. Phagocytosis by monocytes and monocyte-derived macrophages. *J. Immunol.* 146:2783–89
87. Marodi L, Schreiber S, Anderson DC, MacDermott RP, Korchak HM, John-

ston RB Jr. 1993. Enhancement of macrophage candidacidal activity by interferon-gamma. Increased phagocytosis, killing, and calcium signal mediated by a decreased number of mannose receptors. *J. Clin. Invest.* 91: 2596–601
88. Martich GD, Boujoukos AJ, Sufferdini AF. 1993. Response of man to endotoxin. *Immunobiology* 187:403–16
89. Matsumoto T, Mizumoe Y, Sakamoto N, Tanaka M, Kumazaawa J. 1990. Increased renal scarring by bacteria with mannose-sensitive pili. *Urol. Res.* 18: 299–303
90. May AK, Bloch CA, Sawyer RG, Spengler MD, Pruett TL. 1993. Enhanced virulence of *Escherichia coli* bearing a site-targeted mutation in the major subunit of type 1 fimbriae. *Infect. Immun.* 61:1667–73
91. Mork T, Hancock REW. 1993. Mechanisms of nonopsonic phagocytosis of *Pseudomonas aeruginosa. Infect. Immun.* 61:3287–93
92. Mosser DM. 1994. Receptors on phagocytic cells involved in microbial recognition. See Ref. 187, pp. 99–130
93. Mosser DM, Vlassara H, Edelson PJ, Cerami A. 1987. Leishmania promastigotes are recognised by the macrophage receptor for advanced glycosylation end products. *J. Exp. Med.* 165:140–45
94. Muller E, Franco MW, Schauer R. 1981. Involvement of membrane galactose in the *in vivo* and *in vitro* sequestration of desialylated erythrocytes. *Hoppe-Seyler's Z. Physiol. Chem.* 362:1615–20
95. Mulligan MS, Jones ML, Vaporciyan AA, Howard MC, Ward PA. 1993. Protective effects of IL-4 and IL-10 against immune complex-induced lung injury. *J. Immunol.* 151:5666–74
96. Mundi H, Bjorksten B, Svanborg C, Ohman L, Dahlgren C. 1991. Extracellular release of reactive oxygen species from human neutrophils upon interaction with *Escherichia coli* strains causing renal scarring. *Infect. Immun.* 59: 4168–72
97. Murakami Y, Hanazawa S, Nishida K, Iwasaka H, Kitano S. 1993. N-Acetyl-D-galactosamine inhibits TNF-α gene expression induced in mouse peritoneal macrophages by fimbriae of *Porphyromonas (Bacteroides) gingivalis,* and oral anaerobe. *Biochem. Biophys. Res. Commun.* 192:826–32
98. Naids FL, Belisle B, Lee N, Rest RF. 1991. Interaction of *Neisseria gonorrhoea* with human neutrophils: Studies with purified PII (Opa) outer membrane proteins and synthetic Opa peptides. *Infect. Immun.* 59:4628–35
99. Naids FL, Rest RF. 1991. Stimulation of human neutrophil oxidative metabolism by nonopsonized *Neisseria gonorrhoea. Infect. Immun.* 59:4383–90
100. Nathan C, Yoshida R. 1988. Cytokines: interferon-γ. In *Inflammation: Basic Principles and Clinical Correlates,* ed. JI Gallin, IM Goldstein, R Snyderman, pp. 229–51. New York: Raven
101. Ngeleka M, Martineau-Doize B, Fairbrother JM. 1994. Septicemia-inducing *Escherichia coli* O115:K"V165"F165$_1$ resist killing by porcine polymorphonuclear leukocytes *in vitro:* role of F165$_1$ fimbriae and K"V165" O-antigen capsule. *Infect. Immun.* 62:398–404
102. Novak MJ, Cohen HJ. 1991. Depolarization of polymorphonuclear leukocytes by *Porphyromonas (Bacteroides) gingivalis* 381 in the absence of respiratory burst activation. *Infect. Immun.* 59:3134–42
103. Ofek I, Beachey EH. 1979. Lipoteichoic acid–sensitive attachment of group A streptococci to phagocytes. In *Pathogenic Streptococci,* ed. MT Parker, pp. 44–46. Chertsey/Surrey, UK: Redbooks. 296 pp.
104a. Ofek I, Doyle RJ. 1994. *Bacterial Adhesion to Cells and Tissues,* pp. 171–94. London/New York: Chapman & Hall. 578 pp.
104b. Ofek I, Doyle RJ. 1994. See Ref. 104a, pp. 2–3
104c. Ofek I, Doyle RJ. 1994. See Ref. 104a, pp. 321–512
104d. Ofek I, Doyle RJ. 1994. See Ref. 104a, pp. 239–320
104e. Ofek I, Doyle RJ. 1994. See Ref. 104a, pp. 513–61
104f. Ofek I, Doyle RJ. 1994. See Ref. 104a, pp. 94–135
105. Ofek I, Kabha K, Athamna A, Frankel G, Wozniak DJ, et al. 1993. Genetic exchange of determinants for capsular polysaccharide biosynthesis between *Klebsiella pneumoniae* strains expressing serotypes K2 and K21a. *Infect. Immun.* 61:4208–16
106. Ofek I, Rest RF, Sharon N. 1992. Nonopsonic phagocytosis of microorganisms. *ASM News* 58:429–35
107. Ofek I, Sharon N. 1988. Lectinophagocytosis: a molecular mechanism of recognition between cell surface sugars and lectins in the phagocytosis of bacteria. *Infect. Immun.* 56:539–47
108. Ohman L, Hed J, Stendahl O. 1982. Interaction between human polymorphonuclear leukocytes and two different strains of type 1 fimbriae-bearing *Es-*

cherichia coli. *J. Infect. Dis.* 146:751–57

109. Ohman L, Maluszynska G, Magnusson K-E, Stendahl O. 1988. Surface interaction between bacteria and phagocytic cells. *Prog. Drug Res.* 32:131–47
110. Ohshima Y, Beuth J, Yassin A, Ko HL, Pulverer G. 1988. Stimulation of human monocytes chemiluminescence by staphylococcal lipoteichoic acid. *Med. Microbiol. Immunol.* 177:115–21
111. Ørskov I, Ørskov F. 1984. Serotyping of *Klebsiella*. *Methods Microbiol.* 14:143–64
112. Pacheco-Soares C, Gaziri LCJ, Loyola W, Felipe I. 1992. Phagocytosis of enteropathogenic *Escherichia coli* and *Candida albicans* by lectin-like receptors. *Brazil. J. Med. Biol. Res.* 25:1051–24
113. Palatnik CB, Borojevic R, Previato JO, Mendonca-Previato L. 1989. Inhibition of *Leishmania donovani* promastigote internalization into murine macrophages by chemically defined parasite glycoconjugate ligands. *Infect. Immun.* 57:754–63
114. Perry A, Ofek I, Silverblatt FJ. 1983. Enhancement of mannose-mediated stimulation of human granulocytes by type 1 fimbriae aggregated with antibodies on *Escherichia coli* surfaces. *Infect. Immun.* 39:1334–45
115. Pistole TG, Sloan AR. 1992. Direct binding of group B streptococci to mouse peritoneal macrophages. *Abstr. Annu. Meet. Am. Soc. Microbiol.* 1992: D-1
116. Pizzo SV, Lehrman MA, Imber MJ, Guthrow E. 1981. The clearance of glycoproteins in diabetic mice. *Biochem. Biophys. Res. Commun.* 101:704–8
117. Pontow SE, Kery V, Stahl PD. 1992. Mannose receptor. *Int. Rev. Cytol.* 137B:221–44
118. Quaissi MA. 1992. Role of the RGD sequence in parasite adhesion to host cells. *Parasitol. Today* 4:169–73
119. Rao SP, Ogata K, Catanzaro A. 1993. *Mycobacterium avium–M. intracellulare* binds to the integrin receptor $\alpha_v\beta_3$ on human monocytes and monocyte-derived macrophages. *Infect. Immun.* 61:663–70
120. Rautelin H, Blomberg B, Järnerot G, Dnnielsson D. 1994. Nonopsonic activation of neutrophils and cytokin production by *Helicobacter pylori*: ulcerogenic marker. *Scand. J. Gastroenterol.* 29:128–32
121. Relman D, Tuomanen E, Falkow S, Golenbock DT, Saukkonen K, Wright SD. 1990. Recognition of a bacterial adhesin by an integrin: macrophage CR3 ($\alpha_m\beta_2$, CD11b/CD18) binds filamentous hemagglutinin of Bordetella pertussis. *Cell* 61:1375–82
122. Rest RF, Frangipane JV. 1992. Growth of *Neisseria gonorrhoea* in CMP-N-acetylneuraminic acid inhibits nonopsonic (opacity-associated outer membrane protein-mediated) interactions with human neutrophils. *Infect. Immun.* 60:989–97
123. Rice KG, Weiss OA, Barthel T, Lee RT, Lee YC. 1990. Defined geometry of binding between triantennary glycopeptide and an asialoglycoprotein receptor of rat hepatocytes. *J. Biol. Chem.* 265:18429–34
124. Robin G, Markovich S, Athamna A, Keisari Y. 1991. Human recombinant granulocyte-macrophage colony-stimulating factor augments viability and cytotoxic activities of human monocyte-derived macrophages in long-term cultures. *Lymphokine Cytokine Res.* 10:257–63
125. Ross GD, Vetvicka V. 1993. CR3 (CD11b, CD18): a phagocyte and NK cell membrane receptor with multiple ligand specificities and functions. *Clin. Exp. Immunol.* 92:181–84
126. Rumlet S, Metzger Z, Kariv N, Rosenberg M. 1988. Clearance of *Serratia marcescens* from blood of mice: role of hydrophobic versus mannose-sensitive interactions. *Infect. Immun.* 56:1167–70
127. Ruoslahti E, Pierschbacher MD. 1987. New perspectives in cell adhesion: RGD and integrins. *Science* 238:491–97
128. Russell DG, Talamas-Rohana PSD. 1989. *Leishmania* and the macrophage: a marriage of inconvenience. *Immunol. Today* 10:328–33
129. Saiman L, Prince A. 1993. *Pseudomonas aeruginosa* pili bind to asialo GM1 which is increased on the surface of cystic fibrosis epihelial cells. *J. Clin. Invest.* 92:1875–80
130. Salmon JE, Kapur S, Kimberly RP. 1987. Opsonin-independent ligation of Fcγ receptors: The 3G8-bearing receptors on neutrophils mediate the phagocytosis of concanavalin A–treated erythrocytes and nonopsonized *Escherichia coli*. *J. Exp. Med.* 166:1798–813
131. Sandberg AL, Mudrick LL, Cisar JO, Metcalf JA, Malech HL. 1988. Stimulation of superoxide and lactoferrin release from polymorphonuclear leukocytes by the type 2 fimbrial lectin of *Actinomyces viscosus* T14V. *Infect. Immun.* 56:267–69
132. Saraiva EMB, Andrade AFB, de Souza W. 1987. Involvement of the macro-

phage mannose-6-phosphate receptor in the recognition of *Leishmania mexicana amazonensis*. *Parasitol. Res.* 73:411–16
133. Sastry K, Ezekowitz RAB. 1993. Collectins: pattern recognition molecules involved in first line host defense. *Curr. Opin. Immunol.* 5:59–66
134. Saukkonen K, Burnette WN, Mar VL, Masure HR, Tuomanen EI. 1992. Pertussis toxin has eukaryotic-like carbohydrate recognition domains. *Proc. Natl. Acad. Sci. USA* 89:118–22
135. Saukkonen K, Cabellos C, Burroughs M, Prasad S, Tuomanen E. 1991. Integrin-mediated localization of *Bordetella pertussis* within macrophages: role in pulmonary colonization. *J. Exp. Med.* 173:1143–49
136. Sauter SL, Rutherfurd SM, Wagener C, Shively JE, Hefta SA. 1991. Binding of non-specific cross-reacting antigen, a granulocyte membrane glycoprotein, to *Escherichia coli* expressing type 1 fimbriae. *Infect. Immun.* 59:2485–93
137. Sauter SL, Rutherfurd SM, Wagener C, Shively JE, Hefta SA. 1993. Identification of the specific oligosaccharide sites recognized by type-1 fimbriae from *Escherichia coli* on non-specific cross-reacting antigen, a cluster granulocyte glycoprotein. *J. Biol. Chem.* 268:15510–16
138. Sawyer RT, Garner RE, Hudson JA. 1992. Arg-Gly-Asp (RGD) peptides alter hepatic killing of *Candida albicans* in the perfused mouse liver model. *Infect. Immun.* 60:213–18
139. Sawyer RT, Horst MN, Garner RE, Hudson J, Jenkins PR, Richardson AL. 1990. Altered hepatic clearance and killing of *Candida albicans* in the isolated perfused mouse liver model. *Infect. Immun.* 58:2869–74
140. Schlepper-Schafer J, Kolb-Bachofen V. 1988. Red cell aging results in a change of cell surface carbohydrate epitopes allowing for recognition by galactose-specific receptors of rat liver macrophages. *Blood Cells* 14:259–69
141. Schlesinger LS. 1993. Macrophage phagocytosis of virulent but not attenuated strains of *Mycobacterium tuberculosis* is mediated by mannose receptors in addition to complement receptors. *J. Immunol.* 150:2920–30
142. Schlesinger LS, Hull SR, Kaufman TM. 1994. Binding of the terminal mannosyl units of lipoarabinomannan from a virulent strain of *Mycobacterium tuberculosis* to human macropahges. *J. Immunol.* 152:4070–79
143. Schreiber S, Blum JS, Chappel JC, Stenson WF, Stahl PD, et al. 1990. Prostaglandin E specifically upregulates the expression of the mannose-receptor on mouse bone marrow-derived macrophages. *Cell. Regul.* 1:403–13
144. Schreiber S, Perkins SL, Teitelbaum SL, Chappel J, Stahl PD, Blum JS. 1993. Regulation of mouse bone marrow macrophage mannose receptor expression and activation by prostaglandin E and IFN-gamma. *J. Immunol.* 151:4973–81
145. Seow WK, Seymour GJ, Thong YH. 1987. Direct modulation of human neutrophil adherence by coaggregating periodontopathic bacteria. *Int. Arch. Allergy Appl. Immunol.* 83:121–28
146. Shafer WM, Rest RF. 1989. Interactions of gonococci with phagocytic cells. *Annu. Rev. Microbiol.* 43:121–45
147. Sharon N, Lis H. 1993. Carbohydrates in cell recognition. *Sci. Am.* 268(1):82–90
148. Shepherd VL, Abdolrasulnia R, Garrett M, Cowan HB. 1990. Down-regulation of mannose receptor activity in macrophages after treatment with lipopolysaccharide and phorbol esters. *J. Immunol.* 145:1530–36
149. Shepherd VL, Campbell EJ, Senior RM, Stahl PD. 1982. Characterization of the mannose/fucose receptor on human mononuclear phagocytes. *J. Reticuloendothel. Soc.* 32:423–31
150. Shepherd VL, Konish MG, Stahl PD. 1985. Dexamethazone increases expression of mannose receptors and decreases extacellular lysosomal enzyme accumulation by macrophages. *J. Biol. Chem.* 260:160–65
151. Soderstrom T, Ohman L. 1984. The effect of monoclonal antibodies against *Escherichia coli* type 1 pili and capsular polysaccharides on the interaction between bacteria and human granulocytes. *Scand. J. Immunol.* 20:299–305
152. Sokurenko EV, Courtney HS, Ohman DE, Klemm P, Hasty DL. 1994. FimH family of type 1 adhesins: functional heterogeneity due to minor sequence variations among *fimH* genes. *J. Bacteriol.* 176:748–55
153. Soteriadou KP, Remoundos MS, Katsikas MC, Tzinia AK, Tsikaris V, et al. 1992. The Ser-Arg-Tyr-Asp region of the major surface glycoprotein of *Leishmania* mimics the Arg-Gly-Asp-Ser cell attachment region of fibronectin. *J. Biol. Chem.* 267:13980–85
154. Speert DP. 1992. Macrophages in bacterial infection. In *The Macrophage*, ed. CE Lewis, JO'D McGee, pp. 215–63. New York: IRL. 423 pp.
155. Speert DP, Eftekhar F, Puterman ML. 1984. Nonopsonic phagocytosis of

156. strains of *Pseudomonas aeruginosa* from cystic fibrosis patients. *Infect. Immun.* 43:1006–11
156. Speert DP, Loh BA, Cabral DA, Salit IE. 1986. Nonopsonic phagocytosis of nonmucoid *Pseudomonas aeruginosa* by human neutrophils and monocyte-derived macrophages is correlated with bacterial piliation and hydrophobicity. *Infect. Immun.* 53:207–12
157. Speert DP, Thorson L. 1991. Suppression by human recombinant gamma interferon of in vitro macrophage nonopsonic and opsonic phagocytosis and killing. *Infect. Immun.* 59:1893–98
158. Speert DP, Wright SD, Silverstein SC, Mah BA. 1988. Functional characterization of macrophage receptors for in vitro phagocytosis of unopsonized *Pseudomonas aeruginosa. J. Clin. Invest.* 82:872–79
159. Stahl PD. 1992. The mannose receptor and other macrophage lectins. *Curr. Opin. Immunol.* 4:49–52
160. Steadman R, Topley N, Jenner DE, Davis M, Williams JD. 1988. Type 1 fimbriated *Escherichia coli* stimulates a unique pattern of degranulation by human polymorphonuclear leukocytes. *Infect. Immun.* 56:815–22
161. Stein M, Keshav S, Harris N, Gordon S. 1992. Interleukin 4 potentially enhances murine macrophage mannose receptor activity: a marker of alternative immunologic macrophage activation. *J. Exp. Med.* 176:287–92
162. Stewart J, Glass EJ, Weir DM 1982. Macrophage binding of *Staphylococcus albus* is blocked by anti I–region alloantibodies. *Nature* 298:852–54
163. Sung SJ, Nelson RS, Silverstein SC. 1983. Yeast mannans inhibit binding and phagocytosis of *Candida* by mouse peritoneal macrophages *J. Cell. Biol.* 96:160–66
164. Sussman M, ed. 1985. *Virulence of Escherichia coli: Reviews and Methods.* London: Academic. 473 pp.
165. Svanborg-Eden C, Bjursten LM, Hull R, Magnusson K-E, Leffler M. 1984. Influence of adhesins on the interaction of *Escherichia coli* with human phagocytes. *Infect. Immun.* 44:672–80
166. Taichman NS, Tsai CC, Shenker BJ, Boehringer H. 1984. Neutrophil interaction with oral bacteria as a pathogenic mechanism in periodontal diseases. In *Advances in Inflammation Research*, ed. R Weissman, pp. 113–42. New York: Raven
167. Tauber AI, Chernyak L. 1991. Current views of phagocytic functions. In *Metchnikoff and the Origin of Immunology*, ed. AI Tauber, L Chernyak, pp. 191–98. New York: Oxford Univ. Press. 247 pp.
168. Taylor ME, Drickamer K. 1993. Structural requirements for high affinity binding of complex ligands by the macrophage mannose receptor. *J. Biol. Chem.* 268:399–404
169. Te-Velde AA, Klomp JP, Yard BA, de Vries JE, Figdor CG. 1988. Modulation of phenotypic and functional properties of human peripheral blood monocytes by IL-4. *J. Immunol.* 140:1548–54
170. Tewari R, MacGregor JI, Ikeda T, Little JR, Hultgren SJ, Abraham SN. 1993. Neutrophil activation by nascent FimH subunits of type 1 fimbriae purified from periplasm of *Escherichia coli. J. Biol. Chem.* 268:3009–15
171. Tewari R, Ikeda T, Malaviya R, MacGregor JI, Little JR, Hultgren SJ, Abraham SN. 1994. Negatively charged PapG tip adhesin of P fimbriae protects *Escherichia coli* from neutrophil bactericidal activity. *Infect. Immun.* 62:5296–304
172. Topley N, Steadman R, Mackenzie R, Knowlden JM, Williams JD. 1989. Type 1 fimbriae strains of *Escherichia coli* initiate renal parenchymal scarring. *Kidney Int.* 36:609–16
173. Tuomanen E, Towbin H, Rosenfelder G, Braun D, Larson G, et al. 1988. Receptor analogs and monoclonal antibodies that inhibit adherence of *Bordetella pertussis* to human ciliated respiratory epithelial cells. *J. Exp. Med.* 168:267–77
174. Vandenbroucke-Grauls CMJE, Thijssen HMWM, Verhoef J. 1984. Interaction between human polymorphonuclear leucocytes and *Staphylococcus aureus* in the presence and absence of opsonins. *Immunology* 52:427–31
175. Venture Y, Scheffer J, Hacker J, Goebel W, Konig W. 1990. Effects of adhesins from mannose-resistant *Escherichia coli* on mediators release from human lymphocytes, monocytes and basophils and from polymorphonuclear granulocytes. *Infect. Immun.* 58:1500–8
176. Vercautren R, Dom P, Haesebrouck F. 1993. Virulence of bacteria in relation to their hydrophobicity, adhesiveness and phagocytosis. *Vlaams Diergeneeskd. Tijdschr.* 62:29–34
177. Warr GA. 1980. A macrophage receptor (mannose/glucosamine)-glycoproteins of potential importance in phagocytic activity. *Biochem. Biophys. Res. Commun.* 93:737–40
178. Weir DM. 1980. Surface carbohydrates

and lectins in cellular recognition. *Immunol. Today* 1:45–51
179. Weiss SJ. 1989. Tissue destruction by neutrophils. *New Engl. J. Med.* 320:365–76
180. Wells CL, Feltis BA, Hanson DF, Jechorek RP, Erlandsen SL. 1993. Oral infectivity and bacterial interactions with mononuclear phagocytes. *J. Med. Microbiol.* 38:345–53
181. Williams P, Lambert PA, Haigh PA, Brown MRW. 1986. The influence of the O and K antigens of *Klebsiella aerogenes* on surface hydrophobicity and susceptibility to phagocytosis and antimicrobial agents. *J. Med. Microbiol.* 21:125–32
182. Williams P, Tomas JM. 1990. The pathogenicity of *Klebsiella pneumoniae*. *Rev. Med. Microbiol.* 1:196–204
183. Wilson ME, Donelson JE, Pearson RD. 1992. Macrophage receptors and *Leishmania*. In *Molecular Recognition in Host-Parasite Interaction*, ed. TK Korhonen, T Hovi, PH Makela, pp. 17–30. New York: Plenum. 230 pp.
184. Wilson ME, Pearson RD. 1986. Evidence that *Leishmania donovani* utilizes a mannose receptor on human mononuclear phagocytes to establish intracellular parasitism. *J. Immunol.* 136:4681–85
185. Wright SD, Levine SM, Jong MTC, Chad Z, Kabbash LG. 1989. CR3 (CD11b/CD18) expresses one binding site for Arg-Gly-Asp–containing peptides and a second site for bacterial lipopolysaccharide. *J. Exp. Med.* 169:175–79
186. Wright SD, Silverstein SC. 1983. Receptors for C3b and C3bi promote phagocytosis but not the release of toxic oxygen from human phagocytes. *J. Exp. Med.* 158:2016–23
187. Zwilling BS, Eisenstein TK, eds. 1994. *Macrophage-Pathogen Interactions*. New York: Dekker. 643 pp.

PEPTIDES AS WEAPONS AGAINST MICROORGANISMS IN THE CHEMICAL DEFENSE SYSTEM OF VERTEBRATES

Pierre Nicolas and Amram Mor

Laboratoire de Bioactivation des Peptides, Institut Jacques Monod, Université Paris 7, 2 place Jussieu, 75251 Paris Cedex 05, France

KEY WORDS: antimicrobial peptides, molecular immunity, cytotoxicity, cell membrane

CONTENTS

INTRODUCTION	278
DEFENSINS AND β-DEFENSINS	279
Defensins and β-Defensins: Endogenous Antibiotics of Myeloid-Derived Cells	279
Defensins and β-Defensins Are Intrinsic to Mammalian Epithelial Tissues	280
PROLINE- AND ARGININE-RICH ANTIMICROBIAL PEPTIDES	283
AMPHIPATHIC HELICAL PEPTIDES	284
Structure and Antimicrobial Properties	284
Biosynthetic Pathways	288
Tissue-Specific Expression	291
BREVININS, ESCULENTINS, AND RANALEXIN	292
ANTIMICROBIAL MECHANISMS	294
β-Sheet Peptides with Three Disulfide Bonds	294
Amphipathic Helical Peptides	296
EVOLUTIONARY SIGNIFICANCE	298
PERSPECTIVES	298

Abstract

The innate immunity of vertebrates to microbial invasion is arbitrated by a network of host-defense mechanisms involving both the long-lasting highly specific responses of the cell-mediated immune system and a nonspecific chemical defense system based on a series of broad-spectrum antimicrobial peptides that are analogous to those found in insects. Vertebrate antibiotic peptides secreted by nonlymphoid cells of the mucosal surfaces of the respiratory and gastrointestinal tracts as well as by the granular glands of the skin reportedly cause the lysis of numerous pathogenic microorganisms, including viruses, gram-positive and gram-negative bacteria, protozoa, yeasts, and fungi, as well as of cancer cells. Antimicrobial peptides isolated from vertebrates have three characteristic properties: They are relatively small (20–46 amino acid residues), basic (lysine- or arginine-rich), and amphipathic. Although these peptides differ widely in length and amino acid sequences, they may be grouped in four broad families based on characteristic structural features. Although the precise mechanism of action of these peptides remains to be defined, their microbicidal effect very likely results from their capacity to form channels or pores within the microbial membrane in order to permeate the cell and impair its ability to carry out anabolic processes. This secondary, chemical immune system provides vertebrates with a repertoire of small peptides that are promptly synthesized upon induction, easily stored in large amounts, and readily available for antimicrobial warfare.

INTRODUCTION

Vertebrates must often defend themselves against invading microorganisms that can easily penetrate the epithelia of the respiratory, gastrointestinal, and genital systems. In addition, numerous microorganisms live as commensals or saprophytes in internal cavities of most living organisms. Although commensal microbes may be useful to the host, for example, in preventing access of pathogenic organisms to their ecological niches, their growth must be controlled to avoid opportunistic infections and severe pathological disorders. To do so, vertebrates have developed a network of host-defense mechanisms involving both nonspecific physical barriers and highly specific responses of the cell-mediated immune system. Although the activation of the latter system requires multiple signals and a complex cascade of events, it provides the organism with durable and highly specific protection against occasional pathogenic microbes. Though primitive lymphoidal tissues are present in invertebrates, cell-mediated, long-lasting immunity does not seem to play a major role as no immunoglobulin-related molecule has ever been detected in these organisms. Invertebrates nevertheless combat very efficiently the proliferation of microorganisms in their natural flora by synthesis and delivery, during occasional infections, of a fixed repertoire of broad-spectrum antimicrobial peptides.

The repeated discovery in the past decade of various antimicrobial peptides originating from nonmyeloid cells of vertebrates, including those in mammals, revealed that vertebrates are also endowed with a chemical defense system based on families of broad-spectrum microbicidal peptides that are analogous to those found in insects. The analysis of the components and of the mode of action of this chemical-warfare system has led to new perspectives in the treatment of opportunistic infections and has improved our understanding of the means by which vertebrate animals survive in a world laden with microorganisms. So far, more than 100 different antimicrobial peptides have been isolated and characterized in vertebrates. Regardless of their origin—inferior or superior vertebrates—all these peptides are small, polycationic, and membrane active. Unlike the antibiotic peptides from microorganisms, which are produced mostly by multi-enzymatic complexes, the defensive peptides of vertebrates are classically synthesized by the ribosomal route via biosynthetic precursors. Although differing widely in length and amino acid sequences, the vertebrate antimicrobial peptides discussed here may be grouped in four broad families based on characteristic structural features.

DEFENSINS and β-DEFENSINS

Defensins and β-Defensins: Endogenous Antibiotics of Myeloid-Derived Cells

Phagocytosis represents one of the major components of the nonspecific defense system against microorganisms and is found throughout the phylogenies of invertebrates and vertebrates. Among circulating phagocytic cells, macrophages and polymorphonuclear neutrophils play an essential role by engulfing and digesting microorganisms through both oxygen-dependent and oxygen-independent mechanisms. Mammalian defensins (79) and β-defensins (92) are two multimember families of cationic (Arg-rich) trisulfide-containing peptides of 29–42 residues (Tables 1 and 2) that are stored in the azurophil granules of circulating neutrophils and macrophages (see 55 for an authoritative review). These endogenous peptides play a decisive role in the nonoxidative microbicidal mechanisms of their producing cells through delivery to phagocytic vacuoles containing ingested microorganisms. Although the β-defensins resemble classical defensins in their location in cytoplasmic granules of hematopoietic cells, their amino acid composition, and their vitro microbicidal activity, these two peptide classes are distinguished by their unique consensus sequences and by differing tridisulfide motifs (104). Peptides of both classes exhibit broad-range microbicidal spectra encompassing gram-positive and gram-negative bacteria, mycobacteria, and spirochetes as well as some fungi and enveloped viruses (55). Evidence suggests that defensins and β-defensins exert their antimicrobial effect by permeating the cytoplasmic membrane of target cells via a mechanism involving the formation of voltage-regulated ion channels (55). NMR spectroscopy (78) and X-ray crystallography (37) showed that defensin molecules consist of a structurally rigid triple-stranded antiparallel

Table 1 Amino acid sequences of defensins originating from hematopoietic cells[a]

		1	10	20	30
Human	HNP1	A C Y C R I P A C I A G E R R Y G T C I Y Q G R L W A F C C			
	HNP2	C Y C R I P A C I A G E R R Y G T C I Y Q G R L W A F C C			
	HNP3	D C Y C R I P A C I A G E R R Y G T C I Y Q G R L W A F C C			
	HNP4	V C S C R L V F C R R T E L R V G N C L I G G V S F T Y C C T R V			
Rabbit	NP1	V V C A C R R A L C L P R E R R A G F C R I R G R I H P L C C R R			
	NP2	V V C A C R R A L C L P L E R R A G F C R I R G R I H P L C C R R			
	NP3a	G I C A C R R R F C P N S E R F S G Y C R V N G A R Y V R C C S R R			
	NP3b	G R C V C R K Q L C S Y R E R R I G D C K I R G V R F P F C C P R			
	NP4	V S C T C R R F S C G F G E R A S G S C T V N G V R H T L C C R R			
	NP5	V F C T C R G F L C G S G E R A S G S C T I N G V R H T L C C R R			
Rat	RtNP1	T C Y C R R T R C G F R E R L S G A C G Y R G R I Y R L C C R			
	RtNP2	V T C Y C R S T R C G F R E R L S G A C G Y R G R I Y R L C C R			
	RtNP3	C 3 C R T S S C R F G E R L S G A C R L N G R I Y R L C C			
	RtNP4	A C Y C R I G A C V S G E R L T G A C G L N G R I Y R L C C R			
Guinea pig	GNP	R R C I C T T R T C R F P Y R R L G T C I F Q N R V Y T F C C			
Consensus and disulfide bonds		- - C - C - - - - C - - - - - - - G - C - - - - - - - - - C C - - -			

[a] The sequences are aligned to demonstrate the most conserved amino acids. The numbering of residues is indexed at the top to the longest of the defensin peptides.

β-sheet stabilized by intramolecular disulfide bonds. cDNAs for seven myeloid defensins (rabbit NP-1, NP-2, NP-3a, NP-4, and NP-5; human HNP-1 and HNP-3) have been cloned and sequenced (21, 32, 76, 77). In each case, the mature defensin sequence constituted the carboxy terminus of a prepro-peptide (93–95 amino acids) containing a typical amino-terminal signal sequence followed by a 40– to 45–amino acid anionic propiece. This propiece may be responsible for masking the cytotoxic potential of defensins prior to their sequestration in lysosome-like organelles (9, 59).

Defensins and β-Defensins Are Intrinsic to Mammalian Epithelial Tissues

Until recently, only hematopoietic cells were known to contain mammalian defensins. Recently, data demonstrated that epithelial cells, as opposed to leukocytes within the tissues, are also a source of defensin and β-defensin peptides. Five new intestinal defensins, termed cryptidins (Table 3), occur

Table 2 Amino acid sequences of β-defensins from bovine neutrophils and tracheal mucosa[a]

		1	10	20	30	40
Neutrophils	BNBD-1		D F A S C H T N G G I C L P N R C P G H M I Q I G I C F R P R V K C C R S W			
	BNBD-2		V R N H V T C R I N R G F C V P I R C P G R T R Q I G T C F G P R I K C C R S W			
	BNBD-3	pE G V R N H V T C R I N R G F C V P I R C P G R T R Q I G T C F G P R I K C C R S W				
	BNBD-4	pE R V R N P Q S C R W N M G V C I P F L C R V G M R Q I G T C F G P R V P C C R R				
	BNBD-5	pE V V R N P Q S C R W N M G V C I P I S C P G N M R Q I G T C F G P R V P C C R				
	BNBD-6	pE G V R N H V T C R I Y G G F C V P I R C P G R T R Q I G T C F G R P V K C C R R W				
	BNBD-7	pE G V R N F V T C R I N R G F C V P I R C P G H R R Q I G T C L G P R I K C C R				
	BNBD-8		V R N F V T C R I N R G F C V P I R C P G H R R Q I G T C L G P Q I K C C R			
	BNBD-9	pE G V R N F V T C R I N R G F C V P I R C P G H R R Q I G T C L G P Q I K C C R				
	BNBD-10	pE G V R S Y L S C W G N R G I C L L N R C P G R M R Q I G T C L A P R V K C C R				
	BNBD-11		G P L S C R R N G G V C I P I R C P G P M R Q I G T C F G R P V K C C R S W			
	BNBD-12		G P L S C G R N G G V C I P I R C P V P M R Q I G T C F G R P V K C C R S W			
	BNDD-13	S G I S G P L S C G R N G G V C I P I R C P V P M R Q I G T C F G R P V K C C R S W				
Tracheal mucosa	TAP		N P V S C V R N K G I C V P I R C P G S M K Q I G T C V G R A V K C C R K K			
consensus and disulfide pairing		- - - - - - - - C - - - - G - C - - - - C - - - - - Q I G - C - - - - - - C C R - -				

[a] Primary structures are aligned to maximize sequence similarities. The numbering of residues is indexed at the top of the longest of the defensin peptides. Note the presence of a N-terminal pyroglutamyl residue in BNBD 3–7 and BNBD 9–10. The disulfide pairing of β-defensins differs from that of defensins.

exclusively in mouse Paneth cells, granulated epithelial cells that reside at the base of the crypts of Lieberkühn throughout the small intestines and proximal colons of many mammalian species (27, 41, 42, 59, 76, 91). Although enteric defensins have amino termini three to six residues longer than those of leukocyte-derived classical defensins, both of these peptides are clearly homologous and exhibit similar broad spectra of antimicrobial activity. Mouse cryptidin-1 mRNA is predicted to code for a 93–amino acid protein that is similar to the deduced human and rabbit neutrophil defensin precursors. The cryptidin gene, *De fer*, in the proximal region of chromosome 8, shows conserved linkage homology with the human defensin gene(s) *DEFL* on 8p23 (97). In humans, Paneth cells also contain high levels of mRNA encoding the putative prepro-defensin peptides HNP-5 and HNP-6 (Table 3) that are highly homologous to the four HNPs isolated from human neutrophils. Biochemical and immunological analyses of the luminal content in the small intestines suggested that cryptidin peptides are localized in eosinophilic secretory granules of the Paneth

Table 3 Amino acid sequences of murine and human enteric defensins (cryptidins)[a]

		1	10	20	30
Human	HD-5	Q A R A T C Y C R T G R C A T R E S L S G V C E I S G R L Y R L C C R
	HD-6	T R A F T C H C R R - S C Y S T E Y S Y G T C T V M G I N H R F C C L
Mouse	cryptidin-1	L R D L V C Y C R S R G C K G R E R M N G T C R K G H L L Y T L C C R
	cryptidin-2	L R D L V C Y C R T R G C K R R E R M N G T C R K G H L M Y T L C C R
	cryptidin-3	L R D L V C Y C R K R G C K R R E R M N G T C R K G H L M Y T L C C R
	cryptidin-4	G L L C Y C R K G H C K R G E R V R G T C - - G - I R F L Y C C P R
	cryptidin-5	L S K K L I C Y C R I R G C K R R E R V F G T C R N L F L T F V F C C
consensus and disulfide bridges		- - - - - - C - C R - - - C - - - E - - - G - C - - - - - - - - - C C - -

[a] Primary structures are aligned to maximize sequence similarity. Hyphens denote gaps in the cryptidin-4 and HD-6 sequences. The numbering of residues is indexed at the top of the longest of the defensin peptides.

cells (91) and secreted into the lumen, in a pattern similar to the Paneth cell secretion of lysozyme. Active secretion of intestinal defensins would distinguish them from phagocyte defensins, which are not normally secreted and are primarily targeted for intracellular delivery to phagolysosomes.

These observations suggest two possible, nonexclusive physiological roles for enteric defensins (41). First, secretion of defensins into the space above the crypt may contribute to the establishment of a local antibacterial milieu that limits bacterial colonization and invasion of the small bowel. Second, the defensins could be important in mucosal defense against microbial invasion by preserving the integrity of the villus epithelium and thereby maintaining the critical function of nutrient absorption.

The tracheal antimicrobial peptide (TAP) is a new member of the β-defensin family, originally isolated from the bovine tracheal mucosa (23, 24). Like β-defensins of neutrophils, TAP is a basic molecule with a broad-spectrum antimicrobial activity and contains six cysteines, all involved in disulfide bonds (Table 2). In situ hybridization of TAP mRNA indicated that TAP is expressed along the entire length of the conducting airways, from nasal to bronchiolar tissues. TAP mRNA is localized in columnar cells of the pseudostratified epithelium, suggesting its expression in the ciliated cells. The fact that the β-defensins found in circulating phagocytes, and TAP from the tracheal epithelium, are members of the same family of antimicrobial peptides strongly supports the hypothesis that TAP contributes to the host defense of the airways.

PROLINE- AND ARGININE-RICH ANTIMICROBIAL PEPTIDES

Mammalian neutrophils produce a family of Pro- and Arg-rich antibacterial peptides in addition to the defensins. So far, this family includes Bac-5 and Bac-7, which were isolated from bovine neutrophils (31), and PR-39 (2), a 39-residue peptide that was first isolated from pig intestines and later found in pig bone marrow cells (101). These three peptides have a peculiar amino acid composition in that proline (47, 47, and 49%, respectively) and arginine (21, 29, and 26%, respectively) represent more than 60% of the constitutive residues. The other amino acids, with Leu and Ile as the major constituents, are mainly apolar. The sequences of these peptides are highly repetitive, as characterized by several Pro-Arg-Pro and/or Arg-Pro-Pro repeats (Table 4). Although no obvious sequence homology can be delineated among the three peptides, their amino acid composition and their spectra of activity are very similar (2, 31, 101). Although the three peptides are mainly active against gram-negative bacteria, Bac-5 and Bac-7 decrease the ATP content and the transport of amino acids and nucleotides, while PR-39 does not lyse bacteria but may stop both DNA and protein synthesis in *Escherichia coli*. Investigation of the secondary structure of PR-39 through circular dichroism and Fourier-transform infrared spectroscopy suggested a polyproline type II conformation in water (17). No data are available on the intracellular localization and storage form of the intestinal PR-39. The fact however, that PR-39 found in circulating

Table 4 Amino acid sequences of Pro+Arg-rich peptides isolated from bovine neutrophils and pig intestine

	1	10	20	30
Bac 5	R F R P P I R R P P I R P P F Y P P F R P P I R P P I F P P			
		40		
	I R P P F R P P L R F P			
	1	10	20	30
Bac 7	R R I R P R P P R L P R P R P R P L P F P R P G P R P I P R			
		40	50	
	P L P F P R P G P R P I P R P L P F P R P G P R P I P R P			
	1	10	20	30
PR-39	R R R P R P P Y L P R P R P P P F F P P R L P P R I P P G F			
	P P R F P P R F P NH2			

phagocytes and PR-39 from the pig intestinal epithelium are identical supports the hypothesis that enteric PR-39 may be involved in host-defense of the mucosal surface of intestines. The primary structure of the precursor of pig myeloid PR-39 (101) revealed a prepro-sequence highly similar to that of the respective precursors of Bac-5, indolicidin, cyclic dodecapeptides, and prointegrin-2, four structurally unrelated neutrophil antimicrobial peptides. An interesting feature of these precursors is the extensive similarity of the common prepro-region to the prepro-sequence of a porcine inhibitor of the cysteine proteinase cathepsin L, termed cathelin (100) or PLCPI (pig leukocyte cysteine proteinase inhibitor). The structural similarity of the N-terminal regions of all these proteins suggests a common evolutionary origin and identifies a novel protein family characterized by a cathelin-like N-terminal domain.

AMPHIPATHIC HELICAL PEPTIDES

The peptide secretions from amphibian dermal glands are the main source of several potent antibiotic peptides originating from the nonmyeloid cells of vertebrates. The multinucleated granular glands of frog skin contain, aside from numerous mammalian-like bioactive hormones and neuropeptides (12, 29, 48, 98), several families of broad-spectrum microbicidal peptides, large amounts of which are stored in secretory granules. These peptides are thought to be involved in the defense of the naked skin of frogs against microbial invasion. The main families of skin peptides that exhibit antimicrobial properties belong to a large group of linear amphipathic helical peptides, 20–36 residues long, whose overall structure is very similar. Although differing widely in length and amino acid sequence, all these peptides are cationic, containing a variable number of lysine residues that punctuate alternating hydrophobic and hydrophilic segments. Their unique primary structures are thought to endow these membrane-active peptides with the ability to form amphipathic α-helices in an anisotropic environment, such as a membrane interface.

Structure and Antimicrobial Properties

BOMBININS The first report associating a vertebrate nonmyeloid cell and an antimicrobial peptide was in 1969 by Csordas & Michl (20). They isolated two nonapeptides from skin exudate of *Bombina variegata* that were later shown to be part of a cationic carboxamidated 24-residue peptide, designated bombinin. The complete amino acid sequence and activity spectrum of this multimember peptide class have only recently been established. Bombinin peptides from *B. variegata* and *Bombina orientalis* (Discoglossidae family) (33, 95) are highly homologous and have a common origin (Table 5). Bom-

Table 5 Amino acid sequences of amphipapthic helical peptides from the skin and gastrointestinal tract of amphibians

Bombinins

B. orientalis

	1	10	20	
BLP-1	G I G A S I L S A G K S A L K G L A K G L A E H F A N	NH2
BLP-2	G I G S A I L S A G K S A L K G L A K G L A E H F A N	NH2
BLP-3	G I G A A I L S A G K S A L K G L A K G L A E H F	NH2
BLP-4	G I G A A I L S A G K S I L K G L A N G L A E H F	NH2

B. variegata

7	G I G G A L L S A G K S A L K G L A K G L A E H F A N NH2
9	G I G G A L L S A A K V G L K G L A K G L A E H F A N NH2
10	G I G A S I L S A A K V G L K G L A K G L A E H F A N NH2
13	G I G G A L I S A G K S A L K G L A K G L A E H F A N NH2
14	G I G G A L L S D A K V G L K G L A K G L A E H F A N NH2
Consensus	G I G - - - - S - - K - - L K G L A - G L A E H F - -_

Magainins and related peptides from *X. laevis*

Magainin 1	G I G K F L H S A G K F G K A F V G E I M K S
Maigainin 2	G I G K F L H S A K K F G K A F V G E I M N S
PGQ	G V L S N V I G Y L K K L G T G A L N A V L K Q
PGLa	G M A S K A G A I A G K I A K V A L K A L NH2
XPF	G W A S K I G Q T L G K I A K V G L K Q L I Q P K
LPF	G W A S K I G Q T L G K I A K V G L Q G L M Q P K
CPF (1+5)[a]	G F G S F L G K A L K A A L K I G A N A L G G S P Q Q

Dermaseptins

P. sauvagei

S1	A L W K T M L K K L G T M A L H A G K A A L G A A A D T I S Q G T Q
S2	A L W F T M L K K L G T M A L H A G K A A L G A A A N T I S Q G T Q
S3	A L W K N M L K G I G K L A G K A A L G A V K K L V G A E S
S4	A L W M T L L K K V L K A A A K A A L N A V L V G A N A
S5	G L W S K I K T A G K S V A K A A A K A A V K A V T N A V

P. bicolor

| B1 | A M W K D V L K K I G T V A L H A G K A A L G A V A D T I S Q NH2 |
| B2 (adenoregulin)[b] | G L W S K I K E V G K E A A K A A A K A A G K A A L G A V S E A V NH2 |

[a] Several other CPF fragments differing by a few substitutions are derived by processing the four prepro-caerulein precursors (83, 109, 110).
[b] Adenoregulin (dermaseptin B2) was first isolated by Daly and coworkers (22).

binins are 25–27 residues long, with a constant C-terminal region that includes an amidated end (from residues 14–27). They contain three lysine residues, a single histidine residue, and an acidic amino acid—either glutamic or aspartic acid—near the carboxyl terminus. Theoretical prediction and circular dichroism (CD) spectroscopy analysis have shown that all bombinins can be fit to a well-behaved α-helix that delineates a hydrophobic and a hydrophilic face. The spectrum of activity of bombinins is quite narrow: Out of more than 200 colonies tested, these peptides exhibited antimicrobial activity only against gram-negative nonenteric bacteria. They exhibited no or very little activity against several enteric strains from blood isolates, including *Escherichia, Pseudomonas,* and *Klebsiella* strains. Preliminary evidence with synthetic bombinins indicated that this class of peptides is not hemolytic, nor does it lyse the sensitive bacteria, as bacterial growth is resumed after addition of fresh medium.

MAGAININS AND RELATED PEPTIDES   Magainins 1 and 2 (114), also designated [Gly10, Lys22]PGS and PGS (34), respectively, in articles whose publications overlapped, are 23-residue peptides isolated from *Xenopus laevis* (Pipidae family) skin that differ by substitutions in positions 10 and 22 (Table 5). Unlike bombibins, these α-helical amphipathic peptides present a large spectrum of antimicrobial activity (116). At micromolar concentrations, magainins inhibit the growth and/or induce osmotic lysis of several gram-positive and gram-negative bacteria and of some fungi and protozoa. They also induce lysis of hematopoietic tumoral cells and of solid cells but are inactive against differentiated mammalian cells. Several related peptides isolated from the skin of *X. laevis* also display broad-spectrum antimicrobial activity (96). These include PGLa (peptide with amino terminal glycine and carboxyl terminal leucinamide) (6), 12 variants of the caerulein precursor fragment (CPFs) (83, 109, 110), xenopsin precursor fragments (XPFs) (103), the levitide precursor fragment (LPF) (80), and PGQ (peptide with amino terminal glycine and carboxyterminal glutamine) (62). These peptides contain 27 (CPFs), 25 (XPFs and LPF), 24 (PGQ), or 21 (PGLa) amino acids each. The bombinins, the magainins, and the CPFs exhibit clear sequence homology. All these peptides are active against gram-positive and gram-negative bacteria and some fungi (116).

DERMASEPTINS   A new family of linear helical antimicrobial peptides analogous to the bombinins, the magainins, and their related peptides was recently isolated from the skin of *Phyllomedusa sauvagei* and *Phyllomedusa bicolor* (Hylidae family). The dermaseptins (65, 68, 69, 71) are broad-spectrum microbicidal cationic peptides, 28–34 amino acids long, with 3–6 lysine residues (Table 5). Their amino acid sequences exhibit the periodic pattern of an α-helix containing sharply demarcated polar and nonpolar surfaces. Like the *X. laevis* peptides, the dermaseptins are active against a very wide spectrum of pathogenic microorganisms but differ in their potency for killing the various agents. Dermaseptins were shown to exert a lytic effect over numerous bacteria,

Table 6 Growth inhibition activity in vitro of the five dermaseptins from *Phyllomedusa sauvagei* against bacteria, yeasts, and filamentous fungi

Organisms	Minimal inhibitory concentration (μM)[a]				
	S1	S2	S3	S4	S5
Aeromonas caviae (IP67-16T)	0.5	1	1	0.5	1
Escherichia coli (IP76-24)	1	2.5	1	2.5	4
Enterococcus faecalis (IP103214)	5	10	10	20	40
Staphylococcus aureus (IP76-25)	5	20	10	10	2
Nocardia brasiliensis (IP16-80)	35	20	5	10	40
Saccaromyces cerevisiae (IP118079)	5	5	5	20	5
Candida albicans (IP886-65)	10	10	10	20	10
Cryptococcus neoformans (IP962-67)	0.5	1	1	1	1
Microsporum canis (IP1194)	15	15	15	15	15
Tricophyton rubrum (IP2043-92)	35	35	40	40	20
Tricophyton mentagrophytes (IP877-71)	20	20	20	20	20
Arthroderma simii (IP1063-74)	30	30	30	30	30
Aspergillus niger (IP218-53)	30	20	10	30	30
Aspergillus fumigatus (IP1025-70)	30	20	20	20	>70
	Hemolysis (μM)[b]				
Human erythrocytes	>70	70	80	1	>90

[a] The minimal inhibitory concentration (MIC) is defined as the dose at which 100% inhibition of growth was observed after 24-h incubation in culture media (36-h for *M. canis*, *T. rubrum*, and *T. mentagrophytes*).
[b] 100% hemolysis after 1-h incubation.

protozoa, and yeasts as well as against the filamentous fungi responsible for the opportunistic infections that accompany the immunodeficiency syndrome or the use of immunosuppressive agents (Table 6).

Despite the considerable structural similarities between the dermaseptin family members, they differ markedly in their ability to inhibit microbial proliferation (67). For instance, whereas dermaseptins S1, S2, S3, and S4, from *P. sauvagei*, are very potent against *Aeromonas caviae* and *E. coli*, dermaseptin S5 is much less potent against these bacteria (Table 6). Conversely, dermaseptin S5 is the most effective peptide at inhibiting the proliferation of the gram-positive cocci *Staphylococcus aureus*. These differences in potency are also observed against the yeasts *Candida albicans* and *Saccharomyces cerevisiae*. In contrast, all five peptides inhibit the growth of pathogenic fungi with nearly the same efficiency. Another noticeable difference is the ability of the dermaseptins to lyse red blood cells. Although 100 μM concentrations of dermaseptins S1 or S5 do not permeate the erythrocytes, a 1 μM concentration of dermaseptin S4 produces 100% hemolysis. In addition, the dermaseptins show dramatic synergy of action. Various peptide combinations have up to

100-fold greater antibiotic potency than the individual peptides. Such a synergism has been also reported between magainin-2 amide and PGLa (12). Hence, the biological significance of the antimicrobial peptides with similar sequences in frog skin, such as the bombinins in *Bombina* spp., the magainins and their related peptides in *X. laevis,* and the dermaseptins in *Phyllomedusa* spp., may be that they provide the frog with maximum protection against a wider range of potential invading microorganisms at a minimum metabolic cost.

Biosynthetic Pathways

All the helical antimicrobial peptides so far isolated in frog skin are synthesized as prepro-peptides. These biosynthetic precursors (3, 4, 33, 34, 80, 83, 95, 103, 105, 110, 114) may be grouped in two classes depending on whether they are mono- or plurifunctional. For instance, PGLa, the magainins, the bombinins, and the dermaseptins issue from monofunctional precursors. Other helical antimicrobial peptides, such as XPFs, CPFs, and LPF (fragments of proxenopsin, procaerulein, and prolevitide, respectively), result from post-translational activation of multifunctional hormonal precursors. Thus, a peptide with an antimicrobial activity and a peptide with neural or hormonal activity are synthesized together in a prepro-protein and are cosecreted in skin exudates. For instance, besides XPF, proxenopsin yields xenopsin (8), which is the amphibian counterpart of mammalian neurotensin. Procaerulein yields CPFs and a hormone, caerulein (5, 28, 47, 61), which is the counterpart of gastrin/CCK of mammals. Levitide (80), a peptide liberated from prepro-levitide together with LPF, has no known biological activities but its structural resemblance to xenopsin suggests a dedicated hormonal or neuronal function. The xenopsin precursor protein has the simplest organization of the plurifunctional prepro-proteins encoding a signal peptide, XPF, an acidic connecting peptide, and the octapeptide xenopsin. A similar organization is found in prepro-levitide. Prepro-caeruleins are more complex proteins. Each contains multiple copies of caerulein and identical acidic connecting peptides and some slightly different copies of CPF.

Genes that encode xenopsin, levitide, and caerulein precursors were mapped and sequenced (40, 47). The transcribed segments of these genes show similar exon patterns despite considerable variations in size. In the three genes, the second exon codes for the signal peptide and for a segment of the acidic connecting peptide.

Prepro-PGLa, prepro-dermaseptins, prepro-bombinins, and prepro-magainins are composed of three distinct domains arranged in a similar manner, that is, a signal peptide, an acidic leader peptide of variable length, and a progenitor sequence coding for an antimicrobial peptide. Whereas the precursors of dermaseptins and PGLa contain a single copy of an acidic leader peptide and a single antimicrobial progenitor sequence each, prepro-magainins and prepro-bombinins are made of, respectively, two and six highly homologous repeats of this tandem module. Interestingly, the signal peptides of the precursors encoding PGLa, XPF, CPF, LPF, and magainins exhibit significant homology. These *X. laevis* precursors may arise from a common ancestral gene that was subjected to duplication and exons shuffling (40, 47).

The deduced amino acid sequences of prepro-dermaseptins have striking similarities with the precursor proteins prepro-dermorphin/dermenkephalin and prepro-deltorphins (84, 85). D-Amino acid–containing opioid peptides (64) isolated from the skin of *Phyllomedusa* species are highly potent at, and exquisitely selective for, the mu or delta opioid receptors (30, 46, 49, 60, 64, 88): dermorphin, Tyr-D.Ala-Phe-Gly-Tyr-Pro-Ser-NH_2 (60); dermenkephalin, Tyr-D.Met-Phe-His-Leu-Met-Asp-NH_2 (66), also named dermorphin gene-associated peptide (49) or deltorphin A (30, 46); and the deltorphins B and C, Tyr-D.Ala-Phe-X-Val-Val-Gly-NH_2 (where X is either Asp or Glu) (85). These peptides are synthesized as part of two larger precursor proteins containing multiple copies of the bioactive heptapeptide sequences. As shown in Figure 1, prepro-dermaseptins B1 and B2, prepro-dermorphin/dermenkephalin, and prepro-deltorphins are composed of distinct domains arranged in a similar manner. The amino-terminal domain includes a 22-residue signal peptide followed by an acidic leader peptide domain containing 20–24 residues. The third domain consists of a progenitor sequence of variable lengths coding for either a membranolytic or a neuroactive peptide that is flanked by a pair of basic residues at its amino end and a tripeptide (Gly-Glu-Ala/Gln) at its C-terminus. The Gly residue is involved in the formation of the C-terminal peptide amide in all four precursors. Whereas prepro-dermaseptins B1 and B2 each contains only a single copy of an acidic spacer peptide and a bioactive sequence assembled in tandem, prepro-dermorphin/dermenkephalin and prepro-deltorphins are made of five highly homologous repeats of this tandem module.

A comparison of the cDNA and deduced amino acid sequences of the four prepro-proteins (4) shows that the signal peptides (72% identical at the amino acid level) and the acidic spacer peptides (50% identical at the amino acid level) are very similar. This similarity also extends into the 5'-untranslated portions of the respective mRNAs (71% nucleotide positional identity). However, the similarity is not quite as high in the 3' untranslated regions of the mRNAs. The putative processing sites of the precursors are also arranged identically; the paired basic residues Lys-Arg and the C-terminal tripeptide extension Gly-Glu-Gln/Ala that flank the bioactive sequence(s) are the same in the four prepro-proteins. The extensive similarities between the prepro-regions of the four precursors that encode end products with very different biological activities support the suggestion that the genes encoding these

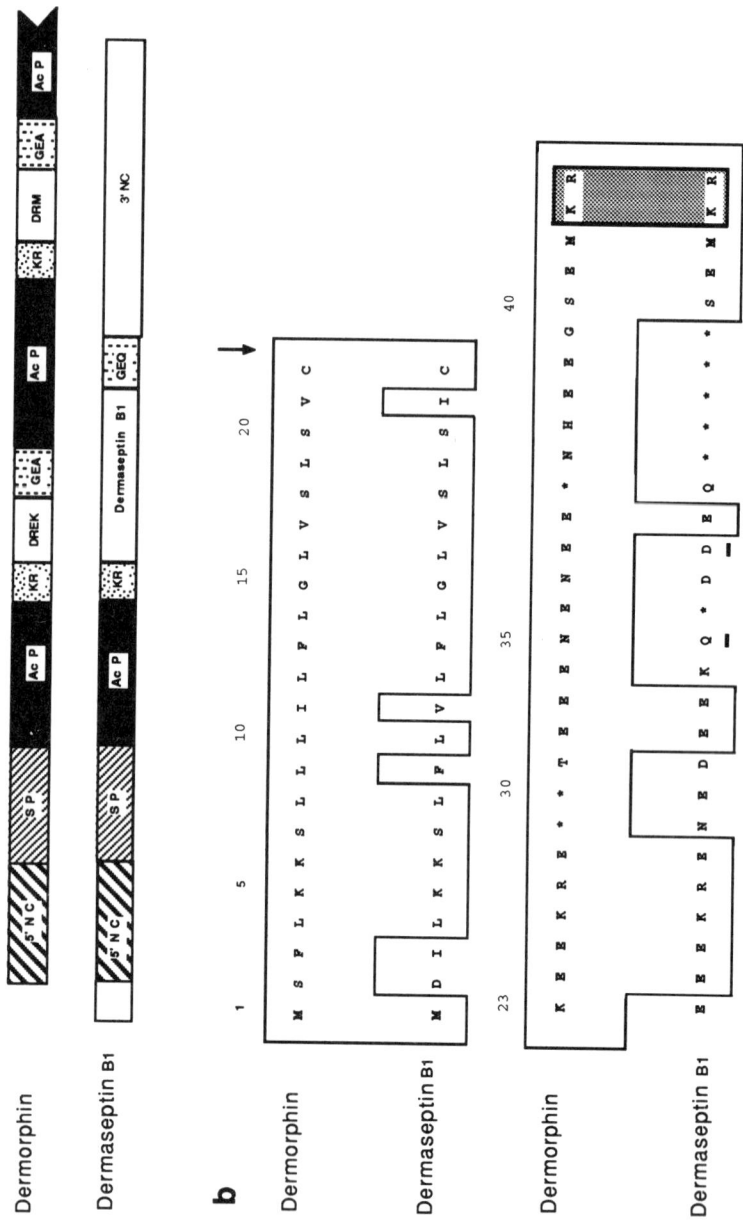

proteins are all members of a multigene family. These genes might contain a homologous exon comprising a portion of the 5′-untranslated region, the 22-residue signal peptide, the 20- to 24-residue acidic spacer peptide, and the prehormone processing site Lys-Arg. This finding suggests that the exon encoding the prepro-region was already present early in the evolution of amphibia.

The ubiquitous presence of acidic connecting peptides encoded within the precursors of hormonal and/or antimicrobial peptides from various frog species strongly suggests that these acidic sequences are of functional significance. Williams' group has identified and sequenced acidic peptides originating from PGLa, XPF, LPF, magainins, and CPF precursor proteins (75). Whether the anionic sequences participate in intracellular routing and processing of the precursor molecules or protect the glandular cells from lysis by preventing the cationic antimicrobial peptides from binding to lipid bilayers is not known.

Tissue-Specific Expression

Most amphibian skin antimicrobial peptides are also present in the gastro-intestinal tracts of frogs. For instance, the gastric mucosa of *X. laevis* contains granular multinucleated syncytial cells adjacent to the lumen that are similar in morphology to granular gland cells of the skin (63, 87). Northern blot analysis and immunohistochemical and immunogold methods demonstrated that *X. laevis* skin antimicrobial peptides are stored together in the rice-shaped granules present in these enteric cells. The morphological similarities between skin and stomach granular multinucleated cells led to the proposal that the enteric cells may release their content onto the stomach lumen by a holocrine mechanism involving the rupture of the plasma membrane and the extrusion of the antimicrobial peptide-containing granules through a duct opening on the surface. Direct isolation of authentic magainins and congeners from extracts of *X. laevis* stomach tissues strongly supports the above proposal (62).

Figure 1 (*a*) Block diagrams of the prepro-dermaseptin B1 (clone C 15), and prepro-dermorphin (clone D-1/2) cDNAs. Regions of high similarity are labeled with the same boxed patterns as follows: 5′ NC, 5′-noncoding region (*left-hatched box*); SP, signal peptide (*right-hatched box*); Ac P, acidic peptide (*black box*); KR, Lys-Arg dipeptide (*stippled box*); GEQ/A, Gly-Glu-Gln/Ala tripeptide (*dashed box*). Open boxes denote sequences with low or no homology: 3′ NC, 3′-noncoding region; DRM, dermorphin; DREK, dermenkephalin. Only partial sequences of the cDNA encode prepro-dermorphin. In D-1/2, the unit corresponding to DRM-GEA-Ac P-KR is repeated four times. (*b*) Maximized multiple sequence alignments of the predicted amino acid sequences of the signal peptides (numbered 1 through 22) and acidic leader peptides (numbered 23 through 44) from prepro-dermorphin (D-1/2) and prepro-dermaseptin B1 (C 15). Regions of identity are boxed. The arrow indicates the common carboxy terminal cleavage site for the signal peptides. Numbers above the prepro-dermorphin sequence indicate positions of the amino acids corresponding only to the dermaseptin B1 precursor. The asterisks indicate gap of one amino acid sequence relative to the others. The putative dibasic cleavage site Lys-Arg is represented in a dark dotted box. A heavy line is drawn below nonidentical but highly homologous residues.

Tissue distribution of magainins and PGLa gene expression have also identified the *X. laevis* gastrointestinal tract as a major site of expression (82). Besides the multinucleated gastric mucosal cells, in situ hybridizations have localized magainins and PGLa expression in eosinophilic granule-filled cells, which are functional analogues of mammalian Paneth cells. The existence of a network of specific enteric cells that may release a fixed repertoire of potent antimicrobial peptides onto the gastrointestinal tract supports a dedicated role for these peptides in maintaining control over bacterial invasions.

BREVININS, ESCULENTINS, AND RANALEXIN

A novel family of antimicrobial peptides has been detected in the skin of *Rana brevipoda* (73) and *Rana esculenta* (18, 93, 94), frogs of the family Ranidae. Different from all of the frog skin peptides mentioned above, the brevinins (73) and the esculentins (93) are characterized by the presence of two cysteine residues in positions 1 and 7 as counted from the carboxyl terminus. These residues are linked in a disulfide bridge. On the basis of size and additional sequence characteristics, four subfamilies can be discerned (94), the brevinins-1 and -2 and the esculentins-1 and -2 (Table 7). The type 1 brevinins are cationic molecules of 24 residues in which six positions are occupied by the same residue in all the peptides. Typical conserved features are a Pro in position 3, an Ala doublet in positions 9–10, and a pair of basic residues at the carboxyl-terminal end. Identities between different members of the brevinins-1 subfamily range from 20 to 96%. An additional member, termed ranalexin (18), has been isolated from the skin of the bullfrog *Rana catesbeiana*. This 20–amino acid peptide is highly homologous to brevinins-1 but lacks both the conserved proline residue in position 3 and the Ala doublet in positions 9–10. Brevinins-1 and ranalexin exhibit strong sequence similarities to the pipinins, three peptides originally isolated from the skin of the frog *Rana pipiens* in 1985 by means of their histamine-releasing activity (38) (Table 7).

Multiple names, such as pipinin, brevinin-1, and ranalexin, for antibacterial compounds with a convincing homology are unfortunate and confuse the literature. For the sake of simplicity and regardless of historical priorities, these peptides should simply be named ranalexin followed by a letter to indicate from which species of the Ranidae family the peptide was isolated. Likewise, in view of the growing confusion surrounding the ever-increasing number of antimicrobial peptides, a simplified nomenclature would be more than welcome at this stage.

The type 2 brevinins are also highly homologous peptides, 29–34 residues in length, in which 13 positions are occupied by the same residue in all the peptides (Table 7). They are characterized, as a subfamily, by a common

Table 7 Amino acid sequences of brevinins, esculentins, and pipinins from amphibian skin[a]

Brevinins and pipinins

	1	10	20
Brevinin-1	F L P V L A G I A A K V V P A L F C K I T K K C
Brevinin-1E	F L P L L A G L A A N F L P K I F C K I T R K C
Brevinin-1Ea	F L P A I F R M A A K V V P T I I C S I T K K C
Brevinin-1Eb	V I P F V A S V A A E M M Q H V Y C A A S R K C
Brevinin-1Ec	F L P L L A G L A A N F F P K I F C K I T R K C
Ranalexin (brevinin-1Ed)	F F G G L I K I V P A M I P K I F C K I T R K C
Pipinin I	F L P I I A G V A A K V F P K I F C A I S K K C
Pipinin II	F L P I I A G I A A K V F P K I F C A I S K K C
Pipinin III	F L P I I A S V A A K V F S K I F C A I S K K C

	1	10	20	30
Brevinin-2	G - L L D S L K G F A A T A G K G V L Q S L L S T A S C K L A K T C
Brevinin-2E	G - I M D T L K N L A K T A G K G A L Q S L L N K A S C K L S G Q C
Brevinin-2Ea	G - I L D T L K L N A I S A A K G A A Q G L V N K A S C K L S G Q C
Brevinin-2Eb	G - I L D T L K N L A K T A G K G A L Q G L V K M A S C K L S G Q C
Brevinin-2Ec	G I L L D K L K N F A K T A G K G V L Q S L L N T A S C K L S G Q C
Brevinin-2Ed	G - I L D S L K N L A K N A G - - - - Q I L L N K A S C K L S G Q C
Brevinin-2Ee	G - I F D K L K N F A K G V A - - - - Q S L L N K A S C K L S G Q C
Brevinin-2Ef	G - I M D T L K N L A K T A G K G A L Q S L V K M A S C K L S G Q C

Esculentins

	1	10	20	30	40
Esculentin-1	G I F S K L G R K K I K N L L I S G L K N V G K E V G M D V V R T G I D I A G C K I K G E C
Esculentin-1a	G I F S K L A G K K I K N L L I S G L K N V G K E V G M D V V R T G I D I A G C K I K G E C
Esculentin-1b	G I F S K L A G K K L K N L L I S G L K N V G K E V G M D V V R T G I D I A C C K I K C E C
Esculentin-2a	G I L S L V K G V A K L A G K G L A K E G G K F G L E L I A C K I A K Q C
Esculentin-2b	G I F S L V K G A A K L A G K G L A K E G G K F G L E L I A C K I A K Q C

[a] The sequences are aligned to demonstrate the most conserved amino acids. The numbering of residues is indexed at the top to the longest peptide of each family.

carboxyl-terminal nonapeptide encompassing the disulfide bridge. Esculentin-1a and -1b are almost identical to esculentin-1. These three peptides have the same chain length (46 residues) and only two and three amino acid substitutions, respectively. Moreover, the linear part of the esculentins shares some similarity with the bombinins, antimicrobial peptides from an evolutionary distant frog species, *B. variegata*. Esculentins and brevinins have a distinct activity spectrum, whereas peptides within each subfamily do not differ significantly in their cell lytic activity. For example, all of these peptides are highly active at micromolar concentrations against *E. coli, Bacillus megaterium,* and *S. aureus*. Esculentins differ from brevinins in their activity against *C. albicans, S. cerevisiae,* and *Pseudomonas aeruginosa*. Whereas brevinins-1 are highly hemolytic, brevinins-2 and esculentins do not lyse erythrocytes.

Nucleotide sequence analysis of cDNAs coding the corresponding precursors (18, 94) revealed that the predicted precursors for the esculentins and the brevinins are quite similar, indicating that they were formed by gene duplication from a common ancestor. They all contain a putative signal sequence of 22 residues at the NH_2 terminus and a spacer sequence of 16–25 residues, of which about half are acidic residues. In each case, the sequence of the single copy of the mature peptide is preceded by a Lys-Arg processing site and followed by a stop codon. As already noted for the dermaseptins, the signal peptides and part of the proregion of the precursors of brevinins, esculentins, and ranalexin are similar (56–72% identity) to the corresponding regions of the precursors of dermorphin and deltorphins, although these preproteins originate from two species of a different family (18, 94). This similarity reinforces the hypothesis that one or more exons encoding the prepro-region were already present in the evolution of amphibia.

ANTIMICROBIAL MECHANISMS

Despite the marked diversity in their primary and tertiary structures, size, and specificity of action, defensins and helical vertebrate antimicrobial peptides have conspicuous membranolytic properties towards their target cells. Inspection of the known three-dimensional structures reveals that they all present an amphipathic character and are also strongly cationic. Whether they adopt α-helical, β-sheet, or less-defined structures, they are all organized in such a way that polar or charged residues are topologically separated from apolar residues. The lytic action of these peptides is not mediated by membrane-anchored specific receptors or enzymes because, for example, synthetic cecropins, magainins, or dermaseptins in which all the constituent amino acids are in the D configuration are as active as the natural parent peptides, although they form left-handed instead of the natural right-handed helices (11, 108). Hence, the explanation for the molecular mechanisms that lead to cell lysis probably lies in the amphipathy and basicity of these structures.

β-*Sheet Peptides with Three Disulfide Bonds*

Defensins are all β-sheet peptides with no helix, and are stabilized by three disulfide bonds. Defensins cause the sequential permeation of the inner and outer membranes of *E. coli* with a simultaneous cessation of nucleic acid and protein synthesis as well as of respiration (52, 55, 89, 107). Bacterial death was attributed to inner-membrane permeation, which allows both the loss of intracellular contents and the entry of normally excluded molecules. Moreover, a membrane potential is required for defensin action because only cells that are metabolically active are killed (52, 54). Treatments that depolarize the target cells render them resistant to the defensins, and artificial phospholipid bilayers are permeated only when a transmembrane potential is applied with a polarity that enables the cationic defensin peptide to insert into the membrane (45). Following the insertion of defensins, the membrane conductance increases as the second to fourth power of the peptide concentration, suggesting that two to four defensin molecules aggregate to form a voltage-regulated channel. X-ray crystallography revealed that human defensin HNP-3 crystallizes as a dimer (37). The triple-stranded antiparallel β-sheet of each monomer is extended across the dimer interface to form a six-stranded continuous sheet stabilized by hydrophobic interactions and hydrogen bonds. Defensin dimers resemble a basket that has an apolar base and a polar top. The amino terminal β-strands of the two monomers are close together in the dimer polar top and are hydrogen bonded through ordered water molecules that form an aqueous mini-channel that traverses the dimer. Six flexible arginine side chains are distributed around the middle of the basket.

Current models (37, 55) envision that the effects of defensins on microorganisms develop in three stages. First, the electrostatic interactions between the defensin's arginine groups and the negatively charged head groups of the target cell membrane concentrate defensin dimers at the surface of the outer leaflet. Second, the amphiphilicity of the dimer basket allows it to be pulled into the target cell membrane. Either the dimer directly distorts lipid-lipid interactions by burying its apolar base into the bilayer while the guanidium groups interact with lipid's phosphate groups (wedge model), or two dimers assemble in the membrane with their polar tops toward each other and apolar bases facing lipid tails. In this configuration, the side chains of the six arginine residues move up and down to bind lipid head groups of both leaflets. Thus, two of the solvent mini-channels seen in the dimer structure completely span the bilayer (dimer-pore model). The third model (general-pore model) also has dimers completely spanning the bilayer but with the polar top surface lining the pore and the apolar surface interacting with lipid acyl chains. The concentration-dependent microbicidal activity of defensins might reflect a transition from wedge to dimer-pore to general-pore mechanism (45). In any case, insertion of defensin dimers into the bilayer disrupts its integrity, probably by forming voltage-dependent pores (stage 3), and allows the entry of normally excluded molecules (ions, peptides) and the leakage of essential minerals and metabolites, thus inducing the dissipation of the membrane potential and the arrest of the respiration.

Amphipathic Helical Peptides

Peptides from this group, when plotted on a Schiffer-Edmunson α-helical wheel, reveal amphipathic structures (90), i.e. the basic and polar residues are aligned on a portion of the helical cylinder whereas the lipophilic side chains occupy the remaining surface, the proportions of each surface varying from one peptide to another. Several lines of experimental evidence demonstrated that the ability to form well-behaved α-helices at membrane interfaces is linked to the antimicrobial properties of these peptides. However, amphipathic helical peptides represent a heterogenous group of antimicrobial molecules that can display multiple modes of action—bacteriostatic, microbicidal, lytic, hemolytic, and other cytolytic properties—that are sequence and concentration dependent. In addition, each family of antimicrobial peptides, and often each peptide within a family, can target specific microorganisms. Although a description of the precise molecular events associated with each one of these multiple modes of action is not yet available, a model picturing the general mechanism by which helical antimicrobial peptides bind to and permeate microbial membranes is emerging.

The general scheme of action expressed by this group of peptides (10, 25, 26, 35, 43, 44, 56–58, 81, 102, 106, 111–113) is exemplified by the magainins and the dermaseptins that are cytotoxic for microbial cells and thought to act primarily by causing an increase in membrane permeability of the victim cell. At low concentrations, these peptides adopt an α-helical amphipathic structure in the presence of acidic phospholipids. They form channels or pores in lipid bilayers; they also enhance the permeability of liposomes, disrupt the membrane potential through alteration of the electrochemical gradients across the cell membrane, and interfere with the free-energy metabolism of the target cell, thus impairing its ability to carry out anabolic processes. The microbicidal activities of these peptides exhibit a strong concentration dependence. At lower concentrations, the peptides are virtually inactive until a critical concentration is reached. Then, additional peptides strongly affect the permeability of the cell membrane in a way consistent with the view that the active form of the peptide is an oligomer. At higher concentrations, these peptides solubilize zwitterionic phosphatidylcholine membranes (113) and arrest the motility and growth of *Leishmania* promastigotes, leading to death through ultrastructurally visible alterations in the cell membrane (36; A Mor, I Vouldoukis, P Nicolas, M Gentilini & L Monjour, unpublished results). Peptide analogues with enhanced helix content, or increased positive charge, exhibit higher microbicidal activity.

Combining these data, a proposed general model stipulates: (*a*) Electrostatic interactions between the lysines and the negatively charged phospholipid head groups of the target cell promote binding to and accumulation of the helical peptide on the outer leaflet of the membrane. The bound peptide lies on the membrane with its long helix axis parallel to the membrane surface. (b) At low concentrations, the peptide adsorbs to the membrane as a monomer. At higher concentrations, the membrane-bound peptide begins to self-aggregate and equilibrates with a multimeric water-filled pore (two to six monomers in a rosette) in which the hydrophilic residues of the individual monomers face inward and the apolar residues interact with the acyl chains of the lipids. This transition occurs at roughly the same critical concentration required for cytotoxic activity, implying that peptide insertion into the membrane is responsible for cytotoxicity. A transmembrane electric potential, negative inside, enhances the pore-forming activity of the peptide favoring insertion by interacting with the permanent dipole moment of the peptide that is oriented along the long axis of the helix. (c) The solvent-filled unspecific pores, by permeating the membrane to various substances including ions, cause the electric potential to dissipate across the membrane, leading to death of the microbial cell.

A length of approximately 18–20 residues may be necessary to provide an α-helix capable of spanning the hydrocarbon moiety of the lipid bilayer (50). However, short peptide analogues 8–12 residues long could kill certain microorganisms and/or form ion-channels (7, 13, 67, 70). In that case, an oligomeric bundle of head-to-tail dimers was proposed as an exploratory channel model (1). Although the above models explain the permeating effect of helical antimicrobial peptides, they do not account for their remarkable specificity of action, which may be partly explained by the difference in lipid composition of the membrane of the target cells and the exact nature of the peptide's side chains.

At low concentrations, the magainins and the dermaseptins act primarily by causing a limited permeability change in the victim cell rather than a complete lysis. At higher concentrations, however, they promote lysis of microorganisms and/or hemolysis. Several studies suggest that the activities described by membrane permeation, lysis, and hemolysis underlie different mechanisms of action resulting in part from differences in the membrane composition and the peptide's net charge, the charge distribution, the hydrophobicity of the nonpolar face, and the average angle subtended by the polar surface of the peptides. For instance (13, 67, 102), numerous contiguous apolar residues in a helix appear to be necessary for a significant hemolysis to occur. Shortening the hydrophobic region or reducing the length of the peptide yields a net decrease in hemolytic activity. Although initial electrostatic interactions similar to those implicated in permeation are involved, lysis and hemolysis may result from a detergent-like action of the peptides through partial penetration of the peptides into the membranes, weakening of the lipid-lipid interactions, and solubilization of the bilayer leaflets.

EVOLUTIONARY SIGNIFICANCE

Antimicrobial peptides, initially thought to be the exclusive attribute of microorganisms, are in fact widely distributed in nature, in both the animal and plant kingdoms (for reviews, see 15, 16, 55, 53, 74, 99, 115). Present in the skin and the lumen of the respiratory and gastrointestinal tracts of vertebrate animals, most of these peptides have identical or closely related counterparts in invertebrates and plants. For instance, cecropins are prototypical amphipathic α-helical cationic peptides initially isolated from insects (15) and later discovered in intestinal pig mucosa (51). Cysteine-rich antimicrobial peptides that resemble mammalian defensins can be detected in the hemolymph of certain insects and plants (14, 19, 39). All vertebrate antimicrobial peptides known to date are small and have a marked cationic character as well as the propensity to form an amphipathic structure in the presence of lipid bilayers. These characteristics allow these peptides to interact with biological membranes, to be inserted into the bilayer, and to perturb its organization, which finally results in membrane permeation, cell-lysis, and death of the target cell.

The antimicrobial peptides present in vertebrates, which may be considered ancestral effectors of immunity, are more than mere remnants: Consider the advantages conferred to vertebrates by this molecular immune system in helping and complementing the classical long-lasting cell-mediated immune system. Indeed, a fixed but large repertoire of small antimicrobial peptides offer a particularly efficient mode of control over commensal microorganisms. If one considers that antimicrobial peptides are induced by simple signals and are produced and can diffuse much faster than antibodies, one realizes that these low-cost products with broad-spectrum activities allow defensive responses much faster than the proliferation rate of microorganisms. Such peptides are thus perfectly adequate to offer a first line of defense against infectious agents and to ensure continuous control over the proliferation of commensal microorganisms. Moreover, this chemical defense system offers partial help in case of temporary or prolonged failures of the classical immune system, as observed in immunodepressed patients.

PERSPECTIVES

Studying the constituents and the mechanisms of this chemical defense system has not only improved our understanding of the means used in nature to survive in a hostile environment laden with pathogenic microorganisms, but has also opened new perspectives in the conception of new therapeutic treatments of infections caused by these pathogens. Indeed, much of the therapeutic armamentarium available today is made of byproducts of fungi (e.g. penicillins) and other unicellular organisms. Unfortunately, resistant strains have evolved, as they will almost inevitably do, given sufficient time of exposure. A promising solution to this resistance problem may reside in new antibiotics as different as possible from those that have lost their effectiveness. Because peptide antibiotics act completely unlike traditional antibiotic agents, these animal-derived products—which actually extend the chemical defenses that animals have been using for millions of years—may just fulfill that requirement and thus be more durable remedies. In support of this supposition, the number of successful in vivo therapeutic attempts appearing in the literature is increasing. For example, when applied to diseased animal models, defensins abrogated or delayed the development of cutaneous lesions caused by *Treponema pallidum* following their intradermal injection into experimental rabbit models (16a). Likewise, magainins reportedly act efficiently in vivo against murine peritoneal ascites tumors (8a). Finally, dermaseptins were revealed to be efficient in curing murine leishmaniasis (A Mor, I Vouldoukis, P Nicolas, M Gentilini & L Monjour, unpublished results) caused by *Leishmania major,* one of the protozoan parasites responsible worldwide for severe muco-visceral infectious diseases for which, to date, no completely satisfactory treatment is available. Overall, these studies demonstrate the potential interest in the direct use of these peptide antibiotics and suggest a potential usefulness as molecular models for the conception of more potent structural analogues.

> Any *Annual Review* chapter, as well as any article cited in an *Annual Review* chapter, may be purchased from the Annual Reviews Preprints and Reprints service.
> 1-800-347-8007; 415-259-5017; email: arpr@class.org

Literature Cited

1. Agawa Y, Lee S, Onon S, Aoyagi H, Ohno M, et al. 1991. Interactions with phospholipid bilayers, ion channel formation and antimicrobial activity of basic amphipathic helical model peptides of various chain lengths. *J. Biol. Chem.* 266:20218–22
2. Agerberth B, Lee JY, Bergman T, Carlquist M, Boman HG, et al. 1991. Amino acid sequence of PR-39, isolation from pig intestine of a new member of the family of Pro, Arg-rich antibacterial peptides. *Eur. J. Biochem.* 202:849–54
3. Amiche M, Ducancel F, Boulain JC, Lajeunesse C, Menez A, Nicolas P. 1993. Molecular cloning of a cDNA encoding the precursor of adenoregulin. *Biochem. Biophys. Res. Commun.* 191: 983–90
4. Amiche M, Ducancel F, Mor A, Boulain JC, Menez A, Nicolas P. 1994. Precursors of vertebrate peptide antibiotics, dermaseptin b and adenoregulin show considerable sequence identities with precursors for the opioid peptides, dermorphin, dermenkephalin and deltorphins. *J. Biol. Chem.* 269:17847–52
5. Anastasi A, Erspamer V, Endean R. 1968. Isolation and sequence of caerulein, the active decapeptide of skin of *Hyla caerulea. Arch. Biochem. Biophys.* 125:57–68
6. Andreu D, Aschauer H, Kreil G, Merrifield RB. 1985. Solid-phase synthesis of PYLa and isolation of its natural counterpart PGLa from skin secretion of *Xenopus laevis. Eur. J. Biochem.* 149:531–35
7. Anzai K, Hamasuna M, Kadono H, Lee S, Aoyagi H, Kirino Y. 1991. Formation of ion-channels in planar

lipid bilayer membranes by synthetic basic peptides. *Biochim. Biophys. Acta* 1064: 256–61
8. Araki T, Tachibana S, Uchiyama M, Nakajima T, Yasuhara T. 1973. Isolation and structure of a new active peptide "Xenopsin" on the smooth muscle, especially on a strip of fundus from a rat stomach, from the skin of *Xenopus laevis. Chem. Pharmac. Bull.* 21:2801–4
8a. Baker AM, Maloy LW, Zasloff M, Jacob SL. 1993. Anticancer efficacy of magainin 2 and analogue peptides. *Cancer Res.* 53:3052–57
9. Barcker RL, Gluck GJ, Pease LR. 1988. Acidic precursor revealed in human eosinophil granules major basic protein cDNA. *J. Exp. Med.* 168:1493–98
10. Bechinger B, Zasloff M, Opella SJ. 1993. Structure and orientation of the antibiotic peptide magainin in membranes by solid-state NMR spectroscopy. *Protein Sci.* 2:2077–84
11. Bessalle R, Kapitkovsky A, Gozea A, Shalit I, Fridkin M. 1990. All-D magainin: chirality, antimicrobial activity and proteolytic resistance. *FEBS Lett.* 274:151–55
12. Bevins CL, Zasloff M. 1990. Peptides from frog skin. *Annu. Rev. Biochem.* 59:395–410
13. Blondelle SE, Houghteen RA. 1992. Design of model peptides having potent antimicrobial activities. *Biochemistry* 31:12688–94
14. Bolhmann H, Clausen S, Behnke S, Giese J, Hiller C, et al. 1988. Leaf-specific thionins of barley: a novel class of cell wall protein toxic to plant-pathogenic fungi and possibly involved in the defense mechanism of plants. *EMBO J.* 7:1559–65
15. Boman HG, Hultmark D. 1987. Cell free immunity in insects. *Annu. Rev. Microbiol.* 41:103–26
16. Boman HG. 1991. Antibacterial peptides: key components needed in immunity. *Cell* 65:205–7
16a. Borenstein LA, Ganz T, Sell S, Lehrer RI, Miller JN. 1991. Contribution of rabbit leukocyte defensins to the host response in experimental syphilis. *Infect. Immun.* 59:1368–77
17. Cabiaux V, Agerberth B, Johansson J, Homblé F, Goormaghtigh E, Ruysschaert JM. 1994. Secondary structure and membrane interaction of PR-39, a Pro + Arg-rich antibacterial peptide. *Eur. J. Biochem.* 224:1019–27
18. Clark DP, Durell S, Maloy WL, Zasloff M. 1994. Ranalexin: a novel antimicrobial peptide from bullfrog *Rana catesbeiana* skin, structurally related to the bacterial antibiotic polymyxin. *J. Biol. Chem.* 269:10849–55
19. Cociancich S, Ghazi A, Hetru C, Hoffmann JA, Letellier L. 1993. Insect defensin, an inducible antibacterial peptide forming voltage-dependent channels in *Micrococcus luteus. J. Biol. Chem.* 286:19239–45
20. Csordas A, Michl H. 1969. Primary structure of two oligopeptides of the toxin of *Bombina variegata. Toxicon* 7:103–8
21. Daher K, Lehrer RI, Ganz T, Kronenberg M. 1988. Isolation and characterization of human defensin cDNA clones. *Proc. Natl. Acad. Sci. USA* 85:7327–31
22. Daly WJ, Caceres J, Moni WR, Gusovski F, Moos M, et al. 1992. Frog secretions and hunting magic in the upper Amazon: identification of a peptide that interacts with an adenosine receptor. *Proc. Natl. Acad. Sci. USA* 89:10960–63
23. Diamond G, Jones DE, Bevins CL. 1993. Airway epithelial cells are the site of expression of a mammalian antimicrobial peptide gene. *Proc. Natl. Acad. Sci. USA* 90:4596–600
24. Diamond G, Zasloff M, Eck H, Brasseur M, Malloy WL, Bevins CL. 1991. Tracheal antibiotic peptide, a novel cysteine-rich peptide from mammalian tracheal mucosa: peptide isolation and cloning of a cDNA. *Proc. Natl. Acad. Sci. USA* 88:3952–57
25. Duclohier H. 1994. Anion pores from magainins and related defensive peptides. *Toxicology* 87:175–88
26. Duclohier H, Molle G, Spach G. 1989. Antimicrobial peptide magainin-1 from *Xenopus* skin forms anion-permeable channels in planar lipid bilayers. *Biophys. J.* 56:1017–21
27. Eisenhauer PB, Harwings SL, Lehrer RI. 1992. Cryptidin: antimicrobial defensins of the small intestine. *Infect. Immun.* 60:3556–65
28. Erspamer V, Melchiorri P. 1973. Active polypeptides in the amphibian skin and their synthetic analogs. *Pure Appl. Chem.* 35:463–94
29. Erspamer V, Melchiorri P, Falconieri-Erspamer G, Montecucchi PC, de Castiglione R. 1985. Phyllomedusa skin: a huge factory and store-house of a variety of active peptides. *Peptides* 6:7–12
30. Erspamer V, Melchiorri P, Falconieri-Erspamer G, Negri L, Corsi R, et al. 1989. Deltorphins: a family of naturally occurring peptides with high affinity and selectivity for delta-opioid binding

sites. *Proc. Natl. Acad. Sci. USA* 86: 5188–92
31. Frank RW, Gennaro R, Schneider K, Przybylski M, Romeo D. 1990. Amino acid sequence of two proline-rich bactenecins. *J. Biol. Chem.* 265:18871–74
32. Ganz T, Rayner JR, Valore EV, Tumolo A, Talmadge K, Fuller F. 1989. The structure of rabbit macrophage defensin genes and their organ-specific expression. *J. Immunol.* 143:1358–65
33. Gibson BW, Tang D, Mandrell R, Kelly M, Spindel E. 1991. Bombinin-like peptides with antimicrobial activity from skin secretions of the Asian toad *Bombina orientalis. J. Biol. Chem.* 266:23103–11
34. Giovannini MG, Poulter L, Gibson BW, Williams DH. 1987. Biosynthesis and degradation of peptides derived from *Xenopus* prehormones. *Biochem. J.* 243:113–20
35. Gomes AV, de Waal A, Berden JA, Werstherhoff HV. 1993. Electric potentiation, cooperativity and synergism of magainin peptides in protein free liposomes. *Biochemistry* 32:5365–72
36. Hernandez C, Mor A, Dagger F, Nicolas P, Hernandez A, et al. 1992. Functional and structural damages in *Leishmania mexicana* exposed to the cationic peptide dermaseptin. *Eur. J. Cell Biol.* 59:414–24
37. Hill CP, Yee J, Selsted ME, Eisenberg D. 1991. Crystal structure of defensin HNP-3, an amphiphilic dimer: mechanisms of membrane permeabilization. *Science* 251:1481–85
38. Horikawa R, Parker DS, Herring PL, Pisano JJ. 1985. Pipinins: a new mast cell degranulating peptides from *Rana pipiens. Fed. Proc.* 44:695–700
39. Hultmark D. 1993. Immune reactions in *Drosophila* and other insects: a model for innate immunity. *Trends Genet.* 9:178–83
40. Hunt LT, Barker WC. 1988. Relationship of promagainin to three other prehormones from the skin of *Xenopus laevis:* a different perspective. *FEBS Lett.* 233:282–85
41. Jones DE, Bevins CL. 1992. Paneth cells of the human small intestine express antimicrobial peptide genes. *J. Biol. Chem.* 267:23216–25
42. Jones DE, Bevins CL. 1993. Defensin-6 mRNA in Paneth cells: implications for antimicrobial peptides in host defense of human bowel. *FEBS Lett.* 315:187–92
43. Juretic D, Chen HC, Brown JH, Morell JL, Hendler RW, Westerhoff HV. 1989. Magainin-2 amide and analogues. Antimicrobial activity, membrane depolarization and susceptibility to proteolysis. *FEBS Lett.* 249:219–23
44. Juretic D, Hendler RW, Kamp F, Xaughey WS, Zasloff M, et al. 1994. Magainin oligomers reversibly dissipate DmH+ cytochrome oxydase liposomes. *Biochemistry* 33:4562–70
45. Kagan BL, Selsted ME, Ganz T, Lehrer RI. 1990. Antimicrobial defensin peptides form voltage-dependent ion-permeable channels in planar lipid bilayer membranes. *Proc. Natl. Acad. Sci. USA* 87:210–14
46. Kreil G, Barra D, Simmaco M, Erspamer V, Falconieri-Erspamer G, et al. 1989. Deltorphin, a novel amphibian skin peptide with high affinity and selectivity for delta-opioid receptors. *Eur. J. Pharmacol.* 162:123–28
47. Kuchler K, Kreil G, Sures I. 1989. The genes for the frog skin peptides PGLa, xenopsin, levitide and caerulein contain a homologous exon encoding a signal sequence and part of an amphipathic peptide. *Eur. J. Biochem.* 179:281–85
48. Lazarus LH, Attila M. 1993. The toad, ugly and venomous, wears yet a precious jewel in his skin. *Prog. Neurobiol.* 41:473–507
49. Lazarus LH, Wilson WE, de Castiglione R, Guglieta A. 1989. Dermorphin gene sequence peptide with high affinity and selectivity for delta-opioid receptors. *Biol. Chem.* 264:3047–50
50. Lear JD, Wasserman ZR, DeGrado WF. 1988. Synthetic amphiphilic peptide models for protein ion channels. *Science* 240:1177–81
51. Lee JY, Boman A, Chuanxin S, Andersson M, Jornvall H, et al. 1989. Antibacterial peptides from pig intestine: isolation of a mammalian cecropin. *Proc. Natl. Acad. Sci. USA* 86:9159–62
52. Lehrer RI, Barton A, Daher KA, Harwig SSL, Ganz T, et al. 1989. Interaction of human defensin with *E. coli.* Mechanism of bactericidal activity. *J. Clin. Invest.* 94:553–61
53. Lehrer RI, Ganz T, Selsted ME. 1991. Defensins: endogenous antibiotic peptides of animal cells. *Cell* 64:229–30
54. Lehrer RI, Ganz T, Szklarek D, Selsted ME. 1988. Modulation of the in vitro candidacidal activity of human neutrophil defensins by target cell metabolism and divalent cations. *J. Clin. Invest.* 81:1829–35
55. Lehrer RI, Lichtenstein AL, Ganz T. 1993. Defensins: antimicrobial and cytotoxic peptides of mammalian cells. *Annu. Rev. Immunol.* 11:105–28

56. Ludtke SJ, He K, Wa Y, Huang HW. 1994. Cooperative membrane insertion of magainin correlated with cytolytic activity. *Biochim. Biophys. Acta* 1190: 181–84
57. Matsuzaki K, Harada M, Handa T, Funahoshi S, Fujii N, et al. 1989. Magainin-1 induced leakage of entrapped calcein out of negatively-charged lipid vesicles. *Biochim. Biophys. Acta* 981: 130–34
58. Matsuzaki K, Murase O, Tokuda H, Funakoshi S, Fujii N, Miyajima K. 1994. Orientational and aggregational states of magainin 2 in phospholipid bilayers. *Biochemistry* 33:3342–49
59. Michaelson D, Rayner J, Coulo M, Valore EV, Ganz T. 1992. Cationic defensins arise from charge-neutralized propeptides: a potential mechanism for avoiding leukocytes autocytotoxicity. *J. Leukocyte Biol.* 51:634–39
60. Montecucchi PC, de Castiglione R, Piani S, Gozzini L, Erspamer V. 1981. Amino acid composition and sequence of dermorphin. A novel opiate-like peptide from the skin of *Phyllomedusa sauvagei*. *Int. J. Peptide Protein Res.* 17: 275–83
61. Montecucchi PC, Falconieri-Erspamer V, Visser J. 1977. Occurrence of Asn[2], Leu[5]-carrulein in skin of African frog *Hylambates maculatus*. *Experientia* 33: 1138–39
62. Moore K, Bevins CL, Brasseur M, Tomassini N, Turner K, et al. 1991. Antimicrobial peptides in stomach of *Xenopus laevis*. *J. Biol. Chem.* 266: 19851–57
63. Moore KS, Bevins CL, Tomassini N, Huttner KM, Sadler K, et al. 1992. A novel peptide-producing cell in *Xenopus*: multinucleated gastric mucosal cells strikingly similar to the granular gland of the skin. *J. Histochem. Cytochem.* 40:367–74
64. Mor A, Amiche M, Nicolas P. 1992. Enter a new post-translational modification: D-amino acids in gene-encoded peptides. *Trends Biochem. Sci.* 204: 481–85
65. Mor A, Amiche M, Nicolas P. 1994. Isolation, synthesis and activity of dermaseptin b, a novel vertebrate defensive peptide from frog skin: relationship with adenoregulin. *Biochemistry* 33: 6642–50
66. Mor A, Delfour A, Sagan S, Amiche M, Pradelles P, et al. 1989. Isolation of dermenkephalin from amphibian skin, a high affinity delta-selective opioid peptide containing a D-amino acid residue. *FEBS Lett.* 255:269–74
67. Mor A, Hani K, Nicolas P. 1994. The vertebrate peptide antibiotics dermaseptins have overlapping structural features but target specific microorganisms. *J. Biol. Chem.* 269:31635–40
68. Mor A, Nguyen VH, Delfour A, Migliore-Samour D, Nicolas P. 1991. Isolation, amino acid sequence and synthesis of dermaseptin, a novel antimicrobial peptide of amphibian skin. *Biochemistry* 30:8824–30
69. Mor A, Nguyen VH, Nicolas P. 1991. Antifungal activity of dermaseptin, a novel vertebrate skin peptide. *J. Mycol. Med.* 1:5–10
70. Mor A, Nicolas P. 1994. The NH_2-terminal helical domain 1–18 of dermaseptin is responsible for antimicrobial activity. *J. Biol. Chem.* 269:1934–39
71. Mor A, Nicolas P. 1994. Isolation and structure of novel vertebrate defensive peptides from frog skin. *Eur. J. Biochem.* 219:145–54
72. Deleted in proof
73. Morikawa N, Hagiwara K, Nakajima T. 1992. Brevins-1 and 2, unique antimicrobial peptides from the skin of the frog, *Rana brevidipora porsa*. *Biochem. Biophys. Res. Commun.* 189:184–90
74. Nicolas P, Mor A, Delfour A. 1992. Les peptides de la défense antimicrobienne des vertébrés. *Med. Sci.* 8:423–31
75. Nutkins JC, Williams DH. 1989. Identification of highly acidic peptides from processing of the skin prepro-peptides of *Xenopus laevis*. *Eur. J. Biochem.* 181:97–102
76. Ouelette AJ, Greco RM, James M, Frederick D, Naftilan J, Fallon JT. 1989. Developmental regulation of cryptidin, a corticostatin/defensin precursor mRNA in mouse small intestinal crypt epithelium. *J. Cell Biol.* 108:1687–95
77. Ouelette AJ, Lualdi JC. 1990. A novel mouse gene family coding for cationic cysteine-rich peptides. Regulation in small intestine and cells of myeloid origin. *Biol. Chem.* 265:9831–37
78. Pardi A, Hare DR, Selsted ME, Morrison RD, Bassolino DA, et al. 1988. Solution structures of the rabbit neutrophil defensin NP-5. *J. Mol. Biol.* 210: 625–36
79. Patterson-Delafield J, Martinez RJ, Lehrer RI. 1980. Microbicidal cationic proteins in rabbit alveolar macrophages: a potential host-defense mechanism. *Infect. Immun.* 30:180–92
80. Poulter L, Terry AS, Williams DH, Giovannini MG, Moore CH, Gibson BW. 1988. Levitide, a new hormone-like peptide from the skin of *Xenopus laevis*. Peptide and peptide precursor

cDNA sequence. *J. Biol. Chem.* 263: 3279–83

81. Pouny Y, Rapaport D, Mor A, Nicolas P, Shai Y. 1992. Interaction of antimicrobial dermaseptin and its fluorescently labeled analogues with phospholipid membranes. *Biochemistry* 31:12416–23

82. Reilly DS, Tomassini N, Bevins CL, Zasloff M. 1994. A Paneth cell analogue in *Xenopus* small intestine expresses antimicrobial peptide genes: conservation of intestinal host-defense system. *J. Histochem. Cytochem.* 42: 697–704

83. Richter K, Egger R, Kreil G. 1986. Sequence of prepro-caerulein cDNAs cloned from skin of *Xenopus laevis*. A small family of precursors containing one, three or four copies of the final product. *J. Biol. Chem.* 261:3676–80

84. Richter K, Egger R, Kreil G. 1987. D-alanine in the frog skin peptide dermorphin is derived from L-alanine in the precursor. *Science* 238:200–2

85. Richter K, Egger R, Negri L, Corsi R, Severini C, Kreil G. 1990. cDNAs encoding [D-Ala^2] deltorphin precursors from skin of *Phyllomedusa bicolor* also contain genetic information for three dermorphin-related opioid peptides. *Proc. Natl. Acad. Sci. USA* 87:4836–39

86. Richter K, Kawashima E, Egger R, Kreil G. 1984. Biosynthesis of TRH in the skin of *Xenopus laevis*: partial sequence of the precursor deduced from cloned cDNA. *EMBO J.* 3:617–21

87. Sadler KC, Bevins CL, Kaltenbach JC. 1992. Localization of xenopsin and XPF immunoreactivities in the skin and gastrointestinal tract of *Xenopus laevis*. *Cell Tissue Res.* 270:257–63

88. Sagan S, Amiche M, Delfour A, Mor A, Camus A, Nicolas P. 1989. Molecular determinants of receptor affinity and selectivity of the natural delta-opioid agonist, dermenkephalin. *J. Biol. Chem.* 264:17100–6

89. Sawyer JG, Martin NL, Hancock REW. 1988. Interaction of macrophage cationic proteins with the outer membrane of *Pseudomonas aeruginosa*. *Infect. Immun.* 56:693–98

90. Segrest JP, Garber DW, Brouillette CG, Harvey SC, Anantharamarah GM. 1994. The amphipathic α-helix. A multifunctional structural motif in plasma apolipoproteins. *Adv. Protein Chem.* 45: 303–69

91. Selsted ME, Miller SI, Henschen AH, Ouelette AJ. 1992. Enteric defensins: antibiotic peptide components of intestinal host defense. *J. Cell Biol.* 118:929–36

92. Selsted ME, Tang YQ, Morris WL, McGuire PA, Novotny MJ, et al. 1993. Purification, primary structures, and antibacterial activities of β-defensins, a new family of antimicrobial peptides from bovine neutrophils. *J. Biol. Chem.* 268:6641–48

93. Simmaco M, Mignogna G, Barra D, Bossa F. 1993. Novel antimicrobial peptides from skin secretions of the European frog *Rana esculenta*. *FEBS Lett.* 324:159–61

94. Simmaco M, Mignogna G, Barra D, Bossa F. 1994. Antimicrobial peptides from skin secretions of *Rana esculenta*. Molecular cloning of cDNAs encoding esculentins and brevinins and isolation of new active peptides. *J. Biol. Chem.* 269:11956–61

95. Simmaco M, Barra D, Chiarini F, Noviello L, Melchiorri P, et al. 1991. A family of bombinin-related peptides from the skin of *Bombina variegate*. *Eur. J. Biochem.* 199:217–22

96. Soravia E, Martini G, Zasloff M. 1988. Antimicrobial properties of peptides from *Xenopus* granular gland secretions. *FEBS Lett.* 228:337–40

97. Sparkes RS, Kronenberg M, Heinzmann C, Daher KA, Klisak I, et al. 1989. Assignment of defensin gene(s) to human chromosome 8p23. *Genomics* 5: 240–44

98. Spencer JH. 1992. Antimicrobial peptides of frog skin. *Adv. Enzyme Regul.* 32:117–29

99. Spitznagel JK. 1990. Antibiotic proteins of human neutrophils. *J. Clin. Invest.* 86:1381–86

100. Storia P, Zanetti M. 1993. A novel cDNA sequence encoding a pig leukocyte antimicrobial peptide with a cathelin-like pro-sequence. *Biochem. Biophys. Res. Commun.* 196:1363–68

101. Storia P, Zanetti M. 1993. A cDNA derived from pig bone marrow cells predicts a sequence identical to the intestinal antibacterial peptide PR-39. *Biochem. Biophys. Res. Commun.* 196: 1058–65

102. Strahilevitz J, Mor A, Nicolas P, Shai Y. 1994. Spectrum of antimicrobial activity and assembly of dermaseptin-b and its precursor form in phospholipid membranes. *Biochemistry* 33: 10951–60

103. Sures I, Crippa M. 1984. Xenopsin: the neurotensin-like octapeptide from *Xenopus* skin at the C-terminus of its precursor. *Proc. Natl. Acad. Sci. USA* 81: 380–84

104. Tang YQ, Selsted E. 1993. Characterization of the disulfide motif in

BNBD-12, an antimicrobial β-defensin peptide from bovine neutrophils. *J. Biol. Chem.* 268:6649–53
105. Terry AS, Poulter L, Willias DH, Nutkins TC, Giovannini MG, et al. 1988. The cDNA sequence coding for prepro-PGS and aspects of the processing of this prepro-peptide. *J. Biol. Chem.* 263:5745–51
106. Urrutia R, Cruciani RA, Barker JL, Kachar B. 1989. Spontaneous polymerization of the antibiotic peptide magainin-2. *FEBS Lett.* 247:17–21
107. Viljanen P, Kosk P, Vaara M. 1988. Effect of small cationic leukocyte peptides (defensins) on the permeability barrier of the outer membrane. *Infect. Immun.* 56:2324–29
108. Wade D, Boman A, Wahlin B, Drain CM, Andreu D, et al. 1990. All D-amino acid-containing channel forming antibiotic peptides. *Proc. Natl. Acad. Sci. USA* 87:4761–65
109. Wakabayashi T, Kato H, Tachibana S. 1984. An unusual repetitive structure of caerulein mRNA from the skin of *Xenopus laevis*. *Gene* 31:295–99
110. Wakabayashi T, Kato H, Tachibana S. 1985. Complete nucleotide sequence of mRNA for caerulein precursor from *Xenopus* skin: the mRNA contains an unusual repetitive structure. *Nucleic Acids Res.* 13:1817–28
111. Westerhoff HV, Hendler RW, Zasloff M, Juretic D. 1989. Interaction between a new class of eukaryotic antimicrobial agents and isolated rat liver mitochondria. *Biochim. Biophys. Acta* 975:361–69
112. Westerhoff HV, Juretic D, Hendler RW, Zasloff M. 1989. Magainins and the disruption of membrane-linked free-energy transduction. *Proc. Natl. Acad. Sci. USA* 86:6597–601
113. Williams RW, Starman R, Taylor KM, Gable K, Beeler T, et al. 1990. Raman spectroscopy of synthetic antimicrobial frog peptide magainin-2a and PGLa. *Biochemistry* 29:4490–96
114. Zasloff M. 1987. Magainins, a class of antimicrobial peptides from *Xenopus laevis* skin: isolation, characterization of two active forms and partial cDNA sequence of a precursor. *Proc. Natl. Acad. Sci. USA* 84:5449–53
115. Zasloff M. 1992. Antibiotic peptides as mediators of innate immunity. *Curr. Opin. Immunol.* 4:3–7
116. Zasloff M, Martin B, Chen HC. 1988. Antimicrobial activity of synthetic magainin peptides and several analogues. *Proc. Natl. Acad. Sci. USA* 85:910–13

CO DEHYDROGENASE

James G. Ferry
Department of Biochemistry, Microbiology, Molecular & Cell Biology, The Pennsylvania State University, University Park, Pennsylvania 16802-4500

KEY WORDS: carbon monoxide, acetate, carbon cycle, aerobic, anaerobic

CONTENTS

INTRODUCTION	306
AEROBES	307
Microbiology and Physiology	307
The CO Dehydrogenase from Pseudomonas carboxydovorans	307
PHOTOTROPHIC ANAEROBES	308
Microbiology and Physiology	308
The CO Dehydrogenase from Rhodospirillum rubrum	309
ACETOGENIC ANAEROBES	311
Microbiology and Physiology	311
The CO Dehydrogenase from Clostridium thermoaceticum	313
ACETOTROPHIC ANAEROBES	318
Microbiology and Physiology	318
The CO Dehydrogenases from Methanosarcina sp. and Methanothrix soehngenii	321
FUTURE RESEARCH	325

ABSTRACT

Structurally and functionally diverse CO dehydrogenases are key components of various energy-yielding pathways in aerobic and anaerobic microbes from the *Bacteria* and *Archaea* domains. Aerobic microbes utilize Mo-Fe-flavin CO dehydrogenases to oxidize CO in respiratory pathways. Phototrophic anaerobes grow by converting CO to H_2, a process initiating with a CO dehydrogenase that contains nickel and iron-sulfur centers. Acetate-producing anaerobes employ a nickel/iron-sulfur CO dehydrogenase to synthesize acetyl-CoA from a methyl group, CO, and CoA. A similar enzyme is responsible for the cleavage of acetyl-CoA by anaerobic *Archaea* that obtain energy by fermenting acetate to CH_4 and CO_2. Acetotrophic sulfate reducers from the *Bacteria* and *Archaea* also utilize CO dehydrogenase to cleave acetyl-CoA yielding methyl and carbonyl groups. These microbes obtain energy for growth via a respiratory pathway in which the methyl and carbonyl groups are oxidized to CO_2, and sulfate is reduced to sulfide.

INTRODUCTION

CO dehydrogenases are present in physiologically and phylogenetically diverse microbes where the enzyme functions to either oxidize CO, synthesize acetyl-CoA, or cleave acetyl-CoA in a variety of energy-yielding pathways. Acetogenic (acetate-producing) anaerobes reduce $2CO_2$ to acetyl-CoA, which is coupled to the oxidation of growth substrates. CO dehydrogenase is also involved in the synthesis of acetyl-CoA from $2CO_2$ for cell carbon in autotrophic anaerobes. Thus, CO dehydrogenase is essential for the dark fixation of CO_2 into organic matter, which joins the Calvin cycle and the reductive tricarboxylic acid cycle as a primary process for CO_2 fixation into organic matter. Acetotrophic (acetate-utilizing) anaerobes, which cleave acetyl-CoA, produce either $2CO_2$ or CO_2 plus CH_4. Acetogenic and acetotrophic anaerobes are links in microbial food chains degrading complex organic matter to CH_4 and CO_2. An undetermined amount of the CH_4 enters the upper atmosphere, where it is converted to CO and CO_2, contributing to the natural biological sources of atmospheric CO. However, most of the 2700 teragrams (1 Tg = 10^{12} g) of CO entering the Earth's atmosphere each year originates from human activities (56a), thereby perturbing the natural levels of CO. Microbial CO dehydrogenases are essential for maintaining the natural level by either oxidizing CO to CO_2 or incorporating CO into cell carbon and metabolic end products (19a). Thus, CO dehydrogenases play pivotal roles in the anaerobic portion of the global carbon cycle and modulation of atmospheric CO in response to human activities.

CO dehydrogenases presumably have an ancient origin and were prevalent in primitive microbes prior to diversification into the *Bacteria* and *Archaea* domains (54). The first cellular energy–conservation mechanism could have been an oxidation of H_2S and FeS_2 to Fe pyrites and $2H^+$ on the outside of the membrane coupled to reduction by CO dehydrogenase of CO_2 on the inside (59a). The resulting proton gradient would then be used to drive ATP synthesis. Furthermore, the synthesis of acetyl-CoA from $2CO_2$ may have predated the Calvin cycle for fixation of CO_2 into cell carbon (96).

CO dehydrogenases are divided into two groups, the Mo-Fe-flavin enzymes (CO oxidases) from aerobes and the Ni-Fe enzymes from anaerobes. The latter are one of four Ni-containing enzymes in nature, the others being hydrogenase, methyl coenzyme M methylreductase, and urease. Research on CO dehydrogenases has grown rapidly in recent years. This review integrates the microbiology and physiology surrounding CO dehydrogenases with recent advances in the biochemistry of the enzyme.

AEROBES

Microbiology and Physiology

Carboxydotrophic microbes are aerobic chemolithoautotrophs that utilize CO as a sole source of carbon and energy. They are taxonomically diverse, encompassing more than 15 described species in 8 genera (82). Conservation of energy is accomplished via proton translocation coupled to the oxidation of CO and electron transport along a CO-insensitive branch of the respiratory pathway (21). Although most species will also grow with H_2 and CO_2, H_2 is not a physiologically important intermediate during growth with CO. The CO-insensitive branch of the electron-transport system is comprised of ubiquinone 10 (E_m = 100 mV) and cytochromes b_{561} (E_m = 40 mV) and b_{563} (E_m = −105 mV), of which the latter serves as the terminal oxidase. The physiological electron acceptor for the CO dehydrogenase from *Pseudomonas carboxydovorans* is thought to be cytochrome b_{561} (47); however, ubiquinone 10 is reportedly reduced by the CO dehydrogenase from *Pseudomonas carboxydohydrogena* (58). The CO insensitivity of this electron-transport pathway does not arise from the activity of the CO dehydrogenase, but possibly from the high affinity of cytochrome b_{563} for O_2, which keeps the cytochrome in the oxidized state and unable to bind CO (82). The CO_2 resulting from oxidation of CO is assimilated into cell carbon via the reductive pentose phosphate cycle.

The CO Dehydrogenase from Pseudomonas carboxydovorans

The CO dehydrogenase (CO:acceptor oxidoreductase) from carboxydotrophs is a molybdenum hydroxylase that catalyzes Reaction 1. CO_2 (not bicarbonate) is the immediate product (32):

$$CO + H_2O + A(ox) \rightarrow CO_2 + AH_2(red). \qquad 1.$$

The enzyme is routinely assayed with artificial electron acceptors (A) such as methylene blue. The CO dehydrogenase from *P. carboxydovorans* is the best characterized. The enzyme resides on the inner aspect of the cytoplasmic membrane (107), which is consistent with membrane-bound cytochrome b_{561} as the physiological electron acceptor. Molybdenum is contained in the cofactor molybdopterin cytosine dinucleotide (MCD) (53) and is structurally more complex than the molybdopterin cofactors of sulfite oxidases and xanthine oxidases from the *Eucarya* domain (60). Since the discovery of MCD, several molybdoenzymes were isolated from microbes in both the *Bacteria* and *Archaea* domains that contain molybdopterin dinucleotides; thus, the name bactopterin was coined for these novel cofactors.

All CO dehydrogenases thus far isolated from carboxydotrophs contain large (L), medium (M), and small (S) subunits in a $(LMS)_2$ configuration (84). The enzyme from *P. carboxydovorans* contains subunits of 86, 34, and 17 kDa. In addition to two MCDs, the enzyme contains two Fe_2S_2 centers and two noncovalently bound FADs (79). Electron paramagnetic resonance (EPR) spectroscopy indicates the presence of two types of Fe-S centers (16) that are similar to the two spectroscopically distinguishable Fe_2S_2 centers of xanthine oxidase. The location of the redox active centers among the subunits and the centers' specific function remain unknown. Because molybdenum is the site of oxidation of substrates in molybdenum hydroxylases, this metal is purportedly directly involved in the oxidation of CO (79). CO dehydrogenase activity is stimulated by aerobic incubation with selenite (83). The selenium is covalently bound between the sulfurs of cysteines that form seleniumtrisulfide (Cys-S-Se-S-Cys); however, the role of selenium is unknown. Recent research showed that selenium coordinates to molybdenum in the nicotinic acid hydroxylase from *Clostridium barkeri* (34) and the formate dehydrogenase from *Escherichia coli* that also contains molybdenum in the molybdopterin guanine dinucleotide (33). In the formate dehydrogenase, selenium occurs in the form of selenocysteine instead of an unidentified labile form as in the hydroxylase. Apparently, selenium is not coordinated to the molybdenum of the CO dehydrogenase from *P. carboxydovorans,* as the selenite-activated form displays Mo(V) EPR spectra identical to the untreated enzyme (81). This enzyme has hydrogenase and nitrate reductase activities of 10–16% and less than 2.5% (respectively) of CO dehydrogenase activity; however, the physiological significance of these ancillary activities is unknown.

In most cases the CO dehydrogenase structural genes reside on plasmids (8, 43, 80). The complexity of CO dehydrogenases from carboxydotrophs predicts that additional genes are necessary to synthesize an active enzyme. The enzyme is inducible (57), which suggests transcriptional regulation and a requirement for regulatory proteins.

PHOTOTROPHIC ANAEROBES

Microbiology and Physiology

Rhodocyclus gelatinosus (formerly *Rhodopseudomonas gelatinosa*) and *Rhodospirillum rubrum* are phototrophs that grow anaerobically in the dark with CO as the sole carbon and energy source (128, 129). They metabolize CO according to Reaction 2, which is exergonic by 20.1 kJ/mol:

$$CO + H_2O \rightarrow CO_2 + H_2. \qquad 2.$$

The CO_2 is assimilated into cell material via the ribulose 1,5-bisphosphate carboxylase cycle (130). The CO dehydrogenase is membrane associated (29, 131). When the enzyme is solubilized from membranes of *R. rubrum* in the presence of low-potential reductants, it is purified as a monomer. However, when solubilized in a redox-independent manner, the CO dehydrogenase is purified in association with a 22-kDa Fe-S protein that can be dissociated from the CO dehydrogenase with acetonitrile. The 22-kDa protein is reconstituted with the CO dehydrogenase in a 1:1 stoichiometry. The Fe-S protein is required to couple electron flow from the CO dehydrogenase to a CO-induced membrane-bound hydrogenase. Whether additional electron carriers are required for electron transport between the 22-kDa protein and the hydrogenase is unknown. CO-grown *R. gelatinosus* synthesizes ATP during CO oxidation to H_2 (18). Most likely, ATP synthesis is driven by a transmembrane proton gradient. The mechanism of gradient formation is unknown, but any hypothesis must address the apparent location of both the CO dehydrogenase and the hydrogenase on the cytoplasmic side of the cell membrane (18).

The synthesis of CO dehydrogenase is regulated in response to CO and light. In *R. rubrum*, both the CO dehydrogenase and a CO-insensitive hydrogenase are induced upon exposure of cells to CO (10). The regulation is at the level of gene expression as evidenced by de novo protein synthesis. In addition to O_2 inactivation of the CO dehydrogenase and hydrogenase, O_2 also represses the synthesis of CO dehydrogenase. The regulation of CO dehydrogenase synthesis is more complex in *R. gelatinosus* (19), which appears to adjust the relative amounts of CO oxidation and photometabolism to support growth in the presence of CO and light. CO concentration and light intensity determine the levels of CO dehydrogenase and bacteriochlorophyll. A diauxic growth response occurs when the CO concentration decreases and cells switch to photosynthesis accompanied by a reduction in CO dehydrogenase activity. The levels of cyclic AMP change during diauxic growth, suggesting that this molecule is involved in intracellular control (87).

Recently, the genes encoding the *R. rubrum* CO dehydrogenase were cloned and sequenced along with several open reading frames that may be required for CO conversion to H_2 and CO_2 (55). Thus, the stage is set for molecular genetic approaches to investigate gene expression and enzyme mechanisms.

The CO Dehydrogenase from Rhodospirillum rubrum

The CO dehydrogenase from *R. rubrum* is a monomer of M_r 61,800 containing Ni, Fe, acid-labile S, and Zn (11, 13). The enzyme cannot evolve H_2, consistent with the requirement for a membrane-bound hydrogenase to reconstitute H_2-evolving activity. The K_m for CO is 32 µM (28), which is in the K_m range of enzymes from carboxydotrophs that also utilize CO as a sole energy source (82). Unlike the enzymes from the aerobic carboxydotrophs, the *R. rubrum* CO dehydrogenase is inactivated by O_2. In this respect, the *R. rubrum* enzyme resembles the Ni-containing O_2-sensitive CO dehydrogenases from chemotrophic anaerobes (see below). On the other hand, antibodies against the *R.*

rubrum enzyme do not cross-react with Ni-containing CO dehydrogenases from chemotrophic anaerobes (11). Furthermore, the *R. rubrum* enzyme cannot cleave or synthesize acetyl-CoA, which is the primary physiological function of the CO dehydrogenases from acetogenic and acetotrophic anaerobes. The *R. rubrum* CO dehydrogenase contains two low-potential Fe_4-S_4 centers and one Ni per monomer (121). The dye-oxidized form elicits an EPR signal (g values = 2.04, 1.90, and 1.71) that broadens when the enzyme contains ^{61}Ni and ^{57}Fe, a result suggesting a mixed Ni-Fe cluster (121). X-ray absorption spectroscopy of the *R. rubrum* enzyme rules out the presence of Ni in the core of a $NiFe_3S_4$ cubane cluster; instead, the data are more consistent with a mononuclear (Ni^{2+}) site bridged by the sulfur of cysteine or sulfide to one or both of the Fe_4S_4 centers (122). The remaining Ni^{2+} coordination sites could potentially be filled by the sulfur of cysteine or N(O) ligands from the polypeptide chain. This model is consistent with observed electronic coupling of the Ni and Fe-S centers (12). A similar Ni-X-[Fe_4S_4] structure is proposed for the CO dehydrogenase (acetyl-CoA synthesizing) from *Clostridium thermoaceticum* (94); however, this does not appear to be the site for CO oxidation by this enzyme. Direct electrochemical experiments indicate that the reduction potential (−418 mV) of the CO dehydrogenase from *R. rubrum* is pH independent and one electron is transferred per redox center (119).

Several lines of evidence suggest a role for Ni in the CO oxidation site of the *R. rubrum* CO dehydrogenase. Cyanide (which binds to transition metals) and dimethyl gloxime (a Ni-specific chelator) inactivates the enzyme; this observation first suggested a role for Ni in catalysis (13). Kinetic studies indicate that cyanide is a slow-binding inhibitor (28). The cyanide inhibition is reversed with CO, which suggests that cyanide and CO bind at the same site. Although CO accelerates the dissociation of cyanide, CO can apparently still bind to an enzyme that contains cyanide at the active site (44). One interpretation of these results is that the CO dehydrogenase has two nonequivalent CO-binding sites. An inactive Ni-deficient form of the CO dehydrogenase can be purified from cells grown photosynthetically in the absence of the metal (12). Activity is restored to the apo-CO dehydrogenase upon in vitro incubation with $NiCl_2$; Ni is incorporated into the reactivated enzyme, which further supports a role for Ni in catalysis. No additional proteins or soluble components are required for Ni insertion. Activation of the Ni-reconstituted enzyme requires a low-potential one-electron reductant and is correlated with reduction of the Fe_4S_4 centers (27). This result is consistent with involvement of reduced Fe-S clusters in forming the mixed Ni-Fe center. The midpoint potential for activation is approximately −475 mV. As expected, the EPR signal attributable to the novel mixed Ni-Fe center is absent in the apo-CO dehydrogenase and the signal is restored upon reconstitution of the apoenzyme with $NiCl_2$ (121). The Fe-S centers of the Ni-deficient CO dehydrogenase remain intact but are no longer reduced by CO, a result that suggests an oxidation of CO at the Ni site and transfer of electrons to the Fe-S centers (26). That cyanide-treated Ni-deficient apoenzyme can still be reconstituted and activated with Ni suggests that the inhibitor binds to Ni and not to the Fe-S centers (28). All of the above results are compatible with a role for Ni in the active site; however, definitive proof that CO binds to Ni is not yet available. Unfortunately, the EPR-active oxidized enzyme is reduced by CO; thus, the influence of CO on the Ni-Fe EPR signal cannot be evaluated. Nonetheless, the CO analogue cyanide is reported to perturb the Ni-Fe EPR signal (121).

The genes encoding the *R. rubrum* CO dehydrogenase (*cooS*) and the 22-kDa Fe-S protein (*cooF*) have been cloned and sequenced (55). The amino acid sequence deduced from *cooF* contains four 4-cysteine motifs also found in formate dehydrogenase, nitrate reductase, dimethyl sulfoxide reductase, and hydrogenase. Except for a single four-cysteine motif, the amino acid sequence deduced from *cooS* has relatively little sequence identity with the sequences of CO dehydrogenases from the acetogenic anaerobe *C. thermoaceticum* and the acetotrophic CH_4-producing anaerobe *Methanothrix soehngenii*. An open reading frame upstream of *cooF* is thought to encode the CO-induced hydrogenase.

ACETOGENIC ANAEROBES

Microbiology and Physiology

Acetogenic anaerobes reduce $2CO_2$ to acetyl-CoA as a terminal electron-accepting process producing acetate as the major end product. They are present in diverse anaerobic habitats, including marine and freshwater sediments, the rumen, the hindgut of termites, and monogastrics. Nearly 40 described species participate in the anaerobic portion of the global carbon cycle by converting various compounds to acetate (24), the chief intermediate in anaerobic microbial food chains that degrade complex organic matter. Acetate in these food chains is produced primarily from acetyl-CoA generated by the oxidation of substrates. This process also yields H_2 or formate. The reductive synthesis of acetyl-CoA by acetogenic microbes is estimated to contribute as much as 10% to the total production of acetate. The acetate produced in anaerobic food chains is consumed by acetotrophic anaerobes that either oxidize acetate to $2CO_2$ or ferment it to CO_2 and CH_4. Acetogenic anaerobes play a larger role in nonmethanogenic environments, such as the hindgut of termites, where H_2 and formate is used for the reduction of $2CO_2$ to acetate rather than CH_4 (17).

Figure 1 shows a generalized pathway for reductive synthesis of acetyl-CoA (the Wood-Ljungdahl pathway). More complete information on the physiology and enzymology of the pathway is available in recent reviews (70, 96, 97).

Central to the pathway is CO dehydrogenase (acetyl-CoA synthase), which catalyzes the synthesis of acetyl-CoA from a methyl group, CO, and CoA. The assembly of acetyl-CoA on the enzyme proceeds via: (*a*) methylation of a reduced metal site, (*b*) carbonylation of the methyl-metal species, (*c*) methyl migration to form an acetyl-metal intermediate, and (*d*) binding of CoA and thiolytic cleavage of the acetyl-metal bond to form acetyl-CoA. The methyl group is donated to the CO dehydrogenase by transfer from methyl-tetrahydrofolate (CH_3-THF) involving a methylated corrinoid/iron-sulfur (C/Fe-S) protein intermediate. The methyl group of CH_3-THF can derive from the reduction of CO_2, which is a significant route of dark CO_2 fixation in nature. The CO_2 is first reduced to formate; then formyl-THF forms, and a six-electron reduction to CH_3-THF occurs. The electron donors are CO, H_2, HCOOH, or various organic compounds (24). The methyl group of CH_3-THF can also originate from methanol or methoxylated compounds. Carbonylation of the methyl-metal species of CO dehydrogenase is accomplished with CO, CO_2 plus two electrons, or the carbonyl group of pyruvate. The energy of the thioester bond of acetyl-CoA is conserved by the phosphorylation of ADP to ATP in reactions catalyzed by phosphotransacetylase and acetate kinase. The acetyl-CoA can also be used for the synthesis of cell carbon. Several acetotrophic anaerobes can utilize CO as a sole energy source (56, 71, 77, 114). In addition to a precursor of the carbonyl group of acetyl-CoA, CO is oxidized to provide electrons for the reduction of CO_2 to CH_3-THF.

$3CO + 3H_2O \rightarrow 3CO_2 + 6e^- + 6H^+$ 3.

$CO + CO_2 + 6e^- + 6H^+ \rightarrow CH_3CO_2^- + H^+ + H_2O$ 4.

$4CO + 2H_2O \rightarrow CH_3CO_2^- + H^+ + 2CO_2$ ($\Delta = -41.4$ kJ/mol CO)   5.

Clostridium thermoaceticum is the most extensively studied acetogenic microbe. In addition to growth on H_2/CO_2 or CO (56), it can grow heterotrophically, producing three moles of acetate from one mole of glucose (70). The glucose is oxidized, yielding four electrons and two pyruvates, which are converted to acetyl-CoA by pyruvate:ferredoxin oxidoreductase. One pyruvate is converted to acetyl-CoA and CO_2 by an oxidative mechanism yielding two electrons. The other pyruvate is converted to acetyl-CoA without oxidation of the carbonyl group, which is donated directly to the CO dehydrogenase. The six electrons generated by oxidation of glucose to acetyl-CoA are used to reduce CO_2 to CH_3-THF. The CO dehydrogenase synthesizes a third acetyl-CoA from CoA, CH_3-THF, and the carbonyl group of pyruvate. The carboxyl of pyruvate does not equilibrate with free CO_2 prior to insertion into the carbonyl of acetyl-CoA, which suggests a mechanism for direct transfer of the carbonyl from pyruvate to the CO dehydrogenase (90). Assuming only one oxidoreductase, this result implies that the enzyme is capable of oxidizing the carbonyl of one pyruvate and directly transferring the carbonyl of the other

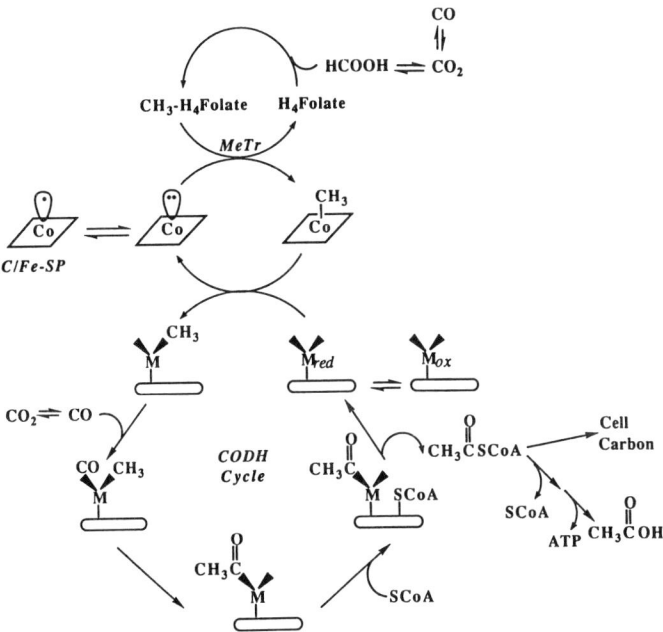

Figure 1 The reductive acetyl-CoA pathway of acetogenesis. The pathway is shown for the reduction of $2CO_2$ to acetate with an emphasis on methyl-transfer reactions and the role of CO dehydrogenase (CODH cycle). See text for variations and additional aspects of the pathway. H_4Folate, tetrahydrofolate; MeTr, methyltransferase; C/Fe-S, corrinoid iron-sulfur protein; M, metal site in center A of the CO dehydrogenase; CoA, coenzyme A. (Reprinted with permission from Reference 35.)

pyruvate to CO dehydrogenase. Thus, the oxidoreductase must have a way of distinguishing between the two activities. A plausible mechanism would be interaction of the oxidoreductase and CO dehydrogenase to channel a direct carbonyl transfer.

The CO Dehydrogenase from Clostridium thermoaceticum

PROPERTIES The CO dehydrogenase (acetyl-CoA synthesizing) from *C. thermoaceticum* is a dimer of $(\alpha\beta)_3$ configuration with α and β subunit molecular masses of 77 and 71 kDa (98). The genes encoding the two subunits are clustered on a 10-kilobase fragment encoding other proteins involved in acetyl-CoA synthesis (105). Each $\alpha\beta$ dimer contains 2 Ni, 12 Fe, 14 acid-labile S, and 1 Zn. The enzyme is oxygen sensitive, as are all other Ni-containing CO dehydrogenases. Ferredoxin is the electron acceptor that forms an electrostatically stabilized complex with the CO dehydrogenase (113). Chemical cross-linking followed by cyanogen bromide cleavage and high-performance liquid chromatography (HPLC) analysis of the resulting polypeptides reveals that ferredoxin binds to residues 229–239 of the 77-kDa subunit.

Characterization of the redox-active metal centers has relied heavily on spectroscopic methods. Two EPR signals attributable to two different Fe-S centers, called center B and center C, have g_{av} values of 1.94 and 1.82 (68, 117). Centers B and C have midpoint potentials of −440 and −220 mV. Fe-S centers with similar EPR properties and midpoint potentials are present in all Ni-containing CO dehydrogenases. The EPR properties of center C correspond to the mixed Ni-Fe center described for the $R.$ $rubrum$ CO dehydrogenase (121). Recent studies showed that Ni is present in center C of the $C.$ $thermoaceticum$ enzyme (D Qiu, M Kumar, SW Ragsdale & TG Spiro, personal communication); however, the structure is unknown. CO oxidation may occur at center C (5, 6, 115, 116), and CO incorporation into acetyl-CoA could occur at a distinct Ni-Fe-C complex called center A. The g_{av} 1.94 EPR signal from center B is clearly assignable to a Fe_4-S_4 cluster (69). A second Fe_4-S_4 center is detected when the redox potential is stabilized below −500 mV. This center has g values similar to those of center B but a midpoint potential of −500 mV. Apparently, the Fe_4-S_4 cluster of center B is converted to another form at low redox potentials. The Fe_4-S_4 cluster is thought to accept electrons from the CO-oxidizing center C (4, 64).

In addition to the low temperature Fe-S signals, a third EPR signal is detected at liquid N_2 temperatures for the CO-reduced CO dehydrogenase from $C.$ $thermoaceticum$ (101). This EPR signal (g_{av} = 2.06; g values of 2.08, 2.07, and 2.03) is attributable to a reduced Ni-Fe center (center A) that binds CO in a spin-coupled complex of Ni, Fe, and CO (the Ni-Fe-C complex). This signal is also observed for the Ni-containing CO dehydrogenase from methanogenic microbes that synthesize or cleave acetyl-CoA, but not for the Ni-containing CO dehydrogenase from $R.$ $rubrum,$ which only oxidizes CO. EPR and electron-nuclear double resonance (ENDOR) spectroscopy have led to the proposal that center A in the CO dehydrogenase from $C.$ $thermoaceticum$ contains three to four Fe atoms (29a). Mössbauer spectra suggest a structure similar to Fe_4S_4 centers (69). These spectroscopic results are the basis for several proposed structures; the most likely is Ni-X-[Fe_4-S_4], in which X is an unknown bridging atom between Ni and a Fe_4-S_4 center (94). X-ray absorption spectroscopy indicates that sulfur atoms are coordinated to Ni (7, 20). Fourier transform infrared spectroscopy of the CO-bound CO dehydrogenase reveals a CO stretching vibration that, when compared to known metal-carbonyl compounds, clearly indicates CO is terminally bound to a metal (65). Because CO is terminally bound, it cannot be the bridging atom X. Resonance Raman spectroscopy (94) has shown that the CO is bound to Fe (Figure 2). CO oxidation

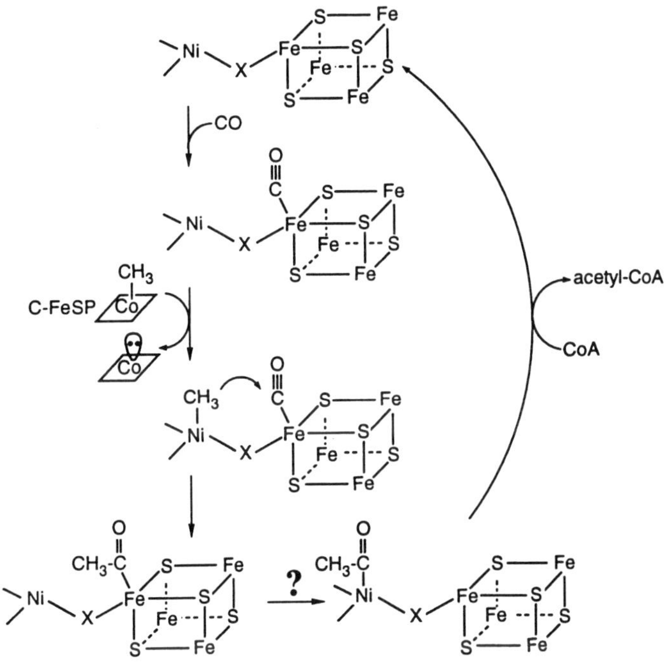

Figure 2 Proposed mechanism for acetyl-CoA synthesis at center A of the CO dehydrogenase from *Clostridium thermoaceticum*. X, unidentified bridging ligand; C-FeSP, corrinoid iron-sulfur protein. (Reprinted with permission from Reference 94. Copyright © 1994, American Association for the Advancement of Science.)

does not occur at center A; rather it is the site proposed for synthesis of acetyl-CoA.

MECHANISM OF ACETYL-CoA SYNTHESIS   Although the CO dehydrogenases from acetogenic anaerobes oxidize CO, the most important physiological function is synthesis of acetyl-CoA. The first evidence suggesting acetyl-CoA synthesis activity was obtained when partially purified (41), and then purified (100, 102), preparations of CO dehydrogenase were shown to catalyze the exchange of CO with the carbonyl of acetyl-CoA:

$$CO + CH_3\text{-}^{14}CO\text{-}SCoA \leftrightarrow {}^{14}CO + CH_3\text{-}CO\text{-}SCoA. \qquad 6.$$

In the reaction, the C-C and C-S bonds of acetyl-CoA must be cleaved and then reformed in order to exchange CO with the carbonyl group of acetyl-CoA, thereby demonstrating the acetyl-CoA synthesis activity. The reaction mixture contains no acceptors for the methyl, carbonyl, or CoA groups; thus, the CO dehydrogenase must bind each group after cleavage and, following equilibration of the bound carbonyl with CO, reassemble the groups to synthesize acetyl-CoA (101). A corrinoid protein may also be required to donate the methyl group to the CO dehydrogenase. Acetyl-CoA synthesis from CH_3-THF, CO, and CoA is obtained in a system reconstituted with purified CO dehydrogenase, ferredoxin, a C/Fe-S protein, and a methyltransferase that catalyzes transfer of the methyl group of CH_3-THF to the C/Fe-S protein (106). The rate of acetyl-CoA synthesis is similar to the in vivo rate of acetate synthesis, which supports the involvement of these enzymes in the pathway. Although not absolutely required for acetyl-CoA synthesis activity, a four-fold reduction occurs when ferredoxin is omitted. Ferredoxin also stimulates the CO/acetyl-CoA exchange reaction (100).

The C/Fe-S protein, which transfers a methyl group directly to the CO dehydrogenase, is an $\alpha\beta$ dimer with subunit molecular weights of 34,000 and 55,000 (42). It contains cobalt in a corrinoid cofactor (5-methoxybenzimidazolylcobamide) and a single Fe_4-S_4 center, both of which are oxidized and reduced during catalysis (99). Recent cloning of the genes, and independent production of the subunits in *E. coli*, reveals that the Fe_4-S_4 center is located in the large subunit and the corrinoid in the smaller subunit (75). The cobalt atom of the corrinoid is methylated with CH_3-THF by the methyltransferase. The methylation requires cobalt in the Co^{1+} redox state, and Co^{1+} functions as a supernucleophile attacking the methyl group of CH_3-THF. Reduction of Co^{2+} to Co^{1+} occurs via an electron transfer from the CO dehydrogenase. The benzimidazole base is not a lower axial ligand to the cobalt atom of the corrinoid (base-off configuration), which renders the midpoint potential of the $Co^{2+/1+}$ couple less negative and puts it within range of physiological electron donors such as CO (40). Transfer of electrons between the CO dehydrogenase and the C/Fe-S protein does not require ferredoxin. The Fe_4-S_4 center of the C/Fe-S protein is probably the initial acceptor of electrons from reduced CO dehydrogenase, and the reduced Fe_4-S_4 center likely donates electrons to Co^{2+}. Methyl transfer from the methylated C/Fe-S protein to the CO dehydrogenase occurs by a nucleophilic attack on the methyl-corrinoid by the CO dehydrogenase. Earlier studies had suggested that methylation occurred on the thiol of a cysteine residue in the smaller subunit of the CO dehydrogenase (91). However, controlled potential studies of the methyl-transfer reaction argues against S-methyl-cysteine and strongly supports a low-potential nucleophilic metal as the initial acceptor of the methyl group on the CO dehydrogenase (73).

Formation of the methylated CO dehydrogenase requires a prior reduction of the metal center. The most likely candidate for this metal site is center A because it appears to also bind the CO in the Ni-Fe-C complex that becomes the precursor to the carbonyl group of acetyl-CoA. The Ni in center A is the most probable site for methylation of the CO dehydrogenase; however, no spectroscopic data support this view. A direct interaction of the CO dehydrogenase and the C/Fe-S protein is required for electron and methyl-group transfer between these proteins. A recently isolated complex containing the CO dehydrogenase and the C/Fe-S protein has enhanced acetyl-CoA synthesis activity over that of the independently isolated and reconstituted protein components (112). The CO dehydrogenase:C/Fe-S complex is apparently held together by a interdisulfide bond involving a cysteine residue in the α subunit of CO dehydrogenase. Interestingly, the CO dehydrogenase and C/Fe-S protein are isolated in a complex from the acetotrophic CH_4-producing anaerobe *Methanosarcina thermophila* (125).

The current body of evidence supports the view that the Ni-Fe-C complex supplies the precursor of the carbonyl group of acetyl-CoA (35). First, when CO dehydrogenase is incubated with [1-^{13}C] acetyl-CoA and CO, the ^{13}C label is completely exchanged into the complex as evidenced by broadening of the Ni-Fe-^{13}C EPR signal owing to magnetic interaction of the ^{13}C nucleus with the unpaired electron. Second, when the CO dehydrogenase is incubated with ^{13}CO and acetyl-CoA, the terminally bound infrared-detectable CO exchanges with the unlabeled carbonyl carbon of acetyl-CoA (65). Finally, the rate of EPR-detectable Ni-Fe-C formation in the presence of CO is severalfold greater than the CO/acetyl-CoA exchange, which demonstrates the kinetic competence of the Ni-Fe-C complex for supplying the carbonyl group of acetyl-CoA. The EPR-detectable Ni-Fe-C complex is not formed upon incubation of the enzyme with acetyl-CoA in the absence of CO unless the enzyme is electrochemically reduced. Formation of the Ni-Fe-C complex requires a one-electron reduction at a midpoint potential of -541 mV (35). Thus, CO apparently binds only to the reduced state of center A to yield the Ni-Fe-C complex. In the cell, electrons for this reduction derive from the oxidation of CO at center C, which transfers electrons to the Fe_4-S_4 cluster in center B and then to an Fe atom in center A (5, 64). Carbonylation of the CO dehydrogenase with CO does not require prior methylation (100), and methylation does not require a prior carbonylation of the enzyme (73), a result which suggests a random sequence for binding the methyl group and CO to center A.

The above collective evidence has led to the proposal of a bimetallic mechanism (Figure 2) for acetyl-CoA synthesis involving a Ni-bound methyl group and a Fe-bound carbonyl group in center A (94). In this mechanism, the methyl group attacks the carbonyl group, forming a Fe-acetyl adduct that could potentially migrate to form a Ni-acetyl species. However, spectroscopic evidence for metal-acetyl intermediates has been elusive. After CoA binds to the enzyme, the catalytic cycle concludes with a thiolysis of the metal-acetyl intermediate that produces acetyl-CoA. The EPR spectrum of the Ni-Fe-C complex is perturbed in the presence of CoA, indicating that the cofactor binds near the complex (101). Arginine and tryptophan residues may reside at or near the CoA binding site (110, 111), probably in the α subunit (86). CoA, but not desulfo-CoA, is a potent inhibitor of the CO/acetyl-CoA exchange activity, which is consistent with binding of CoA to CO dehydrogenase through the sulfur atom of the cofactor (103).

ACETOTROPHIC ANAEROBES

Microbiology and Physiology

METHANE PRODUCERS The anaerobic decomposition of organic matter is a significant component of the global carbon cycle. The process occurs in nearly every conceivable anaerobic environment, including the rumen, the lower intestinal tract of humans, the hindgut of termites, sewage digestors, landfills, rice paddies, and marine and freshwater sediments. At least three interacting metabolic groups of microbes form a food chain in which acetate is the principle intermediate. Fermentative and acetogenic anaerobes degrade complex organic matter to H_2, CO_2, formate, acetate, and higher volatile carboxylic acids. Obligate proton-reducing anaerobes oxidize the higher acids to acetate and either H_2 or formate. In anaerobic freshwater environments, the CH_4-producing *Archaea* are the final link in microbial food chains. About two thirds of the CH_4 produced in nature originate from the methyl group of acetate and about one third from the reduction of CO_2 with electrons derived from the oxidation of H_2 or formate (reviewed in 30). AOR (acetate oxidizing rod), a nonmethanogenic microbe, oxidizes acetate to H_2 and $2CO_2$ (66). The extent to which acetate oxidizers such as AOR occur in anaerobic environments is unknown. In marine environments, sulfate-reducing bacteria out-compete methanogenic microbes for H_2, formate, and acetate.

Although most of the CH_4 produced in nature derives from acetate, only two genera (*Methanosarcina* and *Methanothrix*) and a few species are known to ferment acetate to CO_2 and CH_4. A CO dehydrogenase enzyme complex is central to the pathway of acetate fermentation to CH_4 in the methanosarcina (Figure 3). The Ni/Fe-S component of the complex (CO dehydrogenase) cleaves acetyl-CoA by reversal of the mechanism of the clostridial CO dehydrogenase. The methyl group is transferred to a C/Fe-S protein in the complex and finally to coenzyme M (HS-CoM). This process involves tetrahydrosarcinapterin (H_4SPT) as a methyl-transfer cofactor. In the final step, the CH_3-S-CoM is reductively demethylated to CH_4 with the electron donor 7-mercaptoheptanoylthreonine phosphate (CoB) (89a). The disulfide CoM-S-S-CoB, formed as a consequence of the reaction, is reduced to the active sulfhydryl forms of the cofactors with electrons derived from oxidation of the carbonyl group of acetate to CO_2. No exogenous electron acceptors are required; thus,

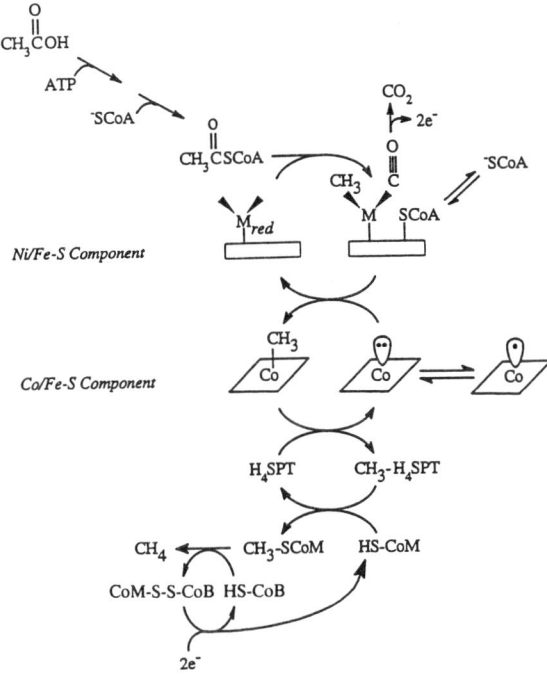

Figure 3 The pathway for the fermentation of acetate in the methanosarcina. The pathway emphasizes methyl transfer steps and the role of the CO dehydrogenase complex. See text for additional aspects of the pathway. CoA, coenzyme A; Ni/FE-S, the CO dehydrogenase component of the complex; M, a metal site in center A of the CO dehydrogenase; Co/Fe-S, the corrinoid iron-sulfur component of the complex; CoM, coenzyme M; CoB, 7-mercaptoheptanoylthreonine phosphate. (Reprinted with permission from Reference 46.)

the formation of CH_4 and CO_2 from acetate is a fermentation. ATP is generated by an electron-transport phosphorylation. A similar pathway operates in *M. soehngenii* (52).

Methanosarcina barkeri can grow with pyruvate, in addition to acetate, as the sole source of carbon and energy (9). Pyruvate is first oxidized to acetyl-CoA, CO_2, and an electron pair. The electrons are consumed in the CO_2-reducing pathway, yielding $1/4CH_4$, while the acetyl-CoA is converted to CH_4 and CO_2 in the acetate fermentation pathway involving CO dehydrogenase. A more in-depth treatment of the physiology and biochemistry of the pathway for the methanogenic fermentation of acetate is available in a recent review (31).

CO dehydrogenase also functions in the pathway of CO_2 fixation for cell carbon in autotrophic methanogenic microbes in which the enzyme catalyzes the reductive synthesis of acetyl-CoA. The methyl group derives from reduction of CO_2 by the same pathway utilized for reduction of CO_2 to CH_4 (reviewed in 30). The carbonyl group is supplied directly from CO or by the two-electron reduction of CO_2 by CO dehydrogenase. The reader should consult a comprehensive review of acetyl-CoA synthesis in autotrophic CH_4 producers (118).

Some species of methanogenic anaerobes can utilize CO as a sole energy source (22); however, growth is slow (89) and the extent to which CO serves as a growth substrate in nature is unknown. Growth on CO implies the involvement of CO dehydrogenase in the oxidation of CO. Apparently, cells convert CO to CO_2 and H_2, which is utilized as a reductant in the energy-yielding CO_2 reduction pathway (89):

$4CO + 4H_2O \rightarrow 4CO_2 + 4H_2$ ($\Delta G^{o\prime} = -20$ kJ/mol H_2)   7.
$CO_2 + 4H_2 \rightarrow CH_4 + 2H_2O$ ($\Delta G^{o\prime} = -131$ kJ/mol CH_4)   8.
$4CO + 2H_2O \rightarrow 3CO_2 + CH_4$ $\Delta G^{o\prime} = -211$ kJ/mol CH_4)   9.

Additional energy for growth could be obtained in the conversion of CO to H_2 (Reaction 7); indeed, proton translocation is coupled to the oxidation of CO to CO_2 and H_2 in acetate-grown *M. barkeri* (14, 15).

SULFATE REDUCERS   Dissimilatory sulfate-reducers out-compete CH_4-producers for acetate in sulfate-rich anaerobic environments. Sulfate reducers oxidize acetate via two pathways, the citric acid cycle and the CO dehydrogenase pathway; additional details are found elsewhere (126, 127). Most acetotrophic sulfate reducers utilize the CO dehydrogenase pathway, in which the enzyme cleaves acetyl-CoA by reversal of the mechanism of the clostridial CO dehydrogenase. The methyl group is transferred to tetrahydrofolate (THF) and then oxidized to CO_2, yielding six electrons. The carbonyl group is oxidized to CO_2 and two electrons. Net ATP synthesis is accomplished via phosphorylation of ADP along with the reduction of sulfate to sulfide; the eight electrons are derived from oxidation of acetate to two CO_2. The fixation of CO_2 into the cell carbon of autotrophic sulfate reducers occurs through a reversal of the CO dehydrogenase pathway (48, 108).

Some sulfate reducers utilize the CO dehydrogenase pathway to oxidize substrates more complex than acetate, including aromatic compounds and saturated hydrocarbons (3, 95). The compounds are first converted to acetyl-CoA, which is completely oxidized to $2CO_2$. Another interesting example is *Archaeoglobus fulgidus,* a sulfate reducer from the *Archaea* domain. This anaerobe does not utilize acetate but oxidizes pyruvate to acetyl-CoA, which is further oxidized to $2CO_2$ by the CO dehydrogenase pathway (85). After cleavage of acetyl-CoA, the methyl group is oxidized by a reversal of the same pathway utilized by CO_2-reducing CH_4 producers (109).

Desulfovibrio vulgaris can utilize CO as a sole source of carbon and energy (76). The CO is first converted to CO_2 and H_2, as is the case for phototrophic and methanogenic anaerobes. Growth is poor, which suggests that the use of CO as a substrate in nature is unlikely. Presumably, CO oxidation primarily functions to recover the reducing potential in CO produced from the carboxyl group of pyruvate during conversion to acetyl-CoA. Although the first demonstration of a CO dehydrogenase was obtained for the enzyme from *Desulfovibrio desulfuricans* (132), comprehensive biochemical characterizations of CO dehydrogenases from sulfate-reducing microbes have not been reported.

OTHER ACETOTROPHIC ANAEROBES   AOR uses protons as an electron acceptor for complete oxidation of acetate to $2CO_2$ and $4H_2$ by the CO dehydrogenase pathway (67). This organism can also synthesize acetate from H_2 and CO_2 in a reversal of the pathway (66). An anaerobe called MPOB dismutates fumarate by oxidation to $4CO_2$ and reduction to succinate (92). The CO dehydrogenase pathway is used for complete oxidation of acetyl-CoA, which is derived from the oxidation of fumarate. Anaerobes utilizing NO_3^-, Fe^{3+}, or Mn^{4+} as electron acceptors for the complete oxidation of acetyl-CoA employ the citric acid cycle, a pathway not involving CO dehydrogenase.

The CO Dehydrogenases from Methanosarcina sp. and Methanothrix soehngenii

PROPERTIES   The most extensively characterized CO dehydrogenases from acetotrophs are those from CH_4 producers. The acetyl-CoA synthesis activity of CO dehydrogenases from acetogenic anaerobes provided the basis for the hypothesis that CO dehydrogenases, abundant in all acetotrophic CH_4 producers, catalyze the reverse reaction (cleavage of acetyl-CoA) in the pathway of acetate fermentation to CH_4 and CO_2. Initial evidence for the involvement of CO dehydrogenase was obtained when antibodies to the enzyme inhibited CH_4 production from acetate in cell extracts of *M. barkeri* (61). The first direct biochemical evidence that supported the hypothesis of an acetyl-CoA cleavage function was obtained by demonstrating an exchange of the carbonyl group of acetyl-CoA with CO catalyzed by purified CO dehydrogenases (49, 104).

Acetate-grown *M. thermophila* contains an enzyme complex that has CO dehydrogenase activity and catalyzes cleavage of the C-C and C-S bonds of acetyl-CoA (104, 125). The complex also synthesizes acetyl-CoA from methyl iodide, CO, and CoA (1). Although the complex also oxidizes CO, the primary function is to cleave acetyl-CoA and oxidize the carbonyl group to CO_2 during growth on acetate. Two enzyme components can be resolved from the five-subunit complex: a Ni/Fe-S enzyme containing 89-kDa and 19-kDa subunits and a C/Fe-S enzyme containing 60-kDa and 58-kDa subunits (2). A fifth

71-kDa subunit has not been characterized. Transcription of the genes encoding subunits of the CO dehydrogenase complex from *M. thermophila* is regulated in response to the growth substrate (120). The Ni/Fe-S enzyme oxidizes CO and reduces ferredoxin (124); thus, it is also referred to as CO dehydrogenase. The CO dehydrogenase from acetate-grown *M. barkeri* was initially isolated as an $\alpha_2\beta_2$ oligomer with subunit molecular masses of approximately 90 and 19 kDa (38, 62). The *M. barkeri* $\alpha_2\beta_2$ CO dehydrogenase was also purified in an enzyme complex that contains a corrinoid protein (36). The complex catalyzes cleavage of acetyl-CoA and transfer of the methyl group to H_4SPT (37). The Ni- and Fe-containing CO dehydrogenase from *M. soehngenii* is an $\alpha_2\beta_2$ oligomer composed of subunits with molecular masses of 79 and 19 kDa (51). It catalyzes the exchange of CO with the carbonyl group of acetyl-CoA, demonstrating C-C and C-S cleavage activity (49). In contrast to the CO dehydrogenases from the methanosarcina, the *M. soehngenii* enzyme is not purified in association with a corrinoid-containing protein. The CO dehydrogenases from each of the methanogenic microbes have EPR properties remarkably similar to the CO dehydrogenase from *C. thermoaceticum*.

The CO dehydrogenase (Ni/Fe-S enzyme) from *M. thermophila* has an EPR-detectable spin-coupled Ni-Fe-C complex with a spectrum nearly identical to the Ni-Fe-C complex of the CO-reduced CO dehydrogenase from *C. thermoaceticum* (74, 123). Thus, the *M. thermophila* CO dehydrogenase has a center A analogous to that in the clostridial enzyme, and this center is the proposed site for cleavage of acetyl-CoA. The intensity of the EPR signal increases upon incubation of the CO-reduced *M. thermophila* enzyme with acetyl-CoA, consistent with the expected consequences of cleavage of acetyl-CoA at center A (123). The Ni-Fe-C EPR signal forms at a rate nearly 100-fold slower than the rate of CO oxidation, which suggests that center A is not involved in the latter activity. These results are consistent with those from previous studies in which the concentration of cyanide required to inhibit acetyl-CoA synthesis was 50-fold greater than that required for the inhibition of CO oxidation (1). Although the CO dehydrogenases from *M. barkeri* and *M. soehngenii* contain Ni and Fe, no EPR signals attributable to Ni have been reported for these enzymes.

The CO dehydrogenase from *M. thermophila* elicits two low-temperature EPR signals (123) similar to the CO dehydrogenase from *C. thermoaceticum* (97). One of the signals from the *M. thermophila* CO dehydrogenase has g values of 2.04, 1.93, and 1.89 ($g_{av} = 1.95$), values that strongly suggest a Fe_4-S_4 center. This center, called center B, has a midpoint potential of -444 mV. At redox potentials less than -500, the EPR signal is replaced by another Fe_4-S_4 EPR signal with g values of 2.05, 1.95, and 1.90 ($g_{av} = 1.97$). The signals of the $g_{av} = 1.95$ and 1.97 species probably originate from a single Fe_4-S_4 cluster that can exist in two states. Another low-temperature EPR signal with g values of 2.02, 1.87, and 1.72 (g_{av} = 1.87) originates from a Fe-S center, called center C, of unknown structure. Center C has a midpoint potential (−154 mV) and EPR properties similar to the Ni-containing Fe-S center C of the *C. thermoaceticum* CO dehydrogenase, which functions in oxidation of CO.

The functions of centers B and C in the *M. thermophila* CO dehydrogenase are unknown. Core extrusion experiments have revealed six Fe_4-S_4 clusters per $\alpha_2\beta_2$ tetramer in the *M. barkeri* CO dehydrogenase (62). Low-temperature EPR spectroscopy of the *M. barkeri* enzyme identifies a Fe_4-S_4 cluster with g values of 2.05, 1.94, and 1.90 (g_{av} 1.96) and a Fe-S center of undetermined composition with g values of 2.005, 1.91, and 1.76 (g_{av} = 1.89), similar to the g values of the CO dehydrogenases from *M. thermophila* and *C. thermoaceticum*. The g_{av} = 1.89 signal is perturbed upon incubation with CO and is observed in whole cells of *M. barkeri* during methanogenesis, suggesting that the cleavage of acetate yields a moiety that CO dehydrogenase recognizes as CO (63).

The genes encoding the subunits of the *M. soehngenii* CO dehydrogenase have been cloned and sequenced (25). The deduced amino acid sequence of the larger α subunit contains eight cysteine residues with spacings that could accommodate Fe-S centers; accordingly, the anaerobically purified (reduced) enzyme exhibits two low-temperature EPR signals (49, 50), one of which has g values of 2.05, 1.93, and 1.87 (g_{av} = 1.95) attributable to a Fe_4-S_4 center. The other low-temperature signal has g values of 2.005, 1.89, and 1.73 (g_{av} = 1.87). Thus, all of the CO dehydrogenases examined from acetotrophic CH_4 producers contain a Fe-S center (g_{av} = 1.87) with EPR properties similar to the Ni-containing Fe-S centers that are the proposed CO oxidation sites in CO dehydrogenases isolated from phototrophic and acetogenic anaerobes. It is not known if this Fe-S center in the CO dehydrogenases from methanogenic microbes contains Ni or if it is the site of CO oxidation. The g_{av} = 1.87 signal from the *M. soehngenii* CO dehydrogenase may elicit from a Fe_6-S_6 prismane-like center; if so, this structure is probably ligated to a four-cysteine motif in the α subunit that is conserved in the CO dehydrogenases from *C. thermoaceticum* and *R. rubrum* (25).

CO dehydrogenases have also been purified and characterized from autotrophic CO_2-reducing methanogenic microbes that synthesize acetyl-CoA from $2CO_2$ by the CO dehydrogenase pathway. Evidence for the presence of Ni in CO dehydrogenases from methanogenic microbes was first obtained with the nonacetotrophic CO_2-reducer *Methanobrevibacter arboriphilicus* (39). The 21-fold purified enzyme from ^{63}Ni-labeled cells comigrated with ^{63}Ni during gel filtration. The CO dehydrogenase purified from *Methanococcus vannielii* (23) is similar to the CO dehydrogenases from the acetotrophs; it has subunits of 89 and 21 kDa in $\alpha_2\beta_2$ configuration and contains 2 Ni; however, no Ni

EPR signal could be detected, and the enzyme was incompetent in the exchange of CO with the carbonyl carbon of acetyl-CoA. None of the CO dehydrogenases from CO_2-reducing species have been purified in a complex with corrinoid proteins.

MECHANISM OF ACETYL-CoA CLEAVAGE   The C/Fe-S component isolated from *M. thermophila* contains factor III {Coα-[α-(5-hydroxybenzimidazolyl)]-cobamide} with the cobalt atom in the Co^{2+} redox state. The cobalt of factor III is reduced to the Co^{1+} state with electrons donated directly by the CO dehydrogenase (2). The Co^{1+} supernucleophile displaces the methyl group bound to the CO dehydrogenase after acetyl-CoA cleavage. EPR spectroscopy of the C/Fe-S protein (46) indicates that factor III is in the base-off configuration, which allows reduction to the methyl-accepting Co^{1+} redox state via a change in the midpoint potential of the $Co^{2+/1+}$ couple to a value less negative than the base-on value. Indeed, redox titration of the $Co^{2+/1+}$ couple reveals a midpoint potential of −486 mV for the C/Fe-S protein from *M. thermophila* (46) and −426 mV for the *M. barkeri* CO dehydrogenase–corrinoid enzyme complex (37). The C/Fe-S enzyme from *M. thermophila* contains a Fe_4-S_4 center with a midpoint potential of −502 mV, which is close to the midpoint potential of the $Co^{2+/1+}$ couple; thus, the Fe_4-S_4 center presumably accepts electrons from the CO dehydrogenase and donates electrons to the cobalt atom of factor III (46).

The genes encoding the two subunits of the C/Fe-S enzyme from *M. thermophila* have been cloned and sequenced (78). The deduced amino acid sequence of the 60-kDa subunit shares strong identity with the sequence deduced from the gene encoding the 55-kDa β subunit of the C/Fe-S protein from *C. thermoaceticum* (75). The greatest identity occurs in the N-terminus, where the sequence CysXXCysXXXXCysX$_{16}$CysPro is perfectly conserved, suggesting an involvement of this sequence in chelation of Fe_4-S_4 centers. The two subunits of the C/Fe-S protein from *M. thermophila* have been independently produced in *E. coli* (78). The larger 60-kDa subunit binds corrinoid in the base-off configuration and contains a Fe-S center. The bound corrinoid is methylated with CH_3-H_4MPT, suggesting that this subunit interacts directly with H_4MPT (JA Maupin-Furlow & JG Ferry, unpublished results).

The properties of the CO dehydrogenase from *M. thermophila* are consistent with a proposed acetyl-CoA cleavage mechanism (Figure 3) analogous to a reversal of that proposed for acetyl-CoA synthesis in *C. thermoaceticum*. In the proposed mechanism, the CO dehydrogenase cleaves the C-C and C-S bonds of acetyl-CoA at center A, which binds the methyl and carbonyl groups at unknown reduced metal sites (M). The methyl group is transferred to the Co^{1+} atom of the C/Fe-S protein and then to H_4MPT. The *M. thermophila* CO dehydrogenase catalyzes an exchange of CoA with acetyl-CoA (104). The rate of exchange is fivefold greater than that of the *C. thermoaceticum* CO dehydrogenase. This increase could reflect the fact that whereas the *M. thermophila* enzyme preferentially cleaves acetyl-CoA, the *C. thermoaceticum* CO dehydrogenase synthesizes it. The CO dehydrogenase from *M. thermophila* also catalyzes an exchange of CO with the carbonyl group of acetyl-CoA (104). The rate of this exchange is considerably less than the *C. thermoaceticum* enzyme, which is consistent with a low rate of exchange of CO with the carbonyl group during methanogenesis from acetate (88).

In addition to participating in C-C and C-S bond-cleavage activity, the CO dehydrogenase from *M. thermophila* may oxidize the bound carbonyl group to CO_2 at a site distinct from center A (74), possibly center C. The electron acceptor for the CO dehydrogenase is a ferredoxin purified from *M. thermophila* (2). The CO dehydrogenase from *M. thermophila* also reduces CO_2 to the carbonyl precursor for synthesis of acetyl-CoA; however, CO is preferentially incorporated into acetyl-CoA without prior oxidation to CO_2 and subsequent reduction to the carbonyl precursor (1). These results are consistent with separate sites for CO oxidation and cleavage of acetyl-CoA.

FUTURE RESEARCH

Although a wealth of spectroscopic information has provided insights into the structure and function of metal centers in CO dehydrogenases, much remains to be learned. For example, direct evidence for the proposed methyl-metal species in the acetyl-CoA synthesis and cleaving enzymes is unavailable. Elucidation of the precise structure of the Ni-Fe center C implicated in CO oxidation will certainly impact an understanding of all Ni-containing CO dehydrogenases. The recent cloning of genes encoding CO dehydrogenases will lead to a deeper understanding of enzyme mechanisms through molecular approaches such as site-directed mutagenesis. The most promising is the CO dehydrogenase from *R. rubrum,* for which a genetic system is currently operable. The assembly of Ni-containing metal clusters has not been investigated. The availability of genetics will surely lead to the discovery of genes involved in metal cluster assembly and regulation of enzyme synthesis. Yet no reliable genetic exchange system is in use for acetotrophic or acetogenic microbes. Although still unreported, production of an active CO dehydrogenase in a heterologous host is the next best approach to study the enzyme mechanism of these CO dehydrogenases.

The apparent ancient origin of CO dehydrogenase raises questions concerning the evolution of this versatile and ubiquitous enzyme. The answers must await the sequence of additional genes encoding CO dehydrogenases and associated proteins from various sources. The extreme phylogenetic difference between the *Archaea* and *Bacteria* offers a unique opportunity for understanding the mechanism and evolution of the functionally similar CO dehydrogenases that synthesize or cleave acetyl-CoA from acetotrophic or acetogenic anaerobes.

CO dehydrogenases and CO dehydrogenase–containing anaerobes hold promise for the bioremediation of toxic waste sites. A strong correlation between the ability of anaerobes to dechlorinate halogenated compounds and the presence of CO dehydrogenase (reviewed in 72) was examined at a biochemical level in *M. thermophila* (45). Although the CO dehydrogenase component of the CO dehydrogenase enzyme complex is inactive, CO and CO dehydrogenase provides the low-potential electrons to the C/Fe-S enzyme component that reductively dechlorinates trichloroethylene; thus, CO-based anaerobic processes for the bioremediation of halogenated compounds is an interesting prospect.

CO is an electron donor for the reductive transformation of 2,4,6,-trinitrotoluene by a sulfate-reducing anaerobe (93). CO also provides electrons for the reductive carboxylation of phenols, which is the first step in the anaerobic degradation of these aromatic compounds (59). The extent to which CO can provide the reducing potential for other degradative and detoxification mechanisms by anaerobic microbes is unknown. CO dehydrogenases have been described from phylogenetically diverse microbes, including species from the *Bacteria* and *Archaea* domains. The enzyme has the assorted functions of CO oxidation, acetyl-CoA synthesis, and acetyl-CoA cleavage. It is present in aerobes and anaerobes with broad metabolic capabilities. This diverse nature of CO dehydrogenase suggests that many more undiscovered microbes utilize the enzyme in novel pathways that could have potential in the biotechnology industry. The recent cloning of CO dehydrogenase genes should provide a source of DNA probes to aid in the isolation and identification of novel CO dehydrogenase-containing microbes.

ACKNOWLEDGMENTS

Research in the author's laboratory was supported by the Department of Energy, Gas Research Institute; National Institutes of Health; National Science Foundation; and Office of Naval Research. Special thanks to Dr. Madeline Rasche for critical reading of the manuscript.

> Any *Annual Review* chapter, as well as any article cited in an *Annual Review* chapter, may be purchased from the Annual Reviews Preprints and Reprints service.
> 1-800-347-8007; 415-259-5017; email: arpr@class.org

Literature Cited

1. Abbanat DR, Ferry JG. 1990. Synthesis of acetyl-CoA by the carbon monoxide dehydrogenase complex from acetate-grown *Methanosarcina thermophila*. *J. Bacteriol.* 172:7145–50
2. Abbanat DR, Ferry JG. 1991. Resolution of component proteins in an enzyme complex from *Methanosarcina thermophila* catalyzing the synthesis or cleavage of acetyl-CoA. *Proc. Natl. Acad. Sci. USA* 88:3272–76
3. Aeckersberg F, Bak F, Widdel F. 1991. Anaerobic oxidation of saturated hydrocarbons to CO_2 by a new type of sulfate-reducing bacterium. *Arch. Microbiol.* 156:5–14
4. Anderson ME, DeRose VJ, Hoffman BM, Lindahl, PA. 1993. Identification of a cyanide binding site in CO dehydrogenase from *Clostridium thermoaceticum* using EPR and ENDOR spectroscopies. *J. Am. Chem. Soc.* 115:12204–5
5. Anderson ME, Lindahl PA. 1994. Organization of clusters and internal electron pathways in CO dehydrogenase from *Clostridium thermoaceticum:* relevance to the mechanism of catalysis and cyanide inhibition. *Biochemistry* 33:8702–11
6. Anderson ME, Lindahl PA. 1994. Organization of clusters and internal electron pathways in CO dehydrogenase from *Clostridium thermoaceticum:* relevance to the mechanism of catalysis and cyanide inhibition. *Biochemistry* 33:8702–11
7. Bastian NR, Diekert G, Neiderhoffer EG, Teo B-K, Walsh CP, Orme-Johnson WH. 1988. Nickel and iron EXAFS of carbon monoxide dehydrogenase from *Clostridium thermoaceticum*. *J. Am. Chem. Soc.* 110:5581–82
8. Black GW, Lyons CM, Williams E, Colby J, Kehoe M, O'Reilly C. 1990. Cloning and expression of the carbon monoxide dehydrogenase genes from *Pseudomonas thermocarboxydovorans* strain C2. *FEMS Microbiol. Lett.* 58:249–54
9. Bock AK, Priegerkraft A, Schonheit P. 1994. Pyruvate—a novel substrate for growth and methane formation in *Methanosarcina barkeri*. *Arch. Microbiol.* 161:33–46
10. Bonam D, Lehman L, Roberts GP, Ludden PW. 1989. Regulation of carbon monoxide dehydrogenase and hydrogenase in *Rhodospirillum rubrum*—effects of CO and oxygen on synthesis and activity. *J. Bacteriol.* 171:3102–7
11. Bonam D, Ludden PW. 1987. Purification and characterization of carbon monoxide dehydrogenase, a nickel, zinc, iron-sulfur protein, from *Rhodospirillum rubrum*. *J. Biol. Chem.* 262:2980–87
12. Bonam D, McKenna MC, Stephens PJ, Ludden PW. 1988. Nickel-deficient carbon monoxide dehydrogenase from *Rhodospirillum rubrum:* in vivo activation by exogenous nickel. *Proc. Natl. Acad. Sci. USA* 85:31–35
13. Bonam D, Murrell SA, Ludden PW. 1984. Carbon monoxide dehydrogenase from *Rhodospirillum rubrum*. *J. Bacteriol.* 159:693–99
14. Bott M, Eikmanns B, Thauer RK. 1986. Coupling of carbon monoxide oxidation to CO_2 and H_2 with the phosphorylation of ADP in acetate-grown *Methanosarcina barkeri*. *Eur. J. Biochem.* 159:393–98
15. Bott M, Thauer RK. 1989. Proton translocation coupled to the oxidation of carbon monoxide to CO_2 and H_2 in *Methanosarcina barkeri*. *Eur. J. Biochem.* 179:469–72
16. Bray RC, George GN, Lange R, Meyer O. 1983. Studies by e.p.r. spectroscopy of carbon monoxide oxidase from *Pseudomonas carboxydovorans* and *Pseudomonas carboxydohydrogena*. *Biochem. J.* 211:687–94
17. Breznak JA, Blum JS. 1991. Mixotrophy in the termite gut acetogen, *Sporomusa termitida*. *Arch. Microbiol.* 156:105–10
18. Champine JE, Uffen RL. 1987. Membrane topography of anaerobic carbon monoxide oxidation in *Rhodocyclus gelatinosus*. *J. Bacteriol.* 169:4784–89
19. Champine JE, Uffen RL. 1987. Regulation of anaerobic carbon monoxide oxidation activity in *Rhodocyclus gelatinosus* (FEM 02895). *FEMS Microbiol. Lett.* 44:307–11
19a. Conrad R, Seiler W. 1980. Role of microorganisms in the consumption and production of atmospheric carbon monoxide by soil. *Appl. Environ. Microbiol.* 40:437–45
20. Cramer SP, Eidsness MK, Pan W-H, Morton TA, Ragsdale SW, et al. 1987. X-ray absorption spectroscopic evidence for a unique nickel site in *Clostridium thermoaceticum* carbon mon-

oxide dehydrogenase. *Inorg. Chem.* 26: 2477–79
21. Cypionka H, Meyer O. 1983. Carbon monoxide-insensitive respiratory chain of *Pseudomonas carboxydovorans*. *J. Bacteriol.* 156:1178–87
22. Daniels L, Fuchs G, Thauer RK, Zeikus JG. 1977. Carbon monoxide oxidation by methanogenic bacteria. *J. Bacteriol.* 132:118–26
23. DeMoll E, Grahame DA, Harnly JM, Tsai L, Stadtman TC. 1987. Purification and properties of carbon monoxide dehydrogenase from *Methanococcus vannielii*. *J. Bacteriol.* 169:3916–20
24. Drake HL. 1994. Acetogenesis, acetogenic bacteria, and the acetyl-CoA "Wood-Ljungdahl" pathway: past and current perspectives. See Ref. 24a, pp. 3–60
24a. Drake HL, ed. 1994. *Acetogenesis*. New York/London: Chapman & Hall. 647 pp.
25. Eggen RIL, Geerling ACM, Jetten MSM, Devos WM. 1991. Cloning, expression, and sequence analysis of the genes for carbon monoxide dehydrogenase of *Methanothrix soehngenii*. *J. Biol. Chem.* 266:6883–87
26. Ensign SA, Bonam D, Ludden PW. 1989. Nickel is required for the transfer of electrons from carbon monoxide to the iron sulfur center(s) of carbon monoxide dehydrogenase from *Rhodospirillum rubrum*. *Biochemistry* 28:4968–73
27. Ensign SA, Campbell MJ, Ludden PW. 1990. Activation of the nickel-deficient carbon monoxide dehydrogenase from *Rhodospirillum rubrum*—kinetic characterization and reductant requirement. *Biochemistry* 29:2162–68
28. Ensign SA, Hyman MR, Ludden PW. 1989. Nickel-specific, slow-binding inhibition of carbon monoxide dehydrogenase from *Rhodospirillum rubrum* by cyanide. *Biochemistry* 8:4973–79
29. Ensign SA, Ludden PW. 1991. Characterization of the CO oxidation/H_2 evolution system of *Rhodospirillum rubrum*—role of a 22-kDa iron-sulfur protein in mediating electron transfer between carbon monoxide dehydrogenase and hydrogenase. *J. Biol. Chem.* 266:18395–403
29a. Fan C, Gorst CM, Ragsdale SW, Hoffman BM. 1991. Characterization of the Ni-Fe-C complex formed by reaction of carbon monoxide with the carbon monoxide dehydrogenase from *Clostridium thermoaceticum* by Q-band ENDOR. *Biochemistry* 30:431–35
30. Ferry JG. 1992. Biochemistry of methanogenesis. *Crit. Rev. Biochem. Mol. Biol.* 27:473–503
31. Ferry JG. 1993. Fermentation of acetate. See Ref. 31a, pp. 304–34
31a. Ferry JG, ed. 1993. *Methanogenesis: Ecology, Physiology, Biochemistry, and Genetics*. New York/London: Chapman & Hall. 536 pp.
32. Futo S, Meyer O. 1986. CO_2 is the first species formed upon CO oxidation by CO dehydrogenase from *Pseudomonas carboxydovorans*. *Arch. Microbiol.* 145:358–60
33. Gladyshev VN, Khangulov SV, Axley MJ, Stadtman TC. 1994. Coordination of selenium to molybdenum in formate dehydrogenase h from *Escherichia coli*. *Proc. Natl. Acad. Sci. USA* 91:7708–11
34. Gladyshev VN, Khangulov SV, Stadtman TC. 1994. Nicotinic acid hydroxylase from *Clostridium barkeri*—electron paramagnetic resonance studies show that selenium is coordinated with molybdenum in the catalytically active selenium-dependent enzyme. *Proc. Natl. Acad. Sci. USA* 91:232–36
35. Gorst CM, Ragsdale SW. 1991. Characterization of the NiFeCO complex of carbon monoxide dehydrogenase as a catalytically competent intermediate in the pathway of acetyl coenzyme A synthesis. *J. Biol. Chem.* 266:20687–93
36. Grahame DA. 1991. Catalysis of acetyl-CoA cleavage and tetrahydrosarcinapterin methylation by a carbon monoxide dehydrogenase-corrinoid enzyme complex. *J. Biol. Chem.* 266:22227–33
37. Grahame DA. 1993. Substrate and cofactor reactivity of a carbon monoxide dehydrogenase corrinoid enzyme complex. Stepwise reduction of iron-sulfur and corrinoid centers, the corrinoid $Co^{2+/1+}$ redox midpoint potential, and overall synthesis of acetyl-CoA. *Biochemistry* 32:10786–93
38. Grahame DA, Stadtman TC. 1987. Carbon monoxide dehydrogenase from *Methanosarcina barkeri*. Disaggregation, purification, and physicochemical properties of the enzyme. *J. Biol. Chem.* 262:3706–12
39. Hammel KE, Cornwell KL, Diekert GB, Thauer RK. 1984. Evidence for a nickel-containing carbon monoxide dehydrogenase in *Methanobrevibacter arboriphilicus*. *J. Bacteriol.* 157:975–78
40. Harder SR, Lu WP, Feinberg BA, Ragsdale SW. 1989. Spectroelectrochemical studies of the corrinoid iron-sulfur protein involved in acetyl coenzyme A synthesis by *Clostridium*

thermoaceticum. Biochemistry 28: 9080–87
41. Hu S-I, Drake HL, Wood HG. 1982. Synthesis of acetyl coenzyme A from carbon monoxide, methyltetrahydrofolate, and coenzyme A by enzymes from *Clostridium thermoaceticum. J. Bacteriol.* 149:440–48
42. Hu S-I, Pezacka E, Wood HG. 1984. Acetate synthesis from carbon monoxide by *Clostridium thermoaceticum.* Purification of the corrinoid protein. *J. Biol. Chem.* 259:8892–97
43. Hugendieck I, Meyer O. 1992. The structural genes encoding CO dehydrogenase subunits (cox L, M and S) in *Pseudomonas carboxydovorans* OM5 reside on plasmid pHCG3 and are, with the exception of *Streptomyces thermoautotrophicus,* conserved in carboxydotrophic bacteria. *Arch. Microbiol.* 157:301–4
44. Hyman MR, Ensign SA, Arp DJ, Ludden PW. 1989. Carbonyl sulfide inhibition of CO dehydrogenase from *Rhodospirillum rubrum. Biochemistry* 28:6821–26
45. Jablonski PE, Ferry JG. 1992. Reductive dechlorination of trichloroethylene by the CO-reduced CO dehydrogenase enzyme complex from *Methanosarcina thermophila. FEMS Microbiol. Lett.* 96:55–60
46. Jablonski PE, Lu WP, Ragsdale SW, Ferry JG. 1993. Characterization of the metal centers of the corrinoid/iron-sulfur component of the CO dehydrogenase enzyme complex from *Methanosarcina thermophila* by EPR spectroscopy and spectroelectrochemistry. *J. Biol. Chem.* 268:325–29
47. Jacobitz S, Meyer O. 1989. Removal of CO dehydrogenase from *Pseudomonas carboxydovorans* cytoplasmic membranes, rebinding of CO dehydrogenase to depleted membranes, and restoration of respiratory activities. *J. Bacteriol.* 171(11):6294–99
48. Jansen K, Fuchs G, Thauer RK. 1985. Autotrophic CO_2 fixation by *Desulfovibrio baarsii:* demonstration of enzyme activities characteristic for the acetyl-CoA pathway. *FEMS Microbiol. Lett.* 28:311–15
49. Jetten MSM, Hagen WR, Pierik AJ, Stams AJM, Zehnder AJB. 1991. Paramagnetic centers and acetyl-coenzyme A/CO exchange activity of carbon monoxide dehydrogenase from *Methanothrix soehngenii. Eur. J. Biochem.* 195:385–91
50. Jetten MSM, Pierik AJ, Hagen WR. 1991. EPR characterization of a high-spin system in carbon monoxide dehydrogenase from *Methanothrix soehngenii. Eur. J. Biochem.* 202:1291–97
51. Jetten MSM, Stams AJM, Zehnder AJB. 1989. Purification and characterization of an oxygen-stable carbon monoxide dehydrogenase of *Methanothrix soehngenii. FEBS Lett.* 181:437–41
52. Jetten MSM, Stams AJM, Zehnder AJB. 1992. Methanogenesis from acetate. A comparison of the acetate metabolism in *Methanothrix soehngenii* and *Methanosarcina* spp. *FEMS Microbiol. Rev.* 88:181–98
53. Johnson JL, Rajagopalan KV, Meyer O. 1990. Isolation and characterization of a second molybdopterin dinucleotide: molybdopterin cytosine dinucleotide. *Arch. Biochem. Biophys.* 283(2): 542–45
54. Kandler O. 1993. The early diversification of life. In *Early Life on Earth. Proc. Nobel Symposium '84,* ed. S Bengtson, pp. 152–60. New York: Columbia Univ. Press
55. Kerby RL, Hong SS, Ensign SA, Coppoc LJ, Ludden PW, Roberts GP. 1992. Genetic and physiological characterization of the *Rhodospirillum rubrum* carbon monoxide dehydrogenase system. *J. Bacteriol.* 174:5284–94
56. Kerby R, Zeikus JG. 1983. Growth of *Clostridium thermoaceticum* on H_2/CO_2 or CO as energy source. *Curr. Microbiol.* 8:27–30
56a. Khalil MAK, Rasmussen RA. 1984. Carbon monoxide in the earth's atmosphere: increasing trend. *Science* 224: 54–56
57. Kim YJ, Kim YM. 1989. Induction of carbon monoxide dehydrogenase during heterotrophic growth of *Acinetobacter* sp. strain JC (DSM 3803) in the presence of carbon monoxide. *FEMS Microbiol. Lett.* 59:207–10
58. Kim YM, Hegeman GD. 1981. Electron transport system of an aerobic carbon monoxide-oxidizing bacterium. *J. Bacteriol.* 148:991–94
59. Knoll G, Winter J. 1988. Anaerobic degradation of phenol in sewage sludge. Benzoate formation from phenol and CO_2 in the presence of hydrogen. *Appl. Microbiol. Biotechnol.* 25:384–91
59a. Koch AL, Schmidt TM. 1991. The first cellular bioenergetic process: primitive generation of a proton-motive force. *J. Mol. Evol.* 33:297–304
60. Kramer SP, Johnson JL, Ribeiro AA, Millington DS, Rajagopalan KV. 1987. The structure of the molybdenum cofactor: characterization of di-(carboxamidomethyl)molybdopterin from

sulfite oxidase and xanthine oxidase. *J. Biol. Chem.* 262:16357–63
61. Krzycki JA, Lehman LJ, Zeikus JG. 1985. Acetate catabolism by *Methanosarcina barkeri:* evidence for involvement of carbon monoxide dehydrogenase, methyl coenzyme M, and methylreductase. *J. Bacteriol.* 163:1000–6
62. Krzycki JA, Mortenson LE, Prince RC. 1989. Paramagnetic centers of carbon monoxide dehydrogenase from aceticlastic *Methanosarcina barkeri. J. Biol. Chem.* 264:7217–21
63. Krzycki JA, Prince RC. 1990. EPR observation of carbon monoxide dehydrogenase, methylreductase and corrinoid in intact *Methanosarcina barkeri* during methanogenesis from acetate. *Biochim. Biophys. Acta* 1015:53–60
64. Kumar M, Lu WP, Liu LF, Ragsdale SW. 1993. Kinetic evidence that carbon monoxide dehydrogenase catalyzes the oxidation of carbon monoxide and the synthesis of acetyl-CoA at separate metal centers. *J. Am. Chem. Soc.* 115:11646–47
65. Kumar M, Ragsdale SW. 1992. Characterization of the CO binding site of carbon monoxide dehydrogenase from *Clostridium thermoaceticum* by infrared spectroscopy. *J. Am. Chem. Soc.* 114:8713–15
66. Lee MJ, Zinder SH. 1988. Isolation and characterization of a thermophilic bacterium which oxidizes acetate in syntrophic association with a methanogen and which grows acetogenically on H_2-CO_2. *Appl. Environ. Microbiol.* 54:124–29
67. Lee MJ, Zinder SH. 1988. Carbon monoxide pathway enzyme activities in a thermophilic anaerobic bacterium grown acetogenically and in syntrophic acetate-oxidizing coculture. *Arch. Microbiol.* 150:513–18
68. Lindahl PA, Munck E, Ragsdale SW. 1990. CO dehydrogenase from *Clostridium thermoaceticum*. EPR and electrochemical studies in CO_2 and argon atmospheres. *J. Biol. Chem.* 265:3873–79
69. Lindahl PA, Ragsdale SW, Munck E. 1990. Mossbauer study of CO dehydrogenase from *Clostridium thermoaceticum*. *J. Biol. Chem.* 265:3880–88
70. Ljungdahl LG. 1994. The acetyl-CoA pathway and the chemiosmotic generation of ATP during acetogenesis. See Ref. 24a, pp. 63–87
71. Lorowitz WH, Bryant MP. 1984. *Peptostreptococcus productus* strain that grows rapidly with CO as the energy source. *Appl. Environ. Microbiol.* 47:961–64
72. Lowe SE, Jain MK, Zeikus JG. 1993. Biology, ecology, and biotechnological applications of anaerobic bacteria adapted to environmental stresses in temperature, pH, salinity, or substrates. *Microbiol. Rev.* 57:451–509
73. Lu WP, Harder SR, Ragsdale SW. 1990. Controlled potential enzymology of methyl transfer reactions involved in acetyl-CoA synthesis by CO dehydrogenase and the corrinoid/iron-sulfur protein from *Clostridium thermoaceticum. J. Biol. Chem.* 265:3124–33
74. Lu WP, Jablonski PE, Rasche M, Ferry JG, Ragsdale SW. 1994. Characterization of the metal centers of the Ni/Fe-S component of the carbon-monoxide dehydrogenase enzyme complex from *Methanosarcina thermophila. J. Biol. Chem.* 269:9736–42
75. Lu WP, Schiau I, Cunningham JR, Ragsdale SW. 1993. Sequence and expression of the gene encoding the corrinoid/iron-sulfur protein from *Clostridium thermoaceticum* and reconstitution of the recombinant protein to full activity. *J. Biol. Chem.* 268:5605–14
76. Lupton FS, Conrad R, Zeikus JG. 1984. CO metabolism of *Desulfovibrio vulgaris* strain madison: physiological function in the absence or presence of exogenous substrates. *FEMS Microbiol. Lett.* 23:263–68
77. Lynd L, Kerby R, Zeikus JG. 1982. Carbon monoxide metabolism of the methylotrophic acidogen *Butyribacterium methylotrophicum. J. Bacteriol.* 149:255–63
78. Maupin JA, Ferry JG. 1993. Corrinoid-containing cobalt/iron-sulfur component of the CO dehydrogenase complex from *Methanosarcina thermophila* strain TM-1: cloning, sequencing, and overexpression in *Escherichia coli. Abstr. General Meeting, Am. Soc. Microbiol., 93rd, 113, 242*
79. Meyer O. 1982. Chemical and spectral properties of carbon monoxide: methylene blue oxidoreductase. *J. Biol. Chem.* 257:1333–41
80. Meyer O, Frunzke K, Gadkari D, Jacobitz S, Hugendieck I, Kraut M. 1990. Utilization of carbon monoxide by aerobes: recent advances. *FEMS Microbiol. Rev.* 87:253–60
81. Meyer O, Frunzke K, Morsdorf G. 1993. Biochemistry of the aerobic utilization of carbon monoxide. In *Microbial Growth on C1 Compounds*, ed. JC Murrell, DP Kelly, pp. 433–59. Andover: Intercept. 520 pp.

82. Meyer O, Jacobitz S, Kruger B. 1986. Biochemistry and physiology of aerobic carbon monoxide-utilizing bacteria. *FEMS Microbiol. Rev.* 39:161–79
83. Meyer O, Rajagopalan KV. 1984. Selenite binding to carbon monoxide oxidase from *Pseudomonas carboxydovorans. J. Biol. Chem.* 259:5612–17
84. Meyer O, Rohde M. 1984. Enzymology and bioenergetics of carbon monoxide-oxidizing bacteria. In *Microbial Growth on C1 Compounds*, ed. RL Crawford, RS Hanson, pp. 26–33. Washington, DC: Am. Soc. Microbiol. 343 pp.
85. Moller-Zinkhan D, Thauer RK. 1990. Anaerobic lactate oxidation to 3 CO_2 by *Archaeoglobus fulgidus* via the carbon monoxide dehydrogenase pathway—demonstration of the acetyl-CoA carbon-carbon cleavage reaction in cell extracts. *Arch. Microbiol.* 153:215–18
86. Morton TA, Runquist JA, Ragsdale SW, Shanmugasundaram T, Wood HG, Ljungdahl LG. 1991. The primary structure of the subunits of carbon monoxide dehydrogenase/acetyl-CoA synthase from *Clostridium thermoaceticum. J. Biol. Chem.* 266:23824–28
87. Murray PA, Uffen RL. 1988. Influence of cyclic AMP on the growth response and anaerobic metabolism of carbon monoxide in *Rhodocyclus gelatinosus. Arch. Microbiol.* 149:312–16
88. Nelson MJK, Terlesky KC, Ferry JG. 1987. Recent developments on the biochemistry of methanogenesis from acetate. In *Microbial Growth on C-1 Compounds*, ed. HW van Verseveld, JA Duine, pp. 70–76. Dordrecht: Nijhoff
89. O'Brien JM, Wolkin RH, Moench TT, Morgan JB, Zeikus JG. 1984. Association of hydrogen metabolism with unitrophic or mixotrophic growth of *Methanosarcina barkeri* on carbon monoxide. *J. Bacteriol.* 158:373–75
89a. Peer CW, Painter MH, Rasche ME, Ferry JG. 1994. Characterization of a CO:heterodisulfide oxidoreductase system from acetate grown *Methanosarcina thermophila. J. Bacteriol.* 176:6974–79
90. Pezacka E, Wood HG. 1984. Role of carbon monoxide dehydrogenase in the autotrophic pathway used by acetogenic bacteria. *Proc. Natl. Acad. Sci. USA* 81:6261–65
91. Pezacka E, Wood HG. 1988. Acetyl-CoA pathway of autotrophic growth. Identification of the methyl-binding site of the CO dehydrogenase. *J. Biol. Chem.* 263:16000–6
92. Plugge CM, Dijkema C, Stams AJM. 1993. Acetyl-CoA cleavage pathway in a syntrophic propionate oxidizing bacterium growing on fumarate in the absence of methanogens. *FEMS Microbiol. Lett.* 110:71–76
93. Preuss A, Fimpel J, Diekert G. 1993. Anaerobic transformation of 2,4,6-trinitrotoluene (TNT). *Arch. Microbiol.* 159:345–53
94. Qiu D, Kumar M, Ragsdale SW, Spiro TG. 1994. Nature's carbonylation catalyst—raman spectroscopic evidence that carbon monoxide binds to iron, not nickel, in CO dehydrogenase. *Science* 264:817–19
95. Rabus R, Nordhaus R, Ludwig W, Widdel F. 1993. Complete oxidation of toluene under strictly anoxic conditions by a new sulfate-reducing bacterium. *Appl. Environ. Microbiol.* 59:1444–51
96. Ragsdale SW. 1991. Enzymology of the acetyl-CoA pathway of CO_2 fixation. *Crit. Rev. Biochem. Mol. Biol.* 26:261–300
97. Ragsdale SW. 1994. CO dehydrogenase and the central role of this enzyme in the fixation of carbon dioxide by anaerobic bacteria. See Ref. 24a, pp. 88–126
98. Ragsdale SW, Clark JE, Ljungdahl LG, Lundie LL, Drake HL. 1983. Properties of purified carbon monoxide dehydrogenase from *Clostridium thermoaceticum*, a nickel, iron-sulfur protein. *J. Biol. Chem.* 258:2364–69
99. Ragsdale SW, Lindahl PA, Munck E. 1987. Mossbauer, EPR, and optical studies of the corrinoid/iron-sulfur protein involved in the synthesis of acetyl-CoA by *Clostridium thermoaceticum. J. Biol. Chem.* 262:14289–97
100. Ragsdale SW, Wood HG. 1985. Acetate biosynthesis by acetogenic bacteria: evidence that carbon monoxide dehydrogenase is the condensing enzyme that catalyzes the final steps of the synthesis. *J. Biol. Chem.* 260:3970–77
101. Ragsdale SW, Wood HG, Antholine WE. 1985. Evidence that an iron-nickel-carbon complex is formed by reaction of CO with the CO dehydrogenase from *Clostridium thermoaceticum. Proc. Natl. Acad. Sci. USA* 82:6811–14
102. Ramer SE, Raybuck SA, Orme-Johnson WH, Walsh CT. 1989. Kinetic characterization of the [3′-^{32}P] coenzyme A/acetyl coenzyme A exchange catalyzed by a three-subunit form of the carbon monoxide dehydrogenase/acetyl-CoA synthase from *Clostridium thermoaceticum. Biochemistry* 28:4675–80
103. Raybuck SA, Bastian NR, Orme-Johnson WH, Walsh CT. 1988. Kinetic

characterization of the carbon monoxide-acetyl-CoA (carbonyl group) exchange activity of the acetyl-CoA synthesizing CO dehydrogenase from *Clostridium thermoaceticum*. *Biochemistry* 27:7698–702

104. Raybuck SA, Ramer SE, Abbanat DR, Peters JW, Orme-Johnson WH, et al. 1991. Demonstration of carbon-carbon bond cleavage of acetyl coenzyme A by using isotopic exchange catalyzed by the CO dehydrogenase complex from acetate-grown *Methanosarcina thermophila*. *J. Bacteriol.* 173:929–32

105. Roberts DL, James-Hagstrom JE, Garvin DK, Gorst CM, Runquist JA, et al. 1989. Cloning and expression of the gene cluster encoding key proteins involved in acetyl-CoA synthesis in *Clostridium thermoaceticum*: CO dehydrogenase, the corrinoid/Fe-S protein, and methyltransferase. *Proc. Natl. Acad. Sci. USA* 86:1–5

106. Roberts JR, Lu WP, Ragsdale SW. 1992. Acetyl coenzyme A synthesis from methyltetrahydrofolate, CO, and coenzyme A by enzymes purified from *Clostridium thermoaceticum*. Attainment of in vivo rates and identification of rate-limiting steps. *J. Bacteriol.* 174:4667–76

107. Rohde M, Mayer F, Meyer O. 1984. Immunocytochemical localization of carbon monoxide oxidase in *Pseudomonas carboxydovorans*. The enzyme is attached to the inner aspect of the cytoplasmic membrane. *J. Biol Chem.* 259: 14788–92

108. Schauder R, Preu BA, Jetten M, Fuchs G. 1989. Oxidative and reductive acetyl CoA/carbon monoxide dehydrogenase pathway in *Desulfobacterium autotrophicum*. 2. Demonstration of the enzymes of the pathway and comparison of CO dehydrogenase. *Arch. Microbiol.* 151:84–89

109. Schmitz RA, Linder D, Stetter KO, Thauer RK. 1991. N_5,N_{10}-methylenetetrahydromethanopterin reductase (coenzyme F_{420}-dependent) and formylmethanofuran dehydrogenase from the hyperthermophile *Archaeoglobus fulgidus*. *Arch. Microbiol.* 156: 427–34

110. Shanmugasundaram T, Kumar GK, Shenoy BC, Wood HG. 1989. Chemical modification of the functional arginine residues of carbon monoxide dehydrogenase from *Clostridium thermoaceticum*. *Biochemistry* 28:7112–16

111. Shanmugasundaram T, Kumar GK, Wood HG. 1988. Involvement of tryptophan residues at the coenzyme A binding site of carbon monoxide dehydrogenase from *Clostridium thermoaceticum*. *Biochemistry* 27:6499–503

112. Shanmugasundaram T, Sundaresh CS, Kumar GK. 1993. Identification of a cysteine involved in the interaction between carbon monoxide dehydrogenase and corrinoid/Fe-S protein from *Clostridium thermoaceticum*. *FEBS Lett.* 326:281–84

113. Shanmugasundaram T, Wood HG. 1992. Interaction of ferredoxin with carbon monoxide dehydrogenase from *Clostridium thermoaceticum*. *J. Biol. Chem.* 267:897–900

114. Sharak-Genthner BR, Bryant MP. 1982. Growth of *Eubacterium limosum* with carbon monoxide as the energy source. *Appl. Environ. Microbiol.* 43:70–74

115. Shin W, Lindahl PA. 1992. Discovery of a labile nickel ion required for CO/acetyl-CoA exchange activity in the NiFe complex of carbon monoxide dehydrogenase from *Clostridium thermoaceticum*. *J. Am. Chem. Soc.* 114: 9718–19

116. Shin WS, Lindahl PA. 1992. Function and CO binding properties of the NiFe complex in carbon monoxide dehydrogenase from *Clostridium thermoaceticum*. *Biochemistry* 31:12870–75

117. Shin W, Stafford PR, Lindahl PA. 1992. Redox titrations of carbon monoxide dehydrogenase from *Clostridium thermoaceticum*. *Biochemistry* 31:6003–11

118. Simpson PG, Whitman WB. 1993. Anabolic pathways in methanogens. See Ref. 31a, pp. 445–72

119. Smith ET, Ensign SA, Ludden PW, Feinberg BA. 1992. Direct electrochemical studies of hydrogenase and CO dehydrogenase. *Biochem. J.* 285: 181–85

120. Sowers KR, Thai TT, Gunsalus RP. 1993. Transcriptional regulation of the carbon monoxide dehydrogenase gene (*cdhA*) in *Methanosarcina thermophila*. *J. Biol. Chem.* 268:23172–78

121. Stephens PJ, McKenna M-C, Ensign SA, Bonam D, Ludden PW. 1989. Identification of a Ni- and Fe-containing cluster in *Rhodospirillum rubrum* carbon monoxide dehydrogenase. *J. Biol. Chem.* 264:16347–50

122. Tan GO, Ensign SA, Ciurli S, Scott MJ, Hedman B, et al. 1992. On the structure of the nickel/iron/sulfur center of the carbon monoxide dehydrogenase from *Rhodospirillum rubrum*. An X-ray absorption spectroscopy study. *Proc. Natl. Acad. Sci. USA* 89:4427–31

123. Terlesky KC, Barber MJ, Aceti DJ, Ferry JG. 1987. EPR properties of the Ni-Fe-C center in an enzyme

complex with carbon monoxide dehydrogenase activity from acetate-grown *Methanosarcina thermophila*. Evidence that acetyl-CoA is a physiological substrate. *J. Biol. Chem.* 262: 15392–95
124. Terlesky KC, Ferry JG. 1988. Purification and characterization of a ferredoxin from acetate-grown *Methanosarcina thermophila*. *J. Biol. Chem.* 263:4080–82
125. Terlesky KC, Nelson MJK, Ferry JG. 1986. Isolation of an enzyme complex with carbon monoxide dehydrogenase activity containing a corrinoid and nickel from acetate-grown *Methanosarcina thermophila*. *J. Bacteriol.* 168: 1053–58
126. Thauer RK. 1988. Citric-acid cycle, 50 years on—modifications and an alternative pathway in anaerobic bacteria. *Eur. J. Biochem.* 176:497–508
127. Thauer RK, Moller-Zinkhan D, Spormann AM. 1989. Biochemistry of acetate catabolism in anaerobic chemo-trophic bacteria. *Annu. Rev. Microbiol.* 43: 43–67
128. Uffen RL. 1976. Anaerobic growth of a *Rhodopseudomonas* species in the dark with carbon monoxide as sole carbon and energy substrate. *Proc. Natl. Acad. Sci. USA* 73:3298–302
129. Uffen RL. 1981. Metabolism of carbon monoxide. *Enzyme Microb. Technol.* 3: 197–206
130. Uffen RL. 1983. Metabolism of carbon monoxide by *Rhodopseudomonas gelatinosa:* cell growth and properties of the oxidation system. *J. Bacteriol.* 155:956–65
131. Wakim BT, Uffen RL. 1983. Membrane association of the carbon monoxide oxidation system in *Rhodopseudomonas gelatinosa*. *J. Bacteriol.* 153:571–73
132. Yagi T. 1959. Enzymic oxidation of carbon monoxide. II. *J. Biochem.* 46: 949–55

NITROGENASE STRUCTURE AND FUNCTION: A Biochemical-Genetic Perspective

John W. Peters, Karl Fisher, and Dennis R. Dean

Department of Biochemistry and Anaerobic Microbiology, The Virginia Polytechnic Institute and State University, Blacksburg, Virginia 24061

KEY WORDS: Fe protein, MoFe protein, P clusters, FeMo cofactor, MgATP, signal transduction, mutagenesis

CONTENTS

INTRODUCTION	336
OVERVIEW	338
ASSIGNMENT OF MoFe PROTEIN METALLOCLUSTER DOMAINS	339
Assignment 1	340
Assignment 2	341
Assignment 3	342
Fe PROTEIN–NUCLEOTIDE BINDING AND HYDROLYSIS	343
COMPONENT PROTEIN INTERACTIONS	347
P-CLUSTER STRUCTURE AND POLYPEPTIDE ENVIRONMENT	352
FeMo COFACTOR STRUCTURE AND POLYPEPTIDE ENVIRONMENT	354
SUMMARY AND COMMENTS	360

ABSTRACT

Biological nitrogen fixation is catalyzed by nitrogenase, an enzyme composed of two component proteins called the Fe protein and the MoFe protein. During catalysis, electrons are delivered one at a time from the Fe protein to the MoFe protein in a process involving component-protein association and dissociation and hydrolysis of at least two MgATP for each electron transfer. The Fe protein contains the sites for MgATP binding and hydrolysis, whereas the site for substrate binding and reduction is located on the MoFe protein. Among the important aspects of nitrogenase enzymology discussed here are (*a*) the structures of the metal centers that participate in electron transfer, (*b*) the organization of the metalloclusters within the polypeptides and their contributions to substrate binding and electron transfer, (c) the nature of the dynamic interactions between the two component proteins that lead to nucleotide hydrolysis and intermolecular electron transfer, (d) the mechanism by which the multiple electrons necessary for substrate reduction are distributed within the MoFe protein, (e) the nature of the intramolecular electron path within the MoFe protein, and (f) where and how substrate and various inhibitors become bound to the substrate-reduction site. This chapter summarizes biochemical-genetic strategies used to address these questions and discusses them in the context of the recently proposed three-dimensional models for both the Fe protein and MoFe protein from *Azotobacter vinelandii*.

Introduction

Nitrogenase is a complex, two-component, metalloenzyme that catalyzes the MgATP-dependent reduction of N_2 to yield two molecules of NH_3. This catalytic reduction of N_2 is called biological nitrogen fixation, and the stoichiometry of the reaction is usually indicated as:

$$N_2 + 8e^- + 8H^+ + 16MgATP \rightarrow 2NH_3 + H_2 + 16MgADP + 16\ Pi.$$

Nitrogen fixation and the processes of nitrification and denitrification comprise the biogeochemical nitrogen cycle. Biological nitrogen fixation research has considerable agronomic and ecologic relevance because the availability of a utilizable, or fixed, form of nitrogen frequently limits plant productivity. In nitrogen-limiting agronomic situations, industrially produced nitrogenous fertilizers are often applied to increase productivity. Production of these fertilizers is expensive because the process necessitates considerable consumption of nonrenewable fossil fuels. Furthermore, the runoff of industrially formed fertilizers represents a source of environmental pollution.

Another aspect of the agronomic significance of biological nitrogen fixation is that some nitrogen-fixing microorganisms, such as the rhizobia, can establish a symbiotic association with certain crop plants. In such a relationship, the microorganism benefits from catabolism of fixed carbon provided by plant photosynthesis, and the plant benefits from utilization of nitrogen fixed by the microorganism.

Within this framework, any improvement in the nitrogen fixation process itself or in the ability of microorganisms to deliver fixed nitrogen to plants could represent economically beneficial and ecologically sound avenues for increasing plant yield. Biological nitrogen fixation is also of general interest to microbiologists because the extreme oxygen lability of the biochemical process has restricted it to microorganisms and to only those capable of anaerobic metabolism or those that have developed mechanisms for protecting the catalytic system from oxygen inactivation. Finally, because nitrogenase

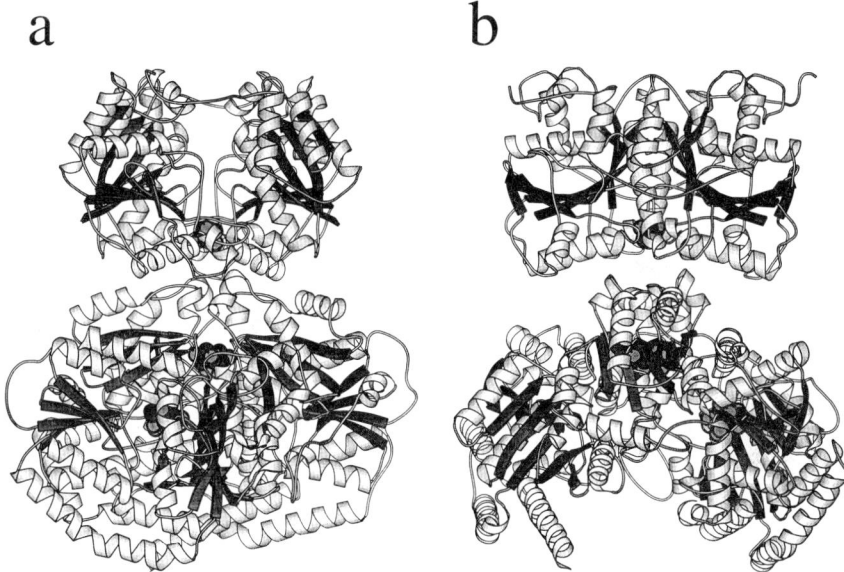

Figure 1 Ribbon diagrams (47) of the *A. vinelandii* Fe protein homodimer and an αβ-unit of the MoFe protein. The two different views (*a, b*) represent 90° rotations about the y axis. The associated metalloclusters are represented by space-filling models. The view in each panel shows the Fe protein (*top*) poised for interaction with the MoFe protein (*bottom*) and is based on the docking model proposed by Rees & Howard (34, 36, 44). Upon docking, the Fe protein's Fe_4S_4 cluster is positioned closest to the MoFe protein's P cluster.

catalysis involves familiar biochemical processes such as protein-protein interactions, signal transduction, and electron-transfer reactions, studies on the enzymology of nitrogenase are of general relevance.

In 1992, three-dimensional models proposed for both of the nitrogenase component proteins and their associated metalloclusters significantly advanced nitrogen fixation research (1–3, 10, 26, 43, 44, 46, 68). Since then, several comprehensive reviews and brief overviews that describe various aspects of the structural features of the nitrogenase component proteins and their mechanistic implications have appeared (14, 24, 36, 45, 58, 67). Also, several reviews published shortly before the availability of the three-dimensional models (7, 9, 59, 79, 92) provide excellent summaries of the biochemical and kinetic features of nitrogenase catalysis. In this chapter, we attempt to avoid extensive overlap with these other reviews by limiting our discussion to a description of several biochemical-genetic strategies that have been used to alter the nitrogenase component proteins to probe their structural and functional properties.

OVERVIEW

The two component proteins of the nitrogenase complex, which can be separately isolated from each other, are often designated the Fe protein and the MoFe protein. These terms reflect the metal compositions of the prosthetic groups contained within the respective component proteins. The Fe protein is a γ_2 homodimer ($M_r \approx 60,000$; encoded by *nifH*) that contains 4 Fe atoms organized into a Fe_4S_4 cluster, and the MoFe protein is an $\alpha_2\beta_2$ heterotetramer ($M_r \approx 250,000$; α encoded by *nifD*, and β encoded by *nifK*) that contains 30 Fe atoms and 2 Mo atoms organized into 2 pairs of metalloclusters, referred to as P clusters (Fe_8S_{7-8}) and FeMo cofactors (Fe_7S_9Mo-homocitrate). A later section describes the structures and proposed functions of the nitrogenase metal clusters in detail. Figure 1 shows ribbon diagrams of two different views representing the current model for the interaction that occurs during catalysis between the *Azotobacter vinelandii* Fe protein and the MoFe protein.

Our discussion of nitrogenase enzymology begins with a summary of several basic mechanistic features of the process, which have generally been accepted but are not necessarily experimentally proven:

1. During catalysis, electrons are delivered one at a time from the Fe protein to the MoFe protein.
2. The path of electron transfer leads primarily to substrate reduction.
3. Intermolecular electron transfer requires the association and dissociation of the component proteins and the hydrolysis of at least two MgATPs for each electron transfer.
4. Dissociation of the component proteins following intermolecular electron transfer is rate limiting.
5. The MgATP-binding sites are located within the Fe protein but no MgATP hydrolysis or intermolecular electron transfer occurs without formation of the Fe protein–MoFe protein complex.
6. The MgATP-binding site and the Fe_4S_4 cluster are separately located within the Fe protein, and they are unlikely to come within intimate contact with each other at any stage of catalysis.
7. Multiple rounds of intermolecular electron transfer must occur before any substrate is reduced.
8. The P cluster and FeMo cofactor are separate entities that do not directly interact with each other.
9. The P cluster is the immediate acceptor in the intermolecular electron-transfer event and probably brokers the intramolecular delivery of electrons to the FeMo cofactor.
10. FeMo cofactor provides the substrate-reduction site.
11. The tetrameric MoFe protein contains two separate, but identical, sub-

strate-reduction sites, each contained within an individual αβ unit of the MoFe protein.
12. Nitrogenase catalyzes evolution of one H_2 for every N_2 reduced.
13. Nitrogenase can reduce a variety of substrates other than N_2, most notably acetylene, which may be reduced by two electrons to yield ethylene.

These considerations show that the essential mechanistic issues associated with nitrogenase catalysis involve (*a*) the role of MgATP in component protein interaction and electron transfer, (*b*) the nature of the interaction between the nitrogenase component proteins, (*c*) how the individual metalloclusters communicate with each other to mediate electron transfer, substrate binding, and substrate reduction, (*d*) where and how multiple electrons are accumulated and stored within the MoFe protein prior to substrate binding and reduction, and (*e*) how and at what redox state(s) are various nitrogenase substrates and inhibitors bound to the active site. One powerful approach to addressing these issues is to specifically alter the polypeptide environments (or structures) of the individual metalloclusters, or to modify the MgATP-binding site, and subsequently determine the spectroscopic, redox, and catalytic consequences that result from such alterations. Several laboratories are vigorously pursuing this approach through alteration of the primary sequences of the nitrogenase component proteins by mutagenesis of the appropriate genes and through structural alteration or elimination of the FeMo cofactor by mutagenesis of genes involved in its assembly. In the following section, we describe strategies that were used to target certain residues or regions for substitution without the benefit of structural models. The results of these site-directed amino acid–substitution studies are discussed in the context of the structural models now available.

ASSIGNMENT OF MoFe PROTEIN METALLOCLUSTER DOMAINS

Once it became established that reasonably facile methods for site-directed mutagenesis and gene replacement could be applied to the nitrogenase system (5, 40, 62, 69), the next challenges became targeting specific amino acids for substitution and deciding which amino acids should be used as substituting residues. Because these studies were initiated without the benefit of structural models, indirect methods were necessary to target individual residues and specific regions for modification. Although the availability of detailed structural information has superseded the need for indirect approaches, a discussion of the logic employed in these initial studies is worthwhile. First, the information and assumptions used to guide early amino acid–substitution studies turned out to be essentially correct, which demonstrates the feasibility of designing informative amino acid–substitution studies without necessarily having a structural model in hand. Second, the fact that conclusions from amino acid–substitution studies and the nitrogenase structural models are in substantial agreement justifies confidence that both techniques provide a correct picture. The same point can be made for spectroscopic studies, such as extended X-ray absorption fine structure, electron nuclear double resonance, and electron spin echo envelope modulation, which were also successfully used to gain many structural insights concerning the nitrogenase-associated metalloclusters and their respective polypeptide environments prior to the emergence of the crystallographically determined models (see discussions in 7, 11, 59). Below we describe criteria that were used to target potential metallocluster environments within the MoFe protein as an example of the logic involved in the development of amino acid–substitution studies for analysis of the nitrogenase system. This chapter discusses results from experiments using both *Klebsiella pneumoniae* and *A. vinelandii,* but for clarity and consistency, all numbers refer only to the *A. vinelandii* primary sequences.

Site-directed mutagenesis programs for study of the nitrogenase MoFe protein were initiated in Blacksburg (*A. vinelandii* model system) and Sussex (*K. pneumoniae* model system) at about the same time, and both groups simultaneously proposed detailed assignments of the polypeptide environments for the metalloclusters contained within the MoFe protein to guide their respective amino acid–substitution strategies (4, 16, 17, 39, 40, 55, 72, 73). Although formulated independently, both models had essentially the same features. The following information was used to target potential metallocluster polypeptide environments and to assign their spatial arrangements within the MoFe protein: (*a*) MoFe protein primary amino acid sequences from various organisms were compared. (*b*) The MoFe protein α- and β-subunit primary sequences were compared with each other. (*c*) The MoFe protein α- and β-subunit primary sequences were respectively compared with the primary sequences of the *nifE* and *nifN* gene products. (*d*) The requirements for chemical extrusion and the spectroscopic features of the metalloclusters were considered. (*e*) The chemical reactivities of the isolated clusters were taken into account. (*f*) Metallocluster-binding motifs from other proteins were considered. (*g*) The results of amino acid–substitution studies were taken into account as such data became available. Described below are salient features that emerged from these considerations and the experimental basis for making the assignments.

Assignment 1

Each 8-Fe-containing P cluster is coordinated to the MoFe protein through cysteine ligands provided by residues α-Cys62, α-Cys88, α-Cys154, β-Cys70, β-Cys95, and β-Cys153 and is solvent exposed or is located close to the polypeptide's surface. The rationale for these assignments is as follows: (*a*)

Fe-S clusters are typically bound to proteins through cysteine ligands and can be quantitatively extruded by unfolding the protein in an organic solvent in the presence of excess thiols (61). Hence, extrusion of P clusters by this method (48) indicated that cysteine ligands coordinate this cluster. (*b*) These Cys residues are strictly conserved in all known MoFe protein sequences and are therefore likely to be functionally important (reviewed in 15). (*c*) Spatial and primary sequence conservations are observed when regions surrounding residues α-Cys62, α-Cys88, and α-Cys154 are compared to the corresponding regions surrounding residues β-Cys70, β-Cys95, and β-Cys153 (30, 49, 85). Such similarities in both spatial arrangement and primary sequence satisfy the requirement for four structurally similar domains within the MoFe protein that, on the basis of Mossbauer studies (23, 56), are needed to accommodate the P clusters. (*d*) Because the Fe protein's Fe_4S_4 cluster forms a bridge between identical subunits (28, 35), during a component-protein interaction, it should contact the MoFe protein at an interface that provides some aspect of two-fold symmetry. Although no motifs are repeated within the primary sequences of either the α-subunit or the β-subunit, sequence conservation between the subunits, including that of the proposed P cluster ligands, could accommodate such an arrangement. (*e*) P clusters should reside near the surface if they are primary acceptors during the intermolecular electron-transfer event.

Assignment 2

The MoFe protein α-subunit Cys275 provides the only thiolate ligand to FeMo cofactor, and it is coordinated to an Fe atom within FeMo cofactor. This assignment has the following rationale: (*a*) Thiols react with isolated FeMo cofactor in a one-to-one stoichiometry, indicating the presence of a single thiol liganded to FeMo cofactor in its protein-bound form (8, 12). (*b*) Nuclear magnetic resonance and extended X-ray absorption spectroscopies indicated that an Fe atom provides the thiol-reactive site on isolated FeMo cofactor (reviewed in 7, 59). (*c*) α-Cys275 is strictly conserved and it is contained in a region exhibiting high primary sequence conservation (15). (*d*) Other cysteines were already assigned as potential P cluster ligands. (*e*) α-Cys275 is the only cysteine residue that is flanked by residues with amide functions that might be displaceable by N-methyl formamide, the chaotropic organic solvent commonly used to extract FeMo cofactor from its polypeptide matrix (77). (*f*) The MoFe protein α-Cys275 has an analogous residue conserved in the *nifE* gene product sequence, E-Cys250 (15). The *nifE* and *nifN* gene products share significant primary sequence identity with the MoFe protein α- and β-subunits, respectively, and they form a heterotetrameric complex that may provide a scaffold for the assembly of FeMo cofactor (6, 64) (see below). (*g*) In the native MoFe protein, α-Cys275 is refractive to alkylation, but in an apo-form of the MoFe protein that lacks FeMo cofactor, this residue is hyperreactive to alkylation (J Howard, personal communication). In other words, when bound to the MoFe protein, FeMo cofactor protects α-Cys275 from alkylation. (*h*) MoFe protein from a mutant strain containing the substitution of α-Ala275 for α-Cys275 has approximately the same electrophoretic mobility as the apo-MoFe protein when electrophoresed under nondenaturing conditions, but a different mobility when compared with the native MoFe protein (39). This result indicated that a thiol group at the α-residue-275 position is required to keep FeMo cofactor attached to the MoFe protein. (*i*) When residues that flank α-Cys275 are replaced with certain other residues, a perturbation in the characteristic $S = 3/2$ EPR signal occurs (16).

Assignment 3

FeMo cofactor is contained entirely within the MoFe protein α-subunit and interacts with domains that include α-Gln191 and α-His195. The following rationale lies behind this assignment:

1. Once FeMo cofactor is assembled upon the NifEN scaffold, it must escape from the biosynthetic complex during maturation of the MoFe protein. Therefore, an FeMo cofactor-binding domain located within the NifEN complex would probably be structurally similar but functionally not equal when compared with the corresponding domain contained within the MoFe protein (6, 73). Namely, certain functional groups contained within the MoFe protein that keep FeMo cofactor attached might not be duplicated in the biosynthetic scaffold. Thus, comparison of the MoFe protein α-subunit and the *nifE* gene product primary sequences identified the MoFe protein residues α-Gln191 and α-His195 as likely to be located within an FeMo cofactor-binding domain (73).
2. α-Gln191 was assumed to interact with the homocitrate moiety because substitution with α-Lys191 results in a biochemical phenotype similar to one in which citrate has replaced homocitrate (29, 72).
3. Electron spin echo envelope modulation (ESEEM) spectroscopic comparison of isolated and MoFe protein–bound FeMo cofactor revealed an N-coordination to the FeMo cofactor present only in the protein-associated species. This modulation was assigned as arising from a deprotonated N-ligand provided by a histidine (84). Because substitution of α-His195 with α-Asn195 eliminated the characteristic ESEEM signal, it was suggested that either FeMo cofactor is covalently bound to the MoFe protein through α-His195 or α-His195 is necessary for N-coordination by some other residue (83).
4. The MoFe protein β-subunit probably does not participate directly in FeMo cofactor binding because comparison of the primary sequences of the *A. vinelandii* MoFe protein β-subunit and the *nifN* gene product revealed

relatively lower sequence identity than did comparison of the MoFe protein α-subunit and the *nifE* gene product (6).

Fe PROTEIN–NUCLEOTIDE BINDING AND HYDROLYSIS

Figure 2 shows a ribbon diagram of the Fe protein structural model. The key function of the Fe protein is coordination of MgATP binding and hydrolysis with electron transfer between its Fe_4S_4 cluster and the MoFe protein's P cluster. Upon binding of MgATP, the Fe protein undergoes a conformational change manifested by a lowering of the redox potential of its Fe_4S_4 cluster, changes in the line shape of its characteristic electron paramagnetic resonance (EPR) spectrum, and an increased susceptibility of its Fe_4S_4 cluster to specific Fe chelators (reviewed in 58). Such changes probably reflect a repositioning of the Fe_4S_4 cluster for the electron-transfer event that is coupled to MgATP hydrolysis, which in turn is triggered by Fe protein–MoFe protein complex formation.

Three important issues concerning the mechanisms of nucleotide binding and hydrolysis and their role in catalysis must be considered. First, because the nucleotide-binding site and the Fe_4S_4 cluster are separated by about 20 Å (26) (see Figure 2), they must share a path for communication. Second, because formation of the Fe protein–MoFe protein complex is an absolute requirement for nucleotide hydrolysis, reciprocal interactions must occur between the two protein partners, ultimately leading to MgATP hydrolysis, electron transfer, and dissociation of the ternary complex. Third, some mechanism must ensure

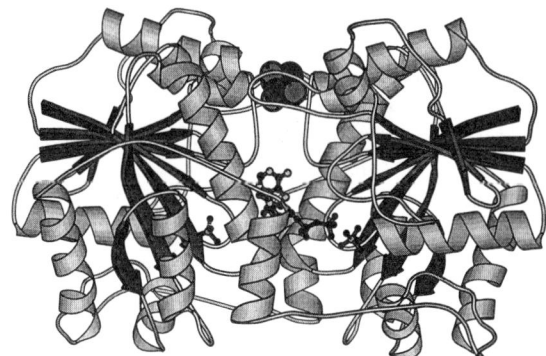

Figure 2 Ribbon diagram (47) of the *A. vinelandii* Fe protein homdimer. A space-filling model of the Fe_4S_4 cluster that bridges the identical subunits is located at the top of the structure. One ball-and-stick ADP molecule is shown bound in the intersubunit configuration. Ball-and-stick molybdate ions also indicate the location of residues that comprise the two Walker A nucleotide-binding motifs, one for each subunit.

that the direction of electron flow occurs mainly toward the substrate-reduction site.

Much of the hypotheses concerning communication between the Fe protein's MgATP-binding site and its Fe_4S_4 cluster have come from the comparison of primary sequence (70, 75) and structural features (26, 36, 45) common to the Fe protein and a functionally diverse class of other nucleotide-binding proteins. The salient functional feature shared among all these proteins is a nucleotide-binding and/or hydrolysis event that is coupled to a protein conformational change. Close inspection of the structural model, described in detail elsewhere (26, 36, 90), reveals that in the absence of MgATP, Lys15 [located within a Walker A–type nucleotide-binding motif (70)] probably forms a salt bridge with Asp125 [located within a Walker B–type nucleotide-binding motif (75, 90)]. Comparison with other nucleotide-binding proteins indicates that, in the MgATP-bound form, Lys15 probably interacts with the β and γ phosphate groups of the nucleotide, whereas Asp125 indirectly interacts with Mg^{2+} through a water molecule. These interactions resemble the rudiments of a signal-transduction mechanism in which upon nucleotide binding disruption of the salt bridge between Lys15 and Asp125 might trigger a cascade of structural rearrangements within the Fe protein that ultimately lead to the observed changes in the Fe_4S_4 cluster environment. In this context, the possibility that Asp125 provides a direct path for communication to the cluster is significant because this residue is connected by a short helix to the cluster-coordinating residue, Cys132. The potential roles for both Lys15 and Asp125 in the signal-transduction mechanism have been examined by replacing these residues with other amino acids.

Substitution of Lys15 with Gln results in an altered Fe protein that cannot undergo the MgATP-induced conformational change (75). MgATP causes neither a perturbation in the EPR spectrum of the altered Fe protein nor an increased susceptibility to chelation of its Fe_4S_4 cluster. However, in the absence of any added nucleotide, the EPR spectrum of the altered Gln15 Fe protein is identical to that of the normal Lys15 Fe protein, indicating that the Gln15 substitution does not cause a direct rearrangement of the polypeptide environment immediately surrounding the Fe_4S_4 cluster but probably compromises the signal-transduction pathway. The Gln15 Fe protein binds MgADP at the normal level but binds MgATP at a level somewhat lower than the unaltered Lys15 Fe protein does. Thus, the observation that Lys15 is needed to stabilize the MgATP complex, but apparently not the MgADP complex, may indicate that, in the MgATP-bound form, Lys15 is normally coordinated to the terminal phosphate group of the nucleotide. This conclusion is in line with expectations based on comparison of the Fe protein structure to other nucleotide-binding proteins. Nevertheless, it is somewhat surprising that, although the altered Gln15 Fe protein can still bind MgATP, there is not even a hint of a nucleotide-induced conformational change, nor is there any detectable MgATP hydrolytic activity when the complete system is assayed under the appropriate catalytic conditions.

These observations, and the fact that the Gln15 Fe protein cannot compete with the Lys15 Fe protein for MoFe protein interactions, suggest that coordination of the terminal phosphate of MgATP by Lys15 is needed for at least two purposes. First, it might anchor the bound nucleotide in an appropriate position to elicit the nucleotide-induced conformational change necessary for complex formation between the Fe protein and the MoFe protein, and second, it might also facilitate MgATP hydrolysis once the complex is formed. Analysis of the altered Gln15 Fe protein clearly shows that the Lys15 plays a critical role in effecting the nucleotide-induced conformational shift. Nevertheless, the signal-transduction mechanism cannot be explained by a simple breaking of a salt bridge between Lys15 and Asp125 because substitution of Lys15 with Gln, which cannot participate in the formation of such a salt bridge, does not render the altered protein's Fe_4S_4 cluster susceptible to chelation. Perhaps a more reasonable scenario is that once the putative salt bridge is eliminated by nucleotide binding, and the MgATP becomes locked into the appropriate position through interaction of its terminal phosphate with Lys15, Asp125 becomes forced to adopt a new position within the nucleotide-binding pocket, which triggers a conformational change in the polypeptide backbone that is propagated down the short loop connecting Asp125 and the cluster-coordinating Cys132. By analogy to *ras* p21, the reorientation of Asp125 within the nucleotide-binding pocket could be induced by indirect coordination of its carboxyl group to Mg^{2+} through an intervening water molecule (90). Thus, in this model the biochemical phenotype of the Gln15 Fe protein can be explained by its inability to lock MgATP into the appropriate position so that either the bound nucleotide remains too remote to interact with Asp125 or the bound MgATP remains flexible enough to interact with Asp125 without demanding a significant reorientation of Asp125 within the nucleotide-binding pocket.

The participation of Asp125 in the signal-transduction mechanism has been probed by replacing this residue with Glu (90). The inactive Glu125 Fe protein shares some similarities with the Gln15 Fe protein because, in the absence of nucleotide, neither substitution appears to directly alter the Fe_4S_4 cluster environment. Also, like the Gln15 Fe protein, the Glu125 Fe protein is ineffective in electron transfer and unable to catalyze MgATP hydrolysis. The distinguishing feature of the Glu125 Fe protein is that, unlike the Gln15 Fe protein, it retains the ability to undergo a conformational shift upon binding MgATP. Moreover, binding of either MgADP or MgATP to the Glu125 Fe protein can effect a conformational change. The requirement for Mg^{2+} has also been eliminated. The observation that MgADP binding induces a conformational change in the Glu125 Fe protein was anticipated in the original design of the experiment, because the goal was to test whether or not the addition of an extra methylene group might extend the primary signaling event to interaction with MgADP in addition to, or instead of, MgATP. Nevertheless, the elimination of the Mg^{2+} requirement was not expected, and a convincing explanation for this effect is not yet available.

Which residues are specifically involved in MgATP hydrolysis? Once again, comparison to other systems leads to speculation that either or both Asp39 and Asp129 (26, 34) are involved. Asp129 is a particularly attractive candidate because it is located between Asp125 and Cys132, and therefore, dynamic changes in the orientation of Asp129 could reflect conformational changes propagated back and forth between Asp125 and Cys132 caused by nucleotide binding or component protein docking. The characterization of an altered Fe protein in which Asp129 was replaced with Glu supports this possibility (W Lanzilotta, M Ryle & L Seefeldt, personal communication). The Glu129 Fe protein can bind nucleotides with normal affinity and undergoes the MgATP-induced conformational shift, but it cannot hydrolyze MgATP or transfer electrons to the MoFe protein.

Characterization of an altered Fe protein in which Ser16, located within the Walker A nucleotide-binding motif, is replaced with Thr indicates that this residue also has an important role in nucleotide hydrolysis (76). The altered Thr16 Fe protein exhibits a higher affinity for MgATP and is dramatically impaired in its ability to utilize MnATP, an alternative form of ATP, during catalysis, compared with such properties of the wild-type Ser16 Fe protein. These results indicate that during MgATP hydrolysis the γ-OH group of Ser16 facilitates the transition of Mg^{2+} from coordination to the γ- and β-phosphates of MgATP to coordination to the β- and α-phosphates of MgADP. This interpretation is in agreement with structural and spectroscopic evidence indicating that the equivalent serine in *ras* p21 not only interacts with Mg^{2+} of MgGTP (discussed in 76), but also has this function. The altered Thr16 Fe protein is the only Fe protein with a substitution at this position that has been characterized at the biochemical level; however, other residues (Ala16, Asp16, Gly16, and Cys16) substituted at this position all lead to inactivation of Fe protein catalytic activity (76). These data provide further evidence for the importance of this residue. In contrast, Fe proteins in which serine or alanine has replaced one of the three threonines located immediately after Ser16 remain active, indicating that these residues are probably not as functionally important as Ser16.

Although a reasonably clear picture of the signal-transduction mechanism leading to MgATP hydrolysis is beginning to emerge, less direct mechanistic information is available concerning how the gate might close following intermolecular electron transfer. Inspection of the structural model has led to some insights. In the as-solved structure, only one of the two available nucleotide-binding sites is occupied (26). This site, which is only partially occupied by MgADP, is not bound in the typical *ras* mode but rather spans a cleft between the subunits as shown in Figure 2. Assuming this binding mode is mechanistically relevant, we can speculate that a nucleotide bound to the Fe protein can assume two configurations (90). For example, upon docking with the MoFe protein, MgATP bound in the *ras*-like mode might be hydrolyzed, which would lead to a transition state at which intermolecular electron transfer can be achieved. Following hydrolysis, the nucleotide, now in the form of MgADP, might adopt the intersubunit binding configuration, which could stabilize a conformation of the Fe protein that cannot accept an electron from the MoFe protein. Thus, the nucleotide can be considered as a swinging gate that helps control the direction of electron flow.

The biochemical properties of the altered Gln15 protein have been interpreted in view of the following model (90): The Gln15 protein only binds nucleotide in the intersubunit configuration to prevent any nucleotide-induced movement of the cluster. This model explains the relatively low level of MgATP binding, but the normal levels of MgADP binding, to the Gln15 Fe protein. Conversely, for the Glu125 Fe protein, the *ras*-like binding mode could be available to both ADP and ATP so that either nucleotide can induce a conformational change yet neither can be hydrolyzed to achieve electron transfer. After the MgATP-induced conformational change, the Glu125 Fe protein apparently cannot transmit the nucleotide hydrolysis signal that normally results from the component protein-docking event, perhaps by preventing Asp129 and/or Asp39 from adopting the correct position for the postulated water-assisted attack on the terminal phosphate of MgATP or MgADP.

COMPONENT PROTEIN INTERACTIONS

How do the Fe protein and MoFe protein interact to couple nucleotide hydrolysis to intermolecular electron transfer? A docking model based on the crystal structures of the separate components that takes into account chemical cross-linking studies and amino acid–substitution studies has been proposed (34, 36, 44). This model pairs the two-fold symmetric surface of the Fe protein homodimer with the exposed surface of a MoFe protein pseudosymmetric $\alpha\beta$-unit interface. Figure 1 shows two different views of this docking model. In this arrangement, the Fe protein's Fe_4S_4 cluster is positioned as close as possible to the MoFe protein's P cluster, which accommodates the view that the primary electron-transfer event involves transient delivery of an electron from the Fe_4S_4 cluster of the Fe protein to a MoFe protein P cluster. Also, the arrangement of several charged groups on the respective surfaces of the Fe and MoFe proteins could permit reciprocal ionic interactions between the component proteins. Studies on the salt sensitivity of nitrogenase catalytic

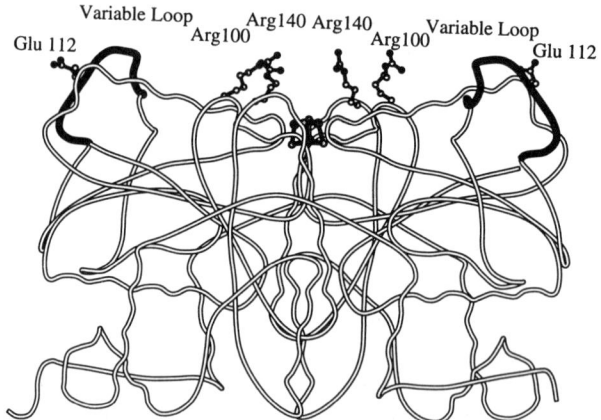

Figure 3 Coil diagram (47) of the *A. vinelandii* Fe protein homodimer highlighting the proposed component protein interaction sites. The variable loop discussed in the text appears in bold. The view of the Fe protein is the same as in Figure 1*b* except that it is rotated 180°.

activity indicate that such ionic interactions are critical to productive complex formation (18, 65, 91).

Inspection of the Fe protein structural model reveals a crown of positively charged residues (Arg100, Arg140, and Lys143) located within the proposed docking interface that surround the Fe protein's Fe_4S_4 cluster (Figure 3). One of these residues, Arg100, corresponds to the reversible ADP-ribosylation site involved in the regulation of Fe protein activity in *Rhodospirillum rubrum* (66). ADP ribosylation exerts its regulatory effect by preventing intermolecular electron transfer. Altered Fe proteins in which His replaces Arg100 (52, 91), Gln replaces Arg140 (74), or Gln replaces Lys143 (74) all share similar biochemical phenotypes, exhibiting various levels of increased salt sensitivity and various degrees of uncoupling of MgATP hydrolysis from intermolecular electron transfer. The biochemical changes indicate that residues Arg100, Arg140, and Arg143 normally provide dominant ionic interactions with the MoFe protein during complex formation. Namely, elimination of any of these ionic interactions seems to render other ionic interaction sites more susceptible to disruption by salt. Also, the close proximity of these residues to the Fe_4S_4 cluster is compatible with a model in which reciprocal ionic interactions are involved in bringing the Fe protein's Fe_4S_4 cluster and a complementary site on the MoFe protein into the proper juxtaposition to effect electron transfer. Nevertheless, the functional role of these residues cannot be assigned simply to ionic interactions, because certain substitutions that eliminate the potential for the contribution of a specific ionic interaction, for example, substitution of

Arg100 with Tyr, only modestly affects catalytic activity, and MgATP hydrolysis remains effectively coupled to electron transfer. Furthermore, because MgATP hydrolysis becomes substantially uncoupled from intermolecular electron transfer in certain altered Fe proteins [e.g. His100 (52, 91) and Gln143 (74)], these residues are not necessarily directly involved in the primary signaling process that leads to MgATP hydrolysis. In other words, if MgATP hydrolysis and electron transfer are obligately coupled to exactly the same signal provided by component protein interaction, they should be equally affected by any perturbation in the pathway.

Inspection of the component docking model shows that a loop (designated the variable loop in Figure 3) (see 34) contained within residues 61–75 is likely to interact with the MoFe protein during component protein interaction. The potential role for this region was tested by construction of an *A. vinelandii* strain that produces a hybrid Fe protein for which a portion of this loop was replaced with the corresponding residues from the Fe protein of *Clostridium pasteurianum* (65). The experimental rationale for the construction was based on the observation that a heterologous mixture of Fe protein from *C. pasteurianum* and MoFe protein from *A. vinelandii* is not catalytically active, but instead forms a tight complex (25). Thus, replacement of this region of the *A. vinelandii* Fe protein primary sequence with the corresponding *C. pasteurianum* Fe protein sequence might result in duplication of the phenomenon of inactive complex formation between the resulting hybrid protein and the MoFe protein.

The hybrid Fe protein constructed in this way exhibits half the maximum specific activity of the normal Fe protein, and its activity is insensitive to inhibition by low levels of salt. Also, the hybrid Fe protein activity is hypersensitive to a molar excess of MoFe protein, which also results in the uncoupling of MgATP hydrolysis from substrate reduction. These results can be explained if the hybrid Fe protein forms a relatively tighter complex with the MoFe protein, a configuration consistent with the original experimental rationale and also in agreement with the docking model. Direct evidence for this conclusion came from stopped flow spectrophotometric experiments showing that the hybrid Fe protein dissociates from the MoFe protein at only half the normal rate (65). These experiments also revealed that moderate salt concentrations slowed the apparent dissociation rate of the wild-type complex, but dissociation of the complex involving the hybrid Fe protein was not affected under the same conditions. One interpretation of these results is that the hybrid Fe protein lacks an ionic interaction involving an electrostatic repulsion that facilitates component protein dissociation. A similar series of experiments, in which the carboxyl terminus of the *A. vinelandii* Fe protein was replaced with the corresponding region from the *C. pasteurianum* Fe protein, led to the conclusion that this region does not play a significant role in component protein

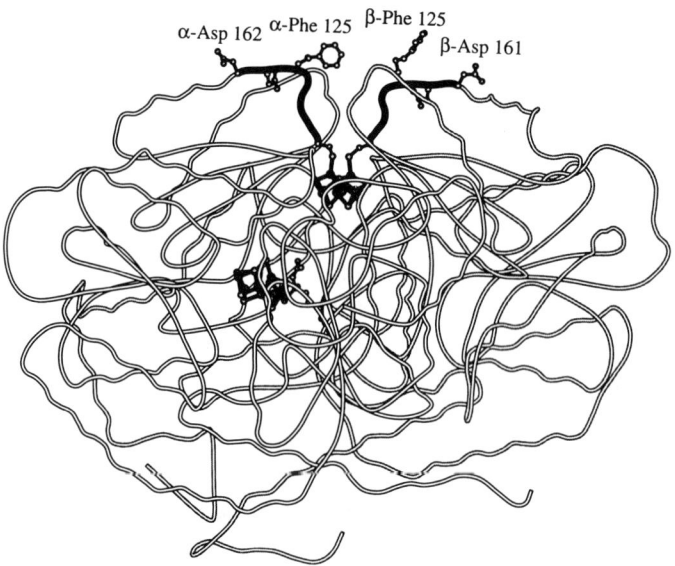

Figure 4 Coil diagram (47) of an αβ unit of the *A. vinelandii* MoFe protein that highlights the proposed component protein interaction sites. The P cluster is located at the pseudosymmetric αβ interface. Pathways from the surface of the molecule to the P cluster coordinating residues α-Cys154 and β-Cys153 appear in bold. This view of the MoFe protein is the same as shown in Figure 1*a*.

interaction (38). This conclusion is also in line with the docking model, which places the carboxyl region of the Fe protein on the opposite side from the proposed docking interface (Figures 1 and 2).

Experiments involving the use of the chemical cross-linking reagent carbodiimide have shown that the Fe protein Glu112 and the MoFe protein β-subunit Lys400 come in close contact during some stage of component protein interaction (89). Such cross-linking is sensitive to high concentrations of salt (88). Glu112 is located on the same face of the Fe protein but at the opposite end of a helix that extends to Arg100 and is, therefore, appropriately positioned for interaction with the MoFe protein. This helix and the variable loop discussed above comprise ridges of an anionic cleft that, in the docking model, probably come in contact with the MoFe protein. Moreover, Glu112 and the adjacent Glu111 from the *A. vinelandii* primary sequence are replaced by leucine and glutamine, respectively, in the corresponding *C. pasteurianum* primary sequence. Thus, future experiments should substitute Glu112 with Leu and Glu111 with Gln, both separately and in combination, to determine if these differences in interspecific primary sequence also contribute to the inactive complex formation observed in heterologous mixtures of *C. pasteurianum* Fe protein and *A. vinelandii* MoFe protein.

Although the docking model places the Fe protein's Fe_4S_4 cluster in the closest possible proximity to the MoFe protein's P cluster, a distance of about 15 Å still separates the clusters because the P cluster resides below the polypeptide surface (see Figures 1 and 4). Hence, either the MoFe protein polypeptide itself provides an adequate path for cluster-to-cluster electron transfer, or the MoFe protein undergoes a conformational change when it interacts with the Fe protein such that its cluster becomes more accessible to the Fe protein's cluster. The possibility that, upon association and dissociation with the Fe protein, the MoFe protein might undergo sequential conformational changes that respectively result in movement of the P cluster toward and away from the polypeptide surface is attractive, because movement of the P cluster toward the polypeptide surface could permit a direct cluster-to-cluster electron jump, and the subsequent sequestering of the P cluster below the polypeptide surface could contribute to the unidirectionality of the electron-transfer pathway. Moreover, conformational changes within the MoFe protein induced by docking and MgATP hydrolysis could lead to a rearrangement in the P cluster structure that alters its ability to accept or donate an electron (36, 67).

Which MoFe protein residues might be involved in transducing such conformational changes? The view of the MoFe protein structure in Figure 4 shows that residues α-Asp162, and α-Phe125, and the corresponding residues β-Asp161, and β-Phe125, reside at or near the mouth of the pseudosymmetric cleft leading to the P cluster. The side chains of α-Asp162 and β-Asp161 are solvent exposed and appear to be appropriately positioned to interact with one of the positively charged residues that form the crown surrounding the Fe protein's Fe_4S_4 cluster. Thus, ionic interactions between positively charged side chains located on the surface of the Fe protein near its Fe_4S_4 cluster and the negatively charged aspartates could result in coordinate transmission of signals down short helices connecting α-Asp162–α-Asp161 and β-Asp161–β-Asp160 to the P cluster coordinating ligands α-Cys154 and β-Cys153, respectively. The parallel α- and β-subunit pathways from the MoFe protein surface to the respective cluster coordinating ligands appear as dark lines in Figure 4. Although not shown in the figure, another interesting feature is that the carboxylate group of α-Asp161 appears to be hydrogen bonded to α-His83 and α-Gly127, and the analogous carboxylate group of β-Asp160 appears to be hydrogen-bonded to β-His90 and β-Gly127. This network of hydrogen bonding allows consideration of potential communication between other surface residues, for example α-Phe125 and β-Phe125, and the residues that provide the ligands bridging the P cluster subcluster fragments, α-Cys88 and β-Cys95.

In the *K. pneumoniae* MoFe protein, the β-Phe125 site is uniquely accessible to cleavage by chymotrypsin, which supports the location of this residue on the polypeptide surface (86). The involvement of β-Phe125 in component protein interaction was subsequently confirmed by kinetic analysis of an altered β-Ile125 MoFe protein that exhibits an approximately 70% reduction in the rate of primary electron transfer from the Fe protein to the MoFe protein. These results indicate that hydrophobic interactions, as well as ionic interactions, are important in the component protein docking process. On the other hand, substitution of asparagine for α-Asp162 and β-Asp161, either singly or in combination, has very little effect on catalytic activity (42). In contrast, substitution of α-Asp161 with asparagine leads to a biochemical phenotype almost identical to the altered Arg100 Fe protein. Namely, productive complex formation becomes hypersensitive to elevated salt concentrations and MgATP hydrolysis becomes substantially uncoupled from electron transfer. The interesting aspect of the biochemical phenotype of the α-Asn161 MoFe protein is that α-Asp161 is not exposed to the surface in the structural model. Thus, during component protein interaction, the hydrogen-bonding pattern mentioned above might indeed be disrupted, effecting a conformational change within the MoFe protein that could ultimately result in an ionic interaction between α-Asp161 and a positively charged residue located on the Fe protein. Such a conformational change, if it does occur, could also cause a perturbation in the P cluster polypeptide environment. Curiously, the analogous substitution within the β-subunit, β-Asn for β-Asp160, has very little effect on catalytic activity. Thus, although the structure of the α- and β-subunit interface in these pseudosymmetric regions is apparently conserved, functionality is not stringently conserved.

P-CLUSTER STRUCTURE AND POLYPEPTIDE ENVIRONMENT

Each P cluster contains eight Fe atoms constructed from two linked, cuboidal Fe_4S_4 subcluster fragments. Four cysteines, two from the α-subunit (α-Cys62 and α-Cys154) and two from the β-subunit (β-Cys70 and β-Cys153), coordinate individual Fe atoms within the P cluster as typical cysteinyl thiolate ligands (10). In addition, two cysteines (α-Cys88 and β-Cys95) link the subclusters by binding two Fe atoms, one from each subcluster. A corner-to-corner disulfide bond may also link the subcluster fragments, and this disulfide bridge has been proposed to be what permits redox reactions to occur within the all-ferrous P cluster (36, 67). In addition to the cysteine ligands that coordinate the P cluster, the *A. vinelandii* MoFe protein structural model indicates that β-Ser188 also appears to coordinate the same iron as β-Cys153. Figure 5 shows two different views of the P cluster model and the organization of its coordinating ligands.

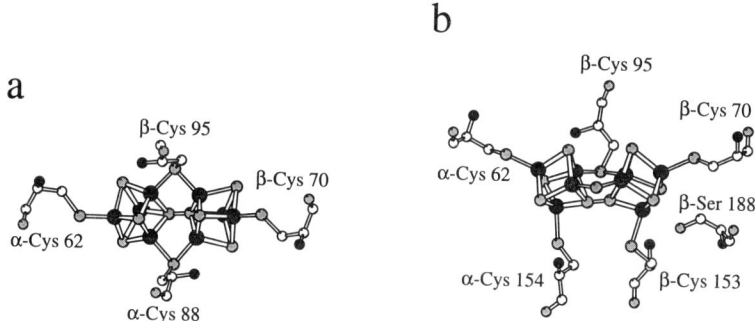

Figure 5 Ball-and-stick models (47) of the P cluster and the organization of its coordinating cysteine ligands. Each P cluster is coordinated by residues α-Cys62, α-Cys88, α-Cys154, β-Cys70, β-Cys95, and β-Cys153. For the sake of clarity, not all of the cluster coordinating ligands are shown in each view, but the location of residues not shown can be ascertained by considering that each view is separated by an approximately 90° rotation about the x-axis.

Although the apparent structural symmetry between the corresponding polypeptide environments of the individual P cluster subfragments is rather striking, amino acid–substitution studies have revealed that counterpart ligands from the two subunits are not necessarily functionally identical (17, 39, 40, 55). Substitution of serine or alanine for α-Cys62, α-Cys154, β-Cys70, or β-Cys95 inactivates MoFe protein activity. In many cases, the loss of enzymatic activity associated with these substitutions, and certain others located in the P cluster environment, is accompanied by failure to form the tetramer (27, 39), indicating that the P clusters are necessary for proper assembly of the MoFe protein. In contrast to the P cluster cysteine ligands mentioned above, certain substitutions can be tolerated for the α-Cys88 and β-Cys153 residues. Why substitutions can be tolerated at these positions, but apparently not at the sites of other P cluster ligands, is not yet known. Furthermore, substitution of alanine for either α-Cys88 or β-Cys95 inactivates the MoFe protein, but some activity is recovered when both residues are replaced with alanine (39). In a similar experiment, individual substitution of asparagine for α-His83 or the analogous β-His90 did not appreciably affect MoFe protein activity, but a combination of both substitutions eliminated this activity (17). These results and the effect of certain substitutions on the assembly of the MoFe tetramer originally led to the hypothesis that P clusters are located at the αβ interface, but again, the mechanistic implications of these findings are not understood.

It is a generally accepted working hypothesis that the catalytic role for the P cluster involves the acceptance and storage of electrons from the Fe protein and their ultimate delivery to the substrate-reduction site provided by FeMo cofactor. The spectroscopic and kinetic evidence along these lines was recently reviewed (51). Although biochemical-genetic studies have not yet contributed to the development of a mechanistic role for the P cluster in nitrogenase catalysis, they have provided some evidence that the P cluster does indeed broker intermolecular and intramolecular electron-transfer events (55). Substitution of β-Ser for the known P cluster ligand β-Cys153 results in an altered MoFe protein that supports only 50% of the maximum activity under conditions of high flux.

Such a decrease in activity can be explained by (a) an alteration in the substrate reduction site, (b) a disruption in the specific component protein interactions, or (c) an alteration in the intramolecular delivery of electrons to the substrate reduction. Possibility a was ruled out by the demonstration that the catalytic and spectroscopic features of the substrate-reduction site remain unchanged in the β-Ser153 MoFe protein. Possibilities b and c were evaluated by comparing the specific activities of the β-Ser153 MoFe protein and the wild-type β-Cys153 MoFe protein under conditions of both high and low flux. Under low-flux conditions, both proteins have approximately the same specific activity, but under high-flux conditions, the β-Ser153 exhibits only 50% of the maximum specific activity compared with the wild-type protein. This result suggests that, under conditions of high electron flux, the maximum specific activity of the β-Ser153 MoFe protein must be limited by the intramolecular delivery of electrons to the substrate-reduction site. On the other hand, if the β-Ser153 MoFe protein was defective in its ability to interact with the Fe protein, it should exhibit a lower specific activity relative to the wild-type MoFe protein under conditions of both low and high flux, which was not observed. More recently, similar results were obtained for an altered MoFe protein in which the P cluster coordinating residue β-Ser188 was replaced with glycine. Spectroscopic studies on both the altered β-Ser153 and β-Gly188 MoFe proteins from *A. vinelandii* reveal marked differences in their magnetic circular dichroic (MCD) spectra compared with the spectra of wild-type MoFe protein (K Fisher, M Finnegan, W Newton, D Dean & M Johnson, unpublished data). These results indicate substantial changes in the electronic properties of the P clusters contained within the altered β-Ser153 and β-Gly188 MoFe proteins, as might be expected from their altered catalytic properties. In contrast, MCD studies on the β-Ser152 MoFe protein from *K. pneumoniae*, which is analogous to the β-Ser153 MoFe protein from *A. vinelandii*, indicate that it has properties similar to those of the wild-type MoFe protein (78).

FeMo COFACTOR STRUCTURE AND POLYPEPTIDE ENVIRONMENT

FeMo cofactor contains a metal-sulfide core (Fe_7S_9Mo) and one molecule of (R)-homocitrate (Figure 6). The metal-sulfide core is constructed from two

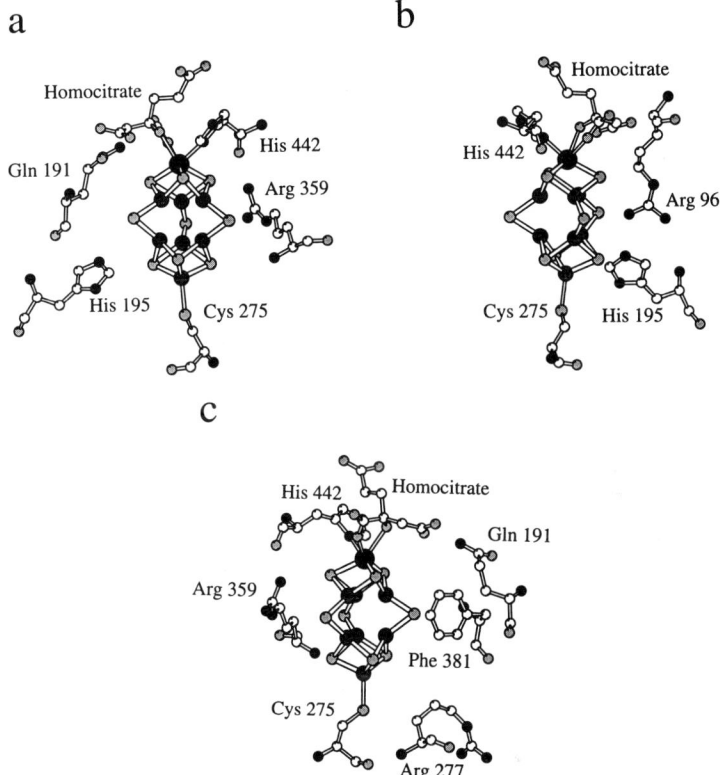

Figure 6 Ball-and-stick models (47) of FeMo cofactor and selected residues contained within its polypeptide environment. (*a–c*) Views related by rotations about the *y*-axis.

partial cubes, one each of $MoFe_3S_3$ and Fe_4S_3 subcluster fragments, that are joined by a ring of three sulfide bridges that connect pairs of opposing Fe atoms. The organic constituent, homocitrate, is coordinated to the Mo atom through its 2-hydroxy and 2-carboxyl groups. FeMo cofactor is covalently attached to the α-subunit through a thiolate ligand provided by α-Cys275 to an Fe atom at one end of the molecule and by the side-chain nitrogen atom of α-His442 to the Mo atom at the opposite end. Although FeMo cofactor is completely contained within the α-subunit, some β-subunit residues, for example β-Tyr98, approach the homocitrate and are indirectly linked to it by water molecules. No amino acids other than α-Cys275 and α-His442 are covalently attached to the FeMo cofactor, although several other residues appear to interact with it through hydrogen bonding. Figure 6 shows not only the structure of FeMo cofactor, but also three different views of its polypeptide environment.

An earlier section of this review outlined the logic used to assign the location of certain residues, for example, α-Gln191, α-His195, and α-Cys275, within the FeMo cofactor–binding pocket. In addition to the structural information, biochemical-genetic evidence that α-Cys275 and α-His442 both participate in anchoring FeMo cofactor within the binding pocket was obtained by nondenaturing electrophoretic studies of altered MoFe proteins having substitutions for or near these residues. For example, replacement of α-Cys275 with α-Ala275 (39) or α-His442 with α-Asn442 (D Govezensky & A Zamir, personal communication) results in altered MoFe proteins that exhibit native electrophoretic mobilities characteristic of the apo-MoFe protein. The studies also showed that the pool of available FeMo cofactor is increased in extracts of the altered α-Ala275 MoFe protein (40). Although such substitution studies were useful in the original assignment of potential FeMo-cofactor ligands and are in agreement with the structural models, they are of little use in mechanistic studies because all substitutions placed at either the α-Cys275 or α-His442 position tested so far lead to the complete loss of all MoFe protein activities. In contrast, substitutions for α-Gln191 and α-His195, residues contained within the binding pocket but not covalently attached to FeMo cofactor, do not necessarily result in a loss of all activities. Certain of these altered MoFe proteins have been studied in detail.

Substitution of α-Gln191 with α-Lys191 results in an altered MoFe protein that cannot reduce N_2 but can still reduce protons and acetylene, although at lower levels (72). The α-Lys191 MoFe protein also differs from the wild-type MoFe protein in that it exhibits proton reduction that is 50% sensitive to CO. This feature is characteristic of MoFe protein produced by *nifV* mutant strains (29), which raises the possibility that an alteration in FeMo cofactor's polypeptide environment might also lead to an alteration in the structure of the bound FeMo cofactor. Inactivation of the *nifV* gene from *K. pneumoniae* results in formation of an altered FeMo cofactor that contains citrate rather than homocitrate (31–33, 50). The possibility that the α-Lys191 MoFe protein results in a structural rearrangement of FeMo cofactor was addressed by using FeMo cofactor isolated from the α-Lys191 MoFe protein to reconstitute an apo-form of an otherwise wild-type α-Gln191 MoFe protein (72). Apo-MoFe protein reconstituted in this way exhibits proton reduction that is insensitive to CO inhibition, the wild-type phenotype. The reverse experiment showed that when FeMo cofactor isolated from the wild-type MoFe protein is used to reconstitute an apo-form of the α-Lys191 MoFe protein, the reconstituted protein does exhibit CO-sensitive proton reduction. Thus, the CO-sensitive phenotype of the α-Lys191 MoFe protein is only a consequence of alterations in the polypeptide environment of FeMo cofactor rather than a structural alteration in FeMo cofactor itself.

The similar biochemical consequences of either substituting citrate for homocitrate or α-Lys for α-Gln191 can now be understood in light of the structural model because α-Gln191 is normally hydrogen bonded to a terminal carboxylate of homocitrate (Figure 6). The fact that either an alteration in the organic acid attached to the Mo atom or the substitution of an amino acid that coordinates the organic acid leads to CO-sensitive proton reduction indicates that at least a portion of hydrogen evolution catalyzed by the MoFe protein must occur at the FeMo cofactor site and that homocitrate likely plays an integral role in the mechanism of hydrogen evolution. Although the chemical basis for substrate binding and reduction is beyond the scope of this review, we draw the attention of the reader to several recent articles on this subject (13, 19, 24, 37, 71).

Another interesting feature of the α-Lys191 MoFe protein is its ability to catalyze the reduction of acetylene by both two and four electrons to give either ethylene or ethane (72, 73), whereas the wild-type MoFe protein is only able to catalyze the two-electron reduction of acetylene to give ethylene (21). Reduction of acetylene to both ethylene and ethane is also a property of the vanadium-dependent nitrogenase (22). The V-dependent nitrogenase is structurally and functionally similar to the Mo-dependent nitrogenase except that its cofactor contains V rather than Mo (reviewed in 79) (note, the V-containing nitrogenase should not be confused with the NifV- phenotype). Reconstitution of an apo-form of the Mo-dependent nitrogenase with the V-containing cofactor results in a species that can reduce acetylene to both ethylene and ethane (80). Thus, the ability of the V-dependent nitrogenase to reduce acetylene by two and four electrons can be attributed to the nature of the metal composition of its cofactor rather than the cofactor's polypeptide environment. This conclusion is supported by the results of the converse experiment showing that replacement of the V-containing cofactor from the V-dependent nitrogenase with FeMo cofactor results in a hybrid species that can reduce acetylene to ethylene but does not catalyze ethane formation. Nevertheless, the FeMo cofactor–containing V-dependent nitrogenase cannot reduce N_2, which suggests specific differences in the polypeptide interactions between the various cofactor types and their respective proteins (80).

Ethane production catalyzed by the α-Lys191 MoFe protein probably occurs by a mechanism different from that of ethane formation catalyzed by the V-dependent enzyme because the α-Lys191 MoFe protein (*a*) does not exhibit a lag before ethane production, (*b*) exhibits an increased sensitivity to inhibition by CO, (*c*) does not show a temperature dependence of ethylene or ethane formation, and (*d*) catalyzes ethane and ethylene formation in a ratio that does not change with variations in electron flux (72). Since these observations were reported, studies have shown that MoFe protein from a *nifV* mutant of *Rhodobacter capsulatus* can also catalyze the reduction of acetylene by two and four electrons to give ethylene and ethane (54), and this phenotype has also been observed for the MoFe protein isolated from a *nifV*-deficient strain of *A. vinelandii* (L Comaratta & D Dean, unpublished observations). The observations that substitution of V for Mo, an alteration in the organic acid attached to Mo, and substitution of a residue coordinated to homocitrate all lead to profound changes in the catalytic properties of the MoFe protein again point to an important role for homocitrate in the catalytic mechanism. Whether or not this role is a structural one that serves to properly orient FeMo cofactor within the polypeptide pocket or a functional one that involves the direct participation of homocitrate in substrate binding and reduction is not yet known. Another approach, involving the use of various homocitrate analogues to probe the functional role of this organic acid in nitrogenase catalysis has been discussed in detail (53).

The MoFe protein α-His195 was originally targeted for amino acid–substitution studies because primary amino acid sequence comparisons indicated that this residue might be located within an FeMo cofactor-binding domain (73), and a pulsed EPR technique (ESEEM) suggested that a histidine might be covalently coordinated to one of FeMo cofactor's metal atoms (84). To test this possibility, α-His195 was replaced with α-Asn195, which indeed resulted in the disappearance of the ESEEM signal characteristic of the N-coordination to FeMo cofactor and a concomitant loss of N_2 reduction activity (83). In contrast, substitutions for certain other histidine residues, for example α-His83, α-His196, α-His274, or β-His90, did not eliminate either the ESEEM signature or the ability to reduce N_2. Although the most logical interpretation of these data was that α-His195 is directly coordinated to one of FeMo cofactor's metal atoms, the structural model revealed that α-His442, rather than α-His195, is covalently attached to the Mo atom of FeMo cofactor (see Figure 6). In retrospect, ESEEM studies could not have determined that α-His442 provides a coordinating ligand to FeMo cofactor, because substitutions at this position lead to accumulation of an apo-MoFe protein that does not exhibit any EPR spectrum (J Peters, W Newton & D Dean, unpublished data). However, one research group did propose α-His442 as a candidate for providing N-coordination to FeMo cofactor; this prediction was based on the native electrophoretic mobility of an altered MoFe protein with a substitution located near α-His442 (27). Once the X-ray crystallographic data revealed that α-His195 is not co- valently attached to a metal atom of FeMo cofactor but is within hydrogen-bonding distance of one of the bridging sulfides, the ESEEM analysis of altered MoFe proteins with substitutions at this position were revisited (20). Although the new studies confirmed the dramatic decrease in the intensity of the ESEEM signal resulting from the α-Asn195 substitution, they revealed that substitution with α-Gln195 caused no detectable change in the modulation. These results indicate that the observed nitrogen modulation is not directly associated with the hydrogen bond provided by α-His195, but rather with the nitrogen moiety of a different residue whose proximity to the FeMo cofactor is sensitive to certain substitutions at the α-His195 position. Thus, the NH-S hydrogen bond normally provided by α-His195 might be important in positioning FeMo cofactor within the polypeptide pocket.

Characterization and comparison of the catalytic features of the altered α-Asn195 and α-Gln195 MoFe proteins has provided some insight into the nature of the interaction of different substrates with the nitrogenase substrate-reduction site (41). Although N_2 is not reduced by the α-Gln195 MoFe protein, it is an effective competitive inhibitor of both acetylene reduction and proton reduction. In contrast, acetylene and proton reduction catalyzed by the α-Asn195 protein is not inhibited by N_2. The most reasonable interpretation of these results is that the α-Gln195 MoFe protein, but not the α-Asn195 protein, retains the ability to bind N_2. The ability of N_2 to bind altered MoFe proteins and the presence or absence of the ESEEM signal appear to be correlated. The observation that N_2 is not reduced by the α-Gln195 MoFe protein but effectively inhibits both proton reduction and acetylene reduction suggests that N_2 can compete with both acetylene and protons for active-site occupancy. Although these results do not address the question whether different substrates bind to the active site in different ways, bind at different subsites within the active site, or bind at different redox states, they suggest that different substrates cannot bind and be reduced at the active site at the same time.

The other important feature of the α-Gln195 MoFe protein is that, although N_2 slows the overall rate of proton reduction, it does not slow the overall rate of MgATP hydrolysis. In other words, in the presence of N_2, MgATP hydrolysis catalyzed by the α-Gln195 MoFe protein becomes uncoupled from proton reduction. One explanation for this result is that because N_2 can occupy the substrate-reduction site, but cannot be reduced, electrons accumulated within the α-Gln195 MoFe protein are ultimately back-donated to the Fe protein (60). A related possibility is that, once the altered MoFe protein becomes saturated with electrons that cannot be captured by substrate reduction, the Fe and MoFe proteins retain their ability to associate and effect MgATP hydrolysis but can no longer achieve intermolecular electron transfer. This possibility is supported by evidence that nucleotide hydrolysis precedes intermolecular electron transfer (87), although MgATP-dependent proton release appears to be slower than electron transfer (57). The observation that N_2 uncouples MgATP hydrolysis from proton reduction catalyzed by the α-Gln195 MoFe protein, but does so without substantially lowering the overall rate of MgATP hydrolysis, indicates that in the unaltered system the flow of electrons from the Fe protein to the MoFe protein must be partly controlled by the substrate, which acts as an effective electron sink.

SUMMARY AND COMMENTS

Prior to the availability of the three-dimensional structures of the nitrogenase component proteins and their associated metal clusters, considerable data had already accumulated from biochemical characterization of nitrogenases altered by site-directed mutagenesis. The experimental basis of much of this work involved the attempted identification of the metal-cluster ligands within the MoFe protein and probes of the importance of individual amino acids identified within the nucleotide-binding motifs of the Fe protein. When the nitrogenase three-dimensional structures became available in 1992, the majority of the predictions made using biochemical, biophysical, and genetic probes turned out to be correct, although incomplete. Thus, both the validity and limitations of using a biochemical-genetic approach to probe functional aspects of the nitrogenase complex have become evident. Currently, perhaps the more profound contribution of the biochemical-genetic approach is that it has established the feasibility of altering individual features of nitrogenase catalysis by substituting amino acids without necessarily compromising other aspects of the process. Moreover, altered nitrogenase components obtained by site-directed mutagenesis are amenable to purification and biochemical characterization. With the detailed structural information now in hand, the biochemical-genetic analysis of nitrogenase will move away from descriptive issues such as the identification and organization of the major players involved in the catalytic process to the analysis of more mechanistic issues. Namely, now that the involvement of certain residues in catalysis has been established, the challenge is to elucidate the molecular mechanisms or structural features associated with their participation in catalysis.

The potential importance of biological nitrogen fixation research and, more specifically, the study of nitrogenase catalysis with the goal of increasing plant productivity is well documented. With the availability of structural models, however, nitrogenase research has achieved broader significance because the system has emerged as a model for more general biochemical processes, such as signal transduction, protein-protein interaction, inter- and intramolecular electron transfer, and metal center–mediated enzyme catalysis. In this regard, the complexity of the nitrogenase system might be viewed as a drawback in its use as a model for examination of these general biochemical processes. However, from our perspective, the very complexity of nitrogenase, together with the numerous biochemical and biophysical probes that can be used to characterize specific aspects of the catalytic process, make it an ideal model system. This complexity also gives researchers the opportunity to ascertain how these general processes act in concert. For example, how MgATP binding and hydrolysis and signal transduction are linked to effect electron transfer from the Fe protein to the MoFe protein in a manner greatly favoring a unidirectional electron flow to substrate reduction remains an intriguing and fundamental biological question. A better understanding of how these aspects of nitrogenase catalysis are intimately linked is clearly relevant to related processes that also utilize conformational changes induced by nucleotide binding and hydrolysis to drive biochemical reactions.

The most striking feature of nitrogenase catalysis is the dynamics of the process. Namely, two proteins associate and dissociate in a MgATP-dependent manner that leads to electron transfer. In our view, another major challenge of future biochemical-genetic analyses will be to dissect the molecular details involved in this dynamic interaction. Of particular interest will be the construction of altered forms of the nitrogenase components that are trapped in intermediate stages of catalysis. In this regard, we emphasize that the current structural view of the nitrogenase component proteins focuses on their resting states. To date, amino acid–substitution studies strongly indicate that such efforts will be successful. For example, altered forms of the Fe protein that are able to bind MgATP but either can or cannot undergo the nucleotide-induced conformational change are already available. Thus, comparison of the structural features of these altered Fe proteins by means of X-ray crystallographic techniques should be very informative. Another valuable approach is the isolation of altered component proteins that can associate but that fail to dissociate. Again, the feasibility of this approach is already indicated by heterologous mixing experiments and amino acid–substitution studies.

The obvious value of obtaining a stable Fe protein–MoFe protein complex is that structural analysis of the complex should provide detailed information concerning the nature of the component protein interaction and could also provide mechanistic insight by revealing the presence or absence of a significant structural rearrangement around any of the nitrogenase-associated metalloclusters during complex formation. Clearly, this information will provide new targets for amino acid substitution. Also, the availability of a mutant strain that produces an inactive complex caused by alteration of one component protein could then be used in genetic experiments to search for suppressor mutations that lead to an alteration in the other component protein. These methods should allow us to identify residues involved in various aspects of the dynamic reciprocal interactions that must occur between the component proteins during catalysis but that are not necessarily obvious from the available structural models. Finally, the structural and functional analysis of altered component proteins that have a combination of different amino acid substitutions could prove extremely valuable. For example, once investigators have obtained an altered Fe protein that forms a tight inactive complex with the MoFe protein, we can determine the biochemical and structural consequences of altering the ability of this complex to bind or hydrolyze MgATP.

Acknowledgments

Our work is supported by the National Science Foundation (MCB9303800) and the National Institutes of Health (DK-37255). We are grateful to numerous colleagues including Drs. J Howard, B Burgess, J Bolin, and B Hoffman, and the Sussex group, all of whom have made intellectual contributions to our work. We also acknowledge the contributions of our previous and current colleagues at Virginia Tech.

Any *Annual Review* chapter, as well as any article cited in an *Annual Review* chapter, may be purchased from the Annual Reviews Preprints and Reprints service.
1-800-347-8007; 415-259-5017; email: arpr-class.org

Literature Cited

1. Bolin JT, Campobasso N, Muchmore SW, Minor W, Morgan TV, Mortenson LE. 1993. Structure of the nitrogenase MoFe protein: spatial distribution of the intrinsic metal atoms determined by X-ray anomalous scattering. See Ref. 63, pp. 89–94
2. Bolin JT, Campobasso N, Muchmore SW, Morgan TV, Mortenson LE. 1993. Structure and environment of the metal clusters in the nitrogenase molybdenum-iron protein from *Clostridium pasteurianum*. See Ref. 82, pp. 186–95
3. Bolin JT, Ronco AE, Morgan TV, Mortenson LE, Xuong N-H. 1993. The unusual metal clusters of nitrogenase: structural features revealed by X-ray anomalous diffraction studies of the MoFe protein from *Clostridium pasteurianum*. *Proc. Natl. Acad. Sci. USA* 90:1078–82
4. Brigle KE, Newton WE, Dean DR. 1985. Complete nucleotide sequence of the *Azotobacter vinelandii* nitrogenase structural gene cluster. *Gene* 37:37–44
5. Brigle KE, Setterquist RA, Dean DR, Cantwell JS, Weiss MC, Newton WE. 1987. Site-directed mutagenesis of the nitrogenase MoFe protein of *Azotobacter vinelandii*. *Proc. Natl. Acad. Sci. USA* 84:7066–69
6. Brigle KE, Weiss MC, Newton WE, Dean DR. 1987. Products of the iron-molybdenum cofactor-specific biosynthetic genes, *nifE* and *nifN*, are structurally homologous to the products of the nitrogenase molybdenum-iron protein genes, *nifD* and *nifK*. *J. Bacteriol.* 169:1547–53
7. Burgess BK. 1990. The iron-molybdenum cofactor of nitrogenase. *Chem. Rev.* 90:1377–406
8. Burgess BK, Stiefel EI, Newton WE. 1980. Oxidation-reduction properties and complexation reactions of the iron-molybdenum cofactor of nitrogenase. *J. Biol. Chem.* 255:353–56
9. Burris RH. 1991. Nitrogenases. *J. Biol. Chem.* 266:9339–42
10. Chan MK, Kim J, Rees DC. 1993. The nitrogenase FeMo-cofactor and P-cluster pair: 2.2 Å resolution structures. *Science* 260:792–94
11. Chen J, Christiansen J, George SJ, van Elp J, Tittsworth R, et al. 1993. Extended X-ray absorption fine structure and L-edge spectroscopy of nitrogenase molybdenum-iron protein. See Ref. 82, pp. 231–42
12. Conradson SD, Burgess BK, Holm RH. 1988. Flourine-19 chemical shifts as probes of the structure and reactivity of the iron-molybdenum cofactor of nitrognease. *J. Biol. Chem.* 263:13743–49
13. Dance IG. 1994. The binding and reduction of dinitrogen at the Fe_4 face of the FeMo cluster of nitrogenase. *Aust. J. Chem.* 47:979–90
14. Dean DR, Bolin JT, Zheng L. 1993. Nitrogenase metalloclusters: structures, organization, and synthesis. *J. Bacteriol.* 175:6737–44
15. Dean DR, Jacobson MR. 1992. Biochemical genetics of nitrogenase. See Ref. 81, pp. 763–834
16. Dean DR, Scott DJ, Newton WE. 1990. Identification of FeMoco domains within the nitrogenase MoFe protein. In

Nitrogen Fixation: Achievements and Objectives, ed. PM Gresshoff, LE Roth, G Stacey, WE Newton, pp. 95–102. New York: Chapman & Hall. 869 pp.
17. Dean DR, Setterquist RA, Brigle KE, Scott DJ, Laird NF, Newton WE. 1990. Evidence that conserved residues Cys-62 and Cys-154 within the *Azotobacter vinelandii* nitrogenase MoFe protein α-subunit are essential for nitrogenase activity but conserved residues His-83 and Cys-88 are not. *Mol. Microbiol.* 4:1505–12
18. Deits TL, Howard JB. 1990. Effect of salts on *Azotobacter vinelandii* nitrogenase activities. *J. Biol. Chem.* 265:3859–67
19. Deng H, Hoffman R. 1993. How N_2 might be activated by the FeMo-cofactor in nitrogenase. *Agnew. Chem. Int. Ed. Engl.* 32:1062–65
20. DeRose VJ, Kim C-H, Newton WE, Dean DR, Hoffman BM. 1994. Electron spin echo envelope modulation analysis of altered nitrogenase MoFe proteins from *Azotobacter vinelandii*. *Biochemistry* 34:2809–14
21. Dilworth MJ. 1966. Acetylene reduction by nitrogen-fixing preparations from *Clostridium pasteurianum*. *Biochim. Biophys. Acta* 127:285–94
22. Dilworth MJ, Eady RR, Robson RL, Miller RW. 1987. Ethane formation from acetylene as a potential test for vanadium nitrogenase in vivo. *Nature* 327:167–68
23. Dunham WR, Hagen WR, Braaksma A, Grande HJ, Haaker H. 1985. The importance of quantitative Mossbauer spectroscopy of the MoFe-protein from *Azotobacter vinelandii*. *Eur. J. Biochem.* 146:497–501
24. Eady RR, Leigh GJ. 1994. Metals in the nitrogenases. *J. Chem. Soc. Dalton Trans.* 19:2739–47
25. Emerich DW, Ljones T, Burris RH. 1978. Nitrogenase: properties of the catalytically inactive complex between *Azotobacter vinelandii* MoFe protein and the *Clostridium pasteurianum* Fe protein. *Biochim. Biophys. Acta* 527:359–69
26. Georgiadis MM, Komiya H, Chakrabarti P, Woo D, Kornuc JJ, Rees DC. 1992. Crystallographic structure of the nitrogenase iron protein from *Azotobacter vinelandii*. *Science* 257:1653–59
27. Govezensky D, Zamir A. 1989. Structure-function relationships in the α-subunit of *Klebsiella pneumoniae* nitrogenase MoFe protein from analysis of *nifD* mutants. *J. Bacteriol.* 171:5729–35
28. Hausinger RP, Howard JB. 1983. Thiol reactivity of the nitrogenase Fe-protein from *Azotobacter vinelandii*. *J. Biol. Chem.* 258:13486–92
29. Hawkes TR, McLean PA, Smith BE. 1984. Nitrogenase from *nifV* mutants of *Klebsiella pneumoniae* contains an altered form of the iron-molybdenum cofactor. *Biochem. J.* 217:317–21
30. Holland D, Zilberstein A, Zamir A, Sussman J. 1987. A quantitative approach to sequence comparisons of nitrogenase MoFe protein α- and β-subunits including the newly sequenced *nifK* gene from *Klebsiella pneumoniae*. *Biochem. J.* 247:277–85
31. Hoover TR, Imperial J, Liang J, Ludden PW, Shah VK. 1988. Dinitrogenase with altered substrate specificity results from the use of homocitrate analogues for in vitro synthesis of the iron-molybdenum cofactor. *Biochemistry* 27:3647–52
32. Hoover TR, Imperial J, Ludden PW, Shah VK. 1988. Homocitrate cures the NifV⁻ phenotype in *Klebsiella pneumoniae*. *J. Bacteriol.* 170:1978–79
33. Hoover TR, Imperial J, Ludden PW, Shah VK. 1989. Homocitrate is a component of the iron-molybdenum cofactor of nitrogenase. *Biochemistry* 28:2768–71
34. Howard JB. 1993. Protein component complex formation and adenosine triphosphate hydrolysis in nitrogenase. See Ref. 82, pp. 271–89
35. Howard JB, Davis R, Moldenhauer B, Cash VL, Dean DR. 1989. Fe-S cluster ligands are the only cysteines required for nitrogenase Fe-protein activities. *J. Biol. Chem.* 264:11270–74
36. Howard JB, Rees DC. 1994. Nitrogenase: a nucleotide-dependent molecular switch. *Annu. Rev. Biochem.* 63:235–64
37. Hughes DL, Ibrahim SK, Querne G, Laouenen A, Talarmin J, et al. 1994. On carboxylate as a leaving group at the active site of Mo nitrogenase: electrochemical reactions of some molybdenum and tungsten carboxylates, formation of mono-, di- and tri-hydrides and the detection of an $MoH_2(N)_2$ intermediate. *Polyhedron*. In press
38. Jacobson MR, Cantwell JS, Dean DR. 1990. A hybrid *Azotobacter vinelandii–Clostridium pasteurianum* iron protein that has in-vivo and in-vitro catalytic activity. *J. Biol. Chem.* 265:19429–33
39. Kent HM, Baines M, Gormal C, Smith BE, Buck M. 1990. Analysis of site-directed mutations in the α- and β-subunits of *Klebsiella pneumoniae*

nitrogenase. *Mol. Microbiol.* 4:1497–504

40. Kent HM, Ioannidis I, Gormal C, Smith BE, Buck M. 1989. Site-directed mutagenesis of *Klebsiella pneumoniae* nitrogenase. *Biochem. J.* 264:257–64
41. Kim C-H, Newton WE, Dean DR. 1994. The role of the MoFe protein α-subunit histidine-195 residue in FeMo-cofactor binding and nitrogenase catalysis. *Biochemistry* 34:2798–808
42. Kim C-H, Zheng L, Newton WE, Dean DR. 1993. Intermolecular electron transfer and substrate reduction properties of MoFe proteins altered by site-specific amino acid substitution. See Ref. 63, pp. 105–10
43. Kim J, Rees DC. 1992. Structural models for the metal centers in the nitrogenase molybdenum-iron protein. *Science* 257:1677–82
44. Kim J, Rees DC. 1992. Crystallographic structure and functional implications of the nitrogenase molybdenum-iron protein from *Azotobacter vinelandii*. *Nature* 360:553–60
45. Kim J, Rees DC. 1994. Nitrogenase and biological nitrogen fixation. *Biochemistry* 33:389–97
46. Kim J, Woo D, Rees DC. 1993. X-ray crystal structure of the nitrogenase molybdenum-iron protein from *Clostridium pasteurianum* at 3.0-Å resolution. *Biochemistry* 32:7104–15
47. Kraulis PJ. 1991. Molscript: A program to produce both detailed and schematic plots of protein structures. *J. Appl. Crystallogr.* 24:946–50
48. Kurtz DM, McMillan RS, Burgess BK, Mortenson LE, Holm RH. 1979. Identification of iron-sulfur centers in the iron-molybdenum proteins of nitrogenase. *Proc. Natl. Acad. Sci. USA* 76:4986–89
49. Lammers PJ, Haselkorn R. 1983. Sequence of the *nifD* gene coding for the α-subunit of dinitrogenase from the cyanobacterium *Anabaena*. *Proc. Natl. Acad. Sci. USA* 80:4723–27
50. Liang J, Madden M, Shah VK, Burris RH. 1990. Citrate substitutes for homocitrate in the nitrogenase of a *nifV* mutant of *Klebsiella pneumoniae*. *Biochemistry* 29:8577–81
51. Lowe DJ, Fisher K, Thorneley RNF. 1993. *Klebsiella pneumoniae* nitrogenase: Pre-steady state absorbance changes show that redox changes occur in the MoFe protein that depend on substrate and component protein ratio; a role for P-centres in reducing nitrogen. *Biochem. J.* 292:93–98
52. Lowery RG, Chang CL, Davis LC, McKenna M-C, Stephens PJ, Ludden PW. 1989. Substitution of histidine for arginine-101 of dinitrogenase reductase disrupts electron transfer to dinitrogenase. *Biochemistry* 28:1206–12
53. Ludden PW, Shah VK, Roberts GP, Homer M, Allen R, et al. 1993. Biosynthesis of the iron-molybdenum cofactor of nitrogenase. See Ref. 82, pp. 196–215
54. Masepohl B, Angermuller S, Hennecke S, Hubner P, Moreno-Vivian C, Klipp W. 1993. Nucleotide sequence and genetic analysis of the *Rhodobacter capsulatus* ORF6-*nifUISVW* gene region: possible role of NifW in homocitrate processing. *Mol. Gen. Genet.* 238:369–82
55. May HD, Dean DR, Newton WE. 1991. Altered nitrogenase MoFe proteins from *Azotobacter vinelandii*. *Biochem. J.* 277:457–64
56. McLean PA, Papaefthymiou V, Orme-Johnson WH, Munck É. 1987. Isotopic hybrids of nitrogenase: Mossbauer study of MoFe protein with selective ^{57}Fe enrichment of the P-cluster. *J. Biol. Chem.* 262:12900–3
57. Mensink RE, Wassink H, Haaker H. 1992. A reinvestigation of the pre-steady-state ATPase activity of the nitrogenase from *Azotobacter vinelandii*. *Eur. J. Biochem.* 208:289–94
58. Mortenson LE, Seefeldt LC, Morgan TV, Bolin J. 1993. The role of metal clusters and MgATP in nitrogenase catalysis. *Adv. Enzymol.* 67:299–373
59. Newton WE. 1992. Isolated iron-molybdenum cofactor of nitrogenase. See Ref. 81, pp. 877–930
60. Orme-Johnson WH, Davis LC. 1977. Current topics and problems in the enzymolgy of nitrogenase. In *Iron-Sulfur Proteins*, ed. W Lovenberg, pp. 15–60. New York: Academic. 443 pp.
61. Orme-Johnson WH, Holm RH. 1978. Identification of Fe-S clusters in proteins. *Methods Enzymol.* 53:268–74
62. Page WJ, von Tigerstrom M. 1979. Optimal conditions for transformation of *Azotobacter vinelandii*. *J. Bacteriol.* 139:1058–61
63. Palacios R, Mora J, Newton WE, eds. 1993. *New Horizons in Nitrogen Fixation*. Dordrecht: Kluwer Academic. 788 pp.
64. Paustian TD, Shah VK, Roberts GP, 1989. Purification and characterization of the *nifN* and *nifE* gene products from *Azotobacter vinelandii* mutant UW45. *Proc. Natl. Acad. Sci. USA* 86:6082–86
65. Peters JW, Fisher K, Dean DR. 1994. Identification of a nitrogenase protein-

protein interaction site defined by residues 59 through 67 within the *Azotobacter vinelandii* Fe protein. *J. Biol. Chem.* 269:28076–83
66. Pope MR, Murrell SA, Ludden PW. 1985. Covalent modification of the iron protein of nitrogenase from *Rhodospirillum rubrum* by adenosine diphosphoribosylation of a specific arginine residue. *Proc. Natl. Acad. Sci. USA* 82:3173–77
67. Rees DC, Chan MK, Kim J. 1993. Structure and function of nitrogenase. *Adv. Inorg. Chem.* 40:89–119
68. Rees DC, Kim J, Georgiadis MM, Komiya H, Chirino AJ, et al. 1993. Crystal structures of the iron protein and the molybdenum-iron protein of nitrogenase. See Ref. 82, pp. 170–85
69. Robinson AC, Burgess BK, Dean DR. 1986. Activity, reconstitution, and accumulation of nitrogenase components in *Azotobacter vinelandii* mutant strains containing defined deletions within the nitrogenase structural gene cluster. *J. Bacteriol.* 166:180–86
70. Robson RL. 1984. Identification of possible adenine nucleotide-binding sites in nitrogenase Fe- and MoFe-proteins by amino acid sequence comparison. *FEBS Lett.* 173:394–98
71. Schrauzer GN, Doemeny PA, Palmer JG. 1993. The chemical evolution of a nitrogenase model. XXII. Reduction of acetylene with catalysts derived from molybdate, homocitric acid and N-methylimidazole and a proposal concerning the active site of functional *Azotobacter* nitrogenases. *Z. Naturforsch.* 48b:1295–98
72. Scott DJ, Dean DR, Newton WE. 1992. Nitrogenase-catalyzed ethane production and CO-sensitive hydrogen evolution from MoFe proteins having amino acid substitutions in an α-subunit FeMo cofactor binding domain. *J. Biol. Chem.* 267:20002–10
73. Scott DJ, May HD, Newton WE, Brigle KE, Dean DR. 1990. Role for the nitrogenase MoFe protein α-subunit in the FeMo-cofactor binding and catalysis. *Nature* 343:188–90
74. Seefeldt LC. 1994. Docking of nitrogenase iron- and molybdenum-iron proteins for electron transfer and MgATP hydrolysis: the role of arginine 140 and lysine 143 of the *Azotobacter vinelandii* iron protein. *Protein Sci.* 3:2073–81
75. Seefeldt LC, Morgan TV, Dean DR, Mortenson LE. 1992. Mapping the sites of MgATP and MgADP interaction with the nitrogenase of *Azotobacter vinelandii*: Lysine 15 of the Fe protein plays a major role in MgATP interaction. *J. Biol. Chem.* 267:6680–88
76. Seefeldt LC, Mortenson LE. 1993. Increasing nitrogenase catalytic efficiency for MgATP by changing serine 16 of its Fe protein to threonine: use of Mn^{++} to show interaction of serine 16 with Mg^{++}. *Protein Sci.* 2:93–102
77. Shah VK, Brill WJ. 1977. Isolation of an iron-molybdenum cofactor from nitrogenase. *Proc. Natl. Acad. Sci. USA* 74:3249–53
78. Smith BE, Buck M, Faridoon KY, Gormal CA, Howes BD, et al. 1993. The metallo-sulphur centres of the nitrogenase MoFe protein from wild-type and mutant strains of *Klebsiella pneumoniae*. *J. Inorg. Biochem.* 51:357
79. Smith BE, Eady RR. 1992. Metalloclusters of the nitrogenases. *Eur. J. Biochem.* 205:1–15
80. Smith BE, Eady RR, Lowe DJ, Gormal C. 1988. The vanadium-iron protein of the vanadium nitrogenase from *Azotobacter chroococcum* contains an iron-vanadium cofactor. *Biochem. J.* 250:299–302
81. Stacey G, Burris RH, Evans HJ, eds. 1992. *Biological Nitrogen Fixation.* New York: Chapman & Hall. 943 pp.
82. Stiefel EI, Coucouvanis D, Newton WE, eds. 1993. *Molybdenum Enzymes, Cofactors and Model Systems.* Washington, DC: Am. Chem. Soc. 387 pp.
83. Thomann H, Bernardo M, Newton WE, Dean DR. 1991. N coordination of the FeMo cofactor requires His-195 of the MoFe protein α subunit and is essential for biological nitrogen fixation. *Proc. Natl. Acad. Sci. USA* 88:6620–23
84. Thomann H, Morgan TV, Jin H, Burgmayer SJN, Bare RE, Stiefel EI. 1987. Protein nitrogen coordination to the FeMo center of nitrogenase from *Clostridium pasteurianum*. *J. Am. Chem. Soc.* 109:7913–14
85. Thöny B, Kaluza K, Hennecke H. 1985. Structural and functional homology between the α and β subunits of the nitrogenase MoFe protein as revealed by sequencing the *Rhizobium japonicum nifK* gene. *Mol. Gen. Genet.* 198:441–48
86. Thorneley RNF, Ashby GA, Fisher K, Lowe DJ. 1993. Electron-transfer reactions associated with nitrogenase from *Klebsiella pneumoniae*. See Ref. 82, pp. 290–302
87. Thorneley RNF, Ashby GA, Howarth JV, Millar NC, Gutfreund H. 1989. A transient-kinetic study of the nitrogenase of *Klebsiella pneumoniae* by stopped-flow calorimetry. *Biochem. J.* 264:657–61

88. Willing AH, Georgiadis MM, Rees DC, Howard JB. 1989. Cross-linking of nitrogenase components. *J. Biol. Chem.* 264:8499–503
89. Willing AH, Howard JB. 1990. Cross-linking site in *Azotobacter vinelandii* complex. *J. Biol. Chem.* 265:6596–99
90. Wolle D, Dean DR, Howard JB. 1992. Nucleotide iron-sulfur cluster signal transduction in the nitrogenase iron-protein: the role of Asp[125]. *Science* 258:992–95
91. Wolle D, Kim C-H, Dean DR, Howard JB. 1992. Ionic interactions in the nitrogenase complex: properties of Fe-protein containing substitutions for Arg-100. *J. Biol. Chem.* 267:3667–73
92. Yates MG. 1992. The enzymology of molybdenum-dependent nitrogen fixation. See Ref. 81, pp. 685–735

CONJUGATIVE TRANSPOSITION

June R. Scott and Gordon G. Churchward
Department of Microbiology and Immunology, Emory University, Atlanta, Georgia 30322

KEY WORDS: conjugation, transposition, recombination, resistance, antibiotic, integrase

CONTENTS

INTRODUCTION	368
General Properties	368
Simple and Compound Transposons	369
Drug Resistance	369
Broad Distribution	370
Use in Genetic Analysis	371
TRANSPOSITION	371
Transposition Is by Excision and Insertion	371
Activity of the Circular Intermediate	373
Polarity of Strand Cleavage	373
Length of the Coupling Sequence	374
Target Sites	375
STRUCTURE	377
Transposon Ends	377
DR-2 Repeats	378
tet(M)	378
Genes to the Right of tet(M)	379
Genes to the Left of tet(M): int and xis	380
Other Open Reading Frames to the Left of tet(M)	381
CONJUGATION	381
Is There Cotransfer of Other DNA?	381
Excision Occurs in the Donor Prior to Conjugation	381
Transfer of a Single Strand During Conjugation	382
Implications for Gene Order During Transfer	384
REGULATION OF CONJUGATIVE TRANSPOSITION	384
Regulation at the Level of Excision	384
Coordinate Regulation of Multiple Transposon Copies	385
Regulation of Expression of Proteins Involved in Transposition and Conjugation	385
Truncated Int Proteins	386
Regulation by Truncated Forms of Int	386
MECHANISM OF RECOMBINATION	386
Lack of Branch Migration During Recombination	386
Binding of Int to Transposon and Target DNA	387
Formation of a Nucleoprotein Complex at the Transposon Ends	389
Host Factors	390

Role of Xis	391
Role of Int in Excision and Integration	391

ABSTRACT

Conjugative transposons are important determinants of antibiotic resistance, especially in gram-positive bacteria. They are remarkably promiscuous and can conjugate between bacteria belonging to different species and genera. Transposon-promoted conjugation may be similar to F plasmid-promoted conjugation, as it appears that only one strand of the transposon DNA is transferred from donor to recipient. The recent determination of the entire nucleotide sequence of Tn916 allowed us to make specific predictions about the possible function of different open reading frames and the position of a (hypothetical) origin of transfer. The mechanism of recombination during conjugative transposition differs from that of other transposons, as shown by the absence of a duplication of the target sequence upon integration. The current model for recombination postulates that staggered double-stranded cleavages occur at each end of the transposon. One DNA strand is cut six bases from the end of the transposon, and the other strand is cut immediately adjacent to the end. The ends of the excised transposon are then ligated to form a circular intermediate with a six-base heteroduplex. Staggered cleavages of the circular intermediate and the target DNA allow the transposon to insert into the target, where it is flanked by heteroduplex regions that are resolved by replication. All hosts examined contain preferential target sites; these are not specific sequences but apparently consist of bent DNA. The site-specific recombinases encoded by conjugative transposons belong to the integrase family. Like phage λ integrase, the integrase of Tn916 has two DNA-binding domains that recognize different sequences, one within the ends of the element and one that includes target DNA. The affinity of Tn916 integrase for target sites correlates with the frequency of integration into a particular site. The similarity between conjugative transposons and phage λ is striking and suggests that both use the same mechanism of recombination. In λ, however, recombining sites must be homologous. Homology may be necessary because of branch migration, which is thought to occur during recombination. In conjugative transposition, the recombining sites are nearly always different, and therefore branch migration probably does not occur. This review presents a speculative model for the alignment of the ends of Tn916 during excision that was adapted from one recently proposed for λ.

INTRODUCTION

General Properties

Conjugative transposons resemble other types of transposable elements because they move from one DNA molecule to another. Unlike most other transposons, however, conjugative transposons do not cause a duplication of target sequences during insertion (18, 48), which indicates an important difference in their mechanism of transposition (see section on transposition, below). They were termed conjugative (32) because the donor molecule is usually in a different cell from the recipient. These elements can cause grampositive bacteria to be conjugational donors, and apparently, the only DNA transferred between the mating bacteria is the transposon (see section on conjugation, below). Recent reviews of conjugative transposons include those by Scott (88–89) and Clewell & Flannagan (17).

Simple and Compound Transposons

The best-studied conjugative transposons are the closely related elements Tn916 (30) and Tn1545 (11, 15, 21). Tn916, which is 18 kilobases (kb) long (29a), appears so far to be the smallest of these conjugative elements. Among clinical isolates of gram-positive bacteria, much larger conjugative transposons—greater in size than 50 kb—have also been identified (40, 42, 43, 54, 55, 93, 94, 104, 105).

In several of the larger conjugative transposons, a Tn916-like element has apparently inserted into a second unrelated transposon, and both are still capable of independent transposition (16, 54). This appears to be the case, for example, for Tn5253, which was originally found in a strain of *Streptococcus pneumoniae* and formerly called omega *cat-tet*. Tn5253 is composed of the Tn916-like element Tn5251 within another element, Tn5252. Both Tn5251 and Tn5252 act independently as transposons (2).

In contrast, when the Tn916-like transposon Tn3703 excises from the large conjugative transposon called Tn3701 (originally found in a strain of *Streptococcus pyogenes*), whether the remaining part of Tn3701 can transpose is unclear. Furthermore, Tn3703 is unable to transfer by conjugation from strains of *Enterococcus faecalis* (56), although it can conjugatively transpose from streptococci to enterococci and other streptococci. A similar large element, Tn3705, was recently identified in the chromosome of *Streptococcus anginosus* and contains within it the Tn916-related element Tn3704 (16). Although Tn3704 conjugatively transposes independently of the larger element, its restriction map is significantly different from that of Tn916. After Tn3704 excises from the larger element, whether the remaining fragment transposes is unknown. Additional variants on the theme of conjugative transposons will probably be detected from gram-positive bacteria associated with disease.

Drug Resistance

Conjugative transposons are very broadly distributed in nature and are common in clinical isolates, especially of gram-positive pathogens. They are often associated with transmissible drug resistance in strains of different groups of streptococci that contain no plasmids (24, 39, 42, 43, 55). All conjugative transposons isolated from pathogenic bacteria carry *tet*(M), (8) or a closely related gene for tetracycline resistance (95). *tet*(M) is expressed in all bacteria for which this property has been assayed.

Although Tn916 carries only *tet*(M), Tn1545 carries in addition the *aphA-3* gene for kanamycin resistance and the *ermAM* gene for erythromycin resistance (21, 72). Larger elements may include genes for resistance to chloramphenicol and the macrolide-lincomycin-streptogramin group (40, 42, 43, 54, 55, 93, 94, 104, 105).

Two elements, named Tn5276 (78) and Tn5301 (41), have been identified in *Lactococcus lactis*. They each carry genes for production of the lantibiotic nisin and for immunity to it. These two transposons, which probably have only minor differences from each other, apparently do not encode tetracycline resistance, and thus are the first exceptions to the rule that all conjugative transposons carry *tet*(M).

Other types of conjugative elements have been identified in species of *Bacteroides* and in *Clostridium difficile* (reviews include 36, 86, 99; see also 66, 83). As too little is currently known about these elements to determine how similar they are to better-studied conjugative transposons, they are not discussed further here.

Broad Distribution

Transfer of conjugative transposons is very promiscuous. They are transferred to strains of different species or even of different genera with the same frequency that they are transferred among isogenic strains that differ only in a single genetic marker (89). Thus, these elements do not seem to be subject to host restriction during conjugation (see section on structure, below). The promiscuous self-transmissibility of these elements and their carriage of the universally expressed *tet*(M) gene, sometimes with additional drug resistance genes, make conjugative transposons a serious clinical problem.

Conjugation mediated by these elements has not been demonstrated convincingly in the laboratory in matings involving gram-negative bacteria. When conjugative transposons are introduced experimentally into *Escherichia coli*, however, they are capable of transposition (19, 22). In addition, the *tet*(M) gene, and sometimes adjacent transposon DNA, has been identified in clinical isolates of some gram-negative pathogens, including *Neisseria gonorrhoeae* and *Neisseria meningiditis* (65, 98). Therefore, although conjugative transposons may be unable to conjugate into gram-negative bacteria independently, they may use other mechanisms (e.g. transduction, transformation of naturally competent cells, or transmission on conjugative plasmids) to enter these bacteria and spread among them.

Use in Genetic Analysis

In some organisms, conjugative transposition is the only form of conjugation that has been described. Because of the ease with which these transposons can be introduced into most bacterial strains, they have been used to generate insertional mutations in bacteria for which there are few if any other types of transposons known. For this purpose, a DNA fragment consisting of the ends of Tn*1545* surrounding a drug resistance marker was inserted into a suicide vector (103). The vector, which replicates in *E. coli* using its pACYC184 origin, has the transfer origin of the IncP broad-host-range conjugative plasmid RK2, and thus can act as a delivery system for the transposon from *E. coli* to gram-positive bacteria.

Conjugative transposons have also been used as shuttle vectors because in some bacteria conjugative transposition is the best method available for introduction of DNA and because it can result in the presence of only a single copy of the introduced gene per chromosome (14, 69). In this process, a cloned gene of interest is inserted into a nonessential site of Tn*916*, which is introduced by transformation, electroporation, or conjugation into the desired host. If necessary, an intermediate gram-positive host that is transformable can be used for this process, and the transposon can then be moved to its final host by conjugative transposition. A resulting transconjugant can be chosen that carries the introduced DNA in a single transposon copy on the bacterial chromosome.

TRANSPOSITION

Transposition Is by Excision and Insertion

The mechanism by which conjugative transposons move from one DNA molecule to another differs from that of all other elements studied. This was first apparent when Clewell et al (18) determined that Tn*916* does not generate duplications at the target site into which it inserts. Instead, conjugative transposons move by a mechanism involving excision from the donor molecule, circularization of the transposon, and insertion into a target site in the recipient molecule (90).

The circular intermediate is formed by staggered nicks at each end of the transposon. One nick of each pair is made in the donor DNA six bases from the transposon end, and the other nick is immediately adjacent to the end of the transposon (13; see Figure 1*a*). The six donor DNA bases at each end of the transposon included in the circular transposon form are called coupling sequences. Because the bases flanking the transposon are not related to each other, the coupling sequences present on the two strands are not complementary. Thus, the circular intermediate contains mismatches at its joint (13).

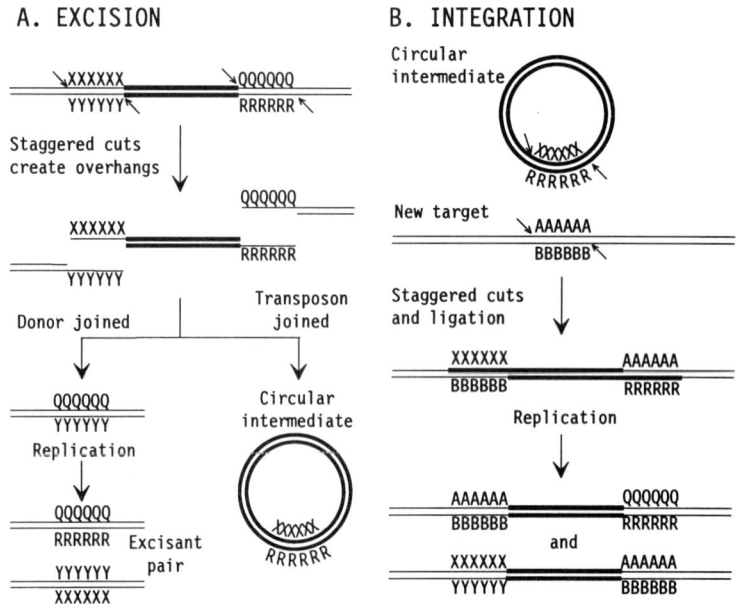

Figure 1 A model for excision and integration of Tn916. The thick lines represent Tn916, and the thin lines represent the DNA adjacent to the transposon. Coupling sequences are indicated by the hypothetical complementary nucleotide pairs X-Y, Q-R, and A-B. (*a*) Staggered cleavages (*arrows*) of the phosphodiester backbone on the 5′ side of the coupling sequence on both strands create molecules shown with 5′ single-stranded ends. The ends are joined to generate an excisant molecule and a circular transposon intermediate, both of which contain heteroduplexes consisting of the base pairs originally present in the coupling sequences. Semiconservative replication resolves the heteroduplex in the excisant and generates a pair of molecules (excisant pair), one of which has the left coupling sequence at the site of excision and the other the right. (*b*) Staggered cleavages in the circular intermediate and the new target, followed by ligation create a new insertion of Tn916 with a heteroduplex at each end. Replication resolves the heteroduplexes and generates a pair of molecules in which each member is flanked by the target sequence at one end and a coupling sequence from the previous insertion at the other end.

The second product of conjugative transposon excision is produced by ligation of the target site from which the transposon has excised. This molecule, called the excisant, contains a six-base region of potentially unpaired bases consisting of the coupling sequences that flanked the transposon. Replication of the excisant molecule resolves this heteroduplex, and both types of excisant products have been identified (13, 76). In one of the two types of excisant, the original DNA sequence present prior to transposon insertion is restored. In the other, a substitution of six base pairs occurs.

Because the excised circular transposon form cannot replicate, the heteroduplex region formed by the unpaired coupling sequences is present in the covalently closed circle form and was directly identified as a mismatch (13). When the circular transposon molecule integrates into a new target site (Figure 1b), it brings with it both six-base coupling sequences. There is no preference for insertion at sites related to either of these coupling sequences, so insertion generates a mismatched region consisting of one coupling sequence strand and one target sequence strand at each end of the transposon. These heteroduplexes are resolved by replication of the recipient molecule to generate two types of insertions: one with the coupling sequence from one strand of the circular intermediate and one with the coupling sequence from the other strand (Figure 1b).

The lantibiotic-producing transposons Tn5276 (78) and Tn5301 (41) differ from Tn916 and Tn1545 in that a duplication of a six-base sequence of the preferred target site flanks the element following insertion (41, 78). This observation suggests important mechanistic differences from Tn916.

Activity of the Circular Intermediate

Excision of Tn916 from some plasmids in *E. coli* occurs at a frequency high enough to detect the circular excised transposon molecule on an agarose gel (90). This circular form of the transposon transforms protoplasts of *Bacillus subtilis* at a frequency 150-fold higher than does the plasmid containing the transposon, which indicates that the circular form of Tn916 can act as an intermediate in transposition in gram-positive hosts. The transformation does not, however, prove that the circular form is an obligate intermediate.

Further evidence that the circular transposon excised from *E. coli* can act as an intermediate in transposition is provided by the observation that both of the expected coupling sequences can be identified among the *B. subtilis* protoplasts transformed with this transposon molecule. As expected, each transformant contains only one of these coupling sequences. Thus, a coupling sequence can be traced through two generations.

Polarity of Strand Cleavage

The excision reaction probably results from staggered endonucleolytic scissions, as shown in Figure 1. No information is available about the actual polarity of strand cleavage. The coupling sequence that was located on the left of the transposon prior to excision can end up on either side of the inserted transposon, suggesting that the nick can be made on either side of the coupling sequence of the circular form. However, the same enzyme can probably not cleave DNA to leave both 5' and 3' protruding ends. A more likely explanation for the observed pattern of inheritance of coupling sequences is that it is caused by mismatch repair of the circular intermediate upon introduction into the recipient and/or repair of the heteroduplex regions present in the target DNA molecule after insertion.

Length of the Coupling Sequence

The number of Ts at the right end of the excised Tn916 varies (12, 13, 18, 76), so defining the transposon ends precisely is difficult. Another factor complicates the determination of the length of the coupling sequences: conjugative transposition occurs naturally and has been studied mainly in bacteria with a high A+T content. Furthermore, conjugative transposons tend to insert at a site containing an A-rich sequence separated by about six bases from a T-rich sequence (91). Therefore, an A or T belonging to the transposon is often difficult to distinguish from one belonging to the coupling sequence or the target. For these reasons, until quite recently, investigators could not determine whether the coupling sequence was five, six, or seven bases long by comparing the sequence flanking the inserted transposon with the independently obtained sequence of the unoccupied target site before insertion (12, 13, 18, 76), although various authors had chosen to report that it was one of these. Experiments performed in *E. coli* using a minitransposon and a defined target provided information that was compatible with a six-base but not a five- or seven-base coupling sequence, although the sequence of the unoccupied target was not presented (102).

To define the coupling sequence length in natural transposition in a gram-positive organism, two conjugative donors were studied in which the composition of the entering coupling sequence was known (85). One case showed that the coupling sequence introduced with Tn916 was six and not five bases, and the other showed it to be six and not seven. Therefore, unless the length of the coupling sequence can vary, it must be six bases long.

All of the published data are consistent with the coupling sequence being six bases long. When the authors have suggested that it is five or seven, the data can be interpreted differently. For example, Ike et al (44) have presented the structure of several insertions into the *cylA* gene. They analyzed this data assuming that all insertions occurred at the same site in the target, following CTTT and preceding AAAATAG, and that the transposon always had five Ts at the right end. In one case (class 3), they show a seven-base coupling sequence adjacent to the right end of Tn916 and a five-base coupling sequence at the left end. However, if one assumes that in this case the transposon has inserted one base to the right compared with the other insertions, then the insertion can be written CTTTA TAGATA Tn916 ATGTAA AAATAG, consistent with a six-base coupling sequence (underlined) at each end. Class 4 sequences have a five-base coupling sequence at the left end. In this case, if the transposon is considered to have inserted at the same site as in the previous case, but with four Ts at the right end, the insertion can be written CTTTA TAGATA Tn916 TAGTTA AAATAG, again consistent with six-base coupling sequences. There is precedent for an insertional hot spot really being a hot region of several bases (91) (see section on target sites, below). Rice & Carias (81) report that the coupling sequence is sometimes five bases and that the joint of the circular transposon molecule they studied is composed of an unequal number of bases on each strand. However, their data on the sequence of the joint in excised Tn*5381* molecules can be explained in the same way as the class 4 example of Ike et al (44).

Thus, the simplest interpretation of a constant length of six bases for the coupling sequence is still consistent with all published information on conjugative transposons. However, coupling sequences might have different lengths, which could possibly arise by two means. Because these sequences result from the location of the staggered nicks made during transposon excision, variable-length coupling sequences might result from imprecision in the measurement by the enzyme that cleaves the DNA. In addition, mutations that change the spacing between the cleavage sites might produce different-length coupling sequences. This kind of behavior has precedents in the phage λ system. Normally, recombination occurs following cleavages that are seven bases apart (61), but for certain mutant sites that have lost or gained a base between the cleavage sites, the cleavages may be separated by six or eight bases, respectively (26, 84).

That excision occurs sometimes with five Ts and sometimes with four Ts at the right end of the transposon can be explained by variability in the site of cleavage. However, other factors may be involved in this phenomenon. Integration always produces a minimum of five Ts at the right end of the integrated transposon (85). When only four Ts have been inherited from the transposon, the fifth T is present in the target or coupling sequence. The meaning of this observation is not clear, but it points to a more complex mechanism than variability in the site of cleavage.

Target Sites

Many different sites for conjugative transposon insertion were compiled previously (91). Probably the most striking aspect of the target sequences for Tn*916* and Tn*1545* insertion is that any six bases can constitute a coupling sequence. This implies that these bases act only as a spacer in the insertion and excision reactions and are not important for direct protein recognition.

Although the degree of homology with sequences near the end of the transposon may determine target preference (102), examination of the collection of target sequences indicates that this is not always the case (91). The only consistent characteristic present in all targets is an A-rich sequence separated by about six bases from a T-rich sequence. Renault et al (80), who looked at introduction of a mini-Tn*1545* into *L. lactis* by electroporation, also failed to find a consensus target sequence. Sequences flanking randomly isolated insertions were determined, and the only consistent aspect of the target sites seemed to be their A+T richness.

In several recent studies, a preferred target site was deduced. For example, insertions into the *cylA* gene, which is thought to regulate expression of hemolysin in the *E. faecalis* plasmid pAD1, had all occurred within one base of each other in a single orientation (44). In another gram-positive bacterial system, a chloramphenicol acetyl transferase (*cat*) gene in the large conjugative plasmid pIP501 was used as a target (91). Tn*916* was introduced by conjugation into *L. lactis* carrying pIP501, and insertions into the plasmid were recovered by conjugational transfer of pIP501 into an *E. faecalis* strain. About half the transferred plasmids contained the transposon in a site within *orfA,* and these transposons were found in both orientations.

Secondary preferred target sites were also identified in this study (91). Independent insertions into the same three different sites of *cat* occurred several times. In one case, the site was actually a region consisting of about four bases. Because these experiments were performed with three different donors, each of which had a different coupling sequence, it could be concluded that, in conjugative transposition in gram-positive bacteria, the coupling sequence composition does not affect the choice of a target site (91). Similar results were obtained when mini-Tn*1545* was introduced by electroporation into *L. lactis.* The same preferred chromosomal target sites were identified using different donors (80).

Trieu-Cuot et al (102) studied a synthetic 32-bp target complementary to the joined ends of the circular intermediate and inserted into a mobilizable plasmid. In *E. coli,* insertions of a minitransposon derivative—consisting of a kanamycin-resistance gene flanked by 203 bp from one end of Tn*1545* and 138 bp from the other—into the plasmid were collected by mating into a naive strain. The minitransposon was complemented by a second plasmid encoding the integrase protein (Int). All but one of the mobilized plasmids selected for the transposon marker had the minitransposon inserted into the target sequence and not elsewhere on the plasmid, and insertions occurred in both orientations. When this 32-bp sequence was placed in the chromosome of *E. faecalis,* however, it did not act as a preferential target site. Therefore, although the authors interpreted their results as indicating that the degree of homology with sequences at the transposon ends determines the preferential use of the target, it seems more likely that the transposon does not recognize a specific sequence, but rather some conformational aspect of the DNA into which it inserts (see section on mechanism of recombination, below).

Therefore, the targets for conjugational transposition do not appear to be random. Furthermore, some targets are highly preferred, although they have no detectable consensus. Recent studies (59a) indicate that some target sites are distinguishable by the presence of an intrinsic bend, even in the absence of protein (see section on mechanism of recombination, below).

STRUCTURE

Transposon Ends

The ends of Tn*1545* and Tn*916* are identical for at least 186 bp at the left and 108 bp at the right. Because of different conventions, the right end of Tn*916* is considered the left of Tn*1545* (12, 18). We use the Tn*916* convention. An imperfect 26-base inverted repeat resides at the ends of these elements (18, 75) (Figure 2). However, the precise definition of the right end of these conjugative transposons is difficult because it includes a variable number of Ts (12, 13, 18, 76). In contrast, the ends of the nisin-encoding transposons are not homologous to those of Tn*916* and do not include any inverted repeats (41, 79).

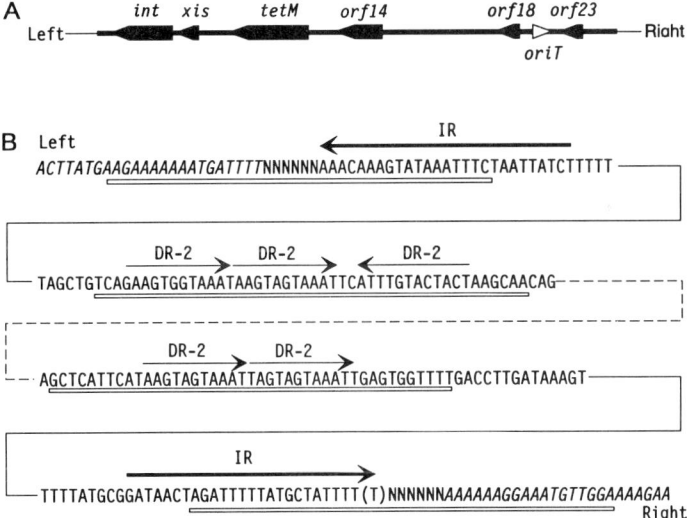

Figure 2 (*a*) Map of Tn*916* showing the relative position and orientation of genes. The gene symbols on the map are not drawn to scale compared with the length of the transposon but indicate the relative length of the genes. The pointed end of the symbols is the 3' end. The predicted position and direction of transfer of an origin of conjugal transfer is shown as an open arrow (*oriT*). (*b*) Structure of the left and right ends of Tn*916*. Italicized characters are flanking host sequences and plain characters represent transposon DNA. The six Ns in each sequence represent the coupling sequences. The T in parentheses in the right transposon end is present when the transposon excises with five Ts, but absent when the transposon excises with four Ts (see section on length of coupling sequence). Arrows indicated inverted (IR) and direct repeats (DR-2). Open bars indicate the extent of DNA protected by Int binding from DNase I cleavage (59).

Eighteen bases of the 26-base inverted repeats at the ends of Tn*916* are protected from DNase I cleavage by Int (Figure 2) and thus serve as specific DNA-binding sites for Int (59). It is not known if the unprotected regions of the inverted repeats play any role in transposition. A role seems possible because in other transposable elements that have inverted repeats at their ends, genetic analysis has revealed domains in the inverted repeats that perform different functions in transposition (e.g. 27, 31).

DR-2 Repeats

At the ends of Tn*916* and Tn*1545*, three different DNA sequences are repeated in direct orientation, termed DR-1, DR-2, and DR-3 (18, 75) (see Figure 2b). No experiments have been performed yet that implicate DR-1 and DR-3 in transposition, but the DR-2 repeat binds Int (59). Two adjacent copies of DR-2 are located 90 bp from the right end of the element, and three copies are located 150 bp from the left end. Pairs of repeats similar in size to the DR-2 repeats of Tn*916* and Tn*1545* are also present near the ends of the nisin element Tn*5276* (78).

tet(M)

The gene encoding tetracycline resistance in conjugative transposons belongs to a class designated M (58). This gene encodes resistance both to tetracycline and, unlike most other *tet* genes, to minocycline. Many of the other classes of *tet* genes that have been studied cause resistance by facilitating export of the antibiotic (57). However, *tet*(M) encodes a protein that causes the process of protein synthesis to become insensitive to tetracycline (9). This protein is active in all gram-positive and gram-negative bacteria in which it has been found or to which it has been transferred, as well as in species of mycoplasma. It also makes in vitro translation systems insensitive to tetracycline. Tet(M) has local sequence similarity with elongation factor G of *E. coli*, EF-Tu of *B. subtilis* (87), and translational initiation factor 2 (97). Apparently Tet(M) binds to tRNA species whose codons begin with U to cause undermodification that, in turn, results in aberrant translation and altered expression of many genes (9).

Expression of *tet*(M) in Tn*916* is increased in the presence of tetracycline. Based on a combination of Northern blot analysis, primer extension, and RNA sequencing, an antitermination mechanism has been proposed to explain this increase (97). Inefficient transcriptional termination from the *tet*(M) promoter might lead to readthrough of the genes required for integration and excision of the transposon (see section on genes to the left of *tet*(M): *int* and *xis*, below), although these genes are at least 3 kb away (96).

All *tet*(M) genes found in natural isolates may have been introduced into new hosts by conjugative trans- posons. In clinical isolates of streptococci and staphylococci, *tet*(M) appears to be present in such elements. In other organisms, including *Ureoplasma urealyticum* (87) and *N. gonorrhoeae* (98), sequences surrounding the *tet*(M) determinant appear to correspond to those of Tn916.

Furthermore, introduction of Tn916 on a suicide vector into *N. meningitidis* by transformation generates an unusual class of Tc^R transformants that carry only the part of the transposon including *tet*(M) (98). The sequences flanking these meningococcal insertions are very different from those of normal Tn916 target sites (see section on target sites, above), which suggests the possibility of an alternative type of recombination for insertion of part of a conjugative transposon in certain hosts.

Genes to the Right of tet(M)

Using Tn5 mutagenesis of a cloned Tn916 in *E. coli*, Senghas et al (92) demonstrated that the right end of Tn916 contains many regions whose inactivation prevents conjugative transfer of the element. Clewell's group recently determined the sequence of the entire 18.3-kb Tn916 element (29a, 97), and homology analyses have revealed possible functions of some of the genes. All the open reading frames from the right of the transposon through *tet*(M) (which is equivalent to *orf11*) read from right to left (see Figure 2a).

One of the predicted proteins encoded in the right side of Tn916 (Orf14) has homology with *iap* of *Listeria monocytogenes*. The *iap* gene encodes a protein called p60 (52), which has bacteriolytic activity and appears to be essential for cell viability (108). It is believed to be necessary for separation of daughter cells at division, presumably because it is involved in autolysis of cell wall material. The similar Tn916 gene may be important in conjugation because it allows DNA to penetrate the cell envelope, either of the bacterium in which it resides and/or of its mating partner.

Another open reading frame in the right half of Tn916 (*orf23*) is predicted to encode a protein with similarity to the MobA family, especially the MbeA version of this protein found in ColE1 (5). This large protein, which is involved in conjugative mobilization of ColE1 and related elements, is found in the relaxosome complex containing the nicked ColE1 DNA. If the Tn916 *orf23*-encoded protein is similar, it might be responsible for nicking the circular transposon DNA at the transfer origin (*oriT*) and transporting it into the recipient cell during conjugation (see section on conjugation, below).

In the right half of Tn916, *orf18* was identified (29a). It shares significant local homology with the restriction protection protein ArdA (alleviation of restriction of DNA) of the unrelated, self-transmissible plasmids pKM101 (IncN) and ColIb-P9 (IncI1) (3, 25). The ArdA protein protects against the five type I restriction systems tested, and it inhibits both restriction and modification activities. The pKM101 and ColIb-P9 Ard proteins are 60% identical, although the former offers some protection against the type II enzyme *Eco*RI while the latter does not.

Conjugative transposons are resistant to DNA restriction upon transfer to a new host (see for example Table 1 in Reference 89). Guild et al (35) demonstrated this resistance directly by using the *Dpn*II system of *S. pneumoniae*. Although the conjugative plasmid pIP501 was restricted when mated into a *Dpn*II⁺ recipient, the complex conjugative transposon Tn*5253* was not restricted in a cross using the same donor and recipient strains. In vitro, this transposon is sensitive to *Dpn*II. Therefore, Orf18 of Tn*916* is probably responsible for the apparent immunity of the transposon to restriction systems following its introduction into a new host and for ensuring that the transposon has a broad host range.

If Orf18 is similar to ArdA, it may also prevent modification of all DNA entering the cell in which it resides. This possibility should probably be taken into account in choosing recipient strains for transformation, transduction, conjugation, or even gene-cloning experiments in which the donor contains a conjugative transposon.

In Tn*916*, *orf20* has homology with an *orf* in a cryptic plasmid of *Bacillus stearothermophilus*. The function of this *orf* has not yet been determined.

Genes to the Left of tet(M): int and xis

The 4 kb on the left of Tn*916* correspond to the sequence obtained for Tn*1545*, as far as that is available, with only a few small differences. Several Tn*5* insertions in Tn*916* that had been isolated in *E. coli* and found to interfere with excision and conjugative transposition are located at the left end of the Tn*916* element (92), within two genes that have been sequenced in Tn*1545* and named *int*(Tn) and *xis*(Tn) (75). These names were based on local homology with the λ integrase (Int) family of proteins (75, 79) and the *xis* gene of P22 and of Pin [which is a Hin-related invertase that catalyzes phage ε15 excision in *E. coli* (74)]. The C-terminal portion of the Int protein contains two sequence domains characteristic of this family of proteins (1), and the domain closest to the C terminus possesses the histidine, arginine, and tyrosine residues important for DNA cleavage and strand-transfer reactions (28, 34, 37, 38, 73). The Int proteins of Tn*916* and Tn*1545* are identical except at position 9, where Tn*916* has a lysine and Tn*1545* an arginine (75, 96). Furthermore, there is extensive similarity between these proteins and the integrase of Tn*5276* throughout their entire lengths (79).

Both *int* and *xis* are transcribed towards the end of the transposon. Although transcription of these genes may begin at the *tet*(M) promoter, several transcriptional terminators have been identified between these sites (97).

Other Open Reading Frames to the Left of tet(M)

Overlapping *xis* (*orf1*) by 17 nucleotides and oriented in the other direction lies *orf5* (97). This reading frame is predicted to produce a 9.2-kDa basic protein. A Tn916 mutation with Tn5 inserted into *orf5* could not conjugatively transpose (92, 97). No complementation experiments have been reported for this mutant, so the possibility remains that the phenotype of the *orf5* insertion mutation may be caused by altered expression of the genes *xis* and/or *int*.

The *orf6* reading frame, which is adjacent to and downstream of *tet*(M), can encode a basic protein and is probably included in the *tet*(M) transcript (97). A possible promoter located upstream of *orf6*, however, might allow *orf6* to be transcribed independently as well. The predicted protein contains a 20–amino acid zinc finger, which indicates that it might bind DNA. No further clues to its function are currently available.

orf7 is located downstream of *tet*(M) and is oriented in the same direction (97). Its predicted product is acidic, and it shares local homology with a σ factor of *Streptomyces coelicolor* and with σ^{28} of *B. subtilis*.

The additional open reading frames identified had no reported similarities with proteins in the database available at the time of the analyses.

CONJUGATION

Is There Cotransfer of Other DNA?

Very little is currently known about the details of DNA transfer during conjugation between gram-positive bacteria. During conjugative transposition, a fairly complete zygote is probably formed (101). The evidence for this zygote comes from analysis of linkage of markers in crosses between a *B. subtilis* donor carrying a conjugative transposon and a multiply marked *B. subtilis* recipient (which cannot be transformed because of a *comG* mutation). Transconjugants inherited linked donor markers even when the markers were far from the inserted transposon and when the transposon was not found in the transconjugants. Unfortunately, no frequency data were presented nor was dependence on the presence of the conjugative transposon shown. Even if cotransfer of chromosomal markers with a conjugative transposon does occur, this does not necessarily imply that there is physical linkage between the transposon and the transferred chromosomal DNA.

Other studies to test for cotransfer of unlinked markers with conjugative transposons have relied upon coresident nonconjugative plasmids. Unfortunately, the results have been plasmid dependent (7, 19, 67), and difficult to interpret.

Excision Occurs in the Donor Prior to Conjugation

Several lines of evidence strongly suggest that prior to conjugation, the transposon is excised in the donor bacterium and is not physically attached to any other DNA at the time of transfer. When Tn916 is located on a nonconjugative plasmid, most of the transconjugants selected for a transposon marker do not contain plasmid markers (7, 19, 29). Furthermore, no transposon-containing transconjugants are obtained when the plasmid marker is selected directly. Thus, the transposon seems to have been excised from the plasmid prior to transfer, and the plasmid DNA does not appear to have been transferred during this mating event. In addition, when a conjugative transposon is present in a self-transmissible plasmid, transfer of the transposon is independent of plasmid transfer (19, 29). On the other hand, when the same plasmid containing the same transposon is introduced into the same recipient strain by protoplast transformation, the expected linkage between the plasmid and transposon is maintained (19). The lack of linkage between these two elements in the mating experiment suggests that the transposon is excised prior to conjugational transfer.

Observations with *L. lactis* subsp. *cremoris* (6) also suggest excision of conjugative transposons prior to transfer in mating. Although *L. lactis* acts normally as a recipient for conjugative transposition, it cannot act as a donor, which suggests that excision of conjugative transposons cannot take place in this organism. If another species is the donor of Tn916 in a mating with *L. lactis*, then the transposon can insert into the *L. lactis* chromosome because it had excised in the donor strain prior to mating.

Mating experiments using *B. subtilis*::Tn916int-1 (an insertion mutation in the *int* gene) were designed to investigate the role of Int in conjugative transposition (7). If the transposon was transferred prior to excision, then Int should not have been needed in the donor, but only in the recipient. Instead, the *int* gene had to be provided only in the donor, but not in the recipient. Furthermore, providing the *int* gene only in the recipient did not allow conjugative transposition to occur. Therefore, excision of Tn916 takes place before conjugational transfer. It seems highly improbable that insertion is Int independent since that would mean it occurs by a mechanism other than excision. Thus, the observation that a functional *int* gene is required only in the donor suggests that Int protein can be transferred with the DNA during conjugation.

Transfer of a Single Strand During Conjugation

In the studies of preferred target sequences in a gram-positive plasmid, Tn916 was introduced into pIP501 in *L. lactis* by mating (91). Three *B. subtilis* strains, each having a single copy of Tn916 in a different place in the chromosome, served as donors. In all cases, the coupling sequences of the donor strains were known. The unexpected finding was that the coupling sequence right of the Tn916 in the donor was never found among the transconjugants, and the coupling sequence left of Tn916 in the donor appeared on either side of the integrated transposon in the recipient (91). This result was in striking contrast to the work in which both coupling sequences were found with approximately

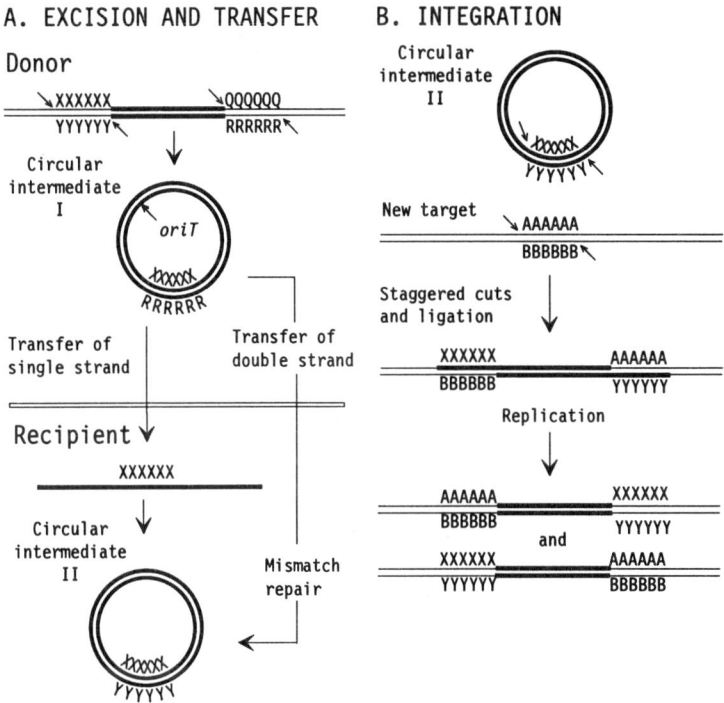

Figure 3 A model for transposition of Tn916 involving conjugation. Symbols as in Figure 1. (*a*) After excision and formation of circular intermediate I in the donor strain, as described in Figure 1, nicking at a putative *oriT* site allows transfer of a single transposon strand into the recipient. Complementary strand synthesis forms circular intermediate II with complementary coupling sequences. Alternatively, double-stranded circular intermediate I transfers to the recipient, and strand-specific mismatch repair corrects the heteroduplex in the intermediate to form the circular intermediate II. (*b*) Cleavage of circular intermediate II and target, followed by ligation, produces a product in which the integrated transposon is flanked by heteroduplex coupling sequences. These are resolved by replication to produce progeny in which the coupling sequence originally present on the left side of the transposon in the donor appears on either the left or the right.

equal frequency after introduction of Tn916 by electroporation or by natural transformation into a gram-positive host (13, 44). Therefore conjugative transfer probably causes the loss of a specific coupling sequence. There are two obvious ways in which this may occur (see Figure 3): mismatch repair, with preferential correction to a specific strand, or transfer of a single specific strand in mating. The latter explanation is more appealing because it is similar to the way in which conjugation occurs in gram-negative bacteria and because it is hard to imagine specific mismatch repair occurring in the many different types of bacterial hosts that can serve as conjugational donors of Tn916 (although only *B. subtilis* was studied in the experiments presented).

If only a single strand is transferred, it is very unlikely that it could be the substrate for integration. First, the observation that following conjugation and integration, a single coupling sequence from the donor can be at either end of the integrated transposon is strong evidence for a circular intermediate. If strand cleavages during recombination occur with a defined polarity, then this intermediate must be double stranded. Second, a double-stranded transposon circle has been shown to act as a substrate for integration (13). Therefore, we believe that complementary strand synthesis occurs in the recipient after transfer of the single strand of Tn916, as is found for conjugative plasmids in *E. coli*. However, conjugative plasmids in gram-negative bacteria carry a DNA primase gene, which is required for this synthesis, and no primase-like gene has yet been reported in Tn916.

Implications for Gene Order During Transfer

The observation that perhaps only a single strand of the transposon is transferred to the recipient implies that the transposon possesses a unique origin of transfer. If conjugative transfer of Tn916 is similar to that of *E. coli* plasmids, the 5' end of the putative *ard* gene should be transferred very early so that Ard protein can be transcribed and translated to protect the entering DNA from restriction. Moreover, a function critical for conjugational transfer, such as *orf23* (related to *mobA*), should enter last. Hence, the location of the predicted origin of transfer of Tn916 and the transfer order of its genes would be expected to occur as indicated by *oriT* in Figure 2a. Further work is needed to test this prediction.

REGULATION OF CONJUGATIVE TRANSPOSITION

Regulation at the Level of Excision

Although no quantitative data have been published, it is generally believed that the excision process is rate limiting for conjugative transposition. If this were so, one would expect that the circular intermediate produced by excision would rapidly undergo conjugation and integration and would not accumulate. In accordance with this expectation, in gram-positive bacteria, the circular transposon cannot be detected in cell lysates by means of gel electrophoresis or, usually, the polymerase chain reaction (PCR) (using primers from within the transposon directed to amplify across the junction).

Recent studies suggest that the amount of circular intermediate is proportional to the frequency of transposition. Jaworski & Clewell (45) studied a set of *E. faecalis* donor strains that differed only in the coupling sequences flanking the integrated transposon and found that transposition frequencies vary by 10^4-fold. High-frequency donors had enough circular transposons to be detected by PCR, while low-frequency donors did not. In similar experiments on *E. faecalis* using a quantitative PCR assay, the frequency of transposition correlated directly with the amount of circular intermediate over a 100-fold range (R Manganelli, L Romano, S Ricci, M Zazzi, PE Valensin, et al, submitted).

The circular transposon is also present in sufficient quantity to be detectable by PCR in the *E. faecalis* transposon Tn*5381* (82). It has been claimed that the presence of a detectable circular transposon is an important difference between Tn*5381* and Tn*916*. However, whether an insertion of different coupling sequences would produce undetectably low quantities of circularized transposon, as is usually the case for Tn*916*, is not clear.

In *E. coli*, the pool of excised circular transposon DNA is much larger than that in gram-positive hosts. Not only is the circularized transposon detectable by PCR, but in some cases it is also visible on ethidium bromide–stained agarose gels (90). Another difference between the behavior of conjugative transposons in gram-positive and gram-negative cells is the frequency of transposon loss, which is very high in *E. coli*. This may reflect differences in the regulation of excision in gram-negative and gram-positive bacteria, but could also be explained by a paucity of good target sites in *E. coli* or by the presence in *E. coli* of a very strongly preferred site in an essential gene.

Coordinate Regulation of Multiple Transposon Copies

During conjugative transposition, the frequency of cotransfer of a second transposon from the donor is much higher than would be expected by chance (19, 29). Therefore, when one transposon excises, the others excise as well. This coordinate excision may be caused by a product of the first excised transposon or by an external factor that stimulates all of the elements to excise. In strains containing more than one copy of a conjugative transposon, the frequency of excision seems to be that of the element with the highest excision frequency.

Regulation of Expression of Proteins Involved in Transposition and Conjugation

Currently, little is known about possible transcriptional and translational regulation of *int* and *xis*, or the effects of altered expression of their proteins on transposition. Transposition is probably not regulated simply by negative control of the level of Int by a repressor, because the presence of a conjugative transposon in a recipient strain (which should produce such a repressor) does not inhibit transfer of a second differently marked copy of the same element into that strain by conjugation (20, 70). Thus, in contrast to phage λ, for conjugative transposons, there is no evidence for immunity to entry of a second element through transposition.

Truncated Int Proteins

Within the large open reading frame for Int [*orf2* of Su et al (97)], three other possible proteins encoded in the same reading frame have plausible ribosome-binding site sequences preceding them. When primed with template DNA containing the *int* gene, in vitro transcription/translation systems produce several protein products (75, 96). Furthermore, Western blot analysis of extracts of *E. coli* cells containing a plasmid carrying the *int* gene shows multiple forms of Int, suggesting that these forms are present in vivo (F Lu & G Churchward, unpublished results).

Comparison of the wild-type Tn*916 int* gene with mutants in which the methionine codon at positions 1, 82, 109, or 138 was altered indicates that these four translation start sites function in vivo and in vitro (F Lu & G Churchward, unpublished results). Therefore, at least in *E. coli,* four different forms of Int, three of which are truncated to different extents at the N-terminal end, are produced. Based on the DNA-binding behavior of Int described in the section on mechanism of recombination below, the truncated protein beginning at codon 82, as well as possibly those beginning at codons 109 and 138, should bind to transposon ends and target DNA.

Regulation by Truncated Forms of Int

A possible function for different forms of Int is in the regulation of transposition. Truncated forms of Int could bind to transposon ends and target sequences and inhibit excision and integration or could combine with full-length Int and produce inactive complexes. In the formation of nucleoprotein complexes at *attL* and *attR* of λ prophage, some λ Int molecules can be recruited into the complex by protein-protein interactions with Int molecules already bound to DNA (49, 50). Similar interactions may occur between Tn*916* Int molecules. In the case of Tn*5,* a truncated form of the transposase lacking the amino-terminal end regulates transposition (47). Experiments with transcriptional and translational gene fusions indicated that the regulation was not at the level of expression of transposase, but that the truncated protein inhibited transposition by interacting with either the ends of the transposon, the transposase protein, or a host factor.

MECHANISM OF RECOMBINATION

Lack of Branch Migration During Recombination

The most striking difference between conjugative transposition and related site-specific recombination systems is the ability of the conjugative transposition mechanism to accommodate regions of nonhomology in recombining

DNA sequences. In the best-studied cases, such as Tn916 and Tn1545, the coupling sequences of recombining sites are almost always different.

In the phage λ system (reviewed in 53), the overlap regions between the sites of strand cleavage (equivalent to the coupling sequences of Tn916) must be identical for normal recombination to occur. Recombination between a wild-type sequence and a sequence carrying a mutation in the overlap region is depressed 100-fold or more, but recombination between two mutant sequences occurs at approximately the wild-type level (106). In biochemical assays, the presence of mismatches between the two recombining λ overlap regions blocks recombination after one round of strand exchange (51). The XerC/XerD recombinase of *E. coli* appears to be less fastidious than λ Int and tolerates mismatches in certain positions of the proposed overlap region, but still requires that certain base pairs be identical in both recombining DNA sequences (60).

Results obtained with λ are consistent with a sequential model for recombination in which one round of strand cleavage and exchange is followed by branch migration and a second round of strand cleavage and exchange (for review, see 53). According to this model, for the second round of cleavage and strand exchange to occur, the crossover must migrate away from the site of the first cleavages to the site of the second cleavages. Thus, mismatches in the overlap region block recombination by preventing branch migration. However, artificial substrates in which the crossover is constrained to fixed positions within the overlap region can still undergo Int-mediated cleavage and ligation at the normal sites (70a). This observation led to the proposal of an alternative strand swapping model, in which branch migration does not occur (70a). Thus, although the branch migration model may not be correct for λ, the requirement for homology in the overlap region still suggests important mechanistic differences between λ recombination and conjugative transposition models.

Binding of Int to Transposon and Target DNA

The Int protein of Tn916, purified as a chimeric protein fused to maltose-binding protein, binds specifically to transposon DNA in vitro (59). A comparison of the binding of chimeric proteins containing different segments of Int showed that this protein contains two independent DNA-binding domains. The C-terminal part of Int is needed for binding to the ends of the transposon and flanking DNA, and the N-terminal part of the protein is needed for binding to the regions of DNA within the transposon containing the DR-2 repeats.

Because the C-terminal end of Int binds to both the ends of the transposon and flanking host DNA, it should bind to target sequences alone. As expected, a chimeric protein containing the C-terminal portion of Int bound to three different target sequences, one within the *cylA* gene of plasmid pAD1, one

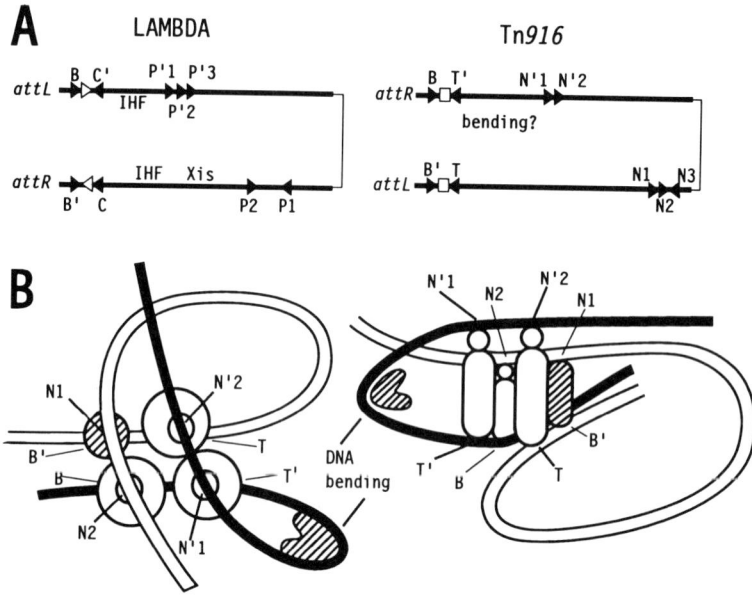

Figure 4 (a) Comparison of the ends of integrated λ prophage and Tn916. For λ, *attL* and *attR* are the left and right ends of the prophage. The open triangles represent the overlap regions at each end of the prophage. Arrows labeled B, C', B', and C are binding sites for the C-terminal domain of λ Int flanking the overlap region. Arrows labeled P'1, P'2, P'3, P1, and P2 represent binding sites for the N-terminal domain of λ Int. Binding sites for integration host factor (IHF) and Xis proteins are labeled. For Tn916, *attL*, and *attR* represent the left and right ends of the integrated transposon; open boxes represent the coupling sequences at each end of the transposon; arrows labeled B, T', B', and T represent binding sites for the C-terminal domain of Tn916 Int; and sites labeled N'1, N'2, N1, N2, and N3 represent the DR-2 repeats shown in Figure 2b that bind the N-terminal domain of Tn916 Int. The region labeled "bending?" indicates the possible binding site for a host factor that bends the DNA between T' and N1' (see section on host factors).

(b) Model for the alignment of the left and right ends of Tn916 during excision (adapted from Reference 50). The left and right parts of the figure show two different views of the complex. The heavy black line represents the DNA of *attR*, and the open line represents the DNA of *attL*. The positions of the Int-binding sites B, T', B', T, N1, N2, N'1, and N'2 are indicated. In the left part of the figure, the large circles represent the C-terminal domain of Int molecules, and the small circles represent the N-terminal domains. On the right, the large cylinders represent the C-terminal domain of Int, and the small circles represent the N-terminal domain. The hatched circle on the left and the hatched cylinder on the right represent an Int molecule, which may have either two DNA-binding domains or only the C-terminal domain (see section on formation of a nucleoprotein complex at the transposon ends). The hatched object labeled "bending" represents a hypothetical host factor that binds and bends the DNA between T' and N'1.

within the *cat* gene, and one within the *orfA* gene of plasmid pIP501 (59a). Competition binding experiments showed that Int binds with a lower affinity to the *cat* target, which was used less frequently in vivo (91) than the other two targets (59a).

In the absence of Tn916 Int protein, all three functional target sites show a pattern of DNase I cleavage characteristic of DNA containing a static bend (59a): They have DNase I–hypersensitive sites approximately 11 bp apart. Circular permutation analysis using gel electrophoresis confirmed the presence of this structural feature in all three target sites (59a). Thus, the shape of the target DNA rather than its particular nucleotide sequence may be the critical factor for target recognition by conjugative transposons.

Formation of a Nucleoprotein Complex at the Transposon Ends

The only other member of the Int family of site-specific recombinases shown to contain two DNA-binding domains is Int of phage λ (63). As shown in Figure 4a, the arrangements of Int-binding sites in Tn916 and in the integrated λ prophage are very similar. We have therefore adopted a variation of the nomenclature for λ recombination sites and Int-binding sites for Tn916. As previously suggested (102), we call the joined transposon ends *attT* and the target sequences *attB*. The left end of the transposon and flanking bacterial DNA is *attL*, while the right end and flanking DNA is *attR*. Int-binding sites within the transposon (equivalent to P and P' in the λ system) are referred to as N and N'.

A detailed model for the formation of a nucleoprotein complex during excisive recombination in the λ system has been proposed (49–50). This model can be adapted for the alignment of the two ends of Tn916 during excision (Figure 4b). We suppose that, as in λ, four Int molecules are involved in bridging the two recombining DNA sites and aligning them in an antiparallel arrangement. One Int molecule, bound to *attR* at site T' by its C-terminal domain, and N'1 by its N-terminal domain, forms an intrastrand loop. A second Int molecule, bound to the T site of *attL* by its C-terminal domain and the N'2 site of *attR* by its N-terminal domain, forms a bridge between *attL* and *attR*. The third Int molecule binds to the B site of *attR* and the N2 site of *attL*, forming a second bridge between *attL* and *attR*. The fourth Int molecule either could bind to the B' and N1 sites of *attL*, forming an intrastrand loop or might be a truncated C-terminal Int protein that binds only to B' because it lacks the N-terminal binding domain. In the λ system, only a single internal binding site, P2, corresponds to N1 and N2 in *attL* of Tn916, and so in the model for the interaction of *attL* and *attR* of λ, the corresponding Int molecule is shown bound only by its C-terminal domain.

This speculative model for interaction between *attL* and *attR* during excision of Tn916 illustrates how the bivalent Int protein may be involved in the formation of a recombination complex during transposon excision. (It has not been shown for Tn916 that the proposed binding sites flanking the coupling sequence and within the transposon are individual binding sites, although they lie within regions protected by Int from DNase I cleavage.) This model also raises two interesting possibilities for further experimentation. The first concerns the role of host factors in Tn916 transposition, and the second is the possibility that different forms of Int are involved in transposition.

Host Factors

In most transposition systems, factors encoded by the host chromosome are required. Three pieces of evidence suggest that this might be true for conjugative transposons as well. First, the frequent loss of Tn916 from *E. coli*, a phenomenon rarely seen in gram-positive bacteria, suggests that host-encoded factors that differ between the two kinds of bacteria affect the excision and insertion process. Second, it appears that *L. lactis* subsp. *cremoris* lacks a factor required for excision of conjugative transposons (6). This organism can act as a recipient for Tn916 transfer and integration, but the transposon is not transferred again once it has inserted into a site on the *L. lactis* chromosome or on a plasmid in *L. lactis*. The host factor that appears to be absent from *L. lactis* may have a direct role in excision or it may be involved indirectly by affecting expression of transposon-encoded functions. Third, F Lu & G Churchward (unpublished results) have detected a host-encoded protein that binds to the ends of Tn916 DNA.

Several groups have noted the presence of binding sequences for the small, basic protein known as integration host factor (IHF) near the ends of conjugative transposons and therefore anticipated that IHF may be involved in excision and/or insertion of these elements (17). IHF was first discovered as a host factor essential for integrative recombination of λ in vitro (68). Because IHF bends DNA sharply (100), it plays an important structural role in the formation of nucleoprotein complexes required for recombination (33). During excision of λ, IHF is required to bend the DNA between the C' and P'1 Int-binding sites in *attL* so that Int can bind both sites (64) (Figure 4). The two sites are too close together for an intrastrand loop to form unless the DNA is actively bent. The equivalent sites in *attR* of Tn916 are farther apart, but still too close to interact without protein-induced bending, unless the DNA has an unusual structure. Therefore a host protein probably binds and bends this region. IHF is also required to bend the DNA between the P2 and C Int-binding sites in *attR* of λ. In *attL* of Tn916, the equivalent sites are far enough apart that a loop should be able to form without active DNA bending.

Excision and integration of Tn916 occur in widely divergent bacteria, including gram-positive and gram-negative hosts. Some of these species, such as *B. subtilis,* are not believed to encode an IHF-like factor, and since homology searches have not revealed the presence of an IHF resembling that of *E. coli,* it seems unlikely that *B. subtilis* has a factor recognizing the same sequence as does IHF of *E. coli.* Therefore, host proteins other than IHF may play a role in Tn*916* transposition. An HU-like protein is a likely candidate. Such proteins have been identified in both gram-positive and gram-negative bacteria, and HU protein plays a role in the Hin and related site-specific inversion systems similar to that envisioned for a host protein in Tn*916* transposition (46).

Role of Xis

In the phage λ system, Xis is required to bend the DNA between the C and P2 Int-binding sites of *attR* to stimulate Int binding (10). Such bending activity is probably not required in the Tn*916* system because the analogous sites T and N1 (Figure 4*a*) are farther apart, and so one would expect that Tn*916* Xis protein may not be required for excision. However, Xis may help stabilize the formation of a nucleoprotein complex that is required for excision and thus stimulate excision, especially from sites where excision occurs at a very low frequency.

In gram-negative bacteria, even when *xis* is not present, the excised minitransposon Tn*1545* can be detected by using PCR (F Lu & G Churchward, unpublished results), and the excisant plasmid molecule produced by excision can be detected using gel electrophoresis (75). In the presence of *xis,* however, more excision product is found (75). In contrast, when Su et al (97) assayed for excision by loss of tetracycline resistance from *E. coli,* they found that both *int* and *xis* are required. Insertion mutations in *orf1* (*xis*) prevented excision, and these mutations were complemented by a plasmid containing only the *xis* gene. Because the tetracycline sensitivity assay is probably less sensitive than the others, the results of Su & Clewell (96) do not rule out the possibility that some excision can occur in the absence of Xis.

Role of Int in Excision and Integration

All members of the Int family of recombinases cut their DNA substrates to leave a 5'-OH group and create a 3' phosphotyrosine linkage to the Int protein (23, 34, 71, 77, 107). Given the sequence similarity between Tn*916* Int and the rest of the Int family of recombinases, the C-terminal domain of Int is probably responsible for strand cleavage and exchange involving the formation of intermediates in which Int monomers are covalently attached to the DNA by 3' phosphotyrosine linkages.

As described above, the occurrence of different forms of Int, some lacking the N-terminal DNA-binding domain required for binding to the transposon ends, suggests that more than one kind of Int protein may function in conjugative transposition. In most of the Int family of site-specific recombination reactions, two molecules of a single kind of recombinase bind to each recombining site. In the case of the XerC/XerD system, however, two different recombinases, encoded respectively by the *xerC* and *xerD* genes, are required both for DNA binding and for recombination (4). In the case of phage Mu, while MuA and host factors are necessary to form a stable synaptic complex between the ends of the phage, a second phage-encoded protein, MuB, which binds relatively nonspecifically to DNA, is necessary for interaction between the stable synaptic complex and target DNA to form a strand transfer complex (for review, see 62). Interaction between a nucleoprotein complex at the joined ends of the circular Tn*916* intermediate and the target may require that the target be bound by C-terminal fragments of Int.

The Tn*7* system (for review, see 22a) contains two separate transposition pathways requiring different combinations of transposon-encoded proteins and host factors that result in either insertion close to a specific chromosomal site or insertion into essentially random targets. In addition, in the Tn*7* system, two different proteins, TnsA and TnsB, are required for cleavage at the 5' and 3' ends of the transposon.

These comparisons with λ and different transposons suggest that integration and excision of conjugative transposons occur by biochemically different pathways that could require the formation of complexes containing different kinds of Int molecules and accessory proteins.

ACKNOWLEDGMENTS

The work in our laboratories is supported by grant GM50376 from the National Institutes of Health. We appreciate the help of all our colleagues who sent us information prior to publication and are grateful to the members of our laboratories and department for many critical discussions.

> Any *Annual Review* chapter, as well as any article cited in an *Annual Review* chapter, may be purchased from the Annual Reviews Preprints and Reprints service.
> 1-800-347-8007; 415-259-5017; email: arpr@class.org

Literature Cited

1. Argos P, Landy A, Abremski K, Egan JB, Haggard-Ljungquist E, et al. 1986. The integrase family of site-specific recombinases: regional similarities and global diversity. *EMBO J.* 5:433–40
2. Ayoubi P, Kilic AO, Vijayakumar MN. 1991. Tn*5253*, the pneumococcal omega (*cat-tet*) BM6001 element, is a composite structure of two conjugative transposons, Tn*5251* and Tn*5252*. *J. Bacteriol.* 173:1617–22
3. Belogurov AA, Delver EP, Rodzevich OV. 1992. IncN Plasmid pKM101 and IncI1 Plasmid ColIb-P9 encode homologous antirestriction proteins in their leading regions. *J. Bacteriol.* 174:5079–85
4. Blakely G, May G, McCulloch R, Arciszewska LK, Burke M, et al. 1993.

Two related combinases are required for site-specific recombination at *dif* and *cer* in *E. coli* K12. *Cell* 75:351–61
5. Boyd AC, Archer JAK, Sherratt DJ. 1989. Characterization of the ColE1 mobilization region and its protein products. *Mol. Gen. Genet.* 217:488–98
6. Bringel F, Van Alstine GL, Scott JR. 1991. A host factor absent from *Lactococcus lactis* subspecies *lactis* MG1363 is required for conjugative transposition. *Mol. Microbiol.* 5:2983–93
7. Bringel F, Van Alstine GL, Scott JR. 1992. Conjugative transposition of Tn*916*: the transposon *int* gene is required only in the donor. *J. Bacteriol.* 174:4036–41
8. Burdett V, Inamine J, Rajagopalan SA. 1982. Heterogeneity of tetracycline resistance determinants in *Streptococcus*. *J. Bacteriol.* 149:995–1004
9. Burdett V. 1993. tRNA modification activity is necessary for Tet(M)-mediated tetracycline resistance. *J. Bacteriol.* 175:7209–15
10. Bushman W, Yin S, Thio LL, Landy A. 1984. Determinants of directionality in lambda site-specific recombination. *Cell* 39:699–706
11. Buu-Hoi A, Horodniceanu T. 1980. Conjugative transfer of multiple antibiotic resistance markers of *Streptococcus pneumoniae*. *J. Bacteriol.* 143:313–20
12. Caillaud F, Courvalin P. 1987. Nucleotide sequence of the ends of the conjugative shuttle transposon Tn*1545*. *Mol. Gen. Genet.* 209:110–15
13. Caparon MG, Scott JR. 1989. Excision and insertion of the conjugative trans-poson Tn*916* involves a novel recombination mechanism. *Cell* 59:1027–34
14. Caparon MG, Scott JR. 1991. Genetic manipulation of pathogenic streptococci. *Methods Enzymol.* 204:556–86
15. Carlier C, Courvalin P. 1982. Resistance of streptococci to aminoglycoside-aminocyclitol antibiotics. See Ref. 87a, pp. 162–66
16. Clermont D, Horaud T. 1994. Genetic and molecular studies of a composite chromosomal element (Tn*3705*) containing a Tn*916*-modified structure (Tn*3704*) in *Streptococcus anginosus* F22. *Plasmid* 31:40–48
17. Clewell DB, Flannagan SE. 1993. The conjugative transposons of Gram-positive bacteria. In *Bacterial Conjugation*, ed. DB Clewell, pp.369–93. New York: Plenum
18. Clewell DB, Flannagan SE, Ike Y, Jones JM, Gawron-Burke C. 1988. Sequence analysis of termini of conjugative transposon Tn*916*. *J. Bacteriol.* 170:3046–52
19. Clewell DB, Flannagan SE, Zitzow LA, Su YA, He P, et al. 1991. Properties of conjugative transposon Tn*916*. See Ref. 27a, pp. 39–44
20. Clewell DB, Pontius LT, Weaver KE, An Y, Ike Y, et al. 1991. *Enterococcus faecalis* hemolysis/bacteriocin plasmid pAD1. See Ref. 27a, pp.3–8
21. Courvalin P, Carlier C. 1986. Transposable multiple antibiotic resistance in *Streptococcus pneumoniae*. *Mol. Gen. Genet.* 205:291–97
22. Courvalin P, Carlier C. 1987. Tn*1545*: a conjugative shuttle transposon. *Mol. Gen. Genet.* 206:259–64
22a. Craig NL. 1995. Transposon Tn*7*. *Curr. Trends Microbiol. Immunol.* In press
23. Craig NL, Nash HA. 1983. The mechanism of phage lambda site-specific recombination: site-specific breakage of DNA by Int topoisomerase. *Cell* 35:795–803
24. De Cespedes G, Derbise A, Trieu-Cuot P, Horaud T. 1994. Mobile chromosomal elements (Tn*3706 Streptococcus agalactiae* B128) resistant to gentamicin and tetracyclines. In *Pathogenic Streptococci: Present and Future*. *Proc. Lancefield Int. Symp. Streptococci and Streptococcal Diseases, 12th, Sept. 6–10, 1993*, ed. A Artem, pp. 272–73. St. Petersburg, Russia: Lancer
25. Delver EP, Kotova VU, Zavilgelsky GB, Belogurov AA. 1991. Nucleotide sequence of the gene (*ard*) encoding the antirestriction protein of plasmid ColIb-P9. *J. Bacteriol.* 173:5887–92
26. De Massey B, Studier FW, Dorgai L, Appelbaum E, Weisberg RA. 1984. Enzymes and sites of genetic recombination: studies with gene-*3* endonuclease of phage T7 and with site-affinity mutants of phage lambda. *Cold Spring Harbor Symp. Quant. Biol.* XLIX:715–26
27. Derbyshire KM, Hwang L, Grindley NDF. 1987. Genetic analysis of the interaction of the insertion sequence IS*903* transposase with its terminal inverted repeats. *Proc. Natl. Acad. Sci. USA* 84:8049–53
27a. Dunny GM, Cleary PP, McKay LL, eds. 1991. *Genetics and Molecular Biology of Streptococci, Lactococci, and Enterococci*. Washington, DC: Am. Soc. Microbiol.
28. Evans BR, Chen J, Parsons RL, Bauer TK, Teplow DB, Jayaram M. 1990. Identification of the active site tyrosine of Flp recombinase: possible relevance of its location to the mechanism of

recombination. *J. Biol. Chem.* 265: 18504–10
29. Flannagan SE, Clewell DB. 1991. Conjugative transfer of Tn916 in *Enterococcus faecalis: trans*-activation of homologous transposons. *J. Bacteriol.* 173:7136–41
29a. Flannagan SE, Zitzow LA, Clewell DB. 1994. Nucleotide sequence of the 18-kb conjugative transposon Tn916 from *Enterococcus faecalis*. *Plasmid* 32:350–54
30. Franke AE, Clewell DB. 1981. Evidence for a chromosome-borne resistance transposon (Tn916) in *Streptococcus faecalis* that is capable of conjugal transfer in the absence of a conjugative plasmid. *J. Bacteriol.* 145:494–502
31. Gamas P, Galas D, Chandler M. 1985. DNA sequence at the end of IS1 required for transposition. *Nature* 317:458–60
32. Gawron-Burke C, Clewell DB. 1982. A transposon in *Streptococcus faecalis* with fertility properties. *Nature* 300:281–84
33. Goodman SD, Nash HA. 1989. Functional replacement of a protein-induced bend in a DNA recombination site. *Nature* 341:251–54
34. Gronostajski RM, Sadowski PD. 1985. The FLP recombinase of the *Saccharomyces cerevisiae* 2μm plasmid attaches covalently to DNA via a phosphotyrosyl linkage. *Mol. Cell. Biol.* 5:3274–79
35. Guild WR, Smith MD, Shoemaker NB. 1982. Conjugative transfer of chromosomal R determinants in *Streptococcus pneumoniae*. See Ref. 87a, pp. 88–92
36. Halula M, Macrina FL. 1990. Tn5030: a conjugative transposon conferring clindamycin resistance in *Bacteriodes* species. *Rev. Infect. Dis.* 12:S235–42
37. Han YW, Gumport RI, Gardner JF. 1993. Complementation of bacteriophage lambda integrase mutants: evidence for an intersubunit active site. *EMBO J.* 12:4577–84
38. Hoess RH, Abremski K. 1985. Mechanism of strand cleavage and exchange in the Cre-*lox* site-specific recombination system. *J. Mol. Biol.* 181:351–62
39. Horaud T, De Cespedes G, Clermont D, David F, Delbos F. 1991. Variability of chromosomal genetic elements in streptococci. See Ref. 27a, pp. 16–20
40. Horaud T, Delbos F, De Cespedes G. 1990. Tn3702, a conjugative transposon in *Enterococcus faecalis*. *FEMS Microbiol. Lett.* 72:189–94
41. Horn N, Swindell S, Dodd H, Gasson M. 1991. Nisin biosynthesis genes are encoded by a novel conjugative transposon. *Mol. Gen. Genet.* 228:129–35
42. Horodniceanu T, Bougueleret L, Bieth G. 1981. Conjugative transfer of multiple-antibiotic resistance markers in beta-hemolytic group A, B, F, and G streptococci in the absence of extrachromosomal deoxyribonucleic acid. *Plasmid* 5:127–87
43. Horodniceanu T, Buu-Hoi A, Delbos F, Bieth G. 1982. High-level aminoglycoside resistance in group A, B, C, D (*Streptococcus bovis*), and viridans streptococci. *Antimicrob. Agents Chemother.* 21:176–79
44. Ike Y, Flannagan SE, Clewell DB. 1992. Hyperhemolytic phenomena associated with insertions of Tn916 into the hemolysin determinant of *Enterococcus faecalis* plasmid pAD1. *J. Bacteriol.* 174:1801–9
45. Jaworski DD, Clewell DB. 1994. Evidence that coupling sequences play a frequency-determining role in conjugative transposition of Tn916 in *Enterococcus faecalis*. *J. Bacteriol.* 176:3328–35
46. Johnson RC, Bruist MF, Simon MI. 1986. Host protein requirements for in vitro site-specific DNA inversion. *Cell* 46:531–39
47. Johnson RC, Yin JCP, Reznikoff WS. 1982. Control of Tn5 transposition in *Escherichia coli* is mediated by protein from the right repeat. *Cell* 30:873–82
48. Jones JM, Gawron-Burke C, Flannagan SE, Yamamoto M, Senghas E, Clewell DB. 1987. Structural and genetic studies of the conjugative transposon Tn916. In *Streptococcal Genetics*, ed. JJ Ferretti, R Curtiss III, pp. 54–60. Washington, DC: Am. Soc. Microbiol.
49. Kim S, De Vargas LM, Nunes-Düby SE, Landy A. 1990. Mapping of a higher order protein–DNA complex: two kinds of long-range interactions in lambda *att*L. *Cell* 63:773–81
50. Kim S, Landy A. 1992. Lambda Int protein bridges between higher order complexes at two distant chromosomal loci *att*L and *att*R. *Science* 256:198–203
51. Kitts PA, Nash HA. 1987. Homology-dependent interactions in phage lambda site-specific recombination. *Nature* 329:346–48
52. Köhler S, Leimeister-Wächter M, Chakraborty T, Lottspeich F, Goebel W. 1990. The gene coding for protein p60 of *Listeria monocytogenes* and its use as a specific probe for *Listeria monocytogenes*. *Infect. Immun.* 58:1943–50
53. Landy A. 1989. Dynamic, structural and regulatory aspect of lambda site-specific

54. Le Bouguenec C, De Cespedes G, Horaud T. 1988. Molecular analysis of a composite chromosomal conjugative element (Tn*3701*) of *Streptococcus pyogenes*. *J. Bacteriol.* 170:3930–36
55. Le Bouguenec C, De Cespedes G, Horaud T. 1990. Presence of chromosomal elements resembling the composite structure Tn*3701* in streptococci. *J. Bacteriol.* 172:727–34
56. Le Bouguenec C, Horaud T, Bieth G, Colimon R, Dauguet C. 1984. Translocation of antibiotic resistance markers of a plasmid-free *Streptococcus pyogenes* (group A) strain into different streptococcal hemolysin plasmids. *Mol. Gen. Genet.* 194:377–87
57. Levy SB. 1992. Active efflux mechanisms for antimicrobial resistance. *Antimicrob. Agents Chemother.* 36:695–703
58. Levy SB, McMurry LM, Burdett V, Courvalin P, Hillen W, et al. 1989. Nomenclature for tetracycline resistance determinants. *Antimicrob. Agents Chemother.* 33:1373–74
59. Lu F, Churchward G. 1994. Conjugative transposition: Tn*916* integrase contains two independent DNA binding domains that recognize different DNA sequences. *EMBO J.* 13:1541–48
59a. Lu F, Churchward G. 1995. Tn*916* target DNA sequences bind the C-terminal domain of integrase protein with different affinities that correlate with transposon insertion frequency. *J. Bacteriol.* 177:1938–46
60. McCulloch R, Coggins LW, Colloms SD, Sherratt DJ. 1994. Xer-mediated site-specific recombination at *cer* generates Holliday junctions in vivo. *EMBO J.* 13:1844–55
61. Mizuuchi K, Weisberg R, Enquist L, Mizuuchi M, Buraczynska M, et al. 1981. Structure and function of the phage lambda *att*: size, *int*-binding sites, and location of the crossover point. *Cold Spring Harbor Symp. Quant. Biol.* 45:429–37
62. Mizuuchi K. 1992. Transpositional recombination: mechanistic insights from studies of Mu and other elements. *Annu. Rev. Biochem.* 61:1011–51
63. Moitoso De Vargas L, Pargellis CA, Hasan NM, Bushman EW, Landy A. 1988. Autonomous DNA binding domains of lambda integrase recognize two different sequence families. *Cell* 54:923–29
64. Moitoso De Vargas L, Kim S, Landy A. 1989. DNA looping generated by recombination. *Annu. Rev. Biochem* 58:913–49 DNA bending protein IHF and the two domains of lambda integrase. *Science* 244:1457–61
65. Morse SA, Johnson SR, Biddle JW, Roberts MC. 1986. High-level tetracycline resistance in *Neisseria gonorrhoeae* is a result of acquisition of streptococcal *tetM* determinant. *Antimicrob. Agents Chemother.* 30:664–70
66. Mullany P, Wilks M, Lamb I, Clayton C, Wren B, Tabaqchali S. 1990. Genetic analysis of a tetracycline resistance element from *Clostridium difficile* and its conjugal transfer to and from *Bacillus subtilis*. *J. Gen. Microbiol.* 136:1343–49
67. Naglich JG, Andrews RE. 1988. Tn*916*-dependent conjugal transfer of PC194 and PUB110 from *Bacillus subtilis* into *Bacillus thuringiensis* subsp. *israelensis*. *Plasmid* 20:113–26
68. Nash HA, Robertson CA. 1981. Purification and properties of the *Escherichia coli* protein factor required for lambda integrative recombination. *J. Biol. Chem.* 256:9246–53
69. Norgren M, Caparon MG, Scott JR. 1989. A method for allelic replacement that uses the conjugative transposon Tn*916*: deletion of the *emm*6.1 allele in *Streptococcus pyogenes* JRS4. *Infect. Immun.* 57:3846–50
70. Norgren MG, Scott JR. 1991. Presence of the conjugative transposon Tn*916* in the recipient strain does not impede transfer of a second copy of the element. *J. Bacteriol.* 173:319–24
70a. Nunes-Düby SE, Azaro MA, Landy A. 1995. Swapping DNA strands and sensing homology without branch migration in lambda site–specific recombination. *Curr. Biol.* 5:139–48
71. Nunes-Düby SE, Matsumoto L, Landy A. 1987. Site-specific recombination intermediates trapped with suicide substrates. *Cell* 50:779–88
72. Ounissi H, Courvalin P. 1982. Heterogeneity of Macrolide-Lincosamide-Streptogramin B-type antibiotic resistance determinants. See Ref. 87a, pp. 167–69
73. Pargellis CA, Nunes-Düby SE, Moitoso De Vargas L, Landy A. 1988. Suicide recombination substrates yield covalent lambda integrase–DNA complexes and lead to identification of the active site tyrosine. *J. Biol. Chem.* 263:7678–85
74. Plasterk RHA, Van De Putte P. 1985. The invertible P-DNA segment in the chromosome of *Escherichia coli*. *EMBO J.* 4:237–42
75. Poyart-Salmeron C, Trieu-Cuot P, Carlier C, Courvalin P. 1989. Molecular characterization of two proteins in-

volved in the excision of the conjugative transposon Tn*1545*: homologies with other site-specific recombinases. *EMBO J.* 8:2425–33

76. Poyart-Salmeron C, Trieu-Cuot P, Carlier C, Courvalin P. 1990. The integration-excision system of the conjugative transposon Tn*1545* is structurally and functionally related to those of lambdoid phages. *Mol. Microbiol.* 4:1513–21

77. Prasad PV, Young L-J, Jayaram M. 1987. Mutations in the 2-µm circle site-specific recombinase that abolish recombination without affecting substrate recognition. *Proc. Natl. Acad. Sci. USA* 84:2189–93

78. Rauch PJG, De Vos WM. 1992. Characterization of the novel nisin-sucrose conjugative transposon Tn*5276* and its insertion in *Lactococcus lactis*. *J. Bacteriol.* 174:1280–87

79. Rauch PJG, De Vos WM. 1994. Identification and characterization of genes involved in excision of the *Lactococcus lactis* conjugative transposon Tn*5276*. *J. Bacteriol.* 176:2165–71

80. Renault P, Nogrette JF, Galleron N, Godon JJ, Ehrlich SD. 1994. *Specificity of insertion of Tn*1545 *transposon family in* Lactococcus lactis *subsp.* lactis. Presented at Am. Soc. Microbiol., Int. Conf. Streptococcal Genetics, 4th

81. Rice LB, Carias LL. 1994. Studies on excision of conjugative transposons in enterococci: evidence for joint sequences composed on strands with unequal numbers of nucleotides. *Plasmid* 31:312–16

82. Rice LB, Marshall SB, Carias LL. 1992. Tn*5381*, a conjugative transposon identifiable as a circular form in *Enterococcus faecalis*. *J. Bacteriol.* 174:7308–15

83. Roberts MC, McFarland LV, Mullany P, Mulligan ME. 1994. Characterization of the genetic basis of antibiotic resistance in *Clostridium difficile*. *J. Antimicrob. Chemother.* 33:419–29

84. Ross W, Shulman M, Landy A. 1982. Biochemical analysis of *att*-defective mutants of the phage lambda site-specific recombination system. *J. Mol. Biol.* 156:505–29

85. Rudy CK, Scott JR. 1994. Length of the coupling sequence of Tn*916*. *J. Bacteriol.* 176:3386–88

86. Salyers AA, Shoemaker NB, Guthrie EP. 1987. Recent advances in *Bacteroides* genetics. *CRC Crit. Rev. Microbiol.* 14:49–71

87. Sanchez-Pescador R, Brown JT, Roberts M, Urdea MS. 1988. Homology of the TetM with translational elongation factors: implication for potential modes of tetM conferred tetracycline resistance. *Nucleic Acids Res.* 3:1218

87a. Schlessinger D, ed. 1982. *Microbiology—1982.* Washington, DC: Am. Soc. Microbiol.

88. Scott JR. 1992. Sex and the single circle: conjugative transposition. *J. Bacteriol.* 174:6005–10

89. Scott JR. 1993. Conjugative transposons. In *Bacillus subtilis and Other Gram-Positive Bacteria,* ed. AL Sonenshein, JA Hoch, R Losick, pp. 597–614. Washington, DC: Am. Soc. Microbiol.

90. Scott JR, Kirchman PA, Caparon MG. 1988. An intermediate in transposition of the conjugative transposon Tn*916*. *Proc. Natl. Acad. Sci. USA* 85:4809–13

91. Scott JR, Bringel F, Marra D, Van Alstine G, Rudy CK. 1994. Conjugative transposition of Tn*916*: preferred targets and evidence for conjugative transfer of a single strand, and for a double-stranded circular intermediate. *Mol. Microbiol.* 11:1099–108

92. Senghas E, Jones JM, Yamamoto M, Gawron-Burke C, Clewell DB. 1988. Genetic organization of the bacterial conjugative transposon Tn*916*. *J. Bacteriol.* 170:245–49

93. Shoemaker NB, Smith MD, Guild WR. 1979. Organization and transfer of heterologous chloramphenicol and tetracycline resistance genes in pneumococcus. *J. Bacteriol.* 139:432–41

94. Shoemaker NB, Smith MD, Guild WR. 1980. DNase-resistant transfer of chromosomal *cat* and *tet* insertions by filter mating in pneumococcus. *Plasmid* 3:80–87

95. Smith MD, Hazum S, Guild WR. 1981. Homology among *tet* determinants in conjugative elements of streptococci. *J. Bacteriol.* 148:232–40

96. Su YA, Clewell DB. 1993. Characterization of the left 4 kb of conjugative transposon Tn*916*: determinants involved in excision. *Plasmid* 30:234–50

97. Su YA, He P, Clewell DB. 1992. Characterization of the *tet*(M) determinant of Tn*916*: evidence for regulation by transcription attenuation. *Antimicrob. Agents Chemother.* 36:769–78

98. Swartley JS, McAllister CF, Hajjeh RA, Heinrich DW, Stephens DS. 1993. Deletions of Tn*916*-like transposons are implicated in *tetM*-mediated resistance in pathogenic *Neisseria*. *Mol. Microbiol.* 10:299–310

99. Tally FP, Shimell MJ, Carson GR, Malamy MH. 1981. Chromosomal and plasmid-mediated transfer of clindamycin resistance in *Bacteroides fragilis*. In

Molecular Biology, Pathogenicity and Ecology of Bacterial Plasmids, ed. SB Levy, RC Clowes, p. 51. New York: Plenum

100. Thompson JF, Landy A. 1988. Empirical estimation of protein-induced DNA bending angles: applications to lambda site-specific recombination complexes. *Nucleic Acids Res.* 16:9687–705
101. Torres OR, Korman RZ, Zahler SA, Dunny GM. 1991. The conjugative transposon tn925: enhancement of conjugal transfer by tetracycline in *Enterococcus faecalis* and mobilization of chromosomal genes in *Bacillus subtilis* and *E. faecalis. Mol. Gen. Genet.* 225: 395–400
102. Trieu-Cuot P, Poyart-Salmeron C, Carlier C, Courvalin P. 1993. Sequence requirements for target activity in site-specific recombination mediated by the Int protein of transposon Tn*1545. Mol. Microbiol.* 8:179–85
103. Trieu-Cuot P, Carlier C, Poyart-Salmeron C, Courvalin P. 1991. An integrative vector exploiting the transposition properties of Tn*1545* for insertional mutagenesis and cloning of genes from Gram-positive bacteria. *Gene* 106: 21–27
104. Vijayakumar MN, Priebe SD, Guild WR. 1986. Structure of a conjugative element in *Streptococcus pneumoniae. J. Bacteriol.* 166:978–84
105. Vijayakumar MN, Priebe SD, Pozzi G, Hageman JM, Guild WR. 1986. Cloning and physical characterization of chromosomal conjugative elements in streptococci. *J. Bacteriol.* 166:972–77
106. Weisberg RA. 1983. Role for DNA homology in site-specific recombination: the isolation and characterization of a site affinity mutant of coliphage lambda. *J. Mol. Biol.* 170:319–42
107. Wierzbicki A, Kendall M, Abremski K, Hoess R. 1987. A mutational analysis of the bacteriophage P1 recombinase Cre. *J. Mol. Biol.* 195:785–94
108. Wuenscher MD, Köhler S, Bubert A, Gerike U, Goebel W. 1993. The *iap* gene of *Listeria monocytogenes* is essential for cell viability, and its gene product, p60, has bacteriolytic activity. *J. Bacteriol.* 175:3491–501

CELLULOSE DEGRADATION IN ANAEROBIC ENVIRONMENTS

Susan B. Leschine

Department of Microbiology, University of Massachusetts, Amherst, Massachusetts 01003-5720

KEY WORDS: cellulase, cellulose fermentation, microbial interactions, cellulolytic microorganisms, carbon cycle

CONTENTS

INTRODUCTION	400
THE SUBSTRATE	400
CELLULOLYTIC ENZYME SYSTEMS OF ANAEROBES	401
CELLULOSE DEGRADATION IN SOILS, SEDIMENTS, AND AQUATIC ENVIRONMENTS	405
The Environments of Free-Living Anaerobic Cellulolytic Bacteria	405
Diversity of Cellulose-Fermenting Microorganisms	407
Interactions Among Microorganisms Engaged in Cellulose Degradation	410
CELLULOSE DEGRADATION IN ASSOCIATION WITH ANIMALS	412
The Rumen	412
Diversity of Cellulolytic Microorganisms in the Rumen	413
Microbial Interactions in Cellulose Degradation in the Rumen	414
Other Animal-Associated Environments	416
CONCLUDING REMARKS	418

ABSTRACT

In anaerobic environments rich in decaying plant material, the decomposition of cellulose is brought about by complex communities of interacting microorganisms. Because the substrate, cellulose, is insoluble, bacterial and fungal degradation occurs exocellularly, either in association with the outer cell envelope layer or extracellularly. Products of cellulose hydrolysis are available as carbon and energy sources for other microbes that inhabit environments in which cellulose is biodegraded, and this availability forms the basis of many microbial interactions that occur in these environments. This review discusses interactions among members of cellulose-decomposing microbial communities in various environments. It considers cellulose decomposing communities in soils, sediments, and aquatic environments, as well as those that degrade cellulose in association with animals. These microbial communities contribute significantly to the cycling of carbon on a global scale.

INTRODUCTION

The development of microbial communities that effect the degradation of cellulose and other abundantly produced plant cell wall polymers in anaerobic environments is one of the hallmarks of evolution. Indeed, these communities play a key role in the cycling of carbon on the planet. They coevolved with organisms that developed light-driven mechanisms to reduce carbon dioxide, and their essential role is to return carbon to the atmosphere. The ecology of cellulose degradation in anaerobic environments is very complex; it involves numerous, varied interactions of metabolically diverse microorganisms whose activities are influenced by a wide range of environmental factors.

Anaerobic environments rich in decaying plant material are prevalent and tremendously varied. Not surprisingly, a wide range of equally varied cellulose-degrading microbial communities has evolved. Rather than serving as a comprehensive review of the ecology of cellulose degradation, this article covers environments and microbes that have been studied most extensively in recent years, focusing on those environments that contribute significantly to the cycling of carbon on a global scale. Included is information on the physiology of cellulolytic microorganisms, the enzyme systems they produce, and their interactions with other microbes in the degradation of cellulose. In addition, the review attempts to draw attention to those topics that, in my opinion, deserve further study.

THE SUBSTRATE

Cellulose is the most abundantly produced biopolymer in terrestrial environments. Each year photosynthetic fixation of CO_2 yields more than 10^{11} tons of dry plant material worldwide (147), and almost half of this material consists of cellulose (39). Cellulose is a homopolymer consisting of glucose units joined by β-1,4 bonds. The disaccharide cellobiose is regarded as the repeating unit in cellulose inasmuch as each glucose unit is rotated by 180° relative to its neighbor. The size of cellulose molecules (degree of polymerization) varies from 7000 to 14,000 glucose moieties per molecule in secondary walls of plants but may be as low as 500 glucose units per molecule in primary walls (93, 139).

Cellulose molecules are strongly associated through inter- and intramolecular hydrogen-bonding and van der Waals forces that result in the formation of microfibrils, which in turn form fibers. Cellulose molecules are oriented in parallel, with reducing ends of adjacent glucan chains located at the same end of a microfibril. These molecules form highly ordered crystalline domains interspersed by more disordered, amorphous regions. The degree of crystallinity in native cellulose is 60–90%. Cellulose can take on at least four different crystalline forms (I to IV), as determined by X-ray crystallography. The predominant native form is referred to as cellulose I. Generally, the percentage and crystalline form of cellulose within a plant cell wall varies according to cell type and developmental stage. For example, cellulose constitutes about 20–40% of wall dry weight in growing primary walls and increases to 40–60% in secondary walls. The secondary walls of cotton seed hairs are nearly 100% cellulose. Secondary cell wall microfibrils have a higher cellulose I crystallinity and may be thicker than primary wall microfibrils (100, 139).

Cellulose almost never occurs alone in nature but is usually associated with other plant substances. This association may affect its natural degradation. Cellulose fibrils are embedded in a matrix of other polymers, primarily including hemicelluloses, pectin, and proteins. Cellulose imparts tensile strength to the wall to resist turgor pressure. High compression strengths are achieved when lignin (a complex aromatic polymer) replaces water in the matrix of cell walls. Lignification greatly increases bonding within the wall and produces rigid, woody tissues able to withstand the compressive force of gravity (139). Hemicelluloses (e.g. xylans, glucomannans) are composed of linear and branched heteropolymers of D-xylose, L-arabinose, D-mannose, D-glucose, D-galactose, and D-glucuronic acid. Xylans, often the most abundant hemicelluloses, have a β-1,4–linked xylopyranose backbone with attached side groups of acetate, arabinofuranose, and O-methyl glucuronic acid (11, 39). Inasmuch as hemicelluloses surround the cellulose microfibrils and occupy spaces between fibrils (39), this polymer must be degraded, at least in part, before cellulose in plant cell walls can be effectively degraded by cellulolytic bacteria (150, 151). Moreover, arabinofuranosyl groups may be esterified by aromatic acids such as ferulic and p-coumaric acid (55) and may participate in lignin-hemicellulose cross-linkages (146), further complicating the microbial degradation of cellulose.

CELLULOLYTIC ENZYME SYSTEMS OF ANAEROBES

One of the most important features of cellulose as a substrate for microorganisms is its insolubility. Bacterial and fungal degradation of cellulose and other insoluble polymers occurs exocellularly, either in association with the outer cell envelope layer or extracellularly. This suggests that the assembly of enzyme systems, which may be extremely complex, also occurs exocellularly.

To function, these enzyme systems must be stable in the exocellular environment; for example, they must be reasonably resistant to proteolytic attack. Also, the products of cellulose hydrolysis may be available as carbon and energy sources for other microbes that inhabit environments in which cellulose is biodegraded, thereby forming the basis of many interactions between microorganisms in these environments, as discussed below.

The enzymology of cellulose degradation has been an area of active research for more than 40 years. The focus of this research has frequently shifted. For example, the initial thrust to understand the mechanism of cellulose degradation was spurred by a fungal attack on the cotton clothing and tents of troops stationed in Southeast Asia during World War II. Research directed toward developing ways to inhibit fungal cellulases was carried out at the US Army Research and Development Command in Natick, Massachusetts. Under the direction of Elwyn Reese and Mary Mandels, this work ultimately led to the development of seminal concepts related to the mechanism of cellulose degradation, including the role of synergism among components of the cellulase system (52, 137, 138). With the energy crises of the 1970s, the focus of research on cellulose biodegradation shifted to developing systems and procedures to use cellulose and other abundant plant polymers as a source of fuels and chemicals that could serve as a potential replacement for fossil hydrocarbons. The voluminous work that followed, aided by the development of recombinant DNA and nucleic acid sequencing techniques, has resulted in our current understanding of the enzyme systems produced by diverse cellulolytic organisms. The enzymology of cellulose decomposition is described in several excellent recent reviews (6–8, 35, 41, 50, 51, 140, 186). It is also briefly considered here, as an understanding of cellulolytic enzyme systems is central to the picture of the overall process of cellulose degradation in natural environments.

The mechanism by which cellulases from anaerobic bacteria catalyze the depolymerization of crystalline cellulose is poorly defined, despite numerous recent investigations. However, this mechanism is clearly fundamentally different from that of the cellulase systems of most aerobic fungi and bacteria (25). Because many of the early studies of bacterial cellulases employed the fungal system as a model, a brief description of the enzymology of that system is presented here for clarity. Details may be found in recent reviews (186, 188).

The cellulase systems of fungi (e.g. *Trichoderma reesei*) comprise three main activities: (*a*) endoglucanases, which randomly hydrolyze 1,4-β bonds within cellulose molecules, thereby producing reducing and nonreducing ends; (*b*) exoglucanases, which cleave cellobiose units from the nonreducing ends of cellulose polymers; and (*c*) β-glucosidases, which hydrolyze cellobiose and low-molecular-weight cellodextrins, thereby yielding glucose. These enzymatic components act synergistically in the hydrolysis of crystalline cellulose.

Synergism has been explained by the proposal that endoglucanases attack amorphous regions of cellulose fibers, forming sites for exoglucanases which can then hydrolyze cellobiose units from more crystalline regions of the fibers. Finally, β-glucosidases, by hydrolyzing cellobiose, prevent the accumulation of this disaccharide, which is an inhibitor of exoglucanase activity. This model for the mechanism of action of the fungal cellulase system may be an oversimplification because it does not explain all types of observed synergism (75).

In contrast to cellulase systems of aerobic fungi described above, the cellulases of most anaerobic microorganisms are organized into large, multiprotein complexes (6–8, 25, 35, 41, 81, 82; and references therein). In general, these complexes efficiently catalyze the hydrolysis of cellulose as long as they retain their integrity, but even partial disassociation of the complexes as occurs under relatively mild conditions causes loss of most activity against crystalline cellulose. Some of the proteins that result from the disassociation of the large complexes have endoglucanase, cellobiohydrolase, or xylanase activity; others do not appear to have enzymatic activity and may have a structural function (e.g. as a scaffolding protein) and/or may be involved in attachment of the complex to the substrate (22, 59, 76, 77, 83, 111, 112, 190). None of the individual proteins in the complexes has been found to have significant activity against crystalline cellulose. However, the cellulase system of the thermophile *Clostridium stercorarium* includes two enzymes designated avicelase I and II. [The avicelase assay measures the ability of cellulase preparations to hydrolyze crystalline cellulose (avicel) and is often considered a measure of exoglucanase activity.] Purified avicelase I is a monomeric 100,000 M_r protein that catalyzes the hydrolysis of crystalline cellulose without requiring additional proteins or cofactors (15). It has been designated an endoglucanase, but differs from endoglucanases produced by other cellulolytic anaerobes by its ability to hydrolyze crystalline cellulose (15).

The most thoroughly investigated cellulase complex, that of *Clostridium thermocellum*, was termed the cellulosome by Lamed and coworkers (81, 82, 87). On the cell surface, these multiprotein, multifunctional enzymes appear as polycellulosomal aggregates and promote adherence of the bacterium to cellulose (5, 85, 104, 105, 177). Cellulosomes of *C. thermocellum* strains range in molecular weight from 2.0 to 6.5 × 10^6 (105, 190) and comprise 14–26 polypeptide subunits (77, 86). Genes encoding several of the polypeptides have been cloned and their nucleotide sequences determined. The largest subunit, the cellulosome integrating protein CipA (also designated S1 or S_L), is a glycoprotein with a mass of 210–250 kDa (87, 189, 190). CipA has a cellulose-binding domain and nine internal repeated sequences that bind the catalytic subunits (45, 47, 132, 145, 162). Catalytic subunits contain a conserved, duplicated segment that is believed to serve as a docking sequence (45, 47, 162). CipA functions as a scaffolding protein of the cellulosome, aligning the catalytic subunits for efficient cellulose hydrolysis. Possibly, the cellulosome catalyzes multiple, nearly simultaneous cuttings of the glucan chain (41, 105). Fujino et al (46) proposed that a protein, ORF3p, encoded by a gene located downstream of the gene encoding CipA, binds to CipA and functions to anchor the cellulosome to the cell surface. However, more recent results (143, 144) argue against this hypothesis and suggest that ORF3p may bind individual cellulases and hemicellulases to the cell surface. Cellulosomes bind bacterial cells to cellulose (5, 81, 177). Furthermore, the products of cellulolysis may pass through the "fibrous contact corridors" (41) that are observed between cells and cellulose. However, as discussed below, noncellulolytic commensal bacteria can grow in cellulose-degrading cocultures with *C. thermocellum*, which indicates that at least some soluble sugars are released from contact corridors or are otherwise available as growth substrate for other microorganisms.

Multiprotein cellulase complexes are produced by many diverse anaerobes, including *Bacteroides cellulosolvens* (84), *Clostridium cellulolyticum* (2, 40, 95), *Clostridium cellulovorans* (35, 149), *Clostridium papyrosolvens* C7 (21, 22), *Fibrobacter* (formerly *Bacteroides*) *succinogenes* (54), *Ruminococcus albus* (187), *Ruminococcus flavefaciens* (34), and the rumen fungus *Neocallimastix frontalis* (180, 181). Several of these cellulolytic anaerobes (*B. cellulosolvens, C. cellulovorans, R. albus*), as well as *Acetivibrio cellulolyticus* and *Clostridium cellobioparum*, have cell-surface localized cellulosome-like structures (85). In the rumen bacteria *R. albus* and *F. succinogenes*, activity against crystalline cellulose apparently results from cell-bound enzymes associated with the capsule (*R. albus*) (155) or outer membrane (*F. succinogenes*) (44). Enzymes associated with capsule or outer-membrane fragments may detach from cells, giving rise to a high-molecular-weight, sedimentable cellulase fraction (44). Generally, *F. succinogenes* seems to need to adhere to cellulose (78), and the properties of its cellulase system differ from those ascribed to cellulosomes (43).

C. papyrosolvens C7, a mesophilic cellulolytic bacterium isolated from a freshwater sediment (89), lacks cellulosome clusters on its surface (21). Also, cells of this clostridium do not adhere to cellulose fibers (21). The cellulase system of this bacterium is found in culture supernatant fluids and comprises at least seven distinct high-molecular-weight multiprotein complexes (M_r 500,000–660,000), each with different polypeptide composition and enzymatic and ultrastructural properties (130; M Pohlschröder, SB Leschine & E Canale-Parola, unpublished data). Two of the complexes have xylanase activity, and three others, avicelase activity. An avicelase-deficient mutant of *C. papyrosolvens* C7, which cannot degrade crystalline cellulose, does not produce the multicomplex system (21, 22). Pohlschröder and coworkers (130) found that hydrolysis of crystalline cellulose by this multicomplex cellulase-xylanase system involves synergistic interactions among its multiprotein components. In some respects, cellulose hydrolysis by the multicomplex system of *C. papyrosolvens* C7 resembles that by the cellulosome of *C. thermocellum*. For example, this system requires Ca^{2+} and a thiol reducing agent for activity (22), and it consists of particles, some of which are similar in ultrastructure to, although smaller than, the cellulosomes of *C. thermocellum* (M Pohlschröder, SB Leschine & E Canale-Parola, unpublished data). However, it differs from cellulosome systems in that it is a multicomplex rather than a unicomplex system. Also, components of the multicomplex system are never found associated with cells. As pointed out by Forsberg et al (44), cell-free cellulase complexes may be able to penetrate small spaces in cellulose fibers that the cell itself cannot enter, and thus the complexes gain access to a greater fraction of the cellulose. Examination of cultures of *C. papyrosolvens* C7 via light microscopy showed that the bacterial cells do not adhere to cellulose fibers; rather, they accumulate near cellulose fibers. Hungate (62) described this same behavior for cellulolytic bacteria he had isolated. Hsing & Canale-Parola (60) proposed that this behavior results from a chemotactic response toward products of cellulose hydrolysis. Chemotactic behavior of this sort could play an important role in the overall process of cellulose degradation in natural environments.

CELLULOSE DEGRADATION IN SOILS, SEDIMENTS, AND AQUATIC ENVIRONMENTS

The Environments of Free-Living Anaerobic Cellulolytic Bacteria

Most cellulose is degraded aerobically, but 5–10% is degraded anaerobically (38, 171). Thus, vast quantities of cellulose are degraded by cellulose-fermenting microorganisms in anaerobic environments. Anaerobic activity starts close to the surface in soils and composts, as well as in freshwater, marine, and estuarine sediments, indicating that aerobic conditions normally prevail only in a thin crust at the atmospheric boundary (93). The anaerobic degradation of soil organic matter plays an extremely important role in the global cycling of carbon. Soils are a huge reservoir of carbon: The top meter of soil contains about twice as much carbon as is found in the atmosphere (133), and some recent studies suggest that the rates of accumulation and turnover of belowground carbon may be underestimated (135). Consequently, understanding the effects of predicted global warming on decomposition of organic matter in soils and sediments is essential for modeling future atmospheric and climatic changes (67).

A community of physiologically diverse microorganisms is responsible for the anaerobic degradation of cellulose. This community structure contrasts with aerobic decomposition, which may be achieved through the activities of single species. For example, both carbohydrate and lignin components of wood are completely decomposed to CO_2 and H_2O by single species of white rot fungi (74). Anaerobic decomposition, on the other hand, requires mixed populations. The metabolic versatility of anaerobes arises largely because they can perform various fermentations and respirations, employing diverse electron acceptors (e.g. carbon dioxide, inorganic sulfur compounds, inorganic nitrogen compounds) in place of oxygen (93, 128, 194; for a review of earlier literature, see 63).

In the absence of oxygen and certain other exogenous inorganic electron acceptors [e.g. nitrate, Mn(IV), Fe(III), sulfate (94)], cellulose is decomposed by the anaerobic community into CH_4, CO_2, and H_2O through a complex microbial food chain (109, 194) shown diagrammatically in Figure 1. The processes are similar in most anaerobic soils and sediments [although perhaps not all (see 14, 80)] and in anaerobic digestors. Cellulolytic microbes produce enzymes that depolymerize cellulose, thereby producing cellobiose, cellodextrins, and some glucose. These sugars are fermented by cellulolytic and other saccharolytic microorganisms. By keeping cellobiose concentrations low, and thus preventing inhibition of the cellulase system by this product of cellulose hydrolysis, noncellulolytic cellobiose-fermenters may play a very important role in this process (93). These fermentations yield CO_2, H_2, organic acids (e.g. acetate, propionate, butyrate), and alcohols. Very little H_2 escapes into the atmosphere because it is immediately consumed by methanogens or homoacetogens. Methanogens use H_2 to reduce CO_2 to CH_4, and homoacetogens use H_2 to reduce CO_2 to acetate. Some methanogenic species use acetate produced by fermenters or by homoacetogens through the acetoclastic cleavage to CH_4 and CO_2 (70, 97). Syntrophic bacteria play a key role in the conversion of cellulose to CH_4 and CO_2. These organisms ferment fatty acids such as propionate and butyrate, or alcohols, and produce acetate, CO_2, and H_2. They grow only in the presence of H_2-consuming organisms through interspecies H_2 transfer. Syntrophic bacteria grow very slowly, and thus the fermentation of fatty acids is usually the rate-limiting step in the anaerobic decomposition of cellulose (109, 185).

Through the combined activities of several major physiological groups of microbes, cellulose is completely dissimilated to CO_2 and CH_4. Thus, as a source of CO_2 and CH_4, the anaerobic decomposition of cellulose plays a major role in carbon cycling on the planet (91, 93). In marine environments, sulfate is plentiful and sulfate-reducing bacteria out-compete methanogens for H_2. Thus, H_2S is a major product of the anaerobic degradation of cellulose in marine systems (93, 128).

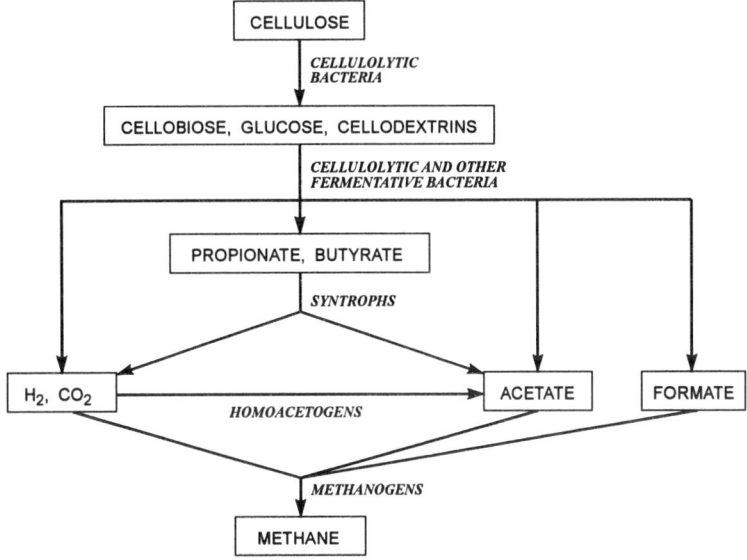

Figure 1 Diagrammatic representation of anaerobic cellulose degradation by microbial communities in soils and freshwater sediments. Lactate, succinate, and ethanol are also produced by fermentative bacteria but usually do not accumulate. In environments where nitrate, Mn(IV), Fe(III), or sulfate are present, the final products of fermentation may differ.

Diversity of Cellulose-Fermenting Microorganisms

Although much of the work on cellulose degradation has involved thermophiles such as *C. thermocellum,* these bacteria probably play a minor role in cellulose decomposition in most natural environments because they require unusually high temperatures for growth and cellulose fermentation (106). Inasmuch as this article aims to review information on cellulose degradation as it occurs in common natural environments, this section focuses primarily on mesophilic cellulose-fermenting microbes. Several mesophilic cellulolytic anaerobes have been isolated from soils, sediments, and composts from geographically widely separated locations. *Clostridium papyrosolvens* NCIB11394 was isolated from estuarine sediments (96), and another strain of this species, strain C7, was isolated from the sediment of a freshwater swamp (89, 130). *Clostridium lentocellum* was isolated from samples of the same estuarine sediments from which *C. papyrosolvens* NCIB11394 was derived (117). *C. lentocellum* differs from other mesophilic cellulolytic bacteria in several ways, but perhaps the most significant is the relatively slow rate of cellulose degradation by cultures of this bacterium. *Clostridium cellulofermen-*

tans inhabits dairy farm soil (192). Cellulolytic anaerobes isolated from wetwood of trees are similar in many respects to isolates from soil (173), and these trees are probably infected by soil bacteria. Although *Clostridium cellobioparum* was originally isolated from bovine rumen contents (62), this clostridium is reportedly also found in soil (152). In addition to the above-mentioned cellulolytic anaerobes, several other strains from soils and sediments have been described (63, 89, 92, 102, 152; SB Leschine, E Canale-Parola, E Monserrate & T Warnick, unpublished results).

All of the mesophilic strains from soils and sediments that have been examined are motile rod-shaped organisms with a similar cell envelope ultrastructure. Also, they have similar G+C contents. Cells of almost all strains stain gram-negative, and most, but not all, form spores. All ferment not only cellulose, cellobiose, and glucose, but also components of the hemicellulosic portion of biomass (e.g. xylan, pentoses). Of those strains examined, all form acetate, ethanol, CO_2, and H_2 as primary products of fermentation. Other free-living cellulose-fermenting mesophiles have been isolated from composts [e.g. *C. cellulolyticum* (129), *Clostridium josui* (161)], sewage sludges [e.g. *Acetivibrio cellulolyticus* (127), *Bacteroides cellulosolvens* (118), strain CM126 (120)], and anaerobic digestors [e.g. *Clostridium aldrichii* (191), *Clostridium celerecrescens* (125), *C. cellulovorans* (154), *Clostridium populeti* (153)]. Table 1 lists some of the properties of these bacteria.

Phylogenetic relationships among species of mesophilic cellulolytic clostridia have been examined via 16S rRNA and rDNA sequence analyses (24, 136). *C. cellulovorans*, along with two rumen cellulolytic bacteria, *Clostridium chartatabidum* (72) and *Clostridium "longisporum"* (64, 168), cluster with members of *Clostridium* group I [as defined by Johnson & Francis (69)], which includes the type species, *Clostridium butyricum* and many other noncellulolytic species (24, 136). *C. lentocellum, C. celerecrescens,* and *C. populeti* are members of *Clostridium* group III (69), which also includes many noncellulolytic species (136). *C. papyrosolvens* and most other mesophilic cellulolytic isolates from soil, *C. cellulolyticum* from composted grass (129), and *A. cellulolyticus* (127) and *B. cellulosolvens* from sewage sludge (118) form phylogenetically coherent clusters within a more deeply branching group of cellulolytic clostridia that includes *C. thermocellum* (136; SB Leschine & R Seward, unpublished results). Thus far, this later group is defined solely by cellulolytic species. A point of interest, *A. cellulolyticus* and *B. cellulosolvens* were not classified as clostridia because they stain gram-negative and spores were not observed (118, 127). However, most cellulolytic clostridia stain gram-negative and are oligosporogenous, and in some strains, spores are observed only rarely. Evidently, *A. cellulolyticus* and *B. cellulosolvens* are related to the cellulolytic clostridia in the above-mentioned group, and they probably share physiological properties (e.g. cellulose fermentation characteristics) common to members of

Table 1 Free-living mesophilic cellulose-fermenting bacteria

Source	Organism	Motility[a]	Growth substrates[b]	Fermentation products[c]	G+C content (mol%)	Reference
Soil and sediment	Clostridium cellulofermentans	+PF	Broad	A, E,	34	192
	Clostridium lentocellum	+F	Broad	A, E	36	117
	Clostridium papyrosolvens	+PF	C, CB, G, ⊕	A, E, L	30–33	89, 96
Compost	Clostridium cellulolyticum	+PF	C, CB, G, ⊕	A, E, L, F	41	129
	Clostridium josui	–	C, CB, G, ⊕ (+ maltose)	A, E, B	40	161
Sewage sludge	Acetivibrio cellulolyticus	+MF	C, CB, salicin	A, E (tr),	38	127
	Bacteroides cellulosolvens	–	C, CB	A, E, L	43	118
Anaerobic digestor	Clostridium aldrichii	+PB	C, CB, xylan	A, P, IB, B, IV, L, S	40	191
	Clostridium celerecrescens	+PF	Broad	A, E, F, B, C, IB, IV, L, S	38	125
	Clostridium cellulovorans	–	Broad	A, B, L, F	26–27	154
	Clostridium populeti	+	Broad	A, B, L	28	153

[a] Abbreviations: +, motile; F, fimbriae; MF, monotrichously flagellated; PB, polar bundle of flagella; PF, peritrichously flagellated; –, nonmotile.
[b] All strains use cellulose (C) and cellobiose (CB) as growth substrates. Some strains use glucose, other 6- and 5-carbon sugars, and xylan (C, CB, G, ⊕). Strains with a "Broad" substrate range use additional substrates such as maltose, sucrose, starch, and/or pectin.
[c] All strains produce CO_2 and H_2. Abbreviations: A, acetate; B, butyrate; C, caproate; E, ethanol; F, formate; IB, isobutyrate; IV, isovalerate; L, lactate; P, propionate; S, succinate; (tr), trace.

this group. Similarities in the cellulase systems of *A. cellulolyticus, B. cellulosolvens,* and *C. thermocellum* (84, 85) were noted above.

Many environments in which cellulose is degraded are deficient in combined nitrogen (for example, peat soils, agricultural and municipal wastes, and composts), and cellulose-fermenting bacteria that satisfy their nitrogen requirements through the fixation of N_2 should have a strong selective advantage over those that require a source of combined nitrogen. Therefore, that several cellulose-fermenting bacteria can satisfy their nitrogen requirement for growth through the fixation of atmospheric N_2 is not surprising (92). Nitrogenase activity has been demonstrated in *C. papyrosolvens, C. cellobioparum,* and in several other mesophilic cellulolytic strains from soils and sediments (12, 92; SB Leschine, T Chen & T Warnick, unpublished results). These findings suggest that nitrogenase may be widespread among cellulose-fermenting bacteria from soils. Thus, in the natural environments they inhabit, these bacteria may contribute significant amounts of combined nitrogen to the growth of other members of cellulose-degrading communities.

Although free-living anaerobic cellulolytic fungi similar to those found in the rumen (124) have not yet been found, Durrant et al (37) reported the isolation, using anaerobic culture conditions, of two strains of morphologically and physiologically distinct cellulose-fermenting fungi from soil. Both strains grow and utilize cellulose more rapidly when incubated under microaerophilic conditions, and one strain degrades cellulose most rapidly in well-aerated cultures (37). Active cellulase and xylanase systems are produced by both strains, and enzymes are present in culture supernatant fluids (36). Clearly, fungi may play a significant role in the anaerobic degradation of cellulose in soils and sediments, and further studies are needed to explore their potential contributions.

Interactions Among Microorganisms Engaged in Cellulose Degradation

Microbial interactions appear to be important processes in the anaerobic degradation of cellulose. The cellulose-fermenting bacterium makes a living by degrading insoluble polymers in a highly competitive, nitrogen-poor environment. To do so, it must export from the cell significant amounts of protein to catalyze the hydrolysis of cellulose, and then it must compete for soluble sugars produced as a result of the activities of these enzymes. Because the products of the enzymatic attack on cellulose are available to other bacteria, the activities of these other bacteria may in turn affect the activity of the cellulose-fermenting bacterium. For example, Cavedon & Canale-Parola (20) found that *C. papyrosolvens* C7 grows in coculture with a noncellulolytic klebsiella in a chemically defined, vitamin-deficient medium containing cellulose as the carbon and energy source. In this coculture, the extracellular cellulase system produced by the clostridium hydrolyzes cellulose to soluble sugars that are utilized as fermentable substrates for the klebsiella, and in turn, the klebsiella excretes vitamins required by the clostridium. In addition, the clostridium fixes N_2, thus allowing growth of the klebsiella, which is not a nitrogen fixer. Mutualistic associations of this sort may take place in natural environments where cellulose is degraded (20). Other mutually beneficial interactions in cellulose-degrading cocultures based on the production of growth factors and the formation of fermentable soluble sugars have been described (e.g. 62, 73, 114, 116).

Spirochetes are frequently observed in anaerobic cellulolytic enrichment cultures, suggesting that free-living spirochetes associate and interact with cellulose-degrading bacteria in anaerobic environments (56). Recently, Pohlschröder et al (131) reported that *Spirochaeta caldaria*, a thermophilic spirochete from a freshwater hot spring, forms stable, cellulose-degrading cocultures with *C. thermocellum*. Cellulose is degraded more rapidly in cocultures than in monocultures of *C. thermocellum*, which suggests that noncellulolytic spirochetes may enhance cellulose breakdown in natural environments. In this study, the factors involved in the increased rate of cellulose degradation in cocultures were not examined; however, a plausible explanation is that the spirochetes keep cellobiose concentrations low, thereby preventing cellobiose inhibition of the cellulase system of *C. thermocellum* (93). Similar results have been obtained in cultures of cellulolytic bacteria with other noncellulolytic saccharolytic bacteria (42, 90, 114, 119).

Several studies have examined the effect of culturing a cellulose-fermenting microorganism with a noncellulolytic saccharolytic bacterium that produces ethanol as the sole or primary nongaseous fermentation product. In all cases, ethanol production from cellulose is enhanced in cocultures as compared to monocultures of the cellulolytic organism grown in the same medium (73, 90, 114, 119, 142, 172). An interesting study carried out by Ng et al (119) demonstrated that stable cocultures of *C. thermocellum* and *Clostridium thermohydrosulfuricum* fermented a variety of cellulosic substrates, including Solka Floc, and that ethanol yields exceeded those of *C. thermocellum* monocultures. The hemicellulosic portion of Solka Floc is not utilized by pure cultures of either bacterium. *C. thermocellum* produces an enzyme system that depolymerizes hemicellulose, but this bacterium cannot ferment pentosan products. In contrast, *C. thermohydrosulfuricum* ferments pentosans but cannot depolymerize hemicellulose. In cocultures, hemicelluloses are fermented through the combined activities of the two coculture members. Thus, both cellulosic and hemicellulosic portions of Solka Floc are fermented by cocultures, and ethanol is produced as a primary product.

The investigations described above suggest that, in natural environments, the soluble sugar products of cellulose hydrolysis are available as growth substrates for noncellulolytic commensal microorganisms, whose activities may affect cellulose degradation in a variety of ways.

Numerous studies have examined the effects on cellulose degradation of coculturing cellulose-fermenting bacteria with syntrophic and/or H_2-consuming microorganisms (reviewed in 93, 128). In general, utilization of H_2 by methanogens in cocultures with cellulose-fermenting bacteria results in an increase in acetate production relative to ethanol. For example, Laube & Martin (88) showed that cellulose-fermenting monocultures of *A. cellulolyticus* produce ethanol, acetate, H_2, and CO_2, and that cocultures of this bacterium with *Methanosarcina barkeri* produce more acetate and less ethanol. *M. barkeri* utilizes acetate, H_2, and CO_2 for methane production, and in cocultures, it utilizes acetate after a lag. In an *A. cellulolyticus–M. barkeri–Desulfovibrio* sp. triculture, the H_2 partial pressure is maintained at a level low enough for *Desulfovibrio* sp. to utilize ethanol. Thus, in tricultures, cellulose is rapidly fermented to CO_2 and CH_4. Furthermore, a greater fraction of cellulose decomposes more rapidly in tricultures than in monocultures of the cellulolytic bacterium (88). In this example, *Desulfovibrio* sp. is a H_2-producing bacterium that apparently grows syntrophically with the methanogen. If sulfate were present, *Desulfovibrio* sp. presumably would function instead as a H_2-consuming bacterium in the reduction of sulfate to H_2S, out-competing the methanogen for H_2.

CELLULOSE DEGRADATION IN ASSOCIATION WITH ANIMALS

The Rumen

Vast amounts of cellulose are degraded anaerobically in the gastrointestinal tracts of herbivorous and wood-eating animals. Herbivores lack the enzymatic capacity needed to degrade plant polysaccharides, particularly cellulose, and instead rely on communities of microorganisms that have this capability. An enlarged section of the gastrointestinal tract of herbivores functions as a fermentation vessel in which slowly degraded plant materials are retained long enough to be degraded by the microbes. This specialized portion of the gastrointestinal tract is separated from the highly acidic pyloric region of the gut and is well buffered. Herbivores have evolved as either pregastric (foregut) or postgastric (hindgut) fermenters, and their anatomy and evolutionary development have been described in detail (29, 61, 167).

Ruminants are cloven-hoofed herbivorous mammals that degrade plant materials in a specialized foregut organ, the rumen. Several recent reviews (30, 58, 98) and the classic work by Hungate (64) have described this extensively studied microbial ecosystem. Plant materials, mixed with saliva containing bicarbonate, enter the rumen and are mechanically ground into smaller particles through the rotary motion of the rumen. Gradually the food and microbial mass passes into the reticulum, another organ of the foregut, where it forms clumps or cuds, which are regurgitated into the mouth for additional chewing. The cud mixed with saliva is swallowed and passes to the omasum and then to the abomasum, an acidic organ that is more like a true stomach. Chemical digestive processes begin in the abomasum and continue in the small and large intestine.

While food remains in the rumen (~9–12 h), plant polysaccharides are hydrolyzed to oligo- and monosaccharides. These carbohydrates are then fermented, producing volatile fatty acids, including primarily acetic, propionic, and butyric acids, that are readily absorbed and serve as essential nutrients for the ruminant. Resultant carbon dioxide and methane are eliminated by eructation. In addition to carrying out this degradative process, the microbial community produces other essential nutrients, such as vitamins, for the animal. Furthermore, when microbial cells pass through the gastrointestinal tract, they are subjected to digestive processes similar to those of nonruminants. Thus, the rumen microbiota serves as a protein source for the animal as well.

In the rumen, cellulose is not completely converted to CO_2 and CH_4 as it is in environments such as soils and sediments. Instead, acetate, propionate, and butyrate are significant products of cellulose fermentation. Longer turnover times in soils, sediments, and other natural environments allow these fatty acids to be converted to CO_2 and CH_4 by the activities of microorganisms, as described above. Thus, in these complete bioconversion systems, acetate, as well as H_2 and CO_2, are primary substrates for methanogens, whereas in the rumen, the predominant substrates for methanogens are H_2 and CO_2 (109).

Diversity of Cellulolytic Microorganisms in the Rumen

The microbiology of the rumen has been the subject of studies for more than 45 years, and the literature describing microbes from this environment has been reviewed many times (e.g. 16, 23a, 30, 31, 57, 58, 64, 65, 98, 124, 158, 179, 182, 184). Bacteria are generally believed to provide most of the cellulolytic activity in the rumen, but rumen fungi, and to a lesser extent rumen protozoa may make a significant contribution. Based on determinations of relative numbers in the rumen and the ability to degrade purified and intact forage cellulose, the principal cellulolytic bacterial species appear to be *F. succinogenes, R. albus,* and *R. flavefaciens* (17–19, 28, 30, 148, 158, 160, 165). Also, some strains of the rumen bacterium *Butyrivibrio fibrisolvens* are cellulolytic (165) and may play an important role in cellulose degradation. Other cellulolytic species found sporadically or in low numbers in the rumen include *Eubacterium* (formerly *Cillobacterium*) *cellulosolvens* (19, 134, 164), *Micromonospora ruminantium* (99), *Micromonospora propionici* (62a), and clostridial species (62, 64, 72, 168). Although these later species are believed to have a limited role in cellulose degradation in the rumen, some may be very actively cellulolytic under certain conditions. For example, Varel (168) found that the rate and extent of alfalfa cell wall degradation by *C. longisporum* exceeded that of principal rumen cellulolytic species. In contrast, barley straw was poorly degraded. These results indicate that rumen cellulolytic microbes may have a high degree of substrate specificity.

Other members of the rumen microbial community are the anaerobic fungi (reviewed in 30, 124). Because they are found in large numbers attached to plant fragments removed from the rumen, and they ferment the major plant polysaccharides, they are believed to take part in decomposing cellulose in the rumen. Five genera of fungi have been described, *Neocallimastix, Caecomyces* (formerly *Sphaeromonas*), *Piromyces* (formerly *Piromonas*), *Orpinomyces,* and *Ruminomyces*. Pure cultures of rumen fungi can solubilize a high proportion of plant fragments, and they seem to have the ability to penetrate plant cell walls and gain access to plant polysaccharides not available to cellulolytic bacteria (124). Thus, it seems likely that the rumen anaerobic fungi play a significant role in fiber degradation in the rumen.

The existence of protozoa in the rumen has been known for more than 150 years (see 64). These anaerobic microbes ferment components of plant material (64). Most are ciliates that, taxonomically, fall into two groups, the holotrich protozoa and the entodiniomorphs (179). The holotrichs are generally believed to utilize nonstructural polysaccharides and soluble sugars (178, 179). On the other hand, all of the entodiniomorphs, except one species, contain cellulase. However, this activity may result from cellulase production by intracellular bacteria (179). Entodiniomorphid protozoa engulf and digest cellulose and use the products for the synthesis of intracellular polysaccharide (64). The ciliate protozoa may be responsible for a third or more of the microbial degradation of fiber in the rumen (32, 179).

Microbial Interactions in Cellulose Degradation in the Rumen

Interactions similar to those described above among free-living cellulose-fermenting bacteria and noncellulolytic bacteria have been observed in studies of rumen bacteria. For example, in cocultures of rumen cellulolytic bacteria and methanogens, interspecies H_2 transfer occurs. Consequently, the oxidative reactions of fermentation pathways are coupled to the reductive reactions of methanogenesis, and the flow of electrons shifts away from reduced products (e.g. lactate, succinate) of cellulolytic members of cocultures, resulting in greater acetate yields (109, 184).

Competitive and cooperative interactions among various cellulolytic and other saccharolytic microorganisms may affect the degradation of plant fiber in the rumen (30). Interactions among different cellulose-fermenting rumen microbes have been examined using 16S rRNA-targeted oligonucleotide probes (121, 122). *R. albus* 8 inhibits the growth of *R. flavefaciens* FD-1 when cocultures are grown with acid-swollen cellulose or with alkaline hydrogen peroxide–treated wheat straw as substrate (122), possibly because *R. albus* 8 produces a bacteriocin-like substance (121). In cocultures of *R. albus* 8 and *F. succinogenes* grown on the above-mentioned substrates, competition apparently does not occur, but *R. flavefaciens* FD-1 out-competes *F. succinogenes* in cocultures grown on acid-swollen cellulose. Complex patterns of competition among microbes occur in tricultures of *R. albus* 8, *R. flavefaciens* FD-1, and *F. succinogenes* (122). This study demonstrates the utility of molecular probes in investigations of competition among cellulolytic microorganisms engaged in cellulose degradation.

Spirochetes are present in the rumen in very large numbers (156). Although cellulose-degrading spirochetes have not been found in the rumen, many spirochetes isolated from rumen fluid can grow at the expense of plant polymers such as xylan, arabinogalactan, pectin, polygalacturonate, and starch (56, 126). Also, many ferment cellobiose (126) and thus may be expected to interact with the cellulolytic members of the rumen. Electron microscopy observations have indicated that spirochetes associate with plant cell wall materials undergoing degradation in the rumen (1, 23, 33, 159). Moreover, Kudo et al (79) observed spirochete cells firmly adhering to fibers of sterilized barley straw in preparations inoculated with pure cultures of *Treponema bryantii*, a cellobiose-fermenting, succinate-producing rumen spirochete (157). In cocultures of *F. succinogenes* and *T. bryantii*, the rod-shaped cells of *F. succinogenes* appeared to colonize and erode the cellulosic substrata, while spirochete cells were observed on the same surface immediately juxtaposed to the cellulolytic cells (79). Stanton & Canale-Parola (157) reported that *T. bryantii* grew in coculture with *F. succinogenes* in a medium that contained cellulose as the fermentable substrate. The spirochete enhanced cellulose breakdown by the rumen cellulolytic bacterium when the two organisms were cocultured on agar media. These investigators suggested that *T. bryantii* is chemotactically attracted by soluble sugars resulting from the cellulolytic activity of *F. succinogenes*, and that as a result of their incessant motion, the spirochetes may randomly push the nonmotile *F. succinogenes* cells toward the cellulose fibers and thus enhance cellulose breakdown. Kudo et al (79) reported that in cocultures of *T. bryantii* with either *F. succinogenes* or *R. albus*, degradation of barley straw, as well as the production of volatile fatty acids, other organic acids, and ethanol were enhanced. These authors concluded that cellulolytic bacteria interact with noncellulolytic spirochetes to promote the digestion of cellulosic materials in the rumen. They further suggested that enhanced cellulose degradation may occur after the cotransportation of cellulolytic and spirochete cells, as described above (157), or through physiological interactions that might be related to the use of cellobiose by the spirochete. If the spirochete utilizes this product of cellulose hydrolysis, such use would relieve cellobiose's inhibition of cellulase activity.

Interactions among rumen bacteria and fungi have been studied extensively. Cellulose degradation and fungal growth are stimulated when fungi are cocultured with methanogenic bacteria (4, 10, 68). Methanogens cause a shift in the fermentation products to more acetate and less lactate, succinate, and ethanol, as would be expected because methanogens maintain low H_2 levels. This methanogenic activity facilitates production of H_2 by fungi, and thus the flow of electrons moves away from the more reduced fungal products (lactate, succinate, ethanol) toward methane (4, 103, 115). Other fungus-bacterium associations, ranging from synergism to antagonism, have been reported (9, 66, 141).

The activities of protozoa apparently have important effects on cellulose-decomposing bacterial and fungal communities. Protozoal predatory activity on rumen bacteria and fungi has been documented (64, 123, 179). In cultures of a rumen fungus, *Piromyces* sp., and of a mixed rumen protozoal population, cellulose degradation was decreased compared with monocultures of the fungus, owing at least in part to protozoal predation (113). However, the amount of cellulose degraded per unit of fungal biomass was larger in mixed cultures (113). Also, formate (which is used as an indicator of fungal growth) was not detected; propionate and butyrate increased; and lactate decreased in mixed cultures compared with production of these organic acids in the fungal monocultures (113). A stimulatory effect of protozoa on bacterial cellulolysis has been reported (193). Although the basis of this stimulation is unclear, hydrolytic enzymes released from protozoal cells may play a role in augmenting plant-fiber degradation (179).

Other Animal-Associated Environments

Besides the ruminants, many other herbivorous and wood-eating animals harbor complex communities of microorganisms and rely, at least to some extent, on the microbes' capacity to degrade the structural polymers of plant cell walls. For example, the hoatzin, a tropical leaf-eating bird from South America, has a foregut fermentation system that resembles the rumen (53). Also, nonruminant mammals such as kangaroos, elephants, and monkeys have complex forestomachs in which plant material is degraded by microbial processes apparently similar to those that occur in the rumen (3, 64, 166). Other mammals such as humans, pigs, horses, and rodents are postgastric (hindgut) fermenters. Host digestion in the stomach precedes microbial fermentation of undigested plant material in the colon or large intestine. These fermentations are carried out by complex microbial communities including cellulolytic bacteria (169, 176), but in contrast to ruminal fermentations, protozoa and fungi are not involved (109). Polysaccharides are degraded by schemes similar to those found in the rumen, producing acetate, propionate, butyrate, H_2, and CO_2 (183). Most of the volatile fatty acids produced in hindgut fermentations are absorbed into the blood and metabolized in tissues, and this process supplies about 5–10% of dietary energy in humans and pigs and about 40% in rabbits and ponies (108, 109).

Production of methane by human intestinal microbial communities occurs only in some individuals (110, 174), even though the kinds and amounts of acids produced by methanogenic and nonmethanogenic intestinal fermentations are very similar (175). Apparently, during nonmethanogenic fermentations in humans, and in other mammals such as rats, guinea pigs, and rabbits, H_2 is used to reduce CO_2 to acetate (14, 109). Similarly, significant amounts of acetate, rather than CH_4, are formed from CO_2 in the guts of various termites (14).

Many insects, such as termites, feed on cellulose, which is degraded by complex microbial communities in the insect gut (reviewed in 13; see also 26). Cellulolytic enzymes in insects may originate not only from the gut microbial community, but also from the insect itself or, in fungus-growing termites, from the insect's feed (101). In phylogenetically lower termites, cellulolytic flagellated protozoa are the major, and possibly the only, source of cellulolytic enzymes, and these protozoa appear to be essential for the survival of termites on wood or native cellulose diets (13). The phylogenetically higher termites lack protozoa and depend largely on the activities of gut bacterial communities for the degradation of cellulose (13). Recently, Gijzen et al (49) reported that cellulase activities in hindgut extracts of the cockroach *Periplaneta americana* are high in cockroaches fed cellulose-rich diets. The number of *Nyctotherus ovalis,* the major hindgut protozoan, is also relatively large in insects on these diets. As *N. ovalis* harbors endosymbiotic methanogens (48), these findings suggest that this protozoan may be involved in producing both cellulolytic and methanogenic activities in the hindgut of cockroaches.

Herbivory in reptiles is most often associated with extinct members of this class, and in terms of biomass consumption and evolutionary longevity, the most successful vertebrates in the history of the Earth were the plant-eating dinosaurs. Although relatively few extant reptiles subsist primarily on plants, some of those that do display adaptations for herbivory that are analogous to those found in nonruminant mammals, with characteristic features of the digestive tract associated with processing high-fiber food and maintaining cellulose-degrading microbial communities (163, and references therein). McBee & McBee (107) have estimated that volatile fatty acid production satisfies 30–40% of the energetic requirement of the green iguana (*Iguana iguana*), indicating that hindgut fermentation of fiber is a significant energy source in this animal. Little information is available on the microorganisms responsible for the hindgut fermentation. Culturing indicated that *Leuconostoc* and *Clostridium* spp. were dominant, and no species of *Bacteroides*, *Fusobacterium*, and *Ruminococcus* were found (107). Variable numbers of ciliated protozoans have been observed in iguana hindguts (163). Testudines (turtles and tortoises) also possess a modified hindgut region in which fibrous plant material is degraded by microbial communities, which include protozoa and bacteria in numbers comparable to those found in the rumen (163).

CONCLUDING REMARKS

This article has reviewed numerous studies that deal with various aspects of the degradation of cellulose in anaerobic environments. Microorganisms that act as the biological catalysts in the anaerobic degradation of cellulose are of enormous biotechnological interest for their hydrolytic enzymes and fermentation products. Great effort has been devoted to studying microbial cellulase systems, resulting in a fair understanding of the enzymology of cellulose degradation. However, fundamental questions remain unanswered. For example, the mechanism of cellulose hydrolysis by the cellulase systems of either bacteria or fungi remains largely unknown, and so it is nearly impossible to discern similarities or differences in the molecular mechanisms employed by different systems. Also, essentially nothing is known about the exo- or extracellular assembly of complex cellulase systems. The biotechnological potential of cellulolytic microorganisms cannot be fully realized without knowledge of their genetics, which is virtually nonexistent at present, and an improved understanding of their physiology. Studies such as those carried out by D'Elia & Chesbro (27) are needed to appreciate the effects of growth rate and energy demands on cellulase production.

Other interest in anaerobic cellulose degradation involves the role of this process in the nutrition of economically important animals, particularly ruminants. Numerous recent publications, including several reviews (e.g. see 71), discuss progress in this area. Although the rumen is one of the best known microbial ecosystems, a greater understanding of interactions among populations of different cellulolytic and noncellulolytic organisms is needed to accurately picture the overall process of plant-fiber degradation in the rumen. In this regard, the above-mentioned studies by Odenyo et al (121, 122) employ molecular tools that may prove generally useful in addressing fundamental questions related to microbial ecology. Finally, on a global scale, the anaerobic degradation of cellulose has a huge impact on the carbon cycle, so there is tremendous interest in learning how this process may be affected by predicted climatic change.

Important aspects of cellulose degradation in anaerobic environments have been addressed infrequently in the literature and hence were given scant attention in this article. For example, facultatively anaerobic microorganisms potentially could play pivotal roles in cellulose decomposition in anaerobic soils, sediments, and other natural environments, and their ecology deserves greater attention. It is hoped that this review will stimulate additional research addressing the many unanswered questions and will lead to a better understanding of the anaerobic degradation of cellulose.

ACKNOWLEDGMENTS

Research in the author's laboratory on the diversity of cellulolytic anaerobes was supported by a grant from the National Science Foundation, and studies of their cellulase systems and interactions in cellulose degradation were supported by a grant from the Department of Energy. I thank all former and present members of the laboratory for their many important contributions, especially Tsute Chen for computer assistance and Ercole Canale-Parola for introducing me to the cellulolytic microbial world and for his untiring support, encouragement, and friendship.

> Any *Annual Review* chapter, as well as any article cited in an *Annual Review* chapter, may be purchased from the Annual Reviews Preprints and Reprints service.
> 1-800-347-8007; 415-259-5017; email: arpr@class.org

Literature Cited

1. Akin DE, Barton FE II. 1983. Rumen microbial attachment and degradation of plant cell walls. *Fed. Proc.* 42:114–21
1a. Aubert J-P, Béguin P, Millet J, eds. 1988. *Biochemistry and Genetics of Cellulose Degradation. FEMS Symp. No. 43.* London: Academic. 428 pp.
2. Bagnara-Tardif C, Gaudin C, Belaich A, Hoest P, Citard T, Belaich J-P. 1992. Sequence analysis of a gene cluster encoding cellulases from *Clostridium cellulolyticum*. *Gene* 119:17–28
3. Bauchop T. 1971. Stomach microbiology of primates. *Annu. Rev. Microbiol.* 25:429–36
4. Bauchop T, Mountfort DO. 1981. Cellulose fermentation by a rumen anaerobic fungus in both the absence and the presence of rumen methanogens. *Appl. Environ. Microbiol.* 42:1103–10
5. Bayer EA, Lamed R. 1986. Ultrastructure of the cell surface cellulosome of *Clostridium thermocellum* and its interaction with cellulose. *J. Bacteriol.* 167:828–36
6. Bayer EA, Morag E, Lamed R. 1994. The cellulosome—a treasure-trove for biotechnology. *Trends Biotechnol.* 12:379–86
7. Béguin P, Aubert J-P. 1994. The biological degradation of cellulose. *FEMS Microbiol. Rev.* 13:25–58
8. Béguin P, Millet J, Chauvaux S, Salamitou S, Tokatlidis K, et al. 1992. Bacterial cellulases. *Biochem. Soc. Trans.* 20:42–46
9. Bernalier A, Fonty G, Bonnemoy F, Gouet P. 1992. Degradation and fermentation of cellulose by the rumen anaerobic fungi in axenic cultures or in association with cellulolytic bacteria. *Curr. Microbiol.* 25:143–48
10. Bernalier A, Fonty G, Gouet P. 1991. Cellulose degradation by two rumen anaerobic fungi in monoculture or in coculture with rumen bacteria. *Anim. Feed Sci. Technol.* 32:131–36
11. Biely P. 1985. Microbial xylanolytic systems. *Trends Biotechnol.* 3:286–90
12. Bogdahn M, Kleiner D. 1986. Inorganic nitrogen metabolism in two cellulose-degrading clostridia. *Arch. Microbiol.* 145:159–61
13. Breznak JA. 1982. Intestinal microbiota

of termites and other xylophagous insects. *Annu. Rev. Microbiol* 36:323–43
14. Breznak JA, Kane MD. 1990. Microbial H_2/CO_2 acetogenesis in animal guts: nature and nutritional significance. *FEMS Microbiol. Rev.* 87:309–14
15. Bronnenmeier K, Staudenbauer WL. 1990. Cellulose hydrolysis by a highly thermostable endo-1,4-β-glucanase (avicelase I) from *Clostridium stercorarum*. *Enzyme Microb. Technol.* 12:431–36
16. Bryant MP. 1973. Nutritional requirements of the predominant rumen cellulolytic bacteria. *Fed. Proc.* 32:1809–13
17. Bryant MP, Burkey LA. 1953. Cultural methods and some characteristics of some of the more numerous groups of bacteria in the bovine rumen. *J. Dairy Sci.* 36:205–17
18. Bryant MP, Doetsch RN. 1954. A study of actively cellulolytic rod-shaped bacteria of the bovine rumen. *J. Dairy Sci.* 37:1176–83
19. Bryant MP, Small N, Bouma C, Robinson IM. 1958. Characteristics of ruminal anaerobic cellulolytic cocci and *Cillobacterium cellulosolvens* n. sp. *J. Bacteriol.* 76:529–37
20. Cavedon K, Canale-Parola E. 1992. Physiological interactions between a mesophilic cellulolytic *Clostridium* and a non-cellulolytic bacterium. *FEMS Microbiol. Ecol.* 86:237–45
21. Cavedon K, Leschine SB, Canale-Parola E. 1990. Cellulase system of a free-living, mesophilic clostridium (strain C7). *J. Bacteriol.* 172:4222–30
22. Cavedon K, Leschine SB, Canale-Parola E. 1990. Characterization of the extracellular cellulase from a mesophilic clostridium (strain C7). *J. Bacteriol.* 172:4231–37
23. Cheng K-J, Stewart CS, Dinsdale D, Costerton JW. 1984. Electron microscopy of bacteria involved in the digestion of plant cell walls. *Feed Sci. Technol.* 10:93–120
23a. Chesson A, Forsberg CW. 1988. Polysaccharide degradation by rumen microorganisms. See Ref. 58, pp. 251–84
24. Collins MD, Lawson PA, Willems A, Cordoba JJ, Fernandez-Garayzabal J, et al. 1994. The phylogeny of the genus *Clostridium*: proposal of five new genera and eleven new species combinations. *Int. J. System. Bacteriol.* 44:812–26
25. Coughlan MP, Ljungdahl LG. 1988. Comparative biochemistry of fungal and bacterial cellulolytic enzyme systems. See Ref. 1a, pp. 11–30
26. Cruden DL, Markovetz AJ. 1987. Microbial ecology of the cockroach gut. *Annu. Rev. Microbiol.* 41:617–43
27. D'Elia J, Chesbro W. 1992. Maintenance energy demand affects biomass synthesis but not cellulase production by a mesophilic *Clostridium*. *J. Ind. Microbiol.* 10:123–33
28. Dehority BA. 1963. Isolation and characterization of several cellulolytic bacteria from in vitro rumen fermentations. *J. Dairy Sci.* 46:217–22
29. Dehority BA. 1986. Protozoa of the digestive tract of herbivorous mammals. *Insect Sci. Appl.* 7:279–96
30. Dehority BA. 1993. Microbial ecology of cell wall fermentation. See Ref. 71, pp. 425–53
31. Dehority BA, Orpin CG. 1988. Development of, and natural fluctuations in, rumen microbial populations. See Ref. 58, pp. 151–83
32. Demeyer DI. 1981. Rumen microbes and digestion of plant cell walls. *Agric. Environ.* 6:295–337
33. Dinsdale E, Morris EJ, Bacon JSD. 1978. Electron microscopy of the microbial populations present and their modes of attack on various cellulosic substrates undergoing digestion in the sheep rumen. *Appl. Environ. Microbiol.* 36:160–68
34. Doerner KC, White BA. 1990. Assessment of the endo-1,4-β-glucanase components of *Ruminococcus flavefaciens* FD-1. *Appl. Environ. Microbiol.* 56:1844–50
35. Doi RH, Goldstein M, Hashida S, Park J-S, Takagi M. 1994. The *Clostridium cellulovorans* cellulosome. *CRC Crit. Rev. Microbiol.* 20:87–93
36. Durrant LR, Canale-Parola E, Leschine SB. 1994. The cellulase system of soil fungi isolated under anaerobic conditions. See Ref. 148a, pp. 531–39
37. Durrant LR, Canale-Parola E, Leschine SB. 1995. Facultatively anaerobic cellulolytic fungi from soil. In *Soil Diversity, Its Function and Regulation*, ed. HP Collins, GP Robertson, MJ Klug. Dordrecht, the Netherlands: Kluwer. In press
38. Ehhalt DH. 1976. The atmospheric cycle of methane. In *Microbial Production and Utilization of Gases*, ed. HG Schlegel, G Gottschalk, N Pfennig, pp. 13–22. Göttingen: Goltze
39. Eriksson K-EL, Blanchette RA, Ander P. 1990. *Microbial and Enzymatic Degradation of Wood and Wood Components*. Berlin: Springer-Verlag. 407 pp.
40. Faure E, Belaich A, Bagnara C, Gaudin C, Belaich J-P. 1989. Sequence analysis of the *Clostridium cellulolyticum* cel-

CCA endoglucanase gene. *Gene* 84:39–46
41. Felix CR, Ljungdahl LG. 1993. The cellulosome: the exocellular organelle of *Clostridium*. *Annu. Rev. Microbiol.* 47:791–819
42. Fond O, Petitdemange E, Petitdemange H, Engasser J-M. 1983. Cellulose fermentation by a coculture of a mesophilic cellulolytic clostridium and *Clostridium acetobutylicum*. *Biotechnol. Bioeng. Symp.* 13:217–24
43. Forsberg CW, Gong J, Malburg JLM, Zhu H, Iyo A, et al. 1994. Cellulases and hemicellulases of *Fibrobacter succinogenes* and their roles in fibre digestion. See Ref. 148a, pp. 125–36
44. Forsberg CW, Beveridge TJ, Hellstrom A. 1981. Cellulase and xylanase release from *Bacteroides succinogenes* and its importance in the rumen environment. *Appl. Environ. Microbiol.* 42:886–96
45. Fujino T, Béguin P, Aubert J-P. 1992. Cloning of a *Clostridium thermocellum* DNA fragment encoding polypeptides that bind the catalytic components of the cellulosome. *FEMS Microbiol. Lett.* 94:165–70
46. Fujino T, Béguin P, Aubert J-P. 1993. Organization of a *Clostridium thermocellum* gene cluster encoding the cellulosomal scaffolding protein CipA and protein possibly involved in attachment of the cellulosome to the cell surface. *J. Bacteriol.* 175:1891–99
47. Gerngross UT, Romaniec MPM, Kobayashi T, Huskisson NS, Demain AL. 1993. Sequencing of a *Clostridium thermocellum* gene (*cipA*) encoding the cellulosomal S_L-protein reveals an unusual degree of internal homology. *Mol. Microbiol.* 8:325–34
48. Gijzen HJ, Broers CAM, Barugahare M, Stumm CK. 1991. Methanogenic bacteria as endosymbionts of the ciliate *Nyctotherus ovalis* in the cockroach hindgut. *Appl. Environ. Microbiol.* 57:1630–34
49. Gijzen HJ, van der Drift C, Barugahare M, op den Camp HJM. 1994. Effect of host diet and hindgut microbial composition on cellulolytic activity in the hindgut of the American cockroach, *Periplaneta americana*. *Appl. Environ. Microbiol.* 60:1822–26
50. Gilbert HJ, Hazlewood GP. 1993. Bacterial cellulases and xylanases. *J. Gen. Microbiol.* 139:187–94
51. Gilkes NR, Henrissat B, Kilburn DG, Miller JRC, Warren RAJ. 1991. Domains in microbial β-1,4-glucanases: sequence conservation, function, and enzyme families. *Microbiol. Rev.* 55:303–15
52. Gilligan W, Reese ET. 1954. Evidence for multiple components in microbial cellulases. *Can. J. Microbiol.* 1:90–107
53. Grajal A, Strahl SD, Parra R, Dominguez MG, Neher A. 1989. Foregut fermentation in the hoatzin, a neotropical leaf-eating bird. *Science* 245:1236–38
54. Groleau D, Forsberg CW. 1981. Cellulolytic activity of the rumen bacterium *Bacteroides succinogenes*. *Can. J. Microbiol.* 27:517–30
54a. Haigler CH, Weimer PJ, eds. 1991. *Biosynthesis and Biodegradation of Cellulose.* New York: Dekker. 694 pp.
55. Hartley RD, Ford CW. 1989. Phenolic constituents of plant cell walls and wall biodegradability. In *Plant Cell Wall Polymers, Biogenesis and Biodegradation, ACS Symp. Ser. 399,* ed. NG Lewis, MG Paice, pp. 135–45. Washington, DC: Am. Chem. Soc.
56. Harwood CS, Canale-Parola E. 1984. Ecology of spirochetes. *Annu. Rev. Microbiol.* 38:161–92
57. Hobson PN. 1971. Rumen micro-organisms. In *Progress in Industrial Microbiology,* ed. DJD Hockenhull, pp. 41–77. London: Churchill
58. Hobson PN, ed. 1988. *The Rumen Microbial Ecosystem.* London: Elsevier Applied Science. 527 pp.
58a. Hollaender A, Rabson R, Rogers P, San Pietro A, Valentine R, Wolfe R, eds. 1981. *Trends in the Biology of Fermentation for Fuels and Chemicals.* New York: Plenum, 591 pp.
59. Hon-nami K, Coughlan MP, Hon-nami H, Ljungdahl LG. 1986. Separation and characterization of the complexes constituting the cellulolytic enzyme system of *Clostridium thermocellum*. *Arch. Microbiol.* 145:13–19
60. Hsing W, Canale-Parola E. 1992. Cellobiose chemotaxis by the cellulolytic bacterium *Cellulomonas gelida*. *J. Bacteriol.* 174:7996–8002
61. Hume ID, Warner ACI. 1980. Evolution of microbial digestion in mammals. In *Digestive Physiology and Metabolism in Ruminants,* ed. Y Ruckebusch, P Thivend, pp. 665–84. Lancaster, UK: MTP
62. Hungate RE. 1944. Studies on cellulose fermentation. I. The culture and physiology of an anaerobic cellulose-digesting bacterium. *J. Bacteriol.* 48:499–513
62a. Hungate RE. 1946. Studies on cellulose fermentation. II. An anaerobic cellulose decomposing *Actinomycete, Micromonospora propionici* N. sp. *J. Bacteriol.* 51:51–56

63. Hungate RE. 1950. The anaerobic mesophilic cellulolytic bacteria. *Bacteriol. Rev.* 14:1–49
64. Hungate RE. 1966. *The Rumen and Its Microbes.* New York: Academic. 533 pp.
65. Hungate RE. 1969. Interrelationships in the rumen microbiota. In *Physiology of Digestion and Metabolism in the Ruminant*, ed. AT Phillipson, pp. 292–305. Newcastle-upon-Tyne: Oriel
66. Irvine HL, Stewart CS. 1991. Interactions between anaerobic cellulolytic bacteria and fungi in the presence of *Methanobrevibacter smithii*. *Lett. Appl. Bacteriol* 12:62–64
67. Jenkinson DS, Adams DE, Wild A. 1991. Model estimates of CO_2 emissions from soil in response to global warming. *Nature* 351:304–6
68. Joblin KN. 1981. Isolation, enumeration, and maintenance of rumen anaerobic fungi in roll tubes. *Appl. Environ. Microbiol.* 42:1119–22
69. Johnson JL, Francis BS. 1975. Taxonomy of the clostridia: ribosomal ribonucleic acid homologies among the species. *J. Gen. Microbiol.* 88:229–44
70. Jones WJ, Nagle JDP, Whitman WB. 1987. Methanogenesis and the diversity of archaebacteria. *Microbiol. Rev.* 51:135–77
71. Jung HG, Buxton DR, Hatfield RD, Ralph J, eds. 1993. *Forage Cell Wall Structure and Digestibility.* Madison, WI: Am. Soc. Agron./Crop Sci. Soc. Am./Soil Sci. Soc. Am. 794 pp.
72. Kelly WJ, Asmundson RV, Hopcroft DH. 1987. Isolation and characterization of a strictly anaerobic, cellulolytic spore former: *Clostridium chartatabidum* sp. nov. *Arch. Microbiol.* 147:169–73
73. Khan AW, Murray WD. 1982. Influence of *Clostridium saccharolyticum* on cellulose degradation by *Acetivibrio cellulolyticus*. *J. Appl. Bacteriol.* 53:379–83
74. Kirk TK. 1971. Effects of microorganisms on lignin. *Annu. Rev. Phytopathol.* 9:185–210
75. Klyosov AA. 1990. Trends in biochemistry and enzymology of cellulose degradation. *Biochemistry* 29:10577–85
76. Kobayashi T, Romaniec MPM, Fauth U, Demain AL. 1990. Subcellulosome preparation with high cellulase activity from *Clostridium thermocellum*. *Appl. Environ. Microbiol.* 56:3040–46
77. Kohring S, Wiegel J, Mayer F. 1990. Subunit composition and glycosidic activities of the cellulase complex from *Clostridium thermocellum* JW20. *Appl. Environ. Microbiol.* 56:3798–804
78. Kudo H, Cheng K-J, Costerton JW. 1987. Electron microscopic study of the methylcellulose-mediated detachment of cellulolytic rumen bacteria from cellulose fibers. *Can. J. Microbiol.* 33:267–72
79. Kudo H, Cheng K-J, Costerton JW. 1987. Interactions between *Treponema bryantii* and cellulolytic bacteria in the in vitro degradation of straw cellulose. *Can. J. Microbiol.* 33:244–48
80. Küsel K, Drake HL. 1994. Acetate synthesis in soil from a Bavarian beech forest. *Appl. Environ. Microbiol.* 60:1370–73
81. Lamed R, Bayer EA. 1988. The cellulosome concept: exocellular/extracellular enzyme reactor centers for efficient binding and cellulolysis. See Ref. 1a, pp. 101–16
82. Lamed R, Bayer EA. 1988. The cellulosome of *Clostridium thermocellum*. *Adv. Appl. Microbiol.* 33:1–46
83. Lamed R, Bayer EA. 1991. Cellulose degradation by thermophilic anaerobic bacteria. See Ref. 54a, pp. 377–410
84. Lamed R, Morag E, Mor-Yosef O, Bayer EA. 1991. Cellulosome-like entities in *Bacteroides cellulosolvens*. *Curr. Microbiol.* 22:27–33
85. Lamed R, Naimark J, Morgenstern E, Bayer EA. 1987. Specialized cell surface structures in cellulolytic bacteria. *J. Bacteriol.* 169:3792–800
86. Lamed R, Setter E, Bayer EA. 1983. Characterization of a cellulose-binding, cellulase-containing complex in *Clostridium thermocellum*. *J. Bacteriol.* 156:828–36
87. Lamed R, Setter E, Kenig R, Bayer EA. 1983. The cellulosome—a discrete cell surface organelle of *Clostridium thermocellum* which exhibits separate antigenic, cellulose-binding and various catalytic activities. *Biotechnol. Bioeng. Symp.* 13:163–81
88. Laube VM, Martin SM. 1981. Conversion of cellulose to methane and carbon dioxide by triculture of *Acetivibrio cellulolyticus, Desulfovibrio* sp., and *Methanosarcina barkeri*. *Appl. Environ. Microbiol.* 42:413–20
89. Leschine SB, Canale-Parola E. 1983. Mesophilic cellulolytic clostridia from freshwater environments. *Appl. Environ. Microbiol.* 46:728–37
90. Leschine SB, Canale-Parola E. 1984. Ethanol production from cellulose by a coculture of *Zymomonas mobilis* and a clostridium. *Current Microbiol.* 11:129–36
91. Leschine SB, Canale-Parola E. 1989. Carbon cycling by cellulose-fermenting

nitrogen-fixing bacteria. *Adv. Space Res.* 9(8):149–52
92. Leschine SB, Holwell K, Canale-Parola E. 1988. Nitrogen fixation by anaerobic cellulolytic bacteria. *Science* 242:1157–59
93. Ljungdahl LG, Eriksson K-E. 1985. Ecology of microbial cellulose degradation. *Adv. Microb. Ecol.* 8:237–99
94. Lovley DR, Goodwin S. 1988. Hydrogen concentrations as an indicator of the predominant terminal electron-accepting reactions in aquatic sediments. *Geochim. Cosmochim. Acta* 52:2993–3003
95. Madarro A, Pena JL, Lequerica JL, Valles S, Gay R, Flors A. 1991. Purification and characterization of the cellulases from *Clostridium cellulolyticum* H10. *J. Chem. Tech. Biotechnol.* 52:393–406
96. Madden RH, Bryder MJ, Poole NJ. 1982. Isolation and characterization of an anaerobic, cellulolytic bacterium, *Clostridium papyrosolvens* sp. nov. *Int. J. Syst. Bacteriol.* 32:87–91
97. Mah RA. 1981. The methanogenic bacteria, their ecology and physiology. See Ref. 58a, pp. 357–74
98. Malburg JLM, Lee JMT, Forsberg CW. 1992. Degradation of cellulose and hemicelluloses by rumen microorganisms. In *Microbial Degradation of Natural Products*, ed. G Winkelmann, pp. 127–59. New York: Weinheim
99. Maluszynska GM, Janota-Bassalik L. 1974. A cellulolytic rumen bacterium, *Micromonospora ruminantium* sp. nov. *J. Gen. Microbiol.* 82:57–65
100. Marchessault RH, Sundararajan PR. 1983. Cellulose. In *The Polysaccharides*, ed. GO Aspinall, 2:11–95. New York: Academic
101. Martin MM. 1983. Cellulose digestion in insects. *Comp. Biochem. Physiol.* 75A:313–24
102. Marty DG. 1986. Description de 23 clostridies cellulolytiques ou non cellulolytiques isolées du milieu marin. *Ann. Inst. Pasteur Microbiol.* 137A:33–43
103. Marvin-Sikkema FD, Richardson AJ, Stewart CS, Gottschal JC, Prins RA. 1990. Influence of hydrogen-consuming bacteria on cellulose degradation by anaerobic fungi. *Appl. Environ. Microbiol.* 56:3793–97
104. Mayer F. 1988. Cellulolysis: ultrastructural aspects of bacterial systems. *Electron Microsc. Rev.* 1:69–85
105. Mayer F, Coughlan MP, Mori Y, Ljungdahl LG. 1987. Macromolecular organization of the cellulolytic enzyme complex of *Clostridium thermocellum* as revealed by electron microscopy. *Appl. Environ. Microbiol.* 53:2785–92
106. McBee RH. 1950. The anaerobic thermophilic cellulolytic bacteria. *Bacteriol. Rev.* 14:51–63
107. McBee RH, McBee VH. 1982. The hindgut fermentation in the green iguana, *Iguana iguana*. In *Iguanas of the World: Their Behavior, Ecology, and Conservation*, ed. GM Burghardt, AS Rand, pp. 77–83. Park Ridge, NJ: Noyes
108. McNeil NI. 1984. The contribution of the large intestine to energy supplies in man. *Am. J. Clin. Nutr.* 39:338–42
109. Miller TL. 1991. Biogenic sources of methane. In *Microbial Production and Consumption of Greenhouse Gases: Methane, Nitrogen Oxides, and Halomethanes*, ed. JE Rogers, WB Whitman, pp. 175–87. Washington, DC: Am. Soc. Microbiol.
110. Miller TL, Wolin MJ. 1986. Methanogens in human and animal intestinal tracts. *Syst. Appl. Microbiol.* 7:223–29
111. Morag E, Bayer EA, Lamed R. 1990. Relationship of cellulosomal and non-cellulosomal xylanases of *Clostridium thermocellum* to cellulose-degrading enzymes. *J. Bacteriol.* 172:6098–105
112. Morag E, Halevy I, Bayer EA, Lamed R. 1991. Isolation and properties of a major cellobiohydrolase from the cellulosome of *Clostridium thermocellum*. *J. Bacteriol.* 173:4155–62
113. Morgavi DP, Sakurada M, Mizokami M, Tomita Y, Onodera R. 1994. Effects of ruminal protozoa on cellulose degradation and the growth of an anaerobic ruminal fungus, *Piromyces* sp. strain OTS1, in vitro. *Appl. Environ. Microbiol.* 60:3718–23
114. Mori Y. 1990. Characterization of a symbiotic coculture of *Clostridium thermohydrosulfuricum* YM3 and *Clostridium thermocellum* YM4. *Appl. Environ. Microbiol.* 56:37–42
115. Mountfort DO, Asher RA, Bauchop T. 1982. Fermentation of cellulose to methane and carbon dioxide by a rumen anaerobic fungus in a triculture with *Methanobrevibacter* sp. strain RA1 and *Methanosarcina barkeri*. *Appl. Environ. Microbiol.* 44:128–34
116. Murray WD. 1986. Symbiotic relationship of *Bacteroides cellulosolvens* and *Clostridium saccharolyticum* in cellulose fermentation. *Appl. Environ. Microbiol.* 51:710–14
117. Murray WD, Hofmann L, Campbell NL, Madden RH. 1986. *Clostridium lentocellum* sp. nov., a cellulolytic species from river sediment containing paper-

mill waste. *Syst. Appl. Microbiol.* 8: 181–84
118. Murray WD, Sowden LC, Colvin JR. 1984. *Bacteroides cellulosolvens* sp. nov., a cellulolytic species from sewage sludge. *Int. J. Syst. Bacteriol.* 34:185–87
119. Ng TK, Ben-Bassat A, Zeikus JG. 1981. Ethanol production by thermophilic bacteria: fermentation of cellulosic substrates by cocultures of *Clostridium thermocellum* and *Clostridium thermohydosulfuricum. Appl. Environ. Microbiol.* 41:1337–43
120. Nitisinprasert S, Temmes A. 1991. The characteristics of a new non-spore-forming cellulolytic mesophilic anaerobe strain CM126 isolated from municipal sewage sludge. *J. Appl. Bacteriol.* 71: 154–61
121. Odenyo AA, Mackie RI, Stahl DA, White BA. 1994. The use of 16S rRNA-targeted oligonucleotide probes to study competition between ruminal fibrolytic bacteria: development of probes for *Ruminococcus* species and evidence for bacteriocin production. *Appl. Environ. Microbiol.* 60:3688–96
122. Odenyo AA, Mackie RI, Stahl DA, White BA. 1994. The use of 16S rRNA-targeted oligonucleotide probes to study competition between ruminal fibrolytic bacteria: pure-culture studies with cellulose and alkaline peroxide-treated wheat straw. *Appl. Environ. Microbiol.* 60:3697–703
123. Orpin CG. 1984. The role of ciliate protozoa and fungi in the rumen digestion of plant cell walls. *Anim. Feed Sci. Technol.* 10:121–43
124. Orpin CG, Joblin KN. 1988. The rumen anaerobic fungi. See Ref. 58, pp. 129–50
125. Palop MLL, Valles S, Piñaga F, Flors A. 1989. Isolation and characterization of an anaerobic, cellulolytic bacterium, *Clostridium celerecrescens* sp. nov. *Int. J. Syst. Bacteriol.* 39:68–71
126. Paster BJ, Canale-Parola E. 1982. Physiological diversity of spirochetes. *Appl. Environ. Microbiol.* 43:686–93
127. Patel GB, Khan AW, Agnew BJ, Colvin JR. 1980. Isolation and characterization of an anaerobic, cellulolytic microorganism, *Acetivibrio cellulolyticus* gen. nov., sp. nov. *Int. J. Syst. Bacteriol.* 30:179–85
128. Peck HD, Odom M. 1981. Anaerobic fermentations of cellulose to methane. See Ref. 58a, pp. 375–95
129. Petitdemange E, Caillet F, Giallo J, Gaudin C. 1984. *Clostridium cellulolyticum* sp. nov., a cellulolytic mesophilic species from decayed grass. *Int. J. Syst. Bacteriol.* 34:155–59

130. Pohlschröder M, Leschine S, Canale-Parola E. 1994. The multicomplex cellulase and xylanase system of *Clostridium papyrosolvens* strain C7. *J. Bacteriol.* 176:70–76
131. Pohlschröder M, Leschine S, Canale-Parola E. 1994. *Spirochaeta caldaria* sp. nov., a thermophilic bacterium that enhances cellulose degradation by *Clostridium thermocellum. Arch. Microbiol.* 161:17–24
132. Poole DM, Morag E, Lamed R, Bayer EA, Hazlewood GP, Gilbert HJ. 1992. Identification of the cellulose binding domain of the cellulosome subunit S1 from *Clostridium thermocellum* YS. *FEMS Microbiol. Lett.* 99:181–86
133. Post WM, Emanuel WR, Zinke PJ, Stangenberger AG. 1982. Soil carbon pools and world life zones. *Nature* 298: 156–59
134. Prins RA, van Vugt F, Hungate RE, van Vorstenbosch CJAHV. 1972. A comparison of strains of *Eubacterium cellulosolvens* from the rumen. *Antonie van Leeuwenhoek J. Microbiol. Serol.* 38: 153–61
135. Raich JW, Nadelhoffer KJ. 1989. Belowground carbon allocation in forest ecosystems: global trends. *Ecology* 70: 1346–54
136. Rainey FA, Stackebrandt E. 1993. 16S rDNA analysis reveals phylogenetic diversity among the polysaccharolytic clostridia. *FEMS Microbiol. Lett.* 113: 125–28
137. Reese ET, Mandels M. 1971. Enzymatic degradation. In *Cellulose and Cellulose Derivatives,* Part 5, ed. NM Bikales, L Segal, pp. 1079–94. New York: Wiley-Interscience
138. Reese ET, Siu RGH, Levinson HS. 1950. The biological degradation of soluble cellulose derivatives and its relationship to the mechanism of cellulose hydrolysis. *J. Bacteriol.* 59:485–97
139. Richmond PA. 1991. Occurrence and functions of native cellulose. See Ref. 54a, pp. 5–23
140. Robson LM, Chambliss GH. 1989. Cellulases of bacterial origin. *Enzyme Microb. Technol.* 11:626–42
141. Roger V, Grenet E, Jamot J, Bernalier A, Fonty G, Gouet P. 1992. Degradation of maize stem by two rumen fungal species, *Piromyces communis* and *Caecomyces communis,* in pure cultures or in association with cellulolytic bacteria. *Reprod. Nutr. Dev.* 32:321–29
142. Saddler JN, Khan AW. 1984. Conversion of pretreated lignocellulosic substrates to ethanol by *Clostridium thermocellum* in mono- and co-culture

with *Clostridium thermosaccharolyticum* and *Clostridium thermohydrosulphuricum*. *Can. J. Microbiol.* 30:212–20
143. Salamitou S, Lemaire M, Fujino T, Ohayon H, Gounon P, et al. 1994. Subcellular localization of *Clostridium thermocellum* ORF3p, a protein carrying a receptor for the docking sequence borne by the catalytic components of the cellulosome. *J. Bacteriol.* 176:2828–34
144. Salamitou S, Raynaud O, Lemaire M, Coughlan M, Béguin P, Aubert J-P. 1994. Recognition specificity of the duplicated segments present in *Clostridium thermocellum* endoglucanase CelD and in the cellulosome-integrating protein CipA. *J. Bacteriol.* 176:2822–27
145. Salamitou S, Tokatlidis K, Béguin P, Aubert J-P. 1992. Involvement of separate domains of the cellulosomal protein S1 of *Clostridium thermocellum* in binding to cellulose and in anchoring of catalytic subunits to the cellulosome. *FEBS Lett.* 304:89–92
146. Scalbert A, Monties B, Lallemand J-Y, Guittet Y, Rolando C. 1985. Ether linkage between phenolic acids and lignin fractions from wheat straw. *Phytochemistry* 24:1359–62
147. Schlesinger WH. 1991. *Biogeochemistry: an Analysis of Global Change*. San Diego: Academic. 443 pp.
148. Shane BS, Gouws L, Kistner A. 1969. Cellulolytic bacteria occurring in the rumen of sheep conditioned to low-protein teff hay. *J. Gen. Microbiol.* 55:445–57
148a. Shimada K, Ohmiya K, Kobayashi Y, Hoshino S, Sakka K, Karita S, eds. 1994. *Genetics, Biochemistry and Ecology of Lignocellulose Degradation*. Tokyo: Uni
149. Shoseyov O, Doi RH. 1990. Essential 170-kDa subunit for degradation of crystalline cellulose by *Clostridium cellulovorans* cellulase. *Proc. Natl. Acad. Sci. USA* 87:2192–95
150. Sinner M, Parameswaran N, Dietrichs HH. 1979. Degradation of delignified sprucewood by purified mannanase, xylanase and cellulases. *Adv. Chem. Ser.* 181:303–29
151. Sinner M, Parameswaran N, Yamazaki N, Liese W, Dietrichs HH. 1976. Specific enzymatic degradation of polysaccharides in delignified wood cell walls. *Appl. Polymer Symp.* 28:993–1024
152. Skinner FA. 1960. The isolation of anaerobic cellulose-decomposing bacteria from soil. *J. Gen Microbiol.* 22:539–54
153. Sleat R, Mah RA. 1985. *Clostridium populeti* sp. nov., a cellulolytic species from a woody-biomass digestor. *Int. J. Syst. Bacteriol.* 35:160–63
154. Sleat R, Mah RA, Robinson R. 1984. Isolation and characterization of an anaerobic, cellulolytic bacterium *Clostridium cellulovorans* sp. nov. *Appl. Environ. Microbiol.* 48:88–93
155. Stack RJ, Hungate RE. 1984. Effect of 3-phenyl-propionic acid on capsule and cellulases of *Ruminococcus albus*. *Appl. Environ. Microbiol.* 48:218–23
156. Stanton TB, Canale-Parola E. 1979. Enumeration and selective isolation of rumen spirochetes. *Appl. Environ. Microbiol.* 38:965–73
157. Stanton TB, Canale-Parola E. 1980. *Treponema bryantii* sp. nov., a rumen spirochete that interacts with cellulolytic bacteria. *Arch. Microbiol.* 127:145–56
158. Stewart CS, Bryant MP. 1988. The rumen bacteria. See Ref. 58, pp. 21–75
159. Stewart CS, Dinsdale D, Cheng K-J, Paniagua C. 1979. The digestion of straw in the rumen. In *Straw Decay and Its Effect on Disposal and Utilization*, ed. E Grossbard, pp. 123–30. Chichester, UK: Wiley & Sons
160. Stewart CS, Paniagua C, Dinsdale D, Cheng K-J. 1981. Selective isolation and characteristics of *Bacteroides succinogenes* from the rumen of a cow. *Appl. Environ. Microbiol.* 41:504–10
161. Sukhumavasi J, Ohmiya K, Shimizu S, Ueno K. 1988. *Clostridium josui* sp. nov., a cellulolytic, moderate thermophilic species from Thai compost. *Int. J. Syst. Bacteriol.* 38:179–82
162. Tokatlidis K, Salamitou S, Béguin P, Dhurjati P, Aubert J-P. 1991. Interaction of the duplicated segment carried by *Clostridium thermocellum* cellulases with cellulosome components. *FEBS Lett.* 291:185–88
163. Troyer K. 1991. Role of microbial cellulose degradation in reptile nutrition. See Ref. 54a, pp. 311–25
164. van Gylswyk NO, Hoffman JPL. 1970. Characteristics of cellulolytic cillobacteria from the rumens of sheep fed teff (*Eragrostis tef*) hay diets. *J. Gen Microbiol.* 60:381–86
165. van Gylswyk NO, Roche CEG. 1970. Characteristics of *Ruminococcus* and cellulolytic *Butyrivibrio* species from the rumens of sheep fed differently supplemented teff (*Eragrostis tef*) hay diets. *J. Gen. Microbiol.* 64:11–17
166. van Gylswyk NO, Schwartz HM. 1984. Microbial ecology of cellulose and hemicellulose metabolism in gastrointestinal ecosystems. In *Current Perspectives in Microbial Ecology*, ed. MJ Klug,

CA Reddy, pp. 588–99. Washington, DC: Am. Soc. Microbiol.
167. Van Soest PJ. 1982. *Nutritional Ecology of the Ruminant.* Corvallis, OR: O & B Books
168. Varel VH. 1989. Reisolation and characterization of *Clostridium longisporum*, a ruminal sporeforming cellulolytic anaerobe. *Arch. Microbiol.* 152:209–14
169. Varel VH, Fryda SJ, Robinson IM. 1984. Cellulolytic bacteria from pig large intestine. *Appl. Environ. Microbiol.* 47:219–21
170. Varel VH, Richardson AJ, Stewart CS. 1989. Degradation of barley straw, ryegrass, and alfalfa cell walls by *Clostridium longisporum* and *Ruminococcus albus. Appl. Environ. Microbiol.* 55:3080–84
171. Vogels GD. 1979. The global cycle of methane. *Antonie van Leeuwenhoek J. Microbiol. Serol.* 45:347–52
172. Wang DIC, Averinos GC, Biocic I, Wang S-D, Fang H-Y. 1983. Ethanol from cellulosic biomass. *Philos. Trans. R. Soc. London Ser. B* 300:323–33
173. Warshaw JE, Leschine SB, Canale-Parola E. 1985. Anaerobic cellulolytic bacteria from wetwood of living trees. *Appl. Environ. Microbiol.* 50:807–11
174. Weaver GA, Krause JA, Miller TL, Wolin MJ. 1986. Incidence of methanogenic bacteria in a sigmoidoscopy population: an association of methanogenic bacteria and diverticulosis. *Gut* 27:698–704
175. Weaver GA, Krause JA, Miller TL, Wolin MJ. 1989. Constancy of glucose and starch fermentations by two different human faecal microbial communities. *Gut* 30:1–19
176. Wedekind KJ, Mansfield HR, Montgomery L. 1988. Enumeration and isolation of cellulolytic and hemicellulolytic bacteria from human feces. *Appl. Environ. Microbiol.* 54:1530–35
177. Wiegel J, Dykstra M. 1984. *Clostridium thermocellum:* adhesion and sporulation while adhered to cellulose and hemicellulose. *Appl. Microbiol. Biotechnol.* 20:59–65
178. Williams AG. 1986. Rumen holotrich ciliate protozoa. *Microbiol. Rev.* 50:25–49
179. Williams AG, Coleman GS. 1988. The rumen protozoa. See Ref. 58, pp. 77–128
180. Wilson CA, Wood TM. 1992. The anaerobic fungus *Neocallimastix frontalis:* isolation and properties of a cellulosome-type enzyme fraction with the capacity to solubilize hydrogen- bond-ordered cellulose. *Appl. Microbiol. Biotechnol.* 37:125–29
181. Wilson CA, Wood TM. 1992. Studies on the cellulase of the rumen anaerobic fungus *Neocallimastix frontalis,* with special reference to the capacity to degrade crystalline cellulose. *Enzyme Microb. Technol.* 14:258–64
182. Wolin MJ. 1979. The rumen fermentation: a model for microbial interactions in anaerobic ecosystems. *Adv. Microb. Ecol.* 3:49–77
183. Wolin MJ, Miller TL. 1983. Carbohydrate fermentation. In *Human Intestinal Microflora in Health and Disease,* ed. DJ Hentges, pp. 147–65. New York: Academic
184. Wolin MJ, Miller TL. 1983. Interactions of microbial populations in cellulose fermentation. *Fed. Proc.* 42:109–13
185. Wolin MJ, Miller TL. 1986. Bioconversion of organic carbon to CH_4 and CO_2. *Geomicrobiol. J.* 5:239–59
186. Wood TM. 1992. Fungal cellulases. *Biochem. Soc. Trans.* 20:46–53
187. Wood TM, Wilson CA, Stewart CS. 1982. Preparation of the cellulase from the cellulolytic anaerobic rumen bacterium *Ruminococcus albus* and its release from the bacterial cell wall. *Biochem. J.* 205:129–37
188. Wood WA, Kellogg ST, eds. 1988. Biomass. Part A: cellulose and hemicellulose. *Methods Enzymol.* 160:1–774
189. Wu JHD, Demain AL. 1988. Proteins of the *Clostridium thermocellum* cellulase complex responsible for degradation of crystalline cellulose. See Ref. 1a, pp. 117–31
190. Wu JHD, Orme-Johnson WH, Demain AL. 1988. Two components of an extracellular protein aggregate of *Clostridium thermocellum* together degrade crystalline cellulose. *Biochemistry* 27:1703–9
191. Yang JC, Chynoweth DP, Williams DS, Li A. 1990. *Clostridium aldrichii* sp. nov., a cellulolytic mesophile inhabiting a wood-fermenting anaerobic digester. *Int. J. Syst. Bacteriol.* 40:268–72
192. Yanling H, Youfang D, Yanquan L. 1991. Two cellulolytic *Clostridium* species: *Clostridium cellulosi* sp. nov. and *Clostridium cellulofermentans* sp. nov. *Int. J. Syst. Bacteriol.* 41:306–9
193. Yoder RD, Trenkle A, Burroughs W. 1966. Influence of rumen protozoa and bacteria upon cellulose digestion *in vitro. J. Anim. Sci.* 25:609–12
194. Zeikus JG. 1983. Metabolic communication between biodegradative populations in nature. *Symp. Soc. Gen. Microbiol.* 34:423–62

NEW MECHANISMS OF DRUG RESISTANCE IN PARASITIC PROTOZOA

P. Borst

Division of Molecular Biology, Netherlands Cancer Institute, 1066 CX Amsterdam, The Netherlands

M. Ouellette

Service d'Infectiologie du Centre Hospitalier de l'Université Laval and Département de Microbiologie, Université Laval, Québec Canada, G1V 4G2

KEY WORDS: P-glycoproteins, drug transport, chemotherapeutic targets, parasitic diseases

CONTENTS

INTRODUCTION	429
AN OVERVIEW OF DRUG RESISTANCE IN PARASITIC PROTOZOA	431
P-GLYCOPROTEINS AND RELATED TRAFFIC ATPases	433
Mammalian P-Glycoproteins	433
P-Glycoproteins and Related Transport Proteins in Parasites	434
RESISTANCE TO CHLOROQUINE, MEFLOQUINE, QUININE, AND HALOFANTRINE	435
RESISTANCE TO ARSENICALS AND ANTIMONIALS	438
Resistance to Arsenicals in Trypanosomes	438
Resistance to Oxyanions in Leishmania spp.	439
A Speculative Model for Oxyanion Resistance in Leishmania spp.	441
RESISTANCE TO ANTIFOLATES	442
Mutations in DHFR and DHPS	443
DHFR Overproduction	444
Transport Mutations in Leishmania spp.	444
Alternative Pathway for Folate Synthesis in Leishmania spp.	445
MISCELLANEOUS RECENTLY DESCRIBED MECHANISMS OF DRUG RESISTANCE	446
Resistance to Purine Analogues	446
Resistance to Metronidazole	448
Resistance to Ornithine Decarboxylase Inhibitors	449
Resistance to Emetine	450
Resistance to Drugs Used in the Laboratory but Not in Patients So Far	450
OUTLOOK	451

ABSTRACT

The main line of defense now available against parasitic protozoa—which are responsible for major diseases of humans and domestic animals—is chemotherapy. This defense is being eroded by drug resistance and, with few new drugs in the pipeline, prevention and circumvention of resistance are medical and veterinary priorities. Although studies of resistance mechanisms in parasites have lagged behind similar studies in bacteria and cancer cells, the tools to tackle this problem are rapidly improving. Transformation with exogenous DNA is now possible with all major parasitic protozoa of humans. Hence, putative resistance genes can be tested in sensitive protozoa, allowing an unambiguous reconstruction of resistance mechanisms. Gene cloning, the polymerase chain reaction, and monoclonal antibodies against resistance-related proteins have made it possible to analyze potential resistance mechanisms in the few parasites that can be obtained from infected people. Hence, the prospect of applying new knowledge about resistance mechanisms to parasites in patients is good, even though today virtually all knowledge pertains to parasites selected for resistance in the laboratory.

Resistance mechanisms highlighted in this review include:

1. Decrease of drug uptake because of the loss of a transporter required for uptake. This decrease contributes to resistance to arsenicals and diamidines in African trypanosomes.
2. The export of drugs from the parasite by P-glycoproteins and other traffic ATPases. This export could potentially be an important mechanism of resistance, as these proteins are richly represented in the few protozoa analyzed. There are indications that such transmembrane transporters can be involved in resistance to emetine in *Entamoeba* spp., to mefloquine in *Plasmodium* spp., and to antimonials in *Leishmania* spp.
3. The possible involvement of the P-glycoprotein encoded by the *Plasmodium falciparum pfmdr1* gene in chloroquine resistance. We present the available data that lead to the conclusion that overproduction of the wild-type version of this protein results in chloroquine hypersensitivity rather than resistance.
4. The involvement of the PgpA P-glycoprotein of *Leishmania* spp. in low-level resistance to arsenite and antimonials. We raise the possibility that this protein transports glutathione conjugates of arsenite and antimonials rather than the compounds themselves.
5. Loss of drug activation as the main mechanism of metronidazole resistance in *Trichomonas* and *Giardia* spp. Recent evidence indicates that a decrease of the proximal cellular electron donor for metronidazole activation, ferredoxin, is the main cause of resistance in *Trichomonas*.

6. Resistance arising through alteration of drug targets. The amino acid substitutions in the dihydrofolate reductase-thymidylate synthase of *Plasmodium* spp. are good examples of this mechanism.

We show here that the field of drug resistance in parasitic protozoa is currently very active and holds considerable future opportunity. With the tools now available, progress should be rapid in the coming years.

INTRODUCTION

Parasitic protozoa are responsible for some of the most devastating and prevalent diseases of humans and domestic animals. Malaria (*Plasmodium* spp.), the various forms of (muco)cutaneous and visceral leishmaniasis (*Leishmania* spp.), African sleeping sickness (*Trypanosoma brucei gambiense, Trypanosoma brucei rhodesiense*), South-American Chagas' disease (*Trypanosoma cruzi*), amoebic dysentery (*Entamoeba* spp.), and toxoplasmosis (*Toxoplasma* spp.) are serious diseases that threaten the lives of nearly one quarter of the human population worldwide. In addition, the parasitic diseases caused by *Trichomonas vaginalis* (vaginitis, urethritis) and *Giardia duodenalis* (diarrhea) are very widespread and unpleasant, even though not life-threatening. Protozoal parasites also result in enormous losses of life and productivity of domesticated animals, both mammals and fowl.

Drugs and prevention are the two major weapons now available against protozoan parasites. There are high hopes that effective vaccines may eventually be available (171), but such vaccines have been slow in coming (30), and one sometimes wonders whether the vaccine aficionados are not underestimating the ability of protozoa to elude the mammalian immune system. Most protozoal diseases are chronic and occur in immunocompetent patients. Is the competence of immunologists really sufficient to boost an immune system that fails to deal with the invader in a natural fashion? We hope the answer is yes, but for the moment drugs are the mainstay in cases when preventive measures fail or prove impractical.

Even though protozoa are eukaryotes and usually contain many of the organelles and metabolic pathways of their hosts, the differences in biochemistry between parasite and host are great enough to leave a large window for the development of parasite-specific drugs. It is not always appreciated that the protozoal parasite differs much more from a human cell than from cells of fungi or plants. On an evolutionary scale deduced from differences in small-subunit ribosomal RNAs (58, 167), protozoa such as *Giardia lamblia* and *Trypanosoma brucei* are nearly as similar to *Escherichia coli* as to humans. Hence, it is not surprising that fairly effective drugs are available for many parasitic diseases. It is even embarrassing that there are still major parasites, such as the American trypanosomes, that cannot be safely and effectively tackled by chemotherapy. Had the same amount of effort and money been invested in the development of antiprotozoal drugs as in the refinement of anticancer drugs, chemotherapy of protozoal diseases would be in a much better position today. It should be easier to develop drugs against trypanosomatids with their exotic biochemistry than against cancer cells, which differ in less than 0.1% of their gene products from normal cells.

As it stands, however, the arsenal of antiprotozoal drugs is limited, and the effectiveness of these drugs is being eroded by drug resistance. For example, there is widespread resistance to some of the most effective drugs ever developed, chloroquine in malaria and metronidazole in anaerobic parasites. With effective vaccines not yet in sight and development of new drugs proceeding slowly, the rising tide of drug resistance is menacing the frail dikes of disease management of protozoal infections.

Studies of resistance mechanisms cannot reverse the tide, but can help to handle the threat more rationally at three levels:

1. by developing tools to recognize resistance early in infection and prevent the loss of time with useless (and often toxic) chemotherapy
2. by pointing the way to more rational use of drugs and drug combinations to minimize development of resistance
3. by pinpointing intracellular drug targets and parasite defense mechanisms allowing the rational development of drug analogues not affected by the most common defenses.

Most of the studies on drug resistance in protozoa have been done with laboratory strains under conditions that do not mimic the normal parasite-host relation. Moreover, most of the studies have been correlative, i.e. the drug-resistance mechanisms have been studied by determining alterations in drug handling or in cellular metabolism of resistant mutants. If an alteration is consistently found in multiple mutants, if it is lost in revertants, and if it provides a plausible mechanism for resistance, then the resistance mechanism is usually thought to be solved. This may be the case, but there are pitfalls in this approach (20). Therefore it is important to reconstruct the resistance mechanism by transforming wild-type cells with the relevant genes. For gain-of-function mutants, such a reconstruction may only require the introduction of additional copies of suitably altered genes. For loss-of-function mutants, the reconstruction may require the directed disruption of endogenous genes. Only if the complete resistance phenotype can be reconstructed in sensitive cells by such directed genetic modifications can one be sure that the resistance mechanism has been solved. This reconstruction requires a convenient procedure for stable transformation, which is now available for a rapidly increasing number of parasitic protozoa. After Bellofatto & Cross (15) and Laban & Wirth (108) showed that electroporation can be used to obtain transient transformation of *Leptomonas seymouri* and *Leishmania enriettii*, respectively, this procedure soon resulted in stable transformation of *Leishmania* spp. (40, 41, 97, 107), *T. brucei* (49, 50, 111, 172), and *L. seymouri* (16). More recently, stable transformation was obtained in *T. cruzi* (35, 79, 100, 131), *Toxoplasma gondii* (48, 102), and *Plasmodium berghei* (MR Van Dijck, AP Waters & CJ Jansen, personal communication), and transient transformation was obtained in *Entamoeba histolytica* (126, 152). In most protozoa studied thus far, exogenous DNA preferentially integrates into host DNA by homologous recombination. Gene disruption experiments, so laborious in mammals, are therefore relatively simple in these protozoa. With the tools in hand it should be possible to reconstruct resistance mechanisms by transformation.

In this review we concentrate on some of the newer mechanisms of resistance that have emerged in the past five years. We indicate those proposed mechanisms that have been proven by transfection, and we emphasize the rare cases in which solid information is available for resistance mechanisms in field isolates. As this review is selective, we refer the reader to other recent overviews for more complete reviews of specific drugs or parasites. A concise overview of modern parasitology with a clear description of the major parasitic protozoa and their metabolic peculiarities that can be targeted by drugs can be found in Cox (39). A special issue of *Science* (June 24, 1994) devoted to parasitology also highlights the impact of parasites on developing countries, the (lack of) funding for parasite control and research, and the cost to society of unchecked parasites. Special issues of *Parasitology Today* (May 1993) and *Acta Tropica* (March 1994) address drug resistance.

AN OVERVIEW OF DRUG RESISTANCE IN PARASITIC PROTOZOA

To interfere with parasite multiplication, a drug must find the parasite (often in a host cell) and reach its target within the parasite. Usually the drug must pass the parasite membrane, and often it must be activated inside. After the drug hits the target, the parasite must be sufficiently incapacitated to be killed by the host defense or to die spontaneously. Each of these steps provides the parasite with opportunities to interfere with drug action, resulting in drug resistance. Resistance mechanisms are most easily studied in parasites made resistant in the laboratory. One can then work with genetically defined, cloned populations, and by a detailed comparison of resistant mutants and the parental strain from which they were derived, one can usually find the gene(s) involved in resistance.

Resistance in clinical samples is less easily defined in biochemical terms. Parasite populations are often heterogeneous; the exact parental strain is often

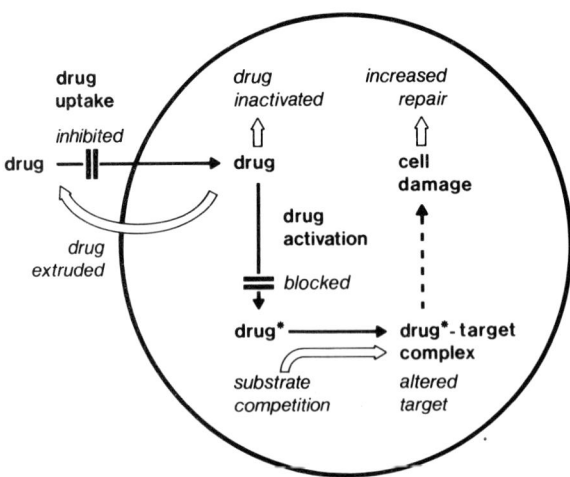

Figure 1 Simplified scheme illustrating the main biochemical mechanisms of drug resistance (20).

not available for comparison, and the culturing of the parasites—often necessary to get sufficient material for analysis—may change resistance. Nevertheless, it should now be possible to verify whether resistance mechanisms defined in the laboratory play a role in the field, as the polymerase chain reaction and monoclonal antibodies allow the study of genes and proteins in a few cells. Studying resistance mechanisms induced under laboratory conditions has been fruitful in the study of clinically relevant forms of drug resistance in cancer (20), and we are optimistic that the same approach will also allow a precise molecular characterization of clinically relevant forms of resistance in protozoal diseases, even if we are still far from that ideal today.

The main biochemical mechanisms responsible for drug resistance are illustrated in Figure 1. Parasites may evade drug action by hiding in sanctuaries such as the brain (e.g. many drugs do not pass the blood-brain barrier); drug uptake may be thwarted by loss of uptake systems or alteration of membrane composition; once inside, drugs may be inactivated, excreted, modified and excreted, or routed into vacuoles (the latter mechanism is still speculative); drug activation mechanisms may be suppressed or lost; the interaction of drug with target may be made less effective by increasing the level of competing substrates or by altering the target to make it less sensitive to the drug; the parasite may learn to live with a blocked target by circumventing the block; and finally the parasite may become more proficient in repairing drug damage. Concrete examples that illustrate the scheme in Figure 1 are discussed in the relevant sections of this review.

P-GLYCOPROTEINS AND RELATED TRAFFIC ATPases

Mammalian P-Glycoproteins

The importance of P-glycoproteins (Pgps) and related transmembrane transporters for drug resistance in organisms ranging from bacteria to human tumor cells has been amply demonstrated since this class of proteins was discovered by Juliano & Ling (96) in multidrug-resistant hamster tumor cells. Pgps belong to the family of Adenine nucleotide Binding Cassette (ABC) transporters (84), also known as traffic ATPases (47). A detailed discussion of the structure and function of these transporters and their possible role in drug resistance can be found in recent reviews (21, 31, 47, 72, 84, 123, 127, 136, 158, 161, 176). The classical Pgps responsible for multidrug resistance (MDR) in mammalian cancer cells are large plasma membrane glycoproteins consisting of two similar halves, each containing six putative transmembrane segments and an ATP-binding site. Drug resistance is caused by the ability of Pgps to extrude drugs against a concentration gradient, resulting in a decrease of the intracellular drug concentration in contact with the drug target.

An interesting new member of the traffic ATPase family is the MDR-associated protein MRP, the second type of mammalian transporter involved in MDR (34, 73). Similar to Pgp, this protein is predominantly present in the plasma membrane and causes drug resistance by lowering the intracellular drug concentration by drug extrusion (195). MRP is not just another member of the Pgp family, however. The sequence identity between human MDR1 Pgp and MRP is only 23%, and MRP seems to have a more asymmetric structure than Pgp (34). Recent experiments indicate that there may also be fundamental differences between MDR1 Pgp and MRP in substrate specificity, even though the two proteins confer resistance to a similar spectrum of natural product drugs. These experiments show that an increase in cellular MRP is associated with an increase in the activity of the elusive GS-X pump (91, 114, 124). This pump, also known as the leukotriene C_4 (LTC_4) transporter (114) or the multispecific anion transporter (132), is known to transport complex molecules with a hydrophobic organic moiety and at least two negative charges (81, 89). It transports a variety of glutathione conjugates out of the cell and it is also responsible for secretion of bilirubin-glucuronides into bile (1, 132). The probable identity of MRP and the GS-X pump explains the puzzling observation that overexpression of MRP may result in cross-resistance to arsenite (124). Arsenite (As) forms a complex with glutathione (GSH), and there is considerable indirect evidence that this $As(SG)_3$ complex is a substrate for the GS-X pump (124).

Drug resistance involving increased conjugation of drug with GSH is a complex affair (89, 124). It requires not only the ability to conjugate the drug (e.g. by a GSH-S-transferase), but also the ability to maintain high levels of GSH and the ability to export the GSH-drug conjugate out of the cell at a sufficient rate to match the continuing influx of drug into the cell. The resistance obtained therefore depends on the activity of at least three enzyme systems: GSH biosynthesis, GSH-S-transferase, and the GS-X pump.

P-Glycoproteins and Related Transport Proteins in Parasites

Table 1 shows that Pgps and other related putative transporters are widely distributed in parasites (136, 177). Only a small minority of these Pgps have been linked to drug resistance, and the functions of most of them are not known. Amplification or overexpression of the gene(s) encoding Pgp has been described in some clinically resistant isolates, and transfection of some parasite *pgp* genes conferred resistance on wild-type parasites (136). The only protozoal

Table 1 P-glycoproteins and other traffic ATPases in parasites

Parasite genus	Gene	Comments	Single (S) or duplicated (D) NBS[a]	References
Entamoeba	Ehpgp1-6	Multigene family with two pseudogenes	D	43
		One gene overexpressed in an emetine-resistant clone		163
	Ehabc1		S	196
Leishmania	pgpA	Involved in resistance to arsenite and antimonials	D	26, 134, 140
	pgpB-E	Gene family, closely related to *pgpA*; more related to *MRP* than to *MDR1*	D	115, 135
	mdr1	Involved in resistance to drugs; part of mammalian *mdr* spectrum	D	32, 82, 83
Plasmodium	pfmdr1	Involved in chloroquine susceptibility and possibly mefloquine resistance	D	36, 178
	pfmdr2	Related to yeast *HMT1*; involved in cadmium resistance	S	160, 194
Schistosoma	smdr1		S	22
	smdr2		D	22
Trichomonas	Tvpgp1-2		S	32, 82, 83
Trypanosoma		Cross-hybridization with *pgpA* probes		26, 134, 140
		Partial sequences of NBS		42

NBS, nucleotide-binding sequence.

Pgp that confers resistance to a spectrum of hydrophobic, natural-product drugs similar to that transported by the mammalian MDR1-type Pgps is the Pgp encoded by the *ldmdr1* gene of *Leishmania donovani* (82, 83). These drugs are not used to treat leishmaniasis, but *Leishmania* spp. may encounter toxic natural products in other stages of its life cycle. The other individual Pgps of parasitic protozoa are discussed in the relevant sections.

RESISTANCE TO CHLOROQUINE, MEFLOQUINE, QUININE, AND HALOFANTRINE

Chloroquine is the drug of choice for treating malaria. The emergence of *Plasmodium falciparum* that is resistant to chloroquine in the late 1950s in Southern Asia and South America has been a major stumbling block in the global control of malaria. Resistance to chloroquine is now distributed widely in all geographical areas where *P. falciparum* is endemic. Despite more than 40 years of usage, the mechanism of action of chloroquine is not completely understood. The most plausible target for the action of chloroquine and related 4-aminoquinolines is the heme polymerase located in the food vacuole of the parasite (166). This polymerase is thought to be essential for the polymerization of ferriprotoporphyrin IX into the malaria pigment of infected erythrocytes, and its inhibition results in the accumulation of toxic heme degradation products, proposed to cause the death of the parasite (61). Other targets remain under investigation, however. For instance, recently two uncharacterized *P. falciparum* proteins of 33 and 42 kDa were found to bind a photoreactive analogue of chloroquine in a specific manner (62). Whether these proteins are related to killing of *P. falciparum* by chloroquine is unknown.

It has long been known that the predominant mechanism of chloroquine resistance results in a decreased accumulation of chloroquine in the resistant parasites (63, 153). In 1987 two observations were reported that seemed to link chloroquine resistance to a Pgp-mediated mechanism. Krogstadt et al (104) reported that chloroquine efflux is increased in resistant parasites, and Martin et al (120) reported that verapamil, the classical agent to reverse MDR in animal cells, is able to restore chloroquine sensitivity to resistant cells. Subsequently, several other agents that reverse MDR were shown to reverse chloroquine resistance as well (63, 153).

It is therefore not surprising that the discovery of a gene encoding Pgp, *pfmdr1*, amplified in some chloroquine-resistant *P. falciparum* strains (65), seemed to settle the mechanism of chloroquine resistance. The Pgh1 Pgp encoded by *pfmdr1* was proposed to be the efflux pump that caused increased efflux of chloroquine from the parasite, and this pump was inhibited by verapamil (65). This simple picture seemed to be supported by the preferential association of chloroquine resistance with specific alleles of *pfmdr1* (64). It came therefore as a surprise that neither resistance nor the rapid chloroquine efflux cosegregated with *pfmdr1* in a genetic cross between a chloroquine-sensitive and a chloroquine-resistant strain of malaria (182). The main resistance locus identified by this genetic approach maps to chromosome 7 (183); the gene(s) involved has not yet been cloned. A further complication was the demonstration by Cowman et al (37) that Pgh1 is not located in the plasma membrane, but in the membrane of the food vacuole of *P. falciparum*. The position of the protein in the membrane strongly indicates that Pgh1 transports its substrate into rather than out of the vacuole. Furthermore, the analysis of the *pfmdr1* gene of recent chloroquine-resistant *P. falciparum* isolates (7, 187) did not reveal any amino acid substitutions that were linked to the resistance phenotype (64). Finally, the association between chloroquine resistance and rapid drug efflux is now also in doubt (24); moreover, the ability of verapamil and other reversal agents to reverse resistance seems not due simply to the block of an efflux pump, as these drugs fail to restore levels of intracellular chloroquine observed in chloroquine-sensitive cells (23).

Recent work seems to have resolved the question of whether overexpression of wild-type *pfmdr1* contributes to chloroquine resistance: It does not. Selection of three cloned lines of *P. falciparum* for increased chloroquine resistance in vitro resulted in deamplification of the *pfmdr1* gene rather than increased amplification. The deamplification was associated with an increase in mefloquine sensitivity (11). Conversely, a study of recent isolates from Thailand indicated an association between level of expression of *pfmdr1* and increased resistance to mefloquine and halofantrine (187). In vitro selection for mefloquine resistance in *P. falciparum* led to *pfmdr1* amplification (145, 186), which can be accompanied by cross-resistance to halofantrine and quinine but increased sensitivity to chloroquine (36). Analogous results were reported in two recent papers (125, 144).

These results have led Cowman et al (36) to suggest that the amplification of the *pfmdr1* gene in multiple isolates from the field may have been caused by selection for quinine resistance, as these isolates were already obtained before mefloquine was introduced in clinical practice. The amplification of the *pfmdr1* gene should have made the parasite more sensitive to chloroquine, which may have led to the selection of the altered *pfmdr1* alleles associated with chloroquine resistance. This speculative picture has received support from experiments in which the *pfmdr1* gene was introduced into Chinese hamster ovary (CHO) cells (178). These cells are hypersensitive to chloroquine, apparently because of an increased accumulation of chloroquine in lysosomes. Presumably, this accumulation is caused by the transport activity of Pgh1 located in the lysosomal membrane, where the Pgp is found within the CHO cells by immunofluorescence. The double mutant *pfmdr1* gene with amino acid replacements at positions 1034 and 1042, i.e. the mutated Pgp associated with chloroquine resistance in some field isolates, gave no chloroquine hypersensitivity in CHO cells (178). How Pgh1 works is still not completely clear. Van Es et al (179) have shown that in CHO cells the protein can lower the pH of lysosomes, suggesting that the increase in chloroquine accumulation in acidic compartments, both in parasites and transfected mammalian cells, might be the passive consequence of the decreased pH, as chloroquine is a weak base. No band of the size of Pgh1 reacted with the photoreactive analogue of chloroquine (62). This suggests that chloroquine does not bind to Pgh1, or at least not with high affinity. Experiments in another heterologous system, however, recently established that Pgh1 can transport bulky molecules. The group of D Wirth (personal communication) has shown that *pfmdr1* can complement the *STE6* mutation in yeast. The *STE6* gene encodes a Pgp that exports the yeast a-type mating factor, a farnesylated peptide of 16 amino acids. Its function can be inefficiently replaced by mammalian Pgps (157), which are known to localize to the plasma membrane, and it is remarkable that Pgh1 is able to do the same, even though its primary cellular location is in the intracellular acidic compartments. It is important to realize, however, that the STE6 Pgp itself appears to recycle between endosomal compartments and plasma membrane and spends most of its time internally rather than at the surface (106). Ligands might also be transported into intracellular compartments and be excreted by exocytosis.

How can these data explain the observation that *pfmdr1* (over)expression results in resistance to mefloquine, quinine, and halofantrine, but in hypersensitivity to chloroquine? In view of the yeast results, it seems likely that Pgh1 can transport drug and that some of the protein can reach the cell membrane even though the bulk of the protein is in the membranes of the endosomal system. The final effect on drug distribution should therefore be determined by the relative rates at which the protein pumps a specific drug into the vacuole or out of the cell and by the ability of the protein to acidify the vacuole contents, an ability that is apparently lacking in some allelic variants. The mechanism of acidification remains unknown. An ion channel through the Pgp itself seems unlikely (see previous section). Rather, acidification could occur from the transport of an endogenous protonated compound that can diffuse back through the vacuolar membrane or by indirect effects on other protonophores. We have even considered the possibility that Pgh1 could be a GS-X pump and transport negatively charged drug complexes, such as drug-glutathione conjugates, but this seems incompatible with the finding that *pfmdr1* can complement *STE6* mutations in yeast. Another possibility to explain the consequences of *pfmdr1* overexpression is that Pgh1 only operates in the vacuolar membrane, but that the outcome of its operation depends on the site of action of the drug transported and the rate at which the drug enters the cell. In the case of chloroquine, which seems to have its main site of action in the vacuole, pumping drug into the vacuole makes matters worse. In the case of halofantrine, which might have DNA as its main target, pumping drug into the vacuole could be beneficial if the drug enters the cell slowly enough to make redistribution of drug between cytosol and vacuole feasible and useful.

Evidence for a second gene with some homology to members of the ABC family of transporters was reported by Wilson et al (186). This gene was called *pfmdr2,* but the complete sequence of the gene revealed that it contains 10 predicted transmembrane domains followed by a region homologous to the nucleotide-binding fold of the ABC transporters (160, 194). The name *pfmdr2* is therefore confusing, and the gene should be renamed once its function is known. Pfmdr2 is related to HMT1, an ABC transporter found in the fission yeast that mediates tolerance to cadmium by compartmentalization of the heavy metal in the vacuole (130). Although *pfmdr2* transcripts were found overexpressed in some chloroquine-resistant parasites (51), more recent work conclusively showed that the Pfmdr2 protein is not usually overexpressed in chloroquine-resistant isolates (160).

The recent development of transient (71, 191) and stable (MR Van Dijck, AP Waters & CJ Janse, personal communication) transformation systems for *Plasmodium* spp. should soon make it possible to study Pgp function in detail in *Plasmodium* spp. This should also allow direct testing of some of the speculative interpretations offered here and characterization of the major putative chloroquine-resistance locus on chromosome 7.

RESISTANCE TO ARSENICALS AND ANTIMONIALS

The arsenical salvarsan, developed by Paul Ehrlich at the beginning of this century, was the first antimicrobial agent designed specifically for chemotherapy. Oxyanions in the form of aromatic arsenicals or drugs containing the related metal antimony are still first-line drugs in the treatment of trypanosomiasis and leishmaniasis.

Resistance to Arsenicals in Trypanosomes

Trypanosoma brucei, a fly-transmitted parasite, is responsible for sleeping sickness in humans. Melamine-based arsenicals such as melarsoprol (MelB), diamidines such as pentamidine, and a few other drugs are used to treat human sleeping sickness, but melarsoprol is the mainstay for treatment of the most advanced cases affecting the central nervous system (8). Resistance to melarsoprol and other arsenical drugs has been encountered both in *Trypanosoma* species selected under laboratory conditions and in clinical isolates refractory to arsenicals (8).

The changing views on the mode of action of arsenicals provide a prime example of the problems encountered in defining drug targets and resistance mechanisms without the help of molecular genetics. As Gutteridge & Coombs (77) put it: "At one time or another, all of the glycolytic kinases of trypanosomes have been suggested as the primary target." More recently, the attention has shifted to trypanothione, a parasite-specific glutathione-spermidine conjugate (56). Trypanothione is the major thiol-containing molecule in the trypanosome and is essential for maintaining an intracellular reducing environment. Secondary targets of arsenicals are thought to include trypanothione reductase (TR) (56), the most extensively studied enzyme involved in trypanothione metabolism (55, 181), and lipoic acid (57).

Resistant cells contain much less lipoic acid, which led to the suggestion that this cofactor of the pyruvate dehydrogenase complex might also be involved in transport of arsenicals (57). Resistance to arsenicals is not associated with changes in the concentration of trypanothione, glutathione, or related molecules (54, 192). The properties of TRs isolated from sensitive and resistant isolates were also indistinguishable (54). Therefore, neither the putative primary target, trypanothione, nor one of the putative secondary targets, TR, seem to be involved in resistance. Trypanosomatid cells transfected with the *Leishmania donovani* TR gene were also not more resistant to drugs thought to act by inducing oxidative damage (99). The recent discovery of substantial amounts of ovothiol, a 4-mercaptohistidine, in *Crithidia fasciculata* (169) and *Leishmania* spp. (168) raises the possibility, however, that there are other thiol-containing targets for arsenicals.

Trypanosomes resistant to melarsoprol are cross-resistant to diamidines (8, 54, 67), and resistant clones accumulate less diamidine (67). This latter finding raised the possibility that the melaminophenyl arsenicals and the diamidines are taken up by a common transporter and that the common structural feature for recognition resides in the melamine and benzamidine moieties of the two classes of drugs (54). To identify the natural function of this transporter, Carter & Fairlamb (29) screened potential substrates for their ability to compete with arsenicals. Only adenine and adenosine were found to prevent lysis of *T. brucei* by melarsen oxide. Wild-type trypanosomes were found to have two high-affinity adenosine transport systems, one of which—the P2 type—also transports melamine-based arsenicals. Trypanosomes resistant to arsenicals lack the P2 adenosine transport system, suggesting that resistance to melamine-based arsenicals is due to loss of uptake (29).

Resistance to Oxyanions in Leishmania spp.

Leishmania spp. are distributed worldwide, with 400,000 new cases of human leishmaniasis each year (6). The clinical manifestations vary, depending on the *Leishmania* spp., from self-healing cutaneous lesions to visceral infections that are usually fatal if left untreated. The treatment of choice for all forms of leishmaniasis is the administration of the pentavalent antimonial-containing drugs (129). The mode of action of antimonials in *Leishmania* spp. is poorly understood but, as with most heavy metals, is thought to be multifactorial. Indeed, antimonials bind to several *Leishmania* proteins (17). It should be noted, however, that multiple sites of action were once popular in explaining the effects of cancer chemotherapy, but that this idea has been discredited by the systematic analysis of resistant mutants. Similarly, the available evidence does not yet exclude that high-affinity binding of antimonials to one key protein is responsible for their therapeutic effect on leishmaniasis.

Unresponsiveness to pentavalent antimonials in *Leishmania* spp. has long been recognized (137), and antimonial-resistant parasites isolated from patients who did not respond to therapy have been described (74, 90). Although most of the work on antimonial resistance in *Leishmania* spp. has been carried out in the insect-stage parasite selected for resistance in vitro, one may expect that the ongoing research on resistance mechanisms in the laboratory will soon yield the tools required to test whether similar mechanisms operate in strains isolated from patients. *Leishmania* spp. mutants selected for resistance to pentavalent antimony but also to the related oxyanions trivalent antimony and arsenite have been described (138). Resistance in these mutants is stable in the absence of drug and predominantly due to decreased accumulation of drug caused by an increase in efflux. This phenomenon was observed in all *L. tarentolae* mutants studied, whether selected for resistance to arsenite, trivalent antimony, or pentavalent antimony (46), and also in some arsenite-resistant mutants of *L. mexicana* (165). The identification of the genes involved in this efflux system seemed simple. Indeed, *Leishmania* spp. often respond to drug selection, at least when applied in vitro, by amplifying specific portions of their genome (18, 137), and several amplicons were observed in *Leishmania* spp. selected for oxyanion resistance (44, 75, 135, 136, 165). The first amplified locus characterized encodes the Pgp protein PgpA (134). This ABC transporter has diverged considerably from the main branch of Pgps; this transporter therefore looked like a good candidate for an oxyanion efflux pump. Attempts to prove this by transfection experiments have given inconclusive results, however. Even with *pgpA* expression levels in transfectants exceeding those observed in arsenite resistant mutants, only a low-level resistance was obtained, and the level of resistance varied depending on the species from which the gene was isolated or in which it was transfected (26, 140). With the *ltpgpA* gene transfected in *L. tarentolae,* only low resistance levels (maximally 2.5-fold for arsenite) were observed, and resistance was not directly proportional to copy number (140).

At these low resistance levels, the steady-state accumulation of radioactive arsenite remained unchanged, and no increased efflux was observed (140). In contrast, the analysis of a *L. major pgpA* transfectant indicates that PgpA reduces active drug uptake (28). As none of these experiments was done with the same gene or in the same recipient strain, it is difficult to deduce how PgpA confers resistance. It is unlikely that high-level resistance requires a mutant form of the transporter, as the amplified *pgpA* allele from a highly resistant strain was used in the experiments of Papadopoulou et al (140). It is therefore more likely that PgpA interacts with other factors to confer high-level resistance and that different *Leishmania* spp. might have different abilities to make these factors.

Candidate genes for these additional factors might reside in a region found in a 50-kilobase linear amplicon unrelated to *ltpgpA* that was found to be amplified in all arsenite-resistant *L. tarentolae* mutants (75). Mutants grown in the absence of the drug concomitantly lost the amplicon and part of their resistance, resulting in partial revertants. Transfection of the linear amplicon into wild-type cells did not lead to an increase in oxyanion resistance, but transfection into the partial revertant restored resistance to the level of the parent mutant (K Grondin, A Haimeur & M Ouellette, unpublished). This result suggests that the resistance gene present in the linear amplicon acts cooperatively with another mutation absent in the wild-type cells but present in the partial revertant to confer resistance. When *ltpgpA* was cotransfected with the linear amplicon in the partial revertant, resistance levels were higher than expected if resistance was due to the simple addition of the contributions of PgpA and the linear amplicon to resistance (K Grondin, A Haimeur & M Ouellete, unpublished). These results indicate that several independent mutations are present in resistant *Leishmania* spp. mutants, and that these mutations act synergistically to confer high-level resistance to oxyanions.

A Speculative Model for Oxyanion Resistance in Leishmania spp.

Recent studies on arsenite metabolism in mammalian cells summarized in the section on mammalian P-glycoproteins indicate that resistance to arsenite is a complicated matter involving formation of a complex between arsenite (As) and GSH and excretion of the complex by the GS-X pump. Although the $As(SG)_3$ complex can form spontaneously, there is circumstantial evidence that complex formation may be accelerated by a glutathione-S-transferase (GST). The GS-X pump appeared identical to MRP, the MDR-associated protein. In the ABC transporter family, MRP is most closely related to PgpA (34). This finding has led us to propose that PgpA is not an oxyanion pump, but a GS-X pump able to extrude oxyanion-GSH complexes—thus providing a plausible explanation for the low and somewhat variable levels of resistance observed in transfection experiments with *ltpgpA* in *Leishmania* spp. (26, 140) and the inability to reproduce the high level of resistance observed in stepwise selected mutants. The high-level resistance obtained by selection of *Leishmania* spp. with arsenite could be due to simultaneous selection for increased

GSH biosynthesis, increased GST activity, and increased export of GSH-arsenite conjugates. (Trypanothione or ovothiol A might also be involved.) If transfection only provides high levels of the export pump (PgpA), only low-level resistance may result, but when other putative GSH metabolizing genes are cotransfected (on the linear amplicon) or are already present (in the partial revertant), higher levels of resistance might be encountered.

Obviously, the interpretation outlined here is speculative, and other interpretations remain to be excluded. Transporters are certainly capable of transporting oxyanions in unconjugated form, as shown by the well-defined, plasmid-mediated, ATP-dependent oxyanion efflux system of *E. coli* (159). Our interpretation also does not account for the alterations in the organization of kinetoplast DNA observed in some arsenite-resistant mutants (113) and, more importantly, for the unexpected observation of Callahan et al (28) that transfection of the *pgpA* gene into *L. major* results in an inhibition of the energy-dependent influx of a trivalent antimonial without any measurable effect on efflux. The authors suggest "that overexpression of PgpA confers a dominant-negative phenotype through interaction with another *Leishmania* spp. protein, possibly belonging to the large and diverse P-glycoprotein superfamily." The availability of convenient *E. coli–Leishmania* spp. shuttle vectors (162) should make it possible to dissect the complex genotype of oxyanion resistance in *Leishmania* spp. in the coming year.

RESISTANCE TO ANTIFOLATES

Dihydrofolate reductase (DHFR) and thymidylate synthase (TS) catalyze consecutive reactions in the de novo synthesis of dTMP. In protozoa, in contrast to most other cells, these two enzymes are fused, resulting in a DHFR-TS protein (59, 69). The enzyme DHFR, which catalyzes the reduction of dihydrofolate to tetrahydrofolate, is an important target for chemotherapy. The anti-DHFR drugs are termed antifolates and are used in the treatment of parasitic infections caused by *Plasmodium falciparum* and *Toxoplasma gondii* in humans. DHFR inhibitors used in the treatment of microbial diseases are often combined with sulfonamides. Sulfonamides are inhibitors of the enzyme dihydropteroate synthase (DHPS), and inhibition of this enzyme blocks the de novo synthesis of dihydrofolates. Antifolates and sulfonamides act synergistically to deplete the pool of reduced folates and eventually to stop DNA synthesis. Many protozoa, however, are incapable of synthesizing their own folates and must obtain them from the environment via specific transporters. Other protozoa supplement the folates they make themselves with imported folates. These folate transporters provide an important means of getting antifolates inside the parasites. Due to their diverse clinical applications and their extensive usage, antifolates and resistance against antifolates have been studied intensively. The most commonly encountered resistance mechanisms are the decreased uptake of the drug due to alterations of the transporter, the overproduction of DHFR, and the production of an altered DHFR with decreased affinity for antifolates (63, 164). Several of these mutations may coexist in the same cell.

Mutations in DHFR and DHPS

Pyrimethamine and cycloguanil are two antifolates used to treat malaria. Pyrimethamine resistance in *P. falciparum* can arise through several mechanisms. The most commonly found is a single point mutation in the DHFR at position 108 changing a serine to an asparagine (38, 148). These parasites are not cross-resistant to cycloguanil. This mutation is also found in field isolates refractory to treatment (146), but in cell lines selected in vitro, other DHFR amino acid residues, in addition to position 108, were altered, and these mutations may confer higher levels of resistance (86, 170, 173). The DHFR of cycloguanil-resistant *P. falciparum* field strains contained a serine-to-threonine change at position 108 along with an alanine-to-valine change at position 16 (64, 147). These isolates are not cross-resistant to pyrimethamine. Strains resistant to both antifolates have an asparagine at position 108 and a few other key point mutations (64, 147). The role of each of these mutations remains to be tested by transfection. The sulfonamide sulfadoxine is frequently used in combination with pyrimethamine, and its target enzyme, DHPS, has recently been analyzed in sulfadoxine-resistant *P. falciparum* isolates. The enzymes from resistant and sensitive strains were found to differ in sequence (25, 175), but the importance of these point mutations for resistance in intact cells must still be tested by reintroduction of the altered DHPS genes in sensitive cells.

Sequence analysis of the *dhfr-ts* gene of pyrimethamine-resistant mutants of *Toxoplasma gondii* failed to reveal any mutation, and the mechanism of resistance is unknown (48). Mutations were introduced into the *T. gondii dhfr-ts* by site-directed mutagenesis at positions corresponding to mutations found in *P. falciparum dhfr-ts* resistant to pyrimethamine. Reintroduction of mutated versions of *dhfr-ts* into *T. gondii* rendered the parasite pyrimethamine resistant (48), but these mutations are apparently not selected for under natural conditions.

A common mechanism by which *Leishmania major* responds to the antifolate methotrexate is by amplification of its *dhfr-ts* gene (33). In at least one case the DHFR overproduced in *L. major* contained a methionine-to-arginine mutation at position 53 (5). This methionine 53 is equivalent to phenylalanine 31 in mammalian DHFRs, and replacement of this phenylalanine is known to result in methotrexate resistance (164). Transfection of the *L. major dhfr* arginine 53 version confirmed that it was associated with higher levels of methotrexate resistance than the wild-type gene (5). Sequence identities between microbial and mammalian DHFRs are low, explaining the efficacy and selectivity of antifolates, but the overall tertiary structures of these enzymes are similar (122, 164). The amino acid substitutions found in parasitic DHFRs and associated with antifolate resistance are aligning with residues known to occur at the active site or involved in antifolate binding in DHFRs of known structure.

DHFR Overproduction

Overproduction of DHFR confers pyrimethamine resistance in *P. falciparum*. This resistance can be achieved either by gene duplication (87) or by other mechanisms resulting in increased expression (173). In *L. major* selected for resistance to methotrexate, *dhfr-ts* amplification is common (18, 33, 52), but *dhfr-ts* amplification has not been observed in other *Leishmania* spp. (138). Interestingly, *dhfr-ts* amplification was recently observed in *L. tarentolae* cells when a preferred resistance mechanism was inactivated by gene targeting (K Grondin, G Roy, E Leblanc, B Papadopoulou, & M Ouellette, unpublished). This observation indicates that *Leishmania* spp. other than *L. major* have the necessary sequences for the generation of *dhfr-ts*-containing amplicons, and do generate such amplicons when the preferred mechanism of resistance is unavailable.

Transport Mutations in Leishmania spp.

Although none of the currently available antifolates is suitable for treatment of leishmaniasis in patients, much experimental work has been done on folate metabolism and resistance to antifolates in *Leishmania* spp. cultured in the laboratory. *Leishmania* spp. appear unable to cover their folate needs by de novo synthesis via DHPS (but see below) and rely heavily on import under standard conditions. *Leishmania* spp. have a common folate/antifolate transporter, and mutations in the gene for this transporter lead to methotrexate resistance (45, 52, 68, 98, 143). As *Leishmania* spp. cells without a folate transporter can still grow, they must either have more than one folate transporter at their disposal or be able to synthesize their own folates via an unconventional route. At least two unrelated transport systems, the reduced folate/methotrexate carrier and a membrane-associated folate-binding protein, are involved in the uptake of antifolates in mammalian cells (184). There is now also evidence for multiple transport systems in *Leishmania* spp. (138).

Under some conditions *Leishmania* spp. cannot synthesize folates de novo. Several *Leishmania* spp. are capable of growing in folate-deficient medium provided it is supplemented with pterins (98, 141, 149). Radioactive biopterin is converted into folates (13). Interestingly, *L. donovani* cells (12) or *L. tarentolae* cells (E Leblanc & M Ouellette, unpublished) with a mutation in their folate transporter gene are also unable to thrive in folate-deficient medium supplemented with pterins. This defect is not due to a decreased uptake of pterins, and revertants regained both their folate transport activity and their capacity to thrive in pterin-supplemented folate deficient medium (12). It is not obvious how a defect in a folate transporter could prevent biosynthesis of folates from biopterin.

Alternative Pathway for Folate Synthesis in Leishmania spp.

Methotrexate selection in *Leishmania* spp. is often associated with amplification of the H locus (18, 19, 85, 143, 150, 185). The gene residing in the H locus responsible for antifolate resistance has been identified by transfection, and characterization of its gene product (termed LTDH, HMTXR, or PTR1) indicated that it belongs to the family of short-chain dehydrogenases (27, 142), a class of enzymes involved in several oxidoreduction reactions. Cells overproducing LTDH are insensitive to methotrexate, whereas *ltdh/ptr1* null mutants are hypersensitive to methotrexate (14, 141). These two phenotypes are consistent with the idea that LTDH, when overproduced, is able to carry out DHFR's job (133, 139). Moreover, a *ptr1* null mutant of *L. major* was able to grow in folate-deficient medium supplemented with reduced pterins, but not oxidized pterins, such as biopterin. This led Beverley and coworkers to propose that LTDH/PTR1 is capable of salvaging oxidized pterins (14), a phenomenon described in Kinetoplastidae 40 years ago. The *ltdh* null mutant of *L. tarentolae* has also lost the ability to grow on biopterin-supplemented folate-deficient medium (141).

Purified LTDH/PTR1 exhibits an NADPH-dependent reductase activity, reduces biopterin and folate, and is inhibited by methotrexate. Activity is greater for the most oxidized form of the two classes of substrate, and biopterin is a much better substrate than folate (14) (J Wang, E Leblanc, B Papadopoulou, T Bray, JM Whiteley, S Linn & M Ouellette, unpublished). LTDH shows significant similarities with dihydropteridine reductase (DHPR), an enzyme that is the source of tetrahydrobiopterin in mammalian cells (180). Despite this sequence similarity, DHPR uses a "quinonoid" dihydropterin, which is not a substrate for LTDH. The role of LTDH/PTR1 in folate metabolism and resistance to antifolates is summarized in Figure 2. The main function of LTDH/PTR1 is to salvage oxidized pterins. The second function, reduction of folates, is less important, as the DHFR activity of LTDH/PTR1 is lower than that of the DHFR-TS present in the same cell. Nevertheless, the contribution of LTDH/PTR1 to folate reduction is not negligible. The *ltdh/ptr1* null mutants are hypersensitive to methotrexate, indicating that LTDH/PTR1 helps to reduce folates even in wild-type cells. In methotrexate-resistant mutants, overproduction of LTDH/PTR1 is sufficient to furnish the cells with reduced folates. It is unlikely that this is only due to sequestration of methotrexate by LTDH/

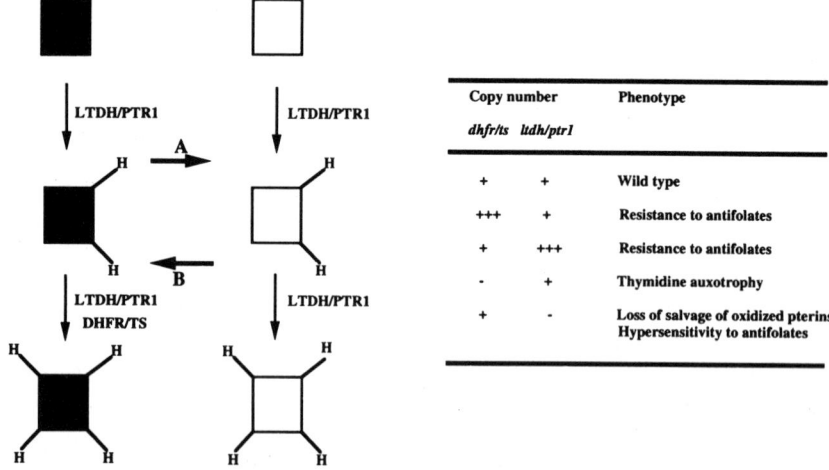

Figure 2 The role of LTDH/PTR1 in *Leishmania* spp. folate metabolism and resistance against antifolates. The filled boxes correspond to folic acid and its reduced dihydro and tetrahydro forms. The open boxes correspond to biopterin and its reduced forms. Reactions labeled A and B have been documented in *Leishmania* spp. A corresponds to a folate-methotrexate hydrolase activity present in *Leishmania* spp. (53, 98), and B represents the conversion of pterins into folates as described by Beck & Ullman (13).

PTR1 allowing DHFR-TS to do its job, as methotrexate has a much higher affinity for DHFR-TS (K_i 0.13 nM) than for LTDH/PTR1 (K_i 8 nM).

The requirement for exogenous pterins was first described in the nonpathogenic parasite *C. fasciculata,* and it is not known yet whether an activity similar to LTDH/PTR1 is present in other organisms besides *Leishmania* spp. As pointed out earlier (14, 141), the novelty and possible uniqueness of the LTDH/PTR1 pathway should make it possible to develop specific inhibitors of this enzyme. As inhibitors of the mammalian DHPR are being developed (156), lead compounds could already be available. LTDH/PTR1 inhibitors, in particular in combination with DHFR-TS inhibitors (see Figure 2), might be highly effective in the treatment of leishmaniasis.

MISCELLANEOUS RECENTLY DESCRIBED MECHANISMS OF DRUG RESISTANCE

Resistance to Purine Analogues

All parasitic protozoa so far studied are auxotrophic for purines, whereas the mammalian cells that they infect are capable of de novo purine synthesis. This clear biochemical difference suggests that purine analogues could be used successfully to treat parasitic disease, and indeed several pyrazolopyrimidine analogues of hypoxanthine and inosine were shown to be toxic to pathogenic parasites (119). The potential for a rational drug target has led to the characterization of several purine salvage enzymes and genes potentially involved in the scavenging of host purines. Notably the enzyme hypoxanthine-guanine phosphoribosyltransferase (HGPRT) of *P. falciparum* (103), *Toxoplasma gondii* (105), *Trypanosoma brucei,* and *Trypanosoma cruzi* (2, 3) has been extensively studied. The purine analogue allopurinol binds to purified HGPRT, and allopurinol was reported to be useful in the treatment of American leishmaniasis (121), although others are less enthusiastic about allopurinol efficacy (129). Recently Pfefferkorn & Borotz (151) mutagenized *Toxoplasma gondii* and selected a clone resistant to the purine analogue 6-thioxanthine. This mutant lacks HGPRT activity, suggesting that in *T. gondii* this pathway is not essential for the salvage of purines. Resistance to purine analogues has not yet been described for parasitic protozoa isolated from the field. Interesting work has been done, however, on resistance to these compounds in the laboratory, and here we summarize the main results. Two separate mutations leading to decreased accumulation of purines were characterized in *Leishmania* spp.; one involved the transport of inosine, guanosine, and their analogues, and the second involved the transport of adenosine and its analogues (4, 88, 155). Mutants lacking both transporters have been obtained by selecting with two toxic analogues, suggesting that other transporters must also be present that allow *Leishmania* spp. to take up purines. Resistance to tubercidin (7-deazaadenosine) in *Trypanosoma cruzi* has also been linked with a mutation in one transporter (60, 128). An extrachromosomal amplified circle was observed in *L. mexicana* cells selected for resistance to inosine dialdehyde or tubercidin (101). Transfection of a fragment derived from this circle conferred tubercidin resistance to wild-type cells by a mechanism apparently related to the decreased uptake of purine analogues (101). This mechanism could either be a novel method of nucleoside base extrusion or, more likely, result from a dominant-negative effect on the transporters involved in uptake.

Mycophenolic acid (MPA) is an effective inhibitor of inosinemonophosphate (IMP) dehydrogenase, an essential enzyme for de novo synthesis of guanine nucleotides. MPA is too toxic for clinical use, but it is used in the laboratory, and interesting resistant mutants have been obtained in several protozoa. As *Leishmania* spp. readily counteract metabolic blocks by amplifying the gene for the enzyme inhibited, it is not surprising that a mutant of *L. donovani,* 100-fold resistant to MPA, proved to have a 15-fold amplification of the IMP dehydrogenase gene and a corresponding increase in enzyme level (190). The amplified gene is not in extrachromosomal circles, however, as in several other *Leishmania* spp. mutants, but mainly in novel 200-kilobase linear chromosomes that are stable in the absence of drug (189). Even more remarkable results were obtained with the analysis of a *T. brucei gambiense* mutant resistant to MPA (188). Whereas all previous attempts to select for DNA amplification in *T. brucei* had failed (18, 109), the *T. brucei gambiense* mutant was found to have amplified the IMP dehydrogenase gene 10-fold. The extra copies resided in 6-Mb chromosomes indistinguishable from the *T. brucei* wild-type chromosome in which the IMP dehydrogenase gene resides. The addition of 20 copies of a 6-Mb chromosome to the *T. brucei* nucleus should increase its DNA content by 70%—a result that was actually found. Although the work of Wilson et al (188) shows that *T. brucei* can increase the copy number of a gene under heavy selective pressure, the impressively clumsy fashion in which this increase is accomplished confirms that trypanosomes lack the pathway(s) that allow *Leishmania* spp. to amplify relatively small gene segments containing single-copy genes efficiently.

The bovine parasite *Tritrichomonas foetus* was also selected for resistance to MPA, but in this parasite no DNA amplification was observed. Instead the resistant parasite had decreases in adenine deaminase activity and in hypoxanthine transport capability, allowing the mutant to salvage xanthine more efficiently (80).

Resistance to Metronidazole

Metronidazole has been used for more than 30 years against anaerobic bacteria and protozoa and is the first choice for treatment of three of the most prevalent human parasites, *Trichomonas vaginalis, Giardia duodenalis,* and *Entamoeba histolytica*. The toxicity of metronidazole and related nitroimidazoles depends on the presence of an electron donor of low redox potential able to convert the nitro group into aggressive free radical and nitroso compounds. Metabolic pathways involving these electron donors only exist in anaerobic organisms, and they can only activate the drug efficiently under anaerobic conditions, as oxygen competes with drug as electron acceptor. This activation pathway explains the highly selective toxicity of nitroimidazoles and related drugs for anaerobic organisms. The electrons responsible for metronidazole activation in protozoa are mainly produced by pyruvate:ferredoxin:oxidoreductase (PFOR) and transferred to the drug by ferredoxin. A decrease in the activity of this pathway appears to be the main mechanism of metronidazole resistance in field isolates of infectious protozoa. In *G. duodenalis* PFOR is predominantly affected; in *T. vaginalis,* ferredoxin is most affected (93, 174, 176). The most extensive molecular studies were done with *T. vaginalis*. Analysis of a resistant strain indicated that metronidazole was converted less efficiently to its cytotoxic reduced form (193). Further analysis showed that resistant strains had a decreased intracellular level of ferredoxin caused by reduced transcription of the ferredoxin gene, which was correlated with point mutations in the 5' flanking sequences of the gene (154).

Analysis of metronidazole-resistant *T. vaginalis* indicated that some have a half-type Pgp overexpressed by a mechanism other than gene amplification (94). No clear correlation was found, however, between levels of expression of the gene for this putative transporter and levels of resistance, and the gene's role in resistance is doubtful. Clearly, a transformation assay for parasitic organisms affected by metronidazole is required to prove unambiguously the resistance mechanisms proposed thus far.

Resistance to Ornithine Decarboxylase Inhibitors

Ornithine decarboxylase is the enzyme catalyzing the conversion of ornithine into the polyamine putrescine. The further conversion of putrescine into spermidine requires S-adenosylmethionine. The conjugation of spermidine and GSH in trypanosomatids results in the formation of trypanothione. A specific inhibitor of ornithine decarboxylase, DL-α-difluoromethylornithine (DFMO, eflornithine) was developed as an antitumor agent, but was also found to be highly effective as an antitrypanosomal agent (8). DFMO is a suicide substrate, and its selectivity resides in the long half-life of trypanosome ornithine decarboxylase, which lacks the C-terminal extension (a PEST sequence), conferring a short half-life on the mammalian enzymes (70). Trypanosome lines selected in vitro in a step-by-step manner for DFMO resistance had a reduced uptake of DFMO with an increase in the intracellular concentration of ornithine (8). The transporter involved has not been identified. *L. donovani* selected for resistance to DFMO in a stepwise manner amplified their ornithine decarboxylase gene, which was correlated with an increase in enzyme activity (78). Neither reduced DFMO uptake nor an increase in ornithine decarboxylase can account for the resistance of the *T. brucei rhodesiense* field strains analyzed by Bacchi et al (9). The most distinctive metabolic alterations in these isolates were related to S-adenosylmethionine (AdoMet) metabolism. Whereas sensitive cells reacted to the inhibition of ornithine decarboxylase with up to a 100-fold increase in AdoMet, the resistant strains showed a more moderate elevation, which correlated with a decrease in AdoMet synthetase. Bacchi et al (9) have suggested that the increased AdoMet may lead to "inappropriate methylation of proteins, nucleic acids or lipids," but this remains to be proven. However, lack of polyamines, essential for trypanothione synthesis, would seem sufficient reason for trypanosomes to die in the presence of DFMO.

As Bacchi has pointed out, the recent research on trypanocidal agents suggests interesting new possibilities for the use of drug combinations. If trypanothione is indeed the main target of arsenicals, DFMO as an inhibitor of trypanothione synthesis should sensitize trypanosomes to arsenicals. Such sensitization has indeed been observed in a mouse model (10). A new AdoMet decarboxylase inhibitor, the deoxyadenosine analogue MDL73811, is an even more effective inhibitor of polyamine synthesis than is DFMO (8). MDL73811 enters trypanosomes via a purine transporter, but which one is not yet known. As melamine-based arsenicals and diamidines appear to enter trypanosomes via the P2 adenosine transporter, purine transporters play a key role in trypanocidal chemotherapy. Trypanosomes may become resistant to arsenicals by the loss of the P2 transporter, but as they have to take up purines to survive, this loss may make them more vulnerable to suitably designed purine analogues that enter via other purine transporters.

Resistance to Emetine

The plant alkaloid toxin emetine used to be the first-line drug for treatment of amebiasis caused by *E. histolytica*, and emetine is still sometimes used for this purpose. Samuelson et al (163) mutagenized *E. histolytica* and selected a clone resistant to emetine. This mutant overexpressed a Pgp homolog and had several characteristics of the MDR phenotype found in cancer cells, including cross-resistance to hydrophobic drugs, increased efflux of emetine, and reversal of resistance by verapamil (43). The role of this Pgp in emetine resistance has not yet been tested by transfection, as transfection of *E. histolytica* was only recently accomplished (126, 152). Involvement of a protozoan Pgp in emetine resistance is plausible, however, as mammalian cells overexpressing Pgp were found to be resistant to emetine (117). Moreover, selection of the nematode *Caenorhabditis elegans* selected for emetine resistance resulted in organisms overexpressing *pgp* genes (116).

Resistance to Drugs Used in the Laboratory but Not in Patients So Far

The advent of gene-transfection techniques has completely transformed the field of molecular parasitology and has permitted conclusive tests of the contribution of a particular gene to drug resistance. The vectors used for transfection contain drug-resistance markers, mainly bacterial antibiotic resistance genes, allowing selection of stable transfectants. A list of these dominant markers is presented in Table 2. It is also possible, by transfecting a dominant-negative selectable marker such as the thymidine kinase of herpes simplex virus, to render parasites more sensitive to a drug than are wild-type cells (110). The parasites transfected with vectors containing antibiotic resistance genes are resistant to drugs not commonly used to treat patients. *Leishmania* spp. cells expressing the neomycin phosphotransferase gene and resistant to the selective drug G418, however, were cross-resistant to paromomycin, a drug now used in the treatment of *Leishmania* spp. (76). The transfer of this marker gene to wild-type *Leishmania* spp. in the field is clearly undesirable.

Table 2 Selectable markers used in stable transfection of parasites

Genes	Selective drugs	References
Kinetoplastidae		
Neomycin phosphotransferase	G-418	97, 107, 111, 172
Hygromycin phosphotransferase	Hygromycin B	41, 112
Phleomycin resistance gene Sh *ble*	Phleomycin	92
Streptothricin acetyl transferase	Nourseothricin	95
Puromycin N-acetyl transferase	Puromycin	66
N-acetyl glucosamine-1-phosphate transferase	Tunicamycin	118
Short-chain dehydrogenase LTDH/PTR1	Methotrexate	142
Herpes simplex thymidine kinase	Ganciclovir	110
Apicomplexa		
Mutated DHFR	Pyrimethamine	48
Chloramphenicol acetyl transferase	Chloramphenicol	102

OUTLOOK

Reviews on drug resistance tend to finish with a grand eulogy on the importance of this topic for humanity (and rightly so). We would like to end here by emphasizing that studies of drug resistance are also great fun. Parasite biochemistry is rarely straightforward and often interesting; hence, the resistance mechanisms are usually ingenious and novel. There is no lack of good projects: Our knowledge of resistance mechanisms in parasites is like a wide-bore fishing net, more hole than thread, as our review has made clear. Even for people who insist on working on Pgps, the parasite field still offers opportunities for plowing virgin soil. It also helps that the tools of the field are rapidly improving. As emphasized in this review, early studies of resistance mechanisms in parasites were often purely correlative and lacked the elegance of forward and reverse genetics, because the parasites refused to mate under controlled conditions and were not transformable with exogenous DNA. This tough situation has drastically changed in the past five years. The power of reverse genetics is now available for many parasites and is being applied with gusto. Moreover, industrious DNA sequencers have discovered parasite genomes, and in the coming years the major ones will get done. With whole genomes in the computer, the identification of genes involved in drug resistance will be simplified and will become a driving force in the improvement of the use of existing drug combinations and in the development of new drugs.

And all that with the possible bonus of discovering something useful in the battle against parasitic disease! So, let's get on with it.

Acknowledgments

We thank Drs. E Schurr, T Wellems, and DF Wirth for their comments on the chloroquine section, our collaborators in Amsterdam and Quebec for constructive critiques, and Bartie van Houten and Anne Helfrich for bearing the secretarial brunt of transatlantic coauthorship. The experimental work in our laboratories is supported by grants of the Netherlands Organization for Scientific Research (NWO/SON) and the Dutch Cancer Society (NKB/KWF) to PB, and by grants of the Medical Research Council, the Natural Sciences and Engineering Research Council, and the United Nations Development Programme/World Bank/World Health Organization Special Programme for Research and Training in Tropical Diseases to MO. MO is also a Fonds Recherche en Santé du Québec research fellow.

> Any *Annual Review* chapter, as well as any article cited in an *Annual Review* chapter, may be purchased from the Annual Reviews Preprints and Reprints service.
> 1-800-347-8007; 415-259-5017; email: arpr@class.org

Literature Cited

1. Akerboom TPM, Narayanaswami V, Kunst M, Sies H. 1991. ATP-dependent S-(2,4-dinitrophenyl)glutathione transport in canalicular plasma membrane vesicles from rat liver. *J. Biol. Chem.* 266:13147–52
2. Allen T, Ullman B. 1993. Cloning and expression of the hypoxanthine-guanine phosphoribosyltransferase gene from *Trypanosoma brucei*. *Nucleic Acids Res.* 21:5431–38
3. Allen T, Ullman B. 1994. Molecular characterization and overexpression of the hypoxanthine-guanine phosphoribosyltransferase gene from *T. cruzi*. *Mol. Biochem. Parasitol.* 65:233–45
4. Aronow B, Kaur K, McCartan K, Ullman B. 1987. Two high affinity nucleoside transporters in *Leishmania donovani*. *Mol. Biochem. Parasitol.* 22:29–37
5. Arrebola R, Olmo A, Reche P, Garvey EP, Santi DV, et al. 1994. Isolation and characterization of a mutant dihydrofolate reductase-thymidylate synthase from methotrexate-resistant *Leishmania* cells. *J. Biol. Chem.* 269:10590–96
6. Ashford RW, Desjeux P, de Raadt P. 1992. Estimation of population at risk of infection and number of cases of leishmaniasis. *Parasitol. Today* 8:104–5
7. Awad-el-Kariem FM, Miles MA, Warhurst DC. 1992. Chloroquine-resistant *Plasmodium falciparum* isolates from the Sudan lack two mutations in the *pfmdr1* gene thought to be associated with chloroquine resistance. *Trans. R. Soc. Trop. Med. Hyg.* 86:587–89
8. Bacchi CJ. 1993. Resistance to clinical drugs in African trypanosomes. *Parasitol. Today* 9:190–93
9. Bacchi CJ, Garofalo J, Ciminelli M, Rattendi D, Goldberg B, et al. 1993. Resistance to DL-α-difluoromethylornithine by clinical isolates of *Trypanosoma brucei rhodesiense*. *Biochem. Pharmacol.* 46:471–81
10. Bacchi CJ, Nathan HC, Yarlett N, Goldberg B, McCann PP, et al. 1994. Combination chemotherapy of drug-resistant *Trypanosoma brucei rhodesiense* infections in mice using DL-α-difluoromethylornithine and standard trypanocides. *Antimicrob. Agents Chemother.* 38:563–69
11. Barnes DA, Foote SJ, Galatis D, Kemp DJ, Cowman AF. 1992. Selection for high level chloroquine resistance results in deamplification of the *pfmdr1* gene and increased sensitivity to mefloquine in *Plasmodium falciparum*. *EMBO J.* 11:3067–75
12. Beck JT, Ullman B. 1990. Nutritional requirements of wild-type and folate transport-deficient *Leishmania donovani* for pterins and folates. *Mol. Biochem. Parasitol.* 43:221–30
13. Beck JT, Ullman B. 1991. Biopterin

conversion to reduced folates by *Leishmania donovani* promastigotes. *Mol. Biochem. Parasitol.* 49:21–28
14. Bello AR, Nare B, Freedman D, Hardy L, Beverley SM. 1994. PTR1: a reductase mediating salvage of oxidized pteridines and methotrexate resistance in the protozoan parasite *Leishmania major. Proc. Natl. Acad. Sci. USA* 91:11442–46
15. Bellofatto V, Cross GAM. 1989. Expression of a bacterial gene in a trypanosomatid protozoan. *Science* 244:1167–69
16. Bellofatto V, Torres-Muoz JE, Cross GAM. 1991. Stable transformation of *Leptomonas seymouri* by circular extrachromosomal elements. *Proc. Natl. Acad. Sci. USA* 88:6711–15
17. Berman JD, Grogl M. 1988. *Leishmania mexicana*: chemistry and biochemistry of sodium stibogluconate (Pentostam). *Exp. Parasitol.* 67:96–103
18. Beverley SM. 1991. Gene amplification in *Leishmania. Annu. Rev. Microbiol.* 45:417–44
19. Beverley SM, Coderre JA, Santi DV, Schimke RT. 1984. Unstable DNA amplifications in methotrexate-resistant *Leishmania* consist of extrachromosomal circles which relocalize during stabilization. *Cell* 38:431–39
20. Borst P. 1991. Genetic mechanisms of drug resistance. *Rev. Oncol.* 4:87–105
21. Borst P, Schinkel AH, Smit JJM, Wagenaar E, Van Deemter L, et al. 1993. Classical and novel forms of multidrug resistance and the physiological functions of P-glycoproteins in mammals. *J. Pharmacol. Ther.* 60:289–99
22. Bosch IB, Wang ZX, Tao LF, Shoemaker CB. 1994. Two *Schistosoma mansoni* cDNAs encoding ATP-binding cassette (ABC) family proteins. *Mol. Biochem. Parasitol.* 65:351–56
23. Bray PG, Boulter MK, Ritchie GY, Howells RE, Ward SA. 1994. Relationship of global chloroquine transport and reversal of resistance in *Plasmodium falciparum. Mol. Biochem. Parasitol.* 63:87–94
24. Bray PG, Howells RE, Ritchie GY, Ward SA. 1992. Rapid chloroquine efflux phenotype in both chloroquine sensitive and chloroquine resistant *Plasmodium falciparum. Biochem. Pharmacol.* 44:1317–24
25. Brooks D, Wang P, Read M, Watkins W, Sims P, Hyde J. 1995. Correlation of sulfadoxine resistance with point mutations located within the bifunctional hydroxymethyldihydrobiopterin pyrophosphokinase-dihydropteroate synthase gene of the human malaria parasite *Plasmodium falciparum. Mol. Biochem. Parasitol.* In press
26. Callahan HL, Beverley SM. 1991. Heavy metal resistance: a new role for P-glycoproteins in *Leishmania. J. Biol. Chem.* 266:18427–30
27. Callahan HL, Beverley SM. 1992. A member of the aldoketo reductase family confers methotrexate resistance in *Leishmania. J. Biol. Chem.* 267:24165–68
28. Callahan HL, Roberts WL, Rainey PM, Beverley SM. 1994. The *pgpA* gene of *Leishmania major* mediates antimony (SbIII) resistance by decreasing influx and not by increasing efflux. *Mol. Biochem. Parasitol.* 68:145–49
29. Carter NS, Fairlamb AH. 1993. Arsenical-resistant trypanosomes lack an unusual adenosine transporter. *Nature* 361:173–75
30. Cerami A, Warren KS. 1994. Drugs. *Parasitol. Today* 10:404–6
31. Childs S, Ling V. 1994. The MDR superfamily of genes and its biological implications. In *Important Advances in Oncology 1994*, ed. J DeVita, S Hellman, SA Rosenberg, pp. 21–37. Philadelphia: Lippincott
32. Chow LMC, Wong AKC, Ullman B, Wirth D. 1993. Cloning and functional analysis of an extrachromosomally amplified multidrug resistance-like gene in *Leishmania enriettii. Mol. Biochem. Parasitol.* 60:195–208
33. Coderre JA, Beverley SM, Schimke RT, Santi DV. 1983. Overproduction of a bifunctional thymidylate synthetase-dihydrofolate reductase and DNA amplification in methotrexate-resistant *Leishmania tropica. Proc. Natl. Acad. Sci. USA* 80:2132–36
34. Cole SPC, Bhardwaj G, Gerlach JH, Mackie JE, Grant CE, et al. 1992. Overexpression of a transporter gene in a multidrug-resistant human lung cancer cell line. *Science* 258:1650–54
35. Cooper R, Ribeiro de Jesus A, Cross GAM. 1993. Deletion of an immunodominant *Trypanosoma cruzi* surface glycoprotein disrupts flagellum-cell adhesion. *J. Cell Biol.* 122:149–56
36. Cowman AF, Galatis D, Thompson JK. 1994. Selection for mefloquine resistance in *Plasmodium falciparum* is linked to amplification of the *pfmdr1* gene and cross-resistance to halofantrine and quinine. *Proc. Natl. Acad. Sci. USA* 91:1143–47
37. Cowman AF, Karz S, Galatis D, Culvenor JG. 1991. A P-glycoprotein homologue of *Plasmodium falciparum*

is localized on the digestive vacuole. *J. Cell Biol.* 113:1033–42
38. Cowman AF, Morry MJ, Biggs BA, Cross GAM, Foote SJ. 1988. Amino acid changes linked to pyrimethamine resistance in the dihydrofolate reductase-thymidylate synthase genes of *Plasmodium falciparum. Proc. Natl. Acad. Sci. USA* 25:9109–13
39. Cox FEG. 1993. *Modern Parasitology.* Oxford: Blackwell Sci. Publ.
40. Cruz A, Beverley SM. 1990. Gene replacement in parasitic protozoa. *Nature* 348:171–73
41. Cruz A, Coburn CM, Beverley SM. 1991. Double targeted gene replacement for creating null mutants. *Proc. Natl. Acad. Sci. USA* 88:7170–74
42. Dallagiovanna B, Castanys S, Gamarro F. 1994. *Trypanosoma cruzi*: sequence of the ATP-binding site of a P-glycoprotein. *Exp. Parasitol.* 79:63–67
43. Descoteaux S, Ayala P, Orozco E, Samuelson J. 1992. Primary sequences of two P-glycoprotein genes of *Entamoeba histolytica. Mol. Biochem. Parasitol.* 54:201–12
44. Detke S, Katakura K, Chang KP. 1989. DNA amplification in arsenite-resistant *Leishmania. Exp. Cell Res.* 180:161–70
45. Dewes H, Ostergaard HK, Simpson L. 1986. Impaired drug uptake in methotrexate resistant *Crithidia fasciculata* without changes in dihydrofolate reductase activity or gene amplification. *Mol. Biochem. Parasitol.* 19:149–61
46. Dey S, Papadopoulou B, Roy G, Grondin K, Dou D, et al. 1994. High level arsenite resistance in *Leishmania tarentolae* is mediated by an active extrusion system. *Mol. Biochem. Parasitol.* 67:49–57
47. Doige CA, Ferro-Luzzi Ames G. 1993. ATP-dependent transport systems in bacteria and humans: relevance to cystic fibrosis and multidrug resistance. *Annu. Rev. Microbiol.* 47:291–319
48. Donald RGK, Roos DS. 1993. Stable molecular transformation of *Toxoplasma gondii*: a selectable dihydrofolate reductase-thymidylate synthase marker based on drug-resistance mutations in malaria. *Proc. Natl. Acad. Sci. USA* 90:11703–7
49. Eid J, Sollner-Webb B. 1991. Stable integrative transformation of *Trypanosoma brucei* that occurs exclusively by homologous recombination. *Proc. Natl. Acad. Sci. USA* 88:2118–21
50. Eid JE, Sollner-Webb B. 1991. Homologous recombination in the tandem calmodulin genes of *Trypanosoma brucei* yields multiple products: compensation for deleterious deletions by gene amplification. *Genes Dev.* 5:2024–32
51. Ekong RM, Robson KJH, Baker DA, Warhurst DC. 1993. Transcripts of the multidrug resistance genes in chloroquine-sensitive and chloroquine-resistant *Plasmodium falciparum. Parasitology* 106:107–15
52. Ellenberger TE, Beverley SM. 1987. Biochemistry and regulation of folate and methotrexate transport in *Leishmania major. J. Biol. Chem.* 262:10053–58
53. Ellenberger TE, Wright JE, Rosowsky A, Beverley SM. 1989. Wild-type and drug-resistant *Leishmania major* hydrolyze methotrexate to N-10-methyl-4-deoxy-4-aminopteroate without accumulation of methotrexate polyglutamates. *J. Biol. Chem.* 264:15960–66
54. Fairlamb AH, Carter NS, Cunningham M, Smith K. 1992. Characterization of melarsen-resistant *Trypanosoma brucei brucei* with respect to cross-resistance to other drugs and trypanothione metabolism. *Mol. Biochem. Parasitol.* 53:213–22
55. Fairlamb AH, Cerami A. 1992. Metabolism and functions of trypanothione in the *Kinetoplastida. Annu. Rev. Microbiol.* 46:695–725
56. Fairlamb AH, Henderson GB, Cerami A. 1989. Trypanothione is the primary target for arsenical drugs against African trypanosomes. *Proc. Natl. Acad. Sci. USA* 86:2607–11
57. Fairlamb AH, Smith K, Hunter KJ. 1992. The interaction of arsenical drugs with dihydrolipoamide and dihydrolipoamide dehydrogenase from arsenical resistant and sensitive strains of *Trypanosoma brucei brucei. Mol. Biochem. Parasitol.* 53:223–32
58. Fernandes AP, Nelson K, Beverley SM. 1993. Evolution of nuclear ribosomal RNAs in kinetoplastid protozoa: perspective on the age and origins of parasitism. *Proc. Natl. Acad. Sci. USA* 90:608–12
59. Ferone R, Roland S. 1980. Dihydrofolate reductase: thymidilate synthetase, a bifunctional polypeptide from *Crithidia fasciculata. Proc. Natl. Acad. Sci. USA* 77:5802–6
60. Finley RW, Cooney DA, Dvorak JA. 1988. Nucleoside uptake in *Trypanosoma cruzi*: analysis of a mutant resistant to tubercidin. *Mol. Biochem. Parasitol.* 31:133–40
61. Fitch CD, Kanjananggulpan P. 1987. The state of ferriprotoporhyrin IX in malaria pigment. *J. Biol. Chem.* 262:15552–55
62. Foley M, Deady LW, Ng K, Cowman

AF, Tilley L. 1994. Photoaffinity labeling of chloroquine-binding proteins in *Plasmodium falciparum*. *J. Biol. Chem.* 269:6955–61
63. Foote SJ, Cowman AF. 1994. The mode of action and the mechanism of resistance to antimalarial drugs. *Acta Trop.* 56:157–71
64. Foote SJ, Galatis D, Cowman AF. 1990. Amino acids in the dihydrofolate reductase-thymidylate synthase gene of *Plasmodium falciparum* involved in cycloguanil resistance differ from those involved in pyrimethamine resistance. *Proc. Natl. Acad. Sci. USA* 87:3014–17
65. Foote SJ, Thompson JK, Cowman AF, Kemp DJ. 1989. Amplification of the multidrug resistance gene in some chloroquine-resistant isolates of *Plasmodium falciparum*. *Cell* 57:921–30
66. Freedman DJ, Beverley SM. 1993. Two more independent selectable markers for stable transfection of *Leishmania*. *Mol. Biochem. Parasitol.* 62:37–44
67. Frommel TO, Balber AE. 1987. Flow cytofluorimetric analysis of drug accumulation by multidrug-resistant *Trypanosoma brucei brucei* and *Trypanosoma brucei rhodesiense*. *Mol. Biochem. Parasitol.* 26:183–92
68. Gamarro F, Amador MV, Chiquero MJ, Legere D, Ouellette M, Castanys S. 1994. Multidrug resistance phenotype and P-glycoprotein overexpression in a methotrexate-resistant *Leishmania infantum*. *Biochem. Pharmacol.* 47:1939–47
69. Garret CE, Coderre JA, Meek TD, Garvey EP, Claman DM, et al. 1984. A bifunctional thymidilate synthetase-dihydrofolate reductase in protozoa. *Mol. Biochem. Parasitol.* 11:257–65
70. Ghoda L, Phillips MA, Bass KE, Wang CC, Coffino P. 1990. Trypanosome ornithine decarboxylase is stable because it lacks sequences found in the carboxyl terminus of the mouse enzyme which target the latter for intracellular degradation. *J. Biol. Chem.* 265:11823–26
71. Goonewarde R, Daily J, Kaslow D, Sullivan TJ, Duffy P, et al. 1993. Transfection of the malaria parasite and expression of firefly luciferase. *Proc. Natl. Acad. Sci. USA* 90:5234–36
72. Gottesman MM, Pastan I. 1993. Biochemistry of multidrug resistance mediated by the multidrug transporter. *Annu. Rev. Biochem.* 62:385–427
73. Grant CE, Valdimarsson G, Hipfner DR, Almquist KC, Cole SP, Deeley RG. 1994. Overexpression of multidrug resistance-associated protein (MRP) increases resistance to natural product drugs. *Cancer Res.* 54:357–61
74. Grogl M, Thomason TN, Franke ED. 1992. Drug resistance in leishmaniasis: its implication in systemic chemotherapy of cutaneous and mucocutaneous disease. *Am. J. Trop. Med. Hyg.* 47:117–26
75. Grondin K, Papadopoulou B, Ouellette M. 1993. Homologous recombination between direct repeat sequences yields P-glycoprotein containing circular amplicons in arsenite resistant *Leishmania*. *Nucleic Acids Res.* 21:1895–901
76. Gueiros-Filho FJ, Beverley SM. 1994. On the introduction of genetically modified *Leishmania* outside the laboratory. *Exp. Parasitol.* 78:425–28
77. Gutteridge WE, Coombs GH, eds. 1977. *Biochemistry of Parasitic Protozoa*, London: MacMillan. 172 pp. 1st ed.
78. Hanson S, Adelman J, Ullman B. 1992. Amplification and molecular cloning of the ornithine decarboxylase gene of *Leishmania donovani*. *J. Biol. Chem.* 267:2350–59
79. Hariharan S, Ajioka J, Swindle J. 1993. Stable transformation of *Trypanosoma cruzi*: inactivation of the *PUB12.5* polyubiquitin gene by targeted gene disruption. *Mol. Biochem. Parasitol.* 57:15–30
80. Hedstrom L, Cheung KS, Wang CC. 1990. A novel mechanism of mycophenolic acid resistance in the protozoan parasite *Tritrichomonas foetus*. *Biochem. Pharmacol.* 39:151–60
81. Heijn M, Oude Elferink RPJ, Jansen PLM. 1992. ATP-dependent multispecific organic anion tranport system in rat erythrocyte membrane vesicles. *Am. J. Physiol.* 262:C104–10
82. Henderson DM, Sifri CD, Rodgers M, Wirth DF, Hendrickson N, Ullman B. 1992. Multidrug resistance in *Leishmania donovani* is conferred by amplification of a gene homologous to the mammalian *mdr1* gene. *Mol. Cell. Biol.* 12:2855–65
83. Hendrickson N, Sifri CD, Henderson DM, Allen T, Wirth DF, Ullman B. 1993. Molecular characterization of the *ldmdr1* multidrug resistance gene from *Leishmania donovani*. *Mol. Biochem. Parasitol.* 60:53–64
84. Higgins CF. 1992. ABC transporters: from microorganisms to man. *Annu. Rev. Cell Biol.* 8:67–113
85. Hightower RC, Ruiz-Perez LM, Lie Wong M, Santi DV. 1988. Extrachromosomal element in the lower eukaryote *Leishmania*. *J. Biol. Chem.* 263:16970–76

86. Hyde JE. 1990. The dihydrofolate reductase-thymidylate synthetase gene in the drug resistance of malaria parasites. *Pharmacol. Ther.* 48:45–49
87. Inselburg J, Bzik DJ, Horii T. 1987. Pyrimethamine resistant *Plasmodium falciparum*: overproduction of dihydrofolate reductase by a gene duplication. *Mol. Biochem. Parasitol.* 26:121–34
88. Iovannisci DM, Kaur K, Young L, Ullman B. 1984. Genetic analysis of nucleoside transport in *Leishmania donovani*. *Mol. Cell Biol.* 4:1013–19
89. Ishikawa T. 1992. The ATP-dependent glutathione S-conjugate export pump. *Trends Biol. Sci.* 17:463–68
90. Jackson JE, Tally JD, Ellis WY, Mebrahtu YB, Lawyer PG, et al. 1990. Quantitative *in vitro* drug potency and drug susceptibility evaluation of *Leishmania spp* from patients unresponsive to pentavalent antimony therapy. *Am. J. Trop. Med. Hyg.* 43:464–80
91. Jedlitschky G, Leier I, Bucholz U, Center M, Keppler D. 1994. ATP-dependent transport of glutathione S-conjugates by the multidrug resistance-associated protein. *Cancer Res.* 54:4833–36
92. Jefferies D, Tebabi P, Le Ray D, Pays E. 1993. The *ble* resistance gene as a new selectable marker for *Trypanosoma brucei*: fly transmission of a stable procyclic transformant to produce antibiotic bloodstream forms. *Nucleic Acids Res.* 21:191–95
93. Johnson PJ. 1993. Metronidazole and drug resistance. *Parasitol. Today* 9:183–86
94. Johnson PJ, Schuck BL, Delgadillo MG. 1994. Analysis of a single-domain P-glycoprotein-like gene in the early-diverging protist *Trichomonas vaginalis*. *Mol. Biochem. Parasitol.* 66:127–37
95. Joshi P, Webb J, Davies J, McMaster WR. 1995. The Streptothricin-Acetyltransferase (*sat*) gene as a selectable marker and the pLEX series of expression vectors for the transfection of *Leishmania. Gene* In press
96. Juliano RL, Ling V. 1976. A surface glycoprotein modulating drug permeability in Chinese hamster ovary cell mutants. *Biochim. Biophys. Acta* 455:152–62
97. Kapler GM, Coburn C, Beverley SM. 1990. Stable transfection of the human parasite *Leishmania* delineates a 30 kb region sufficient for extrachromosomal replication and expression. *Mol. Cell. Biol.* 10:1084–94
98. Kaur K, Coons T, Emmet K, Ullman B. 1988. Methotrexate-resistant *Leishmania donovani* genetically deficient in the folate-methotrexate transporter. *J. Biol. Chem.* 263:7020–28
99. Kelly JM, Taylor MC, Smith K, Hunter KJ, Fairlamb AH. 1993. Phenotype of recombinant *Leishmania donovani* and *Trypanosoma cruzi* which overexpress trypanothione reductase: sensitivity towards agents that are thought to induce oxidative stress. *Eur. J. Biochem.* 218:29–37
100. Kelly JM, Ward HM, Miles MA, Kendall G. 1992. A shuttle vector which facilitates the expression of transfected genes in *Trypanosoma cruzi* and *Leishmania. Nucleic Acids Res.* 20:3963–69
101. Kerby BR, Detke S. 1993. Reduced purine accumulation is encoded on an amplified DNA in *Leishmania mexicana amazonensis* resistant to toxic nucleosides. *Mol. Biochem. Parasitol.* 60:171–86
102. Kim K, Soldati D, Boothroyd JC. 1993. Gene replacement in *Toxoplasma gondii* with chloramphenicol acetyltransferase as selectable marker. *Science* 262:911–14
103. King A, Melton DW. 1987. Characterization of cDNA clones for hypoxanthine-guanine phosphoribosyltransferase from the human malaria parasite, *Plasmodium falciparum*: comparisons to the mam- malian gene and protein. *Nucleic Acids Res.* 15:10469–81
104. Krogstadt DJ, Gluzman IY, Kyle DE, Oduola AM, Martin SK, et al. 1987. Efflux of chloroquine from *Plasmodium falciparum*: mechanism of chloroquine resistance. *Science* 238:1283–85
105. Krug EC, Marr JJ, Berens RL. 1989. Purine metabolism in *Toxoplasma gondii*. *J. Biol. Chem.* 264:10601–7
106. Külling R, Hollenberg CP. 1994. The ABC-transporter Ste6 accumulates in the plasma membrane in a ubiquitinated form in endocytosis mutants. *EMBO J.* 13:3261–71
107. Laban A, Finbarr Tobin JF, Curotto de Lafaille MA, Wirth D. 1990. Stable expression of the bacterial *neo(r)* gene in *Leishmania enriettii*. *Nature* 343:572–74
108. Laban A, Wirth DF. 1989. Transfection of *Leishmania enriettii* and expression of chloramphenicol acetyltransferase gene. *Proc. Natl. Acad. Sci. USA* 86:9119–23
109. Laird PW. 1988. *Transsplicing in Trypanosoma brucei*. PhD thesis. Univ. Amsterdam, Rodopi, Amsterdam. 145 pp.
110. LeBowitz JH, Cruz A, Beverley SM. 1992. Thymidine kinase as a negative

selectable marker in *Leishmania major*. *Mol. Biochem. Parasitol.* 51:321–26

111. Lee MG-S, Van der Ploeg LHT. 1990. Homologous recombination and stable transfection in the parasitic protozoan *Trypanosoma brucei*. *Science* 250:1583–87
112. Lee MGS, Van der Ploeg LHT. 1991. The hygromycin B resistance-encoding gene as a selectable marker for stable transformation of *Trypanosoma brucei*. *Gene* 105:255–57
113. Lee ST, Liu HY, Lee SP, Tarn C. 1994. Selection for arsenite resistance causes reversible changes in minicircle composition and kinetoplast organization in *Leishmania mexicana*. *Mol. Cell. Biol.* 14:587–96
114. Leier I, Jedlitschky G, Buchholz U, Keppler D. 1994. Characterization of the ATP-dependent leukotriene C_4 export carrier in mastocytoma cells. *Eur. J. Biochem.* 220:599–606
115. Légaré D, Hettema E, Ouellette M. 1994. The P-glycoprotein related gene family in *Leishmania*. *Mol. Biochem. Parasitol.* 68:81–91
116. Lincke CR. 1993. *P-glycoprotein genes and multidrug resistance*, PhD thesis. Univ. Amsterdam, Rodopi, Amsterdam. 114 pp.
117. Ling V, Kartner N, Sudo T, Siminovitch L, Riordan JR. 1983. Multidrug-resistance phenotype in Chinese hamster ovary cells. *Cancer Treat. Rep.* 67:869–74
118. Liu X, Chang KP. 1992. The 63-kilobase circular amplicon of tunicamycin-resistant *Leishmania amazonensis* contains a functional N-acetylglucosamine-1-phosphate transferase gene that can be used as a dominant selectable marker in transfection. *Mol. Cell. Biol.* 12:4112–22
119. Marr JJ, Berens RL, Nelson DJ, Krenitsky TA, Spector T, et al. 1982. Antileishmanial action of 4-thiopyrazolo(3,4-d)pyrimidine and its ribonucleoside. *Biochem. Pharmacol.* 31:143–48
120. Martin SK, Oduola AMJ, Milhous WK. 1987. Reversal of chloroquine resistance in *Plasmodium falciparum* by verapamil. *Science* 235:899–901
121. Martinez S, Marr JJ. 1992. Allopurinol in the treatment of American cutaneous leishmaniasis. *New Engl. J. Med.* 326:741–44
122. Matthews DA, Bolin JT, Burridge JM, Filman DJ, Volz KW, et al. 1985. Refined crystal structures of *Escherichia coli* and chicken liver dihydrofolate reductase containing bound trimethoprim. *J. Biol. Chem.* 260:381–91
123. Murren JR, De Vita VT. 1994. Multidrug resistance revisited. *Pezcoller Found. J.* 3:2–9
124. Müller M, Meijer C, Zaman GJR, Borst P, Scheper RJ, et al. 1994. Overexpression of the multidrug resistance associated protein (MRP) gene results in increased ATP-dependent glutathione S-conjugate transport. *Proc. Natl. Acad. Sci. USA* 91:13033–37
125. Nateghpour M, Ward SA, Howells RE. 1993. Development of halofantrine resistance and determination of cross-resistance patterns in *Plasmodium falciparum*. *Antimicrob. Agents Chemother.* 37:2337–43
126. Nickel R, Tannich E. 1994. Transfection and transient expression of chloramphenicol acetyltransferase gene in the protozoan parasite *Entamoeba histolytica*. *Proc. Natl. Acad. Sci. USA* 91:7095–98
127. Nikaido H. 1994. Prevention of drug access to bacterial targets: permeability barriers and active efflux. *Science* 264:382–88
128. Nozaki T, Dvorak JA. 1993. Molecular biology studies of tubercidin resistance in *Trypanosoma cruzi*. *Parasitol. Res.* 79:451–55
129. Olliaro PL, Bryceson ADM. 1993. Practical progress and new drugs for changing patterns of leishmaniasis. *Parasitol. Today* 9:323–28
130. Ortiz DF, Kreppel L, Speiser DM, Scheel G, McDonald G, Ow DW. 1992. Heavy metal tolerance in the fission yeast requires an ATP-binding cassette-type vacuolar membrane transporter. *EMBO J.* 11:3491–99
131. Otsu K, Donelson JE, Kirchhoff LV. 1993. Interruption of *Trypanosoma cruzi* gene encoding a protein containing 14-amino acid repeats by targeted insertion of neomycin phosphotransferase gene. *Mol. Biochem. Parasitol.* 57:317–30
132. Oude Elferink RPJ, Ottenhof R, Liefting W, De Haan J, Jansen PLM. 1989. Hepatobiliary transport of glutathione and glutathione conjugate in rats with hereditary hyperbilirubinemia. *J. Clin. Invest.* 84:476–83
133. Ouellette M, Borst P. 1991. Drug resistance and P-glycoprotein gene amplification in the protozoan parasite *Leishmania*. *Res. Microbiol.* 142:737–46
134. Ouellette M, Fase-Fowler F, Borst P. 1990. The amplified H circle of methotrexate-resistant *Leishmania tarentolae*

contains a novel P-glycoprotein gene. *EMBO J.* 9:1027-33
135. Ouellette M, Hettema E, Wüst D, Fase-Fowler F, Borst P. 1991. Direct and inverted repeats associated with P-glycoprotein gene amplification in drug resistant *Leishmania. EMBO J.* 10:1009-16
136. Ouellette M, Légaré D, Papadopoulou B. 1994. Microbial multidrug resistance ABC transporters. *Trends Microbiol.* 2:407-11
137. Ouellette M, Papadopoulou B. 1993. Mechanisms of drug resistance in *Leishmania. Parasitol. Today* 9:150-53
138. Ouellette M, Papadopoulou B, Haimeur A, Grondin K, Leblanc E, et al. 1995. Transport of antifolates and antimonials in drug-resistant *Leishmania*. In *Drug Transport in Antimicrobial and Anticancer Chemotherapy*, ed. NH Georgopapadakou, pp. 377-402. New York: Dekker
139. Papadopoulou B, Ouellette M. 1993. Frequent amplification of a short chain dehydrogenase gene in methotrexate resistant *Leishmania. Adv. Exp. Med. Biol.* 338:559-62
140. Papadopoulou B, Roy G, Dey S, Rosen BP, Ouellette M. 1994. Contribution of the *Leishmania* P-glycoprotein related gene *ltpgpA* to oxyanion resistance. *J. Biol. Chem.* 269:11980-86
141. Papadopoulou B, Roy G, Mourad W, Leblanc E, Ouellette M. 1994. Changes in folate and pterin metabolism after disruption of the *Leishmania* H locus short-chain dehydrogenase gene. *J. Biol. Chem.* 269:7310-15
142. Papadopoulou B, Roy G, Ouellette M. 1992. A novel antifolate resistance gene on the amplified H circle of *Leishmania. EMBO J.* 11:3601-8
143. Papadopoulou B, Roy G, Ouellette M. 1993. Frequent amplification of a short chain dehydrogenase gene as part of circular and linear amplicons in methotrexate resistant *Leishmania. Nucleic Acids Res.* 21:4305-12
144. Peel SA, Bright P, Yount B, Handy J, Baric RS. 1994. A strong association between mefloquine and halofantrine resistance and amplification, overexpression, and mutation in the P-glycoprotein gene homolog (*pfmdr*) of *Plasmodium falciparum in vitro. Am. J. Trop. Med. Hyg.* 51:648-58
145. Peel SA, Merritt SC, Handy J, Baric RS. 1993. Derivation of highly mefloquine-resistant lines from *Plasmodium falciparum in vitro. Am. J. Trop. Med. Hyg.* 48:385-97
146. Peterson DS, Di Santi SM, Povoa M, Calvosa VS, do Rosario VE, Wellems TE. 1991. Prevalence of the dihydrofolate reductase Asn-108 mutation as the basis for pyrimethamine-resistant falciparum malaria in the Brazilian Amazon. *Am. J. Trop. Med. Hyg.* 45:492-97
147. Peterson DS, Milhous WK, Wellems TE. 1990. Molecular basis of differential resistance to cycloguanil and pyrimethamine in *Plasmodium falciparum. Proc. Natl. Acad. Sci. USA* 87:3018-22
148. Peterson DS, Walliker D, Wellems TE. 1988. Evidence that a point mutation in dihydrofolate reductase-thymidylate synthase confers resistance to pyrimethamine in *falciparum* malaria. *Proc. Natl. Acad. Sci. USA* 85:9114-18
149. Petrillo-Peixoto ML, Beverley SM. 1987. *In vitro* activity of sulfonamides and sulfones against *Leishmania major* promastigotes. *Antimicrob. Agents Chemother.* 31:1575-78
150. Petrillo-Peixoto ML, Beverley SM. 1988. Amplified DNAs in laboratory stocks of *Leishmania tarentolae*: extrachromosomal circles structurally and functionally similar to the inverted H region amplification of methotrexate-resistant *Leishmania major. Mol. Cell. Biol.* 8:5188-99
151. Pfefferkorn ER, Borotz SE. 1994. *Toxoplasma gondii*: characterization of a mutant resistant to 6-thioxanthine. *Exp. Parasitol.* 79:374-82
152. Purdy JE, Mann BJ, Pho LT, Petri JR. WA. 1994. Transient transfection of the enteric parasite *Entamoeba histolytica* and expression of firefly luciferase. *Proc. Natl. Acad. Sci. USA* 91:7099-103
153. Pussard E, Verdier F. 1994. Antimalarial 4-aminoquinolines: mode of action and pharmacokinetics. *Fund. Clin. Pharmacol.* 8:1-17
154. Quon DVK, d'Oliveira CE, Johnson PJ. 1992. Reduced transcription of the ferredoxin gene in metronidazole-resistant *Trichomonas vaginalis. Proc. Natl. Acad. Sci. USA* 89:4402-6
155. Rainey P, Santi DV. 1984. Formycin B resistance in *Leishmania. Biochem. Pharmacol.* 33:1374-77
156. Randles D, Taguchi H, Armarego WLF, Singh SV, He NG, Awasthi YC. 1993. New inhibitors of dihydropteridine reductase (human brain). *Adv. Exp. Med. Biol.* 338:127-30
157. Raymond M, Gros P, Whiteway M, Thomas DY. 1992. Functional complementation of yeast *ste6* by a mammalian multidrug resistance *mdr* gene. *Science* 256:232-34
158. Roninson IB. 1992. From amplification

to function: the case of the *MDR1* gene. *Mutat. Res.* 276:151–61
159. Rosen BP, Dey S, Dou D, Ji G, Kaur P, et al. 1992. Evolution of an iontranslocating ATPase. *Ann. NY. Acad. Sci.* 671:257–72
160. Rubio JP, Cowman AF. 1994. *Plasmodium falciparum*: the pfmdr2 protein is not overexpressed in chloroquine-resistant isolates of the malaria parasite. *Exp. Parasitol.* 79:137–47
161. Ruetz S, Gros P. 1994. A mechanism for P-glycoprotein action in multidrug resistance: are we there yet? *Trends Pharmacol. Sci.* 15:260–63
162. Ryan KA, Dasgupta S, Beverley SM. 1993. Shuttle cosmid vectors for the trypanosomatid parasite *Leishmania*. *Gene* 131:145–59
163. Samuelson J, Ayala P, Orozco E, Wirth D. 1990. Emetine-resistant mutants of *Entamoeba histolytica* overexpress mRNAs for multidrug resistance. *Mol. Biochem. Parasitol.* 38:281–90
164. Schweitzer BI, Dicker AP, Bertino JR. 1990. Dihydrofolate reductase as a therapeutic target. *FASEB J.* 4:2441–52
165. Singh AK, Liu HY, Lee ST. 1994. Atomic absorption spectrophotometric measurement of intracellular arsenite in arsenite-resistant *Leishmania*. *Mol. Biochem. Parasitol.* 66:161–64
166. Slater AFG, Cerami A. 1992. Inhibition by chloroquine of a novel haem polymerase enzyme activity in malaria trophozoites. *Nature* 355:167–69
167. Sogin ML, Gunderson JH, Elwood HJ, Alonso RA, Peattie DA. 1989. Phylogenetic meaning of the kingdom concept: an unusual ribosomal rna from *Giardia lamblia*. *Science* 243:75–77
168. Spies HSC, Steenkamp DJ. 1994. Thiols of intracellular pathogens. *Eur. J. Biochem.* 224:203–13
169. Steenkamp DJ, Spies HSC. 1994. Identification of a major low-molecular-mass thiol of the trypanosomatid *Crithidia fasciculata* as ovothiol A. *Eur. J. Biochem.* 223:43–50
170. Tanaka M, Gu HM, Bzik DJ, Li WB, Inselberg J. 1990. Dihydrofolate reductase mutations and chromosomal changes associates with pyrimethamine resistance of *Plasmodium falciparum*. *Mol. Biochem. Parasitol.* 39:127–34
171. Tanner M, Evans D. 1994. Vaccines or drugs: complementarity is crucial. *Parasitol. Today* 10:406–7
172. Ten Asbroek ALMA, Ouellette M, Borst P. 1990. Targeted insertion of the neomycin phosphotransferase gene into the tubulin gene cluster of *Trypanosoma brucei*. *Nature* 348:174–75
173. Thaitong S, Chan S-W, Songsomboon S, Wilairat P, Seesod N. 1992. Pyremethamine resistant mutations in *Plasmodium falciparum*. *Mol. Biochem. Parasitol.* 52:149–58
174. Townson SM, Boreham PFL, Upcroft P, Upcroft JA. 1994. Resistance to the nitroheterocyclic drugs. *Acta Trop.* 56:173–94
175. Triglia T, Cowman AF. 1994. Primary structure and expression of the dihydropteroate synthetase gene of *Plasmodium falciparum*. *Proc. Natl. Acad. Sci. USA* 91:7149–53
176. Upcroft JA, Upcroft P. 1993. Drug resistance and *Giardia*. *Parasitol. Today* 9:187–90
177. Upcroft P. 1994. Multiple drug resistance in the pathogenic protozoa. *Acta Trop.* 56:195–212
178. Van Es HHG, Karcz S, Chu F, Cowman AF, Vidal S, et al. 1994. Expression of the plasmodial *pfmdr1* gene in mammalian cells is associated with increased susceptibility to chloroquine. *Mol. Cell. Biol.* 14:2419–28
179. Van Es HHG, Renkema H, Aerts H, Schurr E. 1995. Enhanced lysosomal acidification leads to increased chloroquine accumulation in CHO cells expressing the *pfmdr1* gene. *Mol. Biochem. Parasitol.* In press
180. Varughese KI, Xuong NH, Kiefer PM, Matthews DA, Whiteley JM. 1994. Structural and mechanistic characteristics of dihydropteridine reductase: a member of the Tyr-(Xaa)3-Lys-containing family of reductases and dehydrogenases. *Proc. Natl. Acad. Sci. USA* 91:5582–86
181. Walsh C, Bradley M, Nadeau K. 1991. Molecular studies on trypanothione reductase, a target for antiparasitic drugs. *Trends Biochem. Sci.* 16:305–9
182. Wellems TE, Panton LJ, Gluzman IY, do Rosario VE, Gwadz RW, et al. 1990. Chloroquine resistance not linked to *mdr*-like genes in a *Plasmodium falciparum* cross. *Nature* 345:253–55
183. Wellems TE, Walker-Jonah A, Panton LJ. 1991. Genetic mapping of the chloroquine resistance locus on *Plasmodium falciparum* chromosome 7. *Proc. Natl. Acad. Sci. USA* 88:3382–86
184. Westerhof GR, Jansen G, van Emmerik N, Kathmann I, Rijksen G. 1991. Membrane of transport of natural and antifolate compounds in murine L1210 leukemia cells: role of carrier- and receptor-mediated transport systems. *Cancer Res.* 51:5507–13
185. White TC, Fase-Fowler F, van Luenen H, Calafat J, Borst P. 1988. The H

circles of *Leishmania tarentolae* are a unique amplifiable system of oligomeric DNAs associated with drug resistance. *J. Biol. Chem.* 263:16977–83

186. Wilson CW, Serrano AE, Wasley A, Bogenschutz MP, Shankar AH, Wirth DF. 1989. Amplification of a gene related to mammalian *mdr* genes in drug resistant *Plasmodium falciparum*. *Science* 244:1184–86

187. Wilson CW, Volkman SK, Thaithong S, Martin RK, Kyle DE, et al. 1993. Amplification of *pfmdr1* associated with mefloquine and halofantrine resistance in *Plasmodium falciparum* from Thailand. *Mol. Biochem. Parasitol.* 57:151–60

188. Wilson K, Berrens RL, Sifri CD, Ullman B. 1994. Amplification of the inosinate dehydrogenase gene in *Trypanosoma brucei gambiense* due to an increase in chromosome copy number. *J. Biol. Chem.* 269:28979–87

189. Wilson K, Beverley SM, Ullman B. 1992. Stable amplification of a linear extrachromosomal DNA in mycophenolic acid-resistant *Leishmania donovani*. *Mol. Biochem. Parasitol.* 55:197–206

190. Wilson K, Collart FR, Huberman E, Stringer JR, Ullman B. 1991. Amplification and molecular cloning of the IMP dehydrogenase gene of *Leishmania donovani*. *J. Biol. Chem.* 266:1665–71

191. Wu Y, Sifri CD, Lei HH, Su X, Wellems TE. 1994. Transfection of *Plasmodium falciparum* within human red blood cells. *Proc. Natl. Acad. Sci. USA* 92:973–77

192. Yarlett N, Goldberg B, Nathan HC, Garofalo J, Bacchi CJ. 1991. Differential susceptibility of *Trypanosoma brucei rhodesiense* isolates to in vitro lysis by arsenicals. *Exp. Parasitol.* 72:205–15

193. Yarlett N, Yarlet NC, Lloyd D. 1986. Ferredoxin-dependent reduction of nitroimidazole derivatives in drug-resistant and susceptible strains of *Trichomonas vaginalis*. *Biochem. Pharmacol.* 35:1703–8

194. Zalis MG, Wilson CM, Zhang Y, Wirth DF. 1993. Characterization of the *pfmdr2* gene for *Plasmodium falciparum*. *Mol. Biochem. Parasitol.* 62:83–92

195. Zaman GJR, Flens MJ, van Leusden MR, De Haas M, Mülder HS, et al. 1994. The human multidrug resistance-associated protein MRP is a plasma membrane drug-efflux pump. *Proc. Natl. Acad. Sci. USA* 91:8822–26

196. Zhang WW, Samuelson J. 1993. Molecular cloning of the gene for a novel ABC superfamily transporter of *Entamoeba histolytica*. *Mol. Biochem. Parasitol.* 62:131–34

Annu. Rev. Microbiol. 1995. 49:461–87
Copyright © 1995 by Annual Reviews Inc. All rights reserved

ENVIRONMENTAL VIROLOGY: From Detection of Virus in Sewage and Water by Isolation to Identification by Molecular Biology—A Trip of Over 50 Years

T. G. Metcalf, J. L. Melnick, and M. K. Estes

Division of Molecular Virology, Baylor College of Medicine, One Baylor Plaza, Houston, Texas 77030

KEY WORDS: sewage-associated viruses, cell cultures, nucleic acid hybridization, public health concerns, sewage-polluted waters, shellfish

CONTENTS

INTRODUCTION	462
ENVIRONMENTAL VIROLOGY AND PUBLIC HEALTH CONCERNS	462
Poliovirus: Early Studies and Virological Perspectives	462
New Frontiers	465
NUCLEIC ACID–BASED DETECTION METHODS—NEW HORIZONS	470
DETECTION OF ENVIRONMENTALLY TRANSMITTED PATHOGENS	473
Enteroviruses as Indicators of Sewage-Associated Virus Pathogens	473
Hepatitis A Virus	474
Caliciviruses	474
Hepatitis E Virus	476
Rotaviruses	477
Astroviruses	478
SUMMARY AND CONCLUSIONS	479

ABSTRACT

Environmental virology began with efforts to detect poliovirus in sewage and water more than 50 years ago. Since that time, cell-culture methods useful for detection of enteroviruses have been replaced by molecular biology techniques for detection of pathogens (hepatitis A and E viruses, caliciviruses, rotaviruses, and astroviruses) that do not grow in cell culture or grow with great difficulty.

Amplification of viral nucleic acid using the polymerase chain reaction (PCR) is the current preferred method. PCR or RT-PCR (to detect RNA viral genomes) is rapid, sensitive, specific, and quantitative. Method shortcomings include potential inhibition by substances in some environmental samples and an inability of test results to distinguish between infectious and noninfectious virus. Current questions involving use of PCR/RT-PCR tests for public health purposes include: What is the public health significance of a positive test, and should direct tests for viruses replace current public health–monitoring programs?

INTRODUCTION

Methods to detect viruses in sewage and sewage-polluted waters began slightly more than a half century ago. Early efforts prompted by public health concerns and epidemiological research on poliovirus distribution in nature represent the beginning of environmental virology.

Prospects for virus transmission in nature and questions about microbiological safety led to extension of method development to include tests of shellfish, sediments, soils, and potable water sources (treated, renovated, and ground water). Viruses sought were extended to pathogens found in feces, including hepatitis A and E, caliciviruses, rotaviruses, and astroviruses.

Methods developed reflected a transition in procedures for isolation of virus—from ones initially involving production of disease in laboratory animals, to those detecting cytopathic changes in cell cultures, to ones based on immunologic methods, and finally to methods of detecting the viral genome or nucleic acid. Detection of viruses that could not be grown in cell culture or grown only with difficulty combined with public health concerns about safety emphasized the need for new test approaches. New procedures had to be rapid, sensitive, and specific. They also had to be quantitative and preferably able to distinguish between infectious and noninfectious virus.

The advent of molecular biology led to the development of new tools and approaches for meeting current challenges. These new developments stem from our knowledge of virus structures and functions at the molecular level. This article chronicles the odyssey of research in environmental virology over the past half-century toward a new era of virus detection and monitoring that offers the potential for improving public health worldwide.

ENVIRONMENTAL VIROLOGY AND PUBLIC HEALTH CONCERNS

Poliovirus: Early Studies and Virological Perspectives

Environmental virology began over a half-century ago with the detection of poliomyelitis virus in monkeys who were administered environmental samples.

Cell culture for virus isolation was necessary for further development. A brief history is in order.

For 30 years, from the time of its discovery in 1908, poliovirus was believed by most authorities to enter the human body through the nasal route, to multiply in the olfactory bulbs, and then to proceed along nerve fibers directly to the central nervous system (12). Thus, poliomyelitis was considered a strictly neurotropic disease. Neither the spread of virus in nature nor the pathogenesis of the disease was clearly understood. In the late 1930s, investigators rediscovered a finding originally made in 1912 in Sweden: Virus could be isolated not only from the spinal cord but also from the feces (where the virus was present in large quantities over several weeks) of both patients and healthy carriers. In the 1940s, poliomyelitis began to be regarded as an enteric infection (90, 93).

This finding brought into focus the questions: If polio was an enteric infection, was it waterborne? Could it be transmitted by direct or indirect contamination from virus-laden sewage? Pursuit of answers to such questions became one of the key tasks in polio research. During 1940–1945, investigators studied the occurrence of polioviruses in the sewage of New York City, Chicago, and other cities (82). The samples were concentrated, purified, and injected into test animals, one monkey for each sample (82). Each monkey was examined daily for 4 to 5 weeks for muscle weakness, and every day its rectal temperature was taken. At the end of the observation period, the brain and spinal cord were removed and examined in thin sections under the microscope for lesions of poliomyelitis. What did we learn? Although this investigation could not prove waterborne transmission of polio, we learned that, when paralytic cases were prevalent in the community, polioviruses were present in large numbers in sewage and remained infectious in flowing sewage for many weeks. Estimating from the amount of virus excreted daily by a typical carrier, Melnick determined the ratio of inapparent infections to paralytic cases to be well over 100:1 (82).

In the late 1940s, the coxsackieviruses were discovered, the group A viruses by Dalldorf and the group B viruses by Melnick (21). Soon after the echoviruses were identified through the use of tissue cultures (93). The enteroviruses became recognized as a family (84, 86) that included the polio, coxsackie, and echoviruses. All members shared similar chemical and physical properties, as well as the same natural habitat—the human intestinal tract. Although large numbers of enteroviruses may be found in sewage and are isolated regularly from polluted waters, outbreaks attributed to enterovirus-polluted water have been very few, and little direct evidence connects the presence of enteroviruses in sewage with a causal role in the transmission of viral diseases.

However, several waterborne outbreaks of viral hepatitis have occurred. By far the worst epidemic—about 230,000 cases, or a case rate of 2/100—occurred during December 1955 and January 1956 in New Delhi. Over a one-week period six weeks before the epidemic, sewage contaminated the Jumna River, the source of water for the treatment plant (83). During the contamination period, water was coagulated with high doses of alum and disinfected with high doses of chlorine, which prevented the outbreak of bacterial and other enteric viral diseases but failed to prevent the hepatitis epidemic. The causative agent was identified years later as hepatitis E virus, probably a member of the calicivirus family (107).

The development and availability of cell cultures greatly expanded virological research and made it technically more feasible. Such research showed that the resistance of enteroviruses to chlorine is greater than that of the coliform bacteria group used to indicate the sanitary quality of water (125). Enteroviruses have high survival capabilities under various environmental conditions, including wastewater- and water-treatment processes (125). The minimum human infectious dose for enteroviruses is on the order of a single virus particle—two- to sixfold lower than the number of enteric bacteria needed to initiate human infection (108). These findings, coupled with epidemiological studies incriminating water as a vehicle for transmission of infectious hepatitis, led to increased concern about the possible spread of other viral diseases through drinking water (91, 108).

By 1965, interest in the subject of environmental virology had broadened, and the first international conference on the subject was held in Cincinnati (8). The conference, entitled "Transmission of Viruses by the Water Route," highlighted the problem and urged concerted attempts to develop adequate quantitative methods for detecting low-level virus transmission through water, to verify whether viruses can survive undetected under conditions that eliminate coliform bacteria, to determine what constitutes an infective dose of virus for humans, to study virus persistence in the deliberate reuse of wastewater, and to intensify efforts to detect the agent of infectious hepatitis.

During the 10 years following the symposium, virus-monitoring techniques allowing assay of 400-liter samples of water were developed and used in surveys of virus concentrations in water and wastewater throughout the world (132, 133). Research showed that the rate of enteroviruses shed in raw sewage ranges from 100 to 1000 plaque-forming units (PFU)/100 ml. In other studies, varying concentrations of viruses were detected in lakes and seawater used for recreation (108). Investigators rarely detected virus in drinking water that had undergone proper treatment. Many studies advanced our understanding of the virus-removal efficiency of conventional as well as other low-cost processes for water and wastewater treatment (89, 108).

This decade also witnessed several contributions to our understanding of the role played by shellfish in the transmission of viral diseases. Methods developed for detection of enteroviruses in shellfish implicated clams, oysters, and mussels in outbreaks of hepatitis A virus (111, 137). Studies showed that, in water polluted with human feces, shellfish accumulate enteric viruses, and humans can become infected by eating improperly cooked shellfish. The largest epidemic traced to shellfish occurred in Shanghai in 1988; it involved about 300,000 persons (137). Shellfish themselves do not become infected, and most importantly, they eventually free themselves of infectious virus if they are moved to unpolluted water (depurated) (112).

In 1974, another international conference on "Viruses in Water," held in Mexico City (9), reviewed the advances made during the preceding decade. Even though water transmission of enteric viruses was often difficult to trace and although a single virus-monitoring technique had not yet been selected, the time seemed right for setting standards for the permissible virus content of waters. Methods at that time could support testing to demonstrate less than 1 PFU/40 liters of recreational water and less than 1 PFU/400 liters of drinking water. Conference resolutions urged that standards should also be considered for viruses in water used for cultivating shellfish, for agricultural irrigation, and for recycling.

The World Health Organization, recognizing the growing importance of the problem, invited experts from around the world to Geneva for in-depth discussion on human viruses in water, wastewater, and soils and published a report in 1979 for use by those responsible for public health and economic planning in both developing and developed countries (141). This report included an assessment of the public health importance of viruses in water, wastewater, and soils and of the nature of risks for exposed persons. It also evaluated the methods available for monitoring viruses and identified areas for further research.

In 1982, scientists at another symposium on "Enteric Viruses in Water," held in Israel, raised the question of whether the additional effort and cost of rendering water totally virus-free is a justifiable public health expenditure (87). A related concern revolved around the difficulties of trying to set virus standards for drinking water, considering the variety of detection methods then in use. No decision was made at the conference on establishing such standards, and none exists as of this writing.

New Frontiers

A new era in environmental virology began in the 1980s with several significant developments in medical virology. These included (*a*) recognition of hepatitis E virus (HEV) as an enteric virus capable of producing waterborne epidemics; (*b*) adaptation of strains of hepatitis A virus (HAV) to replicate in cell cultures and the development of sensitive assays for HAV antigen; (*c*) development of methods for the concentration of HAV from water; (*d*) recog-

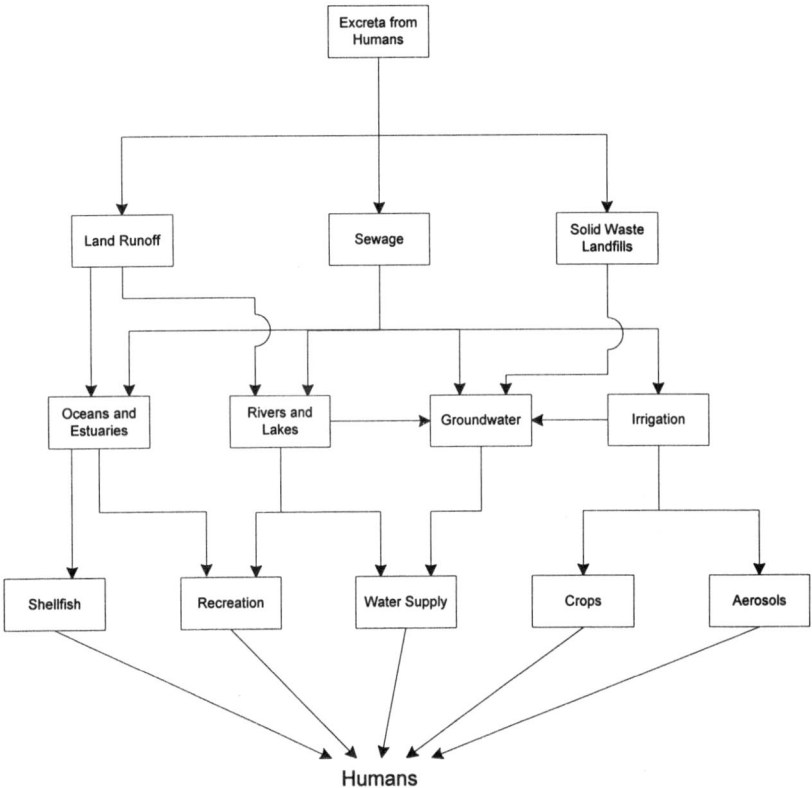

Figure 1 Routes of enteric virus transmission in the environment. (Adapted from Reference 89.)

nition of the involvement of HAV in waterborne epidemics; (*e*) seroepidemiological surveys of outbreaks of nonbacterial gastroenteritis showing causation by small, round structured viruses such as caliciviruses, including Norwalk virus; (*f*) demonstration of the involvement of human rotaviruses in infantile diarrhea; (*g*) recovery of new types of enteric adenoviruses, caliciviruses, and astroviruses from children suffering from acute gastroenteritis; and (*h*) recognition that outbreaks of hepatitis and gastroenteritis could be caused by the transmission of enteric viruses in the environment.

Increasing awareness and follow-up studies led to the recognition that many human populations around the world are exposed to enteric viruses through various routes: shellfish that grow in contaminated ocean or estuary water, food crops grown in land irrigated with wastewater or fertilized and condi-

Table 1 Human virus pathogens transmitted by sewage-polluted water

Virus[a]	Pathogenicity proven		Virus detection by	
	Sewage-polluted water	Sewage-polluted shellfish	Cell culture[d]	Nucleic acid test
Enteroviruses				
Polio	+/−	−	+	+
Coxsackie A	−	−	+	+
Coxsackie B	−	−	+	+
Echo	−	−	+	+
Entero 68–71	−	−	+	+
Hepatitis A[b]	+	+	+/−	+
Caliciviruses				
Human calicivirus, Sapporo[c]	+	+	−	+
Norwalk[c]	+	+	−	+
Norwalk-like[c]	+	+	−	+
Hepatitis E[b]	+	−	−	+
Rotaviruses[c]	+	−	+/−	+
Enteric adenoviruses	−	−	+	+
Astroviruses[c]	+	+	+/−	+
Coronaviruses	−	−	+/−	+

[a] All these viruses contain RNA genomes except enteric adenoviruses, which contain a dsDNA genome. The RNA genome of rotavirus is dsRNA, while the genomes of the other viruses are ssRNA.
[b] Causes hepatitis.
[c] Causes acute gastroenteritis.
[d] Enteroviruses recoverable in primary and continuous-type cultures of primate or human origin. Hepatitis A virus can sometimes be recovered in special cultures of primate or human origin. Coronaviruses are sometimes recoverable in tracheal organ cultures. Enteric adenoviruses are recoverable in Graham 293 cultures. Some rotaviruses and astroviruses can be recovered in cultures of primate or human origin with the help of tryptic enzymes.

tioned with sludge, recreational waters, and even drinking water polluted with viruses. Figure 1 illustrates enteric virus–transmission routes, while Table 1 lists human viral pathogens, together with their transmission potential and method of detection. In addition to the previously mentioned pathogens (e.g. HAV, HEV, calicivirus, and astrovirus), enteric adenoviruses are causally associated with diarrheal illness in children, but water or food transmission routes have not been described. Coronaviruses are also found in diarrheal disease feces, but no causal relationship has been unequivocally established (33, 47, 76).

What circumstances lead to clinical recognition of water-associated outbreaks of disease? The overwhelming majority of reported waterborne outbreaks of hepatitis A and E, Norwalk virus, and rotaviruses can be attributed to untreated or inadequately treated water supplies. In some cases, sewage contamination of a water source that had inadequate or no treatment has caused the dilemma. With large municipal water supplies, the source of the problem can usually be traced to cross-connections or back-siphonage during distribution. These deficiencies have straightforward solutions.

On the other hand, some viral infections can spread insidiously by way of continual low-level transmission through water. The nature of most enteric virus diseases is such that they elude epidemiologists in their quest for definitive data. Many enteroviruses can cause inapparent or silent infections that go unrecognized until secondary person-to-person spread finally leads to overt disease in scattered, virtually untraceable pockets of the population (141, 144). Inadequacies in current water-treatment techniques have been demonstrated on several occasions when viruses were isolated from potable water supplies that had been conventionally treated to remove bacterial pathogens. A question remains concerning the degree to which small amounts of virus that manage to get into drinking water do indeed become foci of infection in a community.

Suspended solids-associated virus in raw sewage is partially transferred into sludge during sewage treatment. Virus concentrations of 5000 to 28,000 PFU/liter are commonly found in raw sludge but are greatly reduced during treatment. An average of about 50 PFU/liter can be expected in effluents from waste-treatment plants. Solids-associated virus in wastewater effluents discharged into an aquatic environment settles through water columns onto bottom sediments, where the concentrations may be 10 to 10,000 times greater than those found in water (87). Sediments represent a reservoir from which viruses can spread when the sediments are disturbed and resuspended in the water column. In estuarine waters, resuspension commonly occurs as a result of, for example, storms, dredging, and tides. Resuspended solids-associated virus can be transported from polluted to unpolluted recreational or shellfish waters and then pose potential hazards (85). Ingestion of virus-polluted water by swimmers or bathers or consumption of virus-polluted shellfish may result in viral disease, as already mentioned (35).

The viral quality of recreational and shellfish waters cannot be determined adequately by reference to fecal coliform indexes. In the 1970s and 1980s, virus pathogens were found repeatedly in such indicator-approved waters (91). Furthermore, numbers of viruses and fecal coliforms in water, shellfish, or sediments do not seem to share any correlation.

Direct application of wastewater or sludge to the land may lead to viral contamination of crops and groundwater. Application of wastewater by sprinkler irrigation introduces a risk of virus aerosols to exposed persons. Treatment of sludge by drying, pasteurization, anaerobic digestion, and composting reduces but does not eliminate enteroviruses, which have survived 30 days of digestion at 50°C. Hepatitis A virus, which is even more thermoresistant, can survive for longer periods of time.

The rate of application of wastewater and soil composition and structure can influence virus contamination of groundwater. Other influential factors include pH level, organic content, and ionic strength of the effluent. Although normal rates of water application for agricultural irrigation are about 1 m^3 water/m^2 land per year, hydraulic loading rates of effluent disposal on land can be as high as 100 m^3 effluent/m^2 land per year. As expected, virus removal levels decline with higher loading rates. Viruses are readily adsorbed onto clays, so the higher the clay content of the soil, the greater the removal. Sandy loams and soils containing organic matter are also favorable for virus removal. Soils containing sand or sand-and-gravel mixtures do not achieve good removal, and fissured limestone aquifers under shallow soil allow virus transport over greater distances and aid in groundwater pollution.

Movement of viruses within soils has also been studied. After application of chlorinated secondary effluent by a sprinkler irrigation system to a sandy soil containing little or no silt or clay, polioviruses and echoviruses were detected in drains well below the surface, demonstrating that viruses survive aeration and sunlight during spraying as well as percolation through 1.5 m of soil (32). Although at first no viruses were detected in wells 3 and 6 m below the surface, they were detected after heavy rainfall because of desorption and migration through the soil. Fecal coliform bacteria were not detected in the water containing viruses (136).

Low pH favors adsorption, whereas high pH results in elution. High concentrations of soluble organic matter in wastewater compete with viruses for adsorption sites on soil particles, resulting in decreased virus adsorption, or even liberation of previously adsorbed viruses. High concentrations of cations tend to enhance virus retention. Viruses retained near soil surfaces may be eluted and washed down to lower strata by heavy rainfall (131).

When wells are located in the vicinity of wastewater irrigation or land-disposal sites, they should not be used as a source of drinking water, pending regular tests of water quality. As factors influencing virus movement in soil are not yet fully elucidated, and effluent and soil conditions vary considerably, careful study of local conditions is required. Reasonable safety measures should include drilling wells suitable distances from likely sources of contamination and viral monitoring of water quality.

Other than sewage and polluted water, where else can viruses transmitted by the fecal-oral route be found in the environment? Sludge and soil, as implicated above, are often sources (85). Flies can also carry such viruses and deposit them on food (88, 135). Recently, however, transmission by this route has received little attention, perhaps because the evidence indicates that virus persists but does not multiply in flies (92).

As the methodology for virus detection improves, the accuracy of the data generated on the amount of virus in the environment and evaluated for appropriate control measures will increase.

NUCLEIC ACID–BASED DETECTION METHODS—NEW HORIZONS

Previously unrecognizable viruses that replicate poorly or not at all in cell cultures became subject to detection with the development of nucleic acid–based methods. A number of difficulties hampered attempts to detect these human pathogens (HAV, HEV, caliciviruses, astroviruses) in samples from environmental sources. First, virus quantities generally are low (94). Second, human caliciviruses and HEV cannot be grown, and wild-type strains of HAV grow slowly if at all, in cell culture. None of these cultured viruses exhibit visible cytopathic effects (10, 106). Growth of astroviruses in cell culture can be slow and uncertain, which also contributes to detection problems (71), although growth of astroviruses in new cell lines has partially overcome this difficulty (142). Radioimmunoassay, enzyme-linked immunosorbent assay, or immuno-electron microscopic methods generally lack the degree of sensitivity required to detect the low quantities of virus expected in samples from environmental sources. An exception is an outbreak of type A hepatitis in Georgetown, Texas, where radioimmunoassays revealed the presence of HAV antigen in heavily contaminated drinking water (43). Initially, investigators used hybridization assays with cDNA or ssRNA probes, but these methods had limited sensitivity and were replaced with amplification [polymerase chain reaction (PCR) for DNA viruses or reverse transcription followed by PCR (RT-PCR) for RNA viruses], which had the best test sensitivity (16, 52, 115–117, 122).

PCR techniques have been applied to the detection of enteric viruses in widely different environmental samples. The degree of detection effectiveness achieved, however, varies among the types of samples examined. The variations are the result of two related factors, the effectiveness of recovery of virus from samples and the degree of final purity of recovered virus. Taken together, these two factors influence the amount and purity of virus nucleic acid recoverable from a sample and determine whether PCR-inhibiting substances that lower test sensitivity are present in the final sample.

Environmental water samples usually are collected in large volumes to increase the chance for detecting expected low quantities of virus. Samples of 400 liters or more may be processed through special charge-modified, virus-retaining filters from which viruses are recovered through elution (132). Eluates of 1 or more liters usually are reduced to final test-volume samples of ≤10 ml, representing a 40,000-fold or greater concentration (133). With such high concentrations, humic compounds (organic compounds of natural origin), which are coconcentrated along with viruses in environmental waters, may cause PCR interference. Sephadex or chelex resin can help remove inhibitory substances from treated samples (1).

The polysaccharide content of shellfish leads to PCR inhibition. Test samples negative upon original testing will become positive after dilution, and after removal of polysaccharides, the samples will be positive without dilution. Because the inhibitory activity persists regardless of the processing method used, treatment to remove inhibitory substances is a constant for all methods currently in use. Cetyltriammonium bromide (CTAB)–based removal of polysaccharides has been an effective treatment for eliminating test inhibition (6, 146). An acid-alcohol polymer, Pro-cipitate, has also been used to remove shellfish-associated inhibitory activity (54). Spin columns, vortex flow filtration, and microconcentration procedures in connection with the above treatments for water and shellfish samples have been used effectively with wastewater and aquatic samples (103, 128).

The PCR-inhibitory activity of some environmental samples has been eliminated through immunologic-based methods that depend on the specificity and affinity of a given antibody for a homologous antigen on a virus surface. Either polyclonal or monoclonal antibody can be used, but the ready availability and unlimited supply of monoclonal antibodies makes these the preferred reagents for specific capture of homologous virus antigen from a sample, providing sufficient affinity is obtained. PCR tests with captured antigen/virus (AC/PCR) avoid competition and eliminate inhibitory substances from clinical samples (53). Judging from detection of virus in sewage (37) and shellfish (24), the technique has a satisfactory degree of sensitivity and specificity with environmental samples. Other procedures analogous to AC/PCR include two versions of an immunomagnetic technique. One version uses antibody-coated magnetic beads to capture virus (97); the second uses biotinylated virus-specific oligonucleotide to capture extracted virus RNA by means of hybridization to virus-specific genomic sequences (98). PCR tests are subsequently performed on the magnetically captured virus or virus RNA. These techniques have the advantages of simplicity and rapid execution, but their effectiveness must still be confirmed.

Alternative PCR formats for enhancing test sensitivity and specificity have been applied to environmental samples. A method called nested PCR uses a second set of internal primers and a second round of amplification. In clinical samples, nested PCR reportedly has sensitivities of close to a single molecule of viral nucleic acid (65, 120). The key to the effectiveness of this method is the careful selection of internal primer sequences closely adjacent upstream and downstream to the virus target genome sequence, and extreme caution must be exercised to ensure that detected sequences do not represent false positives due to contamination. Nested PCR tests have successfully detected enteric viruses in sewage, water, sediment, and shellfish samples (69, 72). The combination of AC/PCR and nested PCR tests is a powerful tool for detecting minimal quantities of enteric viruses in environmental samples. Another test format, called multiplex PCR, has been used for simultaneous detection of up to three virus pathogens in sewage and marine waters (129).

Because of the high sensitivity of PCR methods, a critical evaluation during the development and use of these assays for environmental samples is important. Several issues must be considered as new methods are developed and reported. First, adequate controls must be run to rule out the possibility that any positive result is not due to laboratory contamination and that any negative result is not due to the presence of inhibitory substances in a sample. The problem of contamination of samples cannot be overemphasized: Almost all investigators working with PCR experience this problem, although none expect to initially. One must also evaluate test sensitivity by determining the minimal amount of virus that can be detected in samples. This value should be ascertained by adding dilutions of virus to a sample, processing each sample containing the increasingly smaller amounts of virus, and then performing the final assays and reporting the results. These results should be compared with test sensitivity data obtained by testing nucleic acid directly extracted from diluted virus not added to an environmental sample. These latter comparisons allow one to estimate the efficiency of recovery of lower amounts of virus, such as may be present in environmental samples. Dilutions of final nucleic acid samples also should be tested to monitor for the presence of residual inhibitors. Unfortunately, many of the alternative PCR assay formats have not been rigorously evaluated, so test-performance sensitivity is not precisely known.

With the incorporation of internal standards, quantitative PCR results can be used to assay field samples. A carefully selected or designed nucleic acid sequence is coamplified along with the virus nucleic acid target. Coamplification, in the same reaction tube at the same time using the same primers, provides a basis for quantitation of a PCR test (19, 34, 80). Testing the amplification of an internal control also allows one to identify inhibition of specific target amplification (false negatives).

Interpretation of the public health significance of positive nucleic acid–based tests is a crucial issue involving consumer protection from water- or food-transmitted pathogens. The question that arises is, does a positive result equate with a public health threat? Because any nucleic acid–based test is an expression of total virus with no distinction between infectious and noninfectious qualities, a positive result carries no absolute proof of public health danger. Cell culture amplification prior to probe or PCR testing has been introduced to overcome the problem (116, 121). Amplification of cultivable virus in a sample generates enough virus progeny as a source of nucleic acid to enable ready detection by the nucleic acid–based test. When virus replication does not result in cytopathology, the technique can enhance the likelihood of virus detection or even serve as a means of detection itself, provided the time required for cell culture amplification and subsequent detection of virus is of no consequence. However, the method is not applicable to detection of non-cultivable viruses such as Norwalk or Norwalk-like pathogens.

Results of studies seeking to determine a relationship between infectious, cultivable virus assays (PFU) and nucleic acid–based tests (PCR) are equivocal. Loss of infectious quality is not always accompanied by loss of PCR activity, but loss of PCR activity means a loss of all activity (29, 63).

A novel application of nucleic acid–based tests to issues in environmental virology involves detection of virus in tissue depots of edible shellfish (14, 124). The technique, in situ transcription, has been used to trace the distribution of HAV in shellfish tissues following its bioaccumulation from seawater (114). Individual HAV-carrying cells located in the stomach wall and digestive diverticula (hepatopancreas) were identified. This technique should be useful for evaluation of depuration effectiveness in shellfish purging bioaccumulated virus.

Improvements in PCR-assay methods amenable to environmental laboratory testing include the ability to obtain results without the need for radioisotopes. For detection of PCR-amplification products, current methods involve visual detection of an appropriately sized DNA band on an agarose gel followed by transfer and confirmation of the product's specificity by hybridization with internal probes. In some cases, the PCR products are simply immobilized (dotted) on a membrane and probed directly. Probes labeled with the nonradioactive digoxigenin or with a fluorescent substance are increasingly being used to detect nucleic acids. These methods are essential for environmental testing. New parameters developed for clinical specimens must be carefully evaluated to be sure that the environmental sample does not contain substances that interfere or mimic (autofluoresce) these theoretically useful probes. Again, test sensitivity must be examined.

DETECTION OF ENVIRONMENTALLY TRANSMITTED PATHOGENS

Enteroviruses as Indicators of Sewage-Associated Virus Pathogens

Cross-hybridization studies showing the existence and extent of genome homology among the enteroviruses, followed by molecular sequencing to identify the sequence basis of this homology, formed the foundation of nucleic acid tests for enteroviruses as potential indicators of sewage-associated pollution and the possible presence of virus pathogens. Current practices emphasize identification of any member of the group rather than a specific enterovirus. The tests use pan-enterovirus probes or virus nucleic acid target sequences representing highly conserved 5′ noncoding regions of broad homology (23, 79, 118, 119). Judging from a comparison of enterovirus vs HAV detection in samples from natural sources, the value of enterovirus detection in predicting an existing potential public health threat may be no better than that of fecal coliform indexes. Little consistent correlation between enterovirus and HAV occurrence has been reported. As examples, river water samples negative for enterovirus were positive for HAV (28), and conversely, river water and tap water samples positive for enterovirus were negative for HAV (121). In shellfish and sediment samples from polluted coastal waters, enterovirus and HAV occurrence showed some correlation in the most polluted areas. However, the levels of enterovirus and HAV diverged with decreasing levels of pollution, with either enterovirus or HAV eventually found in the absence of the other (72, 73).

Hepatitis A Virus

Application of molecular hybridization to the detection of HAV in environmental samples initially featured the use of ^{32}P-labeled cDNA probes (57). The specter of false positive results raised by the possible presence of contaminating bacterial plasmid DNA sequences (pBR322 vector) in probes or in sewage-polluted samples threatened to compromise the use of such probes (2). In vitro transcription techniques that generated cRNA (hybridizes with HAV RNA) and vRNA (no reaction with HAV RNA) single-stranded probes solved the vector-contamination problem and simultaneously provided an internal control. The control enabled distinction between true and false-positive results (56). Probe specificity problems were resolved, but the sensitivity was insufficient to detect the minimal quantities of HAV expected in water and shellfish samples. Clearly, sensitivity needed to be improved. The problem was solved by the introduction of RT-PCR tests for detection of HAV based on the sequencing of HAV genomes (18, 74, 113). Highly sensitive and specific tests resulted from the selection of primers focused on HAV target sequences from areas in the 5' noncoding region and the regions coding for the VP3-VP1 capsid protein or the 3D RNA polymerase that shared high sequence homology among all strains of HAV. Natural strains of HAV share a high degree of nucleotide identity and can be expected to react specifically with these target sequences.

RT-PCR-based detection of HAV in polluted natural sources such as sewage, surface- and groundwaters, sediments, and shellfish testify to the credibility of PCR techniques as effective tools for public health monitoring (26, 28, 105, 128). This method has detected virus in shellfish and water samples associated causally with outbreaks of disease (13, 25, 145).

Caliciviruses

Human caliciviruses, of which Norwalk virus is the prototype, are increasingly being associated with water- and shellfish-transmitted gastroenteritis (17, 20, 40, 42, 99, 102, 127). Norwalk virus was originally detected using Norwalk virus antigen and acute and convalescent sera from volunteers (39, 66, 67) in a radioimmunoassay blocking antibody tests and subsequently by ELISA. At first only a few research laboratories could utilize these methods, but this limitation has been overcome since the cloning and expression of the Norwalk virus capsid protein (58–60). This accomplishment and the subsequent sequence analysis and expression of the viral capsid protein have resulted in the development of new diagnostic assays to detect viral antigen, antibody to the virus, and the viral genome (38, 55, 59, 60). Sequence analysis of the cDNAs representing the viral genome showed that Norwalk virus is a calicivirus (61). Expression of the capsid protein using the baculovirus-expression system revealed that the capsid protein self-assembles into virus-like particles, which are produced in high yields (20 mg from 9×10^8 cells) (60). These recombinant Norwalk virus (rNV) particles have been used for ELISA-based antibody assays that are sensitive and able to detect antibody responses in adults given Norwalk virus as well as other Norwalk-like viruses such as the Snow Mountain Agent and Hawaii Agent (126). In this assay, responses in adult volunteers are generally highest to the homologous virus, but broadly reactive serum antibodies can be detected. As almost all adults have preexisting antibody, whether this assay using rNV particles will detect antibodies to heterologous viruses following a primary infection in adults or children remains unclear.

The rNV particles also have been used to produce the first hyperimmune antiserum, and these sera have been used to establish an antigen ELISA (38, 55). The rNV antigen ELISA is very sensitive and also type-specific. Thus, the rNV antigen ELISA does not detect the Snow Mountain or Hawaii agents (38), and it has not detected virus in some samples shown to be positive by the antigen ELISA with human volunteer sera (55). These and other results demonstrate that humans make antibodies to antigens other than the Norwalk virus capsid, and at least one of these other epitopes is present in the nonstructural protein predicted to be similar to the picornavirus 3C protease (61, 81).

To circumvent the problem of not having a broadly reactive antigen ELISA, RT-PCR methods to detect viral nucleic acid were developed, and when the appropriate primer pairs are used, this assay can detect a broad range of caliciviruses (55). Amplification and sequence analysis of more than 50 types of small, round structured viruses have so far confirmed that all are caliciviruses, and genetic analyses have now divided them into at least three genogroups (30). The first genogroup is represented by Norwalk virus; the second genogroup includes both Snow Mountain and the Hawaii agents; and the third genogroup includes the Sapporo strain of human calicivirus. A direct comparison of the sensitivity of the ELISA and RT-PCR for virus detection in the stools of volunteers excreting virus was surprising in that it showed that these two assays are equally sensitive, rather than that RT-PCR was more sensitive (59). This finding apparently reflects the fact that such stools contain large amounts of soluble antigen, which was recently shown to be a cleavage product of soluble virus capsid protein (41).

For virus detection in the environment, RT-PCR is currently the method of choice because the antigen ELISAs are too specific. Use of this assay has identified the virus in numerous outbreaks, and laboratory methods of detecting virus in shellfish and water have been reported (6, 22). These assays have not yet been successfully applied to the detection of virus in environmental samples associated with an outbreak, but such success should come in the future. Tests are still underway to determine which primers would be best for the detection of Norwalk virus and other human caliciviruses. Numerous strains were detected with primers chosen from the genome region that codes for the viral RNA polymerase, but subsequently accumulated sequences have shown that this region is not as well conserved as expected. Other regions of the genome that code for nonstructural proteins are being tested, and some success has been attained by using primers from the region encoding the 2C protease (100, 134). The 5' end of the genome does not appear to be conserved well. Thus, because a primer that can detect all human calicivirus strains remains to be found, development of broadly reactive antigen assays would be useful so that antigen-capture methods such as those developed for HAV detection could be modified for detection of the caliciviruses. The three-dimensional structure of the rNV particles was recently solved. The single capsid protein composing the particles folds into a structure with a shell domain, and arch-like domains protrude from the surface of the particle. This basic information is helping to predict the location of cross-reactive epitopes for the development of broadly reactive assays that can exploit the advantages of antigen capture PCR (104; JM Ball, ME Hardy, LJ White & MK Estes, unpublished data).

Without the ability to cultivate Norwalk virus and other human caliciviruses, we cannot judge the precise sensitivity of any assay method. However, by testing dilutions of stool from volunteers who shed Norwalk virus, we can make some estimates of test assay sensitivity. The stool samples from adult volunteers were estimated to contain approximately 10^5–10^6 virus particles because particles could not be observed directly by electron microscopy (EM). This assumption indicates that the RT-PCR assay for Norwalk virus can detect between 1 and 10 virus particles in stool and between 9 and 90 virus particles in shellfish (5, 6, 59). Other RT-PCR assays for Norwalk virus have not estimated test sensitivity (3, 22, 96, 101, 140).

Hepatitis E Virus

Detection of hepatitis E virus (HEV) first depended upon serologic tests in which HAV and hepatitis B virus (HBV) were excluded [i.e. absence of IgM anti-HAV, HBSAg, IgM anti-HBc, and immunoelectron microscopy (IEM)] (46). New tools from molecular biology studies of the HEV genome now make it possible to detect the virus directly through nucleic acid tests. The complete or nearly so HEV genomes of three strains have been cloned and sequenced. A Burmese isolate (110, 123), a Pakistani isolate (130), and a Mexican isolate (49) are considered closely related at the molecular level. The HEV genome organization (5'- NS-S-3') consists of nonstructural and structural segments, with conserved 5'- and 3'-terminal sequences in all three strains (85% identical). As in the enterovirus detection strategy, nucleotide sequences of broad genomic homology from highly conserved regions of the HEV genome will be useful for developing RT-PCR tests for detection of geographically diverse HEV.

A nested PCR using oligonucleotide primers corresponding to the nucleotide sequence for a presumed polymerase gene allowed the direct detection of HEV in stools of hepatitis patients (109). Six of ten stool samples were positive in tests in which the authenticity of a 239-bp PCR product was confirmed by Southern blot hybridization with a ^{32}P-labeled probe. Direct RT-PCR detection of HEV in raw and treated wastewater has also been reported (64). These results are encouraging, but further data are needed on test sensitivity, and detection capabilities need to be confirmed under varying environmental conditions.

Rotaviruses

Rotaviruses were first detected in 1973 with EM (11), but immunologic and nucleic acid–based procedures are now used. Recently, assays using monoclonal antibodies to detect viral antigen and to serotype viruses directly in stools were developed. These assays generally detect different viral proteins; the most sensitive assay detects common antigens present on the extremely stable protein VP6, which forms the intermediate capsid layer of the rotavirus structure. To detect VP6, samples are treated with chelating agents that are effective in removing the two outer capsid proteins VP4 and VP7. Assays for virus serotypes detect either VP7 or VP4, as both of these contain neutralization antigens. Currently, however, most such assays detect VP7 because well-characterized VP7-specific monoclonals are available, and this outer capsid protein generally remains associated with particles and available for detection if samples are kept in buffer containing calcium chloride. Assays for VP4 have been difficult to develop because this dimeric protein, which makes up the 60 spikes on particles, is present in fewer molecules per virion (120 molecules of VP4 vs 780 for VP6 and VP7), and VP4 often is lost from particles. The advantage of being able to type rotaviruses, in addition to simply detecting them, is that this information is useful in molecular epidemiology studies monitoring the transmission of a particular virus strain.

Methods to detect rotavirus nucleic acid have also been developed and include dot blot hybridization, electrophoresis on polyacrylamide gels, and

RT-PCR. Adaptations of the latter method allow detection of specific virus types based on the genes that code for VP4 and VP7. Consequently, RT-PCR has been useful in typing samples in which the virus is present but the outer capsid proteins have degraded (31, 36). Although RT-PCR is useful for the detection of rotavirus nucleic acids, because denaturation of rotavirus dsRNA is not completely efficient, RT-PCR for rotavirus has less sensitivity than RT-PCR for the detection of the ssRNA genomes of picornaviruses or caliciviruses. Thus, RT-PCR assays can be used to detect a dsRNA isolated from approximately 10–1000 rotavirus particles, depending on the strain (31), in stark contrast to detecting 1–10 copies of calicivirus or picornavirus ssRNA.

Numerous methods for the detection of rotaviruses in environmental samples such as water and shellfish have been developed in the laboratory. However, rotavirus has been detected in food and water only a few times in association with community outbreaks of diarrheal illness (48). A novel rotavirus primarily affecting adults was responsible for a severe form of waterborne gastroenteritis in China in 1982–1983 (51). The surprising fact that rotaviruses have not been associated with shellfish-transmitted illness may reflect some property of the stability of these viruses. Perhaps rotaviruses are taken up by shellfish but do not remain infectious, possibly because the spikes are removed.

Astroviruses

Human astrovirus, which was first described in 1975 (4), is another example of a gastroenteritis-causing pathogen that can be transmitted by fecally polluted water and shellfish, but it is usually transmitted by close personal contact (27, 45, 68, 75, 77, 143). EM detection of astroviruses was replaced partially by methods involving the adaptation of virus to grow with the help of trypsin treatment in human embryonic kidney cultures (71), as well as in CaCo-2 cultures (142). Five serotypes sharing a common group antigen have been identified by enzyme immunoassay (70). Monoclonal antibody prepared against this group antigen forms the basis of an ELISA developed for detection of astrovirus in stool samples (44). Monoclonal antibody is used to capture astrovirus (antigen), and a group-reactive antibody against astrovirus type 2 serves as detector antibody (45). The ELISA is about as sensitive as IEM, which means 10^3 to 10^4 virions/ml are needed for detection. Specificity is reported at about 98%. A modification of this ELISA, a biotin-avidin enzyme immunoassay, is considered an efficient, sensitive, and specific method for routine screening of large numbers of fecal samples (95). The assay can detect 5–15 ng of viral protein. The value of a plaque assay reported for serotypes 1, 2, and 5 for detection of natural virus in feces or feces-polluted samples remains untested (50).

Application of molecular biology approaches to studies of the astrovirus genome led to cloning and sequencing of the 3' terminus of serotype 1, the most commonly found astrovirus (78, 138). A dot hybridization test was developed based on the use of a cDNA probe from the 3′ terminus (sequence of approximately 1000 base pairs) and a second cDNA probe from the internal region (sequence of about 900 base pairs) of the genome (139). A mixture of the two probes maximized cross-serotype reactivity and resulted in greater sensitivity. The probe mixture, when used to detect astrovirus in a panel of stool samples previously classified by means of EM, was more sensitive than EM, and hence the results indicated a higher prevalence of astrovirus in diarrheal stools than that determined by EM. Accordingly, the true incidence of astrovirus-caused gastroenteritis is probably greater than previously reported.

RT-PCR tests for detection of astroviruses focus on 3′ terminus target sequences. A nested RT-PCR of serotype 1 specificity correctly detected 10 serotype 1 positive samples, while failing to react with any of 25 samples of other astrovirus serotypes (62).

Selection of a target sequence and primers of broad serotype reactivity would be more advantageous for detection of all serotypes that might be present in environmental samples. A dependence upon serotype 1–specific tests can prejudice detection of gastroenteritis caused by other astroviruses.

SUMMARY AND CONCLUSIONS

Over the past 50-plus years, starting with the finding of poliovirus in stools of poliomyelitis patients, more than 100 enteric viruses have been identified in human feces and details of their environmental-transmission potential unraveled. Although all these viruses represent a pathogenic potential under certain conditions, the vast majority, at the concentrations usually found in sewage, seem incapable of causing environmentally transmitted illness. The relative few that can include hepatitis A virus, caliciviruses, hepatitis E virus, rotaviruses, and astroviruses. Transmission via sewage-polluted water, sewage, well- or groundwater, shellfish, and sewage- or sludge-amended soils may occur.

Detection of these pathogens has undergone a transition from cell culture to contemporary molecular biology methods because these viruses are either noncultivable in cell culture or poorly so. RT-PCR or PCR techniques have become the sine qua non of modern virus-detection methods. Furthermore, these new molecular methods to directly detect viruses in environmental samples are showing good sensitivity and specificity when performed and analyzed with appropriate controls. These methods have also revealed the presence of inhibitors in environmental samples that can obscure detection of viral nucleic acid and have resulted in the development of additional techniques to reduce the presence of these inhibitors in final test samples (e.g. CTAB to precipitate nucleic acids to separate them from polysaccharide inhibitors). The presence of inhibitors of RT-PCR and the problems they cause in the development and use of sensitive assays to detect pathogens in shellfish are now widely accepted and recognized.

As these new methods are validated and successfully used by laboratories, we must next ask whether these methods are ready to be utilized by regulatory agencies. Several important questions about the reproducibility and reliability of these methods remain to be answered. The following discussion outlines some issues related to the use of molecular methods to detect viral nucleic acid in environmental samples by regulatory agencies:

First, will any method be sensitive enough to detect virus capable of causing disease in environmental samples? A definitive answer to this question will require knowing whether virus detected in any environmental sample is infectious, what is the minimal infectious dose of virus, and how much of the sample (water, shellfish, soils, sediments) needs to be tested to detect contamination that represents a public health hazard. The presently available results suggest that the sensitivity of virus detection is adequate, but this conclusion is currently difficult to prove. In the absence of cultivatable virus, absolute amounts of infectious virus cannot be determined. For example, the question of whether detected Norwalk virus nucleic acid represents infectious virus is an important issue that will not be overcome until the virus can be cultivated. Answers to these questions also can be approached by studying shellfish or water associated with outbreaks of disease. However, to do so effectively, we may need to establish an organized method for obtaining samples, with possible coordination with the state or federal laboratories that monitor outbreaks of gastroenteritis.

Second, will the new molecular methods be restricted to laboratories possessing highly skilled personnel and specialized equipment? A recent collaborative study demonstrated that three federal laboratories in the US could use a molecular method (RT-PCR) to detect Norwalk virus in shellfish. Investigators in each of these federal laboratories felt that the method could be performed in US state laboratories, providing qualified personnel received sufficient training. Personnel trained in molecular biology techniques will be needed so that they can solve problems when they arise. Ultimately, test kits with clear instructions and trouble-shooting information should be available.

Third, should sensitive virus assays be used to determine the pathogen-free status of environmental (water or shellfish) samples? Several other issues are related to this question: (*a*) Should direct tests for water- and shellfish-transmitted pathogens in estuarine environments replace current fecal coliform surveys or public health monitoring programs? That is, should direct tests be used for HAV, Norwalk virus, and other human caliciviruses? Are assays needed to detect astroviruses? (*b*) Should an enterovirus indicator of enteric virus pathogens (assays based on poliovirus or pan-enterovirus genome detection) be used in place of direct tests for HAV, Norwalk virus, and other human caliciviruses or any other virus pathogens? (c) Should shellfish alone be tested as the best and most appropriate target of public health surveys, or should water or surrounding sediment be tested as well as, or instead of, shellfish? (d) How frequently should such tests be performed to assess the public health safety of shellfish beds? (e) Will use of these assays be economically feasible? (f) Would these assays be useful to validate the virus depletion effectiveness of depuration processes?

Current regulations use fecal coliforms to assess shellfish-bed safety. Although the fecal coliform standard is clearly not perfect, it has been reasonably effective in reducing shellfish-associated disease. The major limitation of fecal coliforms as indicators is that they may not accurately reflect viral contamination of shellfish. Additional work using direct methods to detect viral pathogens will be needed to validate their utility. In one method to detect Norwalk virus in shellfish, usable by many laboratories, test results could be obtained within three days (5); test specificity was 100% (no false-positive results); and test sensitivity was 79%. Although the number of samples tested was relatively low, the test sensitivity obtained in this study is similar to that of many assays currently used in clinical laboratories. Therefore, these results are encouraging in that they indicate that new methods could be used to directly detect viral pathogens, at least in shellfish.

Current data indicate that any positive results obtained with such tests would be useful for predicting a public health hazard from shellfish. However, we currently do not know the public health risk of consuming shellfish that contain Norwalk virus nucleic acid detected by RT-PCR because we do not have a way to determine whether this nucleic acid represents infectious virus. Based on the current data, a recommendation for use of this method by state laboratories to regulate shellfish beds seems premature, unless the method is used with full knowledge of its limitations. However, specialized or reference laboratories could provide answers to the questions posed above by comparing results from this Norwalk virus assay with other standard and new indicator tests in field studies.

Evaluations of the magnitude of environmental viral contamination are now within the realm of environmental virology. As we understand the true significance of viruses, appropriate control measures can be introduced to reduce their public health hazards.

Any *Annual Review* chapter, as well as any article cited in an *Annual Review* chapter, may be purchased from the Annual Reviews Preprints and Reprints service. 1-800-347-8007; 415-259-5017; email: arpr@class.org

Literature Cited

1. Abbaszadegan M, Huber MS, Gerba CP, Pepper IL. 1993. Detection of enteroviruses in groundwater with the polymerase chain reaction. *Appl. Environ. Microbiol.* 59:1318–24
2. Ambinder RF, Charache P, Staal S, Wright P, Forman M, et al. 1986. The vector homology problem in diagnostic nucleic acid hybridization of clinical specimens. *J. Clin. Microbiol.* 24:16–20
3. Ando T, Mulders MN, Lewis CD, Estes MK, Monroe SS, Glass RI. 1994. Comparison of the polymerase region of small round structured virus strain previously classified in three antigenic types by solid-phase immune electron microscopy. *Arch. Virol.* 135:217–26
4. Appleton H, Higgins PG. 1975. Letter: viruses and gastroenteritis in infants. *Lancet* 1:1297
5. Atmar RL, Metcalf TG, Neill FH, Estes MK. 1993. Detection of enteric viruses in oysters by using the polymerase chain reaction. *Appl. Environ. Microbiol.* 59:631–35
6. Atmar RL, Neill FH, Romalde JL, Le Guyader F, Woodley CM, et al. 1995. Detection of Norwalk virus and hepatitis A virus in shellfish tissues using the polymerase chain reaction. *Appl. Environ. Microbiol.* In press
7. Deleted in proof
8. Berg G. 1967. *Transmission of Viruses by the Water Route.* New York: Interscience. 484 pp.
9. Berg G, Bodily HL, Lennette EH, Melnick JL, Metcalf TG, eds. 1976. *Viruses in Water.* Washington, DC: Am. Publ. Health Assoc. 256 pp.
10. Binn LN, Lemon SM, Marchwicki RH, Redfield RR, Gates NL, Bancroft WH. 1984. Primary isolation and serial passage of hepatitis A strains in primate cell cultures. *J. Clin. Microbiol.* 20:28–33
11. Bishop RF, Davidson GP, Holmes IH, Ruck BJ. 1973. Virus particles in epithelial cells of duodenal mucosa from children with acute nonbacterial gastroenteritis. *Lancet* 2:1281–83
12. Bodian D. 1959. Poliomyelitis. See Ref. 112a, pp. 47–98
13. Bosch A, Lucena F, Diez JM, Gajardo R, Blasi M, Jofre J. 1991. Waterborne viruses associated with hepatitis outbreaks. *J. Am. Water Works Assoc.* 83:80–83
14. Carstens JM, Tracy S, Chapman NM, Gauntt CJ. 1992. Detection of enteroviruses in cell cultures by using *in situ* transcription. *J. Clin. Microbiol.* 30:25–35
15. Deleted in proof
16. Chapman NM, Tracy S, Gauntt CJ, Fortmueller U. 1990. Molecular detection and identification of enteroviruses using enzymatic amplification and nucleic acid hybridization. *J. Clin. Microbiol.* 28:843–50
17. Chiba S, Sakuma Y, Kogasaka R, Akihara M, Horino K, et al. 1979. An outbreak of gastroenteritis associated with calicivirus in an infant home. *J Med. Virol.* 4:249–54
18. Cohen JI, Ticehurst JR, Purcell RH, Buckler-White A, Baroudy BM. 1987. Complete nucleotide sequence of wild-type hepatitis A virus: comparison with different strains of hepatitis A virus and other picornaviruses. *J. Virol.* 61:50–59
19. Cone RW, Hobson AC, Huang M-LW. 1992. Coamplified positive control detects inhibition of polymerase chain reactions. *J. Clin. Microbiol.* 30:3185–89
20. Cubitt WD, McSwiggan DA, Moore W. 1979. Winter vomiting disease caused by calicivirus. *J. Clin. Pathol.* 32:786–93
21. Dalldorf G, Melnick JL, Curnen EC. 1959. The Coxsackie virus group. See Ref. 112a, pp. 519–46
22. DeLeon R, Matsui SM, Baric RS, Herrman JE, Blacklow NR, et al. 1992. Detection of Norwalk virus in stool specimens by reverse transcription-polymerase chain reaction and nonradioactive oligoprobes. *J. Clin. Microbiol.* 30:3151–57
23. DeLeon R, Shieh C, Baric RS, Sobsey MD. 1990. Detection of enteroviruses and hepatitis A virus in environmental samples by gene probes and polymerase chain reaction. In *Proc. 1990 Water Quality Technology Conf., San Diego, CA.* pp. 833–53. Denver: Am. Water Works Assoc.
24. Deng MY, Day SP, Cliver DO. 1994. Detection of hepatitis A virus in environmental samples by antigen-capture PCR. *Appl. Environ. Microbiol.* 60:1927–33
25. Desenclos J-CA, Klontz KC, Wilder MH, Nainan OV, Margolis HS, et al. 1991. A multistate outbreak of hepatitis A caused by the consumption of raw oysters. *Am. J. Publ. Health* 81:1268–72
26. Divizia M, deFilippio P, DiNapoli A, Venuti A, Perez-Bercoff R, Pana A. 1989. Isolation of wild-type hepatitis A

virus from the environment. *Water Res.* 23:1155–60
27. Donnelli G, Ruggeri FM, Tinari A, Marziano ML, Menichella D, et al. 1988. A three year diagnostic and epidemiological study on viral infantile diarrhoea in Rome. *Epidemiol. Infect.* 100:311–20
28. Dubrou S, Kopecka H, Lopez-Pila JM, Marechal J, Prevot J. 1991. Detection of hepatitis A virus and other enteroviruses in wastewater and surface water samples by gene probe assay. *Water Sci. Technol.* 24:267–72
29. Enriquez CE, Abbaszadegan M, Pepper IL, Richardson KJ, Gerba CP. 1993. Poliovirus detection in water by cell culture and nucleic acid hybridization. *Water Res.* 27:1113–18
30. Estes MK, Hardy ME. 1995. Norwalk virus and other enteric caliciviruses. In *Infections of the Gastrointestinal Tract*, ed. M Blaser, R Smith, J Ravdin, H Greenberg, R Guerrant. New York: Raven. In press
30a. Fields BN, Knipe DM, eds. 1990. *Virology.* New York: Raven. 2nd ed.
31. Gentsch JR, Glass RI, Woods P, Gouvea V, Gorziglia M, et al. 1992. Identification of group A rotavirus gene 4 types by polymerase chain reaction. *J. Clin. Microbiol.* 30:1365–73
32. Gerba CP. 1987. Transport and fate of viruses in soils: field studies. See Ref. 108a, pp. 141–54
33. Gerna G, Passarani N, Battaglia M, Rondanelli EG. 1985. Human enteric coronaviruses: antigenic relatedness to human coronavirus OC43 and possible etiologic role in viral gastroenteritis. *J. Infect. Dis.* 151:796–803
34. Gilliland G, Perrin S, Blanchard K, Bunn HF. 1990. Analysis of cytokine mRNA and DNA: detection and quantitation by competitive polymerase chain reaction. *Proc. Natl. Acad. Sci. USA* 87:2725–29
35. Goldfield M. 1976. Epidemiological indicators for transmission of viruses by water. See Ref. 9, pp. 70–85
36. Gouvea V, Glass RI, Woods P, Taniguichi K, Clark HF, et al. 1990. Polymerase chain reaction amplification and typing of rotavirus nucleic acids from stool specimens. *J. Clin. Microbiol.* 28:276–82
37. Graff J, Ticehurst J, Flehmig B. 1993. Detection of hepatitis A virus in sewage sludge by antigen capture polymerase chain reaction. *Appl. Environ. Microbiol.* 59:3165–70
38. Graham DY, Jiang X, Tanaka T, Opekun AR, Madore HP, Estes MK.

1994. Norwalk virus infection of volunteers: new insights based on improved assays. *J. Infect. Dis.* 169:1364–67
39. Greenberg HB, Valdesuso J, Kapikian AZ, Chanock RM, Wyatt RG, et al. 1979. Prevalence of antibody to the Norwalk virus in various countries. *Infect. Immun.* 26:270–73
40. Grohman GS, Murphy AM, Christopher PJ, Auty E, Greenberg HB. 1981. Norwalk virus gastroenteritis in volunteers consuming depurated oysters. *Aust. J. Exp. Biol. Med. Sci.* 59:219–28
41. Hardy ME, White LJ, Ball JM, Estes MK. 1995. Specific proteolytic cleavage of recombinant Norwalk virus capsid protein. *J. Virol.* 69:1693–98
42. Hedberg CW, Osterholm MT. 1993. Outbreaks of viral gastroenteritis caused by food and water. *Clin. Microbiol. Rev.* 6:199–210
43. Hejkal TW, Keswick B, LaBelle RL, Gerba CP, Sanchez Y, et al. 1982. Viruses in a community water supply associated with an outbreak of gastroenteritis and infectious hepatitis. *J. Am. Water Works Assoc.* 74:318–21
44. Herrmann JE, Nowak NA, Perron-Henry DM, Hudson RW, Cubitt WD, et al. 1990. Diagnosis of astrovirus gastroenteritis by antigen detection with monoclonal antibodies. *J. Infect. Dis.* 161:226–29
45. Herrmann JE, Taylor DN, Echeverria P, Blacklow NR. 1991. Astrovirus as a cause of gastroenteritis in children. *New Engl. J. Med.* 324:1757–60
46. Hollinger FB, Ticehurst J. 1990. Hepatitis A virus. See Ref. 30a, pp. 631–67
47. Holmes KV. 1990. Coronaviridae and their replication. See Ref. 30a, pp. 841–56
48. Hopkins RS, Gaspard GB, Williams FP Jr, Karlin RJ, Cukor G, et al. 1984. A community waterborne gastroenteritis outbreak: evidence for rotavirus as the agent. *Am. J. Publ. Health* 74:263–65
49. Huang C-C, Nguyen D, Fernandez J, Yun KY, Fry KE, et al. 1992. Molecular cloning and sequencing of the Mexico isolate of hepatitis E virus (HEV). *Virology* 191:550–58
50. Hudson RW, Herrmann JE, Blacklow NR. 1989. Plaque quantitation and neutralization assays for human astrovirus. *Arch. Virol.* 108:33–38
51. Hung T, Wang C, Fang Z, Chou Z, Chang X, et al. 1984. Waterborne outbreak of rotavirus diarrhoea in adults in China caused by a novel rotavirus. *Lancet* 1:1139–42
52. Hyypia T, Auvinen P, Maaronen M. 1989. Polymerase chain reaction for hu-

man picornaviruses. *J. Gen. Virol.* 70: 3261–68
53. Jansen RW, Siegl G, Lemon SM. 1990. Molecular epidemiology of human hepatitis A virus defined by an antigen-capture polymerase chain reaction method. *Proc. Natl. Acad. Sci. USA* 87:2867–71
54. Jaykus LA, DeLeon R, Sobsey MD. 1993. Application of RT-PCR for the detection of enteric viruses in oysters. *Water Sci. Technol.* 27:49–53
55. Jiang S, Wang J, Estes MK. 1995. Characterization of SRSVs using RT-PCR and a new antigen ELISA-a short communication. *Arch. Virol.* 140:363–74
56. Jiang X, Estes MK, Metcalf TG. 1987. Detection of hepatitis A virus by hybridization with single-stranded RNA probes. *Appl. Environ. Microbiol.* 53: 2487–95
57. Jiang X, Estes MK, Metcalf TG, Melnick JL. 1986. Detection of hepatitis A virus in seeded estuarine samples by hybridization with cDNA probes. *Appl. Environ. Microbiol.* 52:711–17
58. Jiang X, Graham DY, Wang K, Estes MK. 1990. Norwalk virus genome: cloning and characterization. *Science* 250: 1580–83
59. Jiang X, Wang J, Graham DY, Estes MK. 1992. Detection of Norwalk virus in stool by polymerase chain reaction. *J. Clin. Microbiol.* 30:2529–34
60. Jiang X, Wang M, Graham DY, Estes MK. 1992. Expression, self-assembly and antigenicity of the Norwalk virus capsid protein. *J. Virol.* 66:6527–32
61. Jiang X, Wang M, Wang K, Estes MK. 1993. Sequence and genomic organization of Norwalk virus. *Virology* 195:51–61
62. Jonassen TO, Kjeldsberg E, Grinde B. 1993. Detection of human astroviruses serotype 1 by the polymerase chain reaction. *J. Virol. Methods* 44:83–88
63. Josephson KL, Gerba CP, Pepper IL. 1993. Polymerase chain reaction detection of nonviable bacterial pathogens. *Appl. Environ. Microbiol.* 59:3513–15
64. Jothikumar N, Aparna K, Kamatchiammel S, Paulmurugan R, Saravanadevi S, et al. 1993. Detection of hepatitis E in raw and treated wastewater with the polymerase chain reaction. *Appl. Environ. Microbiol.* 59:2558–62
65. Kammerer U, Kunkel B, Korn K. 1994. Nested PCR for specific detection and rapid identification of human picornaviruses. *J. Clin. Microbiol.* 32:285–91
66. Kapikian AZ, Estes MK, Chanock RM. 1995. Norwalk group of viruses. In *Virology*, ed. BN Fields, DM Knipe, P Howley. New York: Raven. 3rd ed. In press
67. Kapikian AZ, Wyatt RG, Dolin R, Thornhill TS, Kalica AR, et al. 1972. Visualization by immune electron microscopy of a 27nm particle associated with acute infectious non-bacterial gastroenteritis. *J. Virol.* 10:1075–81
68. Konno T, Suzuki H, Ishida N, Chiba R, Mochizuki K, et al. 1982. Astrovirus-associated epidemic gastroenteritis in Japan. *J. Med. Virol.* 9:11–17
69. Kopecka H, Dubrou S, Prevot J, Marechal J, Lopes-Pila JM. 1993. Detection of naturally occurring enteroviruses in waters by reverse transcription, polymerase chain reaction and hybridization. *Appl. Environ. Microbiol.* 59:1213–19
70. Kurtz JB, Lee TW. 1984. Letter: human astrovirus serotypes. *Lancet* 2:1405
71. Lee TW, Kurtz JB. 1977. Letter: astroviruses detected by immunofluorescence. *Lancet* 2:406
72. LeGuyader F, Dubois E, Menard D, Pommepuy M. 1994. Detection of hepatitis A virus, rotavirus and enterovirus in naturally contaminated shellfish and sediment by reverse transcription-seminested PCR. *Appl. Environ. Microbiol.* 60:3665–71
73. LeGuyader F, Apaire-Marchais V, Brillet J, Billaudel S. 1993. Use of genomic probes to detect hepatitis A virus and enterovirus RNAs in wild shellfish and relationships of viral contamination to bacterial contamination. *Appl. Environ. Microbiol.* 59:3963–68
74. Lemon SM, Chao SF, Jansen RW, Binn LN, LeDuc JW. 1987. Genomic heterogeneity of human and non-human strains of hepatitis A virus. *J. Virol.* 61:735–42
75. Lew JF, Glass RI, Petric M, LeBaron CW, Hamond GW, et al. 1990. Six-year retrospective surveillance of gastroenteritis viruses identified at ten electron microscopy centers in the United States and Canada. *Pediatr. Infect. Dis. J.* 9: 709–14
76. Macnaughton MR, Davies HA. 1981. Human enteric coronaviruses, a brief review. *Arch. Virol.* 70:301–13
77. Madeley CR, Cosgrove BP. 1975. Letter: viruses in infantile gastroenteritis. *Lancet* 2:124
78. Major ME, Eglin RP, Easton AJ. 1992. 3′Terminal nucleotide sequence of human astrovirus type 1 and routine detection of astrovirus nucleic acid and antigens. *J. Virol. Methods* 39:217–25
79. Margolin AB, Hewlett MJ, Gerba CP. 1991. The application of a poliovirus cDNA probe for the detection of en-

teroviruses in water. *Water Sci. Technol.* 24:227-80
80. Martino TA, Sole MJ, Penn LZ, Liew C-C, Liu P. 1993. Quantitation of enteroviral RNA by competitive polymerase chain reaction. *J. Clin. Microbiol.* 31:2634-40
81. Matsui SM, Kim JP, Greenberg HB, Su W, Sun Q, et al. 1991. The isolation and characterization of a Norwalk virus-specific cDNA. *J. Clin. Invest.* 87:1456-61
82. Melnick JL. 1947. Poliomyelitis virus in urban sewage in epidemic and nonepidemic times. *Am. J. Hyg.* 45:240-53
83. Melnick JL. 1957. A water-borne urban epidemic of hepatitis. In *Hepatitis Frontiers,* ed. GA LoGrippo, FW Hartman, JG Mateer, J Barron, pp. 211-25. Boston: Little, Brown
84. Melnick JL. 1958. Advances in the study of the enteroviruses. *Progr. Med. Virol.* 1:59-105
85. Melnick JL. 1987. Human enteric viruses in sediments, sludges, and soils: an overview. See Ref. 108a, pp. 1-2
86. Melnick JL. 1993. The discovery of the enteroviruses and the classification of poliovirus among them. *Biologicals* 21:305-9
87. Melnick JL, ed. 1984. *Monographs in Virology,* Vol. 15, *Enteric Viruses in Water.* Basel: Karger. 235 pp.
88. Melnick JL, Emmons J, Coffey JH, Schoof H. 1954. Seasonal distribution of Coxsackie viruses in urban sewage and flies. *Am. J. Hyg.* 59:64-184
89. Melnick JL, Gerba CP, Wallis C. 1978. Viruses in water. *Bull. WHO* 56:499-508
90. Melnick JL, Horstmann DM, Ward R. 1946. The isolation of poliomyelitis virus from human extraneural sources. II. Comparison of virus content of blood, oropharyngeal washings, and stools of contacts. *J. Clin. Investig.* 25:275-77
91. Melnick JL, Metcalf TG. 1985. Distribution of viruses in the water environment. In *Banbury Report 22: Genetically Altered Viruses and the Environment,* pp. 95-102. Cold Spring Harbor, NY: Cold Spring Harbor Lab.
92. Melnick JL, Penner LR. 1952. The survival of poliomyelitis and Coxsackie viruses following their ingestion by flies. *J. Exp. Med.* 96:255-71
93. Melnick JL, Sabin AB. 1959. The ECHO virus group. See Ref. 112a, pp. 547-69
94. Metcalf TG. 1982. Viruses in shellfish-growing waters. *Environ. Int.* 7:21-27
95. Moe CL, Allen JR, Monroe SS, Gary HEJ, Humphrey CD, et al. 1991. Detection of astrovirus in pediatric stool samples by immunoassay and RNA probe. *J. Clin. Microbiol.* 29:2390-95
96. Moe CL, Gentsch J, Grohmann G, Ando T, Monroe SS, et al. 1994. Detection of Norwalk virus by polymerase chain reaction in fecal specimens from outbreaks of gastroenteritis: identification of sequence variability between strains. *J. Clin. Microbiol.* 32:642-48
97. Monceyron C, Grinde B. 1994. Detection of hepatitis A virus in clinical and environmental samples by immunomagnetic separation and PCR. *J. Virol. Methods* 46:157-66
98. Muir P, Nicholson F, Jhetam M, Neogi S, Banatvala JE. 1993. Rapid diagnosis of enterovirus infection by magnetic bead extraction and polymerase chain reaction detection of enterovirus RNA in clinical specimens. *J. Clin. Microbiol.* 31:31-38
99. Murphy AM, Grohman GS, Christopher PJ, Lopes WA, Davey GR, et al. 1979. An Australia-wide outbreak of gastroenteritis from oysters caused by Norwalk virus. *Med. J. Aust.* 2:329-33
100. Neill JD, Meyer R, Seal BS. 1994. Genetic relatedness of the caliciviruses: PCR amplification and sequence analysis of specific regions to the genomic RNAs of San Miguel sea lion and vesicular exanthema of swine viruses. *Annu. Meet. Am. Soc. Virol., 13th, Madison, WI.* Madison: Am. Soc. Virol.
101. Norcott JP, Green J, Lewis D, Estes MK, Brown DWG. 1994. Genomic diversity of small round structured viruses in the UK. *J. Med. Virol.* 44:280-86
102. Oishi I, Maeda A, Yamazaki K, Minekawa Y, Nishimura H, et al. 1980. Calicivirus detected in outbreaks of acute gastroenteritis in school children. *Biken J.* 23:163-68
103. Paul JH, Jiang SC, Rose JB. 1991. Concentration of viruses and dissolved DNA from aquatic environments by vortex flow filtration. *Appl. Environ. Microbiol.* 57:2197-204
104. Prasad BVV, Rothnagel R, Jiang X, Estes MK. 1994. Three-dimensional structure of baculovirus-expressed Norwalk virus capsids. *J. Virol.* 68:5117-25
105. Prevot J, Dubrou S, Marechal J. 1993. Detection of human hepatitis A virus in environmental water by an antigen-capture polymerase chain reaction method. *Water Sci. Technol.* 27:227-33
106. Provost PJ, Hilleman MR. 1979. Propagation of human hepatitis A virus in cell culture *in vitro. Proc. Soc. Exp. Biol. Med.* 160:213-21

107. Purcell RH, Ticehurst JR. 1988. Enterically transmitted non-A, non-B hepatitis: epidemiology and clinical characteristics. In *Viral Hepatitis and Liver Disease*, ed. AJ Zuckerman, pp. 131-37. New York: Liss
108. Rao VC. 1982. Introduction to environmental virology. In *Methods in Environmental Virology*, ed. CP Gerba, SM Goyal, pp. 1-13. New York: Dekker
108a. Rao VC, Melnick JL, eds. 1987. *Human Viruses in Sediments, Sludges, and Soils*. Boca Raton, FL: CRC
109. Ray R, Aggarwal R, Salunke PN, Mehrotra NN, Talwar GP, et al. 1991. Hepatitis E virus genome in stools of hepatitis patients during large epidemic in north India. *Lancet* 2:438-42
110. Reyes GR. 1991. Molecular cloning of the hepatitis E virus. In *Int. Symp. Viral Hepatitis and Liver Disease*, ed. FB Hollinger, SM Lemon, HS Margolis, pp. 514-17. Baltimore, MD: Williams & Williams
111. Richards GP. 1985. Outbreaks of shellfish-associated enteric virus illness in the United States: requisite for development of viral guidelines. *J. Food Prot.* 48:815-23
112. Richards GP. 1988. Microbial purification of shellfish: a review of depuration and relaying. *J. Food Prot.* 51:218-50
112a. Rivers T, Horsfall F, eds. 1959. *Viral and Rickettsial Infections of Man*. Philadelphia: Lippincott. 3rd ed.
113. Robertson BH, Khanna B, Nainan OV, Margolis HS. 1991. Epidemiologic patterns of wild type hepatitis A virus determined by genetic variation. *J. Infect. Dis.* 163:286-92
114. Romalde JL, Estes MK, Szucs G, Atmar RL, Woodley CM, et al. 1994. In situ detection of hepatitis A virus in cell cultures and shellfish tissues. *Appl. Environ. Microbiol.* 60:1921-26
115. Rotbart HA. 1989. Human enterovirus infections: molecular approaches to diagnosis and pathogenesis. In *Molecular Aspects of Picornavirus Infection and Detection*, ed. BL Semler, E Ehrenfeld, pp. 243-64. Washington, DC: Am. Soc. Microbiol.
116. Rotbart HA. 1990. Enzymatic RNA amplification of the enteroviruses. *J. Clin. Microbiol.* 28:438-42
117. Rotbart HA. 1991. Nucleic acid detection systems for enteroviruses. *Clin. Microbiol. Rev.* 4:156-68
118. Schwab KJ, DeLeon R, Baric RS, Sobsey MD. 1991. Detection of rotaviruses, enteroviruses, and hepatitis A virus by reverse transcriptase polymerase chain reaction. In *Advances in Water Analysis and Treatment*, pp. 475-91. Orlando, FL: Am. Water Works Assoc.
119. Schwab KJ, DeLeon R, Sobsey MD. 1993. Development of PCR methods for enteric virus detection in water. *Water Sci. Technol.* 27:211-18
120. Severini GM, Mestroni L, Falaschi A, Camerini F, Giacca M. 1993. Nested polymerase chain reaction for high sensitivity detection of enterovirus RNA in biological samples. *J. Clin. Microbiol.* 31:1345-49
121. Shieh Y-SC, Baric RS, Sobsey MD, Ticehurst J, Miele TA, et al. 1991. Detection of hepatitis A virus and other enteroviruses in water by ssRNA probes. *J. Virol. Methods* 31:119-36
122. Steffan RJ, Atlas RM. 1988. Polymerase chain reaction: applications in environmental microbiology. *Annu. Rev. Microbiol.* 45:137-61
123. Tam AW, Smith MM, Guerra ME, Huang C-C, Bradley DW, et al. 1991. Hepatitis E virus (HEV): molecular cloning and sequencing of the full length viral genome. *Virology* 185:120-31
124. Tecott LH, Barachas JD, Eberwine JH. 1988. In situ transcription: specific synthesis of complementary DNA in fixed tissue sections. *Science* 240:1661-64
125. Trask JD, Melnick JL, Wenner HA. 1945. Chlorination of human, monkey-adapted, and mouse strains of poliomyelitis virus. *Am. J. Hyg.* 41:30-40
126. Treanor JJ, Jiang X, Madore HP, Estes MK. 1993. Subclass specific serum antibody responses to recombinant Norwalk capsid antigen (rNV) in adults infected with Norwalk, Snow Mountain, or Hawaii viruses. *J. Clin. Microbiol.* 31:1630-34
127. Truman BI, Madore HP, Memegus MA, Nitzkin JL, Dolin R. 1987. Snow Mountain agent gastroenteritis from clams. *Am. J. Epidemiol.* 126:516-25
128. Tsai Y-L, Sobsey MD, Sangermano LR, Palmer CJ. 1993. Simple method of concentrating enteroviruses and hepatitis A virus from sewage and ocean water for rapid detection by reverse transcription-polymerase chain reaction. *Appl. Environ. Microbiol.* 59:3488-91
129. Tsai Y-L, Tran B, Sangermano LR, Palmer CJ. 1994. Detection of poliovirus, hepatitis A virus and rotavirus from sewage and ocean water by triplex reverse transcriptase PCR. *Appl. Environ. Microbiol.* 60:2400-7
130. Tsarev SA, Emerson SU, Reyes GR, Tsareva TS, Legters LJ, et al. 1992. Characterization of a prototype strain of hepatitis E virus. *Proc. Natl. Acad. Sci. USA* 89:559-63

131. Vaughn JM, Landry EF. 1983. Viruses in soils and groundwater. In *Viral Pollution of the Environment*, ed. G Berg, pp. 163–210. Boca Raton: CRC
132. Wallis C, Homma A, Melnick JL. 1972. A portable virus concentrator for testing water in the field. *Water Res.* 6:1249–56
133. Wallis C, Melnick JL, Gerba CP. 1979. Concentration of viruses from water by membrane chromatography. *Annu. Rev. Microbiol.* 33:413–37
134. Wang J, Jiang X, Madore HP, Gray J, Desselberger U, et al. 1994. Sequence diversity of small round structured viruses. *J. Virol.* 68:5982–90
135. Ward R, Melnick JL, Horstmann DM. 1945. Poliomyelitis virus in fly-contaminated food collected at an epidemic. *Science* 101:491–93
136. Wellings FM, Lewis AL, Mountain CW. 1974. Virus survival following wastewater spray irrigation of sandy soils. In *Virus Survival in Water and Wastewater Systems*, ed. JF Malina, BP Sagik, pp. 253–60. Austin: Center Res. Water Resources Syst.
137. Wen Y-M, Xu Z-Y, Melnick JL, eds. 1992. *Monographs in Virology*, Vol. 19, *Viral Hepatitis in China: Problems and Control Strategies*. Basel: Karger. 159 pp.
138. Willcocks MM, Carter MJ. 1992. The 3′terminal sequence of a human astrovirus. *Arch. Virol.* 124:279–89
139. Willcocks MM, Carter MJ, Silcock JG, Madeley CR. 1991. A dot-blot hybridization procedure for the detection of astrovirus in stool samples. *Epidemiol. Infect.* 107:405–10
140. Willcocks MM, Silcock JG, Carter MJ. 1993. Detection of Norwalk virus in the UK by the polymerase chain reaction. *FEMS Microbiol. Lett.* 112:7–12
141. World Health Organization. 1979. Report of WHO Scientific Group of Human Viruses in water, wastewater and soil. *Tech. Rep. Ser. WHO.* No. 639
142. Wyn-Jones AP, Herring AJ. 1991. Growth of clinical isolates of astrovirus in a cell line and the preparation of viral RNA. *Water Sci. Technol.* 24:285–90
143. Xu A, Wan X, Qiu F, Pang Q. 1981. Astrovirus in autumn infantile gastroenteritis. *China Med. J.* 94:659–62
144. Xu Z-Y, Hu S-L. 1992. Epidemiology of hepatitis A problems and control strategies. See Ref. 137, pp. 119–25
145. Yang F, Xu X. 1993. A new method of RNA preparation for detection of hepatitis A virus in environmental samples by the polymerase chain reaction. *J. Virol. Methods* 43:77–84
146. Zhou Y-J, Estes MK, Jiang X, Metcalf TG. 1991. Concentration and detection of hepatitis A virus and rotavirus from shellfish by hybridization tests. *Appl. Environ. Microbiol.* 57:2963–68

HOW BACTERIA SENSE AND SWIM

David F. Blair

Department of Biology, University of Utah, Salt Lake City, Utah 84112

KEY WORDS: motility, chemotaxis, protein phosphorylation, membrane receptors, molecular motors

CONTENTS

DESCRIPTION OF THE PHENOMENON	490
THE CHEMOSENSORY PATHWAY	491
Structure and Function of the Protein Components	494
Other Factors in Signaling	503
Unexplained Sensitivity of Attractant Responses	503
HOW BACTERIA SWIM: FLAGELLAR STRUCTURE AND FUNCTION	504
Flagellar Assembly	505
Organization and Regulation of the Flagellar Genes	509
Components Involved in Torque Generation	510
Flagellar Motor Dynamics	512
Models of Torque Generation	513
FUTURE DIRECTIONS	514

ABSTRACT

Cells of *Escherichia coli* or *Salmonella typhimurium* can sense chemicals in their environment and respond by moving toward some and away from others. The ability to sense and swim requires the products of approximately 50 genes, about 10 for detecting and processing sensory cues and the rest for assembly and operation of the flagella. The function of each component in the chemosensory signaling pathway is well understood. Signaling is known to involve phosphorylation of a set of cytoplasmic proteins, but questions remain concerning the protein conformational changes and interactions that take place. Functions have been assigned to almost all of the approximately 40 flagellar proteins, and the sequence of events in flagellar assembly has been largely determined. Flagellar assembly depends on a specialized apparatus for exporting certain flagellar components to their appropriate locations. The structure and mechanism of this apparatus remain a mystery, as does the mechanism by which the flagellar motor generates torque.

DESCRIPTION OF THE PHENOMENON

Bacteria are seldom in a position to control all of the important features of their environment. Many species are equipped, however, to sense what is beneficial in their surroundings and what is not and to then respond by moving toward conditions more likely to promote their survival. Sensory cues include chemicals, which are often nutrients or related compounds, and other variables such as pH, temperature, osmolarity, or viscosity. This review emphasizes the sensing of chemical attractants and repellents. In *Escherichia coli*, attractant chemicals include aspartic acid, serine, dipeptides, and various sugars, while examples of repellents are leucine and Ni or Co ions.

Cells of *E. coli* or *Salmonella typhimurium* swim using flagella, which comprise thin helical filaments, each driven at the base by a rotary motor (12, 157). The motor extends through the cytoplasmic membrane, and the energy for rotation comes directly from the transmembrane proton gradient rather than from ATP (92). The motors can turn either clockwise (CW) or counterclockwise (CCW), which is the basis for controlling the direction of swimming: When the motors turn CCW, the filaments on a cell (typically about five) join into a bundle that propels the cell along a more or less smooth trajectory called a run, whereas if one or more of the motors reverses to the CW direction, the bundle disperses and the uncoordinated action of the filaments causes rapid somersaulting called a tumble (13, 93). Unstimulated cells alternate between the two modes of swimming, running in approximately straight lines for about a second, then tumbling for a fraction of a second and resuming linear motion in a new direction almost uncorrelated with the original trajectory. The result is a random walk, in three dimensions if the cell is swimming in solution. Cells of *E. coli* and *S. typhimurium* can also migrate on moist surfaces, a behavior that might be more relevant to cells in nature (56). The paths of cells swarming on surfaces have not been characterized in detail but are also likely to have some random character.

A purely random walk would be a poor strategy for migration or dispersal because the net distance covered is proportional (on average) to only the square root of the time spent in motion (11). Cells bias their random walk to achieve net progress in a chosen direction by controlling the frequency of motor reversals in response to sensory cues. An attractant stimulus, such as that encountered by a cell swimming up a gradient of aspartic acid, causes momentary suppression of CW rotation and thus prolongs the runs that happen to carry the cell up the gradient. Conversely, a repellent cue (either an increased concentration of repellent or a decreased concentration of attractant) increases the probability of CW rotation, shortening runs in an unfavorable direction (35). In the gradients normally encountered by a cell, attractant stimuli are more important because the cells are more sensitive to them; a larger repellent stimulus is needed to achieve an equivalent change in the probability of motor reversal. In both cases, the relevant cue is the change in concentration of the chemical with time (35, 108). The cell measures the concentration encountered during the past second and compares it with that encountered during the previous three or four seconds, basing decisions to run or tumble on the difference (155). Bacterial chemotaxis thus involves a simple, very short-term memory.

While this strategy for directed movement might appear unusual, it represents a nearly optimal solution to the problems that face an organism as small as *E. coli*. Owing to its small size, a cell of *E. coli* cannot maintain any particular trajectory for more than about 3 s, the time in which thermal (Brownian) motion will impose a new trajectory. Consequently, the cell has a limited amount of time to make measurements of concentration and act upon them. [Berg (11) discusses these and other physical considerations relevant to bacterial chemotaxis.] In addition to the unusual movement pattern employed, two features of the bacterial chemotactic response are remarkable. First, the sensing apparatus is highly adaptable to different effector concentrations. Gradients of α-methyl aspartic acid, for example, can elicit responses over about five orders of magnitude of concentration [1 μM–0.1 M (120)]. Second, the response is very sensitive: The binding of a few, and possibly only a single, attractant molecule(s) to a receptor on the surface of the cell can significantly change the reversal probability of the flagellar motors (25, 155), which implies substantial amplification of the signal at one or more stages.

The products of about 50 genes are needed for motility and chemotaxis in *E. coli* or *S. typhimurium* (106). A dozen or so are involved in detecting and processing chemical sensory cues, and about 40 in the assembly and operation of the flagella. This review first discusses the components involved in detecting and processing chemical signals and then summarizes what is known of the structure, genetics, and function of the flagella. The focus is on recent developments, with most citations of the primary literature limited to work reported in the past few years. The topics of bacterial signal transduction and motility have been reviewed individually in greater detail (30, 72, 106, 107, 139, 152, 159, 174). Most of this chapter deals with the enteric species *E. coli* and *S. typhimurium,* for which motility and sensing are best understood.

THE CHEMOSENSORY PATHWAY

Figure 1 shows the chemosensory pathway of *E. coli*. Detection of a chemoeffector begins when it binds to either a receptor protein within the cytoplasmic

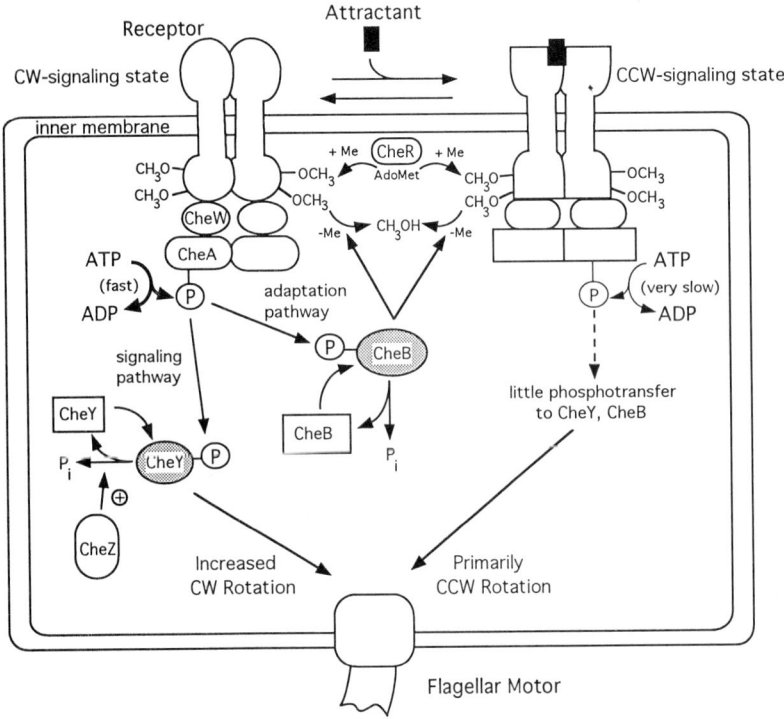

Figure 1 The chemotactic signaling and adaptation pathways of *E. coli*. Chemoeffectors are detected by membrane receptors. The receptors modulate the activity of the autokinase CheA and thus the level of phosphorylation of its substrate CheY. When phosphorylated, CheY promotes CW rotation of the flagellar motors. Dephosphorylation of CheY is accelerated by CheZ. The receptors are hypothesized to exhibit at least two conformations, one that activates CheA (and thus signals CW) and one that does not (and thus signals CCW). Binding of attractant stabilizes the CCW state, and repellent (not shown) the CW state. Adaptation involves the transfer of methyl groups to the receptors by CheR and their removal by CheB. (+Me signifies methylation and –Me demethylation.) CheB can be phosphorylated by CheA, thereby increasing its demethylating activity. Adaptation to an attractant stimulus occurs as follows: Attractant binding reduces the activity of CheB (via the CheW-CheA pathway) and also makes the receptors better substrates for CheR. Thus, over the course of several seconds the receptors become more heavily methylated. Additional methyl groups stabilize the CW state, which causes the equilibrium between CW and CCW states to shift back toward prestimulus levels. For additional details, see the text.

membrane or to a periplasmic protein, which then binds to a transmembrane receptor. The receptors are clustered near the poles of the cell (109), forming in effect a chemical-sensing "nose" (140). *E. coli* contains four transmembrane receptors, also called transducers, which each mediate responses to a specific set of chemoeffectors (30). The receptor Tsr, for example, transduces responses to serine and certain repellents, whereas the receptor Tar transduces responses to aspartate, maltose, and certain repellents. When a chemoeffector (a small molecule, metal ion, or periplasmic receptor occupied by its ligand) binds to the transducer, it undergoes a conformational change that is transmitted across the membrane, signaling this event to components in the cytoplasm. Four cytoplasmic proteins then relay the signal to the flagellar motors. In order of their appearance in the signaling pathway, these proteins are called CheW, CheA, CheY, and CheZ. The Che designation reflects the fact that mutations in any of these components cause a general defect in chemotaxis.

Considerable insight into the signaling process was obtained several years ago in studies of the behavior of mutants with defects in specific chemotaxis genes, or cells deleted of all chemotaxis genes then reconstituted with small subsets (88, 138, 190). These experiments indicated that the CheY protein is involved in signaling the motor to turn CW, that CheW and CheA are needed to make this CW-signaling activity responsive to the membrane receptors, and that the CheZ protein somehow antagonizes the CW signal. The biochemical basis for these effects was then revealed in studies in which many of the signaling steps were reconstituted in vitro using purified proteins (28, 29, 59–61, 126, 132). These studies showed that phosphorylated CheY acts as a CW signal, that CheA is an autokinase that phosphorylates itself and then donates its phosphoryl group to CheY, and that CheW is a coupling factor that makes the autophosphorylation activity of CheA responsive to the state of the receptors. The catalytic activity of CheA explains much of the sensitivity of the chemotactic response; when stimulated by the receptors, CheA can rapidly phosphorylate many molecules of CheY (28, 29, 126). CheZ antagonizes CW signaling because it binds to phospho-CheY and accelerates its dephosphorylation (60).

As bacteria migrate up or down chemical gradients, they must continually adapt to the prevailing average concentrations of attractants and repellents and thus remain responsive to local gradients. Adaptation involves two proteins called CheR and CheB, which together determine the extent of covalent modification of the membrane receptors at specific glutamic acid residues (Figure 1). CheR is a methyltransferase that attaches methyl groups to the receptors, using S-adenosylmethionine as the donor (85, 164). CheB is a methylesterase that removes the methyl groups and also deamidates certain glutamine residues to produce some of the glutamates that act as methyl acceptors (77). Methylation of the receptors enhances their stimulatory effect on CheA and thus augments CW signals, whereas demethylation has the opposite effect (27, 47). By means of a feedback loop that controls the activity of CheB (Figure 1) and additional factors (discussed below) that influence the rate of methylation by CheR, the number of methyl groups on the receptors is maintained at a level that gives an intermediate CW-CCW motor bias. Thus, the cell remains able to respond to changes in attractant or repellent concentration even in the presence of relatively high static concentrations of chemoeffectors. Because they can be methylated and demethylated in this way, the chemoreceptors are often called methyl-accepting chemotaxis proteins, or MCPs.

Structure and Function of the Protein Components

RECEPTORS The transmembrane chemoreceptors of *E. coli* or *S. typhimurium* have the membrane topology shown in Figure 2, with two hydrophobic membrane-spanning segments connecting sizable domains in the periplasm and cytoplasm. The two-spanner membrane topology has been confirmed experimentally (48, 112). The sequences of the different receptors are homologous, most closely in the cytoplasmic domain that interacts with other components of the signaling pathway and less so in the periplasmic domains, which are specialized to bind different ligands (26, 33, 148; 57 is a recent review). In their native state, the receptors are homodimers both with and without a bound ligand, which suggests that signaling does not involve multimer formation or dissociation (122).

The structure of the periplasmic domain of the aspartate receptor Tar has

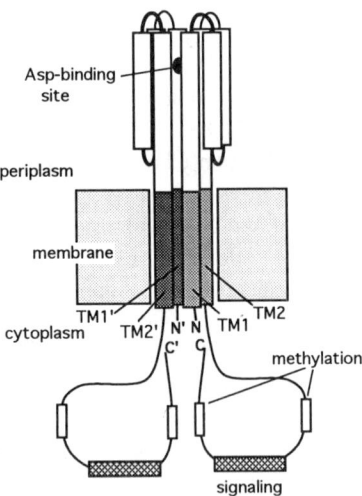

Figure 2 The structure of the aspartate chemoreceptor. Cylinders represent α-helices. The receptor is a dimer of subunits, each with two membrane-spanning segments. The periplasmic domain of each subunit is a four-helix bundle. The helices at the subunit interface form another four-helix bundle that probably continues uninterrupted from the periplasmic domain through the membrane. The tertiary structure of the cytoplasmic domain is not known; it has two segments that contain the sites of methylation (indicated), and between these another segment involved in signaling, which probably contacts the CheA and/or CheW proteins.

been determined by X-ray crystallography, both with and without a bound ligand (121, 154). Like the intact receptor, the periplasmic domain forms dimers. Each subunit is folded into a long four-helix bundle (shown schematically in Figure 2). Two of the helices from one subunit contact their counterparts in the other subunit, forming another, relatively loose four-helix bundle in the center of the dimer. Aspartate binds at the interface between subunits, at a site quite far (>60 Å) from the surface of the membrane. The residues at the binding site are the same as those identified in previous studies of mutants unable to respond to aspartate and include several arginine residues that interact with the negative charges of aspartate (52, 192). A second, nonoverlapping site for aspartate, related to the first by an approximate twofold axis of symmetry, is present in the dimer but is not occupied. Measurements of ligand binding to the intact Tar or Tsr receptors also indicate that each dimer binds tightly to only a single ligand molecule (17, 97). In *E. coli,* the Tar receptor also mediates responses to maltose by binding to the periplasmic receptor for that sugar (52). Mutations in Tar that block this response are clustered together on a surface distinct from the aspartate binding site and also quite far from the membrane (52, 154).

An important question concerning the receptors is how they communicate the binding event in the periplasm to signaling components in the cytoplasm. Contrasting proposals for the mechanism of transmembrane signaling, one involving movements within each subunit, and the other movements between subunits in the dimer, have been discussed by Pakula & Simon (136). The determination of periplasmic domain structures in both unliganded and aspartate-bound forms promised to shed light on this question; the differences between the two structures are slight, however. The conformation of each subunit is essentially unchanged by effector binding (within the precision of the structure determinations). The largest change is a rotation of the subunits relative to each other by approximately 4° (121, 154). This prompted the suggestion that transmembrane signaling might involve a relative movement of the receptor subunits (121). Certain mutations in the periplasmic domain at the interface between subunits can cause the receptor to signal, which also suggests that relative movement of the subunits might be important (193). The periplasmic domain has been studied using ^{19}F NMR, a method sensitive to relatively small conformational changes; those measurements indicate that binding of aspartate triggers significant movements within each subunit, in support of the alternative, intrasubunit hypothesis (42). Evidence concerned with the structure of the membrane-spanning segments, described below, also supports the view that movements within each subunit are most important.

The arrangement of the membrane-spanning segments of several receptors has been deduced using the method of site-directed disulfide cross-linking (48, 49). In this approach, cysteine residues are introduced in various places in the membrane segments, and the rates of disulfide cross-link formation are used to determine which positions are near each other. These experiments indicate that the transmembrane segments are α-helices (95, 135). The four membrane segments in a receptor dimer form a bundle, with the amino-proximal segments of each subunit (called TM1) in close contact near the center and the other segments (called TM2) farther from the center and from each other (48, 95, 135) (Figure 2). The structure of the protein just outside the membrane is not known, but the arrangement of membrane segments is consistent with the suggestion that two of the α-helices in each periplasmic domain extend without interruption through the membrane.

Recent disulfide cross-linking studies suggest that transmembrane signaling involves mainly intrasubunit rather than intersubunit movements. Signaling was not inhibited by the formation of disulfide cross-links between the subunits at several positions in segment TM1 (38, 96, 105, 175). Thus, effector binding evidently does not induce large relative movements of the subunits at this interface. In contrast, cross-links between TM1 and TM2 segments in a single subunit inhibited signaling (96). Also, some fairly conservative amino acid substitutions at the interface between TM1 and TM2 segments of the same subunit can disrupt signaling, suggesting that the precise relationship between membrane spanners within a single subunit is important (71).

In their pure forms, the intrasubunit and intersubunit mechanisms of transmembrane signaling represent somewhat artificial extremes. Although movements within each subunit appear to be most important in the transmembrane domain, those movements will in general cause the relationship between the subunits to change as well, in at least part of the cytoplasmic domains. A full understanding of the mechanism of transmembrane signaling will require determination of the structure of the cytoplasmic domain. The dimensions of the cytoplasmic domain have been estimated via electron microscopy of ordered arrays of intact receptors (9). Improved resolution may allow elements of secondary structure to be identified. Cytoplasmic fragments of the receptor will probably be useful in structural studies; some of these are stable, soluble, and can function in signaling both in vivo and in vitro (4). Some mutant forms are locked in one or another signaling state (3, 4), so it may be possible to determine structures of this domain in both on and off conformations.

Intensive mutational studies have identified a segment in the cytoplasmic domain of the receptor that is closely involved in signaling (3). This signaling domain is flanked by two segments, designated K1 and R1, that contain the sites of methylation. The K1 and R1 segments are probably α-helices (181). One explicit hypothesis for the mechanism of signaling by the receptor suggested that the K1 and R1 segments of both subunits are clustered together, arranged as a four-helix bundle in one signaling state and as a coiled coil in the other (174). The relative stability of those conformations would depend upon the charge of the K1 and R1 segments, which could be adjusted by the level of methylation. Like other less-specific proposals concerning receptor mechanism, this hypothesis assumes that the receptor exists in two states, one that signals CW and the other CCW (6). As discussed below, however, two-state models might prove too simple to explain all aspects of signaling by the receptors.

Methylation of the receptors allows the cell to adapt to prevailing chemoeffector concentrations so that they can continue to respond sensitively to local gradients. Increased concentrations of chemical attractants cause increased receptor methylation, while repellents reduce the level of methylation (164). Methylation of the receptors is correlated in vivo with an increased tendency of the flagellar motors to turn CW (137, 164). The effect of receptor methylation on the signaling process reconstituted in vitro was recently examined using purified receptors in defined states of methylation (27, 47). As expected based on their behavior in vivo, more-methylated receptors produced stronger CW signals in the in vitro assay (i.e. higher levels of phospho-CheY). Conversely, demethylated receptors did not stimulate formation of phospho-CheY. As noted above, the chemotactic response is effective over a remarkably wide range of chemoeffector concentrations (120). Some of this great adaptability could be explained if the affinity of the receptor for ligand depended on the level of methylation, increased methylation causing reduced affinity. Thus, a relevant question is whether methylation of the receptors affects their affinity for ligand. Several studies have addressed this question through measurements of ligand binding to receptors methylated to different degrees, with apparently contradictory results (27, 47, 97, 196). Weis and coworkers (97) have discussed the issue and note that the discrepancies could result from overexpression of the receptors, but not the other signaling components, in most of the studies. The data are consistent with the proposal that methylation of the receptors does alter their affinity for ligands, in some cases by a large factor (196), and that this modulation requires other proteins, probably CheA and CheW, to be present as well.

CheA AND CheW   A significant breakthrough in our understanding of bacterial chemotaxis occurred with the discovery that CheA is a kinase that can phosphorylate itself using ATP and subsequently donate the phosphoryl group to CheY or CheB (60). CheA and CheY each belong to widespread families of bacterial proteins that are involved in detecting various sensory cues and responding to them, most often by modulating gene expression (139, 169). Related proteins were also identified recently in eukaryotic cells (2, 37, 110, 134). With other members of this family that have been studied, the signaling chemistry is analogous to that of CheA and CheY. The members of the CheA family are called sensor kinases, and the phosphoaccepting proteins are called response regulators (139). Sensor kinases have a modular architecture consisting of a transmitter domain involved in autophosphorylation and another domain specialized for sensing some environmental variable. Response regulators are usually composed of a receiver domain containing the phosphoaccepting site and another domain specialized to produce the required output effect (141).

Interactions with other proteins control the rate of autophosphorylation of CheA. Autophosphorylation appears to be the most important control point in the chemotaxis signaling pathway. CheA can phosphorylate itself slowly in the absence of other proteins, but the reaction is accelerated several hundredfold in the presence of membrane receptors and CheW. The stimulatory effect of the receptors is inhibited by binding of an attractant. Thus, the essential features of intracellular signaling can be reconstituted in a system containing only receptors, CheW, CheA, and CheY (28, 29, 126).

CheA, like the receptors, forms dimers (54). CheA, CheW, and receptors together form a stable ternary complex with a stoichiometry of 2:2:2 (55, 117). An excess of CheW disrupts the complex, which suggests that this protein is positioned between CheA and the receptors (55). The in vivo effects of overexpression or mutation of CheW also indicate that it links CheA to the receptors (98, 99). Although stimulation of CheA activity by the receptors requires CheW, its suppression by attractant does not, which indicates that CheA also contacts the receptors directly (98). The ternary complex remains intact through many cycles of CheA autophosphorylation and phosphotransfer to CheY, both in the presence or absence of attractant ligand. Thus, modulation of CheA activity must involve conformational changes within the ternary complex rather than association or dissociation of components (55). The receptor-CheW-CheA complex also can bind to CheY. Formation of the ternary signaling complex on immobilized CheY molecules has been monitored using surface plasmon resonance (153). Phosphorylation of CheY released this protein from the complex, as expected if phospho-CheY acts as a diffusible CW signal.

A more detailed picture of signaling will require a better understanding of the conformational changes that modulate CheA activity in the ternary complex, which will require determination of the component structures. Mutational and biochemical studies of CheA have shown that it is organized into several (at least four) domains with distinct activities. A segment near the middle, spanning approximately residues 260–510, contains the catalytic determinants for phosphotransfer from ATP to the protein (141, 179). The site of autophosphorylation is His48, contained within an amino-terminal segment of ~18 kDa that can be removed via limited proteolysis and yet remain active in phosphotransfer to CheY (59). Between the catalytic and phosphoaccepting segments resides a domain of about 65 residues called P2 that can bind to CheY and is probably important for ensuring that the phosphoryl groups on CheA are donated to that partner rather than one of the many other phosphoaccepting proteins in the cell (123, 179). The P2 domain might also bind to CheB, the other phosphoaccepting partner of CheA (167). The carboxy-terminal segment is necessary for the enhancement of CheA activity by receptors and CheW and for inhibition of its activity by attractant (31). Some of the domains of CheA are joined by protease-sensitive segments that are poorly conserved and probably flexible (123), which suggests the possibility of sizable relative movements of domains.

Autophosphorylation in the CheA dimer is by phosphotransfer from the kinase active site of one subunit to His48 on the other subunit (178, 191). CheA activity might therefore be modulated by the relative movement of subunits (or subunit domains). A second, shorter form of CheA is also synthesized by the cell, in amounts comparable to the full-length version described above (83, 161). This short form of CheA (termed CheA$_s$) lacks the phosphoaccepting residue His48, but when present in a dimer with a full-length partner, it retains the ability to catalyze phosphotransfer to His48 on the partner (178, 191). The role of this second form of CheA is not clear; strains lacking it can carry out efficient chemotaxis (149). The short form of CheA, but not the long form, complexes with CheZ in a stoichiometry of >8 CheZ:1 CheA$_s$ (114). The function of this interaction in signaling is not known, but it might allow modulation of CheZ activity by CheA$_s$ and the receptors, a possibility discussed further below.

CheY When phosphorylated, CheY induces CW rotation of the flagellar motors by binding to a motor protein called FliM (7, 187). Most mutant variants of CheY that cannot be phosphorylated also do not cause CW rotation.

The structure of CheY is known at high resolution from X-ray crystallographic studies of the protein from both *E. coli* (185) and *S. typhimurium* (171). The core of the protein is a five-stranded β-sheet surrounded by five α-helices; this fold resembles the *ras* family of eukaryotic signaling proteins (101). The sequences of 79 members of the response regulator family were recently compared (184). The pattern of residue conservation within the family can be rationalized by reference to the CheY structure. Many of the conserved hydrophobic residues occupy positions in the core of CheY. Several conserved acidic residues are grouped together in a pocket near one end of the molecule; one of these, Asp57, is the site of phosphorylation (32, 150).

Magnesium ion is required for CheY to become phosphorylated; fluorescence (103) and NMR (75) studies indicate a single high-affinity site for Mg^{2+}. The structure of CheY with bound Mg^{2+} was recently reported by two groups (10, 170). One of the reported structures resembles the metal-free protein, differing only in the vicinity of the ion-binding site (170), while the other shows significant changes relative to the apoprotein (10). NMR spectroscopic studies indicate that binding of Mg^{2+} does not grossly alter the structure of the protein in solution (75), suggesting that the former structure is correct. In both structures, the metal ion binds to residues in the acidic pocket, including Asp57. The arrangement of residues in the coordination sphere of Mg^{2+} suggests a plausible mechanism for the phosphotransfer reaction in which the metal ion has a central role, acting as a template to stabilize the transition state (170). Thus CheY itself, rather than CheA, contains the catalytic determinants that promote phosphoryl transfer. Results supporting this notion came from an earlier study that showed that CheY can be phosphorylated by small compounds such as acetylphosphate (102).

Current interest centers on the mechanism by which phosphorylation of CheY makes it active in binding to the flagellar motors and promoting CW rotation. Activation appears to involve a conformational change induced by phosphorylation rather than phosphorylation per se, because some mutant variants of CheY cannot be phosphorylated but still cause CW rotation (32) while others can be phosphorylated yet fail to cause CW rotation (101). The nature of the conformational change has not been determined. ^{19}F NMR studies suggest that phosphorylation affects residues in many parts of the protein (45). A surface on CheY likely to contact the target in the flagellar motor (FliM) was identified by analyzing suppressor mutations in CheY that compensate for defects in motor components involved in switching (147, 162). This surface approaches but does not extensively overlap the site of phosphorylation. Presumably, the activating conformational change affects some or all of the residues on this surface.

Intergenic suppression studies suggested that CheY interacts with one or more of the flagellar proteins FliG, FliM, and FliN, which are involved in controlling the direction of motor rotation (143, 162, 194). Phospho-CheY binds in vitro to FliM, but not FliG or FliN (187). FliM thus appears to be the principal target of phospho-CheY in the motor. Numerous mutations in FliM affect CW-CCW switching (162); some are presumably in parts of the protein that form the phospho-CheY binding site.

CheZ   CheZ is a 24-kDa protein whose structure is not known. It accelerates the dephosphorylation of phospho-CheY (60), acting in effect as an anti-CW signal. CheZ binds to CheY (117), probably recognizing the activated conformation (22). Most mutations in CheZ cause an increased CW bias of the motors, as expected for loss of a CheY-dephosphorylating activity. Two unusual mutations in CheZ, isolated as suppressors of defects in motor switching components, cause excessive CCW rotation. These mutant CheZ proteins promote dephosphorylation of CheY at aberrantly high rates. At normal levels of expression, they do not support efficient chemotaxis, but when they are underexpressed, chemotaxis is restored, as assayed by swarming in soft agar (66). Evidently, the overall activity of CheZ is important for chemotaxis while the precise stoichiometry of the protein is not, at least in the swarming assay. The purpose of CheZ may be to maintain phospho-CheY at a level allowing intermediate CW-CCW motor bias, a prerequisite for chemotaxis because the swimming pattern can be modulated only if the motors spend appreciable time turning each way. As others have noted, however, the need for a separate protein for this function is not obvious (66). Several authors (34, 66, 174) have suggested that the activity of CheZ might not be constant but could be modulated by sensory cues, acting as a separately controllable CCW signal. Reasons for suspecting the existence of a separate CCW signal are discussed below. CheZ might also interact with the motor directly, but none of the current evidence strongly supports this possibility. Intergenic suppression studies suggest that CheZ might interact with components of the flagellum (142), but the suppression could alternatively result from separate, compensating effects on motor bias (162).

CheR AND CheB   CheR and CheB function together to control the level of methylation of the transmembrane receptors (164). CheR has a mass of 33 kDa and CheB 38 kDa; the structures of the proteins are not known. CheR catalyzes the transfer of methyl groups from S-adenosyl methionine to certain Glu residues in the cytoplasmic domains of the receptors, thereby forming glutamyl methyl esters. CheB is an esterase that catalyzes the removal of the methyl groups, producing methanol and restoring the Glu side chain. Each receptor can be methylated at several nonequivalent sites; methyl-group turnover is more rapid at some sites than at others, presumably because they are more accessible to CheR and CheB (165, 173). In newly synthesized receptors, some of the prospective methylation sites are Gln rather than Glu residues. CheB also catalyzes the deamidation of those Gln residues to Glu (77), allowing subsequent methylation.

The level of receptor methylation is controlled both globally and locally. Global control involves the modulation of CheB activity by CheA, CheW, and the receptors (78, 168). The CheB protein has two domains, a C-terminal part that contains the esterase active site and an N-terminal part that is homologous to CheY and the other response regulators (160). The N-terminal domain can be phosphorylated (104), probably at Asp56 (analogous to Asp57 in CheY). In the unphosphorylated CheB molecule, the N-terminal domain is an inhibitor of the C-terminal domain. Phosphorylation releases this inhibition, causing an ~15-fold increase in esterase activity (104). Like CheY, CheB is phosphorylated by CheA. An increase in the autokinase activity of CheA therefore causes both increased CW rotation of the motors and decreased methylation of the receptors, which reduces their stimulatory effect on CheA and counteracts the initial stimulus (Figure 1). This mechanism is global in the sense that a stimulus through any of the receptors can, by affecting CheB activity, influence the methylation level of all of the other receptors.

The level of methylation is also controlled locally by the conformation of the receptors themselves. The binding of attractant to a receptor increases the level of methylation of that receptor more than others, presumably by inducing a conformational change that alters the exposure of methylation sites to CheR and CheB (165, 173). Such a receptor-specific adaptation mechanism would enable a cell swimming through multiple chemoeffector gradients to adapt to each gradient independently. This mechanism also generates negative feedback; attractant binding initially causes CCW signaling (decreased CheA activity) but then leads to increased methylation of the receptor, increasing its stimulatory effect on CheA and restoring prestimulus behavior.

When attractant or repellent binds to a receptor, methylation is readjusted, with a characteristic delay that reflects the relative rates of phosphorylation of CheY and CheB, the rates at which CheR and CheB can catalyze reactions at the receptors, and the size of the stimulus. For small stimuli of both kinds, this delay is a few seconds (24, 155). The level of receptor methylation at a given moment will therefore reflect the chemical environment experienced by the cell a few seconds before. This short-term memory, referred to above, allows the cell to make the temporal comparisons that guide its choices to run or tumble. The fact that cells lacking both CheR and CheB activities can migrate through very steep chemoeffector gradients has led to the suggestion that receptor methylation is not needed for efficient chemotaxis (172). Closer examination has shown that an effective response to relevant (not exceedingly steep) gradients requires CheR and CheB (186) as well as methylatable receptors (58).

To a surprising degree, either the global or the local mechanism for controlling receptor methylation is sufficient for chemotaxis in swarm plates. Two mutant variants of CheB that cannot be phosphorylated by CheA have relatively low, unregulated methylesterase activities. Nevertheless, they support efficient chemotaxis in a soft-agar swarming assay, provided they are moderately overexpressed (167). Regulation of CheB activity by phosphorylation is therefore not necessary for chemotaxis in the gradients encountered in swarm plates. One reason may be that the attractants are continually consumed by the swarming bacteria in this assay, which ensures that the attractant concentration will be in a range that gives intermediate CW-CCW motor bias somewhere in the gradient.

Chemotaxis in swarm plates can also proceed without local control of methylation that is specific to the receptor mediating the response, as shown in studies of a mutant variant of the Trg receptor that lacks the sites of methylation. The wild-type form of this receptor mediates responses to ribose and galactose; when the nonmethylatable Trg variant is the only receptor present in a cell, chemotaxis toward those sugars is not observed. When other receptors specific for serine or aspartate are also present, chemotaxis to ribose and galactose is restored (58). The mutant receptor can therefore mediate effective responses to ribose or galactose while not being methylated itself. Adjustments to the level of methylation of the other receptors, mediated by CheB, evidently suffices for adaptation. The role of the other receptors is presumably to send signals that balance those sent by the nonadapting Trg and thus maintain the CW-CCW bias of the flagellar motors at an intermediate value. The rate of adaptation of these cells to stimuli should be very different from that of the wild-type, which implies that the rate of chemotaxis in swarm plates is not sensitive to the details of adaptation kinetics.

Other Factors in Signaling

Acetate can cause cells of *E. coli* to tumble even when CheY is the only chemotaxis protein present in the cell. The properties of mutants with various defects in acetate metabolism suggested that the metabolite acetyladenylate might be involved in this signal, acting through CheY (189). Acetyladenylate is an intermediate product of acetyl CoA synthetase (ACS). A recent reevaluation of the acetate effect (40) showed that it involves not ACS but the enzyme acetate kinase, which produces acetyl phosphate, a compound that can phosphorylate CheY in vitro (102). When a CheY variant that cannot be phosphorylated is used, the acetate effect is not seen; hence it appears to involve phosphorylation rather than any other modification of CheY.

Changes in the intracellular concentration of free Ca^{2+} can affect the bias of the flagellar motors: increased cytoplasmic $[Ca^{2+}]$ increases tumbling, and decreased $[Ca^{2+}]$ suppresses tumbles (180). The precise target of Ca^{2+} action has not been identified; it requires CheY, CheA, and CheW but not the membrane receptors. Whether changes in Ca^{2+} concentration play an essential role in signaling or occur as a result of other events such as the phosphotransfer reactions remains to be determined.

Fumarate can also influence flagellar reversals. Cell envelopes devoid of all chemotaxis proteins except CheY can rotate their flagellar motors either CW or CCW but do so incessantly, apparently never reversing direction. When fumarate is also included, many of the envelopes can reverse direction (8). Fumarate or a metabolite of fumarate therefore appears to be needed for switching. The relevant target has not been identified but is probably either a component of the flagellar switch (described below) or CheY.

Unexplained Sensitivity of Attractant Responses

In a signaling event as presently conceived, binding of chemoeffector to a membrane receptor causes a conformational change in the receptor-CheW-

CheA ternary complex that modulates the autokinase activity of CheA, either stimulating or inhibiting it depending on whether the ligand is a repellent or an attractant. With a repellent, the signal can be highly amplified because CheA acts catalytically to phosphorylate many CheY molecules. Attractant signals must also be amplified (155), but the mechanism is unknown. In the present scheme, binding of an attractant molecule would inhibit a single CheA dimer among several thousand present in a cell and thus should not significantly change the level of phospho-CheY. Modulation of CheZ activity by the receptors could result in amplification of attractant stimuli because a single CheZ molecule, released from sequestration or inhibition by the receptors when an attractant binds, could act catalytically to dephosphorylate many phospho-CheY molecules.

Apart from the demonstration that CheZ interacts with $CheA_s$ (114), no evidence supports the proposal that CheZ activity is controlled. Defects in the putative CheZ-modulation pathway may not have been identified in mutational studies because they have only subtle effects in swarm plate assays. As noted above, some mutants swarm surprisingly well in soft agar. Defects in CheZ regulation might have more pronounced effects in assays using defined, relatively shallow attractant gradients. An apparatus that might prove useful in such an assay was recently described; it employs an array of capillary tubes to form defined gradients and monitor chemotaxis within them (14).

The same reasoning suggests that receptors might take on more than the two signaling states described above. A system employing catalytic activities for both CW and CCW signals would operate better if most of the receptors in a cell were in a neutral state in which the associated kinase is not highly active and CheZ is also inhibited or sequestered. The steady-state rate of ATP usage would thus be reduced, and the receptors would be poised to signal actively in either direction.

HOW BACTERIA SWIM: FLAGELLAR STRUCTURE AND FUNCTION

The flagellum comprises a thin helical filament that acts as propeller, joined via accessory proteins to a hook-shaped structure that serves as a flexible coupling, in turn attached to a basal structure consisting of several rings centered on a rod (Figure 3). The basal structure lies mostly within the cell envelope and the rings are named according to their locations relative to parts of the envelope. The recently discovered C-ring, for example, is located in the cytoplasm just inside the inner membrane (51, 79). The flagellar motor converts chemical energy stored in the transmembrane proton gradient into the mechanical work of rotation. In some species, the sodium gradient is used (63). Rotation is fast [300 Hz (100)], and the process is efficient in that movement of protons

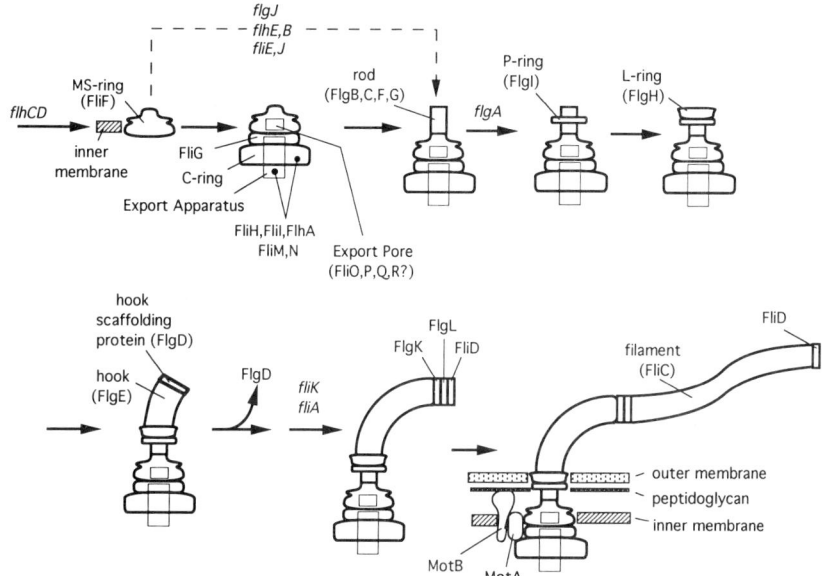

Figure 3 The steps in flagellar assembly. Where component proteins have been identified, they are indicated. The shapes and locations of the export apparatus and export pore are not known; the structures drawn are schematic only. The presence of FliO, P, Q, and R in an export pore is not proven but is suggested by the hydrophobicities and sequence homologies of these proteins (see text). FliM and FliN might be present in the C-ring, but their exact location is not known. Flagellar genes that are needed for assembly but whose exact functions are not known are also indicated at the steps where they are needed. The *flhC* and *flhD* genes encode positive regulators of transcription of all the other flagellar genes and so are needed at the first step. *fliA* encodes a σ factor needed for expression of flagellar genes used late in assembly. *fliK* is needed for control of hook length, but its mechanism of action is unknown. *fliE* encodes a protein of the basal body whose location in the structure is not known. For suggestions concerning the function of the other genes, see the text. Several genes that have regulator roles (*flgM, flgN, fliS,* and *fliT*) do not appear in the scheme.

across the membrane appears to be tightly coupled to rotation, with little slippage (80, 113, 119). The molecular mechanism of energy conversion is not known. Because flagella are relatively complex and depend on a vectorial energy source (the proton gradient), functional reconstitution from purified components has not been achieved, nor even seriously attempted. Still, a great deal is known about the components, their places in the structure, and the sequence of events during assembly.

Flagellar Assembly

The proteins that form most parts of the flagellum have been identified through electron microscopic studies and genetic and biochemical characterization of various flagellar mutants (e.g. 1, 65). The stoichiometry of the component proteins has also been estimated (74, 163). The sequence of events in flagellar assembly (outlined in Figure 3) has been determined by characterizing the partial structures produced in certain mutants (68, 73, 86, 177). The innermost (cell-proximal) structures are assembled first, beginning with the MS-ring found in the cytoplasmic membrane, formed from a single protein called FliF (182). The next step(s)—addition of the basal-body rod—is presently the least well-understood. This step requires, in addition to the four proteins that form the rod, the products of several other genes whose functions are not known precisely but that are likely to be involved in flagellum-specific export.

Most of the flagellum lies outside the cytoplasmic membrane. All except a few of the external component proteins are delivered to their sites of assembly by an export pathway specific to flagella (106). The rod proteins are among those exported via this flagellum-specific pathway, which must partly account for the complexity of the rod-assembly step. Export occurs through a channel in the center of the growing structure (125) and is presumably driven by an apparatus at the base whose shape and exact location are not known. The export apparatus might form part of the C-ring (Figure 3) or might be contained in a separate structure not yet isolated.

Several proteins that function in flagellum-specific export have been identified (183). One of these, called FliI, is especially interesting because it is homologous to the β-subunit of ATP synthases. The protein binds ATP (46) and might use energy from ATP hydrolysis to power steps in export. FliI is quite abundant (~1500 copies per cell), suggesting that it is not a fixed component in the flagellum. Dreyfus et al suggest that it might act as a shuttle, binding to components in the cytoplasm and delivering them to the flagellum (46). Another protein implicated in flagellar export, FlhA, is homologous to several bacterial proteins involved in virulence that probably function in protein secretion [e.g. the LcrD protein of *Yersinia pestis* (144)].

Additional sequence similarities underscore a connection between flagellar assembly and other bacterial secretion systems. The flagellar proteins FliP, FliQ, and FliR are homologous to the Spa24, Spa9, and Spa29 proteins of *Shigella flexneri*, which are involved in the secretion of plasmid-encoded antigens (36, 111, 151). These flagellar proteins are all very hydrophobic, as is the FliO protein encoded in the same operon (111). Given their hydrophobicity and the resemblance of some to proteins involved in secretion, FliO, P, Q, and R may form or stabilize a pore in the membrane through which flagellar components are exported (Figure 3). Several other genes whose precise functions are not known (*flhB, flhE, fliJ*) are cotranscribed with genes implicated in flagellum-specific export, which suggests that they also might encode components of the export apparatus.

The steps following rod assembly are well understood. Two proteins called

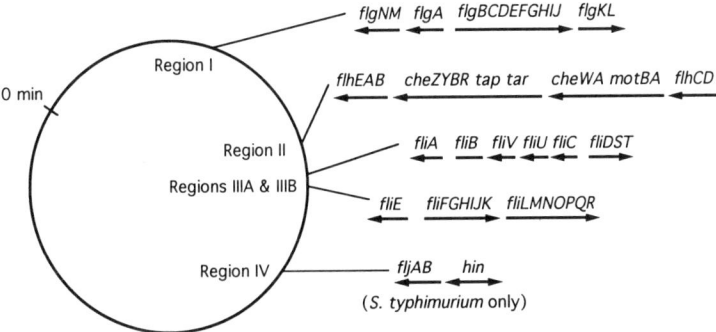

Figure 4 Organization of the genes for flagellation, motility, and chemotaxis in *E. coli* or *S. typhimurium*. The region IV genes, present in *S. typhimurium* only, encode an alternative form of the filament protein and factors needed to control its expression. The *fliB* gene, involved in methylation of the filament protein, has not been demonstrated in *E. coli*. Gene functions are described in the text and in Figure 3.

FlgI and FlgH are secreted via the conventional signal sequence–dependent pathway and assemble on the rod to form the P- and L-rings, respectively (106). Another protein called FlgA is needed for P-ring assembly; it has a signal sequence and is presumably exported into the periplasm (91), but its exact function is not known. Assembly of the P-ring requires in addition the *dsbB* and *dsbA* gene products, which catalyze formation of a disulfide bond in FlgI, the P-ring subunit (41). Subunits of the hook are exported through the basal body and polymerize at the tip of the growing structure. During hook assembly, the FlgD protein resides at the tip, where it presumably promotes polymerization. Upon completion of the hook, it is discarded (130). Hook-filament junction proteins (64), then the filament, are added; all of these components reach their destinations via the flagellum-specific pathway. Two membrane proteins, MotA and MotB, discussed at greater length below, can be added to the flagella last (18, 23).

Salmonella muenchen reportedly contains two new flagellar genes, *fliU* and *fliV*, just downstream of *fliC* (44). In *S. typhimurium*, the *fliB* gene resides in this vicinity (Figure 4). The sequence of *fliB* has not been reported, and its relationship to *fliU* and *fliV* is not clear. The functions of FliU and FliV are not known. Another relatively recent discovery is the *flgN* gene, whose precise function also is not known. Disruption of *flgN* decreases the number of complete flagella and concomitantly increases the number of hook-basal bodies, suggesting that it facilitates filament assembly (91). A gene that seems dis-

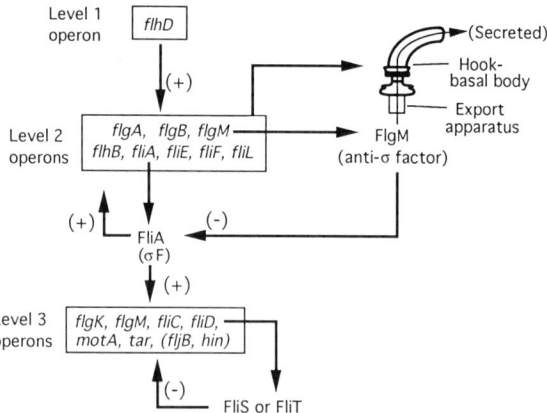

Figure 5 Regulation of transcription of the flagellar genes. The flagellar operons (indicated by their first genes) are organized into a hierarchy with three levels. Both first-level genes (*flhD* and *flhC*) are needed to stimulate transcription of the other genes. FliA is a σ factor needed for transcription of the level 3 genes, which also augments transcription of the level 2 genes. The *flgM* gene was put in both level 2 and 3 because it responds to both FliA and FlhD/FlhC. *flgM* encodes an anti-σ factor that inhibits FliA and that can be pumped out of the cell by the flagellar export apparatus once the hook-basal body is complete. The FliS and/or FliT proteins, products of the *fliD* operon, act as negative regulators of level 3 gene transcription and might function to regulate flagellum length. The *fljB* and *hin* operons are present only in *S. typhimurium*.

pensable in *E. coli* or *S. typhimurium* is *fliL,* which can be deleted with no effect on flagellar structure or function (146).

Many questions remain concerning the specific functions and locations of proteins in the flagellar export apparatus as well as the mechanism by which the apparatus recognizes and transports the appropriate components. Another, probably related, question regards how hook length is determined. The control of hook length involves the FliK protein; mutations in *fliK* give rise to aberrantly long structures called polyhooks (159). A recent analysis of intergenic suppressors also implicated the FlhB protein (62), which as noted above might be involved in flagellar export. As discussed by Macnab (107), a plausible hypothesis is that, when the hook reaches its normal length, the specificity of the export apparatus is switched so that it ceases exporting hook subunits and begins exporting more distal components. The nature of this switch, and the mechanism by which it senses the hook length, are not known. In principal, the hook could be measured by a molecular ruler that contacts it directly. Alternatively, it could be measured indirectly through its effect on export, because as a hook grows longer it could present an increasing impedance to export, influencing events within the export apparatus.

Organization and Regulation of the Flagellar Genes

The flagellar genes are clustered in three (*E. coli*) or four (*S. typhimurium*) regions on the chromosome (Figure 4). [Region III has two subregions (76).] They are further organized into several operons, in many cases according to function. The genes are named *flg, flh, fli,* or *flj* according to the chromosomal region in which they are found. The *fl* designation reflects their involvement in flagellar assembly and function, as distinct from the *che* genes described above, which are essential for chemotaxis but not motility. Two genes called *motA* and *motB* encode products essential for flagellar rotation but not assembly (18, 23, 158).

Expression of the flagellar genes is controlled by a regulatory hierarchy that has three levels (84, 90). Level 1 has two genes (*flhC* and *flhD*) that are both required for the expression of numerous level 2 genes, which are in turn required for expression of the level 3 genes (Figure 5). The level 1 genes are under control of cyclic-AMP (cAMP) levels (in *E. coli*) and other factors probably linked to the cell cycle (127). This regulatory hierarchy roughly parallels the sequence of events in flagellar assembly. Many of the level 2 genes, for example, encode components of the basal body that are used early in assembly, while level 3 genes encode components of the filament that are added later or encode the Che and Mot proteins, which are not needed until flagella are in place.

Expression of the level 3 genes requires almost all of the level 2 genes. Most of the level 2 gene products are components in the flagellum, yet many of them can serve also as regulators of level 3 gene transcription. The mechanism of this regulation was recently elucidated. One of the level 2 genes, *fliA*, encodes an alternative σ factor that is needed for expression of the level 3 genes (128) and also augments the expression of the level 2 genes (89). Another level 2 gene called *flgM* encodes an anti-σ factor that binds to FliA, thus inhibiting level 3 (and to some extent level 2) gene expression (129). Interestingly, the FlgM protein is a substrate of the flagellum-specific export pathway and is actively pumped from the cell upon completion of the basal body and hook (67). Expression of the level 3 genes thus depends upon successful assembly of the basal body and hook and any defect in that structure will inhibit expression of the level 3 genes.

Among the level 3 genes are *fliC*, encoding the protein in the filament, and *fliD*, whose product is needed to direct filament polymerization. Hughes et al (67) suggest that the FlgM export–based regulatory mechanism is an adaptation that allows the cell to adjust gene expression to match the demand for new filaments: When filaments break, as they will if a cell experiences mechanical stress, the shortened filaments should permit more rapid export of FlgM, leading to increased expression of *fliC* and *fliD* and more rapid regrowth of the filament. Given the recent demonstration that level 2 gene expression is also stimulated by FliA (89), both flagellum number and flagellum length may be modulated in this way. FliS and FliT also have been implicated in the control of flagellar length, but their mechanism of action is not known (90).

Components Involved in Torque Generation

Extensive mutational studies have identified five genes that can be mutated to cause paralysis of the flagella with no overt defect in structure. These genes, called *motA, motB, fliG, fliM,* and *fliN,* are the focus of studies directed at understanding the mechanism of torque generation because their products might participate closely in the process.

MotA AND MotB   MotA and MotB are integral membrane proteins that together form a proton-conducting channel across the membrane (19, 20, 53, 176, 188). MotB probably also serves to secure the stationary part of the motor (the stator) to the cell wall so that forces generated by the motor are transmitted effectively to its rotating parts (the rotor) (21, 39). Both MotA and MotB can be incorporated last into otherwise complete flagellar motors. Each motor contains several MotA-MotB complexes (18, 23, 82) that can function independently in torque generation (18, 23).

MotA has four, and MotB a single, hydrophobic segment that could traverse the membrane (43, 166). The topology predicted by the sequence has been confirmed experimentally. MotA crosses the membrane four times; the polar parts of the protein reside mostly in the cytoplasm with only two short segments in the periplasm (20, 197). MotB crosses the membrane once; the bulk of the protein resides in the periplasm (39). A mutational study employing systematic introduction of Trp residues showed that the four hydrophobic segments of MotA are α-helices and identified the parts that are directed toward the channel interior (156). The interior contains very few polar, hydrogen-bonding residues, suggesting that proton conduction occurs largely via water molecules contained in the channel rather than amino acid side chains. An acidic residue in the membrane segment of MotB, Asp32, is conserved in the known sequences and might have an important role in conveying the protons; a mutation that changed it to Asn abolished function (21).

Some bacteria use sodium ions rather than protons to drive flagellar rotation (63). Like proton-driven motors, sodium-driven motors contain several, between 5 and 9, independent torque generators (124). The sodium-conducting membrane components were recently identified in *Vibrio parahaemolyticus* (115, 116). These proteins, called MotX and MotY, do not resemble MotA and MotB except in a short sequence present in both MotB and MotY that is believed to function in attachment to the peptidoglycan layer. The channel component MotX has only a single hydrophobic segment, suggesting that the sodium channel is formed from several copies of the protein (116). Further comparisons of these and other components of the sodium- and proton-driven motors will be informative.

FliG, FliM, AND FliN   Mutations in FliG, FliM, and FliN produce different kinds of defects depending upon the allele. Null mutants have no flagella, implying that these proteins are essential for assembly (69, 162; H Tang, S Lloyd, & DF Blair, unpublished results). Other mutations result in less severe defects, such as paralysis or an increased tendency of the motors to rotate in either the CW or CCW direction (69, 162, 195). Certain mutations in the *fliG, fliM,* or *fliN* genes can be suppressed by mutations in each of the others, suggesting that their products function in a complex, termed the switch complex to reflect its role in controlling the direction of flagellar rotation (194).

The FliG protein has been localized to the cytoplasmic face of the MS-ring (50). A *S. typhimurium* mutant whose FliG protein is translationally fused to FliF (the MS-ring subunit) is motile, implying that these proteins function in close association (50). Since the MS-ring is presumed to be a rotating part of the flagellum, FliG also must rotate. A mutant in which FliM is translationally fused to FliN is also motile (M Kihara & RM Macnab, unpublished results), so FliM and FliN must function in close proximity, either both on the rotor or both on the stator. The weight of evidence favors a location on the rotor. In vitro binding studies suggest an association between FliM and FliF (133). Results from the author's laboratory indicate that FliM, FliN, and FliG form a complex: Isolation of a GST-FliM fusion protein from cell lysates results in copurification of FliG and FliN (H Tang & DF Blair, unpublished results).

FliM and FliN occur in flagellar basal bodies containing a recently discovered feature called the C-ring (51). Preservation of the C-ring requires gentle isolation procedures, which implies that its attachment to the basal body is relatively weak. FliM and FliN might form part of the C-ring; if they were the only components and present in equal stoichiometries, 100 copies of each would be required (51). This is possible, but the C-ring might also contain other, as yet unidentified proteins, including components of the export apparatus.

Because each of the switch-complex proteins can be mutated to give the Mot⁻, or paralyzed, phenotype, all three have been presumed to participate in torque generation (195). Motility can be restored to the *fliM* and *fliN mot* mutants, however, by moderate overexpression of either the mutant proteins or other, evidently interacting, flagellar proteins (H Tang, D Tamir-Elnekave & DF Blair, unpublished results). The *fliM* and *fliN mot* mutations thus appear to affect aspects of flagellar assembly rather than torque-generating activities per se, and they do not support a direct involvement of FliM or FliN in torque generation. In contrast, some *mot* mutants of *fliG* remain immotile when the mutant FliG or other proteins that might interact with it are overexpressed (S Lloyd & DF Blair, unpublished results). Thus, some *fliG* mutations appear to affect torque-generating activities specifically.

The precise function of each switch-complex protein is not known. An extensive analysis of mutants suggests that FliM is involved primarily in direction control (162). This proposal is consistent with the discovery that FliM can bind to phospho-CheY, the CW signal (187). Binding of phospho-CheY presumably triggers a conformational change in FliM that somehow results in clockwise rotation. The FliN protein has significant homology to Spa33, a protein involved in protein secretion in *S. flexneri* (151). A screen for mutants defective in filament growth also identified FliN as having a role in export (183). Thus the primary function of FliN might be in flagellum-specific export rather than torque generation. As noted above, the properties of *fliG* mutants suggest that FliG could be directly involved in torque generation, but its specific role is not known.

Flagellar Motor Dynamics

The dynamics of flagellar rotation have been intensively studied. We can monitor the rotation of individual flagellar motors by tethering cells to coverslips by their flagellar filaments (157), which causes the entire cell body to rotate at easily measured speeds (10–15 Hz). The torque of the motor is balanced by the forces of viscous drag, so torque can be estimated by measuring the rotation speed, cell size, and viscosity of the medium (119). Measurements on tethered cells lead to the following conclusions: The torque of the motor is proportional to the protonmotive force (113), changes little as the temperature varies between 5 and 40°C, and is not affected by substitution of D_2O for H_2O in the medium (80). These results suggest that rotation is tightly coupled to proton flow, with a fixed number of protons (~500) used to drive each revolution (119). At the relatively slow speeds of tethered cells, rotation rate is determined by the energy available and the viscous load, rather than by the rates of internal processes.

The tethering experiment involves unnaturally large loads because the cell body, rather than the thin flagellar filament, rotates. The rotation speed of filament bundles on swimming cells can be measured by the rapid vibrations they induce in the cell bodies (100); the rotation of individual filaments can be monitored directly by darkfield microscopy (87). The filaments on a freely swimming cell rotate at about 300 Hz, which reflects the lighter load. Flagellar motors rotating at this high speed have very different dynamic properties. The torque of the motor depends steeply upon temperature and is significantly greater in H_2O than in D_2O (100). These observations imply that in motors working against small external loads, the speed is limited by the rates of internal processes, probably proton transfers.

The information most useful for understanding the mechanism comes from measurements made with motors rotating at high speeds. Recent studies have used electrorotation to adjust the load on tethered cells rotating at various speeds (15, 70) and have for the first time measured the motor torque across a continuous wide range of speeds (15). The motor torque remained constant at speeds up to about 50 Hz (the exact value depending on the temperature), decreased at higher speeds, and assumed a negative value (i.e. resisting rather than driving rotation) at very high speeds (100–400 Hz, depending upon the temperature). This result contrasts with that of a previous report that torque decreases linearly with increasing speed above zero (100). The torque-speed relationship is useful because it provides information on the rates of processes in the motor and thus constrains mechanisms of torque generation. Specific mechanistic proposals can be developed quantitatively to see if they make predictions in accord with the measured torque-speed relationship. Torque-speed data on various mutants might allow mutational defects to be ascribed to specific steps in torque generation.

Models of Torque Generation

Many models have been proposed to explain torque generation by the flagellar motor (e.g. 16, 80, 94, 118, 131). All models consist essentially of a geometry that describes the trajectories along which the protons and relevant proteins move, together with a mechanism for coupling the proton and rotor movements. Sometimes models give an explicit chemical definition of the coupling mechanism and make quantitative estimates of its efficiency, but most do not.

Because FliG appears to be involved in torque generation and is located in the cytoplasm, the site of torque generation is probably on the cytoplasmic side of the membrane rather than within the bilayer. Measurements of motor torque at various values of pH also suggest that the site of energy conversion is within or near the cytoplasm rather than in the middle of the membrane (81). This result rules out many models that place the site of torque generation within the membrane. Coupling mechanisms can be classed according to whether they are tight (with each revolution coupled to the passage of a fixed number of protons) (118) or loose (with a more flexible stoichiometry) (131). Although loose coupling mechanisms cannot be ruled out conclusively, the available data are most simply interpreted in terms of tight coupling between motor rotation and proton flow.

At least two different classes of tight-coupling models can be imagined. In one, torque generation involves protein conformational changes and cycles of protein binding and release (between rotor and stator), thus resembling the action of ATP-driven motors such as myosin or dynein. A second type of model does not involve large conformational changes or tight interprotein binding but employs more localized, chemically simpler mechanisms (e.g.

electrostatic repulsion) for constraining the movement of motor parts. Currently, no evidence allows us to discriminate between these alternatives. In making analogies to the ATP-driven motors, however, one should note that the flagellar motor is much faster than myosin or dynein. The elementary events in torque generation take place at a rate of about 10,000 s^{-1} (the estimated rate of proton flow through a single MotA-MotB proton channel). For comparison, the dynein turnover rate is of order 100 s^{-1} (145). Mechanisms that do not involve large conformational changes, or cycles of tight protein binding and release, appear more compatible with the rapid pace of events in the flagellar motor.

FUTURE DIRECTIONS

Many questions remain concerning the interactions among chemotaxis proteins and the conformational changes involved in signaling. Progress will depend increasingly upon high-resolution structures of the components, alone or in complexes with each other. A final test of our understanding will be to account for the in vivo signaling process quantitatively in terms of properties of the components (affinities and reaction rates) measured in vitro. Bacterial chemotaxis is sometimes described as a relatively simple model system in which prospects for achieving a quantitative understanding are good. Still, simulations of bacterial chemotaxis by computer require numerous parameters (34), many of which remain unmeasured. Much could be learned by closer examination of the chemotactic response in strains with various defects; the methods that provide the most detailed information, such as iontophoresis or controlled effector ramps, have been used with relatively few of the existing mutants.

As noted above, a major challenge in understanding flagellar assembly will be to elucidate the structure and mechanism of the export apparatus. These studies will overlap and benefit from work on the many related secretion systems in bacteria. A detailed understanding of the mechanism of torque generation will require the combined application of genetic, physiological, and biochemical approaches. Continued in vitro studies of purified motor components should allow their interactions to be characterized. More intensive mutational studies should lead to a more precise definition of the proteins and amino acid residues that participate in torque generation. Eventually, structural studies on the key components could provide a basis for formulating a chemically explicit model of torque generation.

ACKNOWLEDGMENTS

Work in my laboratory was supported by grant GM1-46683 from the National Institute of General Medical Sciences and grant MCB-9117785 from the National Science Foundation. I thank the many colleagues who furnished unpublished information and helpful comments on the manuscript.

> Any *Annual Review* chapter, as well as any article cited in an *Annual Review* chapter, may be purchased from the Annual Reviews Preprints and Reprints service.
> 1-800-347-8007; 415-259-5017; email: arpr@class.org

Literature Cited

1. Aizawa S-I, Dean GE, Jones CJ, Macnab RM, Yamaguchi S. 1985. Purification and characterization of flagellar hook-basal body complex of *Salmonella typhimurium*. *J. Bacteriol.* 161:836–49
2. Alex LA, Simon MI. 1994. Protein histidine kinases and signal transduction in prokaryotes and eukaryotes. *Trends Genet.* 10:133–38
3. Ames P, Parkinson JS. 1988. Transmembrane signaling by bacterial chemoreceptors: *E. coli* transducers with locked signal output. *Cell* 55:817–26
4. Ames P, Parkinson JS. 1994. Constitutively signaling fragments of Tsr, the *Escherichia coli* serine chemoreceptor. *J. Bacteriol.* 176:6340–48
5. Deleted in proof
6. Asakura S, Honda H. 1984. Two-state model for bacterial chemoreceptor proteins. The role of multiple methylation. *J. Mol. Biol.* 176:349–67
7. Barak R, Eisenbach M. 1992. Correlation between phosphorylation of the chemotaxis protein CheY and its activity at the flagellar motor. *Biochemistry* 31:1821–26
8. Barak R, Eisenbach M. 1992. Fumarate or a fumarate metabolite restores switching ability to rotating flagella of bacterial envelopes. *J. Bacteriol.* 174:643–45
9. Barnakov AN, Downing KH, Hazelbauer GL. 1994. Studies of the structural organization of a bacterial chemoreceptor by electron microscopy. *J. Struct. Biol.* 112:117–24
10. Bellsolell L, Prieto J, Serrano L, Coll M. 1994. Magnesium binding to the bacterial chemotaxis protein CheY results in large conformational changes involving its functional surface. *J. Mol. Biol.* 238:489–95
11. Berg HC. 1993. *Random Walks in Biology*. Princeton: Princeton Univ. Press. 152 pp. 2nd ed.
12. Berg HC, Anderson RA. 1973. Bacteria swim by rotating their flagellar filaments. *Nature* 245:380–82
13. Berg HC, Brown DA. 1972. Chemotaxis in *Escherichia coli* analysed by three-dimensional tracking. *Nature* 239:500–4
14. Berg HC, Turner L. 1990. Chemotaxis of bacteria in glass capillary arrays. *Biophys. J.* 58:919–30
15. Berg HC, Turner L. 1993. Torque generated by the flagellar motor of *Escherichia coli*. *Biophys. J.* 65:2201–16
16. Berry RM. 1993. Torque and switching in the bacterial flagellar motor. An electrostatic model. *Biophys J.* 64:961–73
17. Biemann H-P, Koshland DE Jr. 1994. Aspartate receptors of *Escherichia coli* and *Salmonella typhimurium* bind ligand with negative and half-of-the-sites cooperativity. *Biochemistry* 33:629–34
18. Blair DF, Berg HC. 1988. Restoration of torque in defective flagellar motors. *Science* 242:1678–81
19. Blair DF, Berg HC. 1990. The MotA protein of *E. coli* is a proton-conducting component of the flagellar motor. *Cell* 60:439–49
20. Blair DF, Berg HC. 1991. Mutations in the MotA protein of *Escherichia coli* reveal domains critical for proton conduction. *J. Mol. Biol.* 221:1433–42
21. Blair DF, Kim DY, Berg HC. 1991. Mutant MotB proteins in *Escherichia coli*. *J. Bacteriol.* 173:4049–55
22. Blat Y, Eisenbach M. 1994. Phosphorylation-dependent binding of the chemotaxis signal molecule CheY to its phosphatase, CheZ. *Biochemistry* 33:902–6
23. Block SM, Berg HC. 1984. Successive incorporation of force-generating units in the bacterial rotary motor. *Nature* 309:470–72
24. Block SM, Segall JE, Berg HC. 1982. Impulse responses in bacterial chemotaxis. *Cell* 31:215–26
25. Block SM, Segall JE, Berg HC. 1983. Adaptation kinetics in bacterial chemotaxis. *J. Bacteriol.* 154:312–23
26. Bollinger J, Park C, Harayama S, Hazelbauer GL. 1984. Structure of the Trg

protein: homologies with and differences from other sensory transducers of *Escherichia coli*. *Proc. Natl. Acad. Sci. USA* 81:3287–91
27. Borkovich KA, Alex LA, Simon MI. 1992. Attenuation of sensory receptor signaling by covalent modification. *Proc. Natl. Acad. Sci. USA* 89:6756–60
28. Borkovich KA, Kaplan N, Hess JF, Simon MI. 1989. Transmembrane signal transduction in bacterial chemotaxis involves ligand-dependent activation of phosphate group transfer. *Proc. Natl. Acad. Sci. USA* 86:1208–12
29. Borkovich KA, Simon MI. 1990. The dynamics of protein phosphorylation in bacterial chemotaxis. *Cell* 63:1339–48
30. Bourret RB, Borkovich KA, Simon MI. 1991. Signal transduction pathways involving protein phosphorylation in prokaryotes. *Annu. Rev. Biochem.* 60:401–41
31. Bourret RB, Davagnino J, Simon MI. 1993. The carboxy-terminal portion of the CheA kinase mediates regulation of autophosphorylation by transducer and CheW. *J. Bacteriol.* 175:2097–101
32. Bourret RB, Hess JF, Simon MI. 1990. Conserved aspartate residues and phosphorylation in signal transduction by the chemotaxis protein CheY. *Proc. Natl. Acad. Sci. USA* 87:41–45
33. Boyd A, Kendall K, Simon MI. 1983. Structure of the serine chemoreceptor in *Escherichia coli*. *Nature* 301:623–26
34. Bray D, Bourret RB, Simon MI. 1993. Computer simulation of the phosphorylation cascade controlling bacterial chemotaxis. *Mol. Biol. Cell* 4:469–82
35. Brown DA, Berg HC. 1974. Temporal stimulation of chemotaxis in *Escherichia coli*. *Proc. Natl. Acad. Sci. USA* 71:1388–92
36. Carpenter PB, Zuberi AR, Ordal GW. 1993. *Bacillus subtilis* flagellar proteins FliP, FliQ, FliR and FlhB are related to *Shigella flexneri* virulence factors. *Gene* 137:243–45
37. Chang C, Kwok SF, Bleecker AB, Meyerowitz EM. 1993. *Arabidopsis* ethylene-response gene *ETR1*: similarity of product to two-component regulators. *Science* 262:539–44
38. Chervitz SA, Lin CM, Falke JJ. 1995. Engineered disulfides at the subunit interface of the aspartate receptor: implications for the mechanism of transmembrane signaling. *Biochemistry.* Submitted
39. Chun SY, Parkinson JS. 1988. Bacterial motility: membrane topology of the *Escherichia coli* MotB protein. *Science* 239:276–78
40. Dailey FE, Berg HC. 1993. Change in direction of flagellar rotation in *Escherichia coli* mediated by acetate kinase. *J. Bacteriol.* 175:3236–39
41. Dailey FE, Berg HC. 1993. Mutants in disulfide bond formation that disrupt flagellar assembly in *Escherichia coli*. *Proc. Natl. Acad. Sci. USA* 90:1043–47
42. Danielson MA, Biemann H-P, Koshland DE Jr, Falke JJ. 1994. Attractant- and disulfide-induced conformational changes in the ligand binding domain of the chemotaxis aspartate receptor: a ^{19}F NMR study. *Biochemistry* 33:6100–9
43. Dean GE, Macnab RM, Stader J, Matsumura P, Burke C. 1984. Gene sequence and predicted amino-acid sequence of the MotA protein, a membrane-associated protein required for flagellar rotation in *Escherichia coli*. *J. Bacteriol.* 143:991–99
44. Doll L, Frankel G. 1993. Cloning and sequencing of two new *fli* genes, the products of which are essential for *Salmonella* flagellar biosynthesis. *Gene* 126:119–21
45. Drake SK, Bourret RB, Luck LA, Simon MI, Falke JJ. 1993. Activation of the phosphosignaling protein CheY. I. Analysis of the phosphorylated conformation by ^{19}F NMR and protein engineering. *J. Biol. Chem.* 268:13081–88
46. Dreyfus G, Williams AW, Kawagishi I, Macnab RM. 1993. Genetic and biochemical analysis of *Salmonella typhimurium* FliI, a flagellar protein related to the catalytic subunit of the F_0F_1 ATPase and to virulence proteins of mammalian and plant pathogens. *J. Bacteriol.* 175:3131–38
47. Dunten P, Koshland DE Jr. 1991. Tuning the responsiveness of a sensory receptor via covalent modification. *J. Biol. Chem.* 266:1491–96
48. Falke JJ, Dernburg AF, Sternberg DA, Zalkin N, Milligan DL, Koshland DE Jr. 1988. Structure of a bacterial sensory receptor: a site-directed sulfhydryl study. *J. Biol. Chem.* 263:14850–58
49. Falke JJ, Koshland DE Jr. 1987. Global flexibility in a sensory receptor: a site-directed cross-linking approach. *Science* 237:1596–600
50. Francis NR, Irikura VM, Yamaguchi S, DeRosier DJ, Macnab RM. 1992. Localization of the *Salmonella typhimurium* flagellar switch protein FliG to the cytoplasmic M-ring face of the basal body. *Proc. Natl. Acad. Sci. USA* 89:6304–8
51. Francis NR, Sosinsky GE, Thomas D, DeRosier DJ. 1994. Isolation, characterization and structure of bacterial flag-

ellar motors containing the switch complex. *J. Mol. Biol.* 235:1261-70
52. Gardina P, Conway C, Kossman M, Manson M. 1992. Aspartate and maltose-binding protein interact with adjacent sites in the Tar chemotactic signal transducer of *Escherichia coli. J. Bacteriol.* 174:1528-36
53. Garza AG, Harris-Haller LW, Stoebner RA, Manson MD. 1995. Motility protein interactions in the bacterial flagellar motor. *Proc. Natl. Acad. Sci. USA* 92:1970-74
54. Gegner JA, Dahlquist FW. 1991. Signal transduction in bacteria: CheW forms a reversible complex with the protein kinase CheA. *Proc. Natl. Acad. Sci. USA* 88:750-54
55. Gegner JA, Graham DR, Roth AF, Dahlquist FW. 1992. Assembly of an MCP receptor, CheW, and kinase CheA complex in the bacterial chemotaxis signal transduction pathway. *Cell* 70:975-82
56. Harshey RM, Matsuyama T. 1994. Dimorphic transition in *Escherichia coli* and *Salmonella typhimurium:* surface-induced differentiation into hyperflagellate swarmer cells. *Proc. Natl. Acad. Sci. USA* 91:8631-35
57. Hazelbauer GL. 1992. Bacterial chemoreceptors. *Curr. Opin. Struct. Biol.* 2:505-10
58. Hazelbauer GL, Park C, Nowlin DM. 1989. Adaptational "crosstalk" and the crucial role of methylation in chemotactic migration by *Escherichia coli. Proc. Natl. Acad. Sci. USA* 86:1448-52
59. Hess JF, Bourret RB, Simon MI. 1988. Histidine phosphorylation and phosphoryl group transfer in bacterial chemotaxis. *Nature* 336:139-43
60. Hess JF, Oosawa K, Kaplan N, Simon MI. 1988. Phosphorylation of three proteins in the signaling pathway of bacterial chemotaxis. *Cell* 53:79-87
61. Hess JF, Oosawa K, Matsumura P, Simon MI. 1987. Protein phosphorylation is involved in bacterial chemotaxis. *Proc. Natl. Acad. Sci. USA* 84:7609-13
62. Hirano T, Yamaguchi S, Oosawa K, Aizawa S-I. 1994. Roles of FliK and FlhB in determination of flagellar hook length in *Salmonella typhimurium. J. Bacteriol.* 176:5439-49
63. Hirota N, Kitada M, Imae Y. 1981. Flagellar motors of alkalophilic *Bacillus* are powered by an electrochemical potential gradient of Na^+. *FEBS Lett.* 132: 278-80
64. Homma M, Iino T. 1985. Locations of hook-associated proteins in flagellar structures of *Salmonella typhimurium. J. Bacteriol.* 162:183-89
65. Homma M, Ohnishi K, Iino T, Macnab RM. 1987. Identification of flagellar hook and basal body gene products (FlaFV, FlaFVI, FlaFVII, and FlaFVIII) in *Salmonella typhimurium. J. Bacteriol.* 169:3617-24
66. Huang C, Stewart RC. 1993. CheZ mutants with enhanced ability to dephosphorylate CheY, the response regulator in bacterial chemotaxis. *Biochim. Biophys. Acta* 1202:297-304
67. Hughes KT, Gillen KL, Semon MJ, Karlinsey JE. 1993. Sensing structural intermediates in bacterial flagellar assembly by export of a negative regulator. *Science* 262:1277-80
68. Iino T. 1985. Genetic control of flagellar morphogenesis in *Salmonella.* In *Sensing and Response in Microorganisms,* ed. M Eisenbach, M Balaban, pp. 83-92. Amsterdam: Elsevier
69. Irikura VM, Kihara M, Yamaguchi S, Sockett H, Macnab RM. 1993. *Salmonella typhimurium fliG* and *fliN* mutations causing defects in assembly, rotation, and switching of the flagellar motor. *J. Bacteriol.* 175:802-10
70. Iwazawa J, Imae Y, Kobayasi S. 1993. Study of the torque of the bacterial flagellar motor using a rotating electric field. *Biophys. J.* 64:925-33
71. Jeffery CJ, Koshland DE Jr. 1994. A single hydrophobic to hydrophobic substitution in the transmembrane domain impairs aspartate receptor function. *Biochemistry* 33:3457-63
72. Jones CJ, Aizawa S-I. 1991. The bacterial flagellum and flagellar motor: structure, assembly and function. *Adv. Microb. Physiol.* 32:109-72
73. Jones CJ, Macnab RM. 1990. Flagellar assembly in *Salmonella typhimurium:* analysis with temperature-sensitive mutants. *J. Bacteriol.* 172:1327-39
74. Jones CJ, Macnab RM, Okino H, Aizawa S-I. 1990. Stoichiometric analysis of the flagellar hook-(basal body) complex of *Salmonella typhimurium. J. Mol. Biol.* 212:377-87
75. Kar L, Matsumura P, Johnson ME. 1992. Bivalent-metal binding to CheY protein. Effect on protein conformation. *Biochem. J.* 287:521-31
76. Kawagishi I, Müller V, Williams AW, Irikura VM, Macnab RM. 1992. Subdivision of flagellar region III of the *Escherichia coli* and *Salmonella typhimurium* chromosomes and identification of two additional flagellar genes. *J. Gen. Microbiol.* 138:1051-65
77. Kehry MR, Bond MW, Hunkapiller MW, Dahlquist FW. 1983. Enzymatic deamidation of methyl-accepting chem-

otaxis proteins in *Escherichia coli* catalyzed by the *cheB* gene product. *Proc. Natl. Acad. Sci. USA* 80:3599–603
78. Kehry MR, Doak TG, Dahlquist FW. 1984. Stimulus-induced changes in methylesterase activity during chemotaxis in *Escherichia coli*. *J. Biol. Chem.* 259:11828–35
79. Khan IH, Reese TS, Khan S. 1992. The cytoplasmic component of the bacterial flagellar motor. *Proc. Natl. Acad. Sci. USA* 89:5956–60
80. Khan S, Berg HC. 1983. Isotope and thermal effects in chemiosmotic coupling to the flagellar motor of Streptococcus. *Cell* 32:913–19
81. Khan S, Dapice M, Humayun I. 1990. Energy transduction in the bacterial flagellar motor: effects of load and pH. *Biophys. J.* 57:779–96
82. Khan S, Dapice M, Reese TS. 1988. Effects of *mot* gene expression on the structure of the flagellar motor. *J. Mol. Biol.* 202:575–84
83. Kofoid EC, Parkinson JS. 1991. Tandem translation starts in the *cheA* locus of *Escherichia coli*. *J. Bacteriol.* 173:2116–19
84. Komeda Y. 1986. Transcriptional control of flagellar genes in *Escherichia coli* K-12. *J. Bacteriol.* 168:1315–18
85. Kort EN, Goy MF, Larsen SH, Adler J. 1975. Methylation of a membrane protein involved in bacterial chemotaxis. *Proc. Natl. Acad. Sci. USA* 72:3939–43
86. Kubori T, Shimamoto N, Yamaguchi S, Namba K, Aizawa S-I. 1992. Morphological pathway of flagellar assembly in *Salmonella typhimurium*. *J. Mol. Biol.* 226:433–46
87. Kudo S, Magariyama Y, Aizawa S-I. 1990. Abrupt changes in flagellar rotation observed by laser dark-field microscopy. *Nature* 346:677–80
88. Kuo SC, Koshland DE Jr. 1987. Roles of *cheY* and *cheZ* gene products in controlling flagellar rotation in bacterial chemotaxis of *Escherichia coli*. *J. Bacteriol.* 169:1307–14
89. Kutsukake K, Iino T. 1994. Role of the FliA-FlgM regulatory system on the transcriptional control of the flagellar regulon and flagellar formation in *Salmonella typhimurium*. *J. Bacteriol.* 176:3598–605
90. Kutsukake K, Ohya Y, Iino T. 1990. Transcriptional analysis of the flagellar regulon of *Salmonella typhimurium*. *J. Bacteriol.* 172:741–47
91. Kutsukake K, Okada T, Yokoseki T, Iino T. 1994. Sequence analysis of the *flgA* gene and its adjacent region in *Salmonella typhimurium*, and identification of another flagellar gene, *flgN*. *Gene* 143:49–54
92. Larsen SH, Adler J, Gargus JJ, Hogg RW. 1974. Chemomechanical coupling without ATP: the source of energy for motility and chemotaxis in bacteria. *Proc. Natl. Acad. Sci. USA* 71:1239–43
93. Larsen SH, Reader RW, Kort EN, Tso W-W, Adler J. 1974. Change in direction of flagellar rotation is the basis of the chemotactic response in *E. coli*. *Nature* 249:74–77
94. Lauger P. 1988. Torque and rotation rate of the bacterial flagellar motor. *Biophys. J.* 53:53–65
95. Lee GF, Burrows GG, Lebert MR, Dutton DP, Hazelbauer GE. 1994. Deducing the organization of a transmembrane domain by disulfide crosslinking: the bacterial chemoreceptor Trg. *J. Biol. Chem.* 269:29920–27
96. Lee GF, Lebert MR, Lilly AA, Hazelbauer GE. 1995. Transmembrane signaling characterized in bacterial chemoreceptors using sulfhydryl crosslinking in vivo. *Proc. Natl. Acad. Sci. USA* 92:3391–95
97. Lin L-N, Li J, Brandts JF, Weis RM. 1994. The serine receptor of bacterial chemotaxis exhibits half-site saturation for serine binding. *Biochemistry* 33:6564–70
98. Liu J, Parkinson JS. 1989. Role of CheW protein in coupling membrane receptors to the intracellular signalling system of bacterial chemotaxis. *Proc. Natl. Acad. Sci. USA* 86:8703–7
99. Liu JD, Parkinson JS. 1991. Genetic evidence for interaction between the CheW and Tsr proteins during chemoreceptor signaling by *Escherichia coli*. *J. Bacteriol.* 173:4941–51
100. Lowe G, Meister M, Berg HC. 1987. Rapid rotation of flagellar bundles in swimming bacteria. *Nature* 325:637–40
101. Lukat GS, Lee BH, Mottonen JM, Stock AM, Stock JB. 1991. Roles of the highly conserved aspartate and lysine residues in the response regulator of bacterial chemotaxis. *J. Biol. Chem.* 266:8348–54
102. Lukat GS, McCleary WR, Stock AM, Stock JB. 1992. Phosphorylation of bacterial response regulator proteins by low molecular weight phospho-donors. *Proc. Natl. Acad. Sci. USA* 89:718–22
103. Lukat GS, Stock AM, Stock JB. 1990. Divalent metal ion binding to the CheY protein and its significance to phosphotransfer in bacterial chemotaxis. *Biochemistry* 29:5436–42
104. Lupas A, Stock J. 1989. Phosphorylation of an N-terminal regulatory domain ac-

tivates the CheB methylesterase in bacterial chemotaxis. *J. Biol. Chem.* 264: 17337–42
105. Lynch BA, Koshland DE Jr. 1991. Disulfide cross-linking studies of the transmembrane regions of the aspartate sensory receptor of *Escherichia coli*. *Proc. Natl. Acad. Sci. USA* 88:10402–6
106. Macnab R. 1992. Genetics and biogenesis of bacterial flagella. *Annu. Rev. Genet.* 26:129–56
107. Macnab RM. 1990. Genetics, structure, and assembly of the bacterial flagellum. *Symp. Soc. Gen. Microbiol.* 46:77–106
108. Macnab RM, Koshland DE Jr. 1972. The gradient-sensing mechanism in bacterial chemotaxis. *Proc. Natl. Acad. Sci. USA* 69:2509–12
109. Maddock JR, Shapiro L. 1993. Polar location of the chemoreceptor complex in the *Escherichia coli* cell. *Science* 259:1717–23
110. Maeda T, Wurgler-Murphy SM, Saito H. 1994. A two-component system that regulates an osmosensing MAP kinase cascade in yeast. *Nature* 369:242–45
111. Malekooti J, Ely B, Matsumura P. 1994. Molecular characterization, nucleotide sequence, and expression of the *fliO*, *fliP*, *fliQ*, and *fliR* genes of *Escherichia coli*. *J. Bacteriol.* 176:189–97
112. Manoil C, Beckwith J. 1986. A genetic approach to analyzing membrane protein topology. *Science* 233:1403–8
113. Manson MD, Tedesco P, Berg HC. 1980. Energetics of flagellar rotation in bacteria. *J. Mol. Biol.* 138:541–61
114. Matsumura P, Roman S, Volz K, McNally D. 1990. Signalling complexes in bacterial chemotaxis. *Symp. Soc. Gen. Microbiol.* 46:135–54
115. McCarter LL. 1994. MotY, a component of the sodium-type flagellar motor. *J. Bacteriol.* 176:4219–25
116. McCarter LL. 1994. MotX, the channel component of the sodium-type flagellar motor. *J. Bacteriol.* 176:5988–98
117. McNally DF, Matsumura P. 1991. Bacterial chemotaxis signaling complexes: formation of a CheA/CheW complex enhances autophosphorylation and affinity for CheY. *Proc. Natl. Acad. Sci. USA* 88:6269–73
118. Meister M, Caplan SR, Berg HC. 1989. Dynamics of a tightly coupled mechanism for flagellar rotation. *Biophys. J.* 55:905–14
119. Meister M, Lowe G, Berg HC. 1987. The proton flux through the bacterial flagellar motor. *Cell* 49:643–50
120. Mesibov R, Ordal GW, Adler J. 1973. The range of attractant concentrations for bacterial chemotaxis and the threshold and size of response over this range. *J. Gen. Physiol.* 62:203–23
121. Milburn MV, Privé GG, Milligan DL, Scott WG, Yeh J, et al. 1991. Three-dimensional structures of the ligand-binding domain of the bacterial aspartate receptor with and without a ligand. *Science* 254:1342–47
122. Milligan DL, Koshland DE Jr. 1988. Site-directed cross-linking: establishing the dimeric structure of the aspartate receptor of bacterial chemotaxis. *J. Biol. Chem.* 263:6268–75
123. Morrison TB, Parkinson JS. 1994. Liberation of an interaction domain from the phosphotransfer region of CheA, a signaling kinase of *Escherichia coli*. *Proc. Natl. Acad. Sci. USA* 91:5485–89
124. Muramoto K, Sugiyama S, Cragoe EJ Jr, Imae Y. 1994. Successive inactivation of the force-generating units of sodium-driven bacterial flagellar motors by a photoreactive amiloride analog. *J. Biol. Chem.* 269:3374–80
125. Namba K, Yamashita I, Vonderviszt F. 1989. Structure of the core and central channel of bacterial flagella. *Nature* 342:648–54
126. Ninfa EG, Stock A, Mowbray S, Stock J. 1991. Reconstitution of the bacterial chemotaxis signal transduction system from purified components. *J. Biol. Chem.* 266:9764–70
127. Nishimura A, Hirota Y. 1989. A cell division regulatory mechanism controls the flagellar regulon in *Escherichia coli*. *Mol. Gen. Genet.* 216:340–46
128. Ohnishi K, Kutsukake K, Suzuki H, Iino T. 1990. Gene *fliA* encodes an alternative sigma factor specific for flagellar operons in *Salmonella typhimurium*. *Mol. Gen. Genet.* 221:139–47
129. Ohnishi K, Kutsukake K, Suzuki H, Iino T. 1992. A novel transcriptional regulation mechanism in the flagellar regulon of *Salmonella typhimurium*: an anti-sigma factor inhibits the activity of the flagellum-specific sigma factor, sigma F, *Mol. Microbiol.* 6:3149–57
130. Ohnishi K, Ohto Y, Aizawa S-I, Macnab RM, Iino T. 1994. FlgD is a scaffolding protein needed for flagellar hook assembly in *Salmonella typhimurium*. *J. Bacteriol.* 176:2272–81
131. Oosawa F, Masai J. 1982. Mechanism of flagellar motor rotation in bacteria. *J. Phys. Soc. Jpn.* 51:631–64
132. Oosawa K, Hess JF, Simon MI. 1988. Mutants defective in bacterial chemotaxis show modified protein phosphorylation. *Cell* 53:89–96
133. Oosawa K, Ueno T, Aizawa S-I. 1994. Overproduction of the bacterial flagellar

switch proteins and their interactions with the MS ring complex *in vitro*. *J. Bacteriol.* 176:3683–91
134. Ota IM, Varshavsky A. 1993. A yeast protein similar to bacterial two-component regulators. *Science* 262:566–69
135. Pakula A, Simon MI. 1992. Determination of transmembrane protein structure by disulfide cross-linking: the *Escherichia coli* Tar receptor. *Proc. Natl. Acad. Sci. USA* 89:4144–48
136. Pakula A, Simon MI. 1992. Pivots or pistons? *Nature* 355:496–97
137. Park C, Dutton DP, Hazelbauer GL. 1990. Effects of glutamines and glutamates at sites of covalent modification of a methyl-accepting transducer. *J. Bacteriol.* 172:7179–87
138. Parkinson JS. 1978. Complementation analysis and deletion mapping of *Escherichia coli* mutants defective in chemotaxis. *J. Bacteriol.* 135:45–53
139. Parkinson JS. 1993. Signal transduction schemes of bacteria. *Cell* 73:857–71
140. Parkinson JS, Blair DF. 1993. Does *E. coli* have a nose? *Science* 259:1701–2
141. Parkinson JS, Kofoid EC. 1992. Communication modules in bacterial signaling proteins. *Annu. Rev. Genet.* 26:71–112
142. Parkinson JS, Parker SR. 1979. Interaction of the *cheC* and *cheZ* gene products is required for chemotactic behavior in *Escherichia coli*. *Proc. Natl. Acad. Sci. USA* 76:2390–94
143. Parkinson JS, Parker SR, Talbert PB, Houts SE. 1983. Interactions between chemotaxis and flagellar genes in *Escherichia coli*. *J. Bacteriol.* 155:265–74
144. Plano GV, Barve SS, Straley SC. 1991. LcrD, a membrane-bound regulator of the *Yersinia pestis* low-calcium response. *J. Bacteriol.* 173:7293–303
145. Porter ME, Johnson KA. 1989. Dynein structure and function. *Annu. Rev. Cell Biol.* 5:119–51
146. Raha M, Sockett H, Macnab RM. 1994. Characterization of the *fliL* gene in the flagellar regulon of *Escherichia coli* and *Salmonella typhimurium*. *J. Bacteriol.* 176:2308–11
147. Roman SJ, Meyers M, Volz K, Matsumura P. 1992. A chemotactic signaling surface on CheY defined by suppressors of flagellar switch mutations. *J. Bacteriol.* 174:6247–55
148. Russo AF, Koshland DE Jr. 1983. Separation of signal transduction and adaptation functions of the aspartate receptor in bacterial sensing. *Science* 220:1016–20
149. Sanatinia H, Kofoid EC, Morrison TM, Parkinson JS. 1995. The smaller of two overlapping *cheA* gene products is not essential for chemotaxis in *Escherichia coli*. *J. Bacteriol.* 177:2713–20
150. Sanders DA, Gillece-Castro BL, Stock AM, Burlingame AL, Koshland DE Jr. 1989. Identification of the site of phosphorylation of the chemotaxis response regulator protein, CheY. *J. Biol. Chem.* 264:21770–78
151. Sasakawa C, Komatsu K, Tobe T, Fukuda I, Suzuki T, Yoshikawa M. 1993. Eight genes in region 5 that form an operon are essential for invasion of epithelial cells by *Shigella flexneri* 2a. *J. Bacteriol.* 175:2334–46
152. Schuster SC, Khan S. 1994. The bacterial flagellar motor. *Annu. Rev. Biophys. Biomol. Struct.* 23:509–39
153. Schuster SC, Swanson RV, Alex LA, Bourret RB, Simon MI. 1993. Assembly and function of a quaternary signal transduction complex monitored by surface plasmon resonance. *Nature* 365:343–47
154. Scott WG, Milligan DL, Milburn MV, Privé GG, Yeh J, et al. 1993. Refined structures of the ligand-binding domain of the aspartate receptor from *Salmonella typhimurium*. *J. Mol. Biol.* 232:555–73
155. Segall JE, Block SM, Berg HC. 1986. Temporal comparisons in bacterial chemotaxis. *Proc. Natl. Acad. Sci. USA* 83:8987–91
156. Sharp LL, Zhou J, Blair DF. 1995. Features of MotA proton channel structure revealed by tryptophan-scanning mutagenesis. *Proc. Natl. Acad. Sci. USA*. In press
157. Silverman M, Simon M. 1974. Flagellar rotation and the mechanism of bacterial motility. *Nature* 249:73–74
158. Silverman M, Simon M. 1976. Operon controlling motility and chemotaxis in *E. coli*. *Nature* 264:577–80
159. Silverman MR, Simon MI. 1972. Flagellar assembly mutants in *Escherichia coli*. *J. Bacteriol.* 112:986–93
160. Simms SA, Keane MG, Stock J. 1985. Multiple forms of the CheB methylesterase in bacterial chemosensing. *J. Biol. Chem.* 260:10161–68
161. Smith RA, Parkinson JS. 1980. Overlapping genes at the *cheA* locus of *Escherichia coli*. *Proc. Natl. Acad. Sci. USA* 77:5370–74
162. Sockett H, Yamaguchi S, Kihara M, Irikura VM, Macnab RM. 1992. Molecular analysis of the flagellar switch protein FliM of *Salmonella typhimurium*. *J. Bacteriol.* 174:793–806
163. Sosinsky GE, Francis NR, DeRosier DJ, Wall JS, Simon MN, Hainfeld J. 1992.

Mass determination and estimation of subunit stoichiometry of the bacterial hook-basal body flagellar complex of *Salmonella typhimurium* by scanning transmission electron microscopy. *Proc. Natl. Acad. Sci. USA* 89:4801–5
164. Springer MS, Goy MF, Adler J. 1979. Protein methylation in behavioural control mechanisms and in signal transduction. *Nature* 280:279–84
165. Springer MS, Zanolari B, Pierzchala PA. 1982. Ordered methylation of the methyl-accepting chemotaxis proteins of *Escherichia coli*. *J. Biol. Chem.* 257:6861–66
166. Stader J, Matsumura P, Vacante D, Dean GE, Macnab RM. 1986. Nucleotide sequence of the *Escherichia coli motB* gene and site-limited incorporation of its product into the cytoplasmic membrane. *J. Bacteriol.* 166:244–52
167. Stewart RC. 1993. Activating and inhibitory mutations in the regulatory domain of CheB, the methylesterase in bacterial chemotaxis. *J. Biol. Chem.* 268:1921–30
168. Stewart RC, Roth AF, Dahlquist FW. 1990. Mutations that affect control of the methylesterase activity of CheB, a component of the chemotaxis adaptation system in *Escherichia coli*. *J. Bacteriol.* 172:3388–99
169. Stock A, Chen T, Welsh D, Stock J. 1988. CheA protein, a central regulator of bacterial chemotaxis, belongs to a family of proteins that control gene expression in response to changing environmental conditions. *Proc. Natl. Acad. Sci. USA* 85:1403–7
170. Stock AM, Martinez-Hackert E, Rasmussen BF, West AH, Stock JB, et al. 1993. Structure of the Mg^{2+}-bound form of CheY and mechanism of phosphoryl transfer in bacterial chemotaxis. *Biochemistry* 32:13375–80
171. Stock AM, Mottonen JM, Stock JB, Schutt CE. 1989. Three-dimensional structure of CheY, the response regulator of bacterial chemotaxis. *Nature* 337:745–49
172. Stock J, Kersulis G, Koshland DE Jr. 1985. Neither methylating nor demethylating enzymes are required for bacterial chemotaxis. *Cell* 42:683–90
173. Stock JB, Koshland DE Jr. 1981. Changing reactivity of receptor carboxyl groups during bacterial sensing. *J. Biol. Chem.* 256:10826–33
174. Stock JB, Lukat GS, Stock AM. 1991. Bacterial chemotaxis and the molecular logic of intracellular signal transduction networks. *Annu. Rev. Biophys. Biophys. Chem.* 20:109–36
175. Stoddard BL, Bui JD, Koshland DE Jr. 1992. Structure and dynamics of transmembrane signaling by the *Escherichia coli* aspartate receptor. *Biochemistry* 31:11978–83
176. Stolz B, Berg HC. 1991. Evidence for interactions between MotA and MotB, torque-generating elements of the flagellar motor of *Escherichia coli*. *J. Bacteriol.* 173:7033–37
177. Suzuki T, Komeda Y. 1981. Incomplete flagellar structures in *Escherichia coli* mutants. *J. Bacteriol.* 145:1036–41
178. Swanson RV, Bourret RB, Simon MI. 1993. Intermolecular complementation of the kinase activity of CheA. *Mol. Microbiol.* 8:435–41
179. Swanson RV, Schuster SC, Simon MI. 1993. Expression of CheA fragments which define domains encoding kinase, phosphotransfer, and CheY binding activities. *Biochemistry* 32:7623–29
180. Tisa LS, Adler J. 1992. Calcium ions are involved in *Escherichia coli* chemotaxis. *Proc. Natl. Acad. Sci. USA* 89:11804–8
181. Terwilliger TC, Wang JY, Koshland DE Jr. 1986. Surface structure recognized for covalent modification of the aspartate receptor in chemotaxis. *Proc. Natl. Acad. Sci. USA* 83:6707–10
182. Ueno T, Oosawa K, Aizawa S-I. 1992. M ring, S ring and proximal rod of the flagellar basal body of *Salmonella typhimurium* are composed of subunits of a single protein, FliF. *J. Mol. Biol.* 227:672–77
183. Vogler AP, Homma M, Irikura VM, Macnab RM. 1991. *Salmonella typhimurium* mutants defective in flagellar filament regrowth and sequence similarity of FliI to F_oF_1, vacuolar, and archaebacterial ATPase subunits. *J. Bacteriol.* 173:3564–72
184. Volz K. 1993. Structural conservation in the CheY superfamily. *Biochemistry* 32:11741–53
185. Volz K, Matsumura P. 1991. Crystal structure of *Escherichia coli* CheY refined at 1.7-Å resolution. *J. Biol. Chem.* 266:15511–19
186. Weis RM, Koshland DE Jr. 1988. Reversible receptor methylation is essential for normal chemotaxis of *Escherichia coli* in gradients of aspartic acid. *Proc. Natl. Acad. Sci. USA* 85:83–87
187. Welch M, Oosawa K, Aizawa S-I, Eisenbach M. 1993. Phosphorylation-dependent binding of a signal molecule to the flagellar switch of bacteria. *Proc. Natl. Acad. Sci. USA* 90:8787–91
188. Wilson ML, Macnab RM. 1990. Co-

overproduction and localization of the *Escherichia coli* motility proteins MotA and MotB. *J. Bacteriol.* 172:3932–39
189. Wolfe AJ, Conley MP, Berg HC. 1988. Acetyladenylate plays a role in controlling the direction of flagellar rotation. *Proc. Natl. Acad. Sci. USA* 85:6711–15
190. Wolfe AJ, Conley MP, Kramer TJ, Berg HC. 1987. Reconstitution of signaling in bacterial chemotaxis. *J. Bacteriol.* 169:1878–85
191. Wolfe AJ, Stewart RC. 1993. The short form of the CheA protein restores kinase activity and chemotactic ability to kinase-deficient mutants. *Proc. Natl. Acad. Sci. USA* 90:1518–22
192. Wolff C, Parkinson JS. 1988. Aspartate taxis mutants of the *Escherichia coli* Tar chemoreceptor. *J. Bacteriol.* 170:4509–15
193. Yaghmai R, Hazelbauer GL. 1992. Ligand occupancy mimicked by single residue substitutions in a receptor: transmembrane signaling induced by mutation. *Proc. Natl. Acad. Sci. USA* 89:7890–94
194. Yamaguchi S, Aizawa S-I, Kihara M, Isomura M, Jones CJ, Macnab RM. 1986. Genetic evidence for a switching and energy-transducing complex in the flagellar motor of *Salmonella typhimurium*. *J. Bacteriol.* 168:1172–79
195. Yamaguchi S, Fujita H, Ishihara A, Aizawa S-I, Macnab RM. 1986. Subdivision of flagellar genes of *Salmonella typhimurium* into regions responsible for assembly, rotation, and switching. *J. Bacteriol.* 166:187–93
196. Yonekawa H, Hayashi H. 1986. Desensitization by covalent modification of the chemoreceptor of *Escherichia coli*. *FEBS Lett.* 198:21–24
197. Zhou J, Fazzio RT, Blair DF. 1995. Membrane topology of the MotA protein of *Escherichia coli*. *J. Mol. Biol.* In press

BIODEGRADATION OF NITROAROMATIC COMPOUNDS

Jim C. Spain

Armstrong Laboratory, US Air Force, AL/EQC, 139 Barnes Drive, Tyndall AFB, Florida 32403

KEY WORDS: nitroaromatic compounds, TNT, explosives, biodegradation, dioxygenase, biotransformation

CONTENTS

INTRODUCTION	524
CHEMISTRY OF THE AROMATIC NITRO GROUP	526
ANAEROBIC BIODEGRADATION	527
Sulfate-Reducing Bacteria	529
Clostridium	530
BIODEGRADATION BY FUNGI	531
AEROBIC BIODEGRADATION BY BACTERIA	534
Monooxygenase-Catalyzed Initial Reactions	534
Dioxygenase-Catalyzed Initial Reactions	537
Reduction of the Aromatic Ring	541
Partial Reduction of the Nitro Group Under Aerobic Conditions	543
APPLICATIONS IN BIOREMEDIATION	548
CONCLUSIONS	549

Abstract

Nitroaromatic compounds are released into the biosphere almost exclusively from anthropogenic sources. Some compounds are produced by incomplete combustion of fossil fuels; others are used as synthetic intermediates, dyes, pesticides, and explosives. Recent research revealed a number of microbial systems capable of transforming or biodegrading nitroaromatic compounds. Anaerobic bacteria can reduce the nitro group via nitroso and hydroxylamino intermediates to the corresponding amines. Isolates of *Desulfovibrio* spp. can use nitroaromatic compounds as their source of nitrogen. They can also reduce

2,4,6-trinitrotoluene to 2,4,6-triaminotoluene. Several strains of *Clostridium* can catalyze a similar reduction and also seem to be able to degrade the molecule to small aliphatic acids. Anaerobic systems have been demonstrated to destroy munitions and pesticides in soil. Fungi can extensively degrade or mineralize a variety of nitroaromatic compounds. For example, *Phanerochaete chrysosporium* mineralizes 2,4-dinitrotoluene and 2,4,6-trinitrotoluene and shows promise as the basis for bioremediation strategies.

The anaerobic bacteria and the fungi mentioned above mostly transform nitroaromatic compounds via fortuitous reactions. In contrast, a number of nitroaromatic compounds can serve as growth substrates for aerobic bacteria. Removal or productive metabolism of nitro groups can be accomplished by four different strategies. (*a*) Some bacteria can reduce the aromatic ring of dinitro and trinitro compounds by the addition of a hydride ion to form a hydride-Meisenheimer complex, which subsequently rearomatizes with the elimination of nitrite. (*b*) Monooxygenase enzymes can add a single oxygen atom and eliminate the nitro group from nitrophenols. (*c*) Dioxygenase enzymes can insert two hydroxyl groups into the aromatic ring and precipitate the spontaneous elimination of the nitro group from a variety of nitroaromatic compounds. (*d*) Reduction of the nitro group to the corresponding hydroxylamine is the initial reaction in the productive metabolism of nitrobenzene, 4-nitrotoluene, and 4-nitrobenzoate. The hydroxylamines undergo enzyme-catalyzed rearrangements to hydroxylated compounds that are substrates for ring-fission reactions. Potential applications of the above reactions include not only the biodegradation of environmental contaminants, but also biocatalysis and synthesis of valuable organic molecules.

INTRODUCTION

Natural organic compounds are readily biodegradable and can serve as sources of carbon and energy for microorganisms that have evolved over geological time to exploit them. In contrast, xenobiotic compounds synthesized and released into the biosphere only recently by humans can present daunting challenges to heterotrophic microorganisms. The inclusion of unusual chemical bonds or substitution with halogens or other functional groups can render a molecule resistant to microbial degradation. A surprisingly large number of halogenated organic compounds are produced in nature (45, 85). As a result, there are correspondingly large numbers of microorganisms able to degrade halogen-substituted compounds that bear some similarity to natural substances. The mechanisms of enzymatic attack on halogenated compounds have been studied extensively and are relatively well understood (36, 63). In contrast, only a few natural nitro-substituted compounds have been reported (125), and they are mostly antibiotics. The vast majority of the nitroaromatic compounds detected in the environment are anthropogenic and released because of their extensive use in the synthesis of dyes, plasticizers, pesticides, and explosives. They are also produced by incomplete combustion of fossil fuels (93, 123).

The toxicity of nitroarenes and their metabolites has been studied in a variety of systems (9, 38, 59, 130, 132). Both the nitro group and the amino group are relatively stable in biological systems. Interconversion between the two, however, involves the intermediate production of the corresponding nitroso and hydroxylamino derivatives—which are very reactive and, in many instances, more toxic than the parent molecules. Polycyclic nitroaromatic compounds are not very toxic or carcinogenic, but can be activated not only by reduction of the nitro group by intestinal microflora, but also by mammalian cytochrome P-450-mediated oxidation of the aromatic ring (9, 38).

Currently, the most visible environmental problem caused by contamination with nitroaromatic compounds is the widespread contamination of soil by explosives. Sites of former manufacturing, loading, and storage facilities in the United States and Europe are heavily contaminated with munitions from World War II. In addition, the more water-soluble nitroaromatic solvents and pesticides may be appropriate targets for bioremediation. Several barriers must be overcome before biodegradation can provide an effective treatment strategy for nitroaromatic compounds: (*a*) the toxicity of nitroaromatic compounds to microorganisms, (*b*) low bioavailability due to insolubility or sorption of the contaminant, (*c*) complications caused by mixtures of nitroaromatic contaminants, and (*d*) lack of catabolic systems able to degrade the nitroaromatic compound in the microbial community. The discovery and development of appropriate catabolic systems is the subject of this review.

At first glance it would seem that the enzymes involved in the inorganic nitrogen cycle could catalyze transformations of aromatic nitro substituents. Closer examination reveals that the chemistry of the molecules and their reactions are very different and that the enzymes involved in their transformation are unrelated (6). Fortuitous transformation of nitroaromatic compounds has been studied extensively. Many, and perhaps most, living organisms contain enzymes that can catalyze the transformation of the aromatic nitro group. The most common examples are the wide variety of redox enzymes that can serve as nitroreductases (17). Less common are enzymes and complete catabolic pathways that allow bacteria to use nitroaromatic compounds as their source of nitrogen or carbon for growth. A number of such systems were discovered recently, however, and are the focus of intensive research to discover not only their mechanisms and evolutionary origins, but also their potential for biotechnology applications.

The microbial degradation of explosives and nitroaromatic compounds has been reviewed by several authors (43a, 58, 67, 68, 78a, 109, 129); therefore, most of the historical work is not discussed here. In the past four years, the

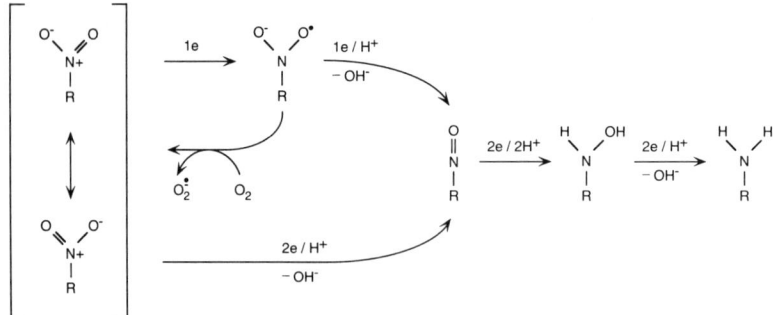

Figure 1 Reduction of nitro groups by one-electron or two-electron mechanisms.

intensive research in this area has led to dramatic progress in understanding the microbial strategies and associated mechanisms for the degradation of nitroaromatic compounds. These recent developments and the questions they have raised are the focus of this review.

CHEMISTRY OF THE AROMATIC NITRO GROUP

The nitro group exists as a resonance hybrid (Figure 1). Because the oxygen atoms are more electronegative than the nitrogen atom, the polarization of the nitrogen-oxygen bond causes the nitrogen atom to carry a partial positive charge and to serve as an electrophile. Therefore, the most common reaction of the nitro group in biological systems is reduction, which can proceed either by one-electron or two-electron mechanisms (Figure 1). In addition, iron(II), other metals and reduced sulfur compounds (31, 44, 56, 94) can serve as reductants for the nonenzymatic reduction of nitroaromatic compounds. Both the nitro group and the amino group are relatively stable. The sequence of reactions involved in reduction of the nitro group to the amine produces highly reactive intermediates, however. The nitroso and hydroxylamino groups are electrophiles that can interact with biomolecules to cause toxic, carcinogenic, and mutagenic effects (9, 59).

The one-electron reduction of the nitro group produces a nitro radical anion, which can be oxidized to the starting material by molecular oxygen with the concomitant production of superoxide. This futile cycle leads to the designation of enzymes that catalyze one-electron reduction of the nitro group as "oxygen sensitive" (17). Enzymes from a variety of sources—including strictly anaerobic bacteria such as *Clostridium* spp. (4), facultative bacteria such as *Es-*

cherichia coli (92) and *Enterobacter* spp. (17), as well as plants and animals (17)—catalyze one-electron reduction of the nitro group.

Reduction of the nitro group by the sequential addition of pairs of electrons is "oxygen insensitive" because no radicals are produced (17). Nitroreductases of this type convert nitro groups to either hydroxylamines or amines by the addition of electron pairs donated by reduced pyridine nucleotides. The reaction pathway includes the nitroso derivatives, but they are difficult to detect because they are so reactive and unstable. The high reactivity of the nitroso and hydroxylamino intermediates is responsible for much of the toxicity and carcinogenicity attributed to nitroaromatic compounds (130). Both structures can react readily with a variety of biological materials and they can also undergo condensation reactions. For example, partial reduction of the nitro group in the presence of oxygen leads to the nonenzymatic production of azoxy compounds as condensation products (79).

The ease of reduction of the aromatic nitro group depends on the nature of the other substituents on the ring and on the reducing potential of the environment. Electron-withdrawing groups activate the molecule for reduction of the nitro group, whereas electron-donating groups make the ring more susceptible to electrophilic attack. In the case of the nitrotoluenes, the probability of reduction increases and the probability of electrophilic attack decreases as the number of nitro groups increases. Therefore, the reduction of one nitro group of TNT is very rapid under a variety of conditions, including those prevalent in growing cultures of aerobic bacteria. In contrast, reduction of 2-amino-4,6-dinitrotoluene (ADNT) requires a lower redox potential, and reduction of 2,4-diamino-6-nitrotoluene (DANT) requires a redox potential below −200 mV (40), because the electron-donating properties of the amino groups lower the electron deficiency of the molecule.

ANAEROBIC BIODEGRADATION

The reactions of nitroaromatic compounds in anaerobic systems almost exclusively involve the reduction of the nitro group. McCormick et al (79) clearly demonstrated that *Viellonella alkalescens* could reduce TNT and described some of the enzymes involved. Subsequently, a variety of other bacteria have been shown to reduce aromatic nitro compounds under anaerobic (4, 79, 90, 97) and even under aerobic (105) conditions (Table 1). The most interesting recent advances are the discoveries that pure cultures of *Desulfovibrio* and *Clostridium* species can extensively degrade TNT. Although a variety of bacteria, soils, and sediments can catalyze the reduction of TNT, the studies with *Desulfovibrio* and *Clostridium* species provide novel insight into the enzymes responsible and the mechanism. Therefore, they are discussed in some detail here.

Table 1 Representative bacteria reported to reduce nitroaromatic compounds[a]

Organism	Nitroaromatic compounds transformed	Reference
Clostridium acetobutylicum	Chloramphenicol, 2-/3-nitrophenol, 2-/3-/4-nitrobenzoate, 2-nitrobenzaldehyde	89
Clostridium pasteurianum, E. coli; *Veillonella alkalescens*	TNT; 40 nitro compounds, including nitrophenols, nitrobenzoates, nitrotoluenes	79
Clostridium kluyveri, Clostridium sp. La1, *C. pasteurianum*	4-Nitrobenzoate, 2-nitroethanol	3
C. propionicum, C. butyricum, C. barkeri, Proteus mirabilis, Micrococcus lactilyticus, Peptostreptococcus anaerobius	4-Nitrobenzoate, 2-nitroethanol	2
Bacteroides fragilis	1-Nitropyrene	72
Clostridium leptum, C. paraputrificum, C. clostridiiforme, Eubacterium sp.	1-Nitropyrene, 1,3-dinitropyrene, 1,6-dinitropyrene	97
Haloanaerobium praevalens, Sporohalobacter marismortui	Nitrobenzene, 2-/3-/4-nitrophenol, 2-/3-/4-nitroaniline, 2,4-dinitrophenol, 2,4-dinitroaniline	90
Desulfovibrio sp. (B. strain)	TNT, 2,4-/2,6-dinitrotoluene, 2,4-dinitrophenol	11, 12
Desulfovibrio sp.	TNT, 2,6-/3,4-dinitrotoluene, 2-/4-nitrotoluene, nitrobenzene	94
Methanobacterium formicicum, Methanobacterium thermoautotrophicum, Methanospirillum hungatei, Methanosarcina sp. KS2002, *Methanosarcina barkeri, Methanosarcina frisia, Methanogenium tationis, Desulfovibrio desulfuricans, D. gigas, Desulfovibrio* sp. AS, *Desulfovibrio* sp. HB, *Desulfotomaculum* sp. GROL	3-/4-Nitrophenol, 2,4-dinitrophenol, 4-nitrobenzoate, 4-nitroaniline	44
Desulfotomaculum orientis, Desulfococcus multivorans	4-Nitrophenol	44
Methanococcus	TNT	13
Pseudomonas	2-/3-/4-Chloronitrobenzene, 2-/3-/4-nitrobenzoate, 2-/3-/4-nitrophenol, 1-chloro-2,4-dinitrobenzene, TNT	105

[a] Modified from Reference 95.

Sulfate-Reducing Bacteria

Two strains of *Desulfovibrio* have been studied extensively because of their ability to transform nitroaromatic compounds. Boopathy & Kulpa studied a strain designated *Desulfovibrio* sp. strain B that uses TNT (11) and a variety of other nitroaromatic compounds (12) as the source of nitrogen for growth and also as the terminal electron acceptor. The authors provided strong evidence that the nitro compounds are reduced to the corresponding amines and proposed that the amino groups are removed from the aromatic ring by a reductive deamination mechanism analogous to that described by Schnell & Schink (106). The evidence (11, 14) that *Desulfovibrio* sp. strain B converts 2,4,6-triaminotoluene (TAT) to toluene is preliminary, and no experimental evidence was provided for the operation of the deamination mechanism, but the proposed reductive deamination would be an exciting discovery if confirmed. Tentative evidence for reductive deamination of aminotoluenes was also provided by studies of bacteria from rumen fluid (25).

Preuss et al (94) isolated a strain of *Desulfovibrio* by selective enrichment with pyruvate as the carbon source, sulfate as the terminal electron acceptor, and TNT as the source of nitrogen. The strain fixes atmospheric nitrogen and can also use ammonia as its nitrogen source. Under growth conditions the sulfide in the medium chemically reduces the TNT to DANT. The DANT is then reduced by the *Desulfovibrio* sp. to TAT, which subsequently disappears from the culture fluid.

Suspensions of cells containing little or no sulfide catalyze the reduction of TNT to TAT when pyruvate is supplied as the electron donor. The rate of reduction of each successive nitro group decreases dramatically because amino groups deactivate the molecule for further reduction. The authors suggested that the facile reduction of TNT to DANT is mediated by nonspecific enzymes that reduce low-potential electron carriers such as ferredoxin. Therefore, the authors focused their subsequent investigations on the mechanism of conversion of DANT to TAT, because it seems to be the rate-determining step in the overall process.

Hydrogen, pyruvate, or carbon monoxide can serve as electron donors for the reduction of DANT by cell suspensions of the *Desulfovibrio* sp. Therefore, the corresponding ferredoxin-reducing enzymes hydrogenase, pyruvate-ferredoxin-oxidoreductase, and carbon monoxide dehydrogenase are probably responsible for the reduction of DANT. When hydrogen or pyruvate serves as the electron donor, DANT is reduced to TAT. In contrast, DANT is only partially reduced, to 2,4-diamino-6-hydroxylaminotoluene (DAHAT), when carbon monoxide serves as the electron donor. Furthermore, with pyruvate as the electron donor, the addition of carbon monoxide inhibits the production of TAT and causes the accumulation of DAHAT. Similarly, inhibition results

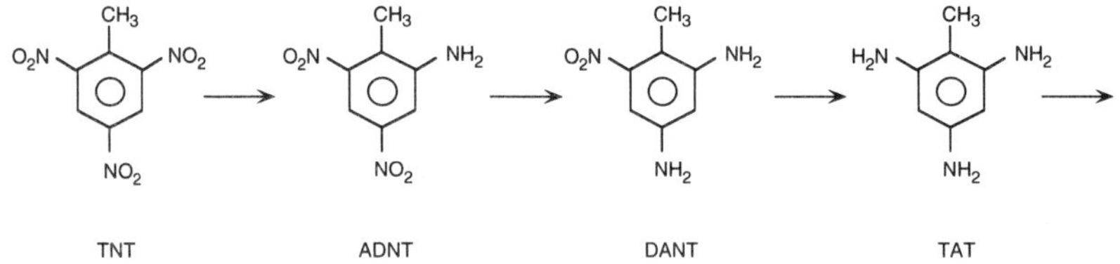

TNT → ADNT → DANT → TAT →

Figure 2 Reduction of TNT by anaerobic bacteria.

when hydroxylamine is included in cell suspensions using hydrogen as the electron donor (94). The above observations led the authors to suggest that sulfite reductase might be responsible for the reduction of DAHAT to TAT, because sulfite reductase is known to be inhibited by carbon monoxide. Subsequent experiments revealed that the reduction of sulfite by cells of the *Desulfovibrio* sp. was inhibited by carbon monoxide, DAHAT, DANT, and hydroxylamine. These results provide strong circumstantial evidence that sulfite reductase is responsible for the reduction of DAHAT to TAT in the *Desulfovibrio* sp. (94). In contrast to the sequence postulated by Boopathy & Kulpa (11), TAT was not further metabolized by the *Desulfovibrio* sp. The overall scheme of TNT reduction by the *Desulfovibrio* sp. is outlined in Figure 2.

Clostridium

Several strains of clostridia have been studied because of their ability to reduce nitroaromatic compounds (4, 79, 94, 97). Angermaier & Simon (4) provided evidence that hydrogenase and ferredoxin in *Clostridium kluyveri* are responsible for a one-electron reduction of nitroaromatic compounds. Rafii et al (97) characterized oxygen-sensitive enzymes from several strains of *Clostridium* isolated from human fecal material. The enzymes reduced 4-nitrobenzoate and several nitropyrenes to the corresponding amines. It is not clear whether the reaction catalyzed by the enzyme is oxygen sensitive because it is a one-electron transfer or because the reaction involves transfer of two electrons catalyzed by an enzyme that is oxygen labile.

Hydrogenase from *Clostridium pasteurianum* and carbon monoxide dehydrogenase from *Clostridium thermoaceticum* reduce DANT to DAHAT when ferredoxin is included in the reaction mixture (94). The reduction also takes place with reduced ferredoxin or methyl viologen in the absence of enzymes, which suggests that the enzymes are not specific for the nitroaromatic compound but only serve to reduce the ferredoxin. Thus the wide distribution of nonspecific, oxygen-sensitive nitroreductases among biological systems seems to reflect the wide distribution of redox enzymes. It follows that much of the reduction under such conditions can be attributed to the ability of many nitroaromatic compounds to be reduced nonenzymatically by a variety of reductants including iron(ll), sulfhydryl compounds, and small electron carriers such as ferredoxin. In contrast, the conversion of DAHAT to TAT is catalyzed more slowly by the clostridia and seems to require the action of specific enzymes such as the ones in *Desulfovibrio* spp. (94).

Kaake et al used an anaerobic mixed culture in the development of a strategy for the biodegradation of dinoseb (2-*sec*-butyl-4,6-dinitrophenol) (65, 66) under methanogenic conditions with starch as the electron donor. Similar enrichment cultures degraded RDX (hexahydro-1,3,5-trinirro-1,3,5-triazine) and TNT to nondetectable levels in contaminated soil (39, 40). Pure cultures of *Clostridium bifermentans* isolated from the consortium (100) and similar strains isolated from other enrichments (26) degrade both RDX and TNT. The pathway for reduction of TNT to TAT by the consortium and the isolated *Clostridium* strains seems to be the same as that proposed by other workers (94). However, the TAT is subsequently degraded extensively not only by the consortium but also by the isolates. The authors provide evidence that TAT is first converted to 2,4,6-trihydroxytoluene by hydrolysis of the amino groups. They propose (26) that reductive dehydroxylation reactions convert 2,4,6-trihydroxytoluene to 4-hydroxytoluene (*p*-cresol), which is then converted to simple organic acids readily assimilated by the culture. The evidence for the proposed pathway is preliminary, and exploration of the mechanism involved will provide an exciting area for future research.

The studies described above indicate that reduction of the nitro group is the major reaction controlling the behavior and fate of nitroaromatic compounds in ecosystems that contain bacteria. The reactions that convert TNT to TAT can be catalyzed by a variety of bacteria, but the subsequent metabolism of TAT is controversial at present.

BIODEGRADATION BY FUNGI

The white rot fungus, *Phanerochaete chrysosporium*, produces a complex system of extracellular peroxidases, small organic molecules, and hydrogen peroxide for the degradation of lignin. The ligninolytic system is nonspecific and can biodegrade a wide range of synthetic chemicals including nitroaromatic compounds. Barr & Aust (8) have provided an excellent review of the mechanisms used by the fungus for transformation of synthetic chemicals. Several groups (18, 35, 80, 114–116, 124) have reported the degradation and even mineralization of nitroaromatic compounds by *P. chrysosporium*.

The initial steps in the fungal degradation of both 2,4-dinitrotoluene and TNT involve reduction of the nitro groups. Valli et al (124) suggested that

2,4-dinitrotoluene is reduced predominantly to 2-amino-4-nitrotoluene by intracellular reductase enzymes that are not part of the ligninolytic system. Evidence for the pathway was provided by transformation experiments with mycelia and purified enzymes. Manganese peroxidase catalyzes the conversion of 2-amino-4-nitrotoluene to 4-nitro-1,2-benzoquinone. The authors suggest that the quinone is reduced to 4-nitrocatechol, but provide no strong evidence for the reaction. 4-Nitrocatechol serves as a substrate for oxidative removal of the nitro group in a reaction catalyzed by manganese peroxidase. Alternatively, methylation reactions and subsequent removal of the nitro group can be catalyzed by lignin peroxidase. The methylated and dihydroxy intermediates were readily degraded by the mycelium, and subsequent reactions in the pathway were demonstrated in cell extracts or with purified enzymes.

Most fungi can catalyze the reduction of at least one nitro group of TNT (91). Mycelia of *P. chrysosporium* reduce TNT to a mixture of 2-amino-4, 6-dinitrotoluene, 4-amino-2,6-dinitrotoluene, and 2,4-diamino-6-nitrotoluene (115). Under ligninolytic conditions the amino compounds disappear, and mineralization can be fairly extensive. Stahl & Aust (116) provided evidence that the reduction of TNT requires live, intact mycelia and that extracellular enzymes and enzymes in cell extracts cannot catalyze the reaction even when supplemented with reduced pyridine nucleotides. They proposed that the reduction reaction is closely coupled to the proton export system used by the fungus to maintain an external pH of 4.5. Generation of protons would produce electrons that could be used in the reduction reaction. In contrast, Valli et al (124) suggested that 2,4-dinitrotoluene is reduced by intracellular enzymes in *P. chrysosporium*. Similarly, Michels & Gottschalk (80) indicated that TNT is reduced by an NADH-dependent nitroreductase in extracts prepared from *P. chrysosporium*. Neither of these latter groups, however, provided data on the reduction reactions. All of the above studies support the conclusion that the ligninolytic system is not involved in the initial reduction of nitroaromatic compounds.

Ligninolytic cultures of *P. chrysosporium* mineralize TNT, whereas nonligninolytic cultures produce primarily the amino derivatives and azoxy condensation products. Nonligninolytic cultures can also mineralize TNT to a lesser extent (80, 94, 114, 115). The mechanism by which the fungi mineralize TNT is not known, but several recent investigations provided preliminary information about the process. The amino derivatives of TNT are not substrates for the ligninolytic system (18), but reduction of the molecule is required prior to mineralization. Michels & Gottschalk (81) provided strong evidence that under nonligninolytic conditions TNT is reduced to 4-amino-2,6-dinitrotoluene via the corresponding hydroxylamino intermediate (Figure 3). The 4-amino-2,6-dinitrotoluene is coverted to 4-formamido-2,6-dinitrotoluene, which is subsequently reduced to 2-amino-4-formamido-6-nitrotoluene and finally con-

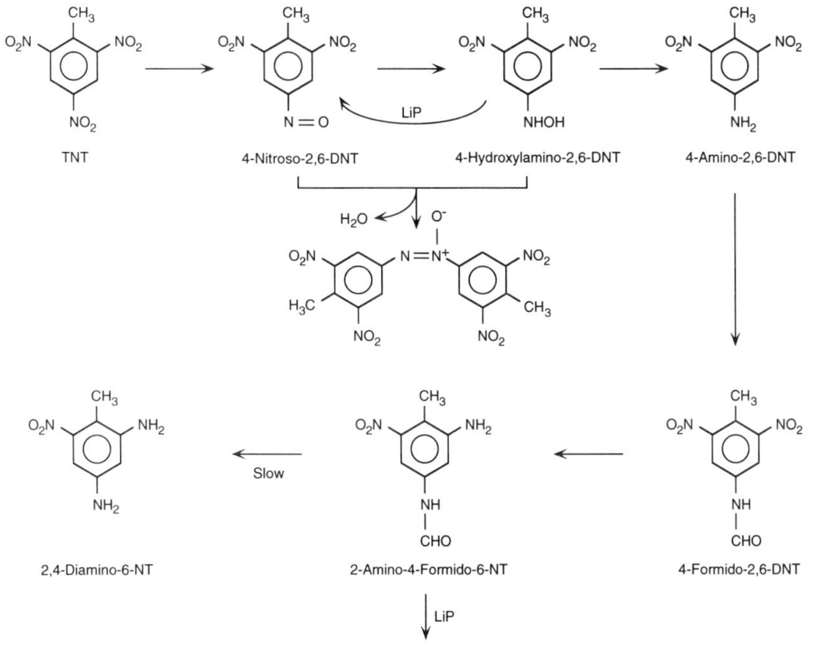

Figure 3 Transformation of TNT by *Phanerochaete chrysosporium*. 4-Hydroxylamino-2,6-DNT and 2-amino-4-formamido-6-nitrotoluene are substrates for lignin peroxidase (LiP) (80, 115).

verted to DANT. They found no evidence for the direct reduction of ADNT to DANT. Both 4-hydroxylamino-2,6-dinitrotoluene and 2-amino-4-formamido-6-nitrotoluene are substrates for oxidation by the ligninolytic system and purified lignin peroxidase. This preliminary work suggests a possible mechanism that would explain previous, seemingly conflicting, observations by several groups. There remain, however, a number of questions concerning the mineralization pathway and the participation of the other isomers of the intermediates produced during reduction of TNT. Even under optimum conditions for mineralization, the majority of the carbon from TNT is not converted to carbon dioxide, and the final products of biodegradation other than carbon dioxide have not been characterized.

The major impediment to the practical application of white rot fungi for biodegradation of TNT has been its toxicity to the fungus. Investigations into the mechanism of toxicity of TNT have revealed that the parent compound and its amino derivatives are not toxic to *P. chrysosporium* (18, 80). In contrast, 4-hydroxylamino-2,6-dinitrotoluene and 2-hydroxylamino-4,6-dinitrotoluene dramatically inhibited veratryl alcohol oxidation by lignin peroxidase. The conversion of veratryl alcohol to veratryl aldehyde is essential for the production of the organic radicals involved in the oxidation of chemicals that are not primary substrates for lignin peroxidase (55). The hydroxylamino compounds are good substrates for the enzyme and competitively inhibit the oxidation of veratryl alcohol. In some studies (80), oxidation of hydroxylaminodinitrotoluenes led to the accumulation of azoxy condensation products; other workers (18) suggest that TNT is the final product of the oxidation. It is clear that development of an approach that avoids the accumulation of hydroxylamino intermediates is the key to practical application of the fungus for biodegradation of TNT.

AEROBIC BIODEGRADATION BY BACTERIA

In contrast to the nonspecific metabolism by fungi and anaerobes, some aerobic bacteria can use nitroaromatic compounds as growth substrates. They can often derive carbon, nitrogen, and energy from degradation of nitroaromatic substrates. Therefore, the isolation and study of such bacteria is much easier than are comparable studies of anaerobes and fungi able to transform nitro compounds. It is remarkable that the capabilities of such aerobic bacteria have only recently come to be appreciated.

Bacteria able to degrade nitrophenols and nitrobenzoates were reported many years ago, but the mechanisms they use for metabolism of the nitro group were not well understood. Recently, bacteria able to degrade a wide range of polar and nonpolar nitroaromatic compounds were reported. Such bacteria use a variety of strategies for the removal or transformation of the nitro group. These include (*a*) monooxygenase-catalyzed elimination of the nitro group as nitrite; (*b*) dioxygenase-catalyzed insertion of two hydroxyl groups with subsequent elimination of the nitro group as nitrite; (*c*) partial reduction of the nitro group to a hydroxylamine, which is the substrate for rearrangement or hydrolytic reactions and elimination of ammonia; and (*d*) partial reduction of the aromatic ring to form a Meisenheimer complex and subsequent elimination of the nitro group as nitrite. Each of these strategies is described in detail below.

Monooxygenase-Catalyzed Initial Reactions

Some of the earliest reports on the biodegradation of nitroaromatic compounds involved studies of bacteria able to grow on nitrophenol. Simpson & Evans (108) provided preliminary evidence in 1953 that a strain of *Pseudomonas* could convert 4-nitrophenol to hydroquinone with concomitant release of nitrite. The details of the pathway (Figure 4) only became clear almost 40 years later (110). Studies with a partially purified enzyme (111) revealed that a strain of *Moraxella* degrades 4-nitrophenol by an initial oxygenase attack that results

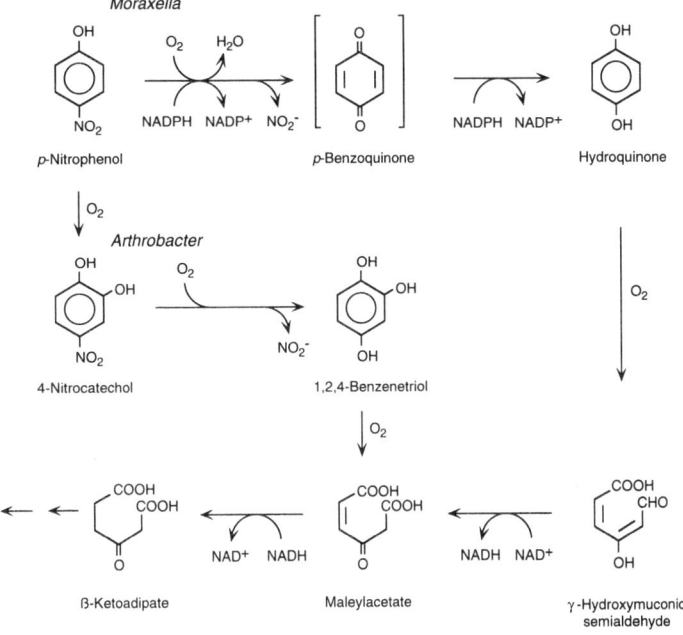

Figure 4 Biodegradation of 4-nitrophenol by *Moraxella* sp. and *Arthrobacter* sp. (62, 110).

in the release of nitrite and the accumulation of hydroquinone. Two moles of NADPH are required to oxidize each mole of 4-nitrophenol. Experiments with $^{18}O_2$ provided rigorous evidence that the mechanism of the reaction is a monohydroxylation (111), and preliminary evidence suggests that the enzyme is a flavoprotein monooxygenase. The stoichiometry of the reaction and the accumulation of hydroquinone as the first detectable intermediate suggest that the initial product of the reaction is 1,4-benzoquinone. An inducible quinone reductase present in cells grown on 4-nitrophenol is not easily separated from the membrane-bound oxygenase that catalyzes the initial reaction. Therefore the quinone could not be detected during the reaction. A similar reaction mechanism has been proposed for the elimination of anionic leaving groups catalyzed by other flavoprotein monooxygenases, including *p*-hydroxybenzoate hydroxylase (60) and pentachlorophenol monooxygenase (134).

The hydroquinone produced by the initial reactions serves as the substrate for a ring-fission reaction catalyzed by a ferrous iron–dependent dioxygenase. The ring-fission product is oxidized to maleylacetic acid, which is then reduced to β-ketoadipate. Only catalytic amounts of NAD^+ are required for the conversion of the ring-fission product to β-ketoadipate in cell extracts because the two reactions recycle the cofactor. Cells of the *Moraxella* sp. grown on 4-nitrophenol also contain an enzyme that catalyzes the oxidation of 1,2,4-benzenetriol. The presence of this enzyme suggests that the nitrophenol could be degraded by an alternate pathway via 4-nitrocatechol, but there is no evidence for the production of 4-nitrocatechol from 4-nitrophenol by this strain. Furthermore, the monooxygenase that initiates the conversion of 4-nitrophenol to hydroquinone is present at levels sufficient to account for the growth of the cells. It is possible that small amounts of 4-nitrocatechol are produced by nonspecific hydroxylases. In support of this hypothesis, 4-nitrocatechol is oxidized slowly by 4-nitrophenol-grown cells with concomitant release of nitrite. It is also possible that a nonspecific hydroxylase converts hydroquinone to 1,2,4-benzenetriol during the operation of the primary pathway. In either case the primary pathway in the *Moraxella* sp. (110) and in an *Arthrobacter* and a *Nocardia* sp. studied by Hanne et al (53) involves the conversion of 4-nitrophenol to hydroquinone. The presence of 1,2,4-benzenetriol dioxygenase in the *Moraxella* remains to be explained.

In contrast to the strains described above, an *Arthrobacter* sp. (62) seems to degrade 4-nitrophenol via 4-nitrocatechol. This pathway (Figure 4) was suggested based on preliminary evidence in an early report (99) on the degradation of 4-nitrophenol by a *Flavobacterium* sp. In cells of the *Arthrobacter* sp., 1,2,4-benzenetriol is produced by the initial reactions and subsequently oxidized to maleylacetic acid. 4-Nitrophenol-grown cells do not contain enzymes capable of oxidizing hydroquinone at detectable rates. The enzyme responsible for the initial attack on 4-nitrophenol has not been detected in cell extracts, so it has not been studied. An enzyme that converts 4-nitrophenol to 4-nitrocatechol has been purified from a *Nocardia* sp. grown on 4-nitrophenol (82), however, and a similar enzyme activity has been demonstrated in another strain of *Nocardia* after growth on phenols (53). The enzyme responsible for the initial attack on 4-nitrophenol in the *Arthrobacter* sp. clearly requires additional study.

The pathway for degradation of 2-nitrophenol, like that of 4-nitrophenol, was suggested based on preliminary evidence in an early report by Simpson & Evans (108). It was confirmed 30 years later by Zeyer & Kearney (135) in a *Pseudomonas putida* isolated from soil. An NADPH-dependent monooxygenase present in extracts of 2-nitrophenol-grown cells catalyzes the conversion of 2-nitrophenol to catechol with the concomitant release of nitrite. Catechol is subsequently oxidized by a 1,2-dioxygenase and degraded via β-ketoadipate. Purified 2-nitrophenol monooxygenase catalyzes the conversion of 2-nitrophenol to catechol with the concomitant oxidation of two moles of NADPH (136). The authors proposed that the enzymatic reaction produces 1,2-benzoquinone by a mechanism analogous to the reaction catalyzed by

4-nitrophenol monooxygenase (111). 2-Nitrophenol monooxygenase is unusual among monooxygenases that catalyze the removal of aromatic nitro groups in that it does not seem to require the participation of a flavin cofactor.

Dioxygenase-Catalyzed Initial Reactions

The catabolism of aromatic hydrocarbons by aerobic bacteria generally requires the activation of the molecule by the addition of two hydroxyl groups to the ring. The reactions are catalyzed by dioxygenase enzymes that introduce two atoms of molecular oxygen on adjacent carbon atoms (42). When the aromatic substrate is a hydrocarbon, the introduction of two hydroxyl groups forms a *cis*-1,2-dihydroxy cyclohexadiene, which is subsequently rearomatized by the action of a dehydrogenase. With substituted aromatic compounds, the introduction of the hydroxyl groups can lead to spontaneous elimination of the substituent and rearomatization of the ring. For example, toluene dioxygenase catalyzes the elimination of hydroxyl groups from phenols (112); naphthalene dioxygenase can remove sulfonyl and hydroxyl groups from substituted naphthalenes (16); 4-sulfobenzoate 3,4-dioxygenase eliminates sulfite from 4-sulfobenzoate (76, 77); and analogous mechanisms have been demonstrated for the removal of chloro (37, 78) and amino (7, 64) substituents.

The removal of aromatic nitro groups by dioxygenase enzymes was first reported by Ecker et al (33) as a result of studies on the transformation of 2,6-dinitrophenol by *Alcaligenes eutrophus*. Additional support for the reaction mechanism came from the observation that the enzyme used by a *Pseudomonas* sp. for the initial oxidation of chlorobenzenes could catalyze the elimination of the nitro group from 2,4,5-trichloronitrobenzene (104). Both of these studies involved fortuitous reactions that resulted in partial transformation of nitroaromatic compounds that could not serve as growth substrates. Subsequently, several groups have isolated aerobic bacteria able to grow on nitroaromatic compounds and remove the nitro group by means of dioxygenase enzymes as the first step in the catabolic pathway.

Nitrobenzene, used extensively as the starting material for synthesis of aniline, is converted to catechol (Figure 5*a*) as the first step in its mineralization by a *Comamonas* sp. isolated from an aerobic waste-treatment plant (88). Experiments with $^{18}O_2$ provided rigorous proof that the initial reaction proceeds by a dioxygenase mechanism. The inducible nitrobenzene dioxygenase is very nonspecific and also catalyzes the oxidation of a variety of nitrophenols, dinitrobenzenes, and nitrotoluenes (SF Nishino & JC Spain, unpublished results). A high activity with naphthalene in intact cells suggests that the dioxygenase might be closely related to naphthalene dioxygenase, but the enzyme was not active in cell extracts and has not been characterized further.

A *Pseudomonas* strain isolated from contaminated soil by selective enrichment grows on 2-nitrotoluene as the sole source of nitrogen and carbon (52).

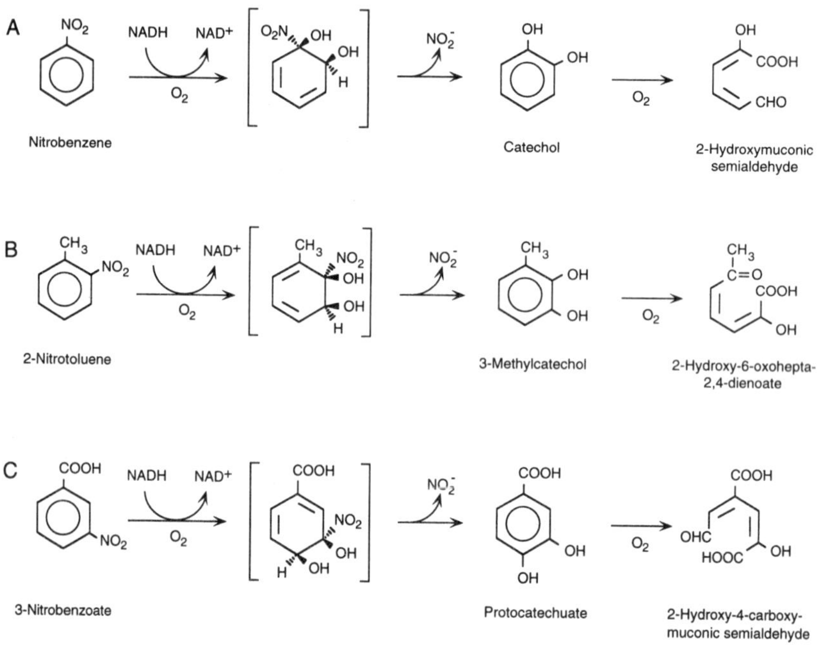

Figure 5 Pathways involving initial reactions catalyzed by dioxygenase enzymes (52, 84, 88).

The catabolic pathway (Figure 5b) involves an initial dioxygenase attack at the 2,3 position of the molecule to form 3-methylcatechol and release nitrite. 3-Methylcatechol is degraded by a typical *meta* cleavage pathway. The enzymes of the 2-nitrotoluene degradative pathway are constitutive, and 2-nitrotoluene dioxygenase has been purified and characterized (1). It consists of three components: an iron-sulfur protein with two subunits similar to the terminal oxygenase component of other dioxygenases, a flavo-iron-sulfur reductase, and a small ferredoxin-like protein, which can be replaced by the ferredoxin obtained from naphthalene dioxygenase. The enzyme system appears to be very similar to naphthalene dioxygenase isolated from another strain of *Pseudomonas* (34). Purification of the enzyme allowed rigorous proof that the insertion of molecular oxygen and release of nitrite involves a dioxygenase mechanism, and that the rearomatization of the ring does not require a separate enzyme.

Early work indicated that protocatechuate is an intermediate in the degradation of 3-nitrobenzoate by a *Nocardia* sp. (22). 3-Hydroxybenzoate was also oxidized by freeze-dried cells but not by resting cells, so the authors proposed

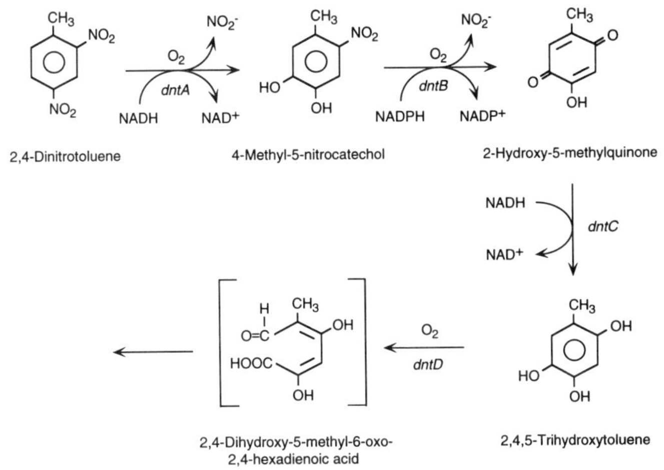

Figure 6 Biodegradation of 2,4-dinitrotoluene by *Pseudomonas* sp. strain DNT (50, 113).

that sequential monooxygenase reactions convert 3-nitrobenzoate to protocatechuate via 3-hydroxybenzoate, and release the nitro group as nitrite. In contrast, strains of *Pseudomonas* and *Comamonas* do not oxidize 3-hydroxybenzoate after growth on 3-nitrobenzoate (84). These strains convert 3-nitrobenzoate to protocatechuate by means of a dioxygenase attack at the 3,4 position with subsequent elimination of nitrite (Figure 5c). Protocatechuate 4,5-dioxygenase catalyzes the oxidation of protocatechuate in the *Pseudomonas* strain.

2,4-Dinitrotoluene (2,4-DNT) is a by-product of the manufacture of TNT and is also used extensively as an intermediate in the synthesis of toluene diisocyanate. It has been released widely into the environment, where it seems to be relatively stable. Bacteria able to mineralize 2,4-DNT have been isolated from a variety of contaminated soils, and the biodegradation of 2,4-DNT has been studied in pure cultures. Mineralization of a nitroaromatic compound via a dioxygenase initial attack was first reported as a result of studies with *Pseudomonas* sp. strain DNT grown on 2,4-DNT. The dioxygenase enzyme that catalyzes the initial reaction is constitutive and has a broad substrate range that is very similar to that of naphthalene dioxygenase (120). It adds hydroxyl groups to the 4 and 5 positions on the ring of 2,4-DNT, and the nitro group is eliminated nonenzymatically as nitrite (Figure 6) (113). The genes that encode 2,4-DNT dioxygenase (*dntA*) are on a 180-kilobase (kb) plasmid in *Pseudomonas* sp. strain DNT (121). Cosmid cloning and restriction analysis led to the isolation of a 6.8 kb fragment that expresses dioxygenase activity in *E. coli*. Sequence analysis (120) revealed four open reading frames whose sequences exhibit a high degree of similarity to the corresponding sequences in the naphthalene dioxygenase genes. The cloned 2,4-DNT dioxygenase expressed in *E. coli* catalyzes the oxidation of naphthalene and the other substrates of naphthalene dioxygenase. The only obvious difference in the substrate ranges of the two enzymes expressed in *E. coli* is the inability of naphthalene dioxygenase to catalyze the release of nitrite from 2,4-DNT.

4-Methyl-5-nitrocatechol (MNC) produced by 2,4-DNT dioxygenase is the substrate for a monooxygenase that catalyzes the replacement of the nitro group and elimination of nitrite (Figure 6). The constitutive enzyme, partially purified from cells of *Pseudomonas* sp. strain DNT, converts MNC to 2-hydroxy-5-methylquinone (50). The reaction mechanism is similar to that described for other enzymes that catalyze the removal of nitro groups from nitrophenols (111, 136) and the removal of other electron-withdrawing groups from substituted phenols (134) or carboxylic acids (60). Detection of the quinone as a product of the reaction provides the first clear evidence of the reaction mechanism that previously has only been postulated based on the stoichiometry of the reaction. In the earlier work, the hypothetical quinone intermediates were not detected because they were reduced rapidly either chemically or by the action of quinone reductases. 2-Hydroxy-5-methylquinone is sufficiently stable under the conditions of the reaction to allow its detection and quantitation.

The gene that encodes MNC monooxygenase (*dntB*) is located on the same large plasmid as *dntA*, but the two genes are not contiguous and seem to be regulated independently (121). It has been cloned on a 2.2 kb fragment and expressed in *E. coli*, but the sequence of the gene and the characteristics of the protein have not been reported.

An inducible NADH-dependent quinone reductase is responsible for the conversion of 2-hydroxy-5-methylquinone to 2,4,5-trihydroxytoluene. The enzyme and the gene involved in its synthesis have not been isolated or studied in any detail. High constitutive levels of quinone reductase in both *Pseudomonas* spp. (50) and *E. coli* (121) have so far precluded determination of whether the enzyme is specific for 2-hydroxy-5-methylquinone.

2,4,5-Trihydroxytoluene serves as the substrate for a ring-fission reaction catalyzed by a dioxygenase enzyme. The mechanism of the reaction is not clear because the ring-fission product is unstable and has not been isolated (50). The gene for 2,4,5-trihydroxytoluene dioxygenase (*dntD*) exhibits a high homology with the catechol 2,3-dioxygenase gene family (54, 122), which suggests the likelihood of an extradiol cleavage as indicated in Figure 6. The evidence is circumstantial, however, and rigorous determination of the reaction mechanism awaits the identification of the ring-fission product.

A strain of *Rhodococcus* (30) and an unidentified bacterium isolated from contaminated soil (86) biodegrade 1,3-dinitrobenzene via an initial dioxygenase mechanism similar to that described above for 2,4-DNT. 4-Nitrocatechol has been identified as the product of the initial reaction, but the details of the subsequent reactions are not known. Both of the isolates studied to date grow very slowly on 1,3-dinitrobenzene, and investigation of the biochemistry will be facilitated by the isolation of more robust strains.

The evolutionary relationships among the dioxygenases that remove nitro groups are a promising area for future study. The available preliminary evidence suggests that some of the enzyme systems are related to naphthalene dioxygenase. The structural changes that would be required to allow naphthalene dioxygenase to accept nitro-substituted molecules as substrates, however, are unknown.

Reduction of the Aromatic Ring

The electron-withdrawing properties of the nitro group cause the aromatic ring of polynitroaromatic compounds to be highly electron deficient and resistant to electrophilic attack. Lenke et al (74) discovered an alternative mechanism of transformation involving reduction of the aromatic ring. They isolated strains of *Rhodococcus erythropolis* that use 2,4-dinitrophenol as the nitrogen source for growth and later discovered that the nitro compound could also serve as a carbon and energy source. Isolation of the *R. erythropolis* strains confirmed earlier work by Hess et al (57), who isolated a *Janthinobacterium* sp. and an actinomycete able to mineralize 2,4-dinitrophenol with stoichiometric release of nitrite. The *R. erythropolis* (74) isolates released nitrite from 2,4-dinitrophenol, and significant amounts of 4,6-dinitrohexanoate transiently accu- mulated during growth. Under anaerobic conditions 2,4-dinitrophenol is converted stoichiometrically to 4,6-dinitrohexanoate and no nitrite is released. The detection of a reduced ring-fission product containing two nitro groups provides clear evidence that the *R. erythropolis* strains contain enzymes able to catalyze reduction of the aromatic ring. It is not clear whether the reduction of the ring leads to a productive catabolic pathway or is a side reaction. The accumulation of 4,6-dinitro-hexanoate suggests that the aliphatic compound is a dead-end metabolite (Figure 7).

Resting cells of *Rhodococcus erythropolis* grown on 2,4-dinitrophenol released nitrite from picric acid, and spontaneous mutants can use picric acid as the nitrogen source (73). The initial reaction in the utilization of picric acid by these strains is the addition of a hydride ion to the aromatic ring to form a hydride-Meisenheimer complex (Figure 7). The reaction has also been demonstrated in cell extracts, and rigorous proof for the structure of the hydride-Meisenheimer complex has been obtained by NMR spectroscopic analysis (103). Addition of a second hydride ion leads to the eventual formation of 2,4,6-trinitrocyclohexanone, which decomposes to form 1,3,5-trinitropentane

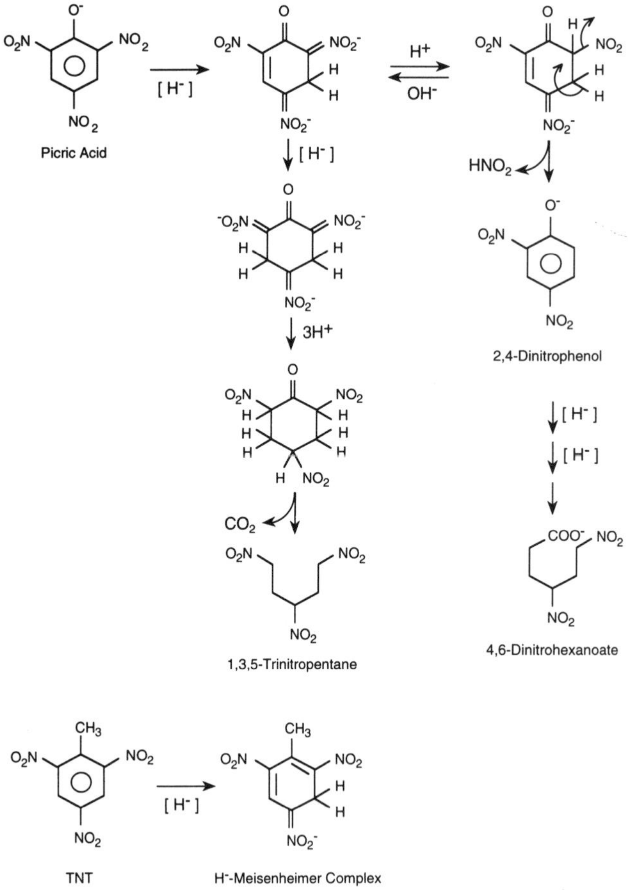

Figure 7 Reduction of the aromatic ring to form an hydride-Meisenheimer complex. Adapted from Lenke & Knackmuss (73), Lenke et al (74), Rieger et al (103), and Vorbeck et al (127).

upon acidification and extraction. This reaction sequence seems to be nonproductive. In contrast, protonation of the hydride-Meisenheimer complex leads to the enzyme-catalyzed rearomatization of the molecule and elimination of nitrite, which can be assimilated by the bacteria. The 2,4-dinitrophenol generated from picric acid is degraded by the bacteria and nitrite is eliminated, but it is not clear whether the mechanism involves formation of a second Meisenheimer complex. A preliminary report that a mixed culture grown on picric acid accumulates 2,4-dinitrophenol transiently (98) also suggests that the elimination of nitrite from the hydride-Meisenheimer complex can lead to productive metabolism of the aromatic ring.

Because the ring of TNT is also very electron deficient, it would be expected to be susceptible to reduction by a hydride ion. Duque et al (32) suggested that a hydride-Meisenheimer complex might be involved in the degradation of TNT by a constructed *Pseudomonas* strain. They initially isolated a strain able to use TNT and several other nitroaromatic compounds as the source of nitrogen. They subsequently selected a more effective strain and, based on preliminary evidence, proposed that it converted TNT to toluene via 2,4-dinitrotoluene, 2,6-dinitrotoluene, and 2-nitrotoluene. Subsequently, rigorous evidence for the conversion of TNT to the corresponding hydride-Meisenheimer complex was provided by Vorbeck et al (127) (Figure 7). Resting cells of a *Mycobacterium* sp. grown on 4-nitrotoluene accumulated several products, and the major metabolite (40%) was the hydride-Meisenheimer complex. A small amount (5%) of 4-ADNT accumulated, as did an additional, unidentified metabolite. A small amount of nitrite was released, but no dinitrotoluenes were detected.

To date, the formation of the hydride-Meisenheimer complexes of polynitroaromatic compounds has only been demonstrated to play a major role in the degradation of picric acid (73, 103). A number of questions remain about the enzymes that catalyze the addition of the hydride ion and those that catalyze the subsequent rearomatization and elimination of the nitro group. Three reactions of the hydride-Meisenheimer complex have been demonstrated in bacteria. The complex can (*a*) spontaneously decompose to the parent compound, (*b*) be reduced to aliphatic compounds, or (*c*) rearomatize by the addition of a proton and the elimination of nitrite. The factors that regulate the relative contributions of these competing reactions in vivo are unknown. Reduction of the nitro groups of polynitroaromatic compounds as described earlier is a competing process in aerobic bacteria that can reduce the aromatic ring by addition of a hydride ion. Thus, amino compounds are also produced from TNT by the strains that accumulate the hydride-Meisenheimer complex (32, 127, 128). Therefore, either the cells must contain competing reductase enzymes, or the same reductases can catalyze both reactions. In contrast, *Rhodococcus erythropolis* does not reduce the nitro groups of picric acid (73).

Partial Reduction of the Nitro Group Under Aerobic Conditions

Very early reports on the biodegradation of 2-nitrobenzoate (19, 20, 71) and 4-nitrobenzoate (21, 22) provided evidence that partial reduction of the nitro group might allow subsequent productive metabolism of nitroaromatic compounds. These studies indicated that the nitro groups are reduced and that the nitrogen is released from the molecule as ammonia. The participation of the

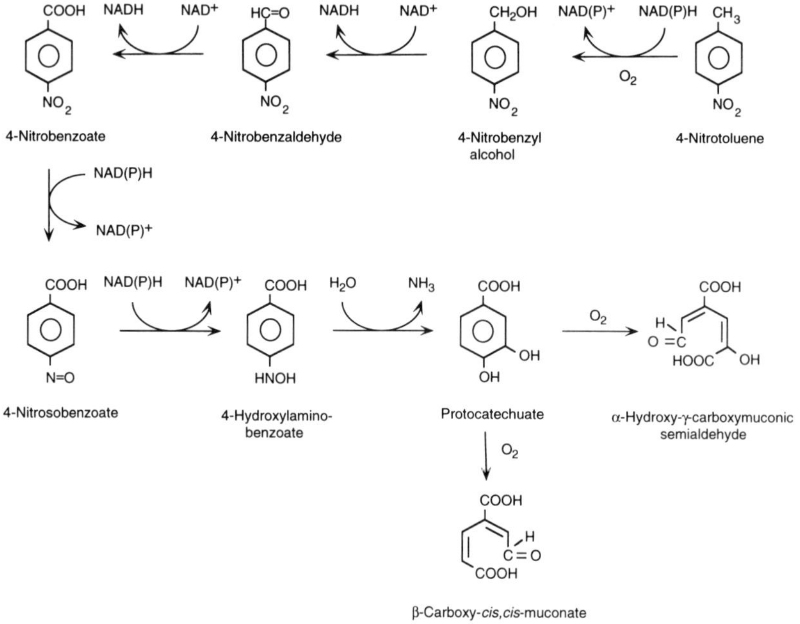

Figure 8 Pathway for catabolism of 4-nitrotoluene and 4-nitrobenzoate. The complete pathway has been found in two strains of *Pseudomonas* (51, 101). The pathway for 4-nitrobenzoate was reported first in *Comamonas acidivorans* (47, 48).

aminobenzoates in the pathway could not be demonstrated, however, and the degradation mechanism remained unclear. The mystery was solved during investigation of the degradation of 4-nitrobenzoate by a *Comamonas acidivorans* isolated from soil (47, 48). A nitroreductase purified from the isolate reduces 4-nitrobenzoate to 4-hydroxylaminobenzoate and does not catalyze further reduction of the molecule to 4-aminobenzoate (Figure 8). Another enzyme purified from 4-nitrobenzoate-grown cells catalyzes the conversion of 4-hydroxylaminobenzoate to protocatechuate without the participation of additional cofactors. The purified enzyme is stimulated by the addition of NADPH and several other reducing agents that seem to function by lowering the redox potential rather than serving as substrates in the reaction (48). The mechanism of the reaction appears to be hydrolytic but has not been studied in detail.

The facile degradation of nitrobenzoates raises the possibility that strains able to degrade the isomeric nitrotoluenes might be constructed if the methyl groups could be oxidized to carboxyl groups. Accordingly, several strains of bacteria have been examined for their ability to transform nitrotoluenes. The enzymes of the TOL pathway involved in the degradation of toluene and xylenes can also oxidize the methyl groups of 3-nitrotoluene and 4-nitrotoluene (28). 2-Nitrotoluene is not a substrate for the initial enzyme in the sequence, toluene monooxygenase. Of the isomeric nitrotoluenes only 4-nitrotoluene can serve as an inducer for the genes of the TOL pathway. The specificity of the effector can be altered by mutagenesis, however, so that all three isomers of nitrotoluene can serve as inducers of the TOL upper pathway (27). This alteration allows pseudomonads containing the TOL plasmid and the mutant regulator to convert 3-nitrotoluene and 4-nitrotoluene to the corresponding nitrobenzoates.

Natural strains of *Pseudomonas* able to grow on 4-nitrotoluene under aerobic conditions have been isolated separately and studied by two groups (51, 101). The catabolic pathways are identical and the initial steps are analogous to those of the TOL pathway. A monooxygenase enzyme oxidizes the methyl group to the corresponding alcohol (Figure 8). Benzyl alcohol dehydrogenase and benzaldehyde dehydrogenase convert 4-nitrobenzyl alcohol to 4-nitrobenzoate. The only obvious difference in the initial steps in the pathways for 4-nitrotoluene degradation of the two *Pseudomonas* isolates is in the properties of their respective 4-nitrobenzyl alcohol dehydrogenases. The strain studied by Haigler & Spain (51) contains an NAD-dependent, membrane-bound enzyme, whereas the strain studied by Rhys-Williams et al (101) contains a soluble enzyme for which the physiological electron acceptor could not be determined. Although the first three reactions in the pathway for 4-nitrotoluene are identical to those in the TOL plasmid-encoded pathway, the genes that encode the enzymes do not seem to be closely related to the TOL genes. Attempts to detect cross-hybridization between a TOL DNA probe and genomic DNA from the 4-nitrotoluene degrader failed (101).

Both of the isolates that grow on 4-nitrotoluene use the pathway described by Groenewegen et al for the degradation of 4-nitrobenzoate (47, 48). The only discernable difference is in the mode of ring fission of protocatechuate. A protocatechuate 3,4-dioxygenase is active in the strain studied by Rhys-Williams et al (101), whereas a 4,5-dioxygenase operates in the strain studied by Haigler & Spain (51). The mechanism of ring fission was not determined for the strain in which the 4-nitrobenzoate pathway was discovered.

Bacteria able to grow on 3-nitrophenol have been isolated (41, 135), and the initial steps in the degradation pathway seem to be reductive rather than oxidative (A Schenzle & H-J Knackmuss, personal communication). Details of the catabolic pathway have not been published, however.

Bacteria containing the catabolic pathway described above for degradation of nitrobenzene via a dioxygenase mechanism (Figure 5a) seem to be unusual. When a range of aerobic nitrobenzene-degrading strains isolated from six

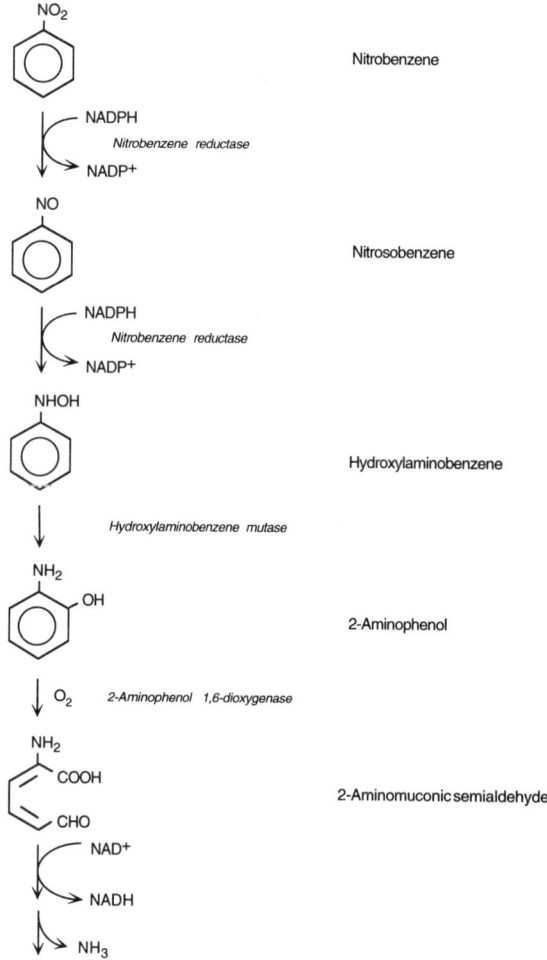

Figure 9 Partial reduction and subsequent oxidation reactions for the degradation of nitrobenzene by *Pseudomonas pseudoalcaligenes* (87).

widely separated ecosystems were examined, only one strain out of nine used the dioxygenase pathway (SF Nishino & JC Spain, unpublished). The other strains use a pathway involving reduction of the nitro group (Figure 9) (87). The enzyme responsible for the NADPH-dependent reduction of nitrobenzene to hydroxylaminobenzene has been purified to homogeneity from a strain of *Pseudomonas pseudoalcaligenes* (108a). It is a flavoprotein with a molecular weight of 30,000, and it is active as a monomer. It is similar to the nitroreductase isolated from *Comamonas acidovorans* (47) in that it does not catalyze the reduction of the hydroxylamino compound to the corresponding amine. Both of these enzymes are unusual in this respect, because most bacterial nitroreductases convert aromatic nitro groups to the amines (5, 10, 17, 43, 72, 94, 97, 126).

A novel enzyme, hydroxylaminobenzene mutase, catalyzes the conversion of hydroxylaminobenzene to 2-aminophenol in the absence of oxygen or any added cofactors. The reaction is analogous to the Bamberger rearrangement (107), in which hydroxylamino aromatic compounds rearrange to aminophenols under mildly acidic conditions. The nonenzymatic rearrangement yields predominantly the 4-aminophenol, whereas the enzyme directs the production of predominantly (>99%) the 2-aminophenol. Analogous enzyme-catalyzed reactions have been reported in animals (117) and in yeast (23) but not in bacteria. The implications of the Bamberger-like rearrangement in biochemistry have been discussed extensively by Corbett & Corbett (24). The relationship between the hydroxylaminobenzene mutase isolated from cells grown on nitrobenzene and the hydroxylaminobenzoate lyase isolated from the *C. acidovorans* grown on 4-nitrobenzoate is not clear. It is possible to postulate similar mechanisms for at least the first steps in both reactions, but the enzyme-catalyzed reactions seem to be specific and show no cross-reactivity (JAM deBont, personal communication). Sequencing of the corresponding genes and additional characterization of the enzymes will be necessary before the relationships can be clarified.

2-Aminophenol produced by the initial steps in the pathway is degraded by a dioxygenase that catalyzes the opening of the ring at the 1,6-position to produce 2-aminomuconic semialdehyde. Only a few examples of enzymes that catalyze ring-fission reactions in the absence of two hydroxyl groups are known. Catechol 1,2-dioxygenases from *Pseudomonas arvilla* (96) and *Pseudomonas aeruginosa* (70) oxidize 2-aminophenol at a rate 1000-fold lower than the rate of oxidation of catechol, their physiological substrate. In contrast, the enzyme from *Pseudomonas pseudoalcaligenes* oxidizes 2-aminophenol, its physiological substrate, 50-fold faster than catechol. An enzyme that catalyzes a similar reaction has also been found in a strain of *Pseudomonas* grown on 5-aminosalicylate (118).

The mechanism for the degradation of 2-aminomuconic semialdehyde by *Pseudomonas pseudoalcaligenes* is not known. Enzymes in crude extracts from cells grown on nitrobenzene catalyze the degradation of the ring-fission product and release of ammonia. The requirement for NAD suggests that the first reaction is an oxidation of the aldehyde, but no clear evidence is available.

The reductive pathway for degradation of nitrobenzene (Figure 9) seems much more complex than the oxidative pathway (Figure 5a). Recruitment of the genes for several unusual enzymes was required for the operation of the pathway. It will be interesting to determine whether the various isolates that use this pathway are derived from a common ancestor or arose separately. In either case, they seem to be more widely distributed than the strains using the simpler oxidative pathway. An explanation for this apparent conundrum lies in the cofactor and oxygen requirements of the pathways. The reductive pathway requires one mole of oxygen and one mole of NADH to convert nitrobenzene to central metabolic intermediates and release ammonia. In contrast, the oxidative pathway requires two moles of oxygen and one mole of NADH that can be regained if the 2-hydroxymuconic semialdehyde undergoes an NAD-dependent oxidation to oxalocrotonate (88). If the isolate is to use the nitrite released by the oxygenolytic reaction as its nitrogen source, three additional moles of NAD(P)H would be required for the reduction of nitrite to ammonia. In nitrobenzene-contaminated subsurface ecosystems, where all of the strains that use the reductive pathway were isolated, oxygen is limiting. In contrast, the strain that uses the oxidative pathway was isolated from an aerobic waste-treatment system. Thus, each pathway seems appropriate to the ecosystem in which it was discovered. The more complex, reductive pathway seems to be well adapted to exploit the conditions of an oxygen-limited ecosystem.

APPLICATIONS IN BIOREMEDIATION

Much of the recent interest in biodegradation of nitroaromatic compounds has been motivated by an increased awareness of the extent of environmental contamination by explosives such as TNT. Therefore, a considerable amount of work has been done on development of treatment systems based on biodegradation. Composting has been used for field-scale cleanup at several TNT-contaminated sites (131) and has been studied on a smaller scale by a number of investigators (61, 69, 83, 133). Composting, however, increases the volume of the waste material and requires a considerable amount of materials handling, and the ultimate products of the process are not well characterized chemically. There is also some evidence of residual toxicity and mutagenicity after composting of explosives-contaminated soil (46). Several groups have worked with aerobic slurry-phase bioreactors (15, 49), and one preliminary report suggests an initial anaerobic conversion of TNT to TAT with subsequent humification under aerobic conditions (75, 102). A similar approach has shown that nitrobenzene can be converted to aniline under anaerobic conditions and that the aniline can be degraded under aerobic conditions (29). Studies with the white rot fungi show considerable promise for use in bioremediation of munitions if problems with toxicity can be overcome (8, 114, 116, 119). Bioremediation under anaerobic conditions seems to be the most favored approach for removal of TNT at present. A process involving methanogenic cultures has been developed and demonstrated for the treatment of soil contaminated with TNT and RDX (26, 40). Other processes employing sulfate-reducing bacteria have been suggested by several groups (14, 94). All of the approaches listed above involve transformation of TNT by microorganisms that use another primary carbon source for growth. Therefore, the processes are more difficult to optimize and control than they would be if the bacteria used TNT as a growth substrate. Unfortunately, reports of microorganisms able to grow on TNT are rare (32).

CONCLUSIONS

Dramatic and rapid progress has been made recently in understanding the biodegradation of nitroaromatic compounds. In addition, a number of exciting questions have been raised as a result of the recent discoveries. The potential for reduction of TNT by anaerobes is well established, yet intriguing areas of uncertainty remain about the metabolism of triaminotoluene. The initial steps in the pathways catalyzed by white rot fungi are clear and the mechanism of toxicity is better understood, but nothing is known about how the reduced metabolites of TNT are mineralized by the fungus. Aerobic bacteria have a hitherto unexpected capacity to convert nitroaromatic compounds into intermediates that can serve as growth substrates. The mechanisms of the reactions, their regulation, and the structure of the enzymes will provide fertile areas for research. Virtually nothing is known about the molecular biology of the systems. Understanding of the molecular basis for the catabolic sequences will allow their capabilities to be enhanced and exploited for practical purposes. Potential applications include not only biodegradation of environmental contaminants, but also the use of the novel enzyme reactions for biocatalysis and synthesis of valuable organic molecules.

ACKNOWLEDGMENTS

This work was supported by the US Air Force Office of Scientific Research. I would like to thank Shirley Nishino for help with editing the manuscript and creating the graphics. I would also like to thank Billy Haigler, Paul-Gerhard Rieger, H.-J. Knackmuss, Chuck Somerville, and Jochen Michels for helpful suggestions.

> Any *Annual Review* chapter, as well as any article cited in an *Annual Review* chapter, may be purchased from the Annual Reviews Preprints and Reprints service.
> 1-800-347-8007; 415-259-5017; email: arpr@class.org

Literature Cited

1. An D, Gibson DT, Spain JC. 1994. Oxidative release of nitrite from 2-nitrotoluene by a three-component enzyme system from *Pseudomonas* sp. Strain JS42. *J. Bacteriol.* 176:7462–67

2. Angermaier L, Hein F, Simon H. 1981. Investigations on the reduction of aliphatic and aromatic nitro compounds by *Clostridium* species and enzyme systems. In *Biology of Inorganic Nitrogen*

and *Sulfur*, ed. H Bothe, A Trebst, pp. 266–75. Berlin: Springer
3. Angermaier L, Simon H. 1983. On nitroaryl reductase activities in several clostridia. *Hoppe-Seyler's Z. Physiol. Chem.* 364:1653–64
4. Angermaier L, Simon H. 1983. On the reduction of aliphatic and aromatic nitro compounds by clostridia, the role of ferredoxin and its stabilization. *Hoppe-Seyler's Z. Physiol. Chem.* 364:961–75
5. Anlezark GM, Melton RG, Sherwood RF, Coles B, Friedlos F, Knox R. 1992. The bioactivation of 5-(aziridin-1-YL)-2,4-dinitrobenzamide (CB1954)-1. Purification and properties of a nitroreductase enzyme from *Escherichia coli*—a potential enzyme for antibody-directed enzyme prodrug therapy (ADEPT). *Biochem. Pharmacol.* 44:2289–95
6. Averill BA. 1995. Transformation of inorganic N-oxides by denitrifying and nitrifying bacteria: pathways, mechanisms, and relevance to transformation of nitroaromatic compounds. See Ref. 109, pp. 183–97
7. Bachofer R, Lingens F, Schäfer W. 1975. Conversion of aniline into pyrocatechol by a *Nocardia* sp.: incorporation of oxygen-18. *FEBS Lett.* 50:288–90
8. Barr DP, Aust SD. 1994. Mechanisms white rot fungi use to degrade pollutants. *Environ. Sci. Technol.* 28:79A–87A
9. Beland FA, Heflich RH, Howard PC, Fu PP. 1985. The in vitro metabolic activation of nitropolycyclic aromatic hydrocarbons. In *Polycyclic Hydrocarbons and Carcinogenesis, ACS Symposium Series*, ed. RG Harvey, pp. 371–96. Washington, DC: Am. Chem. Soc.
10. Blasco R, Castillo F. 1993. Characterization of a nitrophenol reductase from the phototrophic bacterium *Rhodobacter capsulatus* E1F1. *Appl. Environ. Microbiol.* 59:1774–78
11. Boopathy R, Kulpa CF. 1992. Trinitrotoluene (TNT) as a sole nitrogen source for a sulfate reducing bacterium *Desulfovibrio* sp. (B strain) isolated from an anaerobic digester. *Curr. Microbiol.* 25:235–41
12. Boopathy R, Kulpa CF. 1993. Nitroaromatic compounds serve as nitrogen source for *Desulfovibrio* sp. (B strain). *Can. J. Microbiol.* 39:430–33
13. Boopathy R, Kulpa CF. 1994. Biotransformation of 2,4,6-trinitrotoluene (TNT) by a *Methanococcus* sp. (strain B) isolated from a lake sediment. *Can. J. Microbiol.* 40:273–78
14. Boopathy R, Kulpa CF, Wilson M. 1993. Metabolism of 2,4,6-trinitrotoluene (TNT) by *Desulfovibrio* sp. (B strain). *Appl. Microbiol. Biotechnol.* 39:270–75
15. Boopathy R, Manning J, Montemagno C, Kulpa CF. 1994. Evaluation of a soil slurry reactor system for treating soil contaminated with munition compounds. *Abstr. 94th Annu. Meet. Am. Soc. Microbiol.* p. 456
16. Brilon C, Beckmann W, Knackmuss H. 1981. Catabolism of naphthalenesulfonic acids by *Pseudomonas* sp. A3 and *Pseudomonas* sp. C22. *Appl. Environ. Microbiol.* 42:44–55
17. Bryant C, DeLuca M. 1991. Purification and characterization of an oxygen-insensitive NAD(P)H nitroreductase from *Enterobacter cloacae. J. Biol. Chem.* 266:4119–25
18. Bumpus JA, Tatarko M. 1994. Biodegradation of 2,4,6-trinitrotoluene by *Phanerochaete chrysosporium*: identification of initial degradation products and the discovery of a TNT metabolite that inhibits lignin peroxidases. *Curr. Microbiol.* 28:185–90
19. Cain RB. 1966. Induction of anthranilate oxidation system during the metabolism of *ortho*-nitrobenzoate by certain bacteria. *J. Gen. Microbiol.* 42:197–217
20. Cain RB. 1966. Utilization of anthranilic and nitrobenzoic acids by *Nocardia opaca* and a flavobacterium. *J. Gen. Microbiol.* 42:219–35
21. Cartwright NJ, Cain RB. 1959. Bacterial degradation of nitrobenzoic acids. *Biochem. J.* 71:248–61
22. Cartwright NJ, Cain RB. 1959. Bacterial degradation of the nitrobenzoic acids. 2. Reduction of the nitro group. *Biochem. J.* 73:305–14
23. Corbett MD, Corbett BR. 1981. Metabolism of 4-chloronitrobenzene by the yeast *Rhodosporidium* sp. *Appl. Environ. Microbiol.* 41:942–49
24. Corbett MD, Corbett BR. 1995. Bioorganic chemistry of the arylhydroxylamine and nitrosoarene functional groups. See Ref. 109, pp. 151–82
25. Craig AM, Bilich D, Will Y, Lee T, Hovervale J. 1994. Biotransformation of trinitrotoluene by anaerobic ruminal bacteria. *Abstr. 94th Annu. Meet. Am. Soc. Microbiol*, p. 409
26. Crawford RL. 1995. Biodegradation of nitrated munition compounds and herbicides by obligately anaerobic bacteria. See Ref. 109, pp. 87–98
27. Delgado A, Ramos JL. 1994. Genetic evidence for activation of the positive transcriptional regulator XylR, a member of the NtrC family of regulators, by

effector binding. *J. Biol. Chem.* 269: 8059–62
28. Delgado A, Wubbolts MG, Abril M-A, Ramos JL. 1992. Nitroaromatics are substrates for the TOL plasmid upper-pathway enzymes. *Appl. Environ. Microbiol.* 58:415–17
29. Dickel O, Haug W, Knackmuss H-J. 1993. Biodegradation of nitrobenzene by a sequential anaerobic-aerobic process. *Biodegradation* 4:187–94
30. Dickel O, Knackmuss H-J. 1991. Catabolism of 1,3-dinitrobenzene by *Rhodococcus* sp. QT-1. *Arch. Microbiol.* 157:76–79
31. Dunnivant FM, Schwarzenbach RP, Macalady DL. 1992. Reduction of substituted nitrobenzenes in aqueous solutions containing natural organic matter. *Environ. Sci. Technol.* 26:2133–41
32. Duque E, Haidour A, Godoy F, Ramos JL. 1993. Construction of a *Pseudomonas* hybrid strain that mineralizes 2,4,6-trinitrotoluene. *J. Bacteriol.* 175: 2278–83
33. Ecker S, Widmann T, Lenke H, Dickel O, Fischer P, et al. 1992. Catabolism of 2,6-dinitrophenol by *Alcaligenes eutrophus* JMP134 and JMP222. *Arch. Microbiol.* 158:149–54
34. Ensley BD, Gibson DT, Laborde AL. 1982. Oxidation of naphthalene by a multicomponent enzyme system from *Pseudomonas* sp. strain NCIB 9816. *J. Bacteriol.* 149:948–54
35. Fernando T, Bumpus JA, Aust SD. 1990. Biodegradation of TNT (2,4,6-trinitrotoluene) by *Phanerochaete chrysosporium*. *Appl. Environ. Microbiol.* 56: 1666–71
36. Fetzner S, Lingens F. 1994. Bacterial dehalogenases: biochemistry, genetics, and biotechnological applications. *Microbiol. Rev.* 58:641–85
37. Fetzner S, Muller R, Lingens F. 1989. A novel metabolite in the microbial degradation of 2-chlorobenzoate. *Biochem. Biophys. Res. Commun.* 161:700–5
38. Fu PP. 1990. Metabolic activation of nitro-polycyclic aromatic hydrocarbons. *Drug Metab. Rev.* 22:209–68
39. Funk SB, Roberts DJ, Crawford DL, Crawford RL. 1993. Degradation of trinitrotoluene (TNT) and sequential accumulation of metabolic intermediates by an anaerobic bioreactor during its adaptation to a TNT feed. *Abstr. 93rd Annu. Meet. Am. Soc. Microbiol.* p. 421
40. Funk SB, Roberts DJ, Crawford DL, Crawford RL. 1993. Initial-phase optimization for bioremediation of munition compound–contaminated soils. *Appl. Environ. Microbiol.* 59:2171–77
41. Germanier R, Wuhrmann K. 1963. Über den aeroben mikrobiellen Abbau aromatischer Nitroverbindungen. *Pathol. Microbiol.* 26:569–78 (In German)
42. Gibson DT, Subramanian V. 1984. Microbial degradation of aromatic hydrocarbons. In *Microbial Degradation of Organic Compounds*, ed. DT Gibson, pp. 181–252. New York: Dekker
43. Glaus MA, Heijman CG, Schwarzenbach RP, Zeyer J. 1992. Reduction of nitroaromatic compounds mediated by *Streptomyces* sp. exudates. *Appl. Environ. Microbiol.* 58:1945–51
43a. Gorontzy T, Drzyzga O, Kahl MW, Bruns-Nagel D, Breitung J, et al. 1994. Microbial degradation of explosives and related compounds. *Crit. Rev. Microbiol.* 20:265–84
44. Gorontzy T, Kuver J, Blotevogel KH. 1993. Microbial transformation of nitroaromatic compounds under anaerobic conditions. *J. Gen. Microbiol.* 139: 1331–36
45. Gribble GW. 1992. Naturally occurring organohalogen compounds—a survey. *J. Nat. Prod.* 55:1353–95
46. Griest WH, Stewart AJ, Tyndall RL, Caton JE, Ho C-H, et al. 1993. Chemical and toxicological testing of composted explosives-contaminated soil. *Environ. Toxicol. Chem.* 12:1105–16
47. Groenewegen PEJ, Breeuwer P, van Helvoort JMLM, Langenhoff AAM, de Vries FP, de Bont JAM. 1992. Novel degradative pathway of 4-nitrobenzoate in *Comamonas acidovorans* NBA-10. *J. Gen. Microbiol.* 138:1599–605
48. Groenewegen PEJ, de Bont JAM. 1992. Degradation of 4-nitrobenzoate via 4-hydroxylaminobenzoate and 3,4-dihydroxybenzoate in *Comamonas acidovorans* NBA-10. *Arch. Microbiol.* 158: 381–86
49. Gunnison D, Pennington J, Price C, Myrick G, Zappi M, et al. 1994. Characterization of TNT-mineralizing activity. *Abstr. 94th Annu. Meet. Am. Soc. Microbiol.* p. 456
50. Haigler BE, Nishino SF, Spain JC. 1994. Biodegradation of 4-methyl-5-nitrocatechol by *Pseudomonas* sp. strain DNT. *J. Bacteriol.* 176:3433–37
51. Haigler BE, Spain JC. 1993. Biodegradation of 4-nitrotoluene by *Pseudomonas* sp. strain 4NT. *Appl. Environ. Microbiol.* 59:2239–43
52. Haigler BE, Wallace WH, Spain JC. 1994. Biodegradation of 2-nitrotoluene by *Pseudomonas* sp. strain JS42. *Appl. Environ. Microbiol.* 60:3466–69

53. Hanne LF, Kirk LL, Appel SM, Narayan AD, Bains KK. 1993. Degradation and induction specificity in actinomycetes that degrade *p*-nitrophenol. *Appl. Environ. Microbiol.* 59:3505–08
54. Harayama S, Rekik M. 1989. Bacterial aromatic ring-cleavage enzymes are classified into two different gene families. *J. Biol. Chem.* 264:15328–33
55. Harvey PJ, Schoemaker HE, Palmer JM. 1986. Veratryl alcohol as a mediator and the role of radical cations in lignin biodegradation by *Phanerochaete chrysosporium. FEBS Lett.* 195:242–46
56. Heijman CG, Holliger C, Glaus MA, Schwarzenbach RP, Zeyer J. 1993. Abiotic reduction of 4-chloronitrobenzene to 4-chloroaniline in a dissimilatory iron-reducing enrichment culture. *Appl. Environ. Microbiol.* 59:4350–53
57. Hess TF, Schmidt SK, Silverstein J, Howe B. 1990. Supplemental substrate enhancement of 2,4-dinitrophenol mineralization by a bacterial consortium. *Appl. Environ. Microbiol.* 56:1551–58
58. Higson FK. 1992. Microbial degradation of nitroaromatic compounds. *Adv. Appl. Microbiol.* 37:1–19
59. Hlavica P. 1982. Biological oxidation of nitrogen in organic compounds and disposition of N-oxidized products. *CRC Crit. Rev. Biochem.* 12:39–101
60. Husain M, Entsch B, Ballou DP, Massey V, Chapman PJ. 1980. Fluoride elimination from substrates in hydroxylation reactions catalyzed by *p*-hydroxybenzoate hydroxylase. *J. Biol. Chem.* 255: 4189–97
61. Isbister JD, Anspatch GL, Kitchens JF, Doyle RC. 1984. Composting for decontamination of soils containing explosives. *Microbiologica* 7:47–73
62. Jain RK, Dreisbach JH, Spain JC. 1994. Biodegradation of *p*-nitrophenol via 1,2,4-benzenetriol by an *Arthrobacter. Appl. Environ. Microbiol.* 60:3030–32
63. Janssen DB, Pries F, Van der Ploeg J. 1994. Genetics and biochemistry of dehalogenating enzymes. *Annu. Rev. Microbiol.* 48:163–91
64. Junker F, Field JA, Bangerter F, Ramsteiner K, Kohler HP, et al. 1994. Oxygenation and spontaneous deamination of 2-aminobenzenesulphonic acid in *Alcaligenes* sp. strain O-1 with subsequent *meta* ring cleavage and spontaneous desulfonation to 2-hydroxymuconic acid. *Biochem. J.* 300:429–36
65. Kaake RH, Crawford DL, Crawford RL. 1994. Optimization of an anaerobic bioremediation process for soil contaminated with the nitroaromatic herbicide dinoseb (2-*sec*-butyl-4,6-dinitrophenol). In *Applied Biotechnology for Site Remediation*, ed. RE Hinchee, DB Anderson, FB Metting Jr, GD Sayles, pp. 337–41. Boca Raton: Lewis
66. Kaake RH, Roberts DJ, Stevens TO, Crawford RL, Crawford DL. 1992. Bioremediation of soils contaminated with the herbicide 2-*sec*-butyl-4,6-dinitrophenol (Dinoseb). *Appl. Environ. Microbiol.* 58:1683–89
67. Kaplan DL. 1990. Biotransformation pathways of hazardous energetic organo-nitro compounds. In *Biotechnology and Biodegradation*, ed. D Kamely, A Chakrabarty, GS Omenn, pp. 155–80. Houston: Gulf
68. Kaplan DL. 1992. Biological degradation of explosives and chemical agents. *Curr. Opin. Biotechnol.* 3:253–60
69. Kaplan DL, Kaplan AM. 1982. Thermophilic biotransformations of 2,4,6-trinitrotoluene under simulated composting conditions. *Appl. Environ. Microbiol.* 44:757–60
70. Kataeva IA, Golovleva LA. 1984. Extradiol cleavage of 2-aminophenol by pyrocatechase from *Pseudomonas aeruginosa* 2x: reaction mechanism. *Mikrobiol. Zh.* 46:22–26
71. Ke Y-H, Gee LL, Durham NN. 1959. Mechanism involved in the metabolism of nitrophenyl-carboxylic acid compounds by microorganisms. *J. Bacteriol.* 77:593–98
72. Kinouchi T, Ohnishi Y. 1983. Purification and characterization of 1-nitropyrene nitroreductases from *Bacteroides fragilis. Appl. Environ. Microbiol.* 46: 596–604
73. Lenke H, Knackmuss H-J. 1992. Initial hydrogenation during catabolism of picric acid by *Rhodococcus erythropolis* HL 24–2. *Appl. Environ. Microbiol.* 58: 2933–37
74. Lenke H, Pieper DH, Bruhn C, Knackmuss H-J. 1992. Degradation of 2,4-dinitrophenol by two *Rhodococcus erythropolis* strains, HL 24–1 and HL 24–2. *Appl. Environ. Microbiol.* 58: 2928–32
75. Lenke H, Wagener B, Daun G, Knackmuss H-J. 1994. TNT-contaminated soil: a sequential anaerob/aerob process for bioremediation. *Abstr. 94th Annu. Meet. Am. Soc. Microbiol.* p. 456
76. Locher H, Leisinger T, Cook AM. 1991. 4-Sulfobenzoate 3,4-dioxygenase: purification and properties of a desulphonative two-component enzyme system from *Comamonas testosteroni* T-2. *Biochem. J.* 274:833–42
77. Locher HH, Leisinger T, Cook AM. 1989. Degradation of *p*-toluenesul-

phonic acid via sidechain oxidation, desulphonation and *meta* ring cleavage in *Pseudomonas* (*Comamonas*) *testosteroni* T-2. *J. Gen. Microbiol.* 135: 1969–78
78. Marcus A, Klages U, Krauss S, Lingens F. 1984. Oxidation and dehalogenation of 4-chlorophenylacetate by a two component enzyme system from *Pseudomonas* sp. strain CBS3. *J. Bacteriol.* 160:618–21
78a. Marvin-Sikkema FD, de Bont JAM. 1994. Degradation of nitroaromatic compounds by microorganisms. *Appl. Microbiol. Biotechnol.* 42:499–507
79. McCormick NG, Feeherry FF, Levinson HS. 1976. Microbial transformation of 2,4,6-trinitrotoluene and other nitroaromatic compounds. *Appl. Environ. Microbiol.* 31:949–58
80. Michels J, Gottschalk G. 1994. Inhibition of lignin peroxidase of *Phanerochaete chrysosporium* by hydroxylamino-dinitrotoluene, an early intermediate in the degradation of 2,4,6-trinitrotoluene. *Appl. Environ. Microbiol.* 60:187–94
81. Michels J, Gottschalk G. 1995. Pathway of 2,4,6-trinitrotoluene degradation by *Phanerochaete chrysosporium.* See Ref. 109, pp. 19–35
82. Mitra D, Vaidyanathan. 1984. A new 4-nitrophenol 2-hydroxylase from a *Nocardia* sp. *Biochem. Int.* 8:609–15
83. Montemagno CD. 1991. Evaluation of the feasibility of biodegrading explosives-contaminated soils and groundwater at the Newport Army Ammunition Plant (NAAP). *USATHAMA CETHA-TS-CR-9200.* Argonne Natl. Lab., Chicago, Ill.
84. Nadeau LJ, Spain JC. 1995. The bacterial degradation of *m*-nitrobenzoic acid. *Appl. Environ. Microbiol.* 61:840–43
85. Neidleman SL, Geigert J. 1986. *Biohalogenation: Principles, Basic Roles, and Applications.* New York: Halstead
86. Nishino SF, Spain JC. 1992. Initial steps in the bacterial degradation of 1,3-dinitrobenzene. *Abstr. 92nd Annu. Meet. Am. Soc. Microbiol.* p. 358
87. Nishino SF, Spain JC. 1993. Degradation of nitrobenzene by a *Pseudomonas pseudoalcaligenes. Appl. Environ. Microbiol.* 59:2520–25
88. Nishino SF, Spain JC. 1995. Oxidative pathway for the degradation of nitrobenzene by *Comamonas* sp. strain JS765. *Appl. Environ. Microbiol.* 61:2308–13
89. O'Brien RW, Morris JG. 1971. The ferredoxin-dependent reduction of chloramphenicol by *Clostridium acetobutylicum. J. Gen. Microbiol.* 67:265–71
90. Oren A, Gurevich P, Henis Y. 1991. Reduction of nitrosubstituted aromatic compounds by the halophilic anaerobic eubacteria *Haloanaerobium praevalens* and *Sporohalobacter marismortui. Appl. Environ. Microbiol.* 57:3367–70
91. Parrish FW. 1977. Fungal transformation of 2,4-dinitrotoluene and 2,4,6-trinitrotoluene. *Appl. Environ Microbiol.* 34:232–33
92. Peterson FJ, Mason RP, Hovsepian J, Holtzman JL. 1979. Oxygen-sensitive and -insensitive nitroreduction by *Escherichia coli* and rat hepatic microsomes. *J. Biol. Chem.* 254:4009–14
93. Pitts JN, Van Cauwenberghe KA, Grosjean D, Schmid JP, Fitz DR, et al. 1978. Atmospheric reactions of polycyclic aromatic hydrocarbons: facile formation of mutagenic nitro derivatives. *Science* 202:515–18
94. Preuss A, Fimpel J, Diekert G. 1993. Anaerobic transformation of 2,4,6-trinitrotoluene (TNT). *Arch. Microbiol.* 159: 345–53
95. Preuss A, Rieger PG. 1995. Anaerobic transformation of 2,4,6-trinitrotoluene and other nitroaromatic compounds. See Ref. 109, pp. 69–85
96. Que L. 1978. Extradiol cleavage of *o*-aminophenol by pyrocatechase. *Biochem. Biophys. Res. Commun.* 84: 123–29
97. Rafii F, Franklin W, Heflich RH, Cerniglia CE. 1991. Reduction of nitroaromatic compounds by anaerobic bacteria isolated from the human gastrointestinal tract. *Appl. Environ. Microbiol.* 57:962–68
98. Rajan J, Sariaslani S. 1994. Microbial degradation of picric acid. *Abstr. 94th Annu. Meet. Am. Soc. Microbiol.* p. 409
99. Raymond DGM, Alexander M. 1971. Microbial metabolism and cometabolism of nitrophenols. *Pest. Biochem. Physiol.* 1:123–30
100. Regan KM, Crawford RL. 1994. Characterization of *Clostridium bifermentans* and its biotransformation of 2,4,6-trinitrotoluene (TNT) and 1,3,5-triaza-1,3,5-trinitrocyclohexane(RDX). *Biotechnol. Lett.* 16:1081–86
101. Rhys-Williams W, Taylor SC, Williams PA. 1993. A novel pathway for the catabolism of 4-nitrotoluene by *Pseudomonas. J. Gen. Microbiol.* 139:1967–72
102. Rieger P-G, Knackmuss H-J. 1995. Basic knowledge and perspectives on biodegradation of 2,4,6-trinitrotoluene and related nitroaromatic compounds in

contaminated soil. See Ref. 109, pp. 1–18
103. Rieger P-G, Preuss A, Lenke H, Knackmuss H-J. 1994. H⁻-additions as initial steps of aerobic bacterial degradation of 2,4,6-trinitrophenol (picric acid). *Abstr. 94th Annu. Meet. Am. Soc. Microbiol.* p. 409
104. Sander P, Wittaich R-M, Fortnagel P, Wilkes H, Francke W. 1991. Degradation of 1,2,4-trichloro- and 1,2,4,5-tetrachlorobenzene by *Pseudomonas* strains. *Appl. Environ. Microbiol.* 57:1430–40
105. Schackmann A, Müller R. 1991. Reduction of nitroaromatic compounds by different *Pseudomonas* species under aerobic conditions. *Appl. Microbiol. Biotechnol.* 34:809–13
106. Schnell S, Schink B. 1991. Anaerobic aniline degradation via reductive deamination of 4-amino-benzoyl-CoA in *Desulfobacterium anilini*. *Arch. Microbiol.* 155:183–90
107. Shine HJ. 1967. The rearrangement of phenylhydroxylamines. In *Aromatic Rearrangements*, pp. 182–90. Amsterdam/London/New York: Elsevier
108. Simpson JR, Evans WC. 1953. The metabolism of nitrophenols by certain bacteria. *Biochem. J.* 55:XXIV
108a. Somerville CC, Nishino SF, Spain JC. 1995. Purification and characterization of nitrobenzene nitroreductase from *Pseudomonas pseudoalcaligenes* JS45. *J. Bacteriol.* In press
109. Spain JC, ed. 1995. *Biodegradation of Nitroaromatic Compounds*. New York: Plenum
110. Spain JC, Gibson DT. 1991. Pathway for biodegradation of *p*-nitrophenol in a *Moraxella* sp. *Appl. Environ. Microbiol.* 57:812–19
111. Spain JC, Wyss O, Gibson DT. 1979. Enzymatic oxidation of *p*-nitrophenol. *Biochem. Biophys. Res. Commun.* 88:634–41
112. Spain JC, Zylstra GJ, Blake CK, Gibson DT. 1989. Monohydroxylation of phenol and 2,5-dichlorophenol by toluene dioxygenase in *Pseudomonas putida* F1. *Appl. Environ. Microbiol.* 55:2648–52
113. Spanggord RJ, Spain JC, Nishino SF, Mortelmans KE. 1991. Biodegradation of 2,4-dinitrotoluene by a *Pseudomonas* sp. *Appl. Environ. Microbiol.* 57:3200–5
114. Spiker JK, Crawford DL, Crawford RL. 1992. Influence of 2,4,6-trinitrotoluene (TNT) concentration on the degradation of TNT in explosive-contaminated soils by the white rot fungus *Phanerochaete chrysosporium*. *Appl. Environ. Microbiol.* 58:3199–202
115. Stahl JD, Aust SD. 1993. Metabolism and detoxification of TNT by *Phanerochaete chrysosporium*. *Biochem. Biophys. Res. Commun.* 192:477–82
116. Stahl JD, Aust SD. 1993. Plasma membrane dependent reduction of 2,4,6-trinitrotoluene by *Phanerochaete chrysosporium*. *Biochem. Biophys. Res. Commun.* 192:471–76
117. Sternson LA, Gammans RE. 1975. A mechanistic study of aromatic hydroxylamine rearrangement in the rat. *Bioorg. Chem.* 4:58–63
118. Stoltz A, Nörtemann B, Knackmuss H-J. 1992. Bacterial metabolism of 5-aminosalicylic acid. Initial ring cleavage. *Biochem. J.* 282:675–80
119. Sublette KL, Ganapathy EV, Schwartz S. 1992. Degradation of munition wastes by *Phanerochaete chrysosporium*. *Appl. Biochem. Biotechnol.* 34/35:709–23
120. Suen W-C, Haigler BE, Spain JC. 1994. 2,4-Dinitrotoluene dioxygenase genes from *Pseudomonas* sp. strain DNT: homology to naphthalene dioxygenase. *Abstr. 94th Annu. Meet. Am. Soc. Microbiol.* p. 458
121. Suen W-C, Spain JC. 1993. Cloning and characterization of *Pseudomonas* sp. strain DNT genes for 2,4-dinitrotoluene degradation. *J. Bacteriol.* 175:1831–37
122. Suen W-C, Spain JC. 1993. Nucleotide sequence and over-expression of 2,4,5-trihydroxytoluene oxygenase gene from *Pseudomonas* sp. strain DNT in *Escherichia coli*. Abstr. Q-318. In *Abstr. 93rd Annu. Meet. Am. Soc. Microbiol.* p. 404
123. Tokiwa H, Ohnishi Y. 1986. Mutagenicity and carcinogenicity of nitroarenes and their sources in the environment. *CRC Crit. Rev. Toxicol.* 17:23–60
124. Valli K, Brock BJ, Joshi DK, Gold MH. 1992. Degradation of 2,4-dinitrotoluene by the lignin-degrading fungus *Phanerochaete chrysosporium*. *Appl. Environ. Microbiol.* 58:221–28
125. Venulet J, Van Etten RL. 1970. Biochemistry and pharmacology of the nitro and nitroso groups. In *The Chemistry of the Nitro and Nitroso Groups*, ed. H Feuer, pp. 201–89. New York: Interscience
126. Villanueva JR. 1964. Nitro-reductase from a *Nocardia* sp. *Antonie van Leeuwenhoek* 30:17–32
127. Vorbeck C, Lenke H, Fischer P, Knackmuss H-J. 1994. Identification of a hydride-Meisenheimer complex as a metabolite of 2,4,6-trinitrotoluene by a *My-*

128. Vorbeck C, Lenke H, Spain JC, Knackmuss H-J. 1994. Initial steps in the aerobic metabolism of 2,4,6-trinitrotoluene (TNT) by a *Mycobacterium* sp. *Abstr. 94th Annu. Meet. Am. Soc. Microbiol.* p. 408
   cobacterium strain. *J. Bacteriol.* 176: 932–34
129. Walker JE, Kaplan DL. 1992. Biological degradation of explosives and chemical agents. *Biodegradation* 3:369–85
130. Weisburger EK. 1978. Mechanism of chemical carcinogenesis. *Annu. Rev. Pharmacol. Toxicol.* 18:395–415
131. Williams RT, Ziegenfuss PS, Sisk WE. 1992. Composting of explosives and propellant contaminated soils under thermophilic and mesophilic conditions. *J. Ind. Microbiol.* 9:137–44
132. Won WD, Disalvo LH, James NG. 1976. Toxicity and mutagenicity of 2,4,6-trinitrotoluene and its microbial metabolites. *Appl. Environ. Microbiol.* 31: 576–80
133. Woodward RE. 1990. Evaluation of composting implementation: a literature review. *TCN 89363.* US Army Toxic and Hazardous Materials Agency, Aberdeen Proving Ground, MD
134. Xun L, Topp E, Orser CS. 1992. Diverse substrate range of a *Flavobacterium* pentachlorophenol hydroxylase and reaction stoichiometries. *J. Bacteriol.* 174: 2898–902
135. Zeyer J, Kearney PC. 1984. Degradation of o-nitrophenol and m-nitrophenol by a *Pseudomonas putida*. *J. Agric. Food Chem.* 32:238–42
136. Zeyer J, Kocher HP. 1988. Purification and characterization of a bacterial nitrophenol oxygenase which converts *ortho*-nitrophenol to catechol and nitrite. *J. Bacteriol.* 170:1789–94

BIOCATALYTIC SYNTHESES OF AROMATICS FROM D-GLUCOSE: Renewable Microbial Sources of Aromatic Compounds

J. W. Frost and K. M. Draths

Department of Chemistry, Michigan State University, East Lansing, Michigan 48824

KEY WORDS: chemical manufacture, environment, amino acid, biosynthesis, common pathway

CONTENTS

Introduction	558
Traditional Approaches for Increasing Aromatic Biosynthesis	559
Substrate-Limited Aromatic Biosynthesis	561
Impediments to Carbon Flow Through the Common Pathway of Aromatic Amino Acid Biosynthesis	565
Biocatalytic Syntheses	568
Future Horizons	575

ABSTRACT

Chemistry is moving into a new era in which renewable resources and starting materials such as D-glucose will likely be prominent features of industrial chemical manufacture. The keys to this progress are the design, development, and use of microbial biocatalysts. Aromatic biosynthesis serves as a paradigm for how biocatalysts can be manipulated to achieve the yield, rate, and purity criteria central to chemical manufacture. A disproportionate amount of the metabolic carbon flow of the biocatalyst must first be directed into the common pathway of aromatic amino acid biosynthesis. This review describes ways of achieving this goal through the traditional strategy of manipulating the catalytic activity of the first enzyme in the common pathway, as well as the amelioration of limitations in the in vivo availability of common-pathway enzyme substrates. The inability of individual enzymes to convert their substrate to product fast enough to avoid substrate accumulation further impedes carbon flow through the common pathway. This review also discusses identification and removal of these rate-limiting enzymes. Finally, we examine the creation of heterologous biocatalysts and how biocatalysis could be integrated with traditional chemical transformations to expand the number of organic chemicals that can be synthesized from glucose.

Introduction

The concepts and principles that have formed the basis of our understanding of nature's manufacture of chemicals are now being tested by the rapidly burgeoning use of microbial biocatalysis (36). This increasingly important industrial activity differs from traditional chemical manufacture in the type of catalysts used and the use of D-glucose as the starting material. Chemical manufacture (except for within the pharmaceutical industry) typically relies on abiotic, chemical catalysts and starting materials derived from petroleum. As a nonrenewable natural resource, petroleum has several negative environmental and geopolitical problems associated with its use. Plant starch (45) and cellulose (47), by contrast, are abundant, renewable sources of glucose. The comparatively low temperatures, near-atmospheric pressures, and use of water as reaction solvent that characterize microbial biocatalysis are environmentally friendly. Add to this list the avoidance of toxic starting materials, intermediates, reagents, and byproducts, and the ability to synthesize a chemical by microbial biocatalysis presents itself as an appealing alternative to traditional chemical manufacture.

Aromatic amino acids such as phenylalanine and tryptophan figure prominently (36) among the chemicals now being microbially manufactured from glucose. The development of microbial catalysts for synthesis of aromatics exemplifies many of the challenges associated with optimization of microbial metabolism for small-molecule synthesis. These biocatalysts are cultured under conditions, and have been genetically manipulated to achieve metabolic optima, that have little to do with the growth conditions or metabolic demands that confront their wild-type ancestors. Thus, while these biocatalysts are subject to many of the same biochemical tenets that control wild-type microbes, they encounter a new set of growth and metabolic limitations as well.

This review examines aromatic biosynthesis and the unique differences between the metabolism of biocatalytic and wild-type microbes. Substrate availability as a limiting factor in biosynthesis is discussed along with the problems created by rate-limiting enzymes that are encountered when carbon flow surges into the common pathway of aromatic amino acid biosynthesis. We also discuss strategies by which the molecules typically associated with industrial chemistry (e.g. *p*-aminobenzoic acid, *p*-hydroxybenzoic acid, gallic acid, pyrogallol, hydroquinone, benzoquinone, catechol, adipic acid) can be made using approaches typically associated with synthesis of biochemicals (phenylalanine, tyrosine, tryptophan and related secondary metabolites).

Traditional Approaches for Increasing Aromatic Biosynthesis

All carbon flow directed into the synthesis of aromatic amino acids and related secondary metabolites must initially pass through the seven enzymes of the common pathway (30, 31, 35, 57) of aromatic amino acid biosynthesis (Figure 1). After synthesis of chorismic acid, carbon flow is directed through three terminal pathways that respectively result in synthesis (5, 25, 30, 32, 57, 69) of phenylalanine, tyrosine, and tryptophan. The rate of aromatic amino acid biosynthesis is controlled by modulation of the catalytic activity of the first enzyme of the common pathway and the first enzyme in each of the branching terminal pathways. Control of the catalytic activities of the three different isozymes of 3-deoxy-D-*arabino*-heptulosonic acid 7-phosphate (DAHP) is dictated by transcriptional regulation, attenuation, and feedback inhibition. Textbooks often use transcriptional regulation and attenuation of DAHP synthase isozymes as paradigms of biosynthetic regulation. However, the most mundane form of regulation, feedback inhibition, is actually the most important mechanism (72) for control of DAHP synthase isozyme catalytic activities.

Several different strategies, individually or in combination, allow us to increase the percentage of biocatalyst-consumed glucose that is committed to aromatic amino acid biosynthesis. Manipulation of the in vivo catalytic activity of the tyrosine-sensitive isozyme (AroF) of DAHP synthase serves as an example. One option is to introduce a mutation into the locus that encodes the aporepressor (TyrR) for *aroF* transcription. In lieu of TyrR, transcription of *aroF* is derepressed and the number of molecules of AroF increases (57). One can further amplify expression of tyrosine-sensitive DAHP synthase (57) by increasing the copy number of *aroF*, usually by extrachromosomally localizing *aroF* on a plasmid of medium-to-high copy numbers. Increasing the number of *aroF* genes beyond the number of available tyrosine-activated TyrR repressor molecules overrides transcriptional regulation of *aroF*. This approach can expand the number of AroF molecules well beyond that achievable by mutation of the *tyrR* locus.

Amplified expression of DAHP synthase does not necessarily mean that the in vivo activity of DAHP synthase has also been increased, because of the prominent regulatory role played by feedback inhibition (72). Several alleles that encode feedback-resistant DAHP synthase have been obtained by mutation of the *aroF* (76), *aroG* (17), and *aroH* (60) loci. Extensive mutation is not required to make the encoded DAHP synthase feedback resistant. For instance, a single amino acid change (76) can render AroF catalytic activity insensitive to the concentration of Tyr. Feedback-resistant DAHP synthase increases the in vivo catalytic activity of each molecule of DAHP synthase.

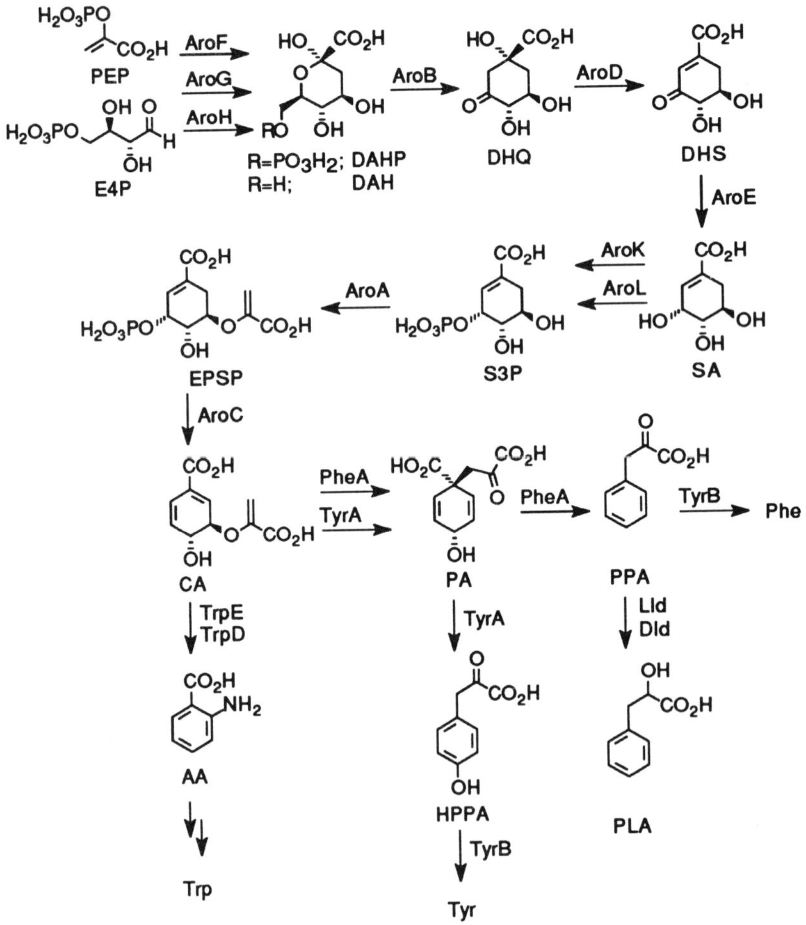

Figure 1 Aromatic amino acid biosynthetic enzymes: AroF, tyrosine-sensitive DAHP synthase; AroG, phenylalanine-sensitive DAHP synthase; AroH, tryptophan-sensitive DAHP synthase; AroB, DHQ synthase; AroD, DHQ dehydratase; AroE, shikimate dehydrogenase; AroL, shikimate kinase II; AroK, shikimate kinase I; AroA, EPSP synthase; AroC, chorismate synthase; PheA, chorismate mutase:prephenate dehydratase; TyrA, chorismate mutase:prephenate dehydrogenase; TyrB, tyrosine aminotransferase; Lld, L-lactate dehydrogenase; Dld, D-lactate dehydrogenase; TrpE TrpD, anthranilate synthase. Metabolites: PEP, phosphoenolpyruvate; E4P, D-erythrose 4-phosphate; DAHP, 3-deoxy-D-*arabino*-heptulosonic acid 7-phosphate; DAH, 3-deoxy-D-*arabino*-heptulosonic acid; DHQ, 3-dehydroquinic acid; DHS, 3-dehydroshikimic acid; SA, shikimic acid; S3P, shikimate 3-phosphate; EPSP, 5-enolpyruvylshikimate 3-phosphate; CA, chorismic acid; PA, prephenic acid; PPA, phenylpyruvic acid; PLA, phenyllactic acid; HPPA, *p*-hydroxyphenylpyruvic acid; AA, anthranilic acid; Phe, L-phenylalanine; Tyr, L-tyrosine; Trp, L-tryptophan.

Substrate-Limited Aromatic Biosynthesis

Ultimately, the catalytic activity of DAHP synthase increases to a point where further amplification of even feedback-resistant DAHP synthase does not increase aromatic amino acid biosynthesis. Researchers have long suspected that phosphoenolpyruvate (PEP) availability might limit the in vivo catalytic activity of amplified DAHP synthase, given the various PEP-dependent enzymes (Figure 2) mediating major cellular processes. For instance, uptake of each molecule of glucose by *Escherichia coli* is mediated by the phosphotransferase system and requires expenditure of one molecule of PEP. Pyruvate kinase-catalyzed reaction of PEP with ADP is essential for regeneration of ATP. Reaction of PEP with carbon dioxide catalyzed by PEP carboxylase determines the levels of oxaloacetate in prokaryotes such as *E. coli*. The concentration of oxaloacetate, in turn, determines the overall rate of reaction of the citric acid cycle.

Attempts at increasing the in vivo supply of PEP include inactivation of PEP carboxylase to avoid loss of PEP as a result of oxaloacetate formation (48). Investigators have examined recycling of pyruvate back to PEP (6) by amplifying expression of PEP synthase (Figure 2). This gluconeogenic enzyme catalyzes the conversion of pyruvate into PEP with expenditure of two of the phosphodiester linkages of ATP. These strategies have not always succeeded

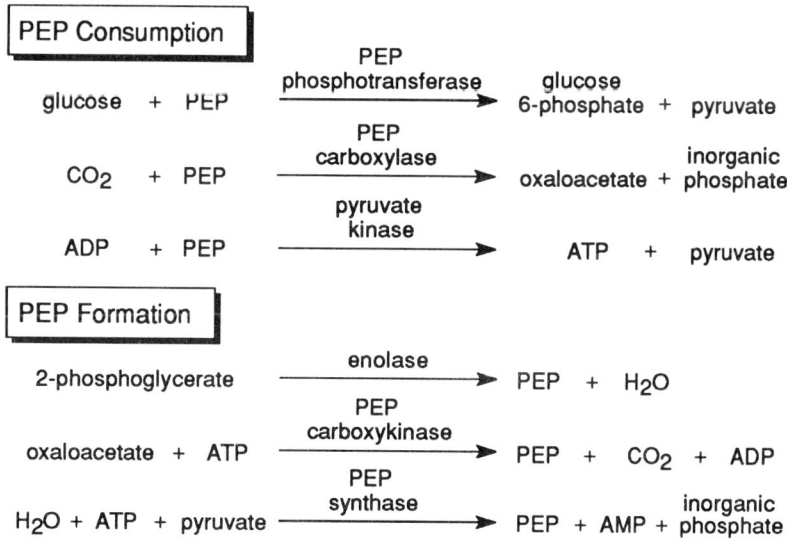

Figure 2 A compilation of enzyme-catalyzed reactions that either consume or form PEP. Although other PEP-consuming and PEP-forming reactions are known, the indicated reactions are associated with major cellular processes.

in increasing the amount of carbon flow directed into aromatic amino acid biosynthesis.

Until recently, the second substrate required for DAHP synthesis, D-erythrose 4-phosphate (E4P), had received almost no attention as a possible limiting factor of in vivo DAHP synthase activity. As an aldose that cannot cyclize, E4P cannot form the stable pyranose and furanose structures in solution that aldose phosphates such as glucose 6-phosphate and ribose 5-phosphate can form. The resulting propensity of E4P to dimerize and polymerize drastically reduces the availability of E4P monomer because of the slow dissociation of the various multimeric forms back to the monomeric form (20). Maintaining E4P in its monomeric form requires careful control of pH, temperature, and concentration.

Nature also must deal with the problematic solution chemistry of E4P.

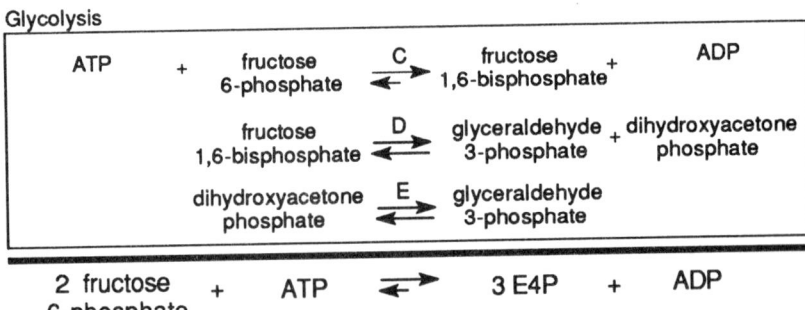

Figure 3 Interplay between the pentose phosphate pathway and glycolysis converts fructose 6-phosphate into erythrose 4-phosphate. Enzymes include: Tkt, transketolase; Tal, transaldolase; A, phosphopentose epimerase; B, phosphopentose isomerase; C, phosphofructokinase; D, aldolase; E, triose phosphate isomerase.

Indeed, E4P has not been detected in extracts obtained from living organisms (78). Thus, the rates of synthesis and utilization of E4P may be closely matched in nature. The maintenance of low concentrations (high dilution) of E4P would favor the monomeric form of this troublesome aldose. From this perspective, intracellular concentrations of fructose 6-phosphate and sedoheptulose 7-phosphate can be viewed as stored forms of E4P that can be released (Figure 3) by the enzymes of the nonoxidative pentose phosphate pathway. E4P is released upon transketolase-catalyzed removal of a ketol from fructose 6-phosphate and transaldolase-catalyzed removal of dihydroxyacetone from sedoheptulose 7-phosphate.

The enzymes of the nonoxidative pentose phosphate pathway and glycolysis mediate (Figure 3) the overall conversion of two molecules of fructose 6-phosphate and one molecule of ATP into three molecules of E4P and one molecule of ADP. Of these pentose phosphate–pathway reactions, conversion of fructose 6-phosphate into E4P and synthesis of sedoheptulose 7-phosphate are catalyzed by transketolase. Transaldolase can then convert the sedoheptulose 7-phophate synthesized by transketolase into E4P. The growth characteristics of *E. coli* BJ502, which carries a leaky mutation resulting in reduced transketolase activity, reflect the presumed importance of transketolase (39, 40). Because of reduced availability of E4P and the associated reduction in the rate of aromatic amino acid biosynthesis, the growth medium must be supplemented with shikimic or aromatic amino acids to achieve normal growth rates of *E. coli* BJ502 (32).

Experiments to ascertain whether amplification of transketolase expression might likewise increase E4P availability began with the use of *E. coli* BJ502 to screen a genomic DNA library. This study led to the isolation (13) of *tkt*, a 5-kb fragment encoding transketolase. Expression of plasmid-localized *tkt* resulted in a tenfold amplification of transketolase levels over those expressed in wild-type *E. coli* (17). Comparison of amplified expression of feedback-resistant DAHP synthase in the presence vs the absence of amplified transketolase indicated that increased in vivo transketolase activity can double the amount of carbon flow into aromatic amino acid biosynthesis (17).

Discovery of the impact of transketolase catalysis on aromatic amino acid biosynthesis has directed research toward the enzymes of the pentose phosphate pathway. Two transketolase isozymes, TktA and TktB, have been cloned, sequenced, and overexpressed; the *tktA* gene has been localized to 63.5 min (70) and *tktB* to 53.0 min (37) on the *E. coli* genome. TktA, which was initially isolated by complementation of *E. coli* BJ502, has been purified (17) to homogeneity and found to be a homodimer with a subunit molecular weight of 72,500. Most of the transketolase activity in *E. coli* is encoded by *tktA* (37). Analysis (37) of the sequence information surrounding *tktB* revealed an open reading frame with a sequence homology to the *tal1* locus (62) of *Saccharomy-*

ces cerevisiae. The close proximity of this locus, designated *talA*, to *tktB* suggests the presence of an operon. Another region of the *E. coli* genome, at 0 min, has sequence homology to the *talA* locus (37).

Elaboration of transketolase's role in primary metabolism has yielded some unexpected dividends: Carbon flow directed into the common pathway can be increased to its theoretical maximum upon overexpression of PEP synthase along with transketolase and feedback-resistant DAHP synthase (56). Amplified expression of PEP synthase should theoretically have recycled pyruvate back to PEP. The fact that PEP synthase amplification alone does not elicit improved carbon flow into the common pathway suggests that the concentration of E4P initially limits the activity of DAHP synthase in *E. coli* (56). Amplified expression of only transketolase increases E4P supplies but leads to a metabolic matrix in which PEP concentrations limit DAHP synthase catalytic activity. This limitation is surmounted by overexpression of PEP synthase.

Another way to increase the supply of PEP is to avoid PEP-dependent phosphotransferase systems for carbon-source uptake. One approach would entail using biocatalysts from microbes that do not require expenditure of PEP for glucose uptake instead of those from *E. coli*. Unfortunately, few microbes have a database as enormous as that amassed for the biochemistry and molecular biology associated with *E. coli*. Carbon sources for *E. coli* biocatalysts could alternatively be switched to carbon sources that do not require a PEP-driven uptake system. These might include five carbon carbohydrates such as xylose or four carbon dicarboxylates such as succinic acid. However, none of the carbon sources for *E. coli* that avoid PEP consumption during uptake are currently as readily available or inexpensive as glucose. Xylose, because of its enzymatic isolation from the xylans typically associated with plants, may ultimately become a viable alternative feedstock. Succinic acid can conveniently be obtained through microbial biocatalysis (81).

Yet another strategy for avoiding PEP expenditure during glucose uptake is to inactivate the phosphotransferase system (58) in *E. coli* and then introduce a different microbe's glucose-uptake system that does not depend on PEP consumption. A successful example (68) is the expression of the *glf* and *glk* genes of *Zymomonas mobilis* in *E. coli* ZSC113, which lacks PEP-driven phosphotransferase systems capable of glucose uptake. *Z. mobilis* utilizes facilitated diffusion for glucose uptake encoded by the *glf* locus. After uptake, phosphorylation of glucose is catalyzed by *glk*-encoded glucokinase. The ability of the *Z. mobilis* to operate in *E. coli* is intriguing given the fundamental differences in the membrane lipids (68) of these two organisms. Whether the expression of the facilitated diffusion system for glucose uptake in *E. coli* results in PEP supplies adequate for amplified DAHP synthase catalytic activity remains to be determined.

Impediments to Carbon Flow Through the Common Pathway of Aromatic Amino Acid Biosynthesis

The carbon flow directed into the common pathway of aromatic amino acid biosynthesis in *E. coli* with amplified expression of feedback-resistant DAHP synthase and transketolase substantially exceeds the carbon flow directed into this pathway in wild-type cells. In the resulting metabolic situation, an individual common-pathway enzyme may become rate limiting. Unable to catalyze the substrate-to-product conversion at an adequate rate, the rate-limiting enzyme begins to accumulate substrate. However, export of this substrate from the cytoplasm to the culture supernatant is usually sufficiently rapid to prevent intracellular accumulation of substantial concentrations. Once deposited in the culture supernatant, the substrate is generally lost to the cell's metabolism. This loss reduces the yield and rate of product formation. Introduced impurities must also be separated from the desired product.

The facile export of substrates provides a mechanism for detection of rate-limiting enzymes. Identification of the rate-limiting enzymes requires use of methodologies such as high-performance liquid chromatography (HPLC) or nuclear magnetic resonance spectroscopy (NMR) to analyze the culture supernatants. HPLC requires that the metabolites found in the culture supernatants be separated to some degree. This is no simple matter given the diverse array of structural elements and functional groups that characterize common-pathway enzyme substrates and related metabolites. Identification by HPLC also requires coinjection with an authentic sample of the tentatively identified substrate. Chemical or enzymatic synthesis of these authentic samples is often a difficult proposition. In contrast, ^{1}H NMR does not require chromatographic separation or access to authentic samples of the metabolites. The NMR spectra of most of the common-pathway enzyme substrates and related metabolites have already been reported in the literature, and comparison of the purported substrate with these NMR structures can often obviate the need for authentic samples. Even complex solution matrices can be analyzed by use of two-dimensional NMR experiments and/or of higher field-strength instruments.

Use of the appropriate auxotrophic mutant along with HPLC (49) or ^{1}H NMR (15) analyses of culture supernatants provides a powerful tool for detection and identification of rate-limiting enzymes. For example, *E. coli* AB2834 is an auxotrophic mutant lacking catalytically active shikimate dehydrogenase (Figure 1) that accumulates 3-dehydroshikimic acid (DHS) in its culture supernatant. Transformation of *E. coli* AB2834 with multicopy plasmid pKD130 encoding *aroF* and *tktA* results in the accumulation (15) of 3-deoxy-D-*arabino*-heptulosonic acid (DAH) in addition to DHS. DAH (Figure 1) is the dephosphorylated substrate of 3-dehydroquinate (DHQ) synthase. Accumulation in the culture supernatant of dephosphorylated substrates is commonly observed, although the identity of the phosphatase(s) and the cellular location of the dephosphorylation process are unknown. Subsequent transformation of *E. coli* AB2834 with pKD136, a multicopy plasmid carrying the *aroB* locus that encodes DHQ synthase in addition to *aroF* and *tktA*, results in the accumulation of increased concentrations of DHS and the absence of DAH in the culture supernatant (15). DHQ synthase appears to be a rate-limiting enzyme in view of the DAH accumulation in the culture supernatant of *E. coli* AB2834/pKD130 whereas the culture supernatant of *E. coli* AB2834/pKD136 lacks DAH but possesses increased DHS concentrations.

To analyze the entire common pathway of aromatic amino acid biosynthesis for rate-limiting enzymes (10), *E. coli* D2704 was used. This *pheA tyrA ΔtrpE-C* auxotroph cannot enzymatically convert chorismic acid (CA) into either prephenic acid (PA) or anthranilic acid (AA) (Figure 1), nor can it enzymatically convert PA into either phenylpyruvic acid (PPA) or *p*-hydroxyphenylpyruvic acid (HPPA). Although in *E. coli* D2704 accumulation of CA might be expected because of the absence of CA-processing enzymes, phenylalanine and phenyllactic acid (PLA) are the products observed in the culture supernatant upon overexpression of DAHP synthase and transketolase. This observation can be attributed to CA's nonenzymatic rearrangement (1, 7), which is followed by nonenzymatic decarboxylation and dehydration (77) of the resulting PA to PPA. Partitioning (Figure 1) of the PPA between transamination and reduction yields Phe and PLA, respectively. TyrB, an enzyme encoded by the *tyrR* regulon, catalyzes the transamination (26), whereas reduction (42) of PPA is likely catalyzed by lactic acid dehydrogenase isozymes.

Along with formation of Phe and PLA, identification of other metabolites in the D2704/pKD130 culture supernatant provided (10) the basis for identifying rate-limiting enzymes. Once these enzymatic impediments to carbon flow were removed, we anticipated that substrates of the rate-limiting enzymes would no longer accumulate in the supernatant and that the concentration of synthesized end-products (Phe and PLA) would increase. Substrates or closely related metabolites exported (10) by D2704/pKD130 included DAH, DHS, shikimic acid (SA), and shikimate 3-phosphate (S3P). Their discovery led to the tentative assignment (10) of DHQ synthase, SA dehydrogenase, SA kinase, and 5-enolpyruvylshikimate (EPSP) synthase as rate-limiting enzymes (Figure 1).

Subsequent overexpression of DHQ synthase in D2704/pKD136 eliminated accumulation (10) of DAH and concomitantly increased the concentration of DHS, SA, and S3P. The other tentatively assigned rate-limiting enzymes were amplified by transforming D2704/pKD136 with a second plasmid encoding the relevant loci (10). Amplified expression of SA dehydrogenase eliminated export of DHS, whereas extracellular accumulation of both DHS and SA ceased when expression levels of SA kinase were increased. The observation that the impediments to carbon flow presented by SA dehydrogenase and SA kinase could be removed by increasing the catalytic activity of only one enzyme led to the discovery of feedback inhibition of SA dehydrogenase by SA. The precedent of such feedback inhibition in plants (3) but not in microbes may reflect the greater flow of carbon into aromatic amino acid biosynthesis in plants relative to that in wild-type microbes lacking amplified DAHP synthase and transketolase.

Although amplified expression of DHQ synthase and SA kinase eliminates accumulation of DAH, DHS, and SA, it does not significantly improve end-product Phe and PLA synthesis. However, transformation of D2704/pKD136 with a second plasmid carrying SA kinase-encoding *aroL* as well as *aroA*, which encodes EPSP synthase, improves end-product synthesis (10). Accumulation of S3P declined but was not eliminated as a result of overexpression of EPSP synthase, which suggested that more than just rate-limiting EPSP synthase activity contributed to S3P accumulation. Because EPSP synthase catalyzes a reversible reaction (4) any EPSP accumulation resulting from rate-limiting CA synthase activity may only be transient. EPSP in the presence of inorganic phosphate would be converted by EPSP synthase back to PEP and S3P.

To assess whether CA synthase was rate-limiting, D2704/pKD136 was transformed with a second plasmid containing *aroC*-encoded CA synthase, *aroL*-encoded SA kinase, and *aroA*-encoded EPSP synthase (10). The resulting biocatalyst, D2704/pKD136/pKAD50, accumulated reduced concentrations of S3P, thereby confirming suspicions that CA synthase was rate limiting. Of greater importance, D2704/pKD136/pKAD50 synthesized twice the amount (10) of end-product Phe and PLA relative to D2704/pKD130.

DHQ dehydratase and SA dehydrogenase are the only common-pathway enzymes (Figure 1) that do not impede the flow of carbon directed into the pathway by amplified expression of DAHP synthase and transketolase. The need to amplify expression of DHQ synthase, SA kinase, EPSP synthase, and CA synthase in addition to DAHP synthase and transketolase raises concerns about the burden these overexpressed enzymes place on the host microbe's metabolism. Resulting instability of the construct under the high-density growth conditions of a fermentor could preclude the utility of the biocatalyst. Fortunately, pronounced overexpression of common-pathway enzymes may not be necessary.

The specific activity of DHQ synthase in D2704/pKD136/pKAD50, although only twofold higher than the level in D2704, is sufficient to remove DHQ synthase's rate-limiting character (10). This observation suggests that the rate-limiting character of enzymes can be circumvented by inserting into the microbe's genome additional copies of rate-limiting common-pathway genes under the control of an appropriately strong promoter. The *aroL* locus may not need to be part of the genomic insertion because SA kinase expression is part of the *tyrR* regulon (57). Derepression of *aroL* expression with a mutant *tyrR* may increase SA kinase's catalytic activity sufficiently to remove the impedance this enzyme poses to carbon flow.

All of the studies examining rate-limiting enzymes in the common pathway utilized D2704. However, are the impediments to carbon flow identified in D2704 the same as those in other *E. coli* hosts? DAH accumulation was observed in both *E. coli* D2704 and *E. coli* AB2834 as well as in other aromatic amino acid–synthesizing *E. coli* constructs that express amplified levels of feedback-resistant DAHP synthase (72). Thus, the rate-limiting character of DHQ synthase may be a general phenomenon in *E. coli* hosts. Future investigations will have to determine whether SA kinase, EPSP synthase, and CA synthase also prove to be, in general, rate-limiting enzymes in *E. coli*. As with the analysis developed for D2704, a single ^{1}H NMR spectrum of the appropriate auxotrophic mutant's culture supernatant would quickly identify the impediments to carbon flow in virtually any microbial host.

Even the common pathway in *E. coli* D2704 will need to be reexamined in light of the aforementioned interplay between transketolase (E4P levels) and PEP synthase (PEP levels). Amplified expression of *aroA* may have surmounted the apparent rate-limiting character of EPSP synthase by facilitating this enzyme's more effective competition with other PEP-utilizing enzymes. Just as availability of PEP is essential to in vivo DAHP synthase activity after amplification of transketolase, enhanced PEP availability with overexpressed PEP synthase may circumvent EPSP synthase's impediment to carbon flow. The common pathway in D2704 will also have to be reexamined as the terminal pathways (Figure 1) are optimized for biocatalytic synthesis of Phe, Trp, Tyr, and related secondary metabolites. The accumulation of substrates or related metabolites as a result of carbon-flow impediments in these terminal and secondary metabolic pathways could lead to unexpected feedback inhibition of common-pathway enzymes.

Biocatalytic Syntheses

The increase in carbon flow directed into and through the common pathway allows the synthesis of a vast array of chemicals. Phe can be enzymatically or chemically converted (55) to aspartame (Figure 4), which has the largest sales volume of all food additives (73). Introduction of naphthalene dioxygenase into a tryptophan-synthesizing microbe that also expresses tryptophanase results in biocatalytic synthesis (21, 50) of indigo (Figure 4), the vat dye that gives blue jeans their faded-blue coloration. More volumes of this dye than of any other are produced worldwide. Tyr can be converted (11) into eumelanin (Figure 4), the mammalian pigmentation that is responsible for hair coloring as well as protection from solar irradiation.

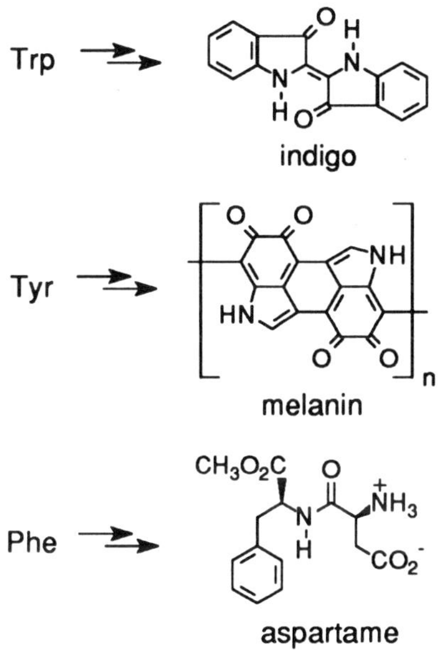

Figure 4 Chemicals synthesized from tryptophan, tyrosine, and phenylalanine.

The substrates of enzymes in the common pathway of aromatic amino acid biosynthesis can also yield valuable chemicals. *p*-Aminobenzoic acid (PABA) is biosynthesized by initial conversion of CA to 4-amino-4-deoxychorismate (ADC) (2) followed by elimination of pyruvate (Figure 5). ADC synthase consists of two subunits encoded by the *pabA* and *pabB* loci. ADC lyase is apparently a homodimer (80) encoded by *pabC*. Each of the genes associated with conversion of CA to PABA has been cloned and sequenced from *E. coli* (28, 29, 41). Although its biological role is as an intermediate in the biosynthesis of folic acid, PABA's chemical use is as an ingredient of sunburn lotions, and it can be esterified to form the local anesthetic known as benzocaine (71a). PABA is currently industrially synthesized from toluene (71a).

CA can also be converted (Figure 5) directly to *p*-hydroxybenzoic acid (PHB) in a reaction catalyzed by chorismate-pyruvate lyase (UbiC). The *ubiC* gene from *E. coli* has been cloned and sequenced (53, 66, 67). Esters of PHB are used as food preservatives (71c). PHB is also a component of thermotropic liquid crystal polymers such as Xydar (43). This class of polymers has attracted considerable attention because of their use as high-performance thermoplastics. PHB can also be synthesized (22) by means of chemical dehydration of SA catalyzed by strong acids (Figure 5). This rather ancient reaction has the

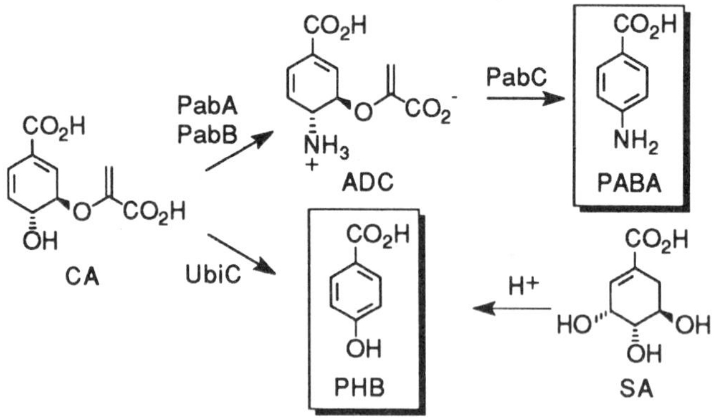

Figure 5 p-Aminobenzoic acid (PABA) can be synthesized from CA via a 4-amino-4-deoxychorismate (ADC) intermediacy using the combined catalytic activities of ADC synthase (PabA PabB) and ADC lyase (PabC). PHB can likewise be derived from CA in a reaction mediated by chorismate pyruvate lyase (UbiC) or obtained from acid-catalyzed dehydration of SA.

advantage that it avoids consumption of the second equivalent of PEP, which is required for CA biosynthesis. Consumption of additional PEP detrimentally affects yields of PHB from glucose, which in turn increases the estimated manufacturing costs. PHB is currently industrially synthesized from phenol via a Kolbe-Schmitt reaction with carbon dioxide (71c).

DHS is another common-pathway substrate that can be chemically converted into industrially useful chemicals. Oxidation (33) of DHS using Fehling solution (Figure 6) yields gallic acid (GA), which is used for applications ranging from tanning to inks and dyes (19). Derivatives of GA include propyl gallate, an important food-grade antioxidant, and pyrogallol, a product of chemical or enzymatic decarboxylation of GA (19). Pyrogallol is one of the strongest known aromatic reducing agents. It is used in photographic developing solutions and is a key building block in the manufacture of bendiocarb, a widely

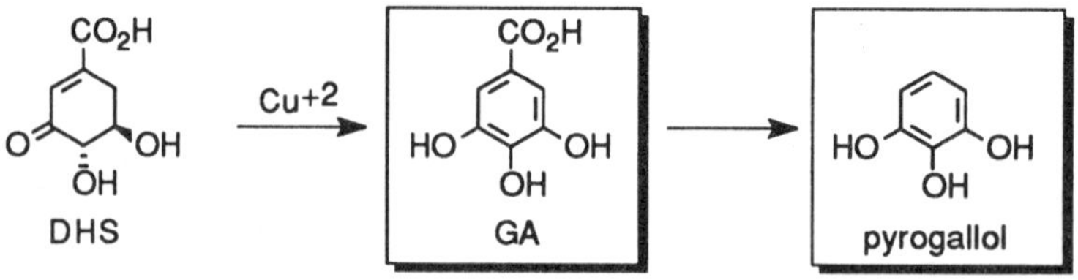

Figure 6 Gallic acid (GA) is the product of the chemical oxidation of DHS. GA decarboxylation affords pyrogallol.

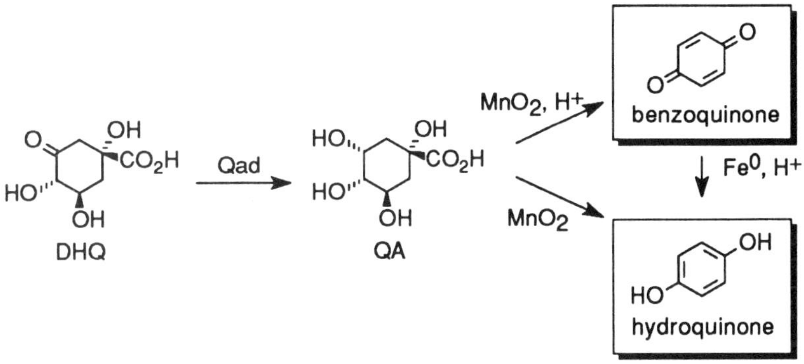

Figure 7 Quinic acid dehydrogenase (Qad)–catalyzed reduction of DHQ yields quinic acid (QA), which can be chemically oxidized to either benzoquinone or hydroquinone.

used insecticide (71d). Both GA and pyrogallol were originally isolated from gall nuts and tara powder (71d). However, the unreliability of these sources prompted a switch to use of petroleum feedstocks for the industrial synthesis of GA and pyrogallol.

The use of heterologous biocatalysts expands even further the number of molecules that can be synthesized from glucose. The manufacture of hydroquinone and benzoquinone (Figure 7) serves as an example (65, 71b, 75). Hydroquinone's selective reduction of photoactivated silver ion is the basis for this organic's widespread use in photography (71b, 75). Benzoquinone is an important chemical precursor in the manufacture of various chemicals. Both hydroquinone and benzoquinone are currently manufactured from benzene, which in turn is derived from nonrenewable fossil fuels. Ironically, the root *quin* in the names hydroquinone and benzoquinone reflects the synthesis (79) of these chemicals in the nineteenth century from a hydroaromatic known as quinic acid (QA) that was isolated from plant sources. This original chemical synthesis suggests that hydroquinone and benzoquinone could be synthesized by using biocatalysts (Figure 7) (18).

Numerous microbes can catabolize hydroaromatics such as QA (Figure 7) and SA (Figure 1) (27, 54). This ability probably imparted an evolutionary advantage to hydroaromatic-degrading organisms, as evidenced by the widespread occurrence of these molecules in plant tissue. Oxidation of QA to DHQ catalyzed by QA dehydrogenase is the first step in QA catabolism. Dehydration of the DHQ catalyzed by DHQ dehydratase yields DHS. Yet another dehydration mediated by DHS dehydratase provides protocatechuic acid (PCA), which can be catabolized via the β-ketoadipate pathway. The reversible reaction catalyzed by QA dehydrogenase thermodynamically favors (8) reduction

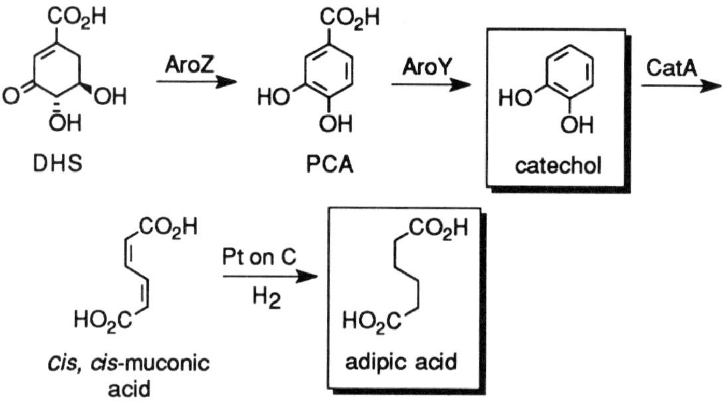

Figure 8 Biocatalytic synthesis of catechol follows from DHS dehydratase (AroZ)–mediated dehydration of DHS to protocatechuic acid (PCA) followed by nonoxidative decarboxylation catalyzed by PCA decarboxylase (AroY). Addition of catechol 1,2-dioxygenase (CatA) activity produces a *cis,cis*-muconic acid–synthesizing biocatalyst. Adipic acid is then produced via hydrogenation of biocatalytically synthesized *cis,cis*-muconic acid.

of DHQ and formation of QA. This equilibrium does not impede QA catabolism because the β-ketoadipate pathway siphons away DHQ as it forms. In contrast, expression of QA dehydrogenase activity in a microbe lacking the β-ketoadipate pathway resulted in a heterologous microbe that synthesized as opposed to catabolized QA (18).

A locus designated *qad* that encoded QA dehydrogenase was isolated from *Klebsiella pneumoniae* (18). Subcloning localized the gene to a 2.9-kb insert in plasmid pTW8090A. Transformation of an *aroD* auxotroph, *E. coli* AB2848, with plasmids pTW8090A and pKD136 yielded *E. coli* AB2848/pKD136/pTW8090A as a QA-synthesizing biocatalyst. In this experiment, QA synthesis catalyzed by *E. coli* AB2848/pKD136/pTW8090A did not have to compete with QA catabolism because *E. coli* could not utilize QA as a sole source of carbon for growth. In vivo catalytic activities of DAHP synthase, transketolase, and DHQ synthase, amplified owing to expression of pKD136, increased the carbon flow into the common pathway and then into DHQ synthesis. DHQ could not be processed through the common pathway because of a mutation in *aroD* that rendered DHQ dehydratase catalytically inactive. QA dehydrogenase catalyzed reduction of DHQ, and export of the product resulted in a culture supernatant in which QA was essentially the only organic in solution (18). The culture supernatant yielded benzoquinone (Figure 7) after cells were removed by centrifugation, sulfuric acid and manganese were added, and the solution was subsequently heated (18). In the absence of added acid, higher yields of benzoquinone resulted when purified QA was oxidized. Heating purified QA with manganese dioxide yielded hydroquinone (Figure 7) (18). High-yielding reduction of benzoquinone provides another route to hydroquinone.

A heterologous biocatalyst has also been constructed for conversion of glucose to catechol (Figure 8) (14). Catechol is currently (23a, 71b, 75) synthesized by reaction of phenol with various formulations of hydrogen peroxide. Phenol in turn is made from fossil fuel–derived benzene. Although it is an essential chemical building block in the manufacture of many chemical products, most catechol is used to make vanillin. Vanillin is responsible for much of the flavor associated with vanilla extracts and is second only to aspartame in terms of sales volume (73).

Biocatalytic synthesis of catechol requires assembly of a pathway in which glucose is converted in high yield into DHS, which is then dehydrated into PCA. Finally, nonoxidative decarboxylation of PCA yields catechol (Figure 8). We have already discussed DHS dehydratase–catalyzed conversion of DHS into PCA in conjunction with the β-ketoadipate pathway. The decarboxylation of PCA to catechol catalyzed by PCA decarboxylase was originally formulated as part of the β-ketoadipate pathway. However, more recent research has demonstrated that this enzyme is not part of aromatic catabolism (12). The true role of PCA decarboxylase in microbial metabolism remains an enigma.

The gene encoding DHS dehydratase has been isolated from *Neurospora crassa* (27, 61, 63, 64) and *Aspergillus nidulans* (34, 44). However, neither gene has been expressed in *E. coli*. Based on the successful expression in *E. coli* of the *K. pneumoniae* gene *qad* (18), the latter bacterium was chosen as the source of the genes for DHS dehydratase and PCA decarboxylase. Expression in *E. coli* of these enzymes from their respective native promoters seemed a reasonable expectation given the close evolutionary similarity (38) between *K. pneumoniae* and *E. coli*. Assembly and screening of a genomic library of *K. pneumoniae* resulted in localization of *aroZ*-encoded DHS dehydratase to a 3.5-kb *Bam*HI fragment and *aroY*-encoded PCA decarboxylase to a 2.3-kb *Hin*dIII fragment (14). *E. coli* AB2834 was chosen as the host for DHS synthesis because of a mutation in its *aroE* locus that encodes SA dehydrogenase. Unable to convert DHS into SA, *E. coli* AB2834 accumulates DHS in its culture supernatant. To ensure increased carbon flow into the common pathway of aromatic amino acid biosynthesis and delivery of this flow into DHS synthesis, the plasmid pKD136, which encodes *aroF, tkt,* and *aroB*, was transformed into *E. coli* AB2834. Plasmid pKD9.069A was then constructed by inserting the 2.3-kb fragment encoding *aroZ* and the 3.5-kb fragment encoding *aroY* into a plasmid with an origin of replication compatible with pKD136. *E. coli* AB2834/pKD136/pKD9.069A converted a minimal salt solution of 56.0 mM glucose into an 18.5 mM solution of catechol (14). Analysis of this crude culture supernatant indicated that acetic acid and trace levels of PCA were the only organic molecules present other than catechol.

With construction of an efficient catechol-synthesizing biocatalyst, creation of a route from glucose to adipic acid (Figure 8) was comparatively simple. Approximately 1.9×10^9 kg of adipic acid are manufactured each year to meet a 4×10^9 kg demand for nylon-6 6, a polymer produced by the condensation of adipic acid with hexamethylenediamine (9, 59). Nearly all adipic acid is synthesized via a route beginning with hydrogenation of benzene (9). The resulting cyclohexane is air oxidized in the presence of metal catalysts to afford cyclohexanone and cyclohexanol. Oxidation of this mixture with concentrated nitric acid yields adipic acid.

The process of industrial synthesis of adipic acid illustrates the problems that can plague traditional, large-volume manufacture of chemicals. Benzene, the starting material, is a carcinogen (46) that must be derived from nonrenewable fossil fuels (23b). Steps in the overall conversion require either elevated temperatures or high-pressure reaction conditions. Of even greater concern is nitrous oxide, a byproduct of the manufacturing process, that has been implicated in both ozone depletion and the greenhouse effect. Adipic acid manufacture may be responsible for 10% of the annual increase in atmospheric nitrous oxide levels (74).

Adipic acid is synthesized from glucose (16) through the assembly of a biocatalyst that converts glucose into *cis,cis*-muconic acid followed by catalytic hydrogenation of the *cis,cis*-muconic acid to adipic acid (Figure 8). The transformation of *E. coli* AB2834 with plasmids pKD136, pKD8.243A, and pKD8.292 led to construction of the *cis,cis*-muconate–synthesizing biocatalyst. Plasmid pKD136 once again ensured an increase in carbon flow to the common pathway and DHS synthesis. DHS dehydratase and PCA decarboxylase activities resulted from expression of the *aroZ* and *aroY* genes carried on pKD8.243A. Expression of pKD8.292 provided *catA*-encoded (51) catechol 1,2-dioxygenase activity. The resulting biocatalyst, *E. coli* AB2834/pKD136/pKD8.243A/pKD8.292, converted a minimal salt solution of 56.0 mM glucose into a solution of 16.8 mM *cis,cis*-muconic acid (15). After the biocatalyst was removed, 10% platinum on carbon was added to the unpurified culture supernatant. Hydrogenation afforded adipic acid in high yield.

The successful interface of biocatalysis and chemical catalysis that enabled the conversion of glucose into adipic acid highlights some of the environmental advantages that accrue from development of fundamentally new routes in the manufacture of industrial chemicals. Manufacturing processes can now begin with renewable plant starch, the current commercial source of glucose, rather than the nonrenewable fossil fuels traditionally required for adipic acid manufacture. The nontoxicity of glucose sharply contrasts with the carcinogenicity of benzene. Mild reaction temperatures and pressures are all that is required for glucose to adipic acid conversion. Finally and perhaps most importantly, synthesis of adipic acid from glucose avoids generation of ozone-depleting gases and greenhouse gases.

Future Horizons

The number of different chemicals that can be synthesized from renewable starting materials such as glucose may be virtually unlimited. This diversity in chemical reactivity is one of the underlying features of biological diversity. What one microbe is unable to synthesize often can be produced by some other microbe. Even if nature does not afford us a biosynthetic pathway for a particular chemical, we can create such a pathway using genes from various sources to construct the appropriate heterologous biocatalyst. The spectrum of chemicals derived from glucose can be further expanded by effectively interfacing biosynthesis with chemical synthesis. Many biosynthetic-pathway intermediates and products that were once available only in small quantities at exorbitant prices are now becoming available in volumes and at prices reminiscent of commodity chemicals. Research aimed at establishing which products can be chemically synthesized from these newly available biochemicals and developing the relevant chemical methodology is a field still in its infancy.

The future of microbial biocatalysis as an integral part of chemical manufacture will have its problems. Expression of genes isolated from one organism and their subsequent expression in evolutionarily distinct microbial hosts will probably always challenge researchers. Potential toxicity of products and intermediates to the synthesizing biocatalyst will also be a recurring dilemma. Investigators must delineate the molecular basis of observed toxicity and devise novel methods to circumvent such toxicity. Advances in bioprocessing must keep pace with advances in biocatalysis. Problematic isolation and purification of biocatalytically synthesized chemicals can negate the utility of even the most exquisitely engineered biocatalyst. Finally, the efficiency with which carbon flow is directed into and through a designated pathway will always be a grave concern.

Despite the favorable environmental characteristics of microbial biocatalysis and glucose utilization, the adoption of this alternate form of chemical manufacture will depend on the marketability of the products at competitive prices. Manufacturing costs estimated (24) for biocatalytic synthesis of several chemicals discussed here are surprisingly competitive with the prices for the same chemicals synthesized by traditional chemical manufacturing routes. Moreover, the costs associated with biocatalytic synthesis of chemicals from glucose should decline steadily. For instance, aromatic amino acid biosynthesis is severely limited in *E. coli* by the availability of the substrates for DAHP synthase. However, increasing the availability of E4P and PEP improves the yields accessible during biocatalytic synthesis, thereby drastically reducing the projected costs of producing chemicals via such a route. This intriguing interplay between research results and product costs contributes to the excitement and the urgency surrounding the biochemical, molecular biological, microbiological, and organic chemical research behind biocatalyst construction and development.

ACKNOWLEDGMENT

Work was supported by grants to JWF from the Environmental Protection Agency and the National Science Foundation.

> Any *Annual Review* chapter, as well as any article cited in an *Annual Review* chapter, may be purchased from the Annual Reviews Preprints and Reprints service.
> 1-800-347-8007; 415-259-5017; email: arpr@class.org

Literature Cited

1. Addadi L, Jaffe EK, Knowles JR. 1983. Secondary tritium isotope effects as probes of the enzymic and nonenzymic conversion of chorismate to prephenate. *Biochemistry* 22:4494–501
2. Anderson KS, Kati WM, Ye Q-Z, Liu J, Walsh CT, et al. 1991. Isolation and structure elucidation of the 4-amino-4-deoxychorismate intermediate in the PABA# enzymatic pathway. *J. Am. Chem. Soc.* 113:3198–200
3. Balinsky D, Dennis AW, Cleland WW. 1971. Kinetic and isotope-exchange studies on shikimate dehydrogenase from *Pisum sativum*. *Biochemistry* 10:1947–52
4. Boocock MR, Coggins JR. 1983. Kinetics of 5-enolpyruvylshikimate-3-phosphate synthase inhibition by glyphosate. *FEBS Lett.* 154:127–33
5. Camakaris H, Pittard J. 1983. Tyrosine biosynthesis. See Ref. 35a, pp. 339–50
6. Chao Y-P, Patnaik R, Roof WD, Young RF, Liao JC. 1993. Control of gluconeogenic growth by *pps* and *pck* in *Escherichia coli*. *J. Bacteriol.* 175:6939–44
7. Copley SD, Knowles JR. 1987. The conformational equilibrium of chorismate in solution: implications for the mechanism of the non-enzymic and enzyme-catalyzed rearrangement of chorismate to prephenate. *J. Am. Chem. Soc.* 109:5008–13
8. Davis BD, Gilvarg C, Mitsuhashi S. 1955. Enzymes of aromatic biosynthesis. *Methods Enzymol.* 2:300–4
9. Davis DD, Kemp DR. 1991. Adipic acid. In *Kirk-Othmer Encyclopedia of Chemical Technology*, ed. J Kroschwitz, M Howe-Grant, M Bickford, L Gray, 1:466–93. New York: Wiley. 4th ed.
10. Dell KA, Frost JW. 1993. Identification and removal of impediments to biocatalytic synthesis of aromatics from D-glucose: rate-limiting enzymes in the common pathway of aromatic amino acid biosynthesis. *J. Am. Chem. Soc.* 115:11581–89
11. della-Cioppa G, Garger SJ, Sverlow GG, Turpen TH, Grill LK. 1990. Melanin production in *Escherichia coli* from a cloned tyrosinase gene. *Biotechnology* 8:634–38
12. Doten RC, Ornston LN. 1987. Protocatechuate is not metabolized via catechol in *Enterobacter aerogenes*. *J. Bacteriol.* 169:5827–30
13. Draths KM, Frost JW. 1990. Synthesis using plasmid-based biocatalysis: plasmid assembly and 3-deoxy-D-*arabino*-heptulosonate production. *J. Am. Chem. Soc.* 112: 1657–59
14. Draths KM, Frost JW. 1995. Environmentally compatible synthesis of catechol from D-glucose. *J. Am. Chem. Soc.* 117:2395–400
15. Draths KM, Frost JW. 1990. Genomic direction of synthesis during plasmid-based biocatalysis. *J. Am. Chem. Soc.* 112:9630–32
16. Draths KM, Frost JW. 1994. Environmentally compatible synthesis of adipic acid from D-glucose. *J. Am. Chem. Soc.* 116:399–400
17. Draths KM, Pompliano DL, Conley DL, Frost JW, Berry A, et al. 1992. Biocatalytic synthesis of aromatics from D-glu-

cose: the role of transketolase. *J. Am. Chem. Soc.* 114:3956-62
18. Draths KM, Ward TL, Frost JW. 1992. Biocatalysis and nineteenth century organic chemistry: conversion of D-glucose into quinoid organics. *J. Am. Chem. Soc.* 114:9725-26
19. Dressler H, Holter SN. 1982. (Polyhydroxy)benzenes. See Ref. 28a, pp. 670-704
20. Duke CC, MacLeod JK, Williams JF. 1981. Nuclear magnetic resonance studies of D-erythrose 4-phosphate in aqueous solution. Structures of the major contributing monomeric and dimeric forms. *Carbohydr. Res.* 95:1-26
21. Ensley BD, Ratzkin BJ, Osslund TD, Simon MJ, Wackett LP, Gibson DT. 1983. Expression of naphthalene oxidation genes in *Escherichia coli* results in the biosynthesis of indigo. *Science* 222:167-69
22. Eykmann JF. 1891. Ueber die shikimisäure. *Ber. Dtsch. Chem. Ges.* 24:1278-303
23a. Franck H-G, Stadelhofer JW. 1987. Production and uses of benzene derivatives. In *Industrial Aromatic Chemistry*, pp. 183-90. New York: Springer-Verlag
23b. Franck H-G, Stadelhofer JW. 1987. Production of benzene, toluene, and xylenes. In *Industrial Aromatic Chemistry*, pp. 99-131. New York: Springer-Verlag
24. Frost JW, Lievense J. 1994. Prospects for biocatalytic synthesis of aromatics in the 21st century. *New J. Chem.* 18:341-48
25. Garner C, Herrmann KM. 1983. Biosynthesis of phenylalanine. See Ref. 35a, pp. 323-38
26. Gelfand DH, Steinberg RA. 1977. *Escherichia coli* mutants deficient in the aspartate and aromatic amino acid aminotransferases. *J. Bacteriol.* 130:429-40
27. Giles NH, Case ME, Baum J, Geever R, Huiet L, et al. 1985. Gene organization and regulation in the *qa* (quinic acid) gene cluster of *Neurospora crassa*. *Microbiol. Rev.* 49:338-58
28. Goncharoff P, Nichols BP. 1984. Nucleotide sequence of *Escherichia coli pabB* indicates a common evolutionary origin of *p*-aminobenzoate synthetase and anthranilate synthetase. *J. Bacteriol.* 159:57-62
28a. Grayson M, Eckroth D, Eastman Cl, Klingsberg A, Spiro L, et al, eds. 1982. *Kirk-Othmer Encyclopedia of Chemical Technology*, Vol. 18. New York: Wiley. 3rd ed.
29. Green JM, Merkel WK, Nichols BP. 1992. Characterization and sequence of *Escherichia coli pabC*, the gene encoding aminodeoxychorismate lyase, a pyridoxal phosphate-containing enzyme. *J. Bacteriol.* 174:5317-23
30. Haslam E. 1974. The shikimate pathway: biosynthesis of the aromatic amino acids. In *The Shikimate Pathway*, pp. 3-48. New York: Wiley
31. Haslam E. 1993. Enzymes and enzymology of the common pathway. In *Shikimic Acid*, pp. 103-50. New York: Wiley
32. Haslam E. 1993. Beyond chorismate-primary essential metabolites. In *Shikimic Acid*, pp. 156-54. New York: Wiley
33. Haslam E, Haworth RD, Knowles PF. 1961. Gallotannins. Part IV. The biosynthesis of gallic acid. *J. Chem. Soc.* pp. 1854-59
34. Hawkins AR, Francisco Da Silva AJ, Roberts CF. 1985. Cloning and characterization of the three enzyme structural genes *QUTB*, *QUTC*, *QUTE*, from the quinic acid utilization gene cluster in *Aspergillus nidulans*. *Curr. Genet.* 9:305-11
35. Herrmann KM. 1983. The common aromatic biosynthetic pathway. See Ref. 35a, pp. 301-22
35a. Herrmann KM, Somerville RL, eds. 1983. *Amino Acids: Biosynthesis and Genetic Regulation*. Reading, MA: Addison-Wesley
36. Hodgson J. 1994. Bulk amino-acid fermentation: technology and commodity trading. *Biotechnology* 12:152-55
37. Iida A, Teshiba S, Mizobuchi K. 1993. Identification and characterization of the *tktB* gene encoding a second transketolase in *Escherichia coli* K-12. *J. Bacteriol.* 175:5375-83
38. Jensen RA. 1985. Biochemical pathways in prokaryotes can be traced backward through evolutionary time. *Mol. Biol. Evol.* 2:92-108
39. Josephson BL, Fraenkel DG. 1969. Transketolase mutants of *Escherichia coli*. *J. Bacteriol.* 100:1289-95
40. Josephson BL, Fraenkel DG. 1974. Sugar metabolism in transketolase mutants of *Escherichia coli*. *J. Bacteriol.* 118:1082-89
41. Kaplan JB, Nichols BP. 1983. Nucleotide sequence of *Escherichia coli pabA* and its evolutionary relationship to *trp(G)D*. *J. Mol. Biol.* 168:451-68
42. Kim M-J, Whitesides GM. 1988. L-Lactate dehydrogenase: substrate specificity and use as a catalyst in the synthesis of homochiral 2-hydroxy acids. *J. Am. Chem. Soc.* 110:2959-64
43. Kirsch MA, Williams DJ. 1994. Under-

standing the thermoplastic polyester business. *CHEMTECH* 24:40–49
44. Lamb HK, Hawkins AR, Smith M, Harvey IJ, Brown J, et al. 1990. Spatial and biological characterisation of the complete quinic acid utilisation gene cluster in *Aspergillus nidulans. Mol. Gen. Genet.* 223:17–23
45. Lee H. 1993. Ethanol's evolving role in the U.S. automobile fuel market. In *Industrial Uses of Agricultural Materials*, pp. 49–54. Washington, DC: US Dept. Agric.
46. Lenga RE, Votoupal KL. 1993. *The Sigma-Aldrich Library of Regulatory and Safety Data.* Milwaukee: Sigma-Aldrich
47. Lynd LR, Cushman JH, Nichols RJ, Wyman CE. 1991. Fuel ethanol from cellulosic biomass. *Science* 251:1318–23
48. Miller JE, Backman KC, O'Connor, MJ, Hatch RT. 1987. Production of phenylalanine and organic acids by phosphoenolpyruvate carboxylase-deficient mutants of *Escherichia coli. J. Ind. Microbiol.* 2:143–49
49. Mousdale DM, Coggins JR. 1985. High-performance chromatography of shikimate pathway intermediates. *J. Chromatogr.* 329:268–72
50. Murdock D, Ensley BD, Serdar C, Thalen M. 1993. Construction of metabolic operons catalyzing de novo biosynthesis of indigo in *Escherichia coli. Biotechnology* 11:381–86
50a. Neidhardt FC, Ingraham JL, Low KB, Magasanik B, Schaechter M, Umbarger HE, eds. 1987. Escherichia coli *and* Salmonella typhimurium: *Cellular and Molecular Biology*, Vols. 1. Washington, DC: Am. Soc. Microbiol.
51. Neidle EL, Ornston LN. 1986. Cloning and expression of *Acinetobacter calcoaceticus* catechol 1,2-dioxygenase structural gene *catA* in *Escherichia coli. J. Bacteriol.* 168:815–20
52. Deleted in proof
53. Nichols BP, Green JM. 1992. Cloning and sequencing of *Escherichia coli ubiC* and purification of chorismate lyase. *J. Bacteriol.* 174:5309–16
54. Ornston LN, Neidle EL. 1991. Evolution of genes for the β-ketoadipate pathway in *Acinetobacter calcoaceticus.* In *The Biology of Acinetobacter*, ed. KJ Towner, E Bergogne-Bérézin, CA Fewson, pp. 201–37. New York: Plenum
55. Oyama K, Irino S, Hagi N. 1987. Production of aspartame by immobilized thermoase. *Methods Enzymol.* 136:503–17
56. Patnaik R, Liao JC. 1994. Engineering of *Escherichia coli* central metabolism for aromatic metabolite production with near theoretical yield. *Appl. Environ. Microbiol.* 60:3903–8
57. Pittard AJ. 1987. Biosynthesis of the aromatic amino acids. See Ref. 50a, pp. 368–94
58. Postma PW. 1987. Phophotransferase system for glucose and other sugars. See Ref. 50a, pp. 127–41
59. Putscher RE. 1982. Polyamides (general). In *Kirk-Othmer Encyclopedia of Chemical Technology*, See Ref. 28a, pp. 328–71
60. Ray JM, Yanofsky C, Bauerle R. 1988. Mutational analysis of the catalytic and feedback sites of the tryptophan-sensitive 3-deoxy-D-arabinoheptulosonate 7-phosphate synthase of *Escherichia coli. J. Bacteriol.* 170:5500–6
61. Rutledge BJ. 1984. Molecular characterization of the *qa-4* gene of *Neurospora crassa. Gene* 32:275–87
62. Schaaff IS, Hohmann S, Zimmermann FK. 1990. Molecular analysis of the structural gene for yeast transaldolase. *Eur. J. Biochem.* 188:597–603
63. Schweizer M, Case ME, Dykstra CC, Giles NH, Kushner SR. 1981. Cloning the quinic acid (*qa*) gene cluster from *Neurospora crassa:* identification of recombinant plasmids containing both *qa-2*$^+$ and *qa-3*$^+$. *Gene* 14:23–32
64. Schweizer M, Case ME, Dykstra CC, Giles NH, Kushner SR. 1981. Identification and characterization of recombinant plasmids carrying the complete *qa* gene cluster from *Neurospora crassa* including the *qa-1*$^+$ regulatory gene. *Proc. Natl. Acad. Sci. USA* 78:5086–90
65. Shearon WH Jr, Davy LG, von Bramer H. 1952. Hydroquinone manufacture. *Ind. Eng. Chem.* 44:1730–35
66. Siebert M, Berchthold A, Melzer M, May U, Berger U, et al. 1992. Cloning of the genes coding for chorismate pyruvate-lyase and 4-hydroxybenzoate octaprenyl transferase from *E. coli. FEBS Lett.* 307:347–50
67. Siebert M, Severin K, Heide L. 1994. Formation of 4-hydroxybenzoate in *Escherichia coli:* characterization of the *ubiC* gene and its encoded enzyme chorismate pyruvate-lyase. *Microbiology* 140:897–904
68. Snoep JL, Arfman N, Yomano LP, Fliege RK, Conway T, Ingram LO. 1994. Reconstitution of glucose uptake and phosphorylation in a glucose-negative mutant of *Escherichia coli* by using *Zymomonas mobilis* genes encoding the

glucose facilitator protein and glucokinase. *J. Bacteriol.* 176:2133–35
69. Somerville RL. 1983. Tryptophan: biosynthesis, regulation, and large-scale production. See Ref. 35a, pp. 351–78
70. Sprenger GA. 1992. Location of the transketolase (*tkt*) gene on the *Escherichia coli* physical map. *J. Bacteriol.* 174:1707–8
71a. Szmant HH. 1989. Benzenoid carboxylic acids and derivatives. In *Organic Building Blocks of the Chemical Industry*, pp. 465–66. New York: Wiley
71b. Szmant HH. 1989. Phenols and derivatives. In *Organic Building Blocks of the Chemical Industry*, pp. 512–18. New York: Wiley
71c. Szmant HH. 1989. Benzenoid carboxylic acids and derivatives. In *Organic Building Blocks of the Chemical Industry*, pp. 467. New York: Wiley
71d. Szmant HH. 1989. Phenols and derivatives. In *Organic Building Blocks of the Chemical Industry*, pp. 519. New York: Wiley
72. Takashi O, Garner C, Markley JL, Herrmann KM. 1982. Biosynthesis of aromatic compounds: ^{13}C NMR spectroscopy of whole *Escherichia coli* cells. *Proc. Natl. Acad. Sci. USA* 79:5828–32
73. Thayer AM. 1991. Use of specialty food additives to continue to grow. *Chem. Eng. News* 69:9–12
74. Thiemens MH, Trogler WC. 1991. Nylon production: an unknown source of atmospheric nitrous oxide. *Science* 251:932–34
75. Varagnat J. 1981. Hydroquinone, resorcinol, and catechol. In *Kirk-Othmer Encyclopedia of Chemical Technology*, ed. M Grayson, D Eckroth, GJ Bushey, CI Eastman, A Klingsberg, et al, 13:39–69. New York: Wiley. 3rd ed.
76. Weaver LM, Herrmann KM. 1990. Cloning of an *aroF* allele encoding a tyrosine-insensitive 3-deoxy-D-*arabino*-heptulosonate 7-phosphate synthase. *J. Bacteriol.* 172:6581–84
77. Weiss U, Gilvarg C, Mingioli ES, Davis BD. 1954. Aromatic biosynthesis. XI. The aromatization step in the synthesis of phenylalanine. *Science* 119:774–75
78. Williams JF, Blackmore PF, Duke CC, MacLeod JK. 1980. Fact uncertainty and speculation concerning the biochemistry of D-erythrose-4-phosphate and its metabolic roles. *J. Biochem.* 12:339–44
79. Woskresensky A. 1838. Ueber die zusammensetzung der chinasäure. *Justus Liebigs Ann. Chem.* 27:257–70
80. Ye Q-Z, Liu J, Walsh CT. 1990. *p*-Aminobenzoate synthesis in *Escherichia coli*: purification and characterization of PabB as aminodeoxychorismate synthase and enzyme X as aminodeoxychorismate lyase. *Proc. Natl. Acad. Sci. USA* 87:9391–95
81. Zeikus JG. 1980. Chemical and fuel production by anaerobic bacteria. *Annu. Rev. Microbiol.* 34:423–64

THE REGULATION OF METHANE OXIDATION IN SOIL

Rocco L. Mancinelli
Mail Stop 239-12, NASA-Ames Research Center, Moffett Field, California 94035-1000

KEY WORDS: methanotrophy, methylotrophy, soil, methanotroph ecology

CONTENTS

INTRODUCTION	582
METHANOTROPH BACTERIA	583
PHYSIOLOGY AND BIOCHEMISTRY OF METHANE OXIDATION	584
Ribulose Monophosphate Pathway	586
Serine Pathway	588
ECOLOGY OF METHANE OXIDATION IN SOILS	589
History	589
Chemical Factors Influencing Soil Methane Oxidation	590
Biological Factors Influencing Soil Methane Oxidation	595
SUMMARY AND FUTURE DIRECTIONS	597

ABSTRACT

The atmospheric concentration of methane, a greenhouse gas, has more than doubled during the past 200 years. Consequently, identifying the factors influencing the flux of methane into the atmosphere is becoming increasingly important. Methanotrophs, microaerophilic organisms widespread in aerobic soils and sediments, oxidize methane to derive energy and carbon for biomass. In so doing, they play an important role in mitigating the flux of methane into the atmosphere. Several physico-chemical factors influence rates of methane oxidation in soil, including soil diffusivity; water potential; and levels of oxygen, methane, ammonium, nitrate, nitrite, and copper. Most of these factors exert their influence through interactions with methane monooxygenase (MMO), the enzyme that catalyzes the reaction converting methane to methanol, the first step in methane oxidation. Although biological factors such as competition and predation undoubtedly play a role in regulating the methanotroph population in soils, and thereby limit the amount of methane consumed by methanotrophs, the significance of these factors is unknown. Obtaining a better understanding of the ecology of methanotrophs will help elucidate the mechanisms that regulate soil methane oxidation.

INTRODUCTION

Methane is second only to carbon dioxide in its importance as a greenhouse gas (8). Consequently, the increase in the atmospheric methane level over the past century has played a significant role in global warming (30, 40, 64, 148, 169, 180). Factors influencing methane flux into the atmosphere clearly affect this trend. One such factor is the oxidation of methane by methane-oxidizing microorganisms (see below).

The ability of microorganisms to oxidize methane has been known since 1906, when Söhngen first isolated an organism capable of growing on methane as a carbon source and named it *Bacillus methanicus* (177). Since that time, methane-oxidizing microorganisms (methanotrophs) have been found in a variety of soil and aquatic environments (2, 30, 35, 50, 54, 71, 77, 78, 80, 81, 91, 92, 102, 108, 111, 117, 125, 126, 135, 150, 169, 186, 187, 198, 199, 206, 213, 214). Because of their wide occurrence in aerobic soils and sediments, methanotrophs play an important role in regulating atmospheric methane content. Indeed, methanotrophs act as important mitigators of methane flux into the atmosphere from aquatic ecosystems (1, 28, 78, 79, 81, 82, 93–97, 119, 124–126, 151–153, 160, 161, 163, 195, 197). Data gathered for savanna, forest, tundra, desert, and agricultural soils show that the soil methanotrophs act as a sink for atmospheric methane (3, 19, 39, 102–104, 114, 133, 170, 184, 186, 199, 214). Soil methanotrophs consume an estimated 1–10% of the atmospheric methane, that is, 5–50 Tg/year (19).

Oxidation of methane to carbon dioxide in soil occurs primarily as part of aerobic metabolism in methanotrophic bacteria (9, 71, 116). Two species of yeast have also been reported to be able to oxidize methane, i.e. *Rhodotorula glutinis* and *Sporobolomyces roseus* (209, 210), but their occurrence and importance in the biosphere are undefined; they are not discussed further in this review. The net reaction of methane oxidation under aerobic conditions can be described as:

$CH_4 + 2O_2 \rightarrow CO_2 + 2H_2O$.

Methane oxidation can also be linked to the reduction of sulfate in anaerobic metabolism (5, 6, 71, 144, 145). The postulated overall reaction for the anaerobic oxidation of methane can be described as follows (11, 127, 154):

$$2CH_4 + SO_4^{2-} + 2H^+ \rightarrow 4H_2 + 2CO_2 + H_2S.$$

This reaction may be performed by a consortium of methane oxidizers and sulfate reducers (162). In environments that have at least 10 mM sulfate, 30 mM bicarbonate, 0.5 mM sulfide, and 0.1 mM methane, the free-energy change for the anaerobic oxidation of methane as given by the above equation is −25 kJ/mol (5). This value certainly should be adequate for organisms to obtain sufficient energy to grow. Anaerobic methane oxidation has not been detected in environments lacking sulfate and appears to be most prevalent in aquatic environments.

METHANOTROPH BACTERIA

Methanotrophs are a subset of the larger group of organisms that utilize one-carbon compounds, the methylotrophs. Prior to 1970, only four methanotrophs had been described. The first, originally named *Bacillus methanicus* (177), was reisolated as *Pseudomonas methanica* by Dworkin & Foster (54) and has now been renamed *Methylomonas methanica*. The three other methanotrophs described are *Pseudomonas methandenitrificans* (51), *Methanomonas methano-oxidans* (21), and *Methylococcus capsulatus* (59). All of these are strict aerobes able to grow only on methane, methanol, or dimethylether and not on multicarbon compounds. In 1970, more than 100 new isolates of methanotrophs were described (206). The first facultative methanotrophs (those that could also grow on multicarbon compounds) were described in 1974 (146, 147).

Classically, the methanotrophs have been divided into a few main genera: *Methylococcus, Methylomonas, Methylobacter, Methylosinus, Methylocystis, Methylobacterium,* and *Methylosporovibrio* (47, 48, 88, 156, 157, 196, 201–203, 205, 206). The discovery of more methanotrophs caused problems in the taxonomic classification. For example, new isolates did not always fit neatly into previously characterized groups, and methanotrophs have relatively few features that can be used to assign organisms to a genus by conventional taxonomic criteria. As a result, the assignment of a methanotroph to a particular genus or species might not reflect its true phylogenetic relatedness to other members of that genus. The use of molecular genetic techniques such as 16S rRNA sequence analysis has helped alleviate the problems associated with phylogenetic classification of these organisms. In fact, results from these analyses reflect the classical taxonomic classification scheme (20, 22, 24, 72, 192, 212). Lidstrom & Stirling (121) recently review the genetics of methanotrophs, so this subject is not addressed in this review.

From a functional standpoint, methanotrophs can be divided into three types based on their carbon-assimilation pathway, or pathways, as well as intracytoplasmic membrane arrangement, cell morphology, and the guanine and cytosine content of their DNA (63, 73, 190, 202, 205). Type I methanotrophs possess the ribulose monophosphate pathway, type II the serine pathway, and type X both pathways (see section on physiology and biochemistry, below, for an explanation of the pathways). Furthermore, phylogenetic classification from 5S and 16S rRNA sequence analyses suggests that the type I methanotrophs cluster within the γ-subgroup of the proteobacteria, and type II seem to cluster with the α-subgroup (20, 22, 24, 192) as defined by Woese (208). The phylogenetic status of the type X methanotrophs has not been well established with respect to these groupings. Examples of type I include *Methylomonas albus* BG8 (183, 202) and the marine *Methylomonas* sp. A4 (120). Type II includes *Methylosinus* sp. 6 (which appears to be the same as *Methylobacterium organophilum* CRL-26) (4, 189) and *Methylosporovibrio methanica* 81Z (13). Finally, type X includes *M. capsulatus* (178, 202).

PHYSIOLOGY AND BIOCHEMISTRY OF METHANE OXIDATION

The majority of methanotrophs isolated to date are gram-negative, microaerophilic rods, vibrios, or cocci. They form a differentiated exospore or cyst that is desiccation and heat resistant. A complex internal arrangement of paired membranes may be involved in oxidation reactions within these bacterial cells (48, 83, 149). They also contain the typical cytochromes *a, b,* and *c* and are catalase and oxidase positive. They can use organic nitrogen compounds, ammonia, nitrate, and nitrite as nitrogen sources (73). Finally, types II and X possess nitrogenase (73).

Many authors have discussed or reviewed the biochemistry of methanotrophs (9, 35, 38, 84, 116), so I present only a summary here. The first step is the oxidation of methane to methanol by methane monooxygenase (MMO), a multicomponent enzyme system (34, 158, 185, 188). This enzyme hydroxylates methane to methanol using molecular oxygen and a reductant [usually NAD(P)H, but in some organisms reduced cytochrome *c*]:

$$CH_4 + O_2 + 2NAD(P)H \rightarrow CH_3OH + H_2O + 2NAD(P)^+.$$

MMO is a copper- and iron-containing enzyme that occurs as a membrane-bound particulate (pMMO) and a soluble (sMMO) form in methanotrophs. The pMMO is universal among all methanotrophs, whereas the sMMO is only found in type II and type X methanotrophs (9, 44, 45). The pMMO and the sMMO are distinguishable not only by their different locations within the cell, but also by differences in substrate specificity, oxygen requirements, copper concentration requirements, kinetics, sensitivity to inhibitors, and NAD(P)H requirements, which all reflect structural differences (25, 36, 43, 44, 60, 67, 158, 179; for a review, see 122).

The differences in the biochemical characteristics of the two MMOs have ecological implications. For example, the sMMO can oxidize aromatic and alicyclic hydrocarbons (25, 36, 43, 134), as well as dehalogenate hydrocarbons (7, 23, 61, 74, 140, 141, 191), whereas the pMMO cannot. The sMMO, however, requires NAD(P)H, which limits its ability to cooxidize substrates other than methane. This attribute makes the pMMO more energy efficient than the sMMO (25, 67, 118). Another distinction between the two enzymes that has particular importance for their ecological role in soil is the apparent difference in their oxygen requirements. The pMMO has a K_m for oxygen of 0.1 µM, whereas the sMMO value is 17 µM (67, 98). This difference suggests that those organisms containing sMMO, that is, types II and X, can more successfully compete for methane under the low-oxygen tension occurring in sediments and in soil at depth.

Hutton & Zobell (92) found that the extensive internal membrane structure of methanotrophs appears to consist of invaginations of the cell membrane. These membranes may be involved in substrate oxidation (174). However, Higgins et al (85), Best & Higgins (17), and Scott et al (168) reported that the internal membrane system of methanotrophs may not be essential for methane oxidation in these organisms.

The MMO enzyme system appears to be responsible for the ability of the methanotrophs to oxidize ammonia to nitrite. Whittenbury & Kelly (204) stated that the oxidation of ammonia by methanotrophs is an example of cooxidation; that is, the organism can oxidize a compound but cannot grow on it as a single source of energy. As stated previously, the MMO is nonspecific and can oxidize many substrates (for a complete discussion, see 36).

In general, methanotrophic bacteria possess both dissimilatory and assimilatory pathways of methane oxidation (Figure 1). In dissimilatory pathways, methane is oxidized completely to carbon dioxide, thereby producing cellular energy, and none of the carbon becomes cellular material, or biomass. The carbon dioxide is given off to the surrounding environment. In assimilatory pathways, methane is oxidized and converted to cellular biomass. In both pathways, methane is first oxidized to methanol, which is then oxidized to formaldehyde. The formaldehyde can be used as reducing power in the electron-transport chain, oxidized to formate, or assimilated by the cell via the ribulose monophosphate pathway and/or the serine pathway. The cell then either oxidizes the formate to carbon dioxide or uses it as reducing power to drive the electron-transport chain. The carbon dioxide can be given off as a gas or be assimilated via the serine pathway. It should be noted, however, that although methane oxidizers consume methane and produce CO_2, another greenhouse gas, their net effect is a reduction in greenhouse gas because most of the methane consumed is converted into biomass and not into CO_2.

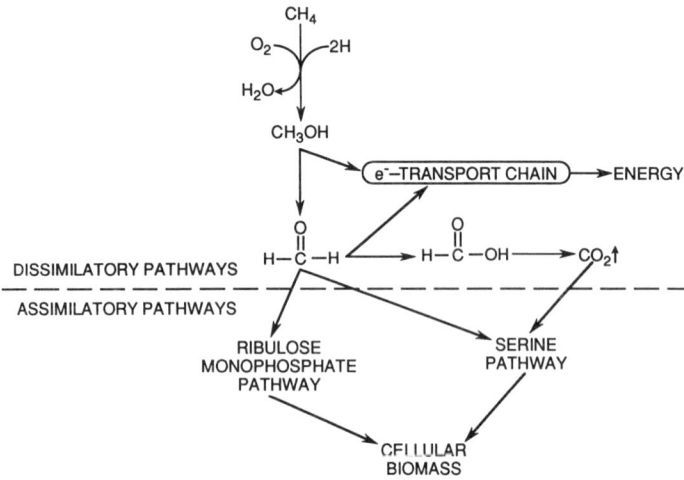

Figure 1 Diagrammatic summary of assimilatory and dissimilatory methane oxidation pathways by methanotrophs.

Ribulose Monophosphate Pathway

The ribulose monophosphate pathway (RuMP) cycle occurs in types I and X methanotrophic bacteria. In this pathway, all the cell's carbon is assimilated at the oxidation level of formaldehyde, produced by the oxidation of methane to formaldehyde:

$$CH_4 + O_2 + 2H^+ \rightarrow CH_3OH + H_2O,$$

$$CH_3OH \rightarrow HCHO + 2H^+.$$

The pathway can be thought of as a cycle consisting of three parts, first fixation, followed by cleavage, and lastly rearrangement reactions (Figure 2). Fixation occurs as a result of an aldol condensation of three molecules of formaldehyde with three molecules of ribulose 5-phosphate (ribulose monophosphate), eventually resulting in the formation of three molecules of fructose 6-phosphate. During cleavage, one molecule of fructose 6-phosphate is converted to 2-keto 3-deoxy 6-phosphogluconate, which is then cleaved to form glyceraldehyde 3-phosphate plus pyruvate (three carbon compounds). Lastly, the remaining two molecules of fructose 6-phosphate and glyceraldehyde phosphate undergo a series of rearrangement reactions resulting in the production of three molecules of ribulose 5-phosphate.

In an alternate pathway (Figure 3), occurring predominantly in facultative methylotrophs, the fructose 6-phosphate in converted to fructose 1,6 diphos-

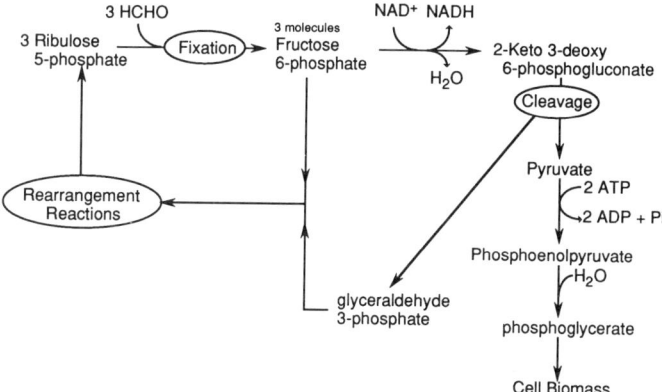

Figure 2 Summary diagram of the predominate ribulose monophosphate cycle of formaldehyde assimilation in methanotrophs.

phate instead of 2-keto 3-deoxy 6-phosphogluconate. The fructose 1,6 diphosphate is cleaved to form glyceraldehyde 3-phosphate and dihydroxyacetone phosphate. The intermediates produced in both variants of this cyclic pathway are used to synthesize cellular biomass.

Formaldehyde can also be oxidized completely to CO_2 via a dissimilatory RuMP cycle, provided the organism possesses the 6-phosphogluconate dehydrogenase. The net reaction is:

$HCHO + 2NADP^+ \rightarrow CO_2 + 2NADPH + 2H^+$.

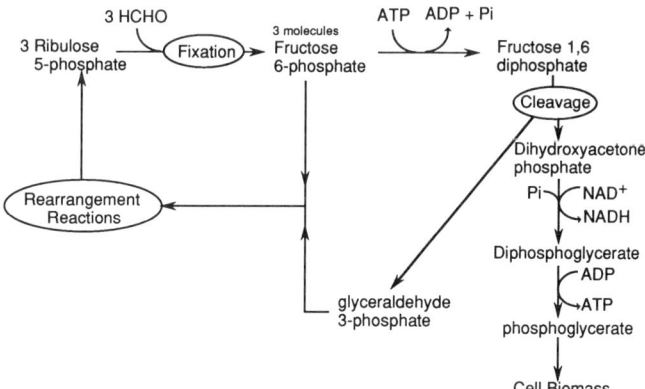

Figure 3 Variant of the ribulose monophosphate cycle of formaldehyde assimilation occurring primarily in facultative methylotrophs.

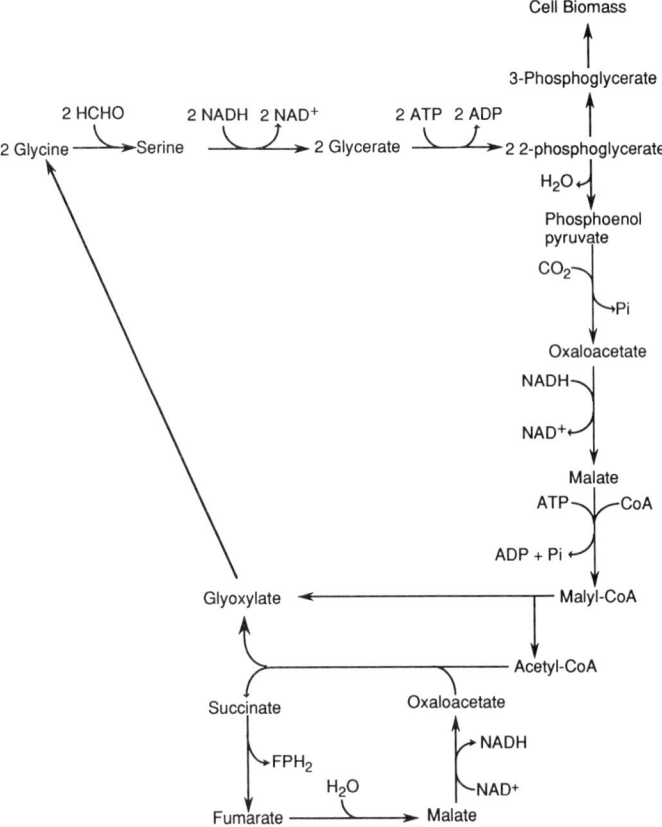

Figure 4 The serine pathway of formaldehyde assimilation in methanotrophs.

Serine Pathway

The serine pathway, depicted in Figure 4, occurs in types II and X methanotrophic bacteria. Like the RuMP pathway, it is cyclic, and cellular carbon is assimilated at the oxidation level of formaldehyde. However, it differs from the RuMP pathway in that the primary intermediates are carboxylic acids and amino acids rather than carbohydrates and that the key formaldehyde-assimilating enzyme is serine transhydroxymethylase. This enzyme catalyzes the addition of formaldehyde to glycine, resulting in the formation of serine:

$$2HCHO + 2CH_2NH_2COOH \rightarrow 2C_2H_4ONH_2COOH.$$

The serine then undergoes a series of reactions to form two molecules of 2-phosphoglycerate, one of which forms one molecule of 3-phosphoglycerate. The 3-phosphoglycerate enters other metabolic pathways and becomes cellular material. The remaining 2-phosphoglycerate molecule forms phosphoenolpyruvate, which then combines with carbon dioxide to form oxaloacetate. The oxaloacetate then proceeds through a series of reactions to eventually form glycine. The net reaction of the pathway can be expressed by the following chemical equation:

$$CO_2 + 2HCHO + 2NADH + 2H^+ + 3ATP \rightarrow C_3H_4O_2PO_3COOH + 2NAD^+ + 3ADP + 2Pi + \text{reduced flavoprotein}.$$

From this pathway, the cell obtains the precursors necessary to synthesize all cellular components.

The ribulose monophosphate pathway is more efficient than the serine pathway for two reasons: (*a*) All of the carbon atoms for constructing biomass are derived from formaldehyde, and (*b*) formaldehyde is at the same oxidation level as cellular biomass, so no reducing power is needed. Because of the different energy requirements of the two pathways, the biomass produced from a given amount of methane is higher for the ribulose monophosphate pathway than for the serine pathway.

ECOLOGY OF METHANE OXIDATION IN SOILS

Methanotrophs are widespread in nature and are found in aquatic and terrestrial ecosystems wherever stable sources of methane are present (107). Methanotrophs are often concentrated in a narrow band where methane diffusing away from its source (an anaerobic zone) meets oxygen from the air. Although nearly all methane oxidizers are obligate aerobes, they are microaerophilic, preferring oxygen levels lower than atmospheric levels. Physiologically, their microaerophily seems related to the K_m of the MMO for oxygen, as stated previously. Methanotrophs play an important role in the global carbon cycle. They convert methane derived from anaerobic decomposition back into organic carbon (biomass) and carbon dioxide.

History

Schollenberger (167) and Harper (76) were among the first to observe that soils surrounding natural gas–pipeline leaks had a higher organic-matter content than surrounding soils. They attributed the increased organic matter to an increased soil microbial population. During the early 1900s, published observations recorded a waxy-looking deposit found on the surface of soils. This deposit, termed paraffin dirt, was used as an indicator of oil and gas deposits along the coast of the Gulf of Mexico (52, 193). Paraffin dirt has a high organic-matter content but a low hydrocarbon content. A few years later, Barton (12) suggested that bacteria living on methane and oxygen were responsible for the production of the organic matter in paraffin dirt. In addition, their study showed that paraffin dirt was closely associated with methane seeps from oil fields along the coast of Louisiana and Texas. The definitive proof of association of paraffin dirt with methanotrophs was not obtained for another 27 years, when Davis (49) conducted an experiment in which he passed a mixture of natural gas and air through soils. After approximately 18 months, paraffin dirt began to form in the soils. The laboratory-formed paraffin dirt consisted primarily of microorganisms and the by-products of their metabolism. Careful field analysis supported these data (50).

Thorough studies coupling hydrocarbon gas transport through soils with microbial methane oxidation were not conducted until the early 1970s. The most complete study occurred in the Netherlands on methane oxidation associated with natural gas–pipeline leaks (2, 86, 87). The data from these studies showed that methane and oxygen levels in the soil surrounding the leak decreased dramatically within a short distance from the leak. Diffusion could not account for this decrease. Coincident with the decreases in methane and oxygen levels in the soil was a sharp increase in the carbon dioxide level. These data suggested that the methane was being oxidized to carbon dioxide by methanotrophs, a hypothesis confirmed by subsequent analyses. The results of this work illustrated that microbial methane oxidation could proceed to completion in soil. Later work in the United States (124–126) showed that total consumption of all of the available methane in soil does not always occur. Various chemical and physical factors that affect methane oxidation by microorganisms prevent complete consumption; these include soil diffusivity, soil water content (water potential), oxygen concentration, and the availability of nutrients (e.g. methane and fixed nitrogen). Temperatures in the range of -1 to $30°C$ have little impact on methane oxidation in forest soils (109).

Chemical Factors Influencing Soil Methane Oxidation

Microaerophilic methane-oxidizing organisms inhabit the aerobic portion of soils (2, 125). The depth and gaseous composition of the aerobic zone of soil, which is primarily governed by soil diffusivity, is an important parameter in determining the population density and growth rate of aerobic methane-oxidizing bacteria. For example, the depth of oxygen penetration into a typical landfill soil has been calculated to be approximately 1 m (123) and measured to depths of 1–2 m using mathematical models of gas diffusion through soils (18). The upward flow of landfill gas displaces the gases diffusing from the air above the landfill through pores in the soil. Only in the region near the soil surface are atmospheric gases found. This region is the one inhabited by the microaerophilic methane oxidizing bacteria (100, 125). The studies of landfill soil point to a dynamic relationship between the methane-oxidizing microorganisms inhabiting the aerobic soil layer covering a sanitary landfill and the gaseous component of the surrounding soil (100, 124, 125). However, the interaction between the microorganisms inhabiting this layer and the gases within the soil have not been well characterized.

METHANE AVAILABILITY The population of methanotrophs inhabiting the aerobic soil layer covering sanitary landfills increases when the methane concentration within that soil also increases (125). The methanotroph population expands until the methane begins to flow hydrodynamically through the soil. For typical clay soils used as landfill cover, such methane flow occurs when the soil gas phase reaches approximately 50% by volume.

Results from laboratory microcosm studies mimicking typical landfill soil conditions and containing high concentrations of methane (i.e. soil gas phase >10% CH_4) showed that ~10–90% of the available methane was oxidized depending on gas-flow rates and oxygen availability (124, 126; RL Mancinelli, unpublished data). Methanotrophs inhabiting certain areas of landfill-cover soil consumed 100% of the available methane (100), and methanotrophs inhabiting lake sediments in northern California consumed more than 90% of the available methane (142). A laboratory study of methane oxidation in landfill topsoil reported that the methane oxidation-rate potential for the soil was high, equaling 45 g CH_4/oxidized m^2 soil (200). This study also showed that the methanotrophs inhabiting the topsoil could oxidize methane from a concentration of <1 ppm to >10 ppt (exhibiting a first-order rate constant $k = -0.54$ h^{-1} to -2.37 h^{-1}). Not all of the methane, however, was consumed by the methanotrophs inhabiting the soil.

Megraw & Knowles (129) conducted laboratory studies on a cultivated humisol containing both methanotrophs and methanogens. Their results led them to suggest that the methanotrophs in this soil consume methane produced by the methanogens in the anaerobic portion of the soil as well as methane from the atmosphere. If this hypothesis is true, soil environments in which methane is produced may serve as a net sink for methane.

Studies of methane oxidation in soils and sediments suggest that the minimum threshold value for methane oxidation in soils is much lower than that in sediments. For example, threshold values of 2–3 ppm have been reported for a sediment, whereas the threshold values for soil are an order of magnitude lower, ranging from <0.1 to 0.4 ppm (19, 199, 200, 214). The cause of the observed differences in the minimum threshold values for methane oxidation is uncertain. These variations may be related to differences in methanotroph population types, in the binding efficiency of the two MMOs to methane at low partial pressures, in the relative expression rates of MMO, and in the availability of oxygen. Nonetheless, the minimum threshold value of a soil system will determine if it will be a sink for atmospheric methane. Furthermore, the kinetics of CH_4 oxidation in soil at ambient levels differs from the kinetics in soils exposed to higher levels of CH_4 (16, 100). Methanotrophs that do not oxidize CH_4 at low concentrations (16), that is, those whose threshold value is too high, may be responsible for the kinetic differences.

WATER POTENTIAL Water potential is the free energy of water in a system, relative to the free energy of a reference pool of pure, free water. Pure free water has, by definition, zero water potential; thus, the potential energy of water in unsaturated soils is negative. Because the resistance posed by a microbial cell wall and membrane is relatively insignificant, the water potential of a microbial cell in soil is likely to be at or near equilibrium with its microenvironment within the soil (56, 207). A system tends to flow from high to low water potential, lowering the availability of water for microbes in the soil as the potential decreases. Changes in soil water content have profound effects on microbial activity (69, 136, 143, 165) that in turn alter the composition of soil microbial populations (31, 33, 68, 132).

The water content of soil regulates methane oxidation by affecting gas transport through soil (i.e. gas-phase molecular diffusion occurring in moist soil is faster than aqueous diffusion) and through the physiological requirement by methanotrophs for water to oxidize methane. When landfill, temperate, and subarctic soils were saturated with water (40–50% wt/vol H_2O), the soil methane oxidation rate decreased. When the soil moisture content was between 5 and 15%, however, the methane oxidation rate was much greater (3, 126, 199). The decrease in methane oxidation in the water-saturated soil probably resulted from slower diffusion of methane and oxygen through the soil to the microorganisms.

The data obtained in the field and laboratory studies discussed above indicate that soil moisture, and thus water potential, are important factors regulating methane oxidation. Unfortunately, none of the studies measured water potential directly. In fact, to date no controlled quantitative studies regarding the relationship of water potential and methane-oxidation rates in soil have been reported.

OXYGEN AVAILABILITY Smirnova (172, 173) and Nesterov et al (138) were among the first to demonstrate that methanotrophic bacteria are sensitive to changes in the oxygen concentration in their environment. Oxygen availability strongly depends upon the soil diffusivity, which in turn affects the utilization of methane by methanotrophs.

Oxygen regulation of methane oxidation is not passive. For example, the photosynthetic alga *Scenedesmus quadricauda* regulates methane oxidation indirectly through photosynthesis by producing the bulk of the oxygen used by the methanotrophs for methane oxidation (194). This phenomenon was confirmed in a series of studies showing that light regimes resulting in benthic photosynthesis and photosynthesis in algal mats affected oxygen penetration into the sediment and therefore methane-oxidation rates (105, 106, 110, 111).

King (106) studied the patterns and controls of methane oxidation in a Danish wetland sediment. The data indicated that shifts in methane emissions coincident with short-term changes in the availability of oxygen resulted from changes in methane oxidation, not availability of methane. Furthermore, even short periods of anoxia result in a loss of methane oxidation (105, 106). In a laboratory microcosm study of methane oxidation in landfill soil permeated with methane and oxygen and contained in gas flow-through chambers (124), methane oxidation shared a significant positive correlation with the oxygen levels in the chambers (RL Mancinelli, unpublished data).

Clearly, soil oxygen levels play a significant role in the rate of methane oxidation in soil. This influence results from the need for oxygen to complete the first step in methane oxidation, the kinetics of oxygen uptake by the cell, and the kinetics of oxygen binding to the MMO. In soil environments with an atmospheric source of methane, methane oxidation is most likely oxygen saturated and methane limited. In systems in which methanogenesis occurs, methane oxidation is probably oxygen limited and methane saturated in all but the very top of the soil exposed to the atmosphere.

NITROGEN COMPOUNDS   Because of the lack of specificity of the sMMO and the pMMO, methanotrophs can oxidize ammonia to nitrite, although they cannot grow lithotrophically using ammonia as a sole electron donor (42, 58, 92, 215; for a recent review, see 15). This ability to oxidize ammonia results in a competitive interaction between methane and ammonia for the MMO. Hence, ammonia is slightly toxic to methanotrophs, and nitrate is the preferred source of fixed nitrogen.

Dalton (42) showed that ammonium oxidation by cell extracts of *M. capsulatus* required NADH + H$^+$. He further showed that the response to both NADH + H$^+$ and ammonium could not be explained by a simple kinetic model. Moreover, substrates such as formate, formaldehyde, and 3-hydroxybutyrate that are linked to NADH + H$^+$ formation stimulate ammonium oxidation (139). Cells and soluble extracts of *M. capsulatus* that oxidize methane to methanol and ammonia to nitrite have a very high apparent K_m for the oxidation of ammonia to nitrite. These data suggest that nitrification would be significant only when either the ammonia concentration was high or the methane concentration was low. Dalton (42) suggested that such conditions exist in aquatic environments where high populations of methane oxidizers occur. If this were true, methanotrophs would be the major nitrifying organisms in such environments. To date, few data support this hypothesis (129–131, 159). Data from kinetic studies, however, show that the affinity and maximal uptake velocity of methanotrophs for ammonium is lower than that of nitrifiers (15), suggesting that nitrifiers in a community would be the primary organisms responsible for ammonia oxidation, even when large populations of methanotrophs occur. Whether methanotrophs play a significant role in nitrification is unclear. Thus, methanotrophs apparently function in the nitrogen cycle not only by assimilating nitrogen, but also by oxidizing ammonia and fixing dinitrogen (150, 211).

Ammonium inhibits methane oxidation in pure cultures of *Methylobacter albus* BG8 and *Methylosinus trichosporium* OB3b (113). The data further suggest that ammonium inhibition results from direct effects of the interaction of ammonia with MMO and indirect effects due to nitrite (113). Methane oxidation in soils appears to be more sensitive to ammonia inhibition than either cell extracts or pure cultures of methanotrophs (3, 133, 137, 184). In addition, the inhibitory effect of ammonia in soils persists for long periods of time, even after the available ammonium dissipates (90, 133, 137). When methane concentration in soils increases to 10 ppm, the inhibition of methane oxidation by ammonia also increases (112). The relative insensitivity of pure cultures of methanotrophs to ammonium inhibition of methane oxidation may result from the cultures' exposure to ammonium under concentrations of methane (1–10%) that are higher than the atmospheric level (42, 113, 139). Recent field data suggest, however, that this inhibition also occurs in landfill soils exposed to concentrations of methane exceeding 10% (RL Mancinelli, unpublished data).

Ammonium inhibition of methane oxidation is not caused solely by substrate competition for the MMO active site. For example, the first step in the oxidation of ammonium by MMO produces hydroxylamine (89). Hydroxylamine reversibly inhibits MMO activity (89). Furthermore, hydroxylamine readily forms nitrite, which inhibits $NADH + H^+$ formation (a key factor for methane oxidation) via formate dehydrogenase (99, 139, 216). Nitrite inhibits methane oxidation in pure cultures of *M. albus* BG8 and *M. trichosporium* OB3b (113). In addition, when soil levels of methane increase, ammonium inhibition of methane oxidation also decreases (164), which precludes simple competition at the enzyme level.

Ammonia oxidation by methanotrophs appears to increase with increasing levels of methane, and the resulting nitrite acts as a toxic agent against the methanotrophs (113, 164). In fact, in soils under ambient levels of methane, nitrite suppressed methane oxidation to a greater degree than ammonium (164). The addition of either methanol, formate, or β-hydroxybutyrate to a forest soil had little effect on methane oxidation. However, in soil treated simultaneously with ammonium plus methanol, formate, or β-hydroxybutyrate, methane oxidation was inhibited to a greater degree than it was in soil treated with ammonium alone; the combination of ammonium and methanol exhibited the greatest inhibitory effect (164). This result may be due to the ammonium inhibition of the MMO, which forces the cell to take advantage of the available methanol, or it may be due to the MMO requirement for a reductant; that is, the oxidation of these organic substrates generates NADH + H$^+$, thereby increasing the available reductant and facilitating ammonium oxidation by MMO (164).

Soil ammonium is classified as an exchangeable ion if it can be displaced by neutral potassium solutions and as fixed (i.e. a nonexchangeable ion) if it cannot be displaced. Ammonium added to soils can be fixed by clay minerals and subsequently released. It can move between exchangeable and fixed sites within soil. However, the range of ammonium release by clays remains unknown. Data from studies in which ^{15}N-labeled ammonium was added to soil indicate that ammonium fixation is much more rapid than its subsequent release (53, 101, 115). For example, in one report, release of ammonium fixed within one week had a half-life of 35–42 weeks (53). Furthermore, the rate of fixed ammonium release is tied to the rate of oxidation of exchangeable ammonium by soil microbes, that is, nitrification of the available (nonfixed) ammonium (66). The fixation of ammonium by clays in soils followed by its slow release probably contributes to the longevity of methane inhibition observed in certain soils. Fixation rates probably also contribute to the different effects of ammonium on methane oxidation found in pure culture studies, laboratory studies, and field studies.

The ecological implications of ammonia inhibition on methane oxidation are broad. For example, the application of ammonia fertilizers along with ammonia from agricultural runoff undoubtedly limits methanotroph activity in agricultural areas. This observation combined with the months-long persistence of ammonium inhibition of methane oxidation (133) suggests that much agricultural land has a significantly reduced capacity for serving as an atmospheric methane sink. Because of the large surface areas involved, atmospheric methane levels increase, which could have a significant impact on the global methane budget.

Data presented in the studies discussed above suggest that water potential, oxygen concentration, and nutrient availability each play a distinct and important role in influencing methane oxidation in soils. However, they are not the sole factors influencing methane oxidation in soils. Biological factors may also mitigate methanotrophic activity.

Biological Factors Influencing Soil Methane Oxidation

Competition and predation are two biological factors known to play important roles in regulating bacterial populations in soil, and they probably affect the soil methanotroph population, thereby influencing methane oxidation.

COMPETITION Each population comprising a community obtains all it needs to survive from its surrounding environment. These resources are finite within a given ecosystem, and often the demand for these nutrients exceeds the supply. When the demand for a critical resource does exceed supply, competition occurs and has a profound impact on natural selection and the composition of microbial communities (for a review, see 62). For example, competition between two populations for a single resource tends to eliminate one of the populations (75). In nature, however, several factors mitigate the severity of competition and allow competitors to coexist (10, 26, 175, 176, 181, 182).

Although no direct competition studies have been reported on methanotrophs in soil, methanotrophs compete with other soil inhabitants for some nutrients. Fixed nitrogen, phosphate, and to a lesser extent oxygen are in finite supply, and the various populations inhabiting the soil must compete for these nutrients. As a result, the competing populations are smaller than would otherwise be expected. Competition for a limiting nutrient among methanotrophs can affect the methanotroph population and rates of methane oxidation. A competition experiment (65) performed on *M. trichosporium* OB3b (a type II methanotroph) and *M. albus* BG8 (a type I methanotroph) in a continuous-flow bioreactor demonstrated these effects. Methane, copper, and nitrate were used as the determinants of microbial selection. Under copper- and nitrate-limiting conditions, *M. trichosporium* dominated, owing to its ability to express sMMO under copper limitation and nitrogenase under nitrate limitation. The greater energy efficiency of the pMMO and the ribulose monophosphate pathway allowed *M. albus* to dominate under methane-limiting conditions. The results of studies done to determine whether nitrifiers and methanotrophs compete for ammonium (164) suggest that these organisms are not in direct competition.

PREDATION During predation, the prey population decreases but is rarely eliminated (27, 70). Nearly all bacterial genera indigenous to soil are susceptible to predation by a multitude of other organisms inhabiting soil. Protozoa are the major predators of bacteria, consuming an estimated 150–900 g bacteria/m^2 per year in arable soil (14, 31, 32, 37, 41, 155, 166). The consumption of bacteria in soil by other soil bacteria also occurs (29, 128, 171).

Predator-prey populations are dependent upon each other (31, 46, 55, 57). In nature, the populations of predators and prey generally remain balanced until the system is disturbed, e.g. by a change in nutrient supply (31). Predator-prey relationships have not been examined for soil methanotrophic populations. A predator-prey relationship may account for the inability of soil methanotroph populations to consume all of the available methane (discussed above).

SUMMARY AND FUTURE DIRECTIONS

Methane is found throughout nature—in the soils, atmosphere, oceans, and lakes. The primary source of methane is bacterial methanogenesis in anaerobic environments, but it also occurs in many coal formations and is a major constituent of natural-gas deposits. As discussed above, an appreciable amount of biogenically produced methane is oxidized by methanotrophs to carbon dioxide before it reaches the atmosphere. Methanotrophs appear to play an important role in maintaining low levels of methane in the atmosphere (~1.7 ppm) and consequently play a role in regulating the Earth's environment. Our recent appreciation of the role of methane and methane oxidation in the global carbon cycle has resulted from the advent of more accurate data on the chemistry and methane content of the atmosphere, as well as an increase in knowledge about the microbiology of methanotrophs in soil and aquatic environments.

The ecology of methane oxidation in soils is complex and not entirely understood. Despite good evidence for anaerobic methane oxidation, the only methanotrophs isolated to date are aerobic. Whether this is a true reflection of the relative abundance of aerobic methanotrophs or an artifact of isolation procedures is unclear. Data from field and laboratory studies indicate that physico-chemical factors (such as water potential; the availability of nutrients, including methane; soil-oxygen concentrations; and soil diffusivity) are important factors regulating methane oxidation in soil. Biological factors (such as competition and predation) probably also influence methane oxidation by affecting the soil methanotroph population.

Many questions regarding the ecology and regulation of methane oxidation in soil remain unanswered. For example, do biological factors (e.g. competition and predation) play a significant and important role in influencing soil methane oxidation rates? Is anaerobic methane oxidation widespread, and does it affect the global carbon budget in any significant way? What is the mechanism underlying the long-term inhibition of methane oxidation by ammonium? To what extent, if any, do nitrifiers contribute to methane oxidation? These are only a few of the questions that must be answered before the true significance of the role of methanotrophs in ecosystem and biosphere dynamics can be understood. Once we have this level of understanding, we can make predictions regarding the susceptibility of methane oxidation to changes in climate and land use. Many more ecological studies of soil methane oxidation must be completed, however, before an accurate picture of the mechanisms that regulate soil methane oxidation in situ can be drawn.

ACKNOWLEDGMENTS

I thank Gerald Livingston and Jane Pretty for commenting on early drafts of the review and Melisa White for help in preparing the manuscript.

Any *Annual Review* chapter, as well as any article cited in an *Annual Review* chapter, may be purchased from the Annual Reviews Preprints and Reprints service.
1-800-347-8007; 415-259-5017; email: arpr@class.org

Literature Cited

1. Abramochkina FN, Bezrukova LV, Koshelev AV, Gal'chenko VF, Ivanov MV. 1987. Microbial oxidation of methane in a body of fresh water. *Mikrobiologiya* 56:464–71
2. Adamse AD, Hoeks J, DeBont JAM, Van Kessel JF. 1972. Microbial activities in soil near natural gas leaks. *Arch. Microbiol.* 83:32–51
3. Adamsen APS, King GM. 1993. Methane consumption in temperate and subarctic forest soils: rates, vertical zonation, and responses to water and nitrogen. *Appl. Environ. Microbiol.* 59:485–90
4. Allen LN, Olstein A, Haber C, Hanson RS. 1984. Genetic and biochemical studies of representative type II methylotrophic bacteria. See Ref. 38, pp. 236–43
5. Alperin MJ, Reeburgh WS. 1984. Geochemical observations supporting anaerobic methane oxidation. See Ref. 38, pp. 282–89
6. Alperin MJ, Reeburgh WS. 1985. Inhibition experiments on anaerobic methane oxidation. *Appl. Environ. Microbiol.* 50:940–45
7. Alvarez-Cohen L, McCarty PL, Boulygina E, Hanson RS, Brusseau GA, Tsien HC. 1992. Characterization of a methane-utilizing bacterium from a bacterial consortium that rapidly degrades trichloroethylene and chloroform. *Appl. Environ. Microbiol.* 58:1886–93
8. Andreae MO, Crutzen PJ. 1985. Atmospheric chemistry. In *Global Change*, ed. TF Malone, JG Roederer, pp. 75–113. Cambridge: ICSU/Cambridge Univ. Press
9. Anthony C. 1982. *The Biochemistry of Methylotrophs*. New York: Academic. 431 pp.
10. Baltzis BC, Fredrickson AG. 1984. Competition of two suspension-feeding protozoan populations for a growing bacterial population in continuous culture. *Microb. Ecol.* 10:61–68
11. Barnes RO, Goldberg RE. 1976. Methane production and consumption in anoxic marine sediments. *Geology* 4:297–300
12. Barton DC. 1925. Review of paper concerning "Paraffin Dirt", by HB Milner. *Am. Assoc. Pet. Geol. Bull.* 9:1118–21
13. Bastien CA, Machlin SM, Zhang Y, Donaldson K, Hanson RS. 1989. The organization of genes required for the oxidation of methanol to formaldehyde in three type II methylotrophs. *Appl. Environ. Microbiol.* 55:3124–30
14. Bazin MJ, Saunders PT, Owen BA, Kilpatrick D. 1978. Predation by slime mould amoebae. In *Microbial Ecology*, ed. MW Loutit, JAR Miles, pp. 21–24. New York: Springer-Verlag
15. Bedard C, Knowles R. 1989. Physiology, biochemistry, and specific inhibitors of CH_4, NH_4^+, and CO oxidation by methanotrophs and nitrifiers. *Microbiol Rev.* 53:68–84
16. Bender M, Conrad R. 1992. Kinetics of CH_4 oxidation in oxic soils exposed to ambient air or high CH_4 mixing ratios. *FEMS Microbiol. Ecol.* 101:261–70
17. Best DJ, Higgins IJ. 1981. Methane oxidizing activity and membrane morphology in a methanol-grown obligate methanotroph, *Methylosinus trichosporium* OB3b. *J. Gen. Microbiol.* 125:73–84
18. Bogner JE, Moore CA. 1986. Gas movement through fractured landfill cover materials. In *Proc. Annu. Madison Waste Conference, 9th*, pp. 288–309. Madison, WI: Univ. Wis. Dep. Eng. Professional Dev.
19. Born M, Dörr H, Ingeborg L. 1990. Methane consumption in aerated soils of the temperate zone. *Tellus B* 42:2–8
20. Bratina BJ, Brusseau GA, Hanson RS. 1992. Use of 16S rRNA analysis to investigate phylogeny of methylotrophic bacteria. *Int. J. Syst. Bacteriol.* 42:645–48
21. Brown LR, Strawinski RJ, McCleskey CS. 1964. The isolation and characterization of *Methanomonas methanooxidans* Brown and Strawinski. *Can. J. Microbiol.* 10:791–99
22. Brusseau GA, Bulygina ES, Hanson RS. 1994. Phylogenetic analysis and development of probes for differentiating methylotrophic bacteria. *Appl. Environ. Microbiol.* 60:626–36
23. Brusseau GA, Tsien HC, Hanson RS, Wackett LP. 1990. Optimization of

trichloroethylene oxidation by methanotrophs and the use of a colorimetric assay to detect soluble methane monooxygenase activity. *Biodegradation* 1: 19–29

24. Bulygina ES, Chumakov KM, Netrusov AI. 1993. Systematics of gram-negative methylotrophic bacteria based on 5S rRNA sequences. See Ref. 134a, pp. 275–84
25. Burrows KJ, Cornish A, Scott D, Higgins IG. 1984. Substrate specificities of the soluble and particulate methane monooxygenases of *Methylosinus trichosporium* OB3b. *J. Gen. Microbiol.* 130:3327–33
26. Butler GJ, Hsu SB, Waltman P. 1983. Coexistence of competing predators in a chemostat. *J. Math. Biol.* 17:133–51
27. Canale RP, Lustig TD, Kehrberger PM, Salo JE. 1973. Experimental and mathematical modeling studies of protozoan predation on bacteria. *Biotech. Bioeng.* 15:707–28
28. Cappenberg TE. 1972. Ecological observations on heterotrophic, methane-oxidizing and sulfate reducing bacteria in a pond. *Hydrobiology* 40:471–85
29. Casida LE. 1988. Minireview: nonobligate bacterial predation of bacteria in soil. *Microb. Ecol.* 15:1–8
30. Cicerone RJ, Oremland RS. 1988. Biochemical aspects of atmospheric methane. *Glob. Biogeochem. Cycles* 2: 299–327
31. Clarholm M. 1981. Protozoan grazing of bacteria in soil—impact and importance. *Microb. Ecol.* 7:343–50
32. Clarholm M. 1984. Heterotrophic, free-living protozoa: neglected microorganisms with an important task in regulating bacteria populations. See Ref. 114a, pp. 321–26
33. Clarholm M, Rosswall T. 1980. Biomass and turnover of bacteria in a forest soil and a peat. *Soil Biol. Biochem.* 12:49–57
34. Colby J, Dalton H. 1978. Resolution of the methane mono-oxygenase of *Methylococcus capsulatus* (Bath) into three components. Purification and properties of component C, a flavoprotein. *Biochem. J.* 171:461–68
35. Colby J, Dalton H, Whittenbury R. 1979. Biological and biochemical aspects of microbial growth on C1 compounds. *Annu. Rev. Microbiol.* 33: 481–87
36. Colby J, Stirling DI, Dalton H. 1977. The soluble methane mono-oxygenase of *Methylococcus capsulatus* (Bath). Its ability to oxygenate *n*-alkanes, *n*-alkenes, ethers and alicyclic, aromatic, and heterocyclic compounds. *Biochem. J.* 165:395–402
37. Coutear M-M, Pussard M. 1982. Nature du régime alimentaire des protozaires du sol. In *New Trends in Soil Biology*, ed. PH Lebrun, M Andre, A De Medts, C Grégoire-Wibo, G Wauthy, pp. 179–94. Ottignies-Lovain-la-Neuve: Dieu-Brichart
38. Crawford RL, Hanson RS, eds. 1984. *Microbial Growth on C1 Compounds.* Washington, DC: Am. Soc. Microbiol. 343 pp.
39. Crill PM. 1991. Seasonal patterns of methane uptake and carbon dioxide release by a temperate woodland soil. *Glob. Biogeochem. Cycles* 5:319–34
40. Crutzen PJ, Graedel TE. 1986. The role of atmospheric chemistry in environment development interactions. In *Sustainable Development of the Biosphere*, ed. WC Clark, RE Muffin, pp. 107–30. Cambridge, MA: Cambridge Univ. Press
41. Curds CR, Bazin MJ. 1977. Protozoan predation in batch and continuous culture. In *Advances in Aquatic Microbiology*, ed. MR Droop, JW Jannasch, pp. 115–76. London: Academic
42. Dalton H. 1977. Ammonia oxidation by the methane oxidizing bacterium *Methylococcus capsulatus* strain Bath. *Arch. Microbiol.* 114:273–79
43. Dalton H. 1980. Oxidation of hydrocarbons by methane monooxygenases from a variety of microbes. *Adv. Appl. Microbiol.* 26:71–87
44. Dalton H, Higgins IJ. 1987. Physiology and biochemistry of methylotrophic bacteria. In *Microbial Growth on C1 Compounds,* ed. HW Van Verseveld, JA Duine, pp. 89–94. Dordrecht: Nijhoff
45. Dalton H, Prior SD, Leak DJ, Stanley SH. 1984. Regulation and control of methane monooxygenase. See Ref. 38, pp. 75–82
46. Darbyshire JF, Greaves MP. 1967. Protozoa and bacteria in the rhizosphere of *Sinapis alba* L., *Trifolium repens* L. and *Lolium perenne* L. *Can. J. Microbiol.* 13:1057–68
47. Davey JF, Whittenbury R, Wilkinson JF. 1972. The distribution in the methylo-bacteria of some key enzymes concerned with intermediary metabolism. *Arch. Mikrobiol.* 87:359–66
48. Davies SL, Whittenbury R. 1970. Fine structure of methane and other utilizing bacteria. *J. Gen. Microbiol.* 61: 227–32
49. Davis JB. 1952. Studies on soil samples from a "paraffin dirt" bed. *Am. Assoc. Pet. Geol. Bull.* 36:2186–88

50. Davis JB. 1967. *Petroleum Microbiology*. London: Elsevier/North Holland
51. Davis JB, Coty VF, Stanley JP. 1964. Atmospheric nitrogen fixation by methane oxidizing bacteria. *J. Bacteriol.* 88:468–72
52. Deussen A. 1918. Review of developments in the Gulf Coast in 1917. *Am. Assoc. Pet. Geol. Bull.* 2:36–37
53. Drury CF, Beauchamp EG. 1991. Ammonium fixation, release, nitrification, and immobilization in high- and low-fixing soils. *Soil Sci. Soc. Am. J.* 55:125–29
54. Dworkin M, Foster JW. 1956. Studies on *Pseudomonas methanicus* (Söhngen) nov. comb. *J. Bacteriol.* 72:646–59
55. Elliott ET, Coleman DC. 1977. Soil protozoan dynamics in a shortgrass prairie. *Soil Biol. Biochem.* 9:113–18
56. Elliott LF, ed. 1981. *Water Potential Relations in Soil Microbiology*. Madison, WI: Soil Sci. Soc. Am. 151 pp.
57. Fenchel T. 1982. Ecology of heterotrophic microflagellates. IV. Quantitative occurrence and importance as consumers of bacteria. *Mar. Ecol. Prog. Ser.* 9:35–42
58. Ferenci T, Strom T, Quayle JR. 1975. Oxidation of carbon monoxide and methane by *Pseudomonas methanica*. *J. Gen. Microbiol.* 91:79–91
59. Foster JW, Davis RH. 1966. A methane-dependent coccus with notes on classification and nomenclature of obligate, methane-utilizing bacteria. *J. Bacteriol.* 91:1924–31
60. Fox BG, Froland WA, Dege JE, Lipscomb JD. 1989. Methane monooxygenase from *Methylosinus trichosporium* OB3b. *J. Biol. Chem.* 264:10023–33
61. Fox BG, Froland WA, Dege JE, Lipscomb JD. 1990. Haloalkene oxidation by the soluble methane monooxygenase from *Methylosinus trichosporium* OB3b: mechanistic and environmental applications. *Biochemistry* 29:6419–27
62. Fredrickson AG, Stephanopoulos G. 1981. Microbial competition. *Science* 213:972–79
63. Gal'chenko VF, Andreev LV. 1984. Taxonomy of obligate methanotrophs. See Ref. 38, pp. 269–75
64. Graedel TE, Crutzen PJ. 1989. The changing atmosphere. *Sci. Am.* 261:136–43
65. Graham DW, Chaudhary JA, Hanson RS, Arnold RG. 1993. Factors affecting competition between type I and type II methanotrophs in two-organism, continuous-flow reactors. *Microb. Ecol.* 25:1–17
66. Green CJ, Blackmer AM, Yang NC. 1994. Release of fixed ammonium during nitrification in soils. *Soil Sci. Soc. Am. J.* 58:1411–15
67. Green J, Dalton H. 1986. Steady-state kinetic analysis of soluble methane monooxygenase from *Methylococcus capsulatus* (Bath). *Biochem. J.* 236:155–62
68. Griffin DM. 1972. *Ecology of Soil Fungi*. London: Chapman & Hall
69. Griffin DM. 1981. Water potential as a selective factor in the microbial ecology of soils. See Ref. 56, pp. 141–51
70. Habte M, Alexander M. 1978. Mechanisms of persistence of low numbers of bacteria preyed upon by protozoa. *Soil Biol. Biochem.* 10:1–6
71. Hanson RS. 1980. Ecology and diversity of methylotrophic organisms. *Adv. Appl. Microbiol.* 11:3–39
72. Hanson RS, Bratina BJ, Brusseau GA. 1993. Phylogeny and ecology of methylotrophic bacteria. See Ref. 134a, pp. 285–302
73. Hanson RS, Netrusov AI, Tsuji K. 1991. The obligate methanotrophic bacteria *Methylococcus*, *Methylomonas*, and *Methylosinus*. In *The Prokaryotes*, ed. A Balows, HG Truper, M Dworkin, K Schliefer, pp. 2350–64. New York: Springer-Verlag
74. Hanson RS, Tsien HC, Tsjui K, Brusseau GA, Wackett LP. 1990. Biodegradation of low-molecular weight halogenated hydrocarbons by methanotrophic bacteria. *FEMS Microbiol. Rev.* 87:273–78
75. Hardin G. 1960. The competitive exclusion principle. *Science* 131:1292–97
76. Harper HJ. 1939. The effect of natural gas on the growth of microorganisms and the accumulation of nitrogen and organic matter. *Soil Sci.* 48:461–66
77. Harriss RC, Gorham E, Sebacher DI, Bartlett KB, Flebbe PA. 1985. Methane flux from northern peatlands. *Nature* 315:652–53
78. Harriss RC, Sebacher DI, Day FP Jr. 1982. Methane flux in the Great Dismal Swamp. *Nature* 297:673–74
79. Harrits SM, Hanson RS. 1980. Stratification of aerobic methane-oxidizing organisms in Lake Mendota, Madison, Wisconsin. *Limnol. Oceanogr.* 25:412–21
80. Heyer J. 1977. Results of enrichment experiments of methane-assimilating organisms from an ecological point of view. In *Microbial Growth on C1 Compounds*, ed. GA Skryabin, MB Ivanov, EN Kondratjeva, GA Zavarzin, YuA

Trotsenko, AI Netrosev, pp. 19–21. Rusching: USSR Acad. Sci.
81. Heyer J, Malaschenko Y, Berger U, Budkova E. 1984. Verbreitung methanotropher Bakterien. *Z. Allg. Mikrobiol.* 24:725–44
82. Heyer J, Suchow R. 1985. Ökologische untersuchungenden methanoxydation in einem sauren Moorsee. *Limnologica* 16:247–66
83. Higgins IJ. 1979. Microbial biochemistry. *Int. Rev. Biochem.* 21:300–53
84. Higgins LJ, Best DJ, Hammond RC, Scott D. 1981. Methane oxidizing microorganisms. *Microbiol. Rev.* 45:556–90
85. Higgins LJ, Best DJ, Scott D. 1981. Hydrocarbon oxidation by *Methylosinus trichosporium*, metabolic implications of the lack of specificity of methane mono-oxygenase. In *Microbial Growth on C1 Compounds*, ed. H Dalton, pp. 11–20. London: Heydon & Son
86. Hoeks J. 1972. Changes in composition of soil air near leaks in natural gas mains. *Soil Sci.* 113:46
87. Hoeks J. 1972. Effect of leaking gas on soil and vegetation in urban areas. *Dutch Agric. Res. Rep. 778.* Wageningen, The Netherlands. 120 pp.
88. Hou CT, Patel R, Laskin AI, Barnabe N, Marczak I. 1979. Microbial oxidation of gaseous hydrocarbons: production of methyl ketones from their corresponding secondary alcohols by methane- and methanol-grown microbes. *Appl. Environ. Microbiol.* 38:135–42
89. Hubley JH, Thomson AW, Wilkinson JF. 1975. Specific inhibitors of methane oxidation in *Methylosinus trichosporium*. *Arch. Microbiol.* 102:199–202
90. Hütsch BW, Webster CP, Powison DS. 1993. Long-term effects of nitrogen fertilization on methane oxidation in soil of the Broadbalk wheat experiment. *Soil Biol. Biochem.* 10:1307–15
91. Hutton WE, Zobell CE. 1949. Occurrence and characteristics of methane-oxidizing bacteria in marine sediments. *J. Bacteriol.* 58:463–73
92. Hutton WE, Zobell CE. 1953. Production of nitrite from ammonia by methane oxidizing bacteria. *J. Bacteriol.* 65:216–19
93. Ivanov MV, Belayev SS, Lauinavichus SS. 1976. Methods of quantitative investigation of microbiological production and utilization of methane. See Ref. 163, pp. 63–67
94. Iversen N, Blackburn TH. 1979. Methane production and oxidation in Santa Barbara Basin sediments. *Estuarine Coastal Mar. Sci.* 8:379–85
95. Iversen N, Jørgenson BB. 1985. Anaerobic methane oxidation rates at the sulfate-methane transition in marine sediments from Kattegat and Skagerrak (Denmark). *Limnol. Oceanogr.* 30:944–55
96. Iversen N, Oremland RS, Klug MJ. 1987. Big Soda Lake (Nevada). 3. Pelagic methanogenesis and anaerobic methane oxidation. *Limnol. Oceanogr.* 32:804–18
97. Jannasch HW. 1975. Methane oxidation in Lake Kivu (Central Africa). *Limnol. Oceanogr.* 20:860–64
98. Joergenson L. 1985. Methane oxidation by *Methylosinus trichosporium* measured by membrane inlet mass spectrometry. In *Microbial Gas Metabolism*, ed. RK Poole, CS Dow, pp. 287–94. New York: Academic
99. Jollie DR, Lipscomb JD. 1991. Formate dehydrogenase from *Methylosinus trichosporium* OB3b: purification and spectroscopic characterization of the cofactors. *J. Biol. Chem.* 266:21853–63
100. Jones HA, Nedwell DB. 1993. Methane emission and methane oxidation in landfill cover soil. *FEMS Microbiol. Ecol.* 102:185–95
101. Juma NG, Paul EA. 1983. Effect of a nitrification inhibitor on N immobilization and release of ^{15}N from nonexchangeable ammonium and microbial biomass. *Can. J. Soil Sci.* 63:167–75
102. Keller M, Goreau TJ, Wofsy SC, Kaplan WA, McElroy MB. 1983. Production of nitrous oxide and consumption of methane by forest soil. *Geophys. Res. Lett.* 10:1156–59
103. Keller M, Kaplan WA, Wofsy SC. 1986. Emission of N_2O, CH_4 and CO_2 from tropical forest soils. *J. Geophys. Res.* 91(D11):11791–802
104. Keller M, Mitre ME, Stallard RF. 1990. Consumption of atmospheric methane in tropical soils of Central Panama. *Glob. Biogeochem. Cycles* 4:21–27
105. King GM. 1990. Regulation by light of methane emissions from a wetland. *Nature* 345:513–15
106. King GM. 1990. Dynamics and controls of methane oxidation in a Danish wetland sediment. *FEMS Microbiol. Ecol.* 74:309–23
107. King GM. 1992. Ecological aspects of methane oxidation, a key determinant of global methane dynamics. In *Advances in Microbial Ecology*, ed. KC Marshall, pp. 431–48. New York: Plenum
108. King GM. 1994. Associations of methanotrophs with the roots and rhizomes

109. of aquatic vegetation. *Appl. Environ. Microbiol.* 60:3220–27
109. King GM, Adamsen APS. 1992. Effects of temperature on methane consumption in a forest soil and in pure cultures of the methanotroph *Methylomonas rubra*. *Appl. Environ. Microbiol.* 58:2758–63
110. King GM, Roslev P, Adamsen APS. 1991. Controls of methane oxidation in a Canadian wetland and forest soils. *Trans. Am. Geophys. Union* 72:79
111. King GM, Roslev P, Skovgaard H. 1990. Distribution and rate of methane oxidation in sediments of the Florida Everglades. *Appl. Environ. Microbiol.* 56:2902–11
112. King GM, Schnell S. 1994. Effect of increasing atmospheric methane concentration on ammonium inhibition of soil methane consumption. *Nature* 370:282–84
113. King GM, Schnell S. 1994. Ammonium and nitrite inhibition of methane oxidation by *Methylobacter albus* BG8 and *Methylosinus trichosporium* OB3b at low methane concentrations. *Appl. Environ. Microbiol.* 60:3508–13
114. King SL, Quay PD, Lansdown JM. 1989. The $^{13}C/^{12}C$ kinetic isotope effect for soil oxidation of methane at ambient atmospheric concentrations. *J. Geophys. Res.* 94(D):18273–77
114a. Klug MJ, Reddy CA, eds. 1984. *Current Perspectives in Microbial Ecology*. Washington, DC: Am. Soc. Microbiol.
115. Kowalenko CG. 1978. Nitrogen transformations and transport over 17 months in field fallow microplots using ^{15}N. *Can. J. Soil Sci.* 58:69–76
116. Large PJ. 1983. *Methylotrophy and Methanogenesis*. Washington, DC: Am. Soc. Microbiol. 88 pp.
117. Leadbetter ER, Foster JW. 1958. Studies on some methane-utilizing bacteria. *Arch. Mikrobiol.* 30:91–118
118. Leak DJ, Dalton H. 1986. Growth yields of methanotrophs. 2. A theoretical analysis. *Appl. Microbiol. Biotechnol.* 23:477–81
119. Lidstrom ME. 1983. Methane consumption in Framvaren, and anoxic marine fjord. *Limnol. Oceanogr.* 28:1247–51
120. Lidstrom ME. 1988. Isolation and characterization of marine methanotrophs. *Antonie van Leeuwenhoek J. Microbiol. Serol.* 54:189–200
121. Lidstrom ME, Stirling DI. 1990. Methylotrophs: genetics and commercial applications. *Annu. Rev. Microbiol.* 44:27–58
122. Lipscomb JD. 1994. Biochemistry of the soluble methane monooxygenase. *Annu. Rev. Microbiol.* 48:371–99

123. Lu A-H, Kunz CO. 1981. Gas-flow model to determine methane production at sanitary landfills. *Environ. Sci. Technol.* 15:436–40
124. Mancinelli RL, McKay CP. 1985. Methane-oxidizing bacteria in sanitary landfills. In *First International Symposium on Biotechnological Advances in Processing Municipal Wastes for Fuels and Chemicals*, ed. A Antonopoulos, pp. 438–50. New York: Noyes
125. Mancinelli RL, Shulls WA, McKay CP. 1981. Methanol-oxidizing bacteria used as an index of soil methane content. *Appl. Environ. Microbiol.* 42:70–73
126. Mancinelli RL, White MR, Bogner J. 1991. Microbial methane oxidation. In *Rio Vista Gas Leak Study, Argonne Natl. Lab. Topical Rep., April 1989–January 1991*, ed. PL Wilkey. Argonne, IL: Argonne Natl. Lab.
127. Martens CS, Berner RA. 1977. Interstitial water chemistry of Long Island Sound sediments. I. Dissolved gases. *Limnol. Oceanogr.* 22:10–25
128. McInerney MJ. 1986. Transient and persistent associations among prokaryotes. In *Bacteria in Nature*, ed. JS Poindexter, ER Leadbetter, 2:334–39. New York: Plenum
129. Megraw SR, Knowles R. 1987. Methane production and consumption in a cultivated humisol. *Biol. Fertil. Soils* 5:56–60
130. Megraw SR, Knowles R. 1989. Methane-dependent nitrate production by a microbial consortium enriched from a cultivated humisol. *FEMS Microbiol. Ecol.* 62:359–66
131. Megraw SR, Knowles R. 1989. Isolation, characterization, and nitrification potential of a methylotroph and two heterotrophic bacteria from a consortium showing methane-dependent nitrification. *FEMS Microbiol. Ecol.* 62:367–74
132. Miller FC. 1989. Matric water potential as an ecological determinant in compost, a substrate dense system. *Microb. Ecol.* 18:59–71
133. Mosier A, Schimel D, Valentine D, Bronson K, Parton W. 1991. Methane and nitrous oxide fluxes in native, fertilized and cultivated grasslands. *Nature* 350:330–32
134. Mountfort DO, White D, Asher RA. 1990. Oxidation of lignin-related aromatic alcohols by cell suspensions of *Methylosinus trichosporium*. *Appl. Environ. Microbiol.* 56:245–49
134a. Murrell JC, Kelly DP, eds. 1993. *Microbial Growth on C1 Compounds*. Andover, UK: Intercept

135. Nagiub M. 1970. On methane-oxidizing bacteria in freshwaters. I. Introduction to the problem and investigations on the presence of obligate methane-oxidizers. *Z. Allg. Mikrobiol.* 10:17–36
136. Nannipieri P, Pedrazzini F, Arcara P, Piovanelli C. 1979. Changes in amino acids, enzyme activities and biomasses during soil microbial growth. *Soil Sci.* 127:24–26
137. Nesbit SP, Breitenbeck GA. 1992. A laboratory study of factors influencing methane uptake by soil. *Agric. Ecosyst. Environ.* 41:39–54
138. Nesterov AI, Mshensky YuN, Gachenko VF, Namsaraev BB, Ilchenko VYu. 1977. A comparative study on parameters of growth of methanotrophic bacteria. *Mikrobiologiya* 46:10–14
139. O'Neil JG, Wilkinson JF. 1977. Oxidation of ammonia by methane-oxidizing bacteria and the effects of ammonia on methane oxidation. *J. Gen. Microbiol.* 100:407–12
140. Oldenhuis R, Oedzes JY, van der Waarde JJ, Janssen DB. 1991. Kinetics of chlorinated hydrocarbon degradation by *Methylosinus trichosporium* OB3b and toxicity of trichloroethylene. *Appl. Environ. Microbiol.* 57:7–14
141. Oldenhuis R, Vink RLJM, Janssen DB, Witholt B. 1989. Degradation of chlorinated aliphatic hydrocarbons by *Methylosinus trichosporium* OB3b expressing soluble methane monooxygenase. *Appl. Environ. Microbiol.* 55:2819–26
142. Ormeland RS, Culbertson CW. 1992. Importance of methane-oxidizing bacteria in the methane budget as revealed by the use of a specific inhibitor. *Nature* 356:421–23
143. Palmer RJ Jr, Friedmann EI. 1990. Water relations and photosynthesis in the cryptoendolithic microbial habitat of hot and cold deserts. *Microb. Ecol.* 19:111–18
144. Panganiban AT, Patt TE, Harr W, Hanson RS. 1979. Oxidation of methane in the absence of oxygen in lake water samples. *Appl. Environ. Microbiol.* 37:303–9
145. Patel RN, Hou CT, Felix A. 1978. Microbial oxidation of methane and methanol: isolation of methane-utilizing bacteria and characterization of a facultative methane-utilizing isolate. *J. Bacteriol.* 136:352–58
146. Patt TE, Cole GC, Bland J, Hanson RS. 1974. Isolation and characterization of bacteria that grow on methane and organic compounds as sources of carbon and energy. *J. Bacteriol.* 120:955–64
147. Patt TE, Cole GC, Hanson RS. 1976. *Methylobacterium*, a new genus of facultatively methylotrophic bacteria. *Int. J. Syst. Bacteriol.* 26:226–29
148. Pearce F. 1989. Methane: the hidden greenhouse gas. *New Sci.* 122:37–41
149. Proctor HM, Norris JR, Ribbons DW. 1969. Fine structure of methane utilizing bacteria. *J Appl. Bacteriol.* 32:118–21
150. Quayle JR. 1972. The metabolism of one carbon compounds by micro-organisms. *Adv. Microb. Physiol.* 7:119–203
151. Reeburgh WS. 1976. Methane consumption in Cariaco Trench waters and sediments. *Earth Planet. Sci. Lett.* 28:337–49
152. Reeburgh WS. 1980. Anaerobic methane oxidation rate depth distribution in Skan Bay sediments. *Earth Planet. Sci. Lett.* 47:345–52
153. Reeburgh WS. 1982. A major sink and flux control for methane in marine sediments: anaerobic consumption. In *The Dynamic Environment of the Ocean Floor*, ed. KA Fanning, FT Manheim, pp. 203–17. Lexington, MA: Heath
154. Reeburgh WS, Heggie DT. 1977. Microbial methane consumption reactions and their effect on methane distributions in fresh water and marine environments. *Limnol. Oceanogr.* 22:1–9
155. Robertson K, Schnürer J, Clarholm M, Bonde TA, Rosswall T. 1988. Microbial biomass in relation to C and N mineralization during laboratory incubations. *Soil Biol. Biochem.* 20:281–86
156. Romanovskoya VA. 1978. Nomenclature of obligate methylotrophs. *Microbiology USSR* 47:1063–72
157. Romanovskoya VA, Sadovmikov YS, Malshenko YR. 1978. Formation of taxons of methane-oxidizing bacteria using numerical analysis methods. *Microbiology USSR* 47:120–30
158. Rosenzwieg AC, Frederick CA, Lippard SJ, Nordlund P. 1993. Crystal structure of a bacterial nonheme iron hydroxylase that catalyzes the biological oxidation of methane. *Nature* 366:537–43
159. Roy R, Knowles R. 1994. Effects of methane metabolism on nitrification and nitrous oxide production in polluted freshwater sediment. *Appl. Environ. Microbiol.* 60:3307–14
160. Rudd JWM, Hamilton RD. 1975. Methane cycling in eutrophic shield lake and its effects on whole lake metosism. *Limnol. Oceanogr.* 23:337–48
161. Rudd JWM, Taylor CD. 1980. Methane cycling in aquatic environments. *Adv. Aquat. Microbiol.* 2:77–150
162. Sahores JJ, Witherspoon PA. 1970. Diffusion of light paraffin hydrocarbons in

water from 2°C–80°C. In *Advances in Organic Geochemistry*, ed. GC Spears, pp. 219–30. New York: Pergamon
163. Schlegel HG, Gottschalk G, Pfennig N, eds. 1976. *Symposium on Microbial Production and Utilization of Gases (H_2, CH_4, CO)*. Gottingen: Akad. Wissenschafter
164. Schnell S, King GM. 1994. Mechanistic analysis of ammonium inhibition of atmospheric methane consumption in forest soils. *Appl. Environ. Microbiol.* 60:3514–21
165. Schnürer J, Clarholm M, Boström S, Rosswall T. 1986. Effects of moisture in soil microorganisms and nematodes: a field experiment. *Microb. Ecol.* 12:217–30
166. Schnürer J, Clarholm M, Rosswall T. 1982. Microorganisms. In *Ecology of Arable Land. The Role of Organisms in Nitrogen Cycling. Progress Report 1981*, ed. T. Rosswall, pp. 77–88. Uppsala: Swedish Univ. Agric.
167. Schollenberger CJ. 1930. Effect of leaking natural gas upon the soil. *Soil. Sci.* 29:261–66
168. Scott D, Brannan J, Higgins IJ. 1981. The effect of growth conditions on intracytoplasmic membranes and methane mono-oxygenase activities in *Methylosinus trichosporium* OB3b. *J. Gen. Microbiol.* 125:63–72
169. Seiler W. 1984. Contributions of biological processes to the global budget of CH_4 in the atmosphere. See Ref. 114a, pp. 468–77
170. Seiler W, Conrad R, Scharffe D. 1984. Field studies of methane emission from termite nests into the atmosphere and measurements of methane uptake by tropical soils. *J. Atmos. Chem.* 1:171–86
171. Shilo M. 1984. *Bdellovibrio* as a predator. See Ref. 114a, pp. 334–39
172. Smirnova ZS. 1971. Efficiency of the utilization of free energy by methane-oxidizing bacteria. *Mikrobiologiya* 40:5–7
173. Smirnova ZS. 1971. The material balance of methane oxidation by microorganisms. *Dokl. Akad. Nauk. SSSR Ser. Biol.* 3:423–27
174. Smith AJ, Hoare DS. 1977. Specialist phototrophs, lithotrophs, and methylotrophs: a unity among a diversity of procaryotes? *Bacteriol. Rev.* 41:419–48
175. Smith HL. 1981. Competitive coexistence in an oscillating chemostat. *SIAM J. Appl. Math.* 42:27–43
176. Smith HL, Waltman B. 1991. The gradostat: a model of competition along a nutrient gradient. *Microb. Ecol.* 22:207–26
177. Söhngen NL. 1906. Ueber Bakterien welche Methan als Koklenstoffnakrung und Energiequelle gebrauchen. *Zentralbl. Bakteriol. Parasitenkd. Infektionskr. Hyg.* 15:513–17
178. Stainthorpe A, Murrell J, Salmond G, Dalton H, Lees V. 1989. Molecular analysis of methane monooxygense from *Methylococcus capsulatus* (Bath). *Arch. Microbiol.* 152:154–59
179. Stanley SH, Prior SD, Leak DJ, Dalton H. 1983. Copper stress underlies the fundamental change in intracellular location of methane monooxygenase in methane-oxidizing organisms: studies in batch and continuous cultures. *Biotechnol. Lett.* 5:487–92
180. Stauffer B, Lochbronner E, Oeschger H, Schwander J. 1988. Methane concentration in the glacial atmosphere was only half that of the preindustrial Holocene. *Nature* 332:812–13
181. Stephanopoulos GN, Aris R, Fredrickson AG. 1979. A stochastic analysis of the growth of competing microbial populations in a continuous biochemical reactor. *Math. Biosci.* 45:99–35
182. Stephanopoulos GN, Fredrickson AG, Aris R. 1979. The growth of competing microbial populations in a CSTR with periodically varying inputs. *AIChE J.* 25:863–72
183. Stephens RL, Haygood MG, Lidstrom ME. 1988. Identification of putative methanol dehydrogenase (*mox F*) structural genes in methylotrophs and cloning of *moxF* genes from *Methylococcus capsulatus* Bath and *Methylomonas albus* BG8. *J. Bacteriol.* 170:2063–69
184. Steudler PA, Bowden RD, Melillo JM, Aber JD. 1989. Influence of nitrogen fertilization on methane uptake in temperate forest soils. *Nature* 341:314–16
185. Stirling DI, Dalton H. 1979. Properties of the methane mono-oxygenase from extracts of *Methylosinus trichosporium* OB3b and evidence for its similarity to the enzyme from *Methylococcus capsulatus* (Bath). *Eur. J. Biochem.* 96:205–12
186. Striegl RG, McConnaughey TA, Thorstenson DC, Woodward JC. 1992. Consumption of atmospheric methane by desert soils. *Nature* 357:145–47
187. Sundh I, Nilsson M, Ganberg G, Svensson BH. 1994. Depth distribution of microbial production and oxidation of methane in northern boreal peatlands. *Microb. Ecol.* 27:253–65
188. Tonge GM, Harrison DEF, Higgins IJ. 1977. Purification and properties of the methane mono-oxygenase enzyme sys-

tem from *Methylosinus trichosporium* OB3b. *Biochem. J.* 161:333–44
189. Toukdarian AE, Lidstrom ME. 1984. DNA hybridization analysis of the *nif* region of *Methylosinus* 6. *J. Bacteriol.* 157:925–30
190. Trotsenko YuA. 1983. Metabolic features of methane and methanol utilizing bacteria. *Acta Biotechnol.* 3:269–77
191. Tsien HC, Brusseau GA, Hanson RS, Wackett LP. 1989. Biodegradation of trichloroethylene by *Methylosinus trichosporium* OB3b. *Appl. Environ. Microbiol.* 55:3155–61
192. Tsuji K, Tsien HC, Hanson RS, De Palma SR, Scholtz R, LaRoche S. 1990. 16S ribosomal RNA sequence analysis for determination of phylogenetic relationship among methylotrophs. *J. Gen. Microbiol.* 136:1–10
193. Udden JA. 1919. Oil-bearing formations in Texas. *Am. Assoc. Pet. Geol. Bull.* 3:82–98
194. Umorin PP, Ermolaev LS. 1986. The interrelations between algae and bacteria in methane oxidation. *Ekologiya* 4:23–27
195. Vogels GD. 1979. The global cycle of methane. *Antonie van Leeuwenhoek J. Microbiol. Serol.* 45:347–52
196. Wake LV, Rickard P, Ralph BJ. 1973. Isolation of methane utilizing micro-organisms: a review. *J. Bacteriol.* 26:92–99
197. Ward BB, Kilpatrick KA, Novelli PC, Scranton MI. 1987. Methane oxidation and methane fluxes in the ocean surface layer and deep anoxic waters. *Nature* 327:226–29
198. Whalen RT, Reeburgh WS. 1988. A methane flux time series for tundra environments. *Glob. Biogeochem. Cycles* 2:399–409
199. Whalen RT, Reeburgh WS. 1990. Consumption of atmospheric methane by tundra soils. *Nature* 346:160–62
200. Whalen SC, Reeburgh WS, Sandbeck KA. 1990. Rapid methane oxidation in a landfill cover soil. *Appl. Environ. Microbiol.* 56:3405–11
201. Whittenbury R, Colby J, Dalton H, Reed HC. 1976. Biology and ecology of methane-oxidizers. See Ref. 163, pp. 281–92
202. Whittenbury R, Dalton H. 1981. The methylotrophic bacteria. In *The Prokaryotes*, ed. MP Starr, H Stolp, H Truper, A Balows, HG Schlegel, pp. 894–902. Berlin: Springer-Verlag
203. Whittenbury R, Davies SL, Davey JF. 1970. Exospores and cysts formed by methane-utilizing bacteria. *J. Gen. Microbiol.* 61:219–26
204. Whittenbury R, Kelly DP. 1977. Autotrophy: a conceptual phoenix. In *Microbial Energetics. Symp. Soc. Gen. Microbiol., 17th*, ed. DA Haddock, WA Hamilton, pp. 121–49. Cambridge: Cambridge Univ. Press
205. Whittenbury R, Krieg NR. 1984. Methanococcaceae. In *Bergey's Manual of Systematic Bacteriology*, ed. NR Krieg, 1:256–61. Baltimore: Williams & Wilkins
206. Whittenbury R, Phillips KC, Wilkinson JF. 1970. Enrichment, isolation and some properties of methane-utilizing bacteria. *J. Gen. Microbiol.* 61:205–18
207. Wiggins PM. 1990. Roll of water in some biological processes. *Microbiol. Rev.* 54:432–49
208. Woese CR. 1987. Bacterial evolution. *Microbiol. Rev.* 51:221–71
209. Wolf HJ, Christiansen M, Hanson RS. 1980. Ultrastructure of methanotrophic yeasts. *J. Bacteriol.* 141:1340–49
210. Wolf HJ, Hanson HS. 1979. Isolation and characterization of methane-utilizing yeasts. *J. Gen. Microbiol.* 114:187–94
211. Wolfe RS, Higgins IJ. 1979. Microbial biochemistry of methane—a study in contrasts. *Int. Rev. Biochem.* 21:267–353
212. Wolfrum T, Stolp H. 1987. Comparative studies on 5S RNA sequences of RuMP-type methylotrophic bacteria. *Syst. Appl. Microbiol.* 9:273–76
213. Yavitt JB, Downey DM, Lancaster E, Lang GE. 1990. Methane consumption in decomposing sphagnum-derived peat. *Soil Biol. Biochem.* 22:441–47
214. Yavitt JB, Downey DM, Lang GE, Sextone AJ. 1990. Methane consumption in two temperature forest soils. *Biogeochemistry* 9:39–52
215. Yoshinari T. 1985. Nitrite and nitrous oxide production by *Methylosinus trischosporium*. *Can. J. Microbiol.* 31:139–44
216. Zahn JA, Duncan C, Dispitito AA. 1994. Oxidation of hydroxylamine by cytochrome P-460 of the obligate methylotroph *Methylococcus capsulatus* (Bath). *J. Bacteriol.* 176:5879–87

DISCOVERY, BIOSYNTHESIS, AND MECHANISM OF ACTION OF THE ZARAGOZIC ACIDS: Potent Inhibitors of Squalene Synthase

James D. Bergstrom, Claude Dufresne, Gerald F. Bills, Mary Nallin-Omstead, and Kevin Byrne

Merck Research Laboratories, Rahway, New Jersey 07065-0900

KEY WORDS: natural products, squalene synthase inhibitors, cholesterol biosynthesis inhibitors, fungi, antifungals

CONTENTS

INTRODUCTION AND DISCOVERY	608
Rationale	608
Screening	609
Discovery of the Zaragozic Acids	610
ZARAGOZIC ACIDS, STRUCTURE AND DIVERSITY	612
Isolation	612
Structural Features, NMR, and Mass Spectroscopy	613
Nomenclature	615
The Zaragozic Acid Family Groups	615
ZARAGOZIC ACID–PRODUCING FUNGI	618
BIOSYNTHESIS OF THE ZARAGOZIC ACIDS	623
Biosynthesis of Zaragozic Acid A	624
Biosynthesis of Zaragozic Acid C	628
Directed Biosynthesis	629
Minor Metabolites in the Biosynthetic Scheme	629
MECHANISM OF ACTION	631
Inhibition of Squalene Synthesis	631
Effects on Other Enzymes	631
Activities in Cells	632
Antifungal Activity	633
In Vivo Activity	634
CONCLUSIONS	635

Abstract

The zaragozic acids (ZAs), a family of fungal metabolites containing a novel 4,6,7-trihydroxy-2,8-dioxobicyclo[3.2.1]octane-3,4,5-tricarboxylic acid core, were discovered independently by two separate groups screening natural product sources to discover inhibitors of squalene synthase. This family of compounds all contain the same core but differ in their 1-alkyl and their 6-acyl side chains. Production of the ZAs is distributed over an extensive taxonomic range of Ascomycotina or their anamorphic states. The zaragozic acids are very potent inhibitors of squalene synthase that inhibit cholesterol synthesis and lower plasma cholesterol levels in primates. They also inhibit fungal ergosterol synthesis and are potent fungicidal compounds. The biosynthesis of the zaragozic acids appears to proceed through alkyl citrate intermediates and new members of the family have been produced through directed biosynthesis. These potent natural product based inhibitors of squalene synthase have potential to be developed either as cholesterol lowering agents and/or as antifungal agents.

INTRODUCTION AND DISCOVERY

Rationale

The discovery of the zaragozic acids (ZAs), a family of fungal metabolite inhibitors of squalene synthase (an enzyme of the sterol biosynthetic pathway), resulted from a concerted effort to screen natural-product sources for novel inhibitors of cholesterol biosynthesis. Inhibitors of cholesterol biosynthesis can be useful therapeutic agents for the treatment of hypercholesterolemia and can also be effective antifungal agents. Lovastatin (3) and other inhibitors of HMG-CoA reductase (2, 41), an early step in the biosynthesis of cholesterol, are cholesterol-lowering agents widely prescribed for human use. Other inhibitors of cholesterol biosynthesis, such as L-659,699 (an inhibitor of HMG-CoA synthase) (47), NB-598 (an inhibitor of squalene epoxidase) (51) and triparinol (an inhibitor of the Δ^{24} reductase involved in the conversion of desmosterol to cholesterol) (22, 23), are also effective cholesterol-lowering agents in humans and/or animals. Inhibitors of squalene epoxidase, oxidosqualene cyclase, and lanosterol demethylation (1, 10) are used as antifungal agents in treating human fungal infections and in crop protection. Therefore, an inhibitor of any step of the cholesterol biosynthetic pathway can be an effective cholesterol-lowering agent and/or an effective antifungal agent.

Sterol synthesis can be divided into three stages: the first or early stage at which acetyl-CoA is converted to mevalonate, the middle stage at which mevalonate is converted to squalene, and the late stage at which squalene is cyclized to lanosterol and then converted to the final sterol product, i.e. cholesterol, ergosterol, or sitosterol. Inhibitors of each of the early-stage enzymes (3, 46, 47) and many of the later-stage enzymes (1, 10, 22, 23, 51) that could block cholesterol synthesis in cultured cells and/or in animals had been described by the late 1980s. However, with the exception of fluoromevalonate (73), inhibitors of middle stages of sterol synthesis that would work in cells were unknown. Squalene synthase, the last enzyme of the middle part of the pathway, is a particularly attractive target for the development of a cholesterol-lowering agent because it is the first enzyme in the pathway that commits its products to sterol biosynthesis. The mammalian isoprenoid biosynthesis pathway produces several important compounds other than sterols, including dolichols, ubiquinones, and the farnesol and geranylgeranyl side chains of prenylated proteins. The synthesis of these essential isoprenoids branches from the sterol synthesis pathway at or before the synthesis of farnesyl-pyrophosphate (F-PP), the substrate for squalene synthase. Thus, inhibitors of squalene synthase should be selective inhibitors of sterol biosynthesis that should not affect the synthesis of these other essential isoprenoids.

The search among natural-product sources for inhibitors of sterol biosynthesis has been very fruitful, yielding potent inhibitors of thiolase (46), HMG-CoA synthase (47), and HMG-CoA reductase (2, 3, 40, 41). Lovastatin, the leading cholesterol-lowering agent worldwide, is a fungal metabolite. Because sterol synthesis inhibitors can have antifungal and growth-regulatory properties, the biosynthesis of these compounds may have evolved to take advantage of these activities. Past discoveries of other inhibitors of sterol synthesis indicated that natural-product screening had a good chance for success in the search to discover inhibitors of the middle steps of cholesterol biosynthesis and in particular inhibitors of squalene synthase.

Screening

The above rationale implicating the importance of squalene synthase as a possible target for the development of cholesterol-lowering agents was apparent to many scientists, but principally those working at pharmaceutical companies. Over the past five years, these scientists have published reports on and patented numerous structurally diverse compounds that inhibit this enzyme (for review, see 1). At least two of these groups, at Merck and Co., Inc., and at Glaxo, Ltd., discovered natural-product inhibitors of this enzyme. Large-scale natural-product screening is best done with a robust biological assay that has a large signal-to-noise ratio and that can be readily automated. Each of these groups developed such an assay.

The assay developed at Merck not only looked for squalene synthase inhibitors, but also would detect inhibitors of any of the enzymatic steps between mevalonate and squalene. This assay used ^{3}H-mevalonate and followed its conversion into heptane-extractable compounds (which would include squalene and all its subsequent metabolites) via a 20,000-g supernatant of a rat liver homogenate supplemented with ATP, Mg^{2+}, and NADPH. The 20,000-g supernatant contains all of the cytosolic enzymes for the conversion of mevalonate to F-PP and the microsomal squalene synthase. An inhibitor of squalene synthase and any of the other five enzymes involved in the conversion of mevalonate to F-PP would prevent the production of heptane-extractable ^{3}H-labeled metabolites. The assay was readily automated for use with a pipetting station (52), and large-scale screening of microbial-fermentation extracts was performed at Centro de Investigaciones Básica de España, a Merck facility in Madrid, Spain. Secondary assays on active cultures were used to determine which of the six enzymes involved were inhibited by the broth extract. The rat-liver enzymes were initially used for screening, but later similar preparations from yeast or HEP-G2 cells were used. Several hundred thousand broth extracts of fungal and bacterial origin were tested in this assay.

RM Tait, a scientist at Glaxo (78), developed a microtiter plate assay using a rat-liver squalene synthase assay based on separating the ^{14}C-squalene produced in the assay from the ^{14}C-F-PP substrate by using a spot wash assay on polyester-backed silica gel TLC sheets. ^{14}C-squalene adheres to the sheets whereas the ^{14}C-F-PP is washed away.

Discovery of the Zaragozic Acids

Within a few months of initiating screening, the Merck scientists discovered four fungi that produced fermentation products that potently and specifically inhibited squalene synthase (12) (see Table 1). Simultaneous purification of the active components from the first three of these broths soon revealed that the active component of the two *Sporormiella intermedia* isolates contained the same major compound and only one of them, MF 5447, was investigated further. The active components of cultures MF 5447 and MF 5453 were then isolated in around 1-mg quantities (isolation procedures are reviewed in the next section), and structural determinations were started. The proton NMR showed that the compounds were closely related. The initial structural characterization of these 1-mg samples gave enough information to resolve the

Table 1 Organisms initially identified as producers of squalene synthase

MF No.	Fungi	Source
5447	Sporormiella intermedia	Cotton tail rabbit dung, Arizona
5446	Sporormiella intermedia	Big horn sheep dung, Arizona
5453	Sterile mycelia	Water sample, Zaragoza, Spain
5465	Leptodontidium elatius	Dead wood, North Carolina

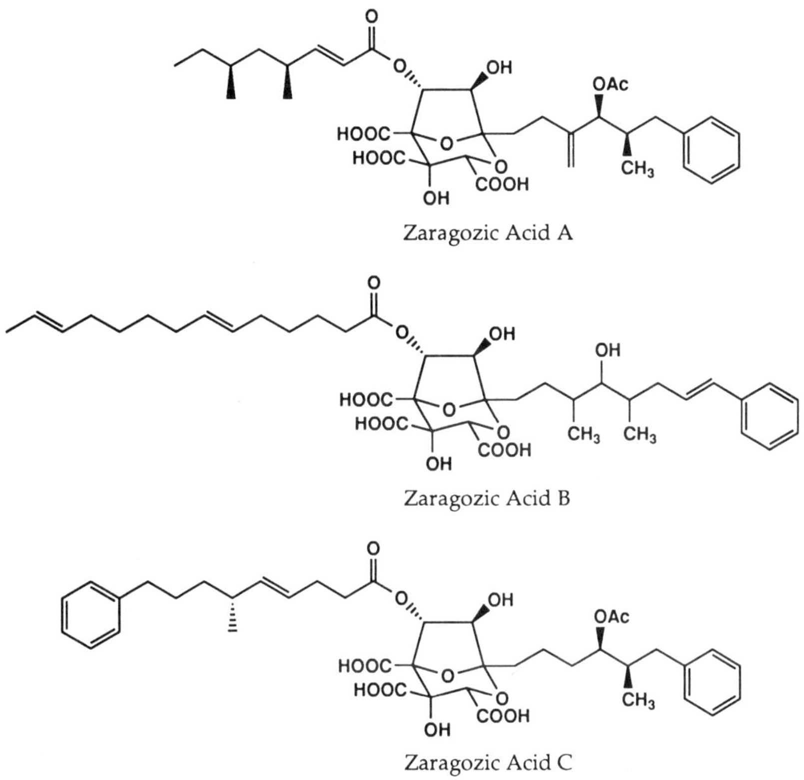

Figure 1 Structures of the first three ZAs discovered.

structure of the side arms, but more material was needed to resolve the core structure.

The original fermentation of culture MF 5453 was in liquid media, whereas the original fermentation culture MF 5447 was in solid media. The liquid fermentation of culture MF 5453 was readily scaled up, allowing the purification of enough of the active component, zaragozic acid A (ZA-A), to fully resolve the structure of the core (see Figure 1). With the core structure known, the structure of the active component of the *S. intermedia* fermentation, zaragozic acid B (ZA-B), (Figure 1) was rapidly resolved. Very soon after that, zaragozic acid C (ZA-C), the active component from a solid fermentation of *Leptodontidium elatius,* was obtained (Figure 1) (15). The isolation of these three compounds marked the discovery of a novel family, the ZAs, of closely related fungal metabolites that were potent inhibitors of squalene synthase.

Names for novel natural products are typically taken from the producing organism, and when the organism is not known, geographic names are often taken from the place the organism was collected. The first ZA whose structure was known came from an unidentified nonsporulating fungal culture, MF 5453, isolated from a water sample taken from the Jalon River in Zaragoza, Spain—hence the name zaragozic acids (12).

Meanwhile, the investigators at Glaxo had discovered a *Phoma* species that produced a potent natural-product inhibitor of squalene synthase (33). Purification of the major active components from this *Phoma* sp. gave three compounds called squalestatin I, II, and III (33). Squalestatin I is identical to ZA-A, and squalestatins II and III are equivalent to 4'-desacetyl-ZA-A (i.e. ZA-A1a) and 6-deacyl ZA-A (i.e. ZA-AX1), respectively.

The publication in the summer of 1991 of a US patent granted to several Merck scientists for the use of ZA-C as an antifungal agent introduced this class of compounds to the public realm (7). Subsequent publication by Merck scientists (6, 8, 11–15, 20, 38, 29, 50, 80) and scientists at Glaxo (9, 24, 33, 61, 72) showed that these groups had independently discovered the same class of novel fungal metabolites.

Work subsequent to the initial discovery of the ZAs has shown that the family contains numerous compounds produced by many members of the fungal kingdom. Subsequent sections in this review detail the diversity of the ZA family of compounds and their producers, the biosynthesis of the ZAs, and the biochemical mechanisms of action of the ZAs.

ZARAGOZIC ACIDS, STRUCTURE AND DIVERSITY

Isolation

SMALL-SCALE METHODS   Several isolation procedures for ZAs have been described (12, 37–39). Fermentation broth extracts were purified using bioassay-guided fractionations. The isolation of the initial small samples of ZA-A, -B, and -C were carried out using standard procedures of extraction, hydrophobic and ion exchange adsorption/elution, and reverse-phase HPLC. The ion-exchange step was performed using Dowex-1 resin (chloride cycle), in which the ZAs were adsorbed from the crude aqueous methanol broth extract at pH 4.5. In more recent isolation efforts, typically the whole fermentation broth is acidified to pH 2 and extracted with a solvent such as methyl ethyl ketone (MEK). The MEK extract is passed through an anion-exchange column equilibrated at pH 4.5, which results in the retention of strong acids. The column is then eluted with a low-pH solution. Active eluate fractions are then fractionated via reverse-phase high-performance liquid chromatography (HPLC).

The fermentation and isolation processes are monitored using HPLC analysis. Samples must be adjusted to pH 2 before injection to obtain reproducible results. Using a Dynamax C8 column (60 Å, 8 μm; 4.6 mm ID × 250 mm, with guard column) and eluting with a mobile phase of 60% CH_3CN/40% (0.1% H_3PO_4 in water) at a flow rate of 1 ml/min, ZA-A, -B, and -C elute with retention times of 13.4, 23.7, and 21.7 min, respectively.

PREPARATIVE METHODS Both the Merck and Glaxo groups developed efficient processes for the isolation of large quantities of ZAs. The Merck process recognizes the strongly acidic nature of the ZAs and makes use of a nonaqueous ion exchange step. The Glaxo process focuses on precipitating calcium salts from a partially purified extract (24).

We felt that the relatively strong acidic nature of the ZAs ($pK_a1 = ~3.5$) should be the key to a successful large-scale isolation process. Nonetheless, the amphipathic nature of the ZAs caused problems in early isolation work—specifically, poor chromatographic behavior and solubility limitations in aqueous solutions that resulted in only modest recoveries. This was most dramatically the case for ZA-B, for which an aqueous anion exchange (Dowex-1 resin) step gave at best modest recoveries and trailing elution, presumably because of backbone interactions. Anion exchange in nonaqueous media can be accomplished using Amberlyst resins. Amberlyst A-21 is a macroreticular weakly basic anion-exchange resin. Adsorption of ZA-B onto the resin (acetate cycle) from an EtOAc solution of the free acid proceeded smoothly. The acetate ion is basic enough to deprotonate ZA-B, a necessary step for ion-exchange retention. The best eluting conditions involved the use of high salt concentrations, specifically 3% ammonium chloride in 90% MeOH/water. The Amberlyst eluate was then desalted using Diaion HP-20. ZA-B, present in the HP-20 eluate as the ammonium salt, was converted to the free acid upon acidification and extraction into EtOAc. As a final purification step, precipitation of ZA-B as its ammonium salt gave a final product of better than 85% purity.

Structural Features, NMR, and Mass Spectroscopy

The very characteristic core bicyclic portion of the ZAs is the novel 4,6,7-trihydroxy-2,8-dioxobicyclo[3.2.1]octane-3,4,5-tricarboxylic acid (see Figure 2). The structure of the core, including relative stereochemistry, was determined using two-dimensional (2D) NMR, specifically heteronuclear mutliple band correlations (HMBC) and selJres experiments (50) and was confirmed by X-ray crystallography of a derivative (80). The absolute stereochemistry was established based on CD measurements on the bis(4-bromobenzoate) (80). Attached to this core are two side chains, termed the alkyl and the acyl side chains. This bicyclic core is common to almost all of the ZAs and thus facilitates their structural elucidation by NMR and mass spectrometry (MS) methods.

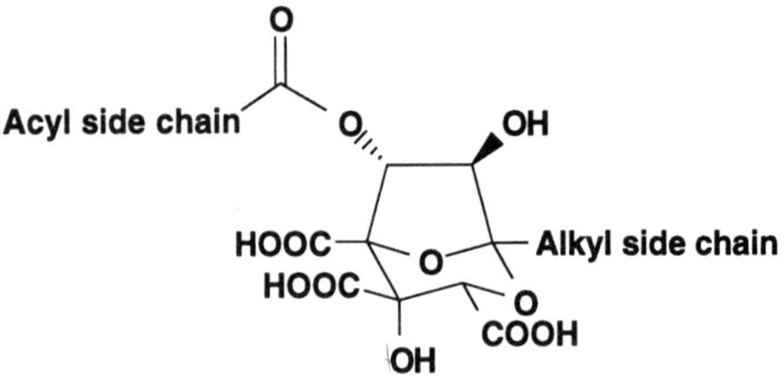

Figure 2 Zaragozic acid core.

NMR ANALYSIS The ^{1}H NMR spectra of the ZAs shows three constant features. The presence of 2-Hz doublets at 4.03 and 6.27 ppm, and of a singlet at 5.25 ppm, is characteristic of H-7, H-6, and H-3, respectively. As shown in Table 2, the ^{13}C NMR spectra also show almost invariable resonances for the carbon atoms of the bicyclic core. The structure of the side chains could be deduced readily from 2D NMR data.

MS ANALYSIS Being strong acids, the ZAs are readily analyzed using negative ion fast atom bombardment (FAB). In addition to the loss of a carboxylate group, loss of the acyl side chain almost always accompanies the loss of ions corresponding to both parting moieties (see Figure 3). Analytical methods for the analysis of ZAs by LC-MS have been described utilizing either dynamic liquid secondary-ion MS (58) or electrospray MS (59) techniques.

Table 2 Comparison of ^{1}H and ^{13}C NMR shifts for the bicyclic core of ZA-A, -B, and -C[a]

Carbon No.	A		B		C	
1	107.2		106.9		107.5	
3	76.6	5.24 s	76.8	5.25 s	76.7	5.26 s
4	75.6		75.8		75.7	
5	90.9		91.3		91.1	
6	81.1	6.23, d, 2 Hz	81.3	6.27, d, 2 Hz	81.0	6.30, d, 2 Hz
7	82.1	4.03, d, 2 Hz	82.7	4.06, d, 2 Hz	82.0	4.03, d, 2 Hz
8	170.1		170.3		170.3	
9	172.4		172.7		172.6	
10	168.5		168.7		168.6	

[a] In CD$_3$OD, 25°C, 300 MHz; s, Singlet, d, doublet.

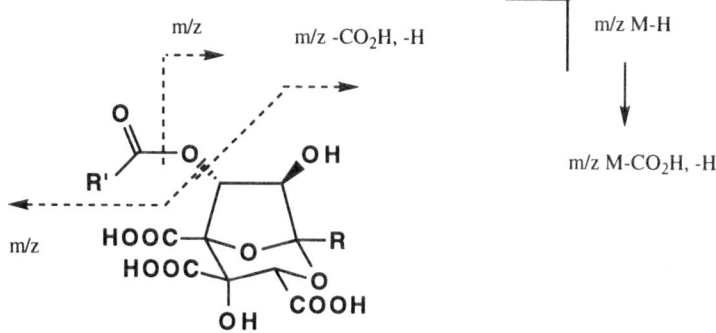

Figure 3 Mass spectral fragmentation observed in negative-ion FAB spectra.

Nomenclature

The first three ZAs isolated were named simply A, B, and C. Subsequent studies revealed that fungi produce numerous structurally similar compounds and indicated the need for a more systematic nomenclature. A logical nomenclature would relate to the biosynthesis of these compounds and hence to their carbon skeleton, i.e. alkyl side chain plus core. Most of the variations observed involve different degrees of oxidation. Variations of the acyl side chain (ester linkage) should not warrant a distinct name. Thus, compounds with the same carbon skeleton as ZA-A, but possessing different ester side chains at C-6, would be named A2, A3, A4, and so on. ZAs lacking an ester substituent at C-6 (no acyl side chain) would be named after the parent ZA but would have an x added to their letter designation. The 6-desacyl analogue of ZA-A would thus be named ZA-Ax, and its various oxidation variants, ZAs Ax2, Ax3, etc. The variations in oxidation patterns are not consistent enough across the various ZAs to warrant a nomenclature for the number attribute. This nomenclature should allow a reader to quickly recognize structural, and probably biosynthetic, similarities between various ZAs (see Figures 4 through 9).

The Zaragozic Acid Family Groups

ZARAGOZIC ACID A GROUP The ZA-A group is characterized by a C-6 alkyl side chain bearing a terminal phenyl ring and methyl substituents at C-3′ and C-5′ (see Figure 4). Variations of the alkyl side chain principally include the desacetyl and desacetoxy analogue at C-4′. The methyl at C-5′ has also been found oxidized to the primary alcohol. Variations of the acyl side chain include different degrees of oxidation and methylation. Some ZA-A analogues also lack the C-6 acyl side chain, with the 6 and 7 positions of the core either

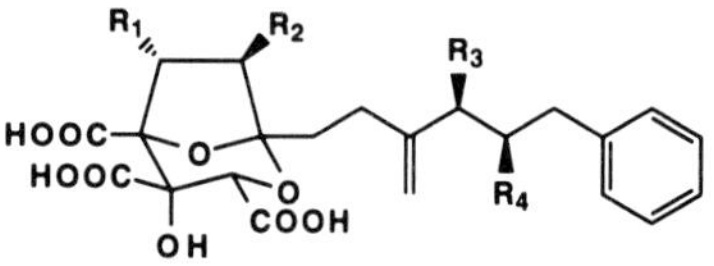

R_1	R_2	R_3	R_4	ZA
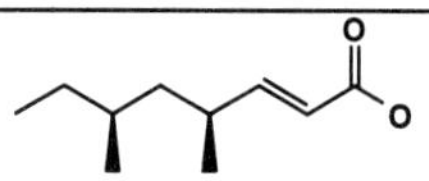 (structure 1)	OH	OAc	CH$_3$	A
	OH	OH	CH$_3$	A1a
	OH	NHAc	CH$_3$	A1b
	OH	OH	CH$_2$OH	A1c
	H	OAc	CH$_3$	A1d
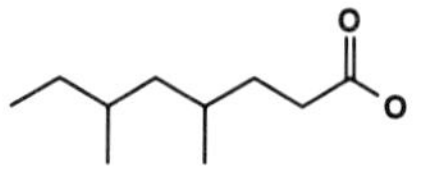 (structure 2)	OH	OAc	CH$_3$	A2
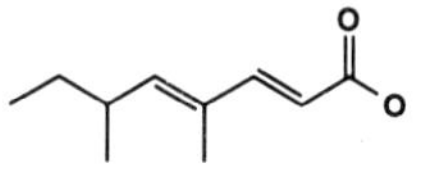 (structure 3)	OH	OAc	CH$_3$	A3
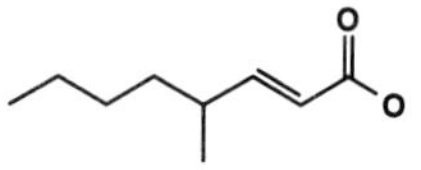 (structure 4)	OH	OAc	CH$_3$	A4
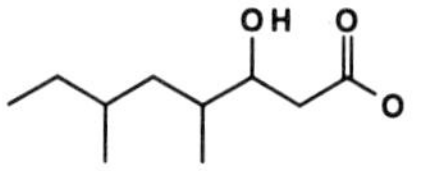 (structure 5)	OH	OAc	CH$_3$	A5
	OH	OH	CH$_3$	A5a
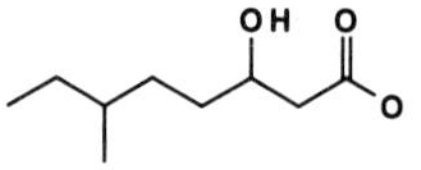 (structure 6)	OH	OAc	CH$_3$	A6
	OH	OH	CH$_3$	A6a
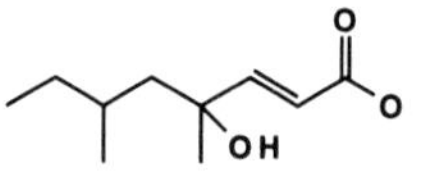 (structure 7)	OH	OAc	CH$_3$	A7
HO	OH	OAc	CH$_3$	Ax1
	OH	OH	CH$_3$	Ax2
H	OH	OAc	CH$_3$	Ax3

Figure 4 Zaragozic acids A group.

oxidized or not (24). Interestingly, L-731,120, an analogue of ZA-A in which the core has not been cyclized, has also been isolated (48) (see Figure 5).

ZARAGOZIC ACID B GROUP The ZA-B group is characterized by a C-8 alkyl side chain bearing a terminal phenyl ring and methyl substituents at C-3′ and C-5′ (see Figure 6). Variation at the alkyl side chain consists mainly of the desoxy analogue at C-4′. Variations of the acyl side chain include different

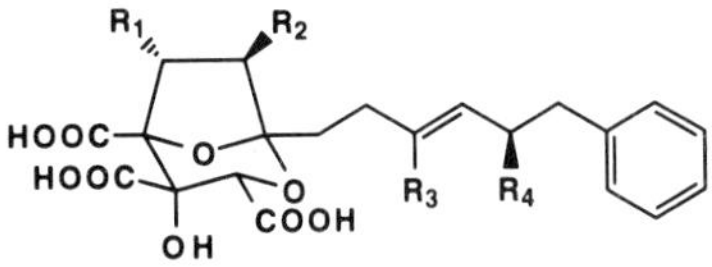

R₁	R₂	R₃	R₄	ZA
[alkenyl chain with ketone]	OH	CH₃	CH₃	A1e
HO	OH	CH₃	CH₃	Ax4
HO	OH	CH₂OH	CH₃	Ax5
HO	OH	CH₃	CH₂OH	Ax6
HO	H	CH₃	CH₃	Ax7
H	OH	CH₃	CH₃	Ax8
H	H	CH₃	CH₃	Ax9

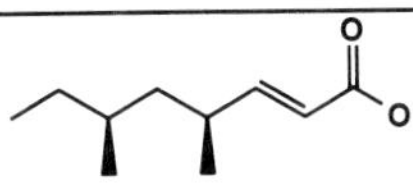

Figure 4 (*Continued*).

degrees of oxidation. ZA-B analogues sometimes lack the C-6 acyl side chain as well, with the 6 and 7 positions of the core either oxidized or not. Analogues of ZA-B, L-731,127 and L-731,128, in which the core has not been cyclized have also been isolated (C Dufresne, unpublished observations) (see Figure 5).

ZARAGOZIC ACID C GROUP   The ZA-C group is characterized by a C-6 alkyl side chain bearing a terminal phenyl ring and methyl substituent at C-5′ (see Figure 7). It differs from the A group by the absence of the olefinic methylene at C-3′. Variation at the alkyl side chain consists mainly of the desacetyl and desacetoxy analogue at C-4′. Variations of the acyl side chain include different degrees of oxidation. The ZA-C was also the first ZA found that possesses a phenyl-terminated acyl side chain (39).

ZARAGOZIC ACID D GROUP   The ZA-D group is characterized by a C-8 alkyl side chain bearing a terminal phenyl ring and methyl substituent at C-5′ (see Figure 8). It differs from the B group by the absence of the methyl group at

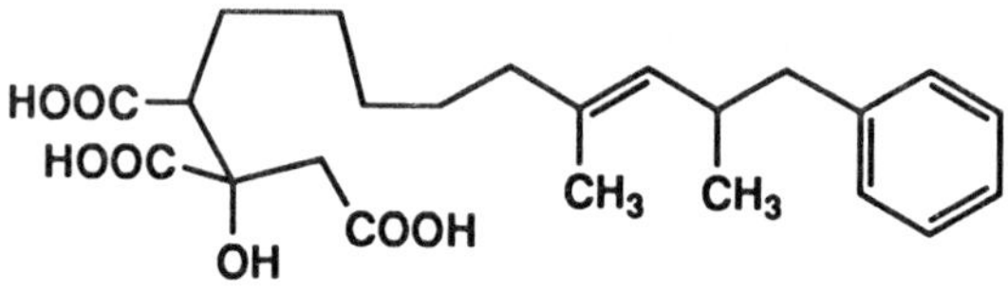

Figure 5 Acyclic compounds related to the ZAs found in fermentation-broth extracts.

L-731,120

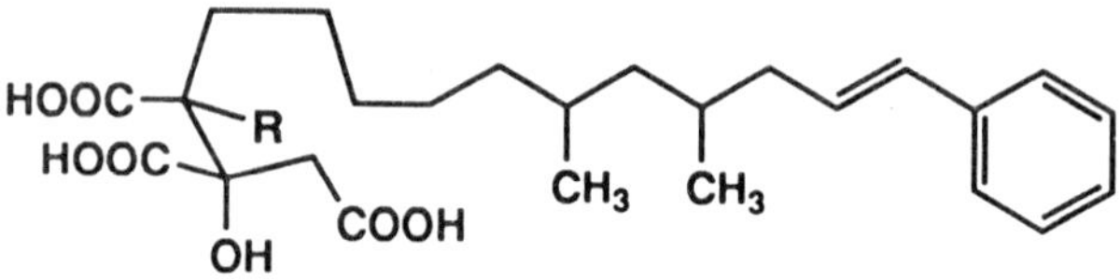

L-731,127 R = OH
L-731,128 R = H

C-3'. No variations of the alkyl side chain have so far been observed. Variations of the acyl side chain include different chain lengths and a phenyl-terminated side chain, as in ZA-C (38).

MISCELLANEOUS ZARAGOZIC ACIDS   ZA-F (Figure 9) represents the first ZA found that possesses a nonaromatic alkyl side chain. This discovery is significant in that biosynthetic studies on ZA-A and -C have indicated that the starter unit is benzoic acid (derived from phenylalanine) and not acetate; the structure of ZA-F implicates an acetate as a starter unit. Furthermore, it shows that a terminal phenyl group is not necessary for squalene synthase activity (37).

ZARAGOZIC ACID–PRODUCING FUNGI

Our high-throughput microbial screening program emphasized filamentous fungi as a chemical source. We believe that discovery of multiple members of the ZA family resulted from the attention given to testing isolates from widespread geographic locations and underutilized, but easily accessible, fungal

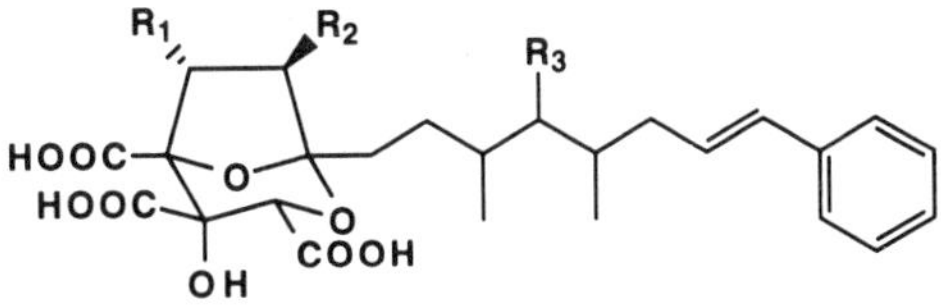

Figure 6 Zaragozic acids B group.

R₁	R₂	R₃	ZA
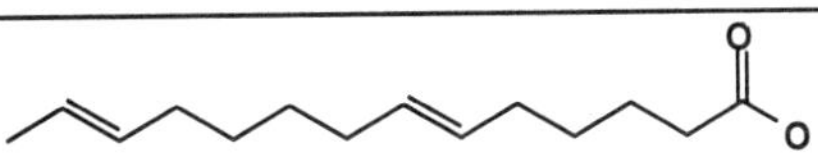 (alkadienoyl chain)	OH OH	OH H	B B1a
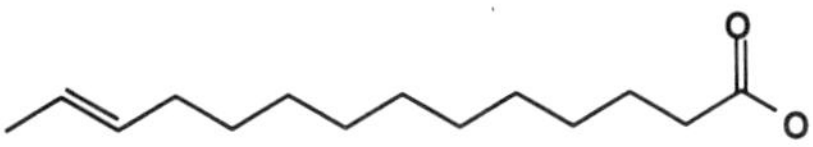 (alkenoyl chain)	OH	OH	B2
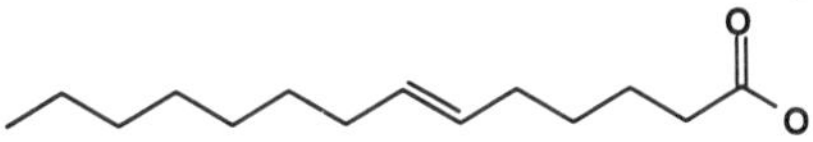 (alkenoyl chain)	OH	OH	B3
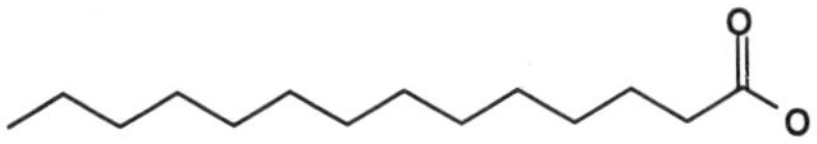 (alkenoyl chain)	OH	OH	B4
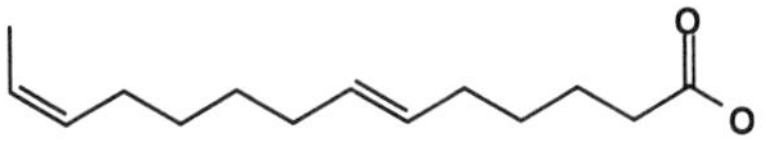 (alkadienoyl chain)	OH	OH	B5
HO	OH H	H H	Bx1 Bx2
H	H	CH_3	Bx3

communities, e.g. coprophilous fungi, endophytes, and wood and plant-litter decomposers.

To date, screening for squalene synthase inhibitors of the ZA family across most major groups of fungi and filamentous bacteria has indicated that this chemical group is absent, or at least undetectable, in prokaryotes, Zygomycotina, and Basidiomycotina. Our results, coupled with those carried out independently in other laboratories, reveal that production of this and a structurally related group of compounds, the cinatrins (53), is distributed over an extensive taxonomic range of Ascomycotina or their anamorphic states (Table 3). Although a clear taxonomic pattern linking the known producers is not evident, many of the taxa, particularly the Pleosporales and Dothideales, have affinities with bitunicate ascomycetes.

ZA-A appears to be most prevalent among the different fungal taxa (Table 3). The strongest taxonomic relationships among the producers belong to group

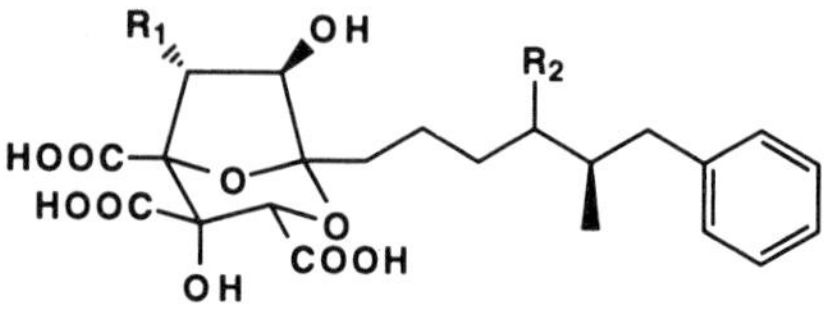

	R_1		R_2	ZA
	(structure: phenyl-CH2CH2-CH=CH-CH(CH3)-CH2-C(=O)-O)		OAc OH H	C C1a C1b
	(structure: phenyl-CH2CH2-CH2-CH2-CH(CH3)-CH2-C(=O)-O)		OAc OH H	C2a C2b C2c
	(structure: phenyl-(CH2)6-C(=O)-O)		OAc OH	C3a C3b
	(structure: phenyl-CH=CH-CH(CH3)-CH=CH-CH2-C(=O)-O)		OAc	C4
	(structure: phenyl-CH2-CH=CH-CH(CH3)-CH=CH-CH2-C(=O)-O)		OH	C5

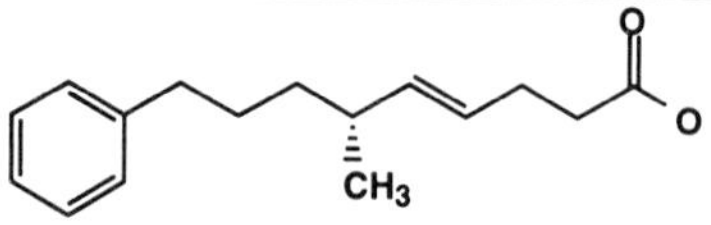

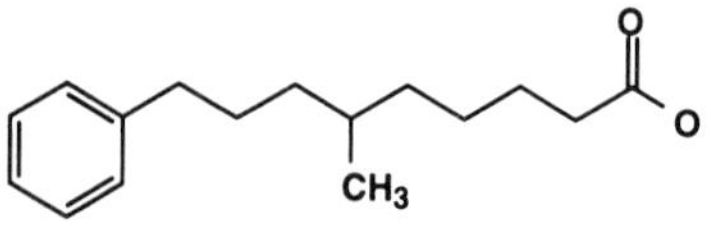

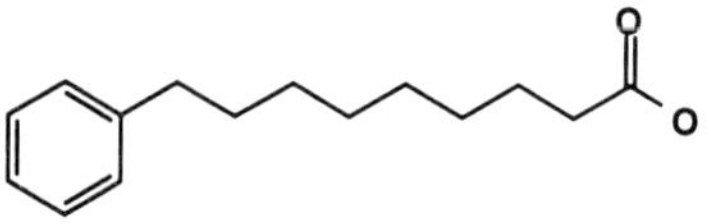

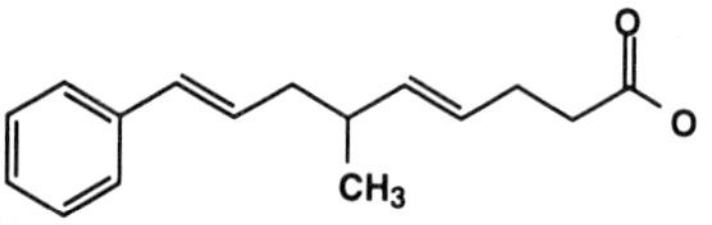

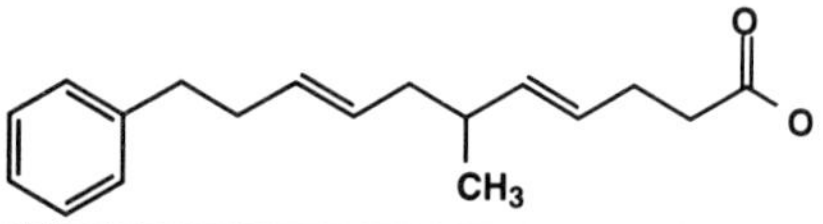

Figure 7 Zaragozic acids C group.

of genera in the family Pleosporaceae, i.e. *Curvularia, Exserohilum,* and *Drechslera* (75). *Setophaeria khartoumensis* has an *Exserohilum* anamorph (*Exserohilum* sp.), whereas *Curvularia lunata* (teleomorph *Cochliobolus lunatus*) and *Exserohilum rostratum* (teleomorph *Setosphaeria rostrata*) are members of the Pleosporaceae. The teleomorph of *Drechslera biseptata* is unknown; however, many *Dreschslera* spp. are conidial states of *Pyrenophora* spp. (75).

At least two different coelomycetes, *Phoma* sp. and *Pseudodiplodia* sp., produce ZA-A. Species of *Phoma* are frequently anamorphs of ascomycetes in the Pleosporales as well as of other bitunicate ascomycetes (74). *Pseudodiplodia* spp. are morphologically similar to *Phoma* sp., differing primarily by having predominantly one-septate conidia (77).

The taxonomic affinities of the sterile strain that produces ZA-A (ATCC

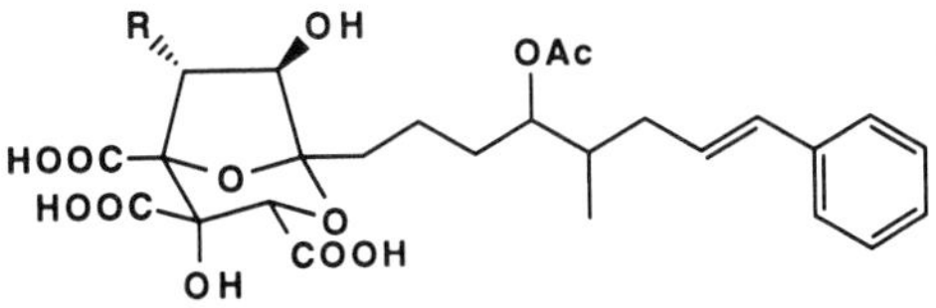

R	ZA
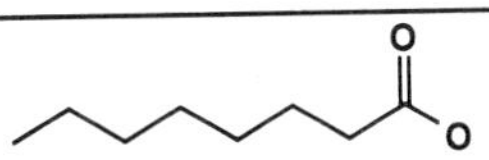	D
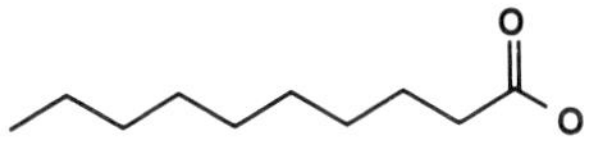	D2
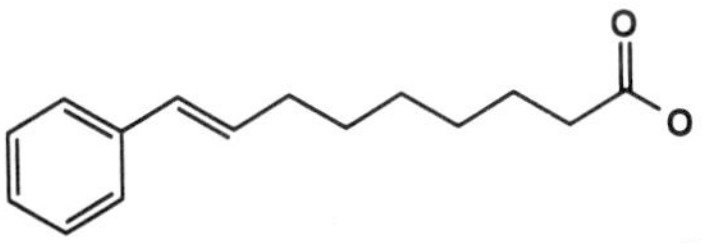	D3
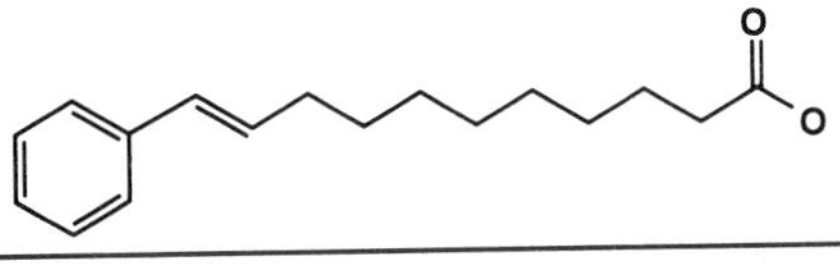	D4

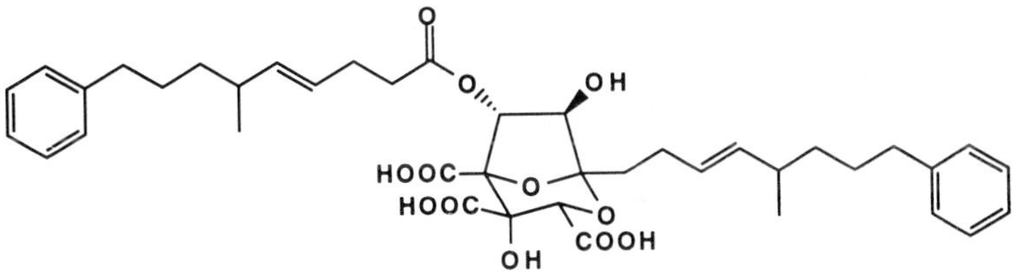

ZA D5

Figure 8 Zaragozic acids D group. ZA D5 was formerly called ZA E.

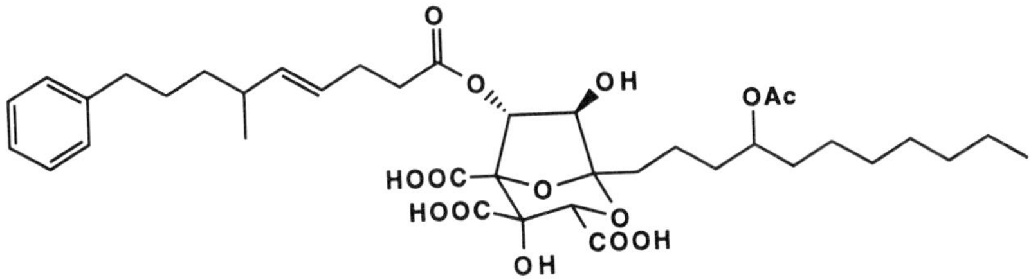

ZA-F

Figure 9 Zaragozic acid F.

Table 3 Zaragozic acid–producing fungi

Organism	Principal product	Source	References
Sterile fungus MF 5453	ZA-A	River filtrates	8, 12, 20
Curvularia lunata var. lunata	ZA-A	Living bark	17, 18, 20
Curvularia lunata var. aeria	ZA-A	Living bark	17, 18, 20
Exserohilum rostratum	ZA-A	Living bark	16, 20
Setosphaeria khartoumensis	ZA-A	Not reported	49
Drechslera biseptata	ZA-A	Living bark	20, 31
Pseudodiplodia sp.	ZA-A	Living bark	19, 20
Phoma sp.	ZA-A and derivatives	Soil	33, 71
Sporormiella intermedia	ZA-B	Herbivore dung	6, 12, 14, 20
Leptodontidium elatius	ZA-C, D5, -F	Dead wood and fungi	7, 12, 20
Amauroascus niger	ZA-D, D_2	Soil baited with hair	20, 38
Libertella sp.	D3	Living bark	Bergstrom et al[a]
Cladosporium cladosporioides	D4	Soil	57

[a] JD Bergstrom, C Dufresne, GF Brills, M Nallin-Omstead & K Byrne, unpublished data.

20986 = MF 5453) to the other species remain uncertain. However, its colony texture and pigmentation and growth rates distinguish it from the other organisms in Table 3. Like other ascomycetes, its hyphal cells are multinucleate, highly branched, simple septate, and darkly pigmented.

Despite intensive screening of a variety of fungi, including many coprophilous ascomycetes (21), ZA-B has so far been detected only in the coprophilous fungus *Sporormiella intermedia. S. intermedia,* likewise, has been classified in the Pleosporaceae (74). Isolates from North America, Europe, and Africa can produce ZA-B; indeed, analysis of dozens of isolates from the southwestern United States indicated that nearly all can produce ZA-B (M Nallin-Omstead & R Jenkins, unpublished data). However isolates vary considerably in the amount produced.

ZA-D4 (TAN-1607A = ZA-D4) was isolated from *Cladosporium cladosporioides* (57). Teleomorphs of *Cladosporium* spp. are often *Mycosphaerella* spp. classified in the Dothideales of the bitunicate ascomycetes.

ZA-D3 was isolated from an endophytic coelomycete that we assigned to the form-genus *Libertella* (Table 3). These fungi are typically anamorphs of ascomycetes in the Diatrypales, an important group of stem-inhabiting fungi. ZA-C, -D5, and -F can be coproduced by strains *Leptodontidium elatius* (20). The teleomorphs of *Leptodontidium* spp. are unknown, but many similar fungi with dematiaceous yeast-like conidial states have teleomorphs in the Dothideales or Sphaeriales (34). *Amauroascus niger,* which produces ZA-D and -D2, is a rare keratinophilic ascomycete and appears to be the most phylogenetically distant species. It is classified in the Onygenales (32).

A related group of fungal natural products, the cinatrins (designated cinatrins

A, B, C_1, C_2, C_3), are fermentation products of *Circinotrichum falcatisporum* (68). The cinatrins are spiro-γ-dilactones and γ-lactones derived from tetra- or trihydroxypentadecane-1,2,3 tricarboxylic acids and share a strong similarity to ZAs devoid of their acyl side chain. *Circinotrichum* spp. are a group of dematiaceous hyphomycetes associated with decaying vegetation (68). Their hyphal characteristics and mode of conidiogenesis indicate that these fungi are ascomycetes, although their teleomorphic states are unknown.

The ZAs, along with the hydroxymethylglutaryl-CoA reductase inhibitors lovastatin (3) and compactin (41), and the hydroxymethylglutaryl-CoA synthetase inhibitor L-659,699 (47), are all fungal metabolites that are potent, specific inhibitors of enzymatic steps early in the sterol biosynthetic pathway, and all have potent antifungal activity. The role of these compounds in fungal metabolism remains uncertain. Because the fungi span a wide array of ecological niches, it is difficult to envision a single ecological function for these specialized metabolites. They might function as fungal antibiotics or antagonists because the compounds have potent antifungal activity. However, the observation that the compounds are associated with the mycelium and not excreted extracellularly complicates this hypothesis. Mycelial-bound antifungal compounds possibly could prevent the invasion and displacement of the producing fungal colony by other fungi. Whatever their natural function may be, the fact that so many fungi make different antifungal compounds that are such specific, potent inhibitors of these early steps in sterol synthesis argues that these metabolites have evolved specifically to impact sterol synthesis.

In recent years, the value of examining fungi as a source of natural products has become better recognized, particularly for metabolites capable of mediating cellular processes in other eukaryotic cells (36, 64, 67). Therefore, expanded screening of fungi from diverse ecological and phylogenetic origins, combined with increased attention to manipulation of fungal-metabolite expression in culture systems, may have contributed significantly to the discovery of the ZAs.

BIOSYNTHESIS OF THE ZARAGOZIC ACIDS

The biosynthesis of the ZAs was an area of intense interest and research by both the Merck and Glaxo groups. Obviously, knowledge of the biosynthetic precursors would be advantageous in efforts to improve the productivity of the fermentations; moreover, the possible alteration of the biosynthetic pathway to include analogues of established precursors also presented an enticing prospect. All of the ZAs discovered to date share the common chemical feature of a novel 2,8-dioxobicyclo[3.2.1]octane-4,6,7-trihydroxy-3,4,5-tricarboxylic acid ring system. The differences among the ZAs arise from variation in the 6-O-acyl and 1-alkyl side chains. The studies anticipated that the biosynthesis of the bicyclic core of the ZAs would remain constant, regardless of the producing fungal species and that the variations in alkyl and acyl side chains were all likely to occur through small alterations in precursor unit substitutions, the level of reduction during chain assembly, and the degree of branch-chain methylation, in an otherwise consistent polyketide chain assembly (35, 56). We initially focused on the biosynthesis of ZA-A (25). The fermentation for ZA-A production was the best developed of the three principal ZA products under investigation in our laboratories in terms of product titer in both production and washed-cell environments. At the same time, the Glaxo group studied the biosynthesis of their primary product, also ZA-A (i.e. squalestatin 1) (33, 55).

Biosynthesis of Zaragozic Acid A

POLYKETIDE BACKBONES OF CHAINS A AND B Feeding of potential ^{14}C-labeled precursors into washed cells of MF 5453 revealed significant incorporation into ZA-A from acetate, succinate, L-phenylalanine, benzoic acid, [ring-^{14}C]-cinnamic acid, and the methyl group of L-methionine (25).

Use of ^{13}C-labeled substrates and carbon NMR analysis of the isolated ZA-A provided evidence for the exact sites of incorporation of the precursors and the direction of assembly of the polyketide skeleton (25). Figure 10 summarizes the incorporation results. Chain A begins with the aromatic ring at C-6' and is extended by addition of five acetate units, presumably in the activated form of malonyl CoA. A sixth acetate unit is used to acetylate the C-4' oxygen. Chain B is formed and extended in a straight-chain polyketide fashion using four acetate/malonate units. The experiment revealed no evidence of specific labeling of ZA-A from [1-^{13}C]propionate, whereas supplementation with L-[methyl-^{13}C]methionine resulted in high levels of enrichment (20- to 25-fold) at all four branched-chain carbon sites, C-9'', 10'', 13', and 14'. Thus, the four branched-chain methyl/methylene groups on the acyl and alkyl side chains of ZA-A are all derived from the methyl group of L-methionine. These results are consistent with reports that fungi possess negligible levels of the enzyme propionyl-CoA carboxylase (70) and cannot metabolize propionyl-CoA to methyl-malonyl-CoA.

THE AROMATIC RING The origin of the aromatic ring at the start of chain A was particularly interesting to us because it represented a segment of the molecule with potential for substitution using directed biosynthetic techniques. The incorporation of [3-^{13}C]phenylalanine, [U-^{14}C]cinnamic acid, and variously labeled ^{13}C–benzoic acid precursors provided convincing evidence that the nine carbons of phenylalanine are metabolized, with the loss of the first two carbon atoms to leave the seven carbons of benzoic acid (25, 33).

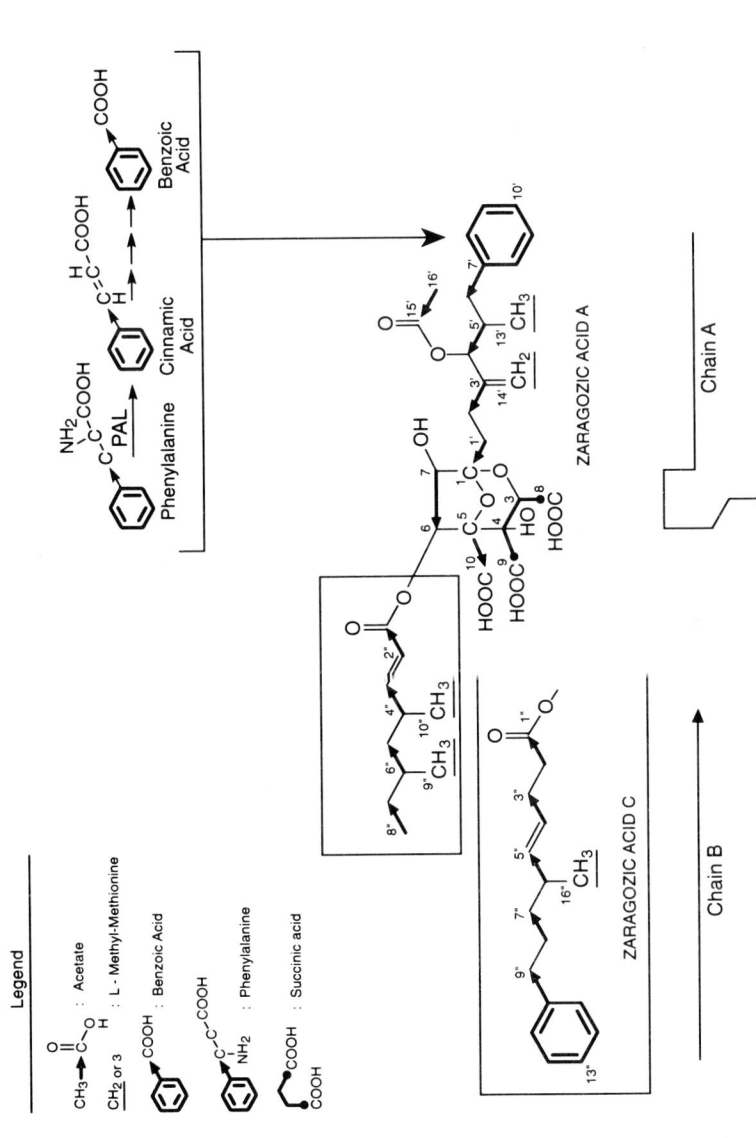

Figure 10 Biosynthetic origin of the carbon skeletons of ZA-A and -C based on the patterns of incorporation of ^{13}C- and ^{14}C-labeled precursors.

This pathway likely involves the common fungal enzyme phenylalanine ammonia lyase (PAL) (43, 60). This enzyme converts phenylalanine to cinnamic acid, which then undergoes cleavage of an acetate unit to generate benzoic acid. The presence of PAL in the producing culture MF 5453 was demonstrated, and inhibitors of PAL, such as D-phenylalanine and phenylpropiolic acid, were found to suppress ZA-A biosynthesis at 1.0 and 0.1 mM, respectively (25, 26).

The actual starter molecule of chain A is probably an activated form of benzoic acid. After addition of the five acetate/malonate units, chain A terminates with the final addition of a succinate molecule, presumably succinyl-CoA.

THE BICYCLIC RING The Merck and Glaxo groups obtained similar data for the biosynthesis of the bicyclic ring from the different fungal producers, and both groups independently arrived at identical conclusions regarding probable precursors (25, 55). The labeling data from ^{13}C precursors (25) indicate that carbons 1, 6, 7, 5, and 10 are derived from acetate, while carbons 3, 4, 8, and 9 arise from succinate. Of particular importance was the coupling observed between carbons C-5 and C-10, the terminal acetate unit of chain A. Weaker enrichment and a lack of significant coupling of bicyclic ring carbons 3, 4, 8, and 9 compared with carbons C-5 and C-10 suggested that the acetate incorporated into the former four carbon atoms was metabolized through the citric acid cycle, leading to dilution and scrambling of the labeled precursor. This result would be consistent with a citric acid cycle precursor for carbons 3, 4, 8, and 9.

With multi-^{13}C-labeled substrates, investigators sought to discern the pathway for the incorporation of succinate and other citric acid–cycle intermediates in the ZA core (25). The experiments showed that (*a*) succinate was incorporated as a single unit; (*b*) incorporation and enrichment from both ^{14}C- and ^{13}C-citrate was less than that from succinate; and (*c*) enrichment from [2,3,4-^{13}C]citrate was heavier at carbons 3, 4, 8, and 9 than at C-5 and C-10, and no significant coupling was observed between carbons 4 and 5. These data strongly indicate that carbons 3, 4, 8, and 9 of ZA-A arise from a four-carbon intermediate or analogue of the citric acid cycle and not as part of a six-carbon precursor. Incorporation of oxaloacetate and aspartic acid was low and nonspecific, which may have resulted in part from poor uptake. Although several studies have demonstrated the incorporation of succinate as a precursor in the biosynthesis of secondary metabolites (4, 63, 79), metabolism to oxaloacetate prior to incorporation is still a possibility. By analogy with the well-established condensation pathway of oxaloacetate with acetyl-CoA, we propose that succinate is a precursor of carbons 3, 4, 8, and 9 but is probably not the direct precursor. The penultimate precursor is more likely to be an oxidized metabolite of succinate, such as oxaloacetate or epoxy-fumarate, that condenses with the α carbon of polyketide chain A.

PRECURSORS OF OXYGEN ATOMS AND BICYCLIC RING FORMATION The Glaxo group investigated the origin of the oxygen atoms of ZA-A by using both [1-^{13}C, ^{18}O$_2$]acetate and ^{18}O$_2$ incubated in the presence of their ZA-A *Phoma* strain (55). Only the ester carbonyls at carbons 1″ and 15′ were definitively derived from acetate oxygens, and oxygen exchange during work-up obviated the results of the ring carboxyl oxygens. Fermentation of the fungus in a closed system containing ^{18}O$_2$ produced ZA-A that displayed characteristic ^{18}O$_2$ carbon NMR shifts at carbons 1, 3, 5, 6, 7, 4′, 15′, and 1″. These results suggested that the five oxygens on carbons 3, 5, 6, 7, and 4′ are all derived from atmospheric oxygen—a conclusion further supported by mass spectrometric analysis of the labeled compound, which indicated the presence of up to five atoms of ^{18}O per molecule of ZA-A. The lack of carbon shift at C-4 under both labeling strategies suggested that the hydroxyl group at this position is probably derived from water.

Certainly one of the most interesting aspects concerning the biosynthesis of ZA-A is the cyclization of the bicyclic ring. The Glaxo group proposed a possible mechanism for the bicyclic ring formation based upon the ^{18}O-labeling results (Figure 11a) (55). This proposal suggests that the cyclization would commence with a nucleophilic attack by a water molecule at C-4, initiating ring formation through C-1. This process in turn initiates a nucleophilic attack by the carbonyl oxygen at C-1 on C-5, opening the second epoxide and leaving a hydroxyl function at C-6 and a methylene at C-7. Implicit in this proposal is a reduction and loss of the acetate-derived carbonyl oxygens at C-1 and C-7, followed by reoxidation with atmospheric oxygen at both carbons. The oxidation at C-1 would have to occur prior to ring closure, generating what the Glaxo authors propose as a trisepoxide that gives rise to the structure in Figure 11a. Despite the precedence for the existence of disepoxides in secondary-product formation (76), stability may preclude the formation of the trisepoxide and disepoxide intermediates.

In contrast, the cyclization may proceed through the intermediate shown in Figure 11b. This compound would be expected to ketalize as shown to the zaragozic acid bicyclic ring. The production of the hydroxyl at C-4 from oxaloacetate would explain why this oxygen was not labeled by either ^{18}O$_2$ or [1-^{13}C, ^{18}O$_2$]acetate. The hydroxyls at C-3 and C-5 could arise by α-oxidation and the hydroxyl at C-6 by β-oxidation. The carbonyl at C-1 must result from aliphatic oxidation and is subsequently lost during bicyclic ring formation from the cyclic hemiketal.

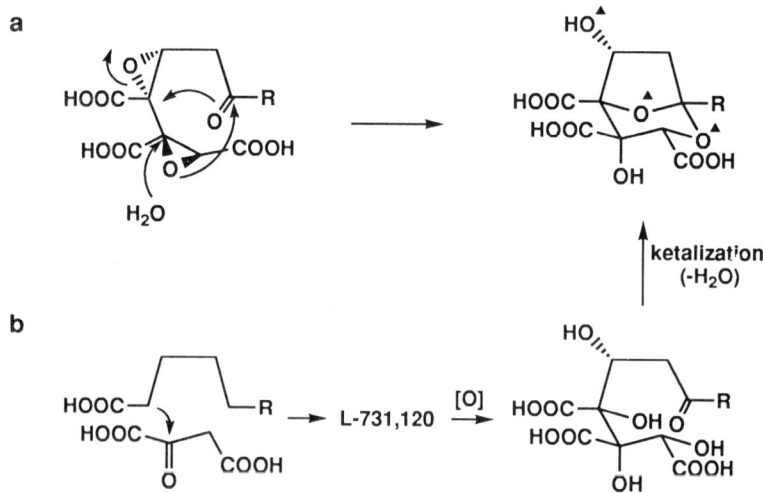

Figure 11 Proposed pathways of bicyclic ring formation. (*a*) Glaxo proposal. The oxygen atoms derived from $^{18}O_2$ (*solid triangles*) are shown in the bicyclic structure. (*b*) Merck proposal originating from minor metabolite L-731,120 or a similar analogue.

Biosynthesis of Zaragozic Acid C

As mentioned previously, variation among ZAs arises principally from changes in the acyl and alkyl side chains. The alkyl side chain of ZA-C is equivalent to that of ZA-A, lacking only the C-methylation from L-methionine at C-3′. The acyl side chain of ZA-C exhibits the greatest diversity among the naturally occurring ZAs identified to date. It contains a single methyl-branched polyketide chain that begins, like the alkyl side chain, with an aromatic ring. Investigation of the biosynthesis of ZA-C was carried out using washed cells of *L. elatius* in a manner similar to that described for ZA-A (25). Not surprisingly, both ^{14}C- and ^{13}C-labeling results were similar to those obtained for ZA-A (20, 25). The only observable differences in ^{13}C-labeling patterns between ZA-A and ZA-C were associated with the aromatic precursors. In ZA-C, [carboxy-^{13}C]benzoic acid labeled both C-6′ and C-9″, whereas [ring-$^{13}C_6$] benzoic acid labeled carbons C-7′ through C-12′ as well as C-10″ through C-15″. Thus, the biosynthesis of both chains A and B of ZA-C probably use an activated form of benzoic acid as a starter group. The ^{14}C- and ^{13}C-labeling patterns of ZA-A and ZA-C were both consistent with metabolism of phenylalanine to benzoic acid, which points to the presence of PAL in all ZA-producing fungi.

Incorporation of [1-^{13}C, $^{18}O_2$]acetate into ZA-C led to only one observable resonance shift (0.038 ppm) corresponding to the acetyl carbonyl at C-15′ (25).

However, the resonance of C-1′ was precisely 0.038 ppm higher than that for C-15′, so a similar shift for the C-1″ resonance would cause it to fall directly under the unshifted C-15′ resonance and therefore be unobservable. Hence, our results of [1-^{13}C, ^{18}O$_2$]acetate incorporation into ZA-C were consistent with the conclusions made by the Glaxo team regarding [1-^{13}C, ^{18}O$_2$]acetate incorporation into ZA-A (squalestatin S1) shown in Figure 11 (55).

Directed Biosynthesis

Knowledge of the biosynthetic precursors of the aromatic moiety of the alkyl side chain of ZA-A permitted the substitution of various analogues using directed biosynthetic techniques. We developed a procedure for directing the biosynthesis of ZA-A utilizing washed cells of MF 5453 (26). Glaxo researchers independently used production-stage fermentations of their *Phoma* strain for directed biosynthesis of ZA-A (27, 28). When advantageous, we also used the PAL inhibitor phenylpropiolic acid to suppress formation of the natural substrate (28, 29). Both research groups explored the ability of their respective fungi to incorporate analogues of benzoic acid into the ZA-A backbone. The resulting analogues of ZA-A produced by the two groups were essentially identical, which suggests that the selectivity of the enzymes required for both activation of the benzoic acid analogues and initiation of chain A biosynthesis in the two fungi was comparable. Approximately 30 substrates were tested, but only the select few listed in Table 4 were incorporated into the ZA-A skeleton to give new analogues. All of the analogues produced via this technique possessed potent squalene synthase–inhibitory activity (26–28).

Minor Metabolites in the Biosynthetic Scheme

As noted earlier and also described by the Glaxo group (24), many of the fungi that produce ZAs also produce minor components that possess many of the

Table 4 Analogues of ZA-A produced by directed biosynthesis

Substrate analogue	Product	References
2-Fluorobenzoic acid	8′-Fluoro-ZA-A	26, 27, 29
3-Fluorobenzoic acid	9′-Fluoro-ZA-A	26, 27, 29
4-Fluorobenzoic acid	10′-Fluoro-ZA-A	26, 27, 29
2,3-Difluorobenzoic acid	8′,9′-Difluoro-ZA-A	28
3,4-Difluorobenzoic acid	9′,10′-Difluoro-ZA-A	28
3,5-Difluorobenzoic acid	9′,11′-Difluoro-ZA-A	28
2-Furylcarboxylic acid	6′-Furyl-ZA-A	26, 29
2-Thiophenecarboxylic acid	6′(8′-Thienyl)-ZA-A	26, 27, 29
3-Thiophenecarboxylic acid	6′(9′-Thienyl)-ZA-A	26, 27, 29
3(-Thienyl)acrylic acid	6′(9′)-Thienyl)-ZA-A	27
3-(Thiophenecarboxaldehyde	6′(9′-Thienyl)-ZA-A	27

structural features of the ZAs. Many of these compounds likely represent degradation products, shunt products, or biosynthetic "mistakes," while others may be true biosynthetic intermediates. Although bioconversion studies using labeled minor components incubated with the producing fungi have not been reported, some proposals regarding possible precursor relationships among these products can be made.

All of the minor components reported for ZA-A vary in the degree of oxidation and methylation of the acyl side chain, as well as the extent of methylation, oxidation, and acetylation of the alkyl side chain (24) (Figure 4). With regard to the biosynthesis, these findings suggest that these modifications probably occur late in biosynthesis, and/or the integrity of the modifications on the alkyl side chain may not be a requirement for acylation at C-6.

The appearance of 4'-desacetyl- and 4'-desacetoxy-ZA-A and -C suggests that the acetylation at C-4' may be one of the final biosynthetic steps. Several ZA-A analogues have been described that lack the oxygen at C-4' and contain a $\Delta^{3',4'}$-vinylic methyl group at C-3' instead of a methylene. The presence of these compounds suggests that, after ring formation, an oxidation of the $\Delta^{3',4'}$ double bond gives rise to the hydroxy at C-4' and a double bond at C-3',14'. ZA-Ax7, a minor product, (Figure 4), lacks the 6-acyl side chain and the hydroxyl C-7, which suggests that oxidation of C-7 likely occurs after ring formation and perhaps before acylation at C-6.

L-731,120, L-731,127, and L-731,128 (Figure 5) are three of the most interesting potential intermediates isolated from the fermentation broths of the ZA-A (L-731,120) and ZA-B (L-731,127 and L-731,128) producers. L-731,120 is an acyclic analogue of ZA-A containing the alkyl side chain, the tricarboxylic acid groups, and the C-4 hydroxyl (48). It also contains the vinylic methyl group observed in several of the cyclized compounds mentioned above. It represents the entire backbone of chain A (Figure 10) prior to extensive oxidation, cyclization, and methylation. Similarly, L-731,128 corresponds to the same chain A structural analogue of ZA-B. L-731,127 is identical to L-731,128, except it is further oxidized at the γ carbon. All three acyclic compounds exhibit squalene synthase activity at the submicromolar level (G Harris, C Dufresne & J Bergstrom, personal communication and unpublished observations). If these compounds were true biosynthetic intermediates of the corresponding zaragozic acids, they would negate the ring-closure mechanism proposed by the Glaxo group. The structures of these compounds suggest that the hydroxy at C-4 exists prior to further oxidation and cyclization and support the proposed cyclization scheme shown in Figure 11*b*. Only further effort to incorporate labeled versions of these acyclic compounds into the zaragozic acids will reveal whether or not they are true biosynthetic intermediates.

MECHANISM OF ACTION

Inhibition of Squalene Synthase

The ZAs are very potent inhibitors of squalene synthase (9, 12, 49). The potency of ZA-A as an inhibitor of the rat liver microsomal squalene synthase varies from an apparent K_i value of 78 pM (12) to 1.6 nM (49) and an IC_{50} value of 12.5 nM (9). Inhibitory values reported for other compounds in this family also vary widely, and laboratories are often at odds as to what structural changes make the compounds more or less active. The large variations in the reported potencies of the ZAs may partly result from the wide variety of protocols used to assay the enzyme (12, 49, 78).

Despite the discrepancies in the reported potencies of the ZAs, these are clearly very potent inhibitors of squalene synthase. Squalene synthase catalyzes a two-step reaction: In the first step, two F-PPs are condensed to form a stable intermediate, presqualene-pyrophosphate (pS-PP) (65), and in the second step the pS-PP goes through a loss of pyrophosphate and a reductive rearrangement to produce squalene. Bergstrom et al (12) and Hasumi et al (49) report that the inhibition of squalene synthase by the ZAs is competitive with respect to F-PP. Hasumi et al (49) showed that ZA-A inhibits both steps of the squalene synthase reaction. The gross structural similarity between the ZAs and pS-PP (12, 49)—both have a highly acidic central core with two long hydrophobic tails attached—has led to the hypothesis that the ZAs inhibit squalene synthase by effectively mimicking the binding of pS-PP to the enzyme. The kinetic mechanism and order of binding of substrates and release of products of squalene synthase have not been satisfactorily resolved, but if F-PP and pS-PP are each capable of binding to the free enzyme and proceeding through the reaction, and if the ZAs bind to only free enzyme in a reversible manner and form a dead-end complex, the ZAs should be competitive inhibitors with respect to both F-PP and pS-PP as substrates. The ZAs are potent inhibitors of squalene synthase from rat liver (9, 12, 33, 49), marmoset liver (9), HEPG-2 cells (a human liver cell line) (9), and *Candida albicans* (9).

Effects on Other Enzymes

F-PP, the substrate for squalene synthase, is at an important metabolic branch point. Because the ZAs are competitive inhibitors of squalene synthase, their effects on these other enzymes that utilize F-PP are of interest. Several ZAs tested against Ras farnesyl transferase (42, 52) have been found to be competitive inhibitors with respect to F-PP. However, the potency of ZA inhibition of Ras farnesyl transferase is 100- to 2000-fold less than that of squalene synthase when measured under comparable conditions (52). Indeed, the ZAs can readily block sterol synthesis of cells in culture (see below), but do not affect protein farnesylation in numerous cell-culture models. Thus, the in vivo relevance of the inhibition of Ras farnesyl transferase by the ZAs is questionable. F-PP pyrophosphatase, an enzyme that can convert F-PP to farnesol, is also inhibited by some ZAs at micromolar concentrations (5). The effects of the ZAs on the prenyl transferases that utilize F-PP to make dolichol, ubiquinone, and geranylgeranyl-pyrophosphate have not been reported.

The synthesis of phytoene in plants and fungi is a two-step process from geranylgeranyl-pyrophosphate via the formation of prephytoene-pyrophosphate, an intermediate analogous to pS-PP (54). The effects of the ZAs on this enzyme have not been reported, but they might well be potent inhibitors of this reaction.

Activities in Cells

The ZAs effectively block cholesterol synthesis from labeled acetate or mevalonate in cultured mammalian cells (Table 5). They are modestly potent inhibitors in HEP-G2 cells and Chinese hamster ovary fibroblasts but are low nanomolar inhibitors in primary cultures of rat hepatocytes. In the nonprimary hepatocyte cells, this level of inhibition was seen after a 2-h preincubation with the ZAs before the label was added (12). Without preincubation, the level of inhibition was much lower (J Bergstrom, unpublished observations), which suggests that the ZAs are poorly taken up by these cell types. However, in primary hepatocytes the low nanomolar inhibition is seen with simultaneous addition of ZA-A and the labeled precursor (9). Pravastatin, an HMG-CoA reductase inhibitor, shows similar behavior with cells in culture and is preferentially taken up by liver cells in vivo, probably via a specific hepatic transport mechanism. Similarly, ZA-A dosed subcutaneously in rats or mice totally blocks hepatic cholesterol synthesis (9, 12) but has little or no effect on cholesterol synthesis in several other tissues (J Bergstrom & M Kurtz, unpublished observations), which suggests that preferential utilization of ZA-A by the primary hepatocytes may be of physiological relevance. Cholesterol synthesis inhibition in cells is measured in serum-free media because the ZAs

Table 5 The activity of the ZAs in inhibiting cholesterol synthesis in cultured cells

Zaragozic acid	HEP-G2	$IC_{50}n, M$ CHO	Hepatocytes
A	6500[a]	1100[b]	14[b], 39[c]
B	600[a]	380[b]	60[b]
C	4000[a]	Not tested	Not tested

[a] Taken from Reference 12.
[b] J Bergstrom, unpublished observations.
[c] Taken from Reference 9.

strongly bind to albumin, and the presence of albumin greatly reduces the efficacy of blocking cholesterol synthesis in cells by the ZAs (J Bergstrom & M Kurtz, unpublished observations).

When cholesterol synthesis is blocked in cells by a ZA, label from ^{3}H-mevalonate accumulates in F-PP (9), farnesol (12), farnesoic acid (12), and dicarboxylic acids derived from farnesol (12). Label in squalene, lanosterol, and cholesterol dramatically decreases in the presence of the ZAs (12). These results show a specific block of cholesterol synthesis at squalene synthase with an accumulation of F-PP. Bansal & Vaidya (5) have shown that a specific F-PP pyrophosphatase activity in liver converts F-PP to farnesol. This enzyme has a high K_m for F-PP, 30 mM (5), whereas the K_m of squalene synthase for F-PP is much lower, 0.5 mM (12). Hence, the pyrophosphatase activity may only become relevant as the F-PP concentrations rise above normal. Farnesol can subsequently be oxidized to farnesoic acid and then be ω-oxidized to a dicarboxylic acid (45). This metabolic pathway, F-PP to farnesol to farnesoic acid to farnesol derived dicarboxylic acids, may be the major metabolic shunt that occurs for mevalonate metabolism in the presence of the ZAs (12).

Antifungal Activity

ZA-A (33, 62) and other ZAs (62) have broad antifungal activity with minimum inhibitory concentrations (MIC) ranging from 0.5 to 128 mg/ml against various yeasts and filamentous fungi. ZAs inhibit squalene synthase from pathogenic fungi (62) and *Saccharomyces cerevisiae* (33, 62). ZA-A can block acetate incorporation into squalene and sterols in fungi (62), and overexpression of the cloned yeast squalene synthetase can produce resistance to the antifungal effects of the ZAs in *S. cerevisiae* (62). Thus, the antifungal activity of the ZAs appears to result from the inhibition of squalene synthase (62). The ZAs show fungicidal activity, leading to a greater than 99% reduction in cell number in 48 h (62). These data suggest that the ZAs may have great potential as antifungal agents; however, the strong binding of the ZAs to albumin and their rapid delivery to liver may limit their utility.

The major enigma of the antifungal activity of the ZAs is that they show fungicidal activity whereas other sterol-synthesis inhibitors have fungistatic activity. The cidal activity cannot be explained merely by inhibition of ergosterol synthesis, because inhibitors of other steps would then also be cidal. However, the antifungal activity can be overcome through the overexpression of squalene synthase, showing that the inhibition of squalene synthase causes the antifungal activity.

The ZA producers, which are all fungi, are resistant to the antifungal activity of the ZAs. Therefore, novel producers might be found by isolating those fungi that can grow in the presence of a ZA.

In Vivo Activity

In vivo activity of cholesterol-synthesis inhibitors can readily be measured in small animals by dosing with the compound and then dosing orally or subcutaneously with radiolabeled acetate or mevalonate and measuring the incorporation of that label into hepatic cholesterol. ZAs dosed i.v. or subcutaneously in mice or rats were very potent, with $ED_{50}s$ of 0.1–0.3 mg/kg (9, 12). When cholesterol synthesis was blocked by the ZAs, labeling of squalene and postsqualene metabolites (12) was dramatically decreased, whereas labeling of F-PP (9) and farnesoic acid–derived metabolites (12) was dramatically increased. These data demonstrate that the ZAs specifically block squalene synthase in vivo. Cholesterol was 15–20% lower in rats dosed subcutaneously at 3 mg/kg for 4 to 7 days with ZA-A (M Kurtz & J Bergstrom, unpublished observations).

In contrast to the i.v. or subcutaneously dosed ZAs, orally dosed ZAs block cholesterol synthesis in rats and mice relatively little, with $ED_{50}s$ greater than 50 mg/kg (30). When ^{3}H-ZA-B was dosed orally in the mouse, less than 1% of the administered dose appeared in plasma or the liver; the bulk of the radioactivity remained in the intestinal contents as unmetabolized ZA-B (J Bergstrom, J Germershausen & R Bostedor, unpublished observations). The difference in activity between oral and subcutaneous dosing, coupled with data on the absorption of ^{3}H-ZA-B, shows that intestinal absorption is the major factor limiting oral activity of the ZAs in these rodents. This conclusion is supported by the finding that certain ester derivatives of the core carboxylic acids of the ZAs have enhanced oral activity in mice and are serving as prodrugs (30). Once past the intestinal absorption barrier, the ZAs were very potent in vivo inhibitors of cholesterol synthesis in rats and mice (9, 12).

Glaxo scientists found that ZA-A was a potent oral cholesterol-lowering agent in marmosets, a primate species with lipoprotein metabolism similar to humans (9), in contrast to the poor oral activity seen in rodents. Baxter et al (9) observed a decrease in cholesterol by ~50% with an oral dose of 10 mg/kg over an 8-week dosing period. They observed a 75% reduction in cholesterol with an oral dose of 100 mg/kg (9). Levels of apolipoprotein B, the major protein component of low-density lipoprotein (LDL) and very low-density lipoprotein (VLDL), were dramatically lowered, whereas levels of apolipoprotein A1, a major protein component of high-density lipoprotein (HDL), were unaffected (9). Thus, the primary cholesterol-lowering effect appeared to be on LDL. Despite the significant reduction in cholesterol levels due to orally dosed ZA-A in marmosets, oral absorption appears, in our opinion, to be very limited. Procopiou et al (69) found that a single i.v. dose of 0.33 mg/kg ZA-A lowered cholesterol 53–63% over a period of 2 to 7 days postdosing. This reduction persisted for over a week (69), which suggests extremely slow metabolism and excretion of the ZAs. The difference in the cholesterol response in rats and marmosets shows that the marmoset's cholesterol levels are much more sensitive than those of the rat to a squalene synthase inhibitor. The cholesterol reduction seen in the marmoset demonstrated that a squalene synthase inhibitor could be an efficacious cholesterol-lowering agent and shows that a ZA, a ZA derivative, or another class of squalene synthase inhibitors could possibly be developed as a cholesterol-lowering agent in humans.

CONCLUSIONS

The ZAs are a large family of fungal metabolites produced by various filamentous fungi that are potent inhibitors of squalene synthase. Although potentially biosynthetically related to other citrate-containing natural products, this family, containing the 4,6,7-trihydroxy-2,8-dioxobicyclo[3.2.1]octane-3,4, 5-tricarboxylic acid core, is a novel group of metabolites.

Given their seemingly ubiquitous distribution among filamentous fungi and their antifungal activity, one might ask why the ZAs were not discovered earlier. The ZAs were not found sooner in part because of the earlier discovery of the compactin-related HMG-CoA reductase inhibitors. As the potential of the latter compounds became apparent, much of the industry shifted its focus to the production and development the HMG-CoA reductase inhibitors. Only after their development was assured did many industrial laboratories interested in cholesterol lowering begin work on other targets. Two other enzymes, thiolase (46) and HMG-CoA synthase (47), were among the first such alternative targets.

Alternatively, antifungal efforts may have been largely directed at the later steps of sterol synthesis because the differences between mammalian and fungal sterol synthesis in the later stages allows specific targeting of fungal sterol biosynthesis. This is the part of the pathway inhibited by the currently marketed antifungals. Thus, industry bias may have precluded a search for inhibitors of squalene synthase and the other steps in the middle of the sterol-synthesis pathway until recently. Another possible reason for the late discovery of ZAs is that, although they are strong antifungal agents, they might not have been detected via traditional antifungal screening. Because ZAs bind strongly to paper discs and diffuse poorly through agar (J Onishi, personal communication), their detection in these assays may have been limited. Producers do give zones of inhibition, but their diameters are so small that they would not normally attract attention in these antifungal screens (J Onishi, personal communication). The discovery of the ZAs came only after the development of the high-throughput biochemical screening assays that were focused on squalene synthase, assays that are more sensitive and specific than the traditional antifungal assays.

In summary, the ZAs are potent and specific inhibitors of squalene synthase (from every species tested) that readily block sterol synthesis of mammalian and fungal cells in culture. In addition to their development for pharmaceutical purposes, the ZAs are being used as agents to specifically block squalene synthase in experiments probing the regulation of HMG-CoA reductase and other aspects of lipoprotein metabolism (44, 66). As have other biologically active natural-product inhibitors, the ZAs will likely become useful tools for biological experimentation. The ZAs have strong broad-spectrum antifungal activity. They are potent in vivo inhibitors of cholesterol synthesis in rodents, and they readily lower cholesterol in the marmoset. Because of these activities, the ZAs or their derivatives have the potential to be developed as antifungal agents and as cholesterol-lowering agents.

> Any *Annual Review* chapter, as well as any article cited in an *Annual Review* chapter, may be purchased from the Annual Reviews Preprints and Reprints service.
> 1-800-347-8007; 415-259-5017; email: arpr@class.org

Literature Cited

1. Abe I, Tomesch JC, Wattanasin S, Prestwich GD. 1994. Inhibitors of squalene biosynthesis and metabolism. *Nat. Prod. Rep.* 11:279–302
2. Alberts A. 1988. HMG-CoA reductase inhibitors—the development. *Atheroscler. Rev.* 18:123–31
3. Alberts AW, Chen J, Kuron G, Hunt V, Huff J, et al. 1980. Mevinolin: a highly potent competitive inhibitor of hydroxymethylglutaryl-coenzyme A reductase and a cholesterol-lowering agent. *Proc. Natl. Acad. Sci. USA* 77:3957–61
4. Ashworth DM, Robinson JA, Turner DL. 1982. Biosynthesis of nonactin from acetate, propionate, and succinate; the assignment of its carbon-13 N.M.R. spectrum by two-dimensional correlation spectroscopy. *J. Chem. Soc. Chem. Commun.* 1982:491–93
5. Bansal VS, Vaidya S. 1994. Characterization of two distinct allyl pyrophosphatase activities from rat liver microsomes. *Arch. Biochem. Biophys.* 315:393–99
6. Bartizal KF, Milligan JA, Rozdilsky W, Onshi JC. 1991. Novel anti-fungal compounds. *US Patent No. 5055487*
7. Bartizal KF, Onishi JC. 1991. Method of inhibiting fungal growth using squalene synthetase inhibitors. *US Patent No. 5026554*
8. Bartizal KF, Rozdilsky W, Onishi JC. 1991. Novel anti-fungal compounds. *US Patent No. 5053425*
9. Baxter A, Fitzgerald BJ, Hutson JL, McCarthy AD, Motteram JM, et al. 1992. Squalestatin 1, a potent inhibitor of squalene synthase, which lowers serum cholesterol in vivo. *J. Biol. Chem.* 267:11705–8
10. Berg D, Plempel M. 1988. *Sterol Biosynthesis Inhibitors. Pharmaceutical and Agrochemical Aspects.* Chichester, UK: Ellis Horwood. 583 pp.
11. Bergstrom JD. 1992. The zaragozic acids: potent natural product derived inhibitors of squalene synthase and cholesterol synthesis. In *Abstr. Int. Symp. on Drugs Affecting Lipid Metabol., 11th.*, p. 120. Houston, TX: Giovanni Lorenzini Med. Found.
12. Bergstrom JD, Kurtz MM, Rew DJ, Amend AM, Karkas JD, et al. 1993. The zaragozic acids. A family of fungal metabolites that are picomolar competitive inhibitors of squalene synthase. *Proc. Natl. Acad. Sci. USA* 90:80–84
13. Bergstrom JD, Liesch JM, Hensens OD, Onishi JC, Huang L, et al. 1991. Antihypercholesterolemics. *Eur. Patent No. 0450812A1*
14. Bergstrom JD, Onishi JC, Hensens OD, Zink DL, Huang L, et al. 1991. Anti-

hypercholesterolemics. *Eur. Patent No. 0448393A1*
15. Bergstrom JD, Onishi JC, Huang L, Bills GF, Nallin M, et al. 1991. Novel squalene synthetase inhibitors. *Eur. Patent No. 0475706A1*
16. Bills GF, Diez MT, Kong YL, Omstead MN, Peláez F. 1993. Novel culture of *Exserohilum rostratum* and processes therefrom. *Eur. Patent No. 0524671A1*
17. Bills GF, Diez MT, Omstead MN, Peláez F. 1993. Processes for preparing novel squalene synthetase inhibitors. *US Patent No. 5250424*
18. Bills GF, Diez MT, Omstead MN, Peláez F. 1993. Squalene synthetase inhibitors and processes therefrom. *US Patent No. 5200342*
19. Bills GF, Omstead MN, Clapp WH, Peláez F. 1994. Process for forming cholesterol lowering compound using *Pseudodiplodia* sp. *US Patent No. 5284758*
20. Bills GF, Peláez F, Polishook JD, Diez-Matas MT, Harris GH, et al. 1994. Distribution of zaragozic acids (squalestatins) among filamentous ascomycetes. *Mycolog. Res.* 98:733–39
21. Bills GF, Polishook JD. 1993. Selective isolation of fungi from dung of *Odocoileus hemionus* (mule deer). *Nova Hedwigia* 57:195–206
22. Blohm TR, Kariya T, Laughlin MW. 1959. Effects of MER-29 cholesterol synthesis inhibitor on mammalian tissue lipides. *Arch. Biochem. Biophys.* 85:250–63
23. Blohm TR, MacKenzie RD. 1959. Specific inhibition of cholesterol biosynthesis by a synthetic compound (MER-29). *Arch. Biochem. Biophys.* 85:245–49
24. Blows WM, Foster G, Lane SJ, Noble D, Piercey JE, et al. 1994. The squalestatins, novel inhibitors of squalene synthase produced by a species of *Phoma*. V. Minor metabolites. *J. Antibiot.* 47:740–54
25. Byrne KM, Arison BH, Nallin-Omstead M, Kaplan L. 1993. Biosynthesis of the zaragozic acids. 1. Zaragozic acid A. *J. Org. Chem.* 58:1019–24
26. Byrne KM, Chen S-ST, Kaplan L, MacConnell JG, Petuch BR, et al. 1994. Cholesterol lowering compounds produced by directed biosynthesis. *US Patent No. 5302604*
27. Cannell RJP, Dawson MJ, Hale RS, Hall RM, Nobel D, et al. 1993. The squalestatins, novel inhibitors of squalene synthase produced by a species of *Phoma*. IV. Preparation of fluorinated squalestatins by directed biosynthesis. *J. Antibiot.* 46:1381–89
28. Cannell RJP, Dawson MJ, Hale RS, Noble D, Lynn S, et al. 1994. Production of additional squalestatin analogues by directed biosynthesis. *J. Antibiot.* 47:247–49
29. Chen TS, Petuch B, MacConnell J, White R, Dezeny G, et al. 1994. The preparation of zaragozic acid A analogues by directed biosynthesis. *J. Antibiot.* 47:1290–94
30. Chiang Y-CP, Biftu T, Doss GA, Plevyak SP, Marquis RW, et al. 1993. Diesters of zaragozic acid A: synthesis and biological activity. *Bioorg. Med. Chem. Lett.* 3:2029–34
31. Clapp WH, Kong YL, Polishook JD. 1993. Fungal microorganism ATCC 74167 capable of producing cholesterol lowering compounds. *US Patent No. 5260215*
32. Currah RS. 1985. Taxonomy of the Onygenales: Arthrodermataceae, Gymnoascaceae, Myxotrichaceae, and Onygenaceae. *Mycotaxon* 24:1–216
33. Dawson MJ, Farthing JE, Marshall PS, Middleton RF, O'Neill MJ, et al. 1992. The squalestatins, novel inhibitors of squalene synthase produced by a species of *Phoma*. I. Taxonomy, fermentation, isolation, physio-chemical properties and biological activity. *J. Antibiot.* 45:639–47
34. De Hoog GS. 1977. *Rhinocladiella* and allied genera. *Stud. Mycol.* 15:1–140
35. Donadio S, Staver MJ, McAlpine JB, Swanson SJ, Katz L. 1991. Modular organization of genes required for complex polyketide biosynthesis. *Science* 252:675–79
36. Dreyfuss MM, Chapela IH. 1994. Potential of fungi in the discovery of novel, low-molecular weight pharmaceuticals. In *The Discovery of Natural Products with Therapeutic Potential*, ed. VP Gullo. pp. 49–80. Boston: Butterworth-Heinmann
37. Dufresne C, Turner-Jones ET, Nallin-Omstead M, Bergstrom JD, Wilson KE. 1995. Novel zaragozic acids from *Leptodontidium elatius*. *J. Nat. Prod.* Submitted
38. Dufresne C, Wilson KE, Singh SB, Zink DL, Bergstrom JD, et al. 1993. Zaragozic acids D and D2: potent inhibitors of squalene synthase and of ras farnesyl protein transferase. *J. Nat. Prod.* 56:1923–29
39. Dufresne C, Wilson KE, Zink D, Smith J, Bergstrom JD, et al. 1992. The isolation and structure elucidation of zaragozic acid C, a novel potent squalene synthase inhibitor. *Tetrahedron* 48:10221–26

40. Endo A. 1979. Monacolin K, a new hypocholesterolemic agent produced by a *Monascus* species. *J. Antibiot.* 32:852–54
41. Endo A, Kuroda M, Tsujita Y. 1976. ML-236A, ML-236B, and ML-236C, new inhibitors of cholesterolgenesis produced by *Penicillium citrinum*. *J. Antibiot.* 29:1346–48
42. Gibbs JB, Pompliano DL, Moser SD, Rands E, Lingham RB, et al. 1993. Selective inhibition of farnesyl-protein transferase blocks ras processing in vivo. *J. Biol. Chem.* 268:7617–20
43. Gilbert HJ, Tully M. 1982. Synthesis and degradation of phenylalanine ammonia-lyase of *Rhodospoidium toruloides*. *J. Bacteriol.* 150:498–505
44. Giron MD, Havel CM, Watson JA. 1994. Mevalonate-mediated suppression of 3-hydroxy-3-methylglutaryl coenzyme A reductase function in α-toxin-perforated cells. *Proc. Natl. Acad. Sci. USA* 91:6398–402
45. Gonzolez-Pacanowska D, Arison B, Havel CM, Watson JA. 1988. Isopentenoid synthesis in isolated embryonic *Drosophila* cells: farnesol catabolism and ω-oxidation. *J. Biol. Chem.* 263:1301–6
46. Greenspan M, Yudkovitz JB, Chen JS, Hanf DP, Chang MN, et al. 1989. The inhibition of cytoplasmic acetoacetyl-CoA thiolase by a triyne carbonate (L-660,631). *Biochem. Biophys. Res. Commun.* 163:548–53
47. Greenspan MD, Yudkovitz JB, Lo CL, Chen JS, Alberts AW, et al. 1987. Inhibition of hydroxymethylglutaryl-coenzyme A synthase by L-659,699. *Proc. Natl. Acad. Sci. USA* 84:7488–92
48. Harris GH, Joshua H, Zink DL. 1994. Cholesterol lowering compounds. *US Patent No. 5286895*
49. Hasumi K, Tachikawa K, Sakai K, Murakawa S, Yoshikawa N, et al. 1993. Competitive inhibition of squalene by squalestatin 1. *J. Antibiot.* 46:689–91
50. Hensens OD, Dufresne C, Liesch JM, Zink DL, Reamer RA, et al. 1993. The zaragozic acids: structure elucidation of a new class of squalene synthase inhibitors. *Tetrahedron Lett.* 34:399–402
51. Horie M, Tsuchiya Y, Hayashi M, Iida Y, Iwasawa Y, et al. 1990. NB-598: a potent competitive inhibitor of squalene expoxidase. *J. Biol. Chem.* 265:18075–78
52. Huang L, Lingham RB, Harris GH, Singh SB, Dufresne C, et al. 1995. New fungal metabolites as potential antihypercholesterolemics and anticancer agents. *Can. J. Bot.* In press
53. Itazaki H, Nagashima K, Kawamura Y, Matsumoto K, Nakai H, et al. 1992. Cinatrins, a novel family of phospholipase A_2 inhibitors. I. Taxonomy and fermentations of the producing culture: isolation and structure of cinatrins. *J. Antibiot.* 45:38–49
54. Jones BL, Porter JW. 1985. Enzymatic synthesis of phytoene. *Methods Enzymol.* 110:209–20
55. Jones CA, Sidebottom PJ, Cannell RJP, Noble D, Rudd BAM. 1992. The squalestains, novel inhibitors of squalene synthase produced by a species of *Phoma*. III. Biosynthesis. *J. Antibiot.* 45:1492–98
56. Katz L, Donadio S. 1993. Polyketide synthesis: propects for hybrid antibiotics. *Annu. Rev. Microbiol.* 47:875–912
57. Kitano K, Tozawa R, Harada S. 1993. Compound TAN-1607A, its derivatives, their production, and use thereof. *Eur. Patent No. 0568946A1*
58. Lane SJ, Brinded KA, Taylor NL. 1993. Analysis of the squalestatins by on-line packed capillary liquid chromatography/dynamic liquid secondary-ion/mass spectrometry. *Rapid Commun. Mass Spectrom.* 7:492–95
59. Lane SJ, Brinded KA, Taylor NL, Watkins PJF, Harrison ME. 1993. Analysis of the squalestatins by on-line packed capillary liquid chromatography combined with electrospray mass spectrometry. *Rapid Commun. Mass Spectrom.* 7:953–56
60. Marusich WC, Jensen RA, Zamir LO. 1981. Induction of L-phenylalanine ammonia-lyase during utilization of phenylalanine as a carbon or nitrogen source in *Rhotorulula glutinis*. *J. Bacteriol.* 146:1013–19
61. McCarthy AD, Fitzgerald BJ, Hutson JL, Motteram JM, Sapra M, et al. 1992. Squalene synthase inhibition: effects of a novel natural product on cholesterol metabolism. In *Abstr. Int. Symp. on Drugs Affecting Lipid Metabol., 11th.*, p. 120. Houston, TX: Giovanni Lorenzini Med. Found.
62. Milligan JA, Onishi JC, Bartizal K, Curotto J, Douglas C, et al. 1992. Zaragozic acids A, B, and C: antifungal activity and mechanism of action. *Abstr. Annu. Meet. Am. Soc. Microbiol.* p. 309
63. Miyakoshi S, Haruyama H, Shioiri T, Takahashi S, Torikata A, et al. 1992. Biosynthesis of griseolic acids: incorporation of ^{13}C-labeled compounds into griseolic acid A. *J. Antibiot.* 45:394–99
64. Monaghan RL, Tkacz JS. 1990. Bioactive microbial products: focus upon

mechanism of action. *Annu. Rev. Microbiol.* 44:271–301
65. Muscio F, Carlson JP, Kuehl L, Rilling HC. 1974. Presqualene pyrophosphate: a normal intermediate in squalene biosynthesis. *J. Biol. Chem.* 249:3746–49
66. Ness GC, Zhao S, Keller RK. 1994. Effect of squalene synthase inhibition on the expression of hepatic cholesterol biosynthetic enzymes, LDL receptor, and cholesterol 7α hydroxylase. *Arch. Biochem. Biophys.* 311:277–85
67. Nisbet LJ, Porter N. 1989. The impact of pharmacology and molecular biology on the exploitation of microbial products. *Bioact. Microb. Prod.* pp. 309–42
68. Pirozynsky KA. 1962. *Circinotrichum* and *Gyrothrix. Mycol. Pap.* 84:1–28
69. Procopiou PA, Bailey EJ, Bamford MJ, Craven AP, Dymock BW, et al. 1994. The squalestatins: novel inhibitors of squalene synthase. Enzyme inhibitory activities and in vivo evaluation of C1-modified analogues. *J. Med. Chem.* 37: 3274–81
70. Pronk JT, van der Linden-Beuman A, Verduyn C, Scheffers WA, van Diijen JP. 1994. Propionate metabolism in *Saccharomyces cerevisiae:* implications for the metabolon hypothesis. *Microbiology* 140:717–22
71. Sidebottom PJ, Hartley CD, Procopiou PA, Lester MG, Watson NS, et al. 1992. Bridged cyclic ketal derivatives. *Eur. Patent No. 0494622A1*
72. Sidebottom PJ, Highcock RM, Laine SJ, Procopiou PA, Watson NS. 1992. The squalestatins, novel inhibitors of squalene synthase produced by a species of *Phoma.* II. Structure elucidation. *J. Antibiot.* 45:648–58
73. Singer FM, Januszka JP, Borman A, Wintersteiner OP. 1959. New inhibitors of in vitro conversion of acetate and mevalonate to cholesterol. *Proc. Soc. Exp. Biol. Med.* 102:370–73
74. Sivanesan A. 1984. *The Bitunicate Ascomycetes and Their Anamorphs.* Vaduz: Cramer. 701 pp.
75. Sivanesan A. 1987. Graminicolous species of *Bipolaris, Curvularia, Drechslera, Exserohilum* and their teleomorphs. *Mycol. Pap.* 158:1–261
76. Staunton J, Sutkowski AC. 1991. ^{17}O NMR in biosynthetic studies: aspyrone, asperlactone and isoasperlactone, metabolites of *Aspergillus melleus. J. Chem. Soc. Chem. Commun.* 1991:1106–8
77. Sutton BC. 1980. *The Coelomycetes.* Kew, UK: Commonw. Mycol. Inst. 696 pp.
78. Tait RM. 1992. Development of a radiometric spot-wash assay for squalene synthase. *Anal. Biochem.* 203:310–16
79. Tanabe M, Hamasaki T, Suzuki Y, Johnson LF. 1973. Biosynthetic studies with carbon-13: Fourier transform magnetic resonance spectra of the metabolite avenaciolide. *J. Chem. Soc. Chem. Commun.* 1973:212–13
80. Wilson KA, Burk RM, Biftu T, Ball RG, Hoogsteen K. 1992. Zaragozic acid A, a potent inhibitor of squalene synthase: initial chemistry and absolute stereochemistry. *J. Org. Chem.* 57: 7151–58

PROSPECTS FOR NEW INTERVENTIONS IN THE TREATMENT AND PREVENTION OF MYCOBACTERIAL DISEASE

Douglas B. Young
Department of Medical Microbiology, St Mary's Hospital Medical School, Imperial College, London W2 1PG, United Kingdom

Kenneth Duncan
Glaxo Research and Development Limited, Gunnels Wood Road, Stevenage, Hertfordshire SG1 2NY, United Kingdom

KEY WORDS: tuberculosis, vaccines, drugs, mycobacteria, immunity

CONTENTS

INTRODUCTION	642
Tuberculosis	642
Other Mycobacterial Diseases	643
BASIC BIOLOGY OF MYCOBACTERIAL INFECTION	643
Stage 1. Initial Uptake of Bacteria	643
Stage 2. T-Cell Responses	646
Stage 3. Dormancy	648
Stage 4. Extracellular Multiplication and Transmission	649
NEW INTERVENTIONS: CURRENT RESEARCH STRATEGIES	650
Treatment	650
Prevention	659
NEW INTERVENTIONS: FUTURE PROSPECTS	665

ABSTRACT

Mycobacterium tuberculosis claims more lives each year than any other single human pathogen. Despite the availability of effective drugs, the incidence of tuberculosis is increasing in much of the developing world and has recently reemerged as a public health problem in industrialized countries. In the first section of this chapter, current understanding of the fundamental biology of mycobacterial infection is reviewed from the perspective of development of new tools for disease control. A second section describes strategies for identification of novel antimycobacterial agents, with particular emphasis on recent progress in defining biosynthetic pathways for unique mycobacterial cell wall components. The third section focuses on current approaches to the development of new vaccine candidates consisting of live attenuated bacteria or individual antigenic subunits.

INTRODUCTION

Tuberculosis

In the last decade of the twentieth century, an estimated 90 million cases of tuberculosis (TB) will occur, resulting in over 30 million deaths—a toll greater than that of any other single human pathogen (20). The vast majority of these cases will occur in the developing world (103), but even industrialized countries, after a century of progressive decline, have seen an increase in TB-case rates in recent years (20, 113). If present trends continue, the incidence of TB will rise by an estimated additional one third over the next decade (49). These statistics are clearly disappointing for a bacterial disease that is fully curable in almost all cases and that until recently was considered one of the major successes of the antibiotic revolution.

Current strategies for control of TB center on treatment with multidrug regimens based on the very effective combination of rifampicin and isoniazid. In endemic areas, the diagnosis and treatment of smear-positive patients—individuals who are secreting large numbers of bacteria—are emphasized in order to interrupt spread of the disease within the community (85). Obstacles to the success of this strategy are the difficulty of early diagnosis and operational problems associated with delivery of a treatment that involves administration of multiple drugs over a period of at least six months (20, 104). Factors that have recently compounded these difficulties include increases in host susceptibility to TB caused by coinfection with human immunodeficiency virus (HIV) and the emergence of *Mycobacterium tuberculosis* strains that are resistant to the front-line drugs (14, 20, 132).

Against this background, described by the World Health Organization (WHO) as a global emergency, stands a need for increased effort to implement the available treatment as widely and efficiently as possible. The estimated mortality of one in three cases underlines the fact that a substantial section of the population is currently excluded from the benefits of optimal therapy. At the same time, however, new strategies for TB control must be developed. In this chapter, we review the current status of mycobacterial research, primarily from the perspective of potential new strategies for prevention and treatment of tuberculosis based on new drugs and immune modulation.

Other Mycobacterial Diseases

Leprosy, caused by *Mycobacterium leprae,* a close relative of *M. tuberculosis,* presents a very different control problem. Multidrug treatment regimens have led to the introduction of defined treatment periods of 6–24 months in place of lifetime treatment mandated by previous monotherapy programs. The number of leprosy patients worldwide has consequently been reduced from an estimated 10–15 million in the 1980s to a current total of less than 5 million (106). WHO has set a goal of reducing the prevalence of leprosy to less than 10 per 100,000 of the population worldwide, which would essentially eliminate leprosy as a public health problem by the year 2000. However, the current success in leprosy control must not lead to the type of complacency in leprosy research that was generated by early successes in TB chemotherapy.

Individuals infected with HIV are very susceptible to *M. tuberculosis,* and synergy between the two infections is a major cause of death in endemic regions (47, 127). In industrialized countries, where exposure to *M. tuberculosis* is less frequent, severely immunodeficient patients are often infected with nontuberculous mycobacteria belonging to the *Mycobacterium avium-intracellulare* complex (MAI). MAI infections, while less contagious than TB, are difficult to treat because these mycobacteria are generally less susceptible than *M. tuberculosis* to the available antimycobacterial agents.

BASIC BIOLOGY OF MYCOBACTERIAL INFECTION

Many mycobacterial species other than *M. tuberculosis* are widespread throughout the environment and are readily isolated from soil and water supplies. The major mycobacterial pathogens characteristically have slow growth rates; the doubling time of *M. tuberculosis* is around 24 h, whereas *M. leprae,* which has yet to be grown in artificial laboratory-culture media, probably divides only once in two weeks. The course of tuberculosis infection has been extensively studied in humans and in experimental animals. Figure 1, adapted from histopathological analysis as described in detail by Dannenberg (41–43), represents a simplified summary of key stages in the disease process. This model provides a framework for identification of stages that may be amenable to novel intervention strategies.

Stage 1. Initial Uptake of Bacteria

M. tuberculosis enters the human host predominantly in the form of aerosol droplets taken up by alveolar macrophages in the lung. These macrophages,

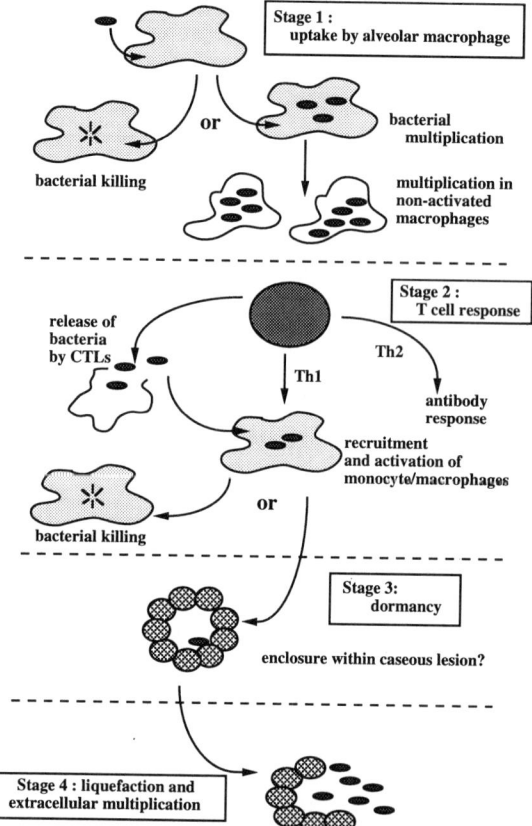

Figure 1 Schematic representation of stages of *M. tuberculosis* infection.

activated by persistent exposure to inhaled particulate material, possess potent microbicidal activities, which, presumably, often arrest TB infection at this initial stage. If the bacteria survive and multiply within the alveolar cells, however, they can then spread to adjacent nonactivated macrophages and set up a focus of infection. *M. tuberculosis* appears to be well-equipped for this initial macrophage encounter and exhibits numerous survival strategies based either on avoidance of exposure to microbicidal mechanisms or on inherent resistance to their toxic effect.

Studying mouse peritoneal macrophages, Hart (9) demonstrated an inhibition of the fusion of lysosomes with phagosomes containing *M. tuberculosis* (9). Phagosome-lysosome fusion was observed when the experiments were repeated in the presence of immune sera (8). With or without fusion, however, the macrophages failed to restrict growth of the phagocytosed *M. tuberculosis*. Survival of *M. tuberculosis* within nonacidic vacuoles has been demonstrated in human macrophages (38), and mycobacteria-containing vacuoles were reported to fuse with lysosomes that lack the proton ATPase pump essential for acidification (139). *M. tuberculosis* can also be found in other intracellular compartments. McDonough et al (93) observed bacilli enclosed within small tightly bound vacuoles and, in the case of virulent but not avirulent strains, found free *M. tuberculosis* within the cell cytoplasm. The diversity of these observations suggests that the initial interaction of *M. tuberculosis* with macrophages can proceed along many pathways. Factors that could influence the course of this interaction include the action of bacterial products released into the cytoplasm or into the phagosome membrane and variations in the pathway of cell entry. Macrophages express several surface receptors available for mycobacterial binding, including complement receptors CR1 and CR3, the mannose receptor, and the antibody Fc receptor (125, 126), and the particular receptor engaged by each individual bacterium might influence subsequent intracellular events.

Other intracellular bacterial pathogens have evolved molecules capable of binding to specific receptors on the eukaryotic cell surface so that the bacteria can effectively regulate their own intracellular fate (52), and the idea that mycobacterial pathogens may have adopted analogous strategies has engendered considerable interest (10). *M. tuberculosis* enters nonprofessional phagocytic cells in vitro, but bacteria occur only within macrophages in vivo (120). Infected cells have enhanced susceptibility to killing mediated by tumor necrosis factor (TNF-α) (53), and phagocytosis of dead cells containing internalized *M. tuberculosis* represents another potentially interesting route for macrophage entry.

In contrast to *M. tuberculosis*, *M. leprae* occurs in many cell types. Leprosy bacilli appear to have a unique ability to enter Schwann cells in the peripheral nervous system, and immune responses triggered by nerve-associated *M. leprae* underlie the pathology of leprosy (78).

Even in the absence of such diversionary tactics, pathogenic mycobacteria are well-equipped to withstand exposure to microbicidal mechanisms. A lipid-rich outer capsule presents a barrier to hydrolytic enzymes and defends the bacterial cell against toxic radicals (25, 96). Superoxide dismutase present in the culture medium of *M. tuberculosis* (6) and MAI strains (91) may provide additional protection against macrophage-derived oxidative radicals. The most consistent evidence for macrophage-mediated killing of *M. tuberculosis* has been obtained using murine cells. Reactive nitrogen intermediates, rather than oxygen radicals, are currently envisaged as the most potent effector molecules (26, 27), and the apparent absence of inducible nitric oxide synthase in human macrophages may contribute to their inability to restrict the growth of *M.*

tuberculosis in in vitro systems (50). Internalized mycobacteria may undergo phenotypic changes to adapt to the intracellular conditions. Transcriptional regulation associated with different stages of infection is a key feature of bacterial pathogenesis (57, 89), and altered gene expression following phagocytosis of *M. avium* has been described (112).

Stage 2. T-Cell Responses

The initial growth of *M. tuberculosis* in macrophages results in formation of small necrotic lesions with a solid caseous center in which mycobacterial multiplication is probably restricted. Cell proliferation and skin test reaction in animal models a few weeks after infection initiation indicate activation of mycobacteria-reactive T lymphocytes at this stage. Vaccination with Bacille Calmette-Guerin (BCG) seems to exert its protective effect at this stage of the infection (94), and the specificity and functional role of mycobacteria-reactive T cells has been the subject of extensive and continuing analysis.

T CELLS AND PROTECTION   An effective T-cell response is essential for protection against mycobacterial disease. Increased risk of TB is a relatively early consequence of diminished T-cell numbers in HIV-infected individuals, while susceptibility to MAI infection is associated with severe immunodeficiency (37). Two functions are currently envisaged for T cells involved in protective immune responses (26). Cytotoxic T cells are thought to be required to release intracellular mycobacteria that have established themselves within macrophages (or other cell types in the case of *M. leprae*); helper T cells are required to recruit and activate new monocytes or macrophages to the site of the lesion. The $CD4^+$ T-cell subset depleted during HIV infection probably plays the central role in macrophage activation, while interferon-γ (IFN-γ) serves as a key cytokine, at least in the mouse model (36, 58). $CD4^+$ T cells isolated from human peripheral blood possess cytotoxic activity in in vitro assays, but this activity is more usually associated with the $CD8^+$ subset (82). A critical role for $CD8^+$ T cells in the murine response to *M. tuberculosis* is indicated by the marked increase in susceptibility of mice that cannot activate such cells because of deletion of the gene encoding β_2-microglobulin, an essential component of antigen recognition by $CD8^+$ cells (59). In addition to their possible cytotoxic function, $CD8^+$ cells could also provide an important source of IFN-γ. Another subset of lymphocytes, characterized by an antigen receptor composed of γ and δ chains, also secrete IFN-γ and may be important triggers at the initial stage of the immune response (76, 100).

T CELLS AND PATHOLOGY   Most patients with mycobacterial disease have an active cell-mediated immune response that appears to contribute to tissue pathology and thus prompts discussion of the relationship between immunopathology and protective immunity. Leprosy, which occurs in various clinical manifestations, provides an important system for studying this relationship. In the model outlined above, a failure to couple cytotoxic activities required for release of intracellular mycobacteria with concomitant macrophage activation could promote unrestricted bacterial growth. Such growth would correspond to the clinical situation in the lepromatous form of leprosy, characterized by very large numbers of bacteria and an absence of IFN-γ production (158). At the other end of the leprosy spectrum, lesions in tuberculoid disease are characterized by activated macrophages and the almost total absence of intact bacilli. Nevertheless, tuberculoid leprosy patients suffer from severe immune-mediated nerve damage.

This immunopathology can be explained in two ways: The immune response in tuberculoid leprosy may have been activated only after bacteria had become established within the lesion, and tissue damage might be an inevitable consequence of an essentially protective response mounted in the presence of high antigen load. TNF-α, for example, has an important role in protective responses (68, 83), but in excess, TNF-α is also a key mediator of pathology (2, 123). Alternatively, the immune response in tuberculoid patients may be inherently pathogenic; tuberculoid and lepromatous disease could each represent a failure to regulate an appropriate balance of T-cell reactivities. Tuberculosis presents a picture intermediate between the polar forms of leprosy, but similar questions related to kinetics and regulation of the responses of different T-cell subsets are also central to understanding of immune protection.

T-CELL SUBSETS   Factors involved in determining the type of immune response generated are under intensive investigation. Activation of CD8$^+$ T-cell responses requires presentation of mycobacterial antigens by proteins belonging to the class I major histocompatibility complex (MHC) (63). Antigens presented by MHC class I are derived from the cytoplasm of the antigen-presenting cell, and the ability of virulent strains of *M. tuberculosis* to escape from phagocytic vacuoles could significantly affect their recognition by CD8$^+$ cells. CD4$^+$ T cells recognize antigens presented by class II MHC molecules, which are preferentially derived from phagocytic vesicles (63).

Extensive evidence for CD4$^+$ T cells [and similar evidence for CD8$^+$ (122)] shows that, during activation, they become committed to expression of one of two distinct sets of cytokines (102). Cytokines that promote macrophage activation (e.g. IFN-γ) characterize the Th1 response, whereas Th2 cytokines [e.g. interleukin (IL)–4] promote antibody formation at the expense of macrophage activation. The balance of Th1 and Th2 responses is influenced by the local cytokine environment at the early stages of T-cell activation (69) and perhaps by additional endocrine factors (118). Key cytokines include IL-12, which promotes Th1 responses, and IL-10, which favors Th2. Both cytokines are secreted by macrophages (as well as other cell types) in response to bacterial components.

Precise details of the interaction of a particular mycobacterial strain and antigen-presenting host cells may therefore be important in determining the type of T-cell response induced. One mycobacterial product known to regulate cytokine expression by macrophages is lipoarabinomannan (LAM) (28). Interestingly, structural differences in LAM isolated from different mycobacterial strains affect its ability to trigger signal transduction pathways in macrophages and to induce expression of TNF-α (15, 29, 114).

T-CELL REPERTOIRE   Protein antigens recognized by mycobacteria-reactive T cells have been extensively characterized (4, 72, 160). Experiments with fractionated and recombinant antigen preparations show that the T-cell repertoire varies considerably between individuals, with no obvious association between disease status and antigen repertoire (66; JER Thole, unpublished data). Attention is currently focused on characterization of protein antigens secreted from actively growing *M. tuberculosis* (1, 6, 7, 108). It is reasoned that secreted proteins may become available for T-cell recognition at an early stage of the infection, and immune responses to such antigens may be preferentially associated with protection.

Stage 3. Dormancy

For most individuals living in endemic areas, infection with *M. tuberculosis* progresses to the T cell–activation stage (as judged by conversion to skin test positivity), but in only a few individuals does the primary infection manifest as clinical disease. Primary, or childhood, TB may occur as meningitis or as a disseminated disease. Among the remaining infected population, one in ten will develop the typical adult form of pulmonary TB at some later stage in life. Progression to clinical disease is dramatically increased to a 3–8% annual risk in HIV-infected individuals (127). The extent to which adult TB results from reinfection as compared with reactivation may vary in different settings, but in a significant number of cases, disease is clearly caused by bacteria that have been carried in an asymptomatic form over many years. The location of such persisting organisms, the mechanisms that allow their continued viability, and the factors responsible for reactivation are questions central to our understanding of mycobacterial pathogenesis.

One attractive location for persistent mycobacteria would be in the caseous center of necrotic lesions that ultimately become walled off by fibrosis. In laboratory culture, *M. tuberculosis* can be converted into a quiescent, or dormant, form by exposure to anaerobic conditions (150); analogous conditions may exist in the center of a caseous lesion. Previous exposure to partial anaerobiosis enhances bacterial survival under anaerobic conditions (151), which is indicative of an active process requiring induction of survival pathways. Understanding of the physiology of dormant mycobacteria could lead to development of novel drug-development strategies using compounds with a direct toxic effect on dormant organisms or compounds that inhibit specific steps involved in entry into the dormant state. The recent report of an effect of metronidazole on anaerobic cultures of *M. tuberculosis* provides an interesting lead in this regard (152).

Alternatively, persisting bacteria may reside intracellularly and be prevented from multiplication by an active immune response. A requirement for continuous suppression by the immune system would account for reactivation disease associated with immunodeficiency. In the extracellular model, reactivation of dormant foci might be considered a periodic event that has clinical significance only if uncontrolled by the appropriate T-cell response. In the absence of immune defects, such localized reactivation may indeed have a beneficial effect in which intermittent boosting effectively sustains T-cell memory. The ability to manipulate the immune response during this dormant stage of the disease could have immense practical benefit in relation to novel approaches to TB control.

Stage 4. Extracellular Multiplication and Transmission

The most common form of TB involves extracellular bacterial multiplication within cavitary lesions in the lung. Solid caseous foci capable of containing the growth of *M. tuberculosis* are converted, by an unknown process, to liquefied capsules ideal for bacterial multiplication (41, 42). Even in the immunocompetent host, release of massive numbers of bacteria from such lesions may be sufficient to overpower the local immune response. As noted above, this form of the disease, in which large numbers of bacteria are released in smear-positive sputum, is the form largely responsible for droplet transmission of TB.

Existing antimycobacterial agents are very effective in killing tubercle bacilli in this stage of the disease, when the bacteria are actively multiplying and mostly extracellular. Currently available treatment regimes are limited by the potential for selection of drug-resistant isolates and by the requirement for prolonged therapy. Treatment with a single drug rapidly selects resistant mutant strains, but combination therapy with at least three drugs very effectively prevents such selection. The recent outbreaks of multidrug-resistant strains (MDR-TB) appear to have resulted from failure to follow prescribed treatment regimens, however, and spread of such strains presents an important threat to existing TB control strategies (67). The requirement of prolonged therapy—six months represents so-called short-course chemotherapy—to prevent recurrence of active disease after cessation of treatment may reflect the fact that some bacteria persist in a dormant, relatively drug-resistant form, as discussed above, or within some sequestered foci out of reach of drugs.

NEW INTERVENTIONS: CURRENT RESEARCH STRATEGIES

Table 1 lists existing control measures in terms of the different stages of TB infection described above along with potential areas for novel interventions. In the remainder of this chapter, we discuss current research activities in relation to these goals. An important aspect underlying much of the current trends in mycobacterial research is the availability of molecular genetic tools for studying mycobacteria. The accumulation of information about the genome structure of mycobacterial pathogens (32, 51), along with the development of vectors and protocols for the introduction of DNA into mycobacteria (74, 86, 130, 138), may lead the way to novel experimental strategies to achieve insights into molecular mechanisms of pathology and immunogenicity.

Treatment

ANTIMYCOBACTERIAL AGENTS The properties of drugs currently recommended for treating mycobacterial infections were reviewed recently (39). The emergence of MDR-TB (14) has highlighted the urgent need for new drug classes to counter the threat posed by these strains. Compliance problems may be alleviated by improving pharmacokinetic properties, thus permitting reduced frequency and/or duration of therapy. Finally, drugs with a broader spectrum of activity are required to combat emerging pathogens such as MAI.

Table 1 Interventions for control of tuberculosis

Stage	Existing interventions		Potential interventions	
	Drugs	Vaccines	Drugs	Vaccines
1. Initial uptake of bacteria	Chemoprophy-laxis		Drugs affecting host cell interactions	Antibody regulation of uptake
2. T-cell response		BCG vaccine		Attenuated vaccines, subunit vaccines
3. Dormancy			Drugs affecting dormancy	Antireactivation vaccine
4. Extracellular multiplication and transmission	Multidrug therapy		New antimycobacterials	Immunotherapy

The desirability of developing agents to be used solely for treating mycobacterial infections is under debate. A novel broad-spectrum antibacterial agent that can also be used to treat mycobacterial infections may at first appear attractive—the range of treatable diseases widens and therefore sales of the agent increase. However, this view does not take into account the fact that the majority of TB patients who must pay for their medications cannot afford the high cost of new medicines that may have been initially developed to treat acute, life-threatening infections. From a clinical viewpoint, an orally active, broad-spectrum therapeutic agent may cause undesirable side effects when taken over the long period specifically needed to completely clear a mycobacterial infection. Disturbances of the normal gut flora may occur, and the risk of developing high-level, transferable drug resistance increases. Moreover, when drug supplies are limited, a broad-spectrum agent will inevitably be diverted to treat acutely ill patients with other bacterial sepsis. Selection of a highly targeted drug has some advantages, not least of which is the optimization of the drug's properties allowing it to act against a very small number of species, rather than the compromises necessary for the broadest spectrum of action. For a specific anti-TB drug, a low overall treatment cost must be maintained so that the drug can benefit the maximum number of patients. However, the volume of sales must still be adequate to recover the high cost of bringing the drug to the market.

NEW DRUGS   New drug therapies may be generated in several ways: First, an existing drug can be chemically modified to improve its antimycobacterial activity and its pharmacokinetic properties and to make it less susceptible to the known mechanisms of resistance. The continuous development in rifampicin analogues exemplifies such changes. Rifabutin, a spiropiperidyl derivative of rifampicin S offers such benefits, and more importantly, its wider spectrum of activity encompasses the nontuberculous mycobacteria (23). Its clinical uses include prophylactic therapy and the treatment of active disseminated *M. avium* infections in immunocompromised patients. A new benzoxazinorifamycin, KRM-1648, has excellent in vitro activity against both *M. tuberculosis* and *M. avium* (121) and significant efficacy in a murine model of *M. intracellulare* infection (144). Furthermore, in a murine TB model, activity of KRM-1648 was superior to rifampicin and to rifabutin; combination experiments with isoniazid or pyrazinamide suggested that very short course regimens of therapy—four months or less—might be feasible (84).

Second, new drugs can be developed from existing lead molecules, the most obvious of which are agents used to treat other bacterial infections. Fluoroquinolones show promising anti-TB activity, particularly those recently developed. Sparfloxacin, clinafloxacin, levofloxacin, ciprofloxacin, and ofloxacin have minimum inhibitory concentrations (MICs) in the range 0.06–0.5 µg ml^{-1}, within achievable serum concentrations, against clinical *M. tuberculosis* isolates (159); *M. avium* is less susceptible. Ofloxacin and ciprofloxacin have been used to treat patients successfully. Quinolones concentrate well within macrophages and distribute into the lungs during therapy, but rapidly emerging resistance is an apparent drawback to their more widespread use. Resistant mycobacteria acquire mutations in the *gyrA* subunit gene of DNA gyrase, the target protein, analogous to those expressed in resistant *Enterobacteriaceae* and staphylococci (24, 142). Therefore, mycobacteria have considerable cross resistance to the entire class of quinolones (159). Animal studies have shown that, although well tolerated, quinolones cause damage to cartilage growth, which precludes their use in children, adolescents, and pregnant women.

Third, completely new lead molecules must be discovered either by random screening or, with a detailed knowledge of a specific target, by rational design. Prospects for both are constantly improving. In the former, technological advances, particularly in the application of robotics, have led to increased screen throughput, and the advent of combinatorial libraries has greatly increased both the number and the diversity of the test samples. Rapid, facile whole-cell assays are available for high-throughput screening, based on bioluminescence (35) or uptake of radiolabeled uracil (GA Chung & K Duncan, unpublished results), to detect antimycobacterial activity. For rational drug design, overexpression of recombinant proteins and better protein-purification methods allow one to obtain sufficient protein for high-resolution structural studies relatively easily. Improvements in X-ray data–collection technology, coupled with increased computing power, have substantially reduced the time taken to resolve structures. With a better understanding of protein-ligand interactions, the concept of rationally designed enzyme inhibitors has become a reality. As an example, the design of potent sialidase inhibitors (148) has led to development of a compound currently in early clinical trials as an anti-influenza virus drug.

Disappointingly few new chemical entities have been reported with significant antimycobacterial activity, although this deficit may simply reflect the lack of effort to find such agents. Two compound classes with promising in vitro and in vivo activity against *M. tuberculosis* are the oxazolidinones, such as DuP721 (12), and the nitroimidazoles, such as CGI 17341 (11). However, no clinical data have been reported, and development of these compounds does not appear to be underway.

NEW TARGETS FOR ANTIMYCOBACTERIAL DRUGS   In this section, we consider new targets in the mycobacterial cell for chemotherapeutic intervention. Certain parameters define the ideal target. For instance, the target's function should be essential for growth, and its inhibition should lead to cell death (i.e. rapid bactericidal activity), rather than just arrest of growth (i.e. bacteriostatic activity). A second attribute would be lack of homologous activity within the mammalian cell, which would eliminate problems involving selectivity (or lack thereof) and potential toxicity. In addition, as set out above, functions specifically found in the mycobacterial cell and absent from other bacteria should be targeted. The identification of new targets in the mycobacterial cell has been hindered by the lack of detailed knowledge of its structural and biochemical features. With our present understanding, listing potential target areas is fairly straightforward, but actually defining the processes involved and then designing inhibitors of those functions is somewhat more challenging.

Clues from existing drugs The mode of action of few TB drugs is known in any detail, yet they indicate essential, bactericidal targets. A knowledge of the target can be exploited to find new inhibitory molecules that do not share common structural features. Hence, resistance will not develop as quickly as when modifications are made to the original drug. Defining a single drug target may reveal other related targets. If the target is an enzyme in a biosynthetic pathway, for example, inhibition of other enzymes in the pathway will probably lead to a similar effect on the cell. Those other pathway enzymes may be more amenable to study or accessible to inhibition. Much of the detailed work characterizing the mode of action of anti-TB drugs began by looking at the effect of the drug on other bacteria and then extended the interpretation to mycobacteria. By default, only agents with at least some broad-spectrum activity can be defined in this way, and consequently, little is known in detail about the mycobacteria-specific agents. Rifampicin binds to RNA polymerase, preventing transcription; resistance emerges because of point mutations over approximately 27 codons near the center of the β-subunit gene, *rpoB*, in both *Escherichia coli* and *M. tuberculosis* (31).

In contrast, both the true mode of action and the reason for the exquisite sensitivity of the pathogenic mycobacteria to isoniazid remain controversial. Several hypotheses have been suggested to explain this drug's action, ranging from inhibition of mycolic acid synthesis to interference with nicotinamide adenine dinucleotide (NAD) metabolism (155). The *katG* product, a catalase-peroxidase, is also implicated. Because strains lacking *katG* are resistant, isoniazid may be altered within the cell before it hits its target (31, 162), perhaps changing into any of several highly reactive species that can oxidize or acylate enzymes (79).

Banerjee et al (13) isolated the *inhA* gene from isoniazid-resistant mycobacteria. A single base change from the gene sequence in isoniazid-sensitive strains presumably renders it insensitive to the activated isoniazid. The InhA protein has significant homology with an *E. coli* protein, EnvM, thought to be involved in fatty acid biosynthesis. Cell-free mycolic acid synthesis assays were less susceptible to inhibition by isoniazid when prepared from organisms harboring the mutant *inhA* gene. Although the definitive role of InhA has not been reported, the structure has been determined to 2.2-Å resolution (163), which will permit design of compounds that can bind to both isoniazid-sensitive and -resistant forms of the protein.

The mycobacterial cell envelope Macromolecules comprising the thick waxy coat of the mycobacterium provide a dense barrier to anti-TB agents and protection from host defenses. This highly complex and well-organized structure, unique to the mycobacteria, represents the best overall target for novel antimycobacterial agents. From the pioneering work of Minniken, Brennan, and their colleagues, much is known about the structure of the cell envelope. Its main elements are peptidoglycan, arabinogalactan, the mycolic acids, and lipoarabinomannan (22, 96, 98). Until recently, our knowledge of the enzymes responsible for biogenesis of the cell wall components was almost negligible. The pathways are now being elucidated in several laboratories, but as yet few specific targets have been clearly identified. We now consider several polymers in turn and highlight some potential targets for new drugs.

Mycobacterial peptidoglycan is similar in structure to that found in other bacteria, except for two notable differences. First, a proportion of the muramic acid is *N*-glycolated, rather than *N*-acetylated. Whether this modification, introduced by oxidation of the *N*-acetyl group in precursor UDP-*N*-acetylmuramic acid, is essential for viability is unclear (62). Second, *Mycobacterium smegmatis* peptidoglycan is cross-linked via both D-alanyl-*meso*-diaminopimelate and *meso*-diaminopimelate-*meso*-diaminopimelate linkages in a 2:1 ratio (154). *Mycobacterium bovis* BCG and *M. tuberculosis* have similar linkages, although their number and distribution were not confirmed. The transpeptidase that catalyzes the cross-linking reaction is uncharacterized. Although the presence of this unusual linkage may explain the poor activity of β-lactam antibiotics against *M. tuberculosis*, the activity of Augmentin (amoxicillin plus the β-lactamase inhibitor clavulanic acid) suggests that β-lactamase activity contributes to resistance (77).

Arabinogalactan consists of ~30 α-D-galactofuranose and ~60 α-D-arabinofuranose residues (40). It is linked to peptidoglycan via a unique diglycosylphosphoryl bridge containing L-rhamnose and *N*-acetylglucosamine (95); agents targeted against formation of this attachment point would dramatically affect integrity of the whole cell wall, yet this point of attack has not been investigated. The galactosyl residues are in the five-membered furanose ring conformation (40) rather than in the thermodynamically favorable six-membered pyranose. Galactofuranose is seldom found in nature; it is restricted to a few microorganisms and is rarely an essential structural component. Inhibitors of its formation and incorporation are likely to be both highly selective and nontoxic. No galactofuranose pathways have been characterized in biological systems. During biosynthesis of galactocaralose [an extracellular β-D-(1→5)-linked poly-galactofuranose found in *Penicillium charlesii*], the pyranose-to-furanose ring contraction takes place on a sugar nucleotide, UDP-galactose (145, 146). A galactofuranosyl transferase activity is probable. However, this activity has not been characterized, and both the donor and acceptor molecules are unknown. No lipid-linked galactose has been reported.

Our knowledge of how arabinose is formed and incorporated into the wall is no better. Experiments in which ^{14}C-glucose was fed to *M. smegmatis* and its fate examined led to the proposal for a pathway that initially follows the pentose phosphate shunt, then proceeds via arabinose-5-phosphate and arabinose-1-phosphate to formation of GDP-arabinose, the sugar phosphate donor of activated arabinose for biosynthesis (141). Ethambutol was thought to act at a stage between formation of D-arabinose and its incorporation into arabinogalactan (141). However, when this proposal was put to the test, key enzymes in the pathway, such as arabinose-5-phosphate isomerase, could not be identified in extracts prepared from *M. smegmatis* (A Weston & K Duncan, unpublished observation). Wolucka et al (156) characterized a family of monoglycosyl polyprenylphosphates in *M. smegmatis* that contained the sugars ribose, arabinose, and mannose. Ethambutol treatment of *M. smegmatis* resulted in accumulation of β-D-arabinofuranosyl-1-monophosphodecaprenol, which indicates that either the transfer step from arabinose donor to wall acceptor, or acceptor synthesis, is inhibited. A novel epimerase interconverting lipid-ribose and lipid-arabinose was postulated, although no evidence for its existence was presented (156). Expression in *M. smegmatis* mc^{2}155 of a DNA fragment cloned from *M. avium* results in resistance to ethambutol. Sequence analysis of this fragment revealed two open reading frames (ORFs). The hydrophobicity profiles of the ORF products suggest membrane association and indicate that they may be arabinosyl transferase(s) (J Inamine, personal communication).

The arabinan portion of arabinogalactan terminates in a unique nonreducing, branched hexa-arabinose motif (97). Ester linkages are formed between the 5-positions of the terminal and penultimate arabinoses, and the mycolic acids are formed (96). Figure 2 shows the arabinan structure, and indicates the potential for multiple targets in the transfer of arabinose sugars. The links between arabinoses differ in type, e.g. (1→3) or (1→5), and in substitutions on neighboring residues. Several enzymes will catalyze the formation of the various bonds. However, each enzyme will recognize a somewhat similar acceptor molecule and probably the same donor molecule. Therefore, a molecule that interferes with the activity of all of the arabinosyl transferases would be lethal to the cell, and a single genetic mutation would almost certainly not result in drug resistance. This situation is analogous to the action of the β-lactam antibiotics in peptidoglycan biosynthesis: Antibiotics hit multiple

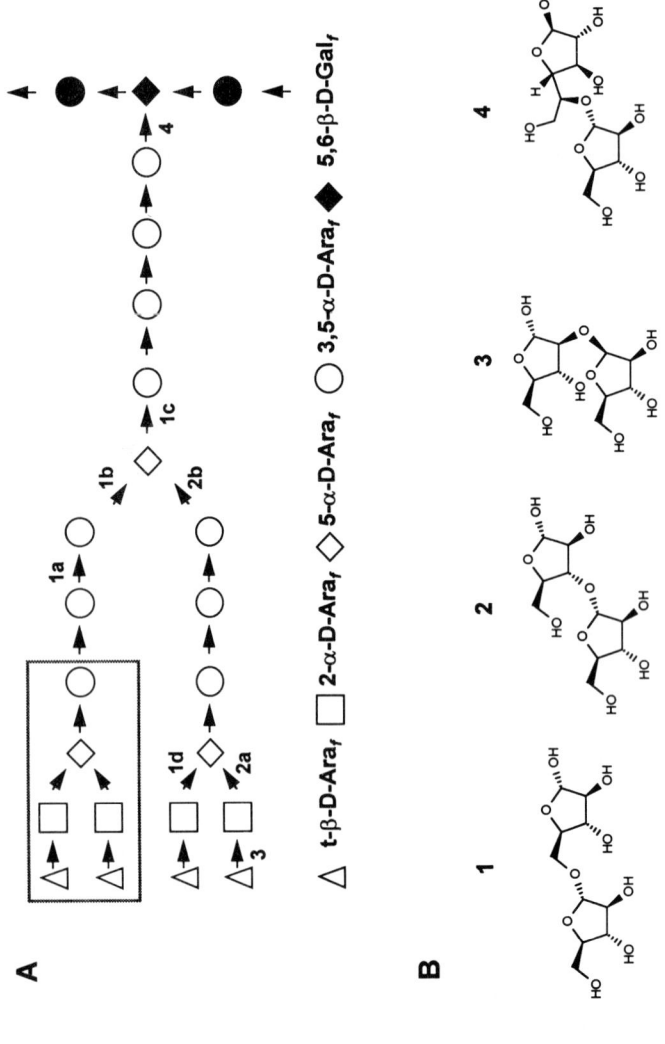

Figure 2 (*A*) Diagrammatic representation of the arabinan portion of arabinogalactan (adapted from Reference 40). Arrows represent bonds between sugar residues; numbers refer to the various bonds seen in arabinan. One of the hexa-arabinose motifs referred to in the text is boxed. (*B*) Structure of four linkage types. Note that, although the same two positions may be linked on neighboring arabinoses, e.g. types 1a–1d are all ($\alpha 1 \rightarrow 5$), the sugars on either side may have different attachments at other positions, and different enzymes may therefore be required for bond formation.

targets, the penicillin-binding proteins, all of which carry out similar transpeptidation reactions. Resistance is mediated by enzymes that inactivate the drug (44, 105), rather than via target modification [although examples of the latter are known (134)].

Mycolic acids, 3-hydroxy,2-alkyl-branched fatty acids containing 60–90 carbon atoms, make up the largest single component of the mycobacterial cell envelope. Mycolic acids have no single structure, but rather a mixture of forms, the nature and composition of which vary among species. Some mycolates are covalently linked to arabinogalactan. Others form an integral part of the wall but can be removed by nondestructive methods. The structure of the mycolates has been reviewed extensively in the past (17, 21, 98, 99). Biosynthetic pathways have been proposed, but little has been published on the nature of the enzymes involved. Pathway intermediates were synthesized (18) and incorporated into mycolates by extracts of *M. smegmatis* (153). Enzymes catalyzing the early biochemical reactions are likely to be located within the mycobacterial cell membrane. Therefore, the enzyme (or enzymes) that catalyzes the covalent attachment of the mycolate onto arabinogalactan is of greater interest for drug targeting, by virtue of accessibility. A recently isolated lipid-linked mycolate, Myc-PL (16) was postulated to be a key intermediate: the final product of the mycolate biosynthetic pathway and substrate for transfer onto arabinose (16). An enzyme that catalyzes the exchange of a mycolyl group between trehalose, trehalose-6-monomycolate, and trehalose-6,6'-dimycolate was purified from *M. smegmatis* and proposed to transfer the mycolate onto arabinogalactan, although no evidence was presented for this proposal (124).

Besra & Chatterjee (17) recently reviewed the composition and structure of lipoarabinomannan (LAM). As discussed above, LAM has been implicated in many of the immunoregulatory activities of mycobacteria that affect expression of cytokines by macrophages. Subtle modifications to its structure have an important impact on its biological activity (28, 29). Curiously, LAM also contains the unusual hexa-arabinose motif described previously, yet no mycolate-capped LAM has ever been described; rather the LAM is capped with mannose. Biosynthesis of LAM has not been explored, but the enzymes that specify its synthesis undoubtedly include good targets for new drugs.

Other targets Several other processes in the cell may represent alternative targets for new drugs. For example, most organisms produce low-molecular-weight thiols that play a crucial role in protection against the effects of molecular oxygen. The major thiol has been isolated from *M. bovis* BCG and identified as 1-D-*myo*-inositol-2-(*N*-acetyl-L-cysteinyl)amino-2-deoxy-α-D-glucopyranoside and given the common name mycothiol (133). Thus far, mycothiol has only been found in mycobacteria; consequently, its synthesis could be targeted if it was shown to be essential for survival in the harsh intracellular environment of the macrophage.

PHARMACOKINETICS AND PERMEABILITY Any new drug development program must be aimed at improving pharmacokinetics, drug delivery, and the ability to surmount the organism's permeability barrier. Few rational routes lead to improved stability or reduced metabolic degradation in vivo; the most common methods employ an empirical approach in which many compound analogues are made and tested. As an alternative to improving a compound per se, the means of delivery can be modified. Artificial depots can be created; in a recent example, implantations provided for slow drug release over a period of several months (80). *M. avium* is not susceptible to most antimycobacterials, yet its cell wall architecture and presumably its biochemical pathways are similar to those found in drug-sensitive organisms. Do the drugs simply not reach their target? Basic research into permeability is urgently needed to both improve upon current drugs and ensure that medicinal chemistry approaches are followed in the future that yield molecules with physico-chemical properties allowing them to breach the barrier.

TESTING NEW DRUGS Clinical trials of new anti-TB therapies pose a particularly intractable problem, which can be summed up by the fact that TB is a readily treatable disease. In treatments of normal pulmonary disease due to fully drug-sensitive organisms, the way to test a new compound is to use it either as a substitute for, or in addition to, the current drugs used. In both cases, with well-monitored therapy (i.e. full compliance), even a drug with much improved in vitro properties will show little improvement over a current therapy that would have worked under these circumstances anyway. In addition, what is a suitable end point for therapy? Clearance from infection would need to be followed for up to two years, and a means of distinguishing between relapse (indicating treatment failure) and reinfection would need to be established. Such a long-term commitment to follow-up increases trial costs substantially. When testing against MDR-TB, the use of historical controls may be necessary as it is unrealistic and unethical to give placebo to patients infected with this bacterium. Finally, new approaches to therapy, as outlined below, may require that we reevaluate the whole question of how to carry out suitable clinical trials.

IMMUNOTHERAPY Although the immune system alone is unable to cope with the vast numbers of bacteria present in active pulmonary disease, some form of immunotherapy may be considered a useful adjunct to chemotherapy. For example, some aspects of immunopathology might be suppressed, and some alteration in the balance of immune activities may be helpful in eliminating persistent intracellular organisms. Robert Koch was the first to attempt immunotherapy of active tuberculosis; he treated patients with mycobacterial extracts, eliciting a "Koch reaction," with fatal consequences in some cases. Immunotherapy continued to attract interest in the treatment of leprosy patients (34, 161), with a view to enhancing clearance of bacterial debris during chemotherapy, and Stanford and colleagues have recently argued strongly in favor of the potential contribution of immunotherapy as an adjunct to drug treatment of TB (135). Patients undergoing standard chemotherapy have been inoculated with an autoclaved preparation of *Mycobacterium vaccae*—a nonpathogenic soil organism. This procedure reportedly improves treatment outcome, perhaps by converting a pathogenic form of immune response to a protective response (119). The mechanism of such an effect is unclear, and further clinical trials are required to substantiate these dramatic observations. A more targeted approach to immunotherapy involves manipulation of local or systemic activities of individual cytokines. Again this approach has attracted attention, particularly for the treatment of leprosy patients (81). An interesting finding is that thalidomide—a drug used to control tissue-damaging immune reactions during leprosy therapy—appears to exert its effect by decreasing expression of TNF-α (101). Thalidomide treatment may also be of use in reducing some of the pathological effects of TB and perhaps in inhibiting progression of HIV replication in dually infected patients (90).

Prevention

The outcome of mycobacterial infection critically depends on the host immune response; disease-control strategies based on immune manipulation are obviously attractive. In spite of intensive research efforts over the past century, however, translation of the theoretical potential for immune manipulation into practical tools for control of mycobacterial disease has met with limited success. Concern over the increasing rates of TB worldwide has focused renewed attention on the search for effective vaccines, and many research groups are investigating the possibility that molecular genetic approaches will lead to novel strategies for mycobacterial vaccine development.

BCG VACCINE   In most individuals, infection with *M. tuberculosis* induces an immune response sufficient for protection against progression to primary disease. Priming such an immune response prior to exposure to virulent challenge might further reduce the incidence of primary disease and perhaps also act against reactivation disease by reducing progression to the dormancy phase of infection. Based on this principle, an optimal vaccine would reproduce as near as possible the initial natural infection but would not impose a disease risk. Calmette & Guerin sought to develop such a vaccine using a strain of *M. bovis* that had been rendered avirulent as a result of serial passage in vitro. The resulting vaccine, BCG, was first used to protect against TB in 1921 and has since been evaluated in many clinical trials that have been reported and interpreted in considerable detail (19, 30, 55). BCG is currently administered to approximately 100 million children each year. A general consensus is that vaccination consistently protects against primary, disseminated forms of TB (115), but protection against pulmonary disease varies widely among different trials (19, 55). Results pooled from a range of trials indicate an overall reduction of 50% in the risk of TB following BCG vaccination (30). BCG also confers consistent protection against leprosy (55). These data are consistent with the idea that BCG vaccination effectively boosts the immune response during primary infection but has limited effect on the subsequent course of dormancy and reactivation. Declining protective efficacy with time after BCG vaccination—little or no protection is seen after 10–15 years (116)—suggests that childhood vaccination will not prevent adult reinfection (140). The demonstration that immune manipulation can indeed alter the risk of mycobacterial disease provides encouragement for vaccine development. In working towards a better BCG, possible strategies would include: (a) enhancement of the ability to prime a protective response to initial infection, perhaps by promoting a sterilizing immunity that precludes formation of dormant foci, and (b) prolongation of protective efficacy with the aim of generating continued protection through the reactivation phase. The latter goal might be achieved by persistence of the vaccine itself, by triggering of a long-lived immunological "memory," or by subsequent boosting vaccination.

RATIONALLY ATTENUATED STRAINS   The nature of the genetic lesion (or lesions) that renders BCG avirulent has not been defined. Loss of virulence may be accompanied by an alteration in immunogenicity from that of the parent virulent strain, and strains carrying alternative attenuating mutations may differ in their potency as vaccines. This principle has been clearly demonstrated in the case of other bacterial pathogens: *Salmonella typhimurium*, an intracellular pathogen capable of macrophage survival, serves as a well-characterized example. Although mutations in numerous *S. typhimurium* genes render the bacteria avirulent, only a subset of such mutants provide protection against subsequent virulent challenge. A clear strategy for mycobacterial vaccine development would therefore be to start with a virulent isolate of *M. tuberculosis* and to generate a series of attenuated mutants with defects in genes required for intracellular survival and virulence. In addition to their potential direct application as vaccines, such strains would provide invaluable research tools for dissecting the fundamental steps involved in mycobacterial immunogenicity and virulence.

Mutagenesis   Procedures for efficient mutagenesis of slow-growing mycobacteria have not yet been developed (92). Transposon mutagenesis has been demonstrated in the fast-growing *M. smegmatis* (64) but has yet to be optimized in slow-growing mycobacteria. Allele exchange, as mediated by homologous recombination, occurs in *M. smegmatis* (71), and analogous, although less efficient, systems have been demonstrated in BCG (3). An unusual structure for the *recA* gene—encoding one of the components essential for homologous recombination—may be responsible for some of the problems associated with gene replacement in pathogenic mycobacteria (45, 46). Development of efficient mutagenesis protocols will be an essential element of vaccine strategies based on rational attenuation.

Targets for attenuation Studies of other bacterial pathogens can lead to the prediction that many genes in *M. tuberculosis* will act as attenuation targets. Genes encoding enzymes involved in metabolic pathways essential for in vitro survival have been targeted in several bacterial species. Disruption of the pathway involved in synthesis of aromatic components has been effectively exploited in vaccine development, for example, and the relevant genes have been identified and characterized in *M. tuberculosis* (60, 61). Transcriptional regulation characterizes bacterial pathogenesis, permitting increased or decreased expression of key genes in response to environmental changes encountered during the process of infection; introduction of mutations in genes subject to in vivo regulation, or genes encoding the regulatory proteins, provides another important strategy for rational attenuation (57). Attempts are being made to identify regulated genes in mycobacterial pathogens in vivo (65, 75, 112), progress in this area should be rapid over the next few years. Transcriptional changes associated with intracellular vs extracellular growth, and with transition between active multiplication and dormancy, are of particular interest in relation to understanding pathogenic mechanisms and development of rationally attenuated vaccines.

SUBUNIT VACCINES The rationale for use of live attenuated vaccine strains is based on the assumption that a stimulus that mimics the natural process of infection will effectively trigger the complex and coordinated set of activities thought to be required for protective immunity. Arguments in favor of simpler subunit vaccine strategies include the possibility that the natural immune response comprises pathogenic as well as protective activities; preferential stimulation of the latter by isolated bacterial components may afford a level of protection greater than that of the natural response. Alternatively, if a rapid response is the key to effective immunity, initial recognition even of a single antigen might provide a sufficient trigger for early activation of the complete set of immune responses. In practical terms of vaccine production, standardization, and delivery, a defined subunit preparation would clearly have many advantages over a live organism.

Protective antigens Considerable effort has been invested over many decades in cataloguing mycobacterial components involved in immune responses, with a view to identification of candidates for subunit vaccines and for diagnostic tests (4, 72, 160). Initial experiments emphasized analysis of proteins containing species-specific epitopes with potential application in diagnostic tests; more recent attention has focused on secreted proteins. Culture filtrate proteins elicit strong immune reactions in animals immunized with live BCG (117), and successful vaccination procedures based on culture filtrate preparations have recently been reported from several laboratories (5, 70, 110). The level of protection obtained in such experiments is comparable to that induced by BCG, and the secreted-antigen approach is attractive as a route toward new vaccines or as a possible strategy for boosting initial BCG responses in older age groups. Knowledge of the physiological function of secreted proteins may assist in selecting appropriate targets; subunit vaccines based on proteins that are not essential for in vivo survival might simply select for growth of strain variants that have deleted the relevant genes, for example. Molecular genetic tools suitable for identification of genes encoding secreted proteins have been adapted for use in mycobacteria (143) and may allow identification of novel antigens, including those produced in vivo. Although recent developments in studying responses to secreted antigens are encouraging, we must remember that the idea that some antigens are inherently protective, while others are detrimental, is a useful working hypothesis that requires further experimental verification.

Delivery systems The way in which an antigen is delivered to the immune system may play a crucial role in its ability to induce a protective response. The secreted antigen examples discussed above used chemical adjuvants as delivery systems, but alternative strategies are available. Genes encoding key antigens can be incorporated into heterologous, live, attenuated bacterial and viral vaccines to take advantage of their known ability to stimulate particular immune pathways. For example, several mycobacterial antigens have been expressed in vaccinia virus (88). BCG itself has been used as a delivery system for recombinant antigens (137, 138) and, while most of the antigens studied to date are shared between BCG and *M. tuberculosis*, strategies could readily be envisaged by which BCG is engineered to overproduce large amounts of an antigen of particular interest. The importance of screening different delivery systems is illustrated by the case of the mycobacterial 65-kDa heat-shock protein. Silva & Lowrie (128) reported that, while immunization of mice with the 65-kDa protein in adjuvant resulted in no significant protection, immunization with the same protein expressed in a transfected cell line did afford protection against subsequent live challenge. The ability to induce mycobacteria-reactive $CD8^+$ T cells could be an important attribute in such a vaccination protocol, and investigations of the related strategy of nucleic acid vaccination may also prove useful (87). The ease of construction of nucleic acid vaccines would allow this strategy to be used for broad screening of numerous genes for protective potential; indeed, the entire mycobacterial genome might be screened for genes encoding protective antigens.

TESTING NEW VACCINES Animal models—principally the mouse and guinea pig (94, 109)—can be used to monitor the ability of vaccines to protect against subsequent challenge with *M. tuberculosis* and are likely to provide the initial screen for novel vaccines. Protection is generally assessed in terms of a reduction in the number of colony-forming units recovered from target tissues at a specified time after challenge, although the effect of vaccination can also be judged by a change in survival times in some systems. BCG affords a significant level of protection in these models, reducing the load of viable bacteria under most circumstances, while not completely eliminating the infection. Various experimental factors can affect results obtained in vaccine models and contribute to marked interlaboratory variation in estimates of relative vaccine efficacy (129). Standard protocols, and comparative testing in independent laboratories, will facilitate progress toward improved vaccines. Convenient model systems for screening potential "anti-reactivation" vaccines are currently unavailable. Prior BCG vaccination had no effect on reactivation in mice that had been infected with *M. tuberculosis* and then incompletely sterilized via chemotherapy (48).

The extent to which a vaccine's protection against experimental challenge in a laboratory animal will parallel its efficacy against the natural disease in humans is very much open to speculation. Continuing debate as to the protective efficacy of BCG highlights the immense practical problems inherent in clinical trials of antimycobacterial vaccines, and an intermediate level of comparative evaluation of new candidates clearly needs to be established. Programs to develop *M. bovis* vaccines in domestic and feral animals could provide useful lessons for design of human vaccines. For example, these programs would allow comparison of strategies based on rational attenuation vs subunit vaccination in control of mycobacterial infection in a naturally susceptible host population. In addition to such model systems, a form of screening procedure is needed that could be used as a surrogate marker for protection in humans and that could be applied in preliminary evaluation of vaccine efficacy.

Immunological correlates of protection The central role of the immune response in determining the outcome of mycobacterial infection suggests that measurement of some immunological parameter might give an indication of disease susceptibility. The only parameter extensively studied at the population level is delayed-type hypersensitivity to a carbohydrate-depleted protein preparation from *M. tuberculosis* known as purified protein derivative (PPD). Following BCG vaccination, most individuals exhibit a positive skin-test reaction to PPD, but the subsequent incidence of TB is the same in PPD$^+$ and PPD$^-$ groups (66). More detailed analysis suggests some correlation between disease susceptibility and the size of the skin-test response: Low to intermediate levels of response are associated with reduced incidence of TB in some populations (56). Immune responses to particular antigens might be associated with protection. For example, tests based on immune recognition of defined secreted antigens might have an advantage over PPD in predicting disease susceptibility, and the feasibility of using such antigens in skin-test assays has been demonstrated in animal model systems (117).

At a further level of molecular definition, synthetic peptides corresponding to individual antigenic determinants from selected mycobacterial proteins could be considered as diagnostic reagents. For example, a peptide corresponding to the C-terminal region of the 38-kDa secreted protein of *M. tuberculosis* induces proliferation of peripheral-blood T cells from PPD$^+$ healthy individuals but induces little or no response in patients with active pulmonary TB (149). This is the only description of a clear difference in the antigen repertoire between protective immunity and disease-associated immunopathology.

Possible markers for protective immunity can include, in addition to antigen repertoire, differences in the nature of the immune response, such as involvement of particular T-cell subsets and patterns of cytokine production. For example, studies that monitor recognition of mycobacterial antigens by CD8$^+$ T cells in different individuals and that measure release of the Th1- and Th2-associated cytokines in response to whole mycobacteria or to particular defined antigens, will be of interest. The feasibility of measuring IFN-γ responses to mycobacterial antigens in whole blood immunoassays has been demonstrated in studies of bovine tuberculosis (157), and this approach could be readily extended to cover additional cytokines.

A third strategy for development of novel tests of immune status would involve use of mycobacteria themselves as a detection system. Novel molecular genetic techniques allow us to introduce reporter genes into mycobacteria and to thereby simply and rapidly assess bacterial viability (73). With increased understanding of transcriptional regulation associated with mycobacterial adaptation to different environments, it may be possible to engineer a series of reporter genes responsive to different physiological signals, and thus provide a means of sensing the intracellular fate of the bacterium, in order to measure the relative intensity and efficacy of the antimycobacterial immune response.

GENETIC SUSCEPTIBILITY   One important consideration in relation to vaccine development is that the vaccine need not necessarily have an identical effect in all individuals. The incidence of TB is strongly influenced by socio-economic factors, but genetic factors also contribute to disease susceptibility (33, 54, 136). A single gene that influences immune responses to BCG infection in inbred strains of mice has been identified (147), but the genes that influence TB in humans remain to be defined. A reassessment of this area in light of recent progress in human genome research may prove useful. High-risk groups could be identified for vaccine evaluation, or perhaps, by working back from some genetic clue, immune parameters could be identified that are aberrant in particularly susceptible individuals.

CHEMOPROPHYLAXIS   The antimycobacterial agents used for treatment of active TB are also effective in preventing disease in asymptomatic individuals exposed to disease (107). Chemoprophylaxis, or preventive therapy, with isoniazid has been extensively analyzed and shown to be cost-effective in the US (131). The high risk of TB in HIV-infected persons suggests an important role for chemoprophylaxis in this group (111), and the potential impact of such a strategy in areas endemic for dual infection is under discussion (104, 107). In light of concern about drug-resistant strains of TB, it would be attractive to have separate drugs available for prophylaxis and for treatment: Perhaps some of the new drug discovery leads discussed above will assist in this area.

NEW INTERVENTIONS: FUTURE PROSPECTS

Current research activities offer reasons to be optimistic about the prospects for identification of novel antimycobacterial agents directed to specific cell wall targets and for a new generation of mycobacterial vaccine candidates based on defined antigens and rationally attenuated mutants. However, will financial incentives for development and marketing of the new drugs be sufficient? If so, will their performance be essentially similar to that of existing drugs, engendering problems of prolonged therapy and emergence of resistant strains? In the case of future vaccine candidates, what are the prospects for their clinical evaluation? Will they follow the same course as BCG, with a 70-year history of uncertainty and debate? In light of the magnitude of the worldwide TB problem, it is important that, while pursuing existing avenues of research, we also look beyond the classical ways of treating mycobacterial diseases.

Novel insights are likely to be achieved through a deeper understanding of the host-pathogen interaction. The current generation of antimycobacterial agents target specific biosynthetic pathways essential for bacterial viability. Although such reagents will obviously remain the central element in treatment of active TB, their use might be optimized by combination with reagents that modify particular aspects of the host-pathogen interaction. Novel reagents that allow a radical reduction in treatment schedules would clearly be of high priority. Disruption of interactions between specific mycobacterial ligands and host-cell receptors might enhance immune-mediated clearance of infection, for example. Reagents capable of interfering with environmental signaling pathways in mycobacteria might similarly shorten treatment times by preventing mycobacterial escape into phenotypically resistant forms. Further understanding of the mechanisms involved in immune-mediated killing of mycobacteria may allow us to develop procedures based on cytokines, or other small molecules, that specifically enhance local antimycobacterial responses. Similarly, modulation of the immune system by means of endocrine signaling could provide a novel therapeutic avenue. If we consider infection a complex host-pathogen interaction, rather than simply a bacterial culture that happens to be physically located inside a mammalian host, we can view control strategies based on antimycobacterials as merged with conventional vaccination strategies. Perhaps the greatest advances in prophylactic vaccines will come from the area of anti-reactivation vaccines, which are designed to boost immune responses in individuals already harboring dormant infections.

Mycobacterial diseases can be seen as a social barometer; TB case rates generally reflect socio-economic conditions. Under optimal conditions, current control measures are very effective against TB, but continued disease outbreaks in underprivileged communities, even in relatively well-off industrialized societies, show that these measures afford a precariously thin veneer of protection. Development of novel scientific interventions to strengthen the barrier against mycobacterial disease presents a formidable conceptual as well as practical challenge. New tools are needed to assist in delivery of existing treatment; efforts to develop and evaluate candidate vaccines and novel drug leads must be intensified; and long-term commitment has to be given to fundamental research aimed at understanding the complex interaction between humans and one of their oldest and most persistent parasites.

> Any *Annual Review* chapter, as well as any article cited in an *Annual Review* chapter, may be purchased from the Annual Reviews Preprints and Reprints service.
> 1-800-347-8007; 415-259-5017; email: arpr@class.org

Literature Cited

1. Abou-Zeid C, Smith I, Grange JM, Ratliff TL, Steele J, Rook GA. 1988. The secreted antigens of *Mycobacterium tuberculosis* and their relationship to those recognized by the available antibodies. *J. Gen. Microbiol.* 134:531–38
2. al-Attiyah R, Moreno C, Rook GA. 1992. TNFα-mediated tissue damage in mouse footpads primed with mycobacterial preparations. *Res. Immunol.* 143:601–10
3. Aldovini A, Husson RN, Young RA. 1993. The *uraA* locus and homologous recombination in *Mycobacterium bovis* BCG. *J. Bacteriol.* 175:7282–89
4. Andersen AB, Brennan P. 1994. Proteins and antigens of *Mycobacterium tuberculosis*. See Ref. 18a, pp. 307–32
5. Andersen P. 1994. Effective vaccination

of mice against *Mycobacterium tuberculosis* infection with a soluble mixture of secreted mycobacterial proteins. *Infect. Immun.* 62:2536–44
6. Andersen P, Askgaard D, Ljungqvist L, Bennedsen J, Heron I. 1991. Proteins released from *Mycobacterium tuberculosis* during growth. *Infect. Immun.* 59:1905–10
7. Andersen P, Heron I. 1993. Specificity of a protective memory immune response against *Mycobacterium tuberculosis*. *Infect. Immun.* 61:844–51
8. Armstrong JA, Hart PD'A. 1975. Phagosome-lysosome interactions in cultured macrophages infected with virulent tubercle bacilli. Reversal of the usual nonfusion pattern and observations on bacterial survival. *J. Exp. Med.* 142:1–16
9. Armstrong JA, Hart PD'A. 1971. Response of cultured macrophages to *Mycobacterium tuberculosis*, with observations on fusion of lysosomes with phagosomes. *J. Exp. Med.* 134:713–40
10. Arruda S, Bomfim G, Knights R, Huima-Byron T, Riley LW. 1993. Cloning of an *M. tuberculosis* DNA fragment associated with entry and survival inside cells. *Science* 261:1454–57
11. Ashtekar DR, Costa-Periera R, Nagrajan K, Vishvanathan N, Bhatt AD, Rittel W. 1993. In vitro and in vivo activities of the nitroimidazole CGI 17341 against *Mycobacterium tuberculosis*. *Antimicrob. Agents Chemother.* 37:183–86
12. Ashtekar DR, Costa-Periera R, Shrinivasan T, Iyyer R, Vishvanathan N, Rittel W. 1991. Oxazolidinones, a new class of synthetic antituberculosis agent. In vitro and in vivo activities of DuP-721 against *Mycobacterium tuberculosis*. *Diagn. Microbiol. Infect. Dis.* 14:465–71
13. Banerjee A, Dubnau E, Quemard A, Balasubramanian V, Um KS, et al. 1994. inhA, a gene encoding a target for isoniazid and ethionamide in *Mycobacterium tuberculosis*. *Science* 263:227–30
14. Barnes PF, Bloch AB, Davidson PT, Snider D Jr. 1991. Tuberculosis in patients with human immunodeficiency virus infection. *N. Engl. J. Med.* 324:1644–50
15. Barnes PF, Chatterjee D, Abrams JS, Lu S, Wang E, et al. 1992. Cytokine production induced by *Mycobacterium tuberculosis* lipoarabinomannan. Relationship to chemical structure. *J. Immunol.* 149:541–47
16. Besra G, Sievert T, Lee R, Slayden R, Brennan P, Takayama K. 1994. Identification of the apparent carrier in mycolic acid synthesis. *Proc. Natl. Acad. Sci. USA* 91:12735–39
17. Besra GS, Chatterjee D. 1994. Lipids and carbohydrates of *Mycobacterium tuberculosis*. In See Ref. 18a, pp. 285–306
18. Besra GS, Minnikin DE, Wheeler PR, Ratledge C. 1993. Synthesis of methyl (Z)-tetracos-5-enoate and both enantiomers of ethyl (E)-6-methyltetracos-4-enoate: possible intermediates in the biosynthesis of mycolic acids in mycobacteria. *Chem. Phys. Lipids* 66:23–34
18a. Bloom BR, ed. 1994. *Tuberculosis. Pathogenesis, Protection, and Control.* Washington, DC: Am. Soc. Microbiol.
19. Bloom BR, Fine PEM. 1994. The BCG experience: implications for future vaccines against tuberculosis. See Ref. 18a, pp. 531–58
20. Bloom BR, Murray CJ. 1992. Tuberculosis: commentary on a reemergent killer. *Science* 257:1055–64
21. Brennan PJ. 1989. Structure of mycobacteria: recent developments in defining cell wall carbohydrates and proteins. *Rev. Infect. Dis.* 11:S420–30
22. Brennan PJ, Draper P. 1994. Ultrastructure of *Mycobacterium tuberculosis*. See Ref. 18a, pp. 271–84
23. Brogden RN, Fitton A. 1994. Rifabutin. A review of its antimicrobial activity, pharmacokinetic properties and therapeutic efficacy. *Drugs* 47:983–1009
24. Cambau E, Sougakoff W, Besson M, Truffot-Pernot C, Grosset J, Jarlier V. 1994. Selection of a gyrA mutant of *Mycobacterium tuberculosis* resistant to fluoroquinolones during treatment with ofloxacin. *J. Infect. Dis.* 170:479–83
25. Chan J, Fujiwara T, Brennan P, McNeil M, Turco SJ, Sibille JC, et al. 1989. Microbial glycolipids: possible virulence factors that scavenge oxygen radicals. *Proc. Natl. Acad. Sci. USA* 86:2453–57
26. Chan J, Kaufmann SHE. 1994. Immune mechanisms of protection. See Ref. 18a, pp. 389–416
27. Chan J, Xing Y, Magliozzo RS, Bloom BR. 1992. Killing of virulent *Mycobacterium tuberculosis* by reactive nitrogen intermediates produced by activated murine macrophages. *J. Exp. Med.* 175:1111–22
28. Chatterjee D, Lowell K, Rivoire B, McNeil MR, Brennan PJ. 1992. Lipoarabinomannan of *Mycobacterium tuberculosis*. Capping with mannosyl residues in some strains. *J. Biol. Chem.* 267:6234–39
29. Chatterjee D, Roberts AD, Lowell K, Brennan PJ, Orme IM. 1992. Structural basis of capacity of lipoarabinomannan

30. Colditz GA, Brewer TF, Berkey CS, Wilson ME, Burdick E, et al. 1994. Efficacy of BCG vaccine in the prevention of tuberculosis. Meta-analysis of the published literature. *J. Am. Med. Assoc.* 271:698–702
31. Cole ST. 1994. *Mycobacterium tuberculosis:* drug-resistance mechanisms. *Trends Microbiol.* 2:411–15
32. Cole ST, Smith DR. 1994. Towards mapping and sequencing the genome of *Mycobacterium tuberculosis.* See Ref. 18a, pp. 227–38
33. Comstock GW. 1978. Tuberculosis in twins: a re-analysis of the Prophit survey. *Am. Rev. Respir. Dis.* 117:621–24
34. Convit J, Aranzazu N, Ulrich M, Pinardi ME, Reyes O, Alvarado J. 1982. Immunotherapy with a mixture of *Mycobacterium leprae* and BCG in different forms of leprosy and in Mitsuda-negative contacts. *Int. J. Lepr. Other Mycobact. Dis.* 50:415–24
35. Cooksey RC, Crawford JT, Jacobs W Jr, Shinnick TM. 1993. A rapid method for screening antimicrobial agents for activities against a strain of *Mycobacterium tuberculosis* expressing firefly luciferase. *Antimicrob. Agents Chemother.* 37:1348–52
36. Cooper AM, Dalton DK, Stewart TA, Griffin JP, Russell DG, Orme IM. 1993. Disseminated tuberculosis in interferon γ gene–disrupted mice. *J. Exp. Med.* 178:2243–47
37. Crowe SM, Carlin JB, Stewart KI, Lucas CR, Hoy JF. 1991. Predictive value of CD4 lymphocyte numbers for the development of opportunistic infections and malignancies in HIV-infected persons. *J. AIDS* 4:770–76
38. Crowle AJ, Dahl R, Ross E, May MH. 1991. Evidence that vesicles containing living, virulent *Mycobacterium tuberculosis* or *Mycobacterium avium* in cultured human macrophages are not acidic. *Infect. Immun.* 59:1823–31
39. Cynamon MH, Klemens SP. 1994. Chemotherapeutic agents for mycobacterial infections. In *Tuberculosis: Current Concepts and Treatment,* ed. LN Freidman, pp. 237–57. Boca Raton, FL: CRC
40. Daffe M, Brennan PJ, McNeil M. 1990. Predominant structural features of the cell wall arabinogalactan of *Mycobacterium tuberculosis* as revealed through characterization of oligoglycosyl alditol fragments by gas chromatography/mass spectrometry and by ^{1}H and ^{13}C NMR analyses. *J. Biol. Chem.* 265:6734–43
41. Dannenberg A Jr. 1989. Immune mechanisms in the pathogenesis of pulmonary tuberculosis. *Rev. Infect. Dis.* 11:S369–78
42. Dannenberg A Jr. 1991. Delayed-type hypersensitivity and cell-mediated immunity in the pathogenesis of tuberculosis. *Immunol. Today* 12:228–33
43. Dannenberg AM, Rook GAW. 1994. Pathogenesis of pulmonary tuberculosis: an interplay of tissue-damaging and macrophage-activating immune responses—dual mechanisms that control bacillary multiplication. See Ref. 18a, pp. 459–84
44. Davies J. 1994. Inactivation of antibiotics and the dissemination of resistance genes. *Science* 264:375–82
45. Davis EO, Jenner PJ, Brooks PC, Colston MJ, Sedgwick SG. 1992. Protein splicing in the maturation of M. tuberculosis RecA protein: a mechanism for tolerating a novel class of intervening sequence. *Cell* 71:201–10
46. Davis EO, Thangaraj HS, Brooks PC, Colston MJ. 1994. Evidence of selection for protein introns in the RecAs of pathogenic mycobacteria. *EMBO J.* 13:699–703
47. de Cock KM. 1994. Impact of interaction with HIV. See Ref. 112a, pp. 35–49
48. Dhillon J, Mitchison DA. 1994. Effect of vaccines in a murine model of dormant tuberculosis. *Tuberc. Lung Dis.* 75:61–64
49. Dolin PJ, Raviglione MC, Kochi A. 1994. Global tuberculosis incidence and mortality during 1990–2000. *Bull. WHO* 72:213–20
50. Douvas GS, Looker DL, Vatter AE, Crowle AJ. 1985. γ-Interferon activates human macrophages to become tumoricidal and leishmanicidal but enhances replication of macrophage-associated mycobacteria. *Infect. Immun.* 50:1–8
51. Eiglmeier K, Honore N, Woods SA, Caudron B, Cole ST. 1993. Use of an ordered cosmid library to deduce the genomic organization of *Mycobacterium leprae. Mol. Microbiol.* 7:197–206
52. Falkow S, Isberg RR, Portnoy DA. 1992. The interaction of bacteria with mammalian cells. *Annu. Rev. Cell Biol.* 8:333–63
53. Filley EA, Rook GA. 1991. Effect of mycobacteria on sensitivity to the cytotoxic effects of tumor necrosis factor. *Infect. Immun.* 59:2567–72
54. Fine PEM. 1981. Immunogenetics of susceptibility to leprosy, tuberculosis, and leishmaniasis. An epidemiological perspective. *Int. J. Lepr. Other Mycobact. Dis.* 49:437–54

55. Fine PEM, Rodrigues LC. 1990. Modern vaccines. Mycobacterial diseases. *Lancet* 335:1016–20
56. Fine PEM, Sterne JAC, Ponnighaus JM, Rees RJW. 1994. Delayed-type hypersensitivity, mycobacterial vaccines and protective immunity. *Lancet* 344:1245–49
57. Finlay BB, Falkow S. 1989. Common themes in microbial pathogenicity. *Microbiol. Rev.* 53:210–30
58. Flynn JL, Chan J, Triebold KJ, Dalton DK, Stewart TA, Bloom BR. 1993. An essential role for interferon-γ in resistance to *Mycobacterium tuberculosis* infection. *J. Exp. Med.* 178:2249–54
59. Flynn JL, Goldstein MM, Triebold KJ, Koller B, Bloom BR. 1992. Major histocompatibility complex class I–restricted T cells are required for resistance to *Mycobacterium tuberculosis* infection. *Proc. Natl. Acad. Sci. USA* 89:12013–17
60. Garbe T, Jones C, Charles I, Dougan G, Young D. 1990. Cloning and characterization of the *aroA* gene from *Mycobacterium tuberculosis*. *J. Bacteriol.* 172:6774–82
61. Garbe T, Servos S, Hawkins A, Dimitriadis G, Young D, et al. 1991. The *Mycobacterium tuberculosis* shikimate pathway genes: evolutionary relationship between biosynthetic and catabolic 3-dehydroquinases. *Mol. Gen. Genet.* 228:385–92
62. Gateau O, Bordet C, Michel G. 1976. Study of the formation of N-glycolylmuramic acid from *Nocardia asteroides*. *Biochim. Biophys. Acta* 421:395–405 (In French)
63. Germain RN, Margulies DH. 1993. The biochemistry and cell biology of antigen processing and presentation. *Annu. Rev. Immunol.* 11:403–50
64. Guilhot C, Otal I, Van-Rompaey I, Martin C, Gicquel B. 1994. Efficient transposition in mycobacteria: construction of *Mycobacterium smegmatis* insertional mutant libraries. *J. Bacteriol.* 176:535–39
65. Gupta S, Tyagi AK. 1993. Sequence of a newly identified *Mycobacterium tuberculosis* gene encoding a protein with sequence homology to virulence-regulating proteins. *Gene* 126:157–58
66. Hart PD'A, Sutherland I, Thomas J. 1967. The immunity conferred by effective BCG and vole bacillus vaccines, in relation to individual variations in induced tuberculin sensitivity and to technical variations in the vaccines. *Tubercle* 48:201–10
66a. Hastings RC, Opromolla DVA, eds. 1994. *Leprosy.* Edinburgh: Churchill Livingstone
67. Heym B, Honore N, Truffot-Pernot C, Banerjee A, Schurra C, et al. 1994. Implications of multidrug resistance for the future of short-course chemotherapy of tuberculosis: a molecular study. *Lancet* 344:293–98
68. Hirsch CS, Ellner JJ, Russell DG, Rich EA. 1994. Complement receptor-mediated uptake and tumor necrosis factor-α-mediated growth inhibition of *Mycobacterium tuberculosis* by human alveolar macrophages. *J. Immunol.* 152:743–53
69. Hsieh CS, Macatonia SE, Tripp CS, Wolf SF, O'Garra A, Murphy KM. 1993. Development of TH1 CD[4+] T cells through IL-12 produced by *Listeria*-induced macrophages. *Science* 260:547–49
70. Hubbard RD, Flory CM, Collins FM. 1992. Immunization of mice with mycobacterial culture filtrate proteins. *Clin. Exp. Immunol.* 87:94–98
71. Husson RN, James BE, Young RA. 1990. Gene replacement and expression of foreign DNA in mycobacteria. *J. Bacteriol.* 172:519–24
72. Ivanyi J, Thole J. 1994. Specificity and function of T- and B-cell recognition in tuberculosis. See Ref. 18a, pp. 437–58
73. Jacobs W Jr, Barletta RG, Udani R, Chan J, Kalkut G, et al. 1993. Rapid assessment of drug susceptibilities of *Mycobacterium tuberculosis* by means of luciferase reporter phages. *Science* 260:819–22
74. Jacobs W Jr, Tuckman M, Bloom BR. 1987. Introduction of foreign DNA into mycobacteria using a shuttle phasmid. *Nature* 327:532–35
75. Jacobs WR, Bloom BR. 1994. Molecular genetic strategies for identifying virulence determinants of *Mycobacterium tuberculosis*. See Ref. 18a, pp. 253–70
76. Janis EM, Kaufmann SH, Schwartz RH, Pardoll DM. 1989. Activation of $\gamma\delta$ T cells in the primary immune response to *Mycobacterium tuberculosis*. *Science* 244:713–16
77. Jarlier V, Gutmann L, Nikaido H. 1991. Interplay of cell wall barrier and β-lactamase activity determines high resistance to β-lactam antibiotics in *Mycobacterium chelonae*. *Antimicrob. Agents Chemother.* 35:1937–39
78. Job CK. 1994. Pathology of leprosy. See Ref. 66a, pp. 193–224
79. Johnsson K, Schultz PG. 1994. Mechanistic studies of the oxidation of isoniazid by the catalase peroxidase

from *Mycobacterium tuberculosis. J. Am. Chem. Soc.* 116:7425–26
80. Kailasam S, Daneluzzi D, Gangadharam PRJ. 1994. Maintenance of therapeutically active levels of isoniazid for prolonged periods in rabbits after a single implant of biodegradable polymer. *Tubercle Lung Dis.* 75:361–65
81. Kaplan G. 1993. Recent advances in cytokine therapy in leprosy. *J. Infect. Dis.* 167:S18–22
82. Kaufmann SH. 1988. CD^{8+} T lymphocytes in intracellular microbial infections. *Immunol. Today* 9:168–74
83. Kindler V, Sappino AP, Grau GE, Piguet PF, Vassalli P. 1989. The inducing role of tumor necrosis factor in the development of bactericidal granulomas during BCG infection. *Cell* 56:731–40
84. Klemens SP, Grossi MA, Cynamon MH. 1994. Activity of KRM-1648, a new benzoxazinorifamycin, against *Mycobacterium tuberculosis* in a murine model. *Antimicrob. Agents Chemother.* 38:2245–48
85. Kochi A. 1991. The global tuberculosis situation and the new control strategy of the World Health Organization. *Tubercle* 72:1–6
86. Lee MH, Pascopella L, Jacobs W Jr, Hatfull GF. 1991. Site-specific integration of mycobacteriophage L5: integration-proficient vectors for *Mycobacterium smegmatis, Mycobacterium tuberculosis,* and bacille Calmette-Guerin. *Proc. Natl. Acad. Sci. USA* 88:3111–15
87. Lowrie D, Tascon R, Colston M, Silva C. 1994. Towards a DNA vaccine against tuberculosis. *Vaccine* 12:1537–40
88. Lyons J, Sinos C, Destree A, Caiazzo T, Havican K, et al. 1990. Expression of *Mycobacterium tuberculosis* and *Mycobacterium leprae* proteins by vaccinia virus. *Infect. Immun.* 58:4089–98
89. Mahan MJ, Slauch JM, Mekalanos JJ. 1993. Selection of bacterial virulence genes that are specifically induced in host tissues. *Science* 259:686–88
90. Makonkawkeyoon S, Limson-Pobre RN, Moreira AL, Schauf V, Kaplan G. 1993. Thalidomide inhibits the replication of human immunodeficiency virus type 1. *Proc. Natl. Acad. Sci. USA* 90:5974–78
91. Mayer BK, Falkinham J. 1986. Superoxide dismutase activity of *Mycobacterium avium, M. intracellulare,* and *M. scrofulaceum. Infect. Immun.* 53:631–35
92. McAdam RA, Guilhot C, Gicquel B. 1994. Transposition in mycobacteria. See Ref. 18a, pp. 199–216
93. McDonough KA, Kress Y, Bloom BR. 1993. Pathogenesis of tuberculosis: interaction of *Mycobacterium tuberculosis* with macrophages. *Infect. Immun.* 61:2763–73
94. McMurray DN. 1994. Guinea pig model of tuberculosis. See Ref. 18a, pp. 135–48
95. McNeil M, Daffe M, Brennan PJ. 1990. Evidence for the nature of the link between the arabinogalactan and peptidoglycan of mycobacterial cell walls. *J. Biol. Chem.* 265:18200–6
96. McNeil MR, Brennan PJ. 1991. Structure, function and biogenesis of the cell envelope of mycobacteria in relation to bacterial physiology, pathogenesis and drug resistance; some thoughts and possibilities arising from recent structural information. *Res. Microbiol.* 142:451–63
97. McNeil MR, Robuck KG, Harter M, Brennan PJ. 1994. Enzymatic evidence for the presence of a critical terminal hexa-arabinose in the cell walls of *Mycobacterium tuberculosis. Glycobiology* 4:165–73
98. Minnikin DE. 1982. Lipids: complex lipids, their chemistry, biosynthesis and roles. See Ref. 112b, pp. 95–184
99. Minnikin DE. 1991. Chemical principles in the organization of lipid components in the mycobacterial cell envelope. *Res. Microbiol.* 142:423–27
100. Modlin RL, Pirmez C, Hofman FM, Torigian V, Uyemura K, et al. 1989. Lymphocytes bearing antigen-specific γδ T-cell receptors accumulate in human infectious disease lesions. *Nature* 339:544–48
101. Moreira AL, Sampaio EP, Zmuidzinas A, Frindt P, Smith KA, Kaplan G. 1993. Thalidomide exerts its inhibitory action on tumor necrosis factor α by enhancing mRNA degradation. *J. Exp. Med.* 177:1675–80
102. Mosmann TR, Coffman RL. 1989. Heterogeneity of cytokine secretion patterns and functions of helper T cells. *Adv. Immunol.* 46:111–47
103. Murray CJ, Styblo K, Rouillon A. 1990. Tuberculosis in developing countries: burden, intervention and cost. *Bull. Int. Union Tuberc. Lung Dis.* 65:6–24
104. Murray CJL. 1994. Issues in operational, social, and economic research on tuberculosis. See Ref. 18a, pp. 583–622
105. Neu HC. 1992. The crisis in antibiotic resistance. *Science* 257:1064–73
106. Noordeen SK. 1994. The epidemiology of leprosy. See Ref. 66a, pp. 29–45
107. O'Brien RJ. 1994. Preventive therapy for tuberculosis. See Ref. 112a, pp. 151–66

108. Orme IM. 1988. Characteristics and specificity of acquired immunologic memory to *Mycobacterium tuberculosis* infection. *J. Immunol.* 140:3589–93
109. Orme IM, Collins FM. 1994. Mouse model of tuberculosis. See Ref. 18a, pp. 113–34
110. Pal PG, Horwitz MA. 1992. Immunization with extracellular proteins of *Mycobacterium tuberculosis* induces cell-mediated immune responses and substantial protective immunity in a guinea pig model of pulmonary tuberculosis. *Infect. Immun.* 60:4781–92
111. Pape JW, Jean SS, Ho JL, Hafner A, Johnson W Jr. 1993. Effect of isoniazid prophylaxis on incidence of active tuberculosis and progression of HIV infection. *Lancet* 342:268–72
112. Plum G, Clark-Curtiss JE. 1994. Induction of *Mycobacterium avium* gene expression following phagocytosis by human macrophages. *Infect. Immun.* 62:476–83
112a. Porter JDH, McAdam KPWJ, eds. 1994. *Tuberculosis: Back to the Future.* Chichester, UK: Wiley & Sons
112b. Ratledge C, Stanford J, eds. 1982. *The Biology of the Mycobacteria.* London: Academic
113. Raviglione MC, Sudre P, Rieder HL, Spinaci S, Kochi A. 1993. Secular trends of tuberculosis in western Europe. *Bull. WHO* 71:297–306
114. Roach TI, Barton CH, Chatterjee D, Blackwell JM. 1993. Macrophage activation: lipoarabinomannan from avirulent and virulent strains of *Mycobacterium tuberculosis* differentially induces the early genes *c-fos, KC, JE,* and tumor necrosis factor-α. *J. Immunol.* 150:1886–96
115. Rodrigues LC, Diwan VK, Wheeler JG. 1993. Protective effect of BCG against tuberculous meningitis and miliary tuberculosis: a meta-analysis. *Int. J. Epidemiol.* 22:1154–58
116. Rodrigues LC, Smith PG. 1990. Tuberculosis in developing countries and methods for its control. *Trans. R. Soc. Trop. Med. Hyg.* 84:739–44
117. Romain F, Augier J, Pescher P, Marchal G. 1993. Isolation of a proline-rich mycobacterial protein eliciting delayed-type hypersensitivity reactions only in guinea pigs immunized with living mycobacteria. *Proc. Natl. Acad. Sci. USA* 90:5322–26
118. Rook GA, Onyebujoh P, Stanford JL. 1993. TH1/TH2 switching and loss of CD^{4+} T cells in chronic infections: an immunoendocrinological hypothesis not exclusive to HIV. *Immunol. Today* 14:568–69
119. Rook GA, Onyebujoh P, Wilkins E, Ly HM, al-Attiyah R, et al. 1994. A longitudinal study of percent agalactosyl IgG in tuberculosis patients receiving chemotherapy, with or without immunotherapy. *Immunology* 81:149–54
120. Rook GAW, Bloom BR. 1994. Mechanisms of pathogenesis in tuberculosis. See Ref. 18a, pp. 485–502
121. Saito H, Tomioka H, Sato K, Emori M, Yamane T, Yamashita K, et al. 1991. In vitro antimycobacterial activities of newly synthesized benzoxazinorifamycins. *Antimicrob. Agents Chemother.* 35:542–47
122. Salgame P, Abrams JS, Clayberger C, Goldstein H, Convit J, et al. 1991. Differing lymphokine profiles of functional subsets of human CD^{4+} and CD^{8+} T cell clones. *Science* 254:279–82
123. Sampaio EP, Kaplan G, Miranda A, Nery JA, Miguel CP, et al. 1993. The influence of thalidomide on the clinical and immunologic manifestation of erythema nodosum leprosum. *J. Infect. Dis.* 168:408–14
124. Sathyamoorthy N, Takayama K. 1987. Purification and characterization of a novel mycolic acid exchange enzyme from *Mycobacterium smegmatis. J. Biol. Chem.* 262:13417–23
125. Schlesinger LS, Bellinger-Kawahara CG, Payne NR, Horwitz MA. 1990. Phagocytosis of *Mycobacterium tuberculosis* is mediated by human monocyte complement receptors and complement component C3. *J. Immunol.* 144:2771–80
126. Schlesinger LS, Hull SR, Kaufman TM. 1994. Binding of the terminal mannosyl units of lipoarabinomannan from a virulent strain of *Mycobacterium tuberculosis* to human macrophages. *J. Immunol.* 152:4070–79
127. Selwyn PA, Hartel D, Lewis VA, Schoenbaum EE, Vermund SH, et al. 1989. A prospective study of the risk of tuberculosis among intravenous drug users with human immunodeficiency virus infection. *New Engl. J. Med.* 320:545–50
128. Silva CL, Lowrie DB. 1994. A single mycobacterial protein (hsp 65) expressed by a transgenic antigen-presenting cell vaccinates mice against tuberculosis. *Immunology* 82:244–48
129. Smith DW, Wiegeshaus EH. 1989. What animal models can teach us about the pathogenesis of tuberculosis in humans. *Rev. Infect. Dis.* 11:S385–93
130. Snapper SB, Lugosi L, Jekkel A, Melton RE, Kieser T, et al. 1988. Lysogeny and

transformation in mycobacteria: stable expression of foreign genes. *Proc. Natl. Acad. Sci. USA* 85:6987–91
131. Snider D Jr, Caras GJ, Koplan JP. 1986. Preventive therapy with isoniazid. Cost-effectiveness of different durations of therapy. *J. Am. Med. Assoc.* 255:1579–83
132. Snider D Jr, Roper WL. 1992. The new tuberculosis. *New Engl. J. Med.* 326:703–5
133. Spies HSC, Steenkamp DJ. 1994. Thiols of intracellular pathogens. *Eur. J. Biochem.* 224:203–13
134. Spratt BG. 1994. Resistance to antibiotics mediated by target alterations. *Science* 264:388–93
135. Stanford JL, Bahr GM, Rook GA, Shaaban MA, Chugh TD, et al. 1990. Immunotherapy with *Mycobacterium vaccae* as an adjunct to chemotherapy in the treatment of pulmonary tuberculosis. *Tubercle* 71:87–93
136. Stead WW, Senner JW, Reddick WT, Lofgren JP. 1990. Racial differences in susceptibility to infection by *Mycobacterium tuberculosis*. *New Engl. J. Med.* 322:422–27
137. Stover CK, Bansal GP, Hanson MS, Burlein JE, Palaszynski SR, et al. 1993. Protective immunity elicited by recombinant bacille Calmette-Guerin (BCG) expressing outer surface protein A (OspA) lipoprotein: a candidate Lyme disease vaccine. *J. Exp. Med.* 178:197–209
138. Stover CK, de-la-Cruz VF, Fuerst TR, Burlein JE, Benson LA, et al. 1991. New use of BCG for recombinant vaccines. *Nature* 351:456–60
139. Sturgill-Koszycki S, Schlesinger PH, Chakraborty P, Haddix PL, Collins HL, et al. 1994. Lack of acidification in *Mycobacterium* phagosomes produced by exclusion of the vesicular proton-ATPase. *Science* 263:678–81
140. Styblo K, Meijer J. 1976. Impact of BCG vaccination programmes in children and young adults on the tuberculosis problem. *Tubercle* 57:17–43
141. Takayama K, Kilburn JO. 1989. Inhibition of synthesis of arabinogalactan by ethambutol in *Mycobacterium smegmatis*. *Antimicrob. Agents Chemother.* 33:1493–99
142. Takiff HE, Salazar L, Guerrero C, Philipp W, Huang WM, et al. 1994. Cloning and nucleotide sequence of *Mycobacterium tuberculosis gyrA* and *gyrB* genes and detection of quinolone resistance mutations. *Antimicrob. Agents Chemother.* 38:773–80
143. Timm J, Perilli MG, Duez C, Trias J, Orefici G, et al. 1994. Transcription and expression analysis, using *lacZ* and *phoA* gene fusions, of *Mycobacterium fortuitum* β-lactamase genes cloned from a natural isolate and a high-level β-lactamase producer. *Mol. Microbiol.* 12:491–504
144. Tomioka H, Saito H, Sato K, Yamane T, Yamashita K, et al. 1992. Chemotherapeutic efficacy of a newly synthesized benzoxazinorifamycin, KRM-1648, against *Mycobacterium avium* complex infection induced in mice. *Antimicrob. Agents Chemother.* 36:387–93
145. Trejo AG, Chittenden GJ, Buchanan JG, Baddiley J. 1970. Uridine diphosphate α-D-galactofuranose, an intermediate in the biosynthesis of galactofuranosyl residues. *Biochem. J.* 117:637–39
146. Trejo AG, Haddock JW, Chittenden GJ, Baddiley J. 1971. The biosynthesis of galactofuranosyl residues in galactocarolose. *Biochem. J.* 122:49–57
147. Vidal SM, Malo D, Vogan K, Skamene E, Gros P. 1993. Natural resistance to infection with intracellular parasites: isolation of a candidate for *Bcg*. *Cell* 73:469–85
148. von Itzstein M, Wu WY, Kok GB, Pegg MS, Dyason JC, et al. 1993. Rational design of potent sialidase-based inhibitors of influenza virus replication. *Nature* 363:418–23
149. Vordermeier HM, Harris DP, Friscia G, Roman E, Surcel HM, et al. 1992. T cell repertoire in tuberculosis: selective anergy to an immunodominant epitope of the 38-kDa antigen in patients with active disease. *Eur. J. Immunol.* 22:2631–37
150. Wayne LG. 1976. Dynamics of submerged growth of *Mycobacterium tuberculosis* under aerobic and microaerophilic conditions. *Am. Rev. Respir. Dis.* 114:807–11
151. Wayne LG, Lin KY. 1982. Glyoxylate metabolism and adaptation of *Mycobacterium tuberculosis* to survival under anaerobic conditions. *Infect. Immun.* 37:1042–49
152. Wayne LG, Sramek HA. 1994. Metronidazole is bactericidal to dormant cells of *Mycobacterium tuberculosis*. *Antimicrob. Agents Chemother.* 38:2054–58
153. Wheeler PR, Besra GS, Minnikin DE, Ratledge D. 1993. Stimulation of mycolic acid biosynthesis by incorporation of cis-tetracos-5-enoic acid in a cell-wall preparation from *Mycobacterium smegmatis*. *Biochim. Biophys. Acta* 1167:182–88
154. Wietzerbin J, Das BC, Petit JF, Lederer E, Leyh-Bouille M, Ghuysen JM. 1974.

Occurrence of D-alanyl-(D)-meso-diaminopimelic acid and meso-diaminopimelyl-meso-diaminopimelic acid interpeptide linkages in the peptidoglycan of mycobacteria. *Biochemistry* 13:3471–76
155. Winder FG. 1982. Mode of action of the antimycobacterial agents and associated aspects of the molecular biology of the mycobacteria. See Ref. 112b, pp. 354–438
156. Wolucka BA, McNeil MR, Hoffmann ED, Chojnacki T, Brennan PJ. 1994. Recognition of the lipid intermediate for arabinogalactan/arabinomannan biosynthesis and its relation to the mode of action of ethambutol on mycobacteria. *J. Biol. Chem.* 269:23328–35
157. Wood PR, Corner LA, Rothel JS, Ripper JL, Fifis T, et al. 1992. A field evaluation of serological and cellular diagnostic tests for bovine tuberculosis. *Vet. Microbiol.* 31:71–79
158. Yamamura M, Uyemura K, Deans RJ, Weinberg K, Rea TH, et al. 1991. Defining protective responses to pathogens: cytokine profiles in leprosy lesions. *Science* 254:277–79
159. Yew WW, Piddock LJV, Li MSK, Lyon D, Chan CY, Cheng AFB. 1994. In vitro activity of quinolones and macrolides against mycobacteria. *J. Antimicrob. Chemother.* 34:343–51
160. Young DB, Kaufmann SH, Hermans PW, Thole JE. 1992. Mycobacterial protein antigens: a compilation. *Mol. Microbiol.* 6:133–45
161. Zaheer SA, Mukherjee R, Ramkumar B, Misra RS, Sharma AK, et al. 1993. Combined multidrug and *Mycobacterium w* vaccine therapy in patients with multibacillary leprosy. *J. Infect. Dis.* 167:401–10
162. Zhang Y, Heym B, Allen B, Young D, Cole S. 1992. The catalase-peroxidase gene and isoniazid resistance of *Mycobacterium tuberculosis*. *Nature* 358:591–31
163. Dessen A, Quemard A, Blanchard JS, Jacobs WR, Sacchettini JC. 1995. Crystal structure and function of the isoniazid target of *Mycobacterium tuberculosis*. *Science* 267:1638–41

DEVELOPMENT AND APPLICATION OF HERPES SIMPLEX VIRUS VECTORS FOR HUMAN GENE THERAPY

J. C. Glorioso*, N. A. DeLuca*, and D. J. Fink**

*Department of Molecular Genetics and Biochemistry and **Department of Neurology, University of Pittsburgh School of Medicine, Pittsburgh, Pennsylvania 15261

KEY WORDS: gene-transfer vector, latency, promoter, cancer, central nervous system

CONTENTS

INTRODUCTION	676
Gene Therapy Vectors: An Overview	676
The Niche of HSV in Gene Transfer	677
THE HSV LIFE CYCLE	678
Molecular Biology of Lytic Infection	678
Molecular Aspects of HSV Latency	682
PROGRESS IN ENGINEERING HSV VECTORS	687
Overview	687
Eliminating Lytic Viral Gene Expression and Cytotoxicity	688
Expression of Therapeutic Genes Using HSV Vectors	690
APPLICATIONS OF HSV VECTORS	696
Gene Therapy of Cancer	696
Gene Therapy of Neurodegenerative Disease	700
CONCLUSIONS AND FUTURE DIRECTIONS	701

ABSTRACT

Advances in understanding the molecular basis of human disease and the development of recombinant DNA methods is rapidly creating new means of disease diagnosis and treatment. Among the most revolutionary developments are technologies for transfer of therapeutic genes to the human body to treat both inherited and acquired disease. Gene therapy offers considerable promise for ameliorating otherwise intractable diseases such as immunopathological conditions, cancer, heart disease, and various metabolic and neurodegenerative syndromes. To fulfill this promise, more efficient and effective methods of gene delivery and appropriate gene expression must be developed. The lack of such techniques is currently the most significant impediment to the use of genetic therapy. Both viral and nonviral delivery systems are under development for specific gene-therapy applications. Herpes simplex virus (HSV) represents a novel vector system for gene delivery to the nervous system and other tissues. HSV is able to establish latency in nondividing neuronal cells in which genomes persist long-term but do not integrate or alter host-cell metabolism and that carry a promoter system uniquely capable of escaping repression that shuts off the expression of HSV-lytic genes during latency. This review examines efforts to create defective HSV vectors that are safe, noncytotoxic, and applicable to the treatment of cancer and diseases affecting peripheral nerves. Perhaps the most important use of HSV vectors will be for the treatment of neurodegenerative diseases of the brain, but additional studies are required to improve the design of promoters to ensure regulatable or effective levels of therapeutic gene expression.

INTRODUCTION

Gene Therapy Vectors: an Overview

Scientific advances in understanding human anatomy in the sixteenth century, physiology in the seventeenth century, and pathology in the nineteenth century each led to revolutionary changes in the treatment of human disease. In the second half of the twentieth century, the practice of medicine has undergone another revolution: an understanding of the molecular genetic basis of inheritance and the development of techniques to manipulate genes.

The underlying principle of gene therapy is the introduction of a functional gene into affected tissue in order to complement a defective gene, which results in a true cure of a recessively inherited disease phenotype. Dominantly inherited diseases will be more difficult to treat by gene transfer, but with the development of methods to inactivate defective gene mRNAs using messenger-specific antisense or ribozyme genes, the treatment of dominantly inherited defects using an approach combining gene inactivation and therapy is also theoretically possible. Specialized gene therapies may include methods to alter mRNA splicing patterns or modification of promoter activities. Diseases of somatic cells such as cancer or degenerative conditions of aging may well be treatable by gene transfer; in those cases, the suitable therapeutic gene would not be based on the identification of a genetic defect but rather on the demonstration of a therapeutic effect of the gene product on the final common pathway leading to the disease phenotype.

For specific applications, introduction of the therapeutic gene directly into cells in the body will be preferable and in some cases necessary. Protein-DNA conjugates, liposomes, and modified viral vectors are all under development for direct in vivo gene transfer, and each will likely have a role in gene therapy applications. In general, retroviruses are less desirable for in vivo delivery for several reasons: First, they are difficult to purify in high enough titers to administer directly. Second, most do not infect nondividing cells in differentiated tissues. Finally, if retroviral vectors could be developed that infect dividing cells and integrate into the host genome in vivo (e.g. lentiviruses), they would nonetheless pose a risk of activating oncogenes or inactivating normal functions via insertional mutagenesis.

Additional viral vector systems are under development that may better meet the requirements for successful in vivo gene delivery. Among these, adenovirus (AV), adeno-associated virus (AAV), and herpes simplex virus type 1 (HSV-1) appear promising. Each of these systems has advantages and disadvantages.

All three of these viruses will enter nondividing cells after direct injection into mature animals, and all have a wide cellular host range. AV and HSV can be propagated to high titers (10^{10}–10^{12} plaque-forming units (pfu)/ml), whereas current methods to produce AAV vectors result in titers that are three to four orders of magnitude lower. The vectors also differ in the amount of foreign DNA, or payload, they can accommodate, in accordance with their respective genome sizes. Each of these vectors will probably have a role in gene transfer for individual applications that will be determined by the target tissue, the size and number of genes to be delivered, and the required duration of expression. For a general review of viral vectors other than HSV, see the chapter by Smith (138a) in this volume.

The Niche of HSV in Gene Transfer

Because HSV is a human neurotropic virus, our laboratories have concentrated on developing HSV vectors for gene transfer to the nervous system, focusing primarily on the central nervous system (CNS). Conventional therapies for diseases of the brain are restricted by the physical constraints imposed by the organ itself. The blood-brain barrier limits the delivery of systemically administered macromolecules to brain parenchyma. The barrier at the ependymal surface is not as restrictive as the blood-brain barrier, but macromolecules delivered directly into the ventricles diffuse only a short distance from the ventricular surface into the substance of the brain. The regional and cellular specialization of the brain means that in some cases the vector may need to be targeted to a defined brain region.

HSV should prove to be a suitable vector for direct gene transfer to the nervous system for several reasons. First, in natural infection of neurons, the virus can enter a latent state characterized by persistence of the HSV genome as an intranuclear episome that does not express viral lytic genes or proteins. Latently infected neurons are not rejected by the host immune response and appear to function normally. Second, the ability of the virus to enter latency does not depend on viral gene expression; replication-defective viruses can readily establish latency. This is an important feature because it allows the safe use of replication-incompetent viruses for gene delivery. Third, HSV can be grown to very high titers, allowing gene delivery with a minimum of disruption to normal brain tissue. Fourth, the size and organization of the HSV genome permits the incorporation of multiple genes or large gene constructs that may include introns and regulatory sequences. Fifth, the presence of few spliced genes makes the genome easy to manipulate, and the virus-recombination machinery allows for foreign gene insertion. Finally, the virus possesses a natural, neuron-specific promoter system that is capable of functioning during latency and thus might be exploited to express foreign genes from the latent viral genome.

This review summarizes the salient features of HSV biology, highlights progress in silencing viral lytic genes, and examines strategies to achieve expression of recombinant foreign genes from the HSV vector genome. We also discuss potential application of HSV vectors to the treatment of human disease via gene therapy.

THE HSV LIFE CYCLE

Molecular Biology of Lytic Infection

The mature HSV-1 virion (Figure 1) consists of an enveloped icosahedral capsid. A layer of proteins, the tegument, lies between the envelope and capsid. The viral genome is a linear double-stranded DNA molecule 152 kilobases (kb) in length composed of two segments, the unique long (U_L) and the unique short (U_S) segment, each flanked by inverted repeats (Figure 1) (91–93). The four viral genes located within these repeat elements are therefore found in two copies each per genome. There are approximately 75 gene products that can be classified as immediate early (IE or α), early (E or β), or late (L or γ) depending on their kinetics of expression during replication, as shown in Figure 2. The viral genes can also be categorized according to whether they are essential or nonessential for virus replication. Essential genes are required to produce new infectious viral particles in highly permissive cell culture infections. Nonessential, or accessory, genes encode products that are not absolutely required in cell culture but are important for optimum lytic replication or affect the natural life cycle of the virus in vivo, contributing to host range, pathogenesis, latency, or spread in postmitotic cells such as neurons.

The viral DNA contains at least 20 essential structural genes (principally L class), which together with the essential IE and E genes represent only approximately half of the viral genome (Figure 1). The right-hand component of

HSV VECTORS FOR GENE THERAPY 679

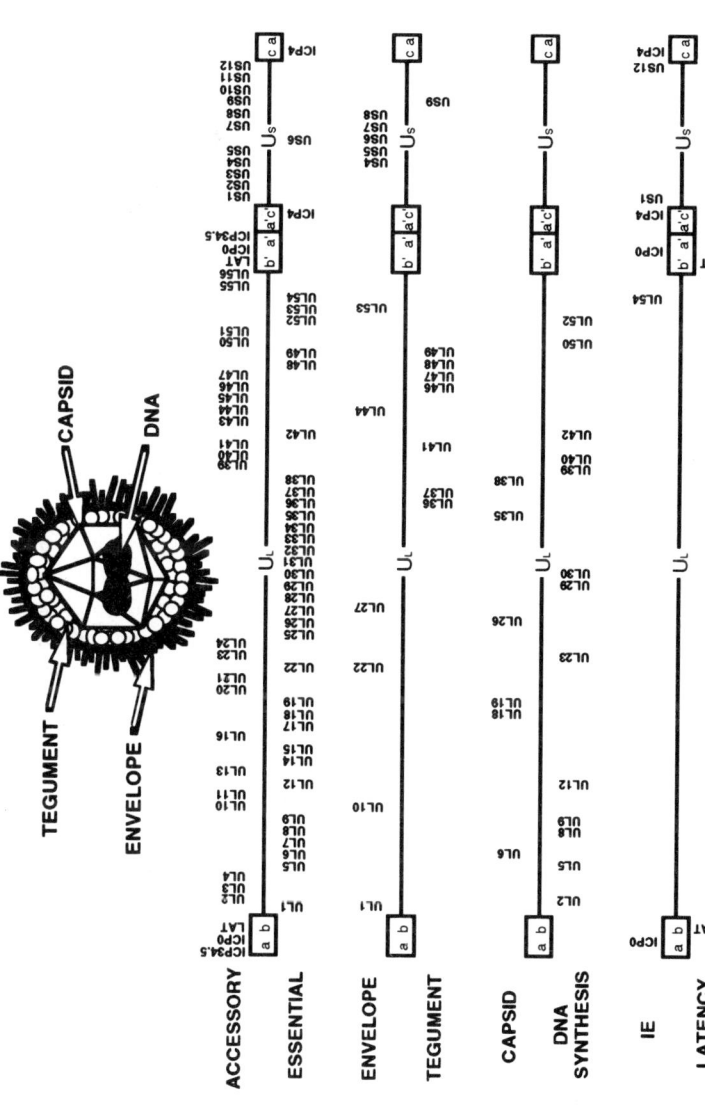

Figure 1 HSV particle structure and gene organization. A schematic of a typical virion appears above a diagram of the HSV prototypic genome displaying the essential and accessory viral genes (127, 165). The genes encoding products that make up specific virus structures or are expressed at specific times during lytic or latent infection are also shown.

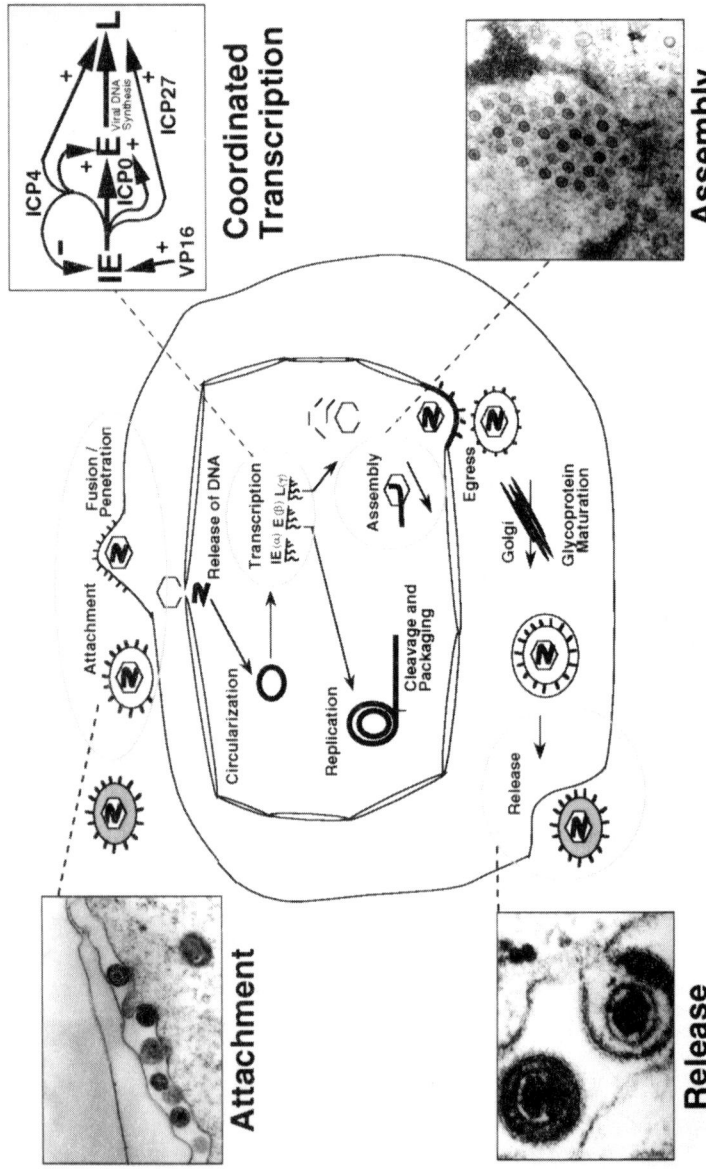

Figure 2 Stages of the HSV life cycle. Electron micrographs display particular events in the HSV life cycle leading up to release of infectious progeny particles and the subsequent death of the host cell. The pattern of the HSV gene regulatory cascade details the role of several key viral gene products.

the genome contains only three essential genes [infected cell polypeptides (ICPs)–4 and –27 and glycoprotein D (gD)], which affords the opportunity to replace large segments of viral sequences with foreign DNA (see 127 for review). Such replacements are possible because inactivation of nonessential genes usually does not greatly affect the virus's ability to propagate in culture. The modified HSV genome should be able to accommodate up to 40–50 kb of foreign sequences.

VIRAL ATTACHMENT AND PENETRATION   In vivo, virus infection is initiated in epithelial cells of the skin or mucosal membranes, and after the initial rounds of replication, the virus is taken up into the axon terminals of neurons innervating the primary site of infection. Some aspects of this initial virus-host interaction have been determined. The viral envelope contains approximately 12 glycoproteins (see Figure 1) and the enveloped particle attaches to the cell surface, presumably by means of a charge interaction between at least two glycoproteins (gB, gC) and heparan sulfate moieties of the plasma membrane. Specific recognition of a second, as yet unidentified, receptor, probably mediated by gD, follows (Figure 2). Attachment triggers fusion of the virus envelope with the cell surface membrane and requires the activities of at least three essential glycoproteins, gB, gD, and gH. Other glycoproteins either modulate immune recognition of the virus or enhance attachment to specific cell types (for reviews of glycoproteins, see 126, 142). The initial phase of virus attachment might be alterable through a combination of glycoprotein gene deletions and modification of essential glycoprotein structures to achieve viral targeting.

After entry, a mechanism (possibly involving specific interactions with the cytoskeleton) transports released virus capsids to the nuclear membrane. In neurons this process results in the retrograde transport of viral particles long distances within axons. The viral DNA is subsequently released through a penton (105) into the nucleus via a nuclear pore to begin the synthetic phase of the replication cycle.

TRANSCRIPTIONAL CONTROL AND DNA SYNTHESIS   Tegument components enter the cytoplasm and nucleus along with the viral genome and assist in the initiation of viral gene expression and host-cell takeover. One of these molecules, referred to as VP16 (also referred to as Vmw65 or αTIF), enhances transactivation of the five IE genes in the nucleus by collaborating with a cellular factor [octamer binding protein one (Oct1)] (7, 15, 76, 94, 112) to recognize an IE gene–specific promoter enhancer that has the consensus sequence TAATGARAT (49). Although the transcriptional activating function of VP16 is not essential for initiation of the replication cycle, the rate of virus gene expression proceeds much more rapidly in the presence of VP16 (1).

Another tegument protein referred to as the virion host shutoff function (vhs) (120) assists viral takeover of the cell machinery by degrading cellular mRNA and interfering with host-cell protein synthesis (77, 108, 120).

The products of the IE genes, which are expressed immediately after viral entry into the nucleus and in the absence of de novo viral protein synthesis, then direct a well-ordered temporal cascade (Figure 2) involving three rounds of coordinated viral gene expression culminating in viral replication (62). Two of the IE genes, ICP4 and -27, are responsible for initiating and controlling expression of E and L genes through both transcriptional and post-transcriptional mechanisms and are essential for viral replication (34, 113, 114, 129, 166). ICP0 is also a transcriptional activator, although unlike ICP4 and -27, it is not essential to viral replication, is not a DNA-binding protein, and can transactivate promoters other than those of HSV (130, 153). The other two IE genes, which encode ICP22 and -47, are not essential for viral replication in cell culture.

The E gene products are primarily responsible for viral DNA synthesis and the production of nucleotide pools in nondividing cells. Viral DNA synthesis occurs through a rolling-circle mechanism producing head-to-tail concatemers of the HSV genome (64). During replication, the U_L and U_S segments can independently reverse their orientation by homologous recombination between the inverted repeat flanking elements, forming four possible isomers (23, 64, 102), all of which appear to be infectious. Isomerization of the genome is not essential for virus replication or the establishment of latency (66, 111).

VIRUS ASSEMBLY, MATURATION, AND ENVELOPMENT   The L genes of the virus are mostly structural genes whose products form the capsid, tegument, and envelope. L gene expression requires viral DNA synthesis in addition to the activities of ICP4 and -27 (61, 89). After L gene expression, the capsid is assembled in the nucleus through several successive stages; the viral DNA is cleaved and packaged through recognition of packaging sequences in the genomic termini referred to as the "a" sequence; and the viral tegument proteins are incorporated through an unknown process (26, 161). The virus particle acquires its envelope by budding through a modified patch of nuclear membrane and is released from the cell through a step involving the Golgi apparatus and perhaps membrane exchange. The lytic cycle is rapid, requiring less than 10 h and invariably results in cell death.

Molecular Aspects of HSV Latency

In the nuclei of sensory neurons of the peripheral nervous system (PNS), the HSV genome can persist throughout the life of the host as a concatemeric or circular molecule in a nonintegrated form (40, 97, 124) bound by nucleosomes (31). Under the influence of certain stresses, HSV genomes can be reactivated from latency and reenter the lytic cycle to produce viral particles, which are then transported in an anterograde direction back along the axons to initiate a secondary infection within cells near the primary site of infection (for review, see 151). Although HSV reactivation can occur in an immune-competent host (9), suppression of effective antiviral immunity can enhance viral spread (98). In rare cases, the virus can infect the CNS, causing encephalitis (46, 74).

During latency, the viral lytic genes, whose expression characterizes the replication cycle, are silenced and only the latency-associated transcript (LAT) gene remains transcriptionally active (21, 24, 125, 143, 152). The diploid LAT sequences map to the repeat regions flanking the U_L component of the genome (Figure 3). The major LAT species is 2.0 kb in length, nonpolyadenylated (162), and intranuclear (24, 25, 152). Current evidence suggests that the major LAT is a highly stable intron spliced from a large polyA$^+$ mRNA (minor or mLAT) that initiates 28 bp downstream of a TATA element and extends to the nearest polyadenylation site 8.7 kb downstream (33, 43). However, the splicing hypothesis awaits confirmation by genetic studies involving the use of viruses containing mutated splice-donor/acceptor sites surrounding the LAT intron. During lytic infection, the 2.0-kb LAT displays L gene kinetics (145). At least two other LAT species of 1.45 and 1.50 kb are detected only during latency (143, 144, 146, 163), although these species are more prominent in certain viral strains than in others (144). The termini of these less-abundant LATs are similar to the 2.0-kb LAT, which indicates that they may be derived from the larger LAT through splicing or RNA processing (146, 163).

The molecular events leading to latency have not been fully determined. Because a replication-deficient mutant virus can establish latency, this state probably does not require replication and likely occurs soon after the viral DNA enters the nucleus. We have observed the expression of LAT in some cells of the rat hippocampus within 24 h postinoculation in neurons that did not coexpress ICP0 mRNA. Neurons destined to harbor latent virus may not express Oct1 (168). Lack of this activity would make VP16 unable to induce the lytic gene transcriptional activators ICP0, -4, and -27 and the cell better able to control the virus. Results using ICP0 or VP16 mutants indicate that deletion or removal of the transactivating function of these genes has a dramatic negative effect on the ability of the virus to replicate in the nervous system; however, these deletions have similar effects on replication in nonneuronal cells (1, 14, 17, 149). Evidence also indicates that ICP0 is required for efficient virus reactivation from latency and that this requirement correlates with the activation function of ICP0 (12, 13, 78). Whether entry into latency is also an active process requiring the specific suppression of the IE promoters is unclear. Transient gene expression assays indicate that such a mechanism is possible. For example, Lillycrop et al (81) have shown that Oct2 expressed in neuronal

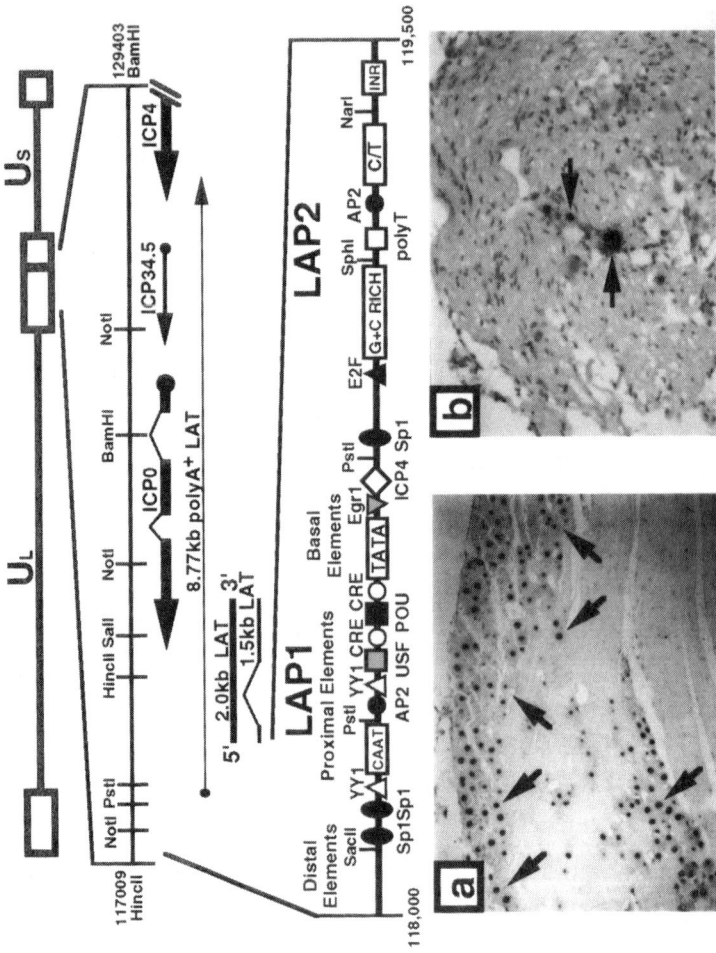

Figure 3 The location of the various HSV latency-associated transcripts (LATs) relative to the ICP0, ICP34.5, and ICP4 lytic gene RNAs within the prototype genome. An enlargement of the latency active promoter (LAP) regions LAP1 and LAP2 shows the position of potentially relevant *cis*-acting elements. (*a*) In situ PCR analysis of latently infected mouse trigeminal ganglia details the numerous dark-staining neuronal cell bodies harboring latent viral genomes. (*b*) In situ hybridization for LAT demonstrates the number of latently infected trigeminal neurons abundantly expressing the LATs.

cells can inhibit the ability of VP16 to activate IE gene promoters, although experiments involving the removal of the Oct2-binding site in IE gene promoters in order to examine the effects on the establishment of latency have not been reported.

The efficiency with which HSV can establish latency in PNS neurons has been estimated using several techniques. The number of neurons that harbor latent virus, determined by the number of cells expressing LATs detectable by in situ hybridization, suggests that only 1–5% of neurons in a peripheral ganglion are latently infected. Estimates of the number of genomes present in the ganglion, determined by Southern blot or by quantitative PCR, vary widely and suggest that latently infected cells can contain from as few as 0.1 to as many as 100,000 genomes (40, 151). We have used in situ DNA PCR to show that latent viral genomes are present in many more neurons (Figure 3a) than those expressing LATs detectable by in situ hybridization using LAT-specific riboprobes (Figure 3b). Other investigators have reported similar results for both HSV-1 and HSV-2 (71). Thus, while LAT expression remains the hallmark of latency, detectable expression occurs in only some of the cells containing genomes, perhaps determined by elements of the neuronal phenotype and the transcriptional environment of the cell. It is possible that LAT expression could occur in all, but at low levels in most, latently infected cells.

Recent studies have suggested that a high percentage of LAT$^+$ neurons in neuronal tissue does not correlate with the likelihood of reactivation (39) and that reactivation after explanation occurs first in cells that do not express detectable LATs (38a). Therefore, the level of LAT expression may play a role in controlling virus reactivation in an individual cell (157). Indeed, LAT has been suggested to function in an antisense manner to block ICP0 activity because it overlaps the 3′ terminus of the ICP0 mRNA (152) and ICP0 expression is an early critical step in virus reactivation. This notion is supported by the observation that the overexpression of LAT from a replication-defective virus using a viral IE promoter inhibits ICP0 protein synthesis and may destabilize the ICP0 mRNA (NA DeLuca, unpublished observation).

No latency-related proteins have been found in vivo (37) despite the presence of two small open reading frames (ORFs) within the 2-kb LAT, and deletion of the LAT locus has no obvious effect on the ability of the virus to establish latency. The majority of studies employing LAT deletion mutants indicate that LAT expression is not required for the establishment or maintenance of latency in mice (42, 58, 59, 65, 78, 150), but the function of the LAT locus is still not completely understood. An intact LAT locus seems to be required for efficient virus reactivation in some animal models (58, 78, 134, 150, 158), whereas in other models, LAT-deleted viruses appear to reactivate with normal kinetics (11, 32, 42, 59, 65, 104, 136). This finding that LAT expression is not required for the establishment of latency is fortuitous because it means the LAT promoter could be used for expression of transgenes during latency.

The transcriptional control of LAT gene expression is complex, although latency active promoters (LAPs) have been identified (5, 6, 33, 36, 54, 79, 106, 116, 174–176; MK Soares, DY Hwang, DJ Fink, MA Schmidt & JC Glorioso, in preparation). The nearest TATA box and basal transcriptional regulatory sequences, which make up LAP1 (36), lie approximately 700–1300 bp upstream of the previously mapped 5' end of the major 2-kb LAT (36). Another latency promoter, LAP2, lies downstream (54) (see Figure 3). Transient gene expression assays in transfected cell cultures have shown LAP1 to be more active than LAP2. Similar analyses have demonstrated that LAP1 contains upstream sequences that confer higher activity in neuronal cells lines (5, 6, 176), and transgenic mice containing a LAP1-CAT construct express chloramphenicol acetyltransferase (CAT) in neuronal cells of sensory ganglia and the limbic system within the brain, which confirms the neuronal specificity of this promoter (A Beckett, unpublished data). LAP2 is highly GC rich, sharing sequence homology with many eukaryotic housekeeping and protooncogene gene promoters (63, 72, 95, 137), and contains sequences that may affect chromatin structure (S French, WF Goins, R Kolluri, A Firulli, AJ Kinniburgh & JC Glorioso, submitted).

The most compelling evidence for the role of LAP1 in LAT gene expression during latency has come from in vivo studies involving a viral mutant (KOS/29) in which sequences surrounding the TATA box promoter and the transcription start site for the 8.7-kb mLAT have been deleted. This mutant establishes latency but does not produce LAT in latently infected mouse trigeminal ganglia (TG) as detected in Northern blot (106) or in situ hybridization (36) experiments. This finding is consistent with the hypothesis that the 2-kb LAT is a processed product. Studies using an HSV-1 recombinant showed that LAP2 juxtaposed to the *lacZ* reporter gene and introduced into the gC locus has long-term expression of β-galactosidase after the establishment of latency in mouse TG (54). In contrast, LAP1 failed to maintain expression of a foreign gene in the gC or native LAT loci (83, 86) in the absence of LAP2 sequences. Therefore, LAP2 sequences must contain an element that can function as a weak promoter in an isolated situation (54). This region may recruit factors or alter overall chromatin structure, thereby influencing LAP1 activity.

Recently, we have attempted to determine the relative roles of LAP1 and LAP2 sequences in expression of LAT in vitro and in vivo using deletion mutants lacking LAP1, LAP2, or both promoters (X Chen, DJ Fink, M Schmidt & JC Glorioso, submitted). In agreement with earlier reports (36, 106), we could not detect LAT by Northern analysis of RNA from latently infected ganglia harboring the LAP1 deletion mutant KOS/29. Nevertheless, LAT was still present, as shown when these RNA samples underwent reverse transcription and amplification by means of the polymerase chain reaction (RT-PCR). The level of LAT expression, as expected, was highly reduced (~500-fold) compared with the level of LAT detected in RNA extracted from TG containing latent wild-type virus. Quantitative RT-PCR analyses of the LAP2 deletion mutant revealed only a two- to threefold reduction in LAT expression during latency. The smaller LAT species (1.50 and 1.45 kb) were also detected in this mutant, indicating that LAP2 was not responsible for their expression.

During lytic infection, however, a different picture emerged. Deletion of LAP1 had little effect on LAT synthesis, while the LAP2 deletion mutant resulted in a significant reduction. Deletion of both regions greatly reduced, but did not completely eliminate, LAT expression. Together, these experiments confirmed that LAP1 is the principal latency promoter, and although latency does not require LAP2 sequences, LAT is not fully expressed in the absence of LAP2. The reduced amounts of LAT seen in cells latently infected with the LAP1 deletion mutants may reflect low-level lytic infection in a few neurons, which would be consistent with previous observations of low-level viral IE gene expression during latency (73). Alternatively, LAP2 may promote a level of LAT expression in all latently infected neurons that is too low to be detected by in situ hybridization or Northern analysis.

PROGRESS IN ENGINEERING HSV VECTORS

Overview

The problems related to HSV vector design fall into two general categories: (*a*) the elimination of lytic viral gene expression and cytotoxicity and (*b*) the engineering of a workable promoter system to achieve appropriate levels of therapeutic gene expression. The solutions to these problems largely depend on the intended use of the vectors. For instance, a vector designed to destroy neoplastic tissue should be cytotoxic with vigorous but short-term gene expression, whereas gene-replacement strategies will require an apathogenic vector, devoid of viral gene expression, that nonetheless is capable of long-term therapeutic gene expression. In some cases, foreign gene expression will need to be specifically regulated, while in others either continuous low-level or prolonged vigorous gene expression will be important. Tissue-specific regulation will be difficult to achieve without a deeper understanding of how the viral genes are regulated during latency. LAP1, for example, promotes expression of specific genes in neuronal cells, but we must learn more about how to exploit other enhancers to alter the tissue specificity of gene expression during latency in order to use HSV for gene therapy in other tissues. The remainder of this review discusses efforts to create appropriate HSV vectors for these different applications.

Eliminating Lytic Viral Gene Expression and Cytotoxicity

The main approach to eliminating cytotoxicity of HSV is to engineer vectors in which virus gene expression is blocked at an early stage, thus aborting the lytic cycle. The natural silencing of the HSV genome and the establishment of latency is thought to involve mechanisms specific to nerve cells. Mutations in the virus that circumvent lytic-gene expression may allow the virus to establish latency in any cell, regardless of the tissue. Because mutant viruses blocked at a very early stage in the lytic regulatory cascade can establish latency (35, 71), the approach of mutationally silencing the viral genome holds considerable promise for the development of HSV as a general gene-transfer vehicle.

The major regulatory protein of HSV is the transactivator ICP4, which is synthesized shortly after the virus enters the nucleus and activates the transcription of nearly all the viral genes. The study of temperature-sensitive (ts) mutants of HSV showed that only the five IE genes and ICP6 are efficiently expressed when ICP4 is inactivated (34). ICP4 also negatively regulates its own promoter through recognition of a specific element near the start of transcription and may play a role in repressing the transcription of other genes (28, 29, 75, 100, 107, 123) through direct interactions with, and possibly modifications of, the cellular transcription preinitiation complex (56, 57, 139). Early studies demonstrated that cell lines could be established that express ICP4 upon infection and that these cell lines could be used to isolate and efficiently propagate viruses lacking both copies of the ICP4 gene (27). ICP4 deletion mutants are less toxic to cells than their conditional lethal counterparts, and in contrast to their ts counterparts, they do not induce the expression of cellular stress proteins (27, 128).

However, infection of cells in vitro with ICP4 mutants alters gross cell morphology and the fragmentation of chromatin (68). Because UV-irradiated virus is not toxic to cells (80), proteins synthesized in the ICP4$^-$ background must be responsible for the toxic effects. The other IE genes, which do not require ICP4 to transactivate their expression, may express proteins that alter host-cell metabolism: ICP6, the large subunit of ribonucleotide reductase and a protein kinase (20) that depends in part on ICP0 for maximal expression (30); ICP27, a regulatory protein that affects gene expression at the transcriptional and post-transcriptional levels (90, 121, 140) as well as the phosphorylation of other viral and cellular proteins (155); ICP22, which affects the phosphorylation of the carboxy-terminal domain of cellular RNA polymerase II (122); ICP47, which interferes with the processing of major histocompatability complex (MHC) I molecules (173) and may protect against immune rejection of neurons during latency; and ICP0, a promiscuous transactivator of gene expression (41, 52, 115). In addition, several protein components of the virion particle [e.g. VP16, which acts to transinduce IE gene expression (7, 112), and the U_L41 gene product, which nonspecifically degrades mRNA molecules (108)] can adversely affect host-cell metabolism. We and others have been systematically deleting these genes to reduce or eliminate the toxicity of ICP4 deletion mutants and thereby improve the efficacy of HSV as a gene-transfer vector.

In the absence of ICP4, deletion of ICP22 and/or ICP47 did not result in the reduction of cytotoxicity (68), nor did the removal of U_L41 (69). Thus, these gene products are not major contributors to the cytotoxicity of the ICP4$^-$ mutant background. Mutations affecting the activation function of VP16, however, did reduce the toxicity of an ICP4$^-$ virus (69), suggesting that expression of IE genes other than ICP4 (either ICP0 or ICP27) must be responsible for the deleterious effects. Because ICP27 is the only IE gene other than ICP4 that is essential for virus growth, we reasoned that if cell lines could be established that simultaneously provide complementing levels of ICP4 and ICP27, then it should be possible to isolate not only viruses deleted for both ICP4 and ICP27, but also viruses simultaneously deleted for the other IE genes as well. Thus, all the genes expressed in the absence of ICP4 could be systematically deleted to determine their relative contribution to virus toxicity.

We have constructed cell lines that host viruses that simultaneously lack ICP4 and ICP27. Approximately 3×10^{10} replication-incompetent infectious progeny can be obtained from 10^8 cells, and more importantly, owing to the construction of the deletions and DNA inserts in the virus and cell line, respectively, replication-competent viruses are not generated during propagation (NA DeLuca, unpublished observation). This feature is a most desirable aspect of the complementing cell line system because it allows the production of vectors safe for in vivo gene-transfer experiments. The ICP4$^-$:ICP27$^-$ recombinant is somewhat less toxic to cells than either of the individual ICP4$^-$ or ICP27$^-$ deletions alone, although viral toxicity remains. The further elimination of ICP6 and U_L41 further increased the survival of infected cells 10- to 30-fold over that seen with an ICP4 deletion mutant (N Wu & NA DeLuca, unpublished data). The residual toxicity is most likely due to ICP0. ICP0 may function in the reactivation of latent virus (55, 78) as well as in the distribution of intranuclear structures (88). Efforts are underway to generate other viruses that lack this function.

Eliminating the toxicity of HSV vectors is a complicated problem requiring manipulation of multiple viral gene products. For example, both ICP27 and U_L41 negatively affect host-cell protein synthesis. One molecule degrades mRNAs and the other affects RNA splicing (108, 132, 140). Therefore, their effects may be somewhat redundant, necessitating the deletion of both, as described above. Moreover, the deletion of one gene may alter intracellular parameters and thereby increase the deleterious effects of one or more remain-

Expression of Therapeutic Genes Using HSV Vectors

STRATEGIES USING TRANSIENTLY ACTIVE PROMOTERS   Early studies demonstrated that eukaryotic genes inserted in the HSV genome were expressed using natural viral promoters during viral replication in cell culture. Examples include the hepatitis B surface antigen (138), hypoxanthine phosphoribosyltransferase (HPRT) (110), and more recently α-interferon (167). These genes were expressed with kinetics appropriate to the class of promoter used. Pallela et al (110) further demonstrated that the HPRT gene under control of the HSV–thymidine kinase (tk) promoter expressed HPRT in neuronal cells deficient for the enzyme, demonstrating that HSV could be used as a vector for gene replacement. However, when the vector was introduced into the brains of mice by intracranial inoculation, the relatively unattenuated virus caused encephalitis (109); hence, despite the production of HPRT mRNA, the HSV vector would not be useful without further attenuation.

Recombinant viruses that have been tested for attenuation of neurovirulence included those with altered expression of accessory functions, such as mutations in the ribonucleotide reductase genes (172); various glycoprotein genes (19, 96, 156); VP16 (159); a neurovirulence gene, ICP34.5 (85, 169); and the Us3 protein kinase gene (45). All of these alterations reduce neurovirulence by affecting the ability of the virus to replicate in the brains of test animals. Experiments even with these recombinants have failed to achieve long-term transgene expression in brain driven by any of the viral lytic gene promoters (45, 109), other viral promoters such as retrovirus LTRs (35, 99), the human cytomegalovirus IE gene promoter (HCMV IEp) (60), various pol III promoters (CA Meaney & JC Glorioso, unpublished data), housekeeping promoters known to be active in brain such as HPRT (KA Lee & JC Glorioso, unpublished data), and neural-specific cellular promoters (2). Prolonged expression was achieved in a few cells by using the nerve-specific enolase promoter in a tk⁻ replication-competent backbone (3), but the promoter failed to remain active in a replication-defective vector (MA Bender, DJ Fink & JC Glorioso, unpublished observation). In general, foreign promoters introduced into the viral genome appear to come under the influence of the viral transcriptional machinery, which results in aberrant behavior.

In contrast to uses in the CNS, the use of non-HSV promoters to express genes in the PNS has met with some success. For example, the Moloney murine leukemia retroviral (MoMuLV) LTR introduced into the ICP4 locus produces long-term expression of the *lacZ* reporter gene in dorsal root ganglion neurons (35), and the neuronal-specific sodium channel promoter results in long-term expression in mouse TG (DA Leib, personal communication). More recently, Feldman and colleagues (83) reported that a cassette containing the MoMuLV LTR and the LAP1 upstream sequence recombined into the gC locus gave long-term expression in TG neurons. This discovery suggests that upstream sequences in LAP1 can enhance the function of the LTR, but the LTR alone does not function during latency (83).

We injected a Us3 deletion mutant that expressed the *lacZ* gene under control of the gC L gene promoter (45) into rat hippocampus and found that after injection of moderate doses of virus, the virus failed to spread from the site of injection and did not cause encephalitis. However, expression limited to the granule cells of the dentate gyrus lasted only a few days, which is consistent with lytic gene inactivation that follows the establishment of latency. LAT could be detected for several weeks by means of in situ hybridization and for up to one year using RT-PCR. High-dose inoculation caused encephalitis, which raised the important question of whether attenuated vectors could be safely exploited for human gene therapy.

Second-generation vectors in which the virus was rendered defective by deletion of essential genes have also been tested. The goal of these experiments was to develop vectors that were free of live virus recombinants yet were still capable of establishing latency and expressing a foreign gene using a promoter system that remained active in the absence of lytic gene expression. The first of these mutants involved introducing the *lacZ* reporter gene into a virus lacking the ICP4 IE gene. The ICP4$^-$ vector (d120) (27) can only replicate in the complementing E5 cell line. Therefore, once the virus was introduced into the brain, only the other IE genes would be expressed, although some low-level E gene expression may occur. The *lacZ* gene was placed under control of a strong non-HSV promoter, the HCMV IEp. This promoter should function independently of ICP4 expression. The experiments revealed that high titers of this vector could be introduced into the rat hippocampus without apparent neuronal cell damage. The vector vigorously expressed the reporter gene for several days, as shown in Figure 4a. The virus readily established latency and maintained high copy numbers over time. These experiments demonstrated that nonreplicating vectors could probably be engineered to establish latency in brain without damage to neurons, although transgene expression driven by HCMV IEp or other promoters shut down upon the establishment of latency. Other investigators have reported similar findings (2, 18, 35).

An alternative approach to the use of modified, whole virus vectors is the use of amplicons, plasmids containing both a bacterial and HSV origin of replication and an HSV-1 packaging site or "a" sequence (141), to deliver foreign genes to neurons. Often referred to as defective HSV-1 vectors, these

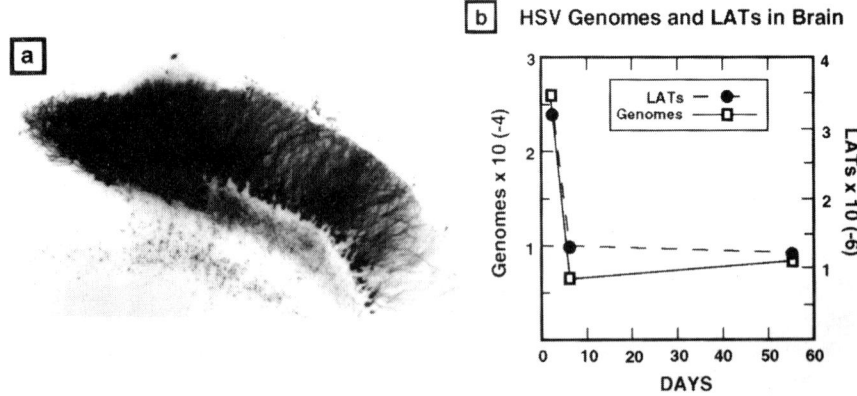

Figure 4 HSV vector-mediated gene expression of genome quantitation in brain. (*a*) X-gal staining detects the *lacZ* reporter gene product expressed in dentate gyrus neurons at 4 days after stereotactic inoculation of the replication-defective recombinant d120 (ICP4⁻), TK⁻::HCMV IEp-*lacZ*, containing the HCMV IEp-β-galactosidase expression cassette. Expression was transient, disappearing by 7 days postinfection. (*b*) Quantitative PCR analyses were performed on DNA isolated at 2, 7, and 56 days postinjection from the hippocampal region of brain slices of animals infected with a defective HSV vector. The number of LAT molecules was also determined and plotted as a function of time.

plasmids can be propagated in bacteria and then cotransfected with a defective HSV recombinant, which results in a mixed population of helper virus particles and vector particles containing concatemers of the plasmid. Geller and colleagues (50, 51) first demonstrated that *lacZ* could be expressed transiently using amplicons in primary cultures of neurons from the PNS or CNS, and similar results have been demonstrated using organotypic hippocampus slice culture (16). In tissue-culture models, appropriate biological effects have been demonstrated with amplicons used to transfer the genes encoding NGF (44) and the human NGFp75 high-affinity receptor (8) to cells.

Amplicons were generated initially from studies of defective interfering HSV particles that arise spontaneously in populations of replicating virus (48, 82). The ratio of amplicon to helper-containing particles is difficult to control, and to favorably increase the amplicon:vector ratio, the preparation must be repeatedly passaged and evaluated because the ratio varies after each round of infection. This method increases the opportunities for recombination either between the vector and the complementing gene used to support the growth of the helper virus in the cell line, or between the amplicon plasmid and the helper-virus genome, which often share homologous sequences.

Gene transfer in vivo has been demonstrated using amplicons. In one of these experiments, after intracranial inoculation, cells positive for expression of β-galactosidase persisted as long as one month (38, 47). One report showed expression of NGF in sympathetic ganglia and what appears to be a biological effect of NGF (44). Amplicon-mediated transfer of the tyrosine hydroxylase (TH) gene into partially denervated striatum of rat brain was recently shown to produce an appropriate biological effect (38). Kaplitt and colleagues reported that amplicons could be used to provide neuronal-specific gene expression in brain (70). However, none of these studies have shown that these preparations actually lack replicating virus produced by recombination during the repeat processing necessary to produce adequate titers of amplicons. The sporadic presence of replication-competent virus and leaky viral gene expression complicates the interpretation of results demonstrating prolonged transgene expression by amplicons.

THERAPEUTIC STRATEGIES USING THE VIRAL LATENCY PROMOTER SYSTEM
One possible explanation for the loss of expression over time after in vivo gene transfer is the clearance of latent viral genomes from neurons. Using quantitative PCR, we have demonstrated that after injection of a defective vector into a single brain region, the number of viral genomes in 10-μm brain slices decreased during an initial period from 2 to 7 days postinoculation but then remained stable to at least 8 weeks (Figure 4b). These data suggest that despite the loss of genomes, adequate vector persists in brain even at 8 weeks. In situ hybridization with riboprobes specific for HSV LAT demonstrated the presence of numerous positive neurons that also remained stable over time; however, long-term LAT expression was too low to detect using in situ procedures. LAT-specific RT-PCR analyses indicated that an average of 10–20 copies of LAT per viral genome persisted (119).

The ability of at least some viral genomes to express LAT suggests that the LAT promoter region must escape the general suppression of HSV promoter activity during latency and therefore might be a candidate to drive transgene expression from the HSV vector. HSV vectors in which a foreign or reporter gene is placed downstream from the TATA box–containing latency promoter (LAP1) in the native LAT site or in an ectopic locus have sometimes failed to express the foreign gene product altogether (83, 86). Alternatively, foreign gene expression in PNS neurons is adequate (36, 170), but prolonged expression in the CNS occurs in only a few cells (170). Using the alternate latency promoter (LAP2), which consists of sequences between LAP1 and the 5′ end of the stable LAT, we have demonstrated long-term expression of the transgene after intracranial inoculation. With the LAP2-*lacZ* cassette placed in either the native locus or the gC locus of the ICP4 deletion mutant (d120), *lacZ* mRNA could be detected using RT-PCR at 2 and 4 weeks after infection into hippocampus (X Chen, DJ Fink, WF Goins & JC Glorioso, unpublished observation).

In contrast, a similar construct in which HCMV IEp controlled *lacZ* expression exhibited no *lacZ* RNA even when observed via RT-PCR.

Low level in situ expression of some gene products, neurotrophic factors for example, might be desirable for some therapeutic applications. However, further work is required to define the localization of the transgene mRNA, determine whether a protein product can be detected, and demonstrate a biological effect of that protein. These experiments are currently underway.

STRATEGIES USING CONSTITUTIVE AND INDUCIBLE TRANSCRIPTIONAL TRANSACTIVATORS   The difficulties in achieving long-term gene expression have prompted us to explore novel ways to express transgenes and still take advantage of the latency promoter system. The first of these strategies is to engineer into the viral genome a novel gene whose product can transactivate a promoter bound by nucleosomes. The target sequence for this transcription factor must also be unique in the vector genome and preferably in the cell harboring the latent virus. One such system has been well described in the literature (131). In another example, a recombinant gene composed of the yeast GAL4 DNA-binding domain and the HSV VP16 acidic transactivation domain specifically transactivates expression of a heterologous promoter containing one or more tandem copies of the GAL4 17-bp recognition element (53). The fusion gene product can activate expression of a gene promoter with the GAL4-binding site even though bound by nucleosomes in yeast and frog oocytes (4, 171). To test whether this system could work in the context of the viral genome, we introduced this fusion gene into the genome of a defective HSV vector at the tk locus either under control of a minimum TATA box promoter with five copies of the GAL4 recognition element or under control of the HCMV IEp modified to contain GAL4 sites. Under control of the minimum promoter, the GAL4-VP16 transactivator (TA) was expressed very poorly. However, GAL4-VP16 expressed by HCMV IEp activated a minimum GAL4-TATA promoter controlling the CAT gene integrated into the cellular chromosome. GAL4-VP16 used to infect the cell line contained the TA virus, and CAT activity was high (Figure 5), while uninfected cells did not express CAT (P Marconi, MA Bender, JC Glorioso unpublished). In a second case, GAL4-VP16 was used to express tyrosine hydroxylase under the control of the GAL4-HCMV IEp embedded in the genome of a second reporter virus. TA-containing GAL4-VP16 and TH reporter viruses were used to coinfect rat neuroma B103 cells in which both DOPA and dopamine synthesis were measured. B103 cells do not express TH but do produce dopamine decarboxylase. DOPA expression when both viruses were present was greater compared with levels observed upon infection with the TH virus alone or those seen upon coinfection with a control virus lacking the GAL4-VP16 TA. These results clearly show that the TA can function in *trans* to up-regulate expression of TH and consequently of

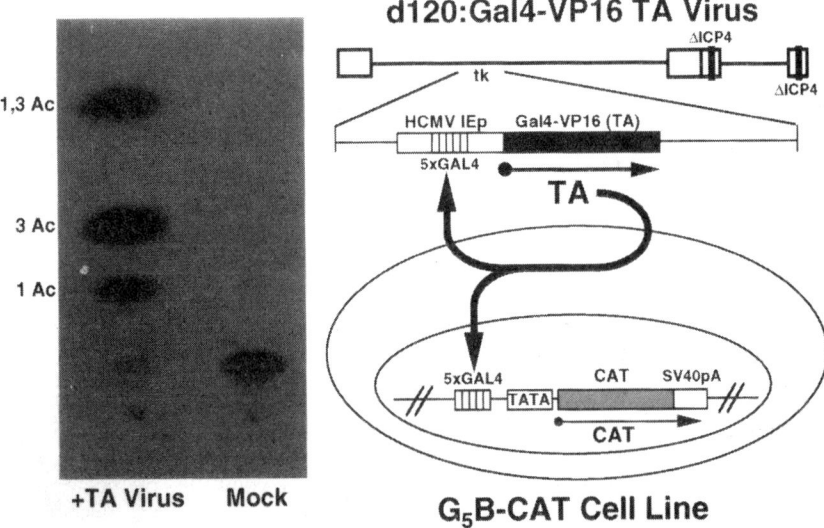

Figure 5 HSV vector-mediated expression of the GAL4-VP16 transactivating function in vitro. A defective HSV vector, which expresses the GAL4-VP16 transactivator (TA) from HCMV IEp and contains five GAl4-binding sites, can activate a CAT reporter gene that is inserted into the cellular genome and is under the control of a minimal TATA box promoter also possessing five GAL4-binding sites (G5B-CAT). Mock-infected G5B-CAT cells failed to synthesize any CAT product in the CAT assays. However, upon infection with the TA-expressing virus, high levels of CAT activity were indicated by the presence of the acetylated forms (1-Ac, 3-Ac, and 1, 3-Ac) of the radioactive chloramphenicol substrate.

DOPA, which leads to a fivefold increase in expression of dopamine. TH virus with and without the TA vector were tested in vivo by stereotaxic injection into rat caudate, but despite the demonstration of autoactivation in vitro, expression did not persist in vivo. The autogene system has yet to be tested in a vector containing both reporter and TA constructs that would guarantee that each infected cell received both gene cassettes. Current experiments aim to introduce the TA gene under control of the LAP in order to achieve low but constitutive expression of the TA product. If such expression is possible, the TA may transactivate a second gene that is engineered into the genome, is controlled by a GAL4 site–containing promoter, and is thus responsive to the TA.

A second approach, taking advantage of the continued transcriptional activity of the LAPs, is to express a TA that can be activated by a drug capable of crossing the blood-brain barrier. The use of a drug to switch the constitutively produced TA product on and off might allow external regulation of the required transgene product. O'Malley and colleagues (164) have reported an example of such a gene switch system. They further modified the GAL4-VP16 TA by adding a third domain derived from the progesterone receptor. Normally, specific progesterone-sensitive promoters become activated once progesterone enters the cytoplasm. The binding domain of the receptor is mutated so that it no longer recognizes progesterone; rather, it is activated by the antiprogestin drug RU486 when GAL4-VP16 is fused to the mutated receptor. Administration of RU486 alters the conformation of the recombinant TA and enables the VP16 acidic transactivation domain to function when the chimeric TA is bound to promoters containing Gal4-binding sites. We have confirmed that this drug-activated TA can stimulate a promoter in the HSV genome in vitro. Constitutive expression of this recombinant, inducible receptor driven by the HSV latency promoter system may be useful in the nervous system. Administration of RU486 could be used to activate the TA in the brain and, in turn, activate a second promoter containing the GAL4-binding sites in order to express a therapeutic gene. In applications requiring intermittent gene expression, the production of therapeutic gene product could then be controlled by the amount and duration of RU486 administration.

APPLICATIONS OF HSV VECTORS

Gene Therapy of Cancer

Various gene therapy strategies are now under evaluation for the treatment of neoplastic disease. Cancer gene therapies involve four general approaches that depend on the nature of the tumor, i.e. whether it is a solid mass or consists of circulating tumor cells of hemopoietic origin. For example, chemotherapy can be used to a greater degree to treat tumors comprised of malignant blood cells. Conventional chemotherapies can lead to bone marrow suppression and loss of the hemopoietic system. However, ex vivo approaches in which bone marrow stem cells are transduced with multiple drug resistance genes using retroviral vectors may allow larger doses of chemotherapeutic agents to be used to destroy the tumor cells without harm to the bone marrow.

Solid tumors are often surgically resected, but surgical intervention may not be possible for tumors that are diffuse or inaccessible. These tumors may develop into large masses that require debulking as part of the treatments involving anticancer drugs and/or radiation. The local transfer of genes whose encoded products produce toxic cytokines (e.g. tumor necrosis factor), induce apoptosis (e.g. p53), or activate prodrugs (e.g. thymidine kinase, cytosine deaminase) could result in reduction in tumor mass.

Some tumors produce substances that suppress the immune response; that lack cell-surface class I and II MHC antigens and costimulatory molecules (e.g. $B7_1$ and $B7_2$) required for immune recognition of tumor cells, or that cause immune tolerance to tumor-specified antigens, such as carcinoembryonic-like proteins, oncogene or mutated oncogene products, or highly expressed normal or alternatively processed products. Thus, another strategy is to present these antigens in an highly enriched cytokine environment in an attempt to break tolerance and promote effective immune surveillance and the development of tumor-rejection mechanisms. An effective treatment for some cancers might be to introduce genes whose products "vaccinate" against the tumors along with clonally expanded, specific immune lymphocytes derived from the patient.

Other vectors can be directly introduced into the tumor that express genes that up-regulate MHC gene expression (e.g. γ-interferon) and recruit professional antigen-processing and -presenting cells (e.g. granulocyte-stimulating factor). Genes whose products may stimulate tumor-specific helper T-cell proliferation (e.g. interleukin 2 and 12), and consequently induce the proliferation of cytotoxic T cells, may provide antitumor immunity capable of destroying metastases elsewhere in the body.

For in vivo methods, viral vectors such as AV can be engineered to carry several genes that induce cytotoxic products following direct tumor injection. However, HSV can be engineered to carry multiple genes that could facilitate direct tumor cell killing, enhance tumor recognition by the immune system, and provide cytokines to promote immune reactivity.

The transfer of the HSV-tk gene combined with ganciclovir (GCV) administration has become a popular method for debulking solid tumors. This general approach, pioneered by Culver and colleagues (22, 117, 118), has been tested in phase I clinical trials for treatment of glioblastoma. The strategy depends on viral tk-mediated phosphorylation of GCV, which in its phosphorylated state is incorporated into newly synthesized cellular DNA. Its incorporation terminates DNA synthesis because the nucleotide analogue cannot participate in chain elongation. Since TK-activated GCV kills only dividing cells, normal brain tissue should be unaffected by the treatment. In animal models in which a cell line capable of producing a retroviral vector carrying the HSV-tk gene was injected into the tumor site, tumor cells infected by the retroviruses subsequently produced in vivo were killed by GCV administration. Uninfected surrounding cells were apparently also destroyed, which suggests a bystander effect. The bystander phenomenon has also been demonstrated in a cell-culture system (10) and may result from transfer of activated drug to neighboring cells across gap junctions.

Several clinical trials using the retroviral HSV-tk producer cell line have begun. In one involving 24 patients, despite demonstrated tumoricidal effects, tumors continued to progress in most patients (KW Culver, personal communication). The producer cell line approach may not be sufficient to provide long-term benefit; few patients survived longer than 15 months and only 1 patient appeared to be free of tumor at 21 months (KW Culver, personal communication). The Recombinant DNA Advisory Committee (RAC) has now approved several protocols involving HSV-tk transfer using adenoviral vectors, and similar herpes virus vectors are under development.

HSV vectors suitable for in vivo gene therapy of cancer have potential advantages over the use of retrovirus-producer cell lines. First, the tumor mass can be directly injected with a vector that is defective for replication and will thus not readily spread to other tissues. Second, short-term gene expression will likely be sufficient to debulk a tumor mass, and currently available HSV vector constructs can be used. The GCV prodrug approach is a good candidate strategy because the drug is safe for human use and is activated by the native HSV-tk gene product. GCV treatment can control HSV replication, thereby adding a built-in safety mechanism. Third, because the HSV genome is large and capable of carrying many foreign genes, the vector can be crafted to express one or more cytokine genes to recruit, activate, and amplify the local immune response to the tumor cells.

HSV vectors can be used that will kill tumor cells by natural cytotoxic mechanisms expressed by the virus, or the virus can be replication defective but still carry the tk gene in an expressible form. Martuza and colleagues showed rapid debulking of brain tumors in experimental animals using a tk deletion mutant virus that replicates poorly in normal brain tissue (87). We have used a Us3$^-$ virus mutant that we previously showed would not replicate and spread in normal brain (45). Using rat 9L cell brain tumors as a target, this mutant virus could effectively kill the tumor cells and had the added advantage that it expressed normal amounts of TK in brain, thereby allowing GCV therapy. Because safety remains a concern with live virus systems, we have explored the possibility of using completely defective viral mutants.

An ICP4 deletion mutant virus was engineered to express the viral tk gene using an HSV IEp, and thus expression is activated upon infection by the natural VP16 IEp activator carried in the virus particle as part of the tegument. As shown in Figure 6a, this vector activates tk expression and produces a bystander effect similar to that observed with the retroviral system. The vector was also effective in debulking tumors in vivo (Figure 6b). Survival studies using this vector in animals are in progress.

The nonreplicating vector system will effectively kill tumor cells through a combination of drug activation and the natural cytotoxicity of this single-deletion mutant virus. However, the entire tumor mass is quite difficult to reach with a single injection. We have therefore introduced the virus at multiple sites in 1-μl amounts, which has resulted in effective tumor destruction, based on histological findings. Whether this strategy alone will be effective in patients

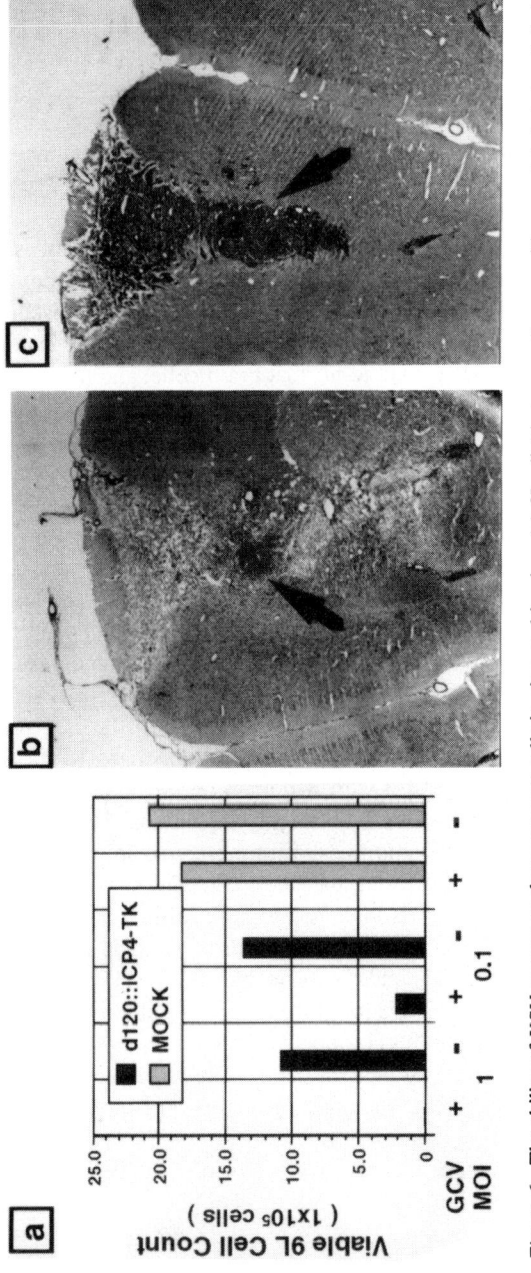

Figure 6 The ability of HSV vectors to destroy tumor cells in vitro and in vivo. (*a*) 9L glioblastoma cells were either mock infected or infected with a defective HSV vector expressing the HSV-tk gene (d120::ICP4-tk) at various multiplicities of infection (MOI) in the presence or absence of ganciclovir (GCV). The cells were then scored for viability. The killing of many more cells than expected based on the MOI demonstrated a bystander-killing effect. For in vivo analyses, (*b*) ra:s were injected with 9L tumor cells and one week later were stereotactically inoculated with replication-defective TK-expressing virus (d120::ICP4-tk). Following GCV treatment for 2 days, brain sections were examined for the presence of tumor and compared with sections from (*c*) mock-infected, tumor-bearing animals.

awaits clinical studies. Although initial clinical trials will probably utilize a nonreplicating tk-expression vector, experiments are underway to test second- and third-generation vectors. These vectors have single and multiple cytokine genes introduced into essential loci in the vector to prevent a fortuitous recombination event with a latent virus that would result in live recombinant virus in vivo. The intention is to induce local antitumor responses that may eliminate tumor cells that escape drug therapy.

Additional prodrug-activating genes are being tested, such as cytosine deaminase, which converts 5-fluoro-cytosine to the cytotoxic compound 5-fluoro-uracil (5-FU). Unlike GCV, 5-FU can be taken up by distal cells and thus may have a more global effect on the tumor. This approach is not without difficulties because 5-FU is toxic to normal brain cells, and thus dose will be become an important issue.

Gene Therapy of Neurodegenerative Disease

The other major class of target diseases for gene transfer to brain are neurodegenerative diseases such as Alzheimer's, Parkinson's, and Huntington's diseases. The pathophysiologies of these complex diseases, despite major advances in recent years, remain poorly understood. Their ultimate treatment in terms of choice of vector and the gene to be delivered will depend on further elucidation of the molecular mechanisms underlying the pathology.

Three types of therapeutic approaches can be envisioned. The first is typified by experimental models of Parkinson's disease, an idiopathic degeneration of dopaminergic cells that project from the substantia nigra to the striatum. Exogenous replacement therapy with a dopamine precursor (L-DOPA) can alleviate the symptoms in patients with early to mid-stage disease. Because tyrosine hydroxylase (TH) is the rate-limiting enzyme in the production of dopamine, one approach to treating the disease would be to transfer the TH gene into nondopaminergic cells in the striatum, resulting in the local production of excess dopamine. Studies using myoblasts transduced with the TH gene transplanted into striatum (67) and direct transfer of the TH gene with amplicons (38) have demonstrated amelioration of apomorphine-induced rotational behavior in 6-hydroxydopamine–lesioned rats, a generally accepted simple animal model of striatal dopamine deficiency. We have shown that an HSV vector containing TH enables cells to produce dopamine in vitro (MA Bender & JC Glorioso, unpublished data).

Nevertheless, the human disease may prove difficult to treat with these measures. First, clinical treatment of patients requires exquisite regulation of the amount of dopamine provided: With too little dopamine, the symptoms persist; with excess dopamine, adventitious choreiform movements supervene. Second, because patients become refractory to dopamine treatment with either L-DOPA or direct dopamine receptor agonist drugs as the disease progresses, they might also become refractory to transgene-produced dopamine. Ultimately, of course, the most desirable therapy would be one that prevented the death of the dopaminergic cells in the substantia nigra.

The second paradigm is represented by Huntington's disease, a dominantly inherited condition causing degeneration of several classes of neurons localized to the caudate and putamen that results in dementia, choreiform movements, and death. The responsible gene was mapped to the short arm of chromosome 4 in 1983 and cloned in 1994 (84). The disease results from expanded trinucleotide repeats in the Huntington gene, although the nature and function of the gene product is not known. Once the molecular basis of the Huntington gene product action is identified, strategies using antisense activities to block its toxic activity could be used to prevent the disease phenotype from developing in patients identified in the presymptomatic state by the presence of the expanded repeats. This preventative treatment contrasts with the therapeutic gene transfer used in the Parkinson's model. The potential pitfalls in this scenario are the untested nature of the antisense approaches and the absence of an animal model in which the vector could be tested.

Alzheimer's disease is a neurodegenerative condition characterized clinically by progressive dementia and pathologically by the accumulation of amyloid, neuritic plaques, and paired helical filaments in the brain. Most cases do not appear to be inherited, and in the familial cases (representing 10% of the total), at least three different chromosomal localizations have been identified (103, 133, 135, 147, 148, 154, 160). Loss of many different types of neurons occurs in this disease; consequently, no single neurotransmitter-replacement therapy is effective. However, were a single trophic factor (nerve growth factor, for instance) to prove useful in preventing degeneration of target cells, or a single enzyme in the final amyloid-processing pathway to prove effective in preventing the accumulation of amyloid and the development of pathology, the genes for such factors could be delivered to at-risk brain regions using a latency-expressing HSV gene-transfer vector.

CONCLUSIONS AND FUTURE DIRECTIONS

The design of HSV vectors has relied on the fundamental understanding of the biology of this virus and its interaction with the host. Considerable progress has been made in muting the cytotoxic features of this virus, and vectors can probably be engineered that are completely devoid of viral gene expression yet are capable of producing therapeutic gene products. An attractive feature of this virus is its ability to infect and persist in nondividing cells, particularly in neurons, and not interfere with normal cell function or induce immunopathologic disease. The virus can be propagated to high titer without the production of wild-type recombinants, making it potentially one of the safest vectors available. The genome is quite large and should be able to accommodate large or multiple gene cassettes. This feature will likely be important because transferring two or more genes to provide a therapeutic benefit will be important in many applications.

Current vectors are suitable for gene therapy of PNS disease using the latency promoter system (e.g. treatment of peripheral neuropathy by transferring the nerve growth factor gene to dorsal root sensory ganglia) or for treatment of solid neoplasms (e.g. glioma) using vigorous but transiently active promoters such as HCMV IEp. For cancer applications, genes that activate prodrugs; recruit, activate, and expand the immune response; and make tumor cells more immunogenic can be engineered into a single vector and used to directly infect the tumor mass. Multiple genes under the control of different promoters have already been incorporated into our HSV vectors.

For applications involving long-term gene expression in brain, however, methods to prevent transcription shutoff must be devised. The LAT promoter system provides an excellent lead in this search, and several laboratories have begun to unravel the molecular details of this system. Whether similar tissue-specific promoter systems can be engineered into the viral genome and whether more vigorous gene expression can be obtained using a modified latency-active promoter remain to be determined. Particularly interesting questions involve resolving how the virus alters the regulation of non-HSV promoters and determining what role methylation and chromatin remodeling play in gene expression during the latent state. Assuming these barriers can be overcome, prospects for exploiting the HSV system for gene therapy for brain disease and for fundamental studies of brain biology appear very promising.

ACKNOWLEDGMENTS

The authors thank William F Goins, Mary Ann Bender, Jim Cavalcoli, Xiaowei Chen, Kevin Lee, Peggy Marconi, David Krisky, Roberto Furlan, Sam French, Karina Soares, Navin Wa, and Ramesh Ramakrishnan for their contributions to this work. This research was sponsored by grants from the National Institute of Health, the National Institutes of Mental Health, and the Veterans Administration.

> Any *Annual Review* chapter, as well as any article cited in an *Annual Review* chapter, may be purchased from the Annual Reviews Preprints and Reprints service.
> 1-800-347-8007; 415-259-5017; email: arpr@class.org

Literature Cited

1. Ace CI, McKee TA, Ryan M, Cameron JM, Preston CM. 1989. Construction and characterization of a herpes simplex virus type 1 mutant unable to transinduce immediate-early gene expression. *J. Virol.* 63:2260–69

2. Andersen JK, Frim DM, Isacson O, Breakefield XO. 1993. Herpesvirus-mediated gene delivery into the rat brain: specificity and efficiency of the neuron-specific enolase promoter. *Cell. Mol. Neurobiol.* 13:503–15
3. Andersen JK, Garber DA, Meaney CA, Breakefield XO. 1992. Gene transfer into mammalian central nervous system using herpes virus vectors: extended expression of bacterial *lacZ* in neurons using the neuron-specific enolase promoter. *Human Gene Ther.* 3:487–99
4. Axelrod JD, Reagan MS, Majors J. 1993. GAL4 disrupts a repressing nucleosome during activation of GAL1 transcription *in vivo*. *Genes Devel.* 7:857–69
5. Batchelor AH, O'Hare PO. 1990. Regulation and cell-type–specific activity of a promoter located upstream of the latency-associated transcript of herpes simplex virus type 1. *J. Virol.* 64:3269–79
6. Batchelor AH, O'Hare PO. 1992. Localization of cis-acting sequence requirements in the promoter of the latency-associated transcript of herpes simplex virus type 1 required for cell-type–specific activity. *J. Virol.* 66:3573–82
7. Batterson W, Roizman B. 1983. Characterization of the herpes simplex virion-associated factor responsible for the induction of α genes. *J. Virol.* 46:371–77
8. Battleman DS, Geller AI. 1993. HSV-1 vector–mediated gene transfer of the human nerve growth factor receptor p75hNGFR defines high-affinity NGF binding. *J. Neurosci.* 13:941–51
9. Berman EJ, Hill JM. 1985. Spontaneous ocular shedding of HSV-1 in latently infected rabbits. *Invest. Ophthalmol. Vis. Sci.* 26:587–90
10. Bi WL, Parysek LM, Warnick R, Stambrook PJ. 1993. In vitro evidence that metabolic cooperation is responsible for the bystander effect observed with HSV tk retroviral gene therapy. *Human Gene Ther.* 4:725–31
11. Block TM, Spivack JG, Steiner I, Deshmane S, McIntosh MT, et al. 1990. A herpes simplex virus type 1 latency-associated transcript mutant reactivates with normal kinetics from latent infection. *J. Virol.* 64:3417–26
12. Cai W, Astor TL, Liptak LM, Cho C, Coen DM, Schaffer PA. 1993. The herpes simplex virus type 1 regulatory protein ICP0 enhances viral replication during acute infection and reactivation from latency. *J. Virol.* 67:7501–12
13. Cai WZ, Schaffer PA. 1989. Herpes simplex virus type 1 ICP0 plays a critical role in the de novo synthesis of infectious virus following transfection of viral DNA. *J. Virol.* 63:4579–89
14. Cai W, Schaffer PA. 1992. Herpes simplex virus type 1 ICP0 regulates expression of immediate-early, early and late genes in productively infected cells. *J. Virol.* 66:2904–15
15. Campbell MEM, Palfreyman JW, Preston CM. 1984. Identification of herpes simplex virus DNA sequences which encode a trans-acting polypeptide responsible for stimulation of immediate early transcription. *J. Mol. Biol.* 180:1–19
16. Casaccia-Bonnefil P, Benedikz E, Shen H, Stelzer A, Edelstein D, et al. 1993. Localized gene transfer into organotypic hippocampal slice cultures and acute hippocampal slices. *J. Neurosci. Methods* 50:341–51
17. Chen J, Silverstein S. 1992. Herpes simplex viruses with mutations in the gene encoding ICP0 are defective in gene expression. *J. Virol.* 66:2916–27
18. Chiocca AE, Choi BB, Cai W, DeLuca NA, Schaffer PA, et al. 1990. Transfer and expression of the *lacZ* gene in rat brain neurons mediated by herpes simplex virus mutants. *New Biol.* 2:739–46
19. Chrisp CE, Sunstrum JC, Averill DR, Levine M, Glorioso JC. 1989. Characterization of encephalitis in adult mice induced by intracerebral inoculation of herpes simplex virus type 1 (KOS) and comparison with mutants showing decreased virulence. *Lab. Invest.* 60:822–30
20. Chung TD, Wymer JP, Smith CC, Kulka M, Aurelian L. 1989. Protein kinase activity associated with the large subunit of herpes simplex virus type 2 ribonucleotide reductase (ICP10). *J. Virol.* 63:3389–98
21. Croen KD, Ostrove JM, Dragovic LJ, Smialek JE, Strause SE. 1987. Latent herpes simplex virus in human trigeminal ganglia. Detection of an immediate early gene "anti-sense" transcript by in situ hybridization. *New Engl. J. Med.* 317:1427–32
22. Culver KW, Ram Z, Walbridge S, Ishii H, Oldfield EH, Blaese RM. 1992. In vivo gene transfer with retroviral vector-producer cells for treatment of experimental brain tumors. *Science* 256:1550–52
23. Davison AJ, Wilkie NM. 1983. Inversion of the two segments of the herpes simplex virus genome in intertypic recombinants. *J. Gen. Virol.* 64:1–18

24. Deatly AM, Spivack JG, Lavi E, Fraser NW. 1987. RNA from an immediate early region of the type 1 herpes simplex virus genome is present in the trigeminal ganglia of latently infected mice. *Proc. Natl. Acad. Sci. USA* 84:3204–8
25. Deatly AM, Spivack JG, Lavi E, O'Boyle D, Fraser NW. 1988. Latent herpes simplex virus type 1 transcripts in peripheral and central nervous system tissues of mice map to similar regions of the viral genome. *J. Virol.* 62:749–56
26. Deiss LP, Chou J, Frenkel N. 1986. Functional domains within the a sequence involved in the cleavage-packaging of herpes simplex virus DNA. *J. Virol.* 57:605–18
27. DeLuca NA, McCarthy A, Schaffer PA. 1985. Isolation and characterization of deletion mutants of herpes simplex virus type 1 in the gene encoding immediate-early regulatory protein ICP4. *J. Virol.* 56:558–70
28. DeLuca NA, Schaffer PA. 1985. Activation of immediate-early, early, and late promoters by temperature-sensitive and wild-type forms of herpes simplex virus type 1 protein ICP4. *Mol. Cell. Biol.* 5:558–70
29. DeLuca NA, Schaffer PA. 1988. Physical and functional domains of the herpes simplex virus transcriptional regulatory protein ICP4. *J. Virol.* 62:732–43
30. Desai P, Ramakrishnan R, Lin ZW, Osak B, Glorioso JC, Levine M. 1993. The RR1 gene of herpes simplex virus type 1 is uniquely transactivated by ICP0 during infection. *J. Virol.* 67:6125–35
31. Deshmane SL, Fraser NW. 1989. During latency, herpes simplex virus type 1 DNA is associated with nucleosomes in a chromatin structure. *J. Virol.* 63:943–47
32. Deshmane SL, Nicosia M, Valyi-Nagy T, Feldman LT, Dillner A, Fraser NW. 1993. An HSV-1 mutant lacking the LAT TATA element reactivates normally in explant cocultivation. *Virology* 196:868–72
33. Devi-Rao GB, Goodart SA, Hecht LM, Rochford R, Rice MK, Wagner EK. 1991. Relationship between polyadenylated and nonpolyadenylated herpes simplex virus type 1 latency-associated transcripts. *J. Virol.* 65:2179–90
34. Dixon RA, Schaffer PA. 1980. Fine-structure mapping and functional analysis of temperature-sensitive mutants in the gene encoding the herpes simplex virus type 1 immediate early protein VP175. *J. Virol.* 36:189–203
35. Dobson AT, Margolis TP, Sederati F, Stevens JG, Feldman LT. 1990. A latent, nonpathogenic HSV-1–derived vector stably expresses β-galactosidase in mouse neurons. *Neuron* 5:353–60
36. Dobson AT, Sederati F, Devi-Rao G, Flanagan WM, Farrell MJ, et al. 1989. Identification of the latency-associated transcript promoter by expression of rabbit β-globin mRNA in mouse sensory nerve ganglia latently infected with a recombinant herpes simplex virus. *J. Virol.* 63:3844–51
37. Doerig C, Pizer LI, Wilcox CL. 1991. An antigen encoded by the latency-associated transcript in neuronal cell cultures latently infected with herpes simplex virus type 1. *J. Virol.* 65:2724–27
38. During MJ, Naegele J, O'Malley KL, Geller AI. 1994. Long-term behavioral recovery in Parkinsonian rats by an HSV vector expressing tyrosine hydroxylase. *Science* 266:1399–403
38a. Ecob-Prince M, Hassan K. 1994. Reactivation of latent herpes simplex virus from explanted dorsal root ganglia. *J. Gen. Virol.* 75:2017–28
39. Ecob-Prince MS, Rixon FJ, Preston CM, Hasson K, Kennedy PG. 1993. Reactivation in vivo and in vitro of herpes simplex virus from mouse dorsal root ganglia which contain different levels of latency-associated transcripts. *J. Gen. Virol.* 74:995–1002
40. Efstathiou S, Minson AC, Field HJ, Anderson JR, Wildy P. 1986. Detection of herpes simplex virus–specific DNA sequences in latently infected mice and in humans. *J. Virol.* 57:446–55
41. Everett RD. 1984. Transactivation of transcription by herpes virus products: requirements for two HSV-1 immediate-early polypeptides for maximum activity. *EMBO J.* 3:3135–41
42. Fareed MU, Spivack JG. 1994. Two open reading frames (ORF1 and ORF2) within the 2.0-kilobase latency-associated transcript of herpes simpelx virus type 1 are not essential for reactivation from latency. *J. Virol.* 68:8071–81
43. Farrell MJ, Dobson AT, Feldman LT. 1991. Herpes simplex virus latency-associated transcript is a stable intron. *Proc. Natl. Acad. Sci. USA* 88:790–94
44. Federoff HJ, Geschwind MD, Geller AI, Kessler JA. 1992. Expression of nerve growth factor in vivo from a defective herpes simplex virus 1 vector prevents effects of axotomy on sympathetic ganglia. *Proc. Natl. Acad. Sci. USA* 89:1636–40
45. Fink DJ, Sternberg R, Weber PC, Mata M, Goins WF, Glorioso JC. 1992. In vivo expression of β-galactosidase in

hippocampal neurons by HSV-mediated gene transfer. *Human Gene Ther.* 3:11–19
46. Fraser NW, Lawrence WC, Wroblewska Z, Gilden DH, Koprowski H. 1981. Herpes simplex virus type 1 DNA in human brain tissue. *Proc. Natl. Acad. Sci. USA* 78:6461–65
47. Freese A, Geller AI, Neve R. 1990. HSV-1 vector mediated neuronal gene delivery. Strategies for molecular neuroscience and neurology. *Biochem. Pharmacol.* 40:2189–99
48. Frenkel N, Jacob RJ, Honess RW, Hayward GS, Locker H, Roizman B. 1975. Anatomy of herpes simplex virus DNA. III. Characterization of defective DNA molecules and biological properties of virus populations containing them. *J. Virol.* 16:153–67
49. Gaffney DF, McLauchlin J, Whitton JL, Clements JB. 1985. A modular system for the assay of transcription regulatory signals: the sequence TAATGARAT is required for herpes simplex virus immediate early gene activation. *Nucleic Acids Res.* 13:7847–63
50. Geller AI, Breakefield XO. 1988. A defective HSV-1 vector expresses *Escherichia coli* β-galactosidase in cultured peripheral neurons. *Science* 241:1667–69
51. Geller AI, Freese A. 1990. Infection of cultured central nervous system neurons with a defective herpes simplex virus 1 vector results in stable expression of *Escherichia coli* β-galactosidase. *Proc. Natl. Acad. Sci. USA* 87:1149 53
52. Gelman IH, Silverstein S. 1985. Identification of immediate early genes from herpes simplex virus that transactivate the virus thymidine kinase gene. *Proc. Natl. Acad. Sci. USA* 82:5265–69
53. Giniger E, Varnum SM, Ptashne M. 1985. Specific DNA binding of GAL4, a positive regulatory protein of yeast. *Cell* 40:767–74
54. Goins WF, Sternberg LR, Croen KD, Krause PR, Hendricks RL, et al. 1994. A novel latency-active promoter is contained within the herpes simplex virus type 1 U_L flanking repeats. *J. Virol.* 68:2239–52
55. Gordon YJ, McKnight JLC, Ostrove JM, Romanowski E, Araullo-Cruz T. 1990. Host species and strain differences affect the ability of an HSV-1 ICP0 deletion mutant to establish latency and spontaneously reactivate in vivo. *Virology* 178:469–77
56. Gu B, DeLuca NA. 1994. Requirements for activation of the herpes simplex virus glycoprotein C promoter in vitro by the viral regulatory protein, ICP4. *J. Virol.* 68:7953–65
57. Gu B, Rivera-Gonzalez R, Smith CA, DeLuca NA. 1993. Herpes simplex virus infected cell polypeptide 4 preferentially represses Sp1-activated over basal transcription from its own promoter. *Proc. Natl. Acad. Sci. USA* 90:9528–32
58. Hill JM, Sedarati F, Javier RT, Wagner EK, Stevens JG. 1990. Herpes simplex virus latent phase transcription facilitates *in vivo* reactivation. *Virology* 174:117–25
59. Ho DY, Mocarski ES. 1989. Herpes simplex virus latent RNA (LAT) is not required for latent infection in the mouse. *Proc. Natl. Acad. Sci. USA* 86:7596–600
60. Ho DY, Mocarski ES, Sapolsky RM. 1993. Altering central nervous system physiology with a defective herpes simplex virus vector expressing the glucose transporter gene. *Proc. Natl. Acad. Sci. USA* 90:3655–59
61. Holland LE, Anderson KP, Shipman C, Wagner EK. 1980. Viral DNA synthesis is required for efficient expression of specific herpes simplex virus type 1 mRNA. *Virology* 101:10–24
62. Honess RW, Roizman B. 1974. Regulation of herpes virus macromolecular synthesis. I. Cascade regulation of the synthesis of three groups of viral proteins. *J. Virol.* 14:8–19
63. Ishii S, Xu Y-H, Stratton RH, Roe BA, Merlino GT, Pastan I. 1985. Characterization and sequence of the promoter region of the human epidermal growth factor receptor gene. *Proc. Natl .Acad. Sci. USA* 82:4920–24
64. Jacob RJ, Morse LS, Roizman B. 1979. Anatomy of herpes simplex virus DNA. XII. Accumulation of head-to-tail concatemers in nuclei of infected cells and their rule in the generation of the four isomeric arrangements of viral DNA. *J. Virol.* 29:448–57
65. Javier RT, Stevens JG, Dissette VB, Wagner EK. 1988. A herpes simplex virus transcript abundant in latently infected neurons is dispensible for establishment of the latent state. *Virology* 166:254–57
66. Jenkins FJ, Roizman B. 1986. Herpes simplex virus 1 recombinants with non-inverting genomes frozen in different isomeric arrangements are capable of independent replication. *J. Virol.* 59:494–99
67. Jiao S, Gurevich V, Wolff JA. 1993. Long-term correction of rat model of Parkinson's disease by gene therapy. *Nature* 362:450–53

68. Johnson P, Miyanohara A, Levine F, Cahill T, Friedman T. 1992. Cytotoxicity of a replication-defective mutant herpes simplex virus type 1. *J. Virol.* 66:2952–65
69. Johnson PA, Wang MJ, Friedman T. 1994. Improved cell survival by the reduction of immediate-early gene expression in replication-defective mutants of herpes simplex virus type 1 but not by mutation of the viron host shutoff function. *J. Virol.* 68:6347–62
70. Kaplitt MG, Kwong AD, Kleopoulos SP, Mobbs CV, Rabkin SD, Pfaff DW. 1994. Preproenkephalin promoter yields region-specific and long-term expression in adult brain after direct in vivo gene transfer via a defective herpes simplex viral vector. *Proc. Natl. Acad. Sci. USA* 91:8979–83
71. Katz JP, Bodin ET, Coen DM. 1990. Quantitative polymerase chain reaction analysis of herpes simplex virus DNA in ganglia of mice infected with replication-incompetent mutants. *J. Virol.* 64:4288–95
72. Kolluri R, Torrey TA, Kinniburgh AJ. 1992. A CT promoter element binding protein: definition of a double-strand and a novel single-strand DNA binding motif. *Nucleic Acids Res.* 20:111–16
73. Kosz-Vnenchak M, Coen DM, Knipe DM. 1990. Restricted expression of herpes simplex virus lytic genes during establishment of latent infection by thymidine kinase-negative mutant viruses. *J. Virol.* 64:5396–402
74. Krause PR, Croen KD, Straus SE, Ostrove JM. 1988. Detection and preliminary characterization of herpes simplex virus type 1 transcripts in latently infected human trigeminal ganglia. *J. Virol.* 62:4819–23
75. Kristie TM, Roizman B. 1986. DNA-binding site of major regulatory protein α4 specifically associated with the promoter-regulatory domains of α genes of herpes simplex virus type 1. *Proc. Natl. Acad. Sci. USA* 83:4700–4
76. Kristie TM, Roizman B. 1987. Host cell proteins bind to the cis-acting site required for virion-mediated induction of herpes simplex virus 1 α genes. *Proc. Natl. Acad. Sci. USA* 84:71–75
77. Kwong AD, Frenkel N. 1987. Herpes simplex virus–infected cells contain a function(s) that destabilizes both host and viral mRNAs. *Proc. Natl. Acad. Sci. USA* 84:1926–30
78. Leib DA, Coen DM, Bogard CL, Hicks KA, Yager DR, et al. 1989. Immediate-early regulatory gene mutants define different stages in the establishment and reactivation of herpes simplex virus latency. *J. Virol.* 63:759–68
79. Leib DA, Nadeau KC, Rundle SA, Schaffer PA. 1991. Promoter of the latency-associated transcripts of herpes simplex virus type 1 contains a functional cAMP-response element: role of the latency-associated transcripts and cAMP in reactivation of viral latency. *Proc. Natl. Acad. Sci. USA* 88:48–52
80. Leiden JM, Frenkel N, Rapp F. 1980. Identification of the herpes simplex virus DNA sequences present in six herpes simplex virus thymidine kinase–transformed mouse cell lines. *J. Virol.* 33:272–85
81. Lillycrop KA, Estridge JK, Latchman DS. 1993. The octamer binding protein Oct-2 inhibits transactivation of the herpes simplex virus immediate-early genes by the virion protein Vmw65. *Virology* 196:888–91
82. Locker H, Frenkel N. 1979. Structure and origin of defective genomes contained in serially passaged herpes simplex virus type 1. *J. Virol.* 29:1065–77
83. Lokensgard JR, Bloom DC, Dobson AT, Feldman LT. 1994. Long-term promoter activity during herpes simplex virus latency. *J. Virol.* 68:7148–58
84. MacDonald ME, Ambrose CM, Duyao MP, Myers RH, Lin C, et al. 1993. A novel gene containing a trinucleotide repeat that is expanded and unstable on Huntington's disease chromosomes. *Cell* 72:971–83
85. MacLean AR, ul-Fareed M, Robertson L, Harland J, Brown SM. 1991. Herpes simplex virus type 1 deletion variants 1714 and 1716 pinpoint neurovirulence-related sequences in Glasgow strain 17⁺ between immediate early gene 1 and the 'a' sequence. *J. Gen. Virol.* 72:631–39
86. Margolis TP, Bloom DC, Dobson AT, Feldman LT, Stevens JG. 1993. Decreased reporter gene expression during latent infection with herpes simplex virus latency-associated transcript promoter constructs. *Virology* 197:585–92
87. Martuza RL, Malick A, Markert JM, Ruffner KL, Coen DM. 1991. Experimental therapy of human glioma by means of a genetically engineered virus mutant. *Science* 252:854–56
88. Maul GG, Everett RD. 1994. The nuclear location of PML, a cellular member of the C3HC4 zinc-binding domain protein family, is rearranged during herpes simplex virus infection by the C3HC4 viral protein ICP0. *J. Gen. Virol.* 75:1223–33
89. Mavromara-Nazos P, Roizman B. 1987.

Activation of herpes simplex virus 1 γ_2 genes by viral DNA replication. *Virology* 161:593–98
90. McCarthy AM, McMahan L, Schaffer PA. 1989. Herpes simplex virus type 1 ICP27 deletion mutants exhibit altered patterns of transcription and are DNA deficient. *J. Virol.* 63:18–27
91. McGeoch DJ, Dalrymple MA, Davison AJ, Dolan A, Frame MC, et al. 1988. The complete DNA sequence of the long unique region in the genome of herpes simplex virus type 1. *J. Gen. Virol.* 69:1531–74
92. McGeoch DJ, Dolan A, Donald S, Brauer DH. 1986. Complete DNA sequence of the short repeat region in the genome of herpes simplex virus type 1. *Nucleic Acids Res.* 14:1727–44
93. McGeoch DJ, Dolan A, Donald S, Rixon FJ. 1985. Sequence determination and genetic content of the short unique region in the genome of herpes simplex virus type 1. *J. Mol. Biol.* 181:1–13
94. McKnight JLC, Kristie TM, Roizman B. 1987. Binding of the virion protein mediating α gene induction in herpes simplex virus 1–infected cells to its cis site requires cellular proteins. *Proc. Natl. Acad. Sci. USA* 84:7061–65
95. Meeker TC, Loeb J, Ayres M, Sellers W. 1990. The human Pim-1 gene is selectively transcribed in different hemato-lymphoid cell lines in spite of a G+C-rich housekeeping promoter. *Mol. Cell. Biol.* 10:1680–88
96. Meignier B, Longnecker R, Mavromara-Nazos P, Sears AE, Roizman B. 1988. Virulence of and establishment of latency by genetically engineered deletion mutants of herpes simplex virus type 1. *Virology* 162:251–54
97. Mellerick DM, Fraser NW. 1987. Physical state of the latent herpes simplex virus genome in a mouse model system: evidence suggesting an episomal state. *Virology* 158:265–75
98. Mester JC, Glorioso JC, Rouse BT. 1991. Protection against zosteriform spread of herpes simplex virus by monoclonal antibodies. *J. Infect. Dis.* 163:263–69
99. Mester JC, Pitha PM, Glorioso JC. 1995. Anti-viral activity of herpes simplex virus vectors expressing murine α interferon. *Gene Ther.* In press
100. Michael N, Roizman B. 1993. Repression of the herpes simplex virus 1 α4 gene by its gene product occurs within the context of the viral genome and is associated with all three identified cognate sites. *Proc. Natl. Acad. Sci. USA* 90:2286–90
101. Deleted in proof
102. Mocarski E, Roizman B. 1982. Structure and role of the herpes simplex virus DNA termini in inversion, circularization and generation of virion DNA. *Cell* 31:89–97
103. Mullan M, Houlden H, Windelspecht M, Fidani L, Lombardi C, et al. 1992. A locus of familial early-onset Alzheimer's disease on the long arm of chromosome 14, proximal to the alpha 1–antichymotrypsin gene. *Nat. Gen.* 2:340–43
104. Natarajan R, Deshmane S, Valyi-Nagy T, Everett R, Fraser NW. 1991. A herpes simplex virus type 1 mutant lacking the ICP0 introns reactivates with normal efficiency. *J. Virol.* 65:5569–73
105. Newcomb WW, Brown JC. 1994. Induced extrusion of DNA from the capsid of herpes simplex virus type 1. *J. Virol.* 68:443–40
106. Nicosia M, Deshmane SL, Zabolotny JM, Valyi-Nagy T, Fraser NW. 1993. Herpes simplex virus type 1 latency-associated transcript (LAT) promoter deletion mutants can express a 2-kilobase transcript mapping to the LAT region. *J. Virol.* 67:7276–83
107. O'Hare P, Hayward GS. 1985. Three trans-acting regulatory proteins of herpes simplex virus modulate immediate-early gene expression in a pathway involving positive and negative feedback regulation. *J. Virol.* 56:723–33
108. Oroskar AA, Read GS. 1989. Control of mRNA stability by the virion host shutoff function of herpes simplex virus. *J. Virol.* 63:1897–906
109. Palella TD, Hidaka Y, Silverman LJ, Levine M, Glorioso J, Kelley WN. 1989. Expression of human HPRT mRNA in brains of mice infected with a recombinant herpes simplex virus type 1 vector. *Gene* 80:137–44
110. Palella TD, Silverman LJ, Schroll CT, Homa FL, Levine M, Kelley WM. 1988. Herpes simplex virus–mediated human hypoxanthine-guanine phosphoribosyltransferase gene transfer into neuronal cells. *Mol. Cell. Biol.* 8:457–60
111. Poffenberger KL, Tabares E, Roizman B. 1983. Characterization of a viable, noninverting herpes simplex virus 1 genome derived by insertion and deletion of sequences at the junction of components L and S. *Proc. Natl. Acad. Sci. USA* 80:2690–94
112. Post LE, Mackem S, Roizman B. 1981. Regulation of α genes of herpes simplex virus: expression of chimeric genes produced by fusion of thymidine kinase with α gene promoters. *Cell* 24:555–65

113. Preston CM. 1979. Control of herpes simplex virus type 1 mRNA synthesis in cells infected with wild-type virus or the temperature-sensitive mutant *tsK*. *J. Virol.* 29:275–84
114. Preston CM. 1979. Abnormal properties of an immediate early polypeptide in cells infected with the herpes simplex virus type 1 mutant *tsK*. *J. Virol.* 32:357–69
115. Quinlan MP, Knipe DM. 1985. Stimulation of expression of a herpes simplex virus DNA-binding protein by two viral functions. *Mol. Cell. Biol.* 5:957–63
116. Rader KA, Acklund-Berglund CE, Miller JK, Pepose JS, Leib DA. 1993. In vivo characterization of site-directed mutants in the promoter of the herpes simplex virus type 1 latency-associated transcripts. *J. Gen. Virol.* 74:1859–69
117. Ram Z, Culver K, Walbridge S, Blaese M, Oldfield EH. 1993. In situ retroviral-mediated gene transfer for the treatment of brain tumors. *Cancer Res.* 53:83–88
118. Ram Z, Walbridge S, Shawker T, Culver KW, Blaese RM, Oldfield EH. 1994. The effect of thymidine kinase transduction and ganciclovir therapy on tumor vasculature and growth of 9L gliomas in rats. *J. Neurosurg.* 81:256–60
119. Ramakrishnan R, Fink DJ, Jiang G, Desai P, Glorioso JC, Levine M. 1994. Competitive quantitatitve PCR analysis of herpes simplex virus type 1 DNA and latency-associated transcript RNA in latently infected cells of the rat brain. *J. Virol.* 68:1864–73
120. Read GS, Frenkel N. 1983. Herpes simplex virus mutants defective in the virion-associated shutoff of host polypeptide synthesis and exhibiting abnormal synthesis of α (immediate-early) viral polypeptides. *J. Virol.* 46:498–512
121. Rice SA, Knipe DM. 1990. Genetic evidence for two distinct transactivation functions of the herpes simplex virus and protein ICP27. *J. Virol.* 64:1704–15
122. Rice SA, Long MC, Lam V, Spencer CA. 1994. RNA polymerase II is aberrantly phosphorylated and localized to viral replication compartments following herpes simplex virus infection. *J. Virol.* 68:988–1001
123. Roberts MS, Boundy A, O'Hare P, Pizzorno MC, Ciufo DM, Hayward GS. 1988. Direct correlation between a negative autoregulatory element at the cap site of the herpes simplex virus type 1 IE175 (α4) promoter and a specific binding site for the IE175 (ICP4) protein. *J. Virol.* 62:4307–20
124. Rock DL, Fraser NW. 1985. Latent herpes simplex virus type 1 DNA contains two copies of the virion DNA joint region. *J. Virol.* 55:849–52
125. Rock DL, Nesburn AB, Ghiasi H, Ong J, Lewis TL, et al. 1987. Detection of latency-related viral RNAs in trigeminal ganglia of rabbits infected with herpes simplex virus type 1. *J. Virol.* 61:3820–26
126. Roizman B, Sears AE. 1990. Herpes simplex viruses and their replication. In *Virology*, ed. BN Fields, pp. 1795–841. New York: Raven
127. Roizman B, Sears AE. 1993. Herpes simplex virus and their replication. In *The Human Herpesviruses*, ed. B Roizman, RJ Whitley, C Lopez, pp. 11–68. New York: Raven
128. Russell J, Stow EC, Stow ND, Preston CM. 1987. Abnormal forms of the herpes simplex virus immediate early polypeptide Vmw175 induce the cellular stress response. *J. Gen. Virol.* 68:2397–406
129. Sacks WR, Greene CC, Aschman DA, Schaffer PA. 1985. Herpes simplex virus type 1 ICP27 is an essential regulatory protein. *J. Virol.* 55:796–805
130. Sacks WR, Schaffer PA. 1987. Deletion mutants in the gene encoding the herpes simplex virus type 1 immediate-early protein ICP0 exhibit impaired growth in cell culture. *J. Virol.* 61:829–39
131. Sadowski I, Ma J, Triezenberg S, Ptashne M. 1988. GAL4-VP16 is an unusually potent transcriptional activator. *Nature* 335:563–64
132. Sandri-Goldin RM, Mendoza GE. 1992. A herpesvirus regulatory protein appears to act post-transcriptionally by affecting mRNA processing. *Genes Dev.* 6:848–63
133. Saunders AM, Strittmatter WJ, Schmechel D, St. George-Hyslop PH, Pericak-Vance MA, et al. 1993. Association of apolipoprotein E allele epsilon 4 with late-onset familial and sporadic Alzheimer's disease. *Neurology* 43:1467–72
134. Sawtell NM, Thompson RL. 1992. Herpes simplex virus type 1 latency-associated transcription unit promotes anatomical site-dependent establishment and reactivation from latency. *J. Virol.* 66:2157–69
135. Schellenberg GD, Bird T, Wijsman EM, Orr HT, Anderson L, et al. 1992. Genetic linkage evidence for a familial Alzheimer's disease locus on chromosome 14. *Science* 258:668–71
136. Sedarati F, Izumi KM, Wagner EK, Stevens JG. 1989. Herpes simplex virus type 1 latency-associated transcript

plays no role in establishment or maintenance of a latent infection in murine sensory neurons. *J. Virol.* 63:4455–58
137. Sehgal A, Patil N, Chao M. 1988. A constitutive promoter directs expression of the nerve growth factor receptor gene. *Mol. Cell. Biol.* 8:3160–67
138. Shih M-F, Arsenakis M, Tiollais P, Roizman B. 1984. Expression of hepatitis B virus S gene by herpes simplex virus type 1 vectors carrying α- and β-regulated gene chimeras. *Proc. Natl. Acad. Sci. USA* 81:5867–70
138a. Smith AE. 1995. Viral vectors in gene therapy. *Annu. Rev. Microbiol.* 49:000–00
139. Smith CA, Bates P, Rivera-Gonzalez R, Gu B, DeLuca NA. 1993. ICP4, the major transcriptional regulatory protein of herpes simplex virus type 1, forms a tripartite complex with TATA-binding protein and TFIIB. *J. Virol.* 67:4676–87
140. Smith IL, Hardwicke MA, Sandri-Goldin RM. 1992. Evidence that herpes simplex virus immediate early protein ICP27 acts post-transcriptionally during infection to regulate gene expression. *Virology* 186:74–86
141. Spaete RR, Frenkel N. 1982. The herpes simplex virus amplicon: a new eucaryotic defective-virus cloning-amplifying vector. *Cell* 30:295–304
142. Spear PG. 1993. Membrane fusion induced by herpes simplex virus. In *Viral Fusion Mechanisms*, ed. J Bentz, pp. 201–32. Boca Raton, FL: CRC
143. Spivack JG, Fraser NW. 1987. Detection of herpes simplex virus type 1 transcripts during latent infection in mice. *J. Virol.* 61:3841–47
144. Spivack JG, Fraser NW. 1988. Expression of herpes simplex virus type 1 latency-associated transcripts in the trigeminal ganglia of mice during acute infection and reactivation of latent infection. *J. Virol.* 62:1479–85
145. Spivack JG, Fraser NW. 1988. Expression of herpes simplex virus type 1 (HSV-1) latency-associated transcripts and transcripts affected by the deletion in avirulent mutant HFEM: evidence of a new class of HSV-1 genes. *J. Virol.* 62:3281–87
146. Spivack JG, Woods GM, Fraser NW. 1991. Identification of a novel latency-specific splice donor signal within herpes simplex virus type 1 2.0-kilobase latency-associated transcript (LAT): translation inhibition of LAT open reading frames by the intron within the 2.0-kilobase LAT. *J. Virol.* 65:6800–10
147. St. George-Hyslop PH, Haines J, Rogaev E, Mortilla M, Vaula G, et al. 1992. Genetic evidence for a novel familial Alzheimer's disease locus on chromosome 14. *Nat. Genet.* 2:330–34
148. St. George-Hyslop PH, Tanzi RE, Polinsky RJ, Haines JL, Nee L, et al. 1987. The genetic defect causing familial Alzheimer's disease maps on chromosome 21. *Science* 235:885–90
149. Steiner I, Spivack JG, Deshmane SL, Ace CI, Preston CM, Fraser NW. 1990. A herpes simplex virus type 1 mutant containing a non-trans-inducing Vmw65 protein establishes latent infection in vivo in the absence of viral replication and reactivates efficiently from explanted trigeminal ganglia. *J. Virol.* 64:1630–38
150. Steiner I, Spivack JG, Lirette RP, Brown SM, MacLean AR, et al. 1989. Herpes simplex virus type 1 latency-associated transcripts are evidently not essential for latent infection. *EMBO J.* 8:505–11
151. Stevens JG. 1989. Human herpesviruses: a consideration of the latent state. *Microbiol. Rev.* 53: 318–32
152. Stevens JG, Wagner EK, Devi-Rao GB, Cook ML, Feldman LT. 1987. RNA complementary to a herpesvirus α gene mRNA is prominent in latently infected neurons. *Science* 235:1056–59
153. Stow ND, Stow EC. 1986. Isolation and characterization of a herpes simplex virus type 1 mutant containing a deletion within the gene encoding the immediate early polypeptide Vmw 110. *J. Gen. Virol.* 67:2571–85
154. Strittmatter WJ, Saunders AM, Schmechel D, Pericak-Vance M, Enghild J, et al. 1993. Apolipoprotein E: high avidity binding to β-amyloid and increased frequency of type 4 allele in late-onset familial Alzheimer's disease. *Proc. Natl. Acad. Sci. USA* 90: 1977–81
155. Su L, Knipe DM. 1989. Herpes simplex virus a protein ICP27 can inhibit or augment viral gene transactivation. *Virology* 170:496–504
156. Sunstrum JC, Chrisp CE, Levine M, Glorioso JC. 1988. Pathogenicity of glycoprotein C negative mutants for the mouse central nervous system. *Virus Res.* 11:17–32
157. Tanaka S, Minagawa H, Yasushi T, Liu Y, Mori R. 1994. Analysis by RNA-PCR of latency and reactivation of herpes simplex virus in multiple neuronal tissues. 75:2691–698
158. Trousdale MD, Steiner I, Spivack JG, Deshmane SL, Brown SM, MacLean AR. 1991. In vivo and in vitro reactivation impairment of a herpes simplex virus type 1 latency-associated tran-

script variant in a rabbit eye model. *J. Virol.* 65:6989–93
159. Valyi-Nagy T, Deshmane SL, Spivack JG, Steiner I, Ace CI, et al. 1991. Investigation of herpes simplex virus type 1 (HSV-1) gene expression and DNA synthesis during the establishment of latent infection by an HSV-1 mutant, in 1814, that does not replicate in mouse trigeminal ganglia. *J. Gen. Virol.* 72:641–49
160. Van Broeckhoven C, Backhovens H, Cruts M, De Winter G, Bruyland M, et al. 1992. Mapping of a gene predisposing to early-onset Alzheimer's disease to chromosome 14q24.3. *Nat. Genet.* 2:335–39
161. Varmuza SL, Smiley JR. 1985. Signals for site-specific cleavage of HSV DNA: maturation involves two separate cleavage events at sites distal to the recognition site. *Cell* 41:792–802
162. Wagner EK, Devi-Rao G, Feldman LT, Dobson AT, Zhang Y, et al. 1988. Physical characterization of the herpes simplex virus latency-associated transcript in neurons. *J. Virol.* 63:1194–2002
163. Wagner EK, Flanagan WM, Devi-Rao GB, Zhang YF, Hill JM, et al. 1988. The herpes simplex virus latency-associated transcript is spliced during the latent phase of infection. *J. Virol.* 62:4577–85
164. Wang Y, O'Malley BW Jr, Tsai SY, O'Malley BW. 1994. A novel regulatory system for gene transfer. *Proc. Natl. Acad. Sci. USA* 91:8180–84
165. Ward PL, Roizman B. 1994. Herpes simplex genes: the blueprint of a successful human pathogen. *Trends Genet.* 10:267–74
166. Watson RJ, Clements JB. 1980. A herpes simplex virus type 1 function continuously required for early and late virus RNA synthesis. *Nature* 285:329–30
167. Weir JP, Elkins KL. 1993. Replication-incompetent herpesvirus vector delivery of an interferon α gene inhibits human immunodeficiency virus replication in human monocytes. *Proc. Natl. Acad. Sci. USA* 90:9140–44
168. Wheatley SC, Dent CL, Wood JN, Latchman DS. 1991. A cellular factor binding to the TAATGARAT DNA sequence prevents the expression of the herpes simplex virus immediate-early genes following infection of nonpermissive cell lines derived from dorsal root ganglion neurons. *Exp. Cell Res.* 194:78–82
169. Whitley RJ, Kern ER, Chatterjee S, Chou J, Roizman B. 1993. Replication establishment of latency, and induced reactivation of herpes simplex virus γ1 34.5 deletion mutants in rodent models. *J. Clin. Invest.* 91:2837–43
170. Wolfe JH, Deshmane SL, Fraser NW. 1992. Herpes virus vector gene transfer and expression of β-glucuronidase in the central nervous system of MPS VII mice. *Nat. Genet.* 1:379–84
171. Xu L, Schaffner W, Rungger D. 1993. Transcription activation by recombinant GAL47-VP16 in the *Xenopus* oocyte. *Nucleic Acids Res.* 21:2775
172. Yamada Y, Kimura H, Morishima T, Daikoku T, Maeno K, Nishiyama K. 1991. The pathogenicity of ribonucleotide reductase–null mutants of herpes simplex virus type 1 in mice. *J. Infect. Dis.* 164:1091–97
173. York IA, Roo C, Andrews DW, Riddell SR, Graham FL, Johnson DC. 1994. A cytosolic herpes simplex virus protein inhibits antigen presentation to CD8⁺ T lymphocytes. *Cell* 77:525–35
174. Zwaagstra JC, Ghiasi H, Nesburn AB, Wechsler SL. 1989. In vitro promoter activity associated with the latency-associated transcript gene of herpes simplex virus type 1. *J. Gen. Virol.* 70:2163–69
175. Zwaagstra JC, Ghiasi H, Nesburn AB, Wechsler SL. 1991. Identification of a major regulatory sequence in the latency-associated transcript (LAT) promoter of herpex simplex virus type 1 (HSV-1). *Virology* 182:287–97
176. Zwaagstra JC, Ghiasi H, Slanina SM, Nesburn AB, Wheatley SC, et al. 1990. Activity of herpes simplex virus type 1 latency-associated transcript (LAT) promoter in neuron-derived cells: evidence for neuron specificity and for a large LAT transcript. *J. Virol.* 64:5019–28

MICROBIAL BIOFILMS

J. William Costerton and Zbigniew Lewandowski
Center for Biofilm Engineering, Montana State University, Bozeman, Montana 59717

Douglas E. Caldwell and Darren R. Korber
Applied Microbiology and Food Science, University of Saskatchewan, Saskatoon, Saskatchewan S7N 5A8, Canada

Hilary M. Lappin-Scott
Biological Sciences, University of Exeter, Hatherley Laboratories, Prince of Wales Road, Exeter, Devon, EX4 4PS, England

KEY WORDS: sessile bacterial communities, σ factor, phenotype change, mass transfer, consortia

CONTENTS

DEFINITION	712
INTRODUCTION	712
DISTRIBUTION AND UBIQUITY OF BIOFILMS	713
THE STRUCTURE OF BIOFILMS	716
As Revealed by Confocal Scanning Laser Microscopy and Fluorescent Probes	717
As Revealed by Microsensors	722
As Revealed by NMR and by Flow Studies	722
As Revealed by Electrochemical Measurements	726
As Revealed by Physicochemical Analyses	728
Summary of Our Current Concept of Biofilm Structure	733
PHENOTYPIC CHANGES IN BIOFILM BACTERIA	735
TELEOLOGICAL THOUGHTS ABOUT MICROBIAL BIOFILMS	736
MICROBIOLOGY AND REALITY	739

ABSTRACT

Direct observations have clearly shown that biofilm bacteria predominate, numerically and metabolically, in virtually all nutrient-sufficient ecosystems. Therefore, these sessile organisms predominate in most of the environmental, industrial, and medical problems and processes of interest to microbiologists. If biofilm bacteria were simply planktonic cells that had adhered to a surface, this revelation would be unimportant, but they are demonstrably and profoundly different. We first noted that biofilm cells are at least 500 times more resistant to antibacterial agents. Now we have discovered that adhesion triggers the expression of a σ factor that derepresses a large number of genes so that biofilm cells are clearly phenotypically distinct from their planktonic counterparts. Each biofilm bacterium lives in a customized microniche in a complex microbial community that has primitive homeostasis, a primitive circulatory system, and metabolic cooperativity, and each of these sessile cells reacts to its special environment so that it differs fundamentally from a planktonic cell of the same species.

DEFINITION

Biofilms are defined as matrix-enclosed bacterial populations adherent to each other and/or to surfaces or interfaces. This definition includes microbial aggregates and floccules and also adherent populations within the pore spaces of porous media.

INTRODUCTION

The real significance of bacterial biofilms has gradually emerged since their first description (126), and the first recognition of their ubiquity (29). It has become increasingly clear that biofilms constitute a distinct growth phase of bacteria that is profoundly different from the planktonic growth phase studied so assiduously during the 15 decades following the discoveries of Louis Pasteur.

During the complex process of adhesion, bacterial cells alter their phenotypes in response to the proximity of a surface (49, 78). During the earliest stages of biofilm formation, sessile bacteria find themselves in a stable juxtaposition with cells of the same species and with those of other species, as single-species and mixed-species microcolonies are formed (82, 91). These cellular juxtapositions, and the exuberant exopolysaccharide matrix production within the developing biofilm, condition the microenvironment of each biofilm bacterium (115). Different biofilm bacteria respond to their specific microenvironmental conditions (59) with different growth patterns, and a structurally complex mature biofilm gradually develops. Physiological cooperativity is a major factor in shaping the structure and in establishing the eventual juxtapositions that make mature biofilms very efficient microbial communities adherent to surfaces (48).

Protein structure and sequential transcription dictate the elaborate structures of enzyme complexes. These complexes—because they are much more efficient than random mixtures of floating enzyme molecules—are predominant in all biological systems. At a higher level of organization, bacteria within biofilms benefit from similar stable juxtaposition and similar physiological cooperativity; such bacteria therefore constitute a coordinated functional community that is much more efficient than mixed populations of floating planktonic organisms (31). In fact, biofilms resemble the tissues formed by eukar- yotic cells, in their physiological cooperativity and in the extent to which they are protected from variations in bulk phase conditions by a primitive homeostasis provided by the biofilm matrix. The analogy with eukaryotic organisms can even be extended to dissemination strategies, in which physiologically efficient, well-protected communities of cells constitute the most successful and most competitive expression of the genome, while less efficient planktonic cells are produced in order to disseminate and to colonize new locations.

As disciples of Koch and Pasteur, we have been taught to extrapolate from single-species laboratory cultures to predict bacterial behavior in actual environments. With modern tools we can now make direct observations of structure and of chemical function in living biofilms growing in specific ecosystems. This perception of functional biofilm communities, reinforced by novel methods for direct observation, will usher in a new golden age of understanding in virtually all fields of microbiology.

DISTRIBUTION AND UBIQUITY OF BIOFILMS

Bacteria constitute the most successful form of life on earth, in terms of total biomass and in terms of the variety and extent of habitats colonized. The pivotal reason for this success is phenotypic plasticity (16). The bacterial genotype is extensive, but it is the ability of this genotype to respond phenotypically to environmental stimuli, rather than the power of its genetic repertoire, that has produced the remarkable success of the bacteria. A general phenotypic strategy has recently become apparent in a majority of bacterial strains, as we have come to understand more of the modes of growth that these organisms can adopt in response to changing growth conditions.

The cells of such ubiquitous bacterial species as *Pseudomonas aeruginosa* respond to favorable nutrient conditions by adhering to available surfaces and by binary fission and exopolymer production to develop mature biofilms. These rod-shaped vegetative cells (0.8×1.2 µm) grow predominantly in this matrix-enclosed sessile mode of growth, in which they are protected from adverse environmental conditions and from biological and chemical antibacterial agents. This protection is sacrificed by the planktonic cells that are periodically shed from established biofilms so that new fresh habitats can be colonized with new biofilms. When nutrient conditions become unfavorable, both sessile and planktonic cells (69) of this species are sharply reduced in size to form very small (± 0.3 µm), spherical ultramicrobacteria by a process that is now well documented as starvation survival (70). In this reversible process the DNA of the bacterial cell is stabilized while its metabolic capabilities are selectively jettisoned to produce dormant cells that can be fully resuscitated many years later. Reduced to its simplest form, this bacterial strategy clearly favors the formation of stationary, metabolically cooperative biofilms in favorable nutrient conditions and of planktonic cells or of dormant ultramicrobacteria when the interests of the species are best served by dissemination and/or by simple survival.

Prior to 1978, biofilms had been described in a certain number of aquatic systems (91, 92, 127), but the proportion of bacteria in a given ecosystem that grew in these adherent populations had not generally been determined. In 1978 Geesey et al (52) adapted a whole series of quantitative recovery methods to allow us to enumerate biofilm bacteria in a pristine mountain stream, and to compare their numbers and their activity with those of planktonic bacteria in the same aquatic system. When biofilm bacteria were clearly shown to predominate in numbers and in metabolic activity, these new methods of quantitative recovery were applied to a large number of aquatic systems in natural (34), industrial (12), and medical (33, 68) ecosystems. We can now make the general statement, based upon detailed analysis of hundreds of aquatic systems, that biofilm populations predominate in virtually all nutrient-sufficient aquatic systems independent of system geometry and of the type of ecosystem involved (78), and that these adherent populations have a very significant metabolic activity (47, 77).

This exhaustive quantitative analysis has now built a sufficient database to allow us to predict the extent of biofilm formation in a particular aquatic system, based on the following principles:

1. Metabolically active (vegetative) bacteria show a remarkable avidity for adhesion to surfaces, and this tendency is especially pronounced in wild-type cells in natural environments.
2. The extent of biofilm accretion on surfaces in any aquatic system is controlled by the amount of nutrient available for cell replication and for exopolysaccharide production.
3. In extremely oligotrophic environments, organic nutrients tend to associate with available surfaces, and to trigger local biofilm development, but bacteria generally do not adhere to surfaces in very nutrient-deficient ecosystems.

Using these principles, we can predict the extent to which biofilm will develop in particular water systems and compare these predictions with direct observations of natural systems within which many local factors may be almost equally important.

It is clear that virtually all body fluids provide sufficient organic nutrients for optimal bacterial growth, and we would therefore predict that most plastic and metal surfaces of medical devices would accrete bacterial biofilms when bacteria are present in these fluids. The development of the biofilms that are the actual etiological agents of infections associated with a wide variety of medical devices attests to the accuracy of this prediction and shows that this accretion process operates in spite of shear forces such as those generated on contact lenses by blinking and on heart valves by blood flow. Once established, the biofilm's natural resistance to natural surfactants (5), phagocytosis (65), and antibiotic therapy (4) allow it to remain as a continuing nidus of living bacteria long after all planktonic organisms have been killed by these host defense factors and by antibacterial agents. Natural endothelial surfaces are more resistant to bacterial adhesion and subsequent biofilm formation, because (*a*) antibacterial components of the tissue-surface milieu (e.g. tissue-associated antibodies) can kill planktonic bacteria as they approach the tissue surface and (*b*) cellular immunity (phagocytic activity) is fully functional on tissue surfaces. When tissue surfaces are covered by thick (>400 µm) biofilms of native (autochthonous) bacteria that condition the tissue-surface milieu by their physiological activity (e.g. lactobacilli in the vagina), the actual tissue surface is well protected from the adhesion of extraneous (allochthonous) organisms. Even though the intestine is equally replete with organic nutrients, allochthonous organisms require very special colonization mechanisms to adhere to the tissue surface, because this surface is covered by a >400 µm thick mucus layer (10) whose viscosity, movement, and autochthonous microbial population make colonization and biofilm development very difficult. The wholesale sloughing of colonized bladder endothelium and the remarkable antibacterial mechanisms of the eye are two additional examples of the many mechanisms that protect healthy tissues from bacterial adhesion, biofilm formation, and subsequent infection.

In natural and industrial aquatic systems that contain sufficient nutrients for bacterial growth and metabolism, we predict, and we note, rapid biofilm accretion on available surfaces (46, 77). This bacterial adhesion is especially rapid and specific if the surface in question is itself a nutrient, as in the case of cellulose and its analogs (32). The adhesion of the bacteria, and their subsequent formation of biofilms, is rapid and is remarkably unaffected by the physical or chemical nature of the surface concerned or by flow regimen in the bulk fluid. Biofilm formation leads to flow modification in linear systems (22) and to plugging in porous media; engineers are thoroughly familiar with the effects of these biofilms in such familiar configurations as heat-exchangers and sand filters. Indeed, sanitary engineers have for decades encouraged the development of biofilm populations in trickling filters and in anaerobic digestors in order to decrease the organic content of treated wastewater by their metabolic activity. While sloughing events and reactions to toxic materials may cause variations in biofilm accretion and metabolic activity, these attached populations usually burgeon until they reach self-generated limitations of space or of sustained flow, and they can actually block conduits of considerable diameter.

In oligotrophic aquatic systems, in both natural and industrial ecosystems, the remarkable plasticity of bacteria is seen most clearly. First principles of surface chemistry predict that organic molecules will concentrate at certain areas of surfaces, and first principles of bacterial adhesion predict that bacteria will adhere to areas of the surface enriched by this process. Adherent bacteria will form biofilms to an extent dictated by nutrient availability in their particular microniche (61), but they may not adhere and they certainly will not form biofilms where nutrients are lacking (102). The ability of bacteria to form stable, essentially dormant ultramicrobacteria under very oligotrophic (starvation) conditions lends a remarkable plasticity to biofilms in these ecosystems. In extremely oligotrophic systems, the bulk fluid will certainly contain ultramicrobacteria (70) that may associate with surfaces but lack the metabolic capability for exopolysaccharide production to accomplish cellular adhesion and biofilm formation. If nutrients become available, then these very small (± 0.3 μm in diameter) ultramicrobacteria will resuscitate to form vegetative cells and even to form biofilms on available surfaces (88). When these nutrients are exhausted, the biofilm of resuscitated bacteria will persist, subsisting on its trapped nutrients, but newly produced cells at the periphery will experience starvation and will produce and shed ultramicrobacteria as their prime starvation-survival response (70). Thus, the overall strategy of bacteria in oligotrophic environments is to grow in biofilms on surfaces where nutrients are locally available and to persist in nutrient-deprived zones as floating, dormant ultramicrobacteria with the full capability of returning to the vegetative state when nutrients again become available.

THE STRUCTURE OF BIOFILMS

The advent of inexpensive high-speed computers and increasingly sophisticated software has transformed traditional microscopic observation into the discipline of analytical imaging. Of the many recent digital imaging devices now available, confocal scanning laser microscopy (CSLM) (108) has proven particularly well suited for the study of microbial biofilms. CSLM allows the nondestructive, in situ analysis of living, fully hydrated biofilms without the need for harsh chemical fixation or embedding techniques. This feature, coupled with the ability to digitally extract optical thin sections from specific biofilm locations, free from out-of-focus optical interference, permits the direct examination of complex xy and xz spatial and chemical relationships that exist between bacteria, their extracellular products, and their environment (118). The episcopic nature of CSLM also enables the examination of biofilms cultivated on nontransparent surfaces (e.g. minerals, metals, gels, synthetics, etc) (23–25, 79), greatly expanding the types of analyses that may be performed. In conjunction with an expanding array of environmentally and chemically sensitive fluorescent molecular probes (2, 40, 60, 110, 114), CSLM may be used to provide detailed information on cell morphology, cellular metabolism, cell phylogeny, microenvironment chemistry, as well as the physical architecture and chemistry of the biofilm matrix and associated polymers.

Direct digital tools of examination, supported by indirect chemical and physical means such as microelectrode probes, have already led to the revision of our early rendering of biofilms, from one of a homogeneous distribution of cells in a uniform exopolysaccharide matrix (106, 116) to a model based upon significant variability and heterogeneity (34, 38, 73, 79). Earlier studies using both light and electron microscopy had sometimes suggested that microbial biofilms were heterogeneous in structure (8). Recent studies show living biofilms to consist of a variable distribution of cells and cellular aggregates, their extracellular polymers, and void spaces or water channels, which may or may not be continuous with the bulk liquid phase (67, 73, 74, 95, 111, 120). The spatially defined pattern of these elements has been termed biofilm "architecture" by Lawrence et al (79), and is often species-specific for pure-culture biofilms, as well as substrate-specific for certain microbial consortia (79, 120).

Notably, heterogeneous biofilm architecture does not appear to be the sole function, or the result, of mixtures of different organisms with variable growth habits inhabiting a similar niche. Pure-culture biofilms also accommodate many of the heterogeneous features found in mixed-cultured biofilms, suggesting a fundamental relationship between biofilm structure and in situ function. Recurrent patterns of various structural elements in both pure and mixed-species biofilms (72–75, 80) further support the premise that basic functional requisites underlie biofilm structure, and that structural diversity actually reflects the adaptation of unicellular organisms to a diverse range of physical, chemical, and communal circumstances on surfaces.

As Revealed by Confocal Scanning Laser Microscopy and Fluorescent Probes

CSLM analysis of microbial biofilms may be performed using a wide range of specific fluorescent probes and nonspecific fluorescent compounds. Nontoxic, nonspecific fluor compounds offer a number of clear benefits for architectural or structural investigations, however. For example, the exclusion by biofilm bacteria of fluorescent compounds added directly to the culture medium (e.g. fluorescein) causes cells to appear negatively stained when viewed with CSLM (24). Bleaching, or fading, of the fluorescent signal following repeated laser scans is precluded using this method, as the fluor is continually replenished at the cell boundary. Furthermore, the nontoxic nature of fluorescein permits the temporal analysis of biofilm development or response without inhibiting normal cell function or growth (24).

Through use of CSLM and negative staining, qualitative differences in pure-culture biofilm architecture have been determined in terms of the amount of biofilm biomass present at different depths or xy-biofilm locations (71, 79, 80). Variability in the distribution of biomass of 24 h *P. fluorescens, P. aeruginosa,* and *Vibrio parahaemolyticus* biofilms has clearly demonstrated the tendency for biofilm bacteria to form aggregates at different horizontal and vertical sites (79), and that area of highest cell density varies among different species. For example, *P. aeruginosa* biofilms were characterized by a dense cell mass located near the biofilm base (27% biomass area at the attachment surface), whereas *V. parahaemolyticus* biofilms exhibited an inverted pyramid structure, with the most biofilm biomass located nearest the biofilm-liquid interface (16% biomass area at the ~3 μm section depth). Significantly, studies of this nature (71, 74, 79) have repeatedly confirmed that living biofilms are highly hydrated, with 50–90% of the total area at each sectioning depth consisting either of polymer and/or void space (liquid).

Recent CSLM examinations of *P. fluorescens* biofilms further detailed the spatial variability that exists within pure-culture biofilm systems. Depth measurements performed at defined xy locations revealed the depth of *P. fluorescens* biofilms after 24 h to be 42±19 μm. The variability in depth measurements was the consequence of adjacent regions where either channels or microcolonies predominated. After 72 h, the average depth of *P. fluorescens* biofilms remained constant (74), but the variability, which correlated with the continued growth of channels and the enlargement of cellular aggregates, increased to 42±28 μm. The actual depth measurements measured for *P. fluorescens* biofilms after 72 h varied from 0 to 90 μm. Similarly, Stewart et al (111) used a light microscopic procedure to characterize the variability in depth of *P. aeruginosa* biofilms. In this case, the mean biofilm depth was reported to be 33 μm, with a range of 13.3 to 60 μm. These data, which may be presented in the form of 3D topographic maps (73) or 2D distance-biofilm thickness profiles (111), have the potential for refining biofilm fluid friction coefficients and models of molecular diffusion. Figure 1 illustrates the clarity of these CSLM images, which can be manipulated to produce a sagittal (xz) or a horizontal (xy) section from the same data. The sagittal section (upper image) is produced at the top edge of the horizontal section (lower image), and this combination clearly shows the microcolonies and the water channels within a living 24 h *P. fluorescens* biofilm positively stained with aeridine orange.

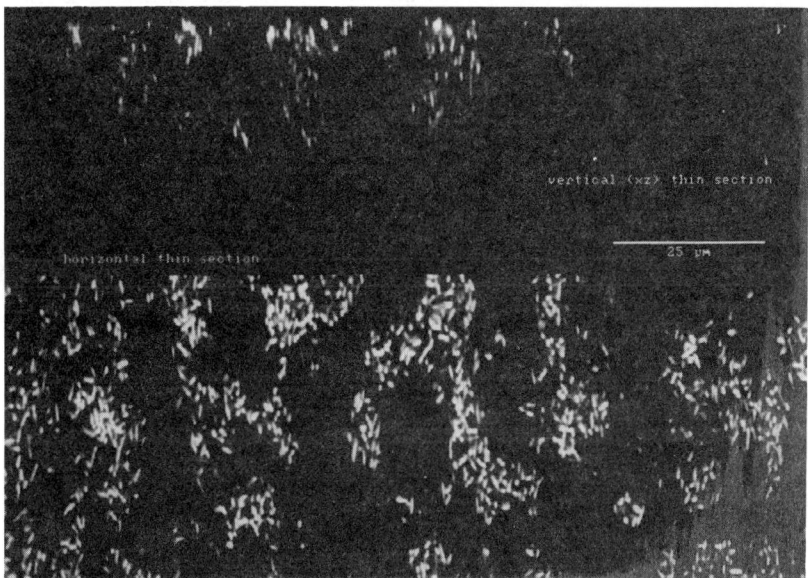

Figure 1 CSLM optical thin sections showing the bacterial microcolonies and the water channels within a living, positively stained 24h *P. fluorescens* biofilm. The upper section shows a sagittal (vertical) xz section, and the lower section shows the corresponding horizontal xy section, whose upper boundary indicates the location of the sagittal section. Bar = 25 μm.

Using high-magnification CSLM imaging, the volumetric significance of structures such as water channels is difficult to determine with certainty. Low-magnification CSLM analyses, performed over large regions of contiguous biofilm (areas ≥ 1 mm^2), lack the cell-level resolution afforded by high-numerical-aperture lenses but can resolve, and be used for quantification of, gross architectural features such as water channels and cell masses. Figure 2 shows a 1×10^6 μm^2 montage consisting of 12 low-magnification optical thin sections of a 48 h *P. fluorescens* biofilm grown under low-flow conditions (~0.1 cm s^{-1}). In this presentation, the quasi-regular array of channels and pores (occurring at 20–40 μm intervals) is clearly evident, and not a rare event. This very regular array of water channels is commonly seen in pure-culture biofilms formed under certain flow regimens, but mixed-species natural biofilms often display a much less regular architecture.

Detailed CSLM architectural studies of a diclofop methyl-degrading microbial consortium (120) linked the structure of biofilms with the nature of the carbon source. When aromatic hydrocarbons were provided as the sole carbon source, biofilm thickness increased, and spatial relationships—including coni-

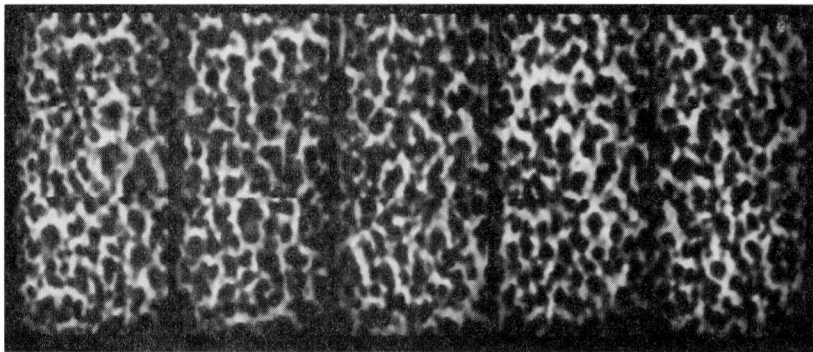

Figure 2 Montage of 12 low-magnification CSLM images of a 48 h *P. fluorescens* biofilm imaged using fluorescein and negative staining. Individual images were digitally positioned in *xy* alignment to create a montage ~ 1×10^6 μm^2 in area, clearly showing a spatial pattern of microcolonies (dark) and water channels (light).

cal bacterial microcolonies, grapelike clusters of cocci, and chemically defined intra- and intergeneric associations—developed after 14–21 days. When the same microbial consortium was grown on a more labile carbon source, biofilms were more homogeneous and less thick, and clear cellular associations were not evident. Furthermore, these architectural patterns could be controlled by alternating substrate. In these cases, biofilms initially grown on 300 µg ml^{-1} trypticase soy broth (TSB) for 21 days regained the attributes of biofilms grown solely on diclofop methyl 2 days after switching to 14 µg ml^{-1} diclofop methyl.

The difference between the architectures of pure and mixed-species biofilms may be rationalized in terms of nutrient limitation in the case of pure-culture biofilms, and niche exploitation in the case of mixed-species biofilms. The regular array of channels and aggregates seen in *P. fluorescens* biofilms would provide an alternate (lateral) route for the transport of nutrients and oxygen to the biofilm base without having to diffuse through the entire vertical length of the film. Trulear (113) previously showed that aerobic *P. aeruginosa* biofilms grew to ~30 to 40 µm in depth as monocultures, but increased in depth to ~130 µm when the culture was amended with anaerobic bacteria. This indirect evidence suggests that depletion of oxygen—not of nutrients—limited the vertical development of the *P. aeruginosa* biofilm.

In terms of penetration of nutrients, cellular wastes, or antimicrobial agents, a quantum model for molecular diffusion may apply (27). Biofilm channels, largely aqueous and under the influence of advective transport in addition to simple diffusion, would represent the primary level, and may consequently facilitate transport of molecules to the depths of the biofilm. Smaller pores or conduits, which permeate cellular aggregates, would represent the secondary level, since (a) advective processes would be minimized and (b) microbial exopolysaccharide (EPS) may occupy the pore space. Pores are not necessarily connected with water channels, and may occur as voids or cavities within an aggregate of bacteria. At the tertiary level, cellular aggregates and associated EPS would hinder diffusion most significantly, as no advection would occur in these zones and molecular transport would be solely diffusion-driven.

A number of fluorescent probes are suitable for the in situ analysis of biofilm population architecture. Thus, the role(s) that specific, or indicator, strains play during these processes may be examined at the phylogenetic, immunological, ultrastructural, or morphological level without the need to isolate the organism from other biofilm community members. This is a significant methodological advance, since many metabolically significant biofilm bacteria are not amenable to growth in pure culture, leading to an underestimation of the natural biofilm biodiversity. Furthermore, it is difficult to find a correlation between reproductive success of an isolate in pure culture and reproductive success of the same organism in a mixed-species biofilm. Fluorescently conjugated 16S rRNA probes are now routinely used for the phylogenetic analysis of microbiological systems. Such molecular probes, in conjunction with either epifluorescence microscopy or CSLM, have already been used to catalog (in some cases, to the subspecies level) the genetic biodiversity that exists within complex biofilm systems. Studies conducted to date include qualitative and quantitative examinations of the bovine intestinal microflora, sewage-degrading consortia, and soil communities (3, 58, 110). Immunological approaches have been used extensively for determinative epifluorescent microscopy applications (15, 62, 87), and have a similar potential for wide application in CSLM studies. Relatively few workers have utilized fluor-conjugated antibodies for determinative CSLM studies to date. One such study (64) used a polyclonal antibody specific for an *Acinetobacter* sp. to examine the distribution of these cells within a 24 h triculture biofilm also containing *P. fluorescens* and *Aeromonas hydrophila*. The *Pseudomonas* and *Aeromonas* spp. were identified based on their morphologies; thus the horizontal and vertical variability in biofilm biomass contributed by each species could readily be determined.

Broader parameters for the identification of bacteria may be established on the basis of cellular morphology, cellular ultrastructure, or cell metabolic capabilities. For example, commercially available fluorescent gram stains detect gram-positive or gram-negative bacteria through a simple, one-step staining procedure. Identification of methanogenic bacteria within mixed-species biofilms is facilitated by the autofluorescence of coenzyme F_{420} (involved in CO_2 reduction) following low-wavelength light excitation (43). *Methanothrix methanobrevibacter* and *Methanosarcina* spp. have been tentatively identified from sludge-digesting biofilms based on their 420-nm excitable fluorescence and unique cell morphologies.

As Revealed by Microsensors

The structural heterogeneity that has been directly observed by CSLM, with and without the use of fluorescent probes, suggests very important ramifications regarding mass transfer within biofilms. Even before this structural heterogeneity was discovered, some workers (109) had hypothesized that there was convective flow through pores in the biofilm because the effective diffusion coefficient in aerobic biofilms was dependent on flow conditions and on biofilm structure. Once these pores were actually visualized using CSLM, we resolved to use microsensors (38) to measure the concentration of substrates directly in various regions of the biofilm during observation by direct microscopy. We used a dissolved oxygen (DO) microelectrode to study the distribution of oxygen in an aerobic biofilm within which water channels and cellular microcolonies could be clearly resolved by CSLM (Figure 3A, C). When this microelectrode was advanced from the bulk fluid through the bulk fluid-biofilm interface (at 100–200 μm) and into a dense, matrix-enclosed microcolony (Figure 3A), the data from the DO sensor showed that the DO values decreased at the interface and reached almost totally anoxic values in the center of the microcolony (Figure 3B) and at the colonized surface (0.0 μm). When the DO microsensor was moved to an adjacent position where it would advance from the bulk fluid through the bulk fluid-biofilm interface and into a much less dense, cell-free water channel (Figure 3C), significant levels of dissolved oxygen were seen at all levels (Figure 3D), including at the colonized surface (0.0 μm). These data clearly show that the structural heterogeneity of the biofilm predicates a corresponding heterogeneity in the distribution of an important physiological substrate: dissolved oxygen. The water channels appear to transport oxygen into the biofilm, but diffusion limitations and oxygen utilization produce very low oxygen levels at the centers of cellular microcolonies, and these direct observations of living biofilms may explain the existence, and even the physiological activity, of fastidious anaerobes within mixed biofilms in aerobic environments.

As Revealed by NMR and by Flow Studies

Another hypothesis that emanates from the conceptual model of biofilm structure presented in Figures 1–3 is the possibility of water movement within the biofilm. Intrabiofilm flow would profoundly affect mass transport within biofilms, and this concept would refute the fundamental assumption (116) of biofilm process modelling that mass transport within the biofilm is entirely due to molecular diffusion. This hypothesis has been examined by two direct methods: nuclear magnetic resonance imaging (NMRI) and particle image velocimetry (PIV).

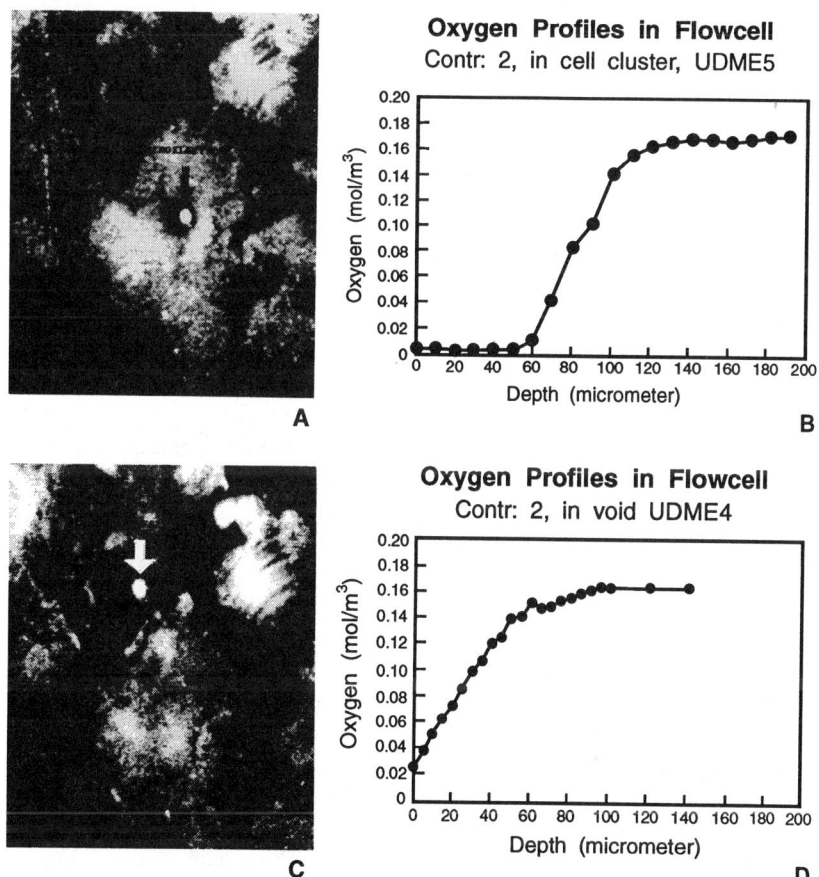

Figure 3 CSLM images (*A* and *C*) of a 24 h mixed-species biofilm showing the location (arrows) of a dissolved oxygen electrode within a microcolony (*A*) and within an adjacent water channel (*C*). *B* shows the oxygen profile as the probe is advanced from the bulk fluid into the microcolony (*A*) and eventually to the colonized surface (depth = 0 mm). *D* shows a corresponding oxygen profile as the probe is advanced into a water channel (*C*). These direct observations of a living biofilm clearly show that the microcolony is anaerobic, while the adjoining water channel contains oxygen throughout its depth.

NMRI provides a direct and non-invasive technique for the study of the chemical and physical properties of small samples and for spatial mapping, or imaging, of larger systems. Morris (97) provides detailed descriptions of current NMRI methods. Our most recent set of experiments (85, 86) used a reactor with a rectangular cross-section, and NMRI velocity measurements were performed at four different flow rates with and without biofilm in the reactor. The flow profiles in spatial cross-sectional coordinates at 15 cm from the inlet indicated that in the reactor with biofilm, increased bulk flow velocity did not affect the character of the velocity profiles. On the other hand, in the reactor without biofilm, bulk flow velocities above 2.4 cm/s had a pronounced effect in these velocity profiles; an increase to 4.0 cm/s resulted in the formation of a jet, and an increase to 4.6 cm/s produced an even more dramatic jet. Such behavior can be explained in terms of the entry length required for development of the velocity profile. The entry length for laminar pipe and channel flows is proportional to the Reynolds number (Equation 1), where ν is the kinematic viscosity, A is the cross-sectional area, and P is the wetted perimeter and V is flow velocity. When the reactor walls are covered with a thick biofilm layer, the flow area does not change much, but the wetted perimeter may increase dramatically, since the biofilm has a very well-developed surface. This effect may stabilize the flow (i.e. decrease the Reynolds number), and consequently the coverage of the surface of a narrow conduit with a biofilm under certain circumstances may stabilize the flow near the surface.

$$Re = \frac{V4A}{\nu P} \qquad 1.$$

We also determined velocity profiles at the same location in identical flat-plate reactors in which thicker bacterial biofilms had been developed for either 3 or 4 days. Small non-uniformities (kinks) were apparent near the reactor walls, indicating that there is flow of the bulk solution in an area that is partially occupied by the biofilm. If flow above the biofilm surface is convective and if only molecular diffusion is operative below the biofilm interface (116), then the biofilm surface should behave as a rigid surface at which flow velocity reaches zero. Because this is clearly not the case, these data strongly suggest that there is convective flow of water within the biofilm layer, but the data are not unequivocal because the effective slice thickness in our NMR apparatus is 5 mm and the biofilm may not be uniform throughout this slice width.

We then turned to an almost equally non-invasive technique (PIV) that has yielded high-resolution data in a number of studies of flow velocity in complex systems. In our laboratory, a Bio-Rad MRC600 confocal scanning laser microscope (CSLM) was used in conjunction with an Olympus BH2 light microscope for simultaneous particle tracking and biofilm visualization. Neutral-density fluorescent latex spheres (Molecular Probes, Eugene Oregon, density 20°C = 1055 kg/m^2, Ex 580 nm/Em 605 nm, diameter = 0.282 μm) were added to the biofilm reactor to achieve a final concentration of 1×10^7 particles/ml, and velocity images were obtained by capturing images at various focal depths using a computer-controlled focus motor. Particles travelling across the field of view could readily be photographed at recorded intervals (Figure 4), so that their velocity could be calculated, and their position within the biofilm could also be

MICROBIAL BIOFILMS 725

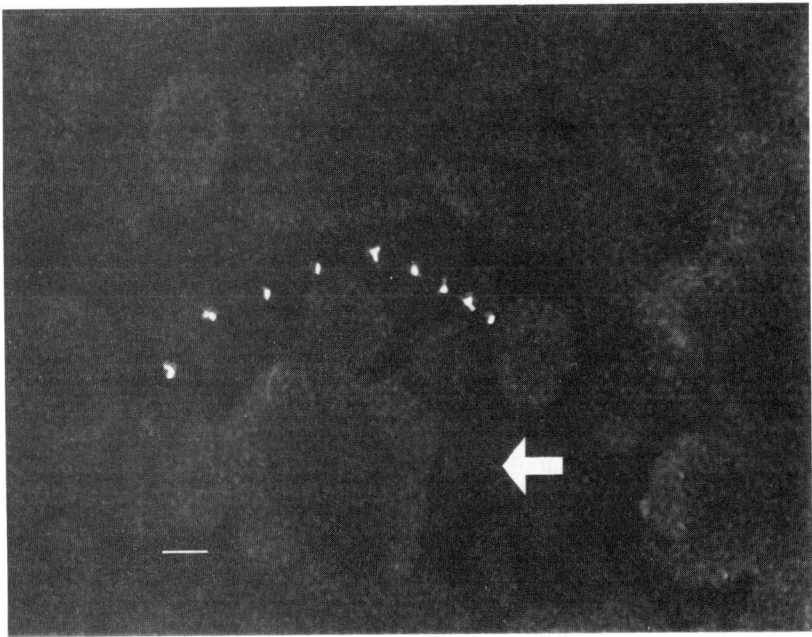

Figure 4 CSLM image showing the movement of a 0.3 μm fluorescent latex sphere, photographed at 2 s intervals, through a water channel within a 24 h *P. aeruginosa* biofilm. The movement of these spheres, which occurs in the same direction as the bulk fluid flow (arrow), is used to quantitate convective flow in the water channels within biofilms.

continuously monitored (112). This direct method of monitoring particle movement within living biofilms produced unequivocal data to show that convective flow occurs in the water channels within bacterial biofilms.

This direct measurement of flow using PIV can also be used in bulk fluids and in zones adjacent to surfaces in flowing systems (39), and we have used it to examine flow in clean systems and in conduits coated with biofilms. Flow velocities, as calculated by PIV, are reduced as bulk fluid approaches a clean surface. The same phenomenon is seen as bulk fluid approaches a biofilm-coated surface, except that flow does not reach zero at the biofilm surfaces and direct evidence of convective flow is seen within the biofilm itself (39). The measurement of velocity profiles in biofilm systems allows the determination of shear stress both at the canal wall and the biofilm surface, as well as the estimation of the thickness of the subviscous laminar boundary layer and of the thickness of the flow regime. These values in turn can be used to estimate mass, heat, and momentum transfer, which are important for understanding and predicting biofilm growth kinetics and (bio)fouling effects on various process operations.

It is clear that hydrodynamic factors influence biofilms in many ways, including their control of the transport of bacteria to the surface during the initial stages of colonization (44). After the biofilm has formed, these hydrodynamic factors—control of the transport of substrates to, and metabolites from, the biofilm and shear stress—are related to biofilm erosion and biofilm sloughing, and to the physical density of the sessile population formed in various flowing systems (72). Biofilm accumulation changes the hydrodynamics of flowing systems, especially as the thickness of the biofilm becomes comparable to that of the boundary layer (104). Biofilm accumulation is essentially an autocatalytic process in that it increases surface roughness (13) which, in turn, provides shelter from shear forces and increases both the surface area and convective mass transport near the surface. Exhaustive studies have shown that biofilm elements may extend outward through the boundary layer that characterizes flow at all surfaces (35), and increase eddy diffusion and external mass transfer (109) as well as erosion (105) due to increased shear forces. Lewandowski & Walser (84) demonstrated that biofilm thickness reaches a maximum within the transition zone between laminar and turbulent flow, at a colonized surface, and they postulated that this thickness is limited by transport of the substrate in the laminar zone and by erosion in the turbulent zone. These irregular extensions of the biofilm into the bulk fluid are regularly seen at the surfaces of mixed-species biofilms formed in natural ecosystems, where they must be presumed to induce significant levels of turbulent flow.

In summary, direct evidence from examinations of living biofilms has clearly shown that these sessile accretions of microbial cells may produce surface roughness that increases turbulence and mass transport at the colonized surface. Mass transport within the biofilm is further enhanced by convective flow of the bulk fluid through the water channels that anastomose throughout the biofilm. This convective flow within the water channels attains a significant rate, and follows the same direction as the flow of the bulk fluid, so that microcolonies within the biofilm are effectively bathed by the bulk fluid even at the deepest levels examined to date.

As Revealed by Electrochemical Measurements

The structural heterogeneity noted above in direct examinations of living mixed-species biofilms, by relatively non-invasive techniques, must necessarily result in a corresponding chemical heterogeneity vis-a-vis matrix composition and the concentration of substrate molecules (e.g. oxygen) within the biofilm. These differences in the structural chemistry of different areas of a mixed-species biofilm would be exacerbated and reinforced by the metabolic activity of microcolonies of specific organisms (e.g. organic acid producers),

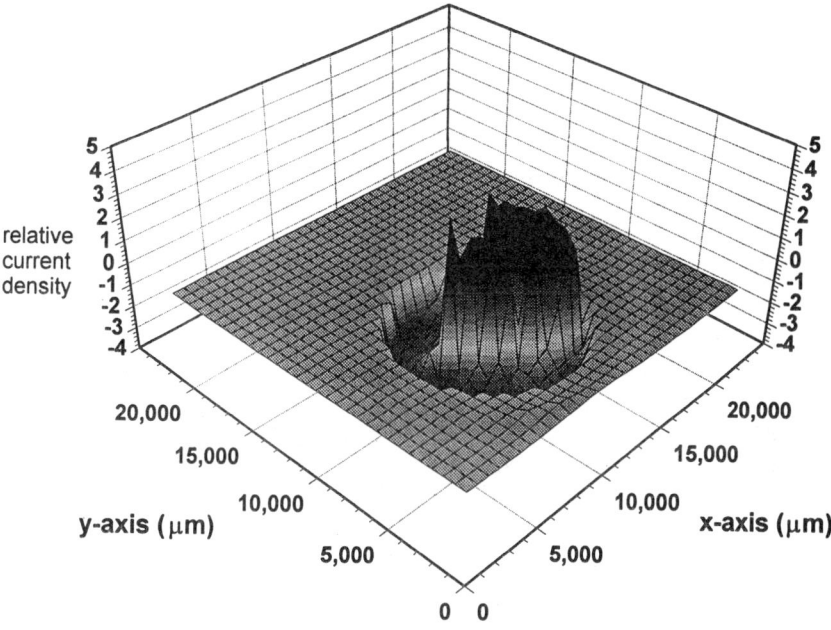

Figure 5 Plot of electrochemical data collected using a scanning vibrating electrode (SVE) when a drop of calcium alginate is placed on the surface of a mild steel coupon. The SVE data detect the development of a local anode (raised area) beneath the alginate deposit with a ringlike cathode (depressed area) on the adjoining area of the steel surface. This non-invasive electrochemical method can be used to detect anodes and cathodes on surfaces colonized by living biofilms.

and we must expect to find electrochemical differences between different loci in a mixed-species biofilm. These electrochemical differences between adjacent loci within mixed-species biofilms may constitute the mechanism of microbially influenced corrosion (MIC), and they must be considered to be a factor in the mass transport of charged ions and molecules within the biofilm. In the most extreme case, a microcolony of matrix-enclosed cells of an acidogenic species within a mixed-species biofilm would constitute a very difficult target for the diffusional attack of a cationic antimicrobial agent.

Because natural mixed-species biofilms are inherently very complex, we have chosen to define the factors operative in electrochemical heterogeneity using single-species biofilms and abiotic systems in which matrix polymers are applied to surfaces. Geesey et al (53) have shown that a measurable degree of electrochemical heterogeneity is established when two different areas of a copper surface are covered with two different acidic polysaccharides representative of biofilm matrix material. More recently, Lewandowski's group used the scanning vibrating electrode (SVE) to map the electrochemistry of metal surfaces covered by single-species biofilms or by abiotic biofilm matrix analogs (polysaccharides). This SVE technique detects anodes and cathodes at the metal surface by detecting gradients of charged species as they stream from the surface into the bulk fluid. Initial studies have indicated that areas of metallic surfaces that are covered by real or by simulated biofilms are anodic, while uncovered areas are generally cathodic (F Roe, unpublished data). Figure 5 shows the electrochemical profile, as detected by the SVE, that results when an area of a mild steel coupon is covered with a drop of calcium alginate. The raised signal pattern indicates that a de facto anode has formed under the alginate deposit; the ringlike depressed trough surrounding this area suggests that a de facto cathode has developed on the immediately adjacent area of the uncoated metal surface. Several mechanisms have been suggested to produce electrochemical heterogeneity in these simple systems, including oxygen concentration differences and metal concentration differences, and we propose to validate these putative mechanisms before proceeding to the direct examination of living mixed-species biofilms using the SVE.

As Revealed by Physicochemical Analyses

Microbial biofilms are characterized in part by the production of an extensive network of highly hydrated exopolysaccharides (31, 117). EPS production by biofilm bacteria serves many functions, including the facilitation of the initial attachment of bacteria to surfaces, the formation and maintenance of microcolony and biofilm structure, enhanced biofilm resistance to environmental stress and antimicrobial agents, protection of the biofilm from protozoan grazing, and biofilm nutrition (1, 20, 26, 93, 103). Biofilm EPS is also highly heterogeneous and has been demonstrated in situ to vary spatially, chemically, and physically (81, 121, 122, 123). In addition, chemically reactive EPS is generally the first biofilm structure to come in contact with potential substrates, predators, antimicrobial agents/antibiotics, and other bacteria, and thus is of considerable applied and ecological importance.

Biofilm bacteria also produce and maintain chemical microenvironments or microzones (23, 34, 38, 73, 83) within the biofilm. Spatial and temporal chemical variation within the biofilm (e.g. gradients of pH, oxygen, etc) permits the involvement of many fastidious bacteria with a limited range of metabolic capabilities. For example, chemical microzones allow the activity of anaerobic, corrosion-causing bacteria within aerobic biofilms, methanogenic syntrophy, reductive dehalogenation of pollutants, fermentative reactions, etc. The direct study of these features in living, fully hydrated biofilms is now possible using analytical CSLM and microprobe electrode methods. Examples of how these tools may be used to elucidate some of the complex physical and chemical relationships present within biofilm systems are provided below.

Physical attributes of the biofilm-EPS matrix, such as effective diffusion coefficients (D_e, cm^2 s^{-1}), help to define the penetration of antimicrobial agents, the transport of nutrients and wastes, and partitioning that may exist between different biofilm regions. Barriers to diffusion imposed by the cell-polymer matrix may be examined using fluorescent compounds and CSLM. For example, fluorescence monitoring and fluorescence recovery after photobleaching (FRAP), using fluorescein and fluorescein-dextran derivatives, have been utilized to determine one-dimensional D_e values for pure-culture and mixed-species biofilm systems (75, 76, 81). Using size-fractionated fluors ranging in molecular weight from 300 to 2×10^6, it was shown that living biofilms and their polymers presented a molecular radius-dependent barrier to diffusion relative to the bulk liquid phase (from ~2 to 13% of D_{aq} values for water), and that D_e values also varied between regions of high and low cell/ polymer density. Regions where channels were directly connected with the bulk aqueous phase correlated with high diffusion coefficients, whereas adjacent areas with dense cell materials and polymers had relatively low diffusion coefficients. The depth of the biofilm material (ranging from ~30 to 100 µm) was not found to influence the magnitude of D_e for pure-culture *P. fluorescens* biofilm systems (76).

Fluorescent probes may also delineate interactions that occur between defined fluors and the microbial matrix. Early models of homogeneous biofilms depicted EPS as possessing a net negative charge. While on average this may hold true, regional variability in EPS charge distribution has been demonstrated for mixed-species biofilm model systems by comparing the binding patterns of polyanionically, cationically, and neutrally charged dextran-fluor conjugates. For example, both sewage- and herbicide-degrading mixed-species biofilms grown in continuous-flow slide culture exhibited considerable spatial variability in the in situ binding patterns of polyanionically and cationically charged fluorescent dextrans (75). In a detailed study involving a nine-member, diclofop methyl–degrading microbial consortium, Wolfaardt et al (122) demonstrated that sites that bound the fluorescent herbicide were predominantly cationic (determined using polyanionic dextrans) and hydrophobic (determined using Nile Red), even though immediately adjacent biofilm sites bound cationic dextrans.

Saccharide-specific fluorescent lectins can also partially characterize the chemical variability of bacteria and their "exopolymers." This approach has previously been utilized for examining polymer footprints left on surfaces by detached or dislodged bacteria (98), and for in situ characterization of exopolymers produced by bacteria present within a marine microcosm (23). Recently, Wolfaardt et al (123) applied a panel of fluorescent lectins (specific for various isomers of mannose, galactose, glucose, fucose, N-acetyl-glucosamine, sialic acid, mannose, etc) to compare the EPS moieties produced by a diclofop-degrading consortium with the same organisms grown on labile substrate. During this study, considerable differences (in terms of the relative abundance and xy-xz lectin-binding patterns of different lectins) were found between those biofilms grown on TSB and those grown on aromatic hydrocarbons. Notably, biofilms grown on a labile carbon source were considerably more uniform (both architecturally and chemically) than biofilms grown on more recalcitrant carbon sources.

Biofilm-EPS chemistry is significant in cases of industrial and medical biofilm control, because the penetration and reactivity of antimicrobial agents vary with the mesh size and chemical characteristics of the microbial EPS (81). Determined indirectly, transport of an antimicrobial agent to the base of a biofilm may be rapid (101); delivery of active molecules to target sites (e.g. the center of a cell-EPS aggregate) may be significantly hindered, however. Combined with the potential for bacterial EPS to act as an absorbent (99), this factor may contribute to the resistance of biofilm bacteria to a wide range of traditional microbial control strategies (33). Identification of unique zones of polymer chemistry using direct CSLM methods may thus be used to provide valuable insight into the mechanisms of resistance, particularly when in situ determinations of cellular viability are performed after inhibitory treatments.

Identification of chemical microenvironments associated with active bacterial cells and cell aggregates may also be accomplished using CSLM and chemically sensitive (e.g. pH-, Eh-, or ion-sensitive) molecular probes. Figure 6 shows an optical thin section of a living *Vibrio parahaemolyticus* biofilm probed with a nontoxic, pH-sensitive fluorescein derivative (5,6-carboxy-fluorescein). The fluorescent signal returned from regions of biofilm channels is much greater than that from areas associated with dense cell aggregates, indicating that the pH in these cell-dense regions is less than that associated with the bulk aqueous phase. Zones of reduced pH resulting from the activity of acid-producing bacteria have previously been inferred from similar studies using pH-sensitive compounds (23). To date, the majority of such studies have been conducted for eukaryotic systems in which the cytoplasmic chemistry remains relatively stable. Within biofilms, in which cell metabolism defines the chemical microenvironment, calibration of environmentally sensitive fluorescent indicators is extremely difficult. Simultaneous use of chemically sensitive fluors with microelectrodes (39, 83), or ratiometric pH indicators, which emit fluorescence in both a pH-dependent and -independent manner (60, 114), may prove helpful in this regard. Overall, there remains a clear need for the development of effective CSLM protocols suitable for routine quantitative analysis of biofilm chemistry.

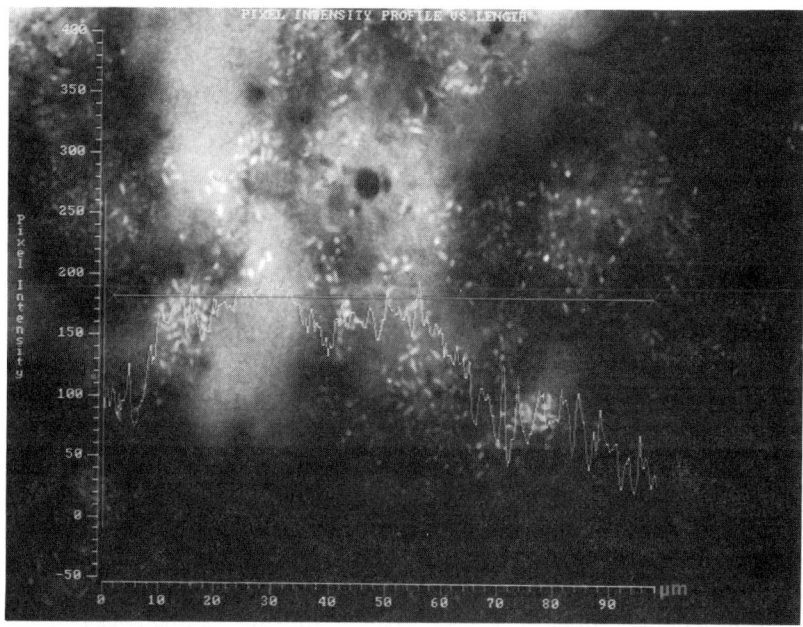

Figure 6 CSLM image of a horizontal (*xy*) section of a living *Vibrio parahaemolyticus* biofilm probed with a pH-sensitive fluorescein derivative: 5,6-carboxy-fluorescein. Areas of high pH return a stronger fluorescent signal, and the water channels are clearly more fluorescent than the bacterial microcolonies, indicating that these cellular aggregates are substantially more acidic. Pixel intensity is quantitated along a straight line that intersects both microcolonies and water channels.

The control of biofilm bacteria has been the focus of a vast amounts of applied and medical research (17, 18, 45, 54, 55, 100). The reason(s) why biofilm bacteria are less susceptible to treatments known to kill their planktonic counterparts remain unclear (18, 31, 100); thus the challenge remains to develop engineered strategies that take into account the features that make biofilms unique. One possibility is that a spatial distribution of cells of different physiological states exists within the biofilm. Such inherent variability would possibly function to maintain a pool of cells at different metabolic states (e.g. rapid or slow growth), which would be important during survival under adverse conditions (17, 54). Eng et al (45) recently demonstrated that increased rates of bacterial growth resulted in the increased efficacy of a range of antimicrobial agents. No antimicrobial agent tested by Eng and coworkers resulted in more than 3 orders of magnitude of killing against slowly growing *Staphylococcus aureus*, and only quinolones were effective against slowly growing or nongrowing gram-negative bacteria. In contrast, rapidly growing cells experienced almost complete killing. A wide range of fluorescent probes (based on cytoplasmic redox potential, electron transport chain activity, enzymatic activity, cell membrane potential, or membrane integrity) have been applied to provide information about the metabolic condition of individual bacteria (7, 107, 125). For example, formazan salts, propidium iodide, ethidium bromide, acridine orange, resazurin, rhodamine 123, and fluorescent substrate analogs have all been applied to various degrees in this regard (96). A wide range of newly developed probes, including RH795, Hoechst 33342, rhodamine 123, calcein AM, and carbocyanine derivatives (60), remain to be examined fully regarding their potential for ecological and applied biofilm research.

Owing to the wide range of different classes of physiological probes available, it is useful to apply the probe most likely to reflect the type of inactivation (or activation) that will occur. For example, during a study of the efficacy of fleroxacin (a fluoroquinolone DNA-gyrase inhibitor) against established (24 h) *P. fluorescens* biofilms, Korber et al (74) used acridine orange (AO) as an indicator of altered nucleic acid conformation. Fleroxacin inhibits the supercoiling of bacterial DNA, thereby altering nucleic acid conformation of affected cells and inhibiting cell replication (actively growing cells become elongated or pleomorphic). The fluorescence emission (36) of AO varies (either in the green or red wavelengths) with the conformation of DNA (the ratio of double- to single-stranded nucleic acids). Although this approach did not attempt to determine whether fleroxacin-affected cells were metabolically crippled or dead, it did identify cells that had undergone some nucleic acid conformational change as a result of fleroxacin activity. The specificity of this approach was verified morphologically by analyzing CSLM images; regions where AO fluorescence indicated cells had been affected by fleroxacin correlated ($P \leq 0.05$) with areas of cell elongation (fluoroquinolones inhibit cell division, and growing cells consequently elongate). This second morphological indicator of fleroxacin efficacy also suggested that a gradient of cell growth rates existed in *P. fluorescens* biofilms, as cells nearest the bulk liquid phase experienced the greatest elongation after 48 h exposure to 2 µg ml^{-1} fleroxacin (average cell length was 8.1±2.4 µm at 15 µm section depth vs 1.7±0.6 µm at the biofilm base). Application of AO to detect the activity of fleroxacin proved much more appropriate than use of an indicator of redox potential, as initial tests revealed that cells affected by fleroxacin continued to be metabolically active for periods greater than 24 h.

CSLM studies using fluorescent dextrans with defined charge (either polyanionic, cationic, or neutral) have confirmed that the reactivity of mixed-species biofilm exopolysaccharide material with these probes is often variable (122). This is generally not the case for pure-culture biofilms, which are much more uniform in chemical and structural terms. The binding intensity of charged dextrans may be modified by specific treatments, however. For ex-

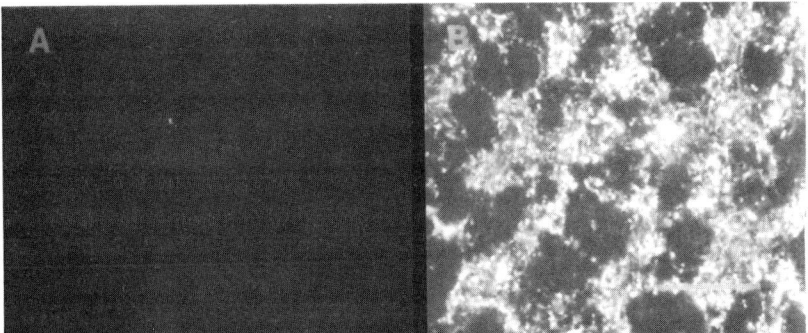

Figure 7 CSLM images of a 48 h *P. fluorescens* biofilm treated with 70 kDa polyanionic TRITC-dextran before (*A*) and after (*B*) the application of a DC electric field sufficient to produce the bioelectric effect (12). The native biofilm shows no detectable affinity for this polyanionic probe, but exposure to the DC electric field produces many cationic sites that react with the probe. These data suggest that DC fields may exert the bioelectric effect by causing profound changes in biofilm electrochemistry that facilitate penetration by antibacterial agents.

ample, 70 kDa polyanionic tetramethylrhodamine isothiocyanate (TRITC)-dextran bound poorly with 48 h *P. fluorescens* biofilm polymer (Figure 7*a*), in agreement with reports that bacterial EPS is generally anionic in nature. Following application of a DC bioelectrical field (11) to an otherwise identical biofilm system, however, strong binding patterns were observed (Figure 7*b*). Such observations may prove valuable for explaining the mechanism by which mild electric current serves to enhance the efficacy of traditional biofilm-control strategies (11), and also explain why biofilms are more resistant to control treatments than are planktonic bacteria. One possibility is that exposure to DC current alters the electrical properties of the EPS by exposing cationically charged polymer sites or Mg^{2+}/Ca^{2+} ions, resulting in greater reactivity with the polyanionically charged dextran.

Summary of Our Current Concept of Biofilm Structure

Direct observations of living biofilms by optical and other non-invasive physical methods have revealed a novel and exciting complexity of structure (Figure 8). Although we have not yet established that this elaborate system of microcolonies and water channels is found in the same form in all biofilms, its occurrence within natural multispecies biofilms on surfaces in a typical river indicates that it may be present in this form in many common and important aquatic environments.

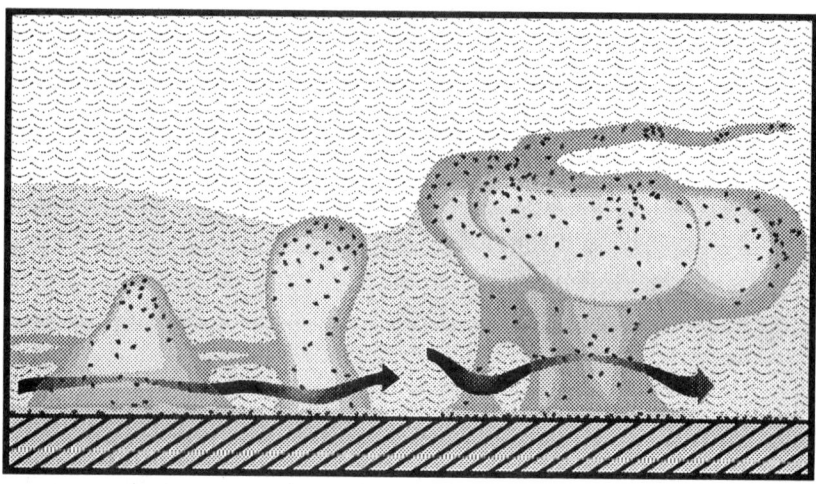

Figure 8 Conceptual model of the architecture of a single-species biofilm based on data collected by CSLM of living biofilms. Some microcolonies are simple conical structures, while others are mushroom shaped. Convective fluid flow is seen in the water channels between and even below the microcolonies within these biofilms.

We find it useful to compare the de facto microenvironment of a bacterial cell in this complex biofilm to that of a planktonic cell of the same organism (34). Following adhesion to a surface, a bacterial cell undergoes a phenotypic change that alters many of its structural molecules (see the next section) and derepresses exopolysaccharide synthesis (37). Cell division coupled with exopolysaccharide synthesis would naturally lead to the development of microcolonies enclosed in dense slime and attached to the colonized surface. Some simple cone-shaped microcolonies are seen within developing biofilms. The persistence of water channels, however, even deep in the biofilm and at the colonized surface, together with the frequent development of mushroom-shaped microcolonies (Figure 8), argues powerfully for some form of cell-cell communication (66) similar to that which mediates the production of fruiting bodies by myxobacteria. Microcolonies would develop as the adherent bacteria that find themselves in the most favorable milieux divide and produce exopolysaccharide locally to provide a dense and coherent matrix to hold the microcolony together and to anchor it to the substratum or to other microcolonies. The microcolony is the basic unit of biofilm growth, just as the tissue is the basic unit of growth of more complex organisms, and each microcolony enjoys a measure of homeostasis in that its internal environment is conditioned by its matrix and by the metabolic activity of its component cells. In mixed microcolonies (9), cells of two physiologically cooperative species may grow together as protected functional consortia. Obviously each microcolony must define its optimal size and must suppress cell division by quorum sensing (51) when a physiologically optimal size has been reached.

The discovery of convective flow within the water channels has revolutionized our concept of bacterial growth within biofilms. We must visualize the water channels as being very open, because they allow the passage of 0.3 µm latex beads (Figure 4), and they have been clearly shown to comprise an anastomosing network that carries bulk fluid throughout the biofilm. Because the convective flow in these water channels maintains the same direction as the bulk fluid flow, and because the demonstrated surface roughness of the biofilm must be expected to enhance mass transport by turbulence, the water channels actually represent a primitive circulatory system analogous to that of higher organisms. Because of this remarkable biofilm architecture, bacterial cells within a microcolony have a degree of homeostasis, optional spatial relationships with cooperative organisms, and an effective means of exchanging nutrients and metabolites with the bulk fluid phase.

PHENOTYPIC CHANGES IN BIOFILM BACTERIA

Direct microscopic observations have been particularly valuable in the use of reporter genes to elucidate the phenotypic changes that bacteria undergo when they adhere to surfaces (37). Reporter gene constructs have been produced by Chakrabarty and by Deretic, and these constructs have been used by Geesey's group (37) and by Costerton's group (63) to show that *algC* and *algD* are upregulated at the time of adhesion. This phenotypic change was intuitive because these genes control the production of phosphomannomutase, and of other enzymes in the alginate synthesis pathway, and alginate is necessary for post-adhesion activities such as biofilm formation. Davies et al (37) have developed a very useful direct visualization technique whereby cells of *P. aeruginosa* that are expressing the *algC* reporter system can be directly visualized by light microscopy, and we can now observe living cells in real time as they express the gene following adhesion and often cease to express it when surrounded by large amounts of alginate. Parallel work with gram-positive organisms, many of which are pathogens, has shown that adhesion triggers the expression of several enzymes that produce exopolysaccharides that are of pivotal importance in adhesion and biofilm formation. These functions are extremely important in the etiology of device-related infections (31).

Biochemical comparisons between biofilm cells and planktonic cells of the same species have shown that at least 30% of the cellular proteins that can be resolved by 2D electrophoresis are expressed to different extents by cells in these two growth phases (HW Yu & JW Costerton, unpublished observations).

Recent studies in Deretic's laboratory (41, 42, 94) indicate that this planktonic-biofilm transformation is controlled by a σ factor that is similar to that which controls sporulation in gram-positive bacteria and to that which controls the rough-smooth transformation in gram-negative organisms. If this fascinating hypothesis (42) stands up to current very intense scrutiny, its impact on the study of bacterial biofilms will be profound. If biofilm bacteria are the product of a σ factor–directed phenotypic change in a large cassette of genes, then they actually constitute a phenotypically distinct expression of the bacterial genome. Adhesion-induced phenotypic changes in proteins in the cell envelope, cell membrane, and cytoplasm may profoundly influence such characteristics as the enigmatic resistance of biofilm bacteria to antibacterial agents (101).

The reversal of this σ factor–directed change would produce cells with the planktonic phenotype, and the expression of alginate lyase (14) would assist in the detachment of these planktonic cells from the biofilm. We have consistently observed that microbial biofilms shed planktonic cells at a steady rate, and that this detachment is sometimes under physiological (6) and even diurnal control. We believe that the controlled shedding of planktonic cells from biofilms is a very important strategy in the bacterial struggle for survival and predominance in aquatic ecosystems.

TELEOLOGICAL THOUGHTS ABOUT MICROBIAL BIOFILMS

If we consider that bacteria thrive in a wide variety of environments throughout the Earth's crust and deep into its oceans, we must concede that they constitute a remarkably successful life form. As new methods of direct observation have been applied in many of these environments, we have seen that bacteria exist predominantly as dormant ultramicrobacteria in a wide variety of harsh oligotrophic systems, but that they actually grow predominantly in exopolysaccharide-enclosed biofilms in those ecosystems that permit growth and replication. Taken together, these data suggest that the planktonic mode of growth is favored for dissemination and for persistence in a dormant form, while the biofilm mode of growth is favored for growth. Because of their obvious success, it is germane to wonder why ultramicrobacteria are such optimal agents of the dissemination of the dormant genome while biofilms are such ideal communities for growth and for the phenotypic expression of metabolic activity.

Because bacteria obviously developed their very successful phenotypic strategy very early in their evolution, it is useful to consider the survival value of this strategy in the milieu of the primitive Earth. Bacterial cells would be attracted to the organic nutrients that concentrate naturally at surfaces in aquatic systems, and the expolysaccharides that mediate their adhesion to surfaces would further concentrate dissolved organic molecules and cations out of the bulk fluid. Primary production by photosynthetic bacteria would be favored by cellular deployment at a surface, and heterotrophic bacteria would thrive because of their close juxtaposition to primary producers and because they can scavenge the biomass when these organisms die. Sustained juxtaposition to the cathodic surface of metallic ores is, in itself, sufficient to allow the growth of heterotrophic bacteria (21) in biofilms. In the primitive Earth we visualize physical (heat) and chemical (acid) threats to aquatic bacteria, and it is clear that the ability to remain essentially stationary in an optimal, or even in a permissive, local environment was one of the most valuable contributions of biofilm growth to bacterial survival. Biofilm bacteria are notably resistant to drying (103, 124), a characteristic that would also enhance bacterial survival as water levels fluctuated. Even in harsh environments in the modern Earth (e.g. hot springs), the growth of bacteria in biofilms allows certain species of bacteria to adapt to challenging conditions gradually as the whole population creeps along the surface towards the very boundary of nonpermissive conditions.

If bacteria can be spoken of as having a strategy, without lapsing into teleology, this set of reactions to environmental conditions is codified in patterns of gene expression triggered by specific environmental stimuli (41). We now know that attachment to a surface induces a de facto σ factor that triggers the expression of a large cassette of genes similar in scope to the sets of genes expressed in sporulation or in starvation survival. This heavily conserved and radically different biofilm phenotype is indicative of the fact that adhesion and biofilm development were positively selected early in the evolution of bacteria. If the elaborately controlled biofilm phenotype allowed bacteria to survive, concentrate nutrients, and grow profusely in the primitive Earth, then we can begin to understand the massive mineralogical monuments to bacterial success that still exist in the form of large ore bodies in which metals have been concentrated (50). The persistence of this positive selection for the biofilm phenotype in the modern Earth is evidenced by the predominance of this sessile mode of growth in virtually all permissive ecosystems.

The tendency of bacteria to grow in protected biofilms proved to be increasingly useful as other life forms evolved. Bacteria within biofilms are notably resistant to bacteriophages, to amoeboid predators, and to vortex-feeding protozoa, and these factors of modern ecosystems tend to reinforce the predominance of biofilms that we note in direct observations of natural systems. The fact that bacteria on plastic and metal surfaces within the human body in persistent device-related infections grow exclusively in the biofilm mode attests to the resistance of this defensive phenotype against host defense factors (antibodies, surfactants, phagocytes) and vigorous antibiotic chemotherapy (56).

Although the aforementioned is remarkable, by far the highest expression of the biofilm phenotype is in the complex, metabolically cooperative, sessile populations that we have begun to study directly in nutrient-sufficient ecosystems. In the simplest case, the colonized surface may itself be a nutrient (e.g. cellulose), and primary colonizers then adhere and initiate the biodegradative process. In the process, these primary colonizers produce surface zones rich in their metabolic products (e.g. butyrate), and secondary and tertiary colonizers may respond to chemotactic cues and build a multilayered and highly functional biofilm community. In more complicated cases (119), many types of bacteria may be present at a surface, in response to a variety of adhesion cues, and each cell in these complex populations then responds to the ecological microniche within which it finds itself (34, 78). These responses may vary from virtual dormancy to wildly accelerated growth, much as the cells of mammalian tissues respond to local hormone and lectin concentrations during embryogenesis, and complex tissuelike biofilms may be formed. Within these tissuelike biofilms there are structural elements (the exopolysaccharide matrix) and a complex array of bacterial cells with specialized functions that are favored by their stable juxtaposition with bacterial cells of complementary metabolic capabilities (89, 121). Within these very complex tissuelike biofilms, the provision of nutritive substrates is important, but the removal of cellular end products is often a more important cooperative function (57). Specific adhesion mechanisms that would tend to initiate or promote physiological cooperation would be positively selected. The characteristic that enables bacteria to construct these elaborate, metabolically linked consortia is their ability to replicate in direct response to nutrient availability. Eukaryotic cells can respond to nutritional or endocrinological stimuli only during certain periods of development, but bacteria can always respond to favorable nutrient conditions very simply and very rapidly, and it is this programmed response that builds functional biofilm consortia.

Biological success can be determined by range or by biomass and, by either criterion, bacteria are eminently successful. Bacteria generally substitute phenotypic plasticity for complexity. In their simplest planktonic forms, bacteria can penetrate into a very wide variety of ecosystems and their range is truly phenomenal. In favorable environments their phenotypic elasticity allows bacteria to develop biofilms, and this mode of growth provides stable cell-to-cell juxtaposition and a measure of primitive homeostasis in that component cells are somewhat protected from toxic changes in the bulk fluid. Recent direct observations established that biofilms have a primitive circulatory system (39) and that cells with specialized metabolic capabilities can form tissuelike cooperative consortia (89). In essence, bacteria can be simple when it suits them to extend their distribution, but they also have the programmed capability of phenotypic change to the biofilm mode of growth, whose complexity and efficiency rivals that of higher organisms and produces prodigious amounts of biomass.

MICROBIOLOGY AND REALITY

Recent advances in quantitative recovery and in the direct observation of microbial populations have led us to the unequivocal conclusion that the biofilm mode of growth is predominant in aquatic ecosystems. These new methods of direct observation, especially the confocal scanning laser microscope, allow us to examine living biofilms in real time, and it is difficult to imagine that they can be misleading. We believe, therefore, that we must now conclude that the biofilms that have assumed considerable importance in medical and industrial areas are actually predominant, numerically and functionally, in virtually all nutrient-sufficient aquatic ecosystems.

This predominance of biofilms was not evident during the several decades during which these same ecosystems were examined by traditional microbiological methods of sampling and culture. It is vitally important to examine the reasons for this very serious error, because the majority of students in microbiology are still taught to examine microbial ecosystems by extrapolation from planktonic samples and from monospecies laboratory cultures. Direct observations of aquatic ecosystems have shown that samples of the bulk fluid typically contain <0.1% of the bacteria in the ecosystem (52) and that swabbing of surfaces recovers a similarly insignificant proportion of the total bacterial population. Because aggregates of bacteria, like those routinely detached from biofilms, produce a single colony on the plates used for determination of colony-forming units (CFU), biofilm organisms have been radically underestimated. For this same reason, the simple rolling of colonized surfaces across agar plates (90) detaches large, slimy aggregates containing thousands of bacterial cells that produce only a single colony unless they are disrupted by special methods. Biofilm bacteria can be accurately quantitated if colonized surfaces are scrapped and if the resultant detached bacterial aggregates are dispersed, by mechanical shaking and/or ultrasonication (30), before dilution and plating by traditional CFU methods. In our experience, quantitative recovery methods must be developed for each ecosystem to be sampled, and success must be assessed by direct microscopic examinations, because some organisms may be difficult to detach and others may be difficult to grow. In many aquatic ecosystems, some of the biofilm population and some of the planktonic bacteria may be viable but nonculturable (28), and cells of many species (e.g. legionella) do not grow on commonly used laboratory media. The quantitative numerical analysis of the microbial ecosystem is not a trivial undertaking, and the best available methods must be constantly tested against the gold standard of direct microscopic examination.

Perhaps the most serious errors have arisen from extrapolation of data obtained from single-species laboratory cultures to explain bacterial behavior in real ecosystems. We now have unequivocal evidence that sessile biofilm bacterial are profoundly phenotypically different from their planktonic counterparts, but liquid cultures of planktonic cells are still routinely used to assess the antibiotic sensitivity of pathogens in device-related infections that have been shown to be caused by biofilm bacteria (19). Similarly, many microbial ecologists still rely on data from single-species planktonic cultures to explain the activity of multispecies biofilms in which the bacterial cells are sessile and are profoundly influenced by their interactions with neighboring organisms. We must begin to use modern methods of direct observation, from microscopy of living biofilms to NMR of whole reactor systems, if we are to understand processes within microbial ecosystems whose complexity is evident from simple, direct observation. We can readily correct the error implicit in our past and current reliance on planktonic sampling as well as our extrapolation from laboratory cultures, and perhaps we can reverse some of the damage to our collective credibility caused by this error. This damage includes the provision by medical laboratories of minimal inhibitory concentration (MIC) data, obtained from planktonic cultures, that erroneously predicted to clinicians that a given concentration of a certain antibiotic would be effective in killing bacteria in device-related infections known to be caused by biofilm bacteria. This damage also includes predictions of the capability of the microbial populations of aquatic ecosystems to degrade particular organic compounds based solely on the degradative capability of planktonic populations, when these planktonic populations have been unequivocally shown to constitute <0.1% of the total microbial community.

Mike Brown and his colleagues have pointed out (16) that the most remarkable characteristic of bacteria is their plasticity in response to changes in their environment. Recent developments have reinforced this contention. Much of what we know of bacteria is derived from their behavior in single-species cultures of planktonic cells in defined media, and this is only a single point on the vast spectrum of what bacteria can do in various environments. It is vitally important that we involve our students in the revision of our discipline, which has been singularly useful to humanity, so that they can lose no time in making microbiology more accurate and even more useful in the solution of modern practical problems.

ACKNOWLEDGMENTS

We are grateful for the sustained support of the Medical Research Council and the Natural Sciences and Engineering Research Council of Canada, and for the support of the Office of Naval Research and the National Science Foundation (EEC-8907039) in the United States. Dr. JR Lawrence has provided excellent advice and generous collaboration throughout this research program.

> Any *Annual Review* chapter, as well as any article cited in an *Annual Review* chapter, may be purchased from the Annual Reviews Preprints and Reprints service.
> 1-800-347-8007; 415-259-5017; email: arpr@class.org

Literature Cited

1. Allison DG, Sutherland IW. 1987. The role of exopolysaccharides in adhesion of freshwater bacteria. *J. Gen Microbiol.* 133:1319–27
2. Amann RI, Kurmholz I, Stahl DA. 1990. Fluorescent-oligonucleotide probing of whole cells for determinative, phylogenetic, and environmental studies in microbiology. *J. Bacteriol.* 172:762–70
3. Amann RI, Stromley J, Devereux R, Key R, Stahl DA. 1992. Molecular and microscopic identification of sulfate-reducing bacteria in multispecies biofilms. *Appl. Environ. Microbiol.* 58:614–23
4. Anwar H, Strap JL, Costerton JW. 1992. Establishment of aging biofilms: a possible mechanism of bacterial resistance to antimicrobial therapy. *Antimicrob. Agents Chemother.* 36:1347–51
5. Anwar H, Strap JL, Costerton JW. 1992. Susceptibility of biofilm cells of *Pseudomonas aeruginosa* to bactericidal actions of whole blood and serum. *FEMS Microbiol. Lett.* 92:235–42
6. Applegate DH, Bryers JD. 1991. Effects of carbon and oxygen limitations and calcium concentrations on biofilm removal processes. *Biotechnol. Bioeng.* 37:17–25
7. Back JP, Kroll RG. 1991. The differential fluorescence of bacteria stained with acridine orange and the effects of heat. *J. Appl. Bacteriol.* 71:51–58
8. Bakke R, Olsson PQ. 1986. Biofilm thickness measurements by light microscopy. *J. Microbiol. Methods* 5:93–98
9. Banks MK, Bryers JD. 1991. Bacterial species dominance within a binary culture biofilm. *Appl. Environ. Microbiol.* 57:1874–79
10. Banwell JG, Howard R, Cooper D, Costerton JW. 1985. Intestinal microflora after feeding phytohemaglutinin (*Phaseolus vulgaris*) lectins to conventional rats. *Appl. Environ. Microbiol.* 50:68–80
11. Blenkinsopp SA, Khoury AE, Costerton JW. 1992. Electrical enhancement of biocide efficacy against *Pseudomonas aeruginosa* biofilms. *Appl. Environ. Microbiol.* 58:3770–73
12. Boivin J, Costerton JW. 1991. Biofilms and biodeterioration. In *Biodeterioration and Biodegradation 8*, ed. HW Rossmore, pp. 53–62. London: Elsevier Appl. Sci.
13. Bouwer EJ. 1987. Theoretical investigation of particle deposition in biofilm systems. *Water Res.* 21:1489–98
14. Boyd A, Chakrabarty AM. 1994. Role of alginate lyase in cell detachment of *Pseudomonas aeruginosa*. *Appl. Environ. Microbiol.* 60:2355–59
15. Brayton PR, Tamplin ML, Huq A, Colwell RR. 1987. Enumeration of *Vibrio cholerae* 01 in Bangladesh waters by fluorescent-antibody direct viable count. *Appl. Environ. Microbiol.* 53:2862–65
16. Brown MRW, Williams P. 1985. The influence of the environment on envelope properties affecting survival of bacteria in infections. *Annu. Rev. Microbiol.* 39:527–56
17. Brown MRW, Allison DG, Gilbert G. 1988. Resistance of bacterial biofilms to antibiotics: a growth-rate related effect? *J. Antimicrob. Chemother.* 22:777–80
18. Brown MRW, Collier PJ, Gilbert P. 1990. Influence of growth rate on susceptibility to antimicrobial agents: modifications of the cell envelope and batch and continuous culture studies. *Antimicrob. Agents Chemother.* 34:1613–23
19. Brown MRW, Gilbert P, Costerton JW. 1991. Extrapolating to bacterial life outside the test tube. *J. Antimicrob. Chemother.* 27:565–67
20. Brown MRW, Gilbert P. 1993. Sensitivity of biofilms to antimicrobial agents. See Ref. 104a, pp. 87S-97S
21. Bryant RD, Laishley EJ. 1989. The role of hydrogenase in anaerobic biocorrosion. *Can. J. Microbiol.* 36:259–64
22. Bryers JD, Characklis WG. 1981. Early fouling biofilm formation in a turbulent

flow system: overall kinetics. *Water Res.* 15:483–91
23. Caldwell DE, Korber DR, Lawrence JR. 1992. Confocal laser microscopy and computer image analysis. In *Advances in Microbial Ecology,* ed. KC Marshall, 12:1–67. New York: Plenum
24. Caldwell DE, Korber DR, Lawrence JR. 1992. Imaging of bacterial cells by fluorescence exclusion using scanning confocal laser microscopy. *J. Microbiol. Methods* 15:249–61
25. Caldwell DE, Korber DR, Lawrence JR. 1993. Analysis of biofilm formation using 2-D versus 3-D digital imaging. See Ref. 104a, pp. 52S-66S
26. Caron DA. 1987 Grazing of attached bacteria by heterotrophic microflagellates. *Microb. Ecol.* 13:203–18
27. Characklis WG, Turakhia MH, Zelver N. 1990. Transfer and interfacial transport phenomena. In *Biofilms,* ed. WG Characklis, KC Marshall, pp. 265–340. New York: Wiley
27a. Characklis WG, Wilderer PA, eds. 1989. *Structure and Function of Biofilms.* New York: Wiley
28. Colwell RR. 1984. *Vibrios in the Environment.* New York: Wiley
29. Costerton JW, Geesey GG, Cheng K-J. 1978. How bacteria stick. *Sci. Am.* 238:86–95
30. Costerton JW, Nickel JC, Ladd TI. 1986. Suitable methods for the comparative study of free-living and surface-associated bacterial populations. In *Bacteria in Nature,* ed. JS Poindexter, ED Ledbetter, 2:49–84. New York: Plenum
31. Costerton JW, Cheng K-J, Geesey GG, Ladd TI, Nickel NC, et al. 1987. Bacterial biofilms in nature and disease. *Annu. Rev. Microbiol.* 41:435–64
32. Costerton JW. 1992. The pivotal role of biofilms in the focused attack of bacteria on soluble substrates. *Int. Biodeter. Biodegrad.* 30:123–33
33. Costerton JW, Anwar H. 1994. *Pseudomonas aeruginosa*: the microbe and pathogen. In Pseudomonas aeruginosa *Infections and Treatment,* ed. A Baltch, P Smith, pp. 1–18. New York: Dekker
34. Costerton JW, Lewandowski Z, DeBeer D, Caldwell DE, Korber DR, James GA. 1994. Biofilms: the customized microniche. *J. Bacteriol.* 176:2137–42
35. Cunningham AB. 1989. Hydrodynamics and solute transport at the fluid–biofilm interface. See Ref. 27a, pp. 19–31
36. Daley RJ. 1970. Direct epifluorescence enumeration of native aquatic bacteria: uses, limitations and comparative accuracy. In *Native Aquatic Bacteria: Enumeration, Activity, and Ecology,* ed. JW Costerton, RR Colwell. pp. 29–45. Philadelphia, PA: Am. Soc. Test. Mat.
37. Davies DG, Chakrabarty AM, Geesey GG. 1993. Exopolysaccharide production in biofilms: substratum activation of alginate gene expression by *Pseudomonas aeruginosa. Appl. Environ. Microbiol.* 59:1181–86
38. DeBeer D, Stoodley P, Roe FL, Lewandowski Z. 1994. Effects of biofilm structures on oxygen distribution and mass transport. *Biotech. Bioeng.* 43:1131–38
39. DeBeer D, Stoodley P, Lewandowski Z. 1994. Liquid flow in heterogeneous biofilms. *Biotech. Bioeng.* 44:636–41
40. DeLong EF, Wickham GS, Pace NR. 1989. Phylogenetic stains: ribosomal RNA–based probes for the identification of single cells. *Science* 243:1360–63
41. Deretic V, Govan JRW, Konyecsni WM, Martin DW. 1990. Mucoid *Pseudomonas aeruginosa* in cystic fibrosis: mutations in the muc loci affect transcription of the *algR* and *algD* genes in response to environmental stimuli. *Mol. Microbiol.* 4:189–96
42. Deretic V, Schurr MJ, Boucher JC, Martin DW. 1994. Conversion of *Pseudomonas aeruginosa* to mucoidy in cystic fibrosis: environmental stress and regulation of bacterial virulence by alternative sigma factors. *J. Bacteriol.* 176:2773–80
43. Dolfing J, Mulder J-W. 1985. Comparison of methane production rate and coenzyme F_{420} content of methanogenic consortia in anaerobic granular sludge. *Appl. Environ. Microbiol.* 49:1142–45
44. Duddridge JE, Kent CA, Laws JF. 1982. Effect of surface shear stress on the attachment of *Pseudomonas fluorescens* to stainless steel under defined flow conditions. *Biotech. Bioeng.* 26:153–64
45. Eng RHK, Padberg FT, Smith SM, Tan EN, Cherubin CE. 1991. Bactericidal effects of antibiotics on slowly growing and nongrowing bacteria. *Antimicrob. Agents Chemother.* 35:1824–28
46. Fletcher M, Loeb GI. 1979. Influence of substratum characteristics on the attachment of a marine pseudomonad to solid surfaces. *Appl. Environ. Microbiol.* 37:67–72
47. Fletcher M. 1986. Measurement of glucose utilization by *Pseudomonas fluorescens* that are free-living and that are attached to surfaces. *Appl. Environ. Microbiol.* 52:672–76
48. Fletcher M. 1987. How do bacteria attach to solid surfaces? *Microbiol. Sci.* 4:133–36
49. Fletcher M. 1991. The physiological

activity of bacteria attached to solid surfaces. *Adv. Microb. Physiol.* 32:53–85
50. Ferris FG, Fyfe WS, Beveridge TJ. 1987. Bacteria as nucleation sites for authigenic minerals in a metal-contaminated lake sediment. *Chem. Geol.* 63:225–32
51. Fuqua WC, Winans SC, Greenberg EP. 1994. Quorum sensing in bacteria: the luxR-luxI family of cell density-responsive transcriptional regulators. *J. Bacteriol.* 176:269–75
52. Geesey GG, Mutch R, Costerton JW, Green RB. 1978. Sessile bacteria: an important component of the microbial population in small mountain streams. *Limnol. Oceanogr.* 23:1214–23
53. Geesey GG, Iwaoka T, Griffiths PR. 1987. Characterization of interfacial phenomena occurring during exposure of a thin copper film to an aqueous suspension of an acidic polysaccharide. *J. Colloid Interface Sci.* 120:370–76
54. Gilbert P, Collier PJ, Brown MRW. 1990. Influence of growth rate on susceptibility to antimicrobial agents: biofilms, cell cycle, dormancy, and stringent response. *Antimicrob. Agents Chemother.* 34:1856–68
55. Gilbert P, Brown MRW. 1995. Mechanisms of the protection of bacterial biofilms from antimicrobial agents. See Ref. 78, In press
56. Gristina AG, Dobbins JJ, Giamara B, Lewis JC, DeVries WC. 1988. Biomaterial-centered sepsis and the total artificial heart: microbial adhesion versus tissue integration. *J. Am. Med. Assoc.* 259:870–77
57. Guiot SR, van den Berg L. 1985. Performance of an upflow anaerobic reactor combining a sludge blanket and a filter treating sugar waste. *Biotech. Bioeng.* 27:800–6
58. Hahn D, Amann RI, Ludwig W, Akkermans ADL, Schleifer K-H. 1992. Detection of micro-organisms in soil after in situ hybridization with rRNA-targeted, fluorescently labelled oligonucleotides. *J. Gen. Microbiol.* 138:879–87
59. Hamilton WA. 1987. Biofilms: microbial interactions and metabolic activities. In *Ecology of Microbial Communities, 41st Symp. Soc. Gen. Microbiol.*, ed. M Fletcher, TRG Gray, JG Jones. Cambridge: Cambridge Univ. Press
60. Haugland RP. 1992. Molecular probes. In *Handbook of Fluorescent Probes and Research Chemicals*, ed. KD Larison, pp. 1–421. Eugene, OR: Molecular Probes
61. Hermansson M, Marshall KC. 1985. Utilization of surface localized substrate by non-adhesive marine bacteria. *Microb. Ecol.* 11:91–105
62. Hoff KA. 1988. Rapid and simple method for double staining of bacteria with 4′,6-diamidino-2-phenylindole and fluorescein isothiocyanate-labeled antibodies. *Appl. Environ. Microbiol.* 54:2949–52
63. Hoyle BD, Williams LJ, Costerton JW. 1993. Production of mucoid exopolysaccharide during development of *Pseudomonas aeruginosa* biofilms. *Infect. Immun.* 61:777–80
64. James GA, Caldwell DE, Costerton JW. 1993. Spatial relationships between bacterial species within biofilms. *Abstr. CSM/SIM Annu. Meet., Toronto, Canada*
65. Jensen ET, Kharazmi A, Lam K, Costerton JW. 1990. Human polymorphonuclear leukocyte response to *Pseudomonas aeruginosa* biofilms. *Infect. Immun.* 58:2383–85
66. Kaiser D, Losick R. 1993. How and why bacteria talk to each other. *Cell* 73:873–85
67. Keevil CW, Walker JT. 1992. Nomarski DIC microscopy and image analysis of biofilms. *Binary* 4:93–95
68. Khoury AE, Lam K, Ellis BD, Costerton JW. 1992. Prevention and control of bacterial infections associated with medical devices. *ASAIO J.* 38:M174–78
69. Kjelleberg S, Humphrey BA, Marshall KC. 1982. The effect of interfaces on small, starved marine bacteria. *Appl. Environ. Microbiol.* 43:1166–72
70. Kjelleberg S. 1993. *Starvation in Bacteria*, pp. 1–277. New York: Plenum
71. Korber DR, Lawrence JR, Hendry MJ, Caldwell DE. 1992. Programs for determining statistically representative areas of microbial biofilms. *Binary* 4:204–10
72. Korber DR, Hanson KG, Lawrence JR, Caldwell DE, Costerton JW. 1993. The effect of environmental laminar flow velocities on the architecture of *Pseudomonas fluorescens* biofilms. *Abstr. ASM Annu. Meet., 94th; Las Vegas, NV*
73. Korber DR, Lawrence JR, Hendry MJ, Caldwell DE. 1993. Analysis of spatial variability within mot⁺ and mot⁻ *Pseudomonas fluorescens* biofilms using representative elements. *Biofouling* 7:339–58
74. Korber DR, James GA, Costerton JW. 1994. Evaluation of fleroxacin activity against established *Pseudomonas fluo-*

rescens biofilms. *Appl. Environ. Microbiol.* 60:1663–69
75. Korber DR, Caldwell DE, Costerton JW. 1994. Structural analysis of native and pure-culture biofilms using scanning confocal laser microscopy. *Natl. Assoc. of Corrosion Engineers (NACE) Canadian Region Western Conf., Calgary, Alberta*
76. Korber DR, Lawrence JR, Lappin-Scott HM, Costerton JW. 1995. The formation of microcolonies and functional consortia within biofilms. See Ref. 78. In press
77. Lappin-Scott HM, Costerton JW. 1989. Bacterial biofilms and surface fouling. *Biofouling* 1:323–42
78. Lappin-Scott HM, Costerton JW, eds. 1995. *Microbial Biofilms*. Cambridge: Cambridge Univ. Press. In press
79. Lawrence JR, Korber DR, Hoyle BD, Costerton JW, Caldwell DE. 1991. Optical sectioning of microbial biofilms. *J. Bacteriol.* 173:6558–67
80. Lawrence JR, Korber DR. 1994. Aspects of microbial surface colonization behavior. In *Trends in Microbial Ecology*, ed. R Guerrero, C Pedros-Alio, pp. 113–18. Barcelona: Span. Soc. Microbiol.
81. Lawrence JR, Wolfaardt GM, Korber DR. 1994. Monitoring diffusion in biofilm matrices using confocal laser microscopy. *Appl. Environ. Microbiol.* 60:1166–73
82. Lawrence JR, Korber DR, Caldwell DE. 1995. Surface colonization strategies of biofilm-forming bacteria. In *Adv. Microb. Ecol.* 14: In press
83. Lens PNL, DeBeer D, Cronenberg CCH, Houwen FP, Ottengraf SPP, Verstraete WH. 1993. Heterogeneous distribution of microbial activity in methanogenic aggregates: pH and glucose microprofiles. *Appl. Environ. Microbiol.* 59:3803–15
84. Lewandowski Z, Walser G. 1991. Influence of hydrodynamics on biofilm accumulation. In *Environ. Eng. Proc., EE Div/ASCE, Reno, NV, July 8–10.* pp. 619–24
85. Lewandowski Z, Altobelli SA, Fukushima E. 1993. NMR and microelectrode studies of hydrodynamics and kinetics in biofilms. *Biotech. Prog.* 9:40–45
86. Lewandowski Z, Altobelli S. 1994. Water flow in a narrow conduit covered with biofilm. *Proc. Int. Assoc. Water Quality*, pp. 19–21. Copenhagen, Denmark
87. Mackie RI, Kreced RC, Els JH, van Niekerk JP, Kirschner LM, Baccker AAW. 1989. Characterization of the microbial community colonizing the anal and vulvar pores of helminths from the hindgut of zebras. *Appl. Environ. Microbiol.* 55:1178–86
88. MacLeod FA, Lappin-Scott HM, Costerton JW. 1988. Plugging of a model rock system using starved bacteria. *Appl. Environ. Microbiol.* 54:1365–72
89. MacLeod FA, Guiot SR, Costerton JW. 1990. Layered structure of bacterial aggregates produced in an upflow anaerobic sludge bed and filter reactor. *Appl. Environ. Microbiol.* 56:1598–607
90. Maki DK, Weise CE, Sarafin HW. 1977. A semi-quantitative method of identifying intravenous catheter-related infection. *New Engl. J. Med.* 296:1305–9
90a. Marshall KC, ed. 1984. *Microbial Adhesion and Aggregation.* New York: Springer Verlag
91. Marshall KC. 1992. Biofilms: an overview of bacterial adhesion, activity, and control at surfaces. *Am. Soc. Microbiol. News* 58:202–7
92. Marshall KC, Stout R, Mitchell R. 1971. Mechanisms of the initial events in the sorption of marine bacteria to surfaces. *J. Gen. Microbiol.* 68:337–48
93. Marshall PA, Loeb GI, Cowan MM, Fletcher M. 1989. Response of microbial adhesives and biofilm matrix polymers to chemical treatments as determined by interference reflection microscopy and light section microscopy. *Appl. Environ. Microbiol.* 55:2827–31
94. Martin DW, Schurr MJ, Yu H, Deretic V. 1994. Analysis of promoters controlled by the putative sigma factor *algU* regulating conversion to mucoidy in *Pseudomonas aeruginosa*: relationship to sigmaE and stress response. *J. Bacteriol.* 176:6688–96
95. McFeters GA, Bazin MJ, Bryers JD, Caldwell DE, Characklis WG, et al. 1984. Biofilm development and its consequences: group report. See Ref. 90a, pp. 109–24
96. McFeters GA, Yu FP, Pyle BH, Stewart PS. 1994. Physiological assessment of bacteria using fluorochromes. *J. Microbiol. Methods* In press
97. Morris PG. 1986. *Nuclear Magnetic Resonance Imaging in Medicine and Biology.* Oxford: Clarendon
98. Neu TR, Marshall KC. 1991. Microbial "footprints"—a new approach to adhesive polymers. *Biofouling* 3:101–12
99. Nichols WW, Dorrington SW, Slack MPE, Walmsley HL. 1988. Inhibition of tobramycin diffusion by binding to alginate. *Antimicrob. Agents Chemother.* 32:518–23

100. Nichols WW. 1989. Susceptibility of biofilms to toxic compounds. See Ref. 27a, pp. 321–31
101. Nickel JC, Ruseska I, Wright JB, Costerton JW. 1985. Tobramycin resistance of cells of *Pseudomonas aeruginosa* growing as a biofilm on urinary catheter material. *Antimicrob. Agents Chemother.* 27:619–24
102. Novitsky JA, Morita RY. 1976. Morphological characterization of small cells resulting from nutrient starvation of a psychrophilic marine *Vibrio*. *Appl. Environ. Microbiol.* 32:617–22
103. Ophir T, Gutnick DL. 1994. A role for exopolysaccharides in the protection of microorganisms from desiccation. *Appl. Environ. Microbiol.* 60:740–45
104. Picologlou BF, Zelver N, Characklis WG. 1980. Biofilm growth and hydraulic performance. *J. Hydraul. Div. Am. Soc. Chem. Eng.* 106(HY5):733–46
104a. Quesnel LB, Gilbert P, Handley PS, eds. 1993. *Microbial Cell Envelopes: Interactions and Biofilms.* Oxford: Blackwell Sci. Publ.
105. Rittman BE. 1982. The effect of shear stress on biofilm loss rate. *Biotech Bioeng.* 24:501–6
106. Rittman BE, Manem JA. 1992. Development and experimental evaluation of a steady-state, multispecies biofilm model. *Biotech. Bioeng.* 39:914–22
107. Rodriguez GG, Phipps D, Ishiguro K, Ridgway HF. 1992. Use of a fluorescent redox probe for direct visualization of actively respiring bacteria. *Appl. Environ. Microbiol.* 58:1801–8
108. Shotton DM. 1989. Confocal scanning optical microscopy and its applications for biological specimens. *J. Cell Sci.* 94:175–206
109. Siegrist H, Gujer W. 1985. Mass transfer mechanisms in a heterotrophic biofilm. *Water Res.* 19(11):1369–78
110. Stahl DA, Flesher B, Mansfeld HR, Montgomery L. 1988. Use of phylogenetically based hybridization probes for studies of ruminal microbial ecology. *Appl. Environ. Microbiol.* 54:1079–84
111. Stewart PS, Peyton BM, Drury WJ, Murga R. 1993. Quantitative observations of heterogeneities in *Pseudomonas aeruginosa* biofilms. *Appl. Environ. Microbiol.* 59(1):327–29
112. Stoodley P, DeBeer D, Lewandowski Z. 1994. Liquid flow in biofilm systems. *Appl. Environ. Microbiol.* 60:2711–16
113. Trulear MG. 1980. *Dynamics of biofilm processes in an annular reactor.* M.S. thesis. Rice Univ., Houston, TX
114. Tsien RY, Waggoner A. 1990. Fluorophores for confocal microscopy: photophysics and photochemistry. In *Handbook of Confocal Microscopy*, ed. JB Pawley, pp. 169–78. New York: Plenum
115. van Loosdrecht MCW, Lyklema J, Norde W, Zehnder AJB. 1990. Influence of interfaces on microbial activity. *Microbiol. Rev.* 54:73–87
116. Wanner O, Gujer W. 1986. A multispecies biofilm model. *Biotech. Bioeng.* 28:314–28
117. White DC. 1984. Chemical characterization of films. See Ref. 90a, pp. 159–76
118. Wilson T. 1990. *Confocal Microscopy*. London: Academic
119. Wimpenny JWT, Kinniment S. 1995. Biochemical reactions and the establishment of gradients within biofilms. See Ref. 78, In press
120. Wolfaardt GM, Lawrence JR, Robarts RD, Caldwell DE. 1994. Multicellular organization in a degradative biofilm community. *Appl. Environ. Microbiol.* 60:434–46
121. Wolfaardt GM, Lawrence JR, Robarts RD, Caldwell DE. 1994. The role of interactions, sessile growth and nutrient amendment on the degradative efficiency of a bacterial consortium. *Can. J. Microbiol.* 40:331–40
122. Wolfaardt GM, Lawrence JR, Headley JV, Robarts RD, Caldwell DE. 1994. Microbial exopolymers provide a mechanism for bioaccumulation of contaminants. *Microb. Ecol.* 27:279–91
123. Wolfaardt GM, Lawrence JR, Robarts RD, Caldwell DE. 1993. In situ visualization of exopolymer chemistry and selective binding of a chlorinated herbicide in microbial biofilms. *Abstr. CSM/SIM Annu. Meet.,* Toronto
124. Wright JB, Ruseska I, Athar MA, Corbett S, Costerton JW. 1989. *Legionella pneumophila* grows adherent to surfaces in vitro and in situ. *Infect. Control* 10:408–15
125. Yu FP, McFeters GA. 1994. Physiological response of bacteria in biofilms to disinfection. *Appl. Environ. Microbiol.* 60:2462–66
126. Zobell CE, Anderson DQ. 1936. Observations on the multiplication of bacteria in different volumes of stored seawater and the influence of oxygen tension and solid surfaces. *Biol. Bull. Woods Hole* 71:324–42
127. Zobell CE. 1943. The effect of solid surfaces upon bacterial activity. *J. Bacteriol.* 46:39–56

LEUCINE-RESPONSIVE REGULATORY PROTEIN: A Global Regulator of Gene Expression in *E. coli*

E. B. Newman

Biology Department, Concordia University, Montreal, Quebec H3G 1M8 Canada

Rongtuan Lin

Lady Davis Institute, 3755 Cote Ste. Catherine, Montreal, Quebec H3T 1E2 Canada

KEY WORDS: transcription regulation, DNA-binding proteins, chromosome structure

CONTENTS

INTRODUCTION	748
THE LEUCINE-RESPONSIVE REGULATORY PROTEIN	750
Lrp AS A REGULATOR	751
Patterns of Regulation by Lrp	751
Regulation of Lrp Synthesis	752
Physiological Consequences of Lrp Action	753
Lrp AS A CHROMOSOME ORGANIZER	758
Interactions of Lrp With DNA: Bending and Macromolecular Organizations	760
Does Lrp Recognize a Particular Sequence?	760
MOLECULAR ASPECTS OF Lrp FUNCTION	762
Definition of Lrp Domains	762
Interactions of Lrp With Specific Promoters	762
KNOWN OPERONS REGULATED BY Lrp	765
Biosynthetic Operons	765
Degradative Operons	766
Other Metabolic Operons	767
Transport Operons	768
Operons Coding for Pili and Fimbriae	770
Other Operons	770

ABSTRACT

The leucine-responsive regulatory protein (Lrp) regulates transcription of the many genes of the Lrp regulon, repressing some and activating others, some in response to L-leucine and some independent of it. The physiology and molecular biology of the regulon in *Escherichia coli* are summarized here. However, the high degree of conservation of the protein suggests that it has an important role in all enterobacteria. We suggest that this role is not only as a transcriptional regulator but also as a determinant of chromosome structure.

INTRODUCTION

A recently described transcriptional regulator called the leucine-responsive regulatory protein, or Lrp, governs expression of a group of genes known as the leucine/Lrp regulon. In its capacity as a leucine-binding protein (67), Lrp integrates the many effects of leucine on gene expression in *Escherichia coli*. However, it also affects transcription at several promoters that are not affected by leucine. Cells deficient in Lrp are so strongly affected as to be physiologically deficient in certain activities such as glycine cleavage (60, 61) and glutamate dehydrogenase (25), neither of which is regulated by leucine.

The remarkable extent to which Lrp is conserved in bacteria other than *E. coli*—over 90% amino acid identity in three other bacteria (see below)—suggests that it plays an important role in organizing metabolism. It functions as a classic regulator with regard to leucine-regulated genes. However, the large number of Lrp molecules in the cell, and the capacity of the Lrp molecule to bend DNA (115) and to organize multimolecular promoter structures (116), led to the suggestion that it also functions as a chromosome organizer (23, 72).

The framework in which Lrp is viewed depends—as often is the case—on the history of experimentation, as well as on the background and interests of the people studying it. The role of Lrp as a regulator of expression of several operons was discovered originally in a search for a mutant defective in regulation of several leucine-regulated genes (60). A mutant in which leucine could not regulate *sdaA* expression turned out to be deficient in expression of other leucine-regulated genes as well. The gene mutated in that strain is now known as *lrp*.

The metabolic description of the regulon was expanded by two random methods: examination of two-dimensional (2D) protein gels from cells grown with and without leucine (25) and screening of random insertions for leucine-regulated genes (61, 108). Both methods provided new examples of leucine-regulated genes. However, identification of the enzymes whose regulation is affected by Lrp did not define a clear metabolic role for Lrp despite various efforts (19, 23, 72).

These random methods require investigators to guess which genes are regulated by leucine and to then test them one by one. However, the realization that Lrp had already been described under different names, including *oppI* (6, 7) and *ilvIH*-binding protein (84), greatly facilitated our understanding of this protein's range of action. Detailed studies of the *ilvIH*-binding protein (83, 84, 114–117, 119, 120) also provided much evidence about the molecular nature of Lrp.

This system of experimentation was based on the beautiful genetics of the *lac* operon, which have influenced a generation of bacterial physiologists. This is a clear case of a protein regulating transcription of a single group of genes whose products affect a limited area of metabolism.

The ideas behind the *lac* system were generalized to much larger groups of genes regulated by a single protein known as a global regulator. These groups of genes were discovered and defined by their physiological and environmental roles. An increase in temperature alters transcription of the heat-shock genes via a global regulator, σ^{32}. Similarly, the catabolite activator protein (Crp) regulates the expression of genes involved in carbohydrate degradation. The RelA protein carries out an even more general role in cell metabolism: By varying the ppGpp concentration, it adjusts the rate of ribosome formation to the availability of aminoacyl-tRNA.

One of the major puzzles of leucine/Lrp regulon is that a similar general statement of its metabolic role cannot be made at this time. Various suggestions as to why exogenous leucine should regulate *E. coli* metabolism are reviewed below. Nevertheless, a complete understanding eludes us.

Lrp is clearly a DNA-binding protein. A typical cell contains some 3000 molecules, which is far more than the number of Lac repressor molecules and a quantity on the order of magnitude of the nonhistone DNA-binding proteins. Such considerable numbers suggest that Lrp functions together with other *E. coli* DNA-binding proteins, such as HU, IHF, Fis, and HN-S (81, 93), to establish the structure of the DNA of the cell (23). Through this influence, Lrp may set the stage for other factors that allow variations in transcription rate (23).

In fact, any of the DNA-binding proteins involved in establishing chromosome structure will appear to be global regulators if transcription in wild-type cells is compared with that in mutants deficient in the particular factor. In *hns* mutants, transcription of many genes is altered (57). However, the Hns level does not appear to vary during metabolism, and this protein is not usually thought of as a regulator. Similarly, although some genes are not transcribed at physiologically significant rates if Lrp is not present (61), there is little evidence that the rate of *lrp* transcription itself varies greatly, except between rich and poor media (61). In these cases, the presence of the protein is clearly required to establish transcription, but variations in Lrp function are not necessarily involved in regulating the actual rate of transcription.

Lrp seems to act both as a global regulator of a group of genes and as a structural element in establishing DNA conformation. Perhaps many other proteins also play a role in establishing the environment of the cell in addition to their currently understood roles as chemical catalysts, membrane and ribosome components, and DNA-binding proteins.

Owing to the lack of clear physiological focus, research on Lrp has had a different orientation from that on other global regulators. Studies have dealt with the biochemistry and molecular biology of Lrp, and with its physiological role. In this review, we survey all the papers published on Lrp up to December 1, 1994, concentrating somewhat less on the molecular biology, which Calvo & Matthews reviewed recently (19). Their paper was perhaps the last review that could cite all Lrp publications, given the increase in interest in this protein. Here, we point to many interesting questions raised by the experiments and to the much fewer conclusions about the nature of Lrp and its influence on the *E. coli* cell.

THE LEUCINE-RESPONSIVE REGULATORY PROTEIN

Lrp is a small basic protein (pI 9.3) composed of two identical subunits of 18,800 Daltons, which form a dimer in solution (19, 120). The mature protein has 163 residues including all natural amino acids except tryptophan. A sequence centered at position 40 may represent a helix-turn-helix motif (120).

Nondescript though Lrp is, it is present in significant quantities—3000 molecules per cell (120). This number is large compared with LacI numbers but is about the same order of magnitude as quantities of Crp, Hns, IHF, and HU. As estimated by an assay of protein reacting with anti-Lrp antibodies, Lrp makes up 0.1% of cellular protein in cells grown in glucose minimal medium (120). This estimate may even be low as it could not take into account molecules bound tightly to DNA. Comparisons with known proteins on 2D gels suggested a similar percentage [0.1–0.2% (25)].

Lrp's importance in cellular function is indicated by its astonishingly high degree of conservation in other microorganisms. The *Enterobacter aerogenes* gene has 91% identity, *Serratia marcescens* 90%, *Klebsiella aerogenes* 89%, and *Salmonella typhimurium* 87%—all coding for proteins of about the same length (19). A symbiotic gene from *Rhizobium* spp. is similar to *lrp*—55% identity over a 241-nucleotide region, but nothing is known about the function of any of these gene products.

The *Pseudomonas bkdR* gene encodes a protein with 36.5% amino acid identity to Lrp (66). The fact that a mutation in *bkdR* is complemented by plasmids carrying *lrp* suggests that even this modest identity is biologically significant and that the identity with other proteins will also prove to be of functional importance. An Lrp function may also be present in Neisseriae (53).

The large number of genes affected by Lrp further supports its importance in cell function. Its absence has profound effects: Estimates from 2D gels suggest that levels of 25 polypeptides are altered in the *lrp* mutant (25). A higher estimate of up to 75 polypeptides with altered levels came from a study of random λplacMu insertions into the *E. coli* genome; these insertions were screened for leucine effects and shown to be Lrp dependent (61). Recent assays of random unscreened insertions suggest that expression of 5–10% of the genes in the *E. coli* chromosome are altered twofold or more in an *lrp* mutant (EB Newman, unpublished results).

The sequence of Lrp gives few clues to its function. It is unlike all the common regulators; indeed, its sequence resembles that of only one other *E. coli* protein, AsnC, a positive regulator of *asnA* (the structural gene of asparagine synthetase A), with which it has 25% amino acid identity. No detailed functional comparison of these two proteins has yet been made.

Purified Lrp binds well to double-stranded DNA containing an operator site, even in the presence of a 1000-fold excess of calf thymus DNA, but it does not bind to single-stranded DNA of the same sequence (115). As might be anticipated from the idea that Lrp is an important element in establishing DNA structure, Lrp binding has major structural effects. A circular permutation assay indicated that binding of a single Lrp molecule bends the DNA an estimated 52°, and two molecules bound to adjacent sites bend DNA 135° (115). Wang & Calvo (115) suggest that Lrp forms an architectural element, which facilitates the assembly of a nucleoprotein complex regulating transcription.

Lrp AS A REGULATOR

Patterns of Regulation by Lrp

Lrp is a transcriptional regulator—activator or repressor—of many operons. Its action is intensified by, and even dependent on, the presence of exogenous leucine at some operators, whereas it is reduced by leucine at others and is independent of leucine at many more.

Whether Lrp acts positively or negatively at a given target operon is usually assessed by comparing expression levels in wild-type and *lrp* mutant cells. Such comparisons do not differentiate between indirect and direct action of Lrp, and indirect effects have been described. Indeed, Lrp acts directly on the *gltBDF* operon and indirectly on a gene of closely related function, *glnA* (25). Lrp-binding studies indicated that Lrp directly affects some 11 operons, and one study showed that Lrp stimulates transcription in vitro (114).

The fact that the presence of leucine in the growth medium often affects Lrp regulation can be understood in terms of the ability of this protein to bind L-leucine (67, 94). Generally, leucine binding reduces the efficiency of Lrp action, leading to weaker activation or repression. The leucine effect is usually less severe than mutational inactivation of Lrp, although for some operons it is quite dramatic, and in one case leucine has a stronger effect than an *lrp* mutation (59). For several operons, Lrp seems to act only in the presence of exogenous leucine, in some cases as activator, in others as repressor. The many physiological effects of leucine on *E. coli* metabolism seem to be mediated by its interaction with Lrp. In the absence of Lrp (i.e. in the *lrp* mutant), the only operon known to be affected by exogenous leucine is the biosynthetic *leuABCD* operon (61), presumably through its attenuator (118). However, even this operon is regulated by Lrp (108).

These results suggest that leucine (or a leucine derivative) is an allosteric effector of Lrp. Some promoters would respond to free Lrp, others to effector-bound Lrp. Some might respond to either. As a result, transcription can decrease or increase, perhaps depending on the gene structure. Promoters might also vary the affinity for free or for leucine-bound Lrp. For example, exogenous leucine causes 3.3-fold higher expression of one operon, whereas mutational loss of Lrp causes only a 1.9-fold increase (59). Possibly in this case, free Lrp represses and the effector-bound form activates transcription. This situation is analogous to regulation at the *araBAD* promoter, which is repressed by free AraC and activated by arabinose-bound AraC (63, 97).

The equilibrium experiments described above examined only leucine binding. However, for many Lrp-regulated operons, exogenous alanine or methionine can have effects similar to those of leucine. Lrp might bind more than one effector. However, the presence of any of these amino acids may cause a particular metabolic change to which Lrp responds. Alternatively, addition of other amino acids may alter the leucine pool, so that the actual effector is L-leucine in all cases.

Regulation of Lrp Synthesis

As indicated above, the question of whether Lrp transcription is regulated is essential to an understanding of its metabolic role. If the level of Lrp in the cell does not vary, it cannot regulate those promoters that do not respond to leucine. One of the factors known to regulate transcription of *lrp* is Lrp itself. Like many transcriptional regulators, Lrp is autogenously regulated: In minimal glucose medium, expression of an *lrp::lacZ* fusion is lowered two- to threefold by the presence of a single chromosomal *lrp*$^+$ gene and tenfold by the presence of a functional *lrp*$^+$ gene on a multicopy plasmid (61, 117). The presence of leucine in the medium has no effect on expression. According to primer extension experiments, the *lrp* promoter resides 267 base pairs (bp) upstream of the translational start codon (117). Lrp binds to a site between 32 and 80 bp upstream of the transcription start, as indicated by gel mobility shifts, DNA footprinting, and deletion studies. Mutations in that area alter the response to

Lrp, strongly suggesting that autogenous regulation is a direct effect of Lrp binding to this region of the DNA (117).

Autogenous regulation is a minor effect compared with the 10-fold reduction in transcription that results from growth in rich medium (61). This Lrp-independent decrease in *lrp* transcription (61) is also observed in medium containing 1% casamino acids, with or without glucose (C Sears, RT Lin & EB Newman, unpublished results). Both the α-ketoglutarate family of amino acids and the oxalacetate family strongly repress *lrp* transcription (C Sears & EB Newman, unpublished results). L-Threonine alone reduced transcription 50%.

The expression of *lrp* also depends heavily on the carbon source supporting growth. In cells grown with more oxidized substrates—acetate, succinate, and pyruvate—transcription was reduced 50%. Cells grown on glycerol or various sugars all showed the same level of *lrp:* about 80% that of cells grown on glucose. The physiological significance of this result seems very limited because no corresponding pattern was seen in the transcription of five *lrp*-regulated target genes in glucose-, succinate-, and acetate-grown cells. Clearly, Lrp is not the major or only effector of transcription of these genes in these media (C Sears & EB Newman, unpublished results).

Even if Lrp in enriched media does decrease 10-fold, Lrp could still be an important metabolic regulator in those media, especially if its binding affinity varies at different promoters. The variations in *lrp* expression may indicate different settings of the cell's metabolic web, at which systems with low affinity for Lrp are significantly regulated by it only during growth in minimal medium (cf 26). The presence or absence of leucine would complicate the picture further.

Physiological Consequences of Lrp Action

Table 1 lists genes known to be regulated by Lrp as of autumn 1994. These genes were identified via several experimental protocols: by assaying leucine-induced genes (those previously known and those identified by screening of random insertions of λplacMu); by identifying proteins whose synthesis rate, as determined on 2D gels, is altered in an *lrp* mutant; and by identifying Lrp-dependent genes through direct determination of the DNA sequences flanking inserts. In 50 recently isolated strains carrying random inserts of λplacMu, 6 were regulated by Lrp, indicating again that Lrp influences transcription of a large proportion of *E. coli* genes (J Zhang & EB Newman, unpublished results).

In the following sections, we attempt to understand the metabolic function(s) of Lrp and consider its overall effects on *E. coli* metabolism. This discussion is based on the individual reactions now known to be regulated by Lrp, and its accuracy is limited by the fact that many reactions may be missing. The individual reactions are reviewed in detail in the final section of this review.

Table 1 *E. coli* operons regulated by Lrp

Operon	Change due to Lrp	Change due to leucine	References
Activated			
ilvIH	30 ↑	20 ↓	87, 114
serA	6 ↑	2 ↓	60, 91
sdaC			61, 99
leuABCD	11 ↑	ND[a]	61, 108
gltBDF	44 ↑	50 ↓	25, 26
gcvTHP	20 ↑	None	61
pntAB	5 ↑	44 ↓	3, 36
ompF			25
malT	1.8 ↑	None	108
malEFG	2 ↑	None	108
malK	6 ↑	None	108
lacZYA	1.5 ↑	None	108
papBA	35 or 430 ↑	None	15, 16, 76
fanABC	76 ↑	10 ↓	15
sfaA	9 ↑	3 ↑	110
daaABCDE	9 or 60	None	12, 112
fimB		13	
Repressed			
sdaA	8 ↓	5 ↑	61
glyA	4 ↓		Unpublished[b]
kbl-tdh	20 ↓	8 ↑	61, 91
op-pABCDF	2.5 ↓		6–8
lysU	22 ↓	4 ↑	30, 62
livJ	85 ↓	105 ↓	41, 61, 79
livKHMG	9 ↓		41, 50
lrp	2 ↓	None	60, 117
ompC			25
fae	3 ↓	None	44
osmY	5 ↓		52

[a] ND, not determined.
[b] M San Martano & EB Newman, unpublished results.

METABOLISM OF A CELL DEVOID OF Lrp   The *lrp* mutant is a well-organized cell: Despite the many changes in expression of its genes, the *lrp* mutant can grow in glucose-minimal medium, albeit slowly. Hence, the biochemical web must still be sufficiently well organized to allow reasonably efficient growth. The *lrp* mutant grows much faster in minimal medium than an *hns* or an unsuppressed *crp* mutant.

However, the reorganization of the *lrp* cell's metabolism makes it highly sensitive to environmental perturbations or to the addition of further mutations. A total loss of Lrp greatly affects several areas of metabolism, including the synthesis of external appendages, the metabolism of serine and one-carbon units, degradation of some amino acids and sugars, nitrogen assimilation, biosynthesis of some amino acids, and several transport capabilities. This list is likely incomplete and in any case reflects the biases of the investigators in the choice of genes to study.

Lrp as a major determinant of fimbrial synthesis As a pathogen, *E. coli* can synthesize several factors that permit it to adhere to host mucosal surfaces, often by using appendages known as fimbriae (65). Many *E. coli* isolated from nature carry several fimbrial operons. Lrp helps modulate the phase variation by which these genes are regulated and helps determine their level of expression. While most studies have focused on the *pap* gene, Lrp controls many other fimbrial operons: *fanABC* (16), *sfaA* (113), *daaABCDE* (12, 113), *fim* (31, 32), and *fae* (44).

The Lrp mutation therefore has a major effect on the cell's pathogenicity. *E. coli* may need one set of appendages for its primary entry into an organism, e.g. in a urinary tract infection, and may then use another to establish itself in the bloodstream. Lrp may take part in that switch as well (75). The molecular details of the regulation of fimbrial genes by Lrp are being studied in elegant detail in the laboratory of D Low, among others (31, 65). The regulation of the *pap* operon is described later in this review.

Decreased availability of L-serine and L-leucine decrease growth rate Growth of the *lrp* mutant at 37°C in glucose minimal medium is markedly slower than that of wild-type strains (84 min doubling time compared with 58 min). This sluggishness is alleviated by adding L-serine and L-leucine to the medium (3, 60). This might be expected from the fact that Lrp is an activator of the first gene in the biosynthetic pathways of both amino acids (60, 61, 108), although a reduced Lrp-independent expression of both genes assures some synthesis in the *lrp* mutant. Because the addition of serine and leucine allows an almost normal growth rate, it is unlikely that other biosynthetic pathways are markedly inhibited in the absence of Lrp.

The *lrp* mutant is probably even more deficient in serine and/or leucine biosynthesis under other environmental conditions. It cannot grow in glucose-minimal medium at 42°C, and it is also unable to grow anaerobically on glucose at 37°C (3). In both cases, growth is restored by the addition of serine to the medium or by increasing the number of copies of the *serA* gene (3).

An assessment of the availability of L-serine in the *lrp* mutant is complicated by its reorganization of other areas of metabolism. In the *lrp* mutant, synthesis of the L-serine–degrading enzyme, L-serine deaminase, is increased sixfold, enough to allow the mutant to use exogenous L-serine as the sole carbon and energy source (60). However L-serine deaminase has a remarkably high K_m (105, 106) and therefore may not decrease the L-serine pool during growth in glucose-minimal medium. In addition, the *lrp* mutant may spare L-serine by producing glycine from threonine via its increased L-threonine dehydrogenase, as discussed below.

The lrp mutant is deficient in glycine cleavage to produce methylene-THF Whereas wild-type *E. coli* has two biosynthetic routes for the synthesis of one-carbon units [methylene-tetrahydrofolic acid (mTHF)], the *lrp* mutant has only one. The mutant relies totally on serine hydroxymethyl transferase (SHMT), the *glyA* gene product, which catalyzes the formation of glycine and mTHF from serine and THF. This is also the major pathway of glycine synthesis during growth on glucose. The mutant is totally deficient in the second route, via the glycine cleavage enzyme complex (Gcv), in which mTHF is formed from the α-carbon of glycine and CO_2 and NH_3 are simultaneously released. However, the *lrp* mutant may produce some glycine from threonine (90).

Similarly, in wild-type cells growing in glucose minimal medium, the glycolysis intermediate 3-phosphoglycerate is the direct precursor of serine, glycine, and one-carbon units. The cell's requirement for one-carbon units is greater than its need for glycine, and the Gcv system permits it to equilibrate these syntheses by overproducing glycine and cleaving the excess (73, 74).

In the *lrp* mutant, this scheme is radically altered because the *gcv* operon absolutely requires Lrp activation for physiologically significant expression (61). Neither the *lrp* mutant nor *gcv* mutants (71) can derive one-carbon units or nitrogen from glycine, and all mTHF must be formed by SHMT. Indeed an *lrp glyA* mutant cannot grow in minimal medium, presumably because it cannot make C1 units for biosynthesis (61).

Nitrogen metabolism The *lrp* mutant grows well with ammonium sulfate as a nitrogen source, at least at the usual high levels. However, it is deficient in glutamate synthase activity (53) and so cannot handle organic nitrogen and does not use arginine or ornithine as the nitrogen source. This deficiency in *gltBDF* expression is probably responsible for the dicarboxylic amino acid requirement of the *lrp pnt* double mutant as well (3).

Transport systems Lrp represses two high-affinity leucine-uptake systems, one of which also transports isoleucine, valine, threonine, and alanine. This is the only case in which leucine intensifies Lrp repression. Lrp also represses an oligopeptide-uptake system, activates a serine transporter, and slightly stimulates expression of the maltose and lactose permeases. No other transporters have as yet been studied. Because the cell has so many alternate transporters, loss of some affects growth little. For example, the *lrp* mutant uses L-serine well as a carbon source. So far, the only phenotype associated with these alterations in transport ability is resistance of the *lrp* mutant to toxic tripeptides (6, 7).

Why does Lrp respond to leucine? The synthesis of Lrp is greatly reduced during growth in Luria broth (LB) (61, 117), which suggests that Lrp is less important to cells that are well provided with metabolites. The fact that some of the genes that Lrp activates are involved in biosynthesis, whereas some of the genes it represses are involved in degradation, supports this view. This has led to the proposal that Lrp is regulated according to a feast-or-famine principle (19, 25, 61).

If *E. coli* evolved first in a rich environment, it must have needed many assimilatory and few biosynthetic systems. Lrp might have evolved as a switch from assimilation to biosynthesis. This particular hypothesis can easily be stated in the other direction. For most of the members of the leucine/Lrp regulon, the reason why exogenous leucine should be taken as a signal is unclear. The choice may be to some degree arbitrary, leucine having been adopted simply as an indicator of active proteolysis in the immediate environment. Other molecules used to signal a general set of conditions via an integrator protein have been described in other organisms. For example, in *Aspergillus nidulans*, β-alanine induces the synthesis of enzymes needed for the degradation of acetamide, in addition to those needed to degrade β-alanine (8). Here, β-alanine is presumably used as a general indicator of rot in the environment, and acetamide is probably rarely if ever encountered without β-alanine. Hence, higher organisms may well have responses similar to the bacterial leucine/Lrp regulon but use a different effector.

It would also be of interest to know whether bacteria have Lrp-like regulons that respond to other signals. *E. coli* exhibits a complex response to exogenous serine (111) and to 2-ketobutyrate and related compounds (21). The possibility that these or other molecules govern global responses has not, to our knowledge, been explored.

Summary: decreased metabolic flexibility of the lrp mutant An *lrp* mutant is totally deficient in the synthesis of glutamate synthetase, glycine cleavage enzymes, and some fimbriae and is limited in the synthesis of serine and glycine. These deficiencies result in only relatively minor problems. The web of reactions the *lrp* mutant expresses permits growth in glucose minimal medium at 37°C: With serine and the three branched-chain amino acids, the *lrp* mutant grows almost as well as its parent (3, 61).

However, wild-type *E. coli* is remarkably flexible—tolerating a wide variety of temperatures, oxygen levels, and secondary mutations, and the *lrp* mutant has lost precisely this metabolic flexibility, a characteristic seen clearly in its reduced tolerance of secondary mutations. For example, an *lrp relA* double mutant has an absolute requirement for leucine. It also requires serine for a normal growth rate, and these requirements suggest that the reduced expression of *serA* and *leuA* no longer suffices. Similarly, an *lrp pnt* double mutant is further hindered in nitrogen metabolism and requires one of glutamate, glutamine, aspartate, and asparagine to grow. However, both *relA* and *pnt* single mutants are prototrophic (3). Moreover, although the cell can tolerate either an *lrp* or a *glyA* mutation, the *lrp glyA* double mutant cannot grow in minimal medium, presumably because it cannot make one-carbon units.

GROWTH OF WILD-TYPE *E. COLI* WITH REDUCED Lrp LEVELS   In LB broth, *E. coli* produces 10-fold less Lrp and is exposed to a high leucine environment, which for many operons lowers the efficiency of Lrp activation or repression (Table 1). During growth in LB then, *E. coli* approaches the state of the *E. coli* mutant. Thus, those operons that depend most strongly on Lrp for activation—*leuDBCA, serA, gcv, pap,* and *gltBDF*—should not be expressed in LB. However, the promoters of these genes might have a particularly high affinity for Lrp, so that a 10-fold decrease might not affect them. The affinity of only one promoter for Lrp has been carefully determined (26). An in vivo assessment of affinities of other promoters, in LB and in minimal medium, is now being carried out and should clarify this issue (L Tao & EB Newman, unpublished results).

AN *lrp* MUTATION FACILITATES GROWTH OF A *metK* MUTANT   The *metK* gene product, S-adenosylmethionine synthetase, catalyzes the formation of S-adenosylmethionine, the direct methyl donor in most methylation reactions. For many years, this synthetase was thought to be dispensable because *metK* mutants were thought to grow in glucose minimal medium. However, it is now known that the *metK* mutant does not in fact grow in minimal medium unless a second mutation is also present. The *metK* strain commonly used in the past was an *lrp metK* double mutant. This observation implies that Lrp makes *metK* mutants almost nonviable: *metK lrp*$^+$ strains grow exceedingly slowly, whereas *metK lrp* double mutants grow well, and indeed *metK* strains tend to accumulate *lrp* mutations rapidly (61).

Lrp AS A CHROMOSOME ORGANIZER

Lrp is known as a global regulator. However, a more profitable way to think of it may be as a determinant of DNA structure, such that the changes in DNA structure when Lrp is reduced or missing are the actual determinants of altered gene expression (23, 72).

One of the first problems cells faced in evolution was the elaboration of a system for packaging the huge and unwieldy DNA molecule. In eukaryotes, packaging involves histones. In bacteria as well, packaging probably involves basic DNA-binding proteins. Until recently, some of these proteins, e.g. Crp, IHF, H-NS, and Lrp, had been considered specific regulators only, rather than chromosome organizers. The idea seems obvious that basic proteins, such as HU and Fis, that bind to DNA with no apparent site specificity, affect its structure (24, 93). However, even proteins that bind specifically to certain sites bend DNA (29, 98).

The conformation of DNA must result from the effects of many protein molecules that are bound to the DNA and thereby bend it. Indeed, cellular DNA might not have any uncoated surfaces in vivo. Moreover, regulation may not involve merely the correct positioning of small proteins on DNA. The cytoplasm is sufficiently crowded that these small, basic molecules must find their way through a collection of other proteins to the DNA. The interactions between these proteins, and the bending they cause once they reach the appropriate site, may have a major effect on gene transcription (17, 80, 121, 122).

Consequently, because of its small, basic, DNA-bending nature and the large number of these molecules per cell (~3000), Lrp must be one of the determinants of chromosome structure. As discussed above, this protein also strongly affects expression of several operons, and the strength of this effect may depend on where Lrp binds in relation to the gene.

We normally detect effects of regulators by the changes they cause in gene regulation. However, these regulators may bind at many sites without causing a change in gene expression, and as a result this binding would be undetectable by methods now in use. In other words, Lrp probably binds DNA at many sites but with varying affinity. If the promoter region of a given gene (e.g. *ilvIH, serA*) contains a high-affinity site, Lrp binding is likely to have a strong effect on expression. If the promoter contains a low-affinity site, Lrp binding might have a lesser effect. Thus, Lrp may play a dual role in cell physiology: as a specific regulator of certain operons whose binding affects expression significantly and as a less-specific DNA-wrapping and -organizing protein. Changes in expression in mutants may not indicate a direct regulatory role of the missing protein, but rather result from structural alterations in less-thoroughly wrapped DNA.

This dual role is probably also the case for Crp, which is usually thought of only as a regulatory protein. Its regulatory role is clearly understood, in part because the known actions of Crp depend on an effector cAMP, whose concentration varies in an intelligible manner according to the physiological conditions of the cell. Likewise, the concentration of Crp varies as well (46). Lrp and Crp differ in that so far no cofactor has been shown for the many non-leucine-dependent Lrp-regulated genes. However no search has been made for Crp-regulated genes that are not affected by cAMP.

The evolution of cells and chromosome structure likely proceeded hand-in-hand. Each new control on gene expression must have affected a structured DNA molecule, albeit one different from that in the present cell. The establishment of a new binding protein during evolution might then have changed the characteristics of the cell greatly—much more than would be expected by addition of a single new factor. In *E. coli*, Lrp purportedly helps convert the cell from metabolism needed inside a host to metabolism needed in independent growth in impoverished media (25, 60, 61). The cell could have acquired this ability quite abruptly with the advent of Lrp during evolution.

Proteins like Lrp may bind in a physiologically significant but nonregulatory manner at DNA sites other than promoters. The observation that Lrp retards 5 of the 11 *Hin*fI fragments from pBR328 (RT Lin & EB Newman, unpublished observations) supports this hypothesis. In vivo DNA-protection studies are generally done with promoter DNA from suspected target operons. Such studies should be done with coding DNA, particularly genes containing sequences similar to the proposed Lrp consensus binding site.

Interactions of Lrp With DNA: Bending and Macromolecular Organizations

A circular permutation assay has shown that Lrp bends the region upstream of *ilvIH* up to 135° depending on the number of molecules bound (115). The region to which it binds is very AT-rich and thus may exhibit an intrinsic bend in the absence of Lrp (19). Several investigators have described long binding sites for Lrp consisting of 100 bp or more, much longer than sites recognized by LacI. Promoters for *ilvIH* (19), *serA* (59), *pap* (16), *lysU* (34, 62), and *fim* (31) contain such sites. The length of these regions might indicate that several molecules of Lrp bind at each promoter or that Lrp is involved in a large complex of proteins that regulates transcription, similar to that seen in TFII-based eukaryotic transcription complexes. Such lengths could also mean that Lrp, alone or with other proteins, forms a structure around which DNA is wrapped (19, 31).

Does Lrp Recognize a Particular Sequence?

Three research groups have proposed closely related consensus sequences for Lrp binding, all of which are extremely AT rich. Rex et al (91) proposed that Lrp binds to an asymmetric 12-bp sequence, TTTATTCtNaAT, upstream of the transcription start of the *kbl-tdh* gene. They found this sequence in both orientations upstream of other Lrp-regulated genes. Deletion of this sequence upstream of the *kbl-tdh* operon resulted in high-level constitutive expression, independent of leucine and Lrp, which strongly suggests that it is indeed part of the operator site recognized by Lrp. However, this attribution is less convincing in light of the sequence's absence from or its unlikely position in other genes (63).

Wang & Calvo (67), in a detailed analysis of the six Lrp-binding sites upstream of the *E. coli ilvIH* operon, found that those with strongest affinity for Lrp have the symmetric, putative consensus sequence AGAATTTTATTCT (with a TTT spacer in the middle). These authors analyzed the sequences at various sites in considerable detail.

Using an efficient method for detection of common motifs in unaligned DNA sequences, H Margalit and coworkers analyzed a set of 23 gene sequences whose transcription is under Lrp control and identified the putative consensus sequence (g/a)(g/c)nnnTTTATtCTgG (53). The core, TTTATtCT, is compatible with the two consensus sequences described above; however the flanking regions differ.

The consensus sequences suggested in these studies are so AT rich to suggest that Lrp recognizes AT richness rather than a particular sequence. Gazeau et al note that periodic AT-rich sequences, by forming regions with a narrowed minor groove, might promote bending (34). They screened 50 sequences with AT-rich regions at the required spacing and found several of the known Lrp-regulated genes in this way. They also cite the case of a *B. subtilis* binding protein, AbrB, a general regulator that also seems to bind to extended AT-rich regions on one face of the helix (104).

Recognition of an AT-rich region may not imply recognition of a sequence but may instead imply recognition of a tertiary structure in the DNA. A DNA-binding protein that selectively recognizes cruciform DNA has been isolated from HeLa cells, although nothing is known about its metabolic function (78). Similarly, HU protein specifically recognizes the type of kinked DNA formed at four-way junctions and therefore was thought to resemble the mammalian protein HMG1 (85). A yeast HMG-like protein shared enough similarity with the bacterial HU protein to serve as an HU substitute in *E. coli* (68). Another eukaryote protein, the lymphoid enhancer binding factor (LEF-1), could replace IHF in bending *E. coli* DNA, provided an appropriate binding sequence was available (35).

Lrp might then be a DNA-wrapping protein of somewhat limited specificity. The suggestions above that Lrp forms a core around which DNA bends are very similar to the suggestion that HMG "has DNA wrapped up" (58). Wrapping might not be its entire role in the cell—by virtue of its interaction with leucine, it may also be a regulatory protein. However, Lrp would also have a less-specific role in forming the DNA structure, and the various DNA-organizing proteins may be to some extent interchangeable in that role. Thus, an *lrp* mutant overproducing Crp might be in better physiological shape than the *lrp* mutant itself: It perhaps could not turn on the *glt* operon, for instance, but it might be able to maintain a more normal chromosome structure. If Lrp, Crp, IHF, and H-NS fill similar DNA organizational functions, overproduction of Lrp in a mutant lacking one of the other proteins might improve the cell's physiology.

MOLECULAR ASPECTS OF Lrp FUNCTION

A recent review examined the biochemistry and molecular biology of Lrp in detail (19). We therefore present only a summary of the main principles of Lrp action as they emerge from two very detailed studies—that of J Calvo on the regulation of *ilvIH* and of D Low's group on fimbrial gene expression—as well as from other less-detailed studies.

Definition of Lrp Domains

The Lrp molecule purportedly consists of three domains: a DNA-binding domain in the N-terminal 40% of the protein; a transcription activation domain in the next 40%; and, overlapping this region, a leucine-response domain in the C-terminal third (83). This composition was suggested in a detailed study in which a plasmid-carried *lrp* gene was subjected to in vitro mutagenesis and introduced into a strain carrying a functional chromosomal *lrp* gene and a chromosomal *ilvIH::lacZ* fusion (83). Mutants affecting expression of the fusion, which were selected for loss of response to exogenous leucine (seven alleles), defined a leucine-response domain in the C-terminal third of the molecule. Other mutants with decreased expression of the *ilvIH::lacZ* fusion (15 alleles) were divided into two classes, according to whether crude extracts could retard *ilvIH* DNA. Mutants whose extracts retarded this DNA were located in the C-terminal half of the molecule and were considered activation mutants. For extracts that did not retard *ilvIH*, the mutations were in a DNA-binding domain in the N-terminal third of the molecule, many from a putative helix-turn-helix region.

Interactions of Lrp With Specific Promoters

COMPLEXITY AND VARIETY OF Lrp-REGULATED PROMOTERS The Lrp-regulated promoters studied in detail have provided us with two specific, beautiful models. These models are very different, and neither of them seems to be generalizable to all Lrp-regulated promoters. However, Lrp is one of several nonhistone general DNA-binding proteins, and it can form complexes with a variety of proteins, some specific to the particular promoter being studied, some specific to a group of promoters, and some much more general. The actual mechanism might then depend on the particular assortment of protein-binding sites that the promoter includes.

The complexity of promoters is indicated by the example of the action of MalT at the *malP* promoter (22). This DNA clearly has four MalT-binding sites and a Crp site. However, only three of the MalT-binding sites are functional—neither the last MalT site nor the Crp site, clearly identified though they are by protection studies, seem to have a physiological role. The first MalT-binding site positions MalT for interaction with RNA polymerase, an understandable position for an activator. The other two functional sites bind MalT with similar affinity, but their function—and that of the other proteins that may form part of the complex—is not known.

These questions are particularly interesting for Lrp function because Lrp interacts with DNA in so many ways—starting with the fact that it activates at some promoters and represses at others. The eukaryote literature describes several cases in which function of a binding protein depends on the other proteins in a complex. Thus, promoter selectivity of mouse and human UBF depends on a second protein, SL1 (11). The demonstration that an HMG-like protein can switch the *Drosophila* protein Dorsal from an activator to a repressor is a particularly elegant example (55), the more so as HMG proteins have been recognized as architectural elements in the assembly of nucleoprotein structures (37, 54). A similar situation was described for prokaryotes by Adhya et al, who showed that defective promoters could become functional through the formation of a multiprotein complex at the promoter (1). We suggest that Lrp cannot be considered on its own, and that studies of Lrp alone at individual promoters, interesting thought they are, tell only a small part of the story.

WHAT IS THE MECHANISM OF LEUCINE ACTION? Lrp was discovered as the mediator of leucine effects, and Lrp is indeed a leucine-binding protein. Equilibrium dialysis experiments show that Lrp in vitro acts as a leucine-binding protein with a dissociation constant of 1.22×10^{-5} and a value of 2 mol leucine bound per mole Lrp (67). Measurements of intracellular concentrations of amino acids are notoriously difficult; nonetheless the level estimated by Quay et al (86) was thought to be in the range of sensitivity needed to affect binding to *ilvIH*, which is affected by leucine in vivo, and to *gltBDF*, which is not sensitive to leucine in vivo (19).

How can we explain leucine-insensitive operons when the cell contains enough leucine to bind to Lrp and to affect transcription in vitro, even of promoters known to be leucine-sensitive in vitro? Ernsting et al suggested that Lrp binds to different promoters with different affinities, and leucine will affect binding only of those operons with low affinity for Lrp (19, 26). They showed that binding of Lrp, even to the *gltBDF* promoter, is clearly modulated by leucine but that, at sufficiently high Lrp concentration, the effect of 30 mM leucine can be counteracted. Whether this would also be true for leucine-sensitive promoters remains to be tested.

Leucine may very well alter the conformation of Lrp. In this case, the altered conformations might bind differently to a multiprotein complex of one type or another. Leucine-insensitive promoters might then be those to which both forms of Lrp can bind, and leucine-sensitive promoters ones that discriminate between the two forms. Binding and mobility-shift studies do not elucidate the binding position. The addition of leucine might displace Lrp by a certain number of base pairs, rather than dislodge it, as is seen in the arabinose operon (97).

In fact, leucine reversed binding only partially, particularly at low Lrp concentration and leucine concentrations well above physiological levels (26, 62, 92, 119, 120). Therefore, the true Lrp effector may not be leucine itself but a leucine derivative, possibly leucyl-tRNA^{leu}, which has been implicated in the repression of *livJ* (87).

MANY Lrp-REGULATED OPERONS HAVE VERY LONG Lrp-BINDING SITES: *ilvIH* AS A PROTOTYPE   The region upstream of *ilvIH* contains six binding sites for Lrp organized as an upstream region between -255 and -215. This area includes two high-affinity Lrp-binding sites and a lower-affinity downstream region (-101 to -56) that encompasses four Lrp-binding sites (92, 114). Lrp is thought to bind cooperatively to several sites and thereby activate transcription (92, 114–116). The presence of this long binding site suggests the possibility that DNA is wrapped around a core of protein, perhaps made of Lrp, but perhaps involving one or more other proteins. Long binding sites of this type have been described also for *lysU* (34, 62), *serA* (59), and *sdaA* (59). These in vitro binding studies all used Lrp as the sole binding protein. However, the promoter initiation complex likely involves several other proteins.

The *ilvIH* promoter has two in vitro transcription start sites, P1, which resides 31 bp upstream of the ATG codon and begins the coding sequence, and P2, some 60 bp farther upstream; both lie within the downstream region defined by binding studies (119). In vitro, P2 is repressed and P1 is activated 2.7-fold by Lrp (half-maximal activation at 15 nM Lrp) (119). In vivo, the parental strain grown in glucose minimal medium transcribes *ilvIH* from P1 only. Adding leucine during growth decreases transcription 4- to 7-fold. In the *lrp* mutant, transcription from P1 is decreased even further (7- to 14-fold).

No in vivo transcription from P2 has been reported. The *serA* promoter also has two transcription start sites, P1, 45 bp from the translation start site, and P2, 93 bp farther upstream (59), but both of these are used in vivo. P2 is repressed by Lrp and is therefore only used in the *lrp* mutant or in growth medium in which Lrp is made at a low level, e.g. in LB broth. P1 is activated by Lrp and therefore used in wild-type cells grown in glucose minimal medium. More fragmentary evidence is available for *gltBDF* (26), *sdaA* (59), and *tdh* (91).

KNOWN OPERONS REGULATED BY Lrp

Several specific members of the leucine/Lrp regulon have been identified. They have been found among genes whose expression is affected by exogenous leucine; among genes coding for proteins whose synthesis rate, as determined on 2D gels, is altered in an *lrp* mutant; and among λplacMu insertions in Lrp-regulated genes identified by means of the insertion mutation phenotype or by direct determination of the flanking DNA sequences. A few more Lrp-regulated operons were established by good guesses (*lrp*, *gcv*THP, *malT*, *lacZYA*). Table 1 lists the genes known to be regulated by Lrp as of November 1994, and these are discussed below. Given the recent interest in Lrp and the availability of an easily transduced *lrp*::Tn*10* mutation, many investigators are beginning to screen for regulation of genes of interest by Lrp.

Biosynthetic Operons

ilvIH (ISOLEUCINE AND VALINE BIOSYNTHESIS) This operon codes for acetohydroxy acid synthase III, one of three isoenzymes catalyzing the first step of branched-chain amino acid synthesis. Lrp activates transcription of *ilvIH* 30-fold, and exogenous leucine almost completely abolishes activation (84, 92, 103, 119). A strain lacking Lrp is physiologically deficient in the enzyme, although this deficiency has little phenotypic effect, presumably because the isoenzyme AHAS I, coded for by *ilvBN*, is expressed at a sufficient level in the absence of Lrp. [AHAS II is not functional in *E. coli* K-12 because of a frame-shift mutation in *ilvG* (54).] A well-characterized protein known as *ilvIH*-binding protein (42, 92) turned out to be Lrp, so the interactions of Lrp at the *ilvIH* promoter is by far the best studied of the Lrp-DNA interactions.

serA (SERINE BIOSYNTHESIS) The *serA* gene codes for phosphoglycerate dehydrogenase, the first enzyme specific to serine biosynthesis. Its transcription is activated sixfold by Lrp, and leucine reduces expression twofold. The sixfold decrease in an *lrp* mutant does not make the strain auxotrophic for serine. However, SerA activity seems to be limiting, because addition of serine to the medium increases the growth rate (3). Furthermore, serine auxotrophy is observed in several double mutants such as *lrp relA* (3).

leuABCD (LEUCINE BIOSYNTHESIS) Leucine synthesis seems to be limiting in *lrp* mutants, because adding leucine to the medium increases the growth rate

(3, 61). Evidence that the *leuABCD* operon is activated by Lrp comes from studies of Lrp-regulated λplacMu insertions that confer leucine auxotrophy and are located in the *leuABCD* operon at 2 min (61, 108). These fusions show 11-fold stimulation by Lrp (61). The *leuABCD* operon also possesses an attenuator that results in higher expression when leucine is limiting (118), and leucine limitation stimulates *leu::lacZ* expression, even in the absence of Lrp (61). As a result, it is difficult to evaluate possible effects of exogenous leucine on the Lrp component of *leu* regulation. This is the only clearly documented case of a transcriptional effect of exogenous leucine in the *lrp* mutant.

glyA (GLYCINE BIOSYNTHESIS)   The *glyA* gene codes for the enzyme serine hydroxymethyl transferase, forming glycine and N5,10-methylene-tetrahydrofolate from L-serine. In wild-type *E. coli* growing in minimal glucose medium, this pathway is the principal source of glycine, as evidenced by the absolute glycine requirement of *glyA* mutants (82). Studies with a *glyA::lacZ* fusion indicate that Lrp represses *glyA* transcription (M San Martano & EB Newman, unpublished results).

gltBDF (GLUTAMATE SYNTHASE)   The *gltBDF* operon codes for a regulatory protein (GltF) and the two subunits of glutamate synthase, an enzyme that, together with glutamine synthetase, is responsible for assimilation of ammonium at low external concentrations. In the *lrp* mutant, little or no glutamate synthase activity is detected, and the region corresponding to GltD is absent in 2D protein gels (25). This deficiency does not cause auxotrophy at high ammonium concentrations, but the *lrp* mutant, like *gltD* mutants, cannot use arginine or ornithine as a nitrogen source. The expression of a *gltBDF::lacZ* fusion was decreased 2.2-fold in the presence of 10 mM leucine and 44-fold by Lrp deficiency (26).

Degradative Operons

sdaA (L-SERINE DEAMINASE)   *E. coli* K-12 makes two quite similar L-serine deaminases that convert L-serine to pyruvate and ammonium (107). The regulation of L-SD I, the product of the *sdaA* gene, is complex. Its expression is stimulated by heat shock, anaerobiosis, ultraviolet irradiation, and leucine and is repressed 7- to 10-fold by Lrp in glucose minimal medium (45, 70). In the *lrp* mutant, the higher level of L-SD permits growth with L-serine as a carbon source. L-SD II is the product of the *sdaB* gene, second gene of the *sdaCB* operon. It is expressed only in rich medium, where it is regulated primarily by Crp (99, 100). Its expression is not significantly affected by exogenous leucine or by an *lrp* mutation. However, transcription of the upstream *sdaC* gene is activated by Lrp (see below). The apparent lack of an effect on *sdaB* may partly result from its very poor ribosome-binding site, with enzyme levels too close to background to detect Lrp activation (100).

kbl-tdh (THREONINE DEGRADATION)   *E. coli* can degrade threonine to acetyl-CoA and glycine in two steps: oxidation to 2-amino-3-ketobutyrate (AKB) followed by cleavage. The two enzymes, threonine dehydrogenase and AKB-CoA lyase, are coded for by the *tdh* and *kbl* genes, respectively, which form an operon. Lrp represses the *kbl-tdh* operon by approximately 20-fold, and leucine partially relieves repression (60, 91). This operon is not normally expressed in minimal medium, as evidenced by direct enzyme assay and by the fact that *glyA* mutants, which lack serine hydroxymethyltransferase and thus cannot form glycine from serine, have an absolute glycine requirement that cannot normally be satisfied by threonine (28). Overexpression of the *kbl-tdh* operon permits formation of glycine and serine from threonine and enables the cell to grow with threonine as a carbon source (20, 90).

lacZ (β-GALACTOSIDASE)   Transcription of the *lacZ* gene is reduced by about 30% in the *lrp* mutant (108). However, the growth rate on 0.2% lactose is not decreased in the absence of Lrp. Whether this modest transcriptional stimulation reflects a direct interaction of Lrp with the *lac* promoter or an indirect effect is unknown.

Other Metabolic Operons

gcvTHP (GLYCINE CLEAVAGE)   The *gcvTHP* operon codes for the glycine-cleavage system, which converts glycine to N5,10-methylene-tetrahydrofolate, NH_3 and CO_2, producing both ammonium and one-carbon units from glycine. The operon is activated by Lrp: An *lrp* mutant makes 20 times less enzyme and is physiologically defective in glycine cleavage, i.e. unable to obtain either one-carbon units or ammonium from glycine (61). Exogenous leucine has no significant effect on *gcv* expression. Lack of glycine cleavage activity in the *lrp* mutant profoundly affects the mutant's physiology.

pntAB (PYRIDINE NUCLEOTIDE TRANSHYDROGENASE)   Pyridine nucleotide transhydrogenase, the product of the *pnt* gene, catalyzes the reaction

$$nH^+_{in} + NADPH + NAD \leftrightarrow nH^+_{out} + NADP + NADH,$$

permitting *E. coli* to replenish its NADPH pool independently of the pentose phosphate shunt or, alternatively, to extrude protons at the expense of NADPH. Transcription of the *pntAB* operon is decreased by leucine four- to fivefold (36) and decreased four- to sixfold in the *lrp* mutant (3). Thus the *pntAB* operon is activated by Lrp. Activation is reversed by leucine and, to a lesser degree, by alanine and methionine (36). The decreased Pnt level in the *lrp* mutant has no known effect on the phenotype. However, lack of both Pnt and Lrp causes unexpected auxotrophies.

lysU (LYSYL-tRNA SYNTHETASE)   *E. coli* has two lysyl-tRNA synthetases, LysU and LysS. This is unusual since each of the 19 other amino acids has a single synthetase. Regulation of the two genes is very different: *lysS* expression is constitutive and Lrp-independent, whereas the *lysU* gene is highly regulated. The latter's expression is stimulated by heat shock, anaerobiosis, low external pH, and leucine and is repressed ninefold by Lrp (33, 47, 56, 62, 69). Leucine relieves Lrp repression and elicits a fivefold increase in *lysU* mRNA (47, 56). Alanine also relieves repression (43). Heat induction of *lysU* is independent of Lrp but requires a small mRNA immediately downstream of the initiation codon (69). Under most laboratory growth conditions, the constitutive synthetase LysS seems to be sufficient for normal growth; LysU alone (in a Δ*lysS* mutant), on the other hand, does not support wild-type growth rates, especially at low temperatures (33).

Transport Operons

livJ AND *livKHMGF* (LEUCINE TRANSPORT)   *E. coli* possesses multiple transport systems for leucine. Two of these have high affinity for leucine, involve periplasmic binding proteins of different specificities, and are regulated by Lrp. The *livJ* gene product binds leucine, isoleucine, and valine, whereas the *livK* gene product is specific to leucine (4, 5, 51). These two systems share a set of membrane components, products of the *livHMGF* genes (in the *livKHMGF* operon). The two operons, adjacent on the genetic map, are both repressed by high leucine concentrations, and this repression requires Lrp. This regulation has been established by direct measurements of transport (46, 47, 79), by analysis of proteins on 2D gels (25), by a study of *livJ::lacZ* and *livK::lacZ* fusions (41), and by identification via inverse PCR and sequencing of the Lrp-regulated insert in strain CP36 (61, 108). This pattern of regulation—repression only in the presence of both leucine and Lrp—suggests that the active repressor is effector-bound Lrp. Biochemical studies are needed to determine whether Lrp action on these two operons is direct. Early work indicated that the LivJ transport system can also be repressed by exogenous methionine (38). Similarly, studies of homoserine transport (apparently via the LivJ system), showed that exogenous methionine or alanine could repress expression of the transport system, albeit to a lesser degree than exogenous leucine (109).

sdaC (SERINE TRANSPORT)   *E. coli* also possesses multiple transport systems for L-serine. One, which is highly specific to serine but is as yet identified only biochemically and not genetically, is induced by leucine (40). On the other hand, *sdaC* has been identified only by genetics, but may code for this serine-uptake system, as judged by the gene's nucleotide sequence and the altered transport characteristics of an *sdaC* mutant (99). Exogenous leucine stimulates expression of *sdaC*, but requires Lrp to do so. As in the case of the *liv* operons, the pattern of regulation is unusual—activation only in the simultaneous presence of leucine and Lrp—and suggests that the activator is effector-bound Lrp. Again, biochemical data are needed to determine whether Lrp action is direct.

oppABCD (OLIGOPEPTIDE TRANSPORT)   *E. coli* and *S. typhimurium* have several oligopeptide permeases that facilitate uptake of di and tripeptides, as well as larger peptides (2). The system specified by the *oppABCD* operon has a periplasmic binding protein and transports a wide range of tripeptides. In *E. coli* (but not in *S. typhimurium*), its expression is increased by the presence of leucine in the growth medium and by the presence of alanine (4). In an *lrp* mutant, the operon has high constitutive expression, suggesting that it is repressed by Lrp (9).

malEFG, malK-lamB-malM, AND *malT* (MALTOSE TRANSPORT)   Maltose uptake in *E. coli* requires four proteins, products of the *malEFGK* genes, which are arranged in two operons, *malEFG* and *malK-lamB-malM,* and transcribed divergently from a 297-bp intergenic region (10). Transcription of the *mal* genes is under the control of several factors, including Crp and a specific transcriptional activator, MalT. Transcription of the two uptake operons is decreased 50–70% in the *lrp* mutant grown in glycerol (108). However, despite the slightly lower expression of the maltose-transport proteins, the *lrp* mutant grows normally in 0.2% maltose. Lrp activation may only be important at low (noninducing) maltose concentrations, where it may help the cell scavenge the sugar and form maltose triose, the endogenous inducer (88). The transcription of *malT* is also decreased 50% in the *lrp* mutant. This decrease alone should reduce expression of the other *mal* operons. However, Lrp probably also affects the individual *malE* and *malK* promoters, as suggested by two observations: First, loss of Lrp does not reduce the expression of the MalT-regulated *malPQ* operon, which codes for maltodextrin phosphorylase and amylomaltase, enzymes involved in maltose catabolism. Second, leucine decreases the expression of the *malE* and *malK* operons by approximately 35% but does not effect *malT* transcription (108). The activation of transcription by MalT involves multiple interactions between proteins and between proteins and DNA—as elucidated in the elegant and detailed studies of Raibaud and his group (88).

ompF AND *ompC* (OUTER MEMBRANE PORINS)   The outer membrane of gram-negative bacteria is relatively permeable to molecules of molecular weight less than approximately 600, allowing their passage into the periplasmic space. This permeability is primarily via porins, outer-membrane proteins that form hydrophilic channels of broad specificity. In *E. coli* and *S. typhimurium,* the principal porins are the *ompC* and *ompF* gene products. Their synthesis is under complex control and includes a response to the osmolarity of the medium: Synthesis of OmpC increases at high osmolarity and that of OmpF decreases (49). A comparison of 2D gels of proteins from wild-type and *lrp* mutant cells suggests that synthesis of OmpC is repressed by Lrp, whereas that of OmpF is stimulated (25). Lrp may also regulate expression of *micF,* which codes for an antisense RNA that regulates *ompF* expression. Regulation of *ompF* by Lrp might therefore be indirect.

Operons Coding for Pili and Fimbriae

E. coli can express many types of external appendages known as fimbriae, a term used interchangeably with pili (57). Which appendages are produced under a given condition depends on an astonishing variety of regulatory mechanisms, which include Lrp in most cases studied. Several common laboratory strains do not express these genes, many of which are carried on plasmids. This work was extensively reviewed recently (19, 65).

Other Operons

osmY Lrp is involved in the regulation of the *osmY* gene in a rather complex manner. Expression of *osmY* is normally induced on entry into stationary phase, particularly in LB broth–grown cells, under the control of a stationary-phase σ factor, σ^S. It is also induced in media containing high salt (101). During growth in LB broth, expression of an *osmY::lacZ* fusion is induced earlier and to a greater extent in the *lrp* mutant than in the wild-type. The effect in minimal medium, where *osmY* expression is not detectable during exponential phase, was even more dramatic: Induction of an *osmY::lacZ* fusion was increased fivefold in an *lrp* mutant (52). Strains defective in the gene coding for the stationary-phase σ factor, *rpoS,* expressed *osmY* to a very small degree. However, an *lrp rpoS* double mutant expressed *osmY* constitutively throughout the growth cycle. This was interpreted as indicating that Lrp represses expression from *osmY* during exponential phase and that σ^S overcomes this repression (52).

o489 Lrp also regulates o489, an unidentified open reading frame (ORF). This sequence resides immediately downstream of the *uspA* gene, which is not regulated by Lrp (77). Inverse PCR experiments, as well as screening for environmentally regulated methylation of GATC sites (39), have shown that this ORF carries the Lrp-regulated insert in strain CP59 (61, 108). The deduced o489 product is exceptionally hydrophobic, suggesting that it may be membrane-bound as a channel or porin.

Any *Annual Review* chapter, as well as any article cited in an *Annual Review* chapter, may be purchased from the Annual Reviews Preprints and Reprints service.
1-800-347-8007; 415-259-5017; email: arpr@class.org

Literature Cited

1. Adhya S, Gottesman M, Garges S, Oppenheim A. 1993. Promoter resurrection by activators—a minireview. *Gene* 132:1–6
2. Alfoldi L, Kerekes E. 1964. Neutralization of the amino acid sensitivity of RCrel *Escherichia coli*. *Biochem. Biophys. Acta* 91:155–57
3. Ambartsoumian G, D'Ari R, Lin RT, Newman EB. 1994. Altered amino acid metabolism in *lrp* mutants of *Escherichia coli* and their derivatives. *Microbiology* 148:1737–44
4. Anderson JJ, Oxender DL. 1977. *Escherichia coli* transport mutants lacking binding protein and other components of the branched-chain amino acid transport systems. *J. Bacteriol.* 130:384–92
5. Anderson JJ, Quay SC, Oxender DL. 1976. Mapping of two loci affecting the regulation of branched-chain amino acid transport in *Escherichia coli* K-12. *J. Bacteriol.* 126:80–90
6. Andrews JC, Blevins TC, Short SA. 1986. Regulation of peptide transport in *Escherichia coli*: induction of the *trp*-linked operon encoding the oligopeptide permease. *J. Bacteriol.* 165:428–33
7. Andrews JC, Short SA. 1986. *opp-lac* operon fusions and transcriptional regulation of the *Escherichia coli trp*-linked oligopeptide permease. *J. Bacteriol.* 165:434–42
8. Arst HN Jr. 1994. Integrator gene in *Aspergillus nidulans*. *Nature* 262:231–34
9. Austin EA, Andrews JC, Short SA. 1989. Selection, characterization and cloning of *oppI*, a regulator of the *E. coli* oligopeptide permease operon. *Abstr. Molecular Genetics. Bacteria Phages*, p. 153. Cold Spring Harbor, NY: Cold Spring Harbor Lab.
10. Bedouelle H, Schmeissner E, Hofnung M, Rosenberg M. 1982. Promoters of the *malEFG* and *malK-lamB* operons in *Escherichia coli* K12. *J. Mol. Biol.* 161:519–31
11. Bell SP, Jantzen HM, Tjian R. 1990. Assembly of alternative multiprotein complexes directs rRNA promoter selectivity. *Genes Dev.* 4:943–54
12. Bilge SS, Apostol JM Jr, Fullner KJ, Moseley SL. 1993. Transcriptional organization of the F1845 fimbrial adhesin determinant of *Escherichia coli*. *Mol. Microbiol.* 7:993–1006
13. Blomfield IC, Calie PJ, Eberhardt KJ, McClain MS, Eisenstein BI. 1993. Lrp stimulates phase variation of type 1 fimbriation in *Escherichia coli* K-12. *J. Bacteriol.* 175:27–36
14. Deleted in proof
15. Braaten BA, Blyn LB, Skinner BS, Low DA. 1991. Evidence for a methylation-blocking factor (*mbf*) locus involved in *pap* pilus expression and phase variation in *Escherichia coli*. *J. Bacteriol.* 173:1789–800
16. Braaten BA, Platko JV, van der Woude MW, Simons WH, de Graaf FK, et al. 1992. Leucine-responsive regulatory protein controls the expression of both the *pap* and *fan* pili operons in *Escherichia coli*. *Proc. Natl. Acad. Sci. USA* 89:4250–54
17. Bracco L, Kotlarz D, Kolb A, Diekmann S, Buc H. 1989. Synthetic curved DNA sequences can act as transcriptional activators in *Escherichia coli*. *EMBO J.* 8:4289–96
18. Deleted in proof
19. Calvo JM, Matthews RG. 1994. Leucine-responsive regulatory protein—a global regulator of metabolism in *Escherichia coli*. *Microbiol. Rev.* 58:466–98
20. Chan TTK, Newman EB. 1981. Threonine as a carbon source for *Escherichia coli*. *J. Bacteriol.* 145:1150–53
21. Daniel J, Joseph E, Danchin A. 1984. Role of 2-ketobutyrate as an alarmone in *E. coli* K12: inhibition of adenylate cyclase activity mediated by the phosphoenolpyruvate:glycose phosphotransfererase transport system. *Mol. Gen. Genet.* 193:467–72
22. Danot O, Raibaud O. 1994. Multiple

protein-NA and protein-protein interactions are involved in transcriptional activation by MalT. *Mol. Microbiol.* 14:335–46
23. D'Ari R, Lin RT, Newman EB. 1993. The leucine-responsive regulatory protein: more than a regulator? *Trends Biochem. Sci.* 18:260–63
24. Drlica K, Rouviere-Yaniv J. 1987. Histonelike proteins of bacteria. *Microbiol. Rev.* 51:301–19
25. Ernsting BR, Atkinson MR, Ninfa AJ, Matthews RG. 1992. Characterization of the regulon controlled by the leucine responsive regulatory protein in *Escherichia coli*. *J. Bacteriol.* 174:1109–18
26. Ernsting BR, Denninger JW, Blumenthal RM, Matthews RG. 1993. Regulation of the *gltBDF* operon of *Escherichia coli*: How is a leucine-insensitive operon regulated by the leucine-responsive regulatory protein? *J. Bacteriol.* 175:7160–69
27. Deleted in proof
28. Fraser J, Newman EB. 1975. Derivation of glycine from threonine in *Escherichia coli* K-12 mutants. *J. Bacteriol.* 122:810–17
29. Freundlich M, Ramani N, Mathew E, Sirko A, Tsui P. 1992. The role of integration host factor in gene expression in *Escherichia coli*. *Mol. Microbiol.* 6:2557–63
30. Friedrich MJ, Kinsey NE, Vila J, Kadner RJ. 1993. Nucleotide sequence of a 13.9 kb sequence of the 90 kb virulence plasmid of *Salmonella typhimurium*: the presence of fimbrial biosynthetic genes. *Mol. Microbiol.* 8:543–58
31. Gally DL, Bucker TJ, Bloomfield IC. 1994. The leucine-responsive regulatory protein binds to the *fim* switch to control phase variation of type I fimbrial expression in *Escherichia coli* K-12. *J. Bacteriol.* 176:5665–72
32. Gally DL, Bogan JA, Eisenstein BI, Blomfield IC. 1993. Environmental regulation of the *fim* switch controlling type 1 fimbrial phase variation in *Escherichia coli* K-12: effects of temperature and media. *J. Bacteriol.* 175:6186–93
33. Gazeau M, Delort F, Dessen P, Blanquet S, Plateau P. 1992. *Escherichia coli* leucine-responsive regulatory protein (Lrp) controls lysyl-tRNA synthetase expression. *FEBS Lett.* 300:254–58
34. Gazeau M, Delort F, Fromant M, Dessen PP, Blanquet S, Plateau P. 1994. Structure-function relationship of the Lrp-binding region up stream of *lysU* in *Escherichia coli*. *J. Mol. Biol.* 241:378–89
35. Geise K, Cox J, Grosschedl R. 1992. The HMG domain of lymphoid enhancer factor 1 bends DNA and facilitates assembly of functional nucleoprotein structures. *Cell* 69:185–95
36. Gerolimatos B, Hanson RL. 1978. Repression of *Escherichia coli* pyridine nucleotide transhydrogenase by leucine. *J. Bacteriol.* 134:394–400
37. Grosschedl R, Giese K, Pagel J. 1994. HMG domain proteins: architectural elements in the assembly of nucleoprotein structures. *Trends Genet.* 10:94–99
38. Guardaiola J, De Felice M, Klopotowski T, Iaccarino M. 1974. Multiplicity of isoleucine, leucine, and valine transport systems in *Escherichia coli* K-12. *J. Bacteriol.* 117:382–92
39. Hale WB, van der Woude MW, Low DA. 1994. Analysis of nonmethylated GATC sites in the *Escherichia coli* chromosome and identification of sites that are differentially methylated in response to environmental stimuli. *J. Bacteriol.* 176:3438–41
40. Hama H, Shimamoto T, Tsuda M, Tsuchiya T. 1988. Characterization of a novel L-serine transport system in *Escherichia coli*. *J. Bacteriol.* 170:2236–39
41. Haney SA, Plakto JV, Oxender DL, Calvo JM. 1992. Lrp, a leucine-responsive protein, regulates branched-chain amino acid transport genes in *Escherichia coli*. *J. Bacteriol.* 174:108–15
42. Haughn GW, Squires CH, De Felice M, Lago CT, Calvo JM. 1985. Unusual organization of the *ilvIH* promoter in *Escherichia coli*. *J. Bacteriol.* 163:186–98
43. Hirshfield IN, Bloch PL, Van Bogelen RA, Neidhardt FC. 1981. Multiple forms of lysyl-transfer ribonucleic acid synthetase in *Escherichia coli*. *J. Bacteriol.* 146:345–51
44. Huisman TT, Bakker D, Klaasen P, de Graaf FK. 1994. Leucine-responsive regulatory protein, IS1 insertions, and the negative regulator FaeA control the expression of the *fae* (K88) operon in *Escherichia coli*. *Mol. Microbiol.* 11:525–36
45. Isenberg S, Newman EB. 1974. Studies on L-serine deaminase in *Escherichia coli* K-12. *J. Bacteriol.* 118:53–58
46. Ishizuka H, Hanamura A, Kunimura T, Aiba H. 1994. A lowered concentration of cAMP receptor protein caused by glucose is an important determinant for catabolite repression in *Escherichia coli*. *Mol. Microbiol.* 6:2489–95

47. Ito K, Kawakami K, Nakamura Y. 1993. Multiple control of *Escherichia coli* lysyl-tRNA synthetase expression involves a transcriptional repressor and a translational enhancer element. *Proc. Natl. Acad. Sci. USA* 90:302-6
48. Deleted in proof
49. Kawaji H, Mizuno T, Mizushima S. 1979. Influence of molecular size and osmolarity of sugars and dextrans on the synthesis of outer membrane proteins O-8 and O-9 of *Escherichia coli* K-12. *J. Bacteriol.* 140:843-47
50. Landick R, Anderson JJ, Mayo MM, Gunsalus RP, Mavromara P, et al. 1980. Regulation of high-affinity leucine transport in *Escherichia coli. J. Supramol. Struct.* 14:527-37
51. Landick R, Oxender DL. 1985. The complete nucleotide sequences of the *Escherichia coli* LIV-BP and LS-BP genes. *J. Biol. Chem.* 260:8257-61
52. Lange R, Barth M, Hengge-Aronis R. 1993. Complex transcriptional control of the σ^S-dependent stationary-phase-induced and osmotically regulated *osmY (cis-5)* gene suggests novel roles for Lrp, cyclic AMP (cAMP) receptor protein–cAMP complex, and integration host factor in the stationary-phase response of *Escherichia coli. J. Bacteriol.* 175:7910-17
53. Lau PCK, Forghani F, Labbe D. 1994. The NlaIV restriction and modification genes of *Neisseria lactamica* are flanked by leucine biosynthesis genes. *Mol. Gen. Genet.* 243:24-31
54. Lawther RP, Calhoun DH, Adams CW, Hauser CA, Gray J, Hatfield GW. 1981. Molecular basis of valine resistance in *Escherichia coli* K-12. *Proc. Natl. Acad. Sci. USA* 78:922-25
55. Lehming N, Thanos D, Brickma JM, Ma J, Maniatis T, Ptashne M. 1994. An HMG-like protein that can switch a transcriptional activator to a repressor. *Nature* 371:175-79
56. Leveque F, Gazeau M, Fromant M, Blanquet S, Plateau P. 1991. Control of *Escherichia coli* lysyl-tRNA synthetase expression by anaerobiosis. *J. Bacteriol.* 173:7903-10
57. Levinthal M, Lejeune P, Danchin A. 1994. The H-NS protein modulates the activation of the *ilvIH* operon of *Escherichia coli* K12 by Lrp, the leucine regulatory protein. *Mol. Gen. Genet.* 242:736-43
58. Lilley DMJ. HMG has DNA all wrapped up. *Nature* 282-83
59. Lin R. 1992. *Characterization of the leucine/Lrp regulon in* Escherichia coli *K-12*. PhD thesis. Concordia Univ., Montreal, Quebec, Canada
60. Lin R, D'Ari R, Newman EB. 1990. The leucine regulon of *Escherichia coli*: a mutation in *rblA* alters expression of leucine-dependent metabolic operons. *J. Bacteriol.* 172:4529-35
61. Lin RT, D'Ari R, Newman EB. 1992. λplacMu insertions in genes of the leucine regulon: extension of the regulon to genes not regulated by leucine. *J. Bacteriol.* 174:1948-55
62. Lin R, Ernsting B, Hirshfield IN, Matthews RG, Neidhardt FC, et al. 1992. The lrp gene product regulates expression of *lysU* in *Escherichia coli* K-12. *J. Bacteriol.* 174:2779-84
63. Lobell RB, Schleif RF. 1991. AraC-DNA looping: orientation and distance-dependent loop breaking by the cyclic AMP receptor protein. *J. Mol. Biol.* 218:45-54
64. Deleted in proof
65. Low DA. 1994. Methylation-dependent and Lrp-dependent fimbrial gene regulation in *Escherichia coli*. In *Molecular Genetics of Bacterial Pathogenesis*, ed. VI Miller, JB Kaper, DA Portnoy, RR Isberg. Washington DC: Am. Soc. Microbiol.
66. Madhusudhan KT, Lorenz D, Sokatch JR. 1993. The *bkdR* gene of *Pseudomonas putida* is required for expression of the *bkd* operon and encodes a protein related to Lrp of *Escherichia coli. J. Bacteriol.* 175:3934-40
67. Marasco R, Varcamonti M, La Cara F, Ricca E, de Felice M, Sacco M. 1994. In vivo footprinting analysis of Lrp binding to the *ilvIH* promoter region of *Escherichia coli. J. Bacteriol.* 176:5197-201
68. McGraw TL, Chae C-B. 1993. Functional complementarity between the HMG1-like yeast mitochondrial histone HMM and the bacterial histone-like protein HU. *J. Biol. Chem.* 268:122758-63
69. Nakamura Y, Ito R. 1993. Control and function of lysyl-tRNA synthetases: diversity and co-ordination. *Mol. Microbiol.* 10:225-31
70. Newman EB, Ahmad D, Walker C. 1982. L-Serine deaminase activity is induced by exposure of *Escherichia coli* K-12 to DNA-damaging agents. *J. Bacteriol.* 152:702-5
71. Newman EB, Batist G, Fraser J, Isenberg S, Weyman P, Kapoor V. 1976. The use of glycine as nitrogen source by *Escherichia coli* K-12. *Biochim. Biophys. Acta* 421:97-105
72. Newman EB, D'Ari R, Lin RT. 1992. The leucine-Lrp regulon in E. coli: a

global response in search of a raison d'etre. *Cell* 68:617–19
73. Newman EB, Magasanik B. 1963. The relation of serine-glycine metabolism to the formation of single carbon units. *Biochim. Biophys. Acta* 78:437–48
74. Newman EB, Miller B, Kapoor V. 1974. Biosynthesis of single-carbon units in *Escherichia coli* K-12. *Biochim. Biophys. Acta* 338:529–39
75. Ngeleka M, Jacques M, Martineau-Doize B, Daigle F, Harel J, Fairbrother JM. 1993. Pathogenicity of an *Escherichia coli* O1115:K"V165" mutant negative for F165 fimbriae in septicemia of gnotobiotic pigs. *Infect. Immun.* 61:836–43
76. Nou X, Skinner B, Braaten B, Blyn L, Hirsch D, Low D. 1993. Regulation of pyelonephritis-associated pili phase-variation in *Escherichia coli:* binding of the PapI and the Lrp regulatory proteins is controlled by DNA methylation. *Mol. Microbiol.* 7:545–53
77. Nystrom T, Neidhardt FC. 1992. Cloning, mapping and nucleotide sequencing of a gene encoding a universal stress protein in *Escherichia coli*. *Mol. Microbiol.* 6:3187–98
78. Pearson CE, Ruiz MT, Price GB, Zannis-Hadjopolous M. 1994. Cruciform DNA binding protein in HeLa cell extracts. *Biochemistry* 33:14185–96
79. Penrose WR, Nichoalds GE, Piperno JR, Oxender DL. 1968. Purification and properties of a leucine-binding protein from *Escherichia coli*. *J. Biol. Chem.* 243:5921–28
80. Perez-Martin J, Rojo F, de Lorenzo V. 1994. Promoters responsive to DNA bending: a common theme in procaryotic gene expression. *Microbiol. Rev.* 58:268–90
81. Petitjohn DE. 1988. Histone-like proteins and bacterial chromosome structure. *J. Biol. Chem.* 263:12793–96
82. Pizer LI. Glycine synthesis and metabolism in *Escherichia coli*. *J. Bacteriol.* 89:1145–50
83. Platko JV, Calvo JM. 1993. Mutations affecting the ability of *Escherichia coli* Lrp to bind DNA, activate transcription, or respond to leucine. *J. Bacteriol.* 175:1110–17
84. Platko JV, Willins DA, Calvo JM. 1990. The *ilvIH* operon of *Escherichia coli* is positively regulated. *J. Bacteriol.* 172:4563–70
85. Pontiggia A, Negri A, Beltrame M, Bianchi ME. 1993. Protein HU binds specifically to kinked DNA. *Mol. Microbiol.* 7:343–56
86. Quay SC, Dick TE, Oxender DL. 1977. Role of transport systems in amino acid metabolism: leucine toxicity and the branched-chain amino acid transport systems. *J. Bacteriol.* 129:1257–65
87. Quay SC, Kline EL, Oxender DL. 1975. Role of leucyl-tRNA synthetase in regulation of branched-chain amino-acid transport. *Proc. Natl. Acad. Sci. USA* 72:3921–24
88. Raibaud O, Richet E. 1987. Maltotriose is the inducer of the maltose regulon of *Escherichia coli.* *J. Bacteriol.* 169:3059–61
89. Deleted in proof
90. Ravnikar PD, Somerville RL. 1987. Genetic characterization of a highly efficient alternative pathway of serine biosynthesis in *Escherichia coli*. *J. Bacteriol.* 169:22611–17
91. Rex JH, Aronson BD, Somerville RL. 1991. The *tdh* and *serA* operons of *Escherichia coli:* mutational analysis of the regulatory elements of leucine-responsive genes. *J. Bacteriol.* 173:5944–53
92. Ricca E, Aker DA, Calvo JM. 1989. A protein that binds to the regulatory region of the *Escherichia coli ilvIH* operon. *J. Bacteriol.* 171:1658–64
93. Rouviere-Yaniv J, Yaniv M, Germond JE. 1979. E. coli DNA-binding protein HU forms nucleosome-like structure with circular double-stranded DNA. *Cell* 17:265–74
94. Sacco M, Ricca E, Marasco R, Paradiso R, De Felice M. 1993. A stereospecific alignment between the promoter and the *cis*-acting sequence is required for Lrp-dependent activation of *ilvIH* transcription in *Escherichia coli*. *FEMS Microbiol. Lett.* 107:331–36
95. Deleted in proof
96. Deleted in proof
97. Schleif R. 1987. The L-arabinose operon. In Escherichia coli *and* Salmonella typhimurium: *Cellular and Molecular Biology*, pp. 1473–81, ed. FC Neidhardt, JL Ingraham, KB Low, B Magasanik, M Schaechter, HE Umbarger. Washington, DC: Am. Soc. Microbiol.
98. Schultz SC, Shields GC, Steitz TA. 1991. Crystal structure of a CAP-DNA complex: The DNA is bent by 90°. *Science* 253:1001–7
99. Shao ZQ, Lin RT, Newman EB. 1994. Sequencing and characterization of the *sdaC* gene and identification of the *sdaCB* operon in *E. coli* K-12. *Eur. J. Biochem.* 222:901–7
100. Shao ZQ, Newman EB. 1993. Sequencing and characterization of the *sdaB* gene from *Escherichia coli* K-12. *Eur. J. Biochem.* 212:777–84

101. Siegele DA, Kolter R. 1992. Life after log. *J. Bacteriol.* 174:345-48
102. Smith DW, Stine WB, Svitil AL, Bakker A, Zyskind JW. 1992. *Escherichia coli* cells lacking methylation-blocking factor (leucine-responsive regulatory protein) have precise timing of initiation of DNA replication in the cell cycle. *J. Bacteriol.* 174:3078-82
103. Squires CH, De Felice M, Wessler SR, Calvo JM. 1981. Physical characterization of the *ilvIH* operon of *Escherichia coli* K-12. *J. Bacteriol.* 147: 797-804
104. Strauch HA, Hoch JA. 1993. Transition state regulators: sentinels of *Bacillus subtilis* post-exponential phase gene expression. *Mol. Microbiol.* 7:337-42
105. Su H, Lang BF, Newman EB. 1989. L-Serine degradation in *Escherichia coli* K-12. Cloning and sequencing of the *sdaA* gene. *J. Bacteriol.* 171:5095-102
106. Su H, Moniakis J, Newman EB. 1993. Use of gene fusions of the structural gene *sdaA* to purify L-serine deaminase 1 from *Escherichia coli* K-12. *Eur. J. Biochem.* 211:521-27
107. Su H, Newman EB. 1991. A novel L-serine deaminase activity in *Escherichia coli* K-12. *J. Bacteriol.* 173: 2473-80
108. Tchetina E, Newman EB. 1995. Identification of Lrp-regulated genes by inverse PCR and sequencing: regulation of two *mal* operons of *E. coli* by leucine-responsive regulatory protein. *J. Bacteriol.* 177:2679-83
109. Templeton BA, Savageau MA. 1974. Transport of biosynthetic intermediates: regulation of homoserine and threonine uptake in *Escherichia coli*. *J. Bacteriol.* 120:114-20
110. Ursini MV, Arcari P, De Felice M. 1981. Acetohydroxy acid synthase isoenzymes of *Escherichia coli* K-12: a *trans*-acting regulatory locus for *ilvIH* gene expression. *Mol. Gen. Genet.* 181:491-96
111. Uzan M, Danchin A. 1978. Correlation between the serine sensitivity and the derepressibility of the *ilv* genes in *Escherichia coli relA⁻* mutants. *Mol. Gen. Genet.* 165:21-30
112. Van der Woude MW, Braaten BA, Low DA. 1992. Evidence for global regulatory control of pilus expression in *Escherichia coli* by Lrp and DNA methylation: model building based on analysis of pap. *Mol. Microbiol.* 6:2429-35
113. Van der Woude MW, Low DA. 1994. Leucine-responsive regulatory protein and deoxyadenosine methylase control the phase variation and expression of the *sfa* and *daa* pili operons in *Escherichia coli*. *Mol. Microbiol.* 11:605-18
114. Wang Q, Calvo JM. 1993. Lrp, a global regulatory protein of *Escherichia coli*, binds co-operatively to multiple sites and activates transcription of *ilvIH*. *J. Mol. Biol.* 229:306-18
115. Wang Q, Calvo JM. 1993. Lrp, a major regulatory protein in *Escherichia coli*, bends DNA and can organize the assembly of a higher-order nucleoprotein structure. *EMBO J.* 12:2495-501
116. Wang Q, Sacco M, Ricca E, Lago CT, De Felice M, Calvo JM. 1993. Organization of Lrp-binding sites upstream of *ilvIH* in *Salmonella typhimurium*. *Mol. Microbiol.* 7:883-91
117. Wang Q, Wu J, Friedberg D, Platko J, Calvo JM. 1994. Regulation of the *Escherichia coli lrp* gene. *J. Bacteriol.* 176:1831-39
118. Wessler SR, Calvo JM. 1981. Control of leu operon expression in *Escherichia coli* by a transcription attenuation mechanism. *J. Mol. Biol.* 149:579-79
119. Willins DA, Calvo JM. 1992. In vitro transcription from the *Escherichia coli ilvIH* promoter. *J. Bacteriol.* 174:7648-55
120. Willins DA, Ryan CW, Platko JV, Calvo JM. 1991. Characterization of Lrp, an *Escherichia coli* regulatory protein that mediates a global response to leucine. *J. Biol. Chem.* 266:10768-74
121. Wu H-M, Crothers DM. 1984. The locus of sequence-directed and protein-induced DNA bending. *Nature* 308:509-13
122. Yamada H, Yoshida T, Tanaka K, Sasakawa C, Mizuno T. 1991. Molecular analysis of the *Escherichia coli hns* gene encoding a DNA-binding protein which preferentially recognizes curved DNA sequences. *Mol. Gen. Gen.* 230: 332-36

MICROBIOLOGY TO 10,500 METERS IN THE DEEP SEA

A. Aristides Yayanos

Scripps Institution of Oceanography, University of California San Diego, La Jolla, California 92093-0202

KEY WORDS: high pressure, microbial ecology, growth physiology, membranes, *PTk* diagram

CONTENTS

INTRODUCTION	778
DEEP-SEA HABITATS	779
METHODS OF SAMPLING FOR MICROBIOLOGY	780
HIGH PRESSURE LABORATORY TECHNIQUES	782
Pressure Vessels	782
Colony Formation at High Pressure	782
High-Pressure Growth Curves	783
Pressurized Temperature-Gradient Instruments	785
THE *PTk* DIAGRAM	785
THE *PT* ENVELOPE OR *PT* REGION	787
NOMENCLATURE	787
CHARACTERISTICS OF DEEP-SEA BACTERIA	788
Sensitivity to Warming, Cooling, and Decompression	788
Physiological Correlates with Habitat Temperature and Pressure	789
Membrane Lipids	791
Membrane Proteins	792
Sensitivity to UV Light and Ionizing Radiation	793
Generation Times in Laboratory Culture and in the Sea	793
PIEZOMESOPHILES AND PIEZOTHERMOPHILES	794
CONCLUSIONS	798

ABSTRACT

Microorganisms in the deep sea live at high pressures, low and high temperatures, and in darkness. These parameters and their food supply govern their lives. The study of these creatures gives us an opportunity to see how life processes work at some of the highest temperatures and pressures of the biosphere. Cultured bacterial isolates can grow to over 100 MPa at 2°C and to over 40 MPa at over 100°C. These cultures comprise the foundation for the study of the molecular biology and biotechnology of these isolates. The *PTk* diagram shows how temperature and pressure affect the growth rate of a bacterium and helps in the search for relationships among bacteria from habitats differing in temperature and pressure.

* * *

Truth is mysterious, elusive, always to be conquered.
A Camus, 1971 (18a)

INTRODUCTION

From the sea-surface microlayer into the depths of sediments in the deepest trenches, bacteria play out their roles as chemical factories. The immense variety of marine microbial processes is vital to the marine ecosystem and affects our food and wastes and our air, offers biotechnological opportunities barely tapped or imagined, and possibly holds clues to the origins of life. The deep sea is principally an oligotrophic, cold, dark, and high-pressure environment, and these properties were not too long ago considered its only attributes. Now we know that deep-sea brines (125), seeps (15, 75), and especially hydrothermal vents also contribute significantly to the variety of deep-sea habitats. The sea pressures that stimulated some of the early investigations in marine microbiology (21) affect all organisms in the deep sea and at the same time keep us from easy access to its inhabitants. The purpose here is to review the status of some of the problems in deep-sea microbiology, chiefly from the standpoint of pressure, a focus demanded by the large scope of deep-sea microbiology.

Pressure is the force acting on an area divided by its size. The force may be gravitational, electromagnetic, mechanical, and so on. In brief, if we imagine a small cube in a liquid or solid, six of the nine conceivable pressures are parallel to the sides of the cube and are shears (18, 95). The other three stresses are normal to the sides of the cube and are the pressures that in a liquid at rest are equal to each other (95) and to the hydrostatic pressure. The hydrostatic pressure is usually positive in a liquid, but liquids in natural and laboratory settings have negative pressures as well (28, 126). The stress is hydrostatic; however, the strain in some crystals under hydrostatic loading is anisotropic. Recent evidence suggests that DNA (23), proteins (79), and membranes (13) also compress anisotropically. We may discover that many of the effects of pressure have their roots in the compressibility, for example, of a hydrogen bond (23).

The hydrostatic pressure increases with depth in an ideal sea, according to the equation of Saunders & Fofonoff (124) that accounts for the latitudinal and depth variation in g, the gravitational acceleration, and the variation in dynamic height. At 10,660 m, for example, the pressure is 110 MPa (~1086 atm). The old rule of thumb that pressure increases by 0.1 atm per meter of depth (0.01 MPa in SI units) in the sea seems good enough for microbiology. (Pressure units are: 1 atm = 1.01325 bars = 1.01325 × 10^5 Pascals.)

The early work on the physics of high pressures is covered in a historical chapter by Bridgman (14a) and the early studies in the biology and microbiology of high pressures are discussed by Cattell (20) and Johnson et al (73). In addition, several reviews cover studies of the past 25 years (5, 33, 65, 67, 68, 91, 93, 100, 101, 104, 112, 150, 155).

DEEP-SEA HABITATS

The largest habitat by volume is the deep ocean psychrosphere (16), which is represented in Figure 1 by the 2°C isotherm extending to beyond 110 MPa. A few deep seas are warmer than 2°C: the Sulu Sea at ~9.8°C and the Mediterranean Sea at ~13.5°C to depths of ~5000 m (44). Hydrothermal vents, the warmest known habitats, are found near the sea surface and to a maximum depth yet to be determined. Although each vent is a spatially and bathymetrically localized environment, vent inhabitants are probably biologically connected, at least by propagules, because the vent fauna of the Atlantic and Pacific share a startling degree of resemblance (137). Figure 1 shows an isopiest at a deep-sea hydrothermal vent. We know that life exists along such isopiests to

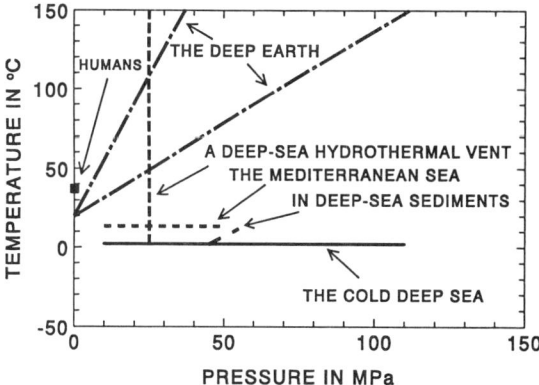

Figure 1 Some of the known and possible subsurface habitats of Earth. Figure shows the temperature (T) and pressure (P). Organisms are extremophiles in direct proportion to the distance in PT-space between their habitat and that of humans (37°C and 0.1 MPa).

a temperature of at least 110°C (34, 109, 128) and along the isotherms to the maximum pressure of ~110 MPa found in the sea. The regions where these values might be surpassed are all subterranean, either in the continents or in the sea floor (50, 52, 77, 122, 128). Figure 1 also shows three possible geothermal gradients. The view is of the contemporary Earth, but in the past, such as during the Cretaceous, the temperature in the abyssal Atlantic Ocean may have been 15°C (11) rather than the 2°C value of today. By linearly extrapolating the trend (9) showing an increase in temperature of 0.12°C over the past 30 years in the deep Mediterranean we can predict that the sea may be 1.2°C warmer in 300 years.

METHODS OF SAMPLING FOR MICROBIOLOGY

This review focuses on collecting samples from the cold deep sea. An ideal microbiological sample would remain in the dark, at the temperature and pressure of the deep sea, and mechanically and chemically undisturbed. The consequences of departing from these conditions depend on the extent and duration of deviation from deep-sea conditions, on the depth sampled, on the level of activity in the sample, and on the captured organisms. Examples of possible sample alteration include the loss of viable deep-sea bacteria, the loss of special relationships on suspended particles or in sediment strata, the unwanted stimulation of growth in the authochthonous bacteria, and the proliferation of the allochthonous bacteria (38, 119, 156). The invader population in a sample from a given depth of the cold deep sea can include upper-ocean bacteria that enter the deep sea on sedimenting particles, bacteria dispersed from hydrothermal vents, bacteria entering a water mass via advective processes from shallower or deeper water masses, propagules, and terrestrial bacteria carried to the sea by eolian processes.

Warming poses the greatest documented threat to the utility of a sample. Examples of commonly used sampling instruments that retain cold temperatures are Niskin bottles, Go-Flo bottles, box corers (1, 61, 120), gravity corers (with PVC core barrels), the Scottish Marine Biological Association (SMBA) multiple corer (2), sediment traps (38), insulated cod ends (25, 144), and insulated animals traps (51, 159). A bacterial isolate autochthonous to the deepest ocean trench came from an insulated animal trap (159). Cold samples serve as sources of cultivable flagellates (141) and live foraminifera (139). These and other studies over the past 12 years support the suggestion that "most deep-ocean microbiology may be possible with thermally insulated equipment for retrieval from the sea and with high-pressure vessels for laboratory incubations" (156, p. 1481).

Pressure-retaining samplers are helping in the search for those microorganisms intolerant of any exposure to atmospheric pressure or partial decompression. More importantly, these samplers help us determine the metabolic activities in natural microbial populations. The first pressure retaining-instrument to recover a water sample kept at its deep-sea pressure is that of Macdonald & Gilchrist (89). A key component of this instrument is a gas-containing accumulator that minimizes pressure loss arising from seal movement and from metal expansion as the sealed vessel rises from the deep sea. The closure on their pressure vessel allows its use as a water sampler, plankton sampler (90), or benthic animal trap (88). Two other pressure-retaining samplers (72, 131) are specifically intended for the study of water-column bacteria. No data obtained with these samplers have been published since a previous review of this field (67). Whereas results with these samplers so far fail to show evidence that continuous pressure retention is needed to determine the microbial activity in a natural sample, recent experiments by Garcin & Bianchi (10) suggest that nondecompressed samples are required to obtain an accurate measure of microbial activity. The criteria of when pressure retention is needed and when it is not remain to be established. Studies assessing the role of bacterial activity in sediments are usually done with decompressed, resuspended, and then recompressed sediments. A significant advance is the development of a device for recovering under pressure a section of a piston core in which sediment stratification is preserved (102).

The first pure culture of a piezophilic bacterium came from a pressure-retaining animal trap (148) wherein decomposition of an animal occurred over several months (158). This trap maintains a pressure close to the sampled habitat pressure by means of a gas-containing accumulator in the fashion described by Macdonald & Gilchrist (51, 89). The trap is deployed in a box having 5.08-cm-thick Lexan sides that provides thermal insulation. With this trap, animals have been recovered alive from the abyssal Pacific Ocean at a pressure of 58 MPa (149) and from the Mariana and Philippine Trenches at over 100 MPa (A Yayanos, unpublished data). Trapped animals have been kept alive without decompression for periods as long as three weeks by maintaining a flow of sea water through the trap at high pressure. Without sea water circulation, animals in the trap die. Microbial degradation appears within three days of death as a visible growth on the animal carcasses (viewed through quartz windows in the trap) (149). Pressure-retaining animal traps hold great promise for deep-sea microbiology. They provide animals in a controlled environment for the study of gut microbiology, symbioses, decomposition, and even the effects of animals on water-column bacteria.

To my knowledge, no reports have been published of pressure-retaining sampling for microbiology at hydrothermal vents. Hyperthermophiles from these deep-sea habitats have been successfully isolated from decompressed samples mostly by means of enrichments at atmospheric pressure (35, 70).

HIGH PRESSURE LABORATORY TECHNIQUES

Several surveys review laboratory high-pressure techniques for chemistry and biology (60, 62, 76, 84, 129, 142). Although an update of safety in high pressure research is long overdue (46), space does not permit it here. This review is restricted to those methods peculiar to laboratory studies in deep-sea microbiology.

Pressure Vessels

Pressure vessels used in deep-sea microbiology transport samples from the field, maintain samples in the laboratory, maintain working cultures, and incubate experiments. Because cells from the cold deep sea grow slowly, experiments should be conducted simultaneously in dozens of pressure vessels. One design, a pin-retained piston closure vessel with a volume of about 340 cc, lends itself to a relatively inexpensive fabrication (151, 154). The piston seal closure is easy to machine and to open quickly without a tool. Further savings in time and effort result from using thin flexible tubing for high pressure connections and a quick-connect coupler to secure a pressure vessel to a high pressure pump (162).

Colony Formation at High Pressure

Bacterial colonies can be grown on agar surfaces at high pressures via two methods. The first is a compressed gas apparatus allowing filtration at high pressure of a high pressure deep-sea water sample (67, 71, 133). The cells on the surface of the filter are sampled at pressure with an inoculating loop used to streak an agar plate at high pressure. The pressure-transmitting medium is an $He-O_2$ gas mixture. Comparison of growth in bacterial cultures compressed with gas mixtures vs with a hydraulic liquid shows that these two means of pressure generation are not equivalent; a high-pressure $He-O_2$ gas mixture yields better growth (58, 106, 132). Marquis (92) and Taylor (133) have reviewed the literature on the effects of high gas pressures on microbial physiology. The compressed gas method is currently the only one that allows growth of colonies of cells from a deep-sea sample with avoidance of decompression. Nevertheless, decompression-intolerant cells have not yet been reported (67). A serious safety concern with this method stems from the amount of energy stored in a gas-filled pressure vessel.

A second method for growing deep-sea colonies on an agar surface does not attempt to maintain constant high pressure (103). The surface of an agar slab is streaked with part of a sample, covered with another sterile slab of agar, placed in a heat-sealed pouch, and then incubated at pressure. Several piezophilic bacteria have been isolated in this manner. However, this method carries three concerns. First, the heat sealing may be cumbersome, as several seals are needed. The second, and potentially more serious, problem is that the space between the two agar surfaces might comprise a thin film of liquid wherein bacteria are free to move, thus contradicting the assumption that each colony arises from a single cell. Third, the agar slabs may require pressure vessels with a larger internal diameter than those normally used in microbiology, which would increase costs. More experiments are clearly needed to further evaluate this method.

The pour-tube method, a well-established technique for growing colonies, can be easily adapted for use in pressure vessels and offers all of the advantages of agar plates except for replica plating (45, 150). Pour tube gels already used with deep-sea bacteria include slush-agar (69), gelatin (157), silicic acid (33, 45, 150), and GELRITE (34). Because silicic acid gels can be formed in ice baths, they have the advantage of precluding thermal shock to psychrophilic cells. The use of gelatin gels subjects the cells to a brief exposure—perhaps 10 s—of warming to about 10°C; these gels are also susceptible to proteolytic digestion. The brief warming causes no detectable effect on the most psychrophilic and pressure-adapted organism with which we have worked, isolate MT41 (22, 157). Thus, if proteolysis is not a concern, easily made gelatin gel pour tubes are preferable. Furthermore, experiments to test colony-forming ability in silicic acid gels vs gelatin gels show that both methods can give the same result. Most significantly, gelatin gels reliably yield plating efficiencies approaching 100%.

In our early work, we used sterile disposable plastic or glass culture tubes to contain the gelled inoculated medium (45, 150). For more than 10 years, however, we have been using sterile polyethylene (PE) (heat-sealable) transfer pipettes made by Samco (31). The pour-tube method with PE transfer pipettes has proven reliable not only for the isolation of new isolates, but also for assays of colony-forming ability to examine survival curves following exposure to heat (156), UV light (85), ionizing radiation (AA Yayanos, Gomez, AS Dietz & R Van Boxtel, unpublished observations), mercury, alkylating agents, and mutagens (85).

High-Pressure Growth Curves

Growth experiments on a culture in a pressure vessel pose problems involving the removal of metabolic wastes, the addition of nutrients, and the removal of samples from the culture for analyses (68, 99). Some sampling methods cannot avoid decompressing the culture during cell removal, whereas others circumvent the need for decompression of the bulk of the culture. Sampling without decompression requires a costly apparatus, which precludes the study of more than a few experimental cultures. For example, if the doubling time of the experimental strain is 10 h, then perhaps one experiment per week will be possible. In contrast, sampling with decompression allows dozens of experiments per week. Until a less-expensive means of sampling cultures without decompression is devised, the best approach seems to be to incorporate both methods into the experimental design. That is, the vessels used in most experiments require decompression, but some of these analyses are repeated by sampling without decompression.

Collecting a sample without decompression requires the high-pressure isopiestic transfer of cells from the culture pressure vessel to a sampling pressure vessel in which the sample is then isolated, decompressed, and removed. During the transfer, a volume of liquid equal to that being removed is injected into the culture vessel. The simultaneous injection of an equal volume of culture is not required if a large hydraulic accumulator is connected to the experimental culture vessel. However, this apparatus greatly increases safety problems. Several instruments now available permit studies without decompression (72, 76, 84, 97, 105, 106, 146).

To sample a culture with periodic decompressions means opening either the vessel itself to remove a sample or a valve through which culture is gently expelled at atmospheric pressure. An example of the latter is an apparatus designed by Herbert Gathrup of Autoclave Engineers, Inc. (Erie, PA) (161). The method now mostly used is the same as that of Certes (21), who more than 100 years ago enclosed cultures in tubes to keep out hydraulic fluid. Today, cultures are kept in heat-sealable containers such as plastic bags (7, 31, 83) and bulbs (31) or in syringes. Samples are taken during brief decompressions or as end-point samples. The pin-retained piston closure vessels and the quick connect coupler greatly facilitate sample collection. In our laboratory, up to 10 pressure vessels sit on a rocking platform in a water bath. With 4 water baths, up to 40 cultures incubate at 4 independently controlled temperatures.

The growth rates of slow-growing, deep-sea psychrophilic bacteria determined by sampling with decompression apparently agree with those determined by sampling without decompression, where the two methods have been compared. However, compression has effects such as cell division stimulation in rapidly growing cultures of *E. coli* (147). Therefore, where fast growth rates are concerned, sampling with decompression may affect the results.

Bacterial cultures containing H_2 pose formidable problems. Hydrogen can enter the walls of some pressure vessels to weaken them through a process called hydrogen embrittlement (14b). Bernhardt et al (8) have tried to circumvent this problem by enclosing the cultures in pressure-transmitting, thin-walled nickel tubes equipped with a port for filling and sampling. These authors report (8, p. 1877) that the "problem of gas leakage cannot be overcome" and that the question is not how to stop the leakage, but rather which metal has a minimal permeability to H_2—a criterion met by nickel. However, if CO is present at high pressures, the formation of nickel carbonyls may also present a safety risk (46). Another attempt to confine H_2 involves a pressure vessel made of sapphire (97), although this vessel is also not perfect (84). Clearly, high-pressure methanogens will continue to rank among the most challenging organisms to study in high-pressure microbiology.

Microorganisms can be directly counted at pressure, using the Coulter principle, with Macdonald's cleverly designed apparatus (87). Reversing the flow through the Coulter aperture tube after each count, so that culture is both inside and outside of the aperture tube, allows one to count culture cells continually. Macdonald studied the effect of pressure and pressure shifts on cell division and mean cell volume in cultures of *Tetrahymena pyriformis.*

Pressurized Temperature-Gradient Instruments

One way to study simultaneously both pressure and temperature effects is to set up temperature gradients in a pressure vessel (78). Such a pressurized temperature-gradient (PTG) apparatus is particularly useful in deep-sea microbiology (164). Long gel-filled tubes inoculated with bacteria incubate in each of eight pressure vessels and each vessel is maintained at a different pressure. At the end of the incubation period, colony growth is evident in the tubes between two temperatures, T_{max} and T_{min}, and these two temperatures are different for each pressure. The *PT* regions determined with a PTG are not perfectly accurate (164) "envelopes of life" (92, p. 1) but are adequate to give a synoptic view of where growth is possible with a given isolate. This instrument has proven useful in defining the *PT* values at which bacteria grow (39, 136, 163, 164) and survive (136) and in revealing the pressures and temperatures at which some of the cultivatable bacteria in a natural population can grow (164).

THE *PTk* DIAGRAM

The conventional methods of displaying growth rate data as a function of temperature or pressure are inadequate for helping us get insights into adaptation in the deep sea, where both temperature and pressure vary. A simple solution outlined here is used throughout the remainder of the review. In a culture containing enough nutrients and lacking cell overcrowding or cell death, the number of bacteria increase exponentially with a characteristic exponential growth rate constant, k (98), where $k = \ln 2$/doubling time. When complications can be avoided, as in controlled laboratory conditions, the value of k indicates how rapidly a bacterial culture can grow as a function of T, P, and the composition of the medium and also serves as a reproducibly measurable trait of the organism. We can express the dependence of k on the medium parameters as

$$k = k(P, T, n_1, n_2, n_3, \ldots, n_N), \qquad 1.$$

where n_i represents the concentration of the ith constituent. In studies of the effects of T on k, investigators have customarily represented the results with a set of Arrhenius plots and recently with the empirical square root function of Ratkowsky and colleagues (96, 114). The effects of P on k lack appropriate expressions. The effects of nutrition on growth rate are often modeled using the Monod equation (17). The medium parameters also include media variables affecting redox conditions and pH, which can influence how a culture responds to a change in pressure (94).

The PTk diagram for a specified set of nutrient conditions is an empirically determined graph of Equation 1 and is simply either a contour plot of lines of constant k in the PT plane or a three-dimensional surface showing how both T and P influence k (152). Figure 2 shows PTk diagrams for bacterial isolates PE36 and MT199. In each case, the highest level contours surround the pressure, $P_{k_{max}}$, and temperature, $T_{k_{max}}$, where the growth rate has its highest value, k_{max}. At pressures somewhat less than $P_{k_{max}}$, k is nearly independent of temperature. $T_{habitat}$ and $P_{habitat}$, denoted by the large cross in Figure 2, represent conditions that are not optimal for growth. The PTk diagram thus illustrates and quantifies several aspects of bacterial growth with respect to temperature and pressure. These features are discussed in the following sections. The terminology $P_{k_{max}}$ and $T_{k_{max}}$ should not be confused with P_{max} and T_{max}, which refer to upper cardinal pressures and temperatures.

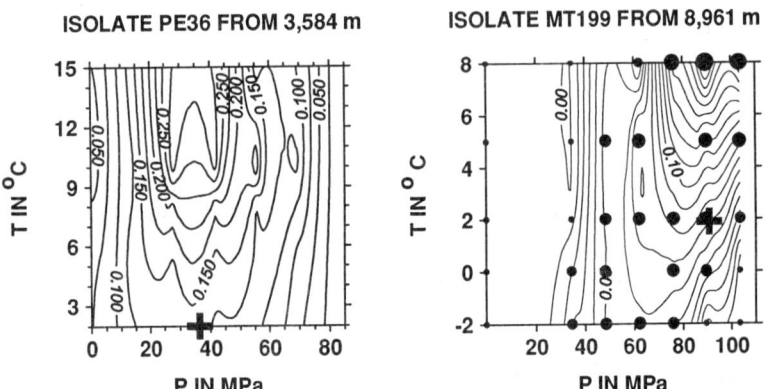

Figure 2 PTk diagrams for two piezopsychrophiles. The large cross denotes the habitat conditions, $P_{habitat}$ and $T_{habitat}$. Each line is drawn at a constant growth rate. The highest contour level surrounds the values of $P_{k_{max}}$ and $T_{k_{max}}$. Piezopsychrophiles typically have $P_{k_{max}} \approx P_{habitat}$ and $T_{k_{max}} \approx T_{habitat}$ + 6–10°C. The black-filled circles indicate the growth rate values used to generate the contours, which are shown only for isolate MT199. The smallest indicates zero growth rate.

THE *PT* ENVELOPE OR *PT* REGION

Marquis (92) has reviewed the origins of the concept of a *PT* envelope or region, the so-called envelope of life. The temperatures and pressures at which an organism can grow form a connected and bounded region on a plot of *T* vs *P*. The pressurized temperature-gradient (PTG) instrument allows one to determine approximate *PT* values defining the envelope for a given bacterial isolate (164). Several features of *PT* envelopes appear noteworthy. First, the values of temperature and pressure where a given bacterial strain grows are bounded by an elliptical curve. Second, judging from the few examples available, the area bounded by a *PT* envelope seems to be the smallest for the deep-sea psychrophiles. Third, the pressure range of growth, $P_{max}-P_{min}$, changes with temperature, and the temperature range of growth, $T_{max}-T_{min}$, changes with pressure. Finally, the *PT* envelopes of bacteria that grow at atmospheric pressure are incomplete elliptical curves with the missing values at negative pressures. Negative pressures in liquids are metastable states. The maximum theoretical negative pressure for water is thought to be approximately −250 MPa (126). In nature, negative pressures occur in fluid inclusions in minerals, in the xylem of tall trees, and in shock waves.

NOMENCLATURE

With respect to the temperature range at which they grow, organisms are classified as psychrophiles, mesophiles, thermophiles, or hyperthermophiles. ZoBell & Johnson (168a) coined the word barophilic to describe "the ability to grow and carry on metabolism as well or better under increased pressure" (p. 183). I propose two nomenclature changes (154). First, instead of the term barophile (literally, a weight lover), we should adopt the term piezophile. The latter term is etymologically an improvement (*piezo* is a verb, to press), and it conforms to the widespread use of the prefix piezo in physics and chemistry to denote pressure. Second, we should classify an organism according to the temperature and pressure intervals wherein lie the values of, respectively, $T_{k_{max}}$ and $P_{k_{max}}$ derived from a *PTk*-diagram obtained from growth experiments at copiotrophic conditions. The current nomenclature for bacteria from atmospheric-pressure habitats would remain intact. Thus, a bacterial strain with a $P_{k_{max}} \approx 0.1$ MPa would be a psychrophile, mesophile, thermophile, or hyperthermophile depending on the temperature interval at which it grows in the conventional context. A bacterial strain with a $P_{k_{max}} > 0.1$ MPa and < 60 MPa (a somewhat arbitrary upper pressure) would be a piezopsychrophile, piezomesophile, piezothermophile, or piezohyperthermophile. Finally, a strain with a $P_{k_{max}} > 60$ MPa would be called a hyperpiezopsychrophile, hyperpiezomesophile, hyperpiezothermophile, or hyperpiezohyperthermophile depending on the temperature interval where it grows. Strains MT199 (Figure 2) and MT41 (152) are examples of hyperpiezopsychrophiles, and strain PE36 (Figure 2) represents a piezopsychrophile. Isolate ES4 shows piezophilic growth at isotherms above its $P_{k_{max}}$. Whether it is a piezohyperthermophile, however, is not certain because its $P_{k_{max}} \approx 0.1$ MPa. Isolate GE5 appears to have a $P_{k_{max}} > 0.1$ MPa, but the data are insufficient to firmly classify it as a piezohyperthermophile. *Pyrococcus* strain GB-D may also barely qualify as a piezohyperthermophile (70). No examples of hyperpiezohyperthermophiles ($P_{k_{max}} \gg 0$), hyperpiezomesophiles ($P_{k_{max}} \gg 0$), nor of hypopiezophiles ($P_{k_{max}} < 0$) have as yet been found.

This terminology is unambiguous, allows us to deal with our expanding view of life shown in Figure 1, and is flexible enough to accommodate improvements in understanding and to name additional piezo-responsive organisms as we discover them.

CHARACTERISTICS OF DEEP-SEA BACTERIA

Sensitivity to Warming, Cooling, and Decompression

Bacterial inhabitants of the cold deep sea are psychrophilic and piezophilic (piezopsychrophilic) (33, 40, 69, 152, 158, 160). The hypothesis that these two traits are so strong as to signify their authentic deep-sea nature (160) would be a tautology were it not for the older literature denying it. An analysis (156) of several deep-sea microbiology papers published between 1952 and 1982 suggests that overlooking the effects of warming may have been the greatest impediment to rapid progress in the field of deep-sea microbiology since ZoBell's 1952 report (166). Cells of isolate CNPT3, the first piezophilic bacterium to be grown in pure culture, have a morphology and colony-forming ability that are particularly sensitive to warming (156). Exposing the cells to 27°C at high pressure results in only a slight reduction in the rate of death (156).

An analysis of the literature indicates that most deep-sea microbiology can be done with samples recovered in thermally insulated devices followed by high-pressure laboratory incubations. We tested this idea in 1981. A colleague on an expedition to the Aleutian Trench returned to us water samples kept on ice. One of the samples had been warmed, and no bacteria showing pressure facilitated growth were recovered from it. All of the other samples yielded bacteria that grow best at high pressure, including some that grow only at high pressure (155). The isolation of a hyperpiezopsychrophile from a warmed sample (40) was possible because all of the cells in a warmed population do not die at once, and the mortality is a function of the duration of warming (156). One way to kill selectively the authochthonous population and thereby evaluate its role (156) is to treat the deep-sea sample with heat, as confirmed by Deming & Colwell (38) with natural populations of abyssal bacteria. Material sedimenting through the water column was collected with sediment traps at water depths of 4270, 4463, and 4830 m and at 200, 20, and 7 m above the sea floor. The first two traps were recovered containing water with a temperature less than 6°C. The third trap was warmed to 24°C for perhaps 6 h because of an 8- to 10-h delay in retrieval. The uptake rate of glutamate in the warmed sample was less at the deep sea than at the atmospheric pressure, whereas the uptake rate in the sample that had been kept cold exhibited the opposite trend. The warmed sample had 42 times more bacteria in it than did the cold sample. However, the warmed population incorporated less labeled glutamate at deep-sea conditions than at atmospheric pressure and 3°C.

There are really not enough examples to fairly judge the importance of habitat depth (pressure) as a determinant of decompression sensitivity. All of the reported piezopsychrophilic bacteria tolerate decompression to atmospheric pressure for some period of time (39, 69, 143, 152). Death by decompression alone (presumably) could be judged from only one example. The hyperpiezopsychrophilic isolate MT41 does not grow at atmospheric pressure, and its ability to form colonies at high pressure diminishes as a function of exposure to atmospheric pressure at 0°C (157). Concurrent with the loss of colony-forming ability, abnormal, plasmolyzed, and lysed cells as well as intracellular vesicles appear (22).

Are there bacteria or other microorganisms that have a strict requirement for pressure in the sense that any exposure to atmospheric pressure would be lethal? The bursting of gas filled vacuoles is currently the only known mechanism leading to death by decompression (59). The presence of gas vacuolate bacteria in marine settings (64) indicates that they may reside in the deep sea. A pressure-retaining sampler would afford a means of detecting them.

Physiological Correlates with Habitat Temperature and Pressure

Inhabitants of the cold deep sea live at a temperature close to 2°C. Yet the $P_{k_{max}}$ from a *PTk* diagram (Figure 2) for a given isolate will be greater than the presumed habitat temperature by 5–10°C (152). The significance, if any, of this difference is open to speculation (39, 152). Unless occasional heat-producing events occur in microniches in the cold deep sea, a piezopsychrophile may never experience at its habitat pressure the temperature where it grows with k_{max}. On the other hand, the deep sea might contain warm microenvironments such as the inside of a decaying carcass or the gut of an animal. Another possibility would be metabolic bursts in bacteria that release heat and thereby require all bacteria to be prepared for an occasional, brief, internal temperature rise (152). Finally, the fact that $T_{k_{max}} > T_{habitat}$ provokes the question of whether an organism should be studied at its physiological optimum temperature ($P_{k_{max}}$) if it never encounters this temperature in nature.

Bacteria of the cold deep sea, the piezopsychrophiles, are stenothermal at their habitat pressure. They reproduce over a temperature range, $T_{max}-T_{min}$, of roughly 10–20°C (39, 152, 164). The breadth of the pressure range, $P_{max}-P_{min}$, for growth is usually ~40 MPa or less for isolates from depths less than 3600 m (69, 160). This pressure range is similar to that found for terrestrial and sea surface bacteria. Isolates of depths greater than 5000 m often have $P_{max}-P_{min}$ ≈ 80 MPa (39, 69, 155, 158, 160). The value of $T_{max}-T_{min}$ changes with pressure, and the value of $P_{max}-P_{min}$ with temperature, reflecting the curvature of the envelope of life (92).

When cells from the cold deep sea are grown at their habitat temperature and over a range of pressures, the maximum growth rate along that isotherm occurs at a pressure less than that at the habitat depth. This pressure at 2°C and the capture depth of an isolate are roughly correlated (160). However, the $P_{k_{max}}$ of a given bacterial isolate and its capture depth pressure are correlated much more strongly (152). Furthermore, the pressure $P_{k_{max}}$ of a given bacterial isolate is within ~5 MPa of its capture depth pressure (152) when grown under copiotrophic conditions. This result is compelling evidence for a role of pressure in the zonation of life in the cold deep sea.

Cell size abnormalities may be sensitive indicators of whether an organism is growing at conditions close to those of its habitat. One of the early observations on the effects of pressure on bacterial cells was their propensity to form filamentous cells at the upper pressure limits of their growth range (168, 169), particularly when grown just below the pressure limiting their growth. Piezopsychrophilic bacteria form large cells and abnormal cell shapes when grown at the lower or the upper limits of their pressure range (69, 155, 158). The greater the habitat depth is of a bacterial isolate, the greater will be the abundance and types of aberrant cells in atmospheric pressure cultures (AA Yayanos & RA Chastain, unpublished data). Isolate MT41 is a hyperpiezophile and does not grow at atmospheric pressure and 0°C. At this temperature and pressure, cells lose colony-forming ability, deteriorate, and disintegrate over a 2-day period (22). In summary, abnormally large cells seem generally indicative of growth at nonhabitat pressures and temperatures, and such sizes could be used to identify the likely habitat of an organism.

The generation times of isolated bacteria tend to increase with increases in the sampled habitat depth (160). The relationship between habitat depth and generation time is difficult to accurately discern because of the unstudied effect of nutrition on the generation time of deep-sea bacteria. The generation time of *Escherichia coli* can be altered more than 10-fold with a change in carbon and nitrogen sources (118). The viscosity, η, of a medium is related to the diffusion constant, D, by $D = kT/6\eta\rho$, where k is Boltzmann's constant, T the absolute temperature, and ρ the diameter of a diffusing molecule. Thus, the general reduction in growth rate with decreasing temperature and with increasing pressure with depth probably results from viscosity increases (and smaller diffusion constants) in membranes and intracellular compartments (154). The effect of pressure on viscosity (and therefore diffusion) is much smaller than is the effect of temperature and likely accounts for the high growth rates found in organisms in the high temperature habitats of hydrothermal vents.

Membrane Lipids

Bacteria of the cold deep sea respond to a change in pressure by altering the composition of their membrane phospholipid fatty acids (30, 31, 66, 74). For example, isolate CNPT3 from a 5782-m depth increases the relative amount of monounsaturated lipids in its membrane phospholipids when the pressure on its culture increases (30). This increase in unsaturation is thought to counter the increase in viscosity caused by increasing pressure, in a fashion analogous to the increase in viscosity caused by decreasing temperature (26). The phospholipids of bacterial isolate PE36 grown at its habitat pressure and temperature have 25% 22:6 fatty acids. The level of 22:6 fatty acids in cultures grown at three different pressures, 0.1, 34.5, and 62.0 MPa, increases with pressure, whereas that of 14:1 and 16:1 fatty acids drops. The unsaturation index shows an increase directly proportional to growth pressure (31).

Two other hyperpiezophiles show a departure from this trend. The unsaturation index increases with decreasing pressure to a maximum value and then decreases with decreasing pressure (31). Furthermore, a comparison of the unsaturation indices of isolates with the same fatty acid profiles but from different depths, when grown at the pressure of their presumed habitat depth (31), shows no trend with depth. The fatty acids occupying the *sn-1* and *sn-2* positions on the glycerol in phospholipids are known to exchange positions in membranes of *Tetrahymena* spp. following temperature shifts and preceding fatty acid compositional change (43, 134). Furthermore, if fluidity is the key property being modulated by the organism, then changes in it need to be confirmed by physical measures of fluidity, as in membranes of fish (26).

The pressure-induced changes in membrane fatty acid composition in piezopsychrophiles are modest compared with the changes following a shift in growth temperature (29). For example, the amount of 22:6 fatty acid changes at 0.1 MPa from a value of 15.7% (w/w) at 2°C to 0% (w/w) at 14°C. The strong homeoviscous response to a temperature change is interesting if these organisms never experience a temperature other than 2°C. It seems that adaptation to an interval of depth—and thereby of pressure—in the isothermal, cold deep sea confers upon an organism an ability to regulate its lipid metabolism over a range of temperatures.

Membrane fatty acid compositional change in bacteria also occurs in response to other factors such as starvation (116) and the phase of growth (54, 113). Rice & Oliver (116) observe a 37% increase in the amount of saturated fatty acids, predominately in 16:0 and 18:0, during a starvation period of 8 days. The changes in unsaturated fatty acids are mostly in 18:1. Rice & Oliver argue that further work will reveal the importance of these changes as part of a starvation strategy.

About 7000 strains of marine bacteria were screened for their ability to produce 20:5 fatty acids (165). The isolates came from fish, invertebrates, water, and sediments, and most were from the upper ocean. Isolates from chum salmon, squid, krill, sea water, and sediments do not produce 20:5 fatty acids. Between 0.2 and 2.9% of the isolates from mackerel, sardine, yellowtail, and other fish do produce 20:5 fatty acids. The amount of 20:5 fatty acids in an *Alteromonas*-like strain varies with temperature from 27.8% (w/w) of the membrane fatty acids at 25°C to 36.3% (w/w) at 4°C. From the standpoint of the deep sea, this work shows that PUFA-synthesizing bacteria are not peculiar to deep-sea bacteria and that they are ubiquitous in the sea. A psychrophilic bacterial isolate producing 22:6 fatty acids, as well as nine other fatty acids, was isolated from sediments at a water depth of 2220 m (54) and showed an altered fatty acid composition with temperature change.

Membrane Proteins

Deep-sea bacteria share many membrane proteins with other bacteria. *Vibrio marinus* and isolates PE36 and CNPT3 share the phosphoenolypyruvate:sugar phosphotransferase system (PTS) for transporting sugars (32). The pressure response of some of the components of the PTS varies from one bacterium to the next, which suggests that the PTS can be adapted to pressure in various strategies (32). Moreover, deep-sea bacteria alter the types and levels of membrane proteins in response to pressure change (4). The mechanism inducing these changes and the role of these proteins is under active investigation (5, 6, 24).

Bartlett and his collaborators are developing the piezomesophile SS9 (29) into a genetic system for the elucidation of pressure effects and the analysis of adaptation to high pressure. Bartlett et al (4) discovered that in SS9 an outer membrane protein designated OmpH (H for high pressure) is produced in greater amounts at high pressure. Analysis of the deduced amino acid sequence of OmpH suggests that it is a porin with some similarity to a porin from *Haemophilus influenzae* (6). Chemical mutagenesis of SS9 yielded a strain designated DB110 deficient in β-galactosidase production (24). Strain DB110 retains the ability to alter its levels of OmpH in response to pressure change. A transcriptional gene fusion of *ompH* and *lacZ* created on plasmid pEC9 and inserted into the genome of strain DB110 resulted in a new strain designated EC10. The *lacZ* gene in strain EC10 is under the control of the *ompH* promoter and thereby allows for tracking of *ompH* expression via assays for β-galactosidase. Mutants in the regulatory and structural parts of the gene fusion in strain EC10 result from chemical mutagenesis. One of these strains, designated EC1002, is highly pressure sensitive, displaying a very long lag phase when grown at 27.6 MPa. Once grown at high pressure, the cells lose their pressure-sensitive response. This and other evidence suggest that strain EC1002 acquires suppressor mutations when grown at high pressure (24). Although the role of OmpH remains to be established, this ongoing work by Chi & Bartlett (24) shows great promise. At the same time, strain SS9 is becoming established as a model genetic system in a piezomesophilic bacterium (5).

Sensitivity to UV Light and Ionizing Radiation

Light in the deep sea is generally at very low intensities, coming largely from bioluminescence and the decay of ^{40}K, a naturally occurring radionuclide. One estimate places the background photon flux from the Cerenkov radiation arising from ^{40}K decay at 120 visible photons cm^{-2} s^{-1} (12). Bioluminescence can result in bursts of light three orders of magnitude greater than the background light levels (12). Four strains of deep-sea bacteria exhibit sensitivity to UV light (85, 153) so high that they rank among organisms usually thought to be defective in DNA repair. Experiments to determine whether these four strains possess photoreactivation were not conclusive. These experiments are difficult because the assay of colony-forming ability done at high pressure requires incubations of up to 5 weeks, and the availability of pressure vessels determines how many experiments can be conducted at any given time. Thus, an experimental result that takes one day with *E. coli* takes one month with deep-sea piezopsychrophiles. The question arises as to the shallowest depth where such exquisite sensitivity to UV light will first become evident. Bacteria such as *Micrococcus euryhalis* and *Photobacterium phosphoreum* are no more sensitive to UV than are most bacteria (AA Yayanos, unpublished data).

The deep sea, although shielded from cosmic rays, contains radioactivity from naturally occurring radionuclides. Some deep-sea organisms accumulate these radionuclides in their tissues, harboring some astoundingly high background levels of ionizing radiation (130). It is not known whether bacteria accumulate natural radionuclides to the same extent as these animals. Results from irradiation of deep-sea bacteria with γ rays from ^{60}Co suggest that deep-sea bacteria have the same sensitivity to ionizing radiation as do shallow-water bacteria (AA Yayanos, LS Gomez, AS Dietz & R Van Boxtel, unpublished data).

Generation Times in Laboratory Culture and in the Sea

Generation times of cold deep-sea piezophiles are roughly between 6 and 35 h under deep-sea temperature, pressure, and copiotrophic conditions (31, 36,

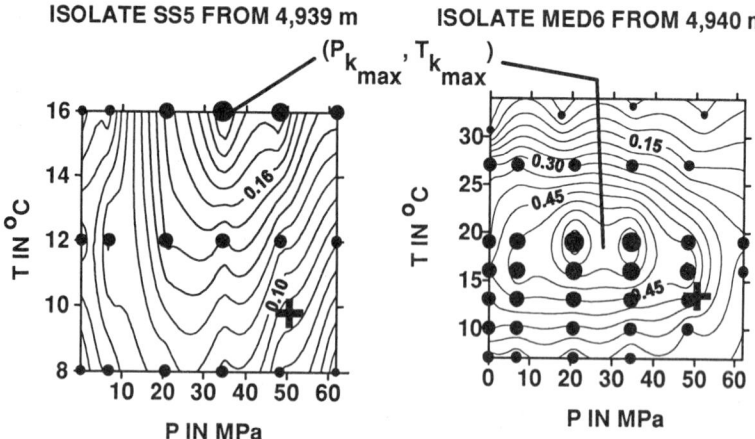

Figure 3 PTk diagrams for two piezomesophiles, SS5 from the Sulu Sea and MED6 from the Mediterranean Sea (EF DeLong & AA Yayanos, unpublished data). The large cross marks the habitat conditions, $P_{habitat}$ and $T_{habitat}$. Each line is drawn at a constant growth rate. The highest contour level surrounds the values of $P_{k_{max}}$ and $T_{k_{max}}$. For these piezomesophiles, $P_{k_{max}} < P_{habitat}$ and $T_{k_{max}} \approx T_{habitat} + 6°C$. Note that MED6 is not piezophilic at its habitat temperature. The filled circles, the smallest indicating zero-growth rate, indicate the growth rate values used to generate the contours.

37, 41, 69, 152, 158–160). ZoBell wrote, but showed no data, that deep-sea bacteria have generation times of about 10–30 h (167). The laboratory growth rates bracket many of the rates determined for natural populations in samples in which copiotrophic conditions prevail (82, 83).

PIEZOMESOPHILES AND PIEZOTHERMOPHILES

Most of the previous discussion is about bacteria of the cold, deep-sea isothermal habitats illustrated in Figure 1 as a straight line drawn at 2°C. Bacteria of the Sulu Sea and Mediterranean Sea can be found at 5000-m depths and at stable temperatures of 9.8 and 13.5°C, respectively, and thus offer a contrast to those of the cold deep ocean. The principal finding is that Mediterranean Sea bacteria are piezophilic but not particularly so at the temperature of their habitat. The piezophilic characteristic, $P_{k_{max}} > 0$, is clear only at temperatures above that in their habitat (29; AA Yayanos & EF DeLong, unpublished data), as shown in Figure 3. Just as with bacteria of the cold deep sea, with these warmer-water bacteria the value of $T_{k_{max}}$ is greater than the value of $T_{habitat}$. Furthermore, these bacteria grow at room temperature and are thus piezomesophiles. The immediate practical significance of this observation is that these bacteria are ideal laboratory organisms for the study of some aspects of pressure adaptation because conventional plating techniques can be employed, unlike the situation with the piezopsychrophiles. The piezomesophile SS9 from the

Sulu Sea is being developed as a genetic system for the study of pressure adaptation (4–6, 24).

The piezopsychrophile pressure relation, $P_{k_{max}} \approx P_{habitat}$ (Figure 2) differs in piezomesophiles, in which $P_{k_{max}} < P_{habitat}$, as can be judged from Figure 3. Comparison of the *PTk* diagrams of SS5 and MED6 suggests that $P_{k_{max}} \to 0$ as $T_{habitat}$ increases. Also noteworthy is that $T_{k_{max}} = T_{habitat} + \sim 6°C$, just as it does in Figure 2. Thus, these piezomesophiles grow fastest at a temperature that they might never experience in their Sulu and Mediterranean Sea habitats.

In the environs of hydrothermal vents, the temperature changes dramatically over spatial scales of centimeters and over time scales measured in seconds. In such dynamic environments, the habitat temperature of a given organism is difficult to judge. Evolution at vents may demand a different strategy than it does in deep seas where the temperature is stable (over millions of years). For the sake of argument only, we can assume that evolution to temperature and pressure is no different at vents than it is in the rest of the environment. If that is the case, then we can look at the *PTk* diagram of a vent bacterium, compare it to *PTk* diagrams of other deep-sea bacteria, and estimate its habitat temperature or pressure. Thus, for ES4, a hyperthermophilic isolate studied by Pledger et al (108), we would estimate a $P_{habitat} \approx 22.3$ MPa based on the water depth above the vent and a $T_{habitat} \approx 87°C$ based on Figure 4 and on $T_{k_{max}} = T_{habitat} +$

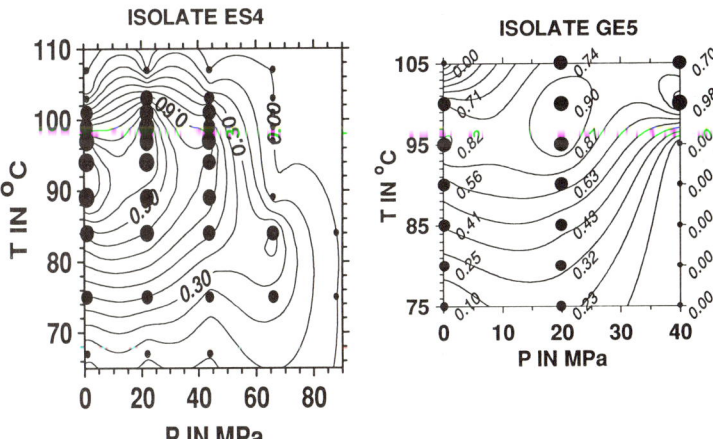

Figure 4 PTk diagrams for two hyperthermophiles, ES4 and GE5, from hydrothermal vents. The diagrams were generated from the published data of Pledger et al (108) for ES4 and of Erauso et al (48) for GE5. It is likely that $P_{habitat} \approx 22$ MPa is for ES4 (108) and $P_{habitat} \approx 20$ MPa for GE5 (48). $T_{habitat}$, however, is difficult to determine because of the steep temperature gradients at vents. Each line is drawn at a constant growth rate. The highest contour level surrounds the values of $P_{k_{max}}$ and $T_{k_{max}}$. For these hyperthermophiles, $P_{k_{max}} < P_{habitat}$ and the relationship between $T_{k_{max}}$ and $T_{habitat}$ is unknown. ES4 is piezophilic at its highest growth temperatures (108), but whether $P_{k_{max}} > 0.1$ MPa is not clear. The numbers next to the black dots for the GE5 plot are the growth rate–constant values. The estimated error in these values is available (48).

~6°C. This estimate, of course, is just an exercise to show the type of comparison and analysis possible with PTk diagrams. Hyperthermophiles may prefer $T_{habitat} > T_{k_{max}}$.

Jannasch et al (70) have reported significant new findings on adaptations in hyperthermophiles that enhance survival during dispersal. Hyperthermophilic isolates were obtained from the Guaymas Basin at a water depth of 2020 m (70). Samples for inocula were scrapings of the outer surface of a black smoker chimney and cores of sediments taken where the temperature in the sediment was 20°C at a 5-cm depth and 125°C at a 75-cm depth. The inoculated standard complex media were incubated anaerobically. The first interesting finding is the survival of hyperthermophiles in samples stored for 5 years. Samples stored for 5 years at ~4°C were used to inoculate 2216 medium with elemental sulfur and TYEG medium. In many tests, hyperthermophiles grew within 24 h. The second noteworthy result is that four of the strict anaerobes tolerate oxygen when exposed at nongrowth temperatures. Jannasch et al (70) postulate that these are essential features enabling the dispersal of these organisms among vents. The survival of these bacteria in the oxic cold deep sea for periods of 5 years or more would seem to allow propagules to go from one vent to another by water mass movements and by particle-transport processes. At the Juan de Fuca Ridge, the pattern of water movement includes two oppositely directed flows parallel to and on either side of the ridge axis and at 2–3 cm s^{-1} (19). Above the ridge, the flow pattern varies and oscillates (19). In 5 years, a bacterium could be transported over a distance greater than 3000 km.

Finally, we come to the question of the upper temperature limit for life and whether pressure plays a pivotal role. The isolation of new and novel hyperthermophilic bacteria is largely due to the efforts of Stetter and his collaborators (109, 127, 128). Because pressure prevents the boiling of water, it clearly contributes to the survival of hyperthermophilic bacteria. The effects of pressure on partial molal heat capacities, partial molal volumes, the dielectric constant of water, viscosity, hydrophobic interactions, hydrogen bonding, phase equilibria, and so on also play roles in bacterial survival. We do not have good data (and sometimes none) for many of these parameters for biochemical reactions. In some instances, an increase in pressure can oppose the effect of an increase in temperature. However, an all-too simplistic picture would be a general principle that pressure will reverse only those high-temperature processes adversely affecting biological systems and thereby allow a dramatic extension of the envelope of life to high temperature.

Since 1983 (3), the highest temperature substantiated for the growth of an organism has been 120°C (34). Colonies grew in GELRITE incubated at 120°C at 26.85 MPa. Unfortunately, this organism was never further studied because it eventually did not subculture and was lost (35). More than ten years have passed since the report that bacteria can live to a temperature of at least 250°C (3).

Virtually all of the evidence in that paper was subsequently found to result from artifacts (135). Even without artifacts, the data do not compose an internally consistent set of results. Nevertheless, these authors steadfastly hold to the view that some bacteria can grow at 250°C or higher (35). A fascinating result derived from the presence of pDNA (particulate DNA) in vent plumes and from an analysis of mixing between vent water and surrounding sea water is that pDNA is highest in the hottest vent waters sampled—350°C (35). Explaining this result is challenging because current evidence on the thermal stability of DNA suggests it cannot exist for very long at this temperature (80, 81) or even at temperatures much above 150°C. The possibility that DNA is stabilized in aqueous systems near the critical point needs to be tested, because pDNA was present at only very low levels in vent waters less than ~310°C (35).

Helical DNA is less susceptible to thermal depurination (81). All studies of the effect of pressure on the helix-coil transition show that high pressure to at least 800 MPa favors the helical structure. The early experiments (56) found that the T_m of DNA increases by 10°C with an increase of pressure to 270 MPa, a pressure 10 times greater than that at currently studied vents. This finding agrees with the observation that the amount of transforming activity remaining in a DNA solution heated to 100°C for 30 min increases from near nil at 150 MPa to 60% at 200 MPa. The stabilization of DNA by pressure also depends on the base composition and on solution variables such the ionic strength (55).

The experimental result that the stabilization of native DNA under 200 MPa, for example, amounts to a T_m that is ~10°C greater than that at 0.1 MPa closely agrees with a theoretical calculation based on a modified self-consistent phonon approximation theory (a phonon is a quantized lattice vibration) that describes how the bonds in a DNA molecule respond to a change in pressure (23). One of the interesting features of the model is that the effect of pressure is ascribed mainly to compression and deformation of the interbase hydrogen bonds in G-C and A-T base pairs. Theoretical estimates (23) for the effect of pressure on the helix-coil transition are $dT_m/dP = 3 \times 10^{-2}$°C MPa^{-1} for poly[d(A-T)] and $dT_m/dP = 4.5 \times 10^{-2}$°C MPa^{-1} for poly[d(G-C)] (23). These values are in remarkable agreement with experiment (55). A tentative conclusion in another study of the DNA-melting problem states that the volume change of the transition, ΔV_{T_m}, probably results from interactions between DNA and surrounding ions. The volume change per base pair was calculated using the Clapeyron equation since the helix-coil transition is a highly cooperative event and can be modeled as a phase transition. The effect of pressure on ΔV_{T_m} is small, decreasing by roughly 20% over a pressure range of 400 MPa. The effect of ionic strength is quite large with ΔV_{T_m} increasing 8-fold between an ionic strength of 0.005m and one of 0.5m. Among other interesting aspects of this study, is that $dT_m/dP=0$

at $T_m \approx 60°C$ and possibly $dT_m/dP < 0$ below this temperature where pressure would then destabilize the DNA helix. The volume change of the helix coil transition has also been determined from the density of DNA solutions (115) and is in agreement to that calculated using the Clapeyron equation (23, 55, 57, 63, 107, 110, 117, 145). The pressure enhancement of DNA stability is thus insufficient on theoretical or experimental grounds to offer significant protection against thermal depurination at vent pressures. It is also an experimental fact that pressure causes damage to DNA in spores and in phage. The mechanism of action is not known (27, 56, 123).

CONCLUSIONS

Papers on newly isolated bacteria from extreme environments appear on a regular basis, but much more could be accomplished. Two additions to our infrastructure would help further these studies. One is the creation of repositories, analogous to the core collections of geologists, wherein samples of seldom-visited environments are properly stored and made available to scientists in academic and industrial settings. Second, we need a culture collection that can maintain and distribute extremophiles even if they have not been named. The literature is replete with work done on isolates that are no longer available. These losses preclude the confirmation of interesting experiments and necessitate expensive field work to resume lines of investigation. Measurement of activities of natural populations of microorganisms has greatly increased over the past few years, and we can look forward to new insights from these studies (42, 111, 138, 140). The Deepstar program in Japan bears close watching as it develops into a major center for the study of the basic biology and biotechnology of extremophiles from the deep sea (53). Many molecular interactions are known to be pressure-sensitive, such as the dissociation of Lac repressor protein subunits (121). The study of this process in piezopsychrophiles and hyperthermophiles may shed light on the essential interactions for the stability of multimeric proteins under pressure (170). Finally, we all know the importance of the properties of water (its dielectric constant, viscosity, heat capacity, hydrogen bonding, its propensity as a medium for hydrophobic interactions, and so on) for the existence of life (47, 49, 86). The new extremophiles already on hand operate under divergent water-property values. They give us an opportunity to find out exactly the significance and role of a given water property to life processes.

> Any *Annual Review* chapter, as well as any article cited in an *Annual Review* chapter, may be purchased from the Annual Reviews Preprints and Reprints service.
> 1-800-347-8007; 415-259-5017; email: arpr@class.org

Literature Cited

1. Alongi DM. 1992. Bathymetric patterns of deep-sea benthic communities from bathyal to abyssal depths in the western South Pacific (Solomon and Coral Seas). *Deep-Sea Res.* 39:549–65
2. Barnett PRO, Watson J, Connelly D. 1984. A multiple corer for taking virtually undisturbed samples from shelf, bathyal and abyssal sediments. *Oceanol. Acta* 7:399–408
3. Baross JA, Deming JW. 1983. Growth of 'black smoker' bacteria at temperatures of at least 250°C. *Nature* 303:423–26
4. Bartlett D, Wright M, Yayanos AA, Siverman M. 1989. Isolation of a gene regulated by hydrostatic pressure in a deep-sea bacterium. *Nature* 342:572–74
5. Bartlett DH. 1992. Microbial life at high pressures. *Sci. Prog. Oxford* 76:479–96
6. Bartlett DH, Chi E, Wright ME. 1993. Sequence of the *ompH* gene from the deep-sea bacterium *Photobacterium* SS9. *Gene* 131:125–28
7. Berger LR, Tam LQ. 1970. A method to grow obligately aerobic bacteria at increased hydrostatic pressure. *Limnol. Oceanogr.* 15:483–85
8. Bernhardt G, Jaenicke R, Lüdemann H-D. 1987. High-pressure equipment for growing methanogenic microorganisms on gaseous substrates at high temperature. *Appl. Environ. Microbiol.* 53:1876–79
9. Bethoux JP, Gentili B, Raunet J, Tailliez D. 1990. Warming trend in the western Mediterranean deep water. *Nature* 347:660–62
10. Bianchi A, Garcin J. 1993. In stratified waters the metabolic rate of deep-sea bacteria decreases with decompression. *Deep-Sea Res.* 40:1703–10
11. Boersma A. 1984. Campanian through Paleocene paleotemperature and carbon isotope sequence and the Cretaceous-Tertiary boundary in the Atlantic Ocean. In *Catastrophes and Earth History. The New Uniformitarianism*, ed. WA Berggren, JA van Couvering, pp. 247–77. Princeton, NJ: Princeton Univ. Press. 464 pp.
12. Bradner H, Bartlett M, Blackinton G, Clem J, Karl D, et al. 1987. Bioluminescence profile in the deep Pacific Ocean. *Deep-Sea Res.* 34:1831–40
13. Braganza LF, Worcester DL. 1986. Structural changes in lipid bilayers and biological membranes caused by hydrostatic pressure. *Biochemistry* 25:7484–88
14a. Bridgman PW. 1931. Historical introduction. In *The Physics of High Pressure*, pp. 1–29. London: Dover Reprint/ Bell & Sons. 398 pp.
14b. Bridgman PW. 1931. Special sorts of rupture peculiar to high pressures. In *The Physics of High Pressure*, pp. 96–97. London: Dover Reprint/Bell & Sons. 398 pp.
15. Brooks JM, Kennicutt MC II, Fisher CR, Macko SA, Cole K, et al. 1987. Deep-sea hydrocarbon seep communities: evidence for energy and nutritional carbon sources. *Science* 238:1138–42
16. Bruun AF. 1957. Deep sea and abyssal depths. *Geol. Soc. Am. Mem.* 67:641–72
17. Button DK. 1985. Kinetics of nutrient-limited transport and microbial growth. *Microbiol. Rev.* 49:270–97
18. Callen HB. 1959. Solid systems—elasticity. In *Thermodynamics*, pp. 213–23. New York: Wiley & Sons. 375 pp.
18a. Camus A. 1971. Acceptance speech (for the 1957 Nobel Prize for Literature). In *Albert Camus, Winston Churchill*, p. 10. New York: Helvetica
19. Cannon GA, Pashinski DJ, Lemon MR. 1991. Middepth flow near hydrothermal venting sites off the Southern Juan de Fuca Ridge. *J. Geophys. Res.* 96:12815–31
20. Cattell M. 1936. The physiological effects of pressure. *Biol. Rev.* 11:441–76
21. Certes A. 1884. De l'action des hautes pressions au les phénomènes de la putréfaction et sur la vitalité des micro-organismes d'eau douce et d'eau de mer. *C. R. Acad. Sci. Paris* 99:385–88
22. Chastain RA, Yayanos AA. 1991. Ultrastructural changes in an obligately barophilic marine bacterium after decompression. *Appl. Environ. Microbiol.* 57:1489–97
23. Chen YZ, Prohofsky EW. 1993. Theory of pressure-dependent melting of the DNA double helix: role of strained hydrogen bonds. *Phys. Rev. E* 47:2100–8
24. Chi E, Bartlett DH. 1993. Use of a reporter gene to follow high-pressure signal transduction in the deep-sea bacterium *Photobacterium* sp. strain SS9. *J. Bacteriol.* 175:7533–40
25. Childress JJ, Barnes AT, Quetin LB, Robison BH. 1978. Thermally protecting cod ends for the recovery of living deep-sea animals. *Deep-Sea Res.* 25:419–22
26. Cossins AR, Macdonald AG. 1989. The adaptation of biological membranes to temperature and pressure: fish from the

deep and cold. *J. Bioenerg. Biomembr.* 21:115–35

27. Da Poian AT, Oliveira AC, Gaspar LP, Silva JL, Weber G. 1993. Reversible pressure dissociation of R17 bacteriophage. The physical individuality of virus particles. *J. Mol. Biol.* 231:999–1008

28. Debenedetti PG, D'Antonio MC. 1988. Stability and tensile strength of liquids exhibiting density maxima. *AIChE J.* 34:447–55

29. DeLong EF. 1986. *Adaptations of deep-sea bacteria to the abyssal environment.* PhD dissertation. Univ. Calif. San Diego, La Jolla. 103 pp.

30. DeLong EF, Yayanos AA. 1985. Adaptation of the membrane lipids of a deep-sea bacterium to changes in hydrostatic pressure. *Science* 228:1101–3

31. DeLong EF, Yayanos AA. 1986. Biochemical function and ecological significance of novel bacterial lipids in deep-sea prokaryotes. *Appl. Environ. Microbiol.* 51:730–37

32. DeLong EF, Yayanos AA. 1987. Properties of the glucose transport system in some deep-sea bacteria. *Appl. Environ. Microbiol.* 53:527–32

33. Deming JW. 1986. Ecological strategies of barophilic bacteria in the deep ocean. *Microbiol. Sci.* 3:205–11

34. Deming JW, Baross JA. 1986. Solid medium for culturing black smoker bacteria at temperatures to 120°C. *Appl. Environ. Microbiol.* 51:238–43

35. Deming JW, Baross JA. 1993. Deep-sea smokers: windows to a subsurface biosphere. *Geochim. Comochim. Acta* 57:3219–30

36. Deming JW, Colwell RR. 1981. Barophilic bacteria associated with deep-sea animals. *BioScience* 31:507–11

37. Deming JW, Colwell RR. 1982. Barophilic bacteria associated with digestive tracts of abyssal holothurians. *Appl. Environ. Microbiol.* 44:1222–30

38. Deming JW, Colwell RR. 1985. Observations of barophilic microbial activity in samples of sediment and intercepted particulates from the Demerara abyssal plain. *Appl. Environ. Microbiol.* 50:1002–6

39. Deming JW, Hada H, Colwell RR, Luehrsen KR, Fox GE. 1984. The ribonucleotide sequence of 5s rRNA from two strains of deep-sea barophilic bacteria. *J. Gen. Microbiol.* 130:1911–20

40. Deming JW, Somers LK, Straube WL, Swartz DG, MacDonell MT. 1988. Isolation of an obligately barophilic bacterium and description of a new genus, *Colwellia* gen. nov. *Syst. Appl. Microbiol.* 10:152–60

41. Deming JW, Tabor PS, Colwell RR. 1981. Barophilic growth of bacteria from intestinal tracts of deep-sea invertebrates. *Microb. Ecol.* 7:85–94

42. Deming JW, Yager PL. 1992. Natural bacterial assemblages in deep-sea sediments: towards a global view. See Ref. 120a, pp. 11–27

43. Dickens BF, Thompson GA Jr. 1982. Phospholipid molecular species alterations in microsomal membranes as an initial key step during cellular acclimation to low temperature. *Biochemistry* 21:3604–11

44. Dietrich G, Kalle K, Krauss W, Siedler G. 1980. Regional oceanography. In *General Oceanography*, pp. 488–504; appendix, p. 569. New York: Wiley & Sons. 626 pp. 3rd ed.

45. Dietz AS, Yayanos AA. 1978. Silica gel media for isolating and studying bacteria under hydrostatic pressure. *Appl. Environ. Microbiol.* 36:966–68

46. Dodge BF. 1950. High pressure technique. In *Chemical Engineers Handbook*, ed. JH Perry, pp. 1233–62. New York: McGraw Hill

47. Edsall JT, Wyman J. 1958. Water and its biological significance. In *Biophysical Chemistry.* Vol. 1. *Thermodynamics, Electrostatics, and the Biological Significance of the Properties of Water*, pp. 27–46. New York: Academic. 699 pp.

48. Erauso G, Reysenbach AL, Godfroy A, Meunier J, Crump B, et al. 1993. *Pyrococcus-abyssi* sp.-nov., a new hyperthermophilic archaeon isolated from a deep-sea hydrothermal vent. *Arch. Microbiol.* 160:338–49

48a. Ernst WG, Morin JG, eds. 1982. *The Environment of the Deep Sea*, Rubey Vol. 2. Englewood Cliffs, NJ: Prentice Hall. 371 pp.

49. Franks F. 1984. *Water.* London: R. Soc. Chem. 96 pp.

50. Ghiorse WC, Wilson JT. 1988. Microbial ecology of the terrestrial subsurface. *Adv. Appl. Microbiol.* 33:107–72

51. Gilchrist I, Macdonald AG. 1983. Techniques for experiments with deep-sea organisms at high pressure. In *Experimental Biology at Sea*, ed. AG Macdonald, IG Priede, pp. 240–76. London: Academic

52. Gold T. 1992. The deep, hot biosphere. *Proc. Natl. Acad. Sci. USA* 89:6045–49

53. Hamamoto T, Horikoshi K. 1993. Deep-sea microbiology research within the Deepstar program. *J. Mar. Biotechnol.* 1:119–22

54. Hamamoto T, Takata N, Kudo T, Horikoshi K. 1994. Effect of temperature and growth phase on fatty acid compo-

sition of the psychrophilic *Vibrio* sp. strain no. 5710. *FEMS Microbiol. Lett.* 119: 77–82
54a. Hattori T, Ishida Y, Maruyama Y, Morita RY, Uchida A, eds. 1989. *Proc. Int. Symp. Microbial Ecology, 5th.* Tokyo: Jpn. Sci. Soc. 724 pp.
55. Hawley SA, Macleod RM. 1977. The effect of base composition on the pressure stability of DNA in neutral salt solution. *Biopolymers* 16:1833–35
56. Hedén C-G. 1964. Effects of hydrostatic pressure on microbial systems. *Bacteriol. Rev.* 28:14–29
57. Hedén C-G, Lindahl T, Toplin I. 1964. The stability of deoxyribonucleic acid solutions under high pressure. *Acta Chem. Scand.* 18:1150–56
58. Hei DJ, Clark DS. 1994. Pressure stabilization of proteins from extreme thermophiles. *Appl. Environ. Microbiol.* 60: 932–39
59. Hemmingsen BB, Hemmingsen EA. 1980. Rupture of the cell envelope by induced intracellular gas phase expansion in gas vacuolate bacteria. *J. Bacteriol.* 143:841–46
60. Heremans K. 1986. Bioinorganic systems. See Ref. 142a, pp. 339–93
61. Hessler RR, Jumars PA. 1974. Abyssal community analysis from replicate box cores in the central North Pacific. *Deep-Sea Res.* 21:185–209
62. Heydemann PLM. 1978. Generation and measurement of pressure. In *High Pressure Chemistry,* ed. H Kelm, pp. 1–49. Dordrecht: Reidel
63. Hughes F, Steiner RF. 1966. Effects of pressure on the helix-coil transitions of the poly A–poly U system. *Biopolymers* 4:1081–90
64. Irgens RL, Suzuki I, Staley JT. 1989. Gas vacuolate bacteria obtained from marine waters of Antarctica. *Curr. Microbiol.* 18:261–65
65. Jannasch HW. 1984. Microbes in the oceanic environment. *Symp. Soc. Gen. Microbiol.* 36(Part II):97–122
66. Jannasch HW. 1987. Effects of hydrostatic pressure on growth of marine bacteria. See Ref. 66a, pp. 1–15
66a. Jannasch HW, Marquis RE, Zimmerman AM, eds. 1987. *Current Perspectives in High Pressure Biology.* New York: Academic. 341 pp.
67. Jannasch HW, Taylor CD. 1984. Deep-sea microbiology. *Annu. Rev. Microbiol.* 38:487–514
68. Jannasch HW, Wirsen CO. 1977. Microbial life in the deep sea. *Sci. Am.* 236:42–52
69. Jannasch HW, Wirsen CO. 1984. Variability of pressure adaptation in deep sea bacteria. *Arch. Microbiol.* 139:281–88
70. Jannasch HW, Wirsen CO, Molyneaux SJ, Langworthy TA. 1992. Comparative physiological studies on hyperthermophilic archaea isolated from deep-sea hot vents with emphasis on *Pyrococcus* strain GB-D. *Appl. Environ. Microbiol.* 58:3472–81
71. Jannasch HW, Wirsen CO, Taylor CD. 1982. Deep-sea bacteria: isolation in the absence of decompression. *Science* 216: 1315–17
72. Jannasch HW, Wirsen CO, Winget CL. 1973. A bacteriological pressure-retaining deep-sea sampler and culture vessel. *Deep-Sea Res.* 20:661–64
73. Johnson FH, Eyring H, Polissar MJ. 1954. Hydrostatic pressure and molecular volume changes. In *The Kinetic Basis of Molecular Biology,* pp. 286–368. New York: Wiley & Sons. 874 pp.
74. Kamimura K, Fuse H, Takimura O, Yamaoka T. 1993. Effects of growth pressure and temperature on fatty acid composition of a barotolerant deep-sea bacterium. *Appl. Environ. Microbiol.* 59:924–26
75. Karl DM. 1987. Bacterial production at deep-sea hydrothermal vents and cold seeps: evidence for chemosynthetic primary production. *Symp. Soc. Gen. Microbiol.* 41:319–60
76. Kelly RM, Deming JW. 1988. Extremely thermophilic archaebacteria: biological and engineering considerations. *Biotechnol. Prog.* 4:47–62
77. Kennedy MJ, Reader SL, Swierczynski LM. 1994. Preservation records of micro-organisms: evidence of the tenacity of life. *Microbiology* 140: 2513–29
78. Kinney M, Jones WR, Royal R, Brauer RW, Sorrel FY. 1981. A gradient tube system for the study of the effect of high hydrostatic pressures on temperature preference behavior in small aquatic animals. *Comp. Biochem. Physiol.* 68A: 501–5
79. Kundrot CE, Richards FM. 1987. Changes in the high resolution structure of crystalline hen egg-white lysozyme produced by a hydrostatic pressure of 1000 atmospheres. See Ref. 66a, pp. 245–55
80. Lindahl T. 1993. Instability and decay of the primary structure of DNA. *Nature* 362:709–15
81. Lindahl T, Nyberg B. 1972. Rate of depurination of native deoxyribonucleic acid. *Biochemistry* 11:3610–18
82. Lochte K. 1992. Bacterial standing stock and consumption of organic carbon in

the benthic boundary layer of the abyssal North Atlantic. See Ref. 120a, pp. 1–10
83. Lochte K, Turley CM. 1988. Bacteria and cyanobacteria associated with phytodetritus in the deep sea. *Nature* 333: 67–69
84. Ludlow JM, Clark DS. 1991. Engineering considerations for the application of extremophiles in biotechnology. *Crit. Rev. Biotechnol.* 10:321–45
85. Lutz LH. 1987. *DNA repair in deep-sea bacteria*. PhD dissertation. Univ. Calif. San Diego, La Jolla. 170 pp.
86. Lüdemann H-D. 1987. Water and aqueous solutions under pressure. See Ref. 66a, pp. 273–85
87. Macdonald AG. 1967. The effect of high hydrostatic pressure on the cell division and growth of *Tetrahymena pyriformis*. *Exp. Cell Res.* 47:569–80
88. Macdonald AG. 1978. Further studies on the pressure tolerance of deep-sea crustacea, with observations using a new high-pressure trap. *Mar. Biol.* 45:9–21
89. Macdonald AG, Gilchrist I. 1969. Recovery of deep seawater at constant pressure. *Nature* 222:71–72
90. Macdonald AG, Gilchrist I. 1972. An apparatus for the recovery and study of deep-sea plankton at constant temperature and pressure. In *Barobiology and the Experimental Biology of the Deep Sea*, ed. RW Brauer, pp. 394–412. Chapel Hill, NC: Univ. N.C. Sea Grant Progr. 428 pp.
91. Marquis RE. 1976. High-pressure microbial physiology. *Adv. Microb. Physiol.* 14:159–241
92. Marquis RE. 1993. Bacteria. In *Advances in Comparative and Environmental Physiology*, Vol. 17, *Effects of High Pressure on Biological Systems*, ed. AG Macdonald, pp. 1–28. New York: Springer-Verlag. 246 pp.
93. Marquis RE, Matsumura P. 1978. Microbial life under pressure. In *Microbial Life in Extreme Environments*, ed. DJ Kushner, pp. 105–58. New York: Academic. 465 pp.
94. Matsumura P, Keller DM, Marquis RE. 1974. Restricted pH ranges and reduced yields for bacterial growth under pressure. *Microb. Ecol.* 1:176–89
95. Maxwell JC. 1872. On the measurement of pressure and other internal forces, and of the effects which they produce. In *Theory of Heat*, pp. 94–107. London: Longmans, Green; 1970. Reprint. Westport, CT: Greenwood. 313 pp. 3rd ed.
96. McMeekin TA, Olley J, Ratkowsky DA. 1988. Temperature effects on bacterial growth rates. In *Physiological Models in Microbiology*, ed. MJ Bazin, JI Prosser, 1:75–89. Boca Raton, FL: CRC
97. Miller JF, Almond EL, Shah NN, Ludlow JM, Zollweg JA, et al. 1988. High-temperature-pressure bioreactor for studying pressure-temperature relationships in bacterial growth and productivity. *Biotechnol. Bioeng.* 31:407–13
98. Monod J. 1949. The growth of bacterial cultures. *Annu. Rev. Microbiol.* 3:371–94
99. Morita RY. 1976. Survival of bacteria in cold and moderate hydrostatic pressure environments with special reference to psychrophilic and barophilic bacteria. In *The Survival of Vegetative Microbes*, ed. RG Gray, JR Postgate, pp. 279–98. New York: Cambridge Univ. Press. 432 pp.
100. Morita RY. 1980. Microbial life in the deep sea. *Can. J. Microbiol.* 26:1375–85
101. Morita RY. 1982. Pressure. Bacteria, fungi and blue-green algae. In *Marine Ecology. Environmental Factors*, ed. O Kinne, pp. 1361–88. Chichester, UK: Wiley & Sons
102. Murray CN, Stanners DA, Jamet M. 1989. Technical note: a piston corer for recovery of deep ocean sediments under pressure. *Mar. Geotech.* 8:69–80
103. Nakayama A, Yano Y, Yoshida K. 1994. New method for isolating barophiles from intestinal contents of deep-sea fishes retrieved from the abyssal zone. *Appl. Environ. Microbiol.* 60: 4210–12
104. Nealson KH. 1982. Bacterial ecology of the deep sea. See Ref. 48a, pp. 179–200
105. Nelson CM, Schuppenhauer MR, Clark DS. 1991. Effects of hyperbaric pressure on a deep-sea archaebacterium in stainless steel and glass-lined vessels. *Appl. Environ. Microbiol.* 57:3576–80
106. Nelson CM, Schuppenhauer MR, Clark DS. 1992. High-pressure, high-temperature bioreactor for comparing effects of hyperbaric and hydrostatic pressure on bacterial growth. *Appl. Environ. Microbiol.* 58:1789–93
107. Nordmeier E. 1992. Effects of pressure on the helix-coil transition of calf thymus DNA. *J. Phys. Chem.* 96:1494–501
108. Pledger RD, Crump BC, Baross JA. 1994. A barophilic response by two hyperthermophilic, hydrothermal vent Archaea: an upward shift in the optimal temperature and acceleration of growth rate at supra-optimal temperatures by elevated pressure. *FEMS Microbiol. Ecol.* 14:233–42
109. Pley U, Schipka J, Gambacorta A, Jannasch HW, Fricke H, et al. 1991. *Py-*

rodictium-abyssi sp-nov. represents a novel heterotrophic marine archaeal hyperthermophile growing at 110°C. *Syst. Appl. Microbiol.* 14:245–53
110. Poon PH, Schumaker VN. 1971. Pressure independence of the rotor speed-induced aggregation of DNA. *Biopolymers* 10:1365–69
111. Poremba K. 1994. Simulated degradation of phytodetritus in deep-sea sediments of the NE Atlantic (47 N, 19 W). *Mar. Ecol. Prog. Ser.* 105:291–99
112. Prieur D. 1992. Physiology and biotechnological potential of deep-sea bacteria. In *Molecular Biology and Biotechnology of Extremophiles*, ed. RA Herbert, RJ Sharp, pp. 163–202. New York: Chapman & Hall
113. Rabinowitch HD, Sklan D, Chace DH, Stevens RD, Fridovich I. 1993. *Escherichia coli* produces linoleic acid during late stationary phase. *J. Bacteriol.* 175:5324–28
114. Ratkowsky DA, Lowry RK, McMeekin TA, Stokes AN, Chandler RE. 1983. Model for bacterial culture growth rate throughout the entire biokinetic temperature range. *J. Bacteriol.* 154:1222–26
115. Rentzeperis D, Kupke DW, Marky LA. 1993. Volume changes correlate with entropies and enthalpies in the formation of nucleic acid homoduplexes: differential hydration of A-conformation and B-conformation. *Biopolymers* 33:117–25
116. Rice SA, Oliver JD. 1992. Starvation response of the marine barophile CNPT-3. *Appl. Environ. Microbiol.* 58:2432–37
117. Robinson CR, Sligar SG. 1994. Hydrostatic pressure reverses osmotic pressure effects on the specificity of *Eco*RI DNA interactions. *Biochemistry* 33:3787–93
118. Rosset R, Julien J, Monier R. 1966. Ribonucleic acid composition of bacteria as a function of growth rate. *J. Mol. Biol.* 18:308–20
119. Rosson RA, Nealson KH. 1982. Manganese bacteria and the marine manganese cycle. See Ref. 48a, pp. 201–16
120. Rowe GT, Deming JW. 1985. The role of bacteria in the turnover of organic carbon in deep-sea sediments. *J. Mar. Res.* 43:925–50
120a. Rowe GT, Pariente V, eds. 1992. *Deep-Sea Food Chains and the Global Carbon Cycle*. Dordrecht: Kluwer Academic. 400 pp.
121. Royer CA, Weber G, Daly TJ, Matthews KS. 1986. Dissociation of the lactose repressor protein tetramer using high hydrostatic pressure. *Biochemistry* 25: 8308–15
122. Russell BF, Phelps TJ, Griffin WT, Sargent KA. 1992. Procedures for sampling deep subsurface microbial communities in unconsolidated sediments. *Ground Water Monit. Rev.* 12:96–104
123. Rutberg L. 1964. On the effects of high hydrostatic pressure on bacteria and bacteriophage. 2. Inactivation of bacteriophages. *Acta Pathol. Microbiol. Scand.* 61:91–97
124. Saunders PM, Fofonoff NP. 1976. Conversion of pressure to depth in the ocean. *Deep-Sea Res.* 23 109–11
125. Sheu DD. 1990. The anoxic Orca Basin (Gulf of Mexico): geochemistry of brines and sediments. *Rev. Aquat. Sci.* 2:491–507
126. Speedy RJ. 1982. Stability-limit conjecture: an interpretation of the properties of water. *J. Phys. Chem.* 86:982–91
127. Stetter KO. 1982. Ultrathin mycelia-forming organisms from submarine volcanic areas having an optimum growth temperature of 105°C. *Nature* 300:258–60
128. Stetter KO, Huber R, Blochl E, Kurr M, Eden Rd, et al. 1993. Hyperthermophilic archaea are thriving in deep North Sea and Alaskan oil reserves. *Nature* 365:743–45
129. Suzuki K. 1973. Measurements at high pressure. *Methods Enzymol.* 26:424–52
130. Swinbanks DD, Shirayama Y. 1986. High levels of natural radionuclides in a deep-sea infaunal xenophyophore. *Nature* 320:354 57
131. Tabor PS, Deming JD, Ohwada K, Davis H, Waxman M, Colwell RR. 1981. A pressure-retaining deep ocean sampler and transfer system for measurement of microbial activity in the deep sea. *Microb. Ecol.* 7:51–65
132. Taylor CD. 1979. Growth of a bacterium under a high-pressure oxy-helium atmosphere. *Appl. Environ. Microbiol.* 37: 42–49
133. Taylor CD. 1987. Solubility properties of oxygen and helium in hyperbaric systems and the influence of high pressure oxy-helium upon bacterial growth, metabolism, and viability. See Ref. 66a, pp. 111–28
134. Thompson GA Jr. 1989. Lipid molecular species retailoring and membrane fluidity. *Biochem. Soc. Trans.* 17:286–89
135. Trent JD, Chastain RA, Yayanos AA. 1984. Possible artefactual basis for apparent bacterial growth at 250°C. *Nature* 307:737–40
136. Trent JD, Yayanos AA. 1985. Pressure effects on the temperature range for

growth and survival of the marine bacterium *Vibrio harveyi*: implications for bacteria attached to sinking particles. *Mar. Biol.* 89:165–72

137. Tunnicliffe V. 1992. The nature and origin of the modern hydrothermal vent fauna. *Palaios* 7:338–50
138. Turley CM. 1993. The effect of pressure on leucine and thymidine incorporation by free-living bacteria and by bacteria attached to sinking oceanic particles. *Deep-Sea Res.* 40:2193–206
139. Turley CM, Gooday AJ, Green JC. 1993. Maintenance of abyssal benthic foraminifera under high pressure and low temperature: some preliminary results. *Deep-Sea Res.* 40:643–52
140. Turley CM, Lochte K. 1990. Microbial response to the input of fresh detritus to the deep-sea bed. *Paleogeogr. Palaeoclimatol. Palaeoecol. Global Planet. Change Sect.* 89:3–23
141. Turley CM, Lochte K, Patterson DJ. 1988. A barophilic flagellate isolated from 4500 m in the mid-North Atlantic. *Deep-Sea Res.* 35:1079–92
142. van Eldik R. 1986. High pressure kinetics: fundamental and experimental aspects. See Ref. 142a, pp. 1–68
142a. van Eldik R, ed. 1986. *Inorganic High Pressure Chemistry. Kinetics and Mechanisms*. New York: Elsevier. 448 pp.
143. Weyland H, Helmke E. 1989. Barophilic and psychrophilic bacteria in the Antarctic Ocean. See Ref. 54a, pp. 43–47
144. Wild RA, Darlington E, Herring PJ. 1985. An acoustically controlled cod-end system for the recovery of deep-sea animals at *in situ* temperatures. *Deep-Sea Res.* 32:1583–89
145. Wu JQ, Macgregor RBJ. 1993. Pressure dependence of the melting temperature of dA•dT polymers. *Biochemistry* 32:12531–37
146. Yayanos AA. 1969. A technique for studying biological reaction rates at high pressure. *Rev. Sci. Instrum.* 40:961–63
147. Yayanos AA. 1975. Stimulatory effect of hydrostatic pressure on cell division in cultures of *Escherichia coli*. *Biochim. Biophys. Acta* 392:271–75
148. Yayanos AA. 1977. Simply actuated closure for a pressure vessel: design for use to trap deep-sea animals. *Rev. Sci. Instrum.* 48:786–89
149. Yayanos AA. 1978. Recovery and maintenance of live amphipods at a pressure of 580 bars from an ocean depth of 5700 meters. *Science* 200:1056–59
150. Yayanos AA. 1980. Measurement and instrument needs identified in a case history of deep-sea amphipod research. In *Advanced Concepts in Ocean Measurements for Marine Biology*, ed. FD Diemer, FJ Vernberg, DZ Mirkes, pp. 307–18. Columbia, SC: Univ. S.C. Press. 572 pp.
151. Yayanos AA. 1982. Deep sea biophysics. In *Subseabed Disposal Program. Annual Rep., Jan. to Sept., 1981*, ed. KR Hinga, 2:407–26. Albuquerque, NM: Sandia Natl. Lab.
152. Yayanos AA. 1986. Evolutional and ecological implications of the properties of deep-sea barophilic bacteria. *Proc. Natl. Acad. Sci. USA* 83:9542–46
153. Yayanos AA. 1989. Physiological and biochemical adaptations to low temperatures, high pressures and radiation in the deep sea. See Ref. 54a, pp. 38–42
154. Yayanos AA. 1996. Empirical and theoretical aspects of life at high pressures in the deep sea. In *Extremophiles*, ed. K Horikoshi, WD Grant. New York: Wiley & Sons. In press
155. Yayanos AA, DeLong EF. 1987. Deep-sea bacterial fitness to environmental temperatures and pressures. See Ref. 66a, pp. 17–32
156. Yayanos AA, Dietz AS. 1982. Thermal inactivation of a deep-sea barophilic bacterium, isolate CNPT-3. *Appl. Environ. Microbiol.* 43:1481–89
157. Yayanos AA, Dietz AS. 1983. Death of a hadal deep-sea bacterium after decompression. *Science* 220:497–98
158. Yayanos AA, Dietz AS, Van Boxtel R. 1979. Isolation of a deep-sea barophilic bacterium and some of its growth characteristics. *Science* 205:808–10
159. Yayanos AA, Dietz AS, Van Boxtel R. 1981. Obligately barophilic bacterium from the Mariana Trench. *Proc. Natl. Acad. Sci. USA* 78:5212–15
160. Yayanos AA, Dietz AS, Van Boxtel R. 1982. Dependence of reproduction rate on pressure as a hallmark of deep-sea bacteria. *Appl. Environ. Microbiol.* 44:1356–61
161. Yayanos AA, Pollard EC. 1969. A study of the effects of hydrostatic pressure on macromolecular synthesis in *Escherichia coli*. *Biophys. J.* 9:1464–82
162. Yayanos AA, Van Boxtel R. 1982. Coupling device for quick high-pressure connections to 100 MPa. *Rev. Sci. Instrum.* 53:704–5
163. Yayanos AA, Van Boxtel R, Dietz AS. 1983. Reproduction of *Bacillus stearothermophilus* as a function of temperature and pressure. *Appl. Environ. Microbiol.* 46:1357–63
164. Yayanos AA, Van Boxtel R, Dietz AS. 1984. High-pressure-temperature gradient instrument: use for determining the temperature and pressure limits of bac-

terial growth. *Appl. Environ. Microbiol.* 48:771–76

165. Yazawa K, Araki K, Watanabe K, Ishikawa C, Inoue A, et al. 1988. Eicosapentaenoic acid productivity of the bacteria isolated from fish intestines. *Nippon Suisan Gakkaishi* 54:1835–38

166. ZoBell CE. 1952. Bacterial life at the bottom of the Philippine Trench. *Science* 115:507–8

167. ZoBell CE. 1954. The occurrence of bacteria in the deep sea and their significance for animal life. In *On the Distribution and Origin of the Deep Sea Bottom Fauna. International Union of Biological Sciences, Ser. B, No. 16*, pp. 20–26. Naples, Italy: Int. Union Biol. Sci. 80 pp.

168. ZoBell CE, Cobet AB. 1962. Growth, reproduction, and death rates of *Escherichia coli* at increased hydrostatis pressures. *J. Bacteriol.* 84:1228–36

168a. Zobell CE, Johnson FH. 1949. The influence of hydrostatic pressure on the growth and viability of terrestrial and marine bacteria. *J. Bacteriol.* 57:179–89

169. ZoBell CE, Oppenheimer CH. 1950. Some effects of hydrostatic pressure on the multiplication and morphology of marine bacteria. *J. Bacteriol.* 60:771–81

170. Zwickl P, Fabry S, Bogedain C, Haas A, Hensel R. 1990. Glyceraldehyde-3-phosphate dehydrogenase from the hyperthermophilic archaebacterium *Pyrococcus woesei*: characterization of the enzyme, cloning and sequencing of the gene, and expression in *Escherichia coli. J. Bacteriol.* 172:4329–38

VIRAL VECTORS IN GENE THERAPY

Alan E. Smith

Genzyme Corporation, One Mountain Road, Framingham, Massachusetts 01701

KEY WORDS: gene therapy, adenovirus vectors, AAV vectors, retrovirus vectors, cationic lipids

CONTENTS

INTRODUCTION	808
OVERALL GOALS OF GENE THERAPY	809
CATEGORIES OF VIRUS CONSIDERED	810
Retroviruses	810
Adenoviruses	812
Adeno-Associated Virus	815
VECTOR DESIGN	816
Retrovirus Vectors	817
Adenovirus Vectors	819
AAV Vectors	821
VECTOR PRODUCTION	821
Retrovirus Vectors	821
Adenovirus Vectors	823
AAV Vectors	824
VECTOR EFFICACY	824
In Vitro Studies	824
Animal Studies	826
ALTERNATIVE VECTOR SYSTEMS	830
Cationic Lipids	830
Molecular Conjugates	830
Naked DNA	830
CONCLUSIONS	830

Abstract

The use of DNA as a drug is both appealing and simple in concept. Indeed in many instances the feasibility of such an approach has been established using model systems. In practical terms, however, the delivery of DNA to human tissues presents a wide variety of problems that differ with each potential therapeutic application. In this review, the design, production, and application of viral vectors for human gene therapy are considered. Although viral vectors are an obvious starting point because viruses have evolved efficient mechanisms to introduce and express their nucleic acid into recipient cells, by the same token the viral hosts have evolved sophisticated mechanisms to rid themselves of such pathogens. The challenge for the therapeutic use of viral vectors is to achieve efficient and often extended expression of the exogenous gene while evading the host defenses. Methodology used and progress towards that goal are reviewed.

INTRODUCTION

Progress in the study of human disorders over the past 25 years has greatly enhanced our ability to describe the molecular basis of many disease states. Molecular genetics techniques have been particularly powerful. They have allowed the isolation of the genes associated with common inherited diseases such as cystic fibrosis (CF), as well as the identification of many other genes that contribute to more complex diseases such as cancers. Because our knowledge of the basic mechanisms underlying many biological processes such as gene expression and protein synthesis has also increased, the challenge today is to use all this information in the development of new treatments for disease.

One very logical application is the use of DNA itself as a drug. The delivery of the appropriate gene to a patient with a recessive inherited disease should correct the genetic defect and potentially cure the disease state. Delivery of genes encoding a toxin might kill cancer cells, whereas other genes might be specifically tailored to kill infectious organisms. The list of potential applications is long, and the rationale behind each application is extremely simple.

However, the very simplicity of the theory carries attendant dangers. It is easy to overlook the myriad of technical details that must be solved before gene therapy can become a practical therapy capable of treating significant numbers of patients with common diseases. Many problems need to be solved in developing any gene-therapy approach: definition of the cells that constitute the target, entry of DNA into those cells, expression of useful levels of gene product over an appropriate time period, avoidance of the almost inevitable response of the host to the introduced agents, and so on. When first confronted with these challenges, investigators turned to the animal viruses as first-generation agents to deliver DNA to cells. These issues closely parallel those facing a virus in going through its life cycle. Viruses have evolved to be extremely efficient not only at delivering nucleic acid to particular cells but also at evading host-defense mechanisms. Fortunately, viruses themselves have been the object of intense study in the recent past, and much is known of their molecular anatomy. This review summarizes progress in developing viruses as gene-therapy vectors. Rather than attempting to be comprehensive, it highlights practical issues and unanswered questions that remain barriers to the widespread application of gene therapy. Several earlier reviews are also recommended (3, 66, 91, 97, 101, 131).

OVERALL GOALS OF GENE THERAPY

Gene therapy consists of the introduction of nucleic acid into cells of a patient in order to use the expression of that nucleic acid for some therapeutic purpose. Though simple, this definition encompasses an extremely wide range of applications such that defining the goals of all gene-therapy applications in more precise terms is difficult.

The following examples illustrate this diversity: CF is a prototypical monogenic, recessive, genetic disease whose treatment would require delivery, primarily to airway epithelial cells, of an integral membrane protein that functions as a chloride channel. Treatment of solid tumors, for example in brain or lung, could require delivery of a gene encoding a toxin or an immune marker or enhancer. A vaccine might need delivery of a gene encoding an immunogen. In some cases, treatment of hemophilia would require delivery of large amounts of a soluble, circulating protein involved in blood clotting, such as factor VIII or IX. Adenosine deaminase (ADA) deficiency treatment requires delivery, ideally to a hemopoietic stem cell, of a gene encoding a housekeeping enzyme.

Bearing in mind these applications, we consider the following issues:

1. What is the nature of the gene product? Is it a nucleic acid, for example an antisense RNA or an RNA decoy? Alternatively, is it a protein? If so, is it secreted, where is it located, how much is required, and over what time period?
2. What is the target tissue? Is it a particular cell type, and is that cell readily accessible? Does the disease state render it more or less accessible?
3. Is the target cell proliferating or nonproliferating?
4. Is the treatment likely to be in vivo or ex vivo? That is, can cells be removed and treated ex vivo, as in all the early protocols, or can, or must, they be treated in situ? If the cells are treated ex vivo, will they be readministered as a separate neo-organ or will they be placed at their normal location? If a neo-organ, will it comprise cells of the same or a different species?
5. Is the requirement for gene expression temporary or permanent? Will the treated cell turn over? Will successful application eliminate the treated cell?
6. Is treatment of all cells in an organ or tissue necessary? Will there be a bystander effect? Will treated cells have a selective advantage?

From these simple considerations, it follows that the requirements for any particular application vary greatly and will profoundly influence the choice of a viral vector to be developed and tested. Factors that will require testing include the efficacy of gene transfer, the efficacy of gene expression, the duration of gene expression, the ability to repeat dosing, and the ability to target appropriate cells and avoid inappropriate cells. Confounding factors that may arise include the inability of virus to enter into or integrate into the chromosomes of particular cells, the shutdown of transcriptional promoters, the loss of input DNA, the destruction of treated cells, and the neutralization of input virus or gene product. All of these factors will strongly depend on the choice of viral vector and on the ability of the host to respond to that virus.

CATEGORIES OF VIRUS CONSIDERED

In this review, we consider vector systems based on three different virus groups: retroviruses (90, 92, 93, 137), adenoviruses (11, 12, 52, 129), and adeno-associated viruses (AAV) (17, 76, 102). The more complex herpes virus–based vectors are reviewed elsewhere in this volume (50a). Clearly, viruses can be used as gene-therapy vectors, at least under ideal circumstances. The question that must be answered is, what barriers at present appear likely to limit the practical application of these virus vectors?

Retroviruses

Retroviruses comprise a large class of enveloped viruses that contain single-stranded RNA as the viral genome (137). During the normal viral life cycle, viral RNA is reverse transcribed to yield double-stranded DNA that integrates into the host genome and is expressed over extended periods. As a result, infected cells shed virus continuously without apparent harm to the host cell. The viral genome is small (approximately 10 kb), and its prototypical organization is extremely simple, comprising three genes encoding *gag*, the group specific antigens or core proteins; *pol*, the reverse transcriptase; and *env*, the viral envelope protein (Figure 1a). The termini of the RNA genome are called long terminal repeats (LTRs) and include promoter and enhancer activities and sequences involved in integration. The genome also includes a sequence required for packaging viral RNA and splice acceptor and donor sites for generation of the separate envelope mRNA (Figure 1b). Importantly, most retroviruses can integrate only into replicating cells, although human immunodeficiency virus (HIV) appears to be an exception (84). This property, possessed by most retroviruses, clearly restricts their use as vectors for gene therapy.

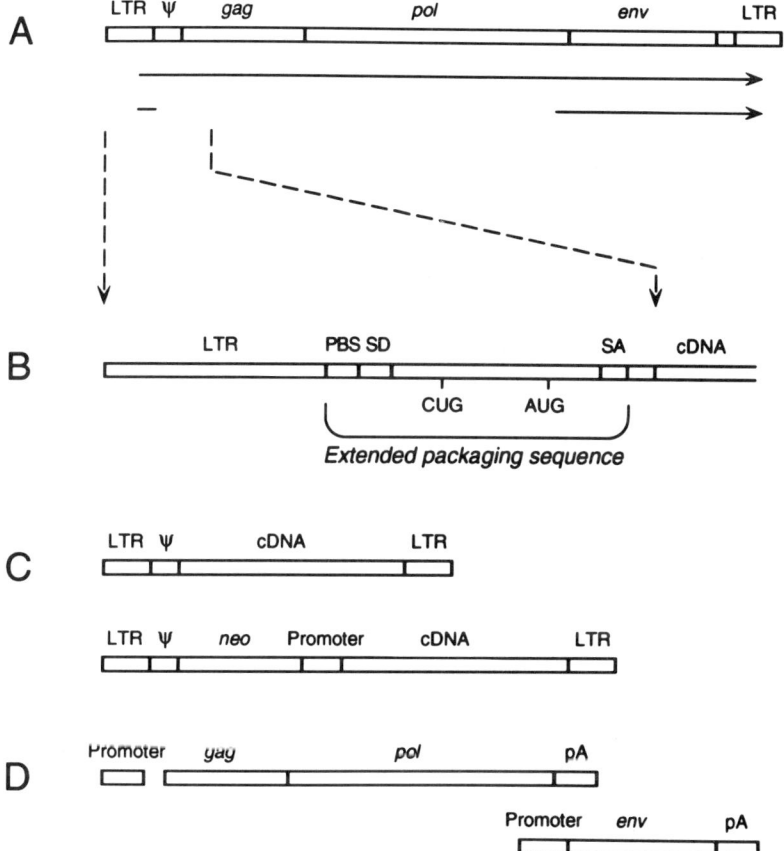

Figure 1 (A) Structure of the integrated form of a prototypical retrovirus. Solid lines with arrows indicate major RNA transcripts. LTR, long terminal repeats; ψ, packaging signal. (B) Detail of 5′ end of integrated retrovirus genome. PBS, primer binding site; CUG and AUG, *gag* initiation codons; SD, splice donor; SA, splice acceptor. (C) Prototypical retrovirus vectors without and with selectable marker. (D) Sequences contained in a prototypical packaging cell line with split genome. Not to scale.

Retroviruses have been studied intensely in the last 20 years, partly because a category of retroviruses causes tumors in animals (137). This ability to cause tumors and to transform the growth properties of cells in culture is now well understood and is based on at least two basic mechanisms. The first is that certain viruses have incorporated activated protooncogenes that upon mutation have acquired the ability to transform cellular growth. The ability of retroviruses to transform cells by this mechanism is not a major concern in gene therapy because the oncogenes are always deleted from the vectors.

The second mechanism of transformation by retroviruses results from insertional mutagenesis upon integration of the viral genome. Because the viral LTR has promoter and enhancer activity and is present at both ends of the genome, insertion of an LTR sequence adjacent to a cellular protooncogene can lead to inappropriate expression of a protein involved in cellular regulation. This mechanism was extensively studied in the activation of c-*myc* by avian leucosis virus (ALV) (61). If retrovirus vectors for gene therapy integrate into the human genome essentially at random, insertional mutagenesis will occur at some frequency. This could lead to protooncogene activation or disruption of a tumor suppressor gene. In practice, however, the frequency of adverse events attributable to insertional mutagenesis is extremely low, perhaps reflecting the observation that much of the human genome is noncoding and that oncogenic transformation usually involves multiple mutagenic events (21, 82).

In essence, retrovirus vectors are relatively simple (66, 92, 93), containing the 5' and 3' LTRs, a packaging sequence, and a transcription unit composed of the gene or genes of interest (Figure 1c). To grow such a vector, one must provide the missing viral functions in *trans* using a so-called packaging cell line (Figure 1d). Such a cell is engineered to contain integrated copies of *gag, pol,* and *env* but to lack a packaging signal so that no helper virus sequences become encapsidated (90). Additional features added to or removed from the vector and packaging cell line reflect attempts to render the vectors more efficacious or reduce the possibility of contamination by helper virus.

The main advantage of retrovirus vectors is that they integrate and are therefore capable, potentially, of long-term expression. They can be grown in relatively large amounts, but care is needed to ensure the absence of helper virus. The host range of the vectors can be manipulated, but their application is limited to replicating cells. Doubts still remain about the possibility of insertional mutagenesis.

Adenoviruses

Adenoviruses comprise a large class of nonenveloped viruses containing linear double-stranded DNA (49, 62, 129). The normal life cycle of the virus does not require dividing cells and involves productive infection in permissive cells during which large amounts of virus accumulates in the nucleus. The productive infection cycle takes about 32–36 hours in cell culture and comprises two phases, the early phase, prior to viral DNA synthesis, and the late phase, during which viral structural proteins and viral DNA are synthesized and assembled into virions. In general, adenovirus infections are associated with mild disease in humans.

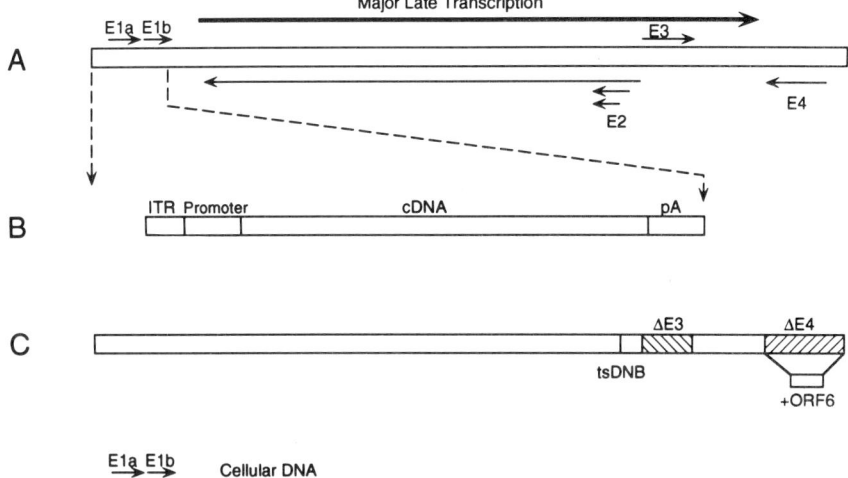

Figure 2 (A) Genome organization of adenovirus. Solid lines with arrows show major early and late (*bold arrow*) transcription units. (B) Detail of 5' end of E1 replacement vector. ITR, inverted terminal repeat; pA, poly A addition sequence/transcriptional terminator. (C) Second-generation vector backbones. tsDNB, vector encoding a temperature-sensitive DNA-binding E2a gene product (39); ΔE3, commonly used E3 deletion; ΔE4, vector lacking E4 with exception of open reading frame 6 (4). (D) Sequences contained within 293 packaging cells. Not to scale.

The adenovirus genome is much larger (about 35 kb), and its organization is much more complex than retroviruses (Figure 2a) (129). Of the four early transcriptional units, each has a separate promoter and encodes several variously spliced mRNAs. The late transcription unit encodes proteins required for virus assembly. Transcription is somewhat sequential in that the E1 genes are expressed early and in part activate transcription of other early genes. The functions of the early proteins fall into various categories; for example, the E1 proteins rescue cells from quiescence, rendering them capable of rapid viral DNA and protein synthesis using predominantly host-coded functions. The E3 genes encode functions to counter host cell defense mechanisms, for example to negate the effects of tumor necrosis factor (TNF) on infected cells and to block the transport of nascent class 1 proteins of the major histocompatibility complex (MHC) so as to reduce presentation of newly synthesized viral peptides to the host immune system (51). The E2 genes encode proteins involved largely in DNA replication. The E4 region is transcriptionally complex and encodes functions regulating the transition between the early and late phases of the viral life cycle.

Adenoviruses have also been studied intensely over the recent past (129). The reasons are twofold: First, the viral genome comprises several interacting transcription units that are complex enough to be interesting, yet simple enough to allow detailed molecular analysis. As a result, many important concepts in eukaryotic molecular biology were first established in studies of adenoviruses, such as mRNA splicing. The second reason is that adenoviruses also can transform the growth properties of cultured cells and form tumors when injected into new-born rodents (129). This process results from adenovirus infection of nonpermissive cells. Under these circumstances, some early gene expression and some DNA synthesis occur, but very little late protein expression and no virus production results. Instead in a very small proportion of the abortively infected cells, viral DNA becomes integrated and early gene expression results. Constitutive expression of E1a and E1b proteins, as in productive infection, negates the tumor-suppressor activity of the Rb and p53 proteins, in this case fortuitously leading to permanent transformation of cellular growth (82). Additional factors are also involved. For example, some serotypes of adenovirus, such as Ad12, are more oncogenic than others, such as Ad2 and Ad5. This ability of adenovirus to transform cells in culture is not regarded as a major problem for gene-therapy applications because despite extensive searches, adenoviruses have never been associated with naturally occurring tumors either in humans (54) or animals and because the E1 transforming genes are absent from most defective adenovirus vectors.

Adenovirus vectors are somewhat larger and more complex than retrovirus or AAV vectors (11, 12, 52), partly because only a small fraction of the viral genome is removed from most current vectors. If additional genes were removed, they would need to be provided in *trans* to produce the vector, which so far has proved difficult. Instead, two general types of adenovirus-based vectors have been studied, E3-deletion and E1-deletion vectors (11, 52). Some viruses in laboratory stocks of wild-type lack the E3 region and can grow in the absence of helper. This ability does not mean that the E3 gene products are not necessary in the wild, only that replication in cultured cells does not require them. Deletion of the E3 region allows insertion of exogenous DNA sequences to yield vectors capable of productive infection and the transient synthesis of relatively large amounts of protein (96, 127).

Deletion of the E1 region disables the adenovirus, but such vectors can still be grown because there exists an established human cell line (called 293) that contains the E1 region of Ad5 and that constitutively expresses the E1 proteins (53). Most recent gene-therapy applications involving adenovirus have utilized E1 replacement vectors grown in 293 cells.

The main advantages of adenovirus vectors are that they are capable of very efficient episomal gene transfer in a wide range of cells and tissues and that

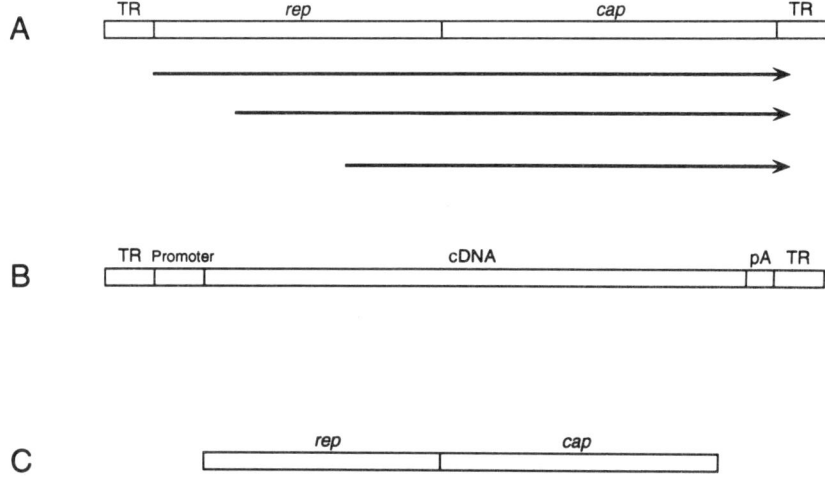

Figure 3 (*A*) Genome organization of AAV. Lines with arrows indicate major transcription units. TR, terminal repeats. (*B*) AAV vector. (*C*) AAV sequences required in packaging cell line. Not to scale.

they are easy to grow in large amounts. The main disadvantage is that the host response to the virus appears to limit the duration of expression and the ability to repeat dosing, at least with high doses of first-generation vectors.

Adeno-Associated Virus

AAV is a very small, very simple, nonautonomous virus containing linear single-stranded DNA (17, 76, 102). The virus requires co-infection with adenovirus or certain other viruses in order to replicate. AAV is widespread in the human population, as evidenced by antibodies to the virus, but it is not associated with any known disease.

The genome organization is extremely simple, comprising just two genes—*rep*, encoding a family of overlapping proteins involved in replication and integration, and *cap*, encoding a family of three viral structural proteins (Figure 3*a*). The termini of the genome comprise terminal repeats (TR) of about 145 nucleotides.

Wild-type AAV can integrate its DNA into the host chromosome of cells in the absence of helper virus. This integration seems to occur most often in a specific region of chromosome 19 and involves the *rep* protein (77).

AAV-based vectors are extremely simple (Figure 3*b*) (17, 76, 102). They contain only the viral TR sequences flanking the transcription unit of interest.

The only constraint appears to be that the length of the vector DNA cannot greatly exceed the viral genome length of 4680 nucleotides. Currently, growth of AAV vectors is cumbersome and involves introducing into the host cell not only the vector itself but also a plasmid encoding *rep* and *cap* to provide helper functions. The helper plasmid lacks TRs and consequently cannot replicate and package in its own right. In addition, helper virus such as adenovirus is required.

The potential advantage of AAV vectors is that they appear capable of long-term expression in nondividing cells, possibly, though not necessarily, because the viral DNA integrates (46). The vectors are structurally simple, and they may therefore provoke less of a host-cell response than adenovirus. Their main limitation at present is that vectors are extremely difficult to grow in large amounts.

VECTOR DESIGN

Viral vectors are made up of at least two components, the modified viral genome and the virion structure surrounding it. Most present-generation vectors comprise virion particles based largely on the wild-type virus structure. This structure packages and protects the viral nucleic acid and provides the means to bind and enter target cells. However, the viral nucleic acid in a vector designed for gene therapy is changed in many ways. The goals of these changes are to disable growth of the virus in target cells while maintaining its ability to grow in vector form in available packaging or helper cells, to provide space within the viral genome for insertion of exogenous DNA sequences, and to incorporate new sequences that encode and enable appropriate expression of the gene of interest. In this way, we may consider vector nucleic acids as comprising two components: those essential viral sequences that remain and the transcription unit for the exogenous gene.

Because any remaining viral coding sequences might be expressed, even by a disabled viral vector, all nonessential viral sequences should be removed. Thus, ideally, a vector will contain only those viral nucleic acid sequences necessarily required in *cis*. However, virtually all other viral functions are still needed to allow replication of the vector genome for production purposes and to provide proteins required to assemble virion particles. Ideally, a specific packaging or helper cell line would be engineered that expresses those viral functions able to function in *trans*. In practice, constitutive expression of viral structural proteins and of proteins involved in replication often proves toxic to a cell line, particularly if the life cycle of the virus normally involves acute, productive infection resulting in cell death rather than chronic infection. Thus, appropriate packaging cell lines required to grow many idealized vectors do not exist. The minimal essential features present in current generations of viral vectors are discussed below.

Retrovirus Vectors

ESSENTIAL VIRAL SEQUENCES Because cells tolerate the presence of constitutive retrovirus proteins, the generation of packaging cell lines expressing all the viral proteins is relatively straightforward. Consequently, retrovirus vectors need only contain nucleic elements required in *cis*. These include the 5' and 3' LTR sequences and the packaging sequence (sometimes referred to as ψ) that lies downstream of the 5' LTR.

The essential 5' region contains numerous viral functions (Figure 1*b*), including many involved in nucleic acid replication and integration, promoter and enhancer sequences, a tRNA primer binding site (PBS), and a splice donor site. This region also encompasses the start codons for the *gag* protein, which in the case of Moloney murine leukemia virus (MoMLV), includes an upstream, in-frame CUG (111) encoding a glycosylated *gag* protein variant with an N terminal extension as well as the more usual AUG.

LTRs from various retroviruses, including those with specificity for different species, have been used successfully to generate retrovirus vectors (60, 74, 93). Even HIV-based vectors have been produced (109). Because murine retroviruses are well characterized and experiments often involve mouse models, vectors commonly utilize the MoMLV LTR and packaging sequence [for example the N2 (5), LNL6 (93), and MFG vectors (105)]. Aside from safety issues involving generation of wild-type recombinant virus in the packing cell line discussed below, for any given application, several features influence the efficacy of an LTR.

Some LTR variants have been especially designed for expression in certain cell types; for example, the murine embryonic stem cell virus (MESV)-based vector is expressed in embryonic cells (55, 56, 60). Changes in the LTR that influence expression include deletions within the U3 region and point mutations that change transcription-factor binding or binding of other cell-specific proteins (56, 72, 130). The PBS immediately adjacent to the 5' LTR also influences expression; for instance, a silencer element from some cells binds to a site for tRNA^{Pro} but not tRNA^{Gln} (19, 55, 72).

The packaging signal currently used is more extensive than that utilized in first-generation vectors and now includes the N-terminal *gag* coding sequences (1, 5). Potential expression from the initiation codon for *gag* synthesis rather than the gene of interest can be prevented by mutation of the AUG sequence or insertion of a downstream termination codon (10). Furthermore, the upstream CUG initiation codon in MoMLV is not present in MoMSV. Consequently for applications such as the LNL6 vector, a hybrid packaging sequence is constructed that consists of MoMSV sequences lacking the upstream CUG and MoMLV sequences containing a mutated AUG (93). This region also contains the naturally occurring splice donor upstream of the *gag* AUG, which normally would interact with the splice acceptor upstream of the *env* coding sequences (Figure 1*a*). However, in the absence of the *env* acceptor site, a cryptic 3' splice acceptor site in the *gag* coding sequences of the extended packaging signal is activated. The use of the splice acceptor-donor combination appears to be important in some vectors (such as N2) but not in others (LNL6) (93). Essential viral sequences at the 3' end of the viral vector include the LTRs but little else.

TRANSCRIPTION UNIT FOR EXOGENOUS GENE EXPRESSION In the process of designing the transcription unit of a retrovirus vector, several questions arise. First, is one protein required or will the vector encode the gene of interest plus a selectable marker? If two proteins are required, will two promoters or a dicistronic mRNA be used? Another important consideration is the amount of the gene product required and the desired duration of expression.

Promoter and enhancer The viral LTR is commonly used as the promoter and enhancer for the gene of interest. Such promoters are usually regarded as relatively strong, and large amounts of gene product can be produced. If a specific cell line is the target, a promoter specially adapted to that cell line might be necessary, as discussed above. When two genes are inserted, a second promoter is often inserted upstream of the second transcription unit (93). A recent systematic study compared different combinations and orientations of the SV40 and CMV promoters with β-galactosidase and neomycin as the test genes and found marked effects, which varied with the precise construct, on both gene expression and vector stability (89). An alternative approach is to incorporate internal ribosome entry sites upstream of the second coding region and rely on a single transcript to generate both gene products (73, 83).

An important consideration for applications requiring long-term expression is promoter shut down. Experiments indicate that in some cases, integrated viral DNA persists but transcription is reduced to undetectable levels (29, 107). This finding has led to the search for long-lasting housekeeping promoters or promoters with greater tissue specificity (59, 122). Also, when a secreted protein is the product, the search has been for an alternative tissue in which the gene can be expressed but shutdown may be less problematic (29, 87, 108, 122).

Splicing Because many vectors are constructed by insertion of an appropriate cDNA into a preformed vector cassette, most rely on the splicing signals provided in the vector backbone. In the N2 vector, for example, a splice donor and acceptor are present within the extended ψ packaging sequence (93). In the MFG vector, the splice acceptor is the naturally occurring sequence utilized in production of *env* mRNA. Placing the AUG of the introduced coding sequence precisely at the site of the *env* AUG has resulted in highly efficient expression (105).

Conclusions—retrovirus vector design   Although the design of retrovirus vectors has reached an advanced stage, it remains largely empirical. This constraint partly results from the very nature of retroviruses themselves. As the viral nucleic acid is RNA, cryptic splice sites present in any given recombinant genome may be functional and, consequently, preclude the packaging of full-length transcripts. Success can only be assured by testing the vector in appropriate packaging cells to determine whether high titer virus is obtained, and whether the vector is stable. Furthermore, its ability to transduce target cells at high efficiency needs to be tested.

In this regard, surprisingly few systematic studies have been done to compare the properties of related vectors. Often constructs used in comparative studies differ by more than one feature, making interpretation difficult. An exception, which can be interpreted, is the systematic study of β-galactosidase- and neomycin-resistance expression in closely related vectors mentioned above (89). A more recent study examined stable expression of human ADA in reconstituted mouse hemopoietic cells by using MFG vectors containing features that reportedly influenced expression in other contexts. The results indicate that of the parameters tested, use of only a PBS mutation or a myeloproliferative sarcoma virus (MPSV) LTR, rather than a MoMLV LTR, yielded significantly improved gene expression (116).

Adenovirus Vectors

E3 deletion-based Ad4 vectors have been designed for use as vaccines (96). In these vectors, the transcription unit encoding the immunogen (for example hepatitis B surface antigen or HIV gp160) is inserted in the E3 region of an otherwise wild-type, viable genome. Tests in animals have been successful, but results from human tests, although limited, have been disappointing (127). This probably reflects the importance of the E3 region in reducing the host cell immune response to adenovirus. In its absence, vector directed antigen synthesis is probably short lived and hence unable to produce immunity.

ESSENTIAL VIRAL GENES   Virtually all adenovirus vectors for gene therapy are E1 replacement vectors, which lack the majority of the E1 coding region and are grown in 293 cells. Vectors retain the immediate 5′ end of the viral genome, including the terminal repeats, sequences required for packaging, and the overlapping E1 enhancer region (Figure 2a). All coding sequences for E1a are deleted, but often some of the 3' ends of the E1b gene and all of the protein IX coding sequences remain. All of the remainder of the viral genome can be left intact (114), but usually some or all of the sequences encoding the E3 region are also removed (118). One report described a vector in which E3 is retained but all sequences within E4 are deleted with the exception of the ORF-6 (4). ORF-6 was retained because it appears to be the only sequence from the E4 region that is essential for growth in tissue culture. The Ad2/ORF-6 vector grows well in 293 cells and is stable.

Attempts to create adenovirus-based vectors with more extensive deletions are limited by the ability to produce appropriate packaging cells. For the present, the deletion of nonessential genes, conditional mutation of other genes, or provision of helper functions by means of co-infection rather than constitutive expression are the more simple options. One report of a so-called pseudo adenovirus vector (PAV) described growth of an Ad2-based vector containing only 5' and 3' viral terminal repeat sequences flanking exogenous DNA encoding β-galactosidase. Growth and packaging of the vector in 293 cells co-infected with wild-type helper is possible, but the difficulty of removing all wild-type helper virus presently limits the practical application of such a vector (25). Another recent report describes an E1- and E3-deleted vector in which the E2a gene product included a temperature-sensitive mutation (39). This construct was produced in the hope that decreased DNA-binding protein activity produced at 37°C would render the remainder of the viral genome more quiescent.

TRANSCRIPTIONAL UNIT FOR EXOGENOUS GENE EXPRESSION *Promoter and enhancer* Various promoters have been used to express exogenous genes in adenovirus vectors, including adenovirus MLP (118), PGK (4), and hybrid CMV/β-actin (38) as well as the endogenous E1a promoter (114). Choice of the appropriate promoter depends on application and tissue. Promoter shutdown as a result of adenovirus vectors has not been reported.

Polyadenylation A variety of signals have been used for polyadenylation, including the endogenous E1b/protein IX sequence. One problem that appears to cause low yield of virus in some instances is promoter occlusion at the protein IX transcription unit, which is caused by read-through from a strong promoter upstream of the introduced gene (7). Reduced expression of protein IX renders packaging less efficient. One can apparently solve this problem by utilizing a strong polyadenylation signal, such as SV40, downstream of the exogenous cDNA or by inverting the transcription unit.

AAV Vectors

ESSENTIAL VIRAL ELEMENTS The only viral sequences essential for an AAV vector are the 145-nucleotide TRs at the 5' and 3' ends of the vector. These provide sequences necessary for replication and packaging of the vector DNA.

TRANSCRIPTION UNIT FOR EXOGENOUS GENE EXPRESSION For some applications, the size of the vector DNA is problematic. DNA larger than ~4800 nucleotides is difficult to replicate and package. Consequently, the promoter-enhancer region and the polyadenylation signal utilized are sometimes selected on the basis of minimal size requirements (44). Indeed, a vector encoding cystic fibrosis transmembrane conductance regulator (CFTR) utilizes a promoter within the TR that had previously not been reported (45).

When the size of the introduced DNA is not problematic, more conventional transcription units can be used. These units can include various exogenous promoters and, in some cases, selectable markers (76). To introduce tissue specificity of expression, segments of locus control regions have also been introduced, but only with limited success (94).

VECTOR PRODUCTION

Retrovirus Vectors

The production of retrovirus-based vectors in a packaging cell has three goals, to produce vector nucleic acid enclosed within a virion particle that has appropriate target specificity, to produce large amounts of vector efficiently, and to avoid the breakout of replication-competent retrovirus (RCR).

PACKAGING CELLS A packaging cell line is produced by means of transfection in a helper virus plasmid encoding *gag, pol,* and *env* and by selecting for cells that express the proteins and can support vector production (86, 90). One can avoid replication of the helper sequences by at least making deletions within the packaging sequence. Similarly, to avoid recombination with the replicating vector, one must remove additional sequences. For example, sequences beyond the envelope-coding sequence, including the 3' LTR, are commonly deleted and replaced with a polyadenylation sequence such as that from SV40 (36). Deletions can be incorporated into the 5' LTR to reduce its ability to replicate, and an exogenous promoter can be used instead (93). Further adaptations designed to ensure that a packaging cell line cannot produce RCR include insertion of point mutations or deletions at key positions in the helper genome and splitting the viral genome into two transcription units, one encoding *gag* and *pol* and a second encoding *env* (88). The combination of such approaches and the transfection of cells at two different times to encourage integration of the split genome at different sites in the host chromosome reduces the possibility of recombination. In spite of this, multiple crossover recombination to generate a breakout of replication-competent vector remains a possibility that must be constantly tested for in production lots.

To ensure the appropriate target specificity, the *env* gene of the helper virus/packaging cell line can be varied. For example, to generate a vector capable of infecting murine cells (a so-called ecotropic vector), one can use a packaging cell line expressing an *env* sequence from a murine retrovirus such as MoMLV, but to produce virus capable of infecting human cells, one can use the *env* sequence of an amphotropic virus such as 4070A instead (9, 23, 30, 98).

Recently, more sophisticated targeting approaches have been taken (121). Initially, these involved covalent modification of virions (104, 119). A more precise approach involves removing sequences from the *env* gene and replacing them with sequences encoding proteins with a different specificity. For example, erythropoietin sequences have been used to target the Epo receptor (70). Another approach is to incorporate a single chain antibody into the *env* sequence (18, 120).

A different approach utilizes the ability of retroviruses to incorporate glycoproteins from other viruses into their envelope to produce so-called pseudotypes. Recently developed methods incorporate the HA protein of influenza virus (35) and the G protein of vesicular stomatitis virus (VSV) as the surface glycoprotein protein of retrovirus vectors (16, 145). Incorporation of VSV has two consequences: First, the vector now has the very broad target range of the VSV G glycoprotein; second, the virion particles appear more stable and can be concentrated by centrifugation without loss of titer.

To produce retrovirus vectors, a given vector in plasmid form is transfected into a selected packaging cell line. The incoming vector sequences then integrate and replicate under the influence of the endogenous helper virus functions, thereby generating progeny virions with the relevant envelope protein as provided by the packaging cell line. Typically, vector production in several producer cell lines is examined to find a stable high producer (75).

REPLICATION-COMPETENT RETROVIRUS Much effort has gone into the development of packaging cell lines because the production of replication-competent virus must be reduced as much as possible (21, 90, 100, 128). Since such virus can replicate, its presence in an otherwise defective population of retrovirus vectors raises serious safety problems. Consider the consequences of injection of helper-free vector and vector containing RCR into primates: Monkeys treated with helper-free virus remained disease free over extended periods of observation (20), whereas those treated with vector containing

RCR developed lymphomas after a few months (34). Presumably, replicating virus is constantly shed and infects additional cells. The resulting viremia leads to multiple rounds of integration, eventually resulting in an accumulation of mutagenic events. The FDA requires rigorous testing of clinical batches of vector meant for use in humans, and although rare, such breakouts of RCR in clinical batches have occurred. Molecular characterization of the resulting RCR shows that minimal regions of sequence homology can give rise to recombinants (106).

PURIFICATION The ability to transduce target cells with retrovirus depends in part on the multiplicity of infection, which in turn depends on the volume and concentration of virus added to cells. Ideally, vectors for clinical use would be purified and concentrated, but this is difficult in practice because retrovirus particles are relatively fragile. New pseudotype viruses may be an exception, but at present, the best that can usually be done is to screen for producer cells yielding a high titer of virus (75). In addition, for human use, the packaging cell line must be tested extensively to exclude the presence of human viruses and other contaminants.

Adenovirus Vectors

PACKAGING CELLS One cell line, 293 cells, is currently used almost exclusively for packaging cell lines for adenovirus vectors. These cells were produced by transfection of sheared Ad5 DNA into human fetal kidney cells (53). After selection and passage, the resulting cell line contains about 11% of the Ad5 genome from the 5' end and is immortalized, presumably under the influence of the E1a and E1b functions. The exact extent of the Ad5 insertion and the exact number of integrated Ad5 E1 sequences is unknown. Because 293 cells contain so little of the adenovirus genome (they encode, at most, one structural protein), their description as a packaging cell line is debatable. Perhaps helper cell line is more appropriate in this case.

REPLICATION-COMPETENT ADENOVIRUS (RCA) Just as with retroviruses, recombination between the Ad5 E1 sequences present in 293 cells and overlapping sequences in the defective partial E1 deleted vectors can generate replication-competent adenovirus (RCA). This problem has been most readily detectable using Ad2-based vectors of greater than wild-type length because once present, the replication-competent progeny rapidly overgrow the input vector, and because the recombinant is an Ad5/Ad2 hybrid that can be distinguished from wild-type (AE Smith, unpublished data). Although less easy to detect and characterize, the generation of RCA with Ad5-based and smaller Ad2-based vectors also occurs. The frequency of such recombination is difficult to measure accurately. In the future, researchers will attempt to minimize sequences that overlap with the 293 cell sequences. Similarly, a valuable development would be a second-generation 293 cell with a better-defined E1 coding region, which might be split between E1a and E1b (as with the retrovirus packaging cells) to discourage recombination.

Although RCA can be generated in the production of adenovirus vectors, whether they are as potentially dangerous as RCR is unclear. Wild-type adenovirus itself may not be serious, and mobilization and widespread dissemination of the vector itself seems unlikely. Nevertheless, the current FDA requirement for clinical lots of adenovirus is one or less RCA per dose.

PURIFICATION   Unlike retroviruses, adenoviruses remain stable when subjected to extensive purification. Procedures usually involve multiple freeze and thaw cycles to disrupt infected cells, and one or more CsCl centrifugation steps followed by dialysis or chromatography. Purified preparations of vector containing 10^{12} particles/ml can be obtained with ease. This is several orders of magnitude greater than is possible with retrovirus or AAV vectors.

Preparations are tested extensively for RCA and for other viruses and contaminants. In addition, preparations must meet an FDA requirement of a particle–to–infectious unit ratio of less than 100.

AAV Vectors

PACKAGING CELLS   AAV vectors lack all viral coding sequences, but no stable packaging cell line has yet been reported. Consequently, the *cap* and *rep* genes must be introduced into cells as plasmids at the same time as the vector plasmid containing the cDNA flanked by TR. The transfection methods used for this purpose are inefficient and difficult to scale up. In addition, transfected cells require superinfection with adenovirus to provide additional helper functions. Efforts to improve the efficiency of production methods are hampered because the *rep* protein is believed to be toxic to cells. Although many ingenious methods have been considered to circumvent this problem, such as engineering a helper adenovirus to express *rep* and *cap,* it remains significant. At present, large-scale efforts to produce material for animal experiments and clinical trials involve brute-force application of the traditional transfection methods.

PURIFICATION   Adenovirus helper can be removed from preparations of AAV by heat treatment and/or by CsCl centrifugation to yield purified virions.

VECTOR EFFICACY

In Vitro Studies

Once one has designed and produced a particular vector for a given application and shown that it can be produced in appropriate amounts, the next question that arises is, does it work? How much of the gene product is made? What is the duration of expression? One can most readily address these questions by using cultured cells.

The first issue that needs to be settled is whether the virus will transfer the gene to the desired cells. For retrovirus vectors, this issue will be largely determined by the envelope protein. For adenovirus-based vectors, knowledge of virus biology does not allow one to predict whether a vector will enter a particular cell or tissue, because the known tropism is defined in terms of the ability of the virus to replicate in a cell. A successful vector requires only that the virus enters a cell and expresses the exogenous gene. Furthermore, the receptor for adenovirus is not well characterized, again making prediction difficult. In practice, adenovirus vectors can enter the cells of a wide variety of tissues and species. The host range of AAV is also difficult to predict from viral-infection data, because the virus is defective in any case. In practice, as with adenovirus, gene-transfer experiments have shown that vectors are capable of entering a wide variety of cells.

Selectable markers are commonly used to aid identification and enrichment of cells into which a vector has transferred a gene, especially with retrovirus and AAV-based vectors. Selection for G418 resistance in cells treated with a vector containing the neomycin-resistance gene is a powerful way to select for cells that likely contain the adjacent gene of interest. Although helpful for initial in vitro studies, such information does not necessarily indicate the degree of gene transfer likely to be achieved in the absence of selection, except perhaps in those rare instances where the transfer of the therapeutic gene itself confers selective advantage. Transfer of ADA cDNA into T cells from ADA deficient patients is one such example (3). Some later generation retrovirus vectors lack selectable markers yet are very efficient, for example the MFG-based vectors (105). In some AAV vectors, the therapeutic gene itself is so large as to preclude inclusion of selectable sequences.

The methods used to detect gene transfer and gene expression should not only be quantitative, but should also indicate the percentage of cells to which the gene is transferred and is expressed. To detect gene transfer, nucleic acid can be examined by Southern, Northern, and PCR-based methods. The PCR methods can be extremely sensitive, but in most configurations, they are not quantitative. None of these methods measure the percentage of cells to which transfer has occurred, but one can gauge this value by using in situ methods to detect nucleic acid in target cells. The integration state of the transferred vector nucleic acid is also often examined. One should remember that just because the behavior of the wild-type virus is well characterized, a related vector will not necessarily behave in the same way. For example, in the absence of *rep*, AAV vectors may not always integrate, and if they do, integration may not always retain wild-type specificity (46).

Detection of the protein product encoded by the vector can utilize many methods. Immunocytochemistry is useful for structural and nonsecreted proteins, although it has limited sensitivity for proteins expressed in only small amounts. Immunoprecipitation methods are sometimes used, but they give little indication of the distribution of protein expression. Secreted proteins are often detected by ELISA or other antibody-based methods, often with very high sensitivity. Some protein products are detected via assay of function, especially several of the genes commonly used for analytical purposes, for example, β-galactosidase, chloramphenicol acetyl transferase (CAT), and luciferase. β-Galactosidase staining is relatively insensitive and subject to artifactual staining in some cells and tissues, although a nuclear localization signal to distinguish exogenous enzyme activity avoids this problem. Luciferase and CAT assays are very sensitive but give no indication of the distribution of expression.

An assay for the function of an expressed therapeutic gene is particularly useful because it allows testing, for example, of complementation of a gene defect or of inhibition of viral growth. Detection of the chloride-channel activity of the CFTR gene product in cells derived from CF patients is a good example of this type of assay (113).

The interpretation of data obtained using cells in culture requires caution. Such cells are often immortalized or transformed or are from the inappropriate tissue or species. Moreover, their growth properties, surface markers and receptors, spectrum of expressed cellular protooncogenes, and tumor suppressors, and even their content of endogenous viruses, may differ from normal. Each of these factors could profoundly influence binding, uptake, and expression of the vector. Furthermore, data obtained in cultured cells necessarily ignores the potential influence of in vivo physiological responses such as the immune response. To ameliorate some of these problems, human primary tissue is desirable, but these cultures are often difficult to obtain and grow and have a limited life span.

Use of human cells has been particularly useful in studying gene transfer for CF. For example, primary respiratory epithelial cells can be grown in monolayers to form tight junctions, thereby allowing assay of electrophysiological properties in Ussing chambers (147). An intermediate step between cell culture and animals is the use of xenografts. For example, denuded rat trachea can be populated with human airway cells and grafted onto nude mice, allowing study of the cell monolayer for several weeks (38).

Animal Studies

Undoubtedly, animal experiments form an essential part of the study of any gene-therapy application and should be initiated at the earliest possible time.

Such experiments have two goals, to study the safety of the vector system and to study the efficacy of gene transfer.

Safety studies are required by the FDA before any human protocol is likely to be permitted. The effect of dose and duration is usually studied in various animals, including in general a nonhuman primate and another animal that is a host of or is permissive to the wild-type virus (for example, cotton rats are often used for adenovirus studies). Spread of viral sequences, especially to the gonads, is usually examined, and replication or partial expression of the supposedly defective viruses is also tested. Any adverse effects such as inflammation following vector administration are studied. The purpose of such experiments is not to show that the vector has no adverse effects, because almost all drugs of any kind have some such properties at some dose. Rather, the aim is to determine the type of adverse event that might be expected if humans were exposed to the vector and to establish the likely doses at which such effects might be predicted. Such data allow an assessment of the likely risk:benefit ratio of any potential trial.

Tests of efficacy require development of relevant models for study. For example, for a genetic disease, a mouse with a gene knockout or an animal with the appropriate phenotype might be available. For hemopoietic cell applications, bone marrow repopulation into lethally irradiated or drug-treated animals can be utilized. For cancer applications, tumor-bearing animals may be used. The animal model is seldom perfect. Knockout mice sometimes do not have the predicted phenotype; mouse hemopoietic cells are much more readily manipulated, sorted, grown, and transduced with vector than primate and especially human cells; and cancer models are notoriously unpredictive of real-life, metastasizing human tumors.

The immune and inflammatory response to either the viral vector or to the exogenous protein are important parameters of study. The use of neonatal animals or nude mice is valuable in assessing effects of vector administration in the absence of a fully functional immune system. Furthermore, strains of mice bearing knockout mutations of important genes in the immune system are available so that any effect can be studied in detail to probe the relative roles of different arms of the immune system. Sensitive assays are available to detect a wide spectrum of mouse cytokines, inflammatory agents, and cell markers.

RETROVIRUSES In most cases, retrovirus vectors are applied ex vivo. Hemopoietic, epithelial, or muscle cells are common targets. Transduction has been demonstrated using murine cells in culture, and repopulation of animals has been achieved using various cDNAs. For example, investigators have reported the transduction of hemopoietic cells with cDNA encoding globin (37), glucocerebrosidase (22, 105), and ADA (68, 85, 139); of fibroblasts with factor

IX (6, 108, 122); and of muscle cells with factor IX (29), growth hormone (33), and ADA (87). Although high-level expression is detected initially, in some cases expression decreases over a period of days to weeks. Analysis of the transferred cells is then necessary. Demonstrating the persistence of integrated DNA in some cases has established that the promoter driving expression of the exogenous gene is shut down (107). Such results have led to the search for more appropriate promoters (59, 122). Despite some success (29, 105), long-lasting expression, especially of high levels of protein, such as abundant blood-clotting factors, remains problematic. Furthermore, extending observations made in mice to other species, especially primates (including humans), can be problematic.

Human subjects have been treated with autologous T cells transduced ex vivo with retrovirus vectors encoding ADA (3) and with liver cells trans-duced ex vivo with low-density lipoprotein receptor (57).

ADENOVIRUS Within the past two years, adenovirus vectors encoding β-galactosidase marker or other proteins have been administered to a remarkable number of tissues in animals. Surprisingly, expression has been detected in numerous tissues, including lung (118, 148), heart (71), liver (40, 64, 65, 78, 123), muscle (126, 132), bone marrow (95), brain (80), CNS (8, 31), endothelial cells (79, 81), kidney (99), retinal cells (67), and solid tumors (14, 32, 58, 125, 138). Human subjects with CF have also been administered adenovirus vectors to the nose (146) and lung (26).

Expression levels in animals depend on the amount of virus added and on the vector used, but in general initial expression is robust. The duration of expression, however, varies widely. In neonate and nude mice, expression is long lasting (125), whereas in adult mice expression declines greatly within two to three weeks (142). The loss of expression is associated, particularly in the liver, with infiltration of inflammatory and immune cells. Earlier studies with wild-type virus in mice that are nonpermissive for viral replication had shown two phases of response characterized by an initial cytokine-mediated infiltration of neutrophils and monocyte cells followed by a T-cell mediated response (50). Much current effort is focused on attempting to understand the relative roles of different cell types in the inflammatory process resulting from administration of adenovirus-based vectors and on the apparent destruction of vector-treated cells. One proposal is that the defective vectors still express significant amounts of viral proteins, which are the target of cytotoxic T lymphocytes (CTLs) that consequently eliminate vector expressing cells (142–144). Protein synthesis by defective vectors might result from leaky gene expression or from complementation of vector growth in target cells by integrated E1 sequences (140), by host cell–coded functions with E1-like activity (15, 124), or by similar proteins encoded by other viruses (110). The CTL-destruction hypothesis, however, runs counter to earlier reports that adenovirus CTLs were restricted, at least in some haplotypes to E1 proteins (112).

Whatever the mechanism in mice, the real questions are, do such mechanisms occur in primates, especially humans, and if so, is it possible to abrogate the effects? For example, expression of the adenovirus E3 gp19 protein might reduce the immune response of the host to the vector because this reduction is its normal function in the wild-type virus life cycle. Other approaches tested include incorporating a temperature-sensitive mutation into the E2 gene for the DNA-binding protein to disfavor viral gene expression (39). Another approach is to delete part of the E4 region for the same purpose (4).

Because adenovirus DNA seldom integrates, many applications of gene therapy using adenovirus vectors will require repeat administration. Repeat-dose studies reveal that animals can become resistant to multiple administrations and subsequent gene expression (96). This phenomenon varies with dose and type of administration and is the subject of intense study. For procedures involving administration via the bloodstream, for example to the liver, a likely explanation is the development of humoral neutralizing antibody. The goal again is to understand the mechanisms involved so as to design improved vectors able to evade the surveillance mechanism.

Careful study of inflammatory and immune responses will define for each application whether there exists a therapeutic window for the use of adenovirus vectors within which efficacy is evident without associated toxicity. This definition is difficult because results obtained in animals seldom predict what will be seen in humans. Indeed, this consideration has prompted the argument for administering gene-therapy vectors to humans at what would otherwise be a very early stage of drug development.

The propensity of some adenovirus vectors to mark treated cells for destruction can be used to advantage in the destruction of cancer cells (14, 58, 138). The use of suicide vectors—encoding cytotoxic agents that often also produce bystander effects (27)—together with the immune response provoked in part by the vector itself might prove an effective antitumor application for adenoviruses.

AAV The difficulties in generating sufficient AAV vector have precluded the availability of much animal data. Ex vivo treatment of fibroblasts (136) and hemopoietic cells (95, 149) has resulted in expression of globin genes and the fanconi anemia gene product (94, 135). In vivo application of tyrosine hydroxylase vectors to brain (69) and of CFTR-encoding vectors in rabbits and monkeys has also been reported (44, 45, 47). To date, expression has been long lasting, and adverse effects from vector administration have not been reported. As more experiments are performed with higher doses, we will need to establish whether AAV vectors avoid some of the inflammatory and immune problems seen with adenovirus vectors, and if so, why.

ALTERNATIVE VECTOR SYSTEMS

Cationic Lipids

Liposomes formed with a variety of cationic lipids and neutral lipids will complex with DNA to form structures capable of transfecting DNA into cultured cells (41, 42, 48). Transfection is inefficient relative to virus transduction in terms of the number of molecules of nucleic acid added per cell. Expression generally is transient. The mechanism by which cationic lipids enhance the ability of DNA to enter cells is unknown, and the rate-limiting steps in the process are not well characterized. Experience in animals is more limited, but generally gene transfer appears less efficient than with viruses (13, 103, 150). Nevertheless, successful corrections of the CF-associated defect in CF knockout mice treated with cationic lipid DNA complexes have been reported (2, 63), and human trials for cancer and CF are in progress.

Molecular Conjugates

DNA:polylysine complexes are stable and can be derivatized to add ligands that aid entry of such complexes into appropriate cells by receptor-mediated endocytosis (24). The asialoglycoprotein receptor and transferin receptors are examples (133, 141). Some experiments involving administration to target rabbit liver cells have shown that duration is transient (141). One of the rate-limiting steps in the entry into cells by this method appears to be escape from the endosome, and one novel approach has been to utilize the ability of conjugated adenovirus to disrupt endosomes and thereby dramatically improve the efficacy of gene transfer, at least in vitro (28). Another receptor that shows promise in delivering molecular conjugates administered systematically to lung and perhaps liver cells is the polymeric IgA receptor (43).

Naked DNA

Surprisingly, naked DNA is taken up into cells in some tissues such as muscle (140). The process is not particularly efficient, but the results can be long lasting. This method is most useful when low-level expression over a protracted period is required, such as in vaccine applications. Here, it may be particularly beneficial in that both humoral and cellular immunity may develop. Naked DNA can also be applied when transient expression of an endothelial cell growth factor might be adequate (115).

CONCLUSIONS

For some applications, such as vaccines, nonviral systems may constitute adequate treatments (24). For other applications, problems with this approach remain in terms of efficacy and duration of expression. A major impetus to continue work on nonviral systems is the widespread, but presently largely unproven, belief that such systems will be safer than viral-based vectors.

Undoubtedly, rapid progress has been made in the application of viral vectors for gene therapy. Cancer applications have led the way and will likely continue to predominate (117). The study of monogenic disease applications progressed with work on ADA, which in retrospect was an excellent early model (3). However, as emphasized throughout, many problems remain in treating more common diseases: control of long-term expression; the biology of the host cell response, in terms of specific and nonspecific effects and of humoral and cellular immunity; the ability to repeat dose; and the efficacy of growth in culture of some ex vivo target cell types, as well as gene transfer to these cells.

Perhaps the best vector to address at least some of these issues would integrate its nucleic acid into any cell type including nondividing cells. This vector would have the efficacy of virus in terms of gene transfer and expression, as well as the supposed safety profile of lipids or molecular conjugates. How best to generate such a vector remains unknown. One approach is to remove undesirable features from particular viruses to improve their safety without compromising their efficacy. The alternative is to add factors, perhaps from viruses (134), to lipids or molecular conjugates to improve their efficacy without compromising their safety. This latter approach requires that we reproduce the evolution of viruses from first principles. This goal is challenging in the absence of biological selection; hence, for the foreseeable future, viral vectors will continue to provide an attractive alternative and will rightly remain the subject of intense widespread research.

ACKNOWLEDGMENTS

I thank Donna Armentano, Sam Wadsworth, Seng Cheng, Johanne Kaplan, and Mary Loeken for helpful comments, and Cathy Janiak for secretarial assistance.

> Any *Annual Review* chapter, as well as any article cited in an *Annual Review* chapter, may be purchased from the Annual Reviews Preprints and Reprints service.
> 1-800-347-8007; 415-259-5017; email: arpr@class.org

Literature Cited

1. Adam MA, Miller AD. 1988. Identification of a signal in a murine retrovirus that is sufficient for packaging of nonretroviral RNA into virions. *J. Virol.* 62:3802–6
2. Alton EW, Middleton PG, Caplen NJ, Smith SN, Steel DM, et al. 1993. Non-invasive liposome-mediated gene delivery can correct the ion transport defect in cystic fibrosis mutant mice. *Nat. Genet.* 5:135–42
3. Anderson WF. 1992. Human gene therapy. *Science* 256:808–13
4. Armentano D, Sookdeo C, Couture L,

Vincent K, Cardoza L, et al. 1993. Second generation adenoviruses for gene therapy of CF. *Pediatr. Pulm. Suppl.* 9:241

5. Armentano D, Yu SF, Kantoff PW, vonRuden T, Anderson WF, Gilboa E. 1987. Effect of internal viral sequences on the utility of retroviral vectors. *J. Virol.* 61:1647–50

6. Axelrod JH, Read M, Brinkhous KM, Verma IM. 1990. Phenotypic correction of factor IX deficiency in skin fibroblasts of hemophilic dogs. *Proc. Natl. Acad. Sci. USA* 87:5173–77

7. Babiss LE, Vaales LD. 1991. Promoter of the adenovirus polypeptide IX gene: similarity to E1B and inactivation by substitution of the simian virus 40 TATA element. *J. Virol.* 65:598–605

8. Bajocchi G, Feldman SH, Crystal RG, Mastrangeli A. 1992. Direct in vivo gene transfer to ependymal cells in the central nervous system using recombinant adenovirus vectors. *Nat. Genet.* 158:39–66

9. Battini J-L, Hard JM, Danos O. 1992. Receptor choice determinants in the envelope glycoproteins of amphotropic, xenotropic and polytropic murine leukemia viruses. *J. Virol.* 66:1468–75

10. Bender MA, Palmer TD, Gelinas RE, Miller AD. 1987. Evidence that the packaging signal of Moloney murine leukemia virus extends into the gag region. *J. Virol.* 61:1639–46

11. Berkner KL. 1988. Development of adenovirus vectors for the expression of heterologous genes. *Biotechniques* 6: 616–29

12. Berkner KL. 1992. Expression of heterologous sequences in adenoviral vectors. *Curr. Top. Microbiol. Immunol.* 158:39–66

13. Brigham KL, Meyrick B, Christman B, Conary JT, King G, et al. 1993. Expression of human growth hormone fusion genes in cultured lung endothelial cells and in the lungs of mice. *Am. J. Respir. Cell Mol. Biol.* 8:209–13

14. Brody SL, Jaffe HA, Han SK, Wersto RP, Crystal RG. 1994. Direct in-vivo gene transfer and expression in malignant cells using adenovirus vectors. *Hum. Gene Ther.* 5:437–47

15. Buck M, Turler H, Chojkier M. 1994. LAP (NF-IL-6), a tissue-specific transcriptional activator, is an inhibitor of hepatoma cell proliferation. *EMBO J.* 13:851–60

16. Burns JC, Friedmann T, Driever W, Burrascano M, Yee JK. 1993. Vesicular stomatitis virus G glycoprotein pseudotyped retroviral vectors: concentration to very high titer and efficient gene transfer into mammalian and nonmammalian cells. *Proc. Natl. Acad. Sci. USA* 90:8033–37

17. Carter BJ. 1994. Adeno-associated virus vectors. *Curr. Opin. Biotechnol.* 3:533–39

18. Chu T-HT, Martinez I, Sheay EC, Dornburg R. 1994. Cell targeting with retroviral vector particles containing antibody-envelope fusion proteins. *Gene Ther.* 1:292–99

19. Colicelli J, Goff SP. 1987. Isolation of a recombinant murine leukemia virus utilizing a new primer tRNA. *J. Virol.* 57:37–45

20. Cornetta K, Moen RC, Culver K, Morgan RA, McLachlin JR, et al. 1990. Amphotropic murine leukemia retrovirus is not an acute pathogen for primates. *Hum. Gene Ther.* 1:15–30

21. Cornetta K, Morgan RA, Anderson WF. 1991. Safety issues related to retrovirus-mediated gene transfer in humans. *Hum. Gene Ther.* 2:5–14

22. Correll PH, Colilla S, Dave HPG, Karlsson S. 1992. High levels of human glucocerebrosidase activity in macrophages of long-term reconstituted mice after retroviral infection of haematopoietic stem cells. *Blood* 80:331–36

23. Cosset F-L, Girod A, Flamant F, Drynda A, Ronfort C, et al. 1993. Use of helper cells with two host ranges to generate high-titer retroviral vectors. *Virology* 193:385–95

24. Cotten M, Wagner E. 1993. Non-viral approaches to gene therapy. *Curr. Opin. Biotechnol.* 4:705–10

25. Couture LA, Armentano D, Souza DW, Cardoza LM, Giuggio VM, et al. 1992. The use of alternative vectors for gene transfer to lung epithelia and tissue culture cells. *Pediatr. Pulm. Suppl.* 8:237

26. Crystal RG, McElvaney NG, Rosenfeld MA, Chu C-S, Mastrangeli A, et al. 1994. Administration of an adenovirus containing the human CFTR cDNA to the respiratory tract of individuals with cystic fibrosis. *Nat. Genet.* 8:42–47

27. Culver KW, Ram Z, Wallbridge S, Ishii H, Oldfield EH, Blaese RM. 1992. In vivo gene transfer with retroviral vector producer cells for treatment of experimental brain tumors. *Science* 256:1550–52

28. Curiel DT, Wagner E, Cotten M, Birnstiel ML, Agarwal S, et al. 1992. High-efficiency gene transfer mediated by adenovirus coupled to DNA-polylysine complexes. *Hum. Gene Ther.* 3:147–54

29. Dai Y, Roman M, Naviaux RK, Verma IM. 1992. Gene therapy via primary

myoblasts: long term expression of factor IX protein following transplantation in vivo. *Proc. Natl. Acad. Sci. USA* 89:10892–95
30. Danos O, Mulligan RC. 1988. Safe and efficient generation of recombinant retroviruses with amphotropic and ecotropic host ranges. *Proc. Natl. Acad. Sci. USA* 85:6460–64
31. Davidson BL, Allen ED, Kozarsky KF, Wilson JM, Roessler BJ. 1993. A model system for in vivo gene transfer into the central nervous system using an adenoviral vector. *Nat. Genet.* 3:219–23
32. Descamps V, Blumenfeld N, Villeval J-L, Vainchenker W, Perricaudet M, Beuzard Y. 1994. Erythropoietin gene transfer and expression in adult normal mice: use of an adenovirus vector. *Hum. Gene Ther.* 5:979–85
33. Dhawan J, Pan LC, Pavlath GK, Travis MA, Lanctot AM, Blau HM. 1991. Systemic delivery of human growth hormone by infection of genetically engineered myoblasts. *Science* 254:1509–12
34. Donahue RE, Kessler SW, Bodine S, McDonagh K, Dunbar C, et al. 1992. Helper virus induced T cell lymphoma in nonhuman primates after retroviral mediated gene transfer. *J. Exp. Med.* 176:1125–35
35. Dong J, Roth MG, Hunter E. 1992. A chimeric avian retrovirus containing the influenza virus hemagglutinin gene has an expanded host range. *J. Virol.* 66:7374–82
36. Dougherty JP, Wisniewski R, Yang S, Rhode BW, Temin HM. 1989. New retrovirus helper cells with almost no nucleotide sequence homology to retroviral vectors. *J. Virol.* 3209:3212
37. Dzierzak EA, Papayannopoulou T, Mulligan RC. 1988. Lineage-specific expression of a human β-globin gene in murine bone marrow transplant recipients reconstituted with retrovirus-transduced stem cells. *Nature* 331:35–41
38. Englehardt JE, Yang Y, Stratford-Perricaudet LD, Allen ED, Kozarsky K, et al. 1993. Direct gene transfer of human CFTR into human bronchial epithelia of xenografts with E1-deleted adenoviruses. *Nat. Genet.* 4:27–34
39. Englehardt JE, Ye X, Doranz B, Wilson JM. 1994. Ablation of E2a in recombinant adenoviruses improves transgene persistence and decreases inflammatory response in mouse liver. *Proc. Natl. Acad. Sci. USA* 91:6196–200
40. Fang B, Eisensmith RC, Li XHC, Finegold MJ, Shedlovsky A, et al. 1994. Gene therapy for phenylketonuria: phenotypic correction in a genetically deficient mouse model by adenovirus-mediated hepatic gene transfer. *Gene Ther.* 1:247–54
41. Felgner JH, Kumar R, Sridhar CN, Wheeler CJ, Tsai YJ, et al. 1994. Enhanced gene delivery and mechanism studies with a novel series of cationic lipid formulations. *J. Biol. Chem.* 269:1–12
42. Felgner PL, Ringold GM. 1989. Cationic liposome-mediated transfection. *Nature* 337:387–88
43. Ferkol T, Kaetzel CS, Davis PB. 1993. Gene transfer into respiratory epithelial cells by targeting the polymeric immunoglobulin receptor. *J. Clin. Invest.* 92:2394–400
44. Flotte TR, Afione SA, Conrad C, McGrath SA, Solow R, et al. 1993. Stable in vivo expression of the cystic fibrosis transmembrane regulator with an adeno-associated virus vector. *Proc. Natl. Acad. Sci. USA* 90:10613–17
45. Flotte TR, Afione SA, Solow R, Drumm ML, Markakis D, et al. 1993. Expression of the cystic fibrosis transmembrane conductance regulator from a novel adeno-associated virus promoter. *J. Biol. Chem.* 268:3781–90
46. Flotte TR, Afione SA, Zeitlin PL. 1994. Adeno-associated virus vector gene expression occurs in nondividing cells in the absence of vector DNA integration. *Am. J. Respir.* 11:517–21
47. Flotte TR, Conrad CK, Afione SA, Reynolds TC, Guggino WB, Carter BJ. 1994. Preclinical trials of adeno-associated virus (AAV) vectors for cystic fibrosis (CF) gene therapy. *Pediatr. Pulm. Suppl.* 10:153–54
48. Gao XA, Huang L. 1991. A novel cationic liposome reagent for efficient transfection of mammalian cells. *Biochem. Biophys. Res. Commun.* 179:280–85
49. Ginsberg HS. 1984. *The Adenoviruses.* New York: Plenum
50. Ginsberg HS, Moldawer LL, Sehgal PB, Redington M, Kilian PL, et al. 1991. A mouse model for investigating the molecular pathogenesis of adenovirus pneumonia. *Proc. Natl. Acad. Sci. USA* 88:1651–55
50a. Glorioso JC, DeLuca NA, Fink DJ. 1995. Development and application of herpes simplex virus vectors for human gene therapy. *Annu. Rev. Microbiol.* 49:675–710
51. Gooding LR, Wold WSM. 1990. Molecular mechanisms by which adeno-

viruses counteract antiviral immune defenses. *Crit. Rev. Immunol.* 10:53–71
52. Graham FL, Prevec L. 1994. Adenovirus-based expression vectors and recombinant vaccines. In *Vaccines: New Approaches to Immunological Problems*, pp. 363–90. Boston: Butterworth-Heinemann
53. Graham FL, Smiley J, Russell WL, Nairn R. 1977. Characterization of a human cell line transformed by DNA from adenovirus 5. *Gen. Virol.* 36:59–72
54. Green M, Wold WSM, Mackey JK, Rigden P. 1979. Analysis of human tonsil and cancer DNAs and RNAs for DNA sequences of group C (serotypes 1, 2, 5, and 6) human adenoviruses. *Proc. Natl. Acad. Sci. USA* 76:6606–10
55. Grez M, Akgun E, Hilberg F, Ostertag W. 1990. Embryonic stem cell virus, a recombinant murine retrovirus with expression in embryonic stem cells. *Proc. Natl. Acad. Sci. USA* 87:9202–6
56. Grez M, Zornig M, Nowock J, Ziegler M. 1991. A single point mutation activates the Moloney murine leukemia virus long terminal repeat in embryonal stem cells. *J. Virol.* 65:4691–98
57. Grossman M, Raper SE, Kozarsky K, Stein EA, Engelhardt JF, et al. 1994. Successful ex vivo gene therapy directed to liver in a patient with familial hypercholesterolaemia. *Nat. Genet.* 6:335–41
58. Haddada H, Ragot T, Cordier L, Duffour MT, Perricaudet M. 1993. Adenoviral interleukin-2 gene transfer into P815 tumor cells abrogates tumorigenicity and induces antitumoral immunity in mice. *Hum. Gene Ther.* 4:703–11
59. Hatzoglou M, Lamers W, Bosch F, Wynshaw-Boris A, Clapp DW, Hanson RW. 1990. Hepatic gene transfer in animals using retroviruses containing the promoter from the gene for phosphoenolpyruvate carboxykinase. *J. Biol. Chem.* 265:17285–93
60. Hawley RG, Lieu FHL, Fong AZC, Hawley TS. 1994. Versatile retroviral vectors for potential use in gene therapy. *Gene Ther.* 1:136–38
61. Hayward WS, Neel BG, Astrin SM. 1981. Activation of a cellular *onc* gene by promoter insertion in ALV-induced lymphoid leukosis. *Nature* 290:475–80
62. Horwitz MS. 1990. Adenoviridae and their replication. In *Virology*, ed. BN Fields, DM Knipe, pp. 1679–721. New York: Raven
63. Hyde SC, Gill DR, Higgins CF, Trezise AE, MacVinish LJ, et al. 1993. Correction of the ion transport defect in cystic fibrosis transgenic mice by gene therapy. *Nature* 362:250–55
64. Ishibashi SM, Brown MS, Goldstein JL, Gerard RD, Hammer RE, Herz J. 1993. Hypercholesterolemia in low density lipoprotein receptor knockout mice and its reversal by adenovirus-mediated gene delivery. *J. Clin. Invest.* 92:883–93
65. Jaffe HA, Danel G, Longenecker G, Metzger M, Setoguchi Y, et al. 1992. Adenovirus-mediated in vivo gene transfer and expression in normal rat liver. *Nat. Genet.* 1:372–78
66. Jolly D. 1994. Viral vector systems for gene therapy. *Cancer Gene Ther.* 1:51–64
67. Jomary C, Piper TA, Dickson G, Couture LA, Smith AE, et al. 1994. Adenovirus-mediated gene transfer to murine retinal cells in vitro and in vivo. *FEBS Lett.* 347:117–22
68. Kantoff PW, Kohn DB, Mitsuya H, Armentano D, Sieberg M, et al. 1986. Correction of adenosine deaminase deficiency in cultured human T and B cells by retrovirus-mediated gene transfer. *Proc. Natl. Acad. Sci. USA* 83:6563–67
69. Kaplitt MG, Leone P, Samulski RJ, Xiao X, Pfaff DW, et al. 1994. Long-term gene expression and phenotypic correction using adeno-associated virus vectors in the mammalian brain. *Nat. Genet.* 8:148–53
70. Kasahara N, Dozy A, Kan YW. 1994. Tissue-specific targeting of retroviral vectors through ligand-receptor interactions. *Science* 266:1373–76
71. Kass-Eisler A, Falck-Pedersen E, Alvira M, Rivera J, Buttrick PM, et al. 1993. Quantitative determination of adenovirus-mediated gene delivery to rat cardiac myocytes in vitro and in vivo. *Proc. Natl. Acad. Sci. USA* 90:11498–502
72. Kempler G, Freitag B, Berwin B, Nanassy O, Barklis E. 1993. Characterization of the Moloney murine leukemia virus stem cell-specific repressor binding site. *Virology* 193:690–99
73. Koo HM, Brown AMC, Kaufman RJ, Prorock CM, Ron Y, Dougherty JP. 1992. A spleen necrosis virus-based retroviral vector which expresses two genes from a dicistronic mRNA. *Virology* 186:669–75
74. Koo HM, Brown AMC, Ron Y, Dougherty JP. 1991. Spleen necrosis virus, an avian retrovirus, can infect primate cells. *J. Virol.* 65:4769–75
75. Kotani H, Newton PB III, Zhang S, Chiang YL, Otto E, et al. 1994. Improved methods of retroviral vector transduction and production for gene therapy. *Hum. Gene Ther.* 5:19–28

76. Kotin RM. 1994. Prospects for the use of adeno-associated virus as a vector for human gene therapy. *Hum. Gene Ther.* 5:793–801
77. Kotin RM, Siniscalco M, Samulski RJ, Zhu X, Hunter L, et al. 1990. Site-specific integration by adeno-associated virus. *Proc. Natl. Acad. Sci. USA* 87:2211–15
78. Kozarsky KF, McKinley DR, Austin LL, Raper SE, Stratford-Perricaudet LD, Wilson JM. 1993. In vivo correction of low density lipoprotein receptor deficiency in the Watanabe Heritable rabbit with recombinant adenoviruses. *J. Biol. Chem.* 269:1–8
79. Lee SW, Trapnell BC, Rade JJ, Virmani R, Dichek DA. 1993. In vivo adenoviral vector-mediated gene transfer in balloon-injured rat carotid arteries. *Circ. Res.* 73:797–807
80. Legal La Salle G, Roberts JJ, Berrard S, Ridoux V, Stratford-Perricaudet LD, et al. 1993. An adenovirus vector for gene transfer into neurons and glia in the brain. *Science* 259:988–90
81. Lemarchand P, Jones M, Yamada I, Crystal RG. 1993. In vivo gene transfer and expression in normal uninjured blood vessels using replication-deficient recombinant adenovirus vectors. *Circ. Res.* 72:1132–38
82. Levine AJ. 1993. The tumor suppressor genes. *Annu. Rev. Biochem.* 62:623–51
83. Levine F, Yee JK, Friedmann T. 1991. Efficient gene expression in mammalian cells from a dicistronic transcriptional unit in an improved retroviral vector. *Gene* 108:167–74
84. Lewis P, Hensel M, Emerman M. 1992. Human immunodeficiency virus infection of cells arrested in the cell cycle. *EMBO J.* 11:3053–58
85. Lim B, Apperley JF, Orkin SH, Williams DA. 1989. Long-term expression of human adenosine deaminase in mice transplanted with retrovirus-infected hematopoietic stem cells. *Proc. Natl. Acad. Sci. USA* 86:8892–96
86. Linial ML, Miller AD. 1990. Retroviral RNA packaging sequence requirements and implications. *Curr. Top. Microbiol. Immunol.* 157:125–52
87. Lynch CM, Clowes MM, Osborne WRA, Clowes AW, Miller AD. 1992. Long-term expression of human adenosine deaminase in vascular smooth muscle cells of rats: a model for gene therapy. *Proc. Natl. Acad. Sci. USA* 89:1138–42
88. Markowitz D, Goff S, Bank A. 1988. A safe packaging line for gene transfer: separating viral genes on two different plasmids. *J. Virol.* 62:1120–24
89. McLachlin JR, Mittereder N, Daucher MB, Kadan M, Eglitis MA. 1993. Factors affecting retroviral vector function and structural integrity. *Virology* 195:1–5
90. Miller AD. 1990. Retrovirus packaging cells. *Hum. Gene Ther.* 1:5–14
91. Miller AD. 1992. Human gene therapy comes of age. *Nature* 357:455–60
92. Miller AD. 1992. Retrovirus vectors. *Curr. Top. Microbiol. Immunol.* 158:1–24
93. Miller AD, Rosman GJ. 1989. Improved retroviral vectors for gene transfer and expression. *Biotechniques* 7.980–90
94. Miller JL, Walsh CE, Ney PA, Samulski RJ, Nienhuis AW. 1993. Single-copy transduction and expression of human gamma-globin in K562 erythroleukemia cells using recombinant adeno-associated virus vectors: the effect of mutations in NF-E2 and GATA-1 binding motifs within the hypersensitivity site 2 enhancer. *Blood* 82:1900–6
95. Mitani K, Graham FL, Caskey CT. 1994. Transduction of human bone marrow by adenoviral vector. *Hum. Gene Ther.* 5:941–48
96. Mittal SK, McDermott MR, Johnson D, Prevec L, Graham FL. 1993. Monitoring foreign gene expression by a human adenovirus-based vector using the firefly luciferase gene as a reporter. *Virus Res.* 28:67–90
97. Morgan RA, Anderson WF. 1993. Human gene therapy. *Annu. Rev. Biochem.* 62:191–217
98. Morgan RA, Nussbaum O, Muenchau DD, Shu L, Couture L, Anderson WF. 1993. Analysis of the functional and the host range-determining regions of the murine ecotropic and amphotropic retrovirus envelope proteins. *J. Virol.* 67:4712–21
99. Moullier P, Friedlander G, Calise D, Ronco P, Perricaudet M, Ferry N. 1994. Adenoviral-mediated gene transfer to renal tubular cells in vivo. *Kidney Int.* 45:1220–25
100. Muenchau DD, Freeman SM, Cornetta K, Zwiebel JA, Anderson WF. 1990. Analysis of retroviral packaging lines for generation of replication competent virus. *Virology* 176:262–65
101. Mulligan RC. 1993. The basic science of gene therapy. *Science* 260:926–32
102. Muzyczka N. 1992. Use of adeno-associated virus as a general transduction vector for mammalian cells. *Curr. Top. Microbiol. Immunol.* 158:97–129
103. Nable EG, Plautz G, Nabel GJ. 1990.

Site-specific gene expression in vivo by direct gene transfer into the arterial wall. *Science* 249:1285–88
104. Neda H, Wu CH, Wu GY. 1991. Chemical modification of an ecotropic murine leukemia virus results in redirection of its target cell specificity. *J. Biol. Chem.* 266:14143–46
105. Ohashi T, Boggs S, Robbins P, Bahnson A, Patrene K, et al. 1992. Efficient transfer and sustained high expression of the human glucocerebrosidase gene in mice and their functional macrophages following transplantation of bone marrow transduced by a retroviral vector. *Proc. Natl. Acad. Sci. USA* 89:11332–36
106. Otto E, Jones-Trower A, Vanin EF, Stambaugh K, Mueller SN, et al. 1994. Characterization of a replication-competent retrovirus resulting from recombination of packaging and vector sequences. *Hum. Gene Ther.* 5:567–75
107. Palmer TD, Rosman GJ, Osbourne WRA, Miller AD. 1991. Genetically modified skin fibroblasts persist long after transplantation but gradually inactivate introduced genes. *Proc. Natl. Acad. Sci. USA* 88:1330–34
108. Palmer TD, Thompson AR, Miller AD. 1989. Production of human factor IX in animals by genetically modified skin fibroblasts: potential therapy for hemophilia B. *Blood* 73:438–45
109. Parolin C, Dorfman T, Palu G, Gottlinger H, Sodroski J. 1994. Analysis in human immunodeficiency virus type 1 vectors of *cis*-acting sequences that affect gene transfer into human lymphocytes. *J. Virol.* 68:3888–95
110. Phelps WC, Yee CL, Munger K, Howley PM. 1988. The human papillomavirus type 16 E7 gene encodes transactivation and transformation functions similar to those of adenovirus E1a. *Cell* 53:539–47
111. Prats A-C, DeBilly G, Wang P, Darlix J-L. 1989. CUG initiation codon used for the synthesis of a cell surface antigen coded by the murine leukemia virus. *J. Mol. Biol.* 205:363–72
112. Rawle FC, Knowles BB, Ricciardi RP, Brahmacheri V, Duerksen-Hughes P, et al. 1994. Specificity of the mouse cytotoxic T lymphocyte response to adenovirus 5. *J. Immunol.* 146:3977–84
113. Rich DP, Anderson MP, Gregory RJ, Cheng SH, Paul S, et al. 1990. Expression of cystic fibrosis transmembrane conductance regulator corrects defective chloride channel regulation in cystic fibrosis airway epithelial cells. *Nature* 347:358–63
114. Rich DP, Couture LA, Cardoza LM, Guiggio VM, Armentano D, et al. 1993. Development and analysis of recombinant adenoviruses for gene therapy of cystic fibrosis. *Hum. Gene Ther.* 4:461–76
115. Rissen R, Rahimizadeh H, Blessing E, Takeshita S, Barry JJ, Isner JM. 1993. Arterial gene transfer using pure DNA applied directly to a hydrogel-coated angioplasty balloon. *Hum. Gene Ther.* 4:749–58
116. Riviere I, Brose K, Mulligan RC. 1994. Effects of retroviral vector design on expression of human adenosine deaminase in murine bone marrow transplant recipients engrafted with genetically modified cells. *Proc. Natl. Acad. Sci. USA.* In press
117. Rosenberg SA. 1992. Gene therapy for cancer. *J. Am. Med. Assoc.* 268:2416–19
118. Rosenfeld MA, Yoshimura K, Trapnell BC, Yoneyama K, Rosenthal ER, et al. 1992. In vivo transfer of the human cystic fibrosis transmembrane conductance regulator gene to the airway epithelium. *Cell* 678:143–55
119. Roux P, Jeanteur P, Piechaczyk M. 1989. A versatile and potentially general approach to the targeting of specific cell types by retroviruses: application to the infection of human cells by means of major histocompatibility complex class I and class II antigens by mouse ecotropic murine leukemia virus-derived viruses. *Proc. Natl. Acad. Sci. USA* 86:9079–83
120. Russell SJ, Hawkins RE, Winter G. 1993. Retroviral vectors displaying functional antibody fragments. *Nucleic Acids Res.* 21:1081–85
121. Salmons B, Gunzburg WH. 1993. Targeting of retroviral vectors for gene therapy. *Hum. Gene Ther.* 4:129–41
122. Scharfmann R, Axelrod JH, Verma IM. 1991. Long-term in vivo expression of retrovirus-mediated gene transfer in mouse fibroblast implants. *Proc. Natl. Acad. Sci. USA* 88:4626–30
123. Smith TAG, Mehaffery MG, Kayda DB, Saunders JM, Yei S, et al. 1993. Adenovirus mediated expression of therapeutic plasma levels of human factor IX in mice. *Nat. Genet.* 5:397–402
124. Spergel JM, Hsu W, Akira S, Thimmappaya B, Kishimoto T, Chen-Kiang S. 1992. NF-IL6, a member of the C/EBP family, regulates E1A-responsive promoters in the absence of E1A. *J. Virol.* 66:1021–30
125. Stratford-Perricaudet LD, Levrero M, Chasse JF, Perricaudet M, Briand P. 1990. Evaluation of the transfer and

expression in mice of an enzyme-encoding gene using a human adenovirus vector. *Hum. Gene Ther.* 1:241–56
126. Stratford-Perricaudet LD, Makeh I, Perricaudet M, Briand P. 1992. Widespread long-term gene transfer to mouse skeletal muscles and heart. *J. Clin. Invest.* 90:626–30
127. Tacket CO, Losonsky G, Lubeck MD, Davis AR, Mizutani S, et al. 1992. Initial safety and immunogenicity studies of an oral recombinant adenohepatitis B vaccine. *Vaccine* 10:673–767
128. Temin HM. 1990. Safety considerations in somatic gene therapy of human disease with retrovirus vectors. *Hum. Gene Ther.* 1:111–23
129. Tooze J. 1980. *DNA Tumour Viruses*. Cold Spring Harbor, NY: Cold Spring Harbor Lab.
130. Tsukiyama T, Ueda H, Hirose S, Niwa O. 1992. Embryonal long terminal repeat–binding protein is a murine homolog of FTZ-Fl, a member of the steroid receptor superfamily. *Mol. Cell. Biol.* 12:1286–91
131. Verma IM. 1994. Human gene therapy. *Sci. Am.* 262:66–84
132. Vincent N, Ragot T, Gilgenkrantz H, Couton D, Chafey P, et al. 1993. Long-term correction of mouse dystrophic muscle degeneration by adenovirus-mediated transfer of a minidystrophin gene. *Nat. Genet.* 5:130–34
133. Wagner E, Cotten M, Foisner R, Birnstiel ML. 1991. Transferrin-polycation-DNA complexes: the effect of polycations on the structure of the complex and DNA delivery to cells. *Proc. Natl. Acad. Sci. USA* 88:4255–59
134. Wagner E, Plank C, Zatloukal K, Cotten M, Birnstiel ML. 1992. Influenza virus hemagglutinin HA-2 N-terminal fusogenic peptides augment gene transfer by transferrin-polylysine-DNA complexes: toward a synthetic virus-like gene-transfer vehicle. *Proc. Natl. Acad. Sci. USA* 89:7934–38
135. Walsh CE, Nienhuis AW, Samulski RJ, Miller JL, Young NS, Liu JM. 1993. Phenotypic correction of fanconi anemia (FACC) in lymphoblasts and CD34+ progenitors with a recombinant adeno-associated virus (rAAV) vector. Am. Soc. Hematol. Abstr. 345a. *Blood* 82: 347A
136. Wei J-F, Wei F-S, Samulski RJ, Barranger JA. 1994. Expression of the human glucocerebrosidase and arylsulfatase A genes in murine and patient primary fibroblasts transduced by an adeno-associated virus vector. *Gene Ther.* 1:261–68
137. Weiss RA, Teich N, Varmus HE, Coffin J. 1984. *RNA Tumor Viruses*. Cold Spring Harbor, NY: Cold Spring Harbor Lab.
138. Wills KN, Manevel DC, Menzel P, Harris MP, Sutjipto S, et al. 1994. Development and characterization of recombinant adenoviruses encoding human p53 for gene therapy of cancer. *Hum. Gene Ther.* 5:1079–88
139. Wilson JM, Danos O, Grossman M, Raulet DH, Mulligan RC. 1990. Expression of human adenosine deaminase in mice reconstituted with retrovirus-transduced hematopoietic stem cells. *Proc. Natl. Acad. Sci. USA* 87:439–43
140. Wolff JA, Malone RW, Williams P, Chong W, Acsadi G, et al. 1990. Direct gene transfer into mouse muscle in vivo. *Science* 247:1465–68
141. Wu CH, Wilson JM, Wu GY. 1989. Targeting genes: delivery and persistent expression of a foreign gene driven by mammalian regulatory elements in vivo. *J. Biol. Chem.* 264:16985–87
142. Yang Y, Ertl HCJ, Wilson JM. 1994. MHC class 1-restricted cytotoxic T lymphocytes to viral antigens destroy hepatocytes in mice infected with E1-deleted recombinant adenoviruses. *Immunity* 1: 433–42
143. Yang Y, Nunes FA, Berencsi K, Furth EE, Gonczol E, Wilson JM. 1994. Cellular immunity to viral antigens limits E1-deleted adenoviruses for gene therapy. *Proc. Natl. Acad. Sci. USA* 91: 4407–11
144. Yang Y, Nunes FA, Berencsi K, Gonczol E, Engelhardt JE, Wilson JM. 1994. Inactivation of E2a in recombinant adenoviruses improves the prospect for gene therapy for cystic fibrosis. *Nat. Genet.* 7:362–69
145. Yee J-K, Miyanohara A, LaPorte P, Bouic K, Burns JC, Friedmann T. 1994. A general method for the generation of high-titer, pantropic retroviral vectors: highly efficient infection of primary hepatocytes. *Proc. Natl. Acad. Sci. USA* 91:9564–68
146. Zabner J, Couture LA, Gregory RJ, Graham SM, Smith AE, Welsh MJ. 1993. Adenovirus-mediated gene transfer transiently corrects the chloride transport defect in nasal epithelia of patients with cystic fibrosis. *Cell* 75:207–16
147. Zabner J, Couture LA, Smith AE, Welsh MJ. 1994. Correction of cAMP-stimulated fluid secretion in cystic fibrosis airway epithelia: efficiency of adenovirus-mediated gene transfer in vitro. *Hum. Gene Ther.* 5:585–93
148. Zabner J, Petersen DM, Puga AP, Graham SM, Couture LA, et al. 1994. Safety

and efficacy of repetitive adenovirus-mediated transfer of CFTR cDNA to airway epithelia of primates and cotton rats. *Nat. Genet.* 6:75–83
149. Zhou SZ, Broxmeyer HE, Cooper S, Harrington MA, Srivastava A. 1993. Adeno-associated virus 2-mediated gene transfer in murine hematopoietic progenitor cells. *Exp. Hematol.* 21:928–33
150. Zhu N, Liggitt D, Liu Y, Debs R. 1993. Systemic gene expression after intravenous DNA delivery into adult mice. *Science* 261:209–11

SUBJECT INDEX

A

Acetate
  bacterial chemotaxis and, 503
  fermentation of
    pathway for, 319
Acetivibrio cellulolyticus
  cellulase complexes produced
    by, 404
  isolation of, 408
Acetoacetyl-coenzyme A thiolase
  ergosterol deprivation and, 101
Acetogenic anaerobes
  microbiology and physiology
    of, 311–13
Acetotrophic anaerobes
  microbiology and physiology
    of, 318–21
Acid-tolerance response (ATR)
  Salmonella typhimurium and,
    152–54
Acinetobacter
  biofilms produced by, 718
Acquired immunodeficiency syndrome
  See AIDS
Acriflavin
  dyskinetoplastic bloodstream
    trypanosomes and, 133
Actinomyces
  type 2 fimbriae of, 252–53
Actinomycin
  biosynthesis of, 214
Actinorhodin
  biosynthesis of, 210
    early pathway intermediates
    of, 218
  structure of, 203
Acyrthosiphon pisum, 58, 61
  DNA sequence of, 64–65
  growth and survival of
    chlortetracycline treatment
    and, 79
  heat-shock response of, 76
  S-endosymbiont of, 64
  tryptophan biosynthesis and,
    81–84
Adeno-associated virus (AAV)
  gene therapy and, 815–16
Adeno-associated virus (AAV)
  vectors
  application of, 829
  design of, 821
  in vivo gene delivery and, 677
  production of, 824
Adenosine deaminase deficiency,
  809

Adenoviruses
  gene therapy and, 812–15
Adenovirus vectors
  application of, 828–29
  design of, 819–20
  in vivo gene delivery and, 677
  production of, 823–24
Adhesins
  of *Escherichia coli*, 249–50
  of *Pseudomonas aeruginosa*,
    251–52
Adipic acid
  biosynthesis of, 574
Aerobactin
  Salmonella typhimurium and,
    150
Aerobes
  biodegradation of nitroaromatic compounds and,
    534–48
  microbiology and physiology
    of, 307
Aeromonas hydrophila
  biofilms produced by, 718
Aflatoxin
  biosynthesis of, 227
    polyketide synthases and,
    208
African sleeping sickness, 429
AIDS, 254
Aklanonic acid
  structure of, 222
Aklanonic acid methyl ester
  structure of, 221
Aklaviketone
  structure of, 221
Alanine transport
  leucine-responsive regulatory
    protein and, 756
Alcaligenes eutrophus
  biodegradation of nitroaromatic compounds and,
    537
Aleyrodoidea
  prokaryotic endosymbionts of,
    63–64
Algae
  thermal, 11
Alkali shift
  GroEL proteins and, 75
Alternaria alternata
  hydroxynaphthalene and melanin biosynthesis in, 207
Alteromonas
  membrane lipids of, 792
ALV
  See Avian leucosis virus

Alzheimer's disease
  gene therapy for
    herpes simplex virus vectors
    and, 700–1
Amastigotes, 177
  invasion of macrophages by,
    179–80
  surface antigens of, 188–89
Amaurouscus nigei
  zaragozic acid production by,
    622
Amebiasis, 450
Amino acid synthesis
  endosymbionts and, 79–80, 84
p-Aminobenzoic acid
  biosynthesis of, 570
Aminolevulinic acid, 100
Aminophenols
  biodegradation of, 547
Ammonium
  methane oxidation and, 594
Amoeba proteus
  GroEL proteins in, 76
Amoebic dysentery, 429
Amphipathic helical peptides,
  284–92
Amphotericin B
  structure of, 203
Anaerobes
  acetogenic
    microbiology and physiology of, 311–13
  acetotrophic
    microbiology and physiology of, 318–21
  cellulolytic
    environments of, 405–6
  cellulolytic enzyme systems
    of, 401–5
  phototrophic
    microbiology and physiology of, 308–9
Anaerobic biodegradation, 527–31
Anaerobic environments
  cellulose degradation in, 399–
    19
Anaerobiosis
  GroEL proteins and, 75
Andrews, N. W., 175–94
Anthracenequinone
  biosynthesis of, 229
Anthranilate synthase
  tryptophan biosynthetic pathway and, 56
Antibiotic resistance
  conjugative transposons and,
    368

839

840   SUBJECT INDEX

Antibiotics
  assembly of
    polyketide synthases and, 205
  endogenous of myeloid-derived cells, 279–82
  GroEL proteins and, 75
Antibodies
  Trypanosoma cruzi invasion and, 191–92
Antifolates
  resistance in parasitic protozoa, 442–46
Antimicrobial peptides, 277–99
  amphipathic helical, 284–92
    biosynthetic pathways of, 288–91
    structure and properties of, 284–88
    tissue-specific expression of, 291–92
  antimicrobial mechanisms of, 294–97
  brevinins, esculentins, ranalexin, 292–94
  defensins, 279–82
  evolutionary significance of, 298
  proline- and arginine-rich, 283–84
Antimonials
  resistance to, 438–42
Antimycin A
  Trypanosoma cruzi invasion of host cells and, 192
Antimycobacterial agents, 650–51
  new targets for, 652–53
Aphids
  endosymbionts of, 55–87
  parthenogenetic reproduction in
    kinetics of growth during, 60–62
    properties of, 57–58
    requirement of *Buchnera* by, 78–79
Apolipoproteins
  cholesterol transport and, 110
Aquatic environments
  cellulose degradation in, 405–12
Arabinogalactan
  in mycobacterial cell envelope, 654–57
Archaebacteria, 22
Archaeoglobus fulgidus, 320
Aromatic nitro group
  chemistry of, 526–27
Aromatics, 557–76
  biosynthesis of
    carbon flow impediments and, 565–68
    substrate-limited, 561–64
    traditional methods of, 559

Arsenicals
  resistance to, 438–42
Arthrobacter
  biodegradation of nitroaromatic compounds and, 536
Ascoquinone A, 208
Aserpgillus nidulans
  sterigmatocystin synthesis in, 208
Aspergillus parasiticus
  aflatoxin biosynthesis in, 208
Aspergillus terreus
  cholesterol biosynthesis inhibitor of, 208
  mevinolin free acid in
    biosynthetic pathway to, 209
Astroviruses
  detection of, 478–79
ATR
  See Acid-tolerance response
Avermectin A
  structure of, 203
Avermectins
  assembly of
    polyketide synthases and, 205
Avian leucosis virus (ALV), 812
Avicelases, 403
Azotobacter vinelandii, 336
  Fe protein homodimer of, 337, 343, 348
  MoFe protein of, 337, 350

B

Bacille Calmette-Guerin (BCG)
  mycobacterial disease and, 646, 659–60
Bacillus
  $\alpha^-\beta^-$ spores of
    DNA damage in, 43–46
  dormant spores of
    α/β-type SASPs of, 41–43
    environment of, 33–35
    oxidizing agents, heat, dessication and, 35–38
    resistance to radiation in, 38–40
    survival of, 30–33
  spore DNA protection in, 29–51
  spore germination of
    DNA repair during, 40–41
Bacillus megaterium
  esculentins and brevinins and, 293
Bacillus methanicus, 582–83
Bacillus stearothermophilus
  dormant spores of
    heat resistance of, 35
Bacillus subtilis
  $\alpha^-\beta^-$ spores of
    DNA damage in, 43–46
  DnaA boxes in, 69

  dormant spores of
    DNA protection in, 33
    environment of, 33–35
    heat resistance of, 35
  origin of replication in, 69
  rRNA operons of, 73
  spore germination of
    DNA repair during, 40–41
Bacteria
  biofilm, 711–40
    phenotypic changes in, 735–36
  in boiling water, 15–16
  chemosensory pathway in, 491–504
  deep-sea, 777–98
    characteristics of, 788–94
  flagellar assembly in, 505–8
  flagellar genes in
    organization and regulation of, 509–10
  flagellar motor dynamics in, 512–13
  gram-positive
    antibiotic resistance in, 368
  high-temperature, 15–16
  mesophilic, 787
  methane-producing, 318–20
  methanotrophic, 583–84
  motility of, 489–514
  polyketide synthases of, 213–25
  sulfate-reducing, 320–21
    biodegradation of nitroaromatic compounds and, 529–30
  surface lectins of
    lectinophagocytosis and, 243–53
  swimming behavior of, 504–14
  thermophilic, 12, 15, 22, 787
  torque generation in
    components involved in, 510–12
    models of, 514
Bacteriocytes, 58–60
  S-endosymbiont and, 58–60
Bacteriomes, 58–60
Bacteriophage infection
  GroEL proteins and, 75
Bacteroides
  conjugative transposons of, 370
Bacteroides cellulosolvens
  cellulase complexes produced by, 404
  isolation of, 408
Baumann, L., 55–87
BCG
  See Bacille Calmette-Guerin
Benzoquinone
  production of, 571
Benzoxazinorifamycin
  mycobacterial disease and, 651

SUBJECT INDEX   841

Bergstrom, J. D., 607–36
Bills, G. F., 607–36
Biofilm bacteria
  phenotypic changes in, 735–36
Biofilms
  See Microbial biofilms
Bioremediation, 524, 548–49
Biosynthetic operons
  leucine-responsive regulatory protein and, 765–66
Blair, D. F., 489–514
Blattaria
  endosymbionts of, 57
Boiling water
  bacteria in, 15–16
Bombinins
  biosynthetic pathways of, 288–91
  structure and antimicrobial properties of, 284–86
Bordetella pertussis
  RGD-containing proteins of, 261
Borrelia burgdorferi, 72
  rRNA genes in, 74
Borst, P., 427–51
Brefeldin A
  trypomastigote cell entry and, 181–82
Brevinins, 292–94
Brock, T. D., 1–26
Buchnera, 55–87
  aromatic amino acid biosynthesis and, 84
  DNA synthesis, transcription, translation in, 65–68
  evolutionary relationships of, 62–64
  genetic characterization of, 64–65
  genome of, 64–65
  role in symbiotic association, 78–85
  rRNA operons of, 73–75
  tryptophan biosynthesis and, 81–84
  ultrastructure, location, transmission of, 58–60
Buchnera aphidicola
  increase in number of aphid growth and, 61
  S-endosymbiont and, 58–60
Burleigh, B. A., 175–94
Butyrivibrio fibrisolvens
  cellulose degradation in rumen and, 413
Byrne, K., 607–36

C

Caecomyces
  cellulose degradation in rumen and, 414

Caerulein, 288
Calcium
  trypomastigote cell entry and, 182–83
Caldwell, D. E., 711–40
Caliciviruses
  detection of, 474–76
Cancer
  gene therapy of
    herpes simplex virus vectors and, 696–700
Candida albicans
  esculentins and, 294
  mannose-inhibitable killing of, 258
  nonopsonic binding to phagocytes, 262
Candidia albicans, 254
Carbohydrates
  Trypanosoma cruzi interaction with host cells and, 190–91
Carbon cycle
  global
    anaerobic decomposition of organic matter and, 318
    methanotrophs and, 589
Carbon monoxide dehydrogenases, 305–26
  of *Clostridium thermoaceticum*, 313–18
  of *Methanosarcina*, 321–25
  of *Methanothrix soehngenii*, 321–25
  of *Pseudomonas carboxydovorans*, 307–8
  of *Rhodospirillum rubrum*, 309–11
Carbon starvation
  GroEL proteins and, 75
  Salmonella typhimurium and, 147–50
Carcinoembryonic antigens (CEAs), 247
Casey, W. M., 95–112
Catechol
  biocatalytic synthesis of, 572
Cationic lipids
  gene therapy and, 830
Cationic peptides
  Salmonella typhimurium and, 158–59
CEAs
  See Carcinoembryonic antigens
Cellulose
  degradation in anaerobic environments, 399–19
  degradation in association with animals, 412–18
  degradation in soils, sediments, aquatic environments, 405–12
Cellulose-fermenting microorganisms

diversity of, 407–10
Cerataphidini, 60, 63
CF
  See Cystic fibrosis
CGI 17341
  for mycobacterial disease, 653
Chagas' disease
  Trypanosoma cruzi and, 176, 191, 429
Chaperonins, 56, 75–76
CheA
  bacterial chemotaxis and, 497–99
CheB
  bacterial chemotaxis and, 501–3
Chemoprophylaxis
  tuberculosis and, 665
CheR
  bacterial chemotaxis and, 501–3
CheW
  bacterial chemotaxis and, 497–99
CheY
  bacterial chemotaxis and, 499–500
CheZ
  bacterial chemotaxis and, 500–1
Chlamydia trachomatis, 64
Chlorine
  enteroviruses resistant to, 464
Chloroflexus, 21
Chloroplasts
  GroEL proteins of, 75
Chloroquine resistance
  in parasitic protozoa, 435–38
Chlortetracycline
  Acyrthosiphon pisum growth and survival and, 79
Cholesterol, 95–96, 110, 608–9
  biosynthesis of
    inhibitors of, 208, 608
    zaragozic acids and, 632–33
  transport of
    apolipoproteins and, 110
Chrysophanol
  biosynthesis of, 228
Churchward, G. G., 367–92
Cinatrins, 622–23
Ciprofloxacin
  for mycobacterial disease, 652–53
Circinotrichum falcatisporum
  fermentation products of, 623
Cladosporium cladosporioides
  zaragozic acid production by, 622
Clark, M. A., 55–87
Clinafloxacin
  for mycobacterial disease, 652

Clostridium
  biodegradation of nitroaromatic compounds and, 530–31
  dormant spores of
    α/β-type SASPs of, 41–43
    DNA protection in, 33
    heat resistance of, 35
    survival of, 30–33
    TNT degradation and, 527
Clostridium aldrichii
  isolation of, 408
Clostridium barkeri
  nicotinic acid hydroxylase of, 308
Clostridium bifermentans
  biodegradation of nitroaromatic compounds and, 531
Clostridium celerecrescens
  isolation of, 408
Clostridium cellobioparum
  cellulase complexes produced by, 404
  isolation of, 408
Clostridium cellulofermentans
  isolation of, 407–8
Clostridium cellulolyticum
  cellulase complexes produced by, 404
  isolation of, 408
Clostridium cellulovorans
  cellulase complexes produced by, 404
  isolation of, 408
Clostridium difficile
  conjugative transposons of, 370
Clostridium josui
  isolation of, 408
Clostridium kluyveri
  biodegradation of nitroaromatic compounds and, 530
Clostridium lentocellum
  isolation of, 407
Clostridium papyrosolvens C7
  cellulase complexes produced by, 404–5
  isolation of, 407
Clostridium papyrosolvens NCIB11394
  isolation of, 407
Clostridium pasteurianum
  biodegradation of nitroaromatic compounds and, 530
  Fe protein of, 349
Clostridium populeti
  isolation of, 408
Clostridium stercorarium
  cellulase system of, 403
Clostridium thermoaceticum
  biodegradation of nitroaromatic compounds and, 530
Clostridium thermocellum
  cellulase system of, 403–4
  cellulose degradation by, 411
Clostridium thermohydrosulfuricum
  cellulose degradation by, 411
Clostridum thermoaceticum
  carbon monoxide dehydrogenase of, 313–18
Cochliobolus miyabeanus
  scytalone dehydratase purified from, 227
Coleoptera
  endosymbionts of, 57
Collectins, 241
Colletortrichum lagenarium
  melanin biosynthesis in, 207
Comamonas
  biodegradation of nitroaromatic compounds and, 537–39
Comamonas acidivorans
  biodegradation of nitroaromatic compounds and, 544
Compactin, 208
Competition
  methane oxidation in soil and, 596
Confocal scanning laser microscopy (CSLM)
  microbial biofilms and, 717–21
Conjugative transposition, 367–92
  circular intermediate in, 373
  cotransfer of DNA in, 381
  by excision and insertion, 371–73
  length of coupling sequence in, 374–75
  polarity of strand cleavage in, 373–74
  recombination and, 386–92
  regulation of, 384–86
  target sites in, 375–77
Conjugative transposons
  antibiotic resistance and, 368, 369–70
  distribution of, 370
  properties of, 368–69
  simple and compound, 369
  structure of, 377–81
  use in genetic analysis, 371
Cooling
  deep-sea bacteria and, 788–89
Costerton, J. W., 711–40
Coxsackieviruses, 463
Crithidia fasciculata
  heat-shock proteins of, 136–37
  kinetoplast DNA of, 118–19
  kinetoplast DNA network in, 121–24
  replication of, 125–28
  maxicircles in
    replication of, 130–31
  minicircle origin-binding protein of, 135–36
  minicircles in, 120–21
    replication of, 129–30
  mitochondrial DNA polymerase activity in, 135
  mitochondrial type II topoisomerases of, 133–35
Cryptidins, 280–81
CSLM
  See Confocal scanning laser microscopy
Cyclobutane-type thymine dimer
  structure of, 34
Cycloguanil, 443
Cysteine biosynthesis
  endosymbionts and, 84–85
Cystic fibrosis (CF), 808
Cytochrome P450 enzymes
  erythromycin biosynthesis and, 228–29
Cytokines
  mannose receptor and, 257–58

D

Daunorubicin
  biosynthesis of, 210
dBcAMP
  trypomastigote cell entry and, 181–82
Dean, D. R., 335–61
DEBS
  See 6-Deoxyerythronolide B synthase
Decompression
  deep-sea bacteria and, 788–89
Deep-sea bacteria
  characteristics of, 788–94
Deep-sea habitats, 779–80
Deep-sea microbiology, 777–98
  high-pressure laboratory techniques in, 782–85
  methods of sampling for, 780–81
  nomenclature in, 787–88
  PT envelope in, 787
  PTk diagram in, 784–86
Deep-sea vents, 18, 22
Defensins, 279–82
  Salmonella typhimurium and, 158
Degradative operons
  leucine-responsive regulatory protein and, 766–67
Deltorphin A, 289
DeLuca, N. A., 675–702
12-Deoxyaklanonic acid
  structure of, 222
6-Deoxyerythronolide B
  biosynthesis of
    intermediates of, 215
6-Deoxyerythronolide B synthase (DEBS)

SUBJECT INDEX   843

erythromycin A formation and, 205
2-Deoxyglucose
  Trypanosoma cruzi invasion of host cells and, 192
Dermaseptans
  biosynthetic pathways of, 288–91
  structure and antimicrobial properties of, 286–88
Dermenkephalin, 289
Dermorphin, 289
Dessication
  α⁻β⁻ spores resistant to, 46
  spore DNA and, 33
  spore resistance to, 38
Desulfovibrio
  biodegradation of nitroaromatic compounds and, 529–30
  TNT degradation and, 527
Desulfovibrio desulfuricans, 321
Desulfovibrio vulgaris, 321
DHFR
  See Dihydrofolate reductase
Dihydrofolate reductase (DHFR)
  mutations in parasitic protozoa and, 443–44
  overproduction of pyrimethamine resistance and, 444
Dihydromonacolin L, 208
Dihydropteroate synthase (DHPS)
  mutations in parasitic protozoa and, 443–44
  sulfonamides and, 442
2,6-Dinitrophenol
  biodegradation of, 537
2,4-Dinitrotoluene
  biodegradation of, 539–40
  fungal degradation of, 531–32
  mineralization of, 524
Dioxygenase enzymes
  biodegradation of nitroaromatic compounds and, 537–41
Dipicolinic acid (DPA)
  structure of, 34
DNA
  leucine-responsive regulatory protein and, 749–50, 760
  protection in *Bacillus* spores, 29–51
  See also Kinetoplast DNA
DNA-damaging agents
  GroEL proteins and, 75
DNA depurination
  spore DNA and, 36–37
DNa polymerases
  kinetoplast DNA structure and replication and, 135

DNA topoisomerases
  kinetoplast DNA structure and replication and, 133–35
Dolichols, 97
Doxorubicin
  structure of, 203
DPA
  See Dipicolinic acid
Draths, K. M., 557–76
Drug resistance
  in parasitic protozoa, 427–51
  P-glycoproteins and, 433–35
Dufresne, C., 607–36
Duncan, K., 641–66
DuP721
  for mycobacterial disease, 653
Dynemicin A
  structure of, 203
Dyskinetoplastidy, 133

E

Electrochemical measurements
  microbial biofilms and, 726–28
Emetine resistance
  in parasitic protozoa, 450
Emodin, 229
  structure of, 222
Emodinanthrone
  structure of, 222
Encephalitis
  herpes simplex virus and, 683
Endosymbionts
  amino acid synthesis and, 79–80, 84
  of aphids, 55–87
  cysteine biosynthesis and, 84–85
  methionine biosynthesis and, 84–85
  primary, 58
  secondary, 58–60
  sulfate reduction and, 84–85
  tryptophan biosynthesis and, 81–84
Englund, P. T., 117–38
Entamoeba histolytica
  emetine resistance in, 450
  metronidazole resistance in, 448
  transient transformation of, 431
Enterobacter aerogenes
  leucine-responsive regulatory protein of, 750
Enterobacteriaceae, 62, 64, 68
  origin of replication in, 69
  type 1 fimbriae of lectinophagocytosis and, 244–49
Enterobactin
  Salmonella typhimurium and, 152

Enterochelin
  Salmonella typhimurium and, 150–51
Enterococcus faecalis
  circular transposon of, 385
  conjugative transposon of, 369
Enteroviruses
  as indicators of sewage-associated pathogens, 473–74
  resistance to chlorine in, 464
Environmentally transmitted pathogens
  detection of, 473–79
Environmental virology, 461–81
  nucleic acid-based detection methods in, 470–73
  public health concerns and, 462–69
Epimastigotes, 177–79
Episterol
  esterification of, 105
Epithelial tissues
  defensins intrinsic to, 280–82
Ergosterol, 95–96, 110–12, 609
  biosynthesis of, 97
  oxygen and, 101
  esterification of, 105
  features and regulation of, 101
  phosphatidylinositol kinase activity and, 107
  respiration and, 102–3
  squalene conversion to biochemical transformations in, 99
Erythromycin
  biosynthesis of, 206, 228–29
Erythromycin A, 205
Escherichia coli, 62
  chemosensory pathway in, 491–504
  DnaA boxes in, 69
  esculentins and brevinins and, 293
  fimbrial and nonfimbrial adhesins of, 249–50
  formate dehydrogenase of, 308
  gene expression in leucine-responsive regulatory protein and, 747–71
  GroEL proteins in, 75
  origin of replication in, 69
  rRNA operons of, 73
  sulfate reduction/cysteine biosynthesis pathways in, 84–85
  transmembrane chemoreceptors of, 494–97
  type 1 fimbriae of lectinophagocytosis and, 244–47
  uropathogenic, 264
Esculentins, 292–94

Esterification
  sterol metabolism and, 103–7
Estes, M. K., 461–81
Ethambutol
  Mycobacterium smegmatis and, 655
Ethanol
  GroEL proteins and, 75
Ethidium bromide
  dyskinetoplastic bloodstream trypanosomes and, 133
Eubacterium cellulosolvens
  cellulose degradation in rumen and, 413
Explosives
  microbial degradation of, 525

F

Farnesylpyrophosphate, 97
Fatty acid synthases
  biosynthesis of
    steps in, 204
Fecosterol, 96
  esterification of, 105
Ferry, J. G., 305–26
Feudomycins, 225
Fibrobacter succinogenes
  cellulase complexes produced by, 404
Fibronectin
  invasion of mammalian cells by trypomastigotes and, 186
Fimbriae
  leucine-responsive regulatory protein and, 770
Fink, D. J., 675–702
Fisher, K., 335–61
FK506
  assembly of
    polyketide synthases and, 205
Fluorescent probes
  microbial biofilms and, 717–21
Fluoroquinolones
  for mycobacterial disease, 651
Forespore, 30, 41
Formaldehyde
  assimilation in methanotrophs
    serine pathway of, 588
Formate
  methane oxidation in soil and, 594–95
Formate dehydrogenase
  of *Escherichia coli*, 308
Foster, J. W., 145–66
Freshwater microbiology, 13
Frost, J. W., 557–76
Fujii, I., 201–32
Fumarate
  bacterial chemotaxis and, 503

Fungi
  biodegradation of nitroaromatic compounds by, 531–34
  cellulase systems of, 402–3
  cellulose degradation in rumen and, 414
  polyketide synthases of, 207–8, 225–26
  sterol synthesis in, 95
  zaragozic acids and, 618–23, 633

G

Galactocaralose
  biosynthesis of, 655
Galactofuranose, 654–55
β-Galactosidase
  leucine-responsive regulatory protein and, 767
Gall formation
  aphids and, 57
Gallic acid
  biosynthesis of, 570
Gamma radiation
  spore resistance to, 40
Ganciclovir
  herpes simplex virus-tk gene combined with
    tumor debulking and, 697
Gastritis
  Helicobacter pylori and, 264
Gastroenteritis
  caliciviruses and, 474
Gene therapy
  adeno-associated virus and, 815–16
  adenoviruses and, 812–15
  alternative vector systems for, 830
  goals of, 809–10
  herpes simplex virus vectors and, 675–702
  retroviruses and, 810–12
  viral vectors and, 807–31
Gene transfer
  herpes simplex virus and, 677–78
Geranylgeranylpyrophosphate, 97
Giardia duodenalis, 429
  metronidazole resistance in, 448
Giardia lamblia, 429
Glorioso, J. C., 675–702
Glucocorticoids
  mannose receptor and, 258
Glutamate
  biosynthesis of
    leucine-responsive regulatory protein and, 766
Glycine
  biosynthesis of

leucine-responsive regulatory protein and, 766
cleavage of
  leucine-responsive regulatory protein and, 767
Goldhar, J., 239–68
Gramicidin
  biosynthesis of, 214
Gram-positive bacteria
  antibiotic resistance in
    conjugative transposons and, 368
Granaticin
  biosynthesis of, 210
Greenbug, 58
Green peach aphid, 58
Griseofulvin
  structure of, 203
GroEL proteins, 75–78
GroES proteins, 75–78

H

Halofantrine resistance
  in parasitic protozoa, 435–38
Hansenula anomala, 4
Heat
  $\alpha^- \beta^-$ spores resistant to, 45–46
  GroEL proteins and, 75
  spore DNA and, 33
  spore killing by
    mechanisms of, 35–37
  spore resistance to, 37–38
Heat-shock proteins
  kinetoplast DNA structure and replication, 136–37
Heat-shock response
  Acyrthosiphon pisum and, 76
  Salmonella typhimurium and, 156
Heavy metals
  GroEL proteins and, 75
Helicobacter pylori
  nonopsonic activation of neutrophils by, 264
Heme compounds
  sterol metabolism and, 102
Hemolysins
  Salmonella typhimurium and, 151
Hepatitis
  See Viral hepatitis
Hepatitis A
  waterborne outbreaks of, 467
Hepatitis A virus, 465–67
  detection of, 474
Hepatitis E
  waterborne outbreaks of, 467
Hepatitis E virus, 464, 465–67
  detection of, 476–77
Herpes simplex virus (HSV)
  gene transfer and, 677–78
  latency of, 682–87

SUBJECT INDEX   845

life cycle of, 678–87
lytic infection due to
molecular biology of, 678–82
Herpes simplex virus (HSV) vectors
applications of, 696–701
expression of therapeutic genes using, 690–96
gene therapy of cancer and, 696–700
gene therapy of neurodegenerative disease and, 700–1
progress in engineering, 687–96
HGPRT
of parasitic protozoa, 447
Histoplasma capsulatum
CD11/CD18 binding and, 262
HIV infection
Mycobacterium tuberculosis and, 642–43
HMG-CoA
isoprenoid biosynthesis and, 99
HMG-CoA synthase
ergosterol deprivation and, 101
HMGR
sterol biosynthesis and, 100
Homoptera
endosymbionts of, 57
HSV
See Herpes simplex virus
Human immunodeficiency virus
See HIV infection
Huntington's disease
gene therapy for
herpes simplex virus vectors and, 700–1
Hutchinson, C. R., 201–32
Hydrogen peroxide
Bacillus subtilis spores and, 36
Hydrophobins, 263
Hydroquinone
production of, 571
3-Hydroxybenzoate
biodegradation of, 538–39
β-Hydroxybutyrate
methane oxidation in soil and, 594–95
3-Hydroxy-3-methylglutaryl coenzyme A
See HMG-CoA
3-Hydroxy-3-methylglutaryl coenzyme A reductase
See HMGR
Hydroxynaphthalene
biosynthesis of
polyketide synthases and, 207
Hydroxystilbamidine
dyskinetoplastic bloodstream trypanosomes and, 133

Hypercholesterolemia
cholesterol biosynthesis inhibitors and, 608
Hyperthermophiles, 12, 787
Hypoxanthine-guanine-phosphoribosyltransferase
See HGPRT

I

IHF
See Integration host factor
Immersion slide technique
use in microbial ecology, 15
Immunoelectron microscopy
α/β-type SASPs and, 42
Immunotherapy
mycobacterial disease and, 658–59
Inflammation
nonopsonic phagocytosis and, 263–65
Insects
associations with intracellular prokaryotes, 57
Integration host factor (IHF)
conjugative transposons and, 390–91
Integrins
as host-cell receptors for *Trypanosoma cruzi* attachment, 189
macrophage
phagocytosis mediated by, 260–62
Interferon-γ
lectinophagocytosis and, 257
Ionizing radiation
deep-sea bacteria and, 793
Iron stress response
Salmonella typhimurium and, 150–52
Isoleucine
biosynthesis of
leucine-responsive regulatory protein and, 765
transport of
leucine-responsive regulatory protein and, 756
Isoniazid, 642
Isopentenoid pathway, 97
Isopentenylpyrophosphate, 97
Isoprenoids
biosynthesis of, 97–99

K

KAP
See Kinetoplast-associated protein
Keisari, Y., 239–68
Ketoreductases

bacteria polyketide synthases and, 227–28
Kinetoplast-associated protein (KAP), 136
Kinetoplast DNA, 117–38
molecular components of, 119–21
proteins condensing, 136
structure and replication of
enzymes and proteins involved with, 133–37
Kinetoplast DNA network
isolated, 121–22
replication of, 125–28
structure of, 121–24
in vivo, 122–24
Kinetoplast DNA replication
model, 131–33
Klebsiella aerogenes
leucine-responsive regulatory protein of, 750
Klebsiella pneumoniae
capsular polysaccharides expressed by, 255
MoFe protein of, 351–52
Korber, D. R., 711–40
KRM-1648
mycobacterial disease and, 651

L

Lactococcus lactis
drug resistance in, 370
Lai, C.-Y., 55–87
Landfills
aerobic soil layer covering methanotrophs inhabiting, 591–92
Lanosterol
demethylation inhibitors of
antifungal properties of, 608
esterification of, 105
Lanosterol synthase, 96
Lappin-Scott, H. M., 711–40
Leaf curling
aphids and, 57
Lectinophagocytosis, 242
bacterial surface lectins and, 243–53
macrophage lectins as receptors in, 253–60
in vitro, 244–47
in vivo, 249
Lectins
bacterial surface
lectinophagocytosis and, 243–53
Gal/GalNAc-specific, 253–54
macrophage
as receptors in lectinophagocytosis, 253–60
sugar specificities of, 260
Man/GlcNAc-specific, 254–59

SUBJECT INDEX

Legionella pneumophila
  survival in macrophages, 161
Leishmania
  folate synthesis in
    alternative pathway for, 445–46
  kinetoplast DNA of, 118
  resistance to oxyanions in, 439–42
  transport mutations in, 444–45
Leishmania donovani
  heat-shock proteins of, 136–37
  kinetoplast DNA network in replication of, 128
Leishmania enriettii
  stable transformation of, 431
Leishmania major
  antifolate resistance in, 443–44
Leishmania mexicana amazonensis
  transkinetoplastidy in, 132–33
Leishmania promastigotes
  interaction of CR3 with, 261–62
Leishmaniasis, 429, 439–40
Leishmania tarentolae
  maxicircles in
    noncoding variable region in, 120
  minicircles in
    guide RNAs encoded by, 120
Lepoteichoic acid
  binding of *Streptococcus pyogenes* to macrophages and, 263
Leprosy, 643, 647, 659
Leptodontidium elatius, 611
  zaragozic acid production by, 622
Leptomonas seymouri
  stable transformation of, 431
Leschine, S. B., 399–19
Leucine
  biosynthesis of
    leucine-responsive regulatory protein and, 765–66
  transport of
    leucine-responsive regulatory protein and, 768
Leucine-responsive regulatory protein, 747–71
  as chromosome organizer, 758–62
  function of, 762–65
  operons regulated by, 765–71
  as regulator, 751–58
Leucothrix mucor, 8–9, 14, 18
Levitide, 288
Levofloxacin
  for mycobacterial disease, 652
Lewandowski, Z., 711–40

Libertella
  zaragozic acid production by, 622
Limnology, 2, 6, 21–22
Lin, T., 747–71
Lipids
  cationic
    gene therapy and, 830
  membrane
    in deep-sea bacteria, 791–92
Lipoarabinomannan
  in mycobacterial cell envelope, 654–57
Lipoteichoic acid
  phagocytic cells stimulated by, 265
Listeria monocytogenes
  escape from vacuole, 193
Lovastatin, 208
  cholesterol biosynthesis and, 608–9
  steryl ester synthase enzyme and, 104
Lysyl-tRNA synthetase
  leucine-responsive regulatory protein and, 768

M

Macrophage integrins
  phagocytosis mediated by, 260–62
Macrophage lectins
  as receptors in lectinophagocytosis, 253–60
  sugar specificities of, 260
Macrophages
  amastigote invasion of, 179–80
  recognition of microorganisms by, 240–42
  Salmonella typhimurium survival in, 161–62
Magainins
  biosynthetic pathways of, 288–91
  structure and antimicrobial properties of, 286
  tissue-specific expression of, 291–92
Magnaporthe grisea
  rice blast disease and, 227
Malaria, 429, 443
Maltose transport
  leucine-responsive regulatory protein and, 769
Mammalian cells
  invasion by *Trypanosoma cruzi*, 175–94
Mancinelli, R. L., 581–97
Mannose receptor, 254–59
  microbial polysaccharides recognized by, 255–57

modulation of, 257–58
role in vivo, 258–59
Marine microbiology, 6–7, 13
See also Deep-sea microbiology
Mass spectroscopy
  zaragozic acids and, 613–15
Maxicircles, 119–20
  replication of, 130–31
Mealybugs
  prokaryotic endosymbionts of, 63–64
Medical microbiology, 5, 7
Mefloquine resistance
  in parasitic protozoa, 435–38
Megoura viciae, 61
Melanin
  biosynthesis of
    polyketide synthases and, 207
Melaphis rhois
  development time of, 83
Melarsoprol resistance
  in trypanosomes, 438–39
Melnick, J. L., 461–81
Membrane lipids
  in deep-sea bacteria, 791–92
Membrane proteins
  in deep-sea bacteria, 792–93
Mesophiles, 787
Metabolic operons
  leucine-responsive regulatory protein and, 767–68
Metcalf, T. G., 461–81
Methane
  oxidation of
    physiology and biochemistry of, 584–89
  oxidation in soils, 581–97
  ecology of, 589–96
Methane monooxygenase (MMO)
  methane oxidation and, 581–82, 584–85
Methane-producing bacteria, 318–20
Methanobrevibacter arboriphilicus, 323
Methanococcus vannielii, 323
Methanomonas methano-oxidans, 583
Methanosarcina
  biofilms produced by, 718
  carbon monoxide dehydrogenases of, 321–25
Methanosarcina barkeri
  pyruvate as source of carbon and energy, 319
Methanosarcina thermophila
  carbon monoxide dehydrogenase of, 317
Methanothrix methanobrevibacter
  biofilms produced by, 718

SUBJECT INDEX 847

Methanothrix soehngenii
  acetate fermentation pathway in, 319
  carbon monoxide dehydrogenase of, 321–25
Methanotrophs, 583–84
  formaldehyde assimilation in serine pathway of, 588
  global carbon cycle and, 589
  paraffin dirt and, 590
Methionine biosynthesis
  endosymbionts and, 84–85
Methotrexate
  Leishmania major and, 443
13-Methylaclacinomycins, 225
Methylation
  products of polyketide synthases and, 229–31
Methylobacter albus BG8
  methane oxidation in ammonium inhibiting, 594
Methylococcus capsulatus, 583
Methylomonas albus, 584
Methylomonas methanica, 583
Methylosinus trichosporium OB3b
  methane oxidation in ammonium inhibiting, 594
6-Methylsalicylic acid
  structure of, 203
6-Methylsalicylic acid synthase (MSAS)
  purification of, 207
Metronidazole
  Mycobacterium tuberculosis and, 649
Metronidazole resistance
  in parasitic protozoa, 448–49
Mevinolin, 208
  steryl ester synthase enzyme and, 104
Microbial biocatalysis, 558
Microbial biofilms, 711–40
  confocal scanning laser microscopy and, 717–21
  distribution and ubiquity of, 713–16
  electrochemical measurements and, 726–28
  flow studies and, 722–26
  fluorescent probes and, 717–21
  microbiology and, 739–40
  microsensors and, 722
  nuclear magnetic resonance imaging and, 722–26
  physicochemical analysis and, 728–33
  structure of, 716–35
  teleology and, 736–39
Microbial ecology, 2, 4, 6–9, 12–13, 15, 23
Micrococcus euryhalis
  ultraviolet light and, 793

Micrococcus luteus
  DnaA boxes in, 69
Micromonospora propionici
  cellulose degradation in rumen and, 413
Micromonospora ruminantium
  cellulose degradation in rumen and, 413
Microorganisms
  phagocytosis of, 239–68
Microsensors
  microbial biofilms and, 722
Minicircle inheritance, 131–32
Minicircle origin-binding protein
  kinetoplast DNA structure and replication and, 135–36
Minicircles, 120–21
  replication of, 129–30
Mitochondria
  GroEL proteins of, 75
MMO
  See Methane monooxygenase
Molecular conjugates
  gene therapy and, 830
Molybdenum
  carbon monoxide oxidation and, 308
Monascus ruber
  compactin from, 208
Monensin A
  structure of, 203
Monooxygenase enzymes
  biodegradation of nitroaromatic compounds and, 534–37
Mor, A., 277–99
Moran, N. A., 55–87
Moraxella
  biodegradation of nitroaromatic compounds and, 534, 536
Motility
  bacterial, 489–514
MPA
  See Mycophenolic acid
MSAS
  See 6-Methylsalicylic acid synthase
Mutualism, 57
Mycetocytes, 58
Mycetomes, 58
Mycobacterial disease, 641–66
  dormancy of, 648–49
  extracellular multiplication and transmission in, 649–50
  initial uptake of bacteria in, 643–46
  prevention of, 659–65
  T-cell responses in, 646–48
  treatment of, 650–59
Mycobacterium avium
  new drugs for, 651–52

Mycobacterium avium-intracellulare complex, 643
Mycobacterium intercellulare
  new drugs for, 651–52
Mycobacterium leprae, 207, 643
Mycobacterium smegmatis
  ethambutol and, 655
  transposon mutagenesis in, 660–61
Mycobacterium tuberculosis
  binding to human macrophages, 255
  extracellular multiplication and transmission, 649–50
  HIV infection and, 642–43
  initial uptake of, 643–46
  T-cell responses to, 646–48
Mycobacterium vaccae, 659
Mycolic acids
  in mycobacterial cell envelope, 654–57
Mycophenolic acid (MPA)
  parasitic protozoa and, 447–48
Mycoplasma capricolum
  DnaA boxes in, 69
Mycoplasma gallisepticum
  rRNA genes in, 74
Myeloid-derived cells
  endogenous antibiotics of, 279–82
Myzus persicae, 58

N

Naked DNA
  gene therapy and, 830
Nallin-Omstead, M., 607–36
Naphthacenequinone
  biosynthesis of, 229
Neisseria
  leucine-responsive regulatory protein and, 750
Neisseria gonorrhoeae
  conjugative transposons of, 370
  outer membrane proteins of, 250–51
Neisseria meningiditis
  conjugative transposons of, 370
Neocallimastix
  cellulose degradation in rumen and, 414
Neocallimastix frontalis
  cellulase complexes produced by, 404
Neurodegenerative disease
  gene therapy of
    herpes simplex virus vectors and, 700–1
Neutrophils
  antimicrobial peptides of, 283–84
  nonopsonic activation of, 264

recognition of microorganisms by, 240–42
stimulated by oral bacteria, 265
Newman, E. B., 747–71
Nicolas, P. N., 277–99
Nicotinic acid hydroxylase
of *Clostridium barkeri*, 308
Nitroarenes
toxicity of, 525
Nitroaromatic compounds, 523–49
biodegradation of
aerobic bacteria and, 534–48
anaerobic, 527–31
fungi and, 531–34
Nitrobenzenes
biodegradation of, 537, 544–45
Nitrobenzoates
biodegradation of, 538, 544–45
Nitrogen
metabolism of
leucine-responsive regulatory protein and, 756
Nitrogenase, 335–61
component protein interactions of, 347–52
FeMo cofactor structure of, 354–59
Fe protein-nucleotide binding and hydrolysis of, 343–47
MoFe protein metallocluster domains of, 339–43
P-cluster structure of, 352–54
Nitrogen compounds
methane oxidation and, 593–95
Nitrogen fixation
nitrogenase and, 335–37
Nitrogen starvation
GroEL proteins and, 75
Salmonella typhimurium and, 147
Nitro groups
aromatic
chemistry of, 526–27
Nitroimidazoles
for mycobacterial disease, 653
2-Nitrophenol
biodegradation of, 536–37
Nitrotoluenes
biodegradation of, 544
NMR
See Nuclear magnetic resonance
Nocardia
biodegradation of nitroaromatic compounds and, 536, 538
Norsolorinic acid
structure of, 203
Norwalk virus
detection of, 474–76
waterborne outbreaks of, 467

Nuclear magnetic resonance (NMR)
microbial biofilms and, 722–26
zaragozic acids and, 613–15

O

Ofek, I., 239–68
Ofloxacin
for mycobacterial disease, 652–53
Oleandomycin
biosynthesis of, 214
Oleic acid
sterol biosynthesis and, 100
Oligomycin
Trypanosoma cruzi invasion of host cells and, 192
Oligopeptide transport
leucine-responsive regulatory protein and, 769
Operons
leucine-responsive regulatory protein and, 765–71
Opsonins, 241
Ornithine decarboxylase inhibitors
resistance in parasitic protozoa, 449–50
Orpinomyces
cellulose degradation in rumen and, 414
Osmotic-shock response
Salmonella typhimurium and, 156–58
Ouellette, M., 427–51
Outer membrane porins
leucine-responsive regulatory protein and, 769–70
Ovothiol, 439
Oxazolidinones
for mycobacterial disease, 653
Oxidation
products of polyketide synthases and, 228–29
Oxidative-stress response
Salmonella typhimurium and, 154–55
Oxidizing agents
$\alpha^-\beta^-$ spores resistant to, 46
spore DNA and, 33
spore killing by
mechanisms of, 35–37
spore resistance to, 37
Oxidosqualene cyclase inhibitors of
antifungal properties of, 608
Oxidosqualenes
synthesis of
HMGR and, 100
Oxyanions
resistance in *Leishmania*, 439–42

Oxygen
ergosterol biosynthesis and, 101
methane oxidation and, 592–93
Oxygen radicals
GroEL proteins and, 75
Oxytetracycline
biosynthesis of, 210
structure of, 203

P

Palmitoleic acid
sterol biosynthesis and, 100
Paraffin dirt, 589–90
Parasitic protozoa
drug resistance in, 427–51
P-glycoproteins and, 433–35
resistance to antifolates in, 442–46
resistance to arsenicals and antimonials in, 438–42
resistance to chloroquine, mefloquine, quinine, halofantrine in, 435–38
resistance to emetine in, 450
resistance to metronidazole in, 448–49
resistance to ornithine decarboxylase inhibitors in, 449–50
resistance to purine analogues in, 446–48
Parasitism, 57
Parasperone A, 208
Parkinson's disease
gene therapy for
herpes simplex virus vectors and, 700–1
Parks, L. W., 95–112
Paromomycin
Leishmania and, 450
Parthenogenetic reproduction, 58
kinetics of growth during, 60–62
Particle image velocimetry (PIV)
microbial biofilms and, 722–26
Paumann, P., 55–87
PAV
See Pseudoadenovirus vector
PCR
See Polymerase chain reaction
Pea aphid, 58
Penetrin
host-cell attachment of trypomastigotes and, 186
Penicillin, 3
Penicillium cyclopium
polyketide synthases of, 207
Penicillium patalum
polyketide synthases of, 207
Penicillium urticae
polyketide synthases of, 207

SUBJECT INDEX 849

Pentamidine resistance
  in trypanosomes, 438
Peptic ulcer disease
  Helicobacter pylori and, 264
Peptide fluoromethyl ketones
  Trypanosoma cruzi invasion of host cells and, 192
Peptides
  See Antimicrobial peptides
Peptidoglycan
  in *Buchnera*, 58
  in mycobacterial cell envelope, 654–57
Peptidyl diazomethane derivatives
  Trypanosoma cruzi invasion of host cells and, 192
Periodontal disease, 265
Peripheral nervous system
  herpes simplex virus latency in, 682–87
Peters, J. W., 335–61
P-glycoproteins
  drug resistance in parasitic protozoa and, 433–35
pH
  spore dormancy and, 33
Phagocytic cells
  hydrophobic interactions of, 262–63
  lipoteichoic acid stimulating, 265
Phagocytosis
  macrophage integrins mediating, 260–62
  nonopsonic, 239–68
    inflammation and tissue injury and, 263–65
Phanerochaete chrysosporium
  biodegradation of nitroaromatic compounds by, 531–32
  bioremediation and, 524
Phenylalanine
  biosynthesis of
    endosymbionts and, 84
  chemicals synthesized from, 569
Phoma
  zaragozic acid production by, 620
Phosphate starvation
  Salmonella typhimurium and, 147
Phosphatidylinositol kinase
  ergosterol and, 107
Phosphorus starvation
  GroEL proteins and, 75
Photobacterium phosphoreum
  ultraviolet light and, 793
Phototrophic anaerobes
  microbiology and physiology of, 308–9

Physicochemical analysis
  microbial biofilms and, 728–33
Phytomonas
  kinetoplast DNA of, 118
Piezomesophiles, 794–98
Piezophiles, 787
Piezothermophiles, 794–98
Pili
  leucine-responsive regulatory protein and, 770
Pirellula marina
  rRNA genes in, 74
Piromyces
  cellulose degradation in rumen and, 414
PIV
  See Particle image velocimetry
Plant ecology, 4
Plant health
  aphids and, 57
Plasmodium berghei
  stable transformation of, 431
Plasmodium falciparum
  antifolates and, 442
  drug resistance in, 435–38
  hypoxanthine-guanine-phosphoribosyltransferase of, 447
  pyrimethamine resistance in, 443
Pneumocystis carinii, 254
Poliovirus, 462–65
Polyadenylation
  adenovirus vectors and, 820
Polyketides
  aromatic
    structures of, 203
    biosynthesis of, 204
    reduction steps in, 221
Polyketide synthases, 201–32
  mechanisms in bacteria, 213–25
  mechanisms in fungi, 225–26
  products of
    enzymes acting on, 226–31
    as source of chemical diversity and drugs, 231–32
  structure and function of, 205–13
Polymerase chain reaction (PCR), 15, 22, 60, 64, 462
  environmental virology and, 470–73
Polymixin E
  Salmonella typhimurium and, 158
Polypeptides
  abnormal
    GroEL proteins and, 75
Porphyromonas gingivalis, 265
PPD
  See Purified protein derivative
Pravastatin, 632

Predation
  methane oxidation in soil and, 596
Pressure vessels, 782
Prokaryotes
  GroEL proteins of, 75
  intracellular
    associations with insects, 57
Prostaglandins
  mannose receptor and, 257–58
Protamine
  Salmonella typhimurium and, 158
Proteins
  membrane
    in deep-sea bacteria, 792–93
Proteobacteria, 62, 64, 74–75
Protozoa
  trypanosomatid
    kinetoplast DNA of, 117–38
  See also Parasitic protozoa
Pseudo adenovirus vector (PAV), 820
Pseudococcidae
  prokaryotic endosymbionts of, 63–64
Pseudodiplodia
  zaragozic acid production by, 620
Pseudomonas
  biodegradation of nitroaromatic compounds and, 534, 537–40
  leucine-responsive regulatory protein and, 750
Pseudomonas aeruginosa
  biodegradation of nitroaromatic compounds and, 547
  biofilms produced by, 713, 718–21
  esculentins and, 294
  fimbriae and adhesins of, 251–52
  killing by monocyte-derived macrophages
    suppression of, 257
Pseudomonas arvilla
  biodegradation of nitroaromatic compounds and, 547
Pseudomonas carboxydohydrogena
  carbon monoxide dehydrogenase of, 307
Pseudomonas carboxydovorans
  carbon monoxide dehydrogenase of, 307–8
Pseudomonas fluorescens
  biofilms produced by, 718–21
Pseudomonas methandenitrificans, 583
Pseudomonas methanica, 583

Pseudomonas pseudoalcaligenes
  biodegradation of nitroaromatic compounds and, 546–47
Pseudomonas putida
  DnaA boxes in, 69
  origin of replication in, 69
Psychrophiles, 787
PT envelope, 787
PTk diagram, 784–86
Purified protein derivative (PPD)
  delayed-type hypersensitivity to, 663–64
Purine analogues
  resistance in parasitic protozoa, 446–48
Pyrenochaete terrestris, 228
Pyridine-2,6-dicarboxylic acid
  structure of, 34
Pyridine nucleotide transhydrogenase
  leucine-responsive regulatory protein and, 767–68
Pyrimethamine resistance
  dihydrofolate reductase overproduction and, 444
  in parasitic protozoa, 443

Q

Quinine resistance
  in parasitic protozoa, 435–38

R

Radiation
  $\alpha^-\beta^-$ spores resistant to, 44–45
  deep-sea bacteria and, 793
  spore DNA and, 33
  spore resistance to, 38–40
Ranalexin, 292–94
Rapamycin
  assembly of
    polyketide synthases and, 205
RCA
  See Replication-competent adenovirus
RCR
  See Replication-competent retrovirus
Recombination
  conjugative transposition and, 386–92
Reduction
  products of polyketide synthases and, 227–28
Renal scarring
  type 1 fimbriated bacteria and, 264
Replication-competent adenovirus (RCA), 823–24

Replication-competent retrovirus (RCR), 821, 822–23
Respiration
  sterols and, 102–3
Retroviruses
  gene therapy and, 810–12
Retrovirus vectors
  application of, 827–28
  design of, 817–19
  production of, 821–23
Rhizobium
  GroEL proteins in, 76
Rhodococcus
  biodegradation of nitroaromatic compounds and, 540–41
Rhodococcus erythropolis
  biodegradation of nitroaromatic compounds and, 541–43
Rhodocyclus gelatinosus
  carbon monoxide dehydrogenase synthesis in, 308–9
Rhodospirillum rubrum, 76
  carbon monoxide dehydrogenase of, 309–11
Rhodotorula glutinis
  methane oxidation in, 582
Ribulose-1,5-biphosphate carboxylase, 76
Ribulose monophosphate pathway (RuMP)
  methane oxidation and, 586–87
Rice blast disease, 227
Rickettsia prowazekii, 64
  ATP/ADP translocase of, 68–69
Rifabutin
  mycobacterial disease and, 651
Rifampicin, 642
Rifampin
  RNA polymerase and, 65
RNA polymerase
  rifampin and, 65
Rotaviruses
  detection of, 477–78
  waterborne outbreaks of, 467
Rouhbakhsh, D., 55–87
rRNA operons
  in *Buchnera*, 73–75
Rumen
  cellulolytic microorganisms in
    diversity of, 413–14
  cellulose degradation in, 412–13
  microbial interactions in, 414–16
Ruminobacter amylophilus, 62
Ruminococcus albus
  cellulase complexes produced by, 404

Ruminococcus flavefaciens
  cellulase complexes produced by, 404
Ruminomyces
  cellulose degradation in rumen and, 414

S

Saccharomyces cerevisiae
  esculentins and, 294
  squalene synthase of
    zaragozic acids and, 633
  sterol alterations in
    physiological effects of, 107–9
  sterol biosynthesis in, 96–99
    regulation of, 99–101
  sterol metabolism in, 101–7
  sterol trafficking in, 109–10
Saccharopolyspora erythraea, 205
  erythromycin biosynthetic pathway in, 206
Salmonella
  type 1 fimbriae expressed by, 247
Salmonella muenchen
  flagellar genes in, 507
Salmonella typhimurium, 145–66
  acid-tolerance response of, 152–54
  cationic peptides and, 158–59
  chemosensory pathway in, 491–504
  fimbrial hemagglutinins expressed by, 250
  heat-shock response of, 156
  iron stress response of, 150–52
  leucine-responsive regulatory protein of, 750
  osmotic-shock response of, 156–58
  oxidative-stress response of, 154–56
  serum resistance of, 164–65
  starvation stress response of, 147–50
  stress survival strategies of, 147–59
  sulfate reduction/cysteine biosynthesis pathways in, 84
  thermotolerance in, 156
  transmembrane chemoreceptors of, 494–97
  virulence of
    stress management and, 159–65
    survival in macrophages and, 161–62
  virulence genes of, 164
  virulence in mice, 165

SUBJECT INDEX 851

voyages through natural environment and animal host, 146–47
Salvarsan, 438
Sanitary landfills
  aerobic soil layer covering methanotrophs inhabiting, 591–92
Scenedesmus quadricauda
  methane oxidation in, 592–93
Schizaphis graminum, 58, 61, 63
  cytoplasmic protein in gene coding for, 68
  DNA fragments in genetic map of, 72, 74
  DNA sequence of, 64–65
  genes found in deduced products of, 70–71
  origin of replication in, 69, 72
  polymerization reactions in genes involved in, 66–67
  tryptophan biosynthesis and, 81–84
Schlechtendalia chinensis
  DNA sequence of, 64
  periplasmic protease in gene coding for, 68
  tryptophan biosynthetic pathway of, 84–85
Scott, J. R., 367–92
Scytalone, 227
  structure of, 221
Sediments
  cellulose degradation in, 405–12
Serine
  biosynthesis of
    leucine-responsive regulatory protein and, 765
  L-Serine deaminase
    leucine-responsive regulatory protein and, 766–67
Serine pathway
  methane oxidation and, 588–89
Serine transport
  leucine-responsive regulatory protein and, 756, 768–69
Serratia marcescens
  leucine-responsive regulatory protein of, 750
Setlow, P., 29–51
Shapiro, T. A., 117–38
Sharon, N., 239–68
Shigella flexneri
  escape from vacuole, 193
  flagellar proteins in, 506
  type 1 fimbriae expressed by, 247
  Sialic acid acceptor molecules
    invasion of mammalian cells by *Trypanosoma cruzi* and, 187–88
Sitosterol, 609

Sleeping sickness, 438
Smith, A. E., 807–31
Sodium azide
  Trypanosoma cruzi invasion of host cells and, 192
Soil mycology, 4
Soils
  cellulose degradation in, 405–12
  methane oxidation in, 581–97
  ecology of, 589–96
  movement of viruses within, 469
Spain, J. C., 523–49
Sparfloxacin
  for mycobacterial disease, 652
Spector, M. P., 145–66
Spirochaeta caldaria
  cellulose-degrading bacteria and, 411
Spiroplasma citri
  DnaA boxes in, 69
Spore DNA, 29–51
  environment of, 33–35
  interaction with α/β-type SASPs, 47–49
  oxidizing agents, heat, dessication and, 35–38
  repair during germination, 40–41
Sporobolomyces roseus
  methane oxidation in, 582
Sporormiella intermedia, 610–11
  zaragozic acid production by, 622
Squalene
  conversion to ergosterol
    biochemical transformations in, 99
  synthesis of
    biochemical transformations in, 98
Squalene epoxidase
  inhibitors of
    antifungal properties of, 608
Squalene synthase
  inhibition of
    zaragozic acids and, 631
  inhibitors of
    screening for, 609–10
Squalene synthetase
  isoprenoid biosynthesis and, 100–1
SSR
  See Starvation stress response
Staphylococcus aureus
  esculentins and brevinins and, 293
  nonopsonized
    recognition by granulocytes, 264
Starvation
  GroEL proteins and, 75

Starvation stress response (SSR)
  Salmonella typhimurium and, 147–50
Sterigmatocystin, 227
  biosynthesis of
    polyketide synthases and, 208
  structure of, 221
Sterol, 95–112
  alterations in yeast
    physiological effects of, 107–9
  biosynthesis of
    stages in, 608–9
  biosynthesis in yeast, 96–99
    regulation of, 99–101
  metabolism of
    cultural conditions and, 101–2
    esterification and, 103–7
    heme compounds and, 102
    respiration and, 102–3
    trafficking in yeast, 109–10
Sterol methyltransferase enzyme, 96
Steryl ester synthase enzyme
  inhibition of, 104
Streptococcus agalactiae
  Gal/GalNAc-specific lectin and, 254
Streptococcus anginosus
  conjugative transposons of, 369
Streptococcus pneumoniae
  conjugative transposons of, 369
Streptococcus pyogenes
  binding to macrophages, 263
  conjugative transposons of, 369
Streptomyces antibioticus
  oleandomycin biosynthesis in, 214
Streptomyces coelicolor
  actinorhodin biosynthesis in, 210
  DnaA boxes in, 69
Streptomyces glaucescens
  polyketide synthase-encoding genes of, 216–17
  tetracenomycin biosynthesis in, 210
Streptomyces violaceoruber
  granaticin biosynthesis genes in, 210
Stress proteins, 75–76
Sulfate-reducing bacteria, 320–21
  biodegradation of nitroaromatic compounds and, 529–30
Sulfate reduction
  endosymbionts and, 84–85
Sulfolobus, 21
Sulfonamides
  dihydropteroate synthase and, 442

SUBJECT INDEX

Sulfur springs
  as steady-state ecosystem, 11
Sybiosomes, 80
Symbionin, 77
Symbiosomes, 58, 78
Syringomycin, 108

T

Tamm-Horsfall protein
  lectinophagocytosis and, 249
TAP
  See Tracheal antimicrobial peptide
Taq polymerase, 15, 19, 26
T cells
  mycobacterial disease and, 646–48
Telescoping of generations, 60
Tetracenomycin
  biosynthesis of, 210
    early pathway intermediates of, 218
Tetracenomycin C
  structure of, 203
Tetracenomycin D3
  structure of, 222
Tetracenomycin F1
  structure of, 221–22
Tetracenomycin F2
  structure of, 221
Tetracycline resistance
  gene encoding, 378–79
Tetrahymena
  membrane lipids of, 791
Tetrahymena pyriformis
  cell division in
    pressure/pressure shifts and, 785
Tetronasin
  assembly of
    polyketide synthases and, 205
Thalidomide
  leprosy therapy and, 659
Thermal algae, 11
Thermal springs
  as model ecosystems, 13
Thermal vents, 18
Thermophiles, 15, 787
  extreme, 12, 22
Thermotolerance
  Salmonella typhimurium and, 156
Thermus aquaticus, 13–15, 19, 26
Thermus thermophilus
  rRNA genes in, 74
Thiothrix, 9–10
Threonine
  degradation of
    leucine-responsive regulatory protein and, 767

transport of
  leucine-responsive regulatory protein and, 756
Thyminyl-thymine adduct
  structure of, 34
Tissue injury
  nonopsonic phagocytosis and, 263–65
TNT
  biodegradation of, 527
  fungal degradation of, 531–34
Topoisomerase II
  kinetoplast DNA network and, 121
Toxoplasma gondii
  antifolates and, 442
  hypoxanthine-guanine-phosphoribosyltransferase of, 447
  pyrimethamine-resistant mutants of, 443
  stable transformation of, 431
Toxoplasmosis, 429
Tracheal antimicrobial peptide (TAP), 282
Transketolase catalysis
  aromatic amino acid biosynthesis and, 563–64
Transkinetoplastidy, 132–33
Transport operons
  leucine-responsive regulatory protein and, 768–70
Transposition
  See Conjugative transposition
Transposon mutagenesis
  in *Mycobacterium smegmatis*, 660–61
Transposons
  See Conjugative transposons
Trans-sialidase
  invasion of mammalian cells by *Trypanosoma cruzi* and, 186–87
Trichoderma reesei
  cellulase system of, 402
Trichomonas vaginalis, 429
  metronidazole resistance in, 448–49
2,4,6-Trinitrotoluene
  mineralization of, 524
Triparinol
  cholesterol biosynthesis and, 608
Trypanosoma brucei
  dyskinetoplastidy in, 133
  hypoxanthine-guanine-phosphoribosyltransferase of, 447
  kinetoplast DNA of, 118
  kinetoplast DNA network in, 121–24
  replication of, 128

mating of
  minicircle exchange during, 132
  maxicircles in
    noncoding variable region in, 120
  replication of, 130–31
  minicircles in
    guide RNAs encoded by, 120
  resistance to arsenicals in, 438–39
  stable transformation of, 431
Trypanosoma cruzi, 175–94
  escape from vacuole, 192–93
  heat-shock proteins of, 136–37
  host-cell attachment by
    parasite surface molecules and, 183–89
  host-cell surface molecules and, 189–91
  hypoxanthine-guanine-phosphoribosyltransferase of, 447
  inhibitors of invasion by, 191–92
  invasion of host cells by, 179–83
  kinetoplast-associated protein of, 136
  kinetoplast DNA of, 118
  kinetoplast DNA network in
    replication of, 128
  life-cycle stages of, 177–79
  nonopsonic binding to phagocytes, 262
  stable transformation of, 431
Trypanosoma equiperdum
  dyskinetoplastidy in, 133
  kinetoplast DNA network in
    replication of, 128
  maxicircles in
    replication of, 131
  minicircles in, 120
    replication of, 129–30
  minicircle homogeneity in, 132
Trypanosoma evansi
  dyskinetoplastidy in, 133
  minicircles in, 120
  minicircle homogeneity in, 132
Trypanosomatid protozoa
  kinetoplast DNA of, 117–38
Trypanosomes
  resistance to arsenicals in, 438–39
Trypanothione, 439
Trypomastigotes, 177–79
  bloodstream surface antigens and, 184–86
  metacyclic surface antigens and, 184
Tryptophan
  biosynthesis of

SUBJECT INDEX   853

anthanilate synthase and, 56
endosymbionts and, 81–84
chemicals synthesized from, 569
TS
  See Thymidylate synthase
Tubercidin resistance
  in parasitic protozoa, 447
Tuberculosis, 641–66
  dormancy of, 648–49
  extracellular multiplication
    and transmission in, 649–50
  initial uptake of bacteria in, 643–46
  prevention of, 659–65
  T-cell responses in, 646–48
  treatment of, 650–59
Tumors
  gene therapy of
    herpes simplex virus vectors and, 696–700
Tylosin
  structure of, 203
Tyrosine
  chemicals synthesized from, 569

U

Ubiquinone, 97
Ultraviolet light
  deep-sea bacteria and, 793
Ureaplasma urealyticum, 379
UV radiation
  $\alpha^- \beta^-$ spores resistant to, 44–45
  spore resistance to, 38–40

V

Vaccines
  mycobacterial disease and, 659–65
Valine
  biosynthesis of
    leucine-responsive regulatory protein and, 765
  transport of
    leucine-responsive regulatory protein and, 756

Verapamil
  chloroquine sensitivity and, 435
Vermelone, 227
  structure of, 221
Versicolorin A
  conversion to sterigmatocystin, 227
  structure of, 221
Vertebrates
  antimicrobial peptides of, 277–99
Vesicular stomatitis virus (VSV), 822
Vibrio marinus
  membrane proteins of, 792
Vibrio parahaemolyticus
  biofilms produced by, 718
Viellonella alkalescens
  TNT degradation and, 527
Viral disease
  transmission of
    aphids and, 57
Viral hepatitis
  waterborne outbreaks of, 463–64
Viral vectors
  design of, 816–21
  efficacy of, 824–29
  in gene therapy, 807–31
Viruses
  environmentally transmitted, 461–81
  detection of, 473–79
Vitamin synthesis
  endosymbionts and, 79
Vitreoscilla, 11
VSV
  See Vesicular stomatitis virus

W

Warming
  deep-sea bacteria and, 788–89
Water
  boiling
    bacteria in, 15–16
Water potential
  methane oxidation and, 592

Wetlands
  methane oxidation in, 593
Whiteflies
  prokaryotic endosymbionts of, 63–64
Wolbachia pipientis, 63
  association with insects, 57

X

Xenopsin, 288

Y

Yayanos, A. A., 777–98
Yeast
  methane oxidation in, 582
  sterol alterations in
    physiological effects of, 107–9
  sterol biosynthesis in, 96–99
    regulation of, 99–101
  sterol metabolism in, 101–7
  sterol trafficking in, 109–10
Yersinia enterocolitica
  survival in macrophages, 161
Young, D. B., 641–66

Z

Zaragozic acids, 607–36
  activities in cells, 632–33
  antifungal activity of, 633
  biosynthesis of, 623–30
  discovery of, 610–12
  effects on enzymes, 631–32
  family groups of, 615–18
  fungi producing, 618–23
  isolation of, 612–13
  mechanism of action of, 631–35
  squalene synthase inhibition and, 631
  structural features of, 613–15
  in vivo activity of, 634–35
Zymomonas mobilis, 564
Zymosterol, 96–97
  esterification of, 105

CUMULATIVE INDEXES

CONTRIBUTING AUTHORS, VOLUMES 45–49

A

Abraham SN, 45:383–415
Adams MWW, 47:627–58
Aharonowitz Y, 46:461–95
Allen BL, 48:585–617
Ames GF-L, 47:291–319
Anderson S, 45:607–35
Andrew PW, 47:89–115
Andrews NW, 49:175–200
Ascher MS, 46:533–64
Atlas RM, 45:137–61

B

Baumann L, 49:55–94
Baumann P, 49:55–94
Beachy RN, 47:739–63
Bej AK, 47:139–66
Beppu T, 46:377–98
Berberof M, 48:25–52
Berens C, 48:345–69
Bergstrom JD, 49:607–39
Beverley SM, 45:417–44
Bills GF, 49:607–39
Blair DF, 49:489–522
Blanchard A, 48:687–712
Boe L, 47:139–66
Borst P, 49:427–60
Boulnois GJ, 47:89–115
Bouvier J, 47:821–53
Brock TD, 49:1–28
Bulawa CE, 47:505–34
Bull AT, 46:219–52
Burlage RS, 48:291–309
Burleigh BA, 49:175–200
Byrne K, 49:607–39

C

Caetano-Anollés G, 45:345–82
Caldwell DE, 49:711–45
Cammack R, 46:277–305
Campbell A, 48:193–222
Campbell WC, 45:445–74
Cardon LR, 48:619–54
Casey WM, 49:95–116
Cerami A, 46:695–729
Chater KF, 47:685–713
Churchward GG, 49:367–97
Citovsky V, 47:167–97
Clark MA, 49:55–94
Cocito CG, 46:95–116
Coene MM, 46:95–116

Cohen G, 46:461–95
Coplin DL, 46:307–46
Costerton JW, 49:711–45
Croen KD, 45:265–82
Cross GAM, 47:385–411
Csonka LN, 45:569–606
Cullen BR, 45:219–50
Cutler JE, 45:187–218

D

Davis BD, 46:1–33
Dean DR, 49:335–66
Debellé F, 46:497–531
Debono M, 48:471–97
DeLuca NA, 49:675–710
Dénarié J, 46:497–531
Descoteaux A, 46:65–94
de Villiers E-M, 48:427–47
Doige CA, 47:291–319
Donachie WD, 47:199–230
Donadio S, 47:875–912
Draths KM, 49:557–79
Dufresne C, 49:607–39
Duncan K, 49:641–73

E

Eichinger D, 48:499–523
Embley TM, 48:257–89
Englund PT, 49:117–43
Ensley BD, 45:283–99
Esko JD, 48:139–62
Estes MK, 49:461–87

F

Fairlamb AH, 46:695–729
Fayet O, 45:301–25
Feagin JE, 48:81–104
Felix CR, 47:791–819
Feng P, 48:401–26
Fenical W, 48:559–84
Ferry JG, 49:305–33
Fink DJ, 49:675–710
Finnerty WR, 46:193–218
Fisher SH, 45:107–35
Fisher K, 49:335–66
Fitchen JH, 47:739–63
Foster PL, 47:467–504
Foster JW, 49:145–74
Friedrich B, 47:351–83
Frost JW, 49:557–79
Fujii I, 49:201–38

G

García-Sastre A, 47:765–90
Georgopoulos C, 45:301–25
Ghuysen J-M, 45:37–67
Givskov M, 47:139–66
Glorioso JC, 49:675–710
Goldhar J, 49:239–76
Gonzalez-Scarano F, 47:117–38
Goodfellow M, 46:219–52
Gordee RS, 48:471–97
Granados RR, 45:69–87
Gresshoff PM, 45:345–82
Griot C, 47:117–38
Guerinot ML, 48:743–72

H

Hagedorn S, 48:773–800
Hager KM, 48:139–62
Hajduk SL, 48:139–62
Hansen JN, 47:535–64
Hanson AD, 45:569–606
Harayama S, 46:565–601
Hengge-Aronis R, 48:53–80
Hill TM, 46:603–33
Hillen W, 48:345–69
Hoch JA, 47:441–65
Hoet PP, 46:95–116
Holloway BW, 47:659–84
Höök M, 48:585–617
Horinouchi S, 46:377–98
Hultgren SJ, 45:383–415
Hutchinson CR, 49:201–38

I

Inouye M, 45:163–86
Inouye S, 45:163–86

J

Janssen DB, 48:163–91
Jensen LB, 47:139–66
Jensen PR, 48:559–84

K

Kaiser D, 46:117–39
Kaphammer B, 48:773–800
Karlin S, 48:619–54
Katz L, 47:875–912
Keisari Y, 49:239–76
Kim SK, 46:117–39

855

Klier AF, 46:429–59
Kok M, 46:565–601
Kolter R, 46:141–63; 47:855–74
Korber DR, 49:711–45
Kristensen CS, 47:139–66
Kuo C-T, 48:291–309
Kuspa A, 46:117–39

L

Lai C-Y, 49:55–94
Lappin-Scott HM, 49:711–45
Leigh JA, 46:307–46
Leschine SB, 49:399–426
Lewandowski Z, 49:711–45
Lin R, 49:747–75
Lindow SE, 47:913–44
Lipscomb JD, 48:371–99
Liu H-w, 48:223–56
Ljungdahl LG, 47:791–819
Loewen PC, 48:53–80
Lory S, 47:565–96
Lovley DR, 47:263–90

M

Magasanik B, 48:1–24
Mancinelli RL, 49:581–605
Marion PL, 45:475–508
Martín JF, 46:461–95
Marzluf GA, 47:31–57
Mason JR, 46:277–305
McGavin MJ, 48:585–617
McKerrow JH, 47:821–53
McKinlay MA, 46:635–54
Melnick JL, 49:461–87
Metcalf TG, 49:461–87
Mitchell TJ, 47:89–115
Molin S, 47:139–66
Montagnier L, 48:687–712
Mor A, 49:277–304
Moran NA, 49:55–94
Moreno F, 46:141–63
Mori H, 47:321–50
Msadek T, 46:429–59
Murphy JW, 45:509–38

N

Nagai H, 47:321–50
Nallin-Omstead M, 49:607–39
Nathanson N, 47:117–38
Nealson KH, 48:311–43
Neidle EL, 46:565–601
Newman EB, 49:747–75
Nicolas P, 49:277–304
Nilsen TW, 47:413–40
Nilsson B, 45:607–35
Normark S, 45:383–415

Nussenzweig V, 48:499–523

O

Ofek I, 49:239–76
Omura S, 47:57–87
Ouellette M, 49:427–60

P

Page BD, 47:231–61
Palese P, 47:765–90
Parks LW, 49:95–116
Paton JC, 47:89–115
Patti JM, 48:585–617
Pays E, 48:25–52
Pereira MEA, 48:499–523
Peters JW, 49:335–66
Pevear DC, 46:635–54
Pfennig N, 47:1–29
Price RW, 46:655–93
Pries F, 48:163–91
Prusiner SB, 48:655–86

R

Ramos JL, 47:139–66
Rapoport G, 46:429–59
Reeve JN, 46:165–91
Reznikoff WS, 47:945–63
Rose MD, 45:539–67
Rosenberg C, 46:497–531
Rosenthal PJ, 47:821–53
Rossmann MG, 46:635–54
Rouhbakhsh D, 49:55–94

S

Sadowsky MJ, 46:399–428
Saffarini D, 48:311–43
Schell MA, 47:597–626
Schenkman S, 48:499–523
Schwartz E, 47:351–83
Scott JR, 49:367–97
Setlow P, 49:29–54
Shannon MJR, 47:715–38
Shapira M, 48:449–70
Shapiro TA, 49:117–43
Sharon N, 49:239–76
Sheppard HW, 46:533–64
Sherker AH, 45:475–508
Siegele DA, 47:855–74
Simons RW, 48:713–42
Slater JH, 46:219–52
Smith AE, 49:807–38
Snyder M, 47:231–61
Sommer JM, 48:105–38
Sonenshein AL, 45:107–35
Spain JC, 49:523–55

Spector MP, 49:145–74
Spencer DC, 46:655–93
Stackebrandt E, 48:257–89
Steele DB, 45:89–106
Steffan RJ, 45:137–61; 48:525–57
Stowers MD, 45:89–106
Straus SE, 45:265–82
Strom MS, 47:565–96
Stuart K, 45:327–44
Sun E, 47:821–53
Sutherland IW, 39:243–70
Swaminathan B, 48:401–26

T

Takle GB, 47:385–411
Tanaka Y, 47:57–87
Taylor DE, 46:35–64
Taylor JM, 46:253–76
Thorson JS, 48:223–56
Timmis KN, 48:525–57
Tormo A, 47:855–74
Triplett EW, 46:399–428
Turco SJ, 46:65–94

U

Unterman R, 47:715–38; 48:525–57

V

van der Ploeg JR, 48:163–91
Vanhamme L, 48:25–52

W

Wagner EGH, 48:713–42
Wang AL, 45:251–63
Wang CC, 45:251–63; 48:105–38
Wickner RB, 46:347–75
Wilson M, 47:913–44
Wolfe RS, 45:1–35
Wood HA, 45:69–87

Y

Yayanos AA, 49:777–805
Young DB, 49:641–73
Yura T, 47:321–50

Z

Zambryski P, 47:167–97
Zeilstra-Ryalls J, 45:301–25
Zilberstein D, 48:449–70
zur Hausen H, 48:427–47

CHAPTER TITLES, VOLUMES 45–49

PREFATORY CHAPTERS

My Kind of Biology	RS Wolfe	45:1–35
Science and Politics: Tensions Between the Head and the Heart	BD Davis	46:1–33
Reflections of a Microbiologist, or How to Learn from the Microbes	N Pfennig	47:1–29
A Charmed Life	B Magasanik	48:1–24
The Road to Yellowstone—and Beyond	TD Brock	49:1–28

ANIMAL PATHOGENS AND DISEASES

Putative Virulence Factors of *Candida albicans*	JE Cutler	45:187–218
Varicella-Zoster Virus Latency	KD Croen, SE Straus	45:265–82
Hepadnaviruses and Hepatocellular Carcinoma	AH Sherker, PL Marion	45:475–508
Mechanisms of Natural Resistance to Human Pathogenic Fungi	JW Murphy	45:509–38
Genetics of *Campylobacter* and *Helicobacter*	DE Taylor	46:35–64
The Lipophosphoglycan of *Leishmania* Parasites	SJ Turco, A Descoteaux	46:65–94
The Structure and Replication of Hepatitis Delta Virus	JM Taylor	46:253–76
The Natural History and Pathogenesis of HIV Infection	HW Sheppard, MS Ascher	46:533–64
Treatment of the Picornavirus Common Cold by Inhibitors of Viral Uncoating and Attachment	MA McKinlay, DC Pevear, MG Rossmann	46:635–54
Human Immunodeficiency Virus and the Central Nervous System	DC Spencer, RW Price	46:655–93
Metabolism and Functions of Trypanothione in the Kinetoplastida	AH Fairlamb, A Cerami	46:695–729
Molecular Analysis of the Pathogenicity of *Streptococcus pneumoniae:* The Role of Pneumococcal Proteins	JC Paton, PW Andrew, GJ Boulnois, TJ Mitchell	47:89–115
The Proteases and Pathogenicity of Parasitic Protozoa	JH McKerrow, E Sun, PJ Rosenthal, J Bouvier	47:821–53
Genetic Controls for the Expression of Surface Antigens in African Trypanosomes	E Pays, L Vanhamme, M Berberof	48:25–52
The Extrachromosomal DNAs of Apicomplexan Parasites	JE Feagin	48:81–104
Targeting Proteins to the Glycosomes of African Trypanosomes	JM Sommer, CC Wang	48:105–38
Human High Density Lipoprotein Killing of African Trypanosomes	SL Hajduk, KM Hager, JD Esko	48:139–62
Rapid Detection of Food-Borne Pathogenic Bacteria	B Swaminathan, P Feng	48:401–26
Human Papillomaviruses	H zur Hausen, E-M de Villiers	48:427–47
The Role of pH and Temperature in the Development of *Leishmania* Parasites	D Zilberstein, M Shapira	48:449–70
Structural and Functional Properties of *Trypanosoma Trans*-Sialidase	S Schenkman, D Eichinger, MEA Pereira, V Nussenzweig	48:499–523

857

MSCRAMM-Mediated Adherence of Microorganisms to Host Tissues	JM Patti, BL Allen, MJ McGavin, M Höök	48:585–617
Biology and Genetics of Prion Diseases	SB Prusiner	48:655–86
AIDS-Associated Mycoplasmas	A Blanchard, L Montagnier	48:687–712
Peptides as Weapons Against Microorganisms in the Chemical Defense System of Vertebrates	P Nicolas, A Mor	49:277–304
New Mechanisms of Drug Resistance in Parasitic Protozoa	P Borst, M Ouellette	49:427–60
Prospects for New Interventions in the Treatment and Prevention of Mycobacterial Disease	DB Young, K Duncan	49:641–73

APPLIED MICROBIOLOGY AND ECOLOGY

Genetically Engineered Baculoviruses as Agents for Pest Control	HA Wood, RR Granados	45:69–87
Techniques for Selection of Industrially Important Microorganisms	DB Steele, MD Stowers	45:89–106
Polymerase Chain Reaction: Applications in Environmental Microbiology	RJ Steffan, RM Atlas	45:137–61
Biochemical Diversity of Trichloroethylene Metabolism	BD Ensley	45:283–99
The Biology and Genetics of the Genus *Rhodococcus*	WR Finnerty	46:193–218
Biodiversity as a Source of Innovation in Biotechnology	AT Bull, M Goodfellow, JH Slater	46:219–52
The Electron Transport Proteins of Hydroxylating Bacterial Dioxygenases	JR Mason, R Cammack	46:277–305
Penicillin and Cephalosporin Biosynthetic Genes: Structure, Organization, Regulation, and Evolution	Y Aharonowitz, G Cohen, JF Martin	46:461–95
Functional and Evolutionary Relationships Among Diverse Oxygenases	S Harayama, M Kok, EL Neidle	46:565–601
Agroactive Compounds of Microbial Origin	Y Tanaka, S Omura	47:57–87
Suicidal Genetic Elements and Their Use in Biological Containment of Bacteria	S Molin, L Boe, LB Jensen, CS Kristensen, M Givskov, JL Ramos, AK Bej	47:139–66
Dissimilatory Metal Reduction	DR Lovley	47:263–90
Molecular Biology of Hydrogen Utilization in Aerobic Chemolithotrophs	B Friedrich, E Schwartz	47:351–83
Evaluating Bioremediation: Distinguishing Fact from Fiction	MJR Shannon, R Unterman	47:715–38
Release of Recombinant Microorganisms	M Wilson, SE Lindow	47:913–44
Pathways and Mechanisms in the Biogenesis of Novel Deoxysugars by Bacteria	H-w Liu, JS Thorson	48:223–56
Living Biosensors for the Management and Manipulation of Microbial Consortia	RS Burlage, C-T Kuo	48:291–309
Iron and Manganese in Anaerobic Respiration: Environmental Significance, Physiology, and Regulation	KH Nealson, D Saffarini	48:311–43
Biochemistry of the Soluble Methane Monooxygenase	JD Lipscomb	48:371–99
Rapid Detection of Food-Borne Pathogenic Bacteria	B Swaminathan, P Feng	48:401–26
Designing Microorganisms for the Treatment of Toxic Wastes	KN Timmis, RJ Steffan, R Unterman	48:525–57
Strategies for the Discovery of Secondary Metabolites from Marine Bacteria: Ecological Perspectives	PR Jensen, W Fenical	48:559–84
Microbial Iron Transport	ML Guerinot	48:743–72

CHAPTER TITLES 859

Microbial Biocatalysis in the Generation of Flavor and Fragrance Chemicals	S Hagedorn, B Kaphammer	48:773–800
Cellulose Degradation in Anaerobic Environments	SB Leschine	49:399–426
Environmental Virology: From Detection of Virus in Sewage and Water by Isolation to Identification by Molecular Biology—A Trip Over 50 Years	TG Metcalf, JL Melnick, MK Estes	49:461–87
Biodegradation of Nitroaromatic Compounds	JC Spain	49:523–55
Biocatalytic Syntheses of Aromatics from D-Glucose: Renewable Microbial Sources of Aromatic Compounds	JW Frost, KM Draths	49:557–79
The Regulation of Methane Oxidation in Soil	RL Mancinelli	49:581–605
Microbial Biofilms	JW Costerton, Z Lewandowski, DE Caldwell, DR Korber, HM Lappin-Scott	49:711–45
Microbiology to 10,500 Meters in the Deep Sea	AA Yayanos	49:777–805

CHEMOTHERAPY AND CHEMOTHERAPEUTIC AGENTS

Ivermectin as an Antiparasitic Agent for Use in Humans	WC Campbell	45:445–74
Genetics of Ribosomally Synthesized Peptide Antibiotics	R Kolter, F Moreno	46:141–63
Penicillin and Cephalosporin Biosynthetic Genes: Structure, Organization, Regulation, and Evolution	Y Aharonowitz, G Cohen, JF Martin	46:461–95
ATP-Dependent Transport Systems in Bacteria and Humans: Relevance to Cystic Fibrosis and Multidrug Resistance	CA Doige, GF-L Ames	47:291–319
Antibiotics Synthesized by Posttranslational Modification	JN Hansen	47:535–64
Polyketide Synthesis: Prospects for Hybrid Antibiotics	L Katz, S Donadio	47:875–912
Mechanisms Underlying Expression of Tn*10*-Encoded Tetracycline Resistance	W Hillen, C Berens	48:345–69
Antibiotics that Inhibit Fungal Cell Wall Development	M Debono, RS Gordee	48:471–97
Strategies for the Discovery of Secondary Metabolites from Marine Bacteria: Ecological Perspectives	PR Jensen, W Fenical	48:559–84
Polyketide Synthase Gene Manipulation: A Structure-Function Approach in Engineering Novel Antibiotics	CR Hutchinson, I Fujii	49:201–38
Discovery, Biosynthesis, and Mechanism of Action of the Zaragozic Acids: Potent Inhibitors of Squalene Synthase	JD Bergstrom, C Dufresne, GF Bills, M Nallin-Omstead, K Byrne	49:607–39

DIVERSITY AND SYSTEMATICS

Biochemical Diversity of Trichloroethylene Metabolism	BD Ensley	45:283–99
The Universally Conserved GroE (Hsp60) Chaperonins	J Zeilstra-Ryalls, O Fayet, C Georgopoulos	45:301–25
Molecular Biology of Methanogens	JN Reeve	46:165–91
Biodiversity as a Source of Innovation in Biotechnology	AT Bull, M Goodfellow, JH Slater	46:219–52
Functional and Evolutionary Relationships Among Diverse Oxygenases	S Harayama, M Kok, EL Neidle	46:565–601
Adaptive Mutation: The Uses of Adversity	PL Foster	47:467–504
Genetics for All Bacteria	BW Holloway	47:659–84
The Molecular Phylogeny and Systematics of the Actinomycetes	TM Embley, E Stackebrandt	48:257–89

Computational DNA Sequence Analysis	S Karlin, LR Cardon	48:619–54
Genetics, Physiology, and Evolutionary Relationships of the Genus *Buchnera:* Intracellular Symbionts of Aphids	P Baumann, L Baumann, C-Y Lai, D Rouhbakhsh, NA Moran, MA Clark	49:55–94

GENETICS

msDNA and Bacterial Reverse Transcriptase	M Inouye, S Inouye	45:163–86
Regulation of Human Immunodeficiency Virus Replication	BR Cullen	45:219–50
RNA Editing in Trypanosomatid Mitochondria	K Stuart	45:327–44
Plant Genetic Control of Nodulation	G Caetano-Anollés, PM Gresshoff	45:345–82
Gene Amplification in *Leishmania*	SM Beverley	45:417–44
Nuclear Fusion in Yeast	MD Rose	45:539–67
Prokaryotic Osmoregulation: Genetics and Physiology	LN Csonka, AD Hanson	45:569–606
Genetics of *Campylobacter* and *Helicobacter*	DE Taylor	46:35–64
Replication Cycle of *Bacillus subtilis* Hydroxymethyluracil-Containing Phages	PP Hoet, MM Coene, CG Cocito	46:95–116
Genetics of Ribosomally Synthesized Peptide Antibiotics	R Kolter, F Moreno	46:141–63
Molecular Biology of Methanogens	JN Reeve	46:165–91
The Biology and Genetics of the Genus *Rhodococcus*	WR Finnerty	46:193–218
Double-Stranded and Single-Stranded RNA Viruses of *Saccharomyces cerevisiae*	RB Wickner	46:347–75
Genetics of Competition for Nodulation of Legumes	EW Triplett, MJ Sadowsky	46:399–428
Positive Regulation in the Gram-Positive Bacterium: *Bacillus subtilis*	A Klier, T Msadek, G Rapoport	46:429–59
Penicillin and Cephalosporin Biosynthetic Genes: Structure, Organization, Regulation, and Evolution	Y Aharonowitz, G Cohen, JF Martin	46:461–95
Signaling and Host Range Variation in Nodulation	J Dénarié, F Debellé, C Rosenberg	46:497–531
Suicidal Genetic Elements and Their Use in Biological Containment of Bacteria	S Molin, L Boe, LB Jensen, CS Kristensen, M Givskov, JL Ramos, AK Bej	47:139–66
Genetics and Molecular Biology of Chitin Synthesis in Fungi	CE Bulawa	47:505–34
Molecular Biology of the LysR Family of Transcriptional Regulators	MA Schell	47:597–626
Genetics for All Bacteria	BW Holloway	47:659–84
Genetics of Differentiation in *Streptomyces*	KF Chater	47:685–713
Genetic Manipulation of Negative-Strand RNA Virus Genomes	A García-Sastre, P Palese	47:765–90
Release of Recombinant Microorganisms	M Wilson, SE Lindow	47:913–44
Genetic Controls for the Expression of Surface Antigens in African Trypanosomes	E Pays, L Vanhamme, M Berberof	48:25–52
The Role of the Sigma Factor σ^S (KatF) in Bacterial Global Regulation	PC Loewen, R Hengge-Aronis	48:53–80
The Extrachromosomal DNAs of Apicomplexan Parasites	JE Feagin	48:81–104
Targeting Proteins to the Glycosomes of African Trypanosomes	JM Sommer, CC Wang	48:105–38
Genetics and Biochemistry of Dehalogenating Enzymes	DB Janssen, F Pries, JR van der Ploeg	48:163–91
Comparative Molecular Biology of Lambdoid Phages	A Campbell	48:193–222
Mechanisms Underlying Expression of Tn*10*-Encoded Tetracycline Resistance	W Hillen, C Berens	48:345–69

Designing Microorganisms for the Treatment of Toxic Wastes	KN Timmis, RJ Steffan, R Unterman	48:525–57
Computational DNA Sequence Analysis	S Karlin, LR Cardon	48:619–54
Biology and Genetics of Prion Diseases	SB Prusiner	48:655–86
Antisense RNA Control in Bacteria, Phages, and Plasmids	EGH Wagner, RW Simons	48:713–42
Mechanisms for the Prevention of Damage to DNA in Spores of *Bacillus* Species	P Setlow	49:29–54
Genetics, Physiology, and Evolutionary Relationships of the Genus *Buchnera*: Intracellular Symbionts of Aphids	P Baumann, L Baumann, C-Y Lai, D Rouhbakhsh, NA Moran, MA Clark	49:55–94
The Structure and Replication of Kinetoplast DNA	TA Shapiro, PT Englund	49:117–43
Polyketide Synthase Gene Manipulation: A Structure-Function Approach in Engineering Novel Antibiotics	CR Hutchinson, I Fujii	49:201–38
Nitrogenase Structure and Function: A Biochemical-Genetic Perspective	JW Peters, K Fisher, DR Dean	49:335–66
Conjugative Transposition	JR Scott, GG Churchward	49:367–97
Leucine-Responsive Regulatory Protein: A Global Regulator of Gene Expression in *E. coli*	EB Newman, R Lin	49:747–75

IMMUNOLOGY

Antibiotics Synthesized by Posttranslational Modification	JN Hansen	47:535–64
Polyketide Synthesis: Prospects for Hybrid Antibiotics	L Katz, S Donadio	47:875–912
Polyketide Synthase Gene Manipulation: A Structure-Function Approach in Engineering Novel Antibiotics	CR Hutchinson, I Fujii	49:201–38

MORPHOLOGY, ULTRASTRUCTURE, AND DIFFERENTIATION

Serine β-Lactamases and Penicillin-Binding Proteins	J-M Ghuysen	45:37–67
msDNA and Bacterial Reverse Transcriptase	M Inouye, S Inouye	45:163–86
The Universally Conserved GroE (Hsp60) Chaperonins	J Zeilstra-Ryalls, O Fayet, C Georgopoulos	45:301–25
Chaperone-Assisted Assembly and Molecular Architecture of Adhesive Pili	SJ Hultgren, S Normark, SN Abraham	45:383–415
Proper and Improper Folding of Proteins in the Cellular Environment	B Nilsson, S Anderson	45:607–35
The Lipophosphoglycan of *Leishmania* Parasites	SJ Turco, A Descoteaux	46:65–94
Control of Cell Density and Pattern by Intercellular Signaling in *Myxococcus* Development	SK Kim, D Kaiser, A Kuspa	46:117–39
Genetics of Competition for Nodulation of Legumes	EW Triplett, MJ Sadowsky	46:399–428
Chromosome Segregation in Yeast	BD Page, M Snyder	47:231–61
The Surface *Trans*-Sialidase Family of *Trypanosoma cruzi*	GAM Cross, GB Takle	47:385–411
Trans-Splicing of Nematode Premessenger RNA	TW Nilsen	47:413–40
Regulation of the Phosphorelay and the Initiation of Sporulation in *Bacillus subtilis*	JA Hoch	47:441–65
Molecular Biology of the LysR Family of Transcriptional Regulators	MA Schell	47:597–626
Genetics of Differentiation in *Streptomyces*	KF Chater	47:685–713
The Cellulosome: The Exocellular Organelle of *Clostridium*	CR Felix, LG Ljungdahl	47:791–819
The Tn5 Transposon	WS Reznikoff	47:945–63

Targeting Proteins to the Glycosomes of African Trypanosomes	JM Sommer, CC Wang	48:105–38
Antibiotics that Inhibit Fungal Cell Wall Development	M Debono, RS Gordee	48:471–97
Structural and Functional Properties of *Trypanosoma Trans*-Sialidase	S Schenkman, D Eichinger, MEA Pereira, V Nussenzweig	48:499–523
MSCRAMM-Mediated Adherence of Microorganisms to Host Tissues	JM Patti, BL Allen, MJ McGavin, M Höök	48:585–617
The Structure and Replication of Kinetoplast DNA	TA Shapiro, PT Englund	49:117–43
The Mechanisms of *Trypanosoma cruzi* Invasion of Mammalian Cells	BA Burleigh, NW Andrews	49:175–200
Nitrogenase Structure and Function: A Biochemical-Genetic Perspective	JW Peters, K Fisher, DR Dean	49:335–66
How Bacteria Sense and Swim	DF Blair	49:489–522
Leucine-Responsive Regulatory Protein: A Global Regulator of Gene Expression in *E. coli*	EB Newman, R Lin	49:747–75

PHYSIOLOGY, GROWTH, AND NUTRITION

Control of Carbon and Nitrogen Metabolism in *Bacillus subtilis*	SH Fisher, AL Sonenshein	45:107–35
RNA Editing in Trypanosomatid Mitochondria	K Stuart	45:327–44
Nuclear Fusion in Yeast	MD Rose	45:539–67
Prokaryotic Osmoregulation: Genetics and Physiology	LN Csonka, AD Hanson	45:569–606
Proper and Improper Folding of Proteins in the Cellular Environment	B Nilsson, S Anderson	45:607–35
The Electron Transport Proteins of Hydroxylating Bacterial Dioxygenases	JR Mason, R Cammack	46:277–305
Autoregulatory Factors and Communication in Actinomycetes	S Horinouchi, T Beppu	46:377–98
Positive Regulation in the Gram-Positive Bacterium: *Bacillus subtilis*	A Klier, T Msadek, G Rapoport	46:429–59
Arrest of Bacterial DNA Replication	TM Hill	46:603–33
Metabolism and Functions of Trypanothione in the Kinetoplastida	AH Fairlamb, A Cerami	46:695–729
Regulation of Sulfur and Nitrogen Metabolism in Filamentous Fungi	GA Marzluf	47:31–55
Transport of Nucleic Acids Through Membrane Channels: Snaking Through Small Holes	V Citovsky, P Zambryski	47:167–97
The Cell Cycle of *Escherichia coli*	WD Donachie	47:199–230
Regulation of the Heat-Shock Response in Bacteria	T Yura, H Nagai, H Mori	47:321–50
Genetics and Molecular Biology of Chitin Synthesis in Fungi	CE Bulawa	47:505–34
Structure-Function and Biogenesis of the Type IV Pili	MS Strom, S Lory	47:565–96
Enzymes and Proteins from Organisms that Grow Near and Above 100°C	MWW Adams	47:627–58
The Stationary Phase of the Bacterial Life Cycle	R Kolter, DA Siegele, A Tormo	47:855–74
The Role of the Sigma Factor σ^S (KatF) in Bacterial Global Regulation	PC Loewen, R Hengge-Aronis	48:53–80
Human High Density Lipoprotein Killing of African Trypanosomes	SL Hajduk, KM Hager, JD Esko	48:139–62
Genetics and Biochemistry of Dehalogenating Enzymes	DB Janssen, F Pries, JR van der Ploeg	48:163–91
Pathways and Mechanisms in the Biogenesis of Novel Deoxysugars by Bacteria	H-w Liu, JS Thorson	48:223–56
Living Biosensors for the Management and Manipulation of Microbial Consortia	RS Burlage, C-T Kuo	48:291–309

CHAPTER TITLES 863

Iron and Manganese in Anaerobic Respiration: Environmental Significance, Physiology, and Regulation	KH Nealson, D Saffarini	48:311–43
Biochemistry of the Soluble Methane Monooxygenase	JD Lipscomb	48:371–99
The Role of pH and Temperature in the Development of *Leishmania* Parasites	D Zilberstein, M Shapira	48:449–70
Designing Microorganisms for the Treatment of Toxic Wastes	KN Timmis, RJ Steffan, R Unterman	48:525–57
Antisense RNA Control in Bacteria, Phages, and Plasmids	EGH Wagner, RW Simons	48:713–42
Microbial Iron Transport	ML Guerinot	48:743–72
Microbial Biocatalysis in the Generation of Flavor and Fragrance Chemicals	S Hagedorn, B Kaphammer	48:773–800
Mechanisms for the Prevention of Damage to DNA in Spores of *Bacillus* Species	P Setlow	49:29–54
Physiological Implications of Sterol Biosynthesis in Yeast	LW Parks, WM Casey	49:95–116
How *Salmonella* Survive Against the Odds	JW Foster, MP Spector	49:145–74
Nonopsonic Phagocytosis of Microorganisms	I Ofek, J Goldhar, Y Keisari, N Sharon	49:239–76
CO Dehydrogenase	JG Ferry	49:305–33
The Regulation of Methane Oxidation in Soil	RL Mancinelli	49:581–605

PLANT-BACTERIA INTERACTIONS

Plant Genetic Control of Nodulation	G Caetano-Anollés, PM Gresshoff	45:345–82
Exopolysaccharides in Plant-Bacterial Interactions	JA Leigh, DL Coplin	46:307–46
Genetics of Competition for Nodulation of Legumes	EW Triplett, MJ Sadowsky	46:399–428
Signaling and Host Range Variation in Nodulation	J Dénarié, F Debellé, C Rosenberg	46:497–531
Genetically Engineered Protection Against Viruses in Transgenic Plants	JH Fitchen, RN Beachy	47:739–63

VIROLOGY

Genetically Engineered Baculoviruses as Agents for Pest Control	HA Wood, RR Granados	45:69–87
Regulation of Human Immunodeficiency Virus Replication	BR Cullen	45:219–50
Viruses of the Protozoa	AL Wang, CC Wang	45:251–63
Varicella-Zoster Virus Latency	KD Croen, SE Straus	45:265–82
Hepadnaviruses and Hepatocellular Carcinoma	AH Sherker, PL Marion	45:475–508
Replication Cycle of *Bacillus subtilis* Hydroxymethyluracil-Containing Phages	PP Hoet, MM Coene, CG Cocito	46:95–116
The Structure and Replication of Hepatitis Delta Virus	JM Taylor	46:253–76
Double-Stranded and Single-Stranded RNA Viruses of *Saccharomyces cerevisiae*	RB Wickner	46:347–75
The Natural History and Pathogenesis of HIV Infection	HW Sheppard, MS Ascher	46:533–64
Treatment of the Picornavirus Common Cold by Inhibitors of Viral Uncoating and Attachment	MA McKinlay, DC Pevear, MG Rossmann	46:635–54
Human Immunodeficiency Virus and the Central Nervous System	DC Spencer, RW Price	46:655–93
Molecular Determinants of the Virulence and Infectivity of California Serogroup Bunyaviruses	C Griot, F Gonzalez-Scarano, N Nathanson	47:117–38
Genetic Manipulation of Negative-Strand RNA Virus Genomes	A García-Sastre, P Palese	47:765–90

Comparative Molecular Biology of Lambdoid Phages	A Campbell	48:193–222
Human Papillomaviruses	H zur Hausen, E-M de Villiers	48:427–47
Environmental Virology: From Detection of Virus in Sewage and Water by Isolation to Identification by Molecular Biology—A Trip Over 50 Years	TG Metcalf, JL Melnick, MK Estes	49:461–87
Development and Application of Herpes Simplex Virus Vectors for Human Gene Therapy	JC Glorioso, NA DeLuca, DJ Fink	49:675–710
Viral Vectors in Gene Therapy	AE Smith	49:807–38

ANNUAL REVIEWS

a nonprofit scientific publisher
4139 El Camino Way
P.O. Box 10139
Palo Alto, CA 94303-0139 • USA

ORDER FORM

ORDER TOLL FREE
1.800.523.8635
from USA and Canada

Fax: 1.415.855.9815

Annual Reviews publications may be ordered directly from our office; through stockists, booksellers and subscription agents, worldwide; and through participating professional societies. **Prices are subject to change without notice. We do not ship on approval.**

- **Individuals:** Prepayment required on new accounts. in US dollars, checks drawn on a US bank.
- **Institutional Buyers:** Include purchase order. Calif. Corp. #161041 • ARI Fed. I.D. #94-1156476
- **Students / Recent Graduates:** $10.00 discount from retail price, per volume. *Requirements:* 1. be a degree candidate at, or a graduate within the past three years from, an accredited institution; 2. present proof of status (photocopy of your student I.D. or proof of date of graduation); 3. Order direct from Annual Reviews; 4. prepay. This discount **does not** apply to standing orders, *Index on Diskette*, Special Publications, ARPR, or institutional buyers.
- **Professional Society Members:** Many Societies offer *Annual Reviews* to members at reduced rates. Check with your society or contact our office for a list of participating societies.
- **California orders** add applicable sales tax. • **Canadian orders** add 7% GST. Registration #R 121 449-029.
- **Postage paid** by Annual Reviews (4th class bookrate/surface mail). UPS ground service is available at $2.00 extra per book within the contiguous 48 states only. UPS air service or US airmail is available to any location at actual cost. UPS requires a street address. P.O. Box, APO, FPO, not acceptable.
- **Standing Orders:** Set up a standing order and the new volume in series is sent automatically each year upon publication. Each year you can save 10% by prepayment of prerelease invoices sent 90 days prior to the publication date. Cancellation may be made at any time.
- **Prepublication Orders:** Advance orders may be placed for any volume and will be charged to your account upon receipt. Volumes not yet published will be shipped during month of publication indicated.

NOTE: For copies of individual articles from any *Annual Review*, or copies of any article cited in an *Annual Review*, call **Annual Reviews Preprints and Reprints (ARPR)** toll free 1-800-347-8007 (fax toll free 1-800-347-8008) from the USA or Canada. From elsewhere call 1-415-259-5017.

ANNUAL REVIEWS SERIES *Volumes not listed are no longer in print*		Prices, postpaid, per volume. USA/other countries	Regular Order Please send Volume(s):	Standing Order Begin with Volume:
☐ Annual Review of	**ANTHROPOLOGY**			
Vols.	1-20	(1972-91) ....$41 / $46		
Vols.	21-22	(1992-93) ....$44 / $49		
Vol.	23-24	(1994 and Oct. 1995) ....$47 / $52	Vol(s). _____	Vol. _____
☐ Annual Review of	**ASTRONOMY AND ASTROPHYSICS**			
Vols.	1, 5-14, 16-29	(1963, 67-76, 78-91) ....$53 / $58		
Vols.	30-31	(1992-93) ....$57 / $62		
Vol.	32-33	(1994 and Sept. 1995) ....$60 / $65	Vol(s). _____	Vol. _____
☐ Annual Review of	**BIOCHEMISTRY**			
Vols.	31-34, 36-60	(1962-65,67-91) ....$41 / $47		
Vols.	61-62	(1992-93) ....$46 / $52		
Vol.	63-64	(1994 and July 1995) ....$49 / $55	Vol(s). _____	Vol. _____
☐ Annual Review of	**BIOPHYSICS AND BIOMOLECULAR STRUCTURE**			
Vols.	1-20	(1972-91) ....$55 / $60		
Vols.	21-22	(1992-93) ....$59 / $64		
Vol.	23-24	(1994 and June 1995) ....$62 / $67	Vol(s). _____	Vol. _____

| **ANNUAL REVIEWS SERIES**
Volumes not listed are no longer in print | **Prices, postpaid, per volume.**
USA/other countries | Regular Order
Please send Volume(s): | Standing Order
Begin with Volume: |

☐ *Annual Review of* **CELL AND DEVELOPMENTAL BIOLOGY** (new title beginning with volume 11)
- Vols. 1-7 (1985-91) .... $41 / $46
- Vols. 8-9 (1992-93) .... $46 / $51
- Vol. 10-11 (1994 and Nov. 1995) .... $49 / $54   Vol(s). _____ Vol. _____

☐ *Annual Review of* **COMPUTER SCIENCE** (Series suspended)
- Vols. 1-2 (1986-87) .... $41 / $46
- Vols. 3-4 (1988-89/90) .... $47 / $52   Vol(s). _____

Special package price for
- Vols. 1-4 (if ordered together) .... $100 / $115   ☐ Send all four volumes.

☐ *Annual Review of* **EARTH AND PLANETARY SCIENCES**
- Vols. 1-6, 8-19 (1973-78, 80-91) .... $55 / $60
- Vols. 20-21 (1992-93) .... $59 / $64
- Vol. 22-23 (1994 and May 1995) .... $62 / $67   Vol(s). _____ Vol. _____

☐ *Annual Review of* **ECOLOGY AND SYSTEMATICS**
- Vols. 2-12, 14-17, 19-22..(1971-81, 83-86, 88-91) ..$40 / $45
- Vols. 23-24 (1992-93) .... $44 / $49
- Vol. 25-26 (1994 and Nov. 1995) .... $47 / $52   Vol(s). _____ Vol. _____

☐ *Annual Review of* **ENERGY AND THE ENVIRONMENT**
- Vols. 1-16 (1976-91) .... $64 / $69
- Vols. 17-18 (1992-93) .... $68 / $73
- Vol. 19-20 (1994 and Oct. 1995) .... $71 / $76   Vol(s). _____ Vol. _____

☐ *Annual Review of* **ENTOMOLOGY**
- Vols. 10-16, 18, 20-36 (1965-71, 73, 75-91) ... $40 / $45
- Vols. 37-38 (1992-93) .... $44 / $49
- Vol. 39-40 (1994 and Jan. 1995) .... $47 / $52   Vol(s). _____ Vol. _____

☐ *Annual Review of* **FLUID MECHANICS**
- Vols. 2-4, 7 (1970-72, 75)
- 9-11, 16-23 (1977-79, 84-91) .... $40 / $45
- Vols. 24-25 (1992-93) .... $44 / $49
- Vol. 26-27 (1994 and Jan. 1995) .... $47 / $52   Vol(s). _____ Vol. _____

☐ *Annual Review of* **GENETICS**
- Vols. 1-12, 14-25 (1967-78, 80-91) ..... $40 / $45
- Vols. 26-27 (1992-93) .... $44 / $49
- Vol. 28-29 (1994 and Dec. 1995) .... $47 / $52   Vol(s). _____ Vol. _____

☐ *Annual Review of* **IMMUNOLOGY**
- Vols. 1-9 (1983-91) .... $41 / $46
- Vols. 10-11 (1992-93) .... $45 / $50
- Vol. 12-13 (1994 and April 1995) .... $48 / $53   Vol(s). _____ Vol. _____

☐ *Annual Review of* **MATERIALS SCIENCE**
- Vols. 1, 3-19 (1971, 73-89) .... $68 / $73
- Vols. 20-23 (1990-93) .... $72 / $77
- Vol. 24-25 (1994 and Aug. 1995) .... $75 / $80   Vol(s). _____ Vol. _____

☐ *Annual Review of* **MEDICINE: Selected Topics in the Clinical Sciences**
- Vols. 9, 11-15, 17-42 (1958, 60-64, 66-42) ... $40 / $45
- Vols. 43-44 (1992-93) .... $44 / $49
- Vol. 45-46 (1994 and April 1995) .... $47 / $52   Vol(s). _____ Vol. _____